ENGINEERING CHEMISTRY

ENGINEERING CHEMISTRY

S.C. BHATIA
B.E. (Chemical), M.B.A.

CBS Publishers & Distributors Pvt. Ltd.

New Delhi • Bengaluru • Chennai • Kochi • Kolkata • Mumbai
Hyderabad • Nagpur • Patna • Pune • Vijayawada

ISBN: 81-239-0766-4

First Edition: 2001
Reprint: 2006, 2008, 2010, 2013, 2016

Published by:
Satish Kumar Jain for CBS Publishers & Distributors Pvt. Ltd.,
4819/XI Prahlad Street, 24 Ansari Road, Daryaganj, New Delhi - 110002
delhi@cbspd.com, cbspubs@airtelmail.in • www.cbspd.com
Ph.: 23289259, 23266861, 23266867 • Fax: 011-23243014

Corporate Office: 204 FIE, Industrial Area, Patparganj, Delhi - 110 092
Ph: 49344934 • Fax: 011-49344935
E-mail: publishing@cbspd.com • publicity@cbspd.com

Branches:
• *Bengaluru:* 2975, 17th Cross, K.R. Road, Bansankari 2nd Stage,
  Bengaluru - 70 • Ph: +91-80-26771678/79 • Fax: +91-80-26771680
  E-mail: cbsbng@gmail.com, bangalore@cbspd.com
• *Chennai:* No. 7, Subbaraya Street, Shenoy Nagar, Chennai - 600030
  Ph: +91-44-26681266, 26680620 • Fax: +91-44-42032115
  E-mail: chennai@cbspd.com
• *Kochi:* Ashana House, 39/1904, A.M. Thomas Road, Valanjambalam,
  Ernakulum, Kochi • Ph: +91-484-4059061-65
  Fax: +91-484-4059065 • E-mail: cochin@cbspd.com
• *Kolkata:* 6-B, Ground Floor, Rameshwar Shaw Road, Kolkata - 700014
  Ph: +91-33-22891126/7/8 • E-mail: kolkata@cbspd.com
• *Mumbai:* 83-C, Dr. E. Moses Road, Worli, Mumbai - 400018
  Ph: +91-9833017933, 022-24902340/41 • E-mail: mumbai@cbspd.com

Representatives:
• Hyderabad: 0-9885175004   • Nagpur: 0-9021734563
• Patna: 0-9334159340   • Pune: 0-9623451994
• Vijayawada: 0-9000660880

Printed at:
J.S. Offset Printers, Delhi

Preface

This text book on Engineering chemistry presents in a lucid manner the fundamental aspects of chemistry for engineers. This invaluable model text book is specifically intended for BE/B.Tech institutions in India, including the Indian Institutes of Technology. The recent technical trends demand sound understanding of chemistry by all engineering students. The rapid strides of modern technology are largely due to several break through in the chemistry of new materials, as many new products of chemical industries are finding increasing application in all field of engineering.

Each chapter covers an important aspects related to basic physico-chemical principles involved, and practical applications and significance. Chapter 1 focuses on basic concepts of chemistry and chapter 2 familiarises students with water and its treatment with special emphasis on environmental protection. Chapter 3 is devoted to structure of atom, and phase equilibria is discussed in chapter 4. Chapter 5 describes the importance and application of solution where as chapter 6 familiarises the students with catalysis. Chapter 7 and 8 focuses on gases, liquids and solids. Chapter 9 focuses on chemical bonding followed by chapter 10 which is devoted to kinetics and chemical equilibrium. Chapter 11 and 12 discusses various aspects of electro chemistry and acids and bases. Chapter 13 deals with complex compounds while chapter 14 describes the importance of various aspects of thermodynamics. Chapter 15 focuses on nuclear chemistry based on radium, uranium and plutonium and emphasis the importance of protection from radioactive, while chapter 16 is devoted to thermochemistry and energetics. Chapter 17 describes the importance of colloids where as chapter 18 is devoted to Stereo-isomerism. Chapter 19 explores explosives and propellants followed by chapter 20 which elaborates on organic reaction mechanism.

Noble gases play an important role as these constitute a group in periodic table is the subject of chapter 21, while chapter 22 is devoted to corrosion and its control. Photochemistry is a science which deals with chemical reaction caused by exposure of reactants to light radiation are discussed in chapter 23. Chapter 24 is devoted to chemical methods of separation and various aspects of analytical chemistry including volumetric and titration aspects are discussed in detail. Chapter 25 covers an important aspects of fuel and combustion, while chapter 26 discusses various high polymers like plastics, rubbers and elastomers. Chapter 27 is devoted to surface coating industries, paints, pigments, varnishes and lacquers –their preparation properties and uses are also discussed in detail. Chapter 28 is devoted to silicate industry and focuses on ceramic, cement and glass.

Chapter 29 describes importance of lubrication and lubricants where as chapter 30 is devoted to preparation, properties and applications of various adhesives. Chapter 31 discusses manufacture of iron, steel and various other ferrous and non-ferrous metals. Finally chapter 32 describes the various aspects of pollution such as air, water, solid wastes, noise and their preventives methods.

This text book also includes a glossary of technical terms that will help the students to comprehend the concepts and ensures an easy grasp of key points. References and index add to the value of the book. The text is throughout supplemented by flow diagrams, figures and tables wherever needed.

While painstacking care has gone into producing a useful and exhaustive textbook, the author would welcome constant constructive criticism and a creative feed back from students and professors, and interaction that will certainly be fruitful.

The author, S. C. Bhatia, is a chemical engineer with management qualifications who has written several books on chemical and allied subjects. At present, he is a renowned consultant in the field of environment, waste heat recovery/energy conservation and petro-chemicals.

Acknowledgments are due to Mr. Keshav Kumar and Harinder Singh Negi, the Computer Operators, who worked long hours to complete the task of bringing out the book on time. Also to Ms. Namita Mohanty Das, M.Sc. (Chemistry), Mr. K. K. Sinha B.E. (Mining) and Mr. Madhuresh Kumar for editing and proof-reading. The author also wishes to express his appreciation for the work of Mr. Harinder Singh Negi and Mr. Aman Bhatia B.E.(Mech) who drew and labelled the flow diagrams.

S. C. Bhatia
Author

CONTENTS AT A GLANCE

Contents In Detail

CHAPTER 3

CHAPTER 7

CHAPTER 8

CHAPTER 9

CHAPTER 10

CHAPTER 11

CHAPTER 14

ELEMENTS OF CHEMICAL THERMODYNAMICS. 350-369

CHAPTER 15

NUCLEAR CHEMISTRY 370-381

CHAPTER 16

THERMO-CHEMISTRY AND ENERGETICS. 382-402

CHAPTER 17

THE COLLOIDAL STATE

CHAPTER 22

CHAPTER 23

CHAPTER 26

CHAPTER 27

CHAPTER 28

CHAPTER 31

CHAPTER 32

Chapter 1

Basic Concepts of Chemistry

INTRODUCTION

It is difficult to give clear cut answers to questions like what chemistry is. There will be various responses to this question depending on the knowledge, experience and interest of the person answering it. A very simple answer that a chemistry student will usually give is: Chemistry is that branch of science dealing with matter, its properties and transformations. If such a definition means that energy is not at all studied in Chemistry, then this definition is definitely incomplete. Chemists normally study the amount of heat evolved or absorbed in a reaction. No study of chemistry can be complete without the study of thermodynamics. Study of the properties of matter can be carried out only after the study of quantum mechanics. Thus the above definition is too narrow. Many other definitions may be given. None of them is really incorrect, but most are incomplete. But none will disagree if it is said that "Chemistry is what chemists do'. And we may safely say that chemists are trying to know the secret about material substances. For this purpose chemists try to purify, identify and estimate substances. They develop and use techniques of purification of a substance from a mixture of substances. Dissolution, filtration, sedimentation, distillation, sublimation, crystallisation, are some such common methods. They identify substances by studying their properties. Properties are measurable attributes of a substance. For studying the properties of a substance, it must be allowed to interact with something. For example, if a substance is kept in the dark, we cannot observe it. When it is brought to light, it interacts with it. Our eye observes it: We note its colour (a physical property). Whenever there is an interaction, there will be a change. The change may be observable, may not be observable. The changes may be physical (where the substances do not really change). The changes may be chemical where the original substance changes to a new one. For the purpose of identification, some characteristic properties (which lead to specific changes) are chosen. Estimation is done by noting the weights and other quantitative parameters of various components in a change. All these are usually studied under the name: "analytical chemistry".

Chemists also try to synthesise more and more substances. This branch of chemistry where a new substance is prepared by the synthesis of two or more substances is termed as synthetic chemistry.

Study of properties is of interest in itself (apart from identification). Chemists engaged in the synthesis and study of a group of substances known as hydrocarbons and their derivatives are organic chemists. This branch of chemistry is termed as organic chemistry. Chemists engaged in the study of substances other than hydrocarbons and their derivatives are inorganic chemists; the branch is inorganic chemistry. Chemists studying living matter only are biochemists.

1

The purpose of science is to interpret, correlate, explain and if possible, predict the phenomenon occurring in bewildering multitude in nature. Naturally chemists like to know why the properties of substances are as they are, how far and how fast can a change go and why. This is the natural tendency of a human being. Man is a thinking and speculative being. He thinks that there is order somewhere in the chaos that is apparent in the nature.

This thought originates from the belief that the nature is uniform. This has led to the concept of cause and effect. We believe that same cause will lead to the same effect. This concept of uniformity of nature has led us to find out the reasons of changes (explanation etc.)—the causes of the effects.

However, we must remember that this is not a fundamental condition of science. After the advent of quantum mechanics, we have come to know that it is not possible to predict exactly what will happen in any fixed set of circumstances. We may predict what the maximum probability is we can find only a statistical average as to what happens. But even then, we would like to predict and find out as to what is most probable. That is, if this thing had led to that (effect) several times, it is most probable that the same thing will happen next time, when the same conditions are repeated.

Therefore, we search for order. We observe certain things, and then we make planned observation (experiment). Observation leads to an idea as to its cause. Then we perform experiments to verify the idea. The sole test of validity of an idea is experiment. We use our imagination in each step of the experiment: we reason. Thus scientific methodology consists of reason and experiment, the first step, of course, being observation.

We develop concepts or fancies from time to time for the purpose of explaining what we observe. Thus we use facts and fancies.

Observations lead to an idea: a hypothesis. Repeated experiments on the idea may lead to a convenient summary of the experimental results. Thus we arrive at a scientific law.

To explain the laws, we again develop hypothesis. And we may test it several times. Ultimately, a theory or model results. We are happy for the time being.

A theory developed in such a way is the atomic theory. An atoms is defined as the smallest particle of an element that can take part in a chemical reaction. All atoms of the same element are identical with each other in all respects. Atoms combine with each other in different ways to form molecules. Molecules are the smallest units of a substance maintaining all its properties. An element is a pure substance containing only one type of atoms. Of course, only the weights of the atoms may be different. Thus the atoms of an element are identical with each other in all respects except in weights. The molecules of an element may be monatomic or diatomic or polyatomic. That is, the molecules of an element may contain two or more than two atoms, e.g., Cl_2, Br_2, S_8, P_4, etc. Atoms of different elements can combine with each other in different ways to give rise to molecules of compounds. Chemists believe that the properties of different substances are due to different atoms and to the different modes of combination of atoms in the formation of their molecules. Therefore, until and unless we know the detailed structure of a molecule, the secret of material substances cannot be known. Consequently a group of chemists study the structure of different molecules. From this study, we may be able to know the distance between different atoms in a molecule (this is defined as bond length or bond distance), and the angle between different atoms (this is known as bond angle). Chemists engaged in such study are structural chemists, and the corresponding branch of chemistry is termed as structural chemistry. However, inorganic chemists study the structure of inorganic compounds and organic chemists study the structure of organic compounds.

We may be able to predict the properties of a substance only after detailed analysis of the substance including its structure. Identification can be made in most cases only after allowing the substance to undergo a chemical change. Chemical change naturally consists of a change of the molecular structure of the original substance which produces a new substance, i.e. a new molecule, with a new molecular structure. Consequently we like to know how these molecular changes take place. This can be known only if a detailed description of reaction pathway is available. This description will, naturally, be imaginary because atoms and molecules are not observable. This involves fancies or concepts. This means that we have to extend our imagination up to the atomic level for explaining the observable changes. The first step in any such study involves the analysis of the rate of a reaction.

The above are, therefore, the principles of the whole of chemistry. And that is what the physical chemists do. They study the rates and therefore naturally the energetics of a chemical change. Then they study the structure of molecules and suggest mechanism of the changes.

Physical chemistry therefore, deals with both the matter and energy. Physical chemists have to study substances in their different states, energetics, that is thermodynamics and thermochemistry, phase transitions, that is conversion of solid to liquid, liquid to gas, etc. Also they have to study the chemistry of surfaces, ions, atoms and molecules.

For the purpose of explanation, correlation and interpretation, a physical chemist will have to find out scientific laws and will have to develop theories and models for explaining them. Now, what is the ultimate result of an experiment? It is nothing but a number. Therefore, the models or theories have to be such that numbers emerge from them. That means that the theories must have a quantitative background and they must be amenable to mathematical analysis. Sometimes a physical chemist or, in general, any scientist has to use completely imaginary ideals or idealisation for the purpose of making the mathematics simpler. These are fictions. And therefore, the world of a physical chemist, consists of facts (experimental results), fancies (concepts), and fictions (idealisation). Because of this mathematical necessity of a theory, Prof. Evans defined a physical chemist as a chemist sitting behind a big mahogany desk with a platinum slide rule in his hand.

The numbers themselves bear no meaning. Theories give them meaning. And a number without unit is in no way amenable to theoretical analysis. Consequently a study of physical chemistry cannot start without a study of units and dimensions. A study of chemistry cannot begin without a knowledge of stoichiometry—the idea of the importance of balanced reactions. Hence we present here a brief outline of stoichiometry, and units and dimensions.

STOICHIOMETRY

Stoichiometry refers to the relationships between quantities of reactants and products in a reaction. Thus stoichiometric analysis is nothing but quantitative analysis of a reaction. Detailed analysis of various reactions led the scientists to the enunciation of the following laws of chemical combination.

1. *Law of conservation of mass was stated by Lomonossoff (Russian scientist) and Lavoisier:* It is the basis of the entire non-nuclear chemistry and states that mass can neither be created nor destroyed.

Thus there must be mass balance in a chemical reaction: the total mass of reactants must be equal to the total mass of products in case of complete conversion. Otherwise, the starting mass must remain the same. Nuclear chemists has changed this concept. We now know that mass can be destroyed and an amount of mass, Δm on destruction releases $\Delta m.C^2$ amount of energy, where 'C' is the velocity of light. This is now generalised as the law of conservation of mass and energy.

2. *Law of definite proportion:* Lavoisier extended the law of conservation of mass to the case of compounds to enunciate law of definite proportion. A chemical compound contains the same proportion by weight of its constituent elements. This is true for majority of compounds. But some compounds like semi-conductor oxides and some sulphides do not strictly obey this law. Iron(II) sulphide has a composition between $Fe_{0.86}S$ and FeS and belongs to this category. These compounds are termed as Berthollides. Other compounds where this law is valid are termed as Daltonides.

3. *Law of multiple proportion:* This was enunciated by Dalton and it states that if two elements form more than one compound, then the weights of one element which combines with a fixed weight of the other are in a simple ratio.

4. *The law of equivalent proportions:* This was propounded by Berzelius and others, states that the weights of the two elements which individually either displace or combine with a fixed weight of a third element are just those weights involved in their combination.

All the four are summary of experimental results and are classified in combination as the laws of stoichiometry. All chemical calculations and equations must follow their rules. This leads to the concept of balancing of equations. That laws of equivalents lead to equivalent weight and along with others led to the determination of atomic weights.

Thus $H_2 + O_2 = H_2O$ is wrong, and $2H_2 + O_2 = 2H_2O$ is a correct balanced equation obeying stoichiometry. The above equation means that two molecules of hydrogen will always (without exception) combine with one molecule of oxygen to produce two molecules of water.

Daltons atomic theory was put forward to explain these laws and correlate these with Gay Lussac's law of gaseous volumes (another conclusion of experimental results and a law of stoichiometry involving gases : Combining volumes of gases bear a fixed simple ratio with each other and also with the products if they are also gaseous), Avogadro's hypothesis defines molecules just as Dalton's theory defines atoms. This hypothesis was initially put forward as a likely consequence of stoichiometric laws. It facilitated the explanation of these laws by Dalton's atomic theory. It has now attained the status of the law. This hypothesis was stated as: equal volumes of all gases under the same conditions of temperature and pressure contain equal number of molecules.

Atoms and molecules are very small. The radius of an atom is of the order of 10^{-8} cm. If we take an orange and can mentally extend its size to that of the earth, then an atom will have the size of the original orange. Thus for carrying out chemical calculations, we need a much larger unit. Such a unit is mole. It is the quantity of a substance whose mass in grams is equal to its atomic or molecular weight. The number of atoms or molecules in one mole has now been very accurately determined. It has been determined to be 6.0228×10^{23} or 6.023×10^{23}. In the present atomic weight scale, one mole is defined as the amount of material containing Avogadro's number of particles, it is the number of atoms in exactly 12 gms of C_{12}.

UNITS AND DIMENSIONS

The result of an experiment is nothing but a number. The numbers are reported either as pure numbers or in terms of units. As for example, the result of several experiments may be 5. In one experiment it may be 5 centimetre, in another 5 millilitre, in yet another 1.5 ergs. The results of the experiments are differentiated by the last portion of the report i.e. their units. When we are reporting 5 centimetre cr 5 cm we could have reported it also as $\frac{5}{2.54}$ inch, $\frac{5}{30.48}$ feet, 0.05 metre etc. All the units quoted are of the same type or dimension, namely, length. We cannot report length in ergs or other. That will be a meaningless report. A particular quality has a particular dimension—it must be expressed in units consistent with the dimension. A quantity may be dimensionless or a pure number. In any physical measurement pure number refers to the ratio of quantities of same dimension.

Length is a dimension. The number of independent dimensions are quite limited. The independent dimensions are termed as basic dimensions. In mechanics we use three basic dimensions—length, mass and time. These three cannot be defined in terms of similar concepts, and hence they are indefinables of mechanics. There are also three other basic dimensions: electric current, temperature and luminous intensity. Another basic quantity is: mole—it is the number of molecules in 1 gm-mole of a substance (Avogadro's number: value 6.023×10^{23}). Other units for quantities like force, energy, velocity, acceleration, electric charge, frequency etc. can all be expressed in terms of these basic dimensions. Thus force has the dimension of mass × length × time^{-2}, volume has the dimension of $(\text{length})^3$ etc.

Quantities on each side of a relation must have the same dimension: we can add or substract quantities of the same dimension. Thus in the relation

$$xy + z^2 + Q = P$$

xy, z^2 and Q must have the same dimension as that of P. When one quantity depends on several others, we may find the nature of dependence by applying the above principle. This method of analysis is termed as dimensional analysis. Naturally, quantities involved in exponential, logarithmic, trigonometric or other functions must be dimensionless numbers. For example $100 = 10^2$. Thus $\log_{10}100 = 2$. Therefore, logarithm is a power—it is a pure number.

When we insert numerical values into equation, it must be checked whether units are consistent or not. It would be inconsistent if we report length in metres, mass in pounds and time in hours, and use them in a calculation. The calculations carried out may be correct but the magnitude of the answer will be meaningless. Thus pressure in the above mixed unit will be force/area=force/$(\text{length})^2$ = mass × $(\text{length})^{-1}$. time^{-2} or pounds. $(\text{metre})^{-1}$. $(\text{hour})^{-2}$. The consistency in the system of units, however, is not very difficult to arrive at. There are a number of systems—the British foot-pound-second system (FPS), the commonly used centimetre-second or C.G.S. system and the modern systeme Internationale d' unite" (S.I.).

A standard length, standard weight and precision time signals are defined as a metre, a kilogram and a second respectively. The standards chosen may change 1 cm is $\frac{1}{100}$th of the standard metre. With the help of dimensional relation the numerical value of a quantity may be changed from one unit to another. But for this, the conversion factors should be known. Here below are given some of the conversion factors.

1 metre	= 39.37 inches
1 inch	= 2.54 cm
1 foot	= 30.48 cm
1 kg	= 2.205 pound (lb)
1 lb	= 435.6 gms
Decimeter (dm)	= 10^{-1} meter
Centimeter (cm)	= 10^{-2} meter
Millimeter (mm)	= 10^{-3} meter
Micrometer (m)	= 10^{-6} meter
Nanometer (nm)	= 10^{-9} meter
Picometer (P)	= 10^{-12} meter
Decameter (dam)	= 10 meter
Hectometer (hm)	= 10^2 meter
Kilometer (hm)	= 10^3 meter
Megameter (mm)	= 10^6 meter
Gigameter (gm)	= 10^9 meter
Terameter (tm)	= 10^{12} meter

We may derive the units of quantities from the fundamental units. Thus density is mass/unit volume. Volume is (length)^3, i.e. L^3. So density is M/L^3 or ML^{-3}. Velocity is length/time or LT^{-1}. Similarly acceleration is LT^{-2}. Force is MLT^{-2}. Energy is force × length, i.e. ML^2T^{-2}. Stress is force/area, $ML^{-1}T^{-2}$.

As we have already outlined dimensional relations may be used to change the numerical values from one unit to another. Thus let a velocity be expressed in C.G.S. system as 20 cm/sec. This may be expressed in meter/hour as follows:

Let its magnitude in the last unit be a

$$\therefore \ \text{a meter / hour} = 20\frac{\text{cm}}{\text{sec}}$$

$$\therefore \ a = 20.\ \frac{\text{cm}}{\text{meter}}.\frac{\text{hour}}{\text{sec}}$$

Using conversion factors (1 meter = 100 cm, 1 hour = 60 × 60 sec)

$$\therefore \ a = 20 \times \frac{1}{100} \times \frac{3600}{1} = 720$$

Thus a velocity of 20 cm/sec is equal to a velocity of 720 meter/hour.

P 1–: Express 2 poundals in dynes.

Poundal is the unit of force in F.P.S. system. It is the force needed to maintain an acceleration of 1 ft/sec^2 for a mass of 1 lb. The dimension of force is MLT^{-2}. Thus it is 2 lb × ft sec^{-2}. In C.G.S. system it will a gm cm sec^{-2}. Thus

$$a \text{ gm cm sec}^{-2} = 2 \text{ lb} \times \text{ft sec}^{-2}$$

$$\therefore \quad a = 2 \times \frac{lb}{gm} \times \frac{ft}{cm}$$

Using conversion factors

$$a = \times \frac{453.6}{1} \times \frac{30.48}{1} = 2.764 \times 10^4$$

The unit of force in C.G.S. system is dyne. Thus 2 poundals is equal to 2.764×10^4 dynes.

M.K.S. system was chosen in preference to the C.G.S. system because in the C.G.S. system the unit of force and hence of energy is too small. In the M.K.S. system the unit of mass is kilogram (kg), the unit of length is metre and the unit of time is second (sec). The unit of force in M.K.S. system is Newtons (N). It is the force required to produce an acceleration of 1 metre/sec^2 in a mass of 1 kg. Let us find out the relation between Newtons and dynes.

$$1 \text{ N} = 1 \text{ kg } 1 \text{ metre} \times \text{sec}^{-2} = a \text{ dynes}$$

$$\therefore \quad a = \frac{1kg}{gm} \cdot \frac{\text{metre}}{cm} = 10^3 \times 10^2 = 10^5$$

$$\therefore \quad 1 \text{ N} = 10^5 \text{ dynes.}$$

Similarly, the relationship between the units of energy in the M.K.S. and C.G.S. system may be found out as follows: The unit of energy in M.K.S. is 1 kg (metre)^2 sec^{-2} and in C.G.S. is 1 gm cm^2 sec^{-2} which is equal to erg. Thus let

$$1 \text{ kg (meter)}^2. \text{ sec}^{-2} = a \text{ erg.}$$

$$\therefore \quad a = 1\frac{g}{cm} \cdot \left(\frac{\text{meter}}{cm} \right)^2 = 10^3 \times 10^4 = 10^7$$

Thus the unit of work in M.K.S. system is 10^7 erg. This is 1 Joule (J). Newton's second law of motion relating mass, length, time and force provides the logical basis for the development of systems of units. In C.G.S., F.P.S., and M.K.S. systems the units of mass, length and time are basic (fundamental) and that of force is derived.

In the gravitational system of units, the units of length, force and time are basic. Thus the metre, kilogram force (or simply kg) and the second are basic in the gravitational metric system—the unit of mass is derived.

1 kg in this system (abbreviated as kg or kgf kilogram force or kilopound (kp) is that force which acting on a mass of 1 kg imparts to it the standard acceleration 9.80665 m/sec^2 (980.665 cm/sec^2 in C.G.S. system). 9.80665 m/sec^2 as the value of standard acceleration has now been internationally accepted. The pound force of gravitational F.P.S. system is similarly defined in terms of the mass of 1 lb (the standard acceleration in F.P.S. rounded off to fourth place of decimal is 32.1740 ft./sec^2).

The units for quantities like velocity, force etc. are derived. Some derived units for quantities such as pressure, certain units of energy and some other quantities are not reported according to any of the systems quoted above. These units are defined in terms of some specific standard: e.g., for pressure, the atmosphere, for energy, the calorie etc.

The pressure is sometimes expressed in mm of Hg. It is a barometric unit and is defined as the pressure exerted by a 1 mm column of liquid of uniform density 13.5951 gm/cc under standard acceleration of 980.665 cm/sec^2.

The standard atmosphere is defined as equal to 760 mm of Hg or 76 cm of Hg. This is often referred :o as 1 atmosphere or 1 atm. Thus

$$1 \text{ atm.} = 76 \times 13.5951 \times 980.665 \text{ gm sec}^{-2}/\text{cm or dyne/cm}^2$$
$$= 76 \times 13.6 \times 980.7 \text{ dyne/cm}^2$$
$$= 1.014 \times 10^6 \text{ dyne/cm}^2$$

(Log table will reproduce values only up to fourth place. It is really 1.013250 dyne/cm^2). In C.G.S. system the unit of pressure is bar, which is 10^6 dynes/cm^2. So 1 atm. = 1.01325 bar. The millibar is defined as 10^3 dynes/cm^2 or 100 N/metre^2. In M.K.S. system the unit of pressure is 1 N/metre^2 or pascal (Pa). Thus 1 Pa = 1 N/metre^2=10^{-5} bar. The torr, which is also in common use as a unit of pressure is $\dfrac{1}{760}$ of 1 standard atm. This is 133.322 N/metre^2, it is also 1 mm of Hg.

P2:— A body weights 1.033 kg. It is placed over a piston of area 1 sq. cm. Express the pressure exerted over the piston in C.G.S. and M.K.S. system.

1.033 kg = 1033 gm.

Thus F = 1033×980.7 dynes = 1.013×10^6 dynes

Thus the pressure is 1.013×10^6 dynes/cm^2.

∴ P = 1.013 bar = 1013 millibar.

Now 10^5 dynes = 1 Newton

1 sq. cm. = 10^{-4} sq. meter

Let 1.013×10^6 dynes/cm^2 = a N/meter^2

$$\therefore \ a = 1.013 \times 10^6 \ \frac{\text{dynes}}{\text{N}} \times \left(\frac{\text{meter}}{\text{cm}} \right)^2$$

$$a = 1.013 \times 10^6 \times 10^{-5} \frac{\text{N}}{\text{N}} \times \left(\frac{\text{meter}}{10^{-2} \text{ meter}} \right)^2$$

$$= 1.013 \times 10^5$$

Therefore the pressure in M.K.S. is

1.013×10^5 N/metre^2 = 1 atm.

P3:— Express 1 atm. in F.P.S. gravitational system.

$$1 \text{ at.} = \frac{1033 \text{ gm}}{\text{cm}^2}$$

$$\text{Let} = \frac{1033 \text{ gm}}{\text{cm}^2} = a \text{ lb} / \text{inch}^2$$

$$\therefore \; a = 1033 \times \frac{1}{453.6} \times (2.54)^2 = 14.69$$

Hence in gravitational F.P.S. system 1 atm. = 14.69 lb/inch². The C.G.S. system can be extended to electric and magnetic measurements. But then there are complications. There was need of introducing at least one non-dynamical unit. The force between two charges q_1 and q_2 at a distance of r is

$$F = \frac{q_1 q_2}{D r^2} \qquad \qquad ...(1.1)$$

Where D is the permittivity. We define unit of electric charge by putting $q_1 = q_2$, F = 1 dyne and D = 1 and r = 1 cm in C.G.S. system. In C.G.S. electromagnetic system, the unit of magnetic pole strength is derived by assigning permeability μ of free space to be 1.

$$F = \frac{m_1 m_2}{\mu r^2} \qquad \qquad ...(1.2)$$

From equation (1.1.), the dimension of electric charge is $M^{1/2}L^{3/2}T^{-1}D^{1/2}$.

$$\left(MLT^{-2} = \frac{q^2}{DL^2} \right)$$

The strength of an electric current is equal to the rate at which an electric charge moves along a conductor. Thus the dimension of current is $M^{1/2}L^{3/2}T^{-2}D^{1/2}$. From equation (1.2), the dimension of pole strength is

$$M^{1/2}L^{3/2}T^{-1/2}\mu^{1/2} \left(MLT^{-2} = \frac{m}{\mu L^2} \right)$$

The work done in carrying a magnetic charge 'm' round a closed path carrying current I is $4\pi mI$, so that the product of pole strength and the current should have the dimensions of work. Thus

$$I = \frac{Work}{Pole \; strength} = \frac{ML^2 T^{-2}}{M^{1/2}L^{3/2}T^{-1}\mu^{1/2}} = M^{1/2}L^{1/2}T^{-1}\mu^{-1/2}.$$

The dimension of current must be the same irrespective of the method of calculation. Thus, let I_s be the measure of current in e.s.u. and I_m be its measure in e.m.u. (electromagnetic unit).

$$I_s[M^{1/2}L^{3/2}T^{-2}D^{1/2} = I_m[M^{1/2}L^{1/2}T^{-1}\mu^{-1/2}]$$

$$\therefore \; \frac{I_s}{I_m}(LT^{-1}) = \frac{1}{\sqrt{D\mu}}$$

$$\therefore \; D^{1/2}\mu^{1/2} = L^{-1}T = \frac{1}{C}$$

Thus $\dfrac{1}{\sqrt{D\mu}}$ represents a velocity of 'c' cm per sec.

Hence $\dfrac{1 \text{ e.m.u. of current}}{1 \text{ e.s.u. of current}} = \dfrac{I_s}{I_m} = C$

The value of 'c' has been experimentally determined by measuring capacitance in the two units. It has been found to be 2.998×10^{10}, i.e., the velocity of light in vacuum.

The dimension of D and μ cannot be expressed alone in terms of M, L and T.

The dimension of some quantities in e.s.u. system are given below :

1. Potential difference V = Charge × potential difference = work.

$$\therefore \quad V = \frac{W}{q} = \frac{ML^2T^{-2}}{M^{1/2}L^{3/2}T^{-1}D^{1/2}} = M^{1/2}L^{1/2}T^{-1}D^{1/2}$$

2. Capacitance $= \dfrac{\text{Charge}}{\text{Potential difference}}$

$$= \frac{M^{1/2}L^{3/2}T^{-1}D^{1/2}}{M^{1/2}L^{1/2}T^{-1}D^{-1/2}} = LD$$

(Ignoring D, capacitance C has the dimension of length in e.s.u.)

3. Resistance $R = \dfrac{\text{Potential difference}}{\text{Current}} = L^{-1}TD^{-1}$

(reciprocal of velocity ignoring D)

4. Electric induction = permittivity × electric inductance

$$= D.\frac{\text{Force}}{\text{Charge}}$$

$$= D.\frac{MLT^{-2}}{M^{1/2}L^{3/2}T^{-1}D^{1/2}} = D^{1/2}M^{1/2}L^{1/2}T^{-1}$$

5. Magnetic pole strength

$$m = \frac{\text{Work}}{\text{Current}} = M^{1/2}L^{1/2}D^{-1/2}$$

6. The dimension of some quantities in e.m.u. system are magnetic intensity (H).

Magnetic intensity × pole strength = force.

$$\therefore \quad H = \frac{\text{Force}}{\text{Pole strength}} = \frac{MLT^{-2}}{M^{1/2}L^{3/2}T^{-1}\mu^{1/2}}$$

$$= M^{1/2}L^{1/2}T^{-1}\mu^{1/2}$$

7. Magnetic induction (B)

   B = Permeability × magnetic intensity (μH)

   $B = M^{1/2}L^{-1/2}T^{-1}\mu^{1/2}$

   (dimension is pole strength/area).

8. Intensity of magnetisation (J)

   J = magnetic moment/Volume

   $$= \frac{M^{1/2}L^{5/2}T^{-1}\mu^{1/2}}{L^8} = M^{1/2}L^{-1/2}T^{-1}\mu^{1/2}$$

   (dimension same as B).

9. Resistance (R)

   $I^2 R$ = Power

   $$\therefore \quad R = \frac{\text{Power}}{I^2} = \frac{\text{Work}/\text{Time}}{I^2} = \frac{ML^2T^{-3}}{MLT^{-2}\mu^{-1}} = LT^{-1}\mu$$

   (Resistance has the dimension of velocity in e.m.u. system).

10. Capacitance (C)

    $C \times V = Q$.

    $$\therefore \quad C = \frac{Q}{V} = \frac{\text{Charge}}{\text{Pd}} = \frac{M^{1/2}L^{1/2}\mu^{-1/2}}{M^{1/2}L^{3/2}T^{-2}\mu^{1/2}} = L^{-1}T^2\mu^{-1}$$

11. Electric intensity (E)

    E = Force/Charge = $M^{1/2}L^{1/2}T^{-2}\mu^{1/2}$

12. Inductance (L)

    1/2 LI^2 = Work

    $$L = \frac{\text{Work}}{I^2} = \frac{ML^2T^{-2}}{MLT^{-2}\mu^{-1}} = L\mu$$

13. Magnetic flux (N) = B × area

    $= M^{1/2}L^{-1/2}T^{-1}\mu \times L^2 = M^{1/2}L^{3/2}T^{-1}\mu.$

It is obvious, therefore, that

$$\frac{1 \text{ e.s.u. of charge}}{1 \text{ e.m.u. of charge}} = \frac{1}{C}$$

Since 1 e.s.u. potential difference (Pd) × 1 e.s.u. of charge = 1 erg.

= 1 e.m.u. of Pd × 1 e.m.u. of charge.

$$\frac{1 \text{ e.s.u. of Pd}}{1 \text{ e.m.u. of Pd}} = \frac{1 \text{ e.m.u. of charge}}{1 \text{ e.s.u. of charge}} = C$$

It can be easily shown that

$$\frac{1 \text{ e.s.u. of capacitance}}{1 \text{ e.m.u. of capacitance}} = \frac{1 \text{ e.m.u. of resistance}}{1 \text{ e.s.u. of resistance}} = \frac{1}{C^2}$$

In most cases the e.s.u. and e.m.u. defined above are inconvenient, being either too small or too big. This led to practical system. The primary practical unit is ampere, the unit of current.

1 ampere = 1/10 e.m.u. of current 3 × 10^9 e.s.u. of current

1 coulomb = 1/10 e.m.u. of charge = 3 × 10^9 e.s.u. of charge

1 volt = 10^6 e.m.u. of Pd = $\frac{1}{300}$ e.s.u. of Pd.

1 Ohm = 10^9 e.m.u. of resistance = $\frac{1}{9 \times 10^{11}}$ e.s.u. of resistance

1 Farad = 10^{-9} e.m.u. of capacitance = 9 × 10^{11} e.s.u. of capacitance

1 Henry = 10^{-9} e.m.u. of inductance $\frac{1}{9 \times 10^{11}}$ e.s.u. of inductance

(We have assumed the velocity of light to be 3 × 10^{10} cm/sec here)

The Oersted (unit of magnetic field strength), the Gauss (unit of magnetic induction) and the Maxwell (unit of magnetic flux) are the only practical units which are same as the corresponding e.m.u. Practical units are, therefore, derived from C.G.S. electromagnetic unit. At first, it was assumed that establishment of units from the definition of the C.G.S. system would tax the resources of many laboratories. It was thus decided by international agreement to define the units retaining as closely as possible the numerical values of the practical system, in a most practical way. Thus ampere was defined in terms of the rate of deposition of silver in a silver coulometer—it is the current which flowing for 1 sec will cause deposition of 0.000118 gm of silver from a solution of silver salt. These units and those derived from them are international units.

In 1948, the small difference between international units and practical units are specified.

International	Practical
1 Ohm	1.00049 Ohm
1 Ampere	0.00985 Ampere
1 Volt	1.0034 Volt
1 Coulomb	0.99985 Coulomb
1 Farad	0.99951 Farad
1 Henry	1.00049 Henry
1 Watt	1.00019 Watt
1 Joule	1.00019 Joule

The practical units are used ever since and they are incorporated in S.I.

In 1901, Giorgi suggested that if we choose the metre and the kilogram as the units of length, and mass, the units in the new system could be made the same as practical units. Apart from time, ampere was chosen as the 4th basic unit.

In M.K.S., the Coulomb's law taking coulomb as the unit of charge becomes as follows.

Two charges of 1 coulomb (3×10^9 e.s.u. each) will exert on each other at a distance of 1 metre

in vacuum a force of $\dfrac{9 \times 10^{18}}{100^2} = 9 \times 10^{14}$ dynes $= 9 \times 10^9$ Newtons.

Thus D has to be given a value as follows

$$9 \times 10^9 \text{ Newton} = \frac{1 \text{ coulomb}^2}{D \text{ meter}^2}$$

Thus a charge of 1 coulomb when placed at a distance of 1 metre from another, will exert a force of 9×10^9 Newton on each other.

The system is now M.K.S.A. The unit of current is defined by the force two parallel currents exert on each other. Using e.m. unit

$$F \text{ (dynes / cm)} = \mu \frac{2 \text{ (current in e.m.u.)}^2}{\text{distance in cm}}$$

where μ is permeability.

In the M.K.S.A. system, we have

$$F \text{ (Newton/metre)} = \mu \frac{2 \text{ (amp)}^2}{\text{metre}}$$

where μ is the permeability of the vacuum.

Thus the force per centimetre between parallel currents 1 e.m.u. each at a distance of 1 cm in vacuum is 2 dynes, since $\mu = 1$.

Thus, two parallel currents 1 ampere each, separated by a distance of 1 metre in vacuum, will exert on each other a force per metre equal to

$$\frac{2 \times (.1)^2}{1 \text{ meter}}$$

$$= 2 \times .01 \text{ dyne}$$

$$= 2 \times .01 \times 10^{-5} \text{ Newton} = 2 \times 10^{-7} \text{ Newton.}$$

$$\therefore \quad \mu \frac{2 \times (1 \text{ cm})^2}{\text{meter}} = 2 \times 10^{-7} \frac{\text{Newton}}{\text{metre}}$$

$$\therefore \quad \mu = 10^{-7} \frac{\text{Newton}}{\text{amp}^2}$$

μ is the permeability of vacuum in M.K.S.A. system.

Thus the ampere is the constant current which, if maintained between two parallel plates of infinite length and negligible radius one metre apart, will exert a force of 2×10^{-7} newton per meterlength.

In M.K.S. units, the unit of work is Joule (Newton $\times$ metre $= 10^5$ dyne $\times 10^2$ cm). But $I^2Rt = $ Work. Hence 1 amp$^2 \times$ unit of resistance = Joule.

Now amp$^2 \times$ ohm $\times$ sec. = Joule. Hence ohm is the unit of resistance. Unit of Pd is ohm $\times$ ampere = Volt.

The factor 4π often appears in equations involving electric and magnetic quantities. This factor is associated with spherical symmetry. To avoid confusion, the units are rationalised. For permittivity of vacuum $4\pi\epsilon_o$ is written and permeability of vacuum $\mu/4\pi$ is written. The equations thus drop factor 4π, but some acquire it. The rationalised C.G.S. system is Heavirside-Lorentz system and MKSA is Giorgi system. By introducing two other basic units—degree, kelvin, and candela (unit of luminous intensity), the Giorgi system is converted to S.I. (system internationale d units).

Thus the S.I. system uses all the six dimensions. The basic units are represented below :

Dimension	Name of the S.I. unit	Symbol
Length (L)	Meter	m
Mass (M)	Kilogramme	kg
Time (T)	Second	S
Electric current (I)	Ampere	A
Temperature (t)	Kelvin	K
Luminous intensity	Candela	Cd

And then there is stoichiometric mass: mole (mol). In the following table some derived S.I. units are given below:

Quantity	Name of S.I. unit	Symbol and basic equivalents
Force	Newton	$N = kg \cdot ms^{-2}$
Energy, work and heat	Joule	$J = Nm = kg\ m^2s^{-2}$
Power	Watt	$W = j\ s^{-1} = kg\ m^2\ s^{-3}$
Electric charge	Coulomb	$C = AS$
Electric potential	Volt	$V = WA^{-1} = kg\ m^2A^{-1}s^{-3}$
Electric capacitance	Farad	$F = ASV^{-1} = A^2S^4\ kg^{-1}m^{-2}$
Electric resistance	Ohm	$\Omega = VA^{-1} = kg\ m^2A^{-2}s^{-2}$
Specific conductance	Siemen/meter (ohm^{-1} meter^{-1})	$\Omega^{-1}\ m^{-1} = kg^{-1}m^{-3}A^2S^3$
Frequency	Hertz (cycle/sec)	$Hz = s^{-1}$
Pressure	Pascal	$Pa = 1\ Nm^{-2} = kg\ m^{-1}s^{-2}$
Viscosity	Poiseuille	$Pl = \dfrac{N.S.}{metre^2} = kg\ m^{-1}s^{-1}$

Normally C.G.S. system is used for the convenience, although the S.I. systems is universally accepted. However, in the following table some conversions are given with the help of which and with the knowledge of this section anybody can convert C.G.S. to S.I.

Quantity	C.G.S. unit	S.I. unit	Conversion factor C.G.S. to M.K.S.
Force	Dyne	Newton	10^5 dyne = 1 N
Pressure	Bar	Pascal	10^{-5} bar = 1Pa
Viscosity	Poise (P)	Poiseuille (Pl)	0.1 or $\dfrac{1}{10}$ P1 = 1P = 1 pascal second
Heat	Calorie	Joule	$\dfrac{1}{4.184}$ calorie = 1 Joule
Heat, work and energy	Erg	Joule	10^7 erg = 1 J
Specific conductance	Ohm^{-1} cm^{-1}	Siemen/meter	10^{-2} ohm^{-1} cm^{-1} = 1 S/m
Charge	e.s.u. & e.m.u.	Coulomb	$\dfrac{1}{10}$ e.m.u. = 1 Coulomb; 1 Coulomb = 3×10^9 e.s.u.

In the C.G.S. system, the force between two electric charges q_1 and q_2 is

$$F = \frac{q_1q_2}{r^2}$$

And in S.I. system it is

$$F = \frac{q_1q_2}{4\pi\varepsilon^\circ r^2}$$

where ε° is 8.85415×10^{-12} Coul^2/N.m^2.

CHEMISTRY AND PHYSICS

Atoms consist of protons, neutrons and electrons. The arrangement of extra nuclear electrons is believed to be the cause of the properties of atoms. This arrangement is assumed to be the prime force of chemical combination. In a molecule, there is a readjustment and the nature of interaction also changes. Naturally, the electronic arrangement of the molecule is responsible for its properties. The structure of a molecule means these electronic arrangement and the distance and angle between the atoms. Modern chemistry studies the structure of molecules, the reasons of chemical changes, the energetics of a reaction and the rates and mechanism of a reaction. Therefore the science of chemistry treats the following groups of phenomena.

1. The structure of the molecules and the properties of substances in relation to their molecular structure and composition.

2. The composition of substances and the changes in the composition and the molecular structure of the substances, and the effect attending such changes.

The science of physics treats the following phenomenon:

1. The characteristic of bodies that are independent of their composition and the properties of substances in their general aspect, without reference primarily to their molecular structure or composition.

2. Changes not involving changes in molecular structure or composition.

The above definitions, although interesting and reasonably complete are made with respect to a theoretical concept: the structure of molecules. A sharp line, however, cannot be drawn between these two upon the basis of molecular structure and composition. The reason is that it is not always possible to distinguish processes that involve changes in molecular structure from process that do not. What is more, physics in dealing with properties in their general aspects take into account also the effect of molecular structure in most of the cases.

Moreover at the basis of both chemistry and physics (called together as physical sciences) lie large bodies of general principles relating to matter and energy. These bodies of knowledge form the basis of both the physical sciences and are often considered as parts of theoretical physics. We may assume these as also parts of physical chemistry. This definition makes theoretical physics a branch of physical chemistry.

According to Feynman, both inorganic and organic chemistry are nothing but physical chemistry and quantum mechanics. Since the above definition includes quantum mechanics as a branch of physical chemistry, inorganic and organic chemistry also becomes branches of physical chemistry. Looked in this way the whole of chemistry is nothing but physical chemistry.

Example 1.1. Calculate the r.m.s. velocity of O_2 at 27°C.

Solution.

$$C_{r.m.s.} = \sqrt{\frac{3RT}{M}}$$

(a) In traditional system if we put $R = 8.31 \times 10^7$ ergs and $M = 32$ gm mole^{-1} we get $C_{r.m.s.}$ in cm/sec.

$$\text{Thus } C_{r.m.s.} = \sqrt{\frac{3 \times 8.31 \times 10^7 \times 300}{32}} = 4.8344 \times 10^4 \text{ cm / sec.}$$

(b) In S. I. system we put R = 8.31 Joule M = 32 × 10⁻³ k.g./mole &

$$Cr.m.s. = \sqrt{\frac{3 \times 8.31 \times 300}{32 \times 10^3}} = 4.8344 \times 10^2 \text{ metre / sec.}$$

Example 1.2 Calculate the potential of the electrode (Pt.) H_2 (1 atom.)/ H^+(a=0.1) at 25°C.

Solution.

$$E_{red}_{H^+/H_2} = \frac{RT}{F} \ln a_{H^+}$$

For this problem traditional system also uses $R = 8.313$ Joule (as 1 Joule = 1 volt-coulomb) to get E in volts. So here both systems use R = 8.313 Joule and

$$E_{red}_{H^+/H_2} = \frac{8.313 \times 298}{96500} \ln 0.1 = -0.0591 \text{ volt.}$$

Chapter 2

Water and its Treatment

INTRODUCTION

Water is one of the abundantly available substances in nature. It is a compound of hydrogen and oxygen. Water is used for industrial and municipal purposes. The largest water requirement is for municipal use but the standard of purity required for this purpose is quite different from that demanded for industrial and commercial use.

The availability of water, both in quantity and quality, is one of the prime factors in deciding the growth of towns and cities as well as industries. For chemical industries, the available water must be as near as possible to the factory site and should also be soft. Otherwise the manufacturing cost will increase.

SOURCES OF WATER SUPPLY

Water is distributed in nature in different forms, some of which are as follows:

Rain Water

Rain water is the purest form of natural occurring water. The water evaporates from sea due to extensive heat. The water vapours thus rising from the surface are drifted by the winds onwards. They rise to high altitudes, where they condense in the form of small droplets. These droplets move in the sky in the form of clouds and go on aggregating continuously till they become heavy enough unable to support their weight when they fall down in the form of rain. Rain water is considered to be very pure being produced by a kind of distillation and is used as distilled water in laboratories. However, it contains dissolved gases such as carbon dioxide, sulphur dioxide, ammonia etc. from the atmosphere.

The rain fall is, however, mainly during monsoon season lasting generally from July to September and a short spell during winter. Water has, therefore, to be conserved in dams or reservoirs to be made available during the rest of the year.

River Water

The rain water in the hilly districts and that from the snows that melt in the mountainous regions flows in the form of rivers. The water is very pure in the hill tracts and takes up suspended impurities as it flows through the plains. During the course water dissolves carbon dioxide from atmosphere which enables it to dissolve carbonates as it passes over the beds.

Spring Water

Some of the rain water percolates underneath the surface of earth, till it reaches an impervious strata which prevents its further penetration. This exudes in the form of springs.

Mineral Water

Water from some springs contains dissolved sulphur compounds. It cures skin diseases, when salts such as magnesium chloride and sulphate are present spring water is known as saline water.

Sea Water

Rivers carry a large volume of water with dissolved impurities into the sea and, as such, the sea contains the maximum amount of dissolved impurities. The total dissolved impurities in sea water are about 3.6 per cent of which 2.6 per cent is common salt; others are magnesium chloride, magnesium sulphate, calcium sulphate, potassium chloride, magnesium bromide and traces of iodides, silica, etc. The quantity of dissolved salts is very much greater in inland seas. Sea water is saline, hence unsuitable for drinking purposes and is hard. Sea water is also faintly alkaline.

Surface Water and Ground Water

As rain falls on the earth, it flows into streams as surface runoff, passes into the air as vapour and goes down into the ground. It is then recovered as surface water or ground water.

The term surface water indicates natural water in rivers, streams, ponds and lakes, although a part of such deposits of water may have percolated more or less through the soil in the course of its flow along the earth's surface. As a result, the water absorbs a part of the materials with which it comes in contact. Surface water, therefore, indicates the chemical constitution and physical conditions of the area in question.

The water obtained from springs and wells in a given locality may be very different from the local surface water. Ground water also indicates the local geological strata and contains a greater percentage of dissolved salts. Ground water is more nearly constant in composition. Some of the ground water may be hot and may contain dissolved gases. It is sometimes used for medicinal baths. Ground water may be different in taste and chemical character—acid, alkaline, bitter, hepatic, ferruginous, iodinous or siliceous.

CHARACTERISTICS OF WATER

Water freezes at 0°C and boils at 100°C at NTP. This is in sharp contrast to the melting and boiling points of its higher analogues, hydrogen sulphide, hydrogen selenide and hydrogen telluride. Moreover, the latter are bad smelling gases at ordinary temperature and pressure. Water also exhibits many other abnormal properties. Compared with hydrogen sulphide, it has a higher surface tension and a higher dielectric constant. Liquid water shows a maximum density at 4°C, unlike all other substances, whereas ice is less dense than water. The specific heat as well as latent heat of fusion and evaporation of water is all abnormally high. All these facts are explained by attributing to water molecules the tendency to associate, giving rise to polymers of the type $(H_2O)_n$ as a result of hydrogen bondings.

Pure water is virtually a non-conductor of electricity. On heating to a very high temperature water molecules are dissociated into the elements. The formation of water from hydrogen and oxygen is an exothermic reaction:

$$H_2 + 1/2\ O_2 = H_2O\ (Steam) \qquad \Delta H = 58,110\ calories.$$

Pure water takes part in a variety of chemical reactions. Pure water at room temperature acts on alkaline metals violently and on alkaline earth metals rapidly with evolution of hydrogen. Magnesium and aluminium amalgams are rapidly attacked. Magnesium, iron, zinc and carbon react with water at high temperature. Iron and lead are also attacked by water in the presence of air. Hard water has less reaction on lead than pure and soft water, as the sulphates and bicarbonates present in hard water form a coating of sulphates and carbonates of lead upon the metal which protect it from further action. Hence, soft water should not be supplied for drinking purposes through lead pipes. A solution containing traces of lead is poisonous to the human system.

Water acts as a catalyst for many physical and chemical changes. Perfectly dry hydrogen and oxygen do not react. Although pure water is a non-conductor, water shows a slight dissociation.

$$2H_2O \rightleftharpoons (H_3O)^+ + (OH)^-$$

At 25°C, the ionic product of water is 1×10^{-14}. Water is an excellent electrolytic solvent, particularly for ionic compounds.

WATER FOR INDUSTRIAL PURPOSES

In industries water is used for various purposes. A large amount of water is used in the industries for cooling purposes. Water used for cooling purpose need not be completely free from the impurities. However, when water is used for solutions, carrying out the reactions, dilution etc. should be reasonably pure. Each industry has its own specifications for water. The specifications of water for some industries are given below:

Water for Boilers

The boiler feed water should be free from soluble salts of calcium and magnesium and should not contain nitrates and organic matter.

Water for cooling system

It should be non-corrosive, and non-scale forming.

Water for beverages

It should not be alkaline.

Water for distilleries

As for as possible, water should be free from micro-organisms and should be pure.

Water for paper mills

It should not contain excess of lime and magnesia and should be free from iron salts.

Water for sugar factories

It should not be rich in sulphates, alkaline carbonates and nitrates. Micro-organisms should be minimum.

Water for dairies and allied industries

Water should be colourless, odourless, tasteless and free from pathogenic bacteria.

water for laundries

It should be soft.

EFFECT OF WATER ON ROCKS AND MINERALS

When water flows through the ground rocks or soil, it comes in contact with various mineral constituents and so gets contaminated because of various physical and chemical changes such as dissolution, hydration or action of dissolved CO_2 etc. For example, some minerals such as $CaSO_4$ (anhydrite), Mg_2SiO_4 (olivine) etc. readily undergo hydration when come in contact with water. As a result, products having increased volume are formed and consequently such mineral bearing rocks undergo disintegration.

$$CaSO_4 + 2H_2O \rightarrow CaSO_4.2H_2O$$
$$\text{(Gypsum)}$$

The resulting gypsum has a volume about 33% more than the volume of the mineral anhydrite, $CaSO_4$. Similarly, Mg_2SiO_4 undergoes hydration according to the reaction.

$$Mg_2SiO_4 + xH_2O \rightarrow Mg_2SiO_4.xH_2O$$
$$\text{(Serpentine)}$$

Mineral constituents of rocks like NaCl (rock salt), gypsum ($CaSO_4.2H_2O$) etc. are readily soluble in water. Dissolved oxygen brings about oxidation as well as hydration reactions. For example:

$$2Fe_3O_4 \xrightarrow[\text{Oxidation}]{1/2\ O_2} 3Fe_2O_3 \xrightarrow[\text{Hydration}]{2H_2O} 3Fe_2O_3.2H_2O$$

$$\text{Magnetite} \qquad\qquad \text{Haematite} \qquad\qquad \text{Limonite}$$

Dissolved CO_2 reacts with insoluble carbonates of Ca, Mg and Fe and forms soluble bicarbonates.

BOILER SCALE AND SLUDGES

We know that a large amount of water is used in boilers for steam generation. The boiler feed water should be as soft as possible and also be free from dissolved gases like oxygen, carbon dioxide and also it should not be alkaline. Therefore it is necessary to treat the boiler feed water suitably otherwise it causes many problems like:

(1) Formation of boiler scales and sludges.

(2) Boiler corrosion.

The formation of boiler scales and their ill effects are discussed in the following section.

In boilers, water is boiled to generate the steam. The steam is continuously withdrawn from the boiler and fresh water is fed into the boiler. The suspended and dissolved matter present in water are left behind in the boiler. With the increase in concentration of these salts, they are separated in the order of their solubility. If the salts thus separated, form hard and sticky coating on the boiler surface it is called *boiler scale*. On the other hand if the deposited material are in the form of soft mud loosely held on the surface, it is called a *sludge*.

Mechanism of Boiler Scale Formation

When temporary hard water is fed into the boiler, due to boiling the calcium and magnesium bicarbonates present decompose and are converted into respective insoluble carbonates. These carbonates deposit on the boiler surface and form the hard scale.

$$Ca(HCO_3)_2 \longrightarrow CaCO_3 \downarrow + H_2O + CO_2 \uparrow$$

$$Mg(HCO_3)_2 \longrightarrow MgCO_3 \downarrow + H_2O + CO_2 \uparrow$$

When hard water is fed into the boiler, on boiling it evaporates leaving behind the suspended matter and dissolved salts. Consequently, the concentration of these materials goes on increasing, ultimately they are separated and form boiler scale or sludge. The solubilities of certain substance, like calcium sulphate decrease at the higher temperature. Hence they are separated and form boiler scale. On the basis of chemical composition, the scales formed in boiler units can be divided into the following four groups:

(1) *Alkali-earth scales:* These scales consist mainly of calcium and magnesium compounds such as $CaCO_3$, $CaSO_4$, $CaSiO_3$, $Ca_3(PO_4)_2$, MgO, $Mg(OH)_2$, $Mg_3(PO_4)_2$ etc. These are further classified as carbonate scale ($CaCO_3$), sulphate scale ($CaSO_4$), phosphate scale [$Ca_3(PO_4)_2$] and other depending upon which of these compounds prevails. If many compounds are present in small amounts, the scale formed is called composite scale.

(2) *Iron oxide scales:* These scales consist of ferric and ferrous compounds such as iron silicates, ferrous phosphate $Fe_3(PO_4)_2$, sodium ferrophosphate ($NaFePO_4$) and iron oxides (Fe_2O_3, Fe_3O_4).

(3) *Copper scales:* Containing considerable amount of copper.

(4) *Silicate scales:* Of various compositions.

The most important property of the scale of any kind is its low thermal conductivity which varies with the nature and porosity of the deposits.

Effects of Boiler Scales and Sludges

Wastage of fuel

The scales formed in the boilers are heat insulators. Due to this, heat supplied to the boiler is not efficiently transferred to water. This results in reduction of the rate of heat transfer. The boiler has to be heated to a higher temperature in order to maintain the supply of the steam and thus more fuel is required.

The boiler may burst

Due to over heating, the boiler metal becomes weak and cannot withstand the pressure developed in the boiler. Also, due to over heating minute cracks are developed in the boiler scale. Water seepages through these cracks are comes in contact with the over heated boiler plates. This is called *bagging*. Chemical reaction takes place between the hot boiler metal and the water in the cracks with the evolution of hydrogen gas. This causes the erosion of the boiler material. The weakened boiler material will not withstand the pressure developed in the boiler and hence the boiler may burst. Large quantities of sludge formed may block the steam outlets. This is called *plugging*. The plugging may result in the bursting of the boiler.

FORMATION OF DEPOSITS IN BOILER UNITS

(1) *Formation of alkali-earth deposits:* These deposits are formed in the presence of sufficiently high concentration of scale forming substances or insoluble substances. If boiler water contains cations such as Ca^{2+} and Mg^{2+}, and anions such as OH^-, CO_3^{2-}, SO_4^{2-}, PO_4^{3-}, or SiO_3^{2-}, then the interaction of these ions give rise to a number of scale forming substances such as $CaCO_3$, $CaSO_4$, $CaSiO_3$, $Ca_3(PO_4)_2$, $Mg(OH)_2$, $Mg_3(PO_4)_2$, $3MgO.2SiO_2.H_2O$, which are not easily soluble.

The rate of formation of alkali earth deposits depends on the concentration of scale forming substances. An increase in the concentration of substances in boiler water is also responsible for the formation of difficult soluble compounds. The solid phase may precipitate or crystallise from a solution and deposit directly on a heating or cooling surface. This process of deposition of hard scales is known as *primary process of scale formation*. In the *secondary process of scale formation*, suspended particles first precipitate in the bulk of water and subsequently form secondary deposits that strongly adhere to the surface.

(2) *Formation of iron oxide deposits:* The formation of iron oxide deposits are most frequently detected in high duty boiler units, in the water wall tubes of saline sections, on the stretches of tubes exposed to high rates of heat flow, at the spots where the burner flame impinges onto heating surface etc. The important components of iron oxide deposits are iron phosphates, depending upon the conditions of scale formation. The cause of such deposits is the introduction of iron into a boiler unit together with feed water. Iron comes into the feed water due to *corrosion processes* taking place in water condensate circuit of electric power stations and industrial boiler houses.

(3) *Formation of copper scales:* Copper scales, which may contain 20–30% copper, also contain iron oxides and impurities such as calcium and magnesium compounds. Copper scales are usually formed in boiler units operated at different pressures and the main factor responsible for their formation is the introduction of copper compounds into boiler units with feedwater. In copper scales, copper is present as metal and metal oxides.

Disadvantages of Scale Formation in Boiler

(1) Scale or sludge is a poor conductor of heat. As a result evaporation capacity of boiler decreases. Hence fuel consumed is much more than usual.

(2) Scale or sludge does not allow the water to come in contact with plates and tubes of the boiler. This causes overheating as well as burning out of these plates and tubes.

(3) Tubes and plates are clogged decreasing the efficiency of the boiler.

(4) Tubes may be corroded by some salts such as $MgCl_2$, if deposited as scales.

Methods of Preventing Deposit Formation

The formation of deposits is usually prevented by using the following measures:

(1) Deep softening of make up water.

(2) Preventing contamination of feed water and its components by the products resulting from the corrosion of structural materials.

(3) By eliminating leakage of cooling water in turbine condensers.

(4) By internal boiler water treatment.

(5) Chemical control of boiler water to meet the required standards.

(6) By eliminating high rates of local heat flow through steam generating tubes.

It is not possible to remove completely the scale forming substances from water by softening. The residual hardness of softened water is generally 3–4 µg-equiv/kg and sometimes more. The boiler water treatment makes the scale forming substances harmless, otherwise they penetrate into the boiler unit. The boiler water treatment thus eliminates their deposition on heating surfaces. For boiler water treatment *orthophosphates* or salts of EDTA are introduced into the boiler units.

PHOSPHATE TREATMENT OF BOILER WATER

Cations such as Ca^{2+} and Mg^{2+} entering a boiler unit are capable of producing scale forming compounds with different anions present in the boiler water. These compounds behave differently in boiler water. For example, some of them as $CaSO_4$, $CaSiO_3$ etc. precipitate from the solution and form scales mainly on the surface of steam generating tubes. Other such compounds as $CaCO_3$, $3Ca_3(PO_4)_2$, $Ca(OH)_2$ precipitate mainly as suspension that turn into sludge and is removed from the boiler unit by blow down.

The object of phosphate treatment is to tie up Ca^{2+} ion so that it becomes harmless, (i.e., to create conditions under which scale formers would precipitate as loose sludge). One way to do this is to form a complex containing Ca^{2+} ions. It should be noted that phosphate treatment prevents only the calcium scale. Triphosphate magnesium $Mg_3(PO_4)_2$ is capable of yielding magnesium scale. Phosphate treatment is based on two methods:

(1) *Phosphate alkaline method* in which a certain concentration of PO_4^{3-} ions and free OH^- ions is continuously maintained and this concentration is not bound by PO_4^{3-} ions.

(2) *Purely phosphate alkalinity method* in which presence of only PO_4^{3-} ions in boiler water is necessary and OH^- ions are bound with PO_4^{3-}. Such conditions are possible only when trisodium phosphate is present in water. Hydrolysis takes place as follows:

$$PO_4^{3-} + H_2O \rightleftarrows HPO_4^{2-} + OH^-$$

If such a solution is concentrated by evaporation, the reaction will proceed to the left. OH^- ions will be bound by HPO_4^{2-} ions and PO_4^{3-} ions are formed. This makes the method of purely phosphate alkalinity highly desirable for boiler units operated at 10.0 MPa or higher. In addition to trisodium phosphate (Na_3PO_4) and hexametaphosphate $(NaPO_3)_6$, tripolyphosphate ($Na_5P_3O_{10}$) has also been used in phosphate treatment.

TREATMENT OF BOILER WATER WITH COMPLEXING AGENTS

The phosphate treatment is not capable of preventing the formation of iron oxide and copper depositions. The conditions of boiler water with complexing agents such as EDTA and disodium salt (trylon B) can secure: (i) scale free and sludge free operation of boiler unit; and (ii) corrosion free operation under certain conditions. The main object of treatment with complexing agents is:

(1) To prevent the deposition of iron oxides in boiler units.
(2) To reduce the carry over of oxides with steam.
(3) To protect the units against steam-water corrosion.

ALKALI TREATMENT OF BOILER WATER

As a result of alkali treatment, the scale forming substances penetrating into the boiler unit precipitate as $CaCO_3$ and $Mg(OH)_2$. In a non-alkaline medium, $CaCO_3$ falls out on heating or cooling surfaces and forms dense scale. With alkaline boiler water, $CaCO_3$ precipitates in the form of a loose sludge not adhering to surfaces. The alkaline treatment of boiler water in boiler units operated at a pressure upto 1.6 MPa is efficient, because stability of CO_3^{2-} ion depends upon temperature, whose increase is followed by decomposition of soda (Na_2CO_3). As a result, the effectiveness of alkaline treatment decreases and steam generated gets contaminated with CO_2. It is to be noted that alkali treatment is useful only when other methods of preventing scale formation prove impracticable.

PREVENTING DEPOSIT FORMATION IN TURBINE CONDENSERS

At electric power stations much water is used as a cooling agent. For the purpose of water economy, closed circuit type cooling water system is used. In this system, cooling water circulates in a closed cycle, i.e., it heats up in the apparatus to be cooled and cools in a supply cooling pond or cooling tower. In such type of water supply systems, various reactions take place in the circulating water. When water is sprayed in a pond or tower, dissolved CO_2 is removed from it. It causes the decomposition of calcium bicarbonate.

$$Ca(HCO_3)_2 \rightarrow CaCO_3 + CO_2 + H_2O$$

This reaction is responsible for the formation of solid carbonate deposits in the tubes of turbine condensers. There is also partial evaporation of water, and thus the concentration of substances in water increases. The formation of carbonate deposits in turbine condensers is prevented by treating make up water in any of the following ways:

(1) *Acidification:* In this process H_2SO_4 is usually introduced into make-up water. It destroys HCO_3^- ions and forms CO_2.

$$HCO_3^- + H^+ \rightarrow H_2O + CO_2$$

Thus carbonate hardness in water decreases and concentration of free CO_2 increases. All this terminates calcium bicarbonate decomposition.

(2) *Lime treatment:* The destruction of HCO_3^- ions can also be carried out by treating water with lime, $Ca(OH)_2$, through

$$HCO_3^- + OH^- \rightarrow CO_3^{2-} + H_2O$$

without formation of CO_2. Lime treatment removes all bicarbonate ions from water.

(3) *Phosphate treatment:* In this process phosphates introduced into water deprive $CaCO_3$ of its scale forming properties.

(4) *Recarbonisation:* Treating water with gases containing CO_2, in particular with the furnace gas from boiler units, is aimed to increase concentration of free CO_2 in water. The process is known as *recarbonisation*, in which CO_2 is blown through water or mixed with water by means of water operated ejector. Thus CO_2 contained in the gas is dissolved in the water. The saturation of water with CO_2 thus prevents the decomposition of bicarbonate.

PRIMING AND FOAMING

The passage of water particles mixed with steam from the boiler is called *priming* which is caused by the presence of large quantity of alkali sulphates and chlorides present in water. Priming is generally associated with *foaming*. When water is associated with suspended fine solid particles, each of these fine particles releases steam bubbles. When these bubbles break, they cause *foam*. The breaking up of bubbles at a certain stage is so quick that the water film enclosing the steam around solid particles, passes out from the boiler along with steam and hence causes *priming*. The dissolved impurities are mainly responsible for priming and foaming and the degree of foaming and priming depends upon the quality and quantity of soluble impurities. As the concentration of water rises in water, the quantity of dissolved impurities increases and hence priming also increases. Priming and foaming can be avoided by: (i) providing sufficient steam space; (ii) removing oil or soap from water; (iii) not allowing accumulation of fine suspended particles in boiler water either by filtration or by efficient softening; and (iv) by blowing off concentrated water for the removal of saline water, at regular intervals. Taking into consideration priming and foaming, water may be graded into three types:

(1) *Non-foaming water:* Which can be used in a locomotive boiler for seven days without any foaming.

(2) *Semi-foaming water:* Which can be used in a locomotive boiler for two days without foaming.

(3) *Foaming water:* Which can not be used for two days without blowing off for preventing foaming.

CAUSTIC EMBRITTLEMENT

This is a form of corrosion which may attack boilers. It is caused by a high concentration of sodium hydroxide, which can react with steels stressed beyond their yield point. This particular form of corrosion is characterised by the formation of irregular intergranular cracks at places of high local stress.

If sodium carbonate is present in water, it is hydrolysed to give sodium hydroxide, which reacts with iron to form magnetic oxide.

Under ordinary conditions in unstressed metal a fairly continuous film of oxide is produced, but when the metal is stressed above its yield point, the oxide coating cracks and the chemical attack continues into the metal. The attack is fairly rapid under these conditions due to the large area of metal exposed. It is considered that the presence of a little sodium sulphate or sodium phosphate inhibit embrittlement either by crystallising out and plugging the capillaries and crevices with solid salts before a dangerously high concentration of sodium hydroxide has been produced.

Hence, by maintenance of sufficient amount of sodium sulphate or sodium phosphate in the boiler water, embrittlement can be prevented. At low boiler pressures, a small amount of sodium sulphate will be sufficient to prevent the corrosion. The use of an excess of sodium carbonate should be avoided in lime-soda process in boiler feed water treatment. This may also prevent caustic embrittlement.

Internal Conditioning of Boiler-Feed Water

Internal conditioning means correcting the water by adding some chemicals to the water in the boiler itself. It is used for supplementing the external treatment such as lime-soda process, demineralisation, deaeration, etc. As regards hardness, the impurities likely to form boiler scales are converted into sludge-forming impurities or they are complexed (sequestered) so that they remain in solution, but yet they are out of harm's way. The compounds used for the internal conditioning of water for boilers are called boiler compounds. Sodium phosphates, tannin and hydrazine are some of the familiar boiler compounds.

The most important step in the internal conditioning of boiler feed water is phosphate treatment. The three phosphates of sodium dihydrogen phosphate (NaH_2PO_4), disodium hydrogen phosphate (Na_2HPO_4) and trisodium phosphate (Na_3PO_4) are used. Any of these phosphates will react with calcium ions to form a sludge of calcium phosphate.

$$3CaCl_2 + 2Na_3PO_4 \rightarrow Ca_3(PO_4)_2 + 6NaCl$$

Depending on the pH of the water to be treated, one of three kinds of phosphate is chosen. If the water is acidic, trisodium phosphate, which on hydrolysis yields alkali, is added. If the water is neutral, disodium hydrogen phosphate is used. The sludge formed is removed by regular blow-down.

Sludge formation can be avoided if calgon, i.e., sodium hexametaphosphate is used. This compound forms a complex with calcium salts in which the calcium forms part of an anion. Hence it no longer forms scales in the boiler.

$$Ca^{2+} + (NaPO_3)_6 \rightarrow Na_2(Ca_2P_6O_{18}) + 4Na^+$$

This complex is water-soluble and hence sludge disposal problem does not arise. A second boiler compound of frequent use of hydrazine. This chemical combines with dissolved oxygen to form water and N_2. Since this treatment does not increase the dissolved salt concentration of the water, it is always preferred to adding ferrous sulphate or sodium sulphite. Tannin is another boiler compound used for the complexing of iron. Some coagulants are also used as boiler additives for removal of oil and silica. Internal conditioning also goes by the name of threshold conditioning because the treatment is done at the gateway to the boiler.

FACTORS PROMOTING CORROSION

The electrochemical corrosion mainly depends upon two factors. These are external factors of corrosion such as nature of the surrounding medium, pH value etc. And internal factors of corrosion such as grades and kind of metal, the presence of stresses, mechanical or heat treatment etc.

Under operating conditions of heat power equipment, the main factor which determines the intensity of corrosion is oxygen dissolved in water. Since oxygen is a depolariser, it contributes to the action of other factors of corrosion effectively. Oxygen corrosion of metal first produces ferrous hydroxide.

$$2Fe + O_2 + 2H_2O \rightarrow 2Fe(OH)_2$$

Corrosion of oxygen in neutral or alkaline medium gives rise to the formation of *pits*, i.e., to sharply pronounced local destruction of metal. At pH < 7, oxygen corrosion usually develops more evenly. Since oxygen comes into water from the atmospheric air, care must be taken to preclude exposure of de-aerated water or condensate to the atmosphere.

The other factor influencing the development of corrosion is the presence of CO_2 in water. With water, it forms carbonic acid, H_2CO_3 that dissociates to form H^+ ions.

$$CO_2 + H_2O \rightleftarrows H_2CO_3 \rightleftarrows H^+ + HCO_3^-$$

The H^+ ions acidify water and cause corrosion attended with hydrogen depolarisation. CO_2 corrosion is most likely to occur in condensate lines, condensate tanks, air heaters and other similar equipments in which internal surfaces are exposed to condensate. The most important characteristic of CO_2 corrosion is that the above reaction does not proceed to completion. Thus H_2CO_3 is also present in water along with H^+ ions and HCO_3^- ions. As neutralisation of H^+ ions on the cathode continues, the equilibrium gets disturbed and again proceeds from left to right. As a result, new H^+ ions are formed and these H^+ ions maintain the corrosion aggressiveness of the solution at the previous level. This process, which makes up for the decrease in H^+ ions, goes on until H_2CO_3 is completely dissociated. When a strong acid such as HCl is added for water acidification, the process of this kind does not occur. Here complete dissociation of HCl is expected to occur and the decrease in H^+ is not compensated by the process of corrosion developed. As a result, concentration of H^+ decrease gradually and hence corrosion slows down. Hence CO_2 solution is more aggressive than HCl solution at the same pH value.

The HCO_3^- ions are bound by Fe^{2+} ions, which originate in the course of *steel corrosion*. Ferrous bicarbonate also decomposes with the formation of $Fe(OH)_2$ and CO_2:

$$Fe(HCO_3)_2 \rightleftarrows Fe(OH)_2 + 2CO_2$$

CO_2 is now able to react again. This process is however, possible when water contains sufficient amount of O_2 required for oxidising Fe^{2+} to Fe^{3+} and eliminating it from the reaction:

$$2Fe(OH)_2 \rightarrow 1/2\ O_2 + H_2O + 2Fe(OH)_3 \downarrow$$

If the corrosion products strongly adhere to the surface of the metal under going corrosion and form a sufficiently dense layer, they may gradually reduce the rate or intensity of corrosion or terminate corrosion altogether.

At electric power stations corrosion occurs in feed lines, steel water economisers, regenerative feed water heaters, heat exchangers and in return condensate lines. The feed line from the deaerator or to the boiler drum is always filled with fresh hot water. When it comes into contact with metal, corrosion takes place if water contains depolarisers. The temperature also increases the rate of corrosion, which also depends upon whether oxygen-containing water is heated in a closed heat exchanger or in a tank open to the atmosphere. In the first case rate of corrosion increases continuously with increase in water temperature. In open system, the rate of corrosion is maximum at 348–358K.

Heaters and evaporators usually suffer corrosion on the side of the heating steam and corrosion is caused by dissolved CO_2 contained in the condensate.

CO_2 finds its way into the steam water circuit of a heat power plant with chemically treated water and due to cooling water leakage in turbine condensers. In both the cases, HCO_3^- and CO_3^{2-} ions are introduced into the system and decomposed in the boiler unit forming CO_2. The most harmful effect of CO_2 corrosion is the contamination of feedwater with iron oxides. This contamination increases with feedwater temperature and time of metal exposure to condensate containing CO_2.

The corrosion of feedwater and steam condensate lines of a power station may be controlled by removing the aggressive gases CO_2 and O_2 from feedwater and condensate. CO_2 corrosion can largely be prevented by carrying out the following measures:

(1) Reducing the amount of free CO_2 in steam by suitable methods of make-up water treatment.

(2) Removing heating steam condensate from regenerative feed water heaters.

(3) Completely venting non-conducting gases from steam space of heaters.

(4) Treating feedwater with ammonium or steam with film forming amines.

CORROSION OF BOILER

Oxygen corrosion occurs in steel water economisers of boiler units. If feedwater is poorly deaerated, steel economisers may fail in 2–3 years after being put in service.

Oxygen corrosion is most commonly encountered in the inlet sections of economisers and iron economisers do not suffer oxygen corrosion. If oxygen concentration in feedwater is sufficiently high, oxygen may pass into the boiler unit too. In boiler units corrosion occurs mainly in drums and down comer tubes. An increase in pressure intensifies oxygen corrosion. The composition of water coming into contact with a metal is also important. Small amount of alkali intensifies local corrosion, while chlorides cause corrosion over the entire surface of metal.

Idling boiler units are subject to *electrochemical corrosion*, which is known as outage or out of service corrosion. The various methods of preventing corrosion when boilers are taken out of service are:

(1) Placing boiler units in dry storage.

(2) Placing boiler units in wet storage.

(3) Coating the internal surfaces of a boiler unit with a film of 10% $NaNO_2$ solution.

(4) Filling the boiler unit with a hydrazine solution.

(5) Connecting the boiler unit to a source of saturated steam at some pressure.

(6) Treating the internal surfaces of a boiler unit with complexing agents.

(7) Preparing for storage by using nitrogen.

Intercrystalline corrosion occurs in reverted and expanded joints of boiler units that are exposed to boiler water. The metal first develops fine cracks which later on grow with time. The cracks run between grains of the metal and hence the corrosion is called *crystalline corrosion*. Disintegration caused by intercrystalline corrosion is also called brittle, stainless. This type of corrosion depends mainly upon the following factors:

(1) High tensile stresses in metal.

(2) Leaks in reverted or expanded joints.

(3) Aggressive properties of boiler water.

Fractures do not occur at all in the absence of any one of the above three factors. Intercrystalline cracks can also develop under local cooling of small sections of a boiler drum.

QUALITY OF WATER

Meaning of Pure Water

The term *pure water* is a relative term and it has to be interpreted in relation to the use of water. The concept of pure water, potable water or wholesome water with relation to various uses of water is understood as follows.

Domestic use

The water required for domestic consumption should posses a high degree of purity and it should be free from suspended impurities, bacteria, etc. A tolerance of small degree of hardness developed due to certain dissolved salts is however permissible. Thus the drinking water and water used in the food industry and some other industries must meet the highest standard of purity. Following are the requirements of *potable* or *wholesome* water for domestic use:

(1) It should be clear, odourless and colourless.

(2) It should be free from harmful and disease producing bacteria.

(3) It should be free from all objectionable substances.

(4) It should be fresh and cool.

(5) It should be palatable i.e., aesthetically attractive.

(6) It should be tasty.

(7) It should not cause corrosion to the pipes and other fittings.

Civic use

For this purpose, a large quantity of water is required to fulfil various civic purposes such as washing of roads, cleaning of sewers, etc. The nature of use of water is such that any degree of impurity can be tolerated. Hence the water containing large amount of suspended and dissolved impurities may be permitted for this purpose. But the water which is considerably mixed up with sewage and other refuse cannot be tolerated for this purpose.

Trade or business use

The water required for a particular trade will depend upon nature of that trade. For instance, the water required for laundry should not be hard as it will result in more consumption of soap. Similarly the water required for bathing cattles and washing floors in case of stables may contain any type of impurities.

Commercial or industrial use

The water required for this purpose should be chemically pure. The various chemical processes involved in the production make it essential to use chemically pure water. A slight amount of impurity may considerably affect the final results of the product.

The nature of water required for various factories is so variable that some of the factories install their own water supply plant to supply water to their industries. Alternatively, a city with well-developed and highly water-consuming industries should have preferably two systems of water supply: (i) supplying potable water for drinking and domestic use; and (ii) providing water for industry. Moscow, Paris and a number of other big cities of the world have now adopted this system in practice and in future, it may become more popular as a means to save the drinking water.

Preparation, Properties and Uses of Water

Preparation

It can be prepared by the following methods: (i) oxidation of hydrogen; (ii) end product of combustion; (iii) end product of acid-base reaction; (iv) end product of condensation reaction. It can be purified by distillation; ion exchange reaction (zeolite); chlorination; filtration.

Properties

Water is a colourless, odourless, tasteless liquid; allotropic forms are ice (solid) and steam (vapour). It is a polar liquid with high dielectric constant (81 at 17°C); this largely accounts for its solvent power. It is a weak electrolyte, ionising as H_3O^+ and OH^-. At atmospheric pressure it has sp. gr. 1.00 (4°C); freezing point 0°C (32°F), and expands about 10% on freezing; viscosity 0.01002 poise (20°C); sp. heat 1 calorie per gram; vapour pressure (100°C) 760 mm Hg. Triple point 273.16°K at 4.6 mm Hg. Surface tension 73 dynes/cm at 20°C. Latent heat of fusion (ice) 80 cal/gram; latent heat of condensation (steam), 540 cal/gram. Wt. per gallon (15°C 8.337 lbs); wt. per cu ft 62.3 lbs. Refractive index 1.333. Water may be superheated by enclosing in an autoclave and increasing pressure; it may be supercooled by adding sodium chloride or other ionising compound. It has definite catalytic activity, especially of metal oxidation. Physiologically water is classed as a nutrient substance.

Uses

It has numerous uses such as suspending agent (paper-making, coal slurries); solvent extraction (extraction, scrubbing); diluent; beer and carbonated beverages; hydration of lime; paper-coatings; textile processing; moderator in nuclear reactors; debarking logs; industrial coolant; filtration; washing and scoring; sulphur mining; hydrolysis; Portland cement; hydraulic system; power source; steam generation; food industry; source of hydrogen by electrolysis and thermochemical decomposition.

Reasons for the Analysis of Water

Following are the reasons or purposes for carrying out the analysis of water:

(1) To ascertain if the supplies maintain the required degree of purity and to find out the extent of any variations which occur.

(2) To ascertain the effect of heavy rainfalls or of long-continued drought on river waters.

(3) To ascertain the quality of the proposed supply to the new consumers.

(4) To decide that the water obtained from some additional source or sources of supply will be pure, wholesome, not too hard and free from the risk of any pollution.

(5) To decide the suitability of water for feeding boilers, hot-water pipes, etc.

(6) To examine the effect of pumping on well waters, especially when wells are situated near the sea or an estuary.

(7) To find out the organisms responsible for the spreading of the water-borne diseases.

(8) To identify the organisms responsible for developing certain effects on water with respect to colour, odour, taste, etc.

(9) To know the characteristics of waters at various depths of deep wells and tube wells.

(10) To know the quality of water submitted to the various purification processes.

(11) To know the quality of water used or proposed to be used for public swimming baths.

(12) To know whether water from a particular supply of water is suitable for specific purpose such as paper making, dyeing, tanning, wool washing, brewing, steam raising, etc.

(13) To study the process of self-purification of streams and rivers.

(14) To suggest the best method of purifying, of softening or of preventing action on mains and supply pipes; etc.

Impurities in Water

It is not possible to find pure water in nature. The rain water as it drops down to the surface of earth absorbs dust and gases from the atmosphere. It is further exposed to the organic matter on the surface of earth and by the time, it reaches the source of water supply, it is found to contain various other impurities also. For the purpose of classification, the impurities present in water may be divided into the following categories: (i) physical impurities; (ii) chemical impurities; and (iii) bacteriological impurities.

Analysis of Water

In order to ascertain the quality of water, it is subjected to the various tests. These tests can be divided into the following categories: (i) physical tests; (ii) chemical tests; and (iii) bacteriological tests.

Physical Tests

Under this category, the tests are carried out to examine water for the following: (i) colour; (ii) taste and odour; (iii) temperature; and (iv) turbidity. Other physical characteristics for which tests are sometimes carried out are density, electrical conductivity, radioactivity and viscosity.

Colour

The pure water is colourless and following are the sources which contribute colour to the water :

(1) Algae metabolism.

(2) End products of degraded organic matter.

(3) Discharge of untreated and partially treated waste-water from various industries like food processing, textile industry, tanneries, paper production, etc.

(4) Divalent species containing iron and manganese; etc.

Taste and odour

The water possesses taste and odour due to various causes and they make the water unpleasant for drinking. The test is carried out by inhaling through two tubes of an osmoscope. One tube is kept in a flask containing diluted water and the other one in a flask containing water to be tested. The taste and odour of water may also be tested by threshold number. In this method, the water to be tested is diluted with odour-free water and the mixture at which odour becomes detectable is determined. It indicates threshold number and other intensities of odour are then worked out. The results of test are greatly affected by the sensitiveness of the observer. For public water supply, the threshold number should not be more than three. If an odour of chlorine or iodoform is found, it should always be recorded. The taste is expressed as brackish, saline, salty, etc. If the taste and odour are suspected to be due to the growth of any kind, the cause may be found out by conducting microscopical and biological examinations.

Temperature

The test for temperature of water has no practical meaning in the sense that it is not possible to give any treatment to control the temperature in any water supply project. The measurement of temperature of water is done with the help of ordinary thermometers. From the study of temperature, the characteristics of water such as density, viscosity, vapour pressure and surface tension can be determined. It also helps in determining the saturation values of solids and gases which can be dissolved in water and also the rates of chemical, biochemical and biological activity.

Turbidity

The colloidal matter present in water interferes with passage of light and thus imparts turbidity to the water. The turbidity in water may also be due to clay and silt particles, discharges of sewage or industrial wastes, presence of large numbers of micro-organisms, etc. and the cloudy appearance developed in water due to turbidity is aesthetically unattractive and it may also be harmful to the consumers. It also disturbs the disinfection process because the solids may partially shield the organisms from the disinfectant.

The measurement of turbidity in the field is done by means of a turbidity rod and it is referred to as the *visual method of turbidity measurement*. For laboratory, the various instruments known as the *turbidimeters* are found out to measure the turbidity of water, the most common being Jackson turbidimeter, Baylis turbidimeter and Nephelometric turbidimeter.

Chemical Tests

Under this category, the tests are carried out to examine water for the following: (i) chlorides; (ii) dissolved gases; (iii) hardness; (iv) hydrogen-ion concentration (pH value); (v) alkalinity; (vi) acidity; (vii) metals and other chemical substances; (viii) nitrogen and its compounds; and (ix) total solids.

Chlorides

The chloride contents, especially of sodium chloride or salt, are worked out for a sample of water. The excess presence of sodium chloride in natural water indicates pollution of water due to sewage, minerals, edible oil mill operations, ice cream plant effluents, chemical industries, sea water intrusion in coastal regions, etc. The water has lower contents of salt than sewage due to the fact that salt consumed in food is excreted by body. For potable water, the highest desirable level of chloride content is 250 mg/litre and its maximum permissible level is 600 mg/litre. The measurement of chloride contents is carried out as follows:

(1) 50 c.c. of sample of water is taken by pipette in a porcelain dish.

(2) Two or three drops of potassium chromate solution are added to the sample of water.)

(3) The chloride contents are then determined by titrating with standard solution of silver nitrate.

The silver reacts first with all chlorides and silver chloride thus formed then reacts with potassium chromate. The silver chromate appears as reddish precipitate and the amount of silver nitrate required to produce such reddish precipitate determines the amount of chlorides present in water. The chemical reactions are as follows :

$$AgNO_3 + NaCl = AgCl + NaNO_3$$
$$2AgCl + K_2Cr_2O_7 = Ag_2Cr_2O_7 + 2KCl$$

The presence of chlorides can corrode and such water cannot be used for boilers because of formation of hydrochloric acid due to presence of magnesium chloride in water. The pH value of sample of water is to be adjusted between 7 and 8 either by adding sulphuric acid or sodium hydroxide solution. Otherwise, the test results are likely to be affected.

Dissolved gases

The water contains various gases from its contact with the atmosphere and ground surfaces. The usual gases are nitrogen, methane, hydrogen sulphide, carbon dioxide and oxygen. The contents of these dissolved gases in a sample of water are suitably worked out.

The nitrogen is not very important. The methane concentration is to be studied for its explosive property. The hydrogen sulphide gives disagreeable odour to the water even if its amount is very small. The carbon dioxide content indicates biological activities, causes corrosion, increases the solubility of many minerals in water and gives taste to the water.

The oxygen in the dissolved state is obtained from atmosphere and pure natural surface water is usually saturated with it. The simple test to determine the amount of dissolved oxygen present in a sample of water is to expose water for 4 hours at a temperature of 27°C with 10% acid solution of potassium permanganate. The quantity of oxygen absorbed can then be calculated. This amount, for potable water, should be about 5 to 10 ppm.

HARDNESS OF WATER

Water which does not easily produce lather with soap is termed as hard water. Soap is manufactured by heating vegetable oils such as palm oil, mahuwa oil or animal fat with caustic soda solution. The complex organic acids such as stearic acid are converted to sodium salt with the liberation of glycerine. Thus, the soap mainly consists of sodium stearate. The soap cleanses by dissolving in the water and loosening the particles of dirt which can be washed off. If the water contains calcium or magnesium salt dissolved in it, that is, it is hard, the soap is converted into calcium stearate which being insoluble separates out without producing lather till all the quantity of calcium or magnesium are removed.

$$2NaSt + CaSO_4 \rightarrow Na_2SO_4 + CaSt_2$$

In this way the valuable stearate group, which loosens the dirt is lost completely. The present day detergent used in laundry work contains a long hydrocarbon alkene of the type C_nH_{2n} where n is between 12 and 20. The alkene is first sulphonated with oleum and then converted into sodium salt. This sodium salt is the detergent.

$$C_{15}H_{30}O + H_2SO_4 = C_{15}H_{31}SO_4H$$

$$C_{15}H_{31}SO_4H + NaOH = C_{15}H_{31}SO_4Na + H_2O$$

They are better than soap inasmuch as, they are not affected by hardness in water.

Hardness is usually expressed in terms of calcium and magnesium salts calculated as calcium carbonate and usually expressed as parts per million parts of water under text. When these constituents are measured by titrating with standard soap solution the hardness is usually termed as soap hardness. The end point is observed when lather is permanent and persists for at least two minutes. The hardness is also measured on Clark's scale, each degree corresponding to one grain of $CaCO_3$ or its equivalent per gallon water. Soft water has less than 10° hardness whereas hard water between 20 to 30°.

Types of Hardness

There are two types of hardness: (i) temporary; and (ii) permanent.

Temporary hardness

Temporary hardness is caused by calcium and magnesium bicarbonates. This can be removed by simply boiling the water to precipitate the insoluble carbonates.

$$Ca(HCO_3) + heat \rightarrow CaCO_3 \downarrow + CO_2 + H_2O$$

$$Mg(HCO_3) + heat \rightarrow MgCO_3 + CO_2 + H_2O$$

It can also be removed by adding lime to precipitate the insoluble carbonates. This method is known as Clark's method.

$$Ca(HCO_3)_2 + Ca(OH)_2 = 2CaCO_2 \downarrow + 2H_2O$$

$$Mg(HCO_3) + Ca(OH)_2 = CaCO_3 \downarrow + MgCO_3 + H_2O$$

Permanent hardness

Permanent hardness is caused by the presence of soluble salt of calcium and magnesium other than bicarbonates such as chlorides and sulphites. This cannot be removed by merely boiling or by adding lime. The term softening is applied to the process whereby we remove or reduce the hardness of water. The principle methods are described below.

Lime Soda Process

In this process, the lime and sodium carbonate or soda ash are used to remove permanent hardness from water. The hardness is brought down to the level of 3 to 4 degrees (Fig. 2.1). Following are the chemical reactions involved in this process:

$$CO_2 + Ca(OH)_2 = CaCO_3 + H_2O \quad \text{...(2.1)}$$

$$Ca(HCO_3)_2 + Ca(OH)_2 = 2CaCO_3 + 2H_2O \quad \text{...(2.2)}$$

$$Mg(HCO_3)_2 + Ca(OH)_2 = CaCO_3 + MgCO_3 + 2H_2O \quad \text{...(2.3)}$$

$$MgSO_4 + Ca(OH)_2 = Mg(OH)_2 + CaSO_4 \quad \text{...(2.4)}$$

$$CaSO_4 + Na_2CO_3 = CaCO_3 + Na_2SO_4 \quad \text{...(2.5)}$$

$$CaCl_2 + Ca(OH)_2 = Ca(OH)_2 + CaCl_2 \quad \text{...(2.6)}$$

$$MgCl_2 + Ca(OH)_2 = Mg(OH)_2 + CaCl_2 \quad \text{...(2.7)}$$

$$CaCl_2 + Na_2CO_3 = CaCO_3 + 2NaCl \quad \text{...(2.8)}$$

$$MgCl_2 + Na_2CO_3 = MgCO_3 + 2NaCl \quad \text{...(2.9)}$$

The compounds calcium carbonate $CaCO_3$ and magnesium hydroxide $Mg(OH)_2$ are insoluble in water and they can therefore be arrested in the sedimentation tanks. The other compounds formed during the chemical reactions are soluble in water and they do not impart the property of hardness to the water.

Equation 2.1 indicates the chemical reaction between lime and carbon dioxide present in water.

Equations 2.2 and 2.3 indicate the removal of temporary hardness by the action of lime on the bicarbonates of calcium and magnesium. Equation 2.4 indicates the chemical reaction between lime and magnesium sulphate. The reaction produces calcium sulphate and hence there is no softening of water as such.

Equation 2.5 indicates the chemical reaction between soda ash and calcium sulphate. Thus calcium sulphate already present in water and also formed by equation 2.4 is removed by this chemical reaction.

Equations 2.6 and 2.7 indicate the chemical reactions between lime and chlorides of calcium and magnesium.

Equation 2.8 and 2.9 indicate the chemical reactions between soda ash and chlorides of calcium and magnesium.

Equipments required: Following units of equipment are required for the working of lime-soda process:

(1) *Feeding and mixing apparatus:* The feeding and mixing of lime and soda ash with water should be done with precision. The lime used in this process may be quick lime or hydrated lime. The former is preferable for large plants as it is cheap and less bulky. The required quantities of both these chemicals should be accurately worked out either volumetrically or gravimetrically.

(2) *Settling tank:* This tank is similar to the coagulation sedimentation tank as adopted in case of rapid sand filters. But in this case, the detention period of tank is longer.

(3) *Recarbonation plant:* It is necessary to remove calcium carbonate formed in this process. Otherwise it will precipitate in sand filters and also cause incrustation in pipes. For this reason, the water is allowed to pass through a recarbonation plant after it has passed through the settling tank. In the settling tank, a dose of alum may be given which will produce carbon dioxide. This carbon dioxide reacts with calcium carbonate in the following way and helps to hold it in solution as bicarbonate:

$$CaCO_3 + CO_2 + H_2O = Ca(HCO_3)_2$$

Alternatively, the carbon dioxide gas may be diffused or blown into water in the recarbonation plant.

(4) *Filters:* This unit is required to finish up the treatment. The filters may be of ordinary type rapid sand filter or pressure filter. If the stage of recarbonation does not proceed filtration, it becomes essential to wash the filters frequently and it also results in the frequent replacement of sand bed.

Lime also removes mineral acids, carbon dioxide, dissolved iron and aluminium salts. Lime soda process is employed in two ways: (i) continuous cold lime soda process; and (ii) hot lime soda process.

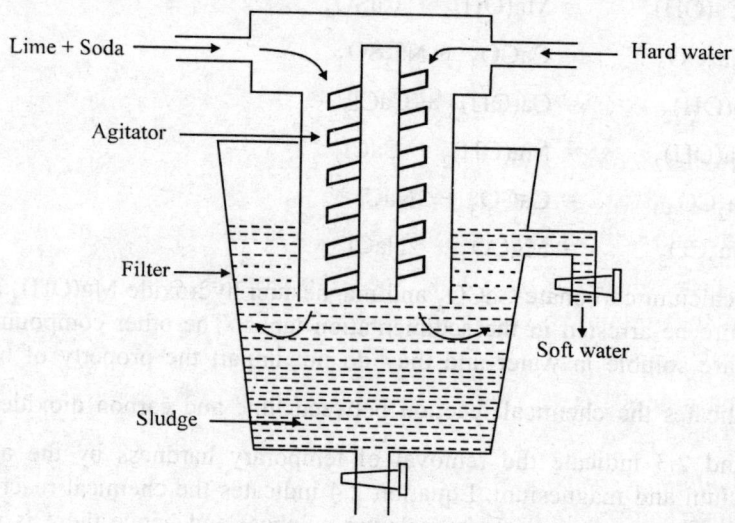

Fig. 2.2. Continuous cold lime soda process.

Continuous cold lime soda process

Hard water is treated with lime and soda ash at room temperature in the treatment tank. The treatment tank is provided with a mechanical agitator. Hard water mixed with the chemicals is agitated thoroughly. Lime and soda ash react with the calcium and magnesium salts present in the hard water. The respective carbonates are precipitated. Coagulating agents like potash alum $Al_2 (SO_4)_3.K_2SO_4.24H_2O$ are added. The suspended particles are coagulated and settle to the bottom of the treatment tank along with the insoluble carbonates of calcium and magnesium. Soft water passes up through the filters and drawn out from the out let.

Hot lime soda process

Hard water lime and soda ash are continuously fed into the treatment tank through which steam is circulated. Lime and soda ash react with the calcium and magnesium salts present in the hard water precipitating respective carbonates. In this process, it is not necessary to add the coagulating agent because at higher temperature, precipitates and suspended particles coagulate and settle to the bottom. Soft water moves upwards and continuously withdrawn. The sludge is removed from a separate outlet periodically (Fig. 2.2).

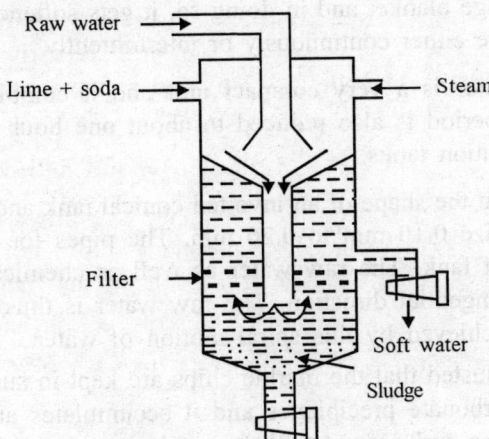

Fig. 2.2. Hot lime soda process.

Advantages of Hot Lime Soda Process

At higher temperature dissolved gases like oxygen and carbon dioxide are expelled off. Since the precipitates coagulate and settle at higher temperature, it is not necessary to add the coagulating agents.

The process is faster than the cold lime soda process. Filtration is easier in hot condition. The hardness is reduced to a minimum level.

Methods adopted in lime-soda process: Following are some of the methods which are employed in the process of lime-soda process: (i) addition of aluminium compounds; (ii) excess lime treatment; (iii) recarbonation; (iv) split method; and (iv) sludge blanket filtration.

(1) *Addition of aluminium compounds:* In this method, the aluminium compounds are added in water in addition to lime and soda ash. The aluminium compounds coagulate finely divided particles. They also convert soluble magnesium salts into insoluble magnesium aluminates which are then arrested and removed in the settling tanks or filters.

(2) *Excess lime treatment:* In this method, slightly more lime is added than necessary. The extra lime completely precipitates magnesium. The soda ash then neutralises the excess lime and it converts all the alkalinity to the sodium alkalinity.

(3) *Recarbonation:* In this method, the raw water is treated with excess lime and the excess lime is then neutralised by the action of carbon dioxide.

(4) *Split method:* In this method, a part of raw water is given excessive treatment and its degree of hardness is lowered to the maximum possible extent. This excessively treated water is then added to the raw water to bring down the hardness of raw water. With the help of this method, it is possible to save a considerable quantity of chemicals.

(5) *Sludge blanket filtration:* In this method, the advantage is taken of the sludge collected at the bottom of coagulation sedimentation tank. The water is softened by filtration through this blanket of sludge. The various devices making use of this process are found out. Following are two such devices.

(a) *Accelerator:* The raw water is mixed with chemicals in the central portion of the tank. The sludge which is previously formed is not allowed to settle down. But it is kept suspension by a mixing fan attached to the bottom of a vertical rotating shaft.

Thus the lower portion of tank is occupied by gently moving sludge. The raw water along with the chemicals rises through this sludge blanket and in doing so, it gets softened. The removal of sludge can be done by sludge removal valve either continuously or intermittently.

The water softening accelerator is a very compact unit and it combines flocculation, settling and sludge removal. The detention period is also reduced to about one hour as compared to 3 or 4 hours in case of coagulation sedimentation tanks.

(b) *Spiractor:* The tank is in the shape of an inverted conical tank and one-half of its depth is filled with marble chips of size 0.10 mm to 0.20 mm. The pipes for raw water and chemicals are placed at the bottom of tank. The raw water as well as chemicals are forced under pressure through nozzles in a tangential direction. The raw water is thus forced upwards spirally. The softening of water is achieved by this spiral motion of water.

The velocity of flow is so adjusted that the marble chips are kept in suspension only and not carried away by water. The calcium carbonate precipitates and it accumulates around the marble chips. This increases the size of marble chips and consequently the volume occupied by them also increases.

The level of marble chips is maintained by periodical removal of marble chips from the bottom and by adding new chips from the top. The marble chips which are removed can be ground, washed and used again. Following are the *advantages* of this device:

(1) The detention period for the tank is only about 5 to 10 minutes.

(2) The wet sludge is not formed. But instead of it, the crystalline and hard granules are formed which are easy to handle.

(3) The tank combines flocculation, settling and sludge removal in one single unit.

Advantages of lime-soda process

Following are the *advantages* of the lime-soda process:

(1) The pH value of water treated by this process is increased which results in decrease in corrosion of the distribution system.

(2) When this process is adopted, less quantity of coagulant will be required.

(3) There is removal of iron and manganese also to some extent.

(4) There is reduction of total mineral content of water.

(5) There is likelihood of killing of pathogenic bacteria. This occurs when causticity caused by calcium hydroxide or sodium hydroxide of 20 to 50 p.p.m. is retained in the treated water for a period of about 4 to 5 hours.

(6) The process is economical.

(7) This process is most suitable for turbid and acidic waters where it will not be possible to adopt zeolite process.

(8) The whole process is easy and simple and it can be accommodated in the existing filter plant of any water supply scheme.

Disadvantages of lime-soda process

Following are the disadvantages of the lime-soda process:

(1) The large quantity of sludge formed during this process is to be disposed off by some suitable method. It can either be discharged directly into river or stream or municipal sewers or it can be used for raising the level of low-lying areas.

(2) The process requires skilled supervision for its successful working.

(3) If recarbonation is omitted, a thick layer of calcium carbonate will be deposited in the filtering media, distribution pipes, etc.

(4) The calcium carbonate is slightly soluble in water to the extent of about 30 mg/l and hence this process can remove hardness upto about 50 mg/l only. Thus the water of zero hardness cannot be produced by this process. But as water of zero hardness is not required for public water supply, this disadvantage does not prove to be serious.

Zeolite Process

This is also known as the *base-exchange* or *ion-exchange* process. The zeolites are compounds of aluminium, silica and soda. They have got the excellent property of interchanging base. They may be obtained from nature or may be prepared synthetically. The naturally available zeolite is green in colour and it is therefore known as the *green sand*. The exchange value of green sand is 7000 to 9000 gm of hardness per m^3 of zeolite (Fig. 2.3).

In this method certain mineral constituents of the soil i.e. hydrated silicates, possessed the property of base-exchange. Thus, in preparing base-exchange materials from solutions of sodium silicate and aluminium sulphate, the reversibility of the process could not be ascertained.

Thus, new methods were adopted in preparing a synthetic hydrated silicate of aluminium and sodium, known as the *Permutit*. Since then the process has steadily advanced and at present, several groups of materials of different character and manufacture having base-exchange properties are available.

The most common artificial zeolite is the *Permutit*. It is manufactured from felspar, kaolin clay and soda. These chemicals are mixed in the required proportion and then the mixture is fused in a furnace. It is allowed to cool after attaining a certain degree of fusing.

The material thus formed is then crushed to form particles of diameter varies from 0.25 mm to 0.50 mm. It is then washed to remove all alkalies and impurities. The Permutit is white in colour and its chemical formula is $2SiO_2Al_2O_3.Na_2O$. The chemical composition is as follows:

Alumina		22.00%
Silica		46.00%
Sodium oxide		13.60%
Water		18.40%
	Total	100.00%

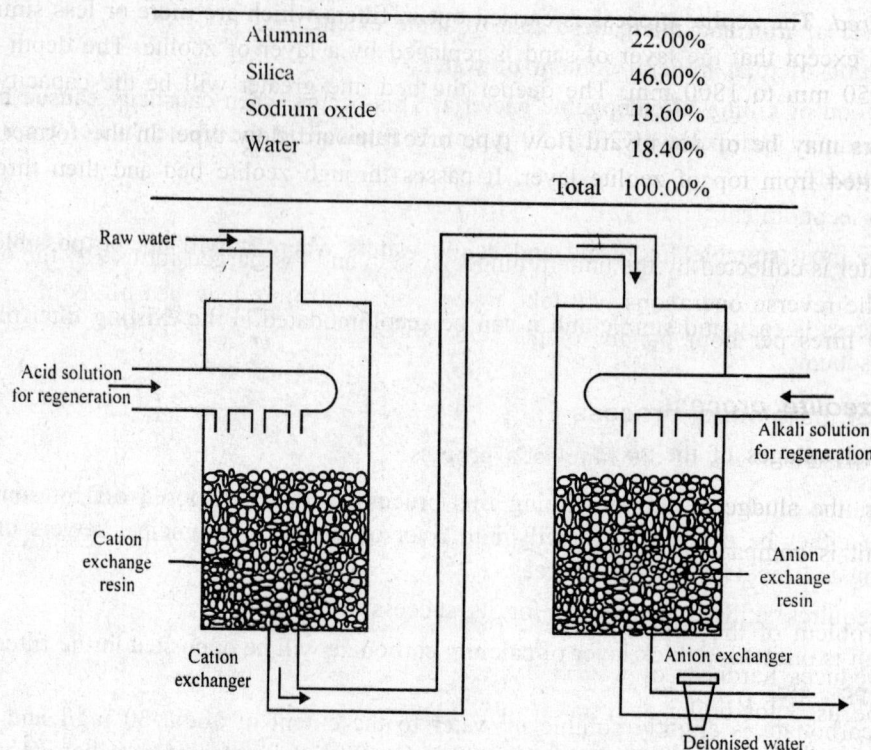

Fig. 2.3. Flow diagram of ion exchange process.

The exchange value of Permutit is 35000 to 41000 gm of hardness per m^3 of zeolite. The chemical reactions involved in the process when permutit is used are as follows:

$$2SiO_2Al_2O_3Na_2O + Ca(HCO_3)_2 = 2SiO_2Al_2O_3CaO + 2NaHCO_3$$

$$2SiO_2Al_2O_3Na_2O + CaSO_4 = 2SiO_2Al_2O_3CaO + Na_2SO_4$$

$$2SiO_2Al_2O_3Na_2O + CaCl_2 = 2SiO_2Al_2O_3CaO + 2NaCl$$

The above reactions are also true for the compounds of magnesium. Thus, when hard water comes into contact with zeolite, the calcium and magnesium are removed and sodium is given in exchange. Thus the hard water is softened and its sodium content is increased as indicated by the above equations.

After some interval of time, the sodium present in zeolite is exhausted. It is very easily regenerated by passing a solution of salt through the zeolite. The chemical reactions take place as follows:

$$2SiO_2Al_2O_3CaO + 2NaCl = 2SiO_2AlO_3Na_2O + CaCl_2$$

$$2SiO_2Al_2O_3MgO + 2NaCl = 2SiO_2Al_2O_3Na_2O + MgCl_2$$

Thus an exchange occurs again and salt of sodium is taken by zeolite. This process is known as the process of regeneration and in order to achieve the results as shown by above equations, the salt solution of required strength is prepared and it is allowed to trickle down over the bed of zeolite for a period of about 8 to 10 hours. The bed is then washed to remove calcium chloride or magnesium chloride for a period of about half an hour or so. The filter bed is then ready for use again.

Equipment required: The zeolite process is carried out in filters which are more or less similar to the pressure type filters except that the layer of sand is replaced by a layer of zeolite. The depth of zeolite layer varies from 750 mm to 1800 mm. The deeper the bed, the greater will be the capacity of filter.

The zeolite filters may be of downward flow type or of upward flow type. In the former case, the hard water is admitted from top of zeolite layer. It passes through zeolite bed and then through base material of gravel.

The softened water is collected by the underdrainage system and discharged for use by the outlet pipe. In the latter case, the reverse operations will take place. For a 750 mm deep bed of zeolite, the rate of flow is about 6000 litres per hour per m^2 of the filter area.

Advantages of zeolite process

Following are the advantages of the zeolite process.

(1) In this process, the sludge is not formed. Hence the problem of sludge disposal does not arise.

(2) The zeolite unit is compact in design. It can be easily operated and does not require any skilled supervision.

(3) There is no problem of the deposition of a layer of calcium carbonate in the distribution system.

(4) This process reduces hardness of water to zero. This advantage has got particular importance for the water to be used for boiler and certain textile industries.

(5) The water of any desired hardness can be prepared by adding softened water of zero hardness to the raw water.

(6) The process is completely automatic and highly skilled labour is not required.

(7) The process proves to be economical where salt is cheaply available.

(8) The load on zeolite process can be reduced by combining it with lime or aeration process. In such a case, the addition of lime or aeration procedes zeolite process. In aerator lime zeolite system, the water is aerated first and it reaches to zeolite softener unit after passing through lime softener and filter.

Disadvantages of zeolite process

Following are the disadvantages of the zeolite process:

(1) The zeolite process cannot be adopted for highly turbid water. The suspended particles get deposited around zeolite particles and hence they cause obstruction to the working of zeolite.

(2) The zeolite process is unsuitable for water containing iron and manganese. This is due to the fact that iron zeolite or manganese zeolite formed during chemical reaction cannot be reconverted into sodium zeolite. The zeolite will thus be wasted in such a case.

(3) The zeolite unit should be carefully operated to avoid injury or damage to the equipment, quality of water and bed of zeolite.

(4) When the ion-exchange capacity of the zeolite has been exhausted, it has to be regenerated and at this time, it gives out a concentrated waste stream of the original contaminant.

Table 2.1 shows the comparison of lime-soda and zeolite processes of softening water supplies.

Table 2.1. Comparison between lime-soda and zeolite processes of softening water supplies.

Item	Lime-soda process	Zeolite process
Economy	Economical	Costly
Effect on bacteria	Can remove pathogenic bacteria	No effect on bacteria
Hardness of treated water	Can produce water upto hardness not less than 50 mg/l.	Water of zero hardness can be obtained
Hardness which can be treated	Excessively hard waters can be treated	Raw waters having hardness more than 800 mg/l cannot be treated
pH of treated water	It increases the pH value of treated water	It does not affect the pH value of treated water
Post treatment	Recarbonation is necessary after sedimentation and filtration	No such treatment is necessary
Removal of colour due to iron and manganese	Can remove such colour to a very small extent	Can remove such colour but proves very costly.
Size of plant	Bulky and large	Compact and small
Skilled supervision	Necessary	Not necessary
Sludge formation	Large quantity of sludge is formed.	No sludge is formed
Turbidity	Highly turbid and acidic waters can be treated.	Difficult to treat highly turbid waters
Use	Public water supplies	Mostly for industrial water supplies

Demineralisation Process

This process is also known as the deionisation process. It is similar to the previous process. But in this process, the hydrogen is exchanged for metallic ions.

The hard water is passed through a bed of resin or carbonaceous material in the hydrogen form. The chemical composition of such materials may be expressed as H_2Y where H represents hydrogen ions and Y represents the organic part of the substance. The chemical reactions involved in the process are as follows:

$$H_2Y + Ca(HCO_3)_2 = CaY + 2CO_2 + 2H_2O$$
$$H_2Y + CaSO_4 = CaY + H_2SO_4$$
$$H_2Y + CaCl_2 = CaY + 2HCl$$

The above reactions are also true for the compounds of magnesium. The treated water contains diluted carbonic acid, sulphuric acid or hydrochloric acid. These acids can be removed by mixing the treated water with the required proportion of alkaline water. After some interval of time, the hydrogen content of the substance is exhausted. It is then regenerated by passing a solution of suitable strength of sulphuric acid or hydrochloric acid. The chemical reactions take place as follows:

$$CaY + H_2SO_4 = H_2Y + CaSO_4$$
$$CaY + 2HCl = H_2Y + CaCl_2$$

The equipment used in this process should be capable of resisting acidity and alkalinity. The demineralisation process is mainly used for preparing water to be used for the industrial purposes.

COAGULATION OF WATER

Purpose

The source of water supply for the most of public water supply projects is surface water. This water is turbid and contains many suspended impurities. It also possesses colour which may be due to colloidal matter and dissolved organic material in water. The turbidity is mainly due to the presence of very fine particles of clay, silt and organic matter.

All these impurities are in a finely divided state and it is not possible to detain them in plain sedimentation tanks unless such tanks are designed for longer detention periods. The other alternative to remove such particles is to increase their size so that they become settle down. The purpose of coagulation is thus to make particles of bigger size by adding certain chemicals known as the *coagulants* to the water. The coagulants react with the impurities in water and convert them in settleable size.

The coagulation is to be adopted when turbidity of water exceeds about 40 p.p.m. It should however be remembered that it is not a complete process by itself. It simply assists plain sedimentation and it is to be followed by the process of filtration. Thus the coagulation is merely a process by which impure water is prepared for successful purification by rapid sand filtration.

Principle of Coagulation

The principle of coagulation can be explained from the following two considerations:

(1) *Floc formation:* When coagulants are dissolved in water and thoroughly mixed with it, they produce a thick gelatinous precipitate. This precipitate is known as the *floc* and this floc has got the property of arresting the suspended impurities in water during its downward travel towards the bottom of tank.

(2) *Electric charges:* The ions of floc are found to possess positive electric charge. Hence they will attract the negatively charged colloidal particles of clay and thus they cause the removal of such particles from water.

Flocculation

The floc produced by the action of coagulants with water is heavy and hence it starts to settle down at the bottom of tank. As it descends, it absorbs and catches more and more suspended impurities present in water. It thus goes slowly on increasing in size. During this process, some amount of bacterial removal also takes place.

The surface of floc is sufficiently wide to arrest colloidal and organic matter present in water. The term *flocculation* is used to denote the process of floc formation and thus the flocculation follows the addition of coagulant and its efficiency depends on the following factors:

(1) *Dosage of coagulant:* The dosage or quantity of coagulant should be carefully determined so as to cause visible floc. The quantity of coagulants should be such that the turbidity of water is brought down to the limit of 10 to 25 p.p.m.

(2) *Feeding:* The feeding of coagulants may be in powder form or in solution form, the latter being more popular.

(3) *Mixing:* The coagulants should be properly mixed with water so as to cause a uniform mass. In the beginning, the mixing may be quick for a period of about 30 to 60 seconds or so.

(4) *pH value:* Depending upon the quality of water and coagulant adopted, suitable pH value should be determined. The pH value should be actually tested in the laboratory at regular intervals. To remove acidity, the lime is added to the water and to remove alkalinity, the sulphuric acid is added to the water.

(5) *Velocity:* The floc should be allowed to more gently after initial quick mixing. The gentle movement of floc results in collision of particles and ultimately, the floc grows in size. The detention period of coagulated sedimentation tanks is about 3 to 4 hours.

The processes of coagulation and flocculation are greatly influenced by the physical characteristics of water, its dissolved constituents and the temperature. The failures in coagulation plant are due to the incorrect dose of the coagulant, inadequate mixing arrangements, improper tank design, etc. Hence the characteristics of water to be submitted to the coagulation plant should be properly studied before deciding the details of the plant.

Usual Coagulants

Following are the usual coagulants which are adopted for coagulation: (i) aluminium sulphate; (ii) chlorinated copperas; (iii) ferrous sulphate and lime; (iv) magnesium carbonate; (v) polyelectrolyte; and (vi) sodium aluminate.

FILTRATION OF WATER

The sedimentation tanks remove a large percentage of the suspended solids and organic matter present in raw water. The process of coagulation of water further assists in the removal of impurities present in the water. But even then, the resultant water is not pure and may contain some very fine suspended particles, bacteria, etc.

In order to remove or to reduce the contents of impurities still further, the water is filtered through the beds of fine granular materials like sand. The process of passing the water through the beds of such granular materials is known as the *filtration*.

The filtered water is potable and palatable and it is free from various undesirable impurities like colour, odour, turbidity, pathogenic bacteria, etc.

Theory of Filtration

The process of filtration forms the most important stage in the purification of water. It usually consists in allowing water to pass through a thick layer of sand. It has been noticed from experience that during the process of filtration, the following effects occur on water:

(1) The suspended and colloidal impurities which are present in water in a finely divided state are removed to a great extent.

(2) The chemical characteristics of water are altered.

(3) The number of bacteria present in water is also considerably reduced.

The theory of filtration to explain why such effects take place is based on the following actions:

(1) Mechanical straining.
(2) Sedimentation.
(3) Biological metabolism.
(4) Electrolytic changes.

Types of filters

(1) Slow sand filters.
(2) Rapid sand filters (gravity type).
(3) Pressure filters..

Slow sand filters

Purpose: In case of slow sand filtration, the water is allowed to pass slowly through a layer of sand placed above the base material and thus the purification process aims at simultaneously improving the biological, chemical and physical characteristics of water. The slow sand filtration is very well suited for rural areas in developing countries because of its simple operation and maintenance procedures. It thus provides safe drinking water at low recurrent cost.

Rapid sand filters (Gravity type)

Purpose: The great *disadvantage* of a slow sand filter is that it requires considerable space for its installation. This requirement makes it uneconomical for places where land values are high.

This difficulty of requiring more space for slow sand filters led the engineers and scientists to find out means to increase the rate of filtration. It was observed that the rate of filtration can be increased in two ways:

(1) By increasing the size of sand so that friction to the water passing through filter media is minimised.
(2) By allowing the water to pass under pressure through the filter media.

The former is achieved in rapid sand filters (gravity type) and it is the most popular method of filtration for public water supply projects. The latter principle is adopted in the working of pressure filters.

Pressure filters

These filters are more or less similar to the rapid sand filters (gravity type). The term pressure filter does not indicate that the water is pumped through the filter under a high pressure loss. But it indicates that a filter is enclosed in space and the water passes under pressure greater than atmospheric pressure. This pressure can be developed by pumping and it may vary from 0.3 to 0.7 N/mm^2.

DISINFECTION OF WATER

When water leaves the filter plant, it is still found to contain some of the impurities. These impurities can be grouped as: (i) bacteria; (ii) dissolved inorganic salts; (iii) colour, odour and taste; and (iv) iron and manganese.

The water should be disinfected before it enters the distribution system. The main purpose of disinfection is to prevent contamination of water during its transit from the treatment plant to the place of its consumption.

The process of disinfection should not be compared with that of sterilisation of water. The destructive effect of disinfection is restricted only to the removal of harmful bacteria. On the other hand, the sterilisation aims at the removal of all sorts of bacteria, whether harmful or harmless to the health.

The materials or substances which are to be used for disinfection are called the disinfectants and the requirements of a good disinfectant are as follows:

(1) Its dose should be such that some residual concentration is obtained to grant protection against contamination in the water during its conveyance and retention.

(2) It should be effective in killing all the harmful pathogenic organisms from the water and make it perfectly safe for use.

(3) It should be harmless, unobjectionable, economical and easily available.

(4) It should be of such a nature that its strength or concentration in the treated water can be quickly determined.

(5) It should not require skilled labour and costly equipment for its application.

(6) It should take only reasonable time in killing the harmful pathogenic organisms at normal temperature.

It has been universally recognised that the chlorine is an ideal material for the disinfection for treating water on a large scale. The disinfection at present therefore is mainly carried out by chlorination. However, there are other minor methods of disinfection.

Minor Methods of Disinfection

Following are the seven minor methods of disinfection: (i) boiling method; (ii) excess lime treatment; (iii) iodine and bromine treatment; (iv) ozone treatment; (v) potassium permanganate treatment; (vi) silver treatment; and (vii) ultra-violet ray treatment. Following factors must be considered in applying the disinfection agents:

(1) concentration and type of chemical agent.

(2) contact time.

(3) intensity and nature of physical agent.

(4) nature of suspending liquid.

(5) number of types of organisms.

(6) temperature.

Chlorination

In this treatment for disinfection, the chlorine (and its compounds) is used as the disinfecting material. For treatment on large scale, the chlorination is invariable used as treatment for disinfection because of the following facts:

(1) It is easy to apply due to relatively high solubility of about 7000 mg per litre.

(2) It is readily available as gas, liquid or powder.

(3) It is very toxic to most of the micro-organisms and thus metabolic activities are stopped.

(4) It leaves harmless residue in solution, but it provides protection in the distribution system.

(5) It produces desired effects which last for a long time.

(6) The treatment by chlorination is cheap and reliable.

However, chlorine is a poisonous gas which requires careful handling and it may also give rise to the problem of taste and odour in water. Some advantages of chlorination is given as under :

(1) It accomplishes greater bacterial purification in minutes than storage achieves in an equal number of days. It thus avoids the construction of costly storage reservoirs.

(2) It is inexpensive and avoids, wholly or in part, the necessity for raw water storage.

(3) It provides extra security to the water against water-borne diseases.

(4) It serves as a convenient accessory to the process of filtration.

Application of chlorine

Chlorine can be applied in water in one of the following ways:

(1) As bleaching powder.

(2) As chloramines and/or.

(3) As free chlorine gas.

WATER SOFTENING

The water to be supplied to the public should not be very hard though there is fear of no health hazard, but it is undesirable as it leads to the following economic disadvantages:

(1) It affects the working of dyeing system and leads to the modification of some of the colours.

(2) It causes corrosion and incrustation of pipes and plumbing fixtures.

(3) It causes more consumption of soap in laundry work and hence proves to be uneconomical for washing processes of textile industries.

(4) It increases the fuel costs.

(5) It makes food tasteless, tough or rubbery.

(6) It provides scales on the boilers and other hot-water heating systems.

It is necessary to understand why soap does not form lather with hard water. The ordinary soap consists mainly of sodium or potassium salts of some of the organic acids which are present in fats and fatty oils. Let us suppose that soap is made from sodium and palmitic acid. It will then be the compound sodium palmitate ($C_{15}H_{31}COONa$). When soap is treated with soft water, it is hydrolysed into caustic soda and palmitic acid. The acid thus liberated unites with a second molecule of the sodium palmitate (soap) and it forms an insoluble substance which, with water, produces the lather. Thus,

Sodium palmitate + H_2O ⇄ NaOH + Palmitic acid

    (soap)

Palmitic acid

    +   + Water = Lather

Sodium palmitate

When soap is treated with hard water, the soluble Ca and Mg salts, present in hard water, react with sodium palmitate forming insoluble Ca and Mg palmitates and they appear as a white precipitate. This process of precipitation and consequent wastage of soap will go on till the whole amount of soluble Ca and Mg salts present in hard water is converted into insoluble Ca and Mg palmitates. The water now becomes soft and will easily form lather with soap.

The hard water is therefore to be softened by some suitable method before it is supplied for consumption. For potable water, it is desirable to bring down the hardness of water to about 5 to 8 degrees. The water having hardness of about 5 degrees is reasonably soft water. But it is found to be tasteless. The water having hardness more than 8 degrees gives undesirable effects as mentioned above. Following are the *advantages* of soft water:

(1) It improves the taste of foods.

(2) It increases the life of textiles which are frequently sent to the laundries.

(3) It leads to overall cleanliness because of the fact that personal washing and domestic cleansing are much more efficient and less laborious with soft water than with hard water.

(4) It makes washing and cleansing easy.

(5) It restricts scale formation and subsequent loss of heat in boilers, hot water pipes, etc. and therefore the economy is achieved in fuel consumption and saving of labour in descaling the affected surfaces.

(6) It results in saving of labour, soap and other detergents.

(7) It undoubtedly proves to be a sound economic proposition to soften public water supplies which are hard in character.

The hard water however possesses the following *advantages* over the soft water.

(1) It is not desirable to have very soft water for the purpose of drinking because some quantities of calcium salts are required for the growth of the human body, especially in the case of children.

(2) The soft water dissolves lead much more readily than the hard water and hence there is less chance of lead poisoning with hard water than with soft water.

DESALINATION OF WATER

If the water contains 1000 to 35,000 mg/lit of dissolved salts, it is called brakish water. Sea water contains on an average 3.5% of dissolved salts. Brakish water has a peculiar salty taste and it is unfit for drinking purposes. If the water contains more dissolved salts than sea water, it is called brine. Desalination is the process of partial or complete removal of the dissolved salts from the sea water and brakish water. Following are some methods of desalination.

1. *Flash evaporation:* It is a process of separating fresh water from the sea water. The sea water is subjected to the multistep flash evaporation. Flash evaporators consists of about 50 units. Each unit is maintained at a slightly lower pressure and temperature than the previous unit. Saline water undergoes is passed through these units and a portion of water undergoes flash distillation, and the distillate is condensed. The residual water is heated under low pressure and subjected to flash distillation. As the saline water as moves from one unit to the other, it is progressively desalinated, the residual water carries the salts removed as brine.

2. *Freezing method:* When the saline water is cooled, ice crystals are formed leaving the salt in the mother liquor. The freezing point of water depends upon the concentration of the salt. The ice crystals thus formed are coated with brine solution. The ice crystals from ice brine slurry are

separated by filtration. The ice crystals are then washed with fresh water to remove the contaminated brine. The ice crystals are then melted by the heat withdrawn from the incoming saline water.

Reverse osmosis: When two solutions of different concentrations are separated by a semi-permeable membrane, the solvent (water) flows from the dilute solution to the concentrated solution till the concentrations of two solution become equal. This process is known as osmosis. This flow of solvent from dilute solution to concentrated solution is due to osmotic pressure. If a hydrostatic pressure in excess of osmotic pressure is applied on the concentrated solution, the flow of solvent is reversed. i.e., the solvent moves from the concentrated side to the dilute side across the membrane. This process is known as reverse osmosis. The process of reverse osmosis can be used to separate pure water from the saline water (Fig. 2.4).

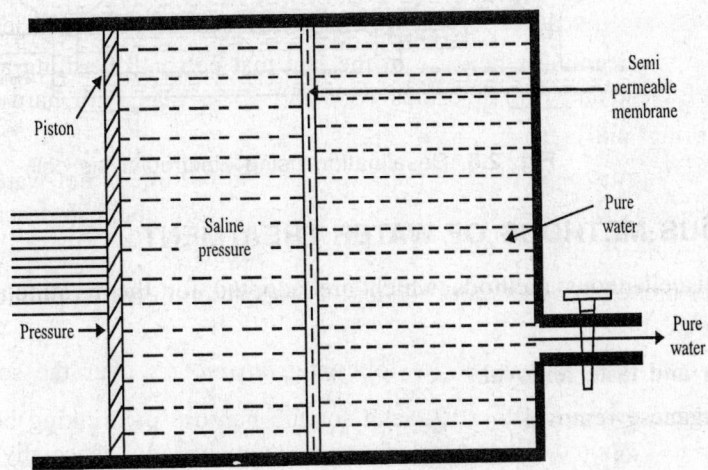

Fig. 2.4. Reverse osmosis.

Saline water is separated from pure water by a membrane which is permeable to water and not to salt. A pressure of about 15 to 40 kg cm^2 is applied on to the saline solution. Pure water from the saline water passes through the membrane due to reverse osmosis, leaving behind the salts. By repeating the process, the water is desalinated.

By this process ionisable, non-ionisable, colloidal and high molecular weight organic compounds present in the water can be separated. The process has low maintenance cost and hence it is more economical.

4. *Electrodialysis process:* The process of migration of ions present in the solution towards the oppositely charged electrodes under the influence of applied electric field is known as electrodialysis. This principle is used to drive away the dissolved minerals present in the saline water. For the electrodialysis process selectively permeable membranes are used (Fig. 2.5).

The dialysis cell consists of alternate cation and anion permeable membranes. The cathode is situated near the cation permeable membrane and the anode is placed near the anion permeable membrane. When emf is applied across the electrodes the anions (Cl$^-$) move towards the anode and the cations (Na$^+$) move towards the cathode. Plastic separators are used to separate the membranes which also direct the flow of water. The salt concentration decreases in one compartment and increases in the other compartment. By using large number of pairs of membranes the brakish water is desalinated. The water of increased salinity i.e. brine solution is rejected. The process is not economical because of high energy requirement. However, if electricity is easily available it is a very good process.

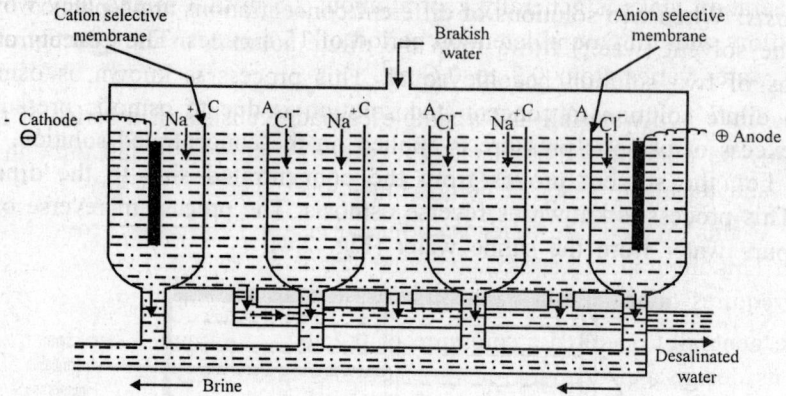

Fig. 2.5. Desalination using electrodialysis cell.

MISCELLANEOUS METHODS OF WATER TREATMENT

The remaining miscellaneous methods which are adopted for the treatment of water under certain circumstances are:

(1) Colour, odour and taste removal.

(2) Iron and manganese removal.

(3) Fluoridation.

Colour, Odour and Taste Removal

It may be mentioned that even pure water is not colourless and it is found to have a pale green-blue tint in large volumes. Some of the treatments methods to remove colour, odour and taste to certain extent are:

(1) Coagulation followed by filtration.

(2) Pre-chlorination.

(3) Super-chlorination followed by dechlorination.

(4) Use of chlorine dioxide.

  Following are however the special methods which can be adopted for the removal of colour, odour and taste: (i) aeration; (ii) treatment by activated carbon; and (iii) use of copper sulphate.

Aeration

This process aims at bringing water in intimate contact with air. Following are the methods of aeration which are generally adopted in the treatment of water: (i) air diffusion; (ii) cascades; (iii) spray nozzles; and (iv) trickling beds.

1. *Air diffusion:* In this method, the perforated pipes are installed at the bottom of tanks. The compressed air is blown through these pipes. The air bubbles while coming up from the bottom of tank come into close contact of water contained in the tank and aeration of water is thus achieved. The depth of aeration tanks is generally kept at about 2.5 m to 3 m and they work on the principle of continuous flow with minimum detention period of 15 minutes. The quantity of air consumed per 1000 litres of water varies from 0.3 to 0.6 m^3.

2. *Cascade:* A cascade is a waterfall and a simple cascade consists of a series of three or four steps. These steps may be of concrete or metal. The water is allowed to fall through a height of about 1 to 3 metres and during this fall, it comes into close contact with air.

3. *Spray nozzles:* In this method, the water is sprinkled in fine jets through nozzles to a height of about 2 m to 2.5 m. This method of aeration removes carbon dioxide to the extent of about 90 per cent or so. But it requires considerable head of water for its working.

The nozzles are generally operated at pressure of 0.7 to 0.14 N/mm^2. The discharge through nozzle will depend upon its design and available head of water.

4. *Trickling beds:* In this method, the beds of coke or slag are prepared. The beds are supported over perforated trays. The size of coke varies from 50 mm to 75 mm. The beds are arranged in vertical series. Usually three beds are placed one above the other. The depth of each bed is about 225 mm and the vertical distance between successive beds is about 100 mm to 150 mm.

The water is discharged through perforated pipes placed at the level of top bed and it is then allowed to trickle down from top bed to bottom bed. During the trickling process, the aeration of water occurs. This method is found to be less effective than that of spray nozzles. But it gives better results than those obtained by the method of cascades.

Treatment by activated carbon

In this treatment, the activated carbon is used to remove colour, odour and taste from water. It is a form of charcoal or vegetable carbon. It is made by heating lignite, charcoal, paper mill waste, sawdust and similar carbonaceous materials in a closed vessel. It is then activated or oxidised by passing air or steam. This removes the hydrocarbons which might interfere with the adsorption of organic matter.

The activated carbon is available in various trade names such as Darco, Nuchar, etc. It is available in granular as well as powder form. The grains are of 6 mm size and below. Its weight is about 4 kN per m^3 and it is highly porous in structure. The usual dose of activated carbon varies from 5 to 20 p.p.m. Following are the *additional advantages* of this treatment:

(1) It aids the process of coagulation, if adopted before filtration of water.

(2) It reduces the chlorine demand of treated water.

(3) It removes the organic matter present in water.

(4) It removes taste due to excess chlorine, hydrogen sulphide, iron, manganese and phenol.

(5) Its overdose is harmless.

Methods of applications

The activated carbon can be applied either as a filter media or in a powder form. As a filter media, the layer of activated carbon is placed instead of sand layer in the rapid sand filters. As time passes, the carbon loses its adsorption quality. It then requires to be renewed or rejuvenated. The interval of time will depend upon the quality of water and may vary from one month to one year. The rejuvenation is carried out by forcing live steam through perforated pipes which are placed in the gravel bed of the filter. The process of rejuvenation of activated carbon can be carried out several times. As a powder form, the activated carbon can be applied in various stages as follows:

(1) It can be added in raw water before any treatment takes place.

(2) It can be added along with the coagulant in mixing tank.

(3) A portion of activated carbon is added in the mixing tank and the remaining portion of it is added in water just before it enters the filter units. This is known as the *split method*.

(4) It can be added just at the point where water enters filters. It can be applied at constant rate or at varying rate. In the latter case, the rate is high when filter is washed and it becomes lower as filter gets clogged.

Use of copper sulphate

In this method, the use is made of copper sulphate to remove colour, odour and taste from water. In addition, it also controls the growth of algae, bacteria, some types of aquatic weeds, etc.

Iron and Manganese Removal

The iron is present in practically all soils, gravels, sands and rocks. The rain, in percolating through soils and rocks, acquires iron in addition to other mineral constituents according to the character of the geological formation and it is observed that practically all waters, whether from surface or underground, contain at least some traces of iron. The manganese, usually in smaller amounts, may accompany iron in water. When their content in water exceeds 0.30 p.p.m., they become objectionable.

Fluoridation

It is found that a fluoride concentration of about 1 p.p.m. in water causes reduction in the cavities of teeth of young children. It also reduces the decaying and missing teeth. For such purposes, the fluoridation is practised.

The fluoride compounds which are usually adopted for the fluoridation of the public water supplies are sodium fluoride NaF, sodium hexafluorosilicate Na_2SiF_6, hexafluorosilicate H_2SiF_6, etc. The fluorides which may occur naturally in water are cryolite $AlF_3 3NaF$, fluorspar of calcium fluoride CaF_2 and fluorapatite $3Ca(PO_4)_2 CaF_2$.

In general, it is found that cryolite and fluorspar occur in volcanic areas and fluorapatite is associated with phosphate deposits. The reaction of fluoride with teeth enamel is ionic and hence any fluoride compound that will give ionised fluoride in a dilute solution will give satisfactory results. It may also be noted that there is no difference in fluorides occurring naturally in water and those artificially added in water.

SEWAGE AND SEWAGE TREATMENT

The waste products of a society including the human excreta had been collected, carried 7 disposal of manually to safe point for disposal by sweepers. This primitive method of collecting and disposing of the society's wastes has now been modernised and replaced by a system in which these wastes are mixed with sufficient quantity of water and carried through closed conduits under the conditions of gravity flow. This mixture of water and waste products, popularly called sewage, thus automatically flow to the place from where it is disposal off, after giving it suitable treatments. The treated sewage effluents may be disposal of either in a running body of water, such as stream, or may be used for irrigation of crops.

Characteristics of Sewage

Sewage although contains more than 99.9 per cent water, the small portion of solids it highly hazardous in nature and may cause pollution of stream, lake or harbour, cause nuisance, endanger water supplies. So it is important to know the characteristics of sewage which will give the idea of degree and method of treatment to be adopted for safe disposal. These characteristics are divided into three classes: (i) physical; (ii) chemical; and (iii) bacteriological.

Physical characteristics

Fresh domestic sewage has a slightly soapy or oily odour and in appearance is cloudy. The solids of sewage are in solution and suspension and include both organic and inorganic matters. Suspended solids are those which can be filtered out on an asbestos mat and dried. They are relatively high inorganic matter. Dissolved solids are obtained by evaporating a filtered sample over water bath. Solids obtained after evaporation when ignited, the loss in weight is ascribed to the volatile solids, which are classed as organic matter; and the remaining or fixed solids are classified as inorganic.

Chemical characteristics

Sewage contains inorganic chemical compounds from the water supply and many complex organic materials derived from faeces, urine and other organic wastes. It is normally alkaline when fresh but tends to acidity as it becomes stale. Complete treatment, however, will restore the alkalinity. It will contain nitrogenous compounds like urea, proteins, amino acids and carbonaceous compounds like soap, fats, carbohydrates etc. By treatment most of the organic matters get stabilised.

Stabilisation

By stabilisation is meant that organic matter has been broken down by bacterial action to simple inorganic substances that will decompose no further. The stabilisation may be brought about by anaerobic or aerobic bacteria. Anaerobic bacteria grow and liberate energy in the absence of free oxygen and obtain it from various compounds which they are to break down. But in anaerobic stabilisation reaction is slow and this process produces unpleasant odours.

Aerobic action takes place when free oxygen is available for bacteria. The oxygen may be available from atmosphere or sewage. In what is known as self-purification of stream, free oxygen is dissolved in the water and becomes available for aerobic bacteria. Nitrogenous compounds by gradual steps of oxidation are converted to nitrates. Those nitrates unlike free nitrogen are usable to the higher plant life. Thus this process is known as nitrogen fixation.

Ether-soluble matter

This is a determination of the fats and greases in a sample of sewage or sludge. The evaporated sample is extracted with ether, and the other is then evaporated to leave the grease behind. Fats and greases when excessive in amount cause scum formation in sedimentation tanks and are injurious to filters.

Disposal of Sewage

Sewage produced, collected, conveyed through sewers finally should be disposed off. It can be treated, or may be disposed of without treatment depending on conditions. This can be done by: (i) land disposal; and (ii) dilution technique.

Land disposal

Mainly there are three methods by which land disposal is done:

(1) *Broad irrigation:* This is also known as surface irrigation. In this process, sewage is allowed to flow over cultivable land, from which a part of sewage evaporates some percolates into the sub-surface and remaining part flows out and collected through surface drainage channel and disposed off.

(2) *Sub-irrigation:* Sewage is allowed to enter the land, through distributors. Sewage by the process of evaporation, filtration, biochemical action etc. gets purified like in previous method. Effluent thus purified can now be suitably disposed off.

(3) *Ridge and furrow method:* This is an improved method of sub-irrigation. Here the sewage is allowed to stand into furrows and crops are grown on the ridges. This way sewage does not get direct touch with the crops.

Sewage farming

It is interesting to note that sewage applying to land by above methods not only solves disposed problems, but also contributes land its fertilising values. This method of growing crops is known as sewage farming. Fertilising elements like nitrogen, phosphate, potash etc. from sewage gets entry to cultivable land through biochemical process.

Sewage sickness

If sewage is continuously applied to some land, the pores of the soil get clogged, preventing free circulation of air. This retards the aerobic action of the sewage also stops filtration of water. So the land looses the purifying capacity of sewage and known as sewage sick land. This condition may be stopped by: (i) applying pretreated sewage instead of raw sewage; (ii) giving rest to the land and using reserved land; and (iii) by rotation of crops.

Advantages of land disposal method

(1) Disposal is accompanied by natural treatment.

(2) In case of irrigation the fertilising value of sewage is returned to the land.

(3) The pollution of natural body of water is prevented.

(4) The method of cheap without requiring any installation of costly machineries involving high initial cost.

(5) Disposal of possible even without a nearby river with enough quantity of water.

Disadvantages of land disposal method

(1) This method requires huge quantity of land both for action as well as for reserve.

(2) In rainy season or wet climate the method is not suitable.

(3) Lands with clayey soil this method is not efficient.

(4) Where crop is grown, this method requires special attention to prevent the spread of diseases.

(5) If not properly supervised the land may become sewage sick.

Dilution technique

Disposal sewage or effluent into a body of water or water course is known as disposal by dilution. In this method we take advantage of the self-purification capacity of reviving water like stream, lake, sea etc. Prior to disposal, the sewage may be treated. The degree of treatment depends on the quantity and concentration of sewage and the dilution ratio.

The physical forces that are helpful in self-purification of water are dilution, dispersion, sedimentation and turbulence of water. By dilution the combined strength of any particular parameter like BOD, COD, chloride content etc. will be lower than that of the effluent quality.

Oxidation ponds

Oxidation ponds are also used for the disposal of sewage. They are shallow basins excavated in the ground which retain the sewage or waste water for a certain desired period. Both settled or raw sewage can be applied and stored to these tanks.

During the process of storage sewage comes in contact with soil bacteria. Again under favourable climatic conditions namely warmth and sunshines there will be growths of algae. Bacterial decomposition of the sewage releases carbon dioxide, ammonia etc. Ammonia and other plant growing substances are used by algae. Algae again by the process of photosynthesis make the pond water to its oxygen saturation value creating favourable condition for decomposition of organic wastes in sewage by aerobic bacteria. The entire process is known as algal-bacterial symbiosis of stabilisation of wastes in oxidation pond. The organic matter in the waste is oxidised by bacteria to carbon dioxide, ammonia, phosphorus compounds and micro nutrients.

Sewage Treatment

Sewage treatment is found necessary for the purpose of converting a raw harmful sewage into an acceptable final effluent and also to dispose of the solids removed in the process. It is therefore, essential first to determine the characteristics of the raw sewage and the required characteristics of the effluent, or the required degree of treatment, before proceeding with the design of the treatment plant.

Once these characteristics treatment requirements have been determined, it should not be assumed that they are fixed and final and not subjected to change. For example, pretreatment or elimination of industrial wastes can have a marked effect on the characteristics of a raw sewage. Even effluent requirement may be subjected to modification under certain conditions. A proper balance of influent characteristics, effluent requirements and degree of treatment should first be ascertained before fixing the design basis. To achieve this balance with optimum economy it will necessitate the use of rational analysis as well as other recognised tools of engineering. Therefore, methods of treatment will depend on several considerations like, the characteristics of influent sewage, condition of available disposal points, effluent standards desired etc.

Classification of sewage treatment

Preliminary treatment

This can be defined as preparatory treatment. This will make sewage fit for further treatment. Screens, shredder, grit chamber and detritus tanks are included in this. This removes about 10 per cent of total solids like floating matters, grit and detritus from domestic sewage.

Primary treatment

This is also known as simple treatment. In the first case the solids removed are mostly inorganic in nature. In this case attempts are made to remove the settleable organic matters. This is done with the help of plain sedimentation tanks. About 40 per cent of total solids gets removed. The BOD reduction varies from 30 to 40 per cent. But the dissolved solids including organic matters escapes along with the effluent.

Final or complete treatment

This is essentially a biological treatment. In this non-settleable and dissolved organic matters get removed. About 90 per cent of total solids and 85 per cent of BOD get removed from sewage after this treatment. Intermittent filters, trickling filter, activated sludge process etc. are included in this treatment.

Screening

Screening of sewage is done for removing floating matters like pieces of wood, charcoal, leaves etc. Sewage admitted to subsequent treatment units or pumping stations should be effectively screened to protect the machinery in the plant and to avoid difficulties and unsightly condition in subsequent stage of treatment.

Chemical precipitation

Chemical precipitation consists adding of such chemicals in sewage which by reaction with one another and also with the sewage constituents will produce a flocculent precipitate and accelerate sedimentation. This is quite similar process as coagulation adopted in water treatment. In this case also suspended solids, both colloidal and unsettleable can be settled easily by chemical precipitation. Sewage containing large quantities of substances that will need only a small amount of chemical to produce the required precipitate is the most favourable for this form of treatment.

The chemicals used for precipitation are lime, ferrous sulphate, ferrous chloride, copperas, sulphuric acid, ferric chloride, ferric sulphate etc. Alum though widely used as a coagulant in water treatment is not normally used in sewage precipitation the reason being it forms lighter floc not suited in sewage treatment. Ferric chloride forms heavy flocs and popularly used in treating sewage although it has the disadvantages of handling and storing difficulties.

Though the chemical reaction may be similar to water treatment process, because of the greater amount of suspended matter in sewage, the dosing usually are considerably greater. In order to keep the cost down, it is, therefore, even more important that optimum pH values be obtained.

In this process dry chemical is fed into a mixing device in which the chemical goes into solution or suspension of calculated strength. This solution is then fed into the sewage which is needed to be precipitated. The mixture flows into the flocculator, where it is stirred gently to induce good coagulation and flocculation. After flocculation the sewage flows to sedimentation tank or clarifier where sludge settles and gets separated.

Mixing period of above is kept about 1 minute, flocculation will take 15 to 20 minutes and for sedimentations detention period is usually kept for 2 hours.

By chemical treatment process dissolved solids are practically remain unaffected. The effluent however becomes quite clear as the turbidity is reduced considerably. With normal dosage of chemicals the suspended solids are removed by 60 to 80 per cent and BOD is reduced by 35 to 60 per cent.

Since chemical precipitation removes most of the suspended matter, it will produce much more sludge then plain sedimentation. The chemical itself forms the floc and add to the quantity of sludge. But as the chemical sludge contains only 90 per cent moisture as against 95 per cent in plain sedimentation the resulting volume may increase only marginally.

The process is particularly suitable where only seasonal treatment is required, as in a summer resort, or as a supplement to other methods of treatment, or for industrial wastes under variable conditions. The process is quickly adjustable to changing conditions and is especially suitable for industrial plants using chemical processes of manufacture. The advantages of the process can be summarised as given below:

(1) Where ordinary plain sedimentation is sufficient but low stream flow at times lowers the dilution factor and necessitates somewhat better treatment. As chemical dosing can be discontinued the process is flexible enough to meet this condition.

(2) As there is forced sedimentation, it is more efficient and space required for treatment unit is less.

(3) Degree of treatment is better than plain sedimentation.

(4) Many industrial wastes do not respond to biological treatment and others require very good clarification as a pre-treatment. In such cases this treatment is very useful.

(5) Colour in waste may be removed by this process.

Disadvantages of chemical precipitation:

(1) Operation cost is high due to high cost of chemicals.

(2) Skilled hands are required for operation of the processes.

(3) Additional provision has to be made for handling and storage of chemicals.

(4) Quantity of sludge accumulated is very high many times the treatment and disposal of them pose a problem.

Skimming tank

The objective of skimming tanks is the separation from the waste-water of the lighter floating substances. The materials collected on the surface of skimming tanks, include oil, grease, soap, pieces of work, wood etc.

Skimming is required to remove the scum which otherwise will find its way to secondary treatment units and will interfere with the biological treatment processes, thus the efficiency of secondary treatment will be adversely affected.

Sludge accumulated gets digested and it is comparatively stable in nature. It is periodically cleaned and may be used as manure after further stabilisation.

Disposal of septic tank effluents

The disposal of septic tank effluent may be done either by soil absorption or by biological filters. The construction of disposal system at the ground surface or below ground level will depend on sub-soil water level condition. When sub-soil water is within 180 cm from ground level the disposal units are constructed partly or fully above ground, otherwise sub-surface construction of this units are suitable.

In case the above mentioned conditions are not feasible and where the effluent has to be discharged into open drain it should be disinfected.

Imhoff tanks

This tank like the septic tank, combines sedimentation along with sludge digestion. There is, however, an important difference, digestion takes place in a different compartment from that where settlement occurs. Second difference is that it is found to be suitable for serving a population much bigger than that of a septic tank, normal range being 5,000 to 30,000 population.

Clarigester

It is a modified type of Imhoff tank. The sewage is fed in a settling tank with an interval cone like bottom. The inside of which acts as the digester hood. These are normally of 5 to 12 metres in diameter suitable for towns having population ranging between 10,000 to 25,000. It is a two storey tank as usual, but instead of hope bottoms, both the upper and lower chambers are flat. The sewage enters the sedimentation chamber through the inlet pipe and flows radially over a circumferential weir and is collected by an effluent channel.

The sludge is deposited on the bottom of the upper chamber. From there a gate permits the entry of sludge into lower chamber, but prevents sewage or gases to move upwards. A central vertical shaft rotates, which has arms attached at four different levels. The upper arms are provided with skimming blades for removal and collection of scum. The next sets of arm move the deposited sludge towards the central sludge door. The third set act as scum breaker arms. The lowest set is for raking and stirring the sludge.

Due to this stirring, mixing of sludge is better. This promotes efficient digestion. Gas collected by the process of digestion, may be utilised in heating the digestion chamber. As the bottom is not hopper, the tank depth is lesser. As the digestion is rapid, lesser space is required. More gas is released due to stirring. As sludge pipe is attached at the bottom for withdrawal of digested sludge.

SLUDGE TREATMENT AND DISPOSAL

Handling and disposal of sludge from biological waste-water treatment plants is an important problem and represents about half the cost of most sewage treatment plants. The concentration of solids in the primary sewage sludge is about 5%; the activated sludge contains less than 1% solids; and the sludge from trickling filters has about 2% solids. This means that the sludge is composed almost entirely of water and volume reduction is the key to economic disposal. In addition to reducing its high water content, the sludge must be stabilised so that its biological activity and tendency towards putrefaction are reduced drastically.

The common unit operations of sludge treatment and disposal involve concentration of thickening, digestion, conditioning, dewatering, oxidation and safe disposal. Fig. 2.6 represents the general flow sheet of sludge treatment techniques in approximately the order in which they could be applied in a treatment scheme.

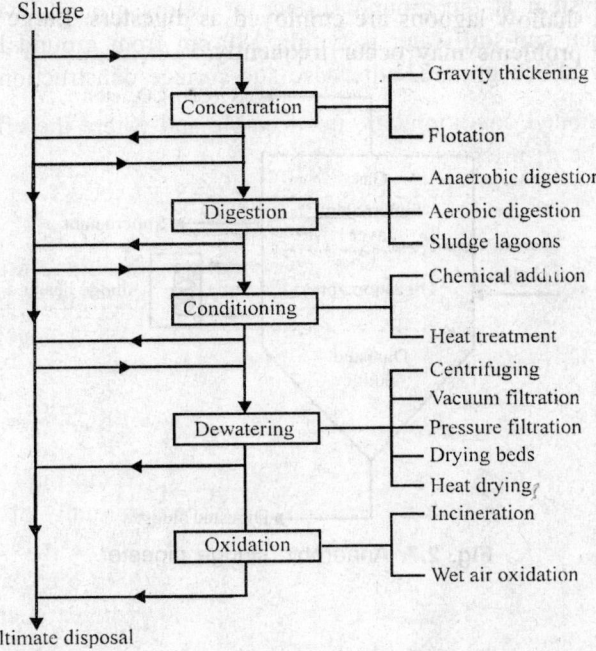

Fig. 2.6. Sequence of operations for sludge treatment (Arrows indicate possible flow paths).

Concentration

The purpose of concentration or thickening is to remove water from the sludge and reduce its volume as much as possible so that the sludge can be handled more efficiently. The common methods for thickening are gravity settling and flotation. In gravity thickeners the sludge is subjected to gentle agitation by means of a slow stirrer which enhances settling. In this manner the combined sludge from primary and secondary settlers can be thickened so as to contain 5–9% solids. Often the thickening of activated sludge is complicated by anaerobic action, particularly under warm conditions when the bacteria in the sludge decompose organic matter and release gases. This also creates settling and odour problems. The sludge can also be thickened by air flotation, particularly the secondary sludge, which keeps the system aerobic. The flotation technique can concentrate the sludge to bring it solids content to 4%.

Digestion

After concentration, the sludge is stabilised by digesting it under aerobic or anaerobic conditions. Anaerobic digestion is the most common method in which the organic content of the sludge decomposes to give mainly methane and carbon dioxide and at the same time the bound water is released from the sludge. Properly digested sludge is black with a faint smell of tar, and is stable. In a typical sludge digester, shown in Fig. 2.7, raw sludge is fed into the active digestion zone and gas lifts the sludge particles and other materials which form a layer on the top of the digestion zone. The gas is collected at the top and the digested sludge is withdrawn from the bottom. The normal detention period in standard digesters varies from 30 to 70 days depending upon the temperature conditions.

Aerobic digestion is also used, and it may be considered similar to extended aeration. The sludge is aerated in a tank for about 20 days at ambient temperatures. During the process the bacterial cells are destroyed and a substantial portion of the sludge is oxidised resulting in the reduction of the solid content by about 30%. Sometimes shallow lagoons are employed as digesters. Large land areas are required for sludge lagoons and odour problems may occur frequently.

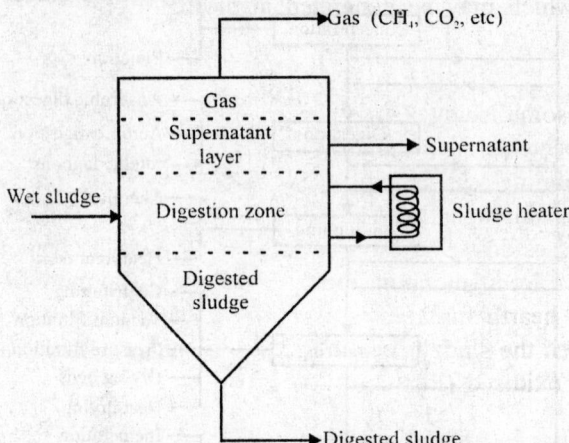

Fig. 2.7. Anaerobic sludge digester.

Conditioning

The sludge after stabilisation may be conditioned to improve its dewatering characteristics. This is done by adding chemicals like iron salts, alum, lime and polyelectrolytes. These chemicals bind the sludge particles together and encourage the release of absorbed water. Physical conditioning methods such as heat treatment are becoming popular. The sludge is heated under pressure and after a period of time the gel structure of the sludge breaks down so that the water as released. Heat treatment has the advantage of sterilising the sludge; at the same time the sludge is partially oxidised and completely stabilised.

Dewatering

The thickened sludge is dewatered for efficient handling and disposal. Dewatering is accomplished by mechanical methods, the most common being centrifugation and filtration, which includes pressure filtration and vacuum filtration. In centrifugation, conditioned sludge is added to a rotating bowl that separates the sludge into a cake and a dilute stream. The solid cake is transported within the bowl and is removed by a screw conveyor at one end of the bowl; the liquid is removed at the opposite end. Centrifugation is a compact method which requires careful control of process variables.

Filtration, using plate-and-frame pressure filters or rotary drum vacuum filters, is widely used for dewatering digested sludge. In pressure filtration, the sludge is pumped slowly with increasing pressure into filter plates supporting a cloth, which retains the solids. It is a bath process, and after the dewatering period the plates are separated and the sludge cake removed. Pressure filtration can produce a cake with a solids content of 25–50%. In contrast to pressure filtration, vacuum filtration is a continuous process where a rotating drum, which is covered with filter cloth, is partially submerged in the sludge. On applying a vacuum of 80–90 kPa inside the drum, the liquid is sucked into the drum. The cake is deposited on the outside of the drum and is removed by a scraping mechanism. Vacuum filtration yields a dewatered sludge with a solids content of about 25%.

Drying beds are also commonly used for dewatering. The bed consists of a filtering medium on which the sludge is applied to a depth of upto 250 mm, depending on the solids content. Dewatering takes place by a combination of drainage and evaporation. Removal of dried sludge is carried out mechanically. Another technique, heat drying, may be utilised in applications where the sludge is to be incinerated or when a saleable commodity can be produced. A major problem associated with this process is the control of gases and ash particles which may be generated in drying.

Oxidation

Before the final disposal, some sludges may be oxidised to reduce the organic content, with the consequent destruction of bacteria and a significant reduction in their volumes. Incineration and wet oxidation are the two common methods employed for sludge oxidation. Incineration is usually performed in a multiple hearth furnace, although fluidised beds or flash dryers may also be used. In the multiple hearth furnace the sludge passes downward through a series of hearths. Vapourisation of moisture occurs in the upper hearths, followed by incineration in the lower ones. The combined efficiency of evaporation and incineration in multiple hearth furnaces is about 55%. The fluidised bed system consists of a bed of sand fluidised by air. When the sludge is introduced, the sludge particles are dried almost instantly as they are dispersed, and are oxidised (Fig. 2.8).

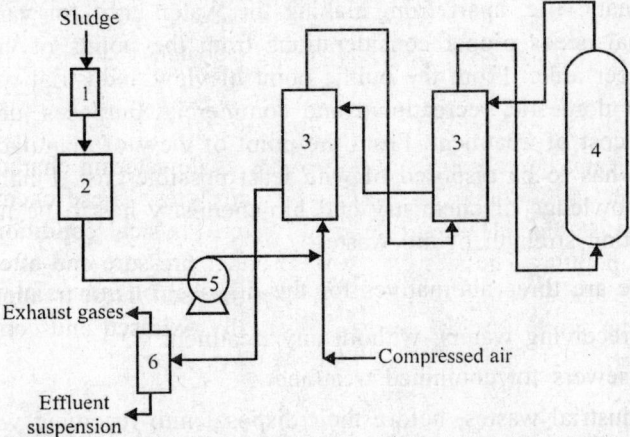

Fig. 2.8. Wet oxidation process flow diagram 1. Grinder; 2. Sludge storage tank; 3. Heat exchangers; 4. Reactor; 5. Sludge pump; 6. Separator.

Wet oxidation is a process in which the sludge is ground, mixed with air in stoichiometric proportions, and then subjected to high temperature reactor. Under the conditions of high temperature (260°C or more) and high pressure (8000–12000 kPa) there is no vapourisation of the sludge liquid. Initially, external heat is applied to start the oxidation of the organic matter in the sludge. The rate of oxidation increases until it reaches an equilibrium value. The products from the top of the reactor are passed through heat exchangers to recover heat which in turn is utilised by the incoming feed.

Ultimate Sludge Disposal

Several methods are employed for the ultimate disposal of sludge. Wet digested sludge may be sprayed onto crop land where it functions as a fertiliser or fertiliser base. It may be lagooned; however, when the lagoons fill they must be abandoned. Dried sludge may be used as a landfill or a soil conditioner. Wet or partially dewatered sludge or ash from incineration may be transported from the shore to dumping grounds at sea.

INDUSTRIAL WASTE-WATER TREATMENT

Industrial wastes are the one which result from industrial processing operation. Though industrial wastes include liquid, solids and gases, we are presently concerned with the liquid part, which is commonly known as industrial waste-water. Industrial waste-waters are either discharged directly into the receiving water (if they meet the necessary standards) or else are discharged into municipal sewers. Some times, industrial waste-waters are treated partially before their discharge into sewers, or else are treated separately through suitable treatment processes so that the treated effluent is safe. From the standpoint of characteristics of their wastes, various industries can be divided into following four groups :

1. Food and food-processing industries like sugar, canning, dairy, fermentation etc.
2. Industries of mineral products, i.e. coal coke and gas, iron and steel, oil, petroleum etc.
3. Industries processing chemicals, chemical manufacturing, metal finishing, paper and pulp, rubber, tanning, textiles etc. and
4. Miscellaneous industries like atomic power, radioactive waste. etc.

Industrial wastes are varied and complex. These wastes affect, in some way or the other, the normal life of a stream or normal functioning of sewerage and sewage treatment plants, unless pretreated at the source point itself. If they are discharged directly in the receiving waters, it may result in discolouring, foul smell and killing of aquatic life, apart from making the water unfit for various other purposes. Industrial waste-water disposal needs proper considerations from the points of view of manufacturer, public and the sanitary engineer alike. From the public point of view, industrial wastes cause pollution to stream making it unfit for domestic, recreational and commercial purposes, deteriorate sewers and treatment plant, and increase cost of treatment. From the point of view of manufacturer or industrialist, the waste is a liability and it has to be disposed of with least possible cost. Finally, from the point of view of sanitary engineer, knowledge of chemistry and bio-chemistry has to be judiciously applied to reduce the volume and polluting strength of the waste.

As mentioned above, there are three alternatives for the disposal of industrial wastes :

1. Direct disposal into the receiving waters, without any treatment.
2. Disposal into municipal sewers for combined treatment.
3. Separate treatment of industrial wastes, before their disposal into receiving water/land.

The factors that affect the choice of any of the above alternatives are: (i) self purification capacity of streams; (ii) tolerance limits for the inland surface waters; (iii) economic aspects of both the manufacturers as well as municipalities; and (iv) technical advantages of mixing the industrial wastes with domestic sewage.

CHARACTERISTICS OF INDUSTRIAL WASTE-WATER

Although the characteristics of the industrial waste-water usually vary from industry to industry, and also vary from process to process even for the same industry, yet industrial effluent as a group differs widely from normal domestic sewage obtained from residences and commercial establishments. They have either too high a proportion of suspended solids, dissolved organic and inorganic solids, BOD, alkalinity or acidity, and their different constituents will not be in the same proportions as they exist in a normal domestic sewage. The pollutants in the industrial waste-water include the raw materials, process chemicals, final products, process intermediates, process by-products and impurities in raw materials and process chemicals. Such industrial waste-waters cannot always be treated easily by the normal methods of treating domestic waste-waters, and certain specially designed methods or sequence of methods may be necessary.

Pollutants in Industrial Waste-water and their Effects

Following are the various types of pollutants that may be present in industrial waste-waters.

1. *Organic substances:* These deplete DO of streams and impose great load on secondary treatment units.

2. *Inorganic substances:* These include carbonates, chlorides, nitrogen, phosphorus etc. They are undesirable for micro-plants of receiving water body. They render the water body unfit for further use.

3. *Acids and alkalies:* These greatly affect the aquatic life of receiving water body. They also cause serious problems in the operation of treatment units.

4. *Toxic substances:* These include cyanides, sulphides, acetylene, alcohol, etc., due to which flora and fauna of receiving waters is greatly affected. They cause problems in the treatment units apart from endangering the life of workmen. Table 2.2 gives the list of toxic substances from some selected industries.

5. *Colour-producing substances:* They impart objectionable colour in the receiving water bodies.

6. *Oil etc.:* These hinder self-purification process of stream. They also cause operational problems in the treatment plants.

Table 2.2. Toxic pollutants in industrial waste-waters.

Industry	Total pollutants
Fertilizers	Ammonia, Arsenic
Coke ovens	Phenols, Cyanide, Thiocyanate, Ammonia
Metallurgicals	Heavy metals (i.e. Copper, Cadmium, Zinc)
Electroplating	Hexa-valent Chromium, Cadmium, Copper, Zinc
Synthetic wool	Acrylonitrile, Acetonitrite, HCN
Petrochemicals	Phenol, heavy metals, Cyanides

METHODS OF TREATMENT

As pointed out earlier, industrial waste-water have either too high a proportion of suspended solids, dissolved organic and inorganic solids, BOD, alkalinity or acidity, and their different constituents will not be in the same proportion throughout, as they exist in domestic waste-water. The important factors which affect the planning for an industrial waste-water treatment plant are: (i) discontinuous flow and sometimes seasonally discharged waste-water; (ii) high concentration of pollutants; and (iii) non-biodegradability and toxicity of some pollution parameters. Treatment of industrial waste-water generally consists of one or more of the following processes: (i) equalisation; (ii) neutralisation; (iii) physical treatment; (iv) chemical treatment; and (v) biological treatment.

Equalisation

When the characteristics of the waste-water varies during the day and also when the discharge is either not uniform or else is discontinuous, equalisation is necessary. The process of equalisation consists of holding the waste-water for some predetermined time, in continuously mixed holding tanks/basins so as to get waste-water of uniform character and at uniform rate. Thus, equalisation needs adequate storage.

Neutralisation

This is necessary when the waste-water contains either excess alkali or excess acid, and is achieved by the addition of either acid or alkali respectively. This may be done either in the equalisation tank, if the conditions so permit, or else in a separate neutralisation tank.

Physical Treatment

Physical treatment is similar to primary treatment of domestic waste-water. Various processes that fall under this head are: (i) screening; (ii) sedimentation; (iii) flotation; and (iv) filtration. Primary sedimentation becomes essential when the waste-water contains high percentage of settleable solids. Floatation is provided to remove the finer particles. In floatation, the rising air bubbles lift the finer particles to the surface from where these are removed by skimming.

Chemical Treatment

Chemical treatment is one of the essential part in the treatment of industrial waste-water, specially for those which are amenable to biological treatment. Sometimes, physico-chemical treatment is applied to the industrial waste-waters, either in absence of biological treatment or in conjunction with biological treatment. Various chemical and physico-chemical processes that are used for industrial waste-water treatment are: (i) coagulation; (ii) chemical precipitation; (iii) hyper-filtration or reverse-osmosis; (iv) chemical oxidation; (v) adsorption; (vi) ion exchange; (vii) air stripping; (viii) electrodialysis; and (ix) thermal reduction.

Most of the above processes have already been described and explained earlier, and hence these are not repeated here.

Biological Treatment

This is resorted to only when the industrial waste-water contains large quantities of biodegradable substances. If the BOD/COD ratio of the industrial waste-water is more than 0.6, it is biologically treatable. If the ratio is less than 0.6 but is upto 0.3, acclimatisation is necessary before biological treatment. If, however, the ratio is less than 0.3, biological treatment is not necessary. Acclimatisation is a process of seeding or raising initial microbial population under a controlled condition, by gradual exposure of the waste-water in increasing concentration. Various biological treatment methods that can be used are (i) activated sludge process; (ii) trickling filter; (iii) oxidation pond; (iv) aerated lagoon; (v) oxidation ditch; and (vi) oxidation lagoon followed by aerated lagoon. Amongst these, trickling filters prove better than activated sludge process, since filters can handle a number of toxic wastes quite satisfactorily. When BOD of waste-water approaches 1000 mg/l, recirculation of effluent will prove more useful. We shall now consider treatment procedures for the effluents from some important industries.

DAIRY INDUSTRY

Large quantities of waste-water originates due to different operations of dairy industry. Various operations in a dairy industry may include pasteurisation, bottling, filling in cans, and preparation of butter, cream, cheese, milk powder etc. Waste-water from dairy industry consists primarily of dilutions of milk and its products which impart a very high BOD, sometimes upto about 1000 mg/l. The waste-water may also contain detergents, germicides and other chemicals. Table 2.3 shows composition of waste-water of a typical dairy. Table 2.4 shows various pollution characteristics of dairy industry waste-water, along with suggested treatments. Fig. 2.9 shows a flow chart for waste-water.

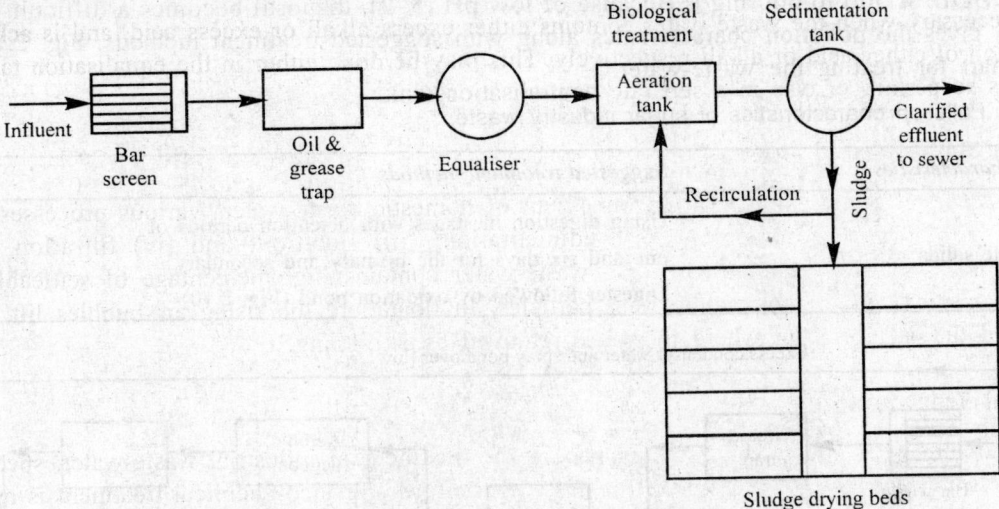

Fig. 2.9. Typical flow chart for treating dairy waste-water.

Table 2.3. Typical composition of dairy waste-water.

Characteristic	Typical values
pH	7.2
Alkalinity	580 mg/l as $CaCO_3$
Suspended solids	720 mg/l
Total dissolved solids	1020 mg/l
BOD	1190 mg/l
COD	90 mg/l
Total nitrogen	90 mg/l
Phosphorous	12 mg/l
Oil and grease	320 mg/l
Chlorides	115 mg/l

Table 2.4. Pollution characteristics of dairy industry waste.

Pollution characteristics	Suggested treatment methods
High dissolved solids High suspended solids High BOD Phosphorous Nitrogen Oil and grease	Physical treatment followed by aerobic biological treatment in the form of high rate trickling filters or activated sludge process. Equalisation tank is commonly provided. Sludge can be dried on sludge drying beds. Oxidation ponds and oxidation ditches can also be provided.

CANE SUGAR INDUSTRY WASTE-WATER

The average quantity of liquid waste produced from a cane sugar industry waste-water is of the order of about 3000 litres/tons of cane crushed. The waste-water is quite strong, having SS of 400 to 500 mg/l and BOD of 500 to 600 mg/l. Because of low pH (5–7), disposal becomes a difficult problem. Table 2.5 gives the pollution characteristics along with suggested treatment methods. Fig. 2.10 shows a flow chart for treating the waste-water.

Table 2.5. Pollution characteristics of sugar industry waste.

Pollution characteristics	Suggested treatment methods
High BOD	Using digestion in stages with detention periods of
High volatile solids	one and six days for the primary and secondary
Low pH	Digester followed by oxidation pond (Fig. 2.10)

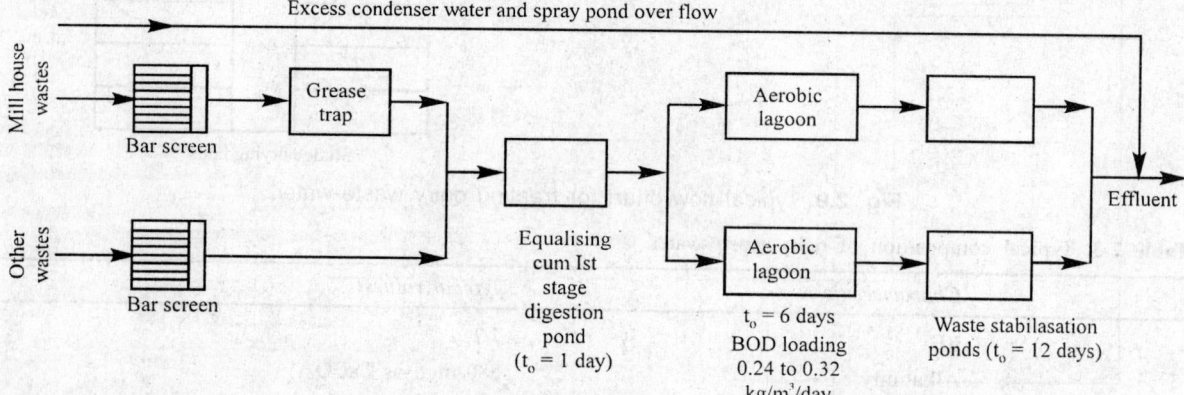

Fig. 2.10. Flow chart for treatment of sugar mill waste.

DISTILLERY AND BREWERY WASTE-WATER

Various products that are manufactured in distilleries include industrial alcohols, rectified spirit, silent spirit, absolute alcohol, beverage alcohol etc. These products are obtained through the high bio-chemical processes of fermentation by yeast, using carbohydrates as raw material. Also, these products contain ethyl alcohol in different proportions.

The unwanted residue obtained at the time of preparation of the medium, contain very high BOD, which makes it non-amenable to aerobic biological treatment. Table 2.6 gives the composition of combined brewery wastes. Table 2.7 gives the pollution characteristics and suggested treatment methods.

Table 2.6. Composition of combined brewery wastes.

Parameter	Values
pH	4.0–7.0
COD (mg/l)	30–1225
BOD (mg/l)	70–3000
Total solids (mg/l)	272–2724
Suspended solids (mg/l)	16–516
Total nitrogen (N) (mg/l)	7–42

Table 2.7. Pollution characteristics of breweries wastes.

Pollution characteristics	Suggested treatment
Very high BOD	Two stage aerobic treatment, using trickling filters.
High COD	
High chlorides and sulphates	Use of anaerobic lagoons and aerated lagoon,
Highly coloured-brownish yellow	followed by polishing ponds

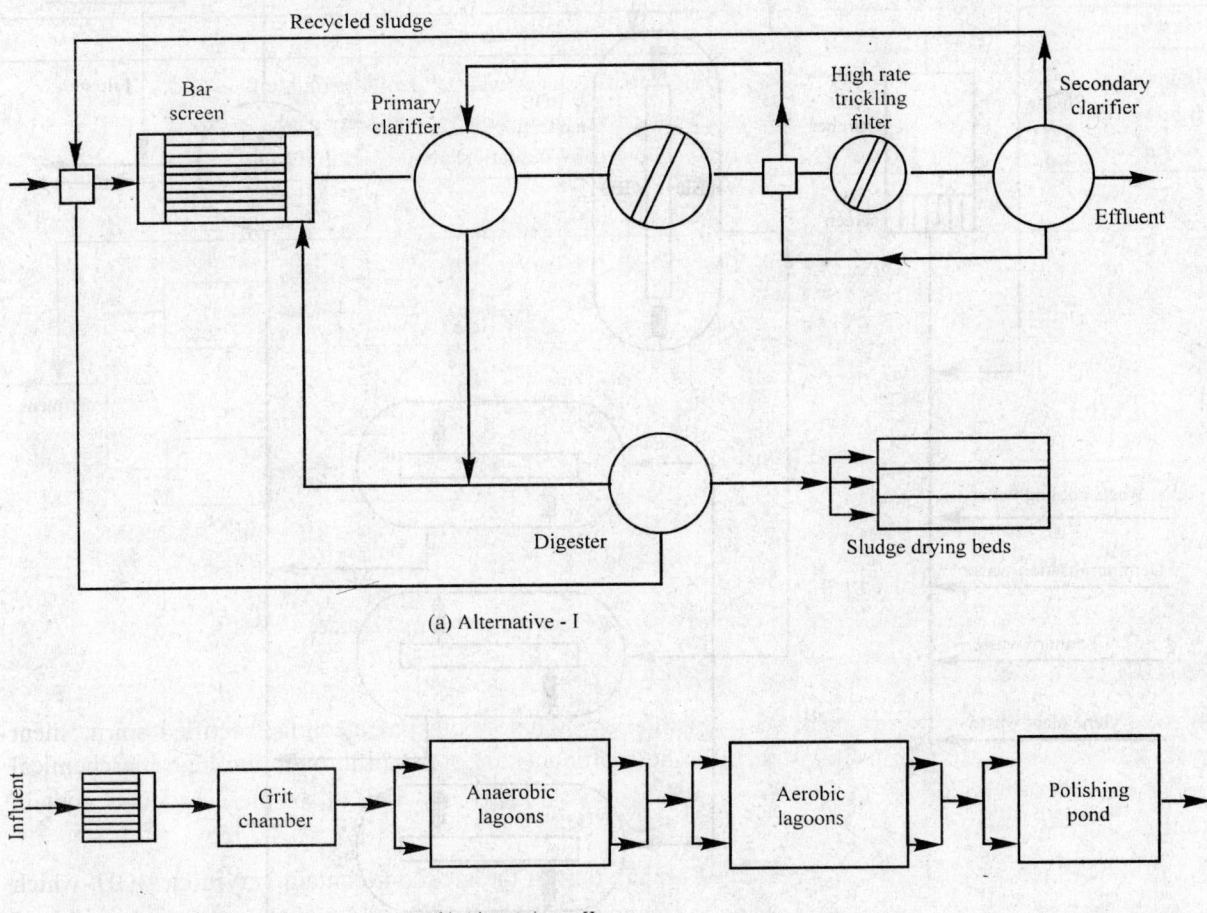

(a) Alternative - I

(b) Alternative - II

Fig. 2.11. Flow chart for treating brewery wastes

PETROCHEMICAL INDUSTRIES WASTE

Petrochemicals are the chemicals which are derived from petroleum, and these products includes both organic as well as inorganic chemicals.

Table 2.8 gives typical characteristics of petro-chemical wastes. Table 2.9 gives pollution characteristics of petro-chemical industries. Fig. 2.12 gives flow chart for treating waste-water from petrochemical industries.

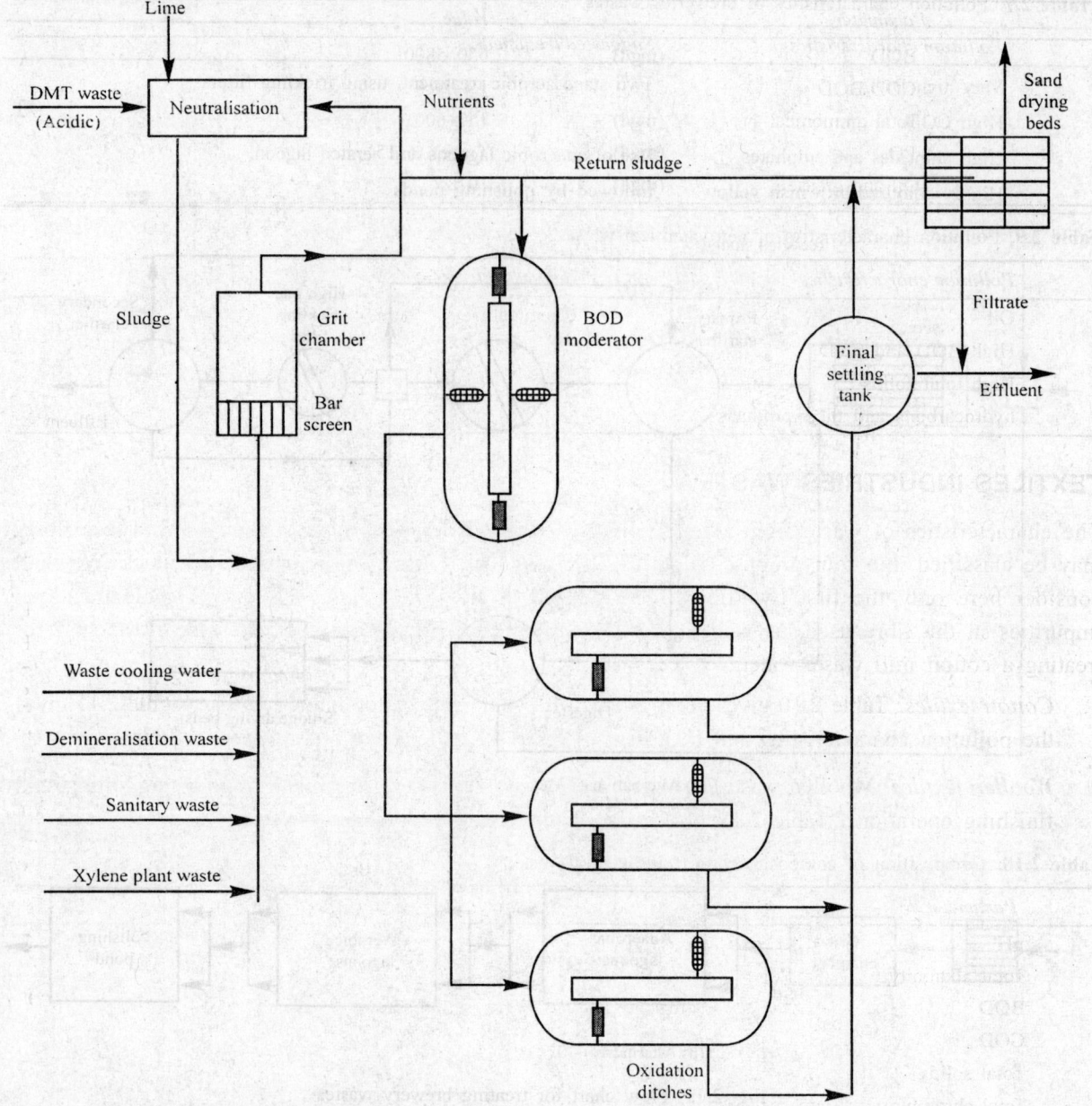

Fig. 2.12. Flow chart for an aromatic waste treatment plant.

Table 2.8. Composition of a typical petrochemical waste.

Parameter		Value
pH	(mg/l)	5.3–10.0
Total solids	(mg/l)	1956–14448
COD	(mg/l)	947–9000

(Cont'd ...)

Parameter		Value
BOD	(mg/l)	650–5800
COD/BOD		1.4–20.0
Total ammonical–N	(mg/l)	150–600
PO_4	(mg/l)	Traces to 15
Oil and grease	(mg/l)	10–85

Table 2.9. Pollution characteristics of petrochemical waste.

Pollution characteristics	Suggested treatment
Oil	Chemical treatment and biological treatment.
High BOD and COD	
High total solids	
hydrocarbons and their products	

TEXTILES INDUSTRIES WASTE

The characteristics of waste from a textile industry depends upto the type of fibre used. These fibres may be classified into four groups: (i) cotton (ii) wool, (iii) regenerated, and (iv) synthetics. We shall consider here only the first two fibres. The pollutants in the textile industry waste include natural impurities in the fibre used, as well as the processing chemicals. Fig. 2.13. gives the flow chart for treating a cotton mill waste-water.

1. *Cotton textiles:* Table 2.10 gives composition of a composite cotton industry waste. Table 2.11 gives the pollution characteristics and treatment methods of cotton textile waste.

2. *Woollen textiles:* Woollen wastes originate from scouring, carbonising, bleaching, dyeing, oiling, and finishing operations. Table 2.12 givers the characteristics of typical wool mill waste

Table 2.10. Composition of composite cotton textile mill waste.

Parameter		Value
pH		9.8–11.8
Total alkalinity	(mg/l)	17.35 as $CaCO_3$
BOD	(mg/l)	760
COD	(mg/l)	1418
Total solids	(mg/l)	6170
Total chromium	(mg/l)	12.5

Table 2.11. Pollution characteristic of cotton textile mill waste.

Pollution characteristics	Suggested treatment
High alkalinity	Chemical and biological treatment
High BOD	
High COD	
High SS	

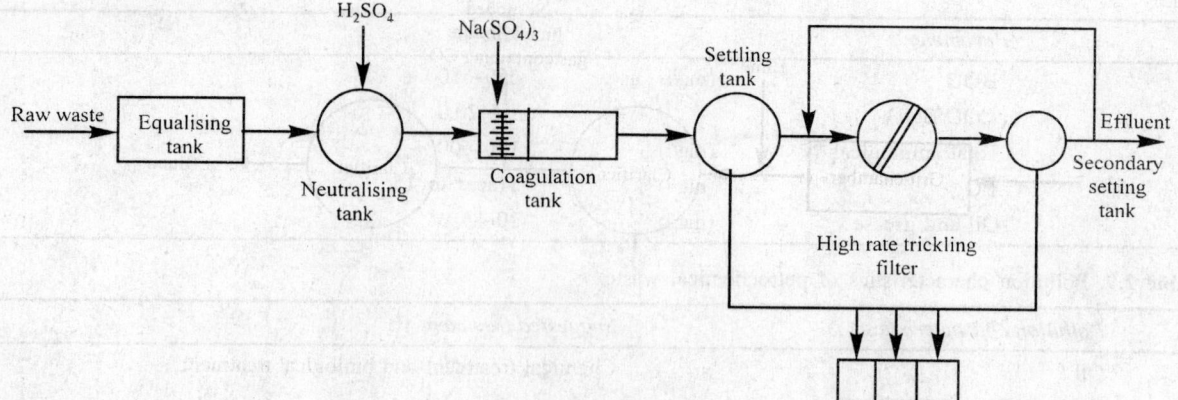

Fig. 2.13. Flow diagram for treatment of cotton textile industries waste.

Table 2.12. Composition of woollen textile mill waste.

Parameter		Value
pH		9–10.5
BOD	(mg/l)	900
Total solids	(mg/l)	3000
Total alkalinity	(mg/l)	600
Total chromium	(mg/l)	4
Suspended solids	(mg/l)	100
Colour	(mg/l)	Brown

PAPER AND PULP MILLS WASTE

A well operated pulp and paper mill, employing kraft process may produce waste-water volume in the range of 200 to 350 m³ per tonne of paper manufactured. The pulp and paper mill waste-water is generally characterised by very strong colour, high BOD, high SS, high COD/BOD ratio. Table 2.13 gives the characteristics of combined effluent of pulp and paper mills. Table 2.14 gives the pollution characteristics along with treatment methods. Fig. 2.14 gives the massive lime treatment for colour removal from pulp and paper mill effluent and Fig. 2.15 gives the flow diagram for the treatment of waste-water from a typical pulp mill.

Table 2.13. Characteristics of combined effluent of pulp and paper Mills.

Parameter		Small mill (production) 20 tons of paper/day	Large mill (Production 2000 tons of paper/day
Flow per day	(m³/t)	330	222
Colour	(units)	–	7800
pH		8.2–8.5	8.2–8.5
Total solids	(mg/l)	–	4410
Suspended solids	(mg/l)	900–2000	3300
COD	(mg/l)	3400–5780	716
BOD	(mg/l)	680–1250	155
COD/BOD		3.9–5	4.6

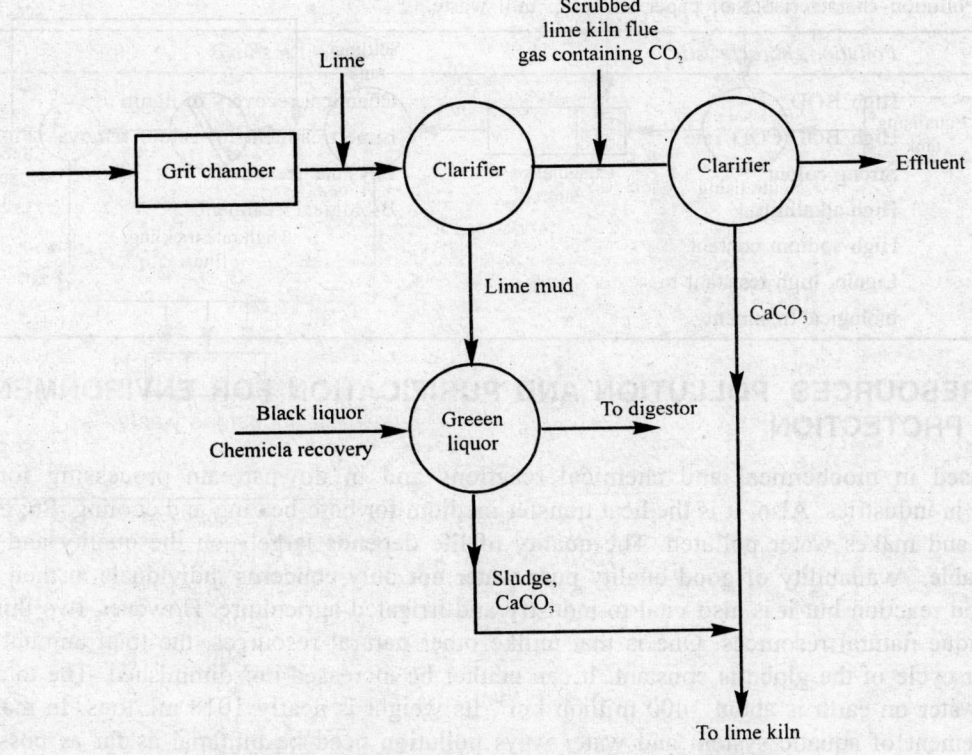

Fig. 2.14. Lime treatment for colour removal.

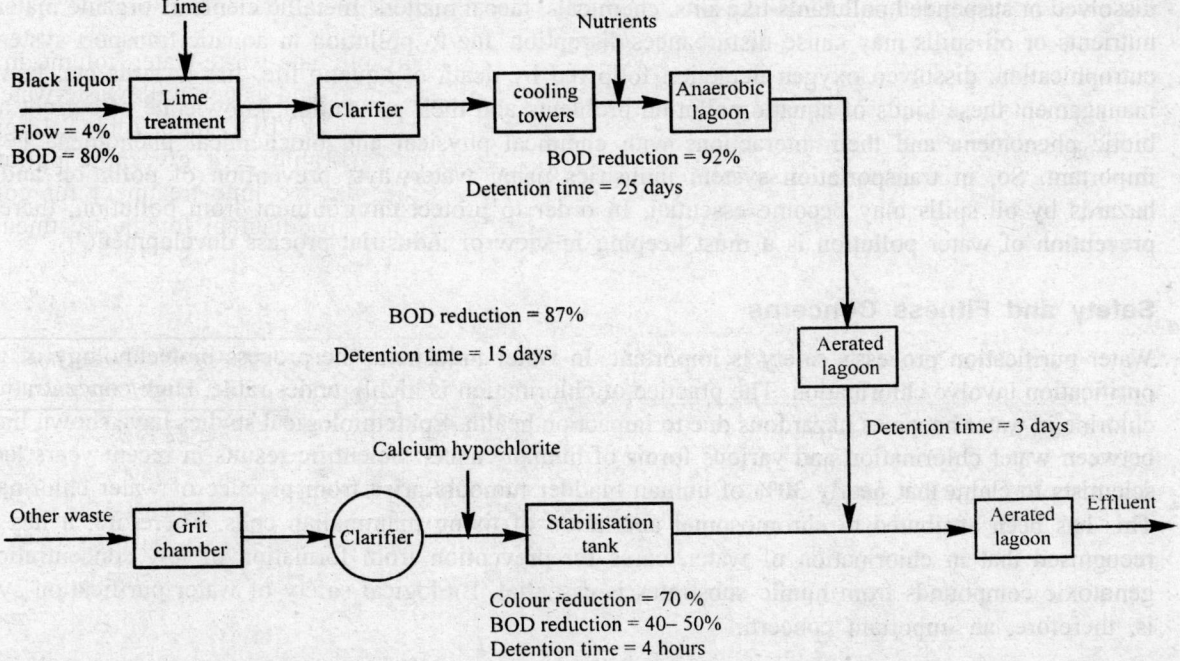

Fig. 2.15. Flow diagram for treatment of waste from a pulp mill.

Table 2.14. Pollution characteristics of paper and pulp mill waste.

Pollution characteristic	Treatment methods
High BOD	Chemical recovery of lignin
High BOD/COD rate	Lime treatment for colour removal (Fig. 2.14)
Strong colour	Physical treatment
High alkalinity	Biological treatment
High sodium content	
Lignin, high resistant to	
biological treatment	

WATER RESOURCES, POLLUTION AND PURIFICATION FOR ENVIRONMENT AND HAZARD PROTECTION

Water is used in biochemical and chemical reactions and in downstream processing for product purification in industries. Also, it is the heat transfer medium for both heating and cooling. So, everybody uses water and makes water polluted. The quality of life depends largely on the quality and purity of water available. Availability of good quality pure water not only concerns individuals in their domestic daily life and reaction but it is also vital to industry and irrigated agriculture. However, two things make water a unique natural resources. One is that unlike other natural resources, the total amount of water in the water cycle of the globe is constant. It can neither be increased nor diminished. The total volume of natural water on earth is about 1400 million km^3. Its weight is nearly 1018 mk tons. In maintenance and management of aquatic system and water ways pollution need be minimal as far as possible. The presence in concentrations higher than normal in natural waterways (lakes, rivers and oceans) of dissolved or suspended pollutants like slits, chemicals, faecal matters, metallic elements organic materials/ nutrients or oil spills may cause disturbances/disruption due to pollution in aquatic transport system by eutrophication, dissolved oxygen depletion followed by death of aquatic life, fire hazards etc. In water management these kinds of aquatic pollution problems and their prevention, knowledge of current aqua-biotic phenomena and their interactions with chemical physical and biochemical phenomena is very important. So, in transportation system industries using waterways, prevention of pollution and fire hazards by oil spills may become essential. In order to protect environment from pollution, therefore, prevention of water pollution is a must keeping in view of industrial process development.

Safety and Fitness Concerns

Water purification process's safety is important. In water industries, the process biotechnology of water purification involve chlorination. The practice of chlorination is highly undesirable. High concentration of chlorine in water becomes hazardous due to impact on health. Epidemiological studies have shown linkage between water chlorination and various forms of human cancer. Scientific results in recent years led the scientists to claim that nearly 30% of human bladder tumours arise from practice of water chlorination. This has been attributed to chromosomal aberration of living mammalian cells. Therefore, it has been recognised that in chlorination of water, cares for prevention from formation of low concentration of genotoxic compounds from humic substrates is essential. Biological safety of water purification system is, therefore, an important concern.

Also, in many cases over chlorinated water has been used in food biotechnology industries like canning for cooling purposes. Cooling water containing more than 100 organisms per ml is usually unfit.

When cooling tanks are used repeatedly this level contamination is usually observed. In such cases water should not be over chlorinated in an attempt to ensure the presence at all times of free chlorine. This is because under certain conditions an excess of chlorine in water may lead to can corrosion with ultimate serious effect on processed food.

Besides hazardous metallic contaminants water and waste usually contain organic nitrogenous material. In waste water purification microbial 'nitrogen rich carbon poor' (or low C/N ratio) environments of denitrification process may serve as key to clean water. When C/N ratio is high causing 'short circulating nitrogen cycle' the objective of waste-water purification process is defeated.

Moreover, in water avenues the concept of bioremediation of hazards and protection from fire hazards due to oil spills and leakage of oil tankers has been suggested by technologically redesigning of biocells with the aid of biomolecular design engineering science. The redesigned biocells would be capable of degradation of multicomponents oil thereby preventing pollution in water.

Oxygen deprivation hazard in water by pollution reduces free oxygen radicals. For water resources to protect aquatic life it is essential that they are served with available oxygen. This is because living biocells use oxygen molecules which are regarded as radicals, each molecule having two unpaired elections. Since urine and faeces of mammals/animals contain large quantities of oxygen scavengers, they are known to be hazardous in terms of oxygen non-availability by pollutants.

Resources and Withdrawal

Water is a multipurpose prime resource. Natural water resources are commonly grouped under four categories (Table 2.15) (a) atmospheric or rain water; (b) surface waters (ocean, sea, river), (c) stored water; and (d) ground waters. Besides these another important source of additional water is the waste-waters generated through human activities for domestic, municipal or industrial purposes. Water is not equally distributed in nature and among people of the world. Nearly 97% of water on earth is sea water. Of the rest, 22% is ground water and 77% is ice locked in the form of glaciers and polar ice caps. It indicates that less than 1% of the supply of fresh water takes part in the hydrological cycle. Nearly half of that 1% is found in rivers, lakes and swamps. In India fresh water resources and withdrawal is very important because it is primarily an agricultural country. It has been estimated that annual internal renewable water amounts to about 1850 Km^3 and 1990 per capita data is 2.17 m^3. Of this resource annual withdrawal percent of water resource is 18 whereas per capita withdrawal is about 612 m^3.

Table 2.15. Water uniqueness and resources.

Uniqueness	Water resource categories
Prime natural resource	Rain water
Essential for life survival	Surface water
In water/hydrological cycle	
total amount of water is constant	Stored water
It can neither be increased nor diminished	Ground water
Essential for process industry	Waste or polluted water

In India, the sectoral withdrawal percent of water in domestic process industries and agriculture is given in Table 2.16(A) comparative information on fresh water resources and withdrawal world wide indicate the largest consumption of fresh water in our country is in agriculture (Table 2.16).

Table 2.16. Comparative information on fresh water withdrawal in India vis-a-vis worldwide and in a modernised region (D-domestic, I-industry, A-agriculture).

Place/Country	Approximate annual withdrawal		Sectoral (%)		
	% of water resources	Per capita (m³)	D	I	A
India	18–20	612	3–5	4–8	93–87
Worldwide	8	660	8	23	69
North and central America	10	1692	9	42	49

Purity, Pollution and Hazard

Amount of natural water resources like ground waters are in general, relatively purer and free from bacteria and other contaminants because of the filtering action of the earth through which the water have penetrated. This filtering action removes not only most of the bacteria but also any suspended particles· and organic materials. Deep wells contain usually fewer microorganisms than water from shallow wells. It is due to the deeper layers of filtering material. Water resources receive microbial load, organic nitrogen and other polluting factors from soil, air, sewerage, organic wastes, dead plants and animals, process industry effluents, municipal sewage etc. Thus to define the purity of water standard, specifications have been used by different nations for safeness of their usability. The important specifying factors and their approximate parametric standard limits have, therefore, been listed (Table 2.17).

Table 2.17. Important pollution specifying factors and their approximate usability specifications in water for different purposes.

Pollution factors	Approximate usability specifications of water for the purpose of		
	Drinking	Process industry	Irrigated agriculture
Physical appearance	Clear, colourless	Clear, colourless	Preferably clear
Odour	Nil	Nil	Nil
pH	7.0	6.9	5.5–9.0
Chemicals (ppm)	500		2000
– Chloride as Cl'	200		600

Pollution factors	Approximate usability specifications of water for the purpose of		
	Drinking	Process industry	Irrigated agriculture
– Sulphate	200	–	1000
– Nitrate	40–50	–	1000
– Phenol	0.002	1.0	1.0
– Lead	Nil	Nil	Nil
– Arsenic	0.005	–	–
– Oil & grease	0.3	1.0	1.0
$BOD_{5(20°C)}$ (ppm)	4.5–5.0	4.5–5.0	6-7
COD (ppm)	7–8.5	–	–

However, ground water is replenished/recharged by hydrological/water cycle. This replenishment may be due to rainfall, less evapotranspiration and surface run off.

Drinking water supplies in almost all urban metropolitan cities in our country have been known to contain pollutants. This is primarily because of either sewerage is absent or partial. In such situations microbial pollutants may predominate sometimes and pose a major health hazard. Not only that relatively minor pollutants relate to leaking sewers. In developed countries as well as in mountaneous health resorts and industrialised areas road deicing and other urban processes also cause pollution in ground water supplies.

Many of the organic contaminants are very site specific. In some urban metropolitan cities in our country two groups of compounds may also be detected in water supplies time to time. These are chlorinated hydrocarbon solvents and aromatic compounds arising from petroleum fraction presence in water during chlorination disinfection.

Industrial organic chemicals in ground water may arise from the same source that create contaminated land e.g. industrial premises, fuel storage transport and pipeline spillages. Also, water resources become polluted and contaminated by sewerage discharge and human excrement. Effluent loads discharged into the natural streams or stored water receptacles at various disposal points cause the water resource polluted. The major effluent load discharging systems may emanate from process industries.

Pollution by sewerage or by human excrement may be greatest danger associated with drinking water. These pollution contributors are usually carriers of such infectious diseases as typhoid fever, dysentery or jaundice. In such cases water contain living microorganisms or viruses which cause the disease thereby causing health hazard. Faecal pollution organisms like E.coli, Streptococci sp. are common resulting exertion of $BOD_{5(20°C)}$ and COD values in potable water as well (Table 2.17). Possibility of five hazards by oil spills in supra-national long oceanic avenues and its prevention scope by technologically redesigned microorganisms has been shown in literature.

Purification for Environment and Hazard Protection

By bioprocessing

For maintaining purity of municipal/corporation water supplies its chlorination is highly effective in purification of water from microbial contamination thereby preventing spread of infectious diseases. Not only municipal/corporation water supplies but also other water supplies must pass through a series of purification steps before pipe line supply to make it safe for either human consumption or for process industry and agriculture requirements. In some cases, however, the water resource may be sufficiently pure to require only treatment with a disinfectant like chlorine, chlorocresol etc. depending on the quality of treatment desired and the purpose. In many instances the condition of water resource is such as to require normally three stages of primary purifications namely, sedimentation, filtration followed by disinfection. However, in recycling of used water in industries or for their disposal in natural water ways stringent treatment methods are needed to make waste-water liquid effluent safe for recycling, for aquatic life and prevention of water basin filling up by metallic or other solids. Reference is, therefore, made to water resources and the danger of their pollution.

Oxic bacterial process biotechnology

It is well reorganised that there are limits to the self purifying ability of a natural body of water. This limit pertains to both capacity and the speed of self purification. In recent industrialised societies, necessity of development of rapid man-made water purification has become obvious.

In a more recent process a mix of various strains of bacteria has been used to assimilate organic matter natural as well as man-made consuming oxygen and producing chiefly CO_2 and H_2O.

The technological success greatly depends on the type of water and pollutants, their amount and concentration, the consistency of these determinants and last but not the least, on the speed of bacterial assimilation and the presence of inhibitory ingredients. Many process biotechnology have been developed in laboratories and pilot plants. Their operational reliability is of utmost importance, both in order to ensure continuous supply and necessary quality of treated clean water after the purification process. Oxic biological nitrification or organic present in polluted water is also one of the major concerns in this bioprocessing to avoid eutrophication.

Anoxic bacterial process biotechnology

Many microbes not only survive but eve grow rapidly when starved of oxygen. For survival these microbes avail gene machines which encode enzymes essential for anoxic growth. The most successful bacterial groups are that can use nitrate as a substitute for oxygen/air. In anoxic stage of water purification/treatment less energy is derived from nitrate reduction than from conventional cell respiration, so genes for nitrate reduction are activated only when nitrate is present but oxygen is unavailable. However, the product of nitrate reduction is very toxic. Actually, nitrate respiring bacteria in water purification does make provision for removal of nitrate, so formed.

Integrated oxic-anoxic process biotechnology consideration

In the water purification bacterial nitrogen metabolism is an important key to clean water. Bacterial metabolism of three molecular species: ammonia, nitrate and nitrogen, dominates the oxic-anoxic integrated biological nitrogen cycle. In oxic biotechnology nitrification plays major role while denitrification fixation governs anoxibiosis in the process biotechnology. A striking balance between oxic-anoxic bioprocessing and water pollution is important so that the objective of waste-water purification is not defeated.

Oxic–anoxic design scheme

In order to overcome the problem of eutrophication in waste water purification nitrogen content in water must be removed otherwise, it will cause BOD increase. The removal of nitrogen from waste-water/water can be done by appropriate schemes of oxic-anoxic design systems for nitrification/denitrification. Operational characteristics of three schemes for designs of nitrification and denitrification have been described below (Fig. 2.16).

Scheme 1: This scheme depends upon the endogenous respiration of the activated sludge to achieve denitrification.

Scheme 2: Here a portion of the influent waste-water is by-passed to the denitrification tank to provide food for the facultative organisms thereby increasing the respiration rate and hence the denitrification rate.

Scheme 3: It uses an influent waste-water which is nitrogen deficient as a food source for denitrifying organisms. The waste-water should contain a readily available carbon source.

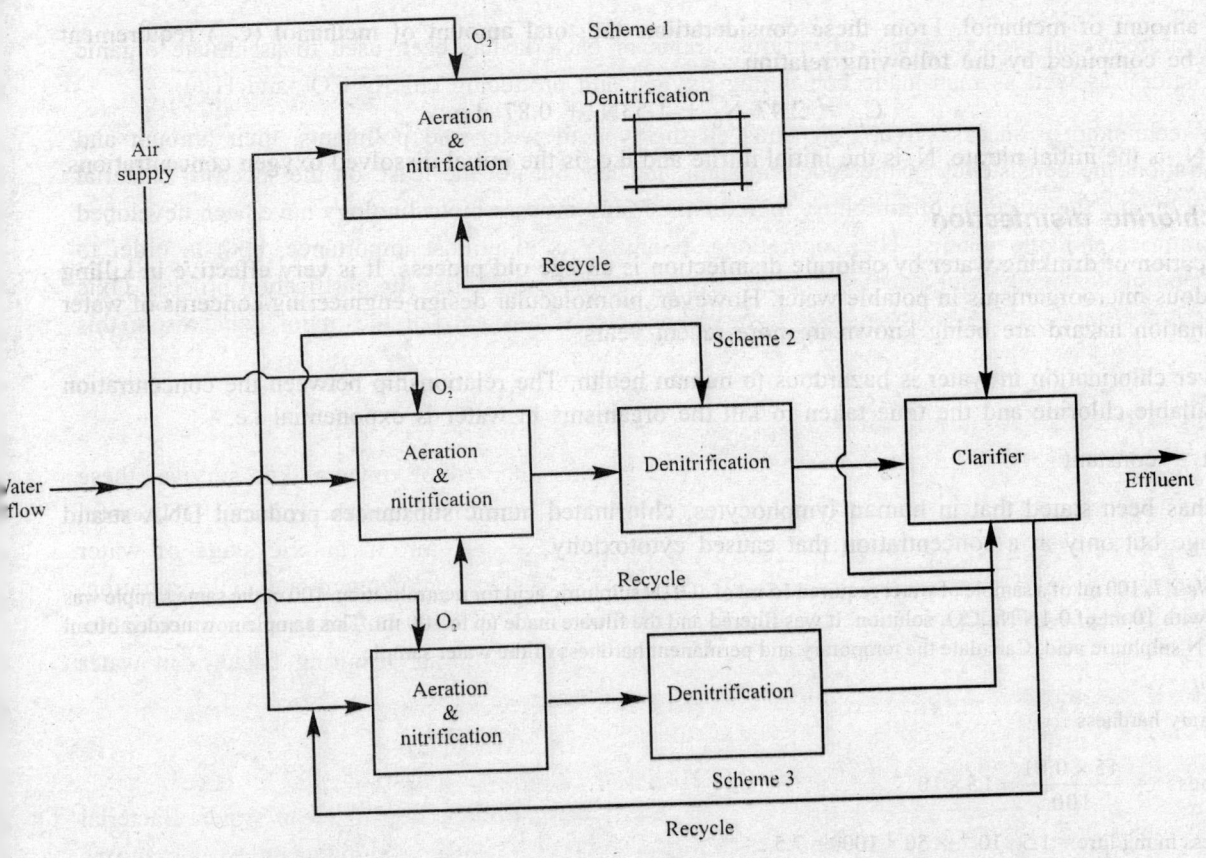

Fig. 2.16. Operational characteristics of three schemes of nitrification and denitrification.

Relative advantages and disadvantages of the Schemes

In Scheme 1 while the bioprocessing achieves a low nitrogen effluent, the slow rate of denitrification under endogenous respiration conditions results in a large denitrification tank. Scheme 2 while some what reduces the required size of the denitrification tank, it has the disadvantage of increasing the unoxidised nitrogen in the treated effluent and in most cases increasing the effluent BOD. While Scheme 2 will not contribute nitrogen to the effluent, careful controlled operation is required to avoid increasing the effluent BOD. Experimental results and experience indicated that economic use of this process scheme necessitates increasing the respiration rate and hence denitrification rate in denitrifying unit. It is also double that required for only carbonaceous BOD removal. If higher O_2 level is maintained in the aeration tank for maximum nitrification rate, the power requirement will be about two and half times that required for conventional activated sludge process operation.

Acceleration of denitrification

Biodenitrification for removal of nitrate and nitrite could be accelerated by adding appropriate amount of methanol. Based on stoichiometric equations of biodenitrification it could be computed that 1 mole NO_3 is equivalent to 5/6 mole methanol. When the waste-water which is to be treated contains dissolved oxygen (d.o) it needs to be removed before denitrification step. This has been accomplished by adding extra amount of methanol. From these considerations the total amount of methanol (C_m) requirement could be computed by the following relation :

$$C_m = 2.47\ N_o + 1.53 N_i + 0.87\ \text{d.o.}$$

Here N_o is the initial nitrate, N_i is the initial nitrite and d.o. is the initial dissolved oxygen concentrations.

By chlorine disinfection

Purification of drinking water by chlorine disinfection is an age old process. It is very effective in killing hazardous microorganisms in potable water. However, biomolecular design engineering concerns of water chlorination hazard are being known in more recent years.

Over chlorination in water is hazardous to human health. The relationship between the concentration of available chlorine and the time taken to kill the organisms in water is exponential i.e.

$c^n t$ = constant

It has been stated that in human lymphocytes, chlorinated humic substances produced DNA strand cleavage but only at a concentration that caused cytotoxicity.

Example 2.1. 100 ml of a sample of water required 15 ml of 0.01N sulphuric acid for neutralisation. 100 of the same sample was boiled with 10 ml of 0.1N Na_2CO_3 solution. It was filtered and the filtrate made up to 100 ml. This sample now needed 30 ml of 0.01N sulphuric acid. Calculate the temporary and permanent hardness of the water sample.

Solution.

Temporary hardness :

$$\text{Hardness} = \frac{15 \times 0.01}{100} = 1.5 \times 10^{-4}$$

Hardness in mg/litre = $1.5 \times 10^{-4} \times 50 \times 1000 = 7.5$

Permanent hardness:

No. of milliequivalents of Na_2CO_3 remainding = No. of milliequivalents of acid needed = $30 \times 0.01 = 0.3$

No. of milliequivalents of Na_2CO_3 reacted with hardness = $1 - 0.3 = 0.7$

This is the same as the hardness in 100 ml of the sample in milliequivalents.

Permanent hardness = $0.7 \times 50 \times 10 = 350$ mg/litre

Example 2.2. 25 ml of 0.1N (i.e., N/10) sodium carbonate was added to 100 ml of a sample of water. After boiling, the filtrate from the above required 20 ml of N/20 sulphuric acid for complete neutralisation, using methyl orange as indicator. Calculate the hardness.

Solution. The hardness is of the non-carbonate type.

No. of meq of Na_2CO_3 added = $25 \times 0.1 = 2.5$

No. of meq of Na_2CO_3 remaining = No. of meq of acid consumed = $20 \times 0.05 = 1$

No. of meq of Na_2CO_3 reacted with hardness = $(2.5-1) = 1.5$

This is the same as the hardness.

Hardness in mg/litre = $1.5 \times 50 \times 10 = 750$.

Chapter 3

Structure of Atom

INTRODUCTION

An atom is the smallest, hard particle of an element that can take part in a chemical reaction. Atoms of most of the elements are very reactive and do not exist in the free state. They exist in combination with the atoms of the same element or another element. Only the atoms of the noble gases like helium, neon, argon, krypton, xenon, etc., are unreactive and exist in the free state. Atoms are the building blocks of all the matter around us. Atoms cannot be subdivided in chemical reactions.

Atom is the number of protons (positively charged mass units) in the nucleus of an atom, upon which its structure and properties depend. This number represents the location of an element in the Periodic Table; it is always the same as the number of negatively charged electrons in the shells. Thus an atom is electrically neutral, except in an ionised state, that is, when one or more electrons have been gained or lost. Atomic numbers range from 1 for hydrogen, to 106, for the most recently discovered element.

The Atomic Hypothesis was first suggested by John Dalton and further developed by Avogadro, Cannizaro and many other Chemists of the 19th century, regarded an atom as the ultimate particle of matter.

DALTON'S ATOMIC THEORY

The main postulates of Dalton's atomic theory were:

(1) An atom cannot be subdivided.

(2) Atoms are neither created nor destroyed during chemical reactions.

(3) Atoms of the same element are all alike; in particular, all atoms of an element have the same mass.

(4) Atoms of different elements are not like; in particular, their masses are different.

This theory provides a satisfactory basis for the law of chemical combination.

Up to the end of 19th century, the Dalton's atomic model remained undisputed. But the new discoveries towards the end of 19th century and early 20th century showed that atom has a complex structure and is not indivisible. The studies also showed that atoms are made up of three fundamental particles, commonly known as *electrons*, *protons* and *neutrons*. This observation led to modification of the Daltonian picture of an atom. It also led to a deeper understanding of how the chemical behaviour of an atom is related to its internal structure. Properties of element are explained on the basis of negatively charged electrons around a positively charged mass, situated at the atom, called nucleus. The positively charged nucleus containing protons and neutrons occupies much less space in an atom compared to the large space in which the electrons are distributed.

ELECTRONS

The electron is a fundamental particle of matter that can exist either as a constituent of an atom or in the free state. It has a negative electric charge (4.8×10^{-10} e.s.u.) and a mass 1/1837th that of a proton, equivalent to 9.1×10^{-28} gram. The number of electrons in an atom of any element is the same as the number of protons in the nucleus, i.e., the atomic number. Thus the range is from 1 electron in hydrogen to 103 in lawrencium. As the negative charge of the electrons equals the positive charge of the protons, all atoms are electrically neutral. Electrons are arranged in from 1 to 7 shells around the nucleus; the maximum number of electrons in each shell is strictly limited by the laws of physics. The tendency of electrons to form complete outer shells accounts for the valence of an element, and they play an essential part in chemical bonding. The outer shells are not always filled: sodium has 2 electrons in the first shell, 8 in the second, and only one in the third. A single electron in the outer shell may be attracted into an incomplete shell of another element, leaving the original atom with a net positive charge. The latter then is called an ion. Valence electrons are those which can be captured by or shared with another atom.

Electrons can be removed from the atoms of some other elements by heat, light, electric energy or bombardment with high-energy particles. In such cases they are totally free from the atomic orbit, and their energy can be utilised by means of a conductor (electricity) or a vacuum tube or semi-conductor. Current is generated by detaching the electrons of a metallic conductor (silver, copper) by means of an electric or magnetic field; the electrons then flow along the conductor to a positively charged terminal. The entire science of electronics is made possible by the tendency of a heated metal cathode to emit a continuous stream of electrons in a vacuum tube.

Free electrons, called beta particles are spontaneously emitted by decaying radioactive nuclei; they have comparatively low energy, but can be accelerated to velocities approaching that of light. The basic nature of the electron has been the subject of much research of the highest order of mathematical rigor. In simplest terms, the electron has the properties of both a particle and a wave, i.e., a standing wave is associated with an electron moving in its orbit. The energy state of any electron is described by four quantum numbers.

Discovery of Electrons

After the Daltons's atomic model, one of the significant advances was made by J.J.Thomson in 1897. This was the discovery of charged particles called electrons. They are extremely small in size. The existence of electron is proved by experiments using discharge tubes.

A common discharge tube is a long glass tube having two metal plates sealed at its two ends (Fig. 3.1) These metal plates are known as electrode. The electrode which is connected to the positive terminal of the battery is known as anode (positive electrode), and the electrode which is connected to the negative terminal of the battery is called cathode (negative electrode). Discharge tube has a side tube through which air (or other gases) can be pumped out by using a vacuum pump, so that experiments can be performed at low pressures. In the following discussion, we will use air as gas in the discharge tube.

1. When air inside the discharge tube is at the atmospheric pressure and a high electric voltage of 10000 volts (or more) is applied to the electrodes, no electricity flows through the air in the discharge tube.

2. If the pressure of air inside the discharge tube is reduced to about 1 mm of mercury and high voltage is applied again, electricity begins to flow through the air and a light is emitted by the air inside the tube. The colour of light changes with the nature of gas taken in the discharge tube.

3. When the pressure of air in the discharge tube is reduced to about 0.001 mm of mercury and a high voltage is applied to the electrodes, the emission of light by air stops. Though the inside of the discharge tube now appears to be dark, the walls of the discharge tube at the end opposite to the cathode begin to glow with greenish light (Fig. 3.1).

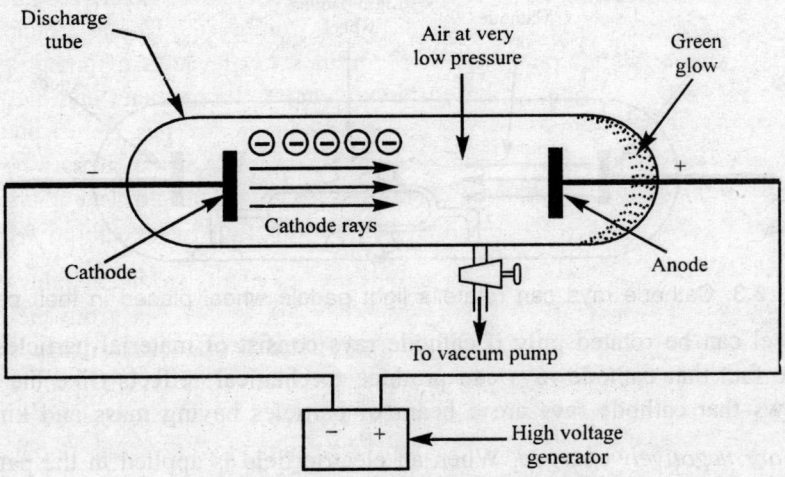

Fig. 3.1. Production of cathode rays.

It is now known that some invisible rays are formed at the cathode and when these rays strike the glass tube, they emit a greenish light. Since these rays are formed at the cathode, they are known as cathode rays.

Properties of Cathode Rays (or Negative Rays)

The important properties of cathode rays are given below:

1. *Cathode rays travel in straight lines.* When an opaque object like a metal cross is placed in the path of cathode rays in a discharge tube, a shadow of the metal cross is formed at the end opposite to the cathode (Fig. 3.2).

    A shadow of the metal cross can be formed only if cathode rays travel in straight lines and cannot bend round the corners of the metal cross. The fact that cathode rays cast shadows of the objects placed in their path shows the cathode rays travel in straight lines.

2. *Cathode rays can produce mechanical effects.* If a light paddle wheel having mica blades is placed in the discharge tube in such a way that cathode rays strike only on the blades in the upper half of the paddle wheel, the paddle wheel rotates (Fig. 3.3).

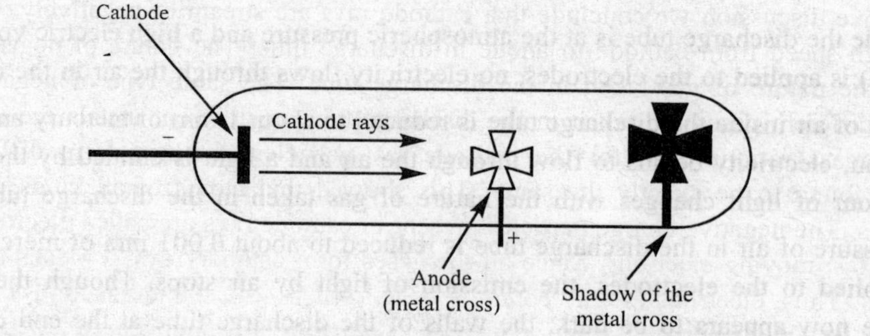

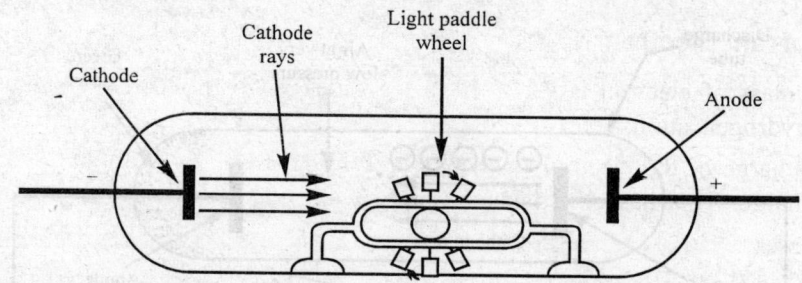

Fig. 3.2. Cathode rays cast shadows of the objects placed in their path.

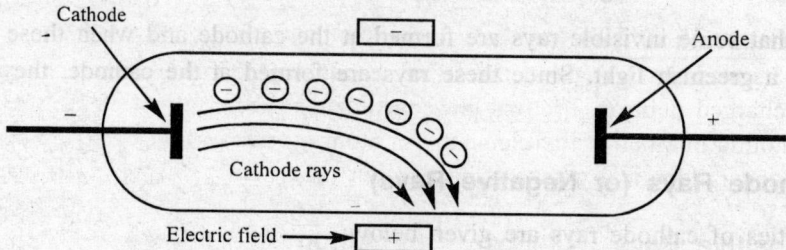

Fig. 3.3. Cathode rays can rotate a light paddle wheel placed in their path.

The paddle wheel can be rotated only if cathode rays consist of material particles having mass and kinetic energy. The fact that cathode rays can produce mechanical effects (like the rotation of a light paddle wheel) shows that cathode rays are a beam of particles having mass and kinetic energy.

3. *Cathode rays are negatively charged.* When an electric field is applied in the path of cathode rays, they are deflected towards the positive plate of electric field (Fig. 3.4).

Fig. 3.4. Effect of electric field on cathode rays. They are deflected towards the positive plate of electric field.

The cathode rays can be deflected towards the positive plate of the electric field only if they have negative charge (because opposite charges attract each other). So, the fact that cathode rays are deflected towards the positive plate of an electric field shows that cathode rays are made up of negatively charged particles. Cathode rays are also deflected by a magnetic field. The direction of deflection of cathode rays in a magnetic field also shows that they consist of negatively charged particles in motion.

4. The nature of cathode rays does not depend on the nature of a gas taken in the discharge tube or material of the cathode.

5. The mass of a cathode ray particle is very, very small as compared to the mass of the atom from which it is formed.

From the above discussion we conclude that cathode rays are stream of negatively charged particles moving with high speed from cathode to anode in discharge tube. The nature of negative particle does not depend on the nature of gas taken in the discharge tube. The same type of negative particles are formed even if different gases are taken in the discharge tube or different metals are used for cathode. The charge to mass ration (e/m) for the cathode ray particles obtained form different gases was determined and found to be exactly the same. This showed that atoms of all kinds contain the same negative particles. The negative charge particles present in cathode rays are called electrons. This cathode rays consists of fast moving electrons. Since all the gases form cathode rays, it means that all the atoms contain electrons. Thus, electrons are the common constituents of all the atoms, and located outside the nucleus in an atom. Only hydrogen atom contains one electron, all other atoms contain more than one electron.

Characteristic of an Electron

1. The absolute mass of electron is 9.1×10^{-31} kg which is approximately 1/1840 of the mass of a proton or a hydrogen atom.

2. The absolute charge of the electron (was measured by Millikan in 1909) is 1.602×10^{-19} coulomb of negative charge. This has been taken to be one unit of negative charge.

PROTONS

The formation of cathode rays has shown that all the atoms contain negatively charged particles called electrons. Now, an atom is electrically neutral, so it must contain some positively charged particles to balance the negative charge of electrons. It has actually been found by experiments that all the atoms contain positively charged particles called protons. The existence of protons in the atoms was shown by Goldstein. We will now discuss the production of anode rays (or positive rays) which led to the discovery of proton.

Or

Proton is a fundamental unit of matter having a positive charge and a mass number of 1, equivalent to 1.67×10^{-24} gram. Its mass is 1837 times that of the negatively charged electron, but is almost identical with that of the uncharged neutron. Protons are constituents of all atomic nuclei, their number in each nucleus being the atomic number of the element. An atom of normal hydrogen contains one proton and one electron. A proton is identical with a hydrogen ion (H^+).

Production of Positive Rays : Discovery of Proton

In the production of positive rays a discharge tube having perforated cathode is used. A perforated cathode is a cathode having holes in it. These perforations or holes are to allow the positive rays to pass through them. When a high voltage of about 10000 volts is applied to a discharge tube having a perforated cathode and containing air at very low pressure of about 0.001 mm of mercury, a faint red glow is observed behind the cathode (Fig. 3.5).

It is now known that some rays are formed at the anode and when these rays strike the walls of the discharge tube, they produce a faint red light. Since these rays are formed at the anode (positive electrode), they are known as anode rays or positive rays. The properties of anode rays have been studied by performing experiments similar to those performed on cathode rays.

Properties of anode rays (or positive rays)

1. Anode rays travel in straight lines. This is shown by the fact that they cast shadows of the objects placed in their path.

2. Anode rays can produce mechanical effects. This is shown by the fact that they can rotate a light paddle wheel placed in their path.

3. Anode rays are positively charged. This is shown by the fact that the anode rays are deflected towards the negative plate of an electric field.

4. The nature of anode rays depends on the gas taken in the discharge tube.

5. The mass of anode rays particle is almost equal to the mass of the atom from which it is formed.

6. The charge to mass (e/m ratio) ratio is not constant for positive ray particles obtained from different gases, it changes with the nature of gas in the discharge tubes.

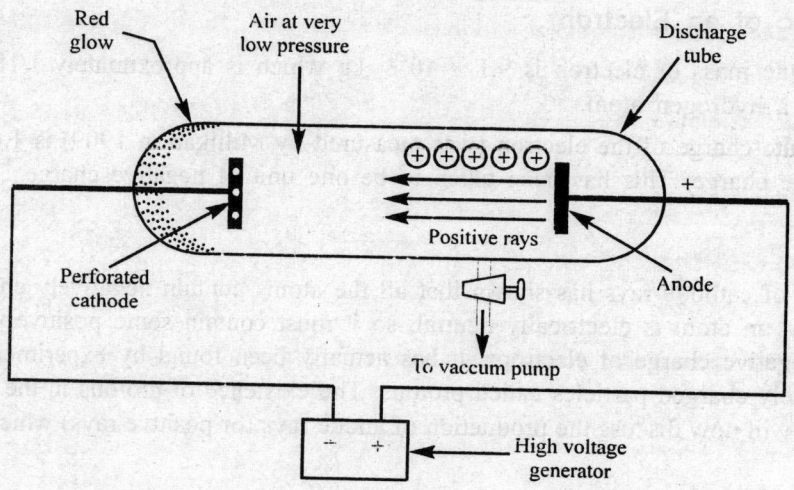

Fig. 3.5. Production of anode rays or positive rays.

Characteristics of proton

1. The absolute mass of a proton is 1.6×10^{-27} kg i.e., the mass of a proton is equal to the mass of one hydrogen atom (i.e., 1840 times that of an electron).

2. The charge of a proton is equal and opposite to the charge of an electron. So, the absolute charge of a proton is 1.6×10^{-19} coulomb of positive charge which has been taken to be one unit of positive charge.

NEUTRON

One atom of hydrogen has one proton and one electron. One atom of helium has two protons and two electrons. Therefore, the relative atomic mass of helium should be twice that of hydrogen. But in fact the relative atomic mass of helium is 4 and not 2. This discrepancy was removed by James Chadwick in 1932. Chadwick's experiments showed that apart from protons, nuclei of all atoms except that of hydrogen atom, contain another type of particles which have the same mass as hydrogen atoms or protons. These particles are known as *neutrons*. The neutron is a fundamental particle of matter having a mass of 1.009 but no electric charge. It is a constituent of the nucleus of all elements except hydrogen, the number of neutrons present being the difference between the mass number and the atomic number of the element. Neutrons may be liberated from the nucleus by fission of uranium-235, plutonium, and a few other elements, each nucleus yielding an average of 2.5 neutrons; they can also be produced by bombardment of their elements, e.g., beryllium, with positively charged particles.

As free neutrons are uncharged they have tremendous penetrating power as a result of their electrical neutrality; hence they have a highly damaging effect on living tissue, requiring the use of a shielding of all equipment in which they are produced. Neutrons directly emitted from atomic nuclei are termed "fast"; it is these that bring about the chain reaction in the atomic bomb. In a nuclear power reactor, where a less rapid reaction is desired, the energy of fast neutrons is partially absorbed by the moderator, and the neutrons so retarded are called "slow" or thermal.

Characteristics of Neutron

1. The mass of a neutron is equal to the mass of a proton i.e., the relative mass of neutron is 1 a.m.u.
2. Neutron has no electric charge i.e. electrically it is neutral.

Table 3.1. Positions, relative masses and relative charges of electrons, protons and neutrons.

Sub-atomic particle	Position in the atom	Mass (relative to the mass of a proton)	Charge
Proton	Nucleus	1	+1
Neutron	Nucleus	1	0
Electron	Energy level around the nucleus (i.e., outside the nucleus)	$\frac{1}{1840}$	−1

MODELS OF ATOM

Thomson's Model

On the basis of the facts known at that time, Thomson proposed the first model. According to this model the positive charge was assumed to be smeared (spread) over a sphere of radius 10^{-10} m with the electrons embedded in the sphere. This was known as plum-pudding model. However, this idea was dropped due to the success of α-scattering experiments performed by Rutherford and Mardson.

Rutherford's Experiment: Discovery of Nucleus

After the discovery of electrons, protons and neutrons, it became clear that an atom is made up of these three fundamental particles. Experiments were then carried out to find out how electrons, protons and neutrons were arranged in an atom. In 1911, Lord Rutherford by his alpha-particle scattering experiment led to the discovery of a small positively charged nucleus in the atom containing all the protons and neutrons.

Rutherford bombarded a gold foil with a stream of alpha-particles (an alpha-particle is a positively charged helium ion, He^{2+}, containing two protons and two neutrons).

When alpha particles (α-particle) are allowed to strike a thin gold foil, it is found that:

1. Most of the alpha particles pass through the gold foil without any deflection (Fig. 3.6).

2. A few of the alpha particles are deflected through small angles, whereas a very few are deflected through large angles.

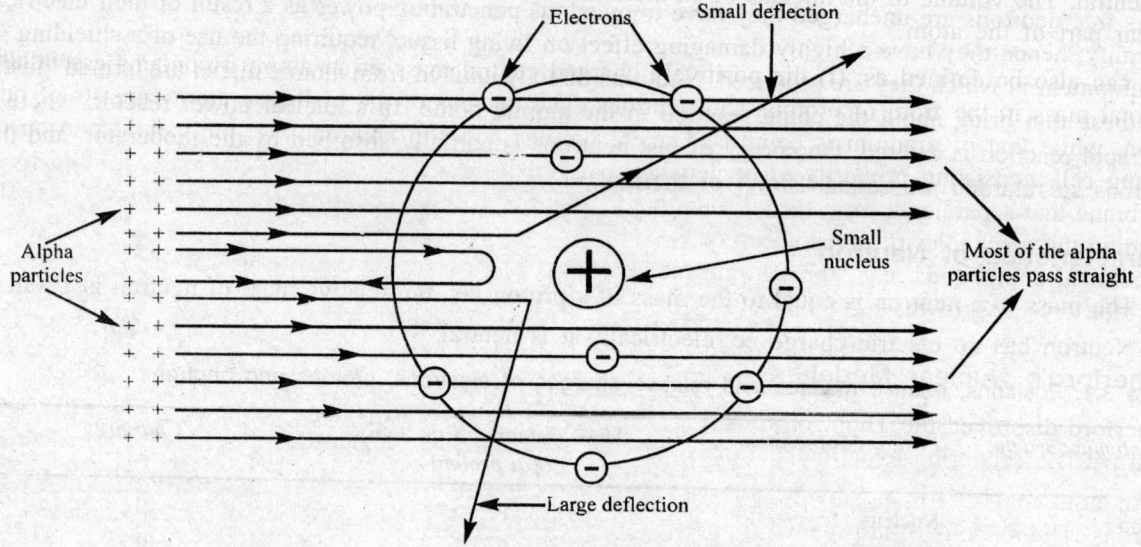

Fig. 3.6. Scattering of alpha particles by the atom of a metal foil.

Rutherford explained these observation in the following way :

1. Gold foil is made up of atoms. If the atoms were solid throughout their volume, then every alpha particle striking them should have changed its path (get deflected). Since most of the alpha particles pass straight without deflection, it shows that there is a lot of empty or hollow space in an atom. Large deflection indicate three things:

(a) The alpha particle which is deflected through large angles is meeting a "centre of high mass" in the atom. This centre does not allow the alpha particles to pass through it.

(b) The centre of high mass should have positive charge, because it repels the positively charged alpha particles. This positive centre of high mass is known as nucleus of the atom.

(c) As the number of alpha particles which are deflected is very small, the volume occupied by the positive centre (known as nucleus) must be very small as compared to the total volume of the atom.

From the above discussion we conclude that Rutherford's α-particle scattering experiment gives the following important information about the nucleus of the atom :

1. Nucleus of an atom is positively charged.

2. Nucleus of an atom is very dense and hard.

3. Nucleus of an atom is very small as compared to size of the atom as a whole.

NUCLEUS

The nucleus is a small positively charged part at the centre of an atom. The nucleus contains all the protons and neutrons, therefore almost the entre mass of an atom is concentrated in the nucleus (the electrons, which are outside the nucleus, have negligible mass). The positive charge on the nucleus is due to the presence of protons in it. The number of protons in the nucleus determines the number of positive charges on the nucleus. The neutrons which are also present in the nucleus have no charge, they are neutral. The volume of the nucleus of an atom is very small as compared to the volume of the extra nuclear part of the atom.

It can also be defined as: (i) the positively charged central mass of an atom; it contains essentially the total mass in the from of protons and neutrons. The nucleus of the hydrogen atom consists of one proton, while that of uranium is comprised of 92 protons and 146 neutrons; (ii) the central portion of a living cell, consisting primarily of nucleoplants in which chromatin is depressed. It is enclosed by a membrane that separates it from the surrounding cytoplasm. All the most important functions of the cell, including the mechanics of division (mitosis) and the nucleus; (iii) the characteristic structure of a group of chemical compounds, e.g., the benzene nucleus; and (iv) any small particle which can serve as the basis for crystal growth.

Rutherford's Nuclear Model of Atom

Rutherford disproved the Thomson model and proposed the nuclear model of the atom. On the basis of alpha particle scattering experiment, Rutherford gave the following picture of the structure of an atom.

An atom consists of a positively charged, dense and very small nucleus containing all the protons and neutrons. The nucleus is surrounded by negatively charged electrons. It was suggested that the electrons are revolving round the nucleus at very high speeds due to which they do not fall into the oppositely charged nucleus. Since an atom is electrically neutral, therefore, the number of electrons in an atom is equal to the number of protons in it. The electrostatic attraction between the positively charged protons and the negatively charged electrons, holds the atom together (Fig. 3.7. and 3.8).

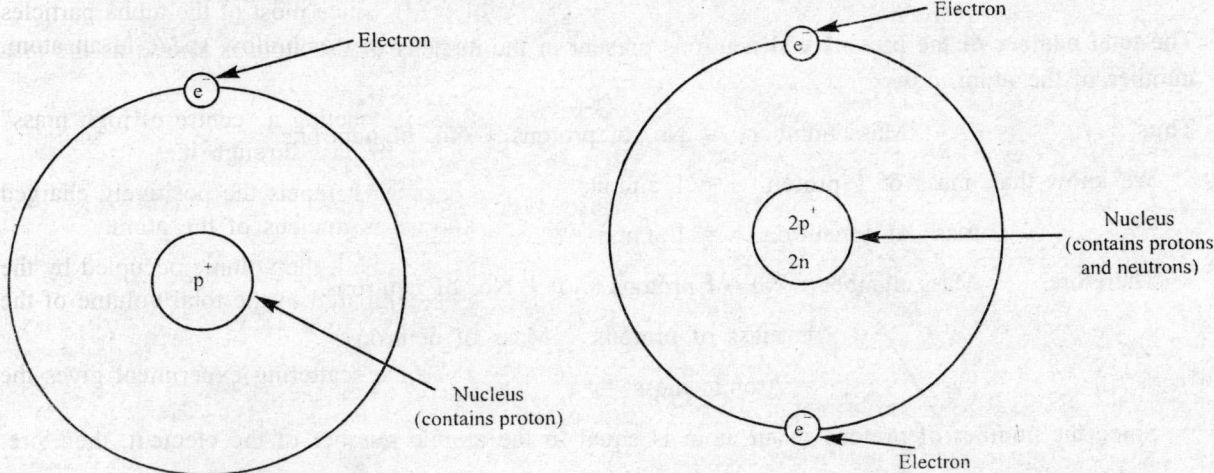

Fig. 3.7. Structure of a hydrogen atom.
Here p^+ = proton; e^- = electron

Fig. 3.8. Structure of a helium atom.
p^+ = proton, n = neutron; e^- = electron

The simplest atom is that of hydrogen. It contains one proton and one electron. According to Rutherford's theory, a hydrogen atom consists of a small nucleus containing one proton, and one electron revolving around it. The nucleus is almost at the centre of the atom. Since the hydrogen atom contains an equal number of protons and electrons (1 each), it is electrically neutral.

Note: The nucleus of an ordinary hydrogen atom does not contain any neutron in it.

The next simplest atom is that of helium. A helium atom consists of a small central nucleus containing 2 protons and 2 neutrons and there are 2 electrons revolving around this nucleus. Since the helium atom contains an equal number of protons and electrons (2 each) therefore, it is electrically neutral.

Atomic Number

Atomic number can be defined as the number of protons (positively charged mass units) in the nucleus of an atom, upon which its structure and properties depend. This number represents the location of an element in the periodic table; it is always the same as the number of negatively charged electrons in the shells. Thus an atom is electrically neutral, except in an ionised state, that is, when one or more electrons have been gained or lost. Atomic numbers range from 1 for hydrogen, to 106, for the most recently discovered element. It was discovered by Moseley in 1913 and it is represented by symbol 'Z'. Since the charge on the nucleus is equal to the number of the protons (p) in it which in turn is equal to the number of electrons (e) in the atom, the atomic number is defined as the number of protons, or the number of electrons for a neutral atom in the atom. It is used to identify an element.

Atomic number of an atom   = Number of protons in one atom.

= Number of electrons in neutral atom.

For example, atomic number 6 tells us that it is carbon element. No other element can have atomic number of 6. Thus, it is the number of protons (or atomic number) which distinguishes the atoms of one element from the atoms of another element.

Mass Number

The total number of the protons and neutrons present in the nucleus of the atom is knows as the mass number of the atom.

Thus,   Mass number   = No. of protons + No. of neutrons

We know that, mass of 1 proton   = 1 a.m.u.

mass of 1 neutron   = 1 a.m.u.

Therefore,   Mass number = No. of protons × 1 + No. of neutrons × 1

= mass of protons + Mass of neutrons

= Atomic mass

Since the number of protons in an atom is equal to the atomic number of the element, therefore,

Mass number   = Atomic number + neutron number.

Thus the mass number of normal helium is 4, of carbon 12, of oxygen 16, and of uranium 238. A given atom is characterised by its atomic number, equivalent to the number of protons, which give its charge and thus determine the kind of element, and by its mass number, which includes the neutrons that make up the remainder of its mass. Helium has two protons and two neutrons, (mass number 4 and atomic number 2). Protons and neutrons each have very close to unit mass, and since the mass change associated with until of the atomic weight of the nuclide.

Example. 3.1. Calculate the atomic number of an element whose atomic nucleus has mass number 23 and neutron number 12. What is the symbol of the element?

Solution. Mass number   = No. of protons + No. of neutrons

or   Mass number   = Atomic number + Neutrons number

                  23   = Atomic number + 12

or   Atomic number   = 23 – 12 = 11

The element with atomic number 11 is sodium and its symbol is Na.

Drawback of Rutherford's Model of an Atom

A major drawback of Rutherford's model of an atom is that is does not explain the stability of the atom. In his model of an atom, the negatively charged electrons are revolving around the positively charged nucleus in circular paths. Now, we know that if an object moves in a circular path, then the motion is said to be accelerated. That means the motion of an election revolving around the nucleus is accelerated.

Bohr's Model of Atom

In order to explain the stability of an atom, Niels Bohr, in 1913, gave a new arrangement of electrons in the atom. According to Bohr's theory:

1. The nucleus of an atom is situated at its centre.
2. The electrons in an atom revolve around the nucleus in definite circular paths known as the energy levels.
3. The energy levels are either designated as K, L, M, N, etc. or numbered n = 1, 2, 3, 4, etc. outwards from the nucleus.
4. Each energy level (or shell) is associated with a fixed amount of energy, the shell nearest to the nucleus having minimum energy and the shell farthest from the nucleus having the maximum energy.
5. There is no change (i.e., neither loss nor gain) in the energy of electrons as long as they keep revolving in the same energy level, and the atom remains stable. The change in the energy of an electrons takes place only when it jumps from a lower energy level to a higher energy level (i.e., gain of energy) or when it comes down from a higher energy level to a lower energy level (i.e., loss of energy).

Distribution of Electrons in Different Energy Levels

The distribution of electrons in different energy levels is governed by a scheme known as Bohr-Bury scheme which states that:

1. The maximum number of electrons that can be accommodated in any energy level (or shell) is $2n^2$ where 'n' is number of that energy level. Thus

$$\text{K-shell } (n = 1) \text{ can have } 2 \times 1^2 = 2 \text{ electrons}$$

$$\text{L-shell } (n = 2) \text{ can have } 2 \times 2^2 = 8 \text{ electrons}$$

$$\text{M-shell } (n = 3) \text{ can have } 2 \times 3^2 = 18 \text{ electrons}$$

$$\text{N-shell } (n = 4) \text{ can have } 2 \times 4^2 = 32 \text{ electrons}$$

The electrons first occupy the shell with the lowest energy.

2. The outermost shell of an atom cannot have more than 8 electrons and next to the outermost shell cannot have more than 18 electrons.

3. The systematic distribution of electrons in different energy shells is called the *electronic configuration* of the atom.

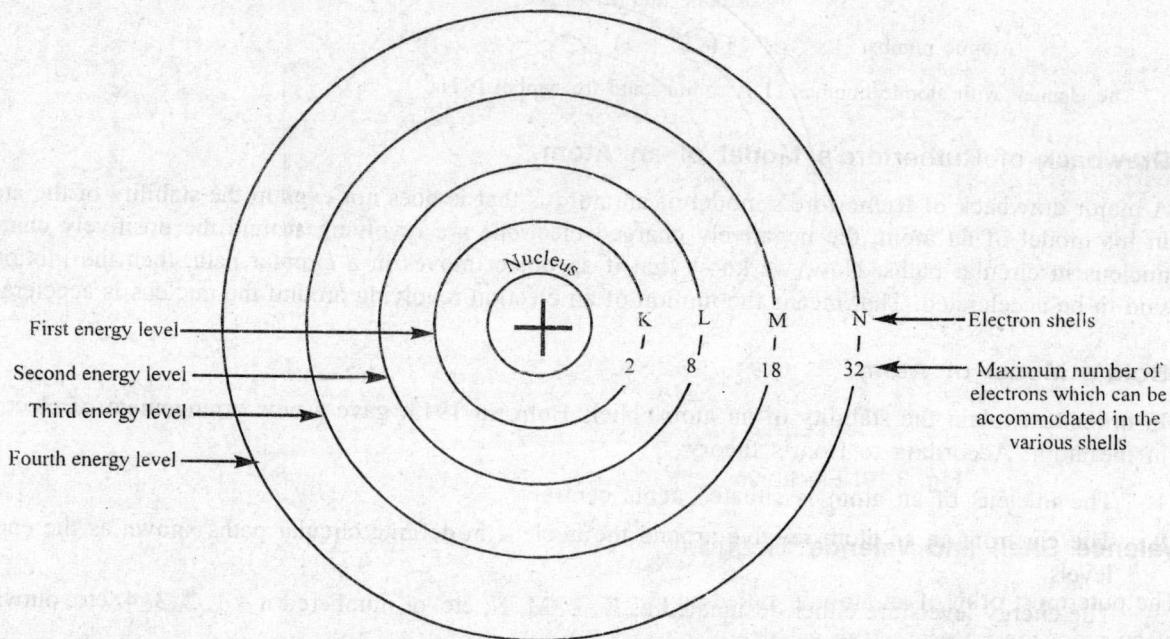

Fig. 3.9. Energy levels or electrons shells in an atom. (This figure shows only first four shells K, L, M and N)

Solved Examples

Example 3.2. Write the electronic configuration of an element X whose atomic number is 12.

Solution. Atomic no. of X is 12 i.e., one atom of X has 12 electrons. First 2 electrons will occupy K shell since it can take maximum of 2 electrons. After filling the 'K' shell the electrons will go to L shell, which can take a maximum of 8 electrons.

In this way 2 + 8 = 10 electrons have been accommodated. The remaining 2 electrons go to M shell (Fig. 3.10). Writing the electron shells together, the electronic configuration of 'X' becomes:

K	L	M
2	8	2

Example 3.3. What would be the electronic configuration of a positively charged sodium ion, Na+? What would be its atomic number? What is its atomic mass?

Solution. The atomic number of sodium is 11. So, a natural sodium atom (Na) has 11 electrons.

1. *A positively charged sodium ion (Na+) is formed by the removal of 1 electron from sodium atom. So a sodium ion has 11–1=10 electrons in it. So, the electronic configuration of sodium ion will be*   K,   L.

   2,   8

2. *The atomic number of an element is equal to the number of protons in its atom. Since a sodium atom as well as a sodium ion contain the same number of protons, therefore, the atomic number of sodium ion is the same as that of a sodium atom, which is 11.*

3. The atomic mass is equal to the mass of protons and neutrons in an atom. Since a sodium atom and a sodium ion contain the same number of protons and neutrons, the atomic mass of a sodium ion is the same as that of sodium atom, which is 23.

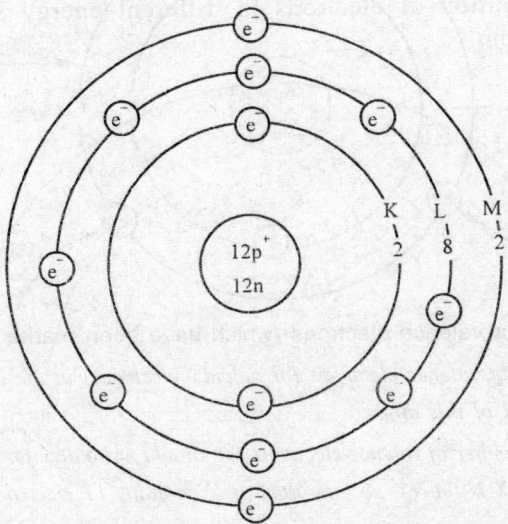

Fig. 3.10. Electronic configuration of element 'X' having 12 electrons.

Valence Shell and Valence Electrons

The outermost orbit of an atom is called as its valence shell. The electrons present in the outermost shell of an atom are known as the valence electrons (or valency electrons) because they decide the valency (combining capacity) of the atom.

Only the valence electrons of an atom take part in chemical reactions because they have more energy than all the inner electrons of the atom.

For example, the atomic number of carbon is 6, which means that one carbon atom has 6 electrons in it. So, the electronic configuration of carbon atom will be K L. In the carbon atom, L shell is the

2 4

outermost shell or valence shell. There are 4 electrons in the outermost shell of carbon atom, therefore, carbon atom has 4 valence electrons. This has been shown clearly in Fig. 3.11 given below.

When a carbon atom combines with other atoms, only the four valence electrons take part in the reaction. The two electrons of the inner shell (K shell) of the carbon atom never take part in chemical reactions. We can now say that those electrons of an atom which take part in chemical reaction are called valence electrons. Valence electrons are located in the outermost shell of an atom. In a chemical reaction, valence electrons of an atom are either transferred to the valence electrons of another atom, or shared with the valence electrons of another atom.

In order to find out the number of valence electrons in an atom of the element, we should write down the electronic configuration of the element by using its atomic number. The outermost shell will be the valence shell and the number of electrons present in it will give us the number of valence electrons.

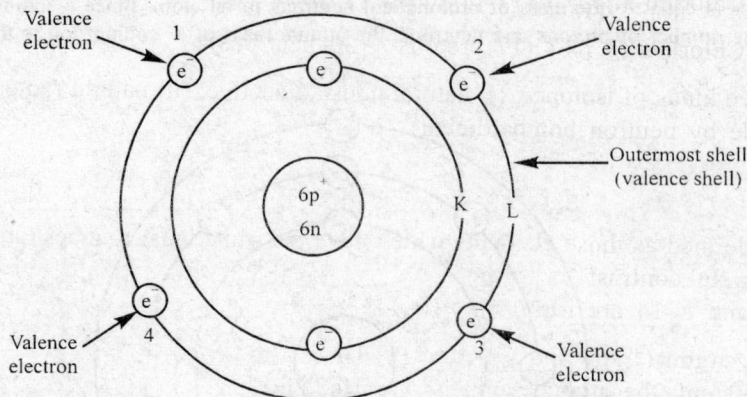

Fig. 3.11. Carbon atom has four valence electrons (which have been marked as 1, 2, 3, and 4 in the figure).

Example 3.4. What is the number of valence electrons (or valency electrons) in the atom of an element X having atomic number 17 Name the valence shell of this atom.

Solution. In order to find out the number of valence electrons, we should write down the electronic configuration of element X. The atomic number of element X is the 17, so one atom of X contains 17 electrons. The electronic configuration will be:

$$
\begin{array}{ccc}
K & L & M \\
2 & 8 & 7
\end{array}
$$

Here M shell is the outermost shell or valence shell of the atom and it has 7 electrons in it. Thus, there are 7 valence electrons in the atom of element X.

Notes:

1. *Elements having the same number of valence electrons in their atoms show similar chemical properties.*

2. *Elements having different number of valence electrons in their atoms show different chemical properties.*

3. *Elements having 1, 2 or 3 valence electrons in their atoms are metals.*

4. *Elements having 4, 5, 6 or 7 valence electrons in their atoms are non-metals.*

Isotopes

In 1919, F.W. Aston discovered that the atoms of some naturally-occurring elements were not exactly alike. He observed that these atoms of the same element had different masses. Such types of atoms of the same element are known as isotopes. All the isotopes of any element have the same atomic number because their nuclei contain the same number of protons, but their mass numbers are different. Since the isotopes of any element contain the same number of electrons, therefore, they have the same chemical properties. However, as the number of neutrons in the nuclei of the atom of different isotopes are different, the physical properties like densities and melting and boiling points are slightly different.

In short, isotopes are atoms of the same element having the same atomic number but different mass number. Examples of some isotopes of some elements are :

Hydrogen   Protium $^{1}_{1}H$ (1 proton only)

            Deuterium $^{2}_{1}H$ (1 proton + 1 neutron)

            Tritium $^{3}_{1}H$ (1 proton + 2 neutrons)

Chlorine	Chlorine-35 $^{35}_{17}Cl$ (17 protons + 18 neutrons)
	Chlorine-37 $^{37}_{17}Cl$ (17 protons + 20 neutrons)

There are three kinds of isotopes: (i) natural non-radioactive; (ii) natural radioactive; and (iii) artificial radioactive (made by neutron bombardment).

Isobars

Isobars may be defined as those elements which have the same mass number (atomic mass) but different atomic numbers. In contrast to isotopes, which have the same atomic number but different mass numbers. C-14 and N-14 are isobars.

For example, argon ($^{40}_{18}Ar$) and Calcium ($^{40}_{20}Ca$), have the same mass number (i.e., 40), and as the elements are different, the atomic numbers are different.

Electronic Structure of Atoms

The chemical behaviour of an atom is mainly controlled by its electronic structure. The term '*electronic structure*' means: (i) number of electrons; (ii) the distribution of these electrons in the space around the nucleus; and (iii) the relative energy of these distributions. We have seen that the number of electrons is determined by the atomic number of the elements. To better *understandable* (ii) and (iii) above consider the hydrogen atom, which is the simplest atom. The hydrogen atom contains only one electron. The study of the emission spectrum of light by the hydrogen atom provided the most important clue to its electronic structure. In order to understand the emission spectrum we shall briefly examine the nature of light.

NATURE OF LIGHT AND ELECTROMAGNETIC WAVES

Corpuscular Theory

The earliest view of light, was put forward by Newton. According to this theory light is a stream of particles more commonly termed as corpuscles of light. While this view explained the experimental laws of reflection and refraction of light, it failed to account for the phenomena of interference and diffraction. The *corpuscular theory* was therefore discarded and replaced by the *wave theory.*

Wave Nature of Light

In 1956, Maxwell proposed that light has wave characteristics. According to him light is form of wave motion and these waves are characterised by wavelength (λ), frequency (v) and speed of propagation (c), which are related by the equation,

$$\lambda v = c$$

The speed of light has been determined and is found to be constant in vacuum. It has the value of $3.00 \times 10^8 m/s$.

Electromagnetic Spectrum

The different colours, e.g., blue, red, green etc. have different wavelengths or different frequencies. Towards the end of the last century, it was shown that light waves are electromagnetic in nature (i.e., they are oscillation of electric and magnetic field in space). It other words, light is electromagnetic radiation. Various types of electromagnetic radiations (EMR) having various wavelengths (or frequencies) are now known. They constitute the so-called *electromagnetic spectrum.*

Different regions of the spectrum are identified by different names. Some examples are : radio frequency region, around 10^6MHz, used for broadcasting; microwave region (around 10^{10}MHz) a component of the sun's radiation. The small portion known as the visible spectrum (around 10^{15}MHz) is what is ordinarily called light. It is the only part which our eyes can detect. Special instruments are necessary to detect non-visible types of electromagnetic radiation.

Example 3.5. The wavelength of a moving body of mass 0.1 mg is 3.312×10^{-29} m. Calculate its kinetic energy (h = 6.625×10^{-34} Js.)

Solution. By de-Broglie's relation, we have:

$$\lambda = \frac{h}{mv} \quad \text{or} \quad v = \frac{h}{m\lambda}$$

Here

$$1 = 3.312 \times 10^{-29} \text{ m};$$

$$m = 0.1 \text{ mg} = 1 \times 10^{-7} \text{ kg};$$

and

$$h = 6.625 \times 10^{-34} \text{ Js}.$$

$$v = \frac{6.625 \times 10^{-34} \text{ Js}}{1 \times 10^{-7} \text{ kg} \times 3.312 \times 10^{-29} \text{ m}}$$

$$= 200 \text{ m s}^{-1}$$

$\therefore$ Kinetic energy

$$= \frac{1}{2} mv^2$$

$$= \frac{1}{2} \times 1 \times 10^{-7} \times (200)^2$$

$$= 2 \times 10^{-3} \text{ J}.$$

Quantum Nature of Light

Another property of light is that it is a form of energy. Albert Einstein, who based his consideration on the work of Max Planck, showed that light energy is carried in packets named photons. The energy of a photon is directly proportional to the frequency (v) of the light wave. The relation between the energy and the frequency is as follows:

$$E = hv$$

where h is a universal constant known as Planck's constant.

The value of h is 6.63×10^{-34} Js, or 3.99×10^{-13} kJs mol^{-1}.

If h is measured in Js, the value of E is in J per photon.

This relation is valid for all forms of EMR's. It shows that higher the frequency the more energetic are the corresponding photons. Sometimes the energy of a photon is also measured in terms of eV known as electron-volt.

$$1 \text{ eV} = 1.6 \times 10^{-19} \text{ J}$$

Photoelectric Effect

It was observed that when a clean metallic surface is bombarded by monochromatic light of proper frequency, electrons are emitted from it. This phenomenon of ejection of the electrons from metal surface was called as *photoelectric effect*. The important experimental observations of this effect were,

1. If the frequency of incident radiation is below a certain minimum value (threshold frequency), no emission takes place, whatsoever the intensity of light may be.

2. The kinetic energy K.E. of the emitted electrons was independent of the intensity of the light. The kinetic energy of the electrons increases linearly with the frequency of incident light radiation. This was highly contrary to the laws of Physics at that time i.e. the energy of the electrons should have been proportional to the intensity of the light, not on the frequency.

These features could not be explained on the basis of wave nature of light. The quantum nature of light could explain the effect satisfactorily.

According to Einstein, when a quantum of light (photon) strikes a metal surface, it transfers its energy to the electrons in the metal. For an electron to escape from the surface of the metal, it must overcome the attractive force of the positive ions in the metal. So some energy has to be supplied to the electron to overcome this attraction, this energy is known as *work function* of the surface and is denoted by ω. A part of the incident energy is utilised in overcoming the work function of the metal and the remaining part of the energy of the photon goes into the kinetic energy of the emitted electron. If E is the energy of the photon, KE is the kinetic energy of the electron and ω be the work function of the metal then Einstein's photoelectric equation is given by :

$$E = K.E + \omega$$

Electrons are emitted only if the frequency of the incident light is equal to or above a threshold frequency. The threshold frequency depends on the work function of the metal. If v_0 is the threshold frequency, and v the frequency of incident light, then we have :

$$\omega = h\, v_0 \text{ and } \qquad E = h\, v$$

$$\Diamond \qquad K.E. = E - \omega \qquad \Diamond \qquad K.E. = h\,(v - v_0)$$

Hence, the quantum theory of light could satisfactorily explain the photoelectric effect.

Note: The photoelectric effect can be explained only by assuming the particle nature of light. The fact that light exhibits diffraction means that light also has a wave character. Thus, light has a dual character.

Example 3.6. Find the ratio of frequencies of violet light ($\lambda = 4.10 \times 10^{-7}$ m) to that of red light ($\lambda = 6.56 \times 10^{-7}$ m). Also determine the ratio of energies carried by them.

Solution. using $c = v\lambda$, we get

$$\Diamond \qquad \frac{v_v}{v_r} = \frac{\lambda_r}{\lambda_v} = \frac{6.56 \times 10^{-7}}{4.10 \times 10^{-7}} = 1.6$$

The energy associated with electromagnetic radiation is given by $E = h\nu$.

$$\frac{E_r}{E_v} = \frac{v_r}{v_v} = \frac{\lambda_v}{\lambda_r} = 1.6$$

The ratio of energies is same as that of frequencies.

Example 3.7. When electromagnetic radiation of wavelength 300 nm falls on the surface of sodium, electrons are emitted with a kinetic energy of 1.68×10^5 J mol^{-1}. What is the minimum energy needed to remove an electron from sodium? What is the maximum wavelength that will cause a photoelectron to be emitted?

Solution. The energy (E) of a 300 nm photon is given by

$$E = h\nu = \frac{hc}{\lambda} = \frac{6.63 \times 10^{-34} \, Js}{300 \times 10^{-9} \, m} \times 3.00 \times 10^8 \, ms^{-1}$$

$$= 6.63 \times 10^{-19} J$$

The K.E. of 1 mol of electron = 1.68×10^5 J mol^{-1}

$$\therefore \quad K.E. \text{ per electron } = \frac{1.68 \times 10^5}{6.023 \times 10^{23}} = 2.789 \times 10^{-19} \text{ J / electron}$$

$\therefore$ The minimum energy needed to remove an electron = E_0

$$E_0 = 6.63 \times 10^{-19} \quad - \quad 2.789 \times 10^{-19}$$

$$= 3.84 \times 10^{-19} \text{ J per electron}$$

For maximum wavelength

$$\lambda_{max} = \frac{hc}{E_0} = \frac{6.63 \times 10^{-34} \times 3 \times 10^8}{3.84 \times 10^{-19}} = 5.179 \times 10^{-7} \text{ m } = 518 \text{ nm}$$

Bohr's Theory of Atomic Structure

Bohr's theory of atomic structure and his equation for hydrogen or hydrogen-like atoms or particles

Neil Bohr, a Danish physicist, proposed the following model of atomic structure on the basis of Planck's Quantum theory.

1. An atom consists of a massive positively charged central core or nucleus with electrons revolving around it in circular orbits and the nucleus is assumed to be at rest.

2. In order to keep the electrons in orbit and prevent them from spiralling towards nucleus the centripetal force being provided by the electrostatic attraction of the electron by the nucleus, the centrifugal force due to its revolution balances it.

3. The electron revolves around the nucleus only in definite, discrete and allowed levels called stationary levels each of which is associated with a definite energy and the electron as long as it moves in stationary levels it does not radiate energy.

4. The size of the allowed stationary orbits is governed by the rule that the angular momentum of the electron therein is an integral multiple of h/2π, i.e., angular momentum (mvr) $= \dfrac{nh}{2\pi}$ where h is Planck's constant and n is the principle quantum number which can have integral values of 1, 2, 3, 4, 5, etc. In other words, the angular momentum is quantised.

5. An electron radiates/absorbs energy when it jumps from one orbit to the other. The energy absorbed or radiated during jumps is equal to

$$\Delta E = E_2 - E_1 = h\nu$$

where h is Plank's constant and v is the frequency of the radiation emitted or absorbed, as the case may be and E_1 is the energy at lower state and E_2 is the energy at higher state.

6. An excited electron has always a tendency to revert to the non-excited or the ground state.

7. On the basis of his theory he worked out an equation to explain the information of different spectral lines of a hydrogen or hydrogen-like particle. The equation runs as follows :

$$\bar{v} = \frac{1}{\lambda} = R_H\left(\frac{1}{n_1^2} - \frac{1}{n_2^2}\right) cm^{-1}$$

where $\bar{v}$ is the wave number, the reciprocal of wavelength (λ) gives the number of waves per cm. and R_H is Rydberg's constant, the value of which is $= 10.97 \times 10^{-6}$ m^{-1}.

8. Energy of the electron in any orbit is given by the expression

$$E_n = -\frac{2.18 \times 10^{-18}}{n^2} \text{ Joules}$$

where n is the number of the orbit. Thus as n increases, i.e., higher the orbit is, less negative is the energy of the electron in it, hence possesses more energy.

9. The radius of nth orbit is given by the expression

$$r_n = \frac{n^2}{Z} \times 5.29 \times 10^{-11} m$$

where n is the orbit no. and Z, the atomic no.

Again, the velocity of the electron in any orbit is given by the expression

$$V_n = \frac{2.185 \times 10^6 Z}{n} \text{ ms}^{-1}$$

10. *Line spectra of hydrogen:* In the production of line spectra, an electron is excited from its original energy level (ground state) to a higher excited state and when it falls back to lower energy level so as to go to its ground state, it can do so in one or more steps and the surplus energy is emitted in the form of line spectrum. The wavelength or frequency of radiation emitted being dependent upon the quantity of energy liberated. Thus the only electron in H-atom may get excited to different levels

in its different atoms depending upon the quantity of the energy absorbed. These electrons when they fall back to lower levels or to the ground state, different series of radiations or lines (emission spectra) are obtained. Five series of lines that are obtained in line spectra are as under :

(a) *Lyman series (Ultraviolet region)* are obtained when electrons from energy levels 2, 3, 4, 5 etc. (n_2) fall to the ground state ($n_1=1$).

(b) *Balmer series (Visible region—4000 A° to 7000 A°)* are obtained when electrons jump from 3, 4, 5, etc. (n_2) to lower level ($n_1=2$).

(c) *Paschen, Brackett and Pfund series* which lie in the infrared region of the spectrum are obtained when n_1 is respectively equal to 3, 4 or 5 and n_2 has correspondingly higher values.

Example 3.8.

1. Calculate the wavelength of a wave of frequency 10^{11} Hertz (Hz) (Cycles per second–cps) travelling with the speed of light (3×10^8 ms^{-1}).

2. Calculate the frequency and wave number of yellow light having wavelength 580 nm (10^{-9} m).

3. Calculate the frequency and wavelength of light radiated when an electron in an excited state falls from the 2nd to the 1st orbit of hydrogen (h=6.62×10^{-34} Js and the energy of H electron in the ground state is -2.18×10^{-18} J atom^{-1} and c=3×10^{-8} ms^{-1}).

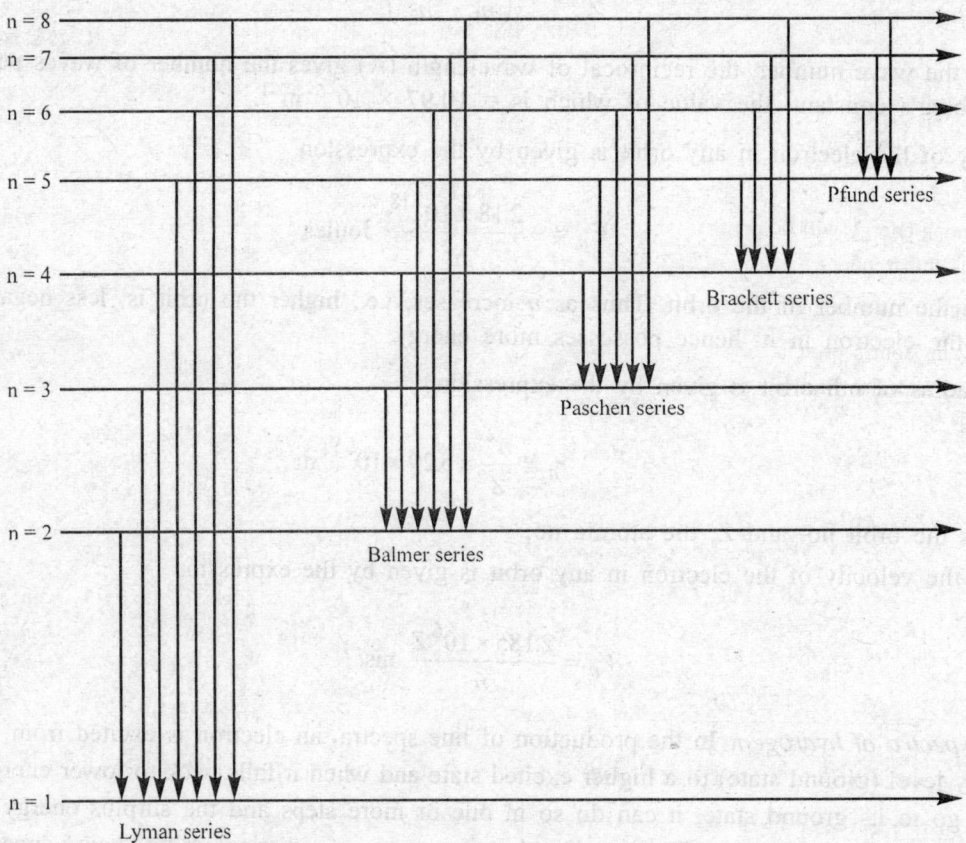

Fig. 3.12. Energy level diagram of hydrogen atom spectrum.

Solution.

(a)
$$\lambda = \frac{c}{v} = \frac{3 \times 10^8}{10^{11}} = 3 \times 10^{-3} \text{ m}$$

(b)
$$v = \frac{c}{\lambda} = \frac{3 \times 10^8}{580 \times 10^{-9}} = 5.17 \times 10^{14} \text{ Hz or cps}$$

$$\bar{v} = \frac{1}{\lambda} = \frac{1}{580 \times 10^{-9}} = 1.72 \times 10^6 \text{ m}^{-1}$$

(c)
$$\Delta E = E_2 - E_1 = -2.18 \times 10^{-18} \left(\frac{1}{n_1^2} - \frac{1}{n_2^2} \right)$$

$$= -2.18 \times 10^{-18} \left(\frac{1}{1} - \frac{1}{4} \right)$$

$$= -2.18 \times 10^{-18} \times \frac{3}{4} = 1.635 \times 10^{-18} \text{ J}$$

$$= h v$$

Example 3.9.

1. How much energy will be emitted when electrons in 1 g. atom of hydrogen fall from the 2nd Bohr orbit to the first? What is the frequency of the radiation?

2. Calculate the wavelength of the first line of the Balmer series (Rydberg's constant = 10.97×10^6 m^{-1}).

3. Energy in Bohr's orbit is given to be equal to $-\dfrac{A}{n^2}$ where A = 2.179×10^{-18} J. Calculate the frequency of the radiation and also find out the wave number when the electron moves from 3rd orbit to the second orbit (h = 6.62 $\times 10^{-34}$ Js).

Solution.

(a)
$$\Delta E = E_2 - E_2 = 2.18 \times 10^{-18} \left(\frac{1}{1} - \frac{1}{4} \right) \text{ J atom}^{-1}$$

$$= 2.18 \times 10^{-18} \times \frac{3}{4} \times 6.02 \times 10^{23} \text{ J / g. atom}$$

$$= 984.25 \text{ kJ}$$

$$v = \frac{\Delta E}{h} = \frac{2.18 \times 10^{-18} \times 3}{6.62 \times 10^{-34} \times 4} = 2.47 \times 10^{15} \text{ Hz}$$

(b) For the first line of the Balmer series

$$\bar{v} = R \left(\frac{1}{2^2} - \frac{1}{3^2} \right) = 10.97 \times 10^6 \left(\frac{1}{4} - \frac{1}{9} \right)$$

$$= 10.97 \times 10^6 \times 5/36 \text{ m}^{-1}$$

$$\lambda = \frac{1}{\bar{v}} = \frac{36}{10.97 \times 10^6 \times 5} = 0.6563 \times 10^{-6} \text{ m}$$

(c) $iE = E_2 - E_1 = hv = 6.62 \times 10^{-34} \times v$

$$= 2.179 \times 10^{-18} \left(\frac{1}{4} - \frac{1}{9} \right)$$

whence $v = \dfrac{2.179 \times 10^{-18} \times 5}{6.62 \times 10^{-34} \times 36} = 4.57 \times 10^{14} \text{ Hz}$

$$\bar{v} = \frac{1}{\lambda} = \frac{4.57 \times 10^{14}}{3.0 \times 10^8} = 1.52 \times 10^6 \text{ m}^{-1}$$

Example 3.10.

Calculate the velocity (cm./sec.) of an electron placed in the third orbit of the hydrogen atom. Also calculate the number of revolutions per second this electron makes around its nucleus.

Solution.

1.   Velocity of the electron being given by the expression

$$\text{Vn} = \frac{2.185 \times 10^8}{n} Z \text{ cms}^{-1}$$

We have, $V_3 = \dfrac{2.185 \times 10^8}{3} \times 1 = 7.28 \times 10^7 \text{ cms}^{-1}$

2.   Radius of the orbit being given by the expression

$$\left(r_n = \frac{n^2}{Z} \times 5.29 \times 10^{-13} \text{ cm} \right)$$

We have, $r_n = \dfrac{9 \times 5.29 \times 10^{-13}}{1} = 4.761 \times 10^{-13} \text{ cm.}$

3.   No. of revolutions per second

$$= \frac{\text{Velocity}}{\text{circumference}} = \frac{7.28 \times 10^7}{2 \times \pi \times r} = \frac{7.28 \times 10^7}{2 \times 3.14 \times 4.76 \times 10^{-13}}$$

$$= 2.44 \times 10^{19}.$$

Drawbacks of Bohr Model

Bohr's model was successful in explaining the hydrogen atom spectra and hence the structure of hydrogen atom. It still left many questions unanswered.

1.   *The theory combined two different concepts:* one from classical physics and second from quantum physics.

2. No justification was given for the quantisation of the angular momentum, though this was a correct assumption.

3. It could not explain the spectrum of atoms or ions having two or more electrons. It accounted only for the spectra of atoms or ions having only one electron.

4. The model could not provide a satisfactory picture for chemical bonding.

5. It could not explain the brightness of the spectral lines, splitting of spectral lines in electric field (Stark Effect) and in magnetic field (Zeeman Effect).

QUANTUM MECHANICAL MODEL OF THE ATOM

Dual Nature of Electrons (Wave Nature and Particles)

Louis de Broglie, a French Physicist, in 1924 introduced the idea that as light is found to have a dual character, i.e. as waves and as particles, it is also possible that matter posses a dual character i.e. every form of matter (electron or proton or any other particle) behaves like waves in some circumstances. According to him a wave can be a particle and a particle can be a wave. These waves are called *matter waves* or *de Broglie's waves*.

Further, he proposed that the wavelength of the matter waves is related to the momentum of the particle. Mathematically it is represented as :

$$\lambda = \frac{h}{mv} = \frac{h}{p}$$

where   h = Planck's constant (=6.63 × 10^{-34} Js)

λ = wavelength of the particle

m = mass of the particle

v = velocity of the particle

p = momentum of the particle

The above equation is known as "de Broglie's" equation.

Example 3.11. Calculate the momentum of a particle which has a de-Broglie wavelength of 0.1 nm.

Solution.

Here wavelength λ = 0.1 nm = 0.1 × 10^{-9} m = 1× 10^{-10} m. But according to de Broglie's equation,

$$\lambda = \frac{h}{p}$$

$$\therefore \text{ Momentum, } p = \frac{h}{\lambda} = \frac{6.6 \times 10^{-34} \text{ kg m}^2\text{s}^{-1}}{1 \times 10^{-10} \text{ m}} = 6.6 \times 10^{-24} \text{ kg ms}^{-1}$$

Example 3.12. What must be the velocity of a beam of electrons, if they are to display a de Broglie wavelength of 100A°? (Mass of electron = 9.1×10^{-31} kg, h = 6.6×10^{-34} Js).

Solution.

Here $\lambda = 100\text{Å} = 1 \times 10^{-8}$ m; m = 9.1×10^{-31} Kg.

According to de Broglie's equation

$$\text{Velocity, } v = \frac{h}{m\lambda} = \frac{6.6 \times 10^{-34}\,\text{Js}}{9.1 \times 10^{-31}\,\text{k} \times 1 \times 10^{-8}} = 7.25 \times 10^{4}\ \text{ms}^{-1}$$

Heisenberg's Uncertainty Principle

The wave nature of electron puts some restriction on how precisely its position can be determined. In answer to this question the famous German physicist, Werner Heisenberg put forward his principle of uncertainty. According to his principle, it is not possible to determine with accuracy both the position and momentum of a particle simultaneously.

This is because the size of the electron is very small and the radiations of high energy with extremely small wavelength are required to detect it. In an attempt to do so, the direction and the speed of the electron is affected. According to Heisenberg we can measure position and momentum of such particles only within certain limits of accuracy. If Δx is the uncertainty in position and Δp be the uncertainty in its momentum, then according to Heisenberg's uncertainty principle,

$$\Delta x \times \Delta p \geq \frac{h}{4\pi}$$

$$\text{or} \quad \Delta x \times m\Delta v \geq \frac{h}{4\pi} [\because \Delta p = \Delta v \times m]$$

$$\text{or} \quad \Delta v \geq \frac{h}{4\pi m \Delta x}$$

where Δv is the uncertainty in velocity and h is the Planck's constant.

Example 3.13. The uncertainty in the position and velocity of a particle are 9.54×10^{-10} m and 5.5×10^{-20} ms^{-1} respectively. Calculate the mass of the particle (h = 6.6×10^{-34} kg m^{2}s^{-1}).

Solution. According to Heisenberg's principle

$$\Delta x . \Delta p \geq \frac{h}{4\pi}$$

$$\text{or} \quad \Delta x . m\Delta v \geq \frac{h}{4\pi}$$

$$\text{or} \quad \Delta v \geq \frac{h}{4\pi \times m \times \Delta x}$$

$$\geq \frac{6.6 \times 10^{-34} \text{ kg m}^2 \text{ s}^{-1}}{4 \times \frac{22}{7} \times 0.150 \text{ kg} \times 1 \times 10^{-10} \text{ m}}$$

or uncertainty in velocity $\geq 3.5 \times 10^{-24}$ ms^{-1}.

Example 3.14. Calculate the product of uncertainty in position/displacement and velocity for an electron of mass 9.1×10^{-31} kg, according to Heisenberg's Uncertainty Principle.

Solution. According to Heisenberg's uncertainty principle, we have,

$$\Delta x . \Delta p \geq \frac{h}{4\pi}$$

or

$$\Delta x . m \Delta v \geq \frac{h}{4\pi}$$

or

$$\Delta x . m \Delta v \geq \frac{h}{4\pi m}$$

$$\geq \frac{6.6 \times 10^{-34} \text{ kg m}^2 \text{ s}^{-1}}{4 \times \frac{22}{7} \times 9.1 \times 10^{-31} \text{ kg}}$$

$$\geq 5.77 \times 10^{-5} \text{m}^2\text{s}^{-1}.$$

Example 3.15. Calculate the uncertainty in the momentum of an electron if it is confined to a linear region of length 1×10^{-10} m. (Planck's constant (h) = 6.6×10^{-34} kg m^{2}s^{-1}).

Solution. According to Uncertainty Principle

$$\Delta x . m \Delta v \geq \frac{h}{4\pi}$$

Substituting the given values we get

$$1 \times 10^{-10} \text{ m} \times \text{momentum} = \frac{6.6 \times 10^{-34} \text{ kg m}^2\text{s}^{-1}}{4 \times \frac{22}{7}}$$

$$\therefore \text{ Momentum} = \frac{6.6 \times 10^{-34} \times 7 \text{ kg m}^2\text{s}^{-1}}{4 \times 22 \times 1 \times 10^{-10} \text{ m}}$$

$$= \frac{2.1}{4} \times 10^{-24} \text{ kg ms}^{-1}$$

$$= \frac{21}{4} \times 10^{-25} \text{ kg ms}^{-1} = 5.25 \times 10^{-25} \text{ kg ms}^{-1}$$

Probability Picture of Electrons

Whenever we find that an exact and precise statement is not possible, we adopt less exact ways for describing the situation. For example, we can say without hesitation that the sun will rise in east tomorrow morning—the weather report even gives the exact time of sunrise—we are unable to predict correctly whether it will rain tomorrow or not. The weather report, therefore, only states whether there is a chance of rain or not. Probabilistic estimate provides the best possible description of a situation which cannot be exactly described.

It is a matter of probability that an electron is more likely to be found in one place or the other. So we can now visualise a region in space (diffused cloud) surrounding the nucleus, where the probability of finding the electron is maximum. Such a region is called as an *orbital*.

An orbital can be defined as the region in space around the nucleus where the probability of finding an electron is greatest. Each orbital can held a maximum of two electrons.

ORBITAL THEORY

The quantum theory of matter applied to the nature and behaviour of the electron, either in single atoms (atomic orbital) or combined atoms (molecular orbital). A combination of Schrodinger's wave mechanics and Heisenberg's uncertainty principle, the orbital theory was formulated in 1926. It has yielded a better understanding of the electron and its critical part in chemical bonding than is possible with Newtonian mechanics. In simple language the orbital theory considers the electron not as a particle, but as a three-dimensional wave that can exist at several energy levels; its exact location and position in the "shell" (which in most elements in a group of orbitals) cannot be precisely determined, but only predicted by the laws of mathematical probability. The orbital levels and the movement of electrons within them are expressed by wave functions and quantum numbers. The probability that an electron will be found in a given volume (i.e., the square of the one-electron wave function) is called the orbital of that electron, and the shape of the orbital is defined by surfaces of constant probability (i.e., spheres and doughnut shaped ellipses). The electron orbital, described in terms of probability, is like a cloud, with indefinite boundaries. The energy state of each electron is given by four quantum numbers which describe its principal level, its angular momentum, its magnetic moment, and its spin. This concept has exerted a profound effect on modern ideas about chemical bonding, transition metal complexes, semi-conductors, and solid state physics.

Main Features of the Orbital Theory

1. In an orbital, there are regions of higher probability where the electron is more likely to be found and there are regions of lower probability where the chances of finding the electron are less.

2. For different orbitals one can say whether, on an average, the electron will be closer or farther from the nucleus, whether it will more likely be along a particular direction and so on.

3. The new theory still defines a definite energy to an orbital in an atom (a remarkable and accepted feature of Bohr's model). The energy is lower if the orbital is concentrated near the nucleus. This is because the electron is more strongly attracted when it is close to the nucleus.

4. The change in energy from one orbital to another is not continuous but discontinuous.

5. The new theory abandons the concept of sharply defined paths.

ORBITALS AND QUANTUM NUMBERS

A large number of electron orbitals are possible in a hydrogen atom. Orbitals can be distinguished in a quantitative manner by their size, shape and orientation. An orbital of smaller size means there is more chance of finding the electron near the nucleus. Similarly, shape and orientation mean that electron distribution has more probability along certain directions and less along certain others. Orbitals are precisely distinguished by what are known as quantum numbers.

Quantum Numbers

An electron in an atom has some location relative to the nucleus and is associated with some energy. These properties taken together describe the state of the electron in the atom. An electron (the state of the electron) within an atom is specified by set of four numbers, (according to the wave mechanical concept of the nature of electron), called quantum numbers. The four quantum numbers are Principal, Azimuthal, Magnetic and Spin quantum numbers. Each of these numbers is associated with a particular characteristic of the electron which they describe.

Principal quantum number

Principal quantum number gives the size of the shell and the relative average distance of the electron, from the nucleus and thereby the energy of the electron. It is denoted by 'n', and 'n' may have any integral value from unity onwards, i.e., 1, 2, 3, 4, etc. Sometimes they are referred to as K, L M, etc. shells. With the increase in the value of n, the distance from the nucleus increases and its energy becomes higher and higher. Large 'n' means a large size. The energy level $n=1$ corresponds to minimum energy and subsequently $n=2, 3, 4, ...$ are arranged in order of increasing energy. It also gives the total number of electrons that can be accommodated in each shell. The capacity of each shell is given by the formula $2n^2$, where 'n' is the principal quantum number.

Azimuthal quantum number

Azimuthal quantum number gives the shape of the orbital for electron and the energy sub-level. It is denoted by letter 'l' and also known as subsidiary or orbital quantum number. Each principal shell is made up of a number of sub-shells and their number is equal to the number of the principal shell. The first shell has only one sub-shell, the second, two sub-shells, the third has three sub-shells and 4th, four sub-shells.

These sub-shells differ from one another in their azimuthal quantum number (l) values. The n sub-shells in the nth principal quantum number are assigned values of $l=0, 1, 2, .. (n=1)$. Thus the first shell or K-shell will have only one sub-shell with the value of $l=0$ and the 2nd shell will have two sub-shells with the values of $l=0$ and $l=l$ and so on. Azimuthal quantum number determines the shape of the sub-shell (or electron cloud) i.e., whether the cloud is spherical $(l=0)$, dumb bell shaped $(l=1)$ or has some other complicated shape.

The values of $l=0, 1, 2$, or 3 are usually referred as s, p, d and f-sub-shells, respectively. These letters are initial letters of the word sharp, principal, diffuse and fundamental series of lines observed in atomic spectra of multiple electron atoms. When more than one electron is present in an atom the energy of the electron is correctly determined by the principal and the azimuthal numbers. For the same principal shell the energy of p sub-shell, more so than p ones and the energy of the f sub-shell of electrons is slightly higher than that of d sub-shell electrons.

Magnetic quantum number

Magnetic quantum number gives the orientation of the orbit, with reference to a fixed direction and is denoted by letter '*m*'.

An electron with angular quantum number can be thought as an electric current circulating in a loop. A magnetic field due to this current is observed. This induced magnetism is determined by the magnetic quantum number. Under the influence of magnetic field, the electrons in a given sub-energy level prefer to orient themselves in certain specific regions in space around the nucleus. The number of possible orientations for a sub-energy level is determined by possible values of *m* (this corresponds to the number of orbitals in a given sub-energy level).

m can have any integral values between $-l$ to $+l$ including 0. In other words a total of $2l+1$ values for given *l*.

In *s* sub-shell ($l = 0$), there is only one orbital.

In a p sub-shell ($l = 1$), there are three orbitals corresponding to three values of *m* $-1, 0, +1$. The three orbitals are represented as P_x; P_y; P_z along X, Y, Z axes perpendicular to each other. In d sub-shell ($l = 2$), there are five orbitals corresponding to $-2, -1, 0, +1, +2$. These five orbitals are represented as d_{xy}, d_{yz}, d_{zx}, d_{x2-y2} and d_{z2}. In f sub-shell ($l=3$), there are seven orbitals corresponding to $-3, -2, -1, 0 +1, +2, +3$.

Shape of the orbitals

The s-orbitals are spherical in shape, i.e., in a s-orbital the electron distribution is symmetric in all directions.

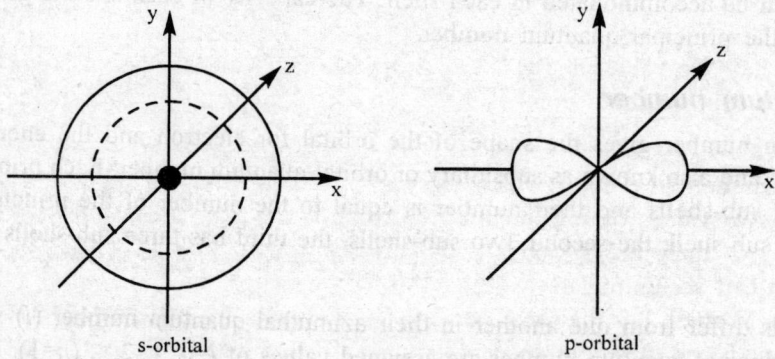

Fig. 3.13. Shape of the orbitals.

The p-orbitals are dumb-bell shaped, i.e., in p orbitals, distribution is symmetrical along the axes.

Note: Apart from the spatial distribution given by the orbital, an electron has an additional characteristic known as the spin.

Spin quantum number

Spin quantum number represents the spin of the electron, i.e. it distinguishes between the clockwise (↑) or anti-clockwise (↓) spin of an electron. It is denoted by letter '*s*'. This number is either ± 1/2 Thus an orbital can accommodate just two electrons provided they have opposite spins.

PAULI'S EXCLUSION PRINCIPLE

This rule was first given by the Austrian scientist, Wolfgang Pauli. It is thus known as Pauli exclusion principle. According to this principle: "No two electrons in the same atom can have all the four quantum numbers alike". If one electron in an atom has some particular value for the quantum numbers, then all the other electrons in that atom are excluded from having the same set of values, hence the name exclusion principle. It follows that any two-electrons in an atom must differ in at least one quantum number.

This principle is very useful in determining the maximum number of electrons in a shell or sub-shell. For K shell ($n=1$), the permitted values of four quantum numbers are as under; $n=1$, $l=0$, $m=0$ and $s = +\frac{1}{2}, -\frac{1}{2}$. Thus, there is only one orbital for the 1st sub-shell, having two electrons of opposite spins.

Again for L shell ($n=2$), the permitted values of four quantum numbers are as under :

$$\text{for } l = 0; \; m = 0 \text{ and} \qquad s = +\frac{1}{2}, -\frac{1}{2}$$

$$n = 2 : l = 0, \text{ and } 1;$$

$$\text{for } l = 1; \; m = -1 \text{ and} \qquad s = +\frac{1}{2}, -\frac{1}{2}$$

$$m = 0 \qquad s = +\frac{1}{2}, -\frac{1}{2}$$

$$m = +1 \qquad s = +\frac{1}{2}, -\frac{1}{2}$$

Thus for $n = 2$, l has two values and accordingly m has four values or orbitals and for each m value there are 2 values for spin quantum number. Eight electrons can thus be accommodated in the 2nd or L-shell.

For the third or M shell ($n = 3$), l will have three values of 0, 1, 2. For $l = 0$, there is only one value of m = 0 and two values of spin quantum number. For 1 = 1, there are 3 values of m, namely $-1, 0, +1$ and for 1 = 2, there are 5 values of m, namely $-2, -1, 0 +1, +2$. Thus there are 9 orbitals in all, each of which can accommodate two electrons of opposite spin. Thus in all 18 electrons can be accommodated in the M shell. For the same reasons, as above, for the 4th or N shell ($n = 4$) l will have 4 values of 0, 1, 2, and 3 and these will respectively have 1, 3, 5 and 7 ($-3, -2, -1, 0 +1, +2, +3$) values of m or there will be 16 orbitals, each orbital accommodating two electrons of opposite spin. There will thus be in all 32 electrons in the 4th shell.

The maximum number of electrons in any sub-shell (s, p d or f) is also given by the formula : (2 ($2l+1$) where $l = 0, 1, 2,$ or 3 for s, p, d or f sub-shell. Thus these sub-shells can have a maximum of 2, 6, 10 and 14 electrons respectively.

ELECTRONIC CONFIGURATION OF ATOMS

Writing the electronic configuration of any atom is now an easy matter. We start filling the orbitals beginning with the lowest energy orbital and keeping in mind the Pauli's exclusion principle.

Aufbau Principle

An atom in its state of lowest energy is said to be in ground state. The ground state is the most stable state in an atom. According to Aufbau principle: "Electrons are added progressively to the various orbitals in the order of their increasing energy starting with the orbital of lowest energy". The order of increasing energy may be determined up as follows :

1. A new electron enters an empty orbital for which the values of $(n + l)$ is minimum.

2. If the value of $(n + l)$ is same for two or more orbitals, the new electron enters on orbital having lower value of n.

According to these two rules the sequence of filling the orbitals is as follows : ls, 2s, 2p, 3s, 3p, 4s, 3d, 4p, 5s, 4d, 5p, 6s, 4f, 5d, 6p, 7s, 5f, 6d, ...

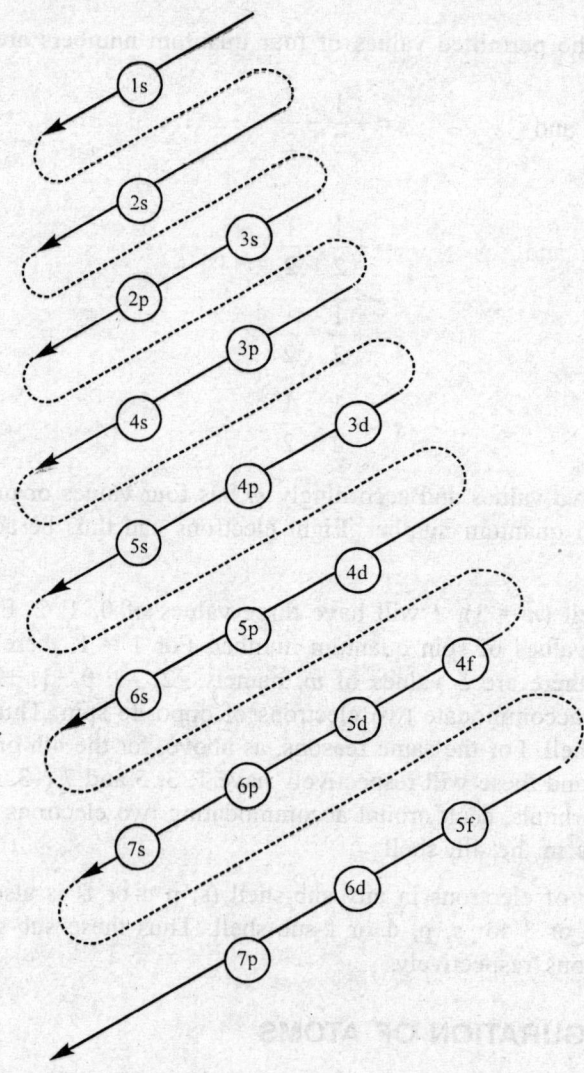

Fig. 3.14. A memory of aid for filling the orbitals according to the Aufbau principle.

Hund's Rule of Maximum Multiplicity

This rule states that "the electrons in any sub-shell (s, p, d, f) occupy orbitals singly to start with and have parallel spins before pairing in any orbital occur with opposite spins". Thus the 4th electron in p orbitals, 6th in d orbitals and 8th in orbitals will be the first to pair in the same sub-shell.

Chapter 4

Phase Equilibria

INTRODUCTION

The term phase is used to describe any part of a system which is itself homogeneous (i.e., uniform throughout so that every sample of it has exactly the same physical properties and chemical composition) and yet is physically distinct from all the other parts of the system. For example, a mixture of a liquid and a gas consists of two different phases, and so does a mixture of a solid and a gas, or a solid and a liquid in equilibrium with each other. On the other hand, a mixture of gases constitutes only a single phase no matter how many gases are present. The same is true of a mixture of liquids which dissolve completely in each other, or of an unsaturated solution of a solid in a liquid, since all these examples are homogeneous throughout. Except when a solid solution is formed, a mixture of solids always consists of as many phases as there are different solids present, and every solid allotropic form is regarded as a separate phase.

Some simple examples of phase equilibria are such as the equilibrium between a liquid and its saturated vapour and that between a saturated solution and excess of solid solute. A systematic study of a wide range of phase equilibria, culminating in the important generalisation known as the phase rule.

PHASE EQUILIBRIUM DIAGRAMS

The relation between the solid, liquid, and vapour states of a given substance can be depicted diagrammatically by what is known as the phase equilibrium diagram of that substance. Fig. 4.1 is a typical example, in which pressure is plotted against temperature for a single substance. Such a diagram is entirely constructed from experimental data, which it summarises in a compact form. For example, the curves OA and BO are the saturated vapour pressure curves of the liquid and solid respectively. They intersect at the point O, which represents, therefore, the conditions of temperature and pressure at which the solid and liquid are in equilibrium with each other under their own vapour pressure. This is not exactly the same as the melting point of the solid, which is the temperature at which solid and liquid are in equilibrium under a pressure of 101325 Nm^{-2}, but the difference is very small. The curve OC shows how this equilibrium point between solid and liquid varies with applied pressure. Since it slopes slightly towards the right in this case, it indicates that the melting point of this substance is raised by the application of pressure.

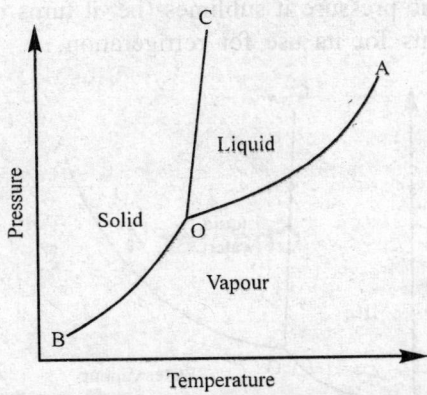

Fig. 4.1. A typical phase equilibrium diagram.

It is important to appreciate the significance of the various areas, lines and points in such a phase diagram. As shown in Fig. 4.1, the diagram is divided into three areas, labelled solid, liquid, and vapour. Thus at a glance it is possible to see what conditions of temperature and pressure correspond to any one of these physical states of the substance. For example, under the conditions prevailing in the area AOC liquid is the only stable phase, whilst under the conditions of the area BOA the substance can only exist as a vapour. The lines in the diagram represent the conditions under which two of the phases are in equilibrium with each other. At any temperature and corresponding pressure along the line OA, for example, the liquid and vapour can coexist indefinitely, but if either the temperature or the pressure is changed, then one phase disappears. The point O is unique, representing the conditions under which all three phases, solid, liquid, and vapour, are in equilibrium with each other. It is known as the triple point for that reason. If the substance is at exactly this temperature and pressure and the conditions are altered, then at least one of the three phases and may be two will disappear.

The line OA ends abruptly at the point A, which corresponds to the critical temperature and pressure of the substance. It is impossible to extend the diagram beyond this point because at higher temperatures and pressures, the liquid and vapour are indistinguishable. The practical difficulties of working at very high and at very low temperatures and pressures limit the extent of OC and OB respectively.

Phase Diagram of Water

This diagram, which is of particular interest and importance, is shown in Fig. 4.2. The triple point, O, occurs at 613 Nm^{-2} pressure and 0.0075°C, and the critical point, A, at 2.20×10^7 Nm^{-2} (217 atm) and 374°C. The line OC inclines very slightly to the left because the melting point of water is lowered by raising the pressure (experiment shows that the change is about 0.0075 K per atmosphere). For most substances OC has a steep positive gradient as shown in Fig. 4.1 (the only other common exceptions are bismuth and type metal). The unusual behaviour of water is in this respect can be related by Le Chatelier's principle to its expansion on freezing, which arises from the very open structure of ice crystals owing to their hydrogen bonding.

Phase Diagram of Carbon Dioxide

Another unusual equilibrium diagram is that of carbon dioxide (Fig. 4.3), because here the triple point O, lies above atmospheric pressure. As the diagram shows, liquid carbon dioxide is not stable at any temperature when the external pressure is below 5.16×10^5 Nm^{-2} Thus when solid carbon dioxide is warmed above −78°C at atmospheric pressure it sublimes (i.e. it turns directly into gas without forming a liquid first), which partly accounts for its use for refrigeration.

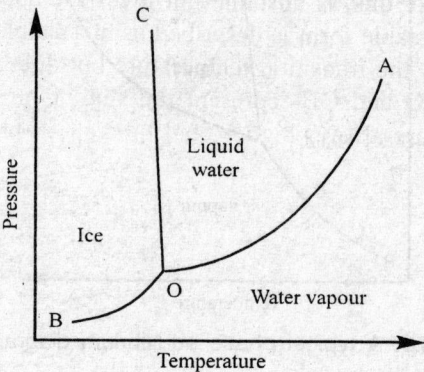

Fig. 4.2. Phase equilibrium diagram of water.

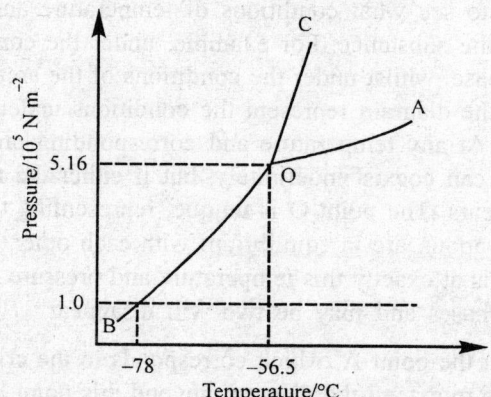

Fig. 4.3. Phase equilibrium diagram of carbon dioxide.

TYPES OF ALLOTROPY

The three main types of allotropy are:

Enantiotropy

This main characteristics of enantiotropy are:

1. Each allotrope is stable over a definite range of temperature.
2. There is a definite transition temperature at which both the solid forms exert the same vapour pressure and can exist in equilibrium with each other.
3. Each allotrope can be converted directly into the other by varying the temperature, i.e. the change from one form into another is reversible.

One of the best examples is provided by sulphur, whose pressure-temperature diagram is shown in Fig. 4.4. The lines AB and BC are the vapour pressure curves of the rhombic and monoclinic allotropes respectively, and B is the transition temperature at which each changes into the other if the temperature is varied slowly. If rhombic sulphur is heated strongly, however, so that its temperature rises rapidly through the transition temperature, it tends to persist for a while even though it is theoretically unstable. Any substance or system which, like this, is unstable but gives the appearance of stability because it is only very slowly changing into the stable form is described as metastable. A similar situation arises when monoclinic sulphur is cooled below the transition temperature but does not change immediately into the rhombic form. The dashed lines BO and GB represent the vapour pressure curves of these metastable forms; it will be noticed that they are always higher than the vapour pressure curves of the more stable forms.

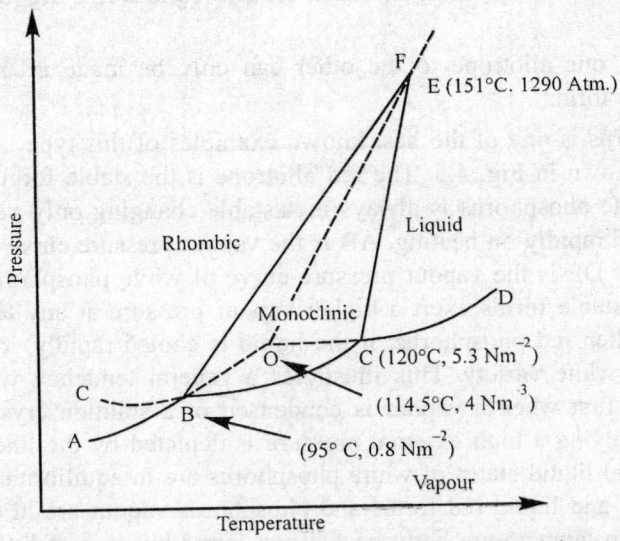

Fig. 4.4. Phase equilibrium diagram of sulphur.

The line CD is the vapour pressure curve of liquid sulphur and C is therefore the melting point of monoclinic sulphur under its own vapour pressure (viz. 5.3 Nm^{-2}). The result of applying a high external pressure is to raise both the transition temperature between the rhombic and monoclinic forms and the melting point of monoclinic sulphur; these changes are represented by the lines BE and CE, though not to scale since the transition temperature rises by only about 0.045 K for each increase of one atmosphere in pressure. Thus the points B and C and the point E are all triple points where three phases are in equilibrium with each other. At B the rhombic and monoclinic allotropes can coexist with sulphur vapour, at C liquid and vapour sulphur are in equilibrium with the monoclinic form, and at the abnormal conditions represented by E the system can contain rhombic, monoclinic, and liquid sulphur together. A fourth triple point, O, is metastable and represents the special conditions when rhombic sulphur is in equilibrium with the liquid and vapour; it is only achieved in practice by heating rhombic sulphur rapidly to its melting point (114.5°C).

The four areas of the diagram represent the various conditions of temperature and pressure under which rhombic, monoclinic, liquid, and vapour sulphur are stable. For example, the monoclinic form can only exist in a stable condition if the temperature and pressure lie somewhere between the limits set by the points B and E. Since no quadruple point exists, it is impossible for all the four phases to be in equilibrium with each other simultaneously.

The transition temperature in enantiotropy is often determined experimentally by observing some property such as colour or volume and finding when it undergoes a sharp change (e.g. by using a dilatometer). Alternatively, the variation of solubility or vapour pressure with temperature can be plotted on a graph and the temperature at which the curves of the two forms intersect can be found.

Monotropy

The chief features of monotropy are:

1. One allotrope is stable and the other metastable at all temperatures up to the melting point.
2. There is no definite transition temperature at which the two solid forms are in equilibrium with each other.
3. The direct change from one allotrope to the other can only be made in one direction, from the metastable to the stable form.

The allotropy of phosphorus is one of the best known examples of this type. A simplified and rather idealised phase diagram is shown in Fig. 4.5. The red allotrope is the stable form at all temperatures up to its melting point, and white phosphorus is always metastable changing only very slowly into the red form at room temperature and rapidly on heating. AB is the vapour pressure curve of the red form, which melts at B (590°C). Similarly DE is the vapour pressure curve of white phosphorus and E is its melting point (44°C) (note that metastable forms exert a higher vapour pressure at any temperature). BC is the vapour pressure curve of molten red phosphorus; if the liquid is cooled rapidly, it follows the path CBE and deposits crystals of the white variety. This illustrates a general tendency whereby the metastable ailotrope is usually obtained first when a vapour is condensed or a solution crystallised. The effect on the two melting points of applying a high external pressure is depicted by the lines EF and BF. Thus at the triple point E the solid and liquid states of white phosphorus are in equilibrium with the vapour, and at the triple point B the solid and liquid red forms and phosphorus vapour are in equilibrium. The point F is theoretically the transition temperature between the two forms but it is of little practical importance because it is above the melting point of either form at atmospheric pressure.

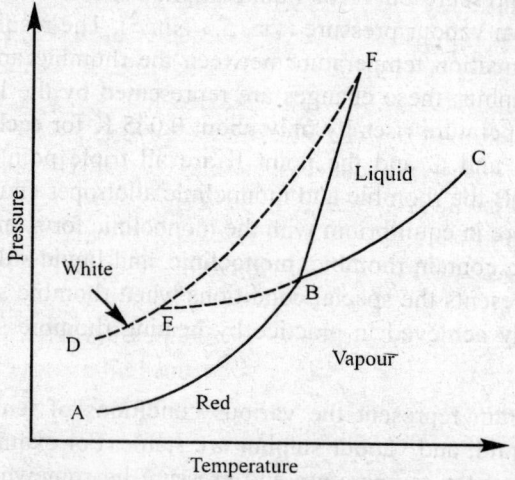

Fig. 4.5. A simplified phase equilibrium diagram of phosphorus.

Dynamic Allotropy

In this type the two allotropes are in dynamic equilibrium with each other over a wide range of temperature, the proportions of each in the gaseous or liquid mixture depending upon the temperature and, sometimes, the pressure. The difference between the allotropes lies not in the way in which the identical particles are packed together in the crystal (as it does not monotropy and enantiotropy) but in the atomicity and shape of the molecules themselves.

For example, the two allotropes oxygen and ozone are continually changing into each other and at any given temperature and pressure there is a fixed proportion of each present in the system when it is at equilibrium. Liquid sulphur provides another example of the same kind.

SOLUTIONS OF GASES IN LIQUIDS

Experiment shows that the mass of a gas which dissolves in a fixed volume of a liquid when the two are in equilibrium depends upon the nature of the gas and the liquid, the temperature, and the pressure. Table 4.1, which gives the volumes of various gases which dissolve in one volume of water at s.t.p., illustrates the wide differences that occur from one gas to another, and Table 4.2 shows how the solubility of one particular gas (oxygen) varies with the solvent.

Table 4.1. Solubilities of various gases in water at s.t.p.

Gas	NH_3	HCl	SO_2	H_2S	CO_2	O_2	N_2	He
Absorption coefficient	1300	500	80	4.6	1.7	0.05	0.024	0.009

Table 4.2. Solubility of oxygen at 20°C.

Solvent	Water	Ethyl alcohol	Benzene	Acetone	Diethyl ether
Absorption coefficient	0.028	0.144	0.163	0.208	0.416

With few exceptions the solubility of a gas in a liquid decreases as the temperature is raised, as might be expected from Le Chatelier's principle, since heat is evolved when gases dissolve and low temperature will therefore favour the exothermic process. For example, the solubility of ammonia in water is 1300 volumes at 0°C, 700 volumes at 20°C, and about 240 volumes at 90°C. This is in direct contrast to the general behaviour of solids, most of which are more soluble at higher temperature. The effect of temperature can be observed very simply by drawing a tumbler of cold water from a tap and leaving it to stand in a warm room, when dissolved air will be expelled from the saturated solution as the temperature of the water rises and will escape as bubbles from the surface. If an aqueous solution of ammonia is boiled for some time in the open air, the dissolved gas is driven off completely and the residual liquid is found to be pure water. This does not show that the solubility of ammonia in water is nil at 100°C, because the gas is not in equilibrium with the solution under these conditions. If the experiment was repeated in such a way that the boiling solution remained in contact with the expelled gas at atmospheric pressure, then the concentration of gas in solution would remain at the equilibrium level for that temperature (195 volumes), however long the boiling was continued. Similarly, if air is bubbled through an aqueous solution of ammonia at room temperature, all the ammonia gas is eventually removed from the solution because the continual flow of air carries away the volatile ammonia and prevents the establishment of equilibrium. These examples emphasise the importance of regarding solutions of gases in liquids in terms of an equilibrium between the two phases.

The solubility of a gas can be expressed in the following ways:

1. As the volume of gas, measured under the conditions of dissolving, which dissolves in unit volume of the solvent. This is known as Ostwald's coefficient of solubility, β.

2. As the volume of gas, reduced to s.t.p., which dissolves in unit volume of solvent at any particular temperature under a partial pressure of one atmosphere. This is known as Bunsen's absorption coefficient, α.

Assuming that the gas laws are obeyed, it follows that $\alpha = \beta \times \dfrac{273}{T}$, where T is the absolute temperature of the experiment.

3. As the mass of gas in gram which dissolves in 100 gram of solvent. This is usually expressed simply as a percentage.

4. As the mole fraction of gas in the solution. This method is particularly useful when considering phase equilibria.

HENRY'S LAW

This states that at constant temperature the mass of gas which dissolves in a given volume of liquid at equilibrium is proportional to the pressure of the gas. If the gas obeys Boyle's law then it follows that the volume which dissolves is independent of the pressure. The law does not hold true for aqueous solutions of very soluble gases such as ammonia or hydrogen chloride which not only dissolve physically in water but also react chemically with it:

$$NH_3 + H_2O \rightleftharpoons NH_4^+ + OH^-$$
$$HCl + H_2O \rightleftharpoons H_3O^+ + Cl^-$$

Deviations are also shown by most gases at very high pressures and low temperatures.

As Dalton showed Henry's law also applies to individual gases in a mixture provided the pressure of each gas is taken as its partial pressure. For example, the partial pressures of oxygen and nitrogen in the air are approximately in the ratio of 1 : 4. Since oxygen is about twice as soluble in water as nitrogen under the same conditions, the weights of oxygen and nitrogen dissolved in water exposed for a long period to the air are in the ratio 1 : 2.

In the Bosch process for manufacturing hydrogen use is made of the effect of pressure upon solubility to remove carbon dioxide from the product by washing; at 50 times atmospheric pressure the dioxide dissolves readily in water and is almost completely removed. Henry's law is also applied in storing acetylene by dissolving it in acetone under a pressure of about 3×10^6 Nm^{-2} (30 atm). If a diver who is wearing a pressurised suit rises to the surface rapidly, the additional gas (mainly nitrogen) which dissolves in his blood at high pressure tends to escape as bubbles, causing acute pain or even death. For this reason gradual decompression is essential.

MEASUREMENT OF THE SOLUBILITY OF GASES

The two main methods in use are:

Pyknometer Method

This is only applicable to gases forming solutions which can be analysed volumetrically, which includes most of the very soluble gases. A pyknometer (Fig. 4.6) is first weighed dry and empty. It is then partially filled with the solvent, as shown in the diagram, and the gas is bubbled through it until a saturated solution is obtained. During this part of the experiment the pyknometer must be kept in a thermostat at the desired temperature. The pyknometer is sealed at both ends and reweighed to give the mass of the solution. It is then immersed in a measured excess of a standard solution of a suitable reagent (one that reacts with the dissolved gas) and one end is broken to allow the liquids to mix. The residual reagent is determined by back titration. From the amount of reagent used up the mass of gas in the solution can be calculated. Taken from the mass of the solution, this gives the mass of solvent. Hence the solubility can be found.

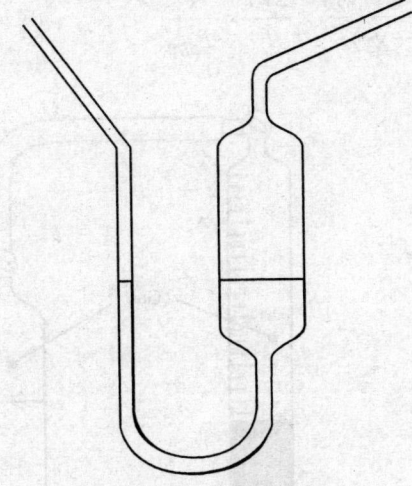

Fig. 4.6. Pyknometer.

Ostwald's Method

The apparatus used is shown in Fig. 4.7. Before starting the experiment the gas burette A is filled with mercury and the pipette B with solvent, and a current of the chosen gas is passed through the flexible tube C to expel the air from it. Then a suitable volume of the gas is drawn into the burette A through the tap T_1 by lowering the other limb, and after equalising the mercury levels its volume at atmospheric pressure is measured. The three-way taps T_1 and T_2 are set so that A and B are connected, and a known mass of the liquid in the pipette is then run out through tap T_3, leaving it about half-full, so that the upper part is filled with gas. After closing all the taps the pipette is shaken vigorously for some minutes to ensure thorough mixing of gas and liquid, and A and B are again connected. The shaking process is repeated until no more gas will dissolve, when the volume of gas (at atmospheric pressure) remaining in A is noted. The temperature and pressure of the atmosphere should be recorded; in accurate work the gas pipette B is completely immersed in a thermostat. The solubility is calculated in the following way:

Total volume of B $= v_1$

Volume of solvent run out $= v_2$

Volume of solvent remaining $= v_1 - v_2$

Volume of gas in A originally $= v_2$

Volume of gas in A finally $= v_4$

$\therefore$ Volume of gas which dissolved $= (v_3 - v_4 - v_2)$

If the temperature is T and the pressure P then the volume of dissolved gas at s.t.p. $= (v_3 - v_4 - v_2) \times \dfrac{273}{T} \times \dfrac{P}{P_{atm}}$

Now this dissolved in $v_1 - v_2$ of solvent.

∴ Absorption coefficient at temperature T and pressure P

$$= \frac{(v_3 - v_4 - v_2)}{(v_1 - v_3)} \times \frac{273}{T} \times \frac{P}{P_{atm}}$$

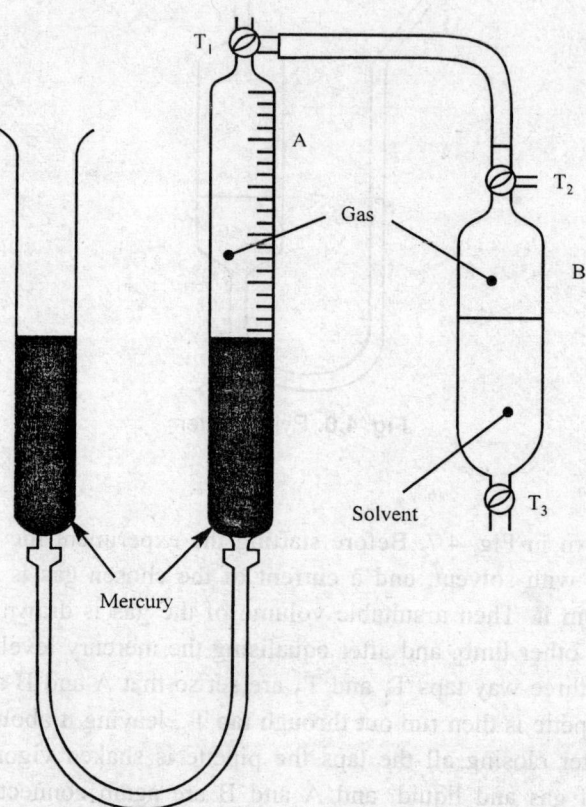

Fig. 4.7. Apparatus for measuring the solubility of sparingly soluble gases by Ostwald's method.

MIXTURES OF MISCIBLE LIQUIDS

Two liquids are said to be miscible when they dissolve completely in each other in all proportions to give a homogeneous mixture. For example, when water is added to ethyl alcohol the two liquids mix intimately forming a solution of uniform composition throughout. The vapour pressure above a mixture of two miscible liquids varies with the composition of the mixture in one of three ways:

1. Whatever the composition of the vapour pressure always lies between the extreme values of the two pure liquids.

2. At a certain composition the vapour pressure has a maximum value which is greater than that of either pure liquid.

3. At a certain composition the vapour pressure has a minimum value which is less than that of either pure liquid.

Types 2 and 3 give rise to constant boiling point mixtures are being considered here.

If Raoult's law is obeyed exactly by two miscible liquids over the whole range of composition then the vapour pressure exerted above the mixture by each constituent is proportional to its mole fraction

i.e.
$$p_A = \frac{n_A}{n_A + n_B} \cdot P \quad \text{and} \quad p_B = \frac{n_B}{n_A + n_B} \cdot P$$

Thus the partial pressure curve of each constituent is a straight line through the origin and the total vapour pressure above the mixture varies regularly with composition, as in Fig. 4.8. Such a mixture is known as an ideal solution; the two liquids forming it do not show any heat change when mixed and the final volume of the mixture is exactly equal to the sum of the volumes of the constituents. In practice most real solutions deviate from this ideal behaviour because their molecules 'interfere' with each other to some extent and exert abnormal attractions for each other, but the concept of an ideal solution is useful, just like that of an ideal gas, because it gives an approximate guide to the behaviour of real substances.

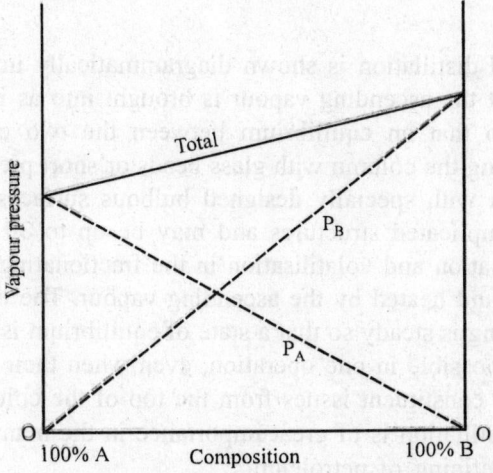

Fig. 4.8. Vapour pressure/composition curve for an ideal solution at constant temperature.

Where deviations from ideal behaviour are only slight and do not lead to a constant boiling point mixture, then vapour pressure and corresponding boiling point curves take the form shown in Fig. 4.9. Mixtures of this type include benzene and toluene, acetone and water, and nitrogen and oxygen in liquid air. At any particular composition of liquid the vapour above the mixture is always richer in the more volatile constituent than is the liquid with which it is in equilibrium. Taking the example shown in Fig. 4.9, if a liquid mixture containing 33% of B is distilled, the vapour which escapes and condenses contains 65% B and 35% A, and the liquid remaining in the flask becomes progressively poorer in B. If this distillate is in turn redistilled it gives a new distillate containing 90% of B. In this way by repeated distillation it is theoretically possible to bring about the complete separation of A and B, but in practice this process is very slow and tedious, particularly if the boiling points of the two constituents are very close together, and the much more efficient fractional distillation is used instead.

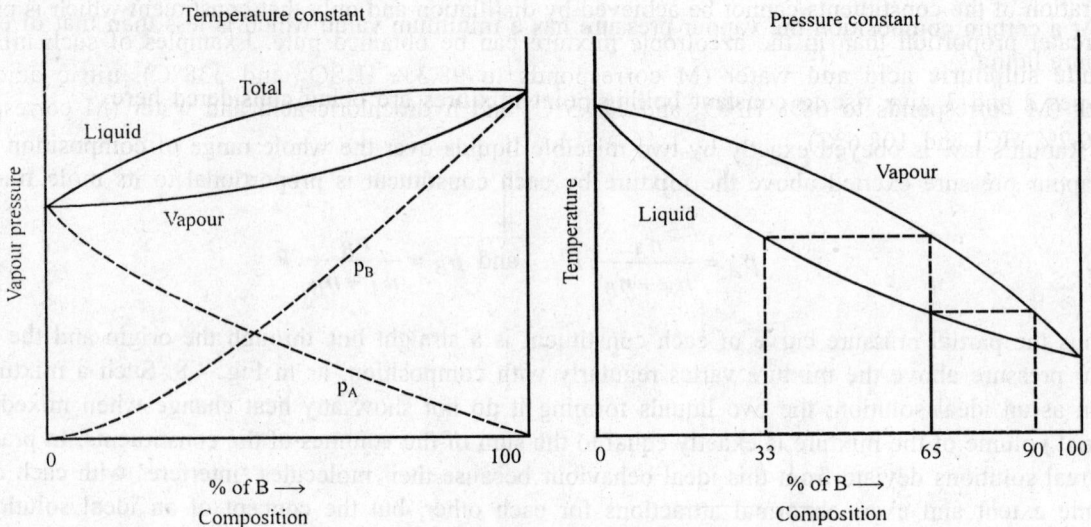

Fig. 4.9. Vapour pressure and boiling point diagrams (Type 1).

Fractional Distillation

The apparatus used for fractional distillation is shown diagrammatically in Fig. 4.10. The fractionating column is designed to ensure that the ascending vapour is brought into as intimate contact with e.q. the descending liquid as possible so that an equilibrium between the two exists at every stage. In the laboratory this is achieved by filling the column with glass beads or short pieces of glass tubing orientated at random or by using a column with specially designed bulbous surfaces, but industrial fractionating columns are often extremely complicated structures and may be up to 30 metre tall. Vapour from the flask undergoes repeated condensation and volatilisation in the fractionating column as it is successively cooled by the descending liquid and heated by the ascending vapour. The effect is the same as repeated distillation and provided the heating is steady so that a state of equilibrium is established, almost complete separation of the constituents is possible in one operation, even when their boiling points differ by only a few degrees. The more volatile constituent issues from the top of the column whilst the other remains behind in the flask. Fractional distillation is of great importance in the manufacture of oxygen, nitrogen, and the inert gases, and in the refining of petroleum.

Constant Boiling Point Mixtures

As already explained in certain cases the vapour pressure above a mixture of miscible liquids does not vary even approximately regularly with composition but reaches a maximum or minimum value at some intermediate composition. Such liquids give rise to constant boiling point or azeotropic mixtures, which are of two types.

Maximum boiling point mixtures

The typical way in which the vapour pressure and the boiling point vary with composition in this type of mixture is shown in Fig. 4.11. On distilling such a mixture the composition of the residue changes gradually and its boiling point rises slowly until the liquid remaining in the flask has the composition M, when vapour of this composition distils over unchanged at constant temperature. Thus complete separation of the constituents cannot be achieved by distillation and only that constituent which is present in greater proportion than in the azeotropic mixture can be obtained pure. Examples of such mixtures include sulphuric acid and water (M corresponds to 98.3% H_2SO_4 and 338°C), nitric acid and water (M corresponds to 68% HNO_3 and 120.5°C) and hydrochloric acid and water (M corresponds to 20.2% HCl and 108.6°C).

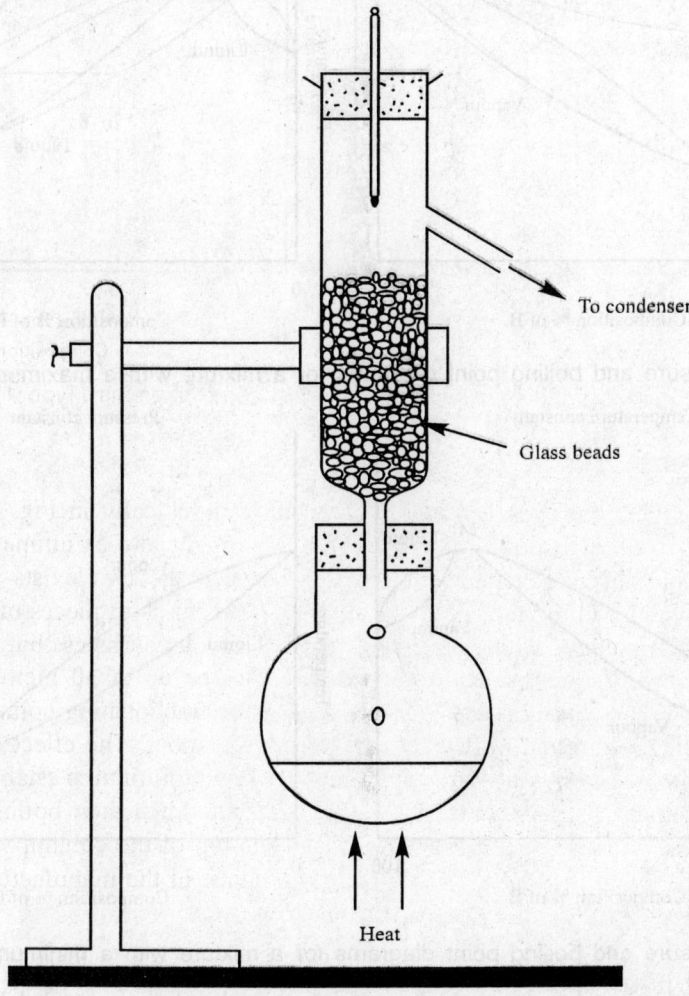

To condenser

Glass beads

Heat

Fig. 4.10. Fractional distillation.

Minimum boiling point mixtures

Typical vapour pressure and boiling point curves for this type of mixture are shown in Fig. 4.12. When such a mixture is fractionally distilled or subjected to repeated redistillation, a distillate of composition M is obtained and this continues to distil over until the whole of one constituent has been vapourised. The pure constituent remaining in the flask then distils over unchanged. Thus it is impossible to separate the two constituents completely by means of distillation. Examples of mixtures of this type include ethyl alcohol and water (M corresponds to 95.6% C_2H_5OH and 78.13°C) and ethyl acetate and water (M corresponds to 91.5% $CH_3.COOC_2H_5$ and 70.4°C).

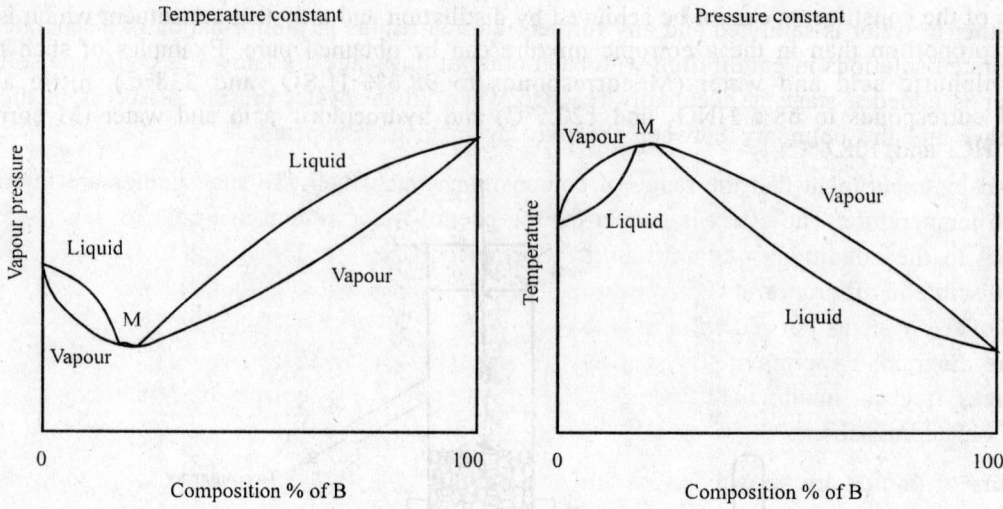

Fig. 4.11. Vapour pressure and boiling point diagrams for a mixture with a maximum boiling point (Type 3).

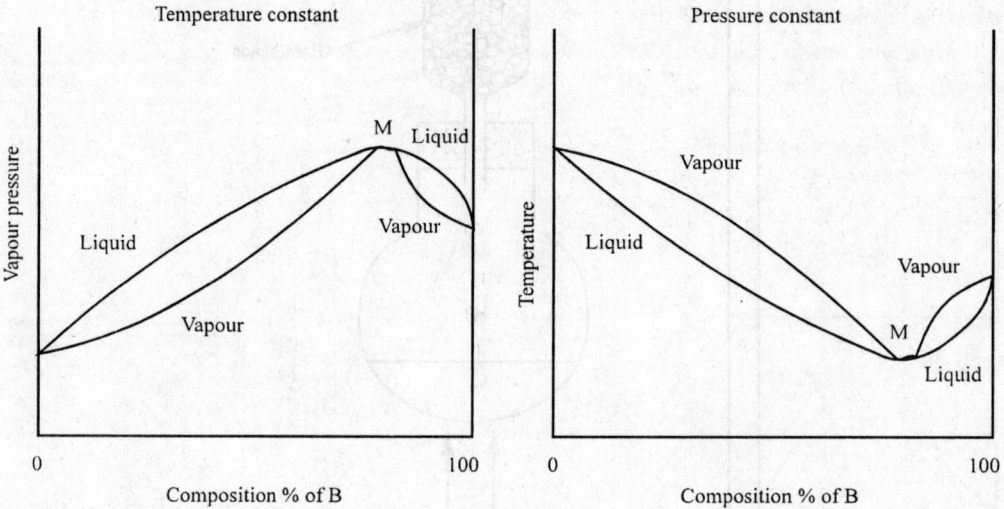

Fig. 4.12. Vapour pressure and boiling point diagrams for a mixture with a minimum boiling point (Type 2).

An azeotropic mixture can be shown to be a mixture and not a compound by varying the external pressure, when its composition varies as well as its boiling point. The constancy of composition at constant pressure is often made use of in preparing a standard solution of hydrochloric acid—the maximum boiling point azeotropic mixture eventually obtained by distilling any sample of hydrochloric acid under 101 325 Nm^{-2} pressure is merely diluted by the required amount.

PARTIALLY MISCIBLE LIQUIDS

Some liquids are only partially miscible, i.e., they dissolve in each other only in certain proportions to form a homogeneous mixture. For example, if diethyl ether is added to water drop by drop, at first it dissolves forming a solution of uniform composition throughout, but there comes a point when the solution of ether in water is saturated and any further addition results in the formation of a separate layer. The two saturated solutions in equilibrium with each other at this stage are known as conjugate solutions. If more ether is added a stage is eventually reached when all the water present dissolves in the large excess of ether and the boundary between the two layers then disappears.

It is found by experiment that the range of composition over which two such liquids are immiscible changes with temperature. The effect is shown for the phenol-water system in Fig. 4.13, where the area corresponding to the conditions of immiscibility is shaded. The lowest temperature at which the two liquids are miscible in all proportions is known as the upper critical solution temperature of the consolute temperature of the system. For phenol and water at atmospheric pressure this temperature is 65.8°C, as shown in the diagram. Experiment shows that it is very sensitive to the presence of impurities, rising several degrees if even small quantities of benzene or naphthalene are present and falling to room temperature on adding about 1% of soap.

Other pairs of liquids are known (e.g. triethylamine and water) which are only completely miscible below a certain temperature and are immiscible in certain proportions above it. The system nicotine-water is unusual in having an upper and a lower critical solution temperature, so that under high pressure its miscibility curve is a closed loop as shown in Fig. 4.14. At the conditions corresponding to the shaded area the two liquids are immiscible and exist as two separate phases, but under all other conditions they form a homogeneous solution.

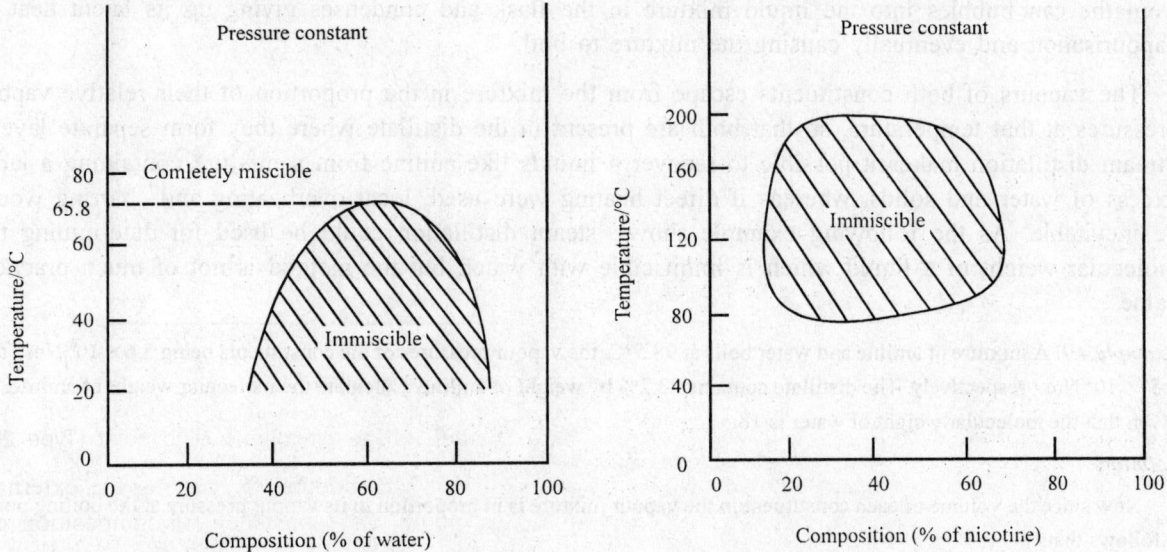

Fig. 4.13. Miscibility of water and phenol.   Fig. 4.14. Miscibility of water and nicotine.

IMMISCIBLE LIQUIDS AND STEAM DISTILLATION

Many pairs of liquids are found to be virtually completely immiscible, i.e. they form separate layers when added to each other in almost any proportions. Examples are mercury and water, carbon disulphide and water, and chlorobenzene and water. It is found by experiment that the vapour pressure above a pair of immiscible liquids of this sort is equal to the sum of the separate vapour pressures of the two pure components and is independent of their relative proportions in the mixture. Since such a mixture will boil when the total vapour pressure is equal to the external pressure, it follows that the mixture always boils at a lower temperature than either pure constituent. This point is illustrated in Fig. 4.15.

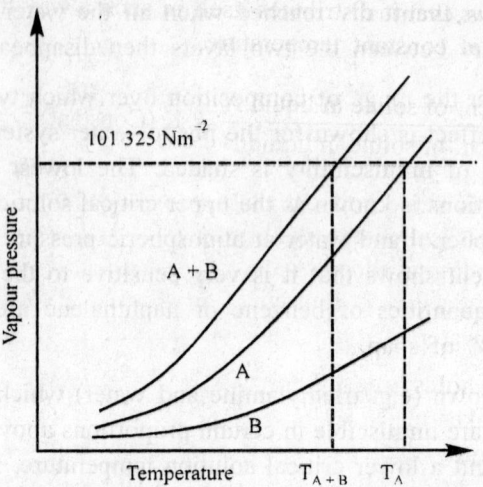

Fig. 4.15. Principle of steam distillation.

Advantage is taken of this behaviour of immiscible liquids in the process of steam distillation. Steam from the can bubbles into the liquid mixture in the flask and condenses giving up its latent heat of vapourisation and eventually causing the mixture to boil.

The vapours of both constituents escape from the mixture in the proportion of their relative vapour pressures at that temperature, so that both are present in the distillate where they form separate layers. Stream distillation makes it possible to recover a liquids like aniline from a mixture containing a large excess of water and solids, whereas if direct heating were used, local overheating and charring would be inevitable. As the following example shows, steam distillation could be used for determining the molecular weight of a liquid which is immiscible with water, but the method is not of much practical value.

Example 4.1. A mixture of aniline and water boils at 98.5°C, the vapour pressures of the constituents being 5.6×10^3 Nm^{-2} and 9.57×10^4 Nm^{-2} respectively. The distillate contains 23.2% by weight of aniline. Calculate the molecular weight of aniline. M, given that the molecular weight of water is 18.

Solution.

Now since the volume of each constituent in the vapour mixture is in proportion to its vapour pressure at the boiling point, it follows that:

$$\frac{\text{Mass of aniline in distillate}}{\text{Mass of water in distillate}} = \frac{\text{Mol. wt. of aniline} \times \text{v.p. of aniline at b.p.}}{\text{Mol. wt. of water} \times \text{v.p. of water at b.p.}}$$

$$\therefore \quad \frac{23.2}{76.8} = \frac{M}{18} \times \frac{5.6}{95.7}$$

so

$$M = \frac{23.2 \times 18 \times 95.7}{76.8 \times 5.6} = 93$$

and the molecular weight of aniline is therefore 93.

DISTRIBUTION OR PARTITION LAW

If to a mixture of two immiscible liquids a substance which is soluble in both is added in varying amounts, then experiment shows that it distributes itself in such a way that the ratio of its concentrations in the two layers is constant at constant temperature,

i.e.
$$\frac{\text{Concn. of solute in liquid A}}{\text{Concn. of solute in liquid B}} = K \text{ (Constant)}$$

The generalisation, which is subject to certain limiting conditions, is known as the distribution or partition law; the constant k is called the distribution or partition coefficient of that system at that temperature. For example, if iodine is added to a mixture of benzene and water and the mixture is shaken until equilibrium is attained, then the concentration of iodine in the benzene layer is always about 400 times as great as that in the water at room temperature.

The distribution law only holds true if the solute is in the same molecular condition in both the solvents. Thus if association takes place in one solvent and not in the other, or if the solute is dissociated to some extent in one solvent only, then the ratio of the concentrations is no longer constant. The quantity of solute added must not be so great that its solubility in either solvent is exceeded and equilibrium must be reached before the concentrations are compared. It is important to notice that the partition coefficient is given by the ratio of the concentrations of the solute and not the ratio of the amounts; this latter ratio varies, of course, with the relative volumes of the two liquids.

It should now be seen that Henry's law is really a special case of the distribution law in which the gaseous solute may be regarded as distributing itself between free space on the one hand and the liquid solvent on the other in a constant ratio at constant temperature.

If the solute has a different molecular weight in the two solvents, then a modified distribution law applies. For example, if the molecular weight of the solute is n times as great in solvent A as in solvent B, then it is found by experiment that at constant temperature

$$\frac{\text{Concn. in liquid A}}{(\text{Concn. in liquid B})^n} = K \text{ (Constant)}$$

For example, acetic acid and benzoic acid are associated by hydrogen bonding into double molecules in benzene but not in water. This follows from the fact that for these systems

$$\frac{\text{Concn. of acid in benzene}}{(\text{Concn. of acid in water})^2} = \text{Constant (at constant temperature)}$$

Thus by evaluating n in a given case the molecular weight of the solute in the two solvents can be compared. The best way of determining n is to plot a graph of the log of the concentration in liquid A against the log of the concentration in liquid B, as in Fig. 4.16 when the gradient of the straight line obtained will be equal to n and the intercept will be equal to log k, since the following relationship applies:

$$\log \text{ concentration in liquid A} = n \times \log \text{ concentration in liquid B} + \log k$$

i.e. y $= m$ x $+ c$

If n is not an integer it indicates that the association or dissociation is only partial.

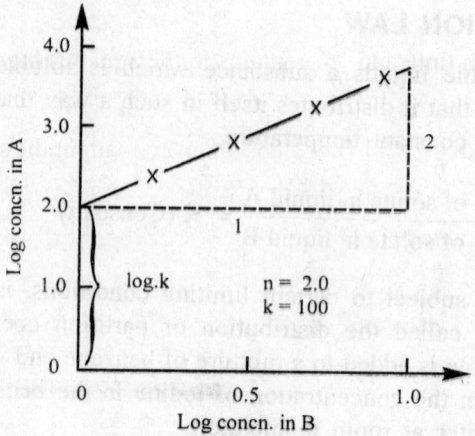

Fig. 4.16. Graphical evaluation of *n* and *k*.

The distribution law is often applied industrially and in the laboratory to extract a solute from water. The general procedure is to shake the aqueous solution with another liquid which is immiscible with it but in which the solute is much more soluble, and then to separate the two layers. For example penicillin is extracted from a dilute aqueous solution with chloroform, and aniline is usually reclaimed from a mixture with water by using ether. The Parks process for recovering silver from lead is another application of the distribution law. In all these cases it is more efficient to shake with several small volumes of the extracting solvent in turn, separating each time, than to use the same total volume all at once, as the following calculation shows:

Example 4.2. Consider the effect of shaking one dm³ of an aqueous solution containing 11 gram of solute X with (a) 100 cm³ of ether at one time, (b) two successive volumes of 50 cm³ of ether, given that the partition coefficient for the system is 100 at that temperature.

Solution.

(a)
$$\frac{\text{Concn. of X in ether}}{\text{Concn. of X in water}} = \frac{100}{1}, \text{ given}$$

$$\therefore \quad \frac{\text{Mass of X in ether layer}}{\text{Mass of X in water layer}} = \frac{100}{1} \times \frac{\text{volume of ether}}{\text{volume of water}}$$

$$= \frac{100}{1} \times \frac{100}{1000} = \frac{10}{1}$$

Thus $\frac{10}{11}$ of X is extracted (i.e. 10 gram) and $\frac{1}{11}$ (i.e. 1 gram) of X remains in the water layer.

(b)
$$\frac{\text{Mass of X in ether layer}}{\text{Mass of X in water layer}} = \frac{\text{Concn. of X in ether}}{\text{Concn. of X in water}} \times \frac{\text{volume of ether}}{\text{volume of water}}$$

$$= \frac{100}{1} \times \frac{50}{1000} = \frac{5}{1}$$

Thus $\frac{5}{6}$ of X is extracted each time and $\frac{1}{6}$ remains in the aqueous layer. In the first extraction, $\frac{5}{6} \times 11$

(i.e. 9.167 gram) are extracted and $\frac{1}{6} \times 11$ (i.e. 1.833 gram) remain in the solution. In the second extraction

$\frac{5}{6}$ e.q. of the remaining X is extracted (i.e. $\frac{5}{6} \times 1.833 = 1.528$ gram) whilst $\frac{1}{6}$ (i.e. $\frac{1}{6} \times 1.833 = 0.305$

gram) remains in the water layer. Thus the mass of X remaining in the water was 1.00 gramme when shaken with a single volume of 100 cm³ of ether and only 0.305 gram when shaken with two successive portions of 50 cm³.

EUTECTIC MIXTURES

On slowly cooling a pure liquid a temperature is eventually reached at which solid begins to form. If the cooling is continued the temperature remains constant until all the liquid has solidified, owing to the evolution of latent heat of fusion, and then begins to fall slowly again. Thus if temperature is plotted against time, a cooling curve like that in Fig. 4.2 is obtained. In practice supercooling often occurs so that the cooling curve shows a dip and hump as shown in the diagram. If the cooling curves of various mixtures of two miscible liquids A and B which do not form solid solutions in each other are plotted in this way, it is possible to determine the temperature at which solid begins to separate from each mixture by the position of the change in the gradient of each curve. Typical curves are shown in Fig. 4.17; from these it is possible to construct an equilibrium diagram of the type shown in Fig. 4.18 in which freezing points measured at atmospheric pressure are plotted against composition. In practice the lines AE and BE are usually curved (not straight as shown in the figure) because the two constituents of the mixture are slightly soluble in each other in the solid state.

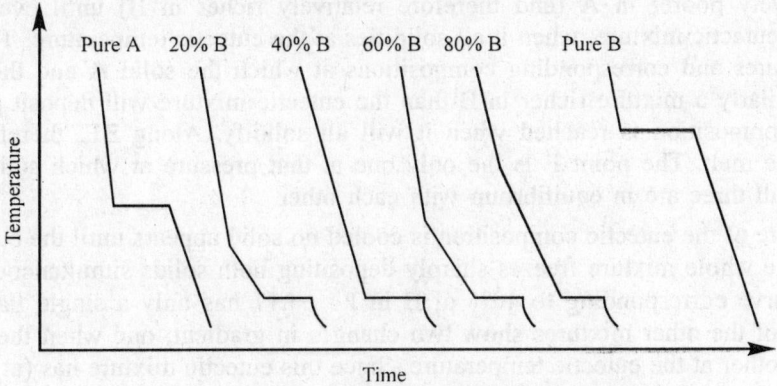

Fig. 4.17. Typical cooling curves of A, B, and A + B.

This diagram shows how the presence of each substance in the mixture lowers the freezing point of the other. For example, the point A is the freezing point of pure A and the line AE represents the progressive lowering of that freezing point as the proportion of B increases. Similarly the freezing point of B follows the course BE as the proportion of A increases. These two lines intersect at the point E which is known as the eutectic point. It is the lowest temperature at that pressure at which a mixture of A and B can exist in the liquid state. The mixture corresponding to the point E is known as the eutectic mixture.

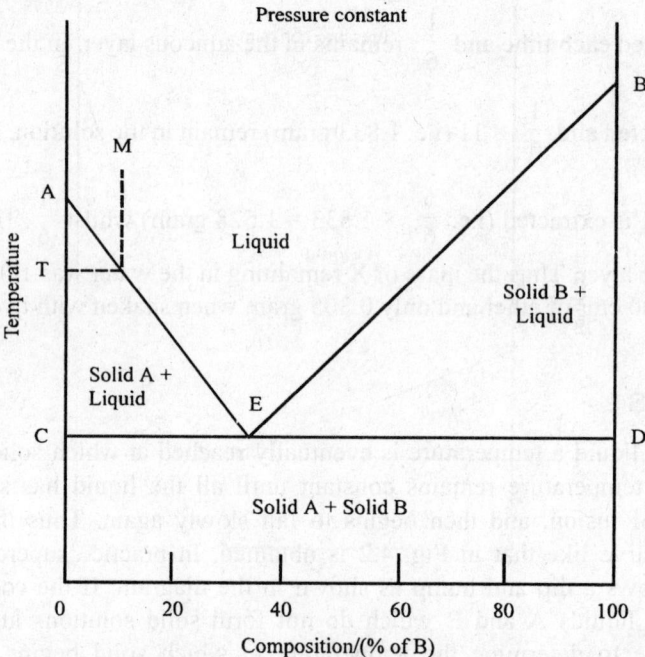

Fig. 4.18. Temperature/composition diagram for A and B.

The lines AE and BE can also be regarded in another way. Consider a mixture M containing a greater proportion of A than the eutectic mixture. When this liquid is cooled it begins to deposit crystals of pure A at temperature T (Fig. 4.18), so that at this temperature solid A and liquid mixture of composition M are in equilibrium. On further cooling more and more pure A is deposited and the remaining liquid becomes progressively poorer in A (and therefore relatively richer in B) until eventually it has the composition of the eutectic mixture, when it all solidifies at the eutectic temperature. Thus AE represents the set of temperatures and corresponding compositions at which the solid A and the liquid phase are in equilibrium. Similarly a mixture richer in B than the eutectic mixture will deposit pure B on cooling until the eutectic composition is reached when it will all solidify. Along BE, therefore, solid B is in equilibrium with the melt. The point E is the only one at that pressure at which solid A, solid B, and the liquid mixture all three are in equilibrium with each other.

If a liquid mixture of the eutectic composition is cooled no solid appears until the eutectic temperature is reached, when the whole mixture freezes sharply depositing both solids simultaneously. This explains why the cooling curve corresponding to 40% of B in Fig. 4.17 has only a single flat portion whereas the cooling curves of the other mixtures show two changes in gradient, one when the first solid begins to separate and the other at the eutectic temperature. Since this eutectic mixture has (at any one pressure) a constant composition and a sharp constant melting point it might, at first sight, be thought to be a compound. That this is not so is clear from the fact that its composition depends upon the pressure, varying when the pressure is altered. Moreover, its heterogeneous structure is often visible under the microscope and its constituents are seldom present in simple whole-number ratios. Thus the eutectic mixture is merely an intimate conglomeration of crystals of the two solids.

Any two substances which are completely miscible as liquids but which do not form solid solutions with each other, so that only pure solid separates out on cooling, can form a eutectic mixture. Many such systems are known, particularly amongst metals. One of the most important is that of tin and lead, portrayed in Fig. 4.19.

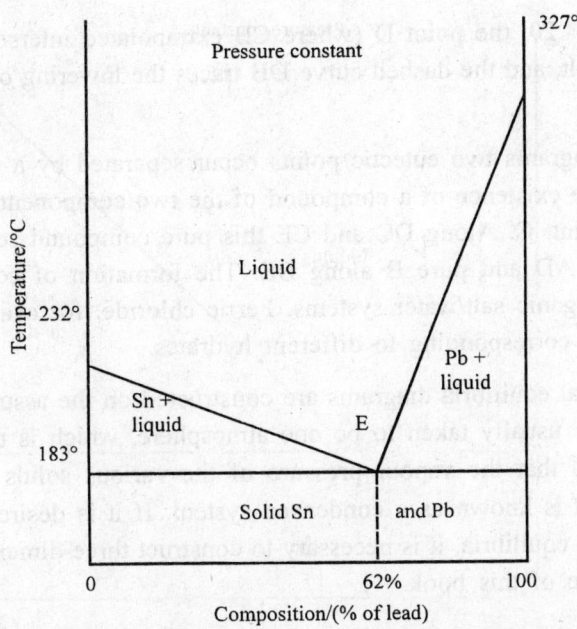

Fig. 4.19. Temperature/composition diagram for tin and lead.

The melting points of pure tin and pure lead are 232°C and 327°C respectively, but they form a eutectic mixture with the composition 62% lead and 38% tin melting at only 183°C. It is the low melting points of mixtures of tin and lead which account for the extensive use of these two metals in solders. Other examples are zinc (m.p. 419°C) and cadmium (m.p. 321°C) which form a eutectic mixture melting at 270°C, o-nitrophenol (m.p. 44°C) and p-toluidine (m.p. 43°C) which give one melting at 15.6°C, and naphthalene (m.p. 80°C) and p-nitrotoluene (m.p. 50°C) which form a eutectic mixture melting at 30°C. The eutectic mixture formed by iron and carbon (melting at 1130°C and containing 4.3% of carbon) is of great industrial importance because it largely determines the carbon content of pig iron and the operating temperature of the blast furnace.

Many inorganic salts (e.g. sodium chloride, potassium chloride, potassium iodide) form eutectic mixtures with water. It is difficult to study these systems over the complete range of composition at atmospheric pressure because the water tends to boil, but a typical equilibrium diagram is shown in Fig. 4.20. The line AC represents the freezing points of progressively stronger solutions of the salt in water, pure ice being deposited at these temperatures. The line BC is the solubility curve of the salt in water, i.e. the temperature at which the pure salt is in equilibrium in the solid state with its saturate solution. The two curves intersect at C, the eutectic point, at which the solid salt, ice, and the solution are all three in equilibrium at atmospheric pressure. This is, of course, the lowest temperature at which the salt can exist in solution. In systems involving a salt and water it is often known as the cryohydric point and the mixture of that particular composition as a cryohydrate.

The use of a mixture of salt and ice as a freezing mixture depends upon the above principle. When the salt is added some dissolves in the water giving a system containing solid salt, ice, and salt solution. Since such a system is only stable at the eutectic point, ice will melt and salt will dissolve (both endothermic processes) until the temperature falls to the eutectic temperature, or until one of the solids is all used up. In this way if sodium chloride, the cheapest salt, is used, a cooling mixture of −21°C can be obtained.

Referring again to Fig. 4.20, the point D (where CB extrapolated intersects the axis) represents the melting point of the pure salt, and the dashed curve DB traces the lowering of this melting point brought about by adding water.

In many equilibrium diagrams two eutectic points occur separated by a distinct hump, as shown in Fig. 4.21. This indicates the existence of a compound of the two components A and B; its composition is that of the maximum point, C. Along DC and CE this pure compound separates from the melt, just as pure A separates along AD and pure B along BE. The formation of compounds is very common amongst metals and in inorganic salt/water systems. Ferric chloride, for example, gives an equilibrium diagram with four maxima corresponding to different hydrates.

All these two-dimensional equilibria diagrams are constructed on the assumption that the pressure of the system is constant. It is usually taken to be one atmosphere, which is the pressure at which most measurements are made, so that the vapour pressure of the various solids can be neglected and the diagram then refers to what is known as a condensed system. If it is desired to incorporate the effect of pressure upon the various equilibria, it is necessary to construct three-dimensional equilibria diagrams, which are beyond the scope of this book.

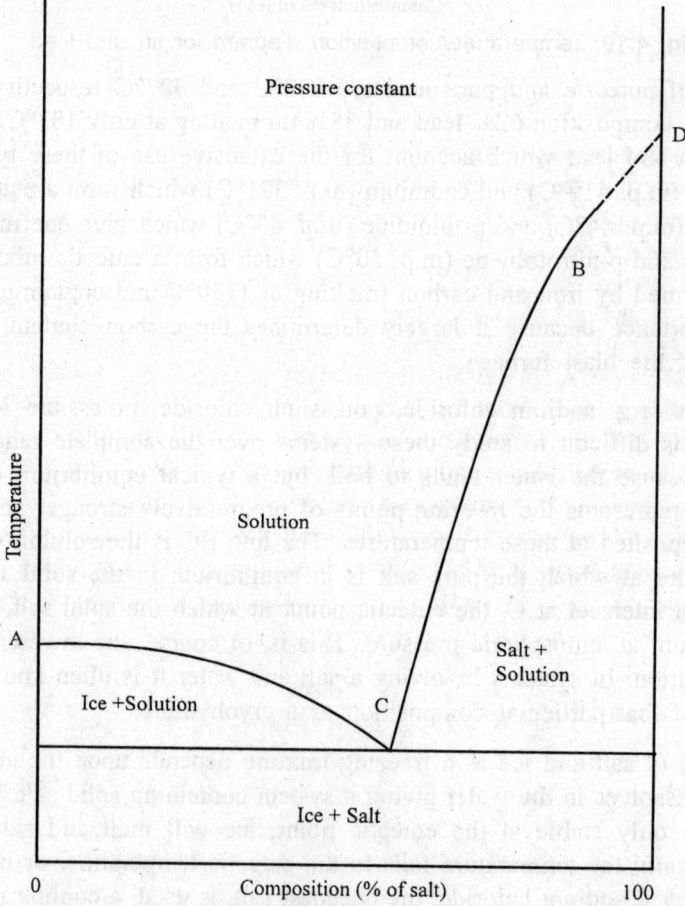

Fig. 4.20. Temperature/composition diagram for a salt and water.

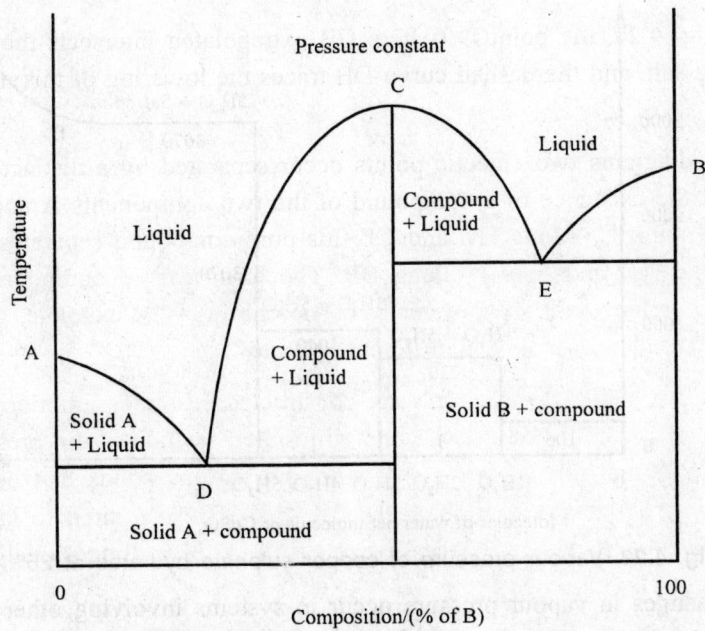

Fig. 4.21. Temperature/composition diagram for a system forming a compound.

VAPOUR PRESSURE OVER SALT HYDRATES

The way in which the pressure of water vapour above salt hydrates varies as their composition changes is well illustrated by considering the example of copper sulphate at 25°C. The results are embodied in Fig. 4.22, which is another condensed system diagram based this time upon measurements at constant temperature. Let us start with a concentrated solution of copper sulphate and consider the changes in water vapour pressure as water is progressively removed from the system. At first the vapour pressure falls along the curve AB as the solution becomes more and more concentrated, but when a saturated solution is obtained the vapour pressure remains at a constant value of 3070 Nm^{-2} despite further removal of water. At this pressure crystals of the pentahydrate are in equilibrium with the saturated solution. Further, dehydration causes a sudden fall in the vapour pressure to 1040 Nm^{-2} when all the saturated solution has gone and the composition corresponds to the pentahydrate. The vapour pressure again remains constant whilst pentahydrate and trihydrate are both present in equilibrium, but falls sharply to 730 Nm^{-2} when the last pentahydrate disappears. After a further constancy whilst the trihydrate and monohydrate exist together, the vapour pressure falls again to about 106 Nm^{-2} when the composition is that of the monohydrate and stays at this value as long as any water remains in the system.

These changes, all of which are reversible, can be represented by the following equations:

$$CuSO_4 \text{ sol.} \rightleftharpoons CuSO_4.5H_2O + nH_2O$$
$$\text{Pentahydrate}$$

$$CuSO_4.5H_2O \rightleftharpoons CuSO_4.3H_2O + 2H_2O$$
$$\text{Trihydrate}$$

$$CuSO_4.3H_2O \rightleftharpoons CuSO_4.H_2O + 2H_2O$$
$$\text{Monohydrate}$$

$$CuSO_4.H_2O \rightleftharpoons CuSO_4 + H_2O$$
$$\text{Anhydrous}$$

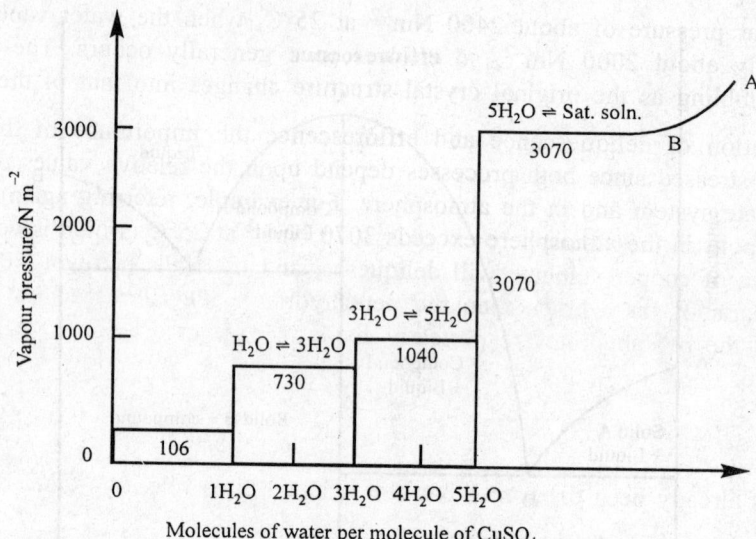

Fig. 4.22. Vapour pressure of copper sulphate hydrates at 25°C.

Similar step-like changes in vapour pressure occur in systems involving other salt hydrates. Indeed these sharp changes in vapour pressure provide one of the best ways of detecting the existence of hydrates. It will be seen that the phrase 'the vapour pressure of a salt hydrate' is really misleading, since any particular hydrate can exist over a range of vapour pressures, but a definite and constant vapour pressure is always associated with the existence of two different hydrates in equilibrium with each other at a particular temperature.

If on exposure to the atmosphere a substance takes up water vapour to such an extent that it dissolves in it forming a solution, that substance is said to be deliquescent. The essential condition for deliquescence is that a saturated solution of the substance should exert a lower water vapour pressure than the atmosphere with which it is in contact. Since the water vapour pressure in the atmosphere at room temperature seldom exceeds $2100 \ Nm^{-2}$ and only a few saturated solutions have a vapour pressure as low as this, the phenomenon is limited to relatively few substances. These are all necessarily very soluble in water since only if their saturated solutions are highly concentrated, are their vapour pressures lowered sufficiently below that of pure water at that temperature. Common examples are calcium chloride, whose saturated solution exerts a water vapour pressure of $10^3 \ Nm^{-2}$ at 20°C, and sodium hydroxide, with the extremely low value of about $130 \ Nm^{-2}$ for the vapour pressure of its saturated solution at room temperature. These substances continue to absorb water from the atmosphere, even in saturated solution, until their solutions are so diluted that they have the same water vapour pressure as the atmosphere. This explains their frequent use as solids in desiccators and for drying gases and liquids, although two other substances, phosphorus pentoxide and magnesium perchlorate, which are hygroscopic (i.e. absorb moisture from the atmosphere) but not deliquescent, are more efficient.

A substance which, on exposure to the atmosphere, loses water and changes into a lower hydrate or an anhydrous form is said to be efflorescent. Efflorescence only occurs when the substance exerts a higher water vapour pressure than the atmosphere around it. Well-known examples of efflorescent substances are sodium carbonate decahydrate, $Na_2CO_3.10H_2O$, and sodium sulphate decahydrate, $Na_2SO_4.10H_2O$, which change into a monohydrate and an anhydrous form respectively on standing in air for some days. For example a mixture of the decahydrate and monohydrate of sodium carbonate exerts a water vapour pressure of about 2400 Nm^{-2} at 25°C, when the water vapour pressure in the atmosphere is usually about 2000 Nm^{-2}, so efflorescence generally occurs. The process is usually accompanied by crumbling as the original crystal structure changes into that of the lower hydrate.

In any consideration of deliquescence and efflorescence the importance of the humidity of the atmosphere must be stressed since both processes depend upon the relative values of the water vapour pressure in the hydrate system and in the atmosphere. For example, referring again to Fig. 4.23, if the pressure of water vapour in the atmosphere exceeds 3070 Nm^{-2} at 25°C (which is very rarely the case), then the pentahydrate of copper sulphate will deliquesce, and if it falls below 1040 Nm^{-2} at 25°C (as it does in certain parts of the world), then the pentahydrate will effloresce, but between these two extremes crystals of the pentahydrate are perfectly stable, showing no tendency to lose or gain water from the atmosphere.

PHASE RULE

The word phase has already been discussed, but before stating this rule, two other technical terms must be explained:

1. *The number of components in a system:* The number of component in a system is the least number of different substances which must be specified to be present in order to describe completely the composition of every phase. For example, in any system consisting of a single substance, only one component is involved even if that substance is present in two or three different phases at the same time. A system containing two different substances which do not react with each other is regarded as containing two components, but a system in which three different compounds are in chemical equilibrium with each other (e.g. A + B ⇌ C) consists of only two components, however, because once any two of the constituents A, B, or C are specified the composition of every phase in the system is settled.

2. *The number of degrees of freedom of a system:* The number of degrees of freedom of a system is the number of factors such as pressure, temperature, and concentration of the various components which can be varied independently of each other without bringing about a change in the number of phases. For example, in a system consisting of a liquid and its vapour in equilibrium with each other, once one of the two variables temperature and pressure is fixed, the other is also settled. It follows that alteration of either of these factors alone will cause one of the phases to disappear and such a system is said to have only one degree of freedom.

The phase rule, which was devised by Gibbs and can be simply stated in the following form:

$$F = C - P + 2$$

where F = the number of degrees of freedom of the system.

C = the number of components in the system at equilibrium.

P = the number of phases present at equilibrium.

As our first example let us consider once again the phase equilibrium diagram shown in Fig. 4.1. This refers to a system containing only one component, so C = 1. At the triple point three phases are present, solid, liquid, and vapour, so under these particular conditions of temperature and pressure, P = 3. Applying the phase rule,

$$F = 1 - 3 + 2 = 0$$

so there are no degrees of freedom in the particular system corresponding to the triple point O.

This means that we can alter neither the temperature nor the pressure without also altering the number of phases present, which is confirmed by experiment. Such a system is said to be invariant.

A system in which a gas is in equilibrium with its saturated solution in a liquid provides our second example. Here we have two components and two phases, since the solution of the gas in the liquid is itself a single phase, so C and P are both equal to 2. Thus, applying the phase rule:

$$F = 2 - 2 + 2 = 2$$

and this system is bivariant, i.e. it has two degrees of freedom—the temperature and the pressure. Both of these factors must be specified in order to define the system completely.

For our third example let us consider a mixture of two immiscible liquids A and B. Here there are two components A and B and namely three phases, liquid A, liquid B, and vapour mixture, all in equilibrium with each other. Thus applying the phase rule to this system, C = 2, P = 3, and

$$F = 2 - 3 + 2 = 1$$

which means that the system is univariant, i.e. it has only one degree of freedom. Thus in this system if we fix the temperature, for example, the vapour pressure exerted above the immiscible liquids and the composition of the vapour phase will both be constant irrespective of the relative amounts of each liquid taken. These predictions are completely fulfilled in practice.

The application of the phase rule to heterogeneous chemical equilibria (i.e. ones in which the reactants and products are not all in the same phase) can be illustrated by considering the thermal dissociation of calcium carbonate:

$$\underset{\text{(Solid)}}{CaCO_3} \rightleftharpoons \underset{\text{(Solid)}}{CaO} + \underset{\text{(Gas)}}{CO_2}$$

In this system there are two solid phases and one gaseous phase, making a total of three. The number of components, however, is only two, because the nature of the third substance present can always be deduced from a knowledge of the other two. Thus P = 3 and C = 2 and

$$F = 2 - 3 + 2 = 1$$

so the system has only one degree of freedom. This means that once the temperature of the system is fixed, the equilibrium pressure will be settled also, and the carbon dioxide will exert a characteristic partial pressure at any particular temperature. This prediction, which is fully borne out by experiment, can also be deduced by applying the law of mass action to the equilibrium. The wide applicability of the phase rule should be apparent from the examples discussed above. Its great importance lies in its ability to predict how many variables will affect a given system in equilibrium. By its help we are able to decide what type of equilibrium can occur in heterogeneous systems and whether it is appropriate in any given case to apply Le Chatelier's principle to the system.

Example 4.3. At 100°C, the specific volumes of water and steam are respectively 1 c.c. and 1673 c.c. Calculate the change in vapour pressure of the system by 1°C change in temperature. The molar heat of vapourisation of water in this range may be taken as 9.70 kcal.

Solution.

Molar volume of liquid water, $V_l = 18 \text{ cm}^3 \text{ mol}^{-1} = 18 \times 10^{-6} \text{ m}^3 \text{ mol}^{-1}$

Molar volume of steam, $V_g = 18 \times 1673 \text{ cm}^3 \text{ mol}^{-1} = 30114 \times 10^{-6} \text{ m}^3 \text{ mol}^{-1}$

Heat of vapourisation, $\Delta H_v = 9700 \text{ cal mol}^{-1} \times 4.184 \text{ J cal}^{-1} = 40584.8 \text{ J mol}^{-1}$

$$\frac{dP}{dT} = \frac{\Delta H_v}{T(V_g - V_l)}$$

In this case,
$$dT = 1 \text{ K}$$
$$T = 273 + 100 = 373 \text{ K}$$

$$\therefore \quad dP = \frac{40584.8 \text{ J mol}^{-1} \times 1 \text{ K}}{373 \text{ K} (30114 - 18) \times 10^{-6} \text{ m}^3 \text{ mol}^{-1}}$$

$$= 0.00361 \times 10^6 \text{ N m}^{-2} \qquad \text{(J = Newton metre)}$$
$$= 0.03561 \text{ atm} = 27.08 \text{ mm of Hg}$$

Thus, the vapour pressure of water increases by about 27 mm of Hg by 1° rise in temperature, at 100°C.

Example 4.4. The vapour pressure of water at 95°C is found to be 634 mm. What would be the vapour pressure at a temperature of 100°C? The heat of vapourisation in this range of temperature may be taken as 40593 J mol^{-1}.

Solution. The integrated form of Clapeyron-Clausius equation for Liquid-Vapour equilibrium is

$$\ln \frac{P_2}{P_1} = \frac{\Delta H_v}{R} \left[\frac{T_2 - T_1}{T_1 T_2} \right]$$

In this case,
$$T_1 = 273 + 95 = 368 \text{ K} \qquad P_1 = 634 \text{ mm}$$
$$T_2 = 273 + 100 = 373 \text{ K} \qquad P_2 = ?$$
$$\Delta H_v = 40593 \text{ J mol}^{-1}$$

Thus,
$$\ln \frac{P_2}{634 \text{ mm}} = \frac{40593 \text{ J mol}^{-1}}{8.314 \text{ J K}^{-1} \text{ mol}^{-1}} \left[\frac{5 \text{ K}}{368 \text{ K} \times 373 \text{ K}} \right]$$
$$P_2 = 759.8 \text{ mm}$$

Chapter 5

Solutions

INTRODUCTION

A solution is regarded as a completely uniform mixture of two or more substances. The term uniform is used to express the fact that it is not possible to discern any inhomogeneity in properties such as density, refractive index, concentration or appearance in different samples drawn from various parts of the solution. At the molecular level, a solution cannot be perfectly homogeneous.

The terms solute and solvent are used in connection with solutions, and it is the solute which is dissolved by the solvent, the latter being the substance which is present in excess. The concentration of a solution is the amount of solute present in a given quantity of solution, and this quantity can be expressed in several ways:

1. Percentage composition
    (a) *by weight (w/w):* This is number of grams of solute contained in 100 g of solution.
    (b) *by volume (v/v):* This is the volume of solute present in 100 volumes of solution.
    (c) *as a weight/volume ratio (w/v):* This is the number of grams of solute present in 100 cm^3 of solution.
2. *Grams per dm^3:* The number of grams of solute per dm^3 of solution.
3. *Moles per dm^3:* The number of moles of solute per dm^3 of solution. This quantity expresses the molarity of a solution. The notation [X] is used to represent the concentration of a species X in moles per dm^3 of solution. Hence, in 0.1_M hydrochloric acid, [H$^+$] = 0.1.
4. *Molality:* This is the number of moles of solute dissolved in 1000 g of solvent.

TYPES OF SOLUTION

There are five general types of solution: (i) gases in gases; (ii) gases in liquids; (iii) liquids in liquids; (iv) solids in liquids; and (v) solids in solids.

Solutions of Gases in Gases

All gases mix, and the relative proportions of each component may be expressed in terms of a partial pressure or mole fraction. Molecules of a liquid may escape into the gaseous phase at temperatures below the boiling point of the liquid producing a vapour. The saturated vapour pressure of a liquid at different temperatures may be measured by a transpiration method, or by using an isoteniscope.

Transpiration method

A measured volume of dry air (at a known pressure) is slowly bubbled through a series of bulbs containing water at a controlled temperature. In this way, the air becomes saturated with water vapour. The saturated air is now passed through a series of calcium chloride tubes where the water vapour in the gas stream is absorbed. From the weight of water vapour taken up by the measured volume of dry air, the vapour pressure of water at the temperature of the experiment may be calculated.

Example 5.1. 15.84 dm³ of dry air (measured at s.t.p.) were bubbled through water at 25°C and then through a system of calcium chloride tubes. A gain in weight of 0.4205 g in the latter was noted. Calculated the saturated vapour pressure of water at 25°C.

Solution.

$$\frac{p_{H_2O}}{P} = \frac{n_{H_2O}}{n_{H_2O} + n_{air}}$$

Number of moles of water $= \dfrac{0.4205}{18} = 0.0233$

Number of moles of air $= \dfrac{15.84}{22.4} = 0.707$

The total pressure is $1.013 \times 10^5 \ Nm^{-2}$.
Hence,

$$p_{H_2O} = \frac{0.0233 \times 1.013 \times 10^5}{(0.0233 + 0.707)} = 3.24 \times 10^3 \ Nm^{-2}$$

Isoteniscope

A common form of the isoteniscope is shown diagrammatically in Fig. 5.1. The bulb is half filled with the liquid under test, which is then heated to boiling several times in order to displace air from the bulb and U-tube. On cooling, some liquid condenses in the U-tube. By tilting the apparatus, liquid can be transferred from the bulb to the U-tube (and vice versa) and in this way the liquid level in the U-tube is adjusted to that shown. It is essential that the vapour space between the bulb and the right-hand limb of the U-tube is completely free from air. The apparatus is placed in a thermostat and a partial vacuum is applied. The pressure on the left-hand side of the system is adjusted by means of the three-way tap until the liquid levels are equal. At this point, the air pressure shown by the manometer is equal to the vapour pressure in the vapour space above the liquid in the bulb. The cold trap prevents the escape of vapour into the manometric and vacuum systems.

Solutions of Gases in Liquids

The solubility of a gas in a liquid depends on the nature of the gas and solvent, and on the temperature and pressure. With water as a solvent, helium is the least soluble and ammonia the most soluble gas.

Usually, heat is liberated when a gas dissolves in water and in accordance with Le Chatelier's principle a decrease in solubility of the gas would be expected following a rise in temperature. This is found to be generally true, although the solubility of hydrogen and the noble gases decreases at first, then increases as the temperature rises. When a beaker of tap water is heated, the dissolved air is expelled as the water warms, and bubbles of air collect on the base and sides of the beaker; hence the use of boiled-out water when making up certain solutions such as potassium permanganate.

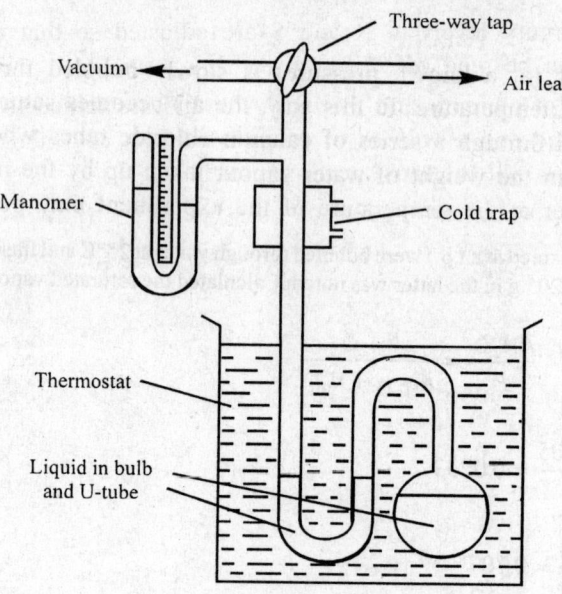

Fig. 5.1. Isoteniscope.

From kinetic considerations, it would be expected that an increase in pressure should increase the solubility of a gas, as on compression the rate of gas molecules striking the liquid interface is increased and the extent of absorption of these molecules into the liquid phase is greater. The effect of pressure on gas solubility is described by Henry's law.

Henry's law: The weight of a gas dissolved by unit volume of liquid is directly proportional to the pressure of the gas with which it is in equilibrium, the temperature being constant.

Thus, the mass of gas dissolved is proportional to pressure, or

$$m = kp$$

where m = mass of the gas

   p = pressure of the gas

   k = constant

The solubility of a gas in a liquid is usually given in terms of an absorption coefficient (α), which is the maximum volume of gas (reduced to s.t.p.) which can be dissolved by 1 cm^3 of liquid at a particular temperature.

Determination of the solubility of sparingly soluble gases

The apparatus used for this determination is shown in Fig. 5.2, and the experiment is carried out in the following stages:

1. The volume of the absorption vessel T is determined by filling with water and either measuring the volume used or by weighing before and after filling.

2. The absorption vessel is filled with boiled-out deionised water, and the apparatus is placed in an air thermostat to attain the desired temperature.

3. R is filled completely with mercury by raising the tube S. Tap A is connected to a flask of gas under test, and, by lowering S, the gas is drawn into R.

4. Tap A is closed and the mercury levels in R and S are adjusted so that the volume of gas at atmospheric pressure in R can be read off.

5. S is raised to provide a head of mercury and the taps A and B are turned to connect R and T. At the same time, tap C is opened and a volume of water is run out of T. In this way a volume of gas is transferred from R to T. All the taps are now closed.

6. The absorption vessel is now shaken for some time, care being taken to ensure that heat from the hands does not change the temperature of T and that the liquid does not splash into the tube below B.

7. R and T are re-connected and the volume in R is read off (the mercury levels in R and S being equal). The gas is transferred back to T and re-shaken.

8. When no further diminution in volume is noticed, the water in T is saturated with gas, and the solubility is calculated as shown in Example 5.2.

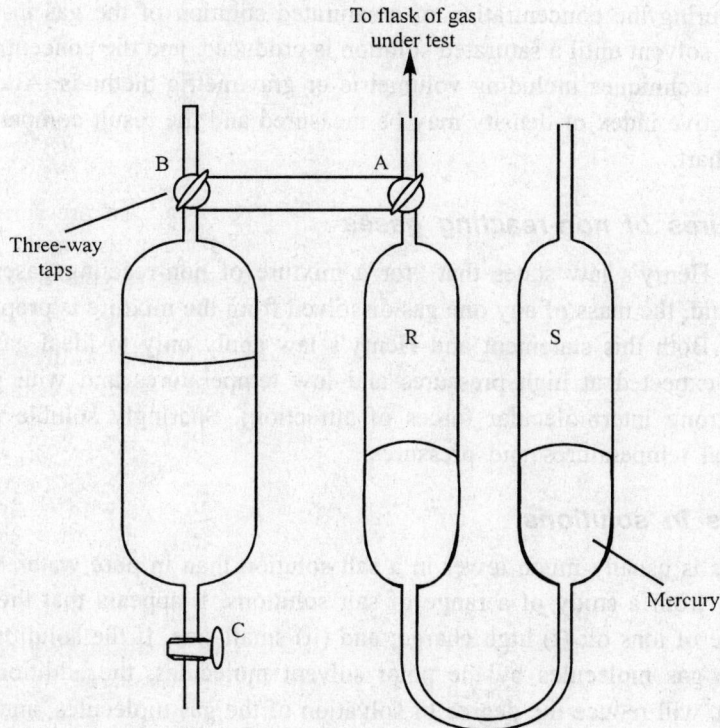

Fig. 5.2. Apparatus for measuring the solubility of a sparingly soluble gas.

Example 5.2. The following results were obtained using the above method to determine the solubility of oxygen in water at 20°C.

Volume of absorption vessel T	= 53.35 cm³
Volume of water displaced	= 21.40 cm³
Initial volume of oxygen in R	= 32.64 cm³
Final volume of oxygen in R after absorption	= 10.15 cm³
Atmospheric pressure	= 9.89×10^4 N m⁻¹

Solution. The volume of water remaining in the absorption vessel in which the oxygen dissolves is

$$53.35 - 21.40 = 31.95 \text{ cm}^3$$

Volume of oxygen originally present was 32.64 cm³, and the volume of oxygen remaining after absorption is

$$21.40 + 10.15 = 31.55 \text{ cm}^3$$

Thus 31.95 cm³ of water dissolve

$$32.64 - 31.55 = 1.09 \text{ cm}^3 \text{ of oxygen}$$

and 1 cm³ of water would dissolve

$$\frac{1.09 \times 273}{31.95 \times 293} \times \frac{9.89 \times 10^4}{1.013 \times 10^5}$$

$$= 0.031 \text{ cm}^3 \text{ of oxygen at s.t.p.}$$

Hence, the absorption coefficient of oxygen at 20°C is 0.031.

Determination of the solubility of highly soluble gases

This is done by measuring the concentration of a saturated solution of the gas in the solvent. The gas is bubbled through the solvent until a saturated solution is produced, and the concentration of this solution is found by analytical techniques including volumetric or gravimetric methods. Alternatively, a physical property such as refractive index or density may be measured and the result compared with a previously obtained calibration chart.

Solutions of mixtures of non-reacting gases

Dalton's extension to Henry's law states that "for a mixture of non-reacting gases (assumed ideal) in equilibrium with a liquid, the mass of any one gas dissolved from the mixture is proportional to the partial pressure of that gas". Both this statement and Henry's law apply only to ideal gases. Deviations from these laws are to be expected at high pressures and low temperatures and with gases that are easily liquefied (i.e. have strong intermolecular forces of attraction). Sparingly soluble gases obey the laws satisfactorily at normal temperatures and pressures.

Solubility of gases in solutions

The solubility of a gas is usually much lower in a salt solution than in pure water. This is known as the *salting out effect* and, from a study of a range of salt solutions, it appears that the salting out effect is greater in the presence of ions of: (i) high charge; and (ii) small size. If the solution of a gas is effected by a solvation of the gas molecules by the polar solvent molecules, the addition of ions, which are preferentially solvated, will reduce the degree of solvation of the gas molecules, and thus lower the mass of gas held in solution. A useful feature of the salting out effect is that it allows moderately soluble gases such as chlorine or carbon dioxide to be collected over saturated sodium chloride solution without too much loss of gas—an important factor when samples are withdrawn from a gas stream for analysis.

Solution of Liquids in Liquids

Liquids which mix with each other completely over all ranges of composition are said to be fully miscible. If one liquid is partially or totally insoluble in the other, the terms partially miscible and immiscible respectively are used.

Vapour pressure of a solution of two miscible liquids

In a mixture of two miscible liquids (say A and B) the ability of molecules of A to escape from the liquid into the gaseous phase is impaired by the presence of B, and vice versa. For an ideal solution, that is one in which no intermolecular forces exist, the vapour pressure of each constituent is given by Raoult's law.

Raoult's law

The vapour pressure of a constituent in an ideal solution is equal to the vapour pressure exerted by the pure constituent at that temperature, multiplied by the mole fraction by which it is present.

Mathematically,

$$p_A = x_A P_A^o$$

where p_A is the vapour pressure of A over the liquid mixture.

x_A is the mole fraction of A in the liquid mixture.

P_A^o is the vapour pressure of pure A at that temperature.

No liquid mixture is ideal, and deviations from Raoult's law are observed, but the law holds well for dilute solutions.

Boiling point–liquid composition curves

It is clear from Fig. 5.3 that at 50°C pure benzene has a vapour pressure greater than any of the benzene-methylbenzene mixtures, and as the temperature rises pure benzene will continue to show the greatest vapour pressure. Consequently, pure benzene will have a vapour pressure of 1 atm (or will equal the external pressure) at a lower temperature than any of the mixtures of benzene and methylbenzene. This means that pure benzene has the lowest boiling point and pure methylbenzene the highest boiling point of all the various compositions.

The boiling points of the various liquid mixtures lie between these two extreme values; mixtures rich in benzene will boil at a lower temperature, while those rich in methylbenzene will have a higher boiling point. This is interpreted by means of a boiling point-liquid composition graph (which is the inverse of the graph of vapour pressures shown in Fig. 5.3), as shown in Fig. 5.4. From this graph, a mixture containing 0.25 mole of benzene and 0.75 mole of methylbenzene boils at 103°C, but it should be noted that the graph is constructed on the assumption that Raoult's law is obeyed over the whole range of compositions.

Composition of the vapour formed by boiling an ideal liquid mixture

Suppose a liquid mixture contains an equal number of moles of two components A and B, and let the vapour pressure of pure A be 1.33×10^5 Nm^{-2}, and that of pure B be 0.696×10^5 Nm^{-2} at a temperature of 120°C. Then the partial pressure of A in the vapour is

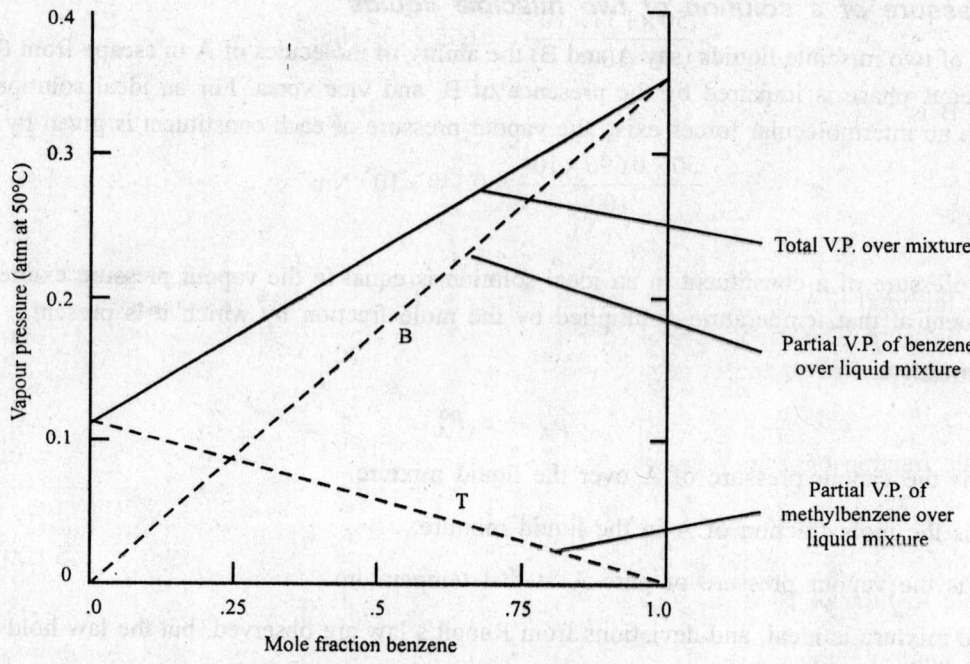

Fig. 5.3. Vapour pressure–composition graph for an ideal liquid mixture at 50°C.

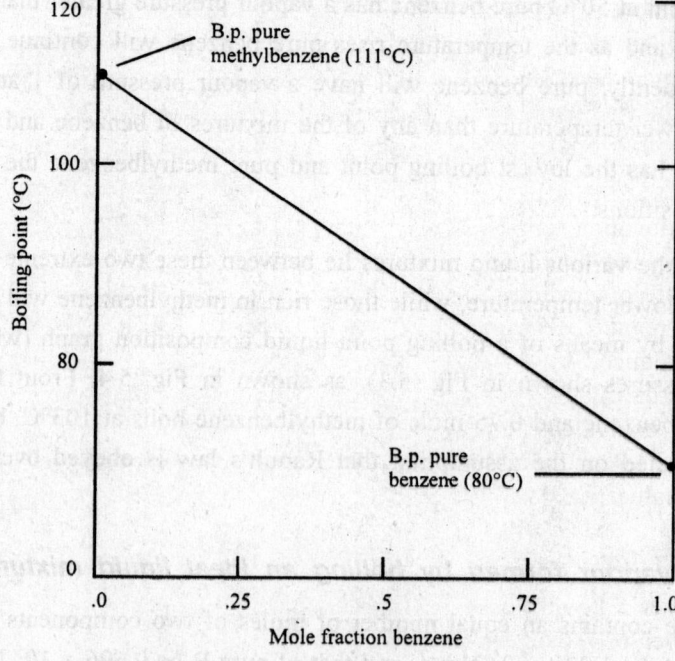

Fig. 5.4. Boiling point-composition graph for mixtures (assumed ideal) of benzene and methylbenzene.

$$\frac{50 \times 1.33 \times 10^5}{100} = 0.655 \times 10^5 \text{ Nm}^{-2}$$

and that of B is

$$\frac{50 \times 0.696 \times 10^5}{100} = 0.348 \times 10^5 \text{ Nm}^{-2}$$

This makes the total pressure 1.013×10^5 Nm^{-2}, and this particular mixture boils at 120°C under one atmosphere pressure. The vapour produced will not have the same composition as the boiling liquid, but will contain amounts of A and B in proportion to their partial vapour pressures.

$$\text{Mole fraction of A in vapour} = \frac{0.665 \times 10^5}{1.013 \times 10^5} = 0.66$$

and

$$\text{Mole fraction of B in vapour} = \frac{0.348 \times 10^5}{1.013 \times 10^5} = 0.34$$

The boiling liquid contains 50 mole % of A, but the vapour produced contains 66 mole % of A. This enrichment on vapourisation of the low boiling component forms the basis of fractional distillation, and this degree of enrichment (produced by boiling a liquid and condensing the vapour in equilibrium with it) is known as that produced by a *theoretical plate*. If the enriched vapour is condensed and reboiled, a further degree of enrichment is produced, and a device for accomplishing this is a fractionation column. Typical columns may be packed with glass beads, glass or metal rings, helices, etc. in order to increase the area of liquid in contact with the vapour. Larger columns contain bubble cap trays in which the vapour from a boiling liquid is made to bubble through the liquid standing on the bubble cap tray. Fig. 5.5 shows diagrams of both types of column. In the column packed with helices the multi-turn helices assist column drainage and help to prevent the formation of liquid locks.

The progress of fractionation can be checked by means of a liquid-vapour composition curve as shown in Fig. 5.6. The boiling liquid has the composition shown by the point X (50 mole % of A) and this produces a vapour of composition Y. This is condensed to give a liquid of the same composition which is represented by point Z. The complete step XYZ is the enrichment produced by a theoretical plate. The enrichment produced by an actual bubble cap plate is less than theoretical; for example, a column containing 20 actual plates may give a separation equal to 14 theoretical plates. For a packed column, H.E.T.P. (height equivalent per theoretical plate) values are quoted; for instance, a laboratory column 100 cm in packed length may produce an enrichment corresponding to 5 theoretical plates, and thus it has an H.E.T.P. value of 20 cm per plate.

Derivation of Raoult's law

If the vapour standing over a liquid behaves as an ideal gas, then the pressure of the vapour is exactly proportional to the number of molecules per unit volume in the vapour phase. In turn, the number of molecules entering the vapour phase is proportional to that fraction of molecules in the liquid which can acquire sufficient energy to escape from the liquid phase.

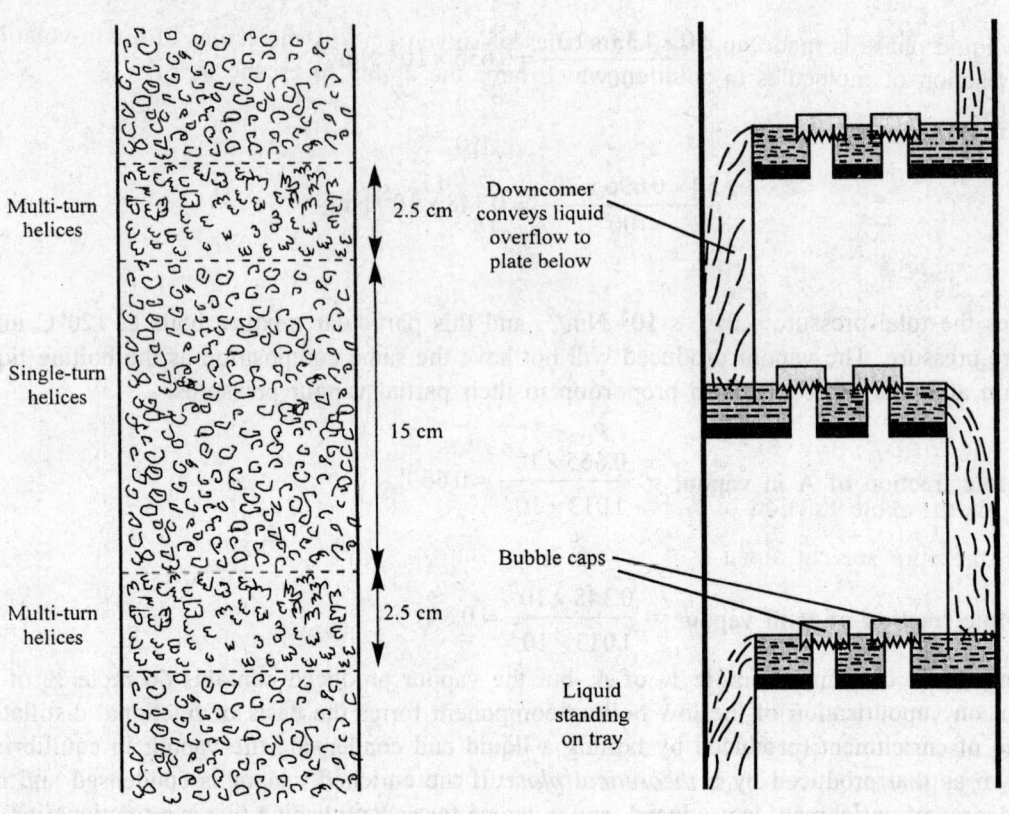

Packed column

Bubble cap column

Fig. 5.5. Fractional column.

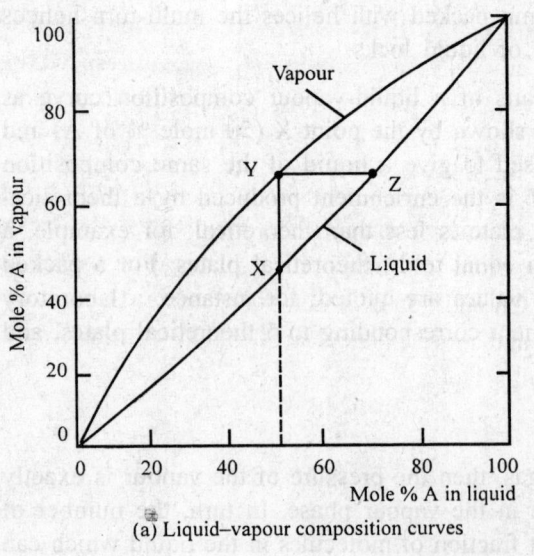

(a) Liquid–vapour composition curves

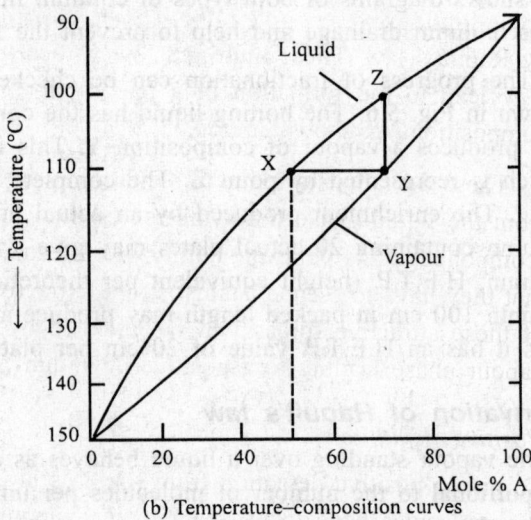

(b) Temperature–composition curves

Fig. 5.6. Liquid-vapour composition curve

If the liquid phase is made up of n_A molecules of solvent and n_B molecules of a non-volatile solute, then the fraction of molecules in solution which have the ability to escape is

$$\frac{n_A}{n_A + n_B}$$

so that

$$p_A \propto \frac{n_A}{n_A + n_B}$$

$$p_A = k\frac{n_A}{n_A + n_B} = kx_A$$

where x_A is the mole fraction of A.

When the pure solvent alone is present, $x_A = 1$ and $p_A = P_A^o$ so that $P_A^o = k$ and

$$p_A/P_A^o = x_A$$

which is Raoult's law.

(The same argument applies to cases in which the solute is volatile, the total vapour pressure over the liquid mixture then being $p = p_A + p_B$). In this derivation, it has been assumed that the ability of a molecule of the solvent to escape remains unaltered on the addition of the second component. (This is implied by assuming the proportionality constant k to be a true constant.) The assumption is not justified, and this leads to the observed deviations from Raoult's law.

Deviations from Raoult's law

Some liquid mixtures give total vapour pressures which are higher than or less than the values predicted by Raoult's law. These high and low values (described as positive or negative deviations respectively) may lead to the formation of a maximum or minimum vapour pressure for the mixture at a certain composition, as shown in Fig. 5.7.

These deviations are due to a change in the ability of the molecules of one component (A) to escape from the liquid phase following the addition of the second component (B). For example, if the solvent (component A) is polar the ability of molecules to escape from this environment is impaired by the fact that they have to escape against strong attractive forces. On addition of a non-polar solute, B, the extent of these attractive forces will be lessened by dilution, and more molecules of A will escape into the vapour phase leading to a positive deviation from Raoult's law.

Positive deviations

To give a positive deviation from Raoult's law, one of the components is normally of the type in which the molecules have to escape into the vapour phase against forces which are larger than usual. Such liquids are usually either those (i) in which strong intermolecular forces of attraction obtain or; (ii) which contain long chain molecules.

An example of the first type is the heptane-ethanol system. Ethanol molecules are attracted by hydrogen bonding in the pure liquid. On mixing, the heptane molecules interpose between the ethanol molecules thus reducing the extent of the hydrogen bonding. This allows the molecules to escape more readily and the vapour pressure rises. The effect of chain length on positive deviations is shown by alcohol-water mixtures. The total vapour pressure curve for methanol and water increases steadily from pure water to pure alcohol, but mixtures of water and propanol or butanol show increasing positive deviations. Systems showing positive deviations are quite common.

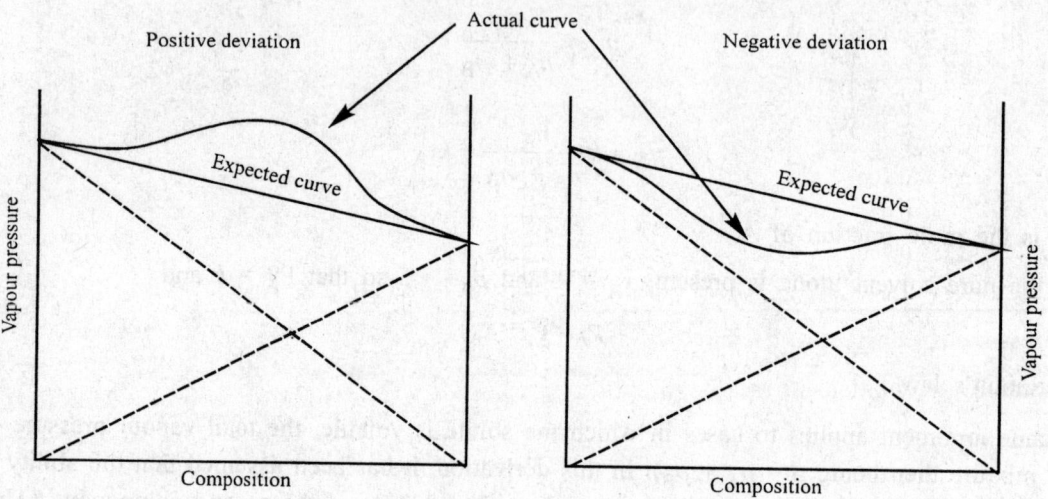

Fig. 5.7. Deviations from Raoult's law.

Negative deviations

Systems which exhibit this type of behaviour do so because extra chemical bonds are formed which prevent the escape of molecules into the vapour phase. This can occur when:

1. There is a strong intermolecular attraction between the two liquids. For example, propanone and trichloromethane attract each other due to the formation of hydrogen bonds.
2. Some type of compound formation can take place between the two liquids. Frequently, such systems contain an acidic and a basic component—pyridine and ethanoic acid for example, or a substance which forms a hydrate with water (e.g. hydrochloric acid and water).

In such systems, heat is evolved when the two liquids (which show negative deviations) are mixed, heat is liberated, corresponding to bond formation. The amount of heat evolved, however, is small (of the order of 20 kJ mol^{-1}) showing that the bonds involved are weak.

In many cases, factors operate to produce both positive and negative deviations from Raoult's law. Occasionally (e.g. in mixtures of benzene and 1,2-dichloroethane) these balance out, giving a pseudo-ideal mixture. The extents of both positive and negative deviations decreases as the dilution of the solution increases, and Roult's law is obeyed in dilute solutions.

Examples of deviations from Raoult's law

A negative deviation from Raoult's law leads to a mixture showing a maximum boiling point—an example of this type of behaviour is shown by the nitric acid and water system. In this mixture, the formation of non-volatile ions in place of volatile molecules together with the existence of hydrates in solution lead to the reduced volatility. The data for this system are:

Mole fraction HNO_3 in vapour	Mole fraction HNO_3 in liquid	Temp. °C
0.000	0.000	100
0.006	0.084	106
0.020	0.120	112
0.160	0.300	121.5
0.380	0.380	122
0.600	0.400	121
0.760	0.460	118
0.890	0.530	112
0.920	0.610	99
0.960	0.930	90
1.000	1.000	86

The pressure is constant at 1 atm.

These values are plotted on the graph shown in Fig. 5.8.

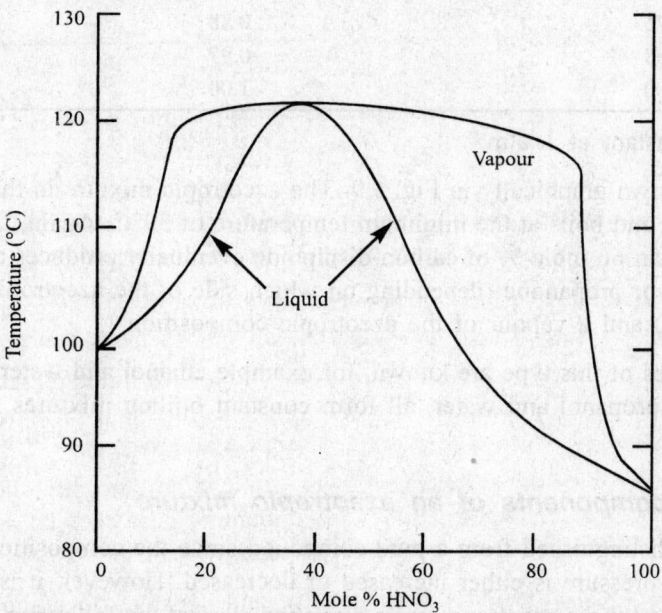

Fig. 5.8. Maximum boiling point mixture nitric acid and water.

When a liquid mixture containing less than 38 mole % of nitric acid is distilled, the vapour produced is richer in water than the boiling liquid. Subsequent fractionation of the condensed vapour will produce pure water. The residual liquid becomes gradually richer in nitric acid and the boiling point rises steadily to 122°C. At this temperature, the liquid contains 38 mole % of nitric acid, and the graph shows that the vapour produced at this composition by the boiling liquid has exactly the same composition as the liquid. This maximum boiling point mixture is referred to as an *azeotrope*. Mixtures containing more than 38 mole % of nitric acid boil at temperatures below 122°C, and again, the liquid composition tends to reach the azeotropic composition on boiling, while the vapour can be fractionated to give pure nitric acid. It is a feature of all mixtures which form an azeotrope, that they cannot be separated into two pure components by fractionation.

Mixtures of hydrochloric acid and water, or trichloromethane and water are other examples of systems which form a constant boiling or azeotropic mixture with a maximum boiling point.

A positive deviation from Raoult's law (Fig. 5.7) leads (if the deviation is large enough) to the mixture showing a minimum boiling point. The carbon disulphide-propanone system is a typical example:

Mole fraction CS_2 in vapour	Mole fraction CS_2 in liquid	Temp. °C
0.00	0.00	56
0.18	0.05	51
0.35	0.13	46
0.44	0.18	44
0.53	0.29	41
0.57	0.38	40
0.60	0.45	39.5
0.66	0.66	39
0.76	0.88	40.5
0.88	0.97	43
1.00	1.00	46

The pressure is constant at 1 atm.

These values are shown graphically in Fig. 5.9. The azeotropic mixture in this case contains 66 mole % of carbon disulphide and boils at the minimum temperature of 39°C. Again, fractionation of a mixture of composition other than 66 mole % of carbon disulphide eventually produces a liquid containing either pure carbon disulphide or propanone (depending on which side of the azeotropic composition the initial liquid composition lies) and a vapour of the azeotropic composition.

Many other examples of this type are known, for example ethanol and water, tetrachloromethane and ethyl ethanoate, and n-propanol and water, all form constant boiling mixtures with a minimum boiling point.

Separation of the components of an azeotropic mixture

An azeotrope may be distinguished from a pure compound since the composition of the constant boiling mixture changes if the pressure is either increased or decreased. However, it is not possible to separate the two components of a binary azeotrope by fractionation, so that the following alternative methods have to be employed.

1. *Extractive distillation with a third component:* For example, the ethanol-water system forms an azeotrope containing 95.5% of alcohol, at 1 atm, boiling at 78.1°C. A small quantity of benzene is added to this mixture, and three separate fractions containing (i) an ethanol, benzene and water three-component azeotrope; (ii) an ethanol and benzene two-component azeotrope; and (iii) pure ethanol, can be distilled off at successively increasing temperatures.

2. *Extraction:* Extraction of one component of the azeotrope using an immiscible solvent.

3. *Adsorption:* Adsorption of one component, using silica gel, for example.

4. *Chemical methods:* For example, the water in the ethanol-water azeotrope can be removed by adding a suitable solid-desiccant (e.g. calcium oxide).

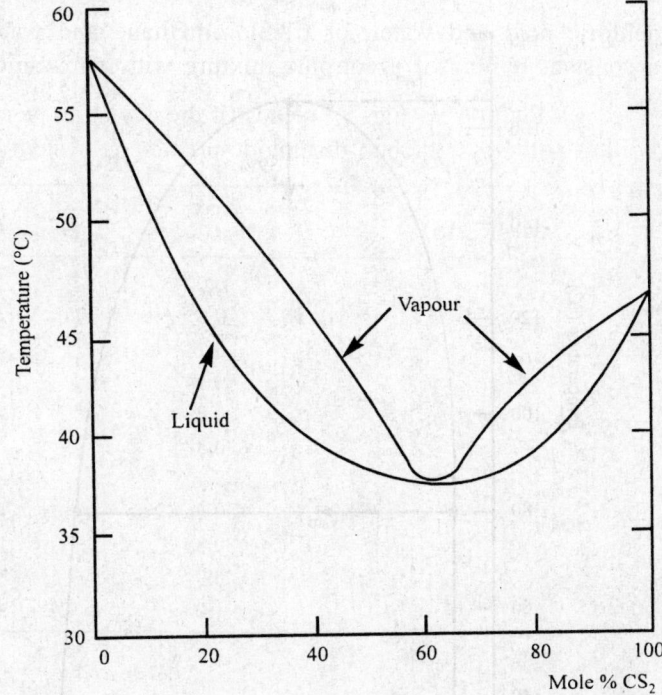

Fig. 5.9. Minimum boiling point mixture: carbon disulphide and propanone.

Partially miscible liquids

Pairs of liquids which dissolve in each other to a limited extent are termed partially miscible. For example, if a little phenol is added to water, the phenol dissolves completely. If a larger quantity of phenol is used, two liquid layers are produced, one being a saturated solution of phenol in water, the other a saturated solution of water in phenol (Fig. 5.10). These solutions are sometimes known as *conjugate solutions*.

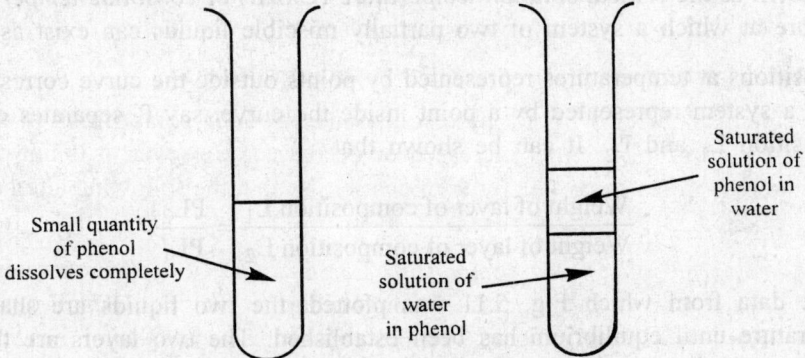

Small quantity of phenol dissolves completely

Saturated solution of water in phenol

Saturated solution of phenol in water

Fig. 5.10. Partially miscible liquids.

Effect of temperature changes on conjugate solutions

Generally, as the temperature is raised, the mutual solubility of the liquids increases, and a graph of the compositions of the liquid layers at various temperatures takes the form of curve shown in Fig. 5.11, which represents the system formed by o-cresol and water.

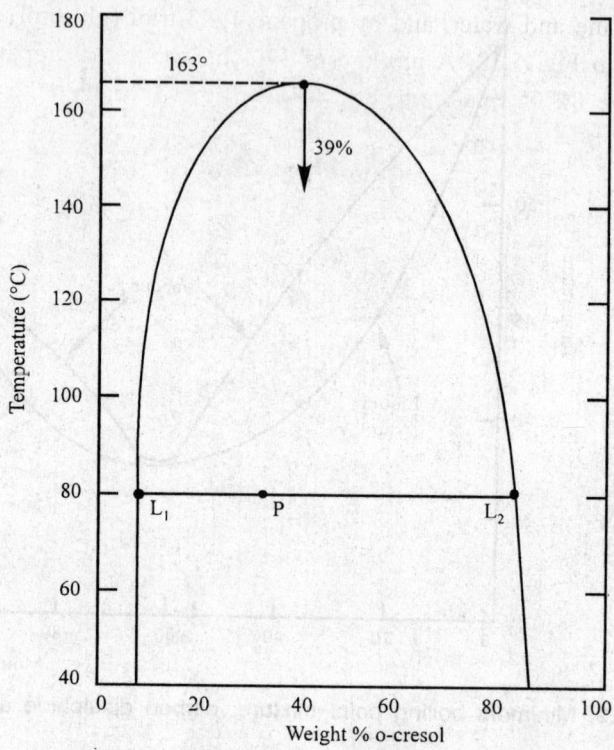

Fig. 5.11. Variation or partial miscibility with temperature: the o-cresol-water system.

As the temperature rises, the two liquid layers become increasingly soluble—the water layer contains a greater proportion of o-cresol, while the organic layer shows an increasing water content. At 163°C, both layers contain 39% by weight of o-cresol, and the two liquid layers mix completely. This temperature is known as the *critical solution temperature* (C.S.T.) or *consolute temperature*, and it is the highest temperature at which a system of two partially miscible liquids can exist as two layers.

Liquid compositions at temperatures represented by points outside the curve correspond to one liquid layer only, while a system represented by a point inside the curve, say P, separates out into two liquid layers, of composition L_1 and L_2. It can be shown that

$$\frac{\text{Weight of layer of composition } L_1}{\text{Weight of layer of composition } L_2} = \frac{PL_2}{PL_1}$$

To obtain the data from which Fig. 5.11 was plotted, the two liquids are shaken together at a controlled temperature until equilibrium has been established. The two layers are then separated and analysed. Alternatively, the temperature at which the two layers just begin to form (shown by the appearance of a turbidity on cooling the clear solution) is determined for a range of mixtures. In both methods, the experiments are repeated to give a large number of points on the graph. In some instances, two partially miscible liquids become more miscible on cooling; for example the system 2:4:6 trimethyl-pyridine and water shows a *lower critical solution temperature*, as shown in Fig. 5.12. Below 5.7°C all mixtures of these two liquids are completely miscible.

A few liquid mixtures exhibit both an upper and a lower critical solution temperature; this type of behaviour is shown by nicotine and water, and by propane 1,2,3-triol (glycerol) and guaiacol, the curve for which system is shown in Fig. 5.13. A mixture of 37% by weight of propane 1,2,3-triol forms one layer below 40°C and above 82°C, but produces two layers between these temperatures.

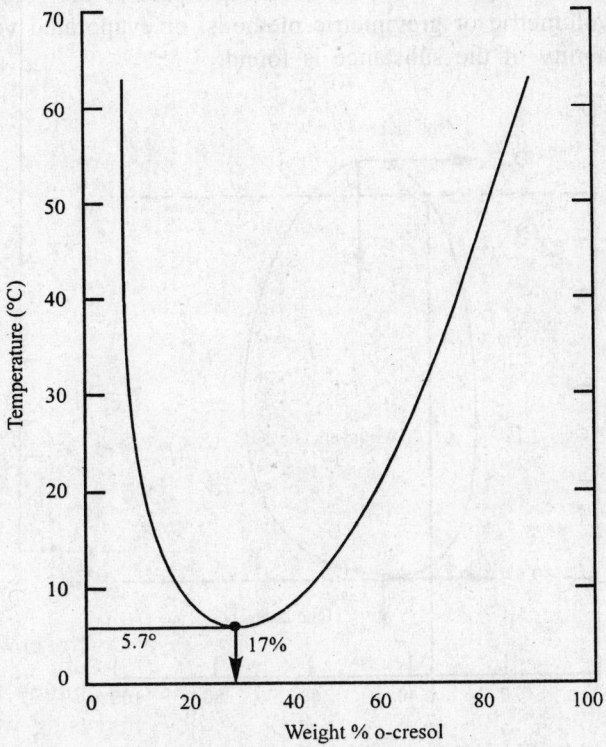

Fig. 5.12. Lower C.S.T. 2:4:6 trimethyl-pyridine-water system.

Solutions of Solids in Liquids

A solid which dissolves in a liquid is said to be soluble, while one which does not is termed insoluble. However, minute amounts of a substance normally regarded as insoluble do dissolve, while there is a limit to the amount of a soluble solid that can dissolve—this maximum quantity dissolved produces a *saturated solution*. Generally, more solid dissolves on warming a solution, and the extra amount crystallises out when a saturated solution is cooled. In some instances, such crystallisation fails to occur spontaneously, and a super-saturated solution results. Vigorous stirring or seeding with a small crystal causes the excess solute to precipitate from a super-saturated solution. In view of this, the solubility of a solid is defined as the maximum weight of solute which will dissolve in 100 g of solvent, at a given temperature, to produce a saturated solution, which is an contact with excess undissolved solute.

Experimental determination of the solubility of a solid

Most methods involve the analysis of a saturated solution, using volumetric or gravimetric method. Since excess solid material has to be present in accordance with the definition of solubility, precautions have to be taken to prevent this excess solid from being removed with the solution withdrawn for analysis.

A saturated solution containing an excess of the undissolved solid is prepared at the temperature required. Some time is allowed for the system to reach equilibrium, after which the solution is decanted through a filter or a sample is withdrawn by means of a pipette (on the end of which is fitted a 'guard' tube containing a glass wool filter). The guard tube is removed before the contents of the pipette are discharged, and all apparatus used must be at the same temperature as the solution. The sample is then weighed, and analysed by volumetric or gravimetric methods, or evaporated very carefully to dryness. From these results, the solubility of the substance is found.

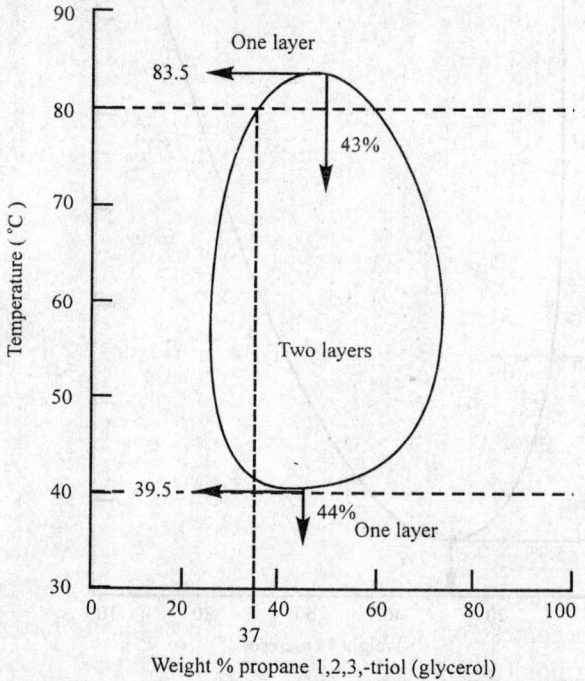

Fig. 5.13. Propane 1,2,3-triol (glycerol)-guaiacol; a system showing upper and lower C.S.T.s.

Solubility curves

The variation of the solubility of a solid with temperature is shown graphically as a solubility curve (Fig. 5.14). Discontinuities in the solubility curve are usually seen when (a) a change in the crystal form of the solid takes place, or (b) a change in the degree of hydration of the salt occurs (Fig. 5.15). The solubility of calcium sulphate decreases with increasing temperature; for example at 40°C the solubility is 0.21, while at 100°C it is 0.17 g/100 g H_2O.

Fractional crystallisation

It is possible to separate the components of a mixture of solids in solution by slowly cooling a hot concentrated solution. The nature of the crystals first deposited depends on the relative concentration of each species and on the solubility of the substances which may be deposited from solution.

For instance, in a solution containing equimolar proportions of sodium, potassium, chloride and nitrate ions, sodium chloride would be deposited first as the solution is concentrated by evaporation (Fig. 5.14). After the removal of these crystals from the hot solution, cooling would give rise to a deposit containing a small quantity of sodium chloride and a large quantity of potassium nitrate. The potassium nitrate could be purified by dissolving the crystals in the minimum quantity of water and recrystallising.

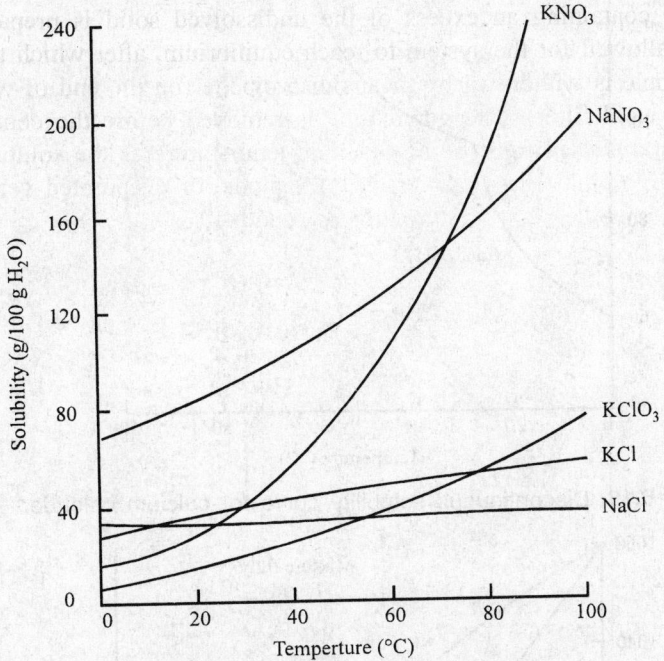

Fig. 5.14. Typical solubility curves.

Solutions of Solids in Solids

A pure substance is, with few exceptions, associated with a sharp melting point. In addition, the temperature interval between the point at which a solid first becomes fully molten (melting point) and the point at which a liquid just becomes completely solid (freezing point) is narrow. When an impurity is added to the solid, the melting point alters (it is usually lowered) and the melting range is extended.

Fig. 5.16 shows the temperature-concentration diagram for mixtures of gold and silver. The upper curve (liquidus) indicates the temperature at which the mixture becomes completely molten, while the lower curve (solidus) shows the temperature to which the molten mixture has to be cooled in order to bring about complete solidification. A molten mixture of gold and silver is a true solution, since the liquid metals are fully miscible. In such circumstances, on cooling, the solid produced does not consist of a mixture of crystals of pure gold and pure silver, but is what is termed an alloy in the case of metals, or more generally, a solid solution. In a solid solution, one component can still be regarded as being dissolved by the other, even in the solid state. Many cases occur where the melting point of the solid solution falls below that of either component to give a system with a minimum melting point. A typical example is fusion mixture (containing sodium carbonate, m.p. 820°C, and potassium carbonate, m.p. 860°C), which has a minimum melting point of 700°C.

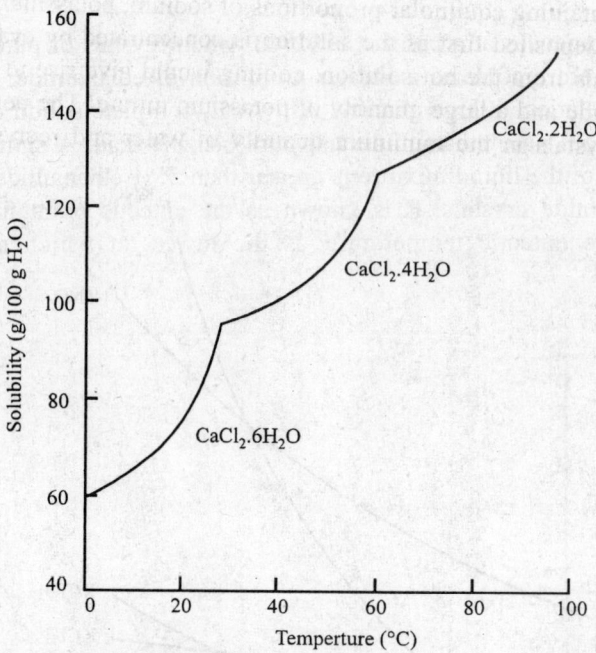

Fig. 5.15. Discontinuous solubility curve for calcium chloride.

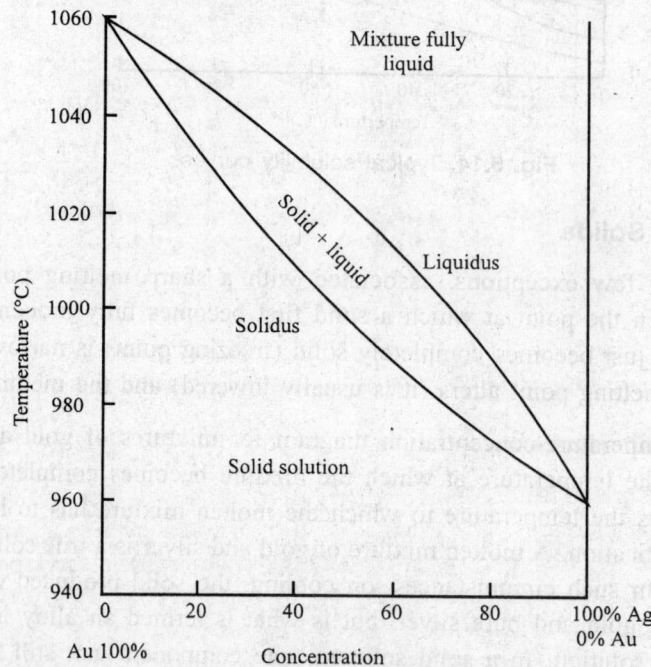

Fig. 5.16. Temperature-concentration diagram for solutions of gold and silver.

Often, the other components of a system which has a minimum melting point are immiscible in the solid state; this leads to the formation of a eutectic. The benzoic acid-ethanamide system shows this behaviour, and the temperature composition graph is given in Fig. 5.17. When a liquid of composition

X is cooled, solidification commences at a point A (102°C), when crystals of pure benzoic acid separate out. The remaining liquid therefore contains a greater proportion of ethanamide, and on further cooling the temperature-composition relation follows the line AE. At E, complete solidification takes place with the deposition of crystals of both pure ethanamide and pure benzoic acid. A similar sequence would be noticed when the composition of the liquid mixture is greater than 55% ethanamide, but the initial deposit would consist of pure ethanamide crystals. E is known as the eutectic point for the system, and the temperature at this point, the eutectic temperature, is the lowest at which mixtures of these two components can remain liquid.

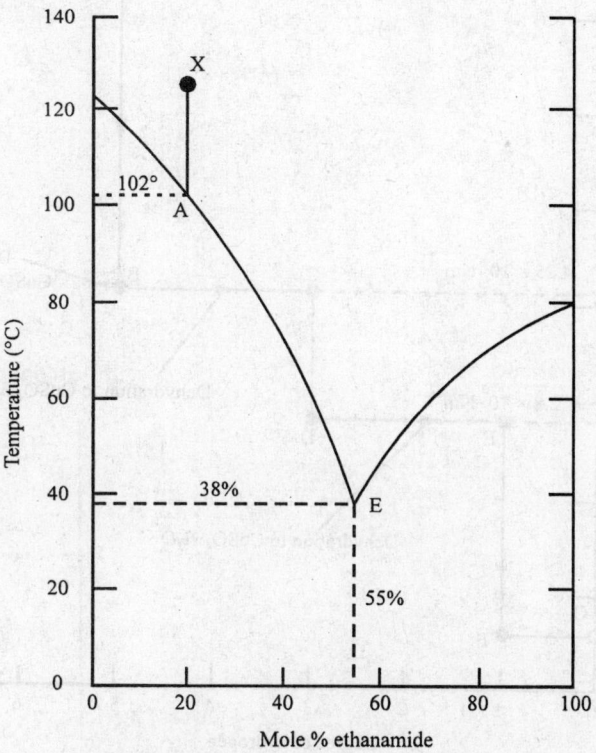

Fig. 5.17. Benzoic acid-ethanamide system.

Vapour Pressure of Salt Hydrates

The saturated vapour pressure of water standing over a solution of a salt gradually decreases as the concentration of the solution increases. The vapour pressure of water over a saturated solution of copper (II) sulphate at 40°C is 8.6×10^3 Nm^{-2}. Removal of this water vapour corresponds to the formation of crystals of the pentahydrate, $CuSO_4.5H_2O$, and the water vapour pressure drops to 4.25×10^3 Nm^{-2} (along the line AB in Fig. 5.18) when crystallisation is complete. (This means that the vapour pressure of water at 40°C in equilibrium with solid $CuSO_4. 5H_2O$ crystals is 4.25×10^3 Nm^{-2}.) From point B, removal of more water vapour leads to the formation, by dehydration, of an increasing amount of trihydrate, but the vapour pressure standing over the solid stays at 4.25×10^3 Nm^{-2} while any pentahydrate remains. At C, conversion to the trihydrate is complete, and the vapour pressure drops to 2.8×10^3 Nm^{-2}. Further dehydration produces the monohydrate and the anhydrous material, the sequence following the lines DE, EF, FG and GO in Fig. 5.18.

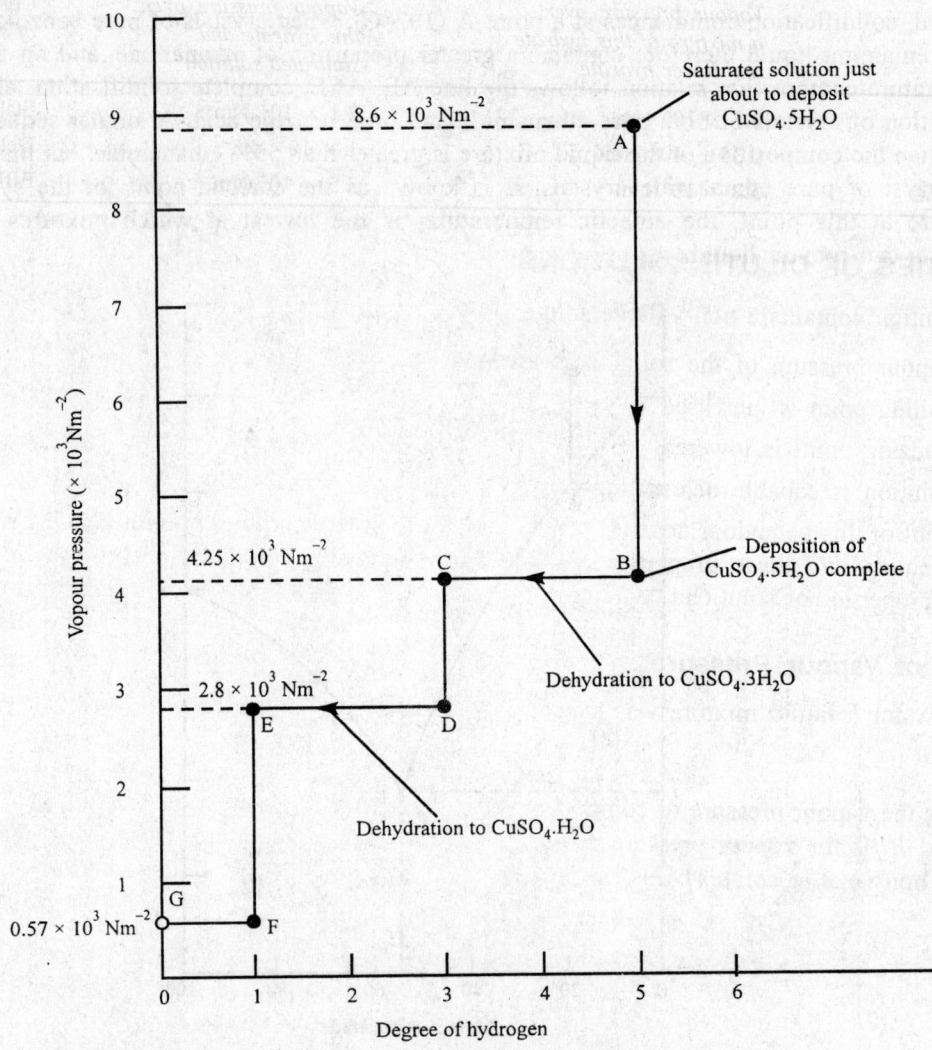

Fig. 5.18. Vapour pressures of the hydrates of copper (II) sulphate at 40°C.

Deliquescence and efflorescence

Suppose a salt can exist either in the anhydrous form or as the monohydrate. At 20°C, the water vapour pressure in this country is about 2×10^3 Nm^{-2}; consequently, if the change from the monohydrate to the anhydrous salt corresponds to a water vapour pressure above 2×10^3 Nm^{-2}, then water will be removed from the monohydrate into the atmosphere and the monohydrate will effloresce. On the other hand, if the monohydrate is deposited from its saturated solution at a vapour pressure below 2×10^3 Nm^{-2}, the monohydrate will take up moisture from the air and the hydrate will be deliquescent. The different cases are summarised below, assuming that the temperature is 20°C and the vapour pressure of water in the atmosphere is 2×10^3 Nm^{-2}.

Salt	Vapour pressure over a mixture of this hydrate and lower hydrate	Vapour pressure over this hydrate and saturated solution	Condition of salt
$CuSO_4.5H_2O$	0.7×10^3 Nm^{-2}	2.1×10^3 Nm^{-2}	Stable
$CaCl_2.6H_2O$	0.35×10^3 Nm^{-2}	1×10^3 Nm^{-2}	Deliquesces
$Na_2CO_3.10H_2O$	3.2×10^3 Nm^{-2}	3.5×10^3 Nm^{-2}	Effloresces

PROPERTIES OF DILUTE SOLUTIONS

When a solution contains a non-volatile solute:

(1) The vapour pressure of the solution is lowered.

(2) The boiling point is increased.

(3) The freezing point is lowered.

(4) The solution is capable of exerting osmotic effects.

The extent of this behaviour depends on the number of solute particles present in solution, rather than on their identity, and these properties are known as colligative properties. The laws relating to the colligative properties of solutions containing a non-volatile solute apply to *dilute solutions only*.

Lowering of Vapour Pressure

Raoult's law for a liquid mixture was stated mathematically in the form:

$$p_A = x_A P_A^0$$

where p_A is the vapour pressure of component A in the vapour, x_A is the mole fraction of A in the liquid mixture and P_A^0 is the vapour pressure of pure A at the given temperature. However, when the solution contains a non-volatile solute, p_A is, in fact, the total vapour pressure of the solution.

Now
$$\frac{p_A}{p_A^0} = x_A = 1 - x_s \text{ (where } x_s \text{ is the mole fraction of the solute)}$$

or
$$\frac{P_A^0 - p_A}{P_A^0} = x_s$$

The factor $(P_A^0 - p_A)/P_A^0$ is termed the relative lowering of vapour pressure.

Thus, Raoult's law may be stated as follows: the relative lowering of vapour pressure of a solution is equal to the mole fraction of the solute in the dilute solution. By measuring the relative lowering of vapour pressure, the weight of one mole of non-volatile solute can be determined. The apparatus used is shown in Fig. 5.19.

Elevation of Boiling Point

It has been shown that a liquid boils when its saturated vapour pressure is equal to the external pressure. Thus, a solution containing a non-volatile solute which has a lower vapour pressure than the pure solvent boils at a higher temperature than the pure solvent. This can be seen from Fig. 5.20, in which the curve A is for the pure solvent, and B and D are the curves for solutions of increasing concentration. From these curves, AC is proportional to $P_A^o - p_A$, which is, in turn, proportional to the concentration of solute in solution (1). Similarly, AE is proportional to the concentration of solute in solution (2). For dilute solutions, the distances AB and BD are small, and CB and ED can be regarded as straight lines, so that ABC and ADE are similar triangles. Thus,

$$\frac{AB}{AD} = \frac{T_1 - T_0}{T_2 - T_0} = \frac{AC}{AE}$$

$$= \frac{\text{Lowering of v.p. for solution (1)}}{\text{Lowering of v.p. for solution (2)}}$$

or, the elevation of boiling point is proportional to the concentration of the solution.

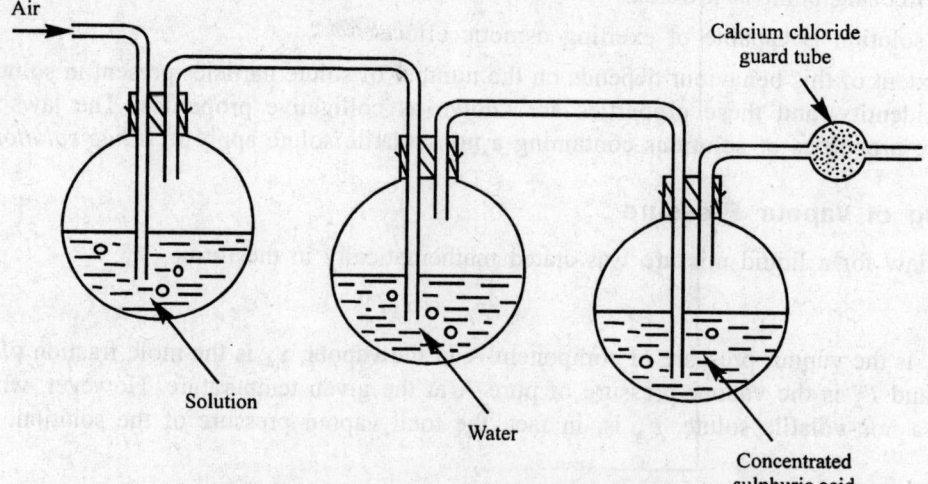

Fig. 5.19. Apparatus for measuring vapour pressure.

The elevation in boiling point (denoted by ΔT) produced by one mole of solute (which does not associate or dissociate) contained in 100 g of a solvent is termed the *elevation of boiling point constant*, or *molar elevation* for the solvent, and this is denoted by K.

Now, if one mole (M grams) of solute dissolved in 100 g of solvent produces an elevation of K degrees, then, one g of solute dissolved in 100 g solvent would produce an elevation of

$$\frac{1}{M} K \text{ degrees}$$

and w grams of solute dissolved in 100 g of solvent would produce an elevation of

$$\frac{w}{M} K \text{ degrees}$$

Thus,

$$\Delta T = \frac{w}{M} K$$

Values of K for several solvents are:

Benzene	26.7	Water	5.15
Propanone	16.7	Ethoxyethane	21.0

Note that, occasionally, the values of K are given for a solution containing one mole of solute in 1000 g of solvent—this is a more dilute solution than the one used in the previous considerations, and so the K values for 1000 g of solvent are ten times smaller than those given above.

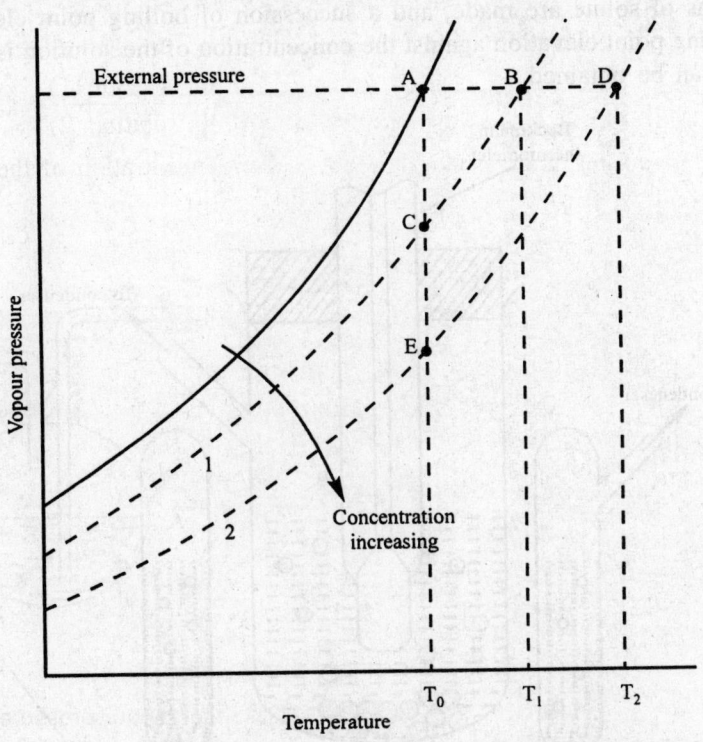

Fig. 5.20. Vapour pressures of solutions of increasing concentration.

Measurement of boiling point elevation

To obtain accurate results, the following precautions must be observed:

1. an accurate large-scale thermometer is required.
2. the solute must be as involatile as possible.
3. superheating of the solution must be avoided.

A diagram of the apparatus used in Beckmann's method is given in Fig. 5.21. The experiment, which takes a considerable time to complete, is carried out in the following stages:

1. The pure solvent is placed in both the inner tube B, and the annular tube A, each of which is connected to a condenser to return any escaping vapour. Some glass beads are placed in B, and a short length of platinum wire is sealed through the base of the tube in order to eliminate superheating.

2. The Beckmann thermometer is adjusted, and fitted into the central tube. Then the whole unit is heated very gently by means of a micro-burner flame directed at the small gauze insert in the asbestos mat C.

3. Heating is continued until the boiling point of the pure solvent is attained, and the reading on the Beckmann is noted.

4. After cooling, a small weighed sample of solute is added to the inner vessel by means of the side arm. Heating is recommenced, and the boiling point of the solution is found.

5. Further additions of solute are made, and a succession of boiling point elevations are determined. A graph of boiling point elevation against the concentration of the solution is plotted, so that the best average value can be obtained.

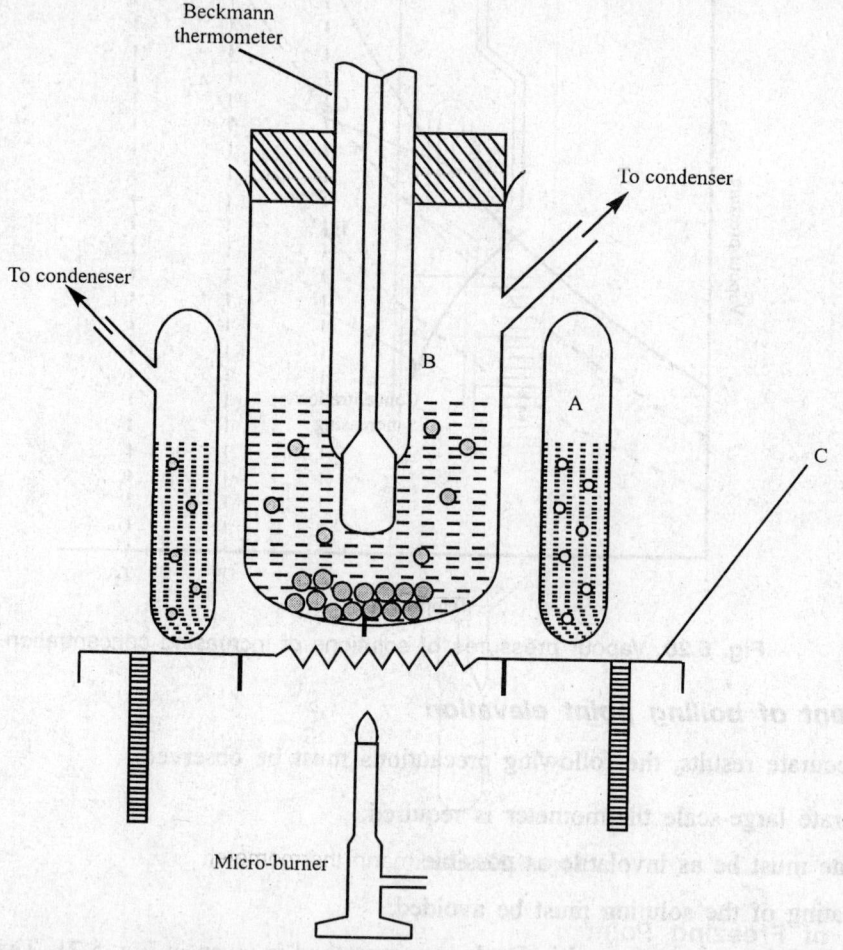

Fig. 5.21. Beckmann's apparatus for determining boiling point elevations.

The *Beckmann thermometer* is shown in Fig. 5.22. This instrument is not intended to give a reading of temperature but to enable temperature differences to be accurately measured. It is used over a very small range of temperature and the scale covers five or six Centigrade degrees. Hence, before use, it must be set so that the mercury level at the temperature of the experiment comes near to the centre of the scale. To do this, the thermometer bulb is warmed until the mercury thread joins up with that in the upper reservoir. A small scale carried on this reservoir shows the point at which the mercury thread should be broken so that the mercury level falls on the main scale at the temperature of the determination; the thread is broken by giving the upper part of the instrument a gentle tap while holding it upright.

The disadvantages of the Beckmann method stem mainly from the problems of eliminating superheating, and the length of time required to complete a determination. In addition, heat is lost from the central tube by radiation, but this is reduced to a minimum by having boiling solvent in the outer annular vessel.

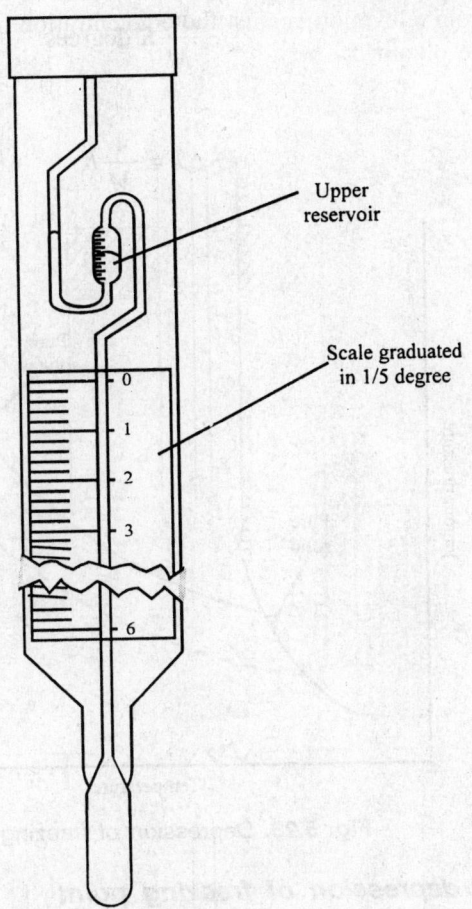

Fig. 5.22. Beckmann thermometer.

Depression of Freezing Point

At the freezing point, the vapour pressure of a liquid and that of the pure solid become equal. It follows that the lowering of vapour pressure caused by the presence of a non-volatile solute in a solution leads to the solution freezing at a temperature below that of the pure solvent (Fig. 5.23).

Again it can be shown that, for dilute solutions, the extent of the freezing point depression is proportional to the concentration of the solution. The depression of freezing point (in Centigrade degrees) brought about by dissolving one mole of non-volatile solute in 100 g of solvent, is termed the depression of freezing point constant, or molar depression, and it is again denoted by K. Some values of K are:

Water	18.6	Ethanoic acid	39.3
Benzene	50.0	Camphor	400

If 1 mole (M g) of solute dissolved in 100 g solvent cause a depression of K degrees, then w g of solute dissolved in 100 g of solvent would cause a depression of

$$\frac{w}{M} K \text{ degrees}$$

and, as before,

$$\Delta T = \frac{w}{M} K$$

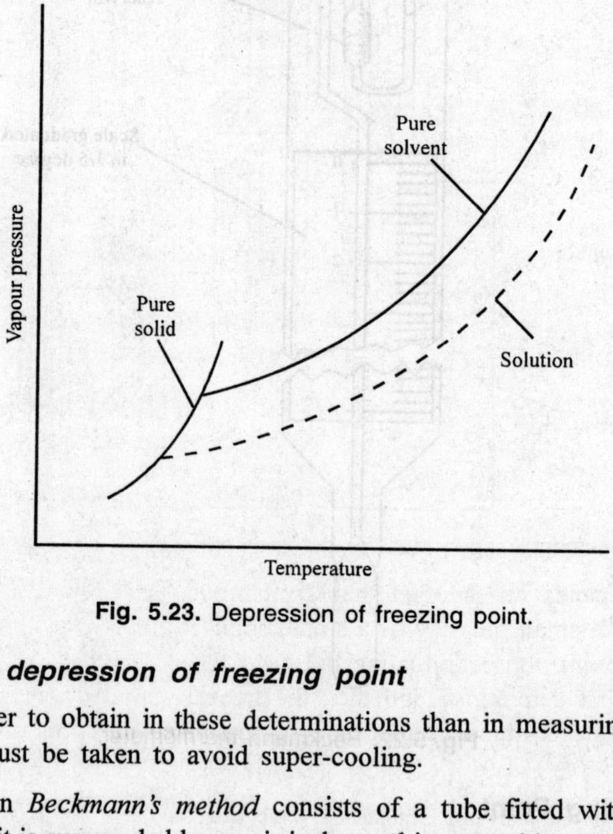

Fig. 5.23. Depression of freezing point.

Measurement of the depression of freezing point

Accurate results are easier to obtain in these determinations than in measuring the elevation of boiling point, but precautions must be taken to avoid super-cooling.

The apparatus used in *Beckmann's method* consists of a tube fitted with a side arm, stirrer and Beckmann thermometer; it is surrounded by an air jacket and immersed in a well-stirred freezing mixture, as shown in Fig. 5.24.

1. A known weight of pure solvent is placed in the inner tube A, which is cooled to a temperature just above the freezing point of the solvent.

2. The apparatus is assembled as shown, and the solvent is thoroughly stirred to eliminate super-cooling.

3. The steady temperature corresponding to the freezing point of the pure solvent is noted. If super-cooling has taken place, the temperature will fall initially below the freezing point of the pure solvent, but will rise to assume the steady value of the freezing point as the solvent crystallises. If a considerable degree of super-cooling has taken place, it is better to repeat the determination.

4. A weighed quantity of solute (in pellet form if possible) is added to the solvent via the side arm in tube A. The freezing point of the solution is then determined.

5. Further weighed portions of solute are added and the freezing point noted after each addition. From a graph of the results, the best average value is taken.

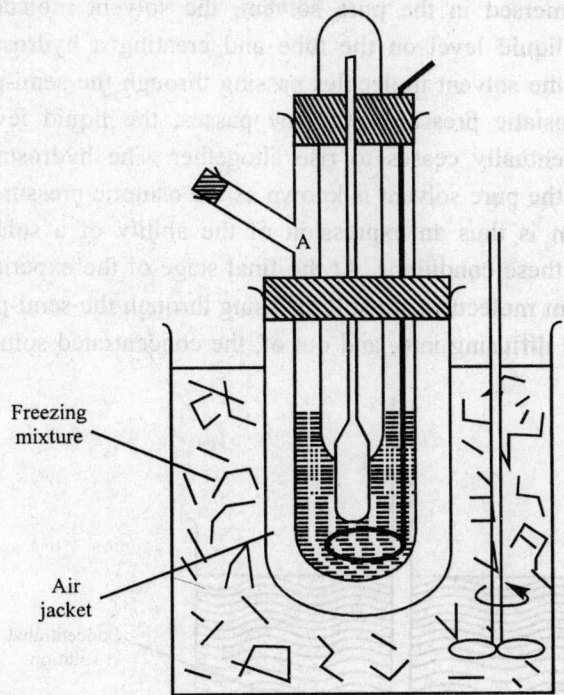

Freezing mixture

Air jacket

Fig. 5.24. Beckmann's apparatus for determining freezing point depressions.

Rast's micro method depends on the fact that camphor has the very high value of 400 for the cryoscopic constant, so that a small quantity of a solute soluble in camphor can produce a considerable depression in the freezing point, thus eliminating the need for a Beckmann thermometer. The freezing point of pure camphor is first determined and then the freezing point of a mixture containing known weights of camphor and solute is found. A thermometer graduated in tenths of a degree is suitable for this determination.

Example 5.3. Deduce the weight of one mole of naphthalene if a mixture of 5 g camphor and 0.16 g naphthalene freezes at 167.3°C. (The freezing point of pure camphor is 177.3°C, and K for 100 g camphor is 400).

Solution. Using $\Delta T = (w/M) K$, we find that the weight w of naphthalene in 100 g camphor is given by

$$w = \frac{100 \times 0.16}{5} = 3.2 \text{ g}$$

Also,

$$M = \frac{3.2 \times 400}{177.3 - 167.3} = 128$$

Osmosis

Osmosis is the term used to describe the diffusion of a solvent from a dilute to a more concentrated solution through a semi-permeable membrane. (A semi-permeable membrane is a barrier which allows the passage of small solvent molecules, but does not allow larger solute molecules to pass through its pores.) In Fig. 5.25, a concentrated solution is contained in a tube sealed with a semi-permeable membrane. When this tube is immersed in the pure solvent, the solvent molecules diffuse into the concentrated solution, raising the liquid level on the tube and creating a hydrostatic head. Once the hydrostatic head has been created, the solvent molecules passing through the semi-permeable membrane do so against an increasing hydrostatic pressure. As time passes, the liquid level in the tube rises progressively more slowly and eventually ceases to rise altogether. The hydrostatic head h which is established between a solution and the pure solvent is known as the osmotic pressure (π) of the solution. The osmotic pressure of a solution is thus an expression of the ability of a solution to produce this hydrostatic head of pressure under these conditions. At the final stage of the experiment, the liquid level does not rise or fall, although solvent molecules are still diffusing through the semi-permeable membrane. This is due to the fact that they are diffusing into, and out of, the concentrated solution at the same rate.

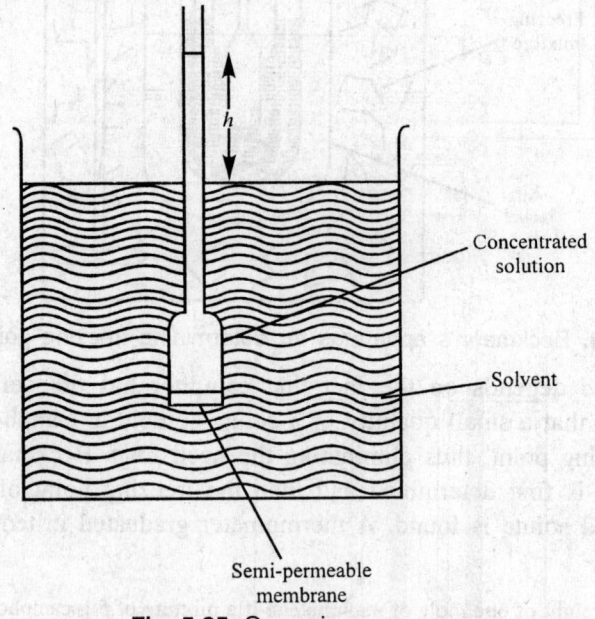

Fig. 5.25. Osmosis.

Alternatively, the osmotic pressure of a solution can be regarded as the minimum external pressure which will prevent the formation of a hydrostatic head between a solution and the pure solvent when they are separated by a semi-permeable membrane. Solutions which have the same osmotic pressure are said to be isotonic.

Measurement of osmotic pressure: Berkeley and Hartley's method

The apparatus is shown in Fig. 5.26, it consists of an inner porous tube carrying a semi-permeable membrane and containing the pure solvent, surrounded by a steel outer tube which contains the solution, and a device for the application and measurement of external pressure.

One side arm attached to the central tube of this apparatus is used for filling purposes, while the other tube serves to indicate any rise or fall in the liquid level. The applied pressure, which may be of the order of several atmospheres, is adjusted until the liquid level in the indicator tube neither rises nor falls; it is then equal to the osmotic pressure of the solution. The semi-permeable membrane is produced within the pores of the inner tube before the start of the experiment. The inner tube is filled with a solution of copper (II) sulphate (either *in situ*, or withdrawn from the apparatus and standing in a beaker; (Fig. 5.27)), and surrounded with a solution of potassium hexacyanoferrate (II). Electrodes are inserted into the solutions (the anode dips into the copper solution) and an electric current is passed, causing the formation of a deposit of copper hexacyanoferrate (II) within the porous tube; this deposit acts as a semi-permeable membrane.

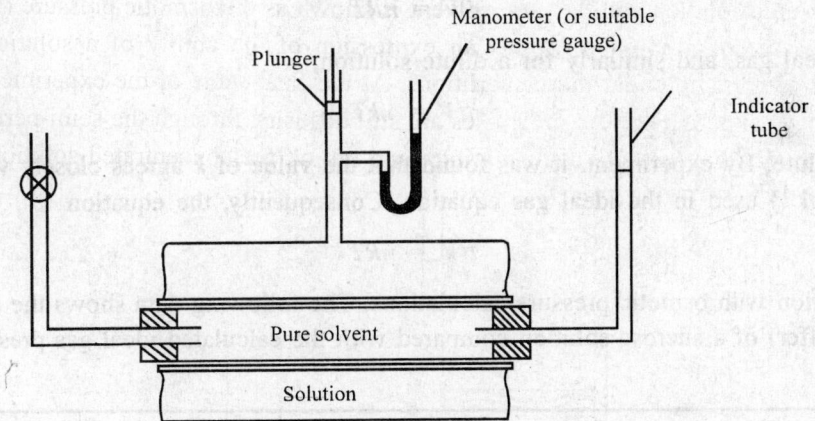

Fig. 5.26. Berkeley and Hartley's apparatus.

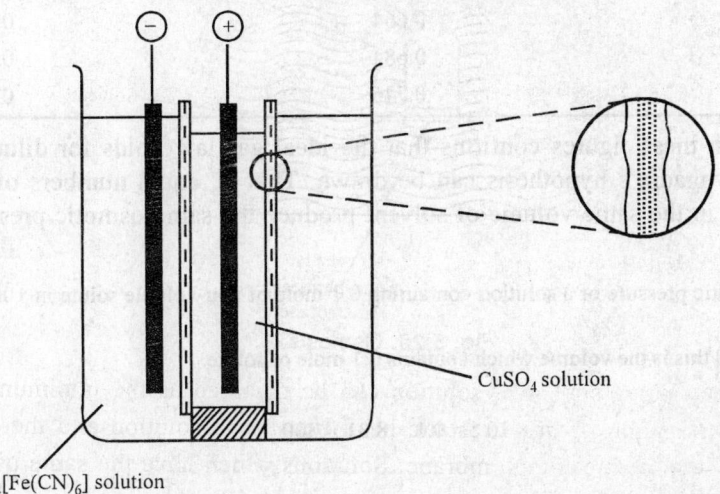

Fig. 5.27. Production of copper hexacyanoferrate (II) membrane.

Laws of osmotic pressure

The osmotic pressure π of a series of sucrose solutions at different concentrations and temperatures has been measured. From these results, it was apparent that:

1. The osmotic pressure of a solution is proportional to the absolute temperature.
2. The osmotic pressure of a solution is proportional to the concentration of the solution.

The analogy between ideal gases and dilute solutions is given as:

$$\pi \propto \text{concentration}$$

and as the concentration in moles dm^{-3} is the reciprocal of the volume, V dm^3, containing one mole, it follows that $\pi \propto 1/V$ which compares with Boyle's law.

The fact that $\pi \propto T$ is the equivalent of Charles' law. Combining Boyle's and Charles' laws leads to the equation

$$PV = nRT$$

for n moles of ideal gas, and similarly for a dilute solution,

$$\pi V = nkT$$

for n moles of solute. By experiment, it was found that the value of k agrees closely with the value of R (8.31 J K^{-1} mol^{-1}) used in the ideal gas equation. Consequently, the equation

$$\pi V = nRT$$

is used in connection with osmotic pressure calculations. The following data shows the osmotic pressure (measured by Pfeffer) of a sucrose solution compared with the calculated ideal gas pressure at the same temperature.

Temperature °C	Osmotic pressure of 0.029 M sucrose	Calculated ideal gas pressure
6.8	0.664	0.666
15.5	0.684	0.687
36.0	0.746	0.737

The agreement between these figures confirms that the ideal gas law holds for dilute solutions, and a conclusion similar to Avogadro's hypothesis can be drawn. That is, equal numbers of different solute molecules when dissolved in the same volume of solvent produce the same osmotic pressure at the same temperature.

Example 5.4. What is the osmotic pressure of a solution containing 0.1 mole of non-volatile solute in 100 cm^3 of solution at 27°C?

Solution. 100 cm^3 = 10^{-4} m^3 and this is the volume which contains 0.1 mole of solute.

Using $\pi V = nRT$ we obtain

$$\pi \times 10^{-4} = 0.1 \times 8.31 \times 300$$

from which

$$\pi = 2.5 \times 10^6 \text{ Nm}^{-2}$$

Association and Dissociation

In common with vapour pressure, boiling point elevation and freezing point depression, the osmotic pressure of a solution depends on the number (and not on the nature) of the solute particles in solution. However, when placed in solution a solute may:

1.. *Associate:* Form aggregates which contain more than one molecule. Consequently, the number of particles in solution is decreased, and the extent of the measured colligative property is reduced. For example, benzoic acid is associated into double molecules in benzene. Therefore, 2 moles of benzoic acid must be dissolved in 100 g of benzene to produce a depression equal to K (50 degrees).

2. *Dissociate:* Either partly or completely. For instance, one mole of sodium chloride which consists of one mole of sodium ions and one mole of chloride ions gives two moles of 'particles' in solution. Hence, 0.5 mole of sodium chloride dissolved in 100 g of water produces a depression of freezing point equal to K (18.6 degrees).

To express this behaviour mathematically, Van't Hoff introduced the factor i, where

$$i = \frac{\text{observed value}}{\text{calculated value}}$$

If i is greater than unity, dissociation has taken place, while an i value of less than one corresponds to association of the solute.

Example 5.6. 15.84 dm^3 of dry air (measured at s.t.p.) were bubbled through water at 25°C and then through a system of calcium chloride tubes. A gain in weight of 0.4205 g in the latter was noted. Calculate the saturated vapour pressure of water at 25°C.

Solution.

$$\frac{p_{H_2O}}{P} = \frac{n_{H_2O}}{n_{H_2O} + n_{air}}$$

Number of moles of water $= \dfrac{0.4205}{18} = 0.0233$

Number of moles of air $= \dfrac{15.84}{22.4} = 0.707$

The total pressure is 1.013×10^5 N m^{-2}. Hence

$$p_{H_2O} = \frac{0.0233 \times 1.013 \times 10^5}{(0.0233 + 0.707)} = 3.24 \times 10^3 \text{ N m}^{-2}$$

Example 5.8. The following resuls were obained using the above method to determine the solubility of oxygen in water at 20°C.

Volume of absorption vessel T	53.35 cm^3
Volume of water displaced	21.40 cm^3
Initial volume of oxygen in R	32.64 cm^3
Final volume of oxygen in R after absorption	10.15 cm^3
Atmospheric pressure	9.89×10^4 N m^{-2}

Solution. The volume of water remaining in the absorption vessel in which the oxygen dissolves is

$$53.35 - 21.40 = 31.95 \text{ cm}^3$$

Volume of oxygen originally present was 32.64 cm^3, and the volume of oxygen remaining after absorption is

$$21.40 + 10.15 = 31.55 \text{ cm}^3$$

Thus 31.95 cm^3 of water dissolve

$$32.64 - 31.55 = 1.09 \text{ cm}^3 \text{ of oxygen}$$

and 1 cm^3 of water would dissolve

$$\frac{1.09 \times 273}{31.95 \times 293} \times \frac{9.89 \times 10^4}{1.013 \times 10^5}$$

$$= 0.031 \text{ cm}^3 \text{ of oxygen at s.t.p.}$$

Hence, the absorption coefficient of oxygen at 20°C is 0.031.

Chapter 6

Catalysis

INTRODUCTION

A catalyst is a substance that increases the rate of a chemical reaction without itself being used up; and can be recovered unchanged chemically at the end of a chemical reaction. A catalyst therefore provides an alternative path, usually of lower activation energy, for the reaction to proceed at an accelerated rate. Consider, for example, the formation of oxygen by heating potassium chlorate. The reaction is very slow and takes place as follows:

$$2KClO_3 \xrightarrow{\text{heating}} 2KCl + 3O_2$$

A small addition of manganese dioxide to potassium chlorate accelerates the rate of the reaction and can be recovered unchanged at the end of the reaction. Thus, manganese dioxide acts as catalyst. Various other examples of catalysts are:

1. In the manufacture of ammonia by Haber's process, iron acts as a catalyst.

$$N_2 + 3H_2 \xrightarrow{\text{Fe}} 2NH_3$$

2. In contact process for the manufacture of sulphuric acid, platinum acts as a catalyst.

$$2SO_2 + O_2 \xrightarrow{\text{Pt}} 2SO_3$$

3. Combination of hydrogen and chlorine takes place in the presence of water vapours.

$$H_2 + Cl_2 \xrightarrow{H_2O} 2HCl$$

In these examples, the catalysts accelerate the rate of a chemical reaction and are, therefore, sometimes termed as *positive catalyst*. On the other hand, there are substances which when added to a chemical reaction retard its reaction rate and are thus called *negative catalysts* or *inhibitors*. Some common examples of negative catalysts are:

1. Auto oxidation of benzaldehyde is strongly inhibited by traces of some sulphur compounds.
2. Decomposition of H_2O_2 is retarded by the presence of a small quantity of sulphuric acid.
3. Oxidation of chloroform is retarded by traces of alcohols.

CHARACTERISTICS OF CATALYSTS

1. A catalyst remains chemically unaffected at the end of a chemical reaction. The catalyst does not undergo any chemical change, although there may be change in its physical state such as particle size or change in the colour of the catalyst etc.

2. Small quantity of a catalyst is usually required to bring about a reaction. A very small amount of a catalyst is sufficient for reactants to combine together. This is because the catalyst is not used up in the reaction. Thus 10^{-4} g of molybdic acid is sufficient for the oxidation of HI by H_2O_2. However, sometimes in many homogeneous catalytic processes, the rate of a catalytic reaction is proportional to the concentration of the catalyst. For example, in the inversion of cane sugar hydrochloric acid acts as a catalyst.

3. Presence of a catalyst does not affect the position of equilibrium in a reversible reaction. This is true when a small amount of the catalyst is used. The catalyst helps in attaining the equilibrium more quickly by increasing the rates of both the forward and the reverse reactions to the same extent. If, however, the catalyst is present in large amount, the same is not true. Some instances are known where the equilibrium constant changes. For example, hydrolysis of ethyl acetate in the presence of varying amounts of HCl, which acts as a catalyst, changes the value of equilibrium constant.

4. A catalyst does not initiate a reaction but only increases or decreases its speed. Generally, a catalyst speeds up the reaction which is already occurring slowly in its absence. However, this is not true in all reactions. Many reactions are known to occur only in the presence of a catalyst.

5. The action of a catalyst is specific. A catalyst can catalyse only a specific reaction and cannot be used for every reaction. For example, manganese dioxide can catalyse the decomposition of potassium chlorate but not potassium nitrate or other substances. Change of a catalyst also changes the nature of the reaction, e.g.,

$$\text{HCOOH} \xrightarrow[\text{Al}_2\text{O}_3 \text{ or TiO}_2]{\text{Cu or ZnO}} \begin{array}{l} \text{H}_2 + \text{CO}_2 \\ \text{H}_2\text{O} + \text{CO} \end{array}$$

$$\text{CO} + \text{H}_2 \xrightarrow[\text{ZnO}]{\text{Ni}} \begin{array}{l} \text{CH}_4 + \text{H}_2\text{O} \\ \text{CH}_3\text{OH} \end{array}$$

6. A catalyst has an optimum temperature at which the action of the catalyst is maximum.

7. A catalyst is poisoned by the presence of traces of certain substances which destroy the catalytic activity and are called *catalytic poisons*. Some of the catalytic poisons are arsenious oxide, carbon monoxide, hydrogen cyanide, etc.

8. The activity of a catalyst is enhanced by the presence of a substance called *promoters*. For example, in the synthesis of ammonia by Haber's process, molybdenum is used as a promoter to the catalyst iron.

TYPES OF CATALYSIS

There are generally two types of catalysis: (i) homogeneous catalysis; and (ii) heterogeneous catalysis.

Homogeneous Catalysis

In homogeneous catalysis, the catalyst is present in the same phase as the reacting substances. Many homogeneous catalysed reactions have been studied in the gas and liquid phases. Some common examples of such catalysis in gas phase are:

1. In the lead chamber process for the manufacture of sulphuric acid, nitric oxide gas catalyses the oxidation of sulphur dioxide.

$$2SO_2 + O_2 \xrightarrow{\ NO\ } 2SO_3$$

2. Decomposition of acetaldehyde is catalysed by iodine vapours.

$$CH_3CHO \xrightarrow[\text{Vapours}]{I_2} CH_4 + CO$$

3. Nitric oxide acts as a catalyst in the combination of carbon monoxide and oxygen.

$$2CO + O_2 \xrightarrow{\ NO\ } 2CO_2$$

Examples of homogeneous catalysts in liquid phase

Important examples of homogeneous catalysis in liquid phase are acid-base catalysis. The most common acid catalyst in water is the hydronium ion and the most common base catalyst is the hydroxyl ion. If an acid catalyses a reaction, the reaction is said to be the subject of acid catalysis. Inversion of cane sugar and hydrolysis of esters are some examples of acid catalysed reactions. However, it was shown that different acids have different catalytic activity; hydrochloric acid has a greater activity than acetic acid. So it is evident that the actual catalysts are H^+ (or H_3O^+) ions. The rates of reaction are found to be proportional to the concentrations of H_3O^+ ions and the concentration of the reacting molecules or ions.

$$C_{12}H_{22}O_{11} + H_2O \xrightarrow{H_3O^+} C_6H_{12}O_6 + C_6H_{12}O_6$$

$$CH_3COOC_2H_5 + H_2O \xrightarrow{H_3O^+} CH_3COOH + C_2H_5OH$$

Such reactions which are catalysed by certain acids (or H_3O^+ ions only) are said to be *specific acid catalysis*. Similarly, there are reactions which are catalysed by OH^- ions only and hence are said to be *specific hydroxyl ion catalysis*. Conversion of acetone into diacetonyl alcohol or the decomposition of nitroso-triacetoneamine are examples of hydroxyl ion catalysis.

$$CH_3COCH_3 + CH_3COCH_3 \xrightarrow{OH^-} CH_5COCH_2C(CH_3)_2OH$$

$$
\begin{array}{ccc}
CH_2\!-\!CMe_2 & & CH=CMe_2 \\
\diagup \qquad \diagdown & & \diagup \\
CO \qquad N\!-\!NO \xrightarrow{\ OH^-\ } N_2 + H_2O + CO & & \\
\diagdown \qquad \diagup & & \diagdown \\
CH_2\!-\!CMe_2 & & CH=CMe_2 \\
& & \text{(Phorone)}
\end{array}
$$

There are many reactions in which both H_3O^+ ions and OH^- ions simultaneously act as catalysts, probably along with water. The mechanisms of hydrolysis of ester can be expressed as follows:

1. With H^+ ions as catalyst

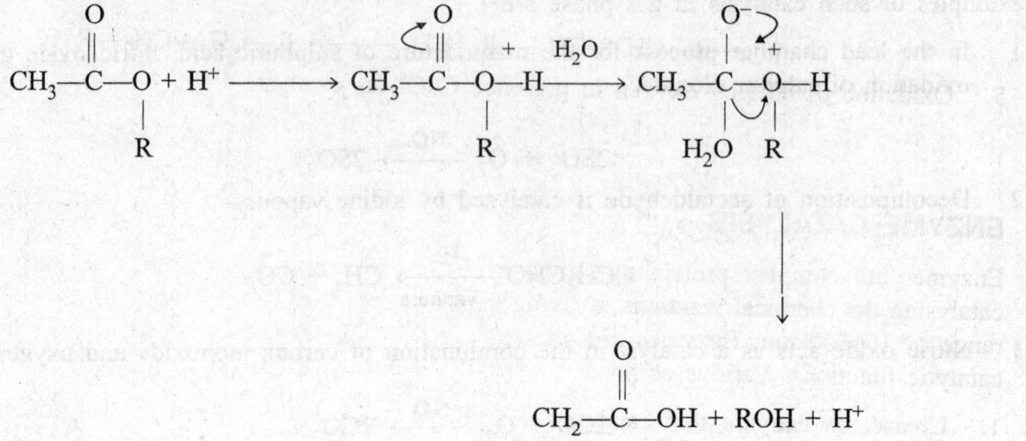

2. With hydroxyl ions as catalyst:

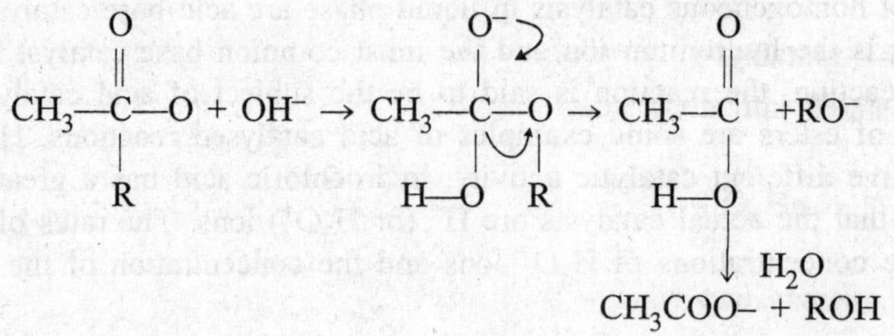

Heterogeneous catalysis

In heterogeneous catalysed reactions, the catalyst is present in a different phase from the reactants. In a number of cases, the catalyst is the solid phase and the reactants are gaseous in most cases or liquids in others. The catalysts which are commonly used are metals like platinum, nickel, copper and iron and certain metal oxides such as ferric oxide, zinc oxide, molybdenum oxide, etc. Some important examples of heterogeneous catalysis are:

1. In contact process for the manufacture of sulphuric acid, sulphur dioxide is directly oxidised into sulphur trioxide by atmospheric oxygen in the presence of platinum as catalyst.

$$2SO_2 + O_2 \xrightarrow{Pt} 2SO_3$$

2. Haber's process for the manufacture of ammonia in which nitrogen and hydrogen in the ratio of 1 : 3 are passed over heated iron catalyst which contains a promoter (molybdenum).

$$N_2 + 3H_2 \xrightarrow{Fe} 2 NH_3$$

3. The oxidation of ammonia to nitric oxide and finally to nitric acid in the presence of a mixture of ferric and bismuth oxide.

$$4NH_3 + 5O_2 \xrightarrow[Bi_2O_3]{Fe_2O_3} 4NO + 6H_2O$$

4. Hydrogenation of unsaturated hydrocarbons in presence of nickel as a catalyst.

$$—R—CH = CH — R' + H_2 \xrightarrow{Ni} R—CH_2—CH_2—R'$$

5. Oxidation of HCl by oxygen in presence $CuCl_2$ as catalyst.

$$4HCl + O_2 \xrightarrow{CuCl_2} 2H_2O + 2Cl_2$$

ENZYME CATALYSIS

Enzymes are complex protein molecules with three-dimensional structures. These are responsible for catalysing the chemical reactions in living organisms. The diameter of the enzyme molecules fall in the range of 10–100 nm. Enzymes are often present in colloidal state and are extremely specific in their catalytic functions. Various enzymes catalysed reactions are known. Some important examples are:

1. Urease, an enzyme that catalyses the hydrolysis of urea but has no effect on the hydrolysis of substituted urea, e.g., methyl urea.

$$NH_2CONH_2 + H_2O \xrightarrow{Urease} 2NH_3 + CO_2$$

2. Peptide, glycyl-L-glutamyl-L-tyrosine, is hydrolysed by an enzyme known as pepsin.

3. Hydrolysis of starch into maltose by diastase.

$$2(C_6H_{10}O_5)_n + nH_2O \xrightarrow{Diastase} nC_{12}H_{22}O_{11}$$
$$\text{Starch} \qquad\qquad\qquad \text{Maltose}$$

4. Conversion of glucose into ethanol by zymase present in yeast.

$$C_6H_{12}O_6 + H_2O \xrightarrow{Zymase} 2C_2H_5OH + 2CO_2$$

5. Conversion of maltose into glucose by maltase.

$$C_{12}H_{22}O_{11} + H_2O \xrightarrow{Maltase} 2C_6H_{12}O_6$$

6. Oxidation of alcohol to acetic acid by micoderma acetic.

$$C_2H_5OH + O_2 \xrightarrow[Aceti]{Micoderma} CH_3COOH + H_2O$$

Almost all enzymes fall into one of the two classes, the *hydrolytic enzymes* and the *oxidation-reduction enzymes*. The hydrolytic enzymes appear to be complex acid-base catalysis which accelerate the ionic reactions mainly due to the transfer of hydrogen ions. The oxidation-reduction enzymes catalyse electron transfer perhaps through the formation of an intermediate radical.

Mechanism of Enzyme Reactions

The mechanism of an enzyme reaction was proposed by Michaelis and Menten and can be represented in the following manner : Let E represent the enzyme and S the substrate it acts on, then the overall reaction is

$$E + S \underset{k_{-1}}{\overset{k_1}{\rightleftharpoons}} [ES] \xrightarrow{k_2} E + P$$

It is to be noted that in the formation of the product P, the enzyme does not undergo any change. The rate of formation of the product depends on the concentration of the enzyme. In the above scheme ES denotes the intermediate between the enzyme and the substrate which decomposes into the product with a first order rate constant k_2. The rate of formation of the product is given by

$$\frac{d[P]}{dt} = k_2[ES] \qquad \qquad ...(6.1)$$

In order to solve the equation (6.1) it is necessary to know the concentration of ES. This can be calculated through the steady-state principle.

$$\frac{d[ES]}{dt} = k_1[E][S] - k_{-1}[ES] - k_2[ES] = 0$$

or

$$[ES] = \frac{k_1[E][S]}{k_{-1} + k_2}$$

$$= \frac{[E][S]}{\dfrac{k_2 + k_{-1}}{k_1}}$$

$$= \frac{[E][S]}{K_m} \qquad \qquad ...(6.2)$$

where K_m is often referred to as Michaelis constant.

In this equation, the quantities $[E]$ and $[S]$ are the concentrations of free enzyme and free substrate. If $[E]_0$ and $[S]_0$ are the initial concentrations of the enzyme and the substrate respectively, then we can write

$$[E]_0 = [E] + [ES]$$

or

$$[E] = [E]_0 - [ES]$$

and

$$[S]_0 = [S] + [ES]$$

Since only a little enzyme is added, hence $[ES]$ is very small in comparison to $[S]$, therefore,

$$[S]_0 \cong [S]$$

Substituting the value of $[E]$ in equation (6.2), we get

$$[ES] = \frac{\{[E]_0 - [ES]\}[S]}{K_m}$$

$$[S + K_m]\ [ES] = [E]_0[S]$$

$$[ES] = \frac{[E]_0[S]}{[S] + K_m}$$

Consequently, the rate of formation of products is

$$\frac{d[P]}{dt} = k_2[ES]$$

$$= \frac{k_2[E]_0[S]}{K_m + [S]} \qquad \qquad ...(6.3)$$

According to equation (6.3) if $[S] \ll K_m$, then the rate of enzymolysis varies linearly with the enzyme and substrate concentrations, i.e., the reaction will be first order in E and S. However, if $[S] \gg K_m$

$$\frac{d[P]}{dt} = \frac{k_2[E]_0[S]}{[S]}$$

$$= k_2\ [E]_0$$

the rate of the reactions will be independent of substrate concentration and will be first order in E.

A plot of $\frac{d[P]}{dt}$ versus $[S]$ for constant enzyme concentration yields a curve (Fig. 6.1) from which it is possible to calculate the value of k_2 and K_m. Further, when the rate is half the maximum value

$$\frac{d[P]}{dt} = \frac{k_2[E]_0}{2} = \frac{k_2[E]_0[S]}{K_m + [S]}$$

or

$$K_m + [S] = 2[S]$$

$$[S] = K_m$$

Rates of enzyme catalysed reactions are slowed down by compounds which are structurally related to the substrate. These compounds combine with the active sites of the enzyme and thus cause inhibition. In cases where the substrate and the inhibitor compete for the active sites, the rate may be increased by taking larger concentrations of substrate. An enzyme reaction has an optimum pH value of which the catalytic activity of the enzyme is maximum. The rate of the reaction decreases as the pH is raised or lowered from the optimum value. At extreme pH they are irreversibly denatured.

Rate versus temperature graph for an enzyme catalysed reaction shows that rate is maximum at a certain temperature. Above this temperature, the enzyme is denatured and hence the rate decreases.

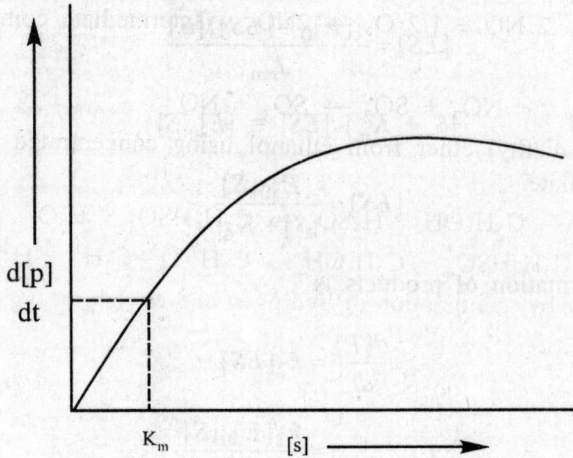

Fig. 6.1. Plot of Michaelis-Menten equation.

THEORY OF CATALYSIS

As stated earlier, the essential requirement for a reaction to occur is that the reacting molecules must acquire sufficient energy. In case a catalyst is added to the reaction, the energy required to activate the molecules is less than in the absence of a catalyst. Due to lower activation energy, more molecules will take part in the reaction and hence the rate of the catalysed reaction would increase. The action of a catalyst can be explained by two different mechanisms, viz. (i) intermediate compound formation theory; and (ii) adsorption theory.

Intermediate Compound Formation Theory

In this theory, essentially two steps are involved:

1. Combination of the catalyst with one or more of the reactants forming intermediate compound.
2. Decomposition of the intermediate compound or its combination with other reactants yielding the product and the catalyst back. Consider a reaction between the reactants A and B giving the product, viz.,

$$A + B \rightarrow AB$$

This reaction is very slow and is catalysed by the presence of a catalyst X. The reaction will therefore proceed as

$$A + X \rightarrow AX \qquad \text{(Intermediate compound)}$$
$$AX + B \rightarrow AB + X$$

The formation of an intermediate compound AX is an easy reaction and needs low energy of activation thereby accelerating the rate of the chemical reaction.

Some examples of intermediate compound formation

1. In lead chamber process for the manufacture of sulphuric acid, the catalyst NO first forms an intermediate compound with oxygen

$$2 \text{ NO} + 1/2 \text{ O}_2 \rightarrow 2\text{NO}_2 \quad \text{(Intermediate compound)}$$

and then

$$\text{NO}_2 + \text{SO}_2 \rightarrow \text{SO}_3 + \text{NO}$$

2. In the preparation of diethyl ether from ethanol using concentrated H_2SO_4, $C_2H_5HSO_4$ is first formed as an intermediate.

$$C_2H_5OH + H_2SO_4 \rightarrow C_2H_5HSO_4 + H_2O$$
$$C_2H_5HSO_4 + C_2H_5OH \rightarrow C_2H_5\text{-O}\text{---}C_2H_5 + H_2SO_4$$

3. The formation of water by combination of hydrogen and oxygen in presence of copper as a catalyst is as follows:

$$2\text{Cu} + 1/2 \text{ O}_2 \rightarrow \text{Cu}_2\text{O}$$
$$\text{Cu}_2\text{O} + \text{H}_2 \rightarrow \text{H}_2\text{O} + 2\text{Cu}$$

Limitations of intermediate compound formation theory

This theory does not explain the cases of heterogeneous catalysis in general and more specifically, the deactivation by a catalytic poison and the activation by a promoter.

Adsorption Theory

A large number of gaseous reactions take place in the presence of solid catalysts. The surface of the catalyst has certain active centres due to the unsaturation of valencies. Appreciable quantities of the reactant molecules are adsorbed or retained by solid surfaces at these active centres and the reactions occur at the surface of the solid. For this reason, this type of catalysis is sometimes referred to as the contact catalysis. The adsorbed molecules form some sort of an activated complex on the surface, which then decomposes forming the products. The products are ultimately desorbed from the surface. A catalytic reaction involves the following steps:

1. Diffusion of the reactants from the bulk on to the surface.
2. Adsorption of the reactants on the surface of the catalyst.
3. Activation of the adsorbed reactants leading to a reaction in the adsorbed phase.
4. Desorption of the products from the surface of the catalyst.
5. Diffusion of the products away from the surface of the catalyst.

Any one of the steps may be slowest and consequently the rate determining but generally step (3) is the rate controlling step.

Due to adsorption, the concentration of the reactants tend to increase on the surface of the catalyst and according to the law of mass action, the rate of the reaction will increase. Furthermore, adsorption being an exothermic process, the heat evolved during adsorption is utilised in the activation of the surface reaction. Adsorption may also lead to proper orientation of the reacting molecules, partial loosening of the bonds in the adsorbed state and thus requiring only small energy to form the activated complex.

Adsorption theory can explain the enhanced catalytic action of a catalyst in the finely divided state. It is due to the larger surface area available for adsorption and also the formation of more active centres.

$$\begin{array}{ccc}
\overset{|}{-}\text{Ni}\overset{|}{-}\text{Ni}\overset{|}{-} & & \overset{|}{-}\text{Ni}\overset{|}{-} \quad \overset{|}{-}\text{Ni}\overset{|}{-} \\
\overset{|}{-}\text{Ni}\overset{|}{-}\text{Ni}\overset{|}{-} & \rightarrow & \overset{|}{-}\text{Ni}\overset{|}{-} \quad + \quad \overset{|}{-}\text{Ni}\overset{|}{-} \\
\overset{|}{} & & \overset{|}{} \qquad \overset{|}{}
\end{array}$$

Action of promoters which enhances the catalytic activity can also be explained in terms of this theory. A promoter generally increases the number of active centres by the adsorption on the surface of a catalyst. Similarly, poisoning of a catalyst results due to the adsorption of the catalytic poisons on the surface and thereby reducing the number of active centres on the surface of a catalyst.

Example 6.1. In the acid hydrolysis reaction

$$A + H_2O + H^+ \longrightarrow \text{Products}$$

where $[H^+] = 0.1$ mol dm^{-3} and H_2O is present in large excess, the apparent rate constant (i.e. the pseudo first order rate constant) is 1.5×10^{-5} s^{-1}. Calculate the true rate constant.

Solution. $r = -d[A]dt = k[A][H_2O][H^+]$

Since $[H^+]$ is essentially constant (because the catalyst is regenerated) and $[H_2O] \gg [A]$, hence

$$r = k_{opp} [A], \text{ where } k_{opp} = k [H_2O] [H^+]$$

$$\therefore \qquad k = k_{opp}/([H_2O][H^+])$$

Since $[H^+] = 0.1$ mol dm^{-3}

and $[H_2O] = 1000$ g dm$^{-3}/18$ g mol$^{-1} = 55.6$ mol dm^{-3} hence,

$$k = \frac{1.5 \times 10^{-5} \, s^{-1}}{(55.6 \text{ mol dm}^{-3})(0.1 \text{ mol dm}^{-3})} = 2.70 \times 10^{-6} \, dm^6 \, mol^{-2} s^{-1}$$

$$= 2.70 \times 10^{-6} \, M^{-2} \, s^{-1} \qquad (\because 1 M = 1 \text{ mol dm}^{-3})$$

Example 6.2. For an enzyme-substrate system obeying the simple Michaelis-Menten mechanism, the rate of product formation when the substrate concentration is very large, has the limiting value 0.02 mol dm^{-3}. At a substrate concentration of 250 mg dm^{-3} the rate is half this value. Calculate k_1/k_{-1} assuming that $k_2 \gg k_{-1}$.

Solution.

$$r = \frac{k_2 [E]_0 [S]}{K_m + [S]}$$

For

$$[S] \gg K_m, \qquad r \equiv V_{max} = k_2[E]_0$$

When r is half of this maximum value, then by definition

$$[S] = K_m = (k_{-1} + k_2)/k_1$$

$$= k_{-1}/k_1 \qquad\qquad\qquad\qquad \text{(assuming that } k_2 \ll k_{-1})$$

$$k_1/k_{-1} = 1/[S] = 1/(250 \text{ mg dm}^{-3}) = 0.004 \text{ dm}^3 \text{ mg}^{-1}.$$

Chapter 7

Gases

INTRODUCTION

A study of the properties of gases leads to the conclusion that gases consist of widely separated molecules in rapid motion. The most familiar gas, air, is actually a mixture of gases. Experiments show that any two (or more) gases can be used in any proportion to prepare a perfectly homogeneous mixture; no such generalisation can be made for liquids. Since the molecules of any gas are thought to be separated by comparatively large distances, it is reasonable that one gas can accommodate molecules of other gases. This molecular model can also be used to explain the compressibility of gases; compression consists of forcing gas molecules closer together.

A gas diffuses throughout any container into which it is introduced. Odourous gases released in a room can soon be detected in all parts of the room. The premise that gas molecules are in constant, rapid motion accounts for the diffusion of gases. Furthermore, gas molecules in their random motion strike the walls of the container, and these myriad impacts explain the fact that gases exert pressure.

STATES OF MATTER

Matter can be classified into three main divisions, namely, gases, liquids and solids.

Gases have neither shape nor size but fill completely any space into which they are introduced. All gases mix completely and a gas exerts pressure equally in all directions. The volume of a gas changes significantly corresponding to changes in temperature and pressure.

Liquids have a definite volume but no shape. They lack rigidity and elasticity but possess a boundary surface which confers many characteristic properties on them. Therefore, when placed in contact they may either diffuse completely (fully miscible) or mix to a limited or negligible extent. The change in the volume of a liquid brought about by an alteration in temperature is much less marked than with gases, while the effect of pressure is almost negligible.

Solids have both size and shape. They resist distortion (i.e. they are rigid) or recover from slight deformations (i.e. they are elastic) when the deforming force is removed. The volume of a solid does not change to any great extent following changes in temperature and pressure. These properties may be explained in terms of the closer packing and increasing intermolecular forces as the physical state changes from gaseous through liquid to solid.

GAS LAWS

The volume of a gas changes considerably as a result of variations in temperature and pressure. The effects of these changes are summarised by the laws of Boyle and Charles.

Boyle's Law

The Boyle's law states that: volume of a given mass of gas is inversely proportional to the pressure, at a constant temperature. Boyle's law states that V to represent the volume and P to represent the total pressure on the gas, the law is expressed by the relation

$$V \propto \frac{1}{P} \quad \text{or} \quad PV = \text{constant (k, say)}$$

The relationship is expressed graphically in Fig. 7.1. Each curve is an isothermal, that is the temperature is the same for all the data lying on that curve.

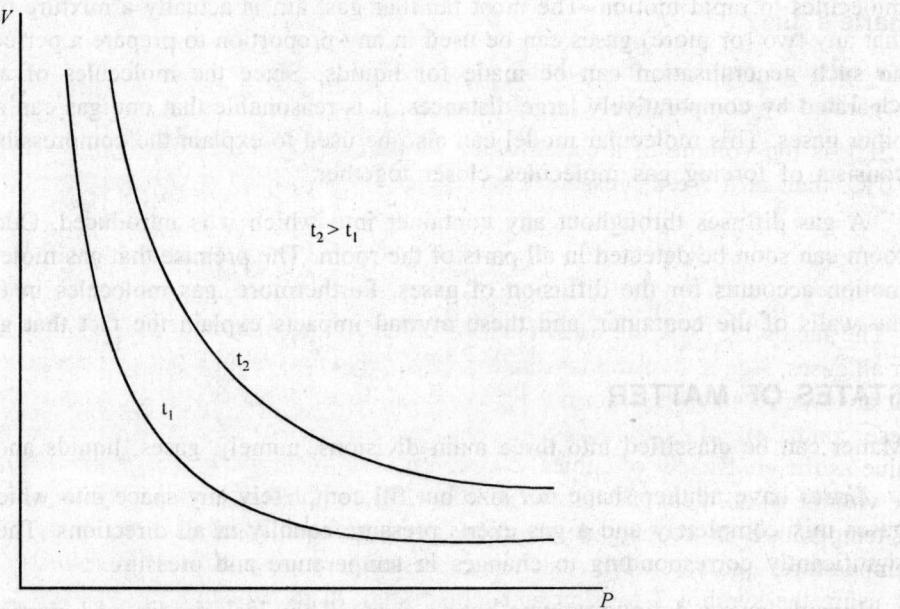

Fig. 7.1. Pressure-volume curve for a gas at a constant temperature.

Alternatively, the product PV may be plotted against P. According to Boyle's law, this product should have a constant value at all pressures, as indicated by the horizontal line (Fig. 7.2). In fact, the actual behaviour of any gas deviates considerably from this requirement; consequently, the evaluation of a constant value for the product PV at all pressures is an attribute of an *ideal* or *perfect gas*.

Boyle's law was based on a restricted number of observations of limited accuracy. Later work by Regnault, and others showed that at high pressures or low temperatures, considerable deviations from ideal behaviour occurred. Some examples are shown in Fig. 7.2, and the reasons underlying this departure from ideality are discussed further in this chapter.

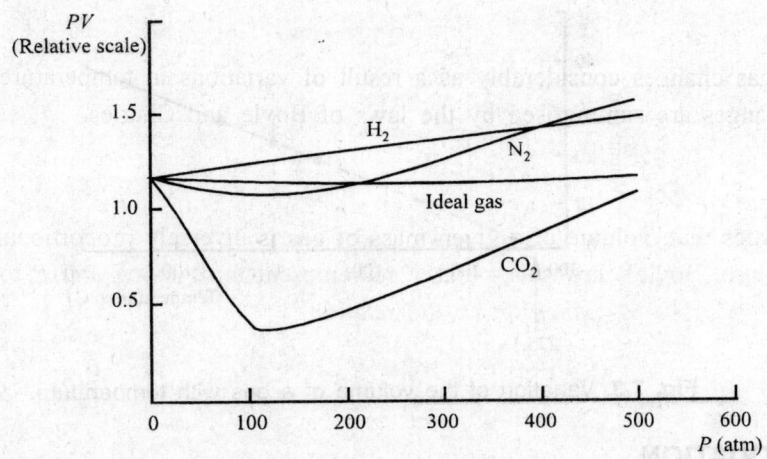

Fig. 7.2. Values of PV for one mole of gas at various pressures and at constant temperature.

Charle's Law

At a constant pressure, the volume of a given mass of gas expands by 1/273 of its volume at 0°C for each 1°C rise in temperature.

If V is the volume of a certain mass of gas at t°C while V_0 is the volume of the same mass of gas at 0°C, then at a fixed pressure Charles' law is expressed by the relation

$$V = V_0\left(1 + \frac{t}{273}\right)$$

The factor 1/273 is the coefficient of cubical expansion of a gas, and this value should be the same for all gases. Again, deviations from the ideal behaviour required by this equation are found with all gases, but at low pressures and normal temperatures most gases obey both Boyle's and Charles' laws. Under these conditions of near ideal behaviour, the coefficient of cubical expansion is 1/273.16. Taking this value as the coefficient of cubical expansion of an ideal gas and extrapolating back as shown in Fig. 7.3, the volume of an ideal gas falls to zero at –273.16°C. This is the absolute zero of temperature. The scale of temperature in centrigrade degrees, starting from absolute zero, is known as the absolute scale. Temperatures quoted on this scale are expressed in degrees Kelvin, K, and the use of this scale is implied by using the symbol T to represent temperature in an equation. Consequently, $t + 273.16 = T$K. Since

$$V = V_0\left(1 + \frac{t}{273.16}\right)$$

then

$$V = \frac{V_0}{273.16}(273.16 + t)$$

For a given mass of a gas at a fixed pressure, the fraction $V_0/273.16$ is a constant, say k'. Hence.

$$V = k'(273.16 + t)$$
$$= k'T$$

which is an alternative form of Charles' law.

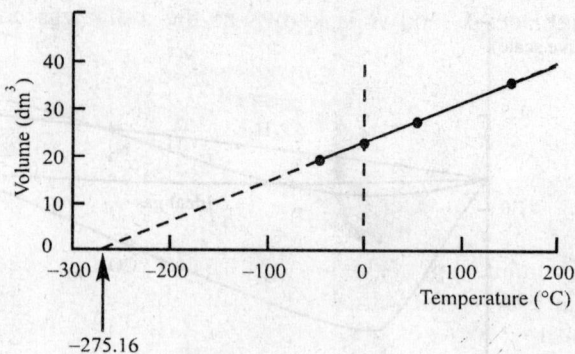

Fig. 7.3. Variation of the volume of a gas with temperature.

IDEAL GAS EQUATION

Boyle's and Charles' laws may be combined to form the ideal gas equation as follows: If

$$V \propto \frac{1}{P} \text{ at a fixed temperature}$$

and

$$V \propto T \text{ at a fixed pressure}$$

then

$$V \propto \frac{T}{P} \text{ when both temperature and pressure vary.}$$

Rearranging

$$PV \propto T$$

$$PV = kT \text{ where } k \text{ is a constant}$$

$$\frac{PV}{T} = k$$

Alternatively, when the volume of a given mass of gas alters from V_1 to V_2 corresponding to changes in pressure from P_1 to P_2 and temperature from T_1 to T_2, the ideal gas equation can take the form

$$\frac{P_1 V_1}{T_1} = \frac{P_2 V_2}{T_2}$$

since each fraction is equal to the constant k.

Avogadro's Law

Avogadro's hypothesis states that equal volumes of gases at the same temperature and pressure contain equal numbers of molecules. Taking the converse of this, one mole (i.e. L molecules) of any gas at the same temperature and pressure occupies the same volume, which means that the value of k in the ideal gas equation, $PV/T = k$ is the same for one mole of any gas assumed to behave ideally. For one mole of any gas, the symbol R replaces k, and R is known as the molar gas constant. Thus

$$\frac{PV}{T} = R$$

or

$$PV = RT$$

This equation, which is a combination of Boyle's law, Charles' law and Avogadro's hypothesis, applies only to, and thus defines, an ideal gas. Consequently, it if referred to as the *ideal gas equation*.

For n moles of any gas,

$$PV = nRT$$

GAS CONSTANT

The value of the molar gas constant, R, depends on the units in which P and V are measured. Pressure is defined as force per unit area or (force) × (length)$^{-2}$ while volume has the dimensions of (length)^{3}. The dimensions of the product PV are therefore (force) × (length) which are those of energy. Hence R, which equal to PV/T, has to be expressed in units of (energy) (degree)$^{-1}$ (mole)$^{-1}$. The volume of 1 mole of any gas at s.t.p. (standard temperature and pressure), which corresponds to 273.16 K and 1.013×10^5 N m^{-2} (or 1 atmosphere pressure), is known as the molar gas volume and it has the value 22.414 dm^3 (or 0.022414 m^3). Using these values in the equation $PV = RT$

$$R = \frac{PV}{T} = \frac{1.013 \times 10^5 \times 0.022414}{273.16}$$

$$= 8.31 \text{ N m K}^{-1} \text{ mol}^{-1}$$

$$= 8.31 \text{ J K}^{-1} \text{ mol}^{-1}$$

(Note, in using the equation, $PV = nRT$ in which the above value of R is used, the units of V are m^3).

CALCULATIONS BASED ON THE GAS LAWS

Errors in these calculations usually arise from:

1. Using inconsistent units on opposite sides of the equation

$$\frac{P_1 V_1}{T_1} = \frac{P_2 V_2}{T_2}$$

(for example, if P_1 is given in N m^{-2}, P_2 must be in the same units).

2. Not converting temperatures to the absolute scale.

3. Either omitting, or using an incorrect value of n in the equations $PV = nRT$.

Example 7.1. 5 dm^3 of air, measured at 27°C and 8×10^4 Nm^{-2} pressure are compressed to 5 atmospheres while the temperature rises to 43°C. Calculate the final volume of the air.

Solution. $P_1 = 8 \times 10^4$ Nm^{-2}

$P_2 = 5 \times 10^5 \text{ Nm}^{-2}$ (taking 1 atmosphere = 10^5 Nm^{-2})

$T_1 = 27 + 273 = 300 \text{ K}$

$T_2 = 43 + 273 = 316 \text{ K}$

$V_1 = 5 \text{ dm}^3$.

Using

$$\frac{P_1 V_1}{T_1} = \frac{P_2 V_2}{T_2}$$

$$V_2 = \frac{P_1 V_1}{T_1} \times \frac{T_2}{P_2}$$

$$= \frac{8 \times 10^4 \times 5 \times 316}{300 \times 5 \times 10^5} = 0.843 \text{ dm}^3$$

Example 7.2. What weight of oxygen would be required to fill a 2 dm³ flask at 20°C and $9 \times 10^4 \text{ N m}^{-2}$ pressure if R has the value 8.31 J K⁻¹ mol⁻¹?

Solution. Using $PV = nRT$,

$$n = \frac{PV}{RT} = \frac{9 \times 10^4 \times 2 \times 10^{-3}}{8.31 \times 293} = 0.074 \text{ mole}$$

Since 1 mole of oxygen weights 32.0 g, the weight required is

$$0.074 \times 32 = 2.37 \text{ g}$$

Example 7.3. What is the pressure of a mixture of 1 g of hydrogen and 1.4 g of nitrogen stored in a 5 dm³ vessel at 127°C?

Solution

Number of moles of hydrogen gas $= \dfrac{1}{2} = 0.5$

Number of moles of nitrogen gas $= \dfrac{1.4}{28} = 0.05$

Total number of moles of gas $= n = 0.55$

Using $PV = nRT$,

$$P = \frac{0.55 \times 8.31 \times 400}{5 \times 10^{-3}}$$

$$= 3.66 \times 10^5 \text{ Nm}^{-2}$$

An alternative method of solving Example 7.3 would be to calculate the pressure exerted by the hydrogen and nitrogen individually :

Pressure exerted by hydrogen, $P_{H_2} = \dfrac{0.5 \times 8.31 \times 400}{5 \times 10^{-3}}$

$$= 3.33 \times 10^5 \text{ N m}^{-2}$$

Pressure exerted by nitrogen, $p_{N_2} = \dfrac{0.05 \times 8.31 \times 400}{5 \times 10^{-3}}$

$$= 3.33 \times 10^5 \text{ Nm}^{-2}.$$

Total pressure, $P = p_{H_2} + p_{N_2} = (3.33 + 0.33) \times 10^5 \text{ Nm}^{-2}$

$$= 3.66 \times 105 \text{ Nm}^{-2}.$$

This is an example of the following law.

Dalton's Law of Partial Pressures

For gases which do not react, the total pressure of a mixture of gases is the sum of the partial pressures of each constituent. The partial pressure of a gas is the pressure it would exert if it alone occupied the total volume. In the above example, p_{H2} and p_{N2} are the partial pressures of hydrogen and nitrogen respectively.

The partial pressure of a gas is proportional to the number of moles of gas present. This fact is important in calculations on steam distillations, as shown in Example 7.4.

Example 7.4. A mixture of nitrobenzene and water boils at 99°C to give a vapour in which the partial pressure of water is 9.65 $\times 10^4$ N m^{-2} and that of nitrobenzene is 0.35 $\times 10^4$ N m^{-2}. Calculate the ratio by weight of nitrobenzene to water in the condensate.

Solution. The number of moles of each constituent is proportional to its partial pressure. Thus,

$$n_{H_2O} \propto p_{H_2O} \quad \text{and} \quad n_{\text{nitrobenzene}} \propto p_{\text{nitrobenzene}}$$

On dividing

$$\frac{n_{\text{nitrobenzene}}}{n_{H_2O}} = \frac{p_{\text{nitrobenzene}}}{p_{H_2O}} = \frac{0.35 \times 10^4}{9.65 \times 10^4}$$

Hence,

$$\frac{\text{Weight of nitrobenzene}}{\text{Weight of water}} = \frac{0.35}{9.65} \times \frac{133}{18} = \frac{1}{3.73}$$

(where 133 and 18 are the relative molecular masses of nitrobenzene and water respectively).

The law of partial pressures is important in connection with the collection of gases over water. Since gases collected this way always contain water vapour, the vapour pressure of water (at the particular temperature) is subtracted from the total pressure to give the true pressure of the gas collected. For example, the partial pressure of a gas collected over water at 1×10^5 N m^{-2} and 15°C is $(1-0.017) \times 10^5 = 0.983 \times 10^5$ N m^{-2} since the vapour pressure of water at 15°C is 0.017×10^5 N m^{-2}.

DALTON'S LAW AND REAL GASES

Dalton's law applies only to ideal gases and for real gases the total pressure is usually slightly in excess of the sum of the partial pressures of the constituents. At very high pressures, inter-molecular attraction becomes significant and this may bring about a negative deviation. At low pressures (up to a few atmospheres) and normal temperatures, the law may be regarded as fairly accurate.

Mole Fraction

The mole fraction x of a species in a mixture is the ratio of the number of moles of that species to the total number of moles present. For example, in a mixture of three components, A, B and C, the mole fraction of A is

$$x_A = \frac{n_A}{n_A + n_B + n_C}$$

where n_A, n_B and n_C are the number of moles of each constituent present.

Example 7.5. What is the mole fraction of nitrogen in a mixture of nitrogen and hydrogen in which the partial pressure of hydrogen is 6.3×10^5 N m^{-2} and the total pressure is 9×10^5 N m^{-2}?

Solution. The partial pressure of nitrogen is

$$(9 - 6.3) \times 10^5 = 2.7 \times 10^5 \text{ N m}^{-2}$$

Hence,

$$x_{N_2} = \frac{n_{N_2}}{n_{N_2} + n_{H_2}} = \frac{p_{N_2}}{p_{N_2} + p_{H_2}} = \frac{2.7 \times 10^5}{9 \times 10^5} = 0.3$$

KINETIC THEORY OF GASES

This theory interprets ideal gas behaviour in terms of molecular motion, and in consequence, the reasons for the behaviour of real gases can be appreciated. The basis of the theory for an ideal gas may be summarised as follows:

1. An ideal gas consists of a large number of particles (representing molecules) moving at random.
2. Each particle is regarded as a point, that is the size of each particle is infinitesimal and the total volume occupied by all the particles is insignificant.
3. The particle are perfectly elastic and no energy is lost when they collide.
4. No forces of attraction obtain between the particles.
5. The particles are in frequent collision with each other and with the walls of the containing vessel. The latter collisions produce the measured pressure exerted by the gas. Between collisions, the particles move in straight lines.
6. A rise in temperature will result in a faster movement of the particles, thus producing an increase in pressure if the volume remains constant.
7. If the volume of the container is decreased, the particles will collide with the walls more frequently and an increase in pressure will result.

DERIVATION OF THE FUNDAMENTAL EQUATION FOR THE PRESSURE OF A GAS

A cube-shaped vessel of side length l cm and of volume V cm^3 contains one mole of ideal gas (represented by N particles each of mass m) at a pressure P and at a fixed temperature. The behaviour of the particles is in accordance with the assumptions detailed above. Taking one particle moving at random, its velocity u may be resolved into three components mutually at right angles such that

$$u^2 = u_x^2 + u_y^2 + u_z^2$$

where u_x is the component of velocity along the x-axis, etc. When the container is placed with one corner at the origin of axes, the sides of the vessel can be made to coincide with the axes as shown in Fig. 7.4.

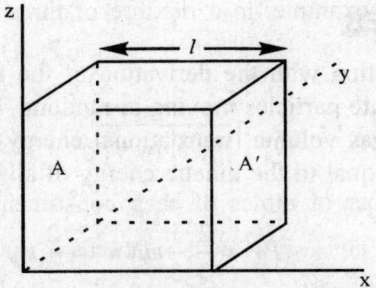

Fig. 7.4. Container and vessel coinciding with axes.

Considering the direction of the x-axis, the particle moves with velocity u_x between the faces A and A′ which are l cm apart, and the time taken for the particle to travel between A and A′ is l/u_x. The number of times per second that the particle collides with either A or A′ will be u_x/l. The particle approaches each face with momentum mu_x and leaves with a momentum $-mu_x$ (or vice versa), since the collisions are perfectly elastic. The change of momentum accompanying each collision is thus $2mu_x$, and the rate of change of momentum per second is

$$2mu_x \frac{u_x}{l} = \frac{2mu_x^2}{l}$$

This is the force on A and A taken together. Similarly, the force exerted by this particle on all six faces of the vessel is

$$\frac{2m}{l}(u_x^2 + u_y^2 + u_z^2)$$

and the total force due to N particles is

$$\frac{2mN}{l}(\overline{u_x^2} + \overline{u_y^2} + \overline{u_z^2})$$

or

$$\frac{2mN}{l}(\overline{u^2})$$

Since pressure is defined as force per unit area, and the total area of the six faces of the container is $6l^2$, the pressure of the gas is

$$P = \frac{2mN\overline{u^2}}{3} \times \frac{1}{6l^2}$$

$$= 1m \frac{N\overline{u^2}}{3l^3}$$

Putting $l^3 = V$, the fundamental gas equation results :

$$PV = \frac{1}{3}mN\overline{u^2}$$

ENERGY OF THE PARTICLES

The system considered in connection with the derivation of the fundamental equation was assumed to be an ideal gas consisting of minute particles moving at random. The energy of these particles is kinetic energy of movement within the gas volume (translational energy). Since kinetic energy is reckoned as $1/2mu^2$, the quantity $1/2mNu^2$ is equal to the kinetic energy of all the particles in one mole of gas. Thus

$$PV = \frac{2}{3} \cdot \frac{1}{3} mN\overline{u^2} = \frac{2}{3} E_k$$

where E_k is the total kinetic energy of the particles in the gas. Heating the gas brings about a rise in temperature corresponding to a conversion of heat energy into kinetic energy. Consequently, if the temperature remains unaltered, the value of E_k is unchanged which is in agreement with Boyle's law that PV is constant.

The particles in a gas cannot all have the same velocity even at a fixed temperature on account of the many collisions that take place. However, J. Clerk Maxwell deduced that the mean kinetic energy per particle, $1/2mu^2$ is the same for all particles irrespective of mass at the same temperature.

For equal volumes of two gases A and B at the same pressure,

$$PV = \frac{1}{3} N_A m_A \overline{u_A^2} = \frac{1}{3} N_B m_B \overline{u_B^2}$$

or

$$N_A m_A \overline{u_A^2} = N_B m_B \overline{u_B^2}$$

If these gases are kept at the same temperature, Maxwell's deduction demands that

$$\frac{1}{2} m_A \overline{u_A^2} = \frac{1}{2} m_B \overline{u_B^2}$$

from which it follows that $N_A = N_B$. This conclusion, that equal volumes of gases at the same temperature and pressure contain equal numbers of molecules, is, of course, Avogadro's hypothesis.

So far, the kinetic theory has been discussed in terms of particles rather than molecules. This distinction has been made to emphasise the fact that the actual behaviour of molecules in real gases will deviate from the assumed behaviour of particles constituting an ideal gas.

TEMPERATURE AND MOLECULAR VELOCITIES

For one mole of gas,

$$PV = RT$$

and from the kinetic theory

$$PV = \frac{1}{3} mN\overline{u^2}$$

$$= \frac{1}{3} M\overline{u^2} = \frac{2}{3} E_k$$

where M represents the weight of one mole of gas.

By comparing these equations, it follows that

$$\frac{1}{3} M \overline{u^2} = \frac{2}{3} E_k = RT$$

From this equation, it is clear that

(1) The kinetic energy of a gas is proportional to the absolute temperature.
(2) The molecular velocity is proportional to the square root of the absolute temperature, and depends also on the molar weight of the gas.

The equation $\frac{1}{3} M \overline{u^2} = RT$ can be re-written to give

$$\overline{u^2} = \frac{3RT}{M}$$

or,

$$\sqrt{(u^2)} = \sqrt{\frac{3RT}{M}}$$

$\sqrt{(u^2)}$ is called the root mean square velocity. If R is in joules or N m K^{-1} mol^{-1} M has to be given in kg as 1 N is 1 kg m s^{-2}. For hydrogen therefore,

$$\sqrt{(u^2)} = \sqrt{\frac{3RT}{M}} = \sqrt{\frac{3 \times 8.31 \times 10^7 \times 273}{2.016 \times 10^{-3}}}$$

$$= 18,4000 \text{ cm s}^{-1} \text{ at s.t.p.}$$

Some other root mean square velocities at s.t.p. are

oxygen, 46 140 cm s^{-1} carbon dioxide, 39 300 cm s^{-1}

helium, 130 600 cm s^{-1} chlorine, 39 950 cm s^{-1}

It can be shown that the average velocity of the molecule $\overline{u}$ is slightly less than the root mean square velocity. The two quantities are related by the equation $\overline{u} = 0.92 \sqrt{(u^2)}$.

The following method for the determination of molecular velocities outlines the chief techniques used by a number of workers investigating this problem. Two discs, each pierced by a slit, are set some distance apart on a common axle. The slits are colinear, and a beam of molecules is directed on to the first disc, while a detecting device is placed behind the second disc. When the discs are stationary, the beam of molecules passes through the two slits and is recorded by the detector. On rotation of the discs, molecules will again enter the detector when the time taken by a molecule to travel between the discs is equal to the time required for the discs to complete one revolution. As a result of these investigations, the majority of molecules were found to have average velocities equal to the values calculated from the kinetic theory. Some scatter of velocities about the mean value was also apparent. Such a scatter is expected, since the molecules are continually colliding and cannot have a steady velocity. The distribution of velocities was expressed in mathematical terms by J. Clerk Maxwell, who showed that in a system containing N molecules, the number n of molecules having a kinetic energy greater than a particular value E, is given by

$$n/N = e^{-E/RT}$$

This relation is expressed graphically in Fig. 7.5, and it is clear that most molecules have a velocity close to the mean value.

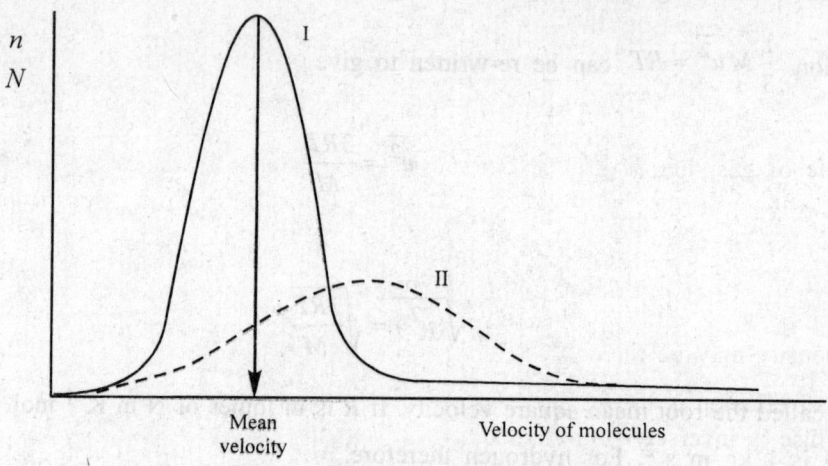

Fig. 7.5. Distribution of molecular velocities.

Curve I represents the distribution of energies (or velocities since $E = \frac{1}{2} M\overline{u^2}$) at a lower temperature, while curve II relates to a higher temperature. From these curves, it can be seen that an increase in temperature causes a general increase in molecular velocities, but a greater scatter of velocities about the mean accompanies this increase. The results of the experimental determination of molecular velocities agree well with the Maxwell distribution law.

EFFUSION AND DIFFUSION

The root mean square velocity of a molecule of ideal gas is given by

$$\sqrt{(u^2)} = \sqrt{\frac{3RT}{M}}$$

For two gases A and B at the same temperature,

$$\frac{\sqrt{(u_A^2)}}{\sqrt{(u_B^2)}} = \sqrt{\frac{3RT}{M_A}} \times \sqrt{\frac{M_B}{3RT}} = \sqrt{\frac{M_B}{M_A}}$$

where M_A and M_B represent the weights of one mole of A and B respectively.

The passage of a gas through a fine hole in the wall of a vessel is known as effusion and the rate of escape of gas should, from the above equations (for the ideal case) be inversely proportional to its mole weight. T. Graham had arrived at this conclusion following his experiments on gaseous effusion and diffusion.

Graham's Law

The rates of effusion (or diffusion) of different gases (under identical conditions) are inversely proportional to the square roots of their densities (or mole weights). For two gases A and B of densities d_A and d_B,

$$\frac{\text{Rate of effusion of A}}{\text{Rate of effusion of B}} = \frac{\sqrt{d_B}}{\sqrt{d_A}} = \frac{\sqrt{M_B}}{\sqrt{M_A}}$$

Since 1 mole of gas, i.e. M grams, occupies 22.4 dm^3 at s.t.p. then the density of d of a gas at s.t.p. is

$$d = \frac{M}{22.4}\,\text{g dm}^{-3}$$

Hence, density may be replaced by the mole weight of the gas.

The rate of effusion is the volume of gas escaping in unit time. The time taken for a given volume of gas to effuse is inversely proportional to the rate of effusion. Hence

$$\frac{\text{Time taken for 1 volume of A to escape}}{\text{Time taken for 1 volume of B to escape}} = \frac{\text{rate of effusion of B}}{\text{rate of effusion of A}}$$

$$= \frac{\sqrt{d_A}}{\sqrt{d_B}} = \frac{\sqrt{M_A}}{\sqrt{M_B}}$$

Example 7.6. The density of methane is four times that of helium. Calculate the time taken for 10 cm^3 of methane to escape through an orifice if, under identical conditions, helium effuses at the rate of 1.5 cm^3 per minute.

Solution. If 1.5 cm^3 of helium effuse in 1 min

then 10 cm^3 of helium would effuse in $1 \times 10/1.5$ min.

Let t be the number of minutes taken for 10 cm^3 of methane to effuse. Then

$$\frac{t}{10/1.5} = \frac{\sqrt{4}}{\sqrt{1}} = 2$$

$$t = \frac{2 \times 10}{1.5} = 13.3 \text{ minutes}$$

Effusiometer

This instrument enables the times required by different gases to escape, under identical conditions, through a small hole in a metal plate to be compared. A diagram of the instrument is shown in Fig. 7.6.

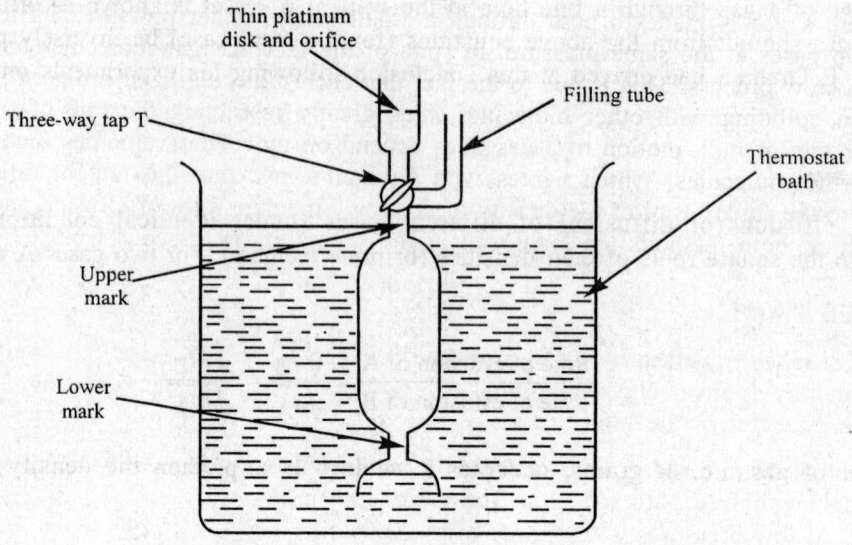

Fig. 7.6. Effusiometer.

The apparatus is filled with the gas under test using the three-way tap, and placed in a thermostatically controlled bath. When a steady temperature has been obtained, the three-way tap T is adjusted to allow the gas under test to escape through the fine hole in the thin platinum disc at the upper end of the apparatus. The time required for the liquid to rise from the lower to the upper mark is noted. The experiment is repeated using a different gas at the same temperature. The liquid used in the thermostat bath should not, of course, react with or dissolve either gas.

Example 7.7. The time taken for a sample of hydrogen containing a small proportion of deuterium to escape from an effusiometer was 3.7 min. If, under the same conditions, an equal volume of pure oxygen effused in 14.1 min, calculate the proportion of deuterium in the sample. (Take the relative atomic masses of oxygen, hydrogen and deuterium as 16.0, 1.0 and 2.0 respectively).

Solution

$$\frac{\text{Time for effusion of oxygen}}{\text{Time for effusion of hydrogen}} = \frac{\sqrt{M_{\text{oxygen}}}}{\sqrt{M_{\text{hydrogen}}}}$$

Thus

$$\frac{14.1}{3.7} = \frac{\sqrt{32}}{\sqrt{M_{\text{hydrogen}}}}$$

$$M_{\text{hydrogen}} = 2.20$$

Let x % of deuterium be present in the hydrogen sample, then the average weight of one mole of hydrogen will be

$$\frac{x \times 4 + (100 - x) \times 2}{100} = 2.20$$

$$x = 10\%$$

In effusion, a gas is made to pass through a small orifice under pressure. Diffusion relates to the mixing of two gases at the same pressure. In spite of the fact that gas molecules have high velocities, diffusion is a slow process. This is due to the fact that each molecule is, at ordinary pressures, involved continually in colliding with other molecules, thus greatly restricting the rate of forward movement. However, the rate of bulk motion of a gas does depend on molecular velocities (and also on the mean free path of the molecules) which agrees with Graham's law, that the rate of diffusion is inversely proportional to the square root of the density of the gas. Since the rates of effusion and diffusion depend on gas densities, both methods are employed in the separation of gaseous isotopes.

MEAN FREE PATH

The average distance traversed by a gas molecule between successive collisions is termed its mean free path. This quantity is inversely proportional to: (i) the number of molecules per unit volume; (ii) the square of the diameter of the molecule.

An estimation of molecular diameters can be made by measuring the rates of diffusion which are related to the mean free path. At s.t.p. the mean free path of a molecule in a gas is of the order of 10^{-5} cm. At low pressures, the number of molecules per unit volume is much less, and at 10^{-5} atm the mean free path is about 1 cm.

REAL GASES AND THE KINETIC THEORY

The kinetic theory explains the behaviour of ideal gases, but modifications of the theory are required to allow for the deviations from ideality shown by real gases. Two of the assumptions made at the outset are clearly untrue for real gases. These are: (i) the gas molecules do not attract each other; and (ii) the gas molecules occupy negligible volume.

Intermolecular Attraction

If the gas molecules do attract each other the gas will be more compressible than the ideal case. The effect of intermolecular attraction is to increase the effective pressure on the gas above the value indicated as the measured external pressure. Consequently, the volume will be less than expected and the value of PV will fall below the ideal case (as shown in the early parts of the curves for carbon dioxide and nitrogen in Fig. 7.2). Moreover, gases can be liquefied, and the liquid state implies a considerable degree of attraction between the molecules; hence it would be reasonable to expect that some degree of attraction between molecules continues in the gaseous state. It also follows that those gases in which the intermolecular attractions are greatest are most easily liquefied. When a gas expands through a porous plug, or an orifice, from a region of high to low pressure, a cooling effect, known as the *Joule-Thomson effect* is noticed. This cooling effect is due to the internal work performed in separating the molecules against their attractions. Hydrogen and helium behave exceptionally as they exhibit the Joule-Thomson effect only at low temperatures. Hydrogen, for example, cools on Joule-Thomson expansion below – 80°C at 120 atm, but warms slightly when expanded above this temperature.

VOLUME OF THE MOLECULES

At low pressures, the volume occupied by the gas molecules is not significant, but at higher pressures the molecules are compacted to a degree such that they tend to behave like a liquid, which is almost incompressible. As a result, at higher pressures the volume of the gas remains larger than the ideal case, and in consequence the product PV rises. This is shown in the later part of the curves for carbon dioxide and nitrogen in Fig. 7.2. Since hydrogen shows an upward trend from ideal behaviour at all pressures, it is concluded that the effect of the volume of the molecules is greater than the effect of intermolecular attraction. In other words, the force of intermolecular attraction between gaseous hydrogen molecules is very small.

VAN DER WAALS EQUATION

The effect of molecular size and intermolecular forces of attraction are taken into account in the van der Waals equation, in which the ideal gas equation $PV = RT$ is changed to

$$(P + a/V^2)(V - b) = RT$$

for one mole of ideal gas. It has already been mentioned that due to the forces of attraction between molecules, the actual pressure experienced by the gas molecules is greater than the indicated external pressure P. The first term is in the van der Waals equation is to allow for this factor, and the fraction a/V^2 is arrived at by considering molecules such as B in Fig. 7.7, which are just about to strike the wall of the container. Such molecules experience an unbalanced force of attraction due to molecules such as A in the bulk of the gas. The force with which a molecule strikes the wall is lessened (i.e. the measured pressure is too low) on account of these unbalanced forces.

Now the loss in pressure due to the inward force of attraction is proportional to the force of attraction per molecule multiplied by the number of molecules moving to the wall, and each of these quantities is proportional to the number of molecules per unit volume. Hence

$$\text{loss in pressure} \propto \frac{n}{V} \times \frac{n}{V} = \frac{a}{V^2}$$

where a is a constant and n is the total number of molecules in the volume V. The term a/V^2 which allows for the intermolecular force of attraction is called the internal pressure. The second term in the van der Waals equation $(V - b)$, contains a correction factor b which is related to the actual volume occupied by the molecules.

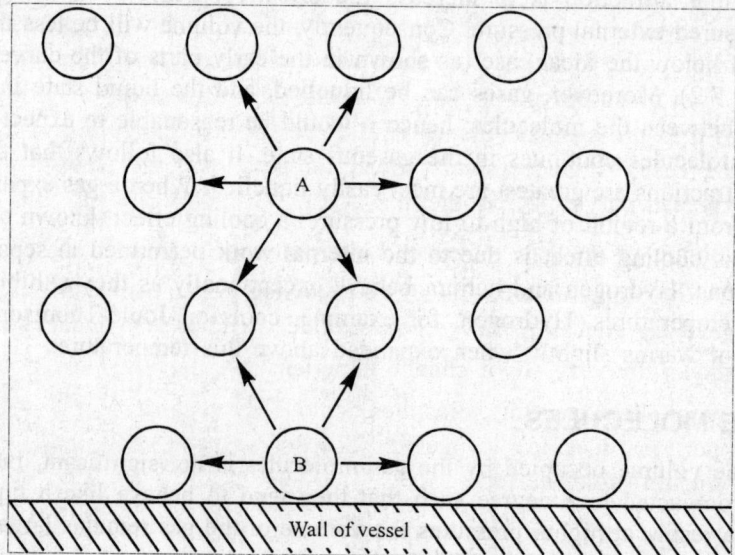

Fig. 7.7. Intermolecular attraction.

The values of a are expressed in N m^4 while b is given in m^3. Some typical values are :

Hydrogen	$a = 0.025$	$b = 2.67 \times 10^{-5}$
Oxygen	$a = 0.152$	$b = 3.12 \times 10^{-5}$
Carbon dioxide	$a = 0.36$	$b = 4.28 \times 10^{-5}$
Sulphur dioxide	$a = 0.67$	$b = 5.6 \times 10^{-5}$

The increasing value of a agrees with increasing ease of liquefaction from hydrogen to sulphur dioxide.

OTHER EQUATIONS OF STATE

Although the van der Waals equation agrees much more closely with the behaviour of real gases than the ideal gas equation, it is found that the values of a and b are not constant but vary with temperature. A number of alternative equations of state have been proposed, such as that of Berthelot:

$$(P + a'/TV^2)\ (V - b) = RT$$

which gives good results at moderate pressures.

LIQUEFACTION OF GASES UNDER PRESSURE

Following the work of Faraday and others, some gases, such as hydrogen chloride, sulphur dioxide, ammonia and chlorine were found to change to the liquid state under the influence of increased pressure. Other gases, even when subjected to pressures of 3000 atmospheres, failed to liquefy, and they were regarded as permanent gases, that is they could not be liquefied. The experiments of Andrews showed that liquefaction of a gas is not possible if the temperature is above the critical value.

ANDREWS' ISOTHERMALS FOR CARBON DIOXIDE

The results of Andrews' investigations on the pressure-volume relations at high pressures and at various temperatures are shown graphically in Fig. 7.8. Each curve is an isothermal, that is the pressure-volume relationship expressed by a given curve applies only to the temperature shown. If carbon dioxide obeys Boyle's law, a graph of pressure against volume should produce a rectangular hyperbola, and at 48.1°C this is found to be the case. At successively lower temperatures, the isothermals show increasing deviations from the expected curve. The isothermal at 13.1°C can be divided into three distinct parts:

(1) Along WX, the volume of the gas decreases corresponding to an increase in pressure.
(2) Between X and Y, the volume decreases rapidly following a slight increase in pressure. Along this line, the gas is liquefying and at Y, the carbon dioxide is completely liquid.
(3) The portion YZ shows the very slight change experienced by a liquid under increasing pressure.

The approximately parabolic dotted line joins the ends of the horizontal portions analogous to XY, and at C, these horizontal portions disappear altogether. The point C is known as the critical point for the gas, and three critical quantities are defined :

(1) The *critical temperature* (T_c) is the temperature above which a gas fails to liquefy under compression.

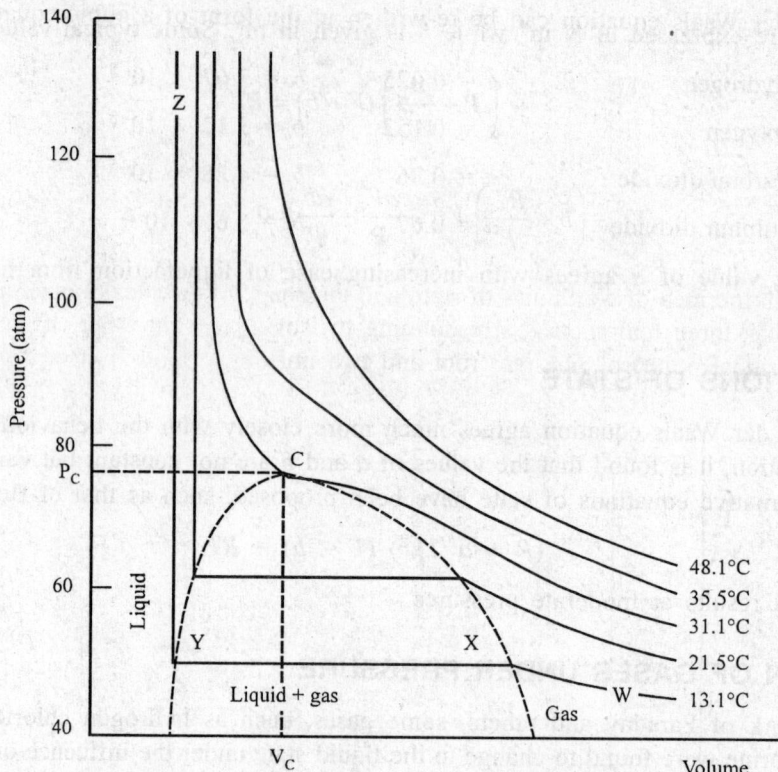

Fig. 7.8. Andrews' isothermals for carbon dioxide.

(2) The *critical pressure* (P_c) is the pressure which is just sufficient to liquefy the gas at its critical temperature.

(3) The *critical volume* (V_c) of a gas is the volume of one mole of gas at the critical temperature and pressure.

All gases show a critical point, and the values of T_c, P_c and V_c are peculiar to each gas. For example, the critical temperature for carbon dioxide is 304.2 K, and the critical pressure is 7.39×10^6 N m^{-2}, while the corresponding values for oxygen are 154.3 K and 5.03×10^6 N m^{-2}.

Usually, when a liquid changes into a gas, a large increase in volume is noted since the gas has a much smaller density than the liquid. As the critical point is approached, the densities of both the liquid and the gas become equal, which implies that there is no difference between a liquid and a gas under these conditions. It is found that no sudden changes in volume or density occur when a liquid changes into a gas near the critical point, but the liquid meniscus suddenly vanishes. The term continuity of state is used to describe this smooth transition from one phase to another.

On compressing a gas (below its critical temperature) in a very clean apparatus with smooth inner surfaces, it is possible to hold the gas at a pressure above that at which liquefaction normally takes place. This is shown by the dotted line XA in Fig. 7.9, in which WXYZ is the normal isothermal.

An increase in pressure beyond A results in a rapid liquefaction of the gas. In a similar way, if the pressure on a pure liquid is smoothly and gradually reduced, the pressure-volume relation follows the dotted curve YB. It has been suggested that joining A and B (by the broken curve shown) would produce a pressure-volume relation relevant to that for an ideal gas.

Now the van der Waals equation can be re-written in the form of a cubic equation:

$$\left(P + \frac{a}{V^2}\right)(V - b) = RT$$

$$V^3 - \left(b + \frac{RT}{P}\right)V^2 + \frac{aV}{P} - \frac{ab}{P} = 0$$

and this agrees with the idea of continuity of state and the shape of the ideal isotherm WABZ. Moreover, a cubic equation has three real roots (corresponding to isotherms below the critical point), a point of inflexion (at the critical point) or one real root and two imaginary roots (corresponding to the gaseous state).

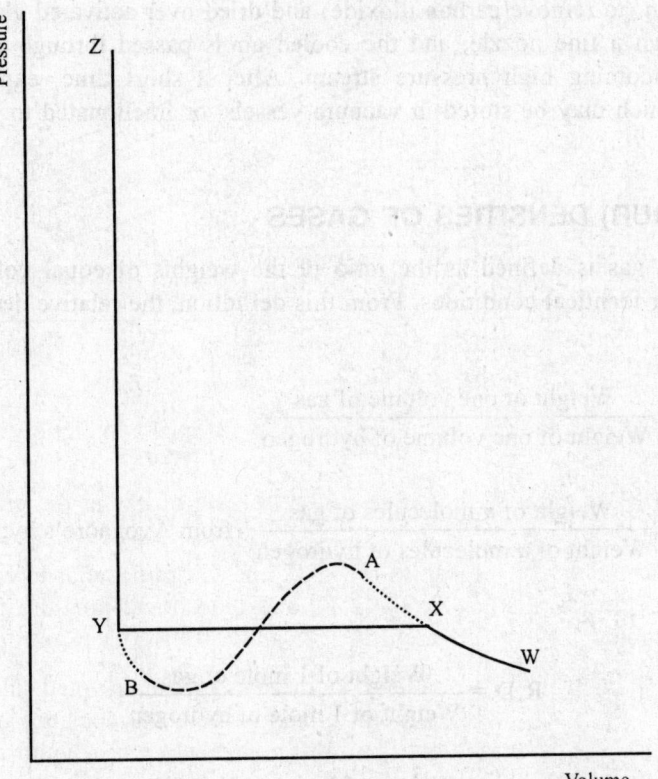

Fig. 7.9. Normal isothermal shown by curve wxyz.

LIQUEFACTION OF GASES

Cooling a gas held under a high pressure is the basis of many methods used for liquefying gases. Pictet used a two-stage process to cool oxygen below its critical temperature of −119°C. Modern methods employ the Joule-Thomson effect or adiabatic expansion to reduce the temperature of the gas below its critical value. The Joule-Thomson effect has already been mentioned, while the adiabatic expansion method relies on the fact that an expanding gas doing work against an external pressure cools down, if the system is insulated against heat gain or loss. The relation between pressure and temperature in an adiabatic expansion is

$$\left(\frac{T_1}{T_2}\right)^{\gamma} = \left(\frac{P_1}{P_2}\right)^{\gamma-1}$$

where γ is the ratio of the molar heat capacities of the gas at constant pressure and at constant volume. Consequently, the higher the initial pressure, the greater the fall in temperature on expansion.

The Joule-Thomson effect has been used for the production of liquid air since 1895 when Linde and Hampson both used this method for obtaining low temperatures. For every gas there is an inversion temperature above which the Joule-Thomson effect produces heating, and below which cooling occurs. With some gases, such as hydrogen, pre-cooling is necessary before the desired results can be obtained.

In the modern process for the production of liquid air, compressed air is bubbled through concentrated sodium hydroxide solution (to remove carbon dioxide) and dried over activated alumina. The air is then allowed to expand through a fine nozzle, and the cooled air is passed through a counter-current heat exchanger against the incoming high pressure stream. After a short time, expansion results in the formation of liquid air which may be stored in vacuum vessels, or fractionated to give oxygen, nitrogen and argon.

RELATIVE (OR VAPOUR) DENSITIES OF GASES

The relative density of a gas is defined as the ratio of the weights of equal volumes of the gas and hydrogen, measured under identical conditions. From this definition, the relative density of a gas is given by:

$$\text{R.D. (or V.D.)} = \frac{\text{Weight of one volume of gas}}{\text{Weight of one volume of hydrogen}}$$

$$= \frac{\text{Weight of } n \text{ molecules of gas}}{\text{Weight of } n \text{ molecules of hydrogen}} \quad \text{(from Avogadro's hypothesis)}$$

Thus

$$\text{R.D.} = \frac{\text{Weight of 1 mole of gas}}{\text{Weight of 1 mole of hydrogen}}$$

$$= \frac{\text{Weight of 1 mole of gas}}{2}$$

Therefore, the weight of 1 mole of gas is twice the relative density :

$$\text{Weight of 1 mole of gas} = 2 \times \text{R.D.}$$

Example 7.8. The densities of oxygen, helium and hydrogen at s.t.p. are 1.427, 0.1784 and 0.0892 g dm^{-3} respectively. Calculate the weights of 1 mole of oxygen and helium.

Solution.

$$\text{R.D. of oxygen} = \frac{1.427}{0.0892} = 16.0$$

Thus

$$1 \text{ mole of oxygen weights } 2 \times 16.0 = 32.0 \text{ g}$$

Similarly

$$\text{R.D. of helium} = \frac{0.1784}{0.0892} = 2.00$$

Thus

$$1 \text{ mole of helium weighs } 2 \times 2.00 = 4.00 \text{ g}$$

DETERMINATION OF THE APPROXIMATE WEIGHTS OF ONE MOLE OF GAS

The following methods are based on measuring the density of a gas or vapour.

Regnault's Method (for Gases)

1. A flask is filled with distilled water at a known temperature and weighed. From this weighing, the volume of the flask can be calculated.

2. The flask is emptied, dried, evacuated and weighed again after filling with gas at a known temperature and pressure.

3. The density of the gas is calculated, and the weight of one mole of gas is found.

   The accuracy of Regnault's method can be improved by correcting for buoyancy effects.

Dumas' Method

In this method (Fig. 7.10) the density of the vapour produced by the evaporation of a volatile liquid is measured. The vapour is contained in a Dumas bulb which is surrounded by a bath of hot liquid kept at a steady temperature which is above the boiling point of the liquid under investigation. The experiment is carried out in the following stages:

1. The Dumas bulb is weighed full of air.

2. A small quantity of the liquid under test is introduced into the bulb.

3. The bulb is maintained at a steady temperature in the heating bath until all the liquid has vapourised and the vapour fills the bulb at atmospheric pressure. The excess vapour escapes into the atmosphere.

4. The tip of the bulb is then sealed by application of a hot flame.

5. The bulb is allowed to cool to room temperature and weighed.

6. The volume of the bulb is found either by weighing when filled with water, or by measuring the volume of water held by the bulb.

   This experiment may be performed at high temperatures using quartz bulbs and liquids of high boiling points (for example, molten metals) in the heating bath. It is also possible to obtain good results with small scale apparatus.

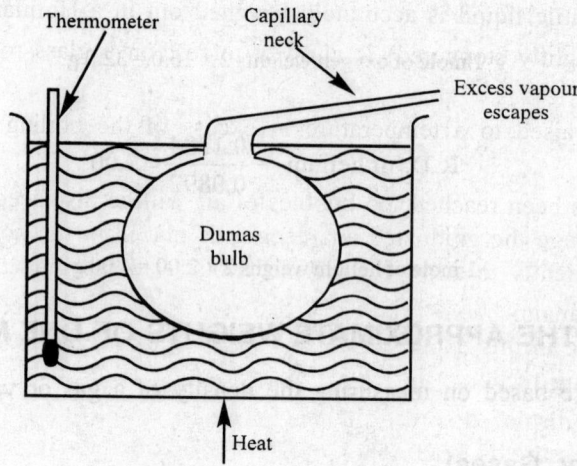

Fig. 7.10. Dumas apparatus.

Victor Meyer's Method

This is another method which measures the volume occupied by a given weight of vapour produced on the evaporation of a volatile liquid. The apparatus is shown in Fig. 7.11.

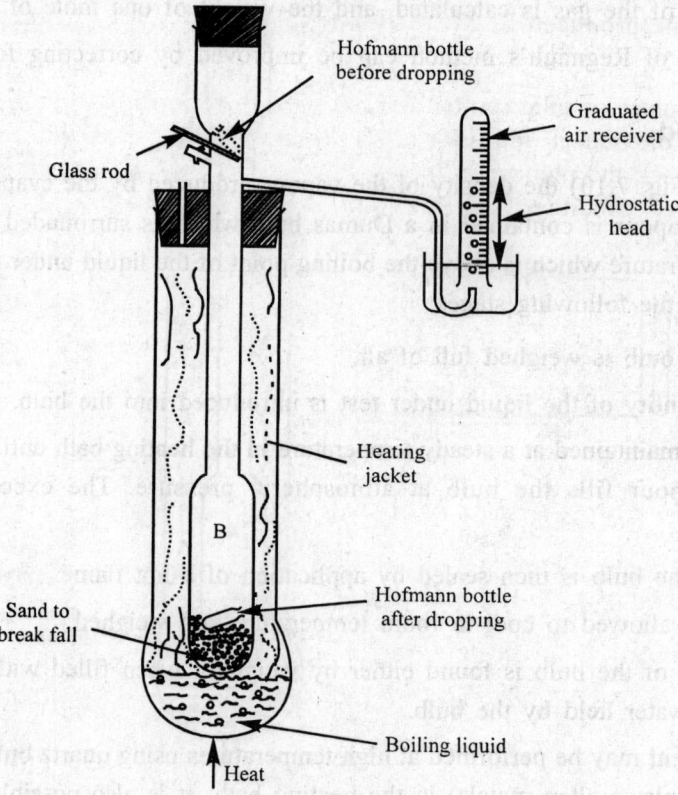

Fig. 7.11. Victor Meyer apparatus.

1. About 0.1 g of the volatile liquid is accurately weighed out in a Hofmann bottle.
2. The Hofmann bottle (lightly stoppered) is allowed to rest on a glass rod in the upper part of the apparatus as shown.
3. The heating jacket is raised to a temperature in excess of the boiling point of the liquid under investigation.
4. When a steady state has been reached, no bubbles of air will be displaced by expansion through the side tube, and at this stage the graduated air receiver is placed in the position shown and the glass rod is withdrawn sufficiently to allow the Hofmann bottle to fall into the bulb B.
5. The stopper of the Hofmann bottle is forced out as the liquid vapourises. (A jammed stopper is a frequent cause of failure in this experiment). The vapour produced displaces an equivalent quantity of air into the measuring receiver.
6. The final volume of air displaced is read off, the temperature of the water in the air receiver is noted and the barometric pressure is read. The pressure of air in the measuring receiver is less than atmospheric by an amount equal to the hydrostatic head of liquid remaining in the receiver, and a correction must also be applied for the presence of water vapour in the displaced air. This method is also capable of being performed at high temperatures and on a small scale.

MEASUREMENT OF GAS DENSITY USING A SYRINGE

A useful laboratory method of determining the density and hence relative molecular mass of a gas, is to measure the volume produced by a suitable chemical or physical process by means of gas syringe (Fig. 7.12).

For example, a piece of solid carbon dioxide is weighed rapidly and transferred immediately to a tube containing carbon dioxide at atmospheric pressure. The solid carbon dioxide evaporates and the gas produced pushes back the plunger in the syringe. In this way the volume of carbon dioxide produced (under the measured conditions of temperature and pressure) from a known weight of solid carbon dioxide is determined.

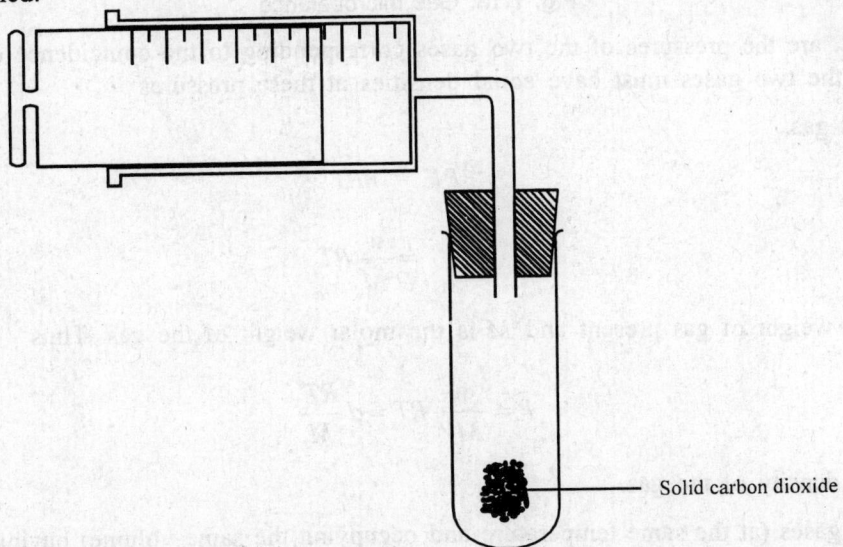

Solid carbon dioxide

Fig. 7.12. Relative molecular mass using a gas syringe.

GAS DENSITY MICROBALANCE

This method, used by R.W. Gray and W. Ramsay to determine the density of radon, is capable of giving very accurate results using very small quantities of gas. The apparatus is shown diagrammatically in Fig. 7.13. The bulb A, of about 8 cm³ capacity, is suspended by a quartz fibre from the balance arm BC, which pivots at C on a transverse quartz fibre. D is a counter-balance plate whose surface area is equal to that of the bulb A. The plate D carries a pointer E which is used in conjunction with a second pointer F attached to the frame (not shown in full) on which the balance is mounted. This part of the apparatus is made entirely of quartz. The balance is enclosed in a glass cylinder, which can be filled with gas at any desired pressure, while the complete unit is maintained at a steady temperature. The apparatus is evacuated and filled with a gas (such as air or nitrogen) of known density. The buoyancy effect of the gas causes the bulb A to rise and the pressure of the gas is adjusted until the pointers E and F (viewed through a microscope) are in exact alignment. The procedure is repeated with the gas of unknown density.

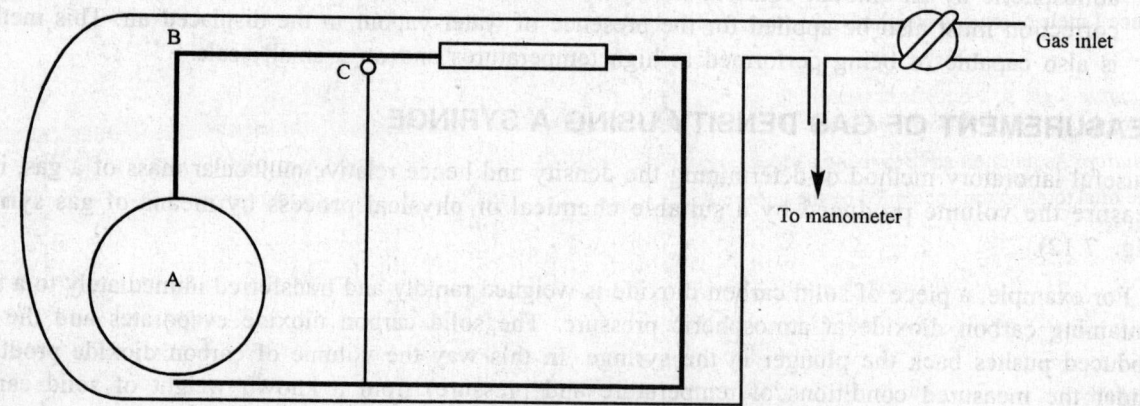

Fig. 7.13. Gas microbalance.

If P_1 and P_2 are the pressures of the two gases corresponding to the coincidence of E and F, then it follows that the two gases must have equal densities at these pressures.

For an ideal gas,

$$PV = nRT$$

$$= \frac{w}{M} RT$$

where w is the weight of gas present and M is the molar weight of the gas. Thus

$$P = \frac{w}{MV} RT = d \frac{RT}{M}$$

where d is the density of the gas.

For the two gases (at the same temperature and occupying the same volume) having equal densities at pressures P_1 and P_2 respectively,

$$d = \frac{M_1 P_1}{RT} = \frac{M_2 P_2}{RT}$$

or

$$\frac{P_1}{P_2} = \frac{M_2}{M_1}$$

A correction is applied in these experiments to allow for the deviation from ideal behaviour.

Example 7.9. What weight of oxygen would be required to fill a 2 dm³ flask at 20°C and 9×10^4 N m⁻² pressure if R has the value 8.31 J K⁻¹ mol⁻¹?

Solution. Using $PV = nRT$,

$$n = \frac{PV}{RT} = \frac{9 \times 10^4 \times 2 \times 10^{-3}}{8.31 \times 293} = 0.074 \text{ mole}$$

Since 1 mole of oxygen weights 32.0 g, the weight required is

$$0.074 \times 32 = 2.37 \text{ g}$$

Example 7.10. A 2 dm³ flask containing carbon dioxide at 6×10^4 N m⁻² pressure is connected to a 4-dm³ flask containing nitrogen at 4.8×10^5 N m⁻² pressure. If the temperature is kept constant, calculate the final pressure of the mixture.

Solution. As an alternative to using a repeated application of Boyle's law, we can proceed as follows :

The final total volume occupied by the gases is 6 dm³, thus, the partial pressure of carbon dioxide falls to 2/6 of its initial value, that is,

$$p_{CO_2} = \frac{2}{6} \times 6 \times 10^5$$

$$= 2 \times 10^5 \text{ N m}^{-2}$$

Similarly,

$$p_{N_2} = \frac{4}{6} \times 4.8 \times 10^5$$

$$= 3.2 \times 10^5 \text{ N m}^{-2}$$

Hence the final pressure is 5.2×10^5 N m⁻².

Chapter 8

Liquids and Solids

INTRODUCTION

The kinetic energies of gas molecules decrease when the temperature is lowered. Consequently the intermolecular attractive forces cause the gas molecules to condense into a liquid when the gas has been cooled sufficiently. The molecules are closer together and the attractive forces exert a greater influence in a liquid than in a gas. Molecular motion, therefore, is more restricted in the liquid state than in the gaseous state. Additional cooling causes the kinetic energies of the molecules to decrease further and ultimately produces a solid. In a crystalline solid, the molecules assume positions in a crystal lattice, and the motion of the molecules is restricted to vibration about these fixed points.

The comparatively high kinetic energies of gas molecules cause the intermolecular attractive forces to assume a role that can be minimised in the development of a satisfactory theory of gases. The comparatively low kinetic energies of molecules (or ions) in crystals are easily overcome by the attractive forces to produce highly ordered, crystalline structures that have been well characterised by diffraction techniques. Our understanding of the intermediate state, the liquid state, however, is not so complete as that of the other two states.

LIQUID STATE

The liquid state is intermediate in character between the complete molecular randomness that characterises gases and the orderly arrangement of molecules typical of crystalline solids. In liquids, the molecules move slowly enough that the intermolecular forces of attraction can hold them together in a definite volume; however, the molecular motion is too rapid for the attractive forces to fix the molecules into the definite positions of a crystal lattice. A liquid, therefore, retains its volume but not its shape; liquids flow to assume the shape of the container.

The space between the molecules of liquids is reduced almost to a minimum by the intermolecular attractions; a change in pressure has almost no effect on the volume of a liquid. An increase in temperature, however, increases the volume of most liquids slightly and consequently decreases the liquid density. As the temperature of a liquid is increased, the kinetic energy of the molecules increases, and this increased molecular motion works against the attractive forces. This expansion is, however, much less than that observed for gases in which the effect of the attractive forces is negligible.

Two miscible (mutually soluble) liquids will diffuse into each other when placed together. If a liquid is carefully overlaid with a second, less dense liquid, a sharp boundary line is observed between the two. This boundary gradually becomes less distinct and, in time, disappears completely as the molecules of the two liquids mix. The diffusion of liquids, however, is a much slower process than the diffusion of gases. Since the molecules of liquids are relatively close together, the mean free path of liquid molecules is relatively short; liquid molecules suffer many more collisions per unit time than do gas molecules.

Any liquid exhibits resistance to flow, a property known as viscosity. One way of determining the viscosity of a liquid is to measure the time that it takes for a definite amount of the liquid to pass through a tube of small diameter under a given pressure. Resistance to flow is largely due to the attractions between molecules, and the measurement of the viscosity of a liquid gives a simple estimate of the strength of these attractions. Allowance should be made, however, for factors such as molecular weight and the structure of the molecules. Liquids with large, irregularly shaped molecules are generally more viscous than liquids with small, spherical molecules. In general, as the temperature of a liquid is increased, the cohesive forces are less able to cope with the increasing molecular motion, and the viscosity decreases. On the other hand, increasing the pressure generally increases the viscosity of a given liquid.

Another property of liquids related to the intermolecular forces of attraction is surface tension. A molecule in the centre of a liquid is attracted equally in all directions by surrounding molecules. Molecules on the surface of a liquid, however, are attracted only toward the interior of the liquid (Fig. 8.1). The surface molecules, therefore, are pulled inward, and the surface area of a liquid tends to be minimised. This behaviour accounts for the spherical shape of liquid drops. Surface tension is a measure of this inward force on the surface of a liquid, the force which must be overcome to expand the surface area. The surface tension of a liquid decreases with increasing temperature since the increased molecular agitation tends to decrease the effect of the intermolecular cohesive forces.

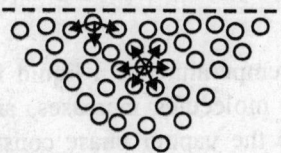

Fig. 8.1. Schematic diagram indicating the unbalanced intermolecular forces on the surface molecules of a liquid.

EVAPORATION

The kinetic energy of the molecules of a liquid follows a Maxwell-Boltzmann distribution similar to the distribution of kinetic energy among gas molecules (Fig. 8.2). The kinetic energy of a given molecule of a liquid is continually changing as the molecule collides with other molecules, but at any given instant, some of the molecules of the total collection have relatively high energies and some have relatively low energies. The molecules with kinetic energies sufficiently high to overcome the attractive forces of surrounding molecules can escape from the liquid and enter the gas phase if they are close to the surface and are moving in the right direction; they use part of their energy to work against the attractive forces when they escape.

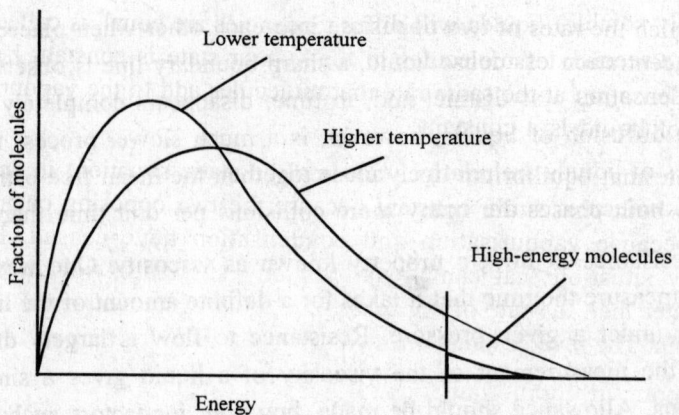

Fig. 8.2. Distribution of kinetic energy among molecules of a liquid.

In time, the loss of a number of high-energy molecules causes the average kinetic energy of the molecules remaining in the liquid to decrease, and the temperature of the liquid falls proportionately. When liquids evaporate from an open container at room temperature, heat flows into the liquid from the surroundings to maintain the temperature of the liquid. Thus, the supply of high-energy molecules is replenished, and the process continues until all of the liquid has evaporated. The total quantity of heat required to vapourise a mole of liquid at a given temperature is called the molar heat of vapourisation of that liquid. For example,

$$H_2O(l) \rightarrow H_2O(g) \qquad \Delta H = +9.7 \text{ kcal}$$

The transfer of heat from the surroundings explains why a swimmer emerging from the water becomes chilled as the water evaporates from his skin. Likewise, the regulation of body temperature is, in part, accomplished by the evaporation of perspiration from the skin. Various cooling devices have made use of this principle. A water cooler of the Middle East consists of a jar of unglazed pottery filled with water. The water saturates the clay of the pottery and evaporates from the outer surface of the jar, thus cooling the water remaining in the jar.

The rate of evaporation increases as the temperature of a liquid is raised. When the temperature is increased, the average kinetic energy of the molecules increases, and the number of molecules with energies high enough for them to escape into the vapour phase constitutes an increased fraction of the total number of molecules (Fig. 8.2).

VAPOUR PRESSURE

When an evaporating liquid is confined in a closed container, the vapour molecules cannot escape from the vicinity of the liquid, and in the course of their random motion, some of the vapour molecules return to the liquid.

$$H_2O(l) \rightleftharpoons H_2O(g)$$

The rate of return depends upon the concentration of molecules in the vapour; the more vapour molecules per unit volume, the greater the chance that some of them will strike the liquid and be recaptured. Thus, a situation develops in which the vapourisation of the liquid is opposed by the condensation of the vapour. Eventually the system reaches a point where the rate of condensation equals the rate of vapourisation.

This condition, in which the rates of two opposite tendencies are equal, is called a state of equilibrium. At equilibrium, the concentration of molecules in the vapour state is constant because molecules leave the vapour through condensation at the same rate that molecules add to the vapour through vapourisation. Similarly, the quantity of liquid is a constant.

It is important to note that equilibrium does not imply a static situation; in any system, the numbers of molecules present in both phases are constant because the two opposing changes are taking place at equal rates and not because vapourisation and condensation have ceased. In a closed container, equilibrium is the only situation that can prevail for any substantial length of time. If the rate of vapourisation were greater than the rate of condensation for an extended period, all the substance would eventually become vapour. The reverse situation is equally improbable; all the material would end as liquid.

The pressure of vapour in equilibrium with a liquid at a given temperature is called the vapour pressure. If the temperature of a liquid-vapour system is constant, the vapour pressure will be constant. The pressure of the vapour depends upon the average kinetic energy of the vapour molecules (which is constant since the temperature is constant) and upon the concentration of vapour molecules. The concentration of vapour molecules, as well as their average kinetic energy, is fixed by temperature since the vapour concentration must be such that the rate of condensation is equal to the rate of vapourisation, and the rate of vapourisation is dictated by the temperature.

For any sample of a liquid at a given temperature, then, the equilibrium concentration of vapour is constant, regardless of the size of the sample or the dimensions of the container. The absolute quantity of vapour may not be the same from system to system, but the vapour concentration is. If the volume of vapour is changed, the consequent alteration of the vapour concentration causes the rate of condensation to increase or decrease until the concentration is re-established at its former (equilibrium) value. Hence, vapour pressure is fixed by temperature alone; volume changes do not alter it.

As the temperature is increased, the vapour pressure of a liquid increases. We have noted that an increase in temperature causes the rate of vapourisation to increase; equilibrium is possible only if the rate of condensation increases proportionately, and this will occur only when the concentration of vapour molecules becomes adjusted to a higher level. This factor and the augmented kinetic energy of the vapour molecules accompanying a rise in temperature account for the observed increase in vapour pressure.

Fig. 8.3 shows the temperature-vapour pressure curves for ethyl ether, ethyl alcohol, and water. The curve for each substance could be extended to the critical temperature of that substance; at this point the vapour pressure would equal the critical pressure. Above the critical temperature only one phase—the gas phase—can exist.

Liquids with weak intermolecular attractive forces have relatively high vapour pressures. At 20°C ethyl ether has a vapour pressure of 0.582 atm (or 422 torr), and water has a vapour pressure of 0.0230 atm (or 17.5 torr). These data indicate that the forces of attraction are stronger in water than in ethyl ether.

BOILING POINT

The temperature at which the vapour pressure of a liquid equals the atmospheric pressure is called the boiling point of the liquid. At this temperature, vapour produced in the interior of a liquid results in the bubble formation and turbulence characteristic of boiling. Bubble formation is impossible at temperatures below the boiling point; the atmospheric pressure on the surface of the liquid prevents the formation of bubbles with internal pressures that are less than atmospheric.

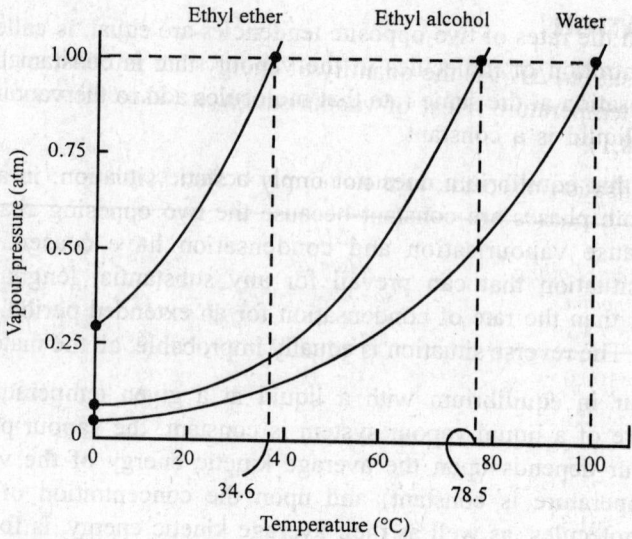

Fig. 8.3. Vapour pressure curves for water, ethyl alcohol, and ethyl ether.

The temperature of a boiling liquid remains constant until all the liquid has been vapourised. In an open container, the maximum vapour pressure that can be attained by any liquid is the atmospheric pressure. We have noted in the preceding section that the vapour pressure of a liquid is determined by temperature alone; hence, if the vapour pressure is fixed, the temperature is fixed. Heat must be added to a boiling liquid to maintain the temperature, because in the boiling process the high-energy molecules are lost by the liquid. If the rate of heat addition is increased above the minimum needed to maintain the temperature of the boiling liquid, the rate of boiling increases, but the temperature of the liquid does not rise.

The boiling point of a liquid changes with changes in external pressure. Water, for example, will boil at 98.6°C at a pressure of 0.950 atm (or 722 torr) and at 101.4°C at a pressure of 1.05 atm (or 798 torr). Only at a pressure of 1.00 atm (or 760 torr) will water boil at 100°C. The normal boiling point of a liquid is defined as the temperature at which the vapour pressure of the liquid equals 1 atm.

The normal boiling points of ethyl ether (34.6°C), ethyl alcohol (78.5°C), and water are indicated on the vapour pressure curves of Fig. 8.3. The boiling point of a liquid can be read from its vapour pressure curve by finding the temperature at which the vapour pressure of the liquid equals the prevailing pressure.

The fluctuations in atmospheric pressure at any one geographic location cause a maximum variation of about 2°C in the boiling point of water. The variations from place to place, however, can be greater than this. The average barometric pressure at sea level is 1.000 atm; at higher elevations, average barometric pressures are less. Thus, at an elevation of 5000 feet above sea level, the average barometric pressure is 0.836 atm; at this pressure, water boils at 95.1°C. Water boils at 90.1°C at 0.695 atm, which is the average atmospheric pressure at 10,000 feet above sea level.

If a liquid has a high normal boiling point or decomposes when heated, it can be made to boil at low temperatures by reducing the pressure. This procedure is followed in vacuum distillation; water can be made to boil at 10°C, which is considerably below room temperature, by adjusting the pressure to 0.0121 atm. Many food products are concentrated by removing unwanted water under reduced pressure; in these procedures, the product is not subjected to temperatures that bring about decomposition or discolouration.

HEAT OF VAPOURISATION

The molar heat of vapourisation, ΔH_v, is the quantity of energy that must be supplied to vapourise a mole of a liquid at a specified temperature. Heat of vapourisation are usually recorded at the normal boiling point in kcal/mol (Table 8.1).

Table 8.1. Heats of vapourisation of liquids at their normal boiling points.

Liquid	t_b Normal boiling point (°C)	ΔH_v Heat of vapourisation (kcal/mol)	ΔH_v Heat of vapourisation (kJ/mol)	$\Delta S_v = \Delta H_v / T_b$ Entropy of vapourisation (cal/°K mol)	$\Delta S_v = \Delta H_v / T_b$ Entropy of vapourisation (J/°K mol)
Water	100.0	9.72	40.7	26.0	109
Benzene	80.1	7.35	30.8	20.8	87
Ethyl alcohol	78.5	9.22	38.6	26.2	110
Carbon tetrachloride	76.7	7.17	30.0	20.5	86
Chloroform	61.3	7.02	29.4	21.0	88
Carbon disulphide	46.3	6.40	26.8	20.0	84
Ethyl ether	34.6	6.21	26.0	20.2	84

Already, a kinetic picture of the vapourisation process has been presented, and the heat of vapourisation was interpreted from this viewpoint. A slightly different interpretation based on a consideration of the heat content (enthalpy, H) of the vapour and of the liquid follows. The temperature of a vapour is the same as the temperature of the liquid with which it is in equilibrium; hence, the molecules of each phase have the same average kinetic energy. The phases, however, differ in total internal energy, which includes potential energy as well as kinetic energy. The molecules of the liquid are held together by cohesive forces; the molecules of the vapour are essentially free. When the liquid is converted into a gas, energy must be supplied to separate the molecules. Thus, the energy of the gas phase is higher than that of the liquid phase by the amount of this difference.

The difference in heat content of the phases (ΔH) takes into account another factor. The volume of a gas is considerably larger than the volume of the liquid from which it is derived (e.g., about 1700 ml of steam is produced by the vapourisation of 1 ml of water at 100°C); energy must be supplied to do the work of pushing back the atmosphere to make room for the vapour. The heat of vapourisation includes both the energy required to overcome the intermolecular cohesive forces and the energy needed to expand the vapour.

When a mole of vapour is condensed to a liquid, the difference in heat content of the phases is released rather than absorbed. In this instance, the heat effect is called the molar heat of condensation; it has a respective sign but is numerically equal to the molar heat of vapourisation at the same temperature.

The heat of vapourisation of a given liquid decreases as the temperature increases and equals zero at the critical temperature of the substance. This parallels an increase in the fraction of high-energy molecules; at the critical temperature all the molecules have sufficient energy to vapourise.

Over a relatively narrow temperature range, the heat of vapourisation can be considered to be constant. Under such conditions, the vapour pressure of a liquid, p (in atm), is related to the temperature at which it is measured, T (in °K), by the equation

$$\log p = -\frac{\Delta H_v}{2.303RT} + C$$

where ΔH_v is the molar heat of vapourisation (in cal/mol, or J/mol), R is the gas constant (1.987 cal/°K mol, or 8.314 J/°K mol), and C is a constant characteristic of the liquid under study. The vapour pressure-temperature curves shown in Fig. 8.3 can be mathematically described by fitting appropriate values into this equation.

If we wish to compare the vapour pressure of a given liquid, p_1, at one temperature, T_1, to the vapour pressure of the same liquid, p_2, at a second temperature, T_2, we can derive a very useful equation.

At T_2: $\log p_2 = -\dfrac{\Delta H_v}{2.303R}\left(\dfrac{1}{T_2}\right) + C$

At T_1: $\log p_1 = -\dfrac{\Delta H_v}{2.303R}\left(\dfrac{1}{T_1}\right) + C$

By substracting the second equation from the first, we get

$$\log p_2 - \log p_1 = -\frac{\Delta H_v}{2.303R}\left(\frac{1}{T_2} - \frac{1}{T_1}\right)$$

which can be rearranged to

$$\log\left(\frac{p_2}{p_1}\right) = \frac{\Delta H_v}{2.303R}\left(\frac{T_2 - T_1}{T_1 T_2}\right)$$

This equation, known as the Clausius-Clapeyron equation, was first proposed by Benoît Clapeyron and later derived from thermo-dynamic theory of Rudolf Clausius.

Example 8.1. The normal boiling point of chloroform is 334°K. At 328°K the vapour pressure of chloroform is 0.823 atm. What is the heat of vapourisation for this temperature range?

Solution.

If we set $T_2 = 334$°K, then $p_2 = 1.000$ atm; $T_1 = 328$°K, and $p_1 = 0.823$ atm.

$$\log\left(\frac{p_2}{p_1}\right) = \frac{\Delta H_v}{2.303R}\left(\frac{T_2 - T_1}{T_1 T_2}\right)$$

$$\log\left(\frac{1.000 \text{ atm}}{0.823 \text{ atm}}\right) = \frac{\Delta H_v}{(2.303)(1.987 \text{ cal/°K mol})}\left[\frac{334°\text{K} - 328°\text{K}}{(328°\text{K})(334°\text{K})}\right]$$

$$\Delta H_v = 7070 \text{ cal/mol}$$

$$= 7.07 \text{ kcal/mol}$$

Example 8.2. The vapour of carbon disulphide at 301°K is 0.526 atm. What is the vapour pressure of CS_2 at 273°K? The heat of vapourisation over this temperature range may be taken as 6.60 kcal/mol.

Solution. We set $T_2 = 301°K$, $p_2 = 0.526$ atm, and $T_1 = 273°K$. Thus,

$$\log\left(\frac{p_2}{p_1}\right) = \frac{\Delta H_v}{2.303R}\left(\frac{T_2 - T_1}{T_1 T_2}\right)$$

$$\log\left(\frac{0.526 \text{ atm}}{p_1}\right) = \frac{6,600 \text{ cal / mol}}{(2.303)(1.987 \text{ cal/°K mol})}\left[\frac{301°K - 273°K}{(273°K)(301°K)}\right]$$

$$= 0.491$$

$$\frac{0.526 \text{ atm}}{p_1} = 3.10$$

$$p_1 = 0.170 \text{ atm}$$

In general, the higher the heat of vapourisation of a substance, the stronger are the intermolecular forces of attraction. Frederick Trouton a scientist discovered that for many liquids the molar heat of vapourisation (at the normal boiling point) divided by the normal boiling point (on the Kelvin scale) is a constant: 21 cal/°K mol (Trouton's rule). Thus,

$$\frac{\Delta H_v}{T_b} = 21 \text{ cal/°K mol}$$

This rule is only approximate, as the values listed in Table 8.2. attest.

The value $(\Delta H_v/T_b)$ of Trouton's rule is called the molar entropy of vapourisation and is given the symbol ΔS_v. The entropy, S, of a system is a measure of the system's randomness, or disorder. A positive change in entropy, ΔS, indicates an increase in disorder; a negative value of ΔS indicates that the system has become more ordered.

The change from liquid to vapour involves an increase in molecular randomness. The extent of this increase is about the same for all non-polar liquids; therefore, the values of ΔS_v are about the same, 21 cal/°K mol. In water and ethyl alcohol, the intermolecular forces of attraction are unusually strong; hence, ΔS_v values for these liquids are comparatively large (Table 8.2), since they represent changes for liquids that are more highly ordered than usual.

Table 8.2. Heats of fusion of solids at the melting point.

Solid	t_f Melting point (C°)	ΔH_f Heat of fusion (kcal/mol)	ΔH_f Heat of fusion (kJ/mol)	$\Delta S_f = \Delta H_f/T_f$ Entropy of fusion (cal/°K mol)	$\Delta S_f = \Delta H_f/T_f$ Entropy of fusion (J/°K mol)
Water	0.0	1.44	6.02	5.26	22.0
Benzene	5.5	2.35	9.83	8.44	35.3
Ethyl alcohol	−117.2	1.10	4.60	7.10	29.7
Carbon tetrachloride	−22.9	0.60	2.51	2.40	10.0
Chloroform	−63.5	2.20	9.20	10.50	43.9
Carbon disulphide	−112.1	1.05	4.39	6.52	27.3

The entropy of vapourisation of a given liquid, like the heat of vapourisation, decreases as the temperature is increased. With increasing temperature, the molecular randomness of the liquid phase increases and approaches that of the gas phase; at the critical temperature they are equal, only one phase exists, $\Delta H_v = 0$, and $\Delta S_v = 0$.

FREEZING POINT

As a liquid is cooled, the molecules move more and more slowly. Eventually a temperature is reached at which some of the molecules have sufficiently low kinetic energies to allow the intermolecular attractions to hold them in a crystal lattice; the substance then starts to freeze. Gradually the low-energy molecules assume positions in the crystal lattice; the molecules remaining in the liquid have a higher temperature because of the loss of these low-energy molecules. Heat must be removed from the liquid to maintain the temperature.

The normal freezing point of a liquid is the temperature at which solid and liquid are in equilibrium under a total pressure of 1 atm. At the freezing point, the temperature of the solid-liquid system remains constant until all of the liquid is frozen. The quantity of heat that must be removed to freeze a mole of a substance at the freezing point is called the molar heat of crystallisation. This quantity represents the difference in the heat content (ΔH) between the liquid and the solid.

At times the molecules of a liquid, as they are cooled, continue the random motion characteristic of the liquid state at temperatures below the freezing point; such liquids are referred to as undercooled or supercooled. These systems can usually be caused to revert to the freezing temperature and the stable solid-liquid equilibrium by scratching the interior walls of the container with a stirring rod or by adding a seed crystal around which crystallisation can occur. The crystallisation process supplies heat, and the temperature is brought back to the freezing point until normal crystallisation is complete.

Some supercooled liquids can exist for long periods, or even permanently, in this state. When these liquids are cooled, molecules solidify in a random arrangement typical of the liquid state rather than in an orderly geometric pattern of a crystal. Substances of this type have relatively high viscosities and generally possess complex molecular forms for which crystallisation would be difficult. They are frequently called amorphous solids, vitreous materials, or glasses; examples include glass, tar, and certain plastics. Some materials (e.g., glass) can be obtained in the crystalline form by careful cooling and change very slowly to the crystalline form from the supercooled state. Amorphous solids have no definite freezing or melting point; rather, these transitions take place over a temperature range. They exhibit conchoidal fracture (the fragments have curved, shell-like surfaces) rather than the cleavage along definite planes at definite angles shown by crystalline materials.

When a crystalline substance is heated, the temperature at which solid-liquid equilibrium is attained under air at 1 atm pressure is called the melting point; it is, of course, the same temperature as the freezing point of the substance. The quantity of heat that must be added to melt a mole of the material at the melting point is called the molar heat of fusion and is numerically equal to the heat of crystallisation.

The molar heats of fusion and molar entropies of fusion of several substances are listed in Table 8.2. The transition from solid to liquid represents an increase in disorder (and consequently in entropy) in the same way that the transition from liquid to gas does. Entropies of fusion, however, are generally of a lower order than entropies of vapourisation because molecular disorder is greater in the gaseous state than in the liquid or solid states. The variation in ΔS_f values from substance to substance is indicative of the differences in the lattice energies of the crystalline solids; ΔS_f values are not nearly so constant as ΔS_t values..

VAPOUR PRESSURE OF A SOLID

Molecules in a crystal vibrate about their lattice positions. A distribution of kinetic energy exists among these molecules similar to that for liquids and gases but on a lower level. Energy is transmitted from molecule to molecule within a crystal; the energy of any one molecule, therefore, is not constant. High-energy molecules on the surface of the crystal can overcome the attractive forces of the crystal and escape into the vapour phase. If the crystal is in a closed container, equilibrium is eventually reached in which the rate of the molecules leaving the solid equals the rate of the vapour molecules returning to the crystal. The vapour pressure of a solid at a given temperature is a measure of the number of molecules in the vapour at equilibrium. Every solid has a vapour pressure, although some pressures are very low. The magnitude of the vapour pressure is inversely proportional to the strength of the attractive forces; thus, ionic crystals have very low vapour pressures.

Since the ability of molecules to escape the intermolecular forces of attraction depends upon their kinetic energies, the vapour pressure of a solid increases as the temperature increases. The temperature-vapour pressure curve for ice is illustrated in Fig. 8.4. This curve intersects the vapour pressure curve for water at the freezing point. In the absence of air, the normal freezing point of water (1 atm total pressure) is 0.0025°C. In air, however, and under a total pressure of 1 atm, the freezing point of water is 0.0000°C, which is the commonly reported value. The difference in freezing point is caused by the presence of dissolved air in the water. The vapour pressures plotted in Fig. 8.4 are the partial pressures of H_2O in air with the total pressure equal to 1 atm. Freezing points are usually determined in air; however, in any event, any change in freezing point of a given substance caused by the presence of air is generally very small.

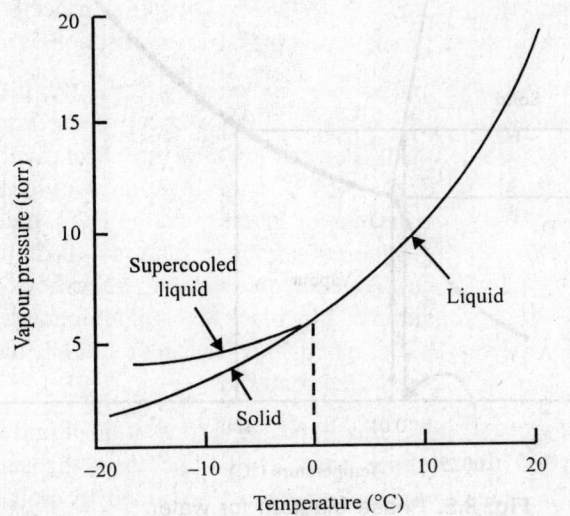

Fig. 8.4. Vapour pressure curves for ice and water near the freezing point. Vapour pressures are partial pressures of H_2O under air at a total pressure of 1 atm.

At the freezing point, the vapour pressures of solid and liquid are equal. If this were not true, one form would gradually be converted into the other. For example, if the vapour pressure of the liquid were higher than that of the solid, the vapour concentration would be higher than the equilibrium value for the solid-vapour system.

$$\text{solid} \rightleftharpoons \text{vapour}$$

In such a case, the rate of crystallisation would increase, the amount of solid would increase, and the concentration of vapour would decrease. This decrease in the concentration of vapour would cause more of the liquid to vapourise in order to re-establish the equilibrium vapour pressure of the liquid.

$$\text{liquid} \rightleftharpoons \text{vapour}$$

The process would continue until all the liquid was converted into solid. At a given temperature, the stable form is the one that has the lower vapour pressure.

PHASE DIAGRAMS

The temperature-pressure phase diagram for water conveniently illustrates the conditions under which water can exist as solid, liquid, or vapour, as well as the conditions that bring about changes in the state of water. Fig. 8.5 is a schematic representation of the water system; it is not drawn to scale, and some of its features are exaggerated to give prominence to important details. Every substance has a unique phase diagram that describes only systems in equilibrium; this diagram must be derived from experimental observations.

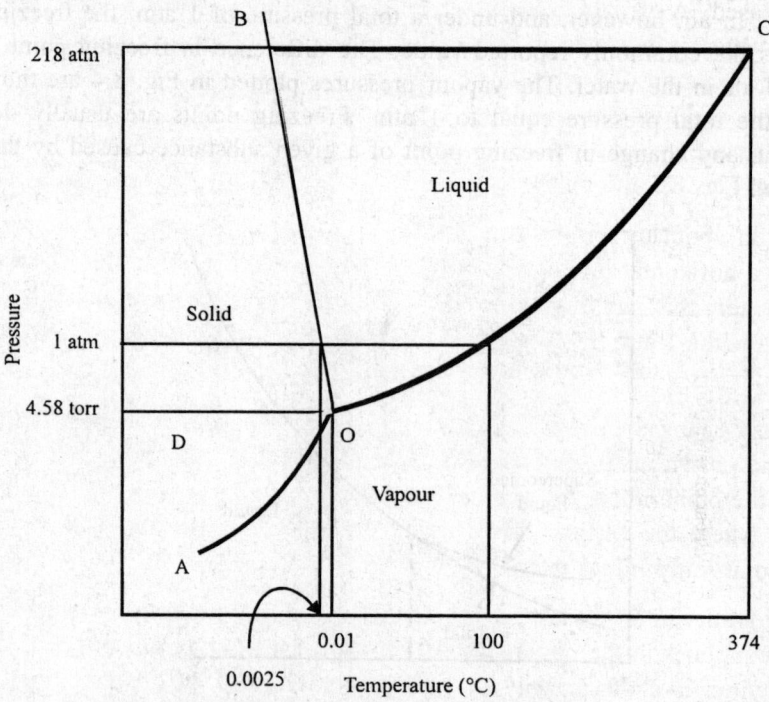

Fig. 8.5. Phase diagram for water.

The diagram of Fig. 8.5 relates to what is called a one-component system; that is, it pertains to the behaviour of water in the absence of any other substance. Thus, the total pressure of any system described by the diagram cannot be due in any part to the pressure of a gas other than water vapour. The vapour pressure curves plotted in Fig. 8.4 (measured in air under a constant total pressure of 1 atm) therefore deviate slightly, but only very slightly, from the vapour pressure curves of Fig. 8.5 (for which the vapour pressure of water is the total pressure). The easiest way to interpret the phase diagram for water is to visualise the total pressure acting on a system in mechanical terms, for example, as a piston acting on the material comprising the system contained in a cylinder.

In Fig. 8.5 curve OC is a vapour pressure curve for liquid and terminates at the critical point, C. Any point on this line describes a set of temperature and pressure conditions under which liquid and vapour can exist in equilibrium. The extension DO is the curve for supercooled liquid; systems between liquid and vapour described by points on this line are metastable. (The term metastable is applied to systems that are not in the most stable possible at the temperature in question). Curve AO is a vapour pressure curve for solid and represents a set of points that describe the possible temperature and pressure conditions for solid-vapour equilibria. The line BO, the melting point curve, represents conditions for equilibria between solid and liquid.

These curves intersect at point O, the triple point, solid, liquid, and vapour can exist together in equilibrium under the conditions represented by this point: 0.01°C (273.16°K) and a pressure of 0.00603 atm (or 4.58 torr) due to the vapour. When a point described by the temperature-pressure coordinates falls in one of the regions labelled solid, liquid, or vapour, a situation is indicated in which only one phase can exist under the conditions described.

The slope of the melting point (or freezing point) curve, BO, shows that the freezing point decreases as the pressure is increased. A variation of this type is observed for only a few substances such as antimony, bismuth, and water; it indicates an unusual situation in which the liquid expands upon freezing. At 0°C a mole of water occupies 18 cm^3, and a mole of ice occupies 19.63 cm^3. Thus, there is an expansion when one mole of liquid water freezes into ice. An increase in pressure on the system would oppose this expansion and the freezing process. Hence, the freezing point of water is lowered as the total pressure is increased. In Fig. 8.5 the slope of the line BO is exaggerated.

Phase changes brought about by temperature changes at constant pressure may be read from a phase diagram by interpreting a horizontal line drawn at the reference pressure (like the line drawn at 1.00 atm in Fig. 8.5). The point where this line intersects curve BO indicates the normal melting point (or freezing point), and the point where the 1 atm line intersects curve CO represents the normal boiling point. Beyond this point, only vapour exists.

In like manner, phase changes brought about by pressure changes at constant temperature may be read from a vertical line drawn at the reference temperature. If the pressure is increased, for example, at 0.0025°C (Fig. 8.5), the point where the vertical line crosses AO is the pressure where vapour changes to solid, and the point where the vertical line crosses BO represents the pressure where solid changes to liquid. Above this point, only liquid exists.

For materials that contract upon freezing (i.e., the solid phase is more dense than the liquid phase), the freezing-point curve inclines in the opposite direction, and the freezing point increases as the pressure is increased. This behaviour is characteristic of most substances, and the freezing-point curves of most phase diagrams slant to the right as is seen in the phase diagram for carbon dioxide (Fig. 8.6).

The process in which a solid goes directly into a vapour without going through the liquid state is known as sublimation; this process is reversible. The phase diagram for carbon dioxide is typical for substances that sublime at ordinary pressures rather than melt and then boil. The triple point of the carbon dioxide system is –55.6 at a pressure of 5.11 atm. Liquid carbon dioxide exists only at pressures greater than 5.11 atm. When solid carbon dioxide (dry ice) is heated at 1 atm pressure, it is converted directly into gas at –78.5°C; the molar heat of sublimation is the heat that must be added to a mole of solid to convert it directly into a gas.

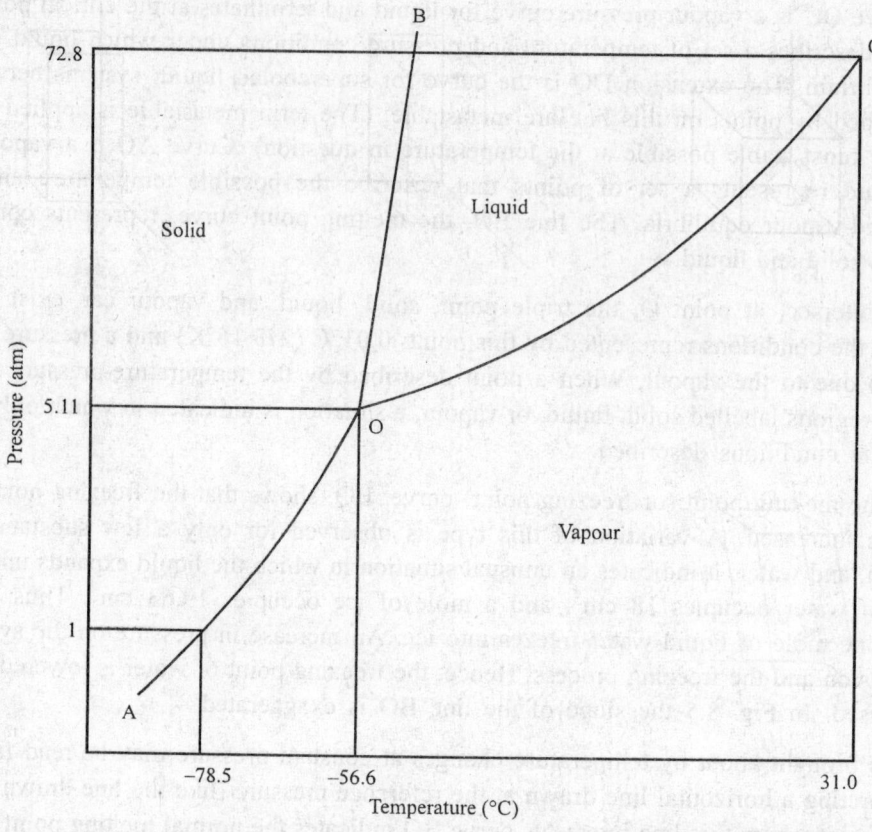

Fig. 8.6. Phase diagram for carbon dioxide.

CRYSTALS

The constituent particles of a crystal are arranged in a repeating three-dimensional pattern called a crystal (or space) lattice. The smallest section of a crystal lattice that can be used to describe the lattice is the unit cell. A crystal can be reproduced, in theory, by stacking its unit cells in three dimensions.

Crystal lattices may be grouped according to symmetry into six crystal systems (Fig. 8.7). A crystal system may be described by the dimensions of a unit cell along its three axes (a, b, c) and the sizes of the three angles between the axes (α, β, γ).

The hexagonal system is sometimes divided into two symmetrically related systems: hexagonal and rhombohedral. Crystals have the same symmetry as their constituent unit cells.

In illustrations of crystal lattices, points are customarily used to indicate the centres of the ions, atoms, or molecules. Six points lattices can be drawn by placing points at the corners of the unit cells shown in Fig. 8.7. However, it is possible to have points at positions other than the corners. Thus, three cubic lattices are known (Fig. 8.8): simple cubic, body-centered cubic, and face-centered cubic. Altogether, there are 14 possible lattices in which each point is surrounded in an identical manner by other points.

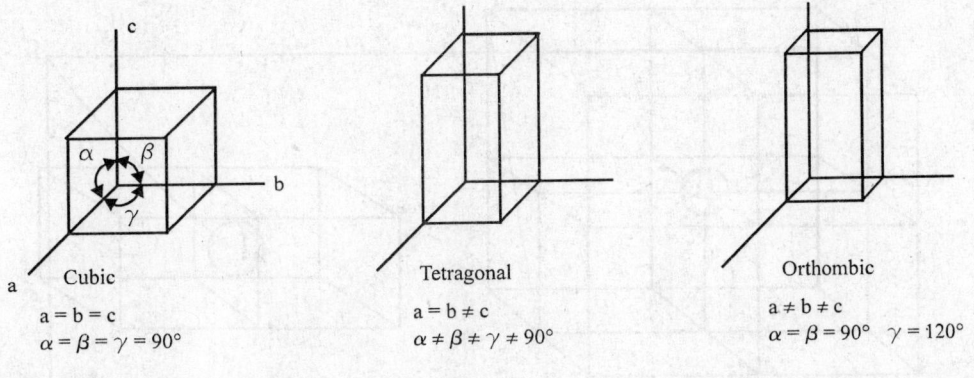

Cubic
a = b = c
$\alpha = \beta = \gamma = 90°$

Tetragonal
a = b ≠ c
$\alpha \neq \beta \neq \gamma \neq 90°$

Orthombic
a ≠ b ≠ c
$\alpha = \beta = 90° \quad \gamma = 120°$

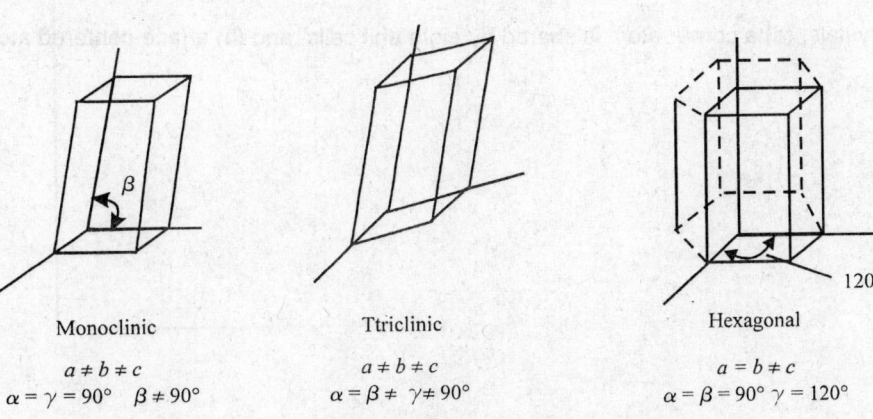

Monoclinic
$a \neq b \neq c$
$\alpha = \gamma = 90° \quad \beta \neq 90°$

Ttriclinic
$a \neq b \neq c$
$\alpha = \beta \neq \gamma \neq 90°$

Hexagonal
$a = b \neq c$
$\alpha = \beta = 90° \quad \gamma = 120°$

Fig. 8.7. Crystal systems.

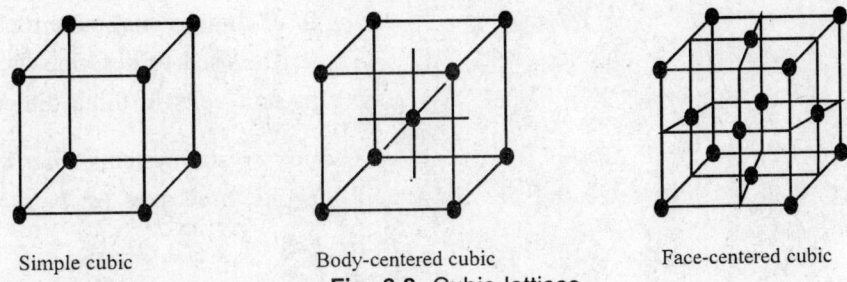

Simple cubic   Body-centered cubic   Face-centered cubic

Fig. 8.8. Cubic lattices.

In counting the number of atoms per unit cell, one must keep in mind that atoms on corners or faces are shared with adjoining cells. Eight unit cells share each corner atom, and two unit cells share each face-centered atom (Fig. 8.9).

The simple cubic unit cell, therefore, contains the equivalent of only one atom (8 corners at 1/8 each). The body-centred unit cell contains two atoms (8 corners at 1/8 each and one unshared atom in the centre). The face-centred unit cell contains the equivalent of four atoms (8 corners at 1/8 each and 6 face-centered atoms at 1/2 each).

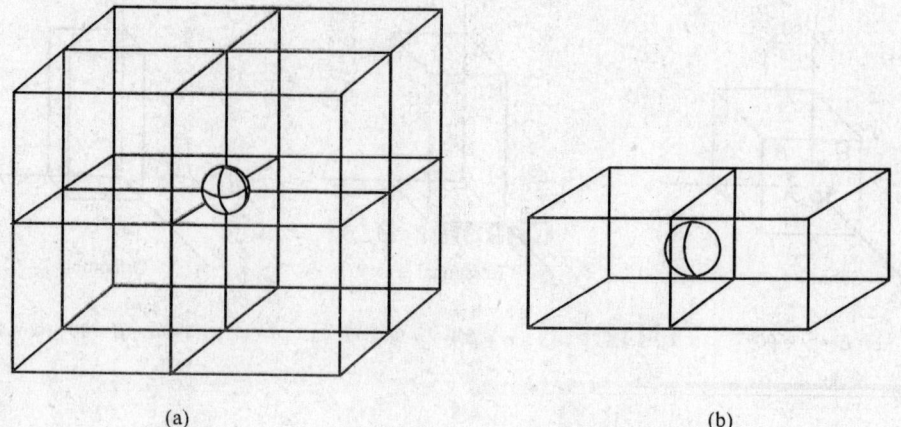

(a)   (b)

Fig. 8.9. In cubic crystals, (a) a corner atom is shared by eight unit cells, and (b) a face-centered atom is shared by two unit cells.

Chapter 9

Chemical Bonding

INTRODUCTION

The electrons of atoms are responsible for chemical bonding. Ionic (also called electrovalent) bonding is characterised by electron transfer; one of the reacting atoms loses one or more electrons, and the other atom gains one or more electrons. In covalent bonding, electrons are not transferred but are shared; a covalent bond consists of a pair of electron shared by two atoms. Pure ionic bonding and pure covalent bonding are seldom encountered. Most bonds have intermediate character, although many bonds are predominantly either ionic or covalent. An understanding of electronic configuration, a principal topic is fundamental to an understanding of chemical bonding. Other properties of the elements, such as atomic size, are also important in discussion of chemical bonding.

ATOMIC SIZES

Determination of atomic sizes poses a problem. If the atom is viewed as a sphere, the radius of the atom should be the distance from the centre of the nucleus to the outer reaches of the last electron. But the electron cloud of an atom has a varying intensity, and the probability of finding an electron extends to infinity. It is impossible to isolate and measure a single atom.

It is possible, however, to measure the distance between the nuclei of two covalently bonded atoms (the so-called bond distance). Atomic (also called covalent) radii are determined from these bond distances by apportioning them between the bonded atoms. For example, one-half the Cl—Cl bond distance, 1.98 Å, gives a value of 0.99 Å for the atomic radius of chlorine. In turn, the atomic radius of chlorine, 0.99 Å, can be subtracted from the C—Cl bond distance, 1.76 Å, to derive the atomic radius of carbon, 0.77 Å. Since changing the environment of an atom may cause its effective size to vary, discrepancies sometimes arise when this method of assigning atomic radii is used. Nevertheless, data of this type are useful in establishing trends and making generalisations.

Atomic radius is plotted against atomic number in Fig. 9.1. Within a group of the periodic table, an increase in atomic radius is generally observed from top to bottom. The alkali metals (group I A) and the halogens (group VII A) are labelled in Fig. 9.1, and the increase in size within each group is clearly evident. This trend is expected since the larger atoms of a group employ more electron levels than the smaller atoms.

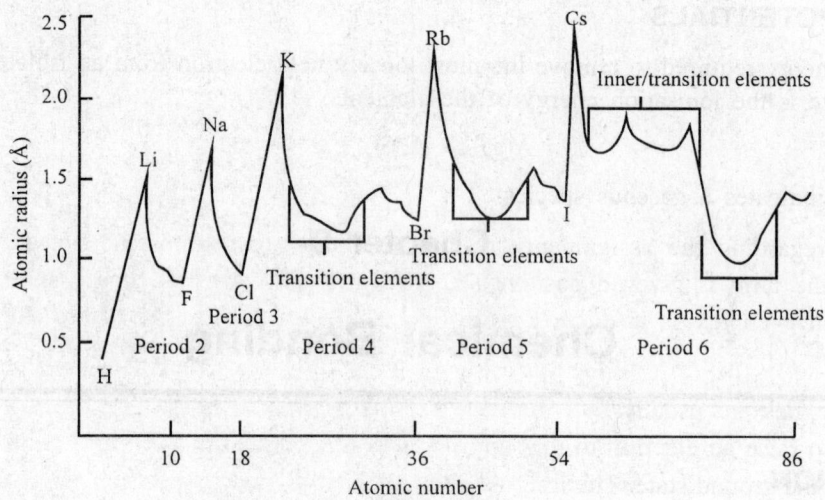

Fig. 9.1. Atomic radii of the elements.

Fig. 9.1. also shows a general decrease in atomic radius from left to right across any period. The outermost electrons of the elements of a given period all have the same principal quantum number. As the atomic number increases from element to element across a period, the additional electron is added to the same level or to an inner level. Since the positive charge on the nucleus also increases (the number of protons increases), the nucleus draws in the electron shells and the atoms decrease in size as a result.

The transition elements and inner-transition elements show some minor variations. For the transition elements the differentiating electrons fill inner d orbitals. The effect of nuclear charge on outer, size-determining electrons is reduced by the screening effect of inner electrons. For a transition series, therefore the gradual build up of electrons in inner d orbitals at first retards the rate of decrease in atomic radius and then, toward the end of the series when the inner d sub-shell nears completion, causes the radius to increase.

In general, the lanthanides (elements 58 to 71) exhibit a slow but significant decrease in atomic radius called the *lanthanide contraction*. The differentiating electrons of these elements are added to the third level from the outside (the $4f$ sub-level). These electrons screen the increasing nuclear charge from the outer $6s$ electrons but are not completely efficient, and therefore the steady reduction in size is observed. The transition elements following the lanthanides exhibit a typical transition-element pattern; however, the effect of the lanthanide contraction causes these elements to have approximately the same size as the corresponding elements of the preceding period. Thus, hafnium (Z = 72) has an atomic radius of 1.44 Å; and zirconium (Z = 40), 1.45 Å. The general trends in atomic radius are summarised in Fig. 9.2.

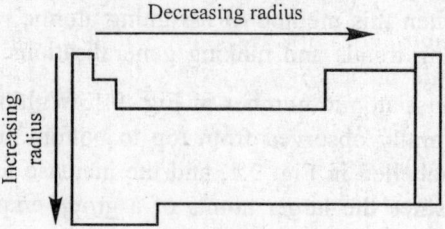

Fig. 9.2. General trends in atomic radius in relation to the periodic classification.

IONISATION POTENTIALS

The amount of energy required to remove the most loosely held electron from an isolated gaseous atom in its ground state is the ionisation energy of the element.

$$A(g) \rightarrow A^+(g) + e^-$$

The symbol (g) indicates a gaseous species.

Conventions regarding the assignment of signs to energy terms must be noted. When a system absorbs energy, the term is given a positive sign. If the system evolves energy, the value is given a negative sign. Since energy is absorbed in the ionisation process described, ionisation energies have positive signs.

Ionisation energies are usually determined by spectroscopic methods. The ionisation energy is the energy required to raise an electron to this limit, or it is the energy emitted when an electron falls from this limit back to the ground state. The limit is readily identified from a spectrum since beyond this point, in the direction of shorter wavelengths, the spectrum no longer consists of lines but is continuous. An electron is not restricted to quantised energy states after it has been removed from the atom and hence produces a continuous spectrum.

The ionisation potential of an element is the minimum potential (measured in volts) required to effect electron removal. The ionisation potential is numerically equal to the ionisation energy expressed in electron volts (eV), which are energy units. One electron volt is the energy acquired by an electron when it falls through a potential difference of 1 V; 1 eV equals 3.829×10^{-20} cal. For one mole (6.022×10^{23}) of electrons, 1 eV is 23.06 kcal/mol. Ionisation energies in electron volts are customarily called ionisation potentials, even though this terminology is not strictly correct.

Certain atoms produce negative ions by the addition of one or more electrons. The terms ionisation energy and ionisation potential pertain only to the production of positive ions by the removal of an electron. Some of the positive ions thus described (such as ions derived from the noble gases or from elements that commonly form only negative ions) are never produced in ordinary chemical reactions.

The factors that influence the magnitude of an ionisation potential are those that control the orbital energies. In Fig. 9.3 the ionisation potentials of the elements are plotted against atomic number. The ionisation potential of helium is larger than that of hydrogen since the nuclear charge of He (2+) is twice that of H (1+) and the electron removed in each case is a 1s electron. The ionisation potential of lithium, however, is much lower than that of He despite the fact that the nuclear charge of Li (3+) is higher than that of He (2+). In He, one electron is not effectively screened from the nuclear charge by the other since they both occupy the same subshell; however, the Li atom is larger than the He atom and has two 1s electrons that strongly shield the nuclear charge from the electron that is removed (a 2s electron).

In the second period (Li to Ne), the nuclear charge increases from 3+ to 10+ and the atomic radius decreases. Since electrons are being added to the same shell (the outer shell), the extent to which they augment the screening effect of the 1s electrons is slight, and the increasing nuclear charge causes the ionisation potential to increase. In general, ionisation potential increases across a period paralleling an increase in nuclear charge and a decrease in atomic radius (Fig. 9.4).

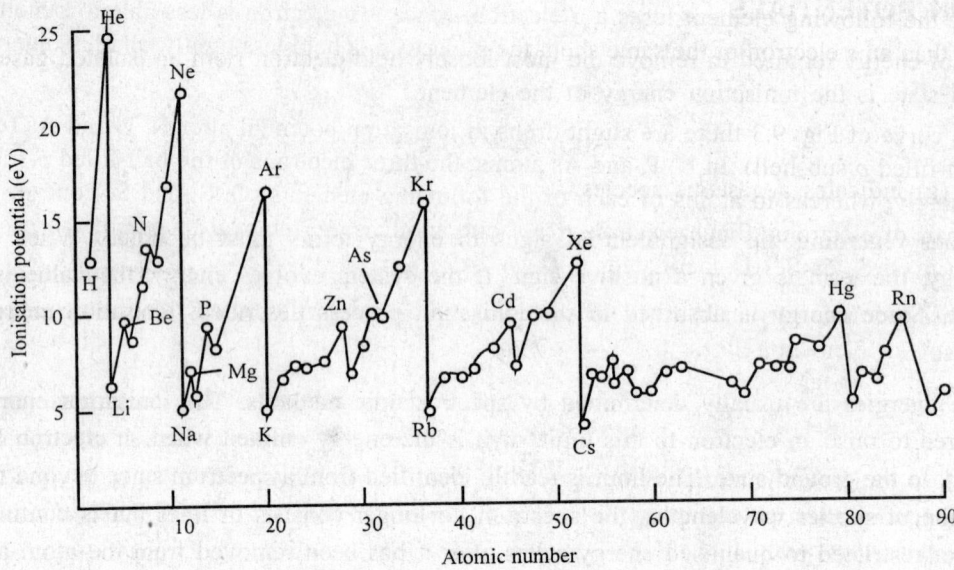

Fig. 9.3. First ionisation potentials of the elements versus atomic number.

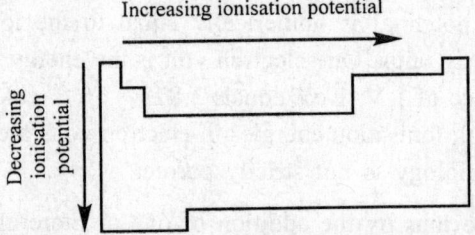

Fig. 9.4. General trends in ionisation potential in relation to the periodic classification.

The ionisation potentials of the transition elements do not increase across a period as rapidly as those of the representative elements; the ionisation potentials of the inner-transition elements remain almost constant. Therefore, these elements have low enough ionisation potentials to react as metals—through electron loss. In these series, electrons are being added to inner shells, and the resulting increase in screening accounts for the effects noted.

For the representative elements, the ionisation potential decreases as we go from the lighter to the heavier members of a group—notice the values for the noble gases (group 0) or the alkali metals (group I A) in Fig. 9.3. The effect of increasing nuclear charge is greatly reduced by the screening effect of the increasing number of inner electrons, and ionisation potential decreases as atomic radius increases (Fig. 9.4).

Several features of the graph of Fig. 9.3 are significant because they give insights into how certain electronic configurations affect the reactivity of atoms. The high ionisation potentials of the noble gases are very stable; it is difficult to remove an electron from these "closed" configurations.

The elements Be, Mg, Zn, Cd, and Hg (which have filled sub-shells) have ionisation potentials that are high in comparison to the ionisation potentials of the elements that follow them in the same period. In the ionisation of each of the elements listed, an electron is removed from a filled s subshell, and in each case the following element loses a p electron. Since a p electron is less able to penetrate screening electrons than an s electron of the same shell, the p electron is held less tightly, and less energy is required to remove it.

In the curve of Fig. 9.3 there are slight drops in ionisation potential after N, P, and As (each of which has a half-filled p subshell). In N, P, and As atoms, the three electrons of the half-filled p subshell occupy orbitals singly, whereas in atoms of each of the following elements (O, S, and Se) one of the p orbitals holds a pair of electrons. Since electrons repel one another, the removal of one of the p electrons from the pair (in O, S, or Se) is facilitated.

The discussion so far has been concerned only with first ionisation potentials. The second ionisation potential of an element relates to the energy required to remove one electron from a 1+ ion of that element.

$$A^+(g) \rightarrow A^{2+}(g) + e^-$$

The third ionisation potential pertains to a process in which one electron is removed from a 2+ ion. Even higher ionisation potentials may be determined. When the term ionisation potential is not further qualified, it is understood that the first ionisation potential is meant.

Predictably, for any given element, the third ionisation potential is higher than the second, and the second is higher than the first. It is more difficult to remove an electron (1–charge) from a 2+ ion than from a 1+ ion; likewise, it is more difficult to remove an electron from a 1+ ion than from a neutral atom. The high values observed for third and fourth ionisation potentials indicate that few highly positive ions exist under ordinary conditions.

In terms of chemical reactivity, remember that ionisation potential relates to the energy required to remove an electron. Thus, the elements most active in losing electrons (metallic behaviour) are those with the lowest ionisation potentials and are found in the lower left of the periodic table. This reactivity in terms of electron loss decreases as one moves upward or to the right from this corner of the chart.

ELECTRON AFFINITIES

The energy effect accompanying the process in which an electron is added to an isolated atom in its ground state is called an electron affinity.

$$e^- + A(g) \rightarrow A^-(g)$$

In this process, energy is usually evolved, and therefore most electron affinities have negative signs. Electron affinity, however, is frequently defined as energy released; in sources that employ such a definition, electron affinity values that relate to the liberation of energy are given positive signs. This sign convention is the reverse of the one that we employ; we shall consistently use a negative sign to indicate that the total energy of the system decreases and a positive sign to indicate that the system absorbs energy.

The direct determination of electron affinity is difficult and has been accomplished for only a few elements. Other values have been determined indirectly by calculation from thermodynamic data. Electron affinity values, therefore, are available for only a limited number of elements, and many of these values are not highly accurate. Some electron affinities are listed in Table 9.1. Notice that some of these values have positive signs indicating that work must be done to force the atom under consideration to accept an additional electron.

Table 9.1. Electron affinities (eV).

H −0.77								He +0.6^a
Li −0.6^a	Be +0.6^a	B −0.2^a	C −1.25	N +0.3^a	O −1.47 (+7.28)^b	F −3.45		Ne +1.0^a
Na −0.2^a	Mg +0.3^a	Al −0.3^a	Si −1.4^a	P −0.6^a	S −2.07 (+3.44)^b	Cl −3.61		
						Br −3.36		
						I −3.06		

^{a} Calculated values.

^{b} Values in parentheses are calculated and pertain to the total energy effects for the addition of two electrons.

A small atom should have a greater tendency to add an electron than a large atom since the added electron is, on the average, closer to the positively charged nucleus in a small atom. This trend is roughly followed from left to right in any period. In the second period, however, exceptions may be noted for beryllium (filled $2s$ subshell), nitrogen (half-filled $2p$ subshell), and neon (all subshells filled). Similar exceptions may be noted for analogous elements of the third period. Elements with relatively stable electronic configurations do not accept additional electrons readily. Since the electronic configuration of each of the group VII A elements is one electron short of a noble gas configuration, each element of group VII A has the greatest tendency toward electron gain of any element in its period.

The values for the halogens illustrate the typical group trend. Electron attracting ability, except for that of fluorine, tends to increase from bottom to top within the group in conformity with decreasing atomic size. The effect of small size, however, may be partially negated by the repulsion of electrons already present in the atom. The exception noted for fluorine, as well as for other second period elements, may be, in part, explained by these repulsions. The concentration of negative charge in a very small shell is greater than the concentration obtained when the same number of electrons is placed in a larger shell.

Some second electron affinities have been determined. These values refer to processes in which an electron is added to a negative ion: for example,

$$e^- + O^-(g) \rightarrow O^{2-}(g)$$

Since there is an electrostatic repulsion between a negative ion and an electron, energy is required, not released. All second electron affinities have positive signs.

IONIC BOND

When a metal reacts with a non-metal, electrons are transferred from the atoms of the metal to the atoms of the non-metal, and an ionic (or electrovalent) compound is produced. For example, a sodium atom loses an electron to a chlorine atom in the reaction of these two elements, and charged particles (called ions) result. Sodium has a comparatively low ionisation potential; relatively little energy is required to remove an electron from a sodium atom. The sodium ion that forms has a 1+ charge since the sodium nucleus contains 11 protons (11+ charge) and the ion has only 10 electrons (one having been lost). On the other hand, the value for the electron affinity of chlorine indicates that the chlorine atom has a strong tendency toward electron gain. The chloride ion that forms has a strong tendency toward electron gain. The chloride ion that forms has a 1– charge since the chlorine nucleus contains 17 protons (17+ charge) and the ion has 18 electrons (one having been gained). Positive ions are called *cations*, and negative ions are called *anions*. These names are derived from the terminology of electrochemistry.

The reactions and compounds of the A family elements are frequently depicted by using the symbols of the elements under consideration together with dots to indicate valence electrons—the only electrons involved in the chemical reactions of these elements.

$$Na\cdot + \cdot\ddot{\underset{\cdot\cdot}{Cl}}: \rightarrow Na^+ + :\ddot{\underset{\cdot\cdot}{Cl}}:^-$$

The complete electronic configurations of the atoms and ions of this reaction are

$$Na\ (1s^2\ 2s^2\ 2p^6\ 3s^1) \rightarrow Na^+\ (1s^2\ 2s^2\ 2p^6) + e^-$$

$$e^- + Cl\ (1s^2\ 2s^2\ 2p^6\ 3s^2\ 3p^5) \rightarrow Cl^-\ (1s^2\ 2s^2\ 2p^6\ 3s^2\ 3p^6)$$

The sodium ion has an electronic configuration identical to that of neon, and the chloride ion has the same configuration as argon. The ions may be said to be isoelectronic (the same in electronic configuration) with neon and argon, respectively.

In ionic reactions most A family elements lose or gain electrons in such a way as to produce ions that are isoelectronic with a noble gas (of the A group elements, only the post-transition metals, such as Ga and Sn, do not). Most of these noble gas ions have an octet of electrons in the outer shell (an s^2p^6 configuration); a few (e.g., Li^+ and H^-) have the $1s^2$ configuration of helium.

In the reaction of sodium and chlorine, the total number of electrons lost by sodium must equal the total number of electrons gained by chlorine. Thus, the number of sodium ions produced is the same as the number of chloride ions produced, and the formula, NaCl, gives the simplest ratio of ions present in the compound (1 to 1). These ions attract one another to form a crystal (Fig. 9.5).

In a sodium chloride crystal, no ion may be considered as belonging exclusively to another. Rather, each sodium ion is surrounded by six chloride ions and each chloride ion is surrounded by six sodium ions. The arrangement of ions in the crystal is such that the repulsion of like-charged ions is more than compensated for by the attraction of oppositely charged ions. This mutual attraction of positive and negative ions holds the crystal together; indeed, the ionic bond may be considered to be just this.

When oxygen undergoes an ionic reaction, each oxygen atom (electron configuration, $1s^2\ 2s^2\ 2p^4$) accepts two electrons and attains the neon configuration ($1s^2\ 2s^2\ 2p^6$); the oxide ion has a charge of 2–. In the reaction of the sodium with oxygen, two atoms of sodium are required for every atom of oxygen since the number of electrons lost must equal the number of electrons gained.

$$2Na\cdot + \ddot{\underset{\cdot\cdot}{O}}: \rightarrow 2Na^+ + :\ddot{\underset{\cdot\cdot}{O}}:^{2-}$$

The simplest ratio of ions present in the product, sodium oxide, is indicated by the formula of the compound, Na_2O. The arrangement of ions in a crystal of sodium oxide, which is different from that of the sodium chloride crystal, in such that the ions are accommodated in the ratio of two sodium ions to one oxide ion.

The electrovalence number of an element is the charge (including sign) of the ion derived from the element after an electrovalent reaction. If we follow the octet principle, the electrovalence number of a group I A element should be 1+; group VII A, 1–; group II A, 2+; group VI A, 2–. These charges are derived from the number of valence electrons that is characteristic of any element of a given group and the number of electrons that must be lost (positive electrovalence) or gained (negative electrovalence) in order for the element to attain a noble-gas configuration.

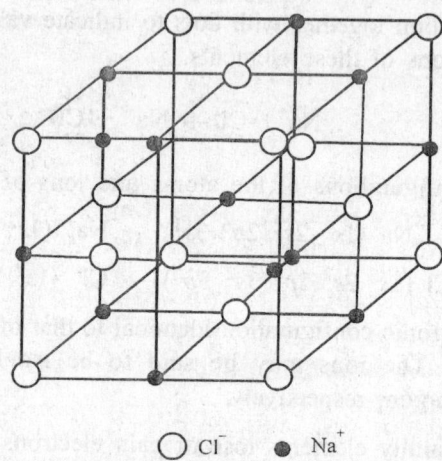

Fig. 9.5. Sodium chloride crystal lattice.

From these ionic charges, formulas may be derived on the basis of the fact that the total positive ionic charge of any compound must equal the total negative ionic charge of that compound. Calcium is a group II A metal, has two valence electrons, and has an electrovalence number of 2+. We have already determined that the electrovalence number of chlorine is 1– and that the electrovalence of oxygen is 2–. Thus, the formula of calcium chloride is $CaCl_2$ and that of calcium oxide is CaO. In a similar manner one can derive the formulae of aluminium oxide (Al_2O_3), sodium sulphide (Na_2S), and potassium nitride (K_3N) from the predicted electrovalence numbers of aluminium (3+), oxygen (2–), sodium (1+), sulphur (2–), potassium (1+), and nitrogen (3–). We shall see later that not every compound predicted by this method exists and that not every one which does not exist is truly ionic in character.

The energy effect associated with the condensation of positive and negative ions into a crystal is called the *crystal energy* or *lattice energy*. Since energy is evolved in these processes, lattice energies have negative signs. The lattice energy (with a positive sign) of a given crystal may also be viewed as the energy required to separate the ions of the crystal.

The driving force of the ionic reaction is the electrostatic attraction of the ions for one another (this attraction results in the liberation of the lattice energy). The ionisation of one mole of sodium (6.02 × 10^{23} atoms) requires +118 kcal (the ionisation energy).

$$Na(g) \rightarrow Na^+(g) + e^-$$

Energy is released when gaseous chlorine atoms gain electrons (–83 kcal for one mole of chlorine atoms),

$$e^- + Cl(g) \rightarrow Cl^-(g)$$

but the amount released is insufficient to supply the energy required for the ionisation of sodium. The deficit is more than compensated for by the lattice energy (−188 kcal/mol),

$$Na^+(g) + Cl^-(g) \rightarrow NaCl(crystal)$$

so that the process as a whole is energetically favourable.

This analysis is not complete since it pertains to the reaction between gaseous atoms of sodium and chlorine, and under ordinary conditions, these elements do not occur in this form. Nevertheless, we can reach some valid conclusions by this approach.

Since electrostatic attraction is greater between multicharged ions than between singly charged ions, the lattice energies of compounds that contain multicharged ions are larger than those that contain only 1+ and 1− ions. The lattice energy of MgO (−948 kcal/mol), for example, is large enough to supply the energy required for the production of the Mg^{2+} ion (+523 kcal/mol, the sum of the first two ionisation energies of Mg) as well as the energy required for the formation of the O^{2-} ion (+168 kcal/mol, the energy absorbed when two electrons are added to each O atom).

If each Mg atom lost only one electron, less energy would be required to produce two moles of Mg^+ ions for Mg_2O (+352 kcal) than is required to produce one mole of Mg^{2+} ions for MgO (+523 kcal). However, the energy released by the formation of the hypothetical Mg_2O crystal (probably about −670 kcal/mol) would be so much lower than the energy released by the formation of the MgO crystal (−948 kcal/mol) that the formation of MgO is favoured.

Why, then, does sodium not form a Na^{2+} ion in the oxide of this element? Table 9.2 lists the first to fourth ionisation energies of some A family metals. For each element, the ionisation energies increase from the first to the fourth, as expected. However, in each case, a jump in the required energy occurs after all the valence electrons have been removed and it is necessary to break into the s^2p^6 noble gas arrangement of the shell underlying the valence shell to remove the next electron (this place is marked with a vertical line in the table). The removal of two electrons from each sodium atom requires +1210 kcal/mol. Since the formation of the O^{2-} ion requires +168 kcal/mol, over 1378 kcal would have to be supplied by the lattice energy in order to form the crystal. The lattice energy can not supply this amount of energy. The lattice energy of the hypothetical $Na^{2+}O^{2-}$ would probably be of the same order of magnitude as that of MgO (−948 kcal/mol).

Table 9.2. Ionisation energies of the third period metals

Metal	Group	Ionisation energies (kcal/mol)			
		First	Second	Third	Fourth
Na	I A	+118	+1091	+1652	+2280
Mg	II A	+176	+347	+1848	+2521
Al	III A	+138	+434	+656	+2767

The foregoing analysis shows why so many metals form cations with s^2p^6 (noble gas) electronic configurations. All anions derived from atoms of the non-metals also have s^2p^6 configurations; the addition of electrons is possible until the s^2p^6 limit is reached, and energy considerations favour the attainment of this limit. Energy is evolved by the formation of the anions of the VII A elements, and under appropriate conditions the lattice energy can supply the energy required for the formation of O^{2-}, S^2, Se^{2-}, and Te^{2-} of group VI A as well as N^{3-} and P^{3-} of group V A.

However, if an electron were to be added beyond the s^2p^6 configuration, the additional electron would be very loosely bound since it would be added to a higher quantum level, screened from the nuclear charge by a closed s^2p^6 shell, and repelled by the electrons of the negatively charged ion to which it was added. Hence, the addition would required a large amount of energy, more than any lattice energy could supply, and the formation of such anions is not observed.

Certain metals engage in ionic reactions but cannot possible produce cations that are isoelectronic with noble gases. The ions Sc^{3+}, Y^{3+}, and La^{3+} have s^2p^6 configurations, but the ions of the other transition metals, as well as the post-transition metals, do not. For these latter metals, a comparatively large number of electrons would have to be lost to achieve a noble gas structure. Zinc (. . . $3s^2\ 3p^6\ 3d^{10}\ 4s^2$), for example, would have to lose 12 electrons. Electron losses of more than three are never observed; the energy required to remove four or more electrons is not obtainable in any ordinary ionic reaction. In its reactions zinc loses two electrons to form the Zn^{2+} ion (. . . $3s^2\ 3p^6\ 3d^{10}$).

In the reactions of the transition elements, inner d electrons can be used as well as outer s electrons. Examples of various types of ions are given in Table 9.3; the configurations listed are those of the outer subshell(s). Notice that a number of ions (including Zn^{2+}) have the configuration $ns^2\ np^6\ nd^{10}$ in their outer shell (listed as d^{10}), and a number of others have a $(n—1)s^2\ (n—1)p^6\ (n—1)d^{10}\ ns^2$ configuration (listed as $d^{10}s^2$). Some metals have a tendency to achieve these two closed configurations in which all sublevels are filled (although it is not so pronounced as the tendency to attain s^2p^6 configuration), but it is impossible for most transition elements to produce ions with any of the "regular" configurations (s^2p^6, d^{10}, or $d^{10}s^2$).

Table 9.3. Types of monatomic ions.

Configuration	Examples
s^2	Li^+, Be^{2+}, H^-
s^2p^6	Na^+, K^+, Rb^+, Cs^+, Mg^{2+}, Ca^{2+}, Sr^{2+}, Ba^{2+}, Al^{3+}, Sc^{3+}, Y^{3+}, La^{3+}, F^-, Cl^-, I^-, O^{2-}, S^{2-}, Se^{2-}, N^{3-}, P^{3-}
d^1	Ti^{3+}
d^2	V^{3+}
d^3	Cr^{3+}, Mo^{3+}
d^4	Cr^{2+}
d^5	Mn^{2+}, Fe^{3+}, Ru^{3+}
d^6	Fe^{2+}, Co^{3+}, Rh^{3+}
d^7	Co^{2+}
d^8	Ni^{2+}, Pd^{2+}, Pt^{2+}, Au^{3+}
d^9	Cu^{2+}
d^{10}	Cu^+, Ag^+, Au^+, Zn^{2+}, Cd^{2+}, Hg^{2+}, Ga^{3+}, In^{3+}, Tl^{3+}
$d^{10}s^2$	Ga^+, In^+, Tl^+, Ge^{2+}, Sn^{2+}, Pb^{2+}, As^{2+}, Sb^{3+}, Br^{3+}

Many transition elements form more than one type of cation; iron, for example, forms either Fe^{3+} or Fe^{2+} depending upon reaction conditions. More energy is required to produce the Fe^{3+} ion than the Fe^{2+} ion, but the lattice energies of Fe^{3+} compounds are larger than those of Fe^{2+} compounds. These factors balance to the point that the preparation of compounds of both ions is feasible.

IONIC RADIUS

The radii of ions have been determined by X-ray diffraction of ionic crystals. Such studies give only the distances between the centres of adjacent ions; apportioning these distances is a problem. To apportion bond distances of covalent molecules to secure covalent radii, one starts with a bond between two like atoms, which can be divided equally. In ionic crystals, however, the sizes of the positive and negative ions are of necessity different.

One solution to this problem is to study a crystal of a compound with a very small cation and a very large anion, such as lithium iodide [Fig. 9.6(a)]. The assumption is then made that the iodide ions touch each other, and the iodide-iodide distance (d in the illustration) is divided in half to obtain the radius of the iodide ion. This method of apportionment of anion-anion distance (d), however, is impossible for a crystal of the more common type illustrated in Fig. 9.6(b) since there is no anion-anion contact. Once the radius of the iodide ion has been fixed as a standard, other ionic radii can be calculated by subtracting the radius of the standard ion from the cation-anion distance (d' in Fig. 9.6 b) of another crystal. For example, the radius of the iodide ion from the K^+ to I^- distance in the potassium iodide crystal. Table 9.4 lists some ionic radii.

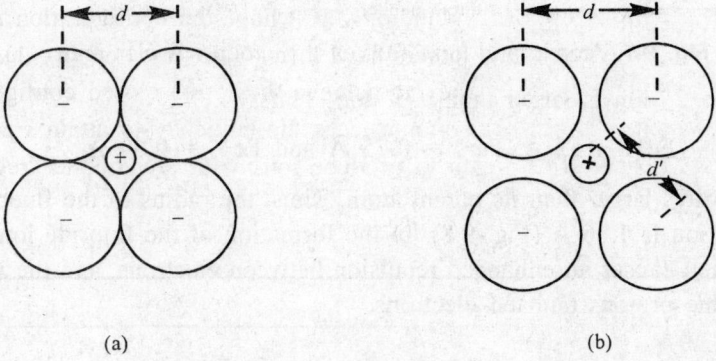

(a)   (b)

Fig. 9.6. Determination of ionic radii.

Table 9.4. Ionic radii (in angström units, Å).

I A	II A	III A	VI A	VII A
Li^+	Be^{2+}		O^{2-}	F^-
0.60	0.31		1.40	1.36
Na^+	Mg^{2+}	Al^{3+}	S^{2-}	Cl^-
0.95	0.65	0.50	1.84	1.81
K^+	Ca^{2+}	Ga^{3+}	Se^{2-}	Br^-
1.33	0.99	0.62	1.98	1.95
Rb^+	Sr^{2+}	In^{3+}	Te^{2-}	I^-
1.48	1.13	0.81	2.21	2.16
Cs^+	Ba^{2+}	Tl^{3+}		
1.69	1.35	0.95		

A positive ion is always smaller than the neutral atom from which it is derived (Fig. 9.7). Thus, the radius of Na is 1.57 Å, and the radius of Na^+ is 0.95 Å. In the formation of the sodium ion, the loss of an electron represents the loss of the entire $n = 3$ shell of the sodium atom. Furthermore, the loss of an electron creates an imbalance in the proton-electron ratio; since the protons outnumber the electrons in the positive ion, the electrons of the ion are drawn in closer to the nucleus.

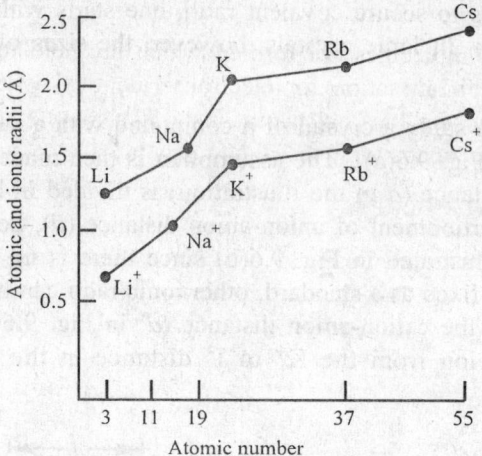

Fig. 9.7. Atomic and ionic radii of the group I A elements.

For similar reasons a 2^+ ion is larger than a 3^+ ion. Thus,

$$\text{Fe} = 1.17 \text{ Å}, \quad \text{Fe}^{2+} = 0.75 \text{ Å and } \text{Fe}^{3+} = 0.60 \text{ Å}$$

A negative ion is always larger than its parent atom. Thus, the radius of the fluorine atom is 0.72 Å, and that of the fluoride ion is 1.36 Å (Fig. 9.8). In the formation of the fluoride ion, the addition of an electron to the $2p$ subshell causes an enhanced repulsion between electrons, and the $n = 2$ shell expands. The fluoride ion has nine protons and ten electrons.

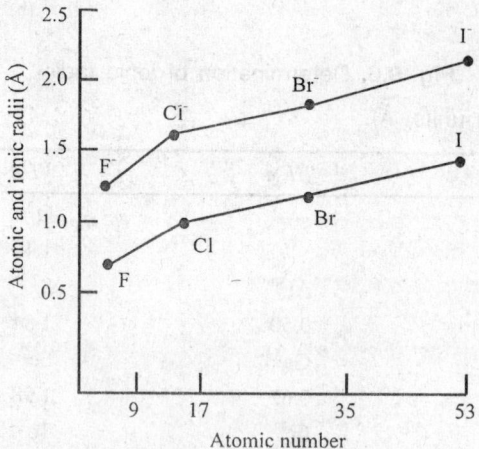

Fig. 9.8. Atomic and ionic radii of the group VII A elements.

Within a period, isoelectronic positive ions show a decrease in ionic radius from left to right because of the increasing nuclear charge. The same trend is observed for the isolelectronic negative ions of a period; ionic radius decreases from left to right.

Within a group of the periodic table, similarly charged ions increase in size from top to bottom. The ions of the heavier members of the group employ more electron shells than those of the lighter members.

COVALENT BOND

When atoms of non-metals interact, molecules are formed which are held together by covalent bonds. Since these atoms are similar in their attraction for electrons (identical when two atoms of the same element are considered), electron transfer does not occur; instead, electrons are shared. A covalent bond consists of a pair of electrons (with opposite spins) that is shared by two atoms.

As an example, consider the bond formed by two hydrogen atoms. An individual hydrogen atom has a single electron that is symmetrically distributed around the nucleus in a 1s orbital. When two hydrogen atoms form a covalent bond, the atomic orbitals overlap in such a way that the electron clouds reinforce each other in the region between the nuclei, and there is an increased probability of finding an electron in this region. According to the Pauli exclusion principle, the two electrons of the bond must have opposite spins. The strength of the covalent bond comes from the attraction of the positively charged nuclei for the negative cloud of the bond (Fig. 9.9).

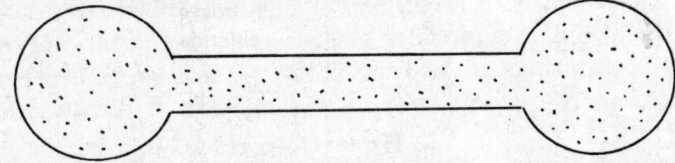

Fig. 9.9. Representation of the electron distribution in a hydrogen molecule.

The hydrogen molecule can be represented by the symbol H:H or H—H. Although the electrons belong to the molecule as a whole, each hydrogen atom can be considered to have the noble gas configuration of helium (two electrons in the $n = 1$ level). This consideration is based on the premise that both shared electrons contribute to the stable configuration of each hydrogen atom.

The formula, H_2, describes a discrete unit—a molecule—and hydrogen gas consists of a collection of such molecules. There are no molecules in strictly ionic materials. The formula Na_2Cl_2 is incorrect because sodium chloride is an ionic compound and the simplest ratio of ions in a crystal of sodium chloride is 1 to 1; a molecule of formula Na_2Cl_2 does not exist. For covalent materials, however, a formula such as H_2O_2 can be correct; this formula describes a molecule containing two hydrogen atoms and two oxygen atoms.

The hydrogen molecule can be described as being diatomic (containing two atoms). Certain other elements also exist as diatomic molecules. An atom of any group VII A element, for example, has seven valence electrons. By the formation of a covalent bond between two of these atoms, each atom attains an octet configuration characteristic of the noble gases. Thus, fluorine gas consists of F_2 molecules.

$$:\ddot{F}\cdot + \cdot\ddot{F}: \rightarrow :\ddot{F}: \quad \ddot{F}:$$

Only the electrons between the two atoms are shared and form a part of the covalent bond (although the molecular orbital theory considers that all of the electrons affect the bonding).

More than one covalent bond may form between two atoms. A nitrogen atom (group V A) has five valence electrons.

$$:\dot{N}\cdot + \cdot\dot{N}: \rightarrow :N : : : N :$$

In the molecule, N_2, six electrons are shared in three covalent bonds (usually called a triple bond). Notice that, as a result of this formulation, each of the nitrogen atoms can be considered to have an octet of electrons.

Non-metallic elements that exist as diatomic molecules are H_2, F_2, Cl_2, Br_2, I_2, N_2, and O_2. These elements are always indicated in this way in chemical equations.

The electron-dot formulae we have been using are called valence-bond structures or Lewis structures, named after Gilbert N. Lewis who proposed this theory of covalent bonding. The Lewis theory emphasises the attainment of noble gas configurations on the part of atoms in covalent molecules. Since the number of valence electrons is the same as the group number for the non-metals, one might predict that VII A elements, such as Cl, would form one covalent bond to attain a stable octet; VI A elements, such as O and S, two covalent bonds; V A elements, such as N and P, three covalent bonds; and IV A elements, such as C, four covalent bonds. These predictions are borne out in many compounds containing only simple covalent bonds.

$$\text{H} \cdot + \cdot \overset{\cdot\cdot}{\underset{\cdot\cdot}{\text{Cl}}} \rightarrow \text{H} \overset{\cdot\cdot}{\underset{\cdot\cdot}{:\text{Cl}}} :$$
hydrogen
chloride

$$2\,\text{H} \cdot + \cdot \text{O} \rightarrow \text{H} \overset{\text{H}}{\underset{\cdot\cdot}{:\text{O}}} :$$
water

$$3\,\text{H} \cdot + \cdot \overset{\cdot}{\text{N}} \cdot \rightarrow \text{H} \overset{\text{H}}{:\overset{\cdot\cdot}{\text{N}}:} \text{H}$$
ammonia

$$4\text{H} \cdot + \cdot \overset{\cdot}{\underset{\cdot}{\text{C}}} \cdot \rightarrow \text{H} \overset{\text{H}}{\underset{\text{H}}{:\text{C}:}} \text{H}$$
methane

Notice that in these molecules, each hydrogen atom can be considered to have a complete $n = 1$ shell; the other atoms have characteristic noble-gas octets.

The covalent bonding of compounds can also be indicated by dashes; each dash represents one bond, a pair of electrons.

$$:\overset{\cdot\cdot}{\underset{}{\text{Cl}}}-\overset{}{\underset{|}{\text{P}}}-\overset{\cdot\cdot}{\underset{\cdot\cdot}{\text{Cl}}}: \qquad \overset{\cdot\cdot}{\underset{\cdot\cdot}{\text{Cl}}}-\overset{\cdot\cdot}{\underset{\cdot\cdot}{\text{O}}}-\overset{\cdot\cdot}{\underset{\cdot\cdot}{\text{Cl}}}: \qquad \overset{\text{H}\ \ \text{H}}{\underset{\text{H}\ \ \text{H}}{\text{H}-\text{C}-\text{C}-\text{H}}}$$

$$:\overset{}{\underset{\cdot\cdot}{\text{Cl}}}:$$

Phorphorus Trichloride   Dichlorine oxide   Ethane

The following are examples of molecules that contain double and triple bonds.

$$:\overset{\cdot\cdot}{O}:+\ :C:+\ :\overset{\cdot\cdot}{O}:\ \rightarrow\ :\overset{\cdot\cdot}{O}::C::\overset{\cdot\cdot}{O}: \qquad \text{(or}\quad :\overset{\cdot\cdot}{O}=C=\overset{\cdot\cdot}{O}:\text{)}$$

Carbon dioxide

$$2H\cdot+\ \cdot\overset{\cdot}{C}:+\ :\overset{\cdot}{C}\cdot+\ 2H\cdot\ \rightarrow\ H:\overset{\cdot\cdot}{C}::\overset{\cdot\cdot}{C}:H \qquad \text{(or}\quad \begin{matrix} H & H \\ | & | \\ H-C & =C-H \end{matrix}\text{)}$$

Ethylene

$$H\cdot+\ \cdot\overset{\cdot}{C}:+\ :\overset{\cdot}{C}\cdot+\ \cdot H\ \rightarrow\ H:C:::C:H \qquad \text{(or}\quad H-C\equiv C-H\text{)}$$

Acetylene

Notice that in each compound that number of covalent bonds on each atom agrees with the number predicted.

FORMAL CHARGE

In the formation of certain covalent bonds, both of the shared electrons are furnished by one of the bonded atoms. For example, in the reaction of ammonia with a proton (a hydrogen atom stripped of its electron), the unshared electron pair of the nitrogen atom of NH_3 is used to form a new covalent bond.

$$H:\overset{\cdot\cdot}{\underset{H}{N}}:H\ +\ H^+\ \rightarrow\ \left[H:\overset{\overset{\displaystyle H}{\cdot\cdot}}{\underset{H}{N}}:H \right]^+$$

A bond formed in this way is frequently called a "coordinate covalent" bond, but it is probably unwise to do so. The coordination process of covalent bond formation is a useful distinction—particularly in the study of coordination complexes and in the interpretation of certain acid-base reactions. However, labeling a specific bond as a "coordinate covalent" bond implies that it is different from other covalent bonds and has little justification.

All electrons are alike no matter what their source. All the bonds in NH_4^+ are identical; it is impossible to distinguish between them. Furthermore, the mode of formation of a covalent bond does not affect its nature. We can imagine any covalent bond as having been formed by electron pairing

$$2H\cdot+\ \cdot\overset{\cdot\cdot}{O}:\ \rightarrow\ H_2O$$

or electron-pair donation

$$2H^+\cdot+\left[:\overset{\cdot\cdot}{\underset{\cdot\cdot}{O}}:\right]^{2-}\rightarrow H_2O$$

Notice, however, that the number of covalent bonds on the N atom of NH_4^+ does not agree with the number predicted in the preceding section. Since a nitrogen atom has five valence electrons (group V A), it would be expected to satisfy the octet principle through the formation of three covalent bonds. This prediction is correct for NH_3; it is not correct for NH_4^+.

An answer to this question may be obtained by calculating the formal charges of the atoms in NH_4^+. A formal charge is calculated by apportioning the bonding electrons equally between the bonded atoms, one electron to each atom for each covalent bond, and then comparing the number of electrons that a given atom has in the structure with the number of valence electrons that this atom would have when electrically neutral.

In NH_4^+ the N atom may be considered to have four electrons—one from the division of each of the covalent bonds. Since a neutral N atom has five valence electrons, the N atom in NH_4^+ is assigned a formal charge of 1+. Each H atom of the NH_4^+ ion has the same number of electrons in the structure that it has as a neutral atom and hence carries no formal charge. We can indicate the formal charge on the N atom of NH_4^+ as

$$
\begin{array}{c}
\text{H} \\
| \\
\text{H—N}^{\oplus}\text{—H} \\
| \\
\text{H}
\end{array}
$$

Thus, the discrepancy between the predicted and actual number of bonds on the N atom in NH_4^+ may be explained on the basis of formal charge since the hypothetical N^+ would have four unpaired electrons and be able to form four covalent bonds.

In the assignment of formal charges, the assumption is made that the electron pair of any covalent bond is shared equally by the bond atoms. Such an assumption is usually not true, and formal charges must be interpreted carefully. The electron density on the N atom in NH_4^+ is less than that on the N atom in NH_3, but the actual charge is not a full positive charge because the bonding electrons are not equally shared. In addition, molecular polarity occurs in some molecules in which each atom has a formal charge of zero.

Lewis structures may be assigned without reference to the source of the bonding electrons. For example, we could proceed in the following way to diagram the structure of the sulphur dioxide molecule; SO_2. This theory of covalent bonding offers no way to predict the arrangement of atoms in a molecule; the arrangement must be derived from experimental evidence. In the SO_2 molecule the O atoms are joined to a central S atom, and the molecule is angular.

$$
\begin{array}{c}
\text{S} \\
\\
\text{O} \qquad \text{O}
\end{array}
$$

We next determine the number of valence electrons to be used in writing the structure. For a molecule this number is simply the sum of the numbers of the valence electrons of the atoms present. If the structure of an ion is being considered, this sum is increased (for an anion) or decreased (for a cation) to take into account the charge of the ion. For SO_2 there is a total of eighteen valence electrons to be distributed (twelve from the two O atoms plus six from the S atom). If we place single bonds between the atoms, four of the eighteen electrons are used. The fourteen electrons that are left must be distributed so that they give an octet to each of the atoms.

In an attempt to accomplish this division, we derive the structure

$$
\ddot{\underset{..}{\text{O}}}\!:\quad \overset{\displaystyle \ddot{\text{S}}}{}\quad :\!\ddot{\underset{..}{\text{O}}}
$$

Both the O atoms in this structure have the proper number of electrons associated with them, but the S atom has only six. We can remedy this situation by switching an electron pair from one of the O atoms in such a way as to make a double bond between it and the S atom.

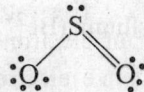

For the calculation of formal charge, the S atom may be considered to have five electrons–two that are not engaged in bonding and three from the division of the three covalent bonds on that atom. Since a neutral S atom has six valence electrons (group VI A), the S atom is assigned a formal charge of 1+. The right-hand O atom has the same number of electrons in the structure that it has as a neutral atom, and hence it carries no formal charge. The O atom on the left has seven valence electrons in the structure (six unshared electrons plus one from the covalent bond). Since the neutral O atom has six valence electrons, this O atom is assigned a formal charge of 1–. Notice that the sum of the formal charges of all the atoms in the SO_2 molecule equals zero. The sum of the foral charges of the atoms of a molecule always equals zero; the sum of the formal charges of the atoms of an ion, however, equals the charge on the ion.

A structure for SO_2 may be diagramed as

The bonding in SO_2, however, is not adequately represented by a single structure of this type.

TRANSITION BETWEEN IONIC AND COVALENT BONDING

The bonding in most compounds is intermediate in character between purely ionic and purely covalent. The best examples of ionic bonding are found in compounds formed by a metal with a low ionisation potential (for example, Cs) and non-metal with a strong tendency toward electron gain (for example, F). In a compound such as CsF, electron transfer is definite and complete; the bonding of the Cs^+ and F^- ions in a crystal is solely the result of electrostatic attraction between the ions.

A purely covalent bond is found only in molecules formed from two identical atoms, such as Cl_2. The electron-attracting ability of one chlorine atom is exactly the same as the electron-attracting ability of another; the bonding results from an electron cloud that is distributed symmetrically with respect to the two atoms. In other words, the bonding electrons are shared equally by the two identical atoms.

The bonding in most compounds lies somewhere between these two extremes. The bonding of intermediate character that occurs in compounds that contain a metal and a non-metal can be interpreted in terms of the interactions between the ions. The positively-charged ion of such a compound is believed to attract the deform the electron cloud of the anion; the electron cloud of the negative ion is drawn toward the cation. In extreme cases, such ion deformation leads to the formation of molecules and predominantly covalent compounds (Fig. 9.10). The degree of covalent character that a compound exhibits can be considered to correspond to the extent of ion distortion in the compound.

The ability of a cation to distort the electron cloud of a neighbouring anion depends upon the charge concentration of the cation; a small cation with a high positive charge is most effective. On the other hand, large multicharged anions are the most readily distorted since the outer electrons are relatively far from the nucleus in ions of this type.

Thus, in any group of metals in the periodic table, the member that forms the smallest cation (for example, Li in group I A) has the greatest tendency toward the formation of compounds with a high degree of covalent character. All compounds of beryllium (Be^{2+} is the smallest cation of group II A element) are significantly covalent.

Within a group of non-metals, the tendency toward increasing covalence parallels increasing atomic number (or increasing anion size). Thus, in the case of the halides of aluminium, AlF_3 is completely ionic; $AlCl_3$ is intermediate in character; and both $AlBr_3$ and AlI_3 are essentially covalent.

In the series formed by the fourth period metals, KCl, $CaCl_2$, $ScCl_3$, and $TiCl_4$, the degree of covalent character increases with increasing charge and decreasing size of the "cation", each of which is isoelectronic with Ar; KCl is strongly ionic and $TiCl_4$ is definitely covalent. Truly ionic compounds that contain cations with a charge of 3+ or higher are rare and exist only when the cation is a large one. Thus, such compounds as $SnCl_4$, $PbCl_4$, $SbCl_5$, and BiF_5 are covalent. Boron forms only covalent compounds; the hypothetical B^{3+} ion would have a comparatively high charge combined with a very small size and would cause extensive anion distortion leading to covalent bonding. Certain highly-charged cations exist, however as hydrated species in water solution.

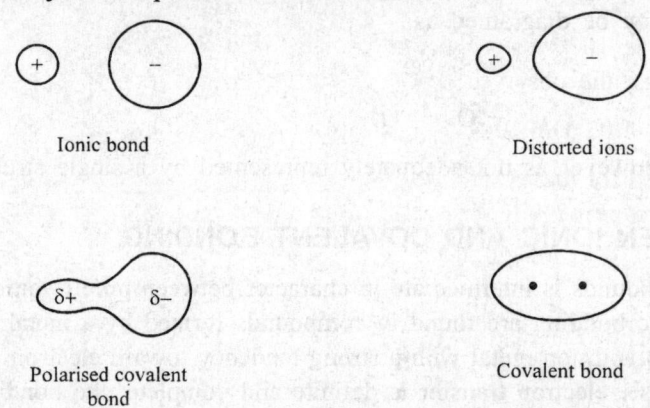

Fig. 9.10. Transition between ionic and covalent bonding.

Another approach to bonds of intermediate character considers the polarisation of covalent bonds. A purely covalent bond results only when two identical atoms are bonded. Whenever two different atoms are joined by a covalent bond, the electron density of the bond is not symmetrically distributed with respect to the two bonded atoms—the electrons of the bond are not shared equally. No matter how similar the atoms may be, there will be some difference in electron-attracting ability.

For example, chlorine has a greater attraction for electrons than bromine has. In the BrCl molecule the electrons of the covalent bond are more strongly attracted by the chlorine atom than by the bromine atom; the electron cloud of the bond is denser in the vicinity of he chlorine atom. Therefore, the chlorine end of the bond has a partial negative charge, and since the molecule as a whole is electrically neutral, the bromine end is left with a partial positive charge of equal magnitude. Such a bond, with positive and negative poles, is called a *polar covalent bond*. The partial charges of the bond are indicated by the symbols δ^+ and δ^- to distinguish them from full ionic charges.

The greater the difference between the electron-attracting ability of two atoms joined by a covalent bond, the more polar the bond, and the larger the magnitude of the partial charges. If the unequal sharing of electrons were carried to an extreme, one of the bonded atoms would have all of the bonding electrons, and separate ions would result.

A polar covalent molecule orients itself in the electric field between the plates of a condenser in such a way that the negative end is toward the positive plate and the positive end is toward the negative plate

(Fig. 9.11). Polar molecules arranged in this way affect the amount of charge that a pair of electrically-charged plates can hold; as a result, measurements can be made that allow the calculation of a value called the dipole moment.

Dipole moment, μ, is defined as the product of the distance separating equal charges of opposite sign and the magnitude of the charge. The dipole moments of non-polar molecules, such as H_2, Cl_2, and Br_2, are zero. The more polar the bond of a diatomic molecule, the larger the dipole moment of the molecule. Linus Pauling has used the dipole moment of a compound to calculate the partial ionic character of its covalent bond. If hydrogen chloride were completely ionic, the H^+ and Cl^- ions would each bear a unit charge (4.80×10^{-10} esu). Since the bond distance between the H and Cl atoms of hydrogen chloride is 1.27 Å, the dipole moment of the hypothetical, completely ionic HCl would be

$$\begin{aligned} \mu \quad &= (4.80 \times 10^{-10} \text{ esu})(1.27 \times 10^{-8} \text{ cm}) \\ &= 6.10 \times 10^{-18} \text{ esu cm} = 6.10 \text{ D} \end{aligned}$$

The debye unit, D, is 10^{-18} esu cm. The experimentally determined dipole moment of hydrogen chloride is 1.03 D. Thus, the observed dipole moment is

$$1.03 \text{ D}/6.10 \text{ D} = 0.17$$

times the value calculated for the hypothetical, ionic compound. Based on dipole moment measurements, the HCl bond appears to be 17% ionic.

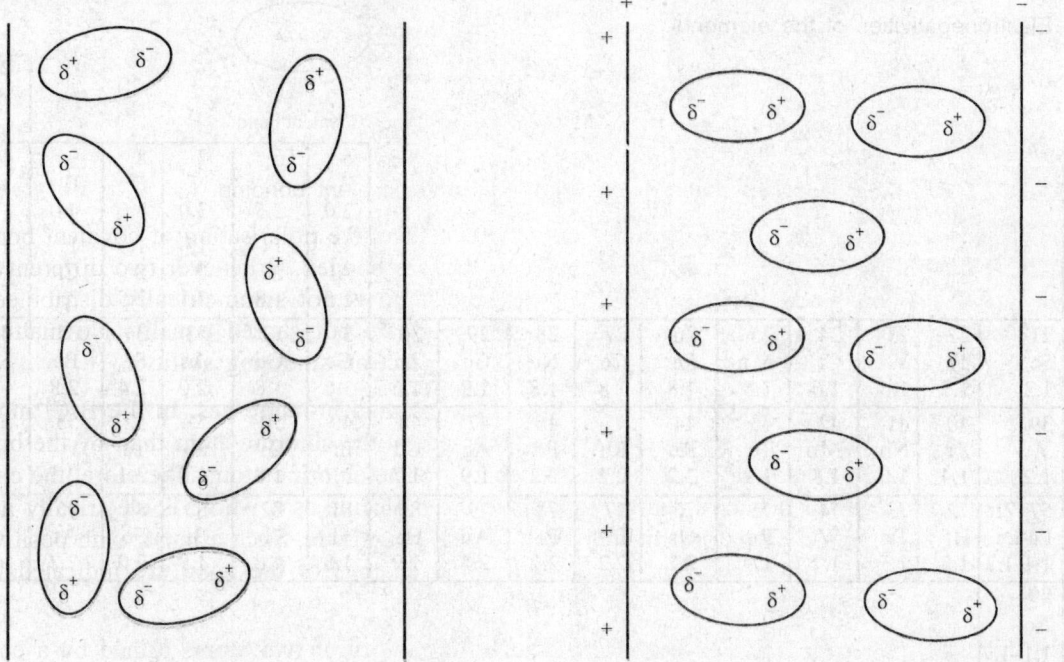

Fig. 9.11. Effect of an electrostatic field on the orientation of polar molecules.

There is a question as to whether the relationship between dipole moment and bond polarity is this simple. It has been demonstrated that the dipole moment of some molecules depends not only upon the bonding in the molecule but also upon the arrangement of the electrons not involved in the bonding.

ELECTRONEGATIVITY

Electronegativity is a measure of the ability of an atom in a molecule to attract electrons to itself. Thus, the polarity of the HCl molecule may be said to arise from the difference between the electronegativity of chlorine and that of hydrogen. Since chlorine is more electronegative than hydrogen, the chlorine end of the molecule is the negative end of the dipole.

The concept of electronegativity is useful but inexact. There is no simple, direct way to measure electronegativities, and various methods of determining them have been proposed. Indeed, since this property depends not only upon the structure of the atom under consideration, but also upon the number and nature of the atoms to which it is bonded, the electronegativity of an atom is not invariable. The electronegativity of phosphorus would be expected to be different in PCl_3 than in PF_5. This concept, therefore, should be regarded as only semi-quantitative.

The electronegativity scale of Pauling is the one most frequently used (Table 9.5) and is based upon experimentally derived values of bond energies. The bond energy of the Br—Br bond, for example, is the energy required to separate the Br_2 molecule into Br atoms; for one mole of Br_2 molecules (6.022×10^{23} molecules), +46 kcal is required. The bond energy of the H—H bond is +104 kcal/mol. Both the Br—Br and H—H bonds are non-polar, and the arithmetic mean of the bond energies

$$\frac{46 \text{ kcal } + 104 \text{ kcal}}{2} = 75 \text{ kcal}$$

may be considered to describe a "non-polar H—Br bond" in which the electrons are shared equally.

Table 9.5. Electronegativities of the elements.

1 H 2.1																	2 He –
3 Li 1.0	4 Be 1.5											5 B 2.0	6 C 2.5	7 N 3.0	8 O 3.5	9 F 4.0	10 Ne –
11 Na 0.9	12 Mg 1.2											13 Al 1.5	14 Si 1.8	15 P 2.1	16 S 2.5	17 Cl 3.0	18 Ar –
19 K 0.8	20 Ca 1.0	21 Sc 1.3	22 Ti 1.5	23 V 1.6	24 Cr 1.6	25 Mn 1.5	26 Fe 1.8	27 Co 1.8	28 Ni 1.8	29 Cu 1.9	30 Zn 1.6	31 Ga 1.6	32 Ge 1.8	33 As 2.0	34 Se 2.4	35 Br 2.8	36 Kr –
37 Rb 0.8	38 Sr 1.0	39 Y 1.2	40 Zr 1.4	41 Nb 1.6	42 Mo 1.8	43 Tc 1.9	44 Ru 2.2	45 Rh 2.2	46 Pd 2.2	47 Ag 1.9	48 Cd 1.7	49 In 1.7	50 Sn 1.8	51 Sb 1.9	52 Te 2.1	53 I 2.5	54 Xe –
55 Cs 0.7	56 Ba 0.9	57–71 La-Lu 1.1-1.2	72 Hf 1.3	73 Ta 1.5	74 W 1.7	75 Re 1.9	76 Os 2.2	77 Ir 2.2	78 Pt 2.2	79 Au 2.4	80 Hg 1.9	81 Tl 1.8	82 Pb 1.8	83 Bi 1.9	84 Po 2.0	85 At 2.2	86 Rn –
87 Fr 0.7	88 Ra 0.9	89– Ac– 1.1–1.7															

The actual HBr bond is polar, and the measured bond energy is +88 kcal/mol. The difference between the measured and calculated values, Δ,

$$\Delta = 88 \text{ kcal } - 75 \text{ kcal } = 13 \text{ kcal}$$

may be considered to represent the portion of the measured bond energy that is required to separate the partial charges produced by the unequal sharing of the bonding electrons. Hence, a bond for which the value of Δ is large would be expected to be highly polar and to be formed from atoms that have a large difference in electronegativity. Consequently, the Δ values for a number of bonds can be used to derive relative electronegativity values for the elements.

In general, electronegativity increases from left to right across any period (with increasing number of valence electrons) and from bottom to top in any group (with decreasing size). Thus, the most highly electronegative elements are found in the upper right corner of the periodic table (ignoring the noble gases) and the least electronegative elements are found in the lower left corner of the chart. These trends parallel those noted for ionisation potentials and electron affinities.

Metals are elements that have small attractions for valence electrons (low electronegativities); non-metals, except for the noble gases, have large attractions (high electronegativities). Thus, electronegativities (as well as ionisation potentials and electron affinities) can be used to rate metallic reactivities and non-metallic reactivities. The positions of the elements in the periodic table are helpful in making predictions concerning chemical reactivity (Fig. 9.12).

Electronegativities can be used to predict the type of bonding that a compound will have. When two elements of widely different electronegativity combine, an ionic compound results; thus, the electronegativity difference between sodium and chlorine is 2.1, and NaCl is an ionic compound.

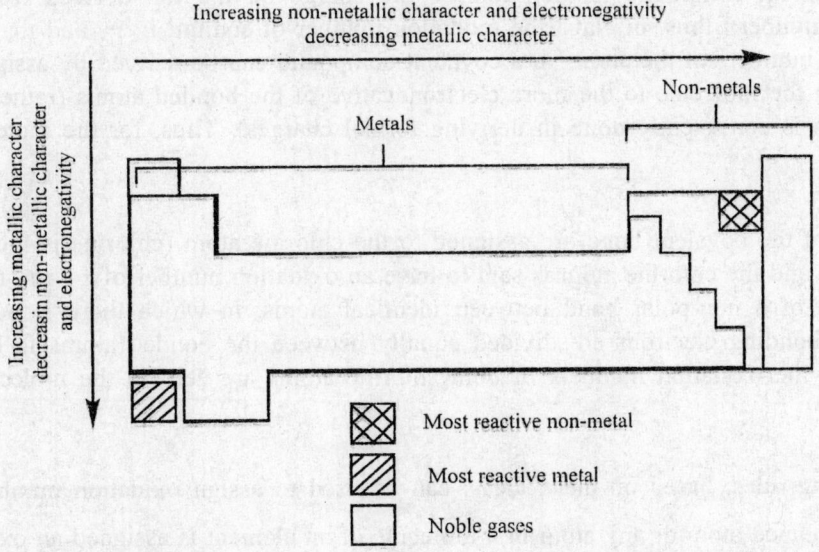

Fig. 9.12. Relation between position in the periodic classification and metallic or non-metallic reactivity.

Covalent bonding occurs between non-metals when the electronegativity differences are not very large. In such cases the electronegativity differences give an indication of the degree of polarity of the covalent bonds. If the electronegativity difference is zero or very small, an essentially non-polar bonds with equal or almost equal sharing of electrons can be assumed. The larger the electronegativity difference, the more polar is the covalent bond (the bond being polarised in the direction of the atom with the larger electronegativity). Thus, from electronegativities we can predict that HF is the most polar, and has the largest bond energy, of any hydrogen halide (Table 9.6).

Table 9.6. Some properties of the hydrogen halides.

Hydrogen halide	Dipole moment (D)	Bond energy (kcal/mol)	Electronegativity of halogen	Electronegativity difference between hydrogen and halogen
HF	1.91	135	F = 4.0	1.9
HCl	1.03	103	Cl = 3.0	0.9
HBr	0.78	87	Br = 2.8	0.7
HI	0.38	71	I = 2.5	0.4

OXIDATION NUMBERS

In the assignment of formal charges to the atoms of a covalent molecule, the bonding electrons are apportioned equally between the bonded atoms, and any bond polarity caused by unequal sharing of electrons is ignored. The charges thus derived are useful in interpreting the structure and some of the properties of covalent molecules, but the concept itself is merely a convention.

Another arbitrary but useful convention is the concept of oxidation numbers. Oxidation numbers are charges (fictitious charges in the case of covalent species) assigned to the atoms of a compound according to arbitrary rules that take into account bond polarity. The oxidation number of an atom in a binary, electrovalent compound is the same as the charge on the ion derived from that atom (the electrovalence number); thus, in NaCl the oxidation number of sodium is 1+ and that of chlorine is 1−. The oxidation numbers of the atoms in a covalent compound can be derived by assigning the electrons of each bond in the molecule to the more electronegative of the bonded atoms (rather than by dividing them equally as is consistently done in deriving formal charges). Thus, for the molecule

$$H : \overset{..}{\underset{..}{Cl}} :$$

both electrons of the covalent bond are assigned to the chlorine atom (chlorine is more electronegative than hydrogen), and the chlorine atom is said to have an oxidation number of 1− and the hydrogen atom 1−. In the case of a non-polar bond between identical atoms, in which there is no electronegativity difference, the bonding electrons are divided equally between the bonded atoms in deriving oxidation numbers. Thus, the oxidation numbers of both chlorine atoms are zero in the molecule

$$Cl : \overset{..}{\underset{..}{Cl}} :$$

The following rules, based on these ideas, can be used to assign oxidation numbers.

1. Any uncombined atom or any atom in a molecule of an element is assigned an oxidation number of zero.
2. The sum of the oxidation numbers of the atoms in a compound is zero since compounds are electrically neutral.
3. The oxidation number of a monatomic ion is the same as the charge on the ion (electrovalence number). The sum of the oxidation numbers of the atoms that constitute a polyatomic ion equals the charge on the ion.
4. The oxidation number of fluorine, the most electronegative element, is 1− in all fluorine-containing compounds.

5. In most oxygen-containing compounds, the oxidation number of oxygen is 2–. There are, however, a few exceptions. In peroxides each oxygen has an oxidation number of 1–. For example, the two oxygens of the peroxide ion, O_2^{2-}, are equivalent; each must be assigned an oxidation number of 1– so that the sum equals the charge on the ion. In the superoxide ion, O_2^-, each oxygen has an oxidation number of 1/2–. In OF_2 the oxygen has an oxidation number of 2+.

6. The oxidation number of hydrogen is 1+ in all its compounds except the metallic hydrides (e.g., CaH_2 and NaH) in which hydrogen is in the 1– oxidation state.

The oxidation states of the constituent elements of most compounds can be determined through the use of these rules. The oxidation numbers of the atoms of H_3PO_4, for example, must add up to zero. If each hydrogen is assigned 1+ (total, 3+) and each oxygen is assigned 2– (total, 8–), the phosphorus must have an oxidation number of 5+. The same conclusion can be reached by examining the ion derived from phosphoric acid, the phosphate ion, PO_4^{3-}. The atoms of a polyatomic ion, such as the phosphate ion, are covalently bonded; the ion exists as an entity and the sum of the oxidation numbers of the atoms that constitute the ion equals the charge of the ion. Since the charge on the PO_4^{3-} ion is 3–, and since each oxygen atom has an oxidation number of 2– (a total of 8– for the four oxygen atoms), the phosphorus atom must have an oxidation number of 5+.

In the dichromate ion, $Cr_2O_7^{2-}$, the seven oxygen atoms have a combined oxidation number of 14–. Since the charge on the ion is 2–, the oxidation numbers of the two chromium atoms must add up to 12+. The oxidation number of chromium in this ion, therefore, is 6+. Usual practice is to report the oxidation state of an element in a compound on the basis of the oxidation number of a single atom. It would be misleading to say that oxygen is in an oxidation state of 2– in H_2O and 4– in SO_2; in both compounds, oxygen has an oxidation number of 2–.

What are the oxidation states of the elements in calcium perchlorate, $Ca(ClO_4)_2$? Since calcium is a member of group II A, its oxidation number (which is the same as its electrovalence number) is 2+. Since there are two perchlorate ions for every Ca^{2+} ion in the compound, the charge on the perchlorate ion must be 1–. Each oxygen atom of a ClO_4^- ion has an oxidation number of 2– (a total of 8– for the four oxygen atoms), and therefore the oxidation number of the chlorine atom is 7+. These values must be interpreted with care. The ionic charges have physical significance; the oxidation numbers of oxygen and chlorine are merely conventions.

Fractional oxidation numbers are possible. Calculation of the oxidation number of iron in Fe_3O_4 gives a value of $2\frac{2}{3}+$. (Since the oxygens add up to 8–, the irons must add up to 8+; 8+ divided by 3 is $2\frac{2}{3}+$). This compound can be considered to be formed from FeO and Fe_2O_3. Hence two iron atoms at 3+ each and one iron atom at 2+ average out to $2\frac{2}{3}+$. When an element has a fractional oxidation number in a particular compound, this often means that the compound contains two (or more) atoms of the element that are not perfectly equivalent.

Frequently an element displays a range of oxidation states in its compounds. For example, in the compounds of nitrogen, this element exhibits oxidation numbers from 3– (e.g., in NH_3) to 5+ (e.g. in HNO_3). Since the number of valence electrons of an A family element is the same as its group number, the highest positive charge (even a hypothetical one) that can logically be assigned to an A family element is the same as its group number. The highest oxidation number of an A family element is, therefore, its group number.

The lowest oxidation number of an A family element is its electro-valence number. Thus, the highest oxidation number of sulphur (group VI A) is 6+ (e.g., in H_2SO_4) and the lowest oxidation number of sulphur is 2– (e.g., in H_2S). The highest oxidation number of sodium (group I A) is the same as the lowest oxidation number, 1+. There are, however, exceptions to these generalisations (e.g., fluorine and oxygen).

NOMENCLATURE OF INORGANIC COMPOUNDS

The name of a binary compound, a compound formed from only two elements, is usually derived from the names of the elements of which it is composed. Ionic binary compounds are named with the name of the metal first and the non-metal second; covalent binaries are named with the name of the less electronegative element first. For all binary compounds, the ending *–ide* is substituted for the usual ending of the name of the element that appears last in the name. All compounds ending in *–ide*, however, are not binaries; there are a few exceptions, such as the cyanides (e.g., NaCN) and the hydroxides (e.g., NaOH).

If two elements form only one compound with each other, the preceding guidelines are sufficient to name that compound. Thus, the name of Na_2O is sodium oxide; $AlCl_3$, aluminium chloride; AgBr, silver bromide; ZnS, zinc sulphide; CaH_2, calcium hydride; and HF, hydrogen fluoride.

Some pairs of elements, however, form several compounds with each other (e.g., $FeCl_2$ and $FeCl_3$). Each compound must have a name that identifies it. In a system of nomenclature, the distinction between such compounds is made by indicating the oxidation state of the first element in Roman numerals immediately after the English name of that element. Thus, $FeCl_2$ is iron(II) chloride [Fe^{2+} is called the iron(II) ion]. Notice that the Roman numeral applies to the oxidation state of the first element of the name (the one in a positive oxidation state) and not to numbers that appear as subscripts in the formula.

Another method to distinguish between the compounds formed by an element that exhibits more than one positive oxidation state is to change the suffix of the name of the element (Latin names are employed when the symbol drives from the Latin). The ending *–ous* is used to indicate the lower oxidation state of a pair, and *–ic* is used for the higher oxidation state. Thus, $FeCl_2$ is ferrous chloride (and Fe^{2+} is the ferrous ion); $FeCl_3$ is ferric chloride (and Fe^{3+} is the ferric ion). Difficulties arise in the use of this method since some pairs of elements form more than two compounds. The two systems of nomenclature that have been presented are compared in Table 9.6.

Table 9.6. Names of some binary compounds.

Formula	Name	
FeO	Iron(II) oxide	Ferrous oxide
Fe_2O_3	Iron(III) oxide	Ferric oxide
Cu_2S	Copper(I) sulphide	Cuprous sulphide
CuS	Copper(II) sulphide	Cupric sulphide
SnF_2	Tin(II) fluoride	Stannous fluoride
SnF_4	Tin(IV) fluoride	Stannic fluoride
Hg_2Br_2[a]	Mercury(I) bromide	Mercurous bromide
$HgBr_2$	Mercury(II) bromide	Mercuric bromide

[a] The mercury(I) ion (or the mercurous ion) is unusual since it is diatomic (Hg_2^{2+}). The two atoms of mercury are covalently bonded. Since the charge on the ion is 2+, each atom has an oxidation number of 1+.

A system of nomenclature that is particularly useful for series of compounds formed from two non-metals employs Greek prefixes: mono-(1), di- (2), tri- (3), tetra- (4), penta (5), hexa- (6), hepta- (7) octa- (8), and so forth. One of the prefixes is added to the name of each element to indicate the number of atoms of that element in the molecule that is being named. The prefix mono- is frequently omitted. The names of the oxides of nitrogen, according to this system as well as to the Stock system, appear in Table 9.7.

Table 9.7. Names of the nitrogen oxides.

Formula	Name	
N_2O	Dinitrogen oxide	Nitrogen(I) oxide
NO	Nitrogen oxide	Nitrogen(II) oxie
N_2O_3	Dinitrogen trioxide	Nitrogen(III) oxide
NO_2	Nitrogen dioxide	Nitrogen(IV) oxide
N_2O_4	Dinitrogen tetroxide	Dimer of nitrogen(IV) oxide[a]
N_2O_5	Dinitrogen pentoxide	Nitrogen(V) oxide

[a] A dimer is a molecule formed from two identical, simpler molecules (monomers). In this case an N_2O_4 molecule is formed from two NO_2 molecules.

Certain binary compounds have acquired non-systematic names by which they are known exclusively. The list of such substances includes water (H_2O), ammonia (NH_3), hydrazine (N_2H_4), and phosphine (PH_3). Notice that the formulae of the last three compounds are customarily inverted; the symbol H should appear first in the formula since hydrogen is the element that occurs in a positive oxidation state.

Acids are covalent compounds of hydrogen that dissociate in water to produce H^+ (aq) ions, in which each H^- is bonded to at least one water molecule. This type of ion is frequently indicated H_3O^+ (which is $H^+.H_2O$) and is called the *hydronium ion*. When hydrogen chloride dissolves in water, an acidic solution is produced that is called hydrochloric acid.

$$HCl(g) + H_2O \rightarrow H_3O^+(aq) + Cl^-(aq)$$

Aqueous solutions of binary compounds that functions as acids are named by adding the prefix hydro- and the suffix –*ic* to the root of the name of the element that is combined with hydrogen and then adding the word acid. Hence, hydrogen fluoride (HF) forms hydrofluoric acid, and hydrogen sulphide ((H_2S) forms hydrosulphuric acid. Although hydrogen cyanide (HCN) is not a binary compound, it is named as though it were; an aqueous solution of HCN is called hydrocyanic acid.

Alkalies are compounds that contain the hydroxide ion (OH^-) and are named accordingly (e.g., NaOH is sodium hydroxide and $Ca(OH)_2$ is calcium hydroxide). Since they contain three elements, the hydroxides are classed as ternary compounds. When an acid and an alkali are mixed, a neutralisation reaction occurs between the H_3O^+ ion from the acid and the OH^- ion from the alkali.

$$H_3O^+(aq) + OH^-(aq) \rightarrow 2H_2O$$

A *salt* is a compound derived from the combination of an alkali and an acid; it consists of the cation of the alkali and the anion of the acid. The salts of binary acids are themselves binary compounds, and the names of these salts have the customary -*ide* ending. Thus, sodium chloride (NaCl) is the salt derived from sodium hydroxide and hydrochloric acid.

Oxygen is one of the three constituent elements of most ternary acids. If a given element (sometimes called the central element) forms only one oxyacid, the acid is named by changing the ending of the name of the element to –*ic* and adding the word acid. Thus, H_3BO_3 is boric acid. If there are two common acids in which the same central element occurs in different oxidation states, the ending –*ous* is used in naming the acid of the element in its lower oxidation state; the ending –*ic* is used to denote the higher oxidation state (Table 9.8).

There are a few series of oxyacids for which two names are not enough. The prefix *hypo-* may be added to the name of an *-ous* acid to indicate an oxidaton state lower than that of the *-ous* acid. The prefix *per-* may be added to the name of an *-ic* acid to indicate an oxidation state higher than that of the *-ic* (see the names of the oxyacids of chlorine in Table 9.8).

Table 9.8. Names of some ternary acids and salts.

Oxidation state of central element	Acid Formula	Name	Sodium salt Formula	Name
5+	HNO_3	Nitric acid	$NaNO_3$	Sodium nitrate
3+	HNO_2	Nitrous acid	$NaNO_2$	Sodium nitrite
6+	H_2SO_4	Sulphuric acid	Na_2SO_4	Sodium sulphate
4+	H_2SO_3	Sulphurous acid	Na_2SO_3	Sodium sulphite
7+	$HClO_4$	Perchloric acid	$NaClO_4$	Sodium perchlorate
5+	$HClO_3$	Chloric acid	$NaClO_3$	Sodium chlorate
3+	$HClO_2$	Chlorous acid	$NaClO_2$	Sodium chlorite
1+	$HOCl$	Hypochlorous acid	$NaOCl$	Sodium hypochlorite

The oxyacids are covalent molecules that produce covalently bonded polyatomic anions through the loss of protons (H^+ ions) in salt formation. The names of the anions produced by the loss of all possible protons are derived from the name of the parent acid by retaining the prefix (if any) and changing the suffix *-ic* to *-ate* or the suffix *-ous* to *-ite*. The salts of these anions are named by combining the name of the cation present (using Stock nomenclature or the *-ous–/ic* convention if necessary) with the name of the anion (Table 9.8). Thus, $Fe(ClO_4)_3$ is iron(III) perchlorate or ferric perchlorate. The names and formulae of some common polyatomic anions are given in Table 9.9.

Table 9.9. Common polyatomic anions.

Formula	Name	Formula	Name
$C_2H_3O_2^-$	Acetate	ClO^-	Hypochlorite
AsO_4^{3-}	Arsenate	OH^-	Hydroxide
AsO_3^{3-}	Arsenite	NO_3^-	Nitrate
CO_3^{2-}	Carbonate	NO_2^-	Nitrite
ClO_3^-	Chlorate	ClO_4^-	Perchlorate
ClO_2^-	Chlorite	MnO_4^-	Permanganate
CrO_4^{2-}	Chromate	PO_4^{3-}	Phosphate
CN^-	Cyanide	SO_2^-	Sulphate
$Cr_2O_7^{2-}$	Dichromate	SO_3^{2-}	Sulphite

Acids that can lose more than one proton per molecule are called polyprotic acids. Sulphuric acid (H_2SO_4, specifically a diprotic acid) can lose one or two protons. Phosphoric acid (H_3PO_4, specifically a triprotic acid) can lose one, two, or three protons. The salts formed by the loss of all possible protons are called *normal salts*. Salts formed by the incomplete neutralisation of polyprotic acids are called *acid salts*; the anions of these salts retain one or more hydrogen atoms of the parent acid. Acid salts are named so as to indicate the number of hydrogen atoms present (Table 9.10). In practice, the prefix mono- is usually omitted. In addition, the monohydrogen salt of a diprotic acid, such as $NaHSO_3$, may also be named by use of the prefix bi-. Thus, $NaHSO_3$ is sodium monohydrogen sulphite, sodium hydrogen sulphite, or sodium bisulphite.

Table 9.10. Names of the sodium salts of some polyprotic acids.

Formula	Name
Na_3PO_4	Sodium phosphate
Na_2HPO_4	Sodium monohydrogen phosphate (sodium hydrogen phosphate)
NaH_2PO_4	Sodium dihydrogen phosphate
Na_2CO_3	Sodium carbonate
$NaHCO_3$	Sodium monohydrogen carbonate (sodium hydrogen carbonate or sodium bicarbonate)

A large number of complex polyatomic cations, anions, and neutral molecules exists. There are not many simple polyatomic cations of importance, but the ammonium ion (NH_4^+) and the mercury(I) ion (Hg_2^{2+}) deserve mention.

Example 9.1. Write the electronic configurations of H_2, H_2^+ and the hypothetical species H_2^- in terms of molecular orbital theory.

Solution. H_2, H_2^+ and H_2^- have 2, 1 and 3 electrons, respectively. Hence, their electronic configurations would be as follows:

$$H_2 \quad : \quad \sigma(1s)^2$$
$$H_2^+ \quad : \quad \sigma(1s)^1$$
$$H_2^- \quad : \quad \sigma(1s)^2 \, \sigma(1s)^1$$

Example 9.2. Why does He_2^+ exist whereas He_2 does not ?

Solution. Electronic configuration of He_2 is : $\sigma(1s)^2 \, \sigma(1s)^2$

∴   Bond order = 1/2 (2 – 2) = 0

Electronic configuration of He_2^+ is : $\sigma(1s)^2 \, \sigma(1s)^1$

∴   Bond order = 1/2 (2 – 1) = 1/2

Thus, while the bond order of He_2 is zero, that of He_2^+ is 1/2. Hence, He_2^+ exists while He_2 does not not.

Chapter 10

Kinetics and Chemical Equilibrium

INTRODUCTION

A chemical equation provides information on the ultimate products of a reaction and on the masses obtained from given amounts of reactants, but it conveys nothing about the speed of the reaction. This may vary widely from very slow, as in corrosion, for example, to very fast as in explosive reactions. The study of the rate of a chemical reaction and the factors upon which it depends is called chemical kinetics. These factors are temperature, presence or absence of a catalyst, intimacy of mixing of the reactants, intensity of ultraviolet light, and the concentration of the reactants.

TEMPERATURE AND RATE OF REACTION

Reaction rates generally increase with rising temperature, approximately doubling for each rise of 10 K. This means that a reaction proceeds about a million times faster if the temperature is raised by 200 K. This is the principal reason for heating substances undergoing reaction and for operating industrial processes at high temperature. The reverse effect of cooling a reaction in order to slow it down is sometimes useful.

CATALYSIS

A substance which alters the speed of a chemical reaction but remains unchanged chemically in mass at the end of the reaction is called a catalyst (positive or negative according to whether it speeds or slows it). Most examples of catalysis, the term used to describe the operation of a catalyst, are positive, but negative catalysts include organic compounds added to solutions of hydrogen peroxide to retard their decomposition during storage and lead tetraethyl which is added to petrol to prevent 'knocking'.

Characteristics of Catalysts

Catalytic reactions are very numerous and whilst the catalysts themselves vary widely in type and chemical composition, most have certain general features in common. They are usually specific, i.e. efficient only for one particular reaction. The most striking examples of this are the enzymes, complicated organic substances which catalyse many of the chemical reactions occurring in digestion. Nevertheless certain metals such as platinum and nickel catalyse a wide range of reactions and water vapour appears to act as a catalyst in many gaseous reactions. Again, benzoyl peroxide is used as a catalyst in the polymerisation of several unsaturated organic compounds.

Another characteristic of catalysts is that every minute amounts may enormously increase the rate of reaction. For example, mere traces of copper ions greatly increase the rate at which a bisulphite is oxidised in solution and extremely low concentrations of cobalt ions accelerate the decomposition of hypochlorites. On the other hand, in the Friedel-Crafts reaction the anhydrous aluminium chloride catalyst must be present in amounts comparable to the reactants to be effective.

Catalysts do not affect the position of equilibrium in a reversible reaction and hence do not affect the yield of product obtained under given conditions. In the presence of a catalyst the forward and backward reactions are accelerated in the same proportion, so that although the position of equilibrium is unchanged, the state of equilibrium is achieved much more quickly.

Experiments have shown that it is often possible to increase the activity of a catalyst considerably by adding small amounts of another substance, e.g. in the Haber process for synthesising ammonia, where molybdenum and the oxides of aluminium and potassium are added to the catalyst of iron. Such substances are called promoters. Conversely traces of certain substances destroy the activity of some catalysts; these are known as catalyst poisons. For example, arsenic poisons the platinum catalyst used in the contact process for making sulphuric acid and consequently all traces of arsenic compounds have to be removed from the reacting gases before use.

Where two substances can react together in more than one way, the choice of catalyst may decide the course of the reaction, presumably by so accelerating one particular reaction that it predominates over the others. For example, carbon monoxide and hydrogen react together at high temperatures and pressures to give methyl alcohol in the presence of zinc oxide catalyst, methane and other volatile hydrocarbons with a catalyst of nickel, and a mixture of higher paraffins with cobalt.

Occasionally one of the products is capable of catalysing the reaction by which it is made. This is called *auto-catalysis*. When an oxalate reacts with acidified potassium permanganate solution, the manganese(II) ions resulting from the reduction of the permanganate catalyse this reaction so that it proceeds at a convenient speed at a lower temperature than the 60°C needed to start it.

Many hydrolyses are catalysed by hydrogen ions or hydroxyl ions in solution, the rate of the reaction being directly proportional to the concentration of the ion concerned. For this reason such hydrolyses are often brought about by solutions of acids or alkalies rather than by water alone. Saponification is a good example.

Theories of Catalysis

Two main explanations of catalysis have been put forward. Heterogeneous catalysis, where reactants and catalyst are in different physical states, as, for example, in the catalysis of gaseous reactions by metals, has been regarded as an example of adsorption, i.e. condensation of the reacting gases upon the surface of the metal in a layer perhaps only a few molecules thick. Presumably the greatly increased concentration of the reactants leads to rapid reaction, the products escaping and being replaced by more reactants. This theory has been elaborated by suggesting that the catalyst has on its surface certain active points where adsorption occurs most readily, and that in this process the molecules of the reactants are activated in some way so that they react together more rapidly.

The other principal theory of catalysis postulates the continuous formation and decomposition of unstable intermediate compounds. This provides a route whereby the final products are obtained more quickly than by direct combination, presumably because the intermediate reactions are comparatively rapid. The theory finds support from the discovery of traces of intermediate compounds remaining in the mixture at the end of the reaction, e.g. small amounts of purple permanganate have been found when manganese dioxide has been used to catalyse the thermal decomposition of potassium chlorate, and traces of chlorine have been detected in the oxygen produced. The lead chamber process for making sulphuric acid is another example where an intermediate compound, nitrosyl sulphuric acid, has been isolated from the reaction mixture.

INTIMACY OF REACTANTS

The rate of a reaction is increased by bringing the reactants into more intimate contact. This is particularly marked in reactions between solids where the state of sub-division and the thoroughness of mixing are important factors. When metals are finely powdered they are often much more reactive owing to their greatly increased surface area, e.g. powdered aluminium, used in the thermite reaction, is a powerful reducing agent, and finely powdered lead inflames spontaneously in air.

Liquids of different densities often react together only slowly, especially when they are immiscible, unless continually stirred or shaken. Similarly, it is found worth while in some industrial processes to use elaborate devices to ensure thorough mixing of gases with liquids, e.g. in the Solvay process for making sodium carbonate.

The catalysts used in heterogeneous reactions, such as hydrogenations, are often prepared by chemical means in a special condition of extreme sub-division, so that the largest possible surface area of the metal is exposed to the reacting gases. Similarly, it is the large surface area of metals in the colloidal condition which accounts for their efficiency as catalysts.

Intensity of Ultra-Violet Light

Certain reactions are greatly accelerated by intense ultra-violet light. For example, hydrogen and chlorine combine only slowly in diffuse light at room temperature, even in the presence of a charcoal catalyst, but react together explosively when exposed to ultra-violet light. Molecules of the reactants absorb the light energy, thereby becoming activated, and react together rapidly in a series of chain reactions. This effect is considered in more detail in this chapter.

Concentration of Reactants

The effect of concentration upon reaction rate is summarised in the *law of mass action* put forward by Guldberg and Waage.

This law states that: "At constant temperature the rate of a reaction is proportional to the active masses of each of the reactants".

The term active mass is a special one best interpreted as the concentration of a substance raised to the appropriate power, which is the number of molecules of it written in the equation for that reaction. The concentration can be expressed in mole dm^{-3} or, where gaseous, as the partial pressure of the substance in the reaction mixture.

Suppose m molecules of A react with n molecules of substance B and let the rate of reaction at any given temperature be r, thus

$$mA + nB \xrightarrow{r}$$

then $r \propto$ (concentration of A)^m

and $r \propto$ (concentration of B)^n

so $r \propto \{$(concentration of A)$^m \times$ (concentration of B)$^n\}$

i.e. $r \propto [A]^m[B]^n$, where [X] is a symbol standing for the molar concentration of the substance X.

or $r \propto p_A{}^m \, p_B{}^n$, where p_x is a symbol standing for the partial pressure of X.

$\langle \; r = k.[A]^m[B]^n$, where k is a constant, called the *velocity constant* of that reaction at that temperature. It embodies all the factors, other than concentration, which affect reaction rate and it is only constant if these factors are not changed. The velocity constant at any temperature is equal to the rate at which the reaction proceeds when the concentrations of each of the reactants is unity, since under these conditions r is equal to k.

Reversible Reactions and Chemical Equilibrium

Many reactions are reversible, i.e. they can proceed in both directions under suitable conditions. If substances A and B are brought together they will react to give products C and D, for example, and if we start with C and D under suitable conditions they will react together to form A and B. In practice, if the reactants in a reversible reaction are left in contact an eqiuilibrium will eventually be set up between the two opposing reactions giving a mixture of reactants and products.

It is instructive to consider the steps by which this state of equilibrium is reached. If we start by mixing A and B, then reactions between them will produce quantities of C and D at a rate proportional to the active masses of A and B, which will, at first, be high. Similarly the C and D formed will combine to give A and B at a rate proportional to the active masses of C and D, which will, at first, be very low. So at the beginning the forward reaction will be much faster than the backward reaction and there will be a net shift from the left side to the right side of the equation representing the reaction. As this process continues, the concentrations of A and B will fall, and those of C and D will rise, and gradually the disparity between the rates of the two opposing reactions will diminish, until eventually when equilibrium is reached the reactions proceed in opposite directions at exactly the same speed.

This condition of equilibrium appears, at first sight, to be one of stagnation because no change occurs in the concentrations of the substances in the reaction mixture once it has been reached. But consideration will show that the condition is really a dynamic one, and no net change is noticed because the products of each reaction are being used up as fast as they are being formed.

Now let us apply the law of mass action to the reversible reaction represented by the following equation, assuming that the temperature is constant throughout :

$$mA + nB \underset{r_2}{\overset{r_1}{\rightleftharpoons}} pC + qD$$

Rate of left-to-right reaction $r_1 = k_1[A]^m[B]^n$

Rate of right-to-left reaction, $r_2 = k_2[C]^p[D]^q$

where k_1 and k_2 are the velocity constants of the forward and backward reactions respectively.

At equilibrium, $r_1 = r_2$

$$\therefore k_1[A]^m[B]^n = k_2[C]^p[D]^q$$

$$\therefore \frac{[C]^p[D]^q}{[A]^m[B]^n} = \frac{k_1}{k_2} = K$$

where K is a constant called the *equilibrium constant* of that reaction at that temperature.

It is important to notice that K is given by the ratio of the two velocity constants k_1 and k_2. Since these usually alter to a different degree when temperature is varied, it follows that K will vary with temperature also, but the addition of a catalyst to the system will affect the values of k_1 and k_2 in the same proportion and so not alter the value of K. This agrees with experimental observation.

To avoid confusion it is conventional to write the expression for K with the concentrations of the products (the substances on the right-hand side of the equation as written) in the numerator and with the concentrations of the reactants in the denominator. If this convention is ignored the value of K obtained is the reciprocal of the accepted value.

APPLICATION OF THE LAW OF MASS ACTION TO CHEMICAL EQUILIBRIA

This is illustrated by the following cases.

Reactants and Products in the Liquid State

$$C_2H_5OH + CH_3.COOH \rightleftharpoons CH_3.COOC_2H_5 + H_2O$$
Ethyl alcohol   Acetic acid   Ethyl acetate   Water

Let a mole of the alcohol react with b mole of the acid in volume V, and let x mole of ethyl acetate be present when the state of equilibrium is reached.

$$C_2H_5OH + CH_3.COOH \rightleftharpoons CH_2.COOC_2H_5 + H_2O$$

Initial amounts (mole):	a	b	0	0
Equilibrium amounts (mole):	$a - x$	$b - x$	x	x
Equilibrium concn. (mole dm^{-3}):	$\dfrac{(a-x)}{V}$	$\dfrac{(b-x)}{V}$	$\dfrac{x}{V}$	$\dfrac{x}{V}$

Now

$$\frac{[CH_3.COOC_2H_5].[H_2O]}{[C_2H_5OH].[CH_3COOH]} = K$$

Taking the molar concentrations at equilibrium as the active masses, we get

$$\frac{\frac{x}{V} \times \frac{x}{V}}{\frac{(a-x)}{V} \times \frac{(b-x)}{V}} = K = \frac{x^2}{(a-x)(b-x)}$$

If a, b, and x are given, then K can be calculated for that system at that temperature. Once we know K we can calculate the concentrations present in the equilibrium mixture at that temperature starting from any given amounts of reactants and products.

Example 10.1. One mole of ethyl alcohol was added to one mole of acetic acid and kept at 25°C; at equilibrium there were 1/3 mole of each of these reactants remaining and 2/3 mole of each of the products. How many gram of ethyl acetate will be present in the equilibrium mixture when 138 gram of ethyl alcohol are added to 120 gram of acetic acid and left to reach equilibrium at 25°C?

Solution.

Now

$$K = \frac{\frac{2}{3} \times \frac{2}{3}}{(1 - \frac{2}{3})(1 - \frac{2}{3})} = 4$$

Mol. wt. of C_3H_5OH = 46   ‹ a = 3 mole

Mol. wt. of CH_3COOH = 60   ‹ b = 2 mole

$$K = \frac{x^2}{(a - x)(b - x)} \qquad \therefore 4 = \frac{x^2}{(3 - x)(2 - x)}$$

$$\therefore \qquad 4(6 - 5x + x^2) = x^2$$

$$\therefore \qquad 3x^2 - 20x + 24 = 0$$

Solving this quadratic equation we get : $x = 5.1$ or $x = 1.57$.

The higher value is rejected as it is clearly impossible to produce over 5 mole of ethyl acetate from the quantities of reactants provided.

‹ There are 1.57 mole of ethyl acetate present at equilibrium. Mol. wt. of $CH_3.COOC_2H_5$ = 88.

‹ There are 1.57×88, i.e. 138.2 gram of ethyl acetate in the equilibrium mixture.

This example demonstrates several points often overlooked by students approaching this type of problem for the first time. Molar concentrations are used in the expression for K, so data quoted in gram has to be converted to mole first. Two values of the unknown are provided by the quadratic equation, but usually one of these can be rejected outright as being incompatible with the original data. The final result should always be stated in words and should provided the information required by the question, not just left in the form $x = 1.57$.

Reactants and Products in Gaseous State

	H_2	+	I_2	$\rightleftharpoons$	$2HI$
	Hydrogen		Iodine		Hydrogen iodide
Initial amounts (mole):	a		b		0
Equilibrium amounts (mole):	$(a - x)$		$(b - x)$		$2x$
Equilibrium concn. (mole dm^{-3}):	$\dfrac{(a - x)}{V}$		$\dfrac{(b - x)}{V}$		$\dfrac{2x}{V}$

Applying the law of mass action

$$\frac{[HI]^2}{[H_2][I_2]} = K_c = \frac{\left(\frac{2x}{V}\right)^2}{\frac{(a-x)}{V} \times \frac{(b-x)}{V}} = \frac{4x^2}{(a-x)(b-x)}$$

It will be noticed that in this example, as in the previous one, the factor V cancelled and did not appear in the final expression for K. It must not be assumed that this happens in every case, but it does occur whenever the total number of molecules of reactants and products are equal, as in these two equations. In such reactions the position of equilibrium is independent of pressure and the composition of the equilibrium mixture is not affected by changes in pressure, a fact confirmed by experimental observation.

Alternatively, using partial pressures in this example instead of molar concentrations, we get

$$H_2 \quad + \quad I_2 \quad \rightleftharpoons \quad 2HI$$

Initial amounts (mole): $\quad a \qquad\qquad b \qquad\qquad 0$

Equilibrium amounts (mole): $\quad a - x \qquad b - x \qquad 2x$

Now the partial pressure of a substance, X, is given by the expression

$$p_X = \frac{\text{Number of mole of X in system}}{\text{Total number of mole in system}} \times P$$

where P is the external or total pressure of the system.

$$\therefore \ p_{HI} = \frac{2x}{(a+b)} \times P \quad p_{H_2} = \frac{(a-x)}{(a-b)} \times P \quad p_{I_2} = \frac{(b-x)}{(a+b)} \times P$$

because the total number of mole in the system

$$= a - x + b - x + 2x$$
$$= a + b$$

Applying the law of mass action

$$K_p = \frac{(p_{HI})^2}{p_{H_2} \times p_{I_2}} = \frac{\left(\frac{2xP}{(a+b)}\right)^2}{\frac{(a-x)P}{(a+b)} \times \frac{(b-x)P}{(a+b)}} = \frac{4x^2}{(a-x)(b-x)}$$

The best way of verifying the law of mass action experimentally is to bring together varied but known concentrations of reactants at a given temperature, determine the composition of the equilibrium mixture in each case, and show the constancy in the value of K. This particular reaction is well suited to this treatment because the amount of iodine present can be determined volumetrically. The system is cooled suddenly to room temperature, which so lowers the rates of the forward and backward reactions that it prevents any substantial readjustment of the position of equilibrium before the analysis is complete.

Another Gaseous Reaction

$$N_2 \quad + \quad 3H_2 \quad \rightleftharpoons \quad 2NH_3$$
$$\text{Nitrogen} \qquad \text{Hydrogen} \qquad \text{Ammonia}$$

Applying the law of mass action to this equilibrium, we get

$$\frac{[NH_3]^2}{[N_2][H_2]^3} = K$$

Suppose that 3 moles of hydrogen are mixed with 1 mole of nitrogen in volume V, and that x mole of nitrogen are used up at equilibrium.

$$N_2 \quad + \quad 3H_2 \quad \rightleftharpoons \quad 2NH_3$$

Initial amounts (mole):	1	3	0
Equilibrium amounts (mole):	$1 - x$	$3 - 3x$	$2x$
Equilibrium concns. (mole dm^{-3}):	$\dfrac{(1-x)}{V}$	$\dfrac{(3-3x)}{V}$	$\dfrac{2x}{V}$

$$\therefore K_6 = \frac{\left(\frac{2x}{V}\right)^2}{\left(\frac{1-x}{V}\right)\left(\frac{3-3x}{V}\right)^3} = \frac{4x^2 V^2}{27(1-x)^4} = \frac{4x^2}{27(1-x)^4 P^2}$$

where P = pressure of the system at equilibrium.

$$\therefore \frac{4x^2}{27(1-x)^4 p^2} \text{ is constant at constant temperature.}$$

If x is small, which, in fact, it is under normal experimental conditions, then $(1 - x)^4$ is approximately equal to 1.

$$\therefore \frac{4x^2}{27P^2} \text{ is approximately constant.}$$

$$\therefore x^2 \text{ is approximately proportional to } P^2.$$

$$\therefore x \text{ is approximately proportional to } P.$$

So we conclude that the yield of ammonia is approximately proportional to the applied pressure, which result is confirmed by experiment.

Thermal Dissociation

As already explained, nitrogen tetroxide when heated undergoes reversible dissociation into nitrogen dioxide in accordance with the equation

$$N_2O_4 \rightleftharpoons 2NO_2$$

Suppose there are originally a mole of the tetroxide in volume V, and let x be the number of mole which are dissociated at any chosen temperature.

	N_2O_4	NO_2
Initial amounts (mole):	a	0
Equilibrium amounts (mole):	$a - x$	$2x$
Equilibrium concns. (mole dm^{-3}):	$\dfrac{a - x}{V}$	$\dfrac{2x}{V}$

Applying the laws of mass action to this equilibrium; we get :

$$\frac{[NO_2]^2}{[N_2O_4]} = K_c$$

$$\therefore \frac{\left(\frac{2x}{V}\right)^2}{\left(\frac{a-x}{V}\right)} = K_c = \frac{4x^2}{(a-x)V}$$

Thus at a constant temperature the extent of the dissociation x varies with the volume V and therefore with the pressure of the system. Since the term V appears in the denominator of the expression for the equilibrium constant, we conclude that the greater the pressure the smaller will be the dissociation, so a low pressure will encourage the dissociation and a high pressure reduce it.

Heterogeneous Equilibrium

Each reaction considered so far has been an example of homogeneous equilibrium in which the reactants and products are all gaseous or all in the liquid state at the equilibrium temperature. In the following reaction, which is a heterogeneous equilibrium, the substances involved are not all in the same phase or physical state:

$$CaCO_3 \quad \rightleftharpoons \quad CaO \quad + \quad CO_2$$

Calcium carbonate   Calcium oxide   Carbon dioxide

(solid)   (solid)   (gas)

Strictly speaking the law of mass action should not be applied to heterogeneous equilibria, but this difficulty is overcome by considering only the vapour phase here. The solids present do exert very small but finite vapour pressures, and these are taken as their active masses in the expression for K. Now the active mass of a solid at constant temperature must be constant because the vapour pressure of that solid is constant, being independent of the amount of solid present.

Applying the law of mass action to this equilibrium and using partial pressures, we get

$$\frac{p_{CaO} \times p_{CO_2}}{p_{CaCO_3}} = K_v$$

But, as explained above, p_{cao} and p_{CaCO_3} are themselves constant at constant temperature.

$$\text{So } p_{CO_2} = K_v \times \frac{\text{a constant}}{\text{a constant}} = \text{a constant.}$$

This means that the partial pressure of carbon dioxide present at equilibrium is constant at constant temperature and does not depend upon the amounts of the two solids present. This remarkable prediction is fully confirmed by experiment, the characteristic equilibrium pressure of carbon dioxide at any particular temperature, which is known as its dissociation pressure, being completely independent of the amounts of carbonate and oxide in the system. The same conclusion can be reached by applying the phase rule to this system. (Fig. 10.1) shows how the extent of the dissociation varies with the temperature; above 900°C the partial pressure of carbon dioxide exceeds 101 325 N m^{-2} and the gas escapes rapidly into the atmosphere.

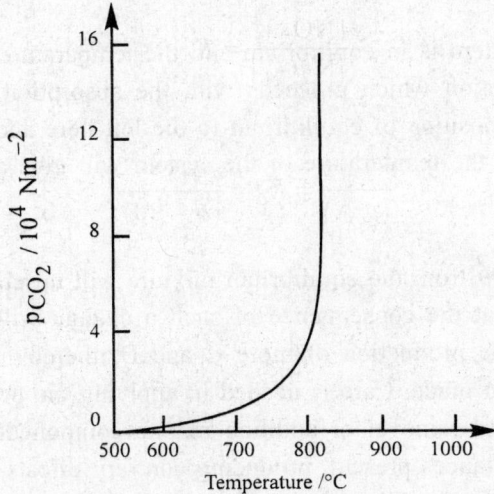

Fig. 10.1. Dissociation of calcium carbonate.

LE CHATELIER'S PRINCIPLE

This was originally stated thus: "If a system is in equilibrium and a constraint be applied, the system will respond in such a way as to tend to annul the constraint".

In this form the principle can be applied generally not only to chemical equilibria, but also to physical equilibria such as those existing between an undissolved solid and its saturated solution, or between a liquid and its vapour. Great care must be taken to ensure that the system to which the principle is applied really is in equilibrium. Here we are concerned with the application of the principle to chemical equilibria only, for which its meaning will be clearer if it is reworded thus: "If a system is in equilibrium and one of the factors pressure, temperature, or the concentration of a component, is altered, then the system responds in such a way as to oppose, or tend to oppose, the change that has been made".

Let us apply this principle to a hypothetical example

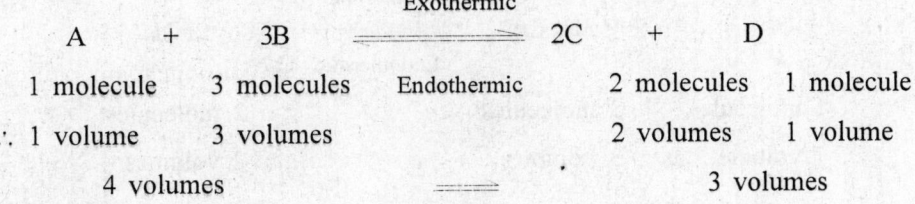

Change of Pressure

If, in a system in equilibrium, the pressure is raised, then Le Chatelier's principle predicts that the system will so respond to this external change as to oppose its effect, i.e. the position of equilibrium will alter by promoting that reaction which results in lower pressure because it proceeds with a reduction in volume. So in this case a higher pressure will cause a shift in the position of equilibrium from left to right and lead to a higher proportion of C and D in the reaction mixture. Lowering the pressure of the system will have the opposite effect.

Change in Temperature

By similar reasoning, if the system is in equilibrium and the temperature is raised, this change will be opposed by promoting the reaction which proceeds with the absorption of heat, i.e. the endothermic reaction, causing a shift in the position of equilibrium to the left here and increasing the proportions of A and B. Conversely, lowering the temperature of the system will give a higher yield of C and D.

Changes in Concentration

Partial or complete removal of D from the equilibrium mixture will upset the system in equilibrium. Le Chatelier's principle predicts that the consequence of such a change will be a shift in the position of equilibrium to the right with the production of more C and D, thereby opposing the reduction in the concentration of D that has been made. Care is needed in applying the principle to predict the effect of changes in concentration because removal or addition of one component of the system may alter the concentrations of the other substances present, producing contrary effects upon the equilibrium, and Le Chatelier's principle, being only a qualitative guide, cannot be used to decide which of these will predominate.

Applications of Le Chatelier's principle

From the above treatment it should be clear that the application of Le Chatelier's principle to chemical equilibria leads to these two generalisations :

1. When an increase in pressure is applied to a system in equilibrium it causes the equilibrium to be displaced in that direction which leads to a contraction in volume.

2. When the temperature is raised in a system in equilibrium, it causes the equilibrium to be displaced in that direction which leads to an absorption of heat, i.e. it favours the endothermic reaction.

The value of Le Chatelier's principle in predicting the conditions for maximum yield of a particular product can be demonstrated by applying these general conclusions to some important reversible reactions :

$$(1) \quad N_2 \ + \ 3H_2 \ \underset{\text{Endothermic}}{\overset{\text{Exothermic}}{\rightleftharpoons}} \ 2NH_3$$

1 molecule	3 molecules	2 molecules
∴ 1 volume	3 volumes	2 volumes
4 volumes		2 volumes

High pressure will favour the left-to-right reaction because it results in a decrease in volume. High temperature will favour the endothermic reaction, the dissociation of ammonia into its elements. We conclude that the conditions necessary for the highest possible yield of ammonia are high pressure and low temperature.

$$(2) \quad 2SO_2 \ + \ O_2 \ \underset{\text{Endothermic}}{\overset{\text{Exothermic}}{\rightleftharpoons}} \ 2SO_3$$

2 molecules	1 molecule		2 molecules
2 volumes	1 volume		2 volumes
	3 volumes	$=\!\!=\!\!=\!\!\rightharpoonup$	2 volumes

Again application of a higher pressure will drive the equilibrium to the right because there is a reduction in volume as this reaction proceeds. Higher temperature will promote the endothermic right-to-left reaction, as in (1). If a large excess of oxygen is added to the system in equilibrium this will drive the equilibrium further to the right. Thus the optimum conditions for a high yield of sulphur trioxide from a given amount of sulphur dioxide would appear to be high pressure, low temperature, and a large excess of air or oxygen.

$$(3) \qquad N_2 \quad + \quad O_2 \quad \overset{\text{Exothermic}}{\underset{\text{Endothermic}}{\rightleftharpoons}} \quad 2NO$$

1 molecule	1 molecule		2 molecules
1 volume	1 volume		2 volumes
	2 volumes	$=\!\!=\!\!=\!\!=$	2 volumes

Altering the pressure will not affect the position of equilibrium because there is no change in volume when the reactions take place. Raising the temperature will displace the equilibrium in the direction of the endothermic reaction, i.e. to the right. Thus the yield of nitric oxide is not affected by pressure and is highest at high temperature.

$$(4) \qquad CO \quad + \quad H_2O \quad \overset{\text{Exothermic}}{\underset{\text{Endothermic}}{\rightleftharpoons}} \quad CO_2 \quad + \quad H_2$$

1 molecule	1 molecule		1 volume	1 molecule
1 volume	1 volume		1 volume	1 volume
	2 volumes	$=\!\!=\!\!=\!\!=$		2 volumes

Again, changing the pressure has no effect upon the position of equilibrium. Raising the temperature of the system will favour the endothermic right-to-left reaction, so the highest yield of hydrogen will be obtained by using as low a temperature as possible.

It must be stressed that Le Chatelier's principle, valuable as it is as a means of predicting the conditions which provide the highest yield of a desired product, is not concerned at all with the factors governing the rate at which that yield is obtained. For instance, when ammonia is synthesised by the Haber process the operating temperature is about 500°C, despite the prediction in (1) above that a low temperature is desirable for the highest possible yield. This is because at lower temperatures nitrogen and hydrogen combine too slowly, even in the presence of a catalyst. The process is operated, therefore, at the lowest temperature consistent with a convenient rate of reaction, the chosen temperature representing a compromise between the conflicting demands of high yield and high rate. The same considerations apply to the oxidation of sulphur dioxide (reaction 2 above), the operating temperature in the contact process being about 500°C, and to the Bosch process (reaction 4 above), which is normally operated at 450°C.

ORDER OF REACTION

The rate of a reaction can be found experimentally by determining the concentration of one of the reactants or products at suitable intervals and calculating the rate of change. This may be done by removing portions from the reaction mixture from time to time and analysing them volumetrically, usually by back-titration methods to prevent further reaction occurring during the estimation. It is often preferable, however, to use a physical method such as observation of changes in volume, pressure, refractive index, colour, or optical rotation, since there is then no danger of the measurements interfering with the speed or course of the reaction.

As a result of these experiments it is found that a reaction tends to get slower and slower as the reactants are used up. Fig. 10.2. shows a typical fall in reaction rate with time. It is also found that reactions differ in the way in which their rates depend upon the concentration of reactants, this difference being expressed in terms of order of reactions.

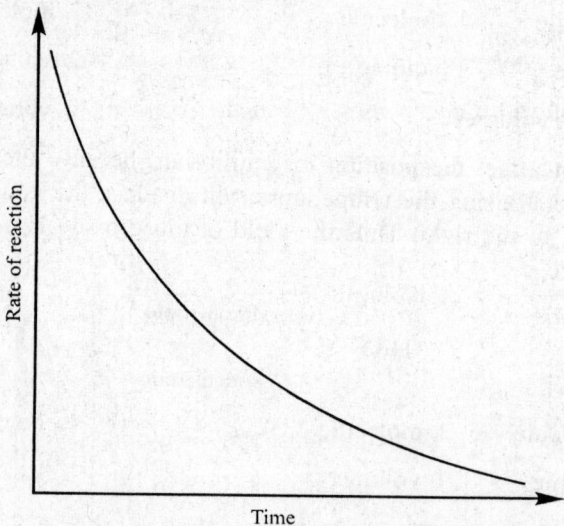

Fig. 10.2. Variation of reaction rate with time.

First Order Reactions

These are reactions where experimental investigation shows that the rate is proportional to the concentration of only one substance. Most of them are reactions in which a substance undergoes chemical decomposition or radioactive decay, as in the following examples.

$$C_4H_9OH \quad = \quad C_4H_8 \ + H_2O$$
Tertiary butyl alcohol $\qquad$ Cyclobutane

$$2N_2O_5 \quad = \quad 4NO_2 + O_2$$
Nitrogen pentoxide

In some reactions between two substances one of the reactants is present in such great excess that its concentration remains virtually unchanged throughout the reaction, so that the rate varies only with the concentration of the other reactant and the reaction is, in practice, of the first order. The hydrolysis of cane sugar is a good example:

$$C_{12}H_{22}O_{11} + H_2O = C_6H_{12}O_6 + C_6H_{12}O_6$$

$$\text{Sucrose} \qquad\qquad \text{Glucose} \qquad \text{Fructose}$$

Let a represent the initial concentration of reactant and $(a - x)$ the concentration of reactant remaining at time t. Then in a first order reaction the rate of reaction $\dfrac{dx}{dt}$ will be proportional to the concentration of reactant at time t, i.e. $(a - x)$, and we can write

$$\text{rate of reaction} = \frac{dx}{dt} = k\,(a - x) \text{ (where } k \text{ is a constant).}$$

Thus

$$\frac{dx}{(a - x)} = k\,dt$$

and on integrating, we get

$$-\log_e (a - x) = kt + \text{integration constant}$$

Now $x = 0$ when $t = 0$, so substituting these values in the equation reveals the integration constant to be $-\log_e a$.

$$\therefore \ \log_e a - \log_e (a - x) = kt$$

$$\therefore kt = \log_e \frac{a}{(a - x)} = 2.303 \log_{10} \frac{a}{(a - x)}$$

Thus in a first order reaction if $(a - x)$, the concentration of reactant at time t, is determined at intervals and a graph of $\log e \dfrac{a}{a - x}$ is plotted against time, we obtain a straight line through the origin as in Fig. 10.3, with a gradient equal to k, the reaction rate constant.

We can obtain an expression for the half-life, i.e. the time taken for 50% of the reactant to be consumed, by substituting $x = \dfrac{a}{2}$ in the equation for a first order reaction.

$$kt_{\frac{1}{2}} = \log_e \frac{a}{a/2} = \log_e 2$$

$$\therefore t_{\frac{1}{2}} = \frac{\log_e 2}{k} = \frac{0.693}{k}$$

Thus in a reaction of the first order the half-life is completely independent of the initial concentration of the reactants, and the reaction takes exactly the same length of time to reach halfway to completion whatever the concentration originally taken. This is a matter which can easily be put to the test experimentally and can therefore be used as a means of recognising first order reactions.

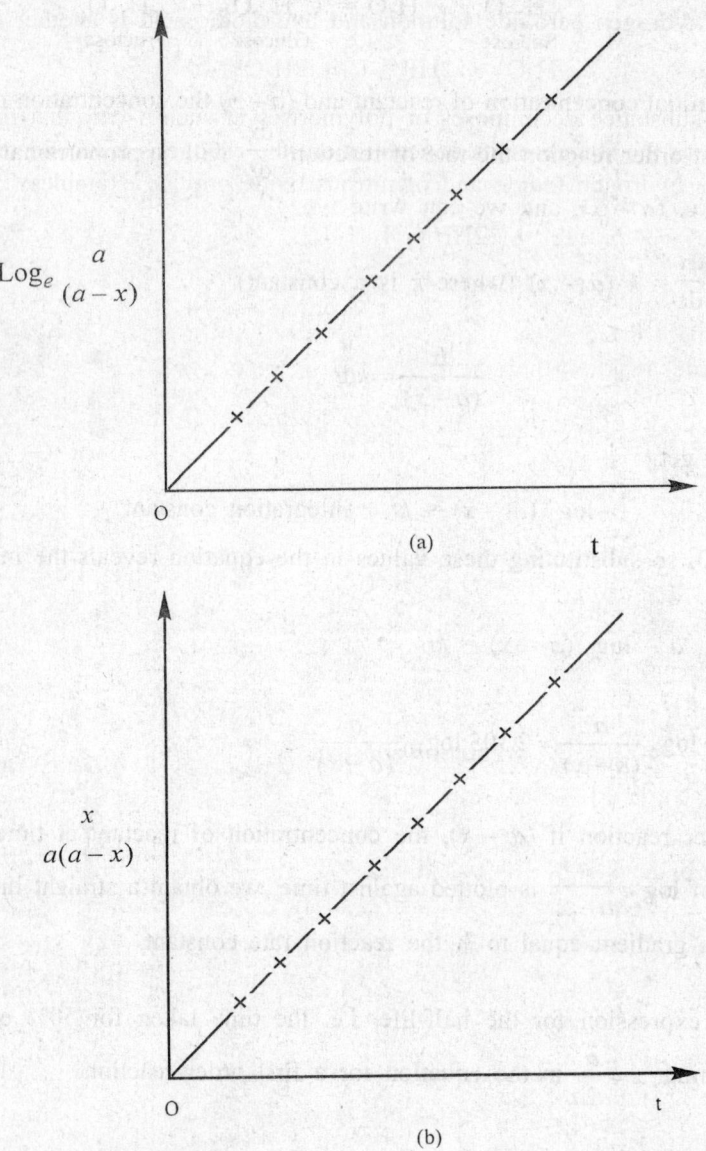

Fig. 10.3. Graphs distinguishing the order of reaction (a) First order reaction; (b) Second order reaction. In each case the gradient is equal to the rate constant, k.

Second Order Reactions

In these reactions the rate is proportional to the concentrations of two substances or to the square of the concentration of one reactant. For example, if in a reaction between two substances it is found by experiment that the rate of reaction at any instant is proportional to the product of their concentrations, then that reaction is of the second order. The hydrolysis of ethyl acetate with sodium hydroxide solution is of this kind:

$$CH_3.COOC_2H_5 + NaOH = CH_3.COONa + C_2H_5OH$$

and the reaction between hydrogen peroxide solution and hydroiodic acid is another example:

$$H_2O_2 + 2HI = I_2 + 2H_2O$$

Alternatively, where a substance decomposes or polymerises in such a way that the reaction rate at any instant is proportional to the square of its concentration, the change is a second order reaction. The decompositions of gaseous hydrogen iodide and of nitrous oxide provide examples:

$$2HI = H_2 + I_2$$

$$2N_2O = 2N_2 + O_2$$

In the simplest case where the concentrations of two reactants are both a initially and $(a - x)$ after time t, then in a second order reaction the rate at time t is proportional to $(a - x)^2$ and the rate equation is written

$$\text{Rate of reaction} = \frac{dx}{dt} = k.(a-x)^2$$

$$\therefore \frac{dx}{(a-x)^2} = k dt$$

which on integration gives

$$\frac{1}{(a-x)} = kt + \text{integration constant}$$

Now $x = 0$ when $t = 0$, so the integration constant is $\frac{1}{a}$ and

$$\frac{1}{(a-x)} = kt + \frac{1}{a}$$

or

$$\frac{x}{a(a-x)} = kt$$

Thus when $\frac{x}{a(a-x)}$ is plotted against time, a second order reaction of this simple type gives a straight line through the origin, as in Fig. 10.3(b), with a gradient equal to the rate constant k.

Again, an expression for the half-life can be obtained by substituting $x = \frac{a}{2}$ when $t = t_{\frac{1}{2}}$:

$$\frac{a/2}{a(a-a/2)} = kt$$

$$\therefore t_{\frac{1}{2}} = \frac{1}{ak}$$

It is thus a characteristic of second order reactions in which the initial concentrations of the reactants are equal that the half-life is inversely proportional to that initial concentration, which enables us to distinguish such reactions experimentally from those of the first order.

MOLECULARITY

Many chemicals reactions, although normally represented by a single chemical equation, take place in fact in a number of consecutive stages or steps, some fast and some relatively slow. Very often the intermediate products formed during these individual stages are free radicals which are too unstable to be isolated, but a great deal of experimental evidence has been accumulated in recent years on the complicated mechanisms of such reactions. Just as a chain is only as strong as its weakest link, so the overall rate of a reaction is limited by its slowest stage, which is known as the rate-determining stage for that reason. The molecularity of a reaction is defined as the number of molecules taking part in a single-step reaction or in the rate-determining stage of a composite reaction. Reactions with a molecularity of one and two are known as unimolecular and bimolecular respectively.

The thermal decomposition of nitrogen dioxide, for example, is bimolecular

$$2NO_2 = 2NO + O_2$$

whereas the decomposition of nitrous oxide which is normally represented by the equation

$$2N_2O = 2N_2 + O_2$$

has a unimolecular rate-determining step as its first stage:

$$N_2O = N_2 + O$$

Unlike order of reaction, which can be determined without any knowledge of mechanism merely from experimental measurements of the rate of a reaction, molecularity is a theoretical quantity which can only be deduced when details of the rate-determining stage of a reaction are known. When a reaction is a simple one proceeding in only one stage, then the order of reaction and molecularity are usually the same; a good example is the decomposition of hydrogen iodide which is both bimolecular (because two HI molecules are involved in the reaction) and of the second order (because the rate of the reaction is proportional to the square of the hydrogen iodide concentration)

$$2HI = H_2 + I_2$$

A warning must be given about chemical equations. Most of them are composite ones embodying the various stages of a reaction in one overall relationship. They indicate the reactants and final products and the relative quantities of each of them involved in any reaction but they convey no information about the mechanism or actual course of the reaction. The student will soon be convinced that this is true if he considers the following equations and reflects for a moment on the remoteness of the chance that these large numbers of reactant molecules might collide together simultaneously in order to react.

$$3Cu + 8HNO_3 = 3Cu(NO_3)_2 + 2NO \uparrow + 4H_2O$$

$$2KMnO_4 + 10FeSO_4 + 8H_2SO_4 = K_2SO_4 + 2MnSO_4 + 5Fe_2(SO_4)_3 + 8H_2O$$

$$4P + 3NaOH + 3H_2O = 3NaH_2PO_2 + PH_3 \uparrow$$

$$4HN_3 + 5O_2 = 4NO + 6H_2O$$

CHAIN REACTIONS

When methane and chlorine react together under suitable conditions research has shown that the first step consists of reaction between a methane molecule and an individual atom of chlorine which is present in the mixture :

$$CH_4 + Cl \cdot = CH_3 \cdot + HCl$$

The methyl radical so formed then reacts with a chlorine molecule producing methyl chloride and another atom of chlorine thus:

$$CH_3 \cdot + Cl_2 = CH_3Cl + Cl \cdot$$

This series of reactions then repeats itself over and over again, each cycle regenerating an active chlorine atom.

A reaction of this kind involving a complete cycle of steps which is capable of unlimited repetition is called a *chain reaction*. In such reactions free radicals or individual atoms are usually formed, as in the above example; they are very active chemically and react rapidly causing the chain to spread through the mixture of gases. Sometimes several million cycles are repeated before such a chain of reactions is terminated. Termination usually happens when the radicals or atoms which initiate the cycle are removed from the mixture by combination with each other thus :

$$CH_3 \cdot + CH_3 \cdot = CH_3.CH_3$$

or
$$Cl \cdot + Cl \cdot = Cl_2$$

or
$$Cl \cdot + CH_3 \cdot = CH_3Cl$$

The reaction of hydrogen and chlorine is another good example of a chain reaction. The stages of the cycle are believed to be :

$$H_2 + Cl \cdot = HCl + H \cdot$$

$$H + Cl_2 = HCl + Cl \cdot$$

In both of these examples, the mechanism suggested for the reaction presupposes the existence in the gaseous mixture of highly reactive individual atoms of chlorine. These atoms, which are responsible for initiating the chain reactions, are produced when a chlorine molecule undergoes dissociation

$$Cl_2 = Cl \cdot + Cl \cdot$$

which it does when heated to a high temperature or exposed to ultra-violet light. It follows that although these reactions are infinitely slow at room temperature in the dark, once the conditions of light or temperature needed to initiate them are provided, they proceed at great speed and can be explosive in nature. Many polymerisations are chain reactions, some of them being of considerable industrial importance. In reactions of this kind a substance such as benzoyl peroxide is frequently used as a catalyst to initiate the reactions by producing a supply of free radicals. In the same way, chain reactions are notoriously sensitive to substances which act as negative catalysts or inhibitors; these are usually impurities or specially chosen substances capable of uniting with and removing free radicals from the reaction mixture so that the propagating cycle is terminated.

ENERGY OF ACTIVATION

As already explained the rate of most chemical reactions is approximately doubled for each 10 K rise in temperature. This is an experimental fact which demands an explanation. Now calculations based upon the kinetic theory show that the number of collisions per second between molecules of a gas only increases a few per cent when the temperature of the system rises by 10 K, so clearly the sharp increase in reaction rate cannot be explained in that way. Moreover, these calculations indicate that only a small proportion of the molecular collisions do, in fact, result in reaction and that in some cases the reaction would be many millions of times faster if every collision was fruitful. This led Arrhenius to propose that two colliding molecules must possess a certain minimum amount of energy, known as the energy of activation, before they could react together. Now it can be calculated that the fraction of the molecules in a gaseous system possessing this critical energy at any instant is approximately $e^{-E/RT}$, where E is the energy of activation of the reaction, R is the universal gas constant, and T is the absolute temperature. Thus in a reaction with a high energy of activation the proportion of collisions leading to reaction would be very small and that reaction would be slow. This is generally true of reactions between covalent molecules, such as those taking place in organic chemistry, whereas reactions occuring between ions in solution involve only small energies of activation and are therefore usually extremely fast.

The activation theory also accounts very satisfactorily for the marked effect of temperature upon reaction rate, since temperature appears in the index of the exponential function which means that even a small rise in temperature causes a rapid increase in the proportion of activated molecules. This point can be made clear by considering briefly the way in which the kinetic energy of a gas is believed to be distributed between its molecules. It can be shown that provided the number of molecules considered is extremely large, which of course is always the case in practice, their velocities at any instant will be as shown in Fig. 10.4.

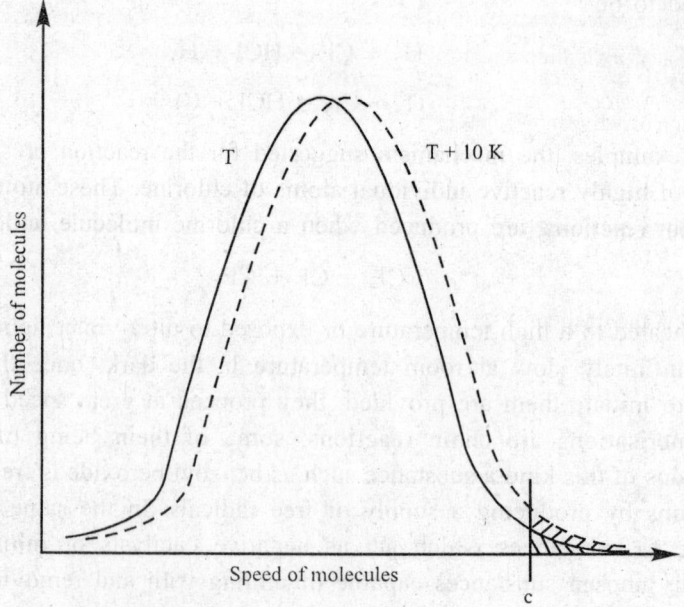

Fig. 10.4. Variation of molecular speeds with temperature.

As a result of random collisions some molecules will be virtually stationary whilst others will be moving at enormous speeds, but the bulk of the molecules will have a velocity between these two extremes. The smooth curve shows the velocities at temperature T and the dashed curve the different distribution when the temperature is 10 K higher. It will be seen that the whole curve has been displaced a little to the right and its shape slightly altered by the rise in temperature, but the important thing to notice is that the proportion of molecules with sufficient energy to react (those with a velocity greater than c) has approximately doubled, as shown by the shaded areas.

The energy of activation idea also helps to explain the working of some types of catalyst. It is suggested that catalysis often brings about an increase in the rate of reaction by making it possible to reach the product by an alternative route each stage of which has a lower energy of activation than the uncatalysed reaction and is therefore relatively fast. It is rather like a person trying to get from one point to another in a mountainous area who finds it quicker to climb over a number of small peaks and follow a roundabout route than to scale the much larger mountain which lies directly between the two points.

Example 10.1. At 25°C the specific rate constant for the hydrolysis of ethyl acetate by NaOH is 6.36 litre mole^{-1} min^{-1}. Starting with concentrations of base and ester of 0.02 moles/litre, what propotion of ester will be hydrolysed in 10 mins.

Solution. $k = 6.36$ litre mole^{-1} min^{-1}. Unit of k indicates that the reaction is second order. So

$$k = \frac{1}{t} \cdot \frac{x}{a(a-x)}$$

Given $6.36 = \frac{1}{10} \cdot \frac{x}{0.02(0.02-x)}$ or $x - 0.011$ moles / litre.

So, % of ester hydrolysed $\times \frac{0.011}{0.02} \times 100 = 55\%$.

Example 10.2. The energy of activation of a bimolecular gaseous decomposition is 20,000 cal/mol. Calculate the fraction of the molecules having sufficient energy to decompose at 27°C and at 227°C.

Solution. We know that if n_E is the number of active molecules having energy $\geq E$ (activation energy) and n_0 is the total molecules; their fraction of active molecules $\dfrac{n_e}{n_0} = e^{-E/RT}$

So at 27°C i.e. 300°A the fraction $= e^{-\frac{20,000}{1.987 \times 300}} = 2.7 \times 10^{-15}$

and at 227°C i.e. 500°A the fraction $= e^{-\frac{20,000}{1.987 \times 500}} = 1.8 \times 10^{-9}$

Chapter 11

Electrochemistry

INTRODUCTION

Electrochemistry is the study of the various chemical effects of electricity. An *electrolyte* is a substance which conducts electricity when dissolved in water or when molten; other substances are known as *non-electrolytes*. For example, all acids, bases, and salts are electrolytes, but substances like sugar and alcohol are not. This distinction can easily be demonstrated by setting up the apparatus shown in Fig. 11.1. When a potential difference is applied to the electrodes, no detectable current flows if the beaker contains pure water or a solution of sugar or alcohol, but as soon as an electrolyte such as dilute sulphuric acid or sodium chloride is added the lamp lights, showing that the solution is now conducting. Similarly, when a potential difference is applied to two carbon electrodes dipping into a U-tube containing molten lead bromide, a current flows through the circuit and the lamp lights, showing the molten salt to be a conductor, and after a short while globules of molten lead appear at the cathode and bromine vapour is given off at the anode.

The conduction of electricity by an electrolyte, which is known as *electrolysis*, differs from metallic conduction in two main respects. Firstly electrolysis is invariably accompanied by chemical changes resulting, for example, in the liberation of a gas, the deposition of a metal, or the dissolution of an electrode, whereas only physical changes occur when an electric current passes through a piece of metal. The second difference concerns the way in which conductivity varies with temperature; an electrolyte conducts electricity more readily at higher temperatures whereas in metals the reverse is the case.

IONIC THEORY

Faraday's Laws

Faraday summarised his experimental results in his laws of electrolysis:

1. The mass (m) of substance dissolved or liberated in electrolysis is proportional to the quantity of electricity which passes through the electrolyte. The quantity of electricity is given by the steady current I multiplied by the time t.

    i.e. $\qquad\qquad m \propto I \times t \qquad\qquad$ for a constant current

    or $\qquad\qquad m \propto \int I dt \qquad\qquad$ for a fluctuating current

    $\qquad\qquad \therefore\ m = e \times I \times t$

266

where e is a constant known as the electrochemical equivalent of the substance and is the mass set free by the passage of one coulomb of electricity (e.g. 1 ampere for 1 second).

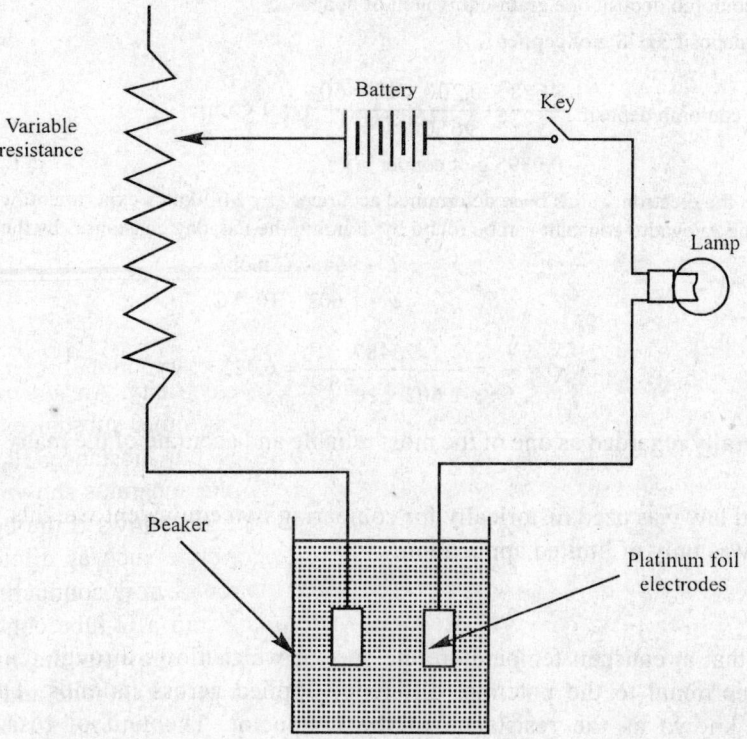

Fig. 11.1. Apparatus to demonstrate electrolytic conduction.

2. The masses of different products set free by passing a given quantity of electricity through different electrolytes are proportional to the chemical equivalents of the substances concerned,

i.e.

$$\frac{m_1}{m_2} = \frac{e_1}{e_2} = \frac{E_1}{E_2}$$

where m_1 and m_2 are the masses liberated by the same quantity of electricity from substances with chemical equivalents E_1 and E_2 respectively. It follows from this law that E/e is a constant and that the same quantity of electricity will liberate one gram-equivalent of any product, i.e. 1.008 g of hydrogen, or 107.88 g of silver, or 31.785 g of copper. This quantity, which is 96487 coulomb mol^{-1}, is known as the Faraday constant, and is denoted by the symbol F.

3. The products of electrolysis appear only at the electrodes, i.e. the wires or plates leading the current into and out of the electrolyte. Faraday called the positive electrode the anode and the other the cathode, terms which have grown into general use.

These laws are obeyed exactly over very wide ranges of temperature, concentration, and current. Advantage was taken of their exactness to define the unit of current, the ampere, in terms of its electrolytic effect, but with the advent of SI units this definition has now been superseded. A silver coulometer which involves a silver anode, a platinum cathode, and a solution of silver nitrate as electrolyte, is used for measuring quantities of electricity very accurately.

Example 11.1. Calculate the mass of copper deposited on the cathode when a current of 0.200 ampere is passed through a solution of copper sulphate for ten minutes. The equivalent weight of copper is 31.78.

Solution. Now 96500 coulomb deposit one gram-equivalent of copper.

$\therefore$ 96500 coulomb deposit 31.78 g of copper.

$\therefore$ 0.200 × 10 × 60 coulomb deposit $\dfrac{31.78 \times 0.200 \times 10 \times 60}{96500}$ g of copper

$$= 0.0395 \text{ g of copper}$$

Since the charge on the electron, e, has been determined accurately by Millikan's experiment, which is described in most textbooks of physics, the Avogadro constant can be found by dividing the Faraday constant F by the electronic charge:

$$F = 96487 \text{ C mol}^{-1}$$
$$e = 1.602 \times 10^{-19} \text{ C}$$

$$\therefore \quad N_A = \frac{F}{e} = \frac{96487}{1.602 \times 10^{-19}} = 6.023 \times 10^{23} \text{ mol}^{-1}$$

This method is generally regarded as one of the most reliable and accurate of the many methods available for determining N_A.

Faraday's second law was used historically for comparing two equivalent weights. Although this was an accurate method it was only of limited application.

Ohm's Law

Experiment shows that at constant temperature the current which flows through a conductor (electrolytic or metallic) is proportional to the potential difference applied across its ends. The ratio of current to applied voltage is known as the resistance of the conductor. The unit of resistance is the ohm; a conductor has a resistance of one ohm when it allows a current of one ampere to pass through it when subjected to a potential difference of one volt. At constant temperature the resistance of a given conductor is directly proportional to its length and inversely proportional to its cross-sectional area.

i.e. $\quad R = \rho . \dfrac{l}{A}$ where R = resistance

ρ = resistivity (rho)
l = length
A = area of cross-section.

Thus the resistivity of a conductor, which is a constant for any given substance, is the resistance between two opposite faces of a unit cube of that substance since l/A is then equal to unity.

Conductivity

In electrochemistry we are rarely concerned with the resistivity of an electrolyte, preferring to use the conductivity instead. This is the reciprocal of the resistivity and is defined as the conductance of a one centimetre cube of the substance or solution. The greater the conductivity the more readily does the electrolyte allow the passage of an electric current through it. Conductivity κ (kappa), is usually expressed in reciprocal ohm per centimetre (Ω^{-1} cm^{-1}), one Ω^{-1} corresponding to the flow of a current of one ampere when the potential difference applied to the conductor is one volt, i.e. $\kappa = 1/\rho$. The conductivity of an electrolyte depends not only upon its nature but also upon the concentration of the solution. The typical way in which it varies with concentration is shown in Fig. 11.2.

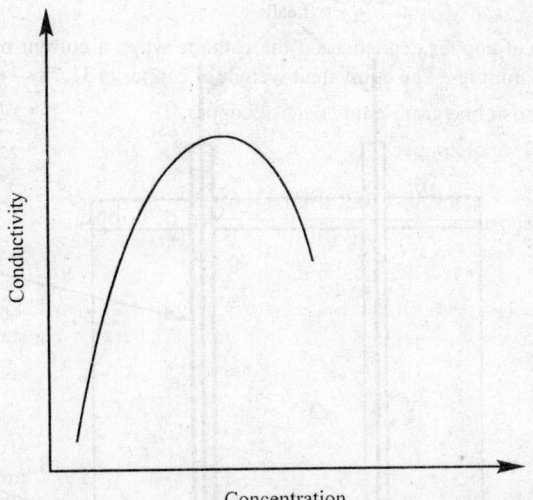

Fig. 11.2. Variation of conductivity with concentration.

Measurement of Conductivity

This is normally carried out by the Kohlrausch method which uses the Wheatstone bridge circuit shown in Fig. 11.3 to compare the resistance of a cell containing the electrolyte with a known but variable resistance R_2. The cell used is shown in Fig. 11.4.

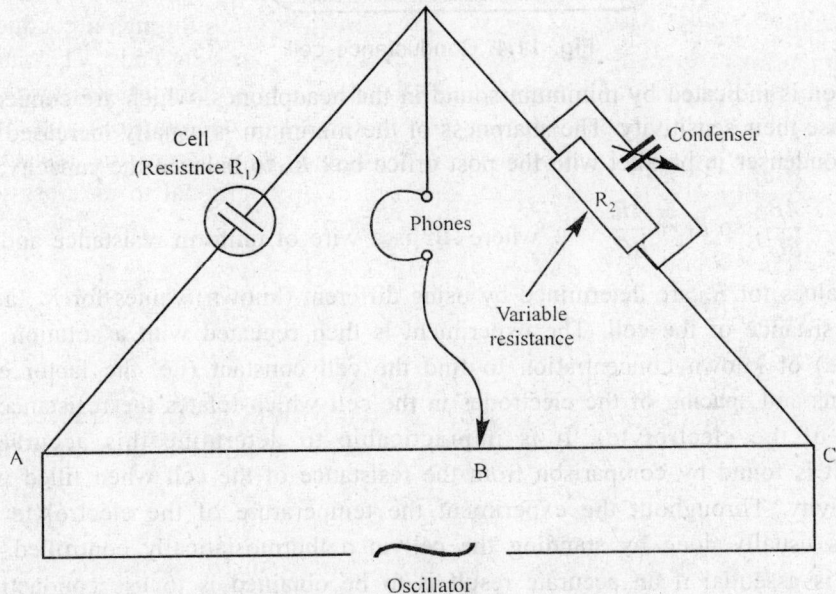

Fig. 11.3. Measurement of conductivity.

It is made of silica or specially resistant glass and is fitted with two platinum electrodes (usually coated with platinum black) which are fixed rigidly in position so that their spacing apart is constant. Alternating current from an induction coil or valve oscillator is used (about 1000 Hz) to avoid polarisation effects (i.e. changes in the resistance of the cell due to the formation of bubbles of gas on the surfaces of the electrodes as a result of electrolysis).

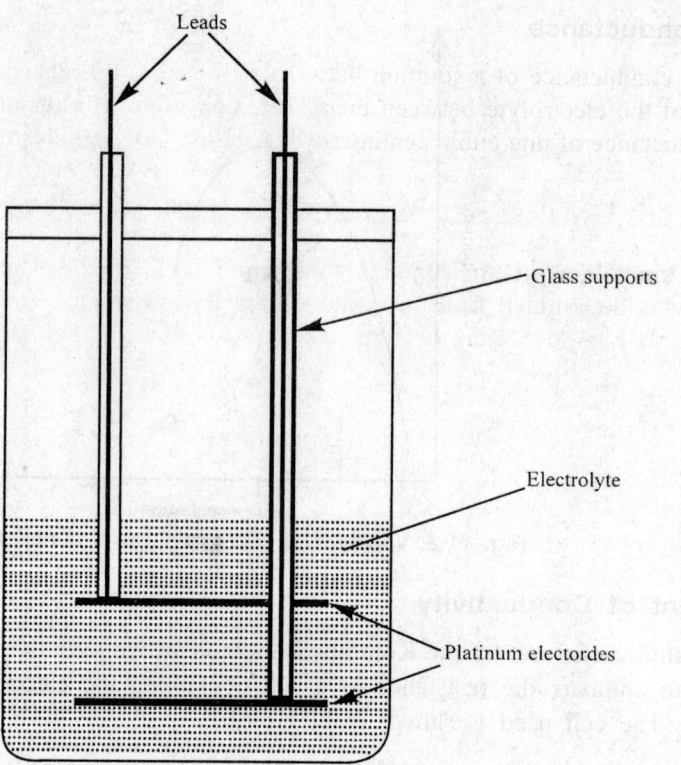

Leads

Glass supports

Electrolyte

Platinum electordes

Fig. 11.4. Conductance cell.

The null position is indicated by minimum sound in the headphones, which are connected to a valve amplifier to increase their sensitivity. The sharpness of the minimum is usually increased by connecting a small variable condenser in parallel with the post office box R_2 to balance the capacity of the cell. At the minimum $\dfrac{R_1}{R_2} = \dfrac{AB}{BC}$, so $R_1 = \dfrac{AB}{BC} \times R_2$ where AC is a wire of uniform resistance and B is a sliding contact. Several values for R_1 are determined by using different (known) values for R_2, and the average is taken as the resistance of the cell. The experiment is then repeated with a solution of a salt (e.g. potassium chloride) of known concentration to find the cell constant (i.e. the factor embodying the physical dimensions and spacing of the electrodes in the cell which relates the resistance of the cell to the conductivity of the electrolyte). It is impracticable to determine this accurately by direct measurement, so it is found by comparison from the resistance of the cell when filled with a solution of know conductivity. Throughout the experiment the temperature of the electrolyte must be kept constant, which is usually done by standing the cell in a thermostatically controlled bath. Another precaution which is essential if an accurate result is to be obtained is to use conductivity water for making the solution of the electrolyte. This is water which has been specially purified by deionisation or by repeated distillation in tin vessels until its conductivity has a constant and extremely low value (about $5 \times 10^{-7}\ \Omega^{-1}\ \text{cm}^{-1}$ at 18°C) which is negligible compared with that of the electrolyte to be dissolved in it. Ordinary distilled water, which often contains alkali dissolved in it from the glass vessel in which it is stored, is not pure enough for this purpose.

Molar Conductance

This is the conductance of a solution between two electrodes spaced one centimetre apart which enclose one mole of the electrolyte between them. It is conventionally denoted by Λ. Since the conductivity, κ, is the conductance of one cubic centimetre of solution between electrodes one centimetre apart, it follows that

$$\Lambda = \kappa \times V$$

where V = the volume in cm^3 containing one mole of electrolyte. The molar conductance is not measured directly but is determined from the conductivity by using this relationship. A physical picture, such as that shown in Fig. 11.5, may help to make clear the distinction between the two.

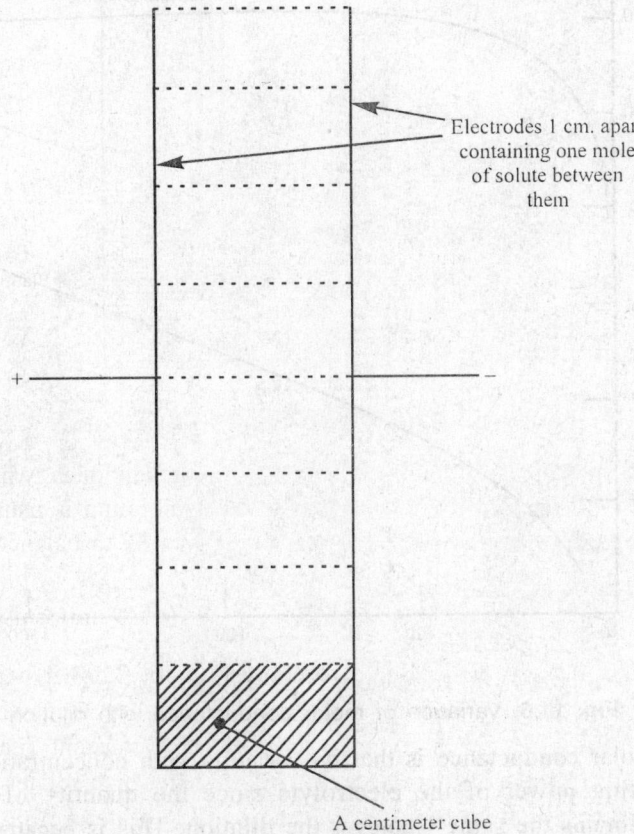

Electrodes 1 cm. apart containing one mole of solute between them

A centimeter cube

Fig. 11.5. Molar conductance.

The molar conductance of a solution of any given electrolyte depends upon the temperature and the concentration. Experiment shows that at room temperature there is an increase of about 2% in conductance for each 1 K rise in temperature. It is significant that this figure agrees with the decrease in viscosity of the solvent with rising temperature, suggesting that there may be a connection between the two properties.

The two general ways in which the molar conductance varies with concentration are shown in Fig. 11.6, where it is plotted against the dilution, i.e. the volume of solution in dm^3 which contains one mole of solute. Curve A is typical of most salts, and of hydrochloric, sulphuric, and nitric acids, and also of sodium and potassium hydroxides. In these cases the conductance is always high and increases only slightly on dilution, soon reaching a maximum value known as the molar conductance at infinite dilution and denoted by Λ_∞. Substances whose molar conductance varies in this way are known as strong electrolytes. Curve B is typical of most organic acids, and of ammonium hydroxide, and of certain mercury salts. In concentrated solutions the molar conductance is very low but it increases steadily with dilution. Substances of this type are known as weak electrolytes. No sharp distinction exists between the two types and some electrolytes of intermediate character are known.

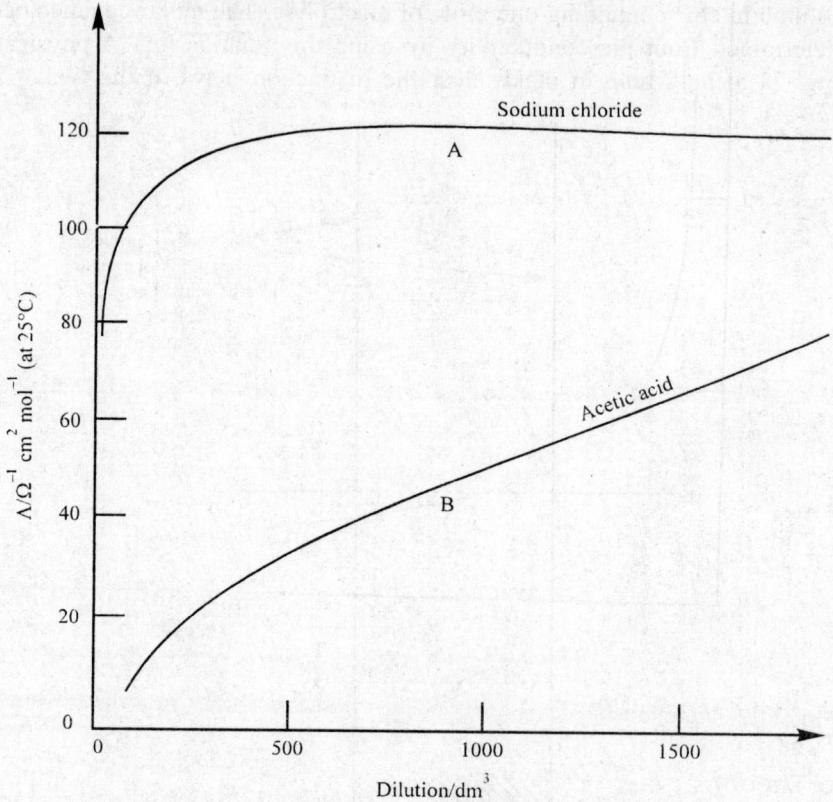

Fig. 11.6. Variation of molar conductance with dilution

The importance of molar conductance is that its variation with concentration faithfully portrays the variation in the conducting power of the electrolyte since the quantity of electrolyte between the electrodes in Fig. 11.5 remains the same whatever the dilution. This is because on dilution the volume of solution taken changes so that one mole of electrolyte is always present. This is not so with conductivity where the variation with concentration represents the resultant of two different changes— one in the conducting ability of the electrolyte and the other in the quantity of electrolyte contained in the cubic centimetre of solution between the two electrodes. Thus changes caused by dilution in the effectiveness of an electrolyte as a conductor of electricity are obscured in the case of conductivity, κ.

For a strong electrolyte the molar conductance at infinite dilution is best determined by plotting the molar conductance against the square root of the concentration. For very dilute solutions this gives a straight line graph (Fig. 11.7) which can then be extrapolated to cut the axis, the intercept being taken as Λ_∞. Weak electrolytes also have a characteristic value for the molar conductance at infinite dilution, but in their case it cannot be determined directly because, as Fig. 11.7 shows, the graph is not a straight line. It is calculated therefore by applying Kohlrausch's law of independent migration of ions as described in the end of this chapter.

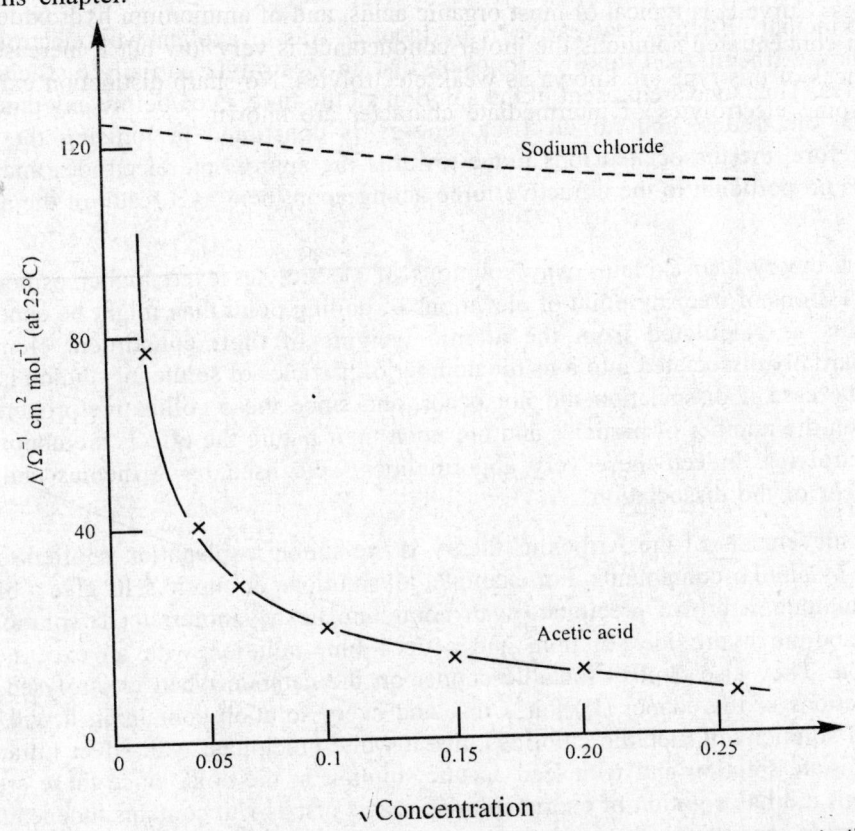

Fig. 11.7. Variation of molar conductance with the square root of concentration.

Simple Ionic Theory

The theory postulates that in solution all electrolytes are spontaneously dissociated to some extent into positive and negative ions. These ions are atoms or groups of atoms carrying a net charge which is numerically equal to the valency number of the element or radical; they have completely different properties from the atoms from which they are derived. The dissociation of an electrolyte into ions in solution is regarded as a reversible process, the extent of dissociation depending upon the nature of the electrolyte, the concentration of the solution, and the temperature. Arrhenius explained the variation in the molar conductance with dilution by proposing that in a solution of a strong electrolyte the proportion dissociated is high at all concentrations and increases when the solution is diluted, soon reaching a maximum value when dissociation is complete and all the electrolyte exists as separate ions.

Electrolytic conduction is explained as being due to the bodily movement of these dissociated ions through the solution towards the electrodes of opposite signs under the influence of the electric field. At the electrodes a process of discharge occurs giving rise to a metallic deposit, the liberation of a gas, or some other chemical change. According to the Arrhenius theory the amount of product liberated at the electrodes during electrolysis is proportional to the number of ions discharged per second and the time, i.e. to the total quantity of electricity which passes through the electrolyte, which explains Faraday's first law. Similarly, the masses of different products liberated by a given quantity of electricity are related to their chemical equivalents because the masses of the different ions per unit of charge are proportional to their equivalent weights.

One of the principal difficulties of earlier ionic theories was to explain why electrolytic conduction obeyed Ohm's law. Arrhenius did this by proposing that an electrolyte partially dissociates into ions as soon as it dissolves. Thus ions are present in the solution all the time, even before any potential difference is applied to the electrodes, and no electrical energy is consumed in ionising the solute. During electrolysis, therefore, the dissociated ions move towards the appropriate electrodes and are discharged at a rate which is proportional to the attractive force acting upon them as a result of the applied potential difference.

The Arrhenius theory also explains why solutions of electrolytes exert higher osmotic pressures or show larger depressions of freezing point or elevations of boiling point than might be expected from their molecular weights as calculated from the atomic weights of their constituent elements. Because electrolytes are partially dissociated into ions the number of particles of solute in solution is always greater than would be the case if dissociation did not occur, and since these colligative properties depend for their magnitude on the number of particles and not upon their nature the effects are abnormally large for solutions of electrolytes. Indeed these very abnormalities were used by Arrhenius and van't Hoff to calculate the extent of the dissociation.

One of the achievements of the Arrhenius theory is the simple explanation it offers of the common properties shown by similar compounds. For example, all solutions of cupric salts give a black precipitate with hydrogen sulphide, a brown precipitate with potassium hexacyanoferrate(II) solution, a light blue precipitate with sodium hydroxide solution, and a deep blue solution with an excess of ammonium hydroxide solution. They also deposit metallic copper on the cathode when electrolysed. These are all characteristic reactions of the copper (II) ion, Cu^{2+}, and every solution containing it will behave in this way. Similarly all solutions of metallic chlorides give a white precipitate with silver nitrate solution and with mercurous nitrate solution and with lead acetate solution in the cold, since these are the reactions of the Cl^- ion. Thus a dilute solution of cupric chloride reacts just as if it contains independent collections of cupric and chloride ions. Similarly Arrhenius realised that the properties of all electrolytes, at least as far as their dilute solutions are concerned, are the sum of the properties of the dissociated ions they contain. Kohlrausch had already shown that this was true of conductance in his law of independent migration of ions.

It is shown that the heat of neutralisation of all strong acids and bases is approximately the same, amounting to about 57300 joule per mole. According to Arrhenius the salts formed by such neutralisations are strong electrolytes which remain almost completely dissociated into ions in solution so the process is effectively the combination of hydrogen and hydroxyl ions to form undissociated water molecules:

e.g.
$$NaOH + HCl = NaCl + H_2O$$

becomes
$$Na^+ + OH^- + H^+ + Cl^- = Na^+ + Cl^- + H_2O$$

or
$$OH^- + H^+ = H_2O$$

Now since these strong acids and bases are almost completely dissociated into ions in solution the same number of ions are present from each mole and hence the same quantity of heat is evolved each time. Support for this view comes from a study of the equilibrium:

$$H^+ + OH^- \rightleftharpoons H_2O$$

which indicates from the change in the position of equilibrium with temperature that the heat evolved when a mole of water is formed from its ions is, in fact, about 57300 joule.

The simple ionic theory can be used in like manner to explain many experimental observations including the hydrolysis of salts, the common-ion effect, the precipitation of insoluble compounds by double decomposition, the action of indicators, the effectiveness of buffer solutions, the solubility of salts of weak acids in solutions of strong acids, and so on. Indirectly they provide impressive confirmation of the existence of ions, since it is very unlikely that any other idea could offer such a satisfactory explanation of such diverse phenomena.

The most powerful evidence of all for the ionic theory was not available at the time the theory was put forward, but only became known in the twentieth century as a result of X-ray analysis of solids. As described many salts are now believed to be composed solely of ions in the solid state, their crystalline structures consisting of ionic lattices and not molecules. When such solids dissolve in water many of these constituent ions are separated by the solvent, which therefore plays an active part in the process of dissolving, and the dissociated ions then move about freely in the solution. Thus a theory of electrolytic conduction no longer has to explain the presence of ions in the solution of an electrolyte and it would be difficult to imagine any modern theory which did not take their presence for granted. The origin of these ions has also been satisfactorily explained so that many of the difficulties encountered by Arrhenius in justifying the existence of ions to his contemporaries do not now arise.

Degree of Dissociation

The proportion of an electrolyte which is dissociated into ions in any given solution is known as the degree of dissociation, α. It is expressed either as a decimal or as a percentage. Arrhenius proposed that since all electrolytes are completely dissociated in an infinitely dilute solution, the degree of dissociation at any particular dilution is given by the ratio of its equivalent conductance at that dilution to its equivalent conductance at infinite dilution, i.e. $\alpha = \Lambda / \Lambda_\infty$.

Now about that time van't Hoff, who had been investigating the abnormally large colligative effects exhibited by solutions of electrolytes, suggested the use of a term i, known as the van't Hoff factor, to relate the observed colligative value to the value calculated on the basis of no dissociation.

Thus
$$i = \frac{\text{Observed osmotic pressure}}{\text{Calculated osmotic pressure}}$$

$$= \frac{\text{Observed freezing point depression}}{\text{Calculated freezing point depression}}$$

$$= \frac{\text{Observed boiling point elevation}}{\text{Calculated boiling point elevation}}$$

and $\pi v = iRT$ for a given solution.

Let us consider an electrolyte each molecule of which dissociates into n ions in solution and let its degree of dissociation at a particular concentration be α (expressed as a decimal). Then if 1 gram-molecule of the electrolyte is present in that solution, α gram-molecules will dissociate forming $n\alpha$ gram-ions and leaving $(1 - \alpha)$ gram-molecules undissociated. Thus the total of solute particles in solution, gram-molecules and gramme-ions together, is $1 - \alpha + n\alpha$, i.e. $1 + (n - 1)\alpha$. This compares with only molecule when no dissociation occurs.

$$\therefore \quad i = \frac{1+(n-1)\alpha}{1} \quad \text{and} \quad \alpha = \frac{(i-1)}{(n-1)}.$$

Arrhenius used this deduction to calculate the degree of dissociation of various electrolytes at particular concentrations from the colligative properties of their solutions and compared the values of α so obtained with those found from measurements of equivalent conductance. In most cases there was remarkably close agreement between the two values of α, which did much to strengthen confidence in the ionic theory. It is now known, however, that in the case of strong electrolytes this agreement was largely fortuitous and was not as significant as it appeared.

Example 11.2. Calculate the degree of dissociation of sodium chloride from the following data which were obtained with a solution of the same concentration :

      1. The molar conductance at 25°C is 96.2 Ω^{-1} cm^2 mol^{-1}, whereas the corresponding value at infinite dilution is 126.5 Ω^{-1} cm^2 mol^{-1}.

      2. The freezing point of the solution (containing 12.5 g in 1000 g of water) is -0.70°C. The cryoscopic constant for 1000 g is 1.86 K and the calculated molecular weight, assuming no dissociation into ions, is 58.5.

Solution.

1. $\alpha = \dfrac{\Lambda}{\Lambda\infty} = \dfrac{96.2}{126.5} = 0.761$ or 76.1% .

2. $\Delta T = \dfrac{Km}{M}$

where ΔT = freezing point depression = 0.70 K

        K = cryoscopic constant for 1000 g = 1.86 K

       m = mass of solute in 1000 g solvent = 12.5 g

       M = apparent mol. wt. of solute

$$\therefore \quad M = \frac{Km}{\Delta T} = \frac{1.86 \times 12.5}{0.70} = 33.2$$

Now $i = \dfrac{\text{calculated mol. wt.}}{\text{apparent mol. wt.}} = \dfrac{58.5}{33.2}$

and $\alpha = \dfrac{i-1}{n-1}$ where n is here (NaCl $\rightleftharpoons$ Na$^+$ + Cl$^-$)

$$\therefore \quad \alpha = \frac{58.5 - 33.2}{33.2} = \frac{25.3}{33.2} = 0.762 \text{ or } 76.2\%$$

It would appear from the data, therefore, that sodium chloride is about 76% dissociated into ions in a solution of this concentration. As we shall see in later sections, this conclusion is no longer accepted, a different interpretation being put upon the result.

Kohlrausch's Law of Independent Migration of Ions

Kohlrausch noticed that pairs of sodium and potassium salts with the same anion showed a constant difference (21 Ω^{-1} cm^2 mol^{-1}) in molar conductance at infinite dilution (Table 11.1). Similar results were observed with other pairs of salts which had a cation or anion in common. Kohlrausch concluded that at infinite dilution each ion shows a characteristic conductance which is independent of the other ions present in the solution. Thus the molar conductance at infinite dilution for any electrolyte is equal to the sum of the molar conductances of its cation and anion, since each ion contributes a definite amount to the total conductance of the electrolyte:

i.e. $$\Lambda_\infty = \Lambda_+ + \Lambda_-$$

where Λ_- and Λ_+ are the ionic conductances at infinite dilution of the cation and anion respectively.

Table 11.1. Molar conductances at infinite dilution.

Salt	$\dfrac{\Lambda_\infty \text{ at } 18°C}{\Omega^{-1}\text{cm}^2\text{mol}^{-1}}$	
KCl	130.1 ⎫	Difference = 21.1 Ω^{-1} cm^2 mol^{-1}
NaCl	109.0 ⎭	
KNO$_3$	126.3 ⎫	Difference = 21.0 Ω^{-1} cm^2 mol^{-1}
NaNO$_3$	105.3 ⎭	

This importance of this law is that it can be used to calculate molar conductances at infinite dilution (such as those of weak electrolytes which cannot be determined by extrapolation), provided the values for three suitable strong electrolytes are known. The calculation is illustrated by the following example:

Example 11.3. Calculate the molar conductance at infinite dilution of acetic acid at 25°C, given that at that temperature the molar conductances at infinite dilution of hydrochloric acid, sodium chloride, and sodium acetate are 426, 126, and 91 Ω^{-1} cm^2 mol^{-1} respectively.

Solution. Now

$$\Lambda_{CH_3COOH} = \Lambda_{CH_2COO^-} + \Lambda_{H^+}$$
$$\Lambda_{HCl} = \Lambda_{H^+} + \Lambda_{Cl^-}$$
$$\Lambda_{NaCl} = \Lambda_{Na^+} + \Lambda_{Cl^-}$$
$$\Lambda_{CH_3COONa} = \Lambda_{CH_3COO^-} + \Lambda_{Na^+}$$
$$\Lambda_{CH_3COOH} = \Lambda_{HCl} + \Lambda_{CH_3COONa} - \Lambda_{NaCl}$$

$$= 426 + 91 - 126$$
$$= 391 \ \Omega^{-1} \text{ cm}^2 \text{ mol}^{-1}$$

Now the fraction of the total current which is carried through the solution by each type of ion depends upon the relative velocities of the cation and anion. This fraction, which is known as the transport number of the ion, is given for a simple binary electrolyte by the expression:

$$t_+ = \frac{\text{current carried by anion}}{\text{total current}} = \frac{v_+}{v_+ + v_-}$$

where v_+ and v_- are the absolute velocities of the anion and cation respectively. The transport number can be measured experimentally by various methods. From a determination of the transport number of an ion, its ionic conductance at infinite dilution can easily be calculated since $t_+ = \Lambda_+/\Lambda_\infty$ and therefore $\Lambda_+ = t_+ \times \Lambda_\infty$. It follows that a list of ionic conductances at infinite dilution can be drawn up for the various ions from which that of compounds can be calculated by straightforward addition. Table 11.2. gives some typical values at 25°C. The conductance of an ion depends upon the speed at which it moves through the solution. The main retarding force which limits the movement of ions when a potential difference is applied is the frictional drag experienced by the ion and its attendant solvent molecules owing to the viscosity of the solvent. Thus the speed of an ion depends upon a number of factors including the nature of the solvent, the temperature, the potential gradient applied, the charge on the ion, the size of the ion, and the degree to which it is solvated. The velocity of an ion under a potential gradient of one volt per centimetre is known as the ionic mobility of that ion in that solvent at that temperature. For most ions in aqueous solution its value lies between 3×10^{-6} metre per second and 8×10^{-6} metre per second at room temperature, which corresponds to a speed of about 2 cm per hour, but H^+ and OH^- ions move considerably faster. Absolute ionic velocities can be determined experimentally, or they can be calculated by dividing the molar conductance of an ion at infinite dilution by 96500.

Table 11.2. Ionic conductances at infinite dilution.

Cation	Λ_∞ at 25°C $\overline{\Omega^{-1}cm^2mol^{-1}}$	Anion	Λ_∞ at 25°C $\overline{\Omega^{-1}cm^2mol^{-1}}$
H^+	349.8	OH^-	198
K^+	73.5	$1/2\ SO_4^{2-a}$	79.7
Ag^+	61.9	Cl^-	76.3
$1/2\ Mg^{2+a}$	53.2	NO_2^-	71.4
Na^+	50.1	CH_3COO^-	40.9

^{a} The 1/2 is necessary here because one gram-equivalent is concerned, not one gram-ion.

The relatively high conductance of the H^+ ion is a point of special interest. Not only does it explain the high conductance of strong acids, which yield these ions on dissociation, but it suggests that this ion moves through the solution in a different way from the other ions. It is thought that the H^+ ion, which is only a proton when not hydrated, is able to transfer one water molecule to another during its passage through the solution instead of clinging only to the same water molecule all the time. The effect of this is much the same as that of a child who whilst taking part in a race by paddling rafts, jumps from time to time from one raft to another ahead of it, using the rafts almost as stepping stones. Naturally his progress will be much faster as a result than it would have been by staying on the same raft all the time. A similar mechanism is believed to take place, but to a lesser extent, with the OH^- ion. The conductometric titration of acids and alkalies is based upon the exceptional conductances of the H^+ and OH^- ions.

The solubility of a sparingly soluble salt such as silver chloride or barium sulphate can be determined from the conductance of its saturated solution provided the ionic conductance of its component ions are known, as the following calculation illustrates.

Example 11.4. The conductivity of a saturated solution of silver chloride at 25°C is found to be 1.50×10^{-6} Ω^{-1} cm^{-1}, after allowing for the conductivity of water. What is the solubility of silver chloride in gramme per dm^3 if

$$\Lambda_{Ag^+} = 62 \quad \text{and} \quad \Lambda_{Cl^-} = 76?$$

Solution. Now a sparingly soluble salt like this can be assumed to be completely dissociated into ions even in saturated solution. Consequently, by Kohlrausch's law, its equivalent conductivity is equal to the sum of the ionic conductances,

i.e.
$$\Lambda_\infty = \Lambda_{Ag^+} + \Lambda_{Cl^-}$$
$$= 62 + 76 = 138$$

But
$$\Lambda_\infty = \kappa . V,$$

where κ = conductivity of saturated solution

V = the volume in cm³ containing 1 gram-equivalent

$\therefore$ Concentration in gram-equivalents per dm³ $= \dfrac{1000}{V}$

$$= \dfrac{1000\kappa}{\Lambda_\infty}$$

$$= \dfrac{1000 \times 1.50 \times 10^{-6}}{138}$$

$\therefore$ Solubility in gram per dm³
$$= \dfrac{1000 \times 1.50 \times 10^{-6} \times 143.5}{138}$$

$$= 1.56 \times 10^{-3}$$

Modern Ionic Theories

In his theory of electrolytic dissociation Arrhenius proposed that the degree of dissociation of an electrolyte at any dilution was equal to the ratio of its molar conductance at that dilution to its molar conductance at infinite dilution, i.e. a = Λ/Λ_∞. When the degree of dissociation of a strong electrolyte is calculated in this way, the electrolyte is found not to obey Ostwald's dilution law, as explained later. Several other pieces of experimental evidence weight against the Arrhenius theory as it applies to strong electrolytes. Studies of absorption spectra show no evidence of undissociated molecules in solutions of salts, and the various heats of neutralisation of strong acids and strong bases are so concordant that they suggest, contrary to the Arrhenius theory, that all these electrolytes are dissociated to exactly the same extent in solution and that this extent does not depend upon their concentrations. Moreover, measurements show that transport numbers vary with the concentration of the solution, which would certainly not be expected from the Arrhenius theory.

These findings have necessitated an important modification of the original Arrhenius theory as far as strong electrolytes are concerned. In relating the degree of dissociation to the ratio of the molar conductances, Arrhenius made the assumption that the velocities of the ions did not change on dilution, so that the number of ions present in the solution was the only variable. This view is no longer accepted and strong electrolytes are now regarded as being completely dissociated into ions in aqueous solution at all dilutions. The variation in their conductance and colligative effects on dilution is attributed solely, therefore, to changes in the velocities of their ions as a result of interionic attraction, and the ratio Λ/Λ_∞, now known as the conductance ratio or the apparent degree of dissociation, is really a measure of the way in which the velocities of ions from strong electrolytes change with concentration.

The modern view, devised by Debye and Hückel and developed further by Onsager, is that in a concentrated solution of a strong electrolyte each ion is surrounded in its immediate vicinity by a cloud or atmosphere of ions which are mostly of opposite charge. In the absence of a potential gradient in the solution this arrangement of ions is symmetrical, but once a potential difference is applied to the electrodes asymmetry arises, as in Fig. 11.8, because the positive and negative ions are attracted in opposite directions. Thus as each ion moves through the solution its ionic atmosphere exerts a drag upon it owing to the excess of ions of opposite charge in its wake and this has the effect of slowing it down appreciably. The situation is rather like a man trying to force his way through a dense crowd, many members of which are moving in the opposite directions. The analogy is all the more realistic if we imagine each person in the crowd to be carrying a couple of large cases (adhering molecules of solvent) which continually get caught in those of other people. Just as the denser the crowd the more the man's progress is impeded, so the more concentrated the solution the greater is the retarding effect of interionic attraction.

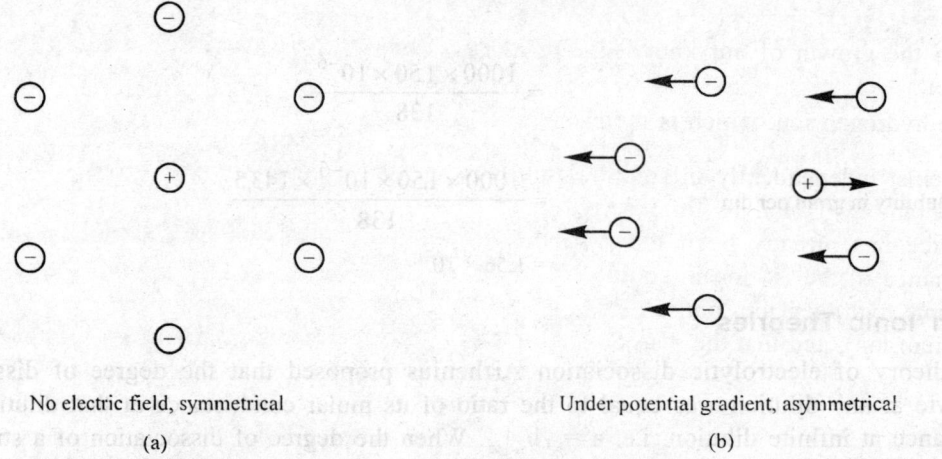

No electric field, symmetrical   Under potential gradient, asymmetrical

(a)   (b)

Fig. 11.8. Diagram to illustrate ionic atmosphere.

It is possible that the attraction between oppositely charged ions is so great in concentrated non-aqueous solutions of strong electrolytes, particularly when the ions are small and highly charged, that a small proportion of the ions may momentarily form pairs or groups. In such cases the electrolyte is said to be completely ionised in solution but not completely dissociated. This is not, of course, the same as Arrhenius proposed in his original theory, since he imagined an equilibrium between unionised molecules and dissociated ions, whereas in this modern theory the solute exists as ions under all conditions.

In their original form these modern theories were highly mathematical but experimental support for them is obtained by applying a very large potential gradient indeed to the solution, when the conductance is appreciably increased. Presumably the ions move so fast under these conditions that they shed their solvating molecules and their retarding atmosphere of oppositely charged ions entirely and stream towards the attracting electrode. Similarly, if an alternating potential difference of very high frequency is applied, the conductance is also raised; in this case the ions change direction so frequently per second that the ionic atmosphere gets no time to build up asymmetrically around them and impede their movements.

For weak electrolytes interionic attraction is slight, since the concentration of ions in solution is never large, and the velocity of the ions does not change very much, therefore, with dilution. Consequently only a small error is incurred in taking the conductance ratio as the degree of dissociation for weak electrolytes and in general the Arrhenius theory accounts satisfactorily for their behaviour. A clear distinction must be drawn here between the terms strong and weak as applied to electrolytes and the terms concentrated and dilute as applied to solutions. An electrolyte is strong or weak depending upon the extent to which it is dissociated into ions in, say, a molar solution, and the way in which its molar conductance changes with dilution, whereas whether a solution is concentrated or dilute depends upon how much solute it contains in unit volume. For example, sulphuric acid is always a strong acid whether dilute or concentrated, and acetic acid is a weak acid even in concentrated solution.

Hydroxonium Ion

In his original theory Arrhenius proposed that when acids dissociated in solution they yielded hydrogen ions, H^+, thus:

$$HCl \rightleftharpoons H^+ + Cl^-$$

$$CH_3COOH \rightleftharpoons H^+ + CH_3COO^-$$

With the growth of our knowledge of atomic structure we now realise that it is highly improbable that the hydrogen ion, which is merely a proton and therefore only about $\dfrac{1}{10000}$ the size of other ions, would exist independently in solution. Any ion carrying such a high density of charge is likely to be unstable and will tend to form a hydrate by combining loosely with one or more water molecules. Reference has already been made to this tendency in explaining the abnormally high equivalent conductance of the H^+ ion in solution. Thus the modern view is that solutions of acids contain not just the simple hydrogen ion, but rather the hydroxonium ion, H_3O^+ (also known as the oxonium ion or the hydronium ion), and that the dissociation of acids in solution is best represented by equilibria of the type:

$$HCl + H_2O \rightleftharpoons H_3O^+ + Cl^-$$

$$CH_3COOH + H_2O \rightleftharpoons H_3O^+ + CH_3COO^-$$

One advantage of this is that it recognised the active part played by the solvent in the ionisation of these acids. Indeed in the completely anhydrous form hydrogen chloride and acetic acid are covalen substances which do not conduct electricity; they do not turn litmus red, corrode metals, or attack carbonates releasing carbon dioxide. The same is true of a solution of hydrogen chloride in benzene or toluene. It is only in the presence of water that these substances show their usual acidic properties. In the same way the dissociation of water may be expressed in terms of the equilibrium:

$$H_2O + H_2O \rightleftharpoons H_3O^+ + OH^-$$

Although the existence of the hydroxonium ion in aqueous solutions of acids is now universally accepted, for the sake of simplicity it is still common practice to use the symbol H^+ in describing ionic equilibria. The hydrogen ion is not the only one, of course, to be hydrated in this way and experimental evidence provided by a study of transport numbers suggests that most other ions are also linked loosely to one or more water molecules in aqueous solution.

Modern Ideas on Acids and Bases

Another consequence of this modern view of dissociation is a wider conception of the terms acid and base. Brönsted and Lowry put forward independently the definition of an acid as any substance which donates protons (i.e. hydrogen ions) to other substances, so that acid salts and the ammonium ion are included with the usual range of compounds. Similarly, a base is defined as any substance which acts as a proton acceptor; this includes substances such as water, amines, and ammonia which contain a lone pair of electrons and all acid radicals, as well as the usual compounds which yield hydroxyl ions. Thus acids and bases are related in the following way:

$$Acid \rightleftharpoons H^+ + Base$$

It will be seen, therefore, that water can act both as a base and as an acid, and does in fact do both simultaneously during its dissociation:

$$H_2O \rightleftharpoons H^+ + OH^-$$

$$H^+ + H_2O \rightleftharpoons H_3O^+$$

These definitions are preferred to the narrower ones formerly in use (acids and bases are substances yielding H^+ and OH^- ions respectively in aqueous solution) because they are also applicable to non-aqueous solvents and because they account for the way in which the strength of an acid varies with the nature of the solvent.

The views of Brönsted and Lowry have been extended further by Lewis, who has defined a base as a substance possessing a lone pair of electrons capable of being used to link it to another atom, and an acid as a substance willing to accept these electrons from a base.

IONIC EQUILIBRIA

In studying ionic equilibria we shall indicate when one side of the equation greatly preponderates over the other by using special arrows as in the following examples :

$$NaCl \rightleftharpoons Na^+ + Cl^-$$

(virtually all the sodium chloride is dissociated into separate ions)

$$H_2O \rightleftharpoons H^+ + OH^-$$

(most water molecules are undissociated, with only a small proportion of ions). Where no great disparity exists in the position of equilibrium, or where the disparity is unpredictable or immaterial, the usual reversible arrows will be used.

Ostwald's Dilution Law

Let us consider a solution containing one mole of a typical simple binary weak electrolyte AB (i.e. one in which each molecule dissociates into two ions in solution) in volume V of solution. Let α be the degree of dissociation under these conditions. Then provided the solution is kept at constant temperature we can apply the law of mass action to the equilibrium between AB and its ions by taking the active mass of each ion to be its ionic concentration:

	AB	$\rightleftharpoons$	A^+	$+$	B^-
Amount present at equilibrium	$(1 - \alpha)$ mole		α mole		α mole
Concentration present at equilibrium	$\dfrac{(1-\alpha)}{V}$ mol dm^{-3}		$\dfrac{\alpha}{V}$ mol dm^{-3}		$\dfrac{\alpha}{V}$ mol dm^{-3}

Applying the law of mass action:

$$\frac{[A^+][B^-]}{[AB]} = \text{constant, at constant temperature.}$$

But

$$[A^+] = \frac{\alpha}{V} \text{ mol dm}^{-3} = [B^-]$$

and

$$[AB] = \frac{(1-\alpha)}{V} \text{ mol dm}^{-3}$$

$\therefore$

$$\frac{\frac{\alpha}{V} \times \frac{\alpha}{V}}{\frac{(1-\alpha)}{V}} = \frac{\alpha^2}{(1-\alpha)V} = \text{constant, } K \text{ (at constant temperature)}$$

This is a statement in symbols of Ostwald's dilution law, the constant K is known as the dissociation constant of the weak electrolyte at that temperature. The law is easily verified by determining the degree of dissociation, α, at various dilutions and substituting corresponding values of V and α in the expression

$\dfrac{\alpha^2}{(1-\alpha)V}$. It is found that for weak electrolytes K is fairly constant, but that for strong electrolytes there

is a wide variation in its value with changing concentration. In fact a weak electrolyte is often defined as one which obeys Ostwald's law and it was the inapplicability of the law to strong electrolytes which first raised doubts about the Arrhenius theory. The term dissociation constant clearly has no meaning, therefore, when applied to strong electrolytes, which explains why when we meet it K always has a very low value, usually less than 10^{-2} mol dm^{-3}. Since K varies with temperature the latter should always be stated when K is quoted unless it is 25°C, the arbitrary standard.

Strength of Acids and Bases

The dissociation constant, K, is a very important index of the strength of an acid or base and is generally preferred to the degree of dissociation, α, for this purpose because unlike the latter it does not vary with concentration. Thus the relative strengths of weak electrolytes can best be determined by comparing their dissociation constants at the same temperature as follows:

$$\text{Acetic acid : } CH_3COOH \rightleftharpoons CH_3COO^- + H^+$$

$$\frac{[CH_3.COO^-][H^+]}{[CH_3COOH]} = K_{CH_3COOH} = 1.85 \times 10^{-5} \text{ mol dm}^{-3} \text{ at } 25°C$$

$$\text{Prussic acid : } HCN \rightleftharpoons H^+ + CN^-$$

$$\frac{[H^+][CN^-]}{[HCN]} = K_{HCN} = 7.1 \times 10^{-10} \text{ mol dm}^{-3} \text{ at } 25°C$$

$$\text{Ammonium hydroxide : } NH_4OH \rightleftharpoons NH_4^+ + OH^-$$

$$\frac{[NH_4^+][OH^-]}{[NH_4OH]} = K_{NH_4OH} = 1.80 \times 10^{-5} \text{ mol dm}^{-3} \text{ at } 25°C$$

Thus acetic acid and ammonium hydroxide are electrolytes of closely comparable strength and both are much stronger than prussic acid, which is a very weak electrolyte.

In calculations of this sort, where α is known to be small, it is often permissible to approximate by calling $(1 - \alpha)$ unity. In this case the error in the value of α by doing so would be less than 1%, since α^2 the equals 1.85×10^{-4} and $\alpha \simeq \sqrt{1.85 \times 10^{-4}} \simeq 1.36 \times 10^{-2}$. When this approximation is made,

Ostwald's dilution law becomes $\frac{\alpha^2}{V} \simeq K$ so $\alpha^2 \simeq KV$ and $\alpha \simeq \sqrt{KV}$. Thus when $V = 1$, $\alpha \simeq \sqrt{K}$, i.e.,

the degree of dissociation of a weak electrolyte in normal solution is approximately equal to the square root of its dissociation constant. It also means that the degree of dissociation of a weak electrolyte is proportional to the square toot of the dilution.

The dissociation constant of a weak electrolyte remains unchanged not only on dilution but also on addition of other electrolytes to the solution, even when the same ion is common to both equilibria.

According to the modern view, in aqueous solution strong electrolytes are completely dissociated into ions at all concentrations. It follows, therefore, that the concentration of H^+ ions in a solution of a strong acid is effectively the same as the concentration of the acid. For example, in decimolar hydrochloric acid, which by definition contains 0.1 mole of acid per dm^3, the concentration of hydrogen ions will be 0.1 mole per dm^3. Frequent use of this fact will be made later in this chapter.

Polybasic acids (i.e. acids containing more than one hydrogen atom in each molecule replaceable by a metal) dissociate into ions in solution in more than one stage and have, therefore, more than one dissociation constant.

e.g.
$$H_2S \rightleftharpoons H^+ + HS^-$$

$$\therefore \qquad \frac{[H^+][HS^-]}{[H_2S]} = K_1$$

$$HS^- \rightleftharpoons H^+ + S^{2-}$$

$$\therefore \qquad \frac{[H^+][S^{2-}]}{[HS^-]} = K_2$$

Multiplying these two expressions together we get:

$$\frac{[H^+]^2[S^{2-}]}{[H_2S]} = K_1 \times K_2 = K_{H_2S}$$

The approximate values of K_1 and K_2 are 10^{-7} and 10^{-15} respectively, so the dissociation constant K_{H_2S} has a value of about 10^{-22} at 25°C. Similarly, tribasic orthophosphoric acid dissociates in three stages as follows:

$$H_3PO_4 \rightleftharpoons H^+ + H_2PO_4^- \rightleftharpoons 2H^+ + HPO_4^{2-} \rightleftharpoons 3H^+ + PO_4^{3-}$$

giving rise to three dissociation constants $K_1(10^{-2})$, $K_2(10^{-7})$, and $K_3(10^{-12})$. Thus a solution of acid contains all three types of orthophosphate ion in equilibrium with H^+ ions and a relatively large concentration of undissociated molecules.

Dissociation of Water

If water is purified by repeated distillation or by using a mixture of ion-exchange resins, its molar conductance falls to a steady, very low value. That this slight conductance remains, no matter how many times the water is distilled or treated with resins, shows that however pure it is water always contains some ions, which are believed to be formed as a result of its own dissociation:

$$H_2O + H_2O \rightleftharpoons H_3O^+ + OH^-$$

or, more simply:

$$H_2O \rightleftharpoons H^+ + OH^-$$

Applying the law of mass action to this equilibrium at constant temperature and taking the concentrations of the ions expressed in mole per dm^3 as their active masses, we get:

$$\frac{[H^+][OH^-]}{[H_2O]} = \text{constant, } K$$

$$\therefore \; [H^+].[OH^-] = K \times [H_2O]$$

Now $[H_2O]$, which is the molar concentration of undissociated water molecules in water, is effectively constant since water is an extremely weak electrolyte only very slightly dissociated into ions. For example, even if the dissociation of water is increased a thousandfold, there will be negligible percentage change in the value of $[H_2O]$ and we can therefore regard it as constant under all conditions, i.e. $[H_2O]$ = constant.

$$\therefore \qquad [H^+].\,[OH^-] = K \times \text{constant} = K_W \text{ (at constant temperature)}$$

where the constant K_W is known as the ionic product for water.

Since each molecule of water gives rise to one hydrogen ion and one hydroxyl ion when it dissociates, the concentration of these two ions must necessarily be the same in pure water. Experiment shows that this concentration is about 10^{-7} mol dm^{-3} at 25°C, i.e. $[H^+] = [OH^-] = 10^{-7}$ mol dm^{-3}. K_W has the value of approximately 1×10^{-14} mol^2 dm^{-6} at 25°C, therefore, and increases to about 50×10^{-14} mol^2 dm^{-6} at 100°C. From this change in K_W we can deduce that the heat of the reaction $H^+ + OH^- \rightleftharpoons H_2O$ is about 57300 joule. A solution in which there is a preponderance of hydrogen ions over hydroxyl ions, i.e. one in which the hydrogen ion concentration exceeds 10^{-7} mol dm^{-3} at 25°C, is said to be *acidic*, whilst a solution in which the hydrogen ion concentration is less than 10^{-7} mol dm^{-3} is called an *alkaline* solution. Only at *neutrality* are the concentrations of H^+ and OH^- ions exactly equal.

It is important to realise that however much the concentration of H^+ and OH^- ions is altered by the addition of acids or alkalies, the ionic product remains constant at constant temperature. For example, the H^+ ion concentration may vary very widely from 1 mol dm^{-3} in a molar solution of hydrochloric acid to 10^{-14} mol dm^{-3} in a molar solution of sodium hydroxide, yet the product $[H^+].\,[OH^-]$ always has the value 10^{-14} mol^2 dm^{-6} at 25°C. Thus it is the product of the two ionic concentrations which is constant, not the concentrations of the individual ions.

Hydrogen Ion Concentration and pH

In describing the acidity or alkalinity of a solution it is adequate to quote the hydrogen ion concentration alone, since the hydroxyl ion concentration can always be deduced from the relationship $[OH^-] = \dfrac{10^{-14}}{[H^+]}$ at 25°C. Now in view of the wide variation which is possible in the H^+ ion concentration, it is convenient to express it on a logarithmic scale, and the one suggested by Sorensen is now in general use. The negative logarithm to the base 10 of the H^+ ion concentration (in mol dm^{-3}) is called the pH of a solution,

i.e.
$$pH = -\log_{10}[H^+] = \log_{10}\frac{1}{[H^+]} \text{ and } [H^+] = 10^{-pH}$$

The sign is reversed in order to obtain a scale on which all the numbers normally encountered are positive. The reverse calculation is sometimes needed.

Example 11.5. A solution has a pH of 4.1. What is its H^+ ion concentration?

Solution. Now
$$4.1 = -\log_{10}[H^+]$$
so
$$\log_{10}[H^+] = -4.1$$
$$= -5.0 + 0.9 = 5.9$$
∴
$$[H^+] = \text{antilog. } 5.9 = 8 \times 10^{-5} \text{ mol dm}^{-3}.$$

The relationship between the acidity or alkalinity of a solution and its pH is shown diagrammatically in Fig. 11.9. An acidic solution, it should be noted, has a high concentration of H^+ ions but a low pH value, i.e. its pH is less than 7. It must also be remembered that the pH scale is a logarithmic one and that a change of only one unit corresponds to a tenfold change in the hydrogen ion concentration.

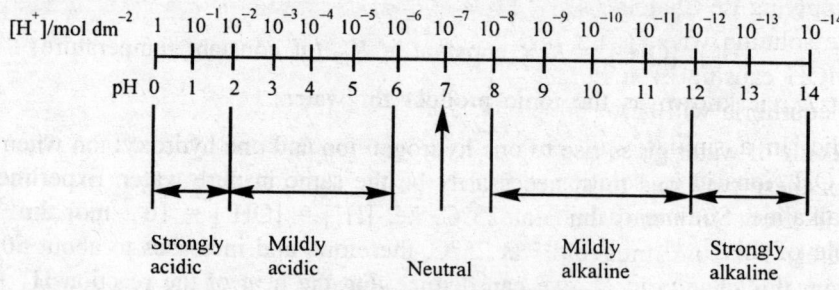

Fig. 11.9. The pH scale.

The hydrogen ion concentration of a solution may be found in various ways. For a weak electrolyte it can be calculated from the degree of dissociation as determined from the conductivity ratio or the van't Hoff factor. The hydrogen ion concentration can also be deduced from its catalytic effect upon the hydrolysis of esters or, more simply, by a comparative colorimetric method. Alternatively the pH can be found by measuring the potential difference developed when a glass electrode and a standard electrode are in contact with the solution (this is the principle of the pH meter, which is popular in industry because it gives direct readings of pH quickly and accurately and can be used by unskilled labour for routine measurements).

Hydrolysis of Salts

An aqueous solution of a salt of a strong acid and a strong base (e.g. sodium chloride or potassium nitrate) has a pH of about 7 at room temperature and is therefore neutral. However, salts formed from a weak acid and a strong base (e.g. sodium phosphate or potassium acetate) give an alkaline solution when dissolved in water, and those obtained from a strong acid and a weak base (e.g. ammonium chloride) give acidic solutions. These facts can readily be explained by examining the ionic equilibria involved in a typical case. Let us consider a salt MA which dissociates completely in aqueous solution as follows :

$$MA \rightarrow M^+ + A^-$$

The water present is also dissociated, but only, of course, to a very small extent :

$$H_2O \rightleftharpoons H^+ + OH^-$$

Thus the solution of a salt contains four types of ions and these will give rise to two further equilibria:

$$M^+ + OH^- \rightleftharpoons MOH$$

and
$$H^+ + A^- \rightleftharpoons HA$$

The extent to which these equilibria proceed to the right depends upon the strength of MOH and HA as electrolytes. If MA is the salt of a strong acid and a weak base, for example, then few, if any, of H^+ and A^- ions will unite since the acid HA will be completely dissociated in solution, but most of the

M^+ and OH^- ions present will combine to form undissociated MOH until the expression $\dfrac{[M^+][OH^-]}{[MOH]}$

has the low value of the dissociation constant K_{MOH}. This removal of OH^- ions from the system upsets the equilibrium existing between water molecules and H^+ and OH^- ions, and since the ionic product of water must be maintained at its constant value of 10^{-14}, further dissociation of water occurs, as might be predicted by applying Le Chatelier's principle. The H^+ ions resulting from this further dissociation will accumulate in the solution whereas the OH^- ions will again be removed as undissociated molecules of the weak base MOH causing even further dissociation of the water. Thus the result of this interaction between the two equilibria will be to produce a large preponderance of H^+ ions in the solution, making the solution acidic. In a similar way, if MA is the salt of a weak acid and a strong base then the concentration of OH^- ions in its solution will be much higher than the concentration of H^+ ions and its solution will be alkaline. Sodium carbonate, which is often titrated against acids in volumetric analysis, is a good example of this.

It is the relative strength of MOH and HA which determines whether the solution will show any marked acidity or alkalinity. Salts formed from acids and bases which are equally strong (e.g. sodium chloride) or equally weak (e.g. ammonium acetate) will be approximately neutral since their solutions will contain equal concentrations of H^+ and OH^- ions. The phenomenon is known as hydrolysis because it involves the interaction of the ions of the salt with those of water. Effectively the process is the opposite of neutralisation and can be represented by the general equation:

$$\text{salt} + \text{water} \rightleftharpoons \text{acid} + \text{base}.$$

For a strong acid and a weak base the overall change is:

$$M^+ + A^- + H_2O \rightleftharpoons MOH + H^+ + A^-$$

Hence, applying the law of mass action of this equilibrium and assuming that the ionic concentrations can be taken as the active masses:

$$\frac{[MOH][H^+][A^-]}{[M^+][A^-][H_2O]} = K_H = \frac{K_W}{K_{MOH}}$$

where K_H is the hydrolysis constant of the salt

K_W is the ionic product of water

K_{MOH} is the dissociation constant of the weak base.

Similarly the equation for the hydrolysis of a salt of a weak acid HA and a strong base MOH is:

$$M^+ + A^- + H_2O \rightleftharpoons M^+ + OH^- + HA$$

and the hydrolysis constant is given by:

$$\frac{[M^+[OH^-][HA]}{[M^+][A^-][H_2O]} = K_H = \frac{K_W}{K_{HA}}$$

Two other aspects of the hydrolysis of salts must be mentioned. Certain salts (e.g. the trichlorides of bismuth and antimony) undergo hydrolysis to give a precipitate of an oxysalt when their solution in hydrochloric acid is diluted:

$$BiCl_3 + H_2O \rightleftharpoons BiOCl \downarrow + 2HCl$$
(Bismuth oxychloride or
bismuthyl chloride)

These hydrolyses are reversible and the precipitate redissolves on adding concentrated hydrochloric acid. Secondly, some salts of weak acids and weak bases are so completely and rapidly hydrolysed that they cannot exist in aqueous solution. For example, aluminium sulphide gives an immediate precipitate of aluminium hydroxide when added to water, and magnesium nitride and calcium phosphide are instantly hydrolysed to their respective hydride and hydroxide :

$$Al_2S_3 + 6H_2O = 2Al(OH)_3 \downarrow + 3H_2S \uparrow$$
$$Mg_3N_2 + 6H_2O = 3Mg(OH)_2 \downarrow + 2NH_3 \uparrow$$
$$Ca_3P_2 + 6H_2O = 3Ca(OH)_2 \downarrow + 2PH_3 \uparrow$$

Common Ion Effect

If a strong electrolyte such as a salt is added to a solution of a weak electrolyte so chosen because one of the ions into which it dissociates in solution is the same as one of the ions from the salt, the degree of dissociation of the weak electrolyte is decreased as a result of the common ion effect. The strong electrolyte is, of course, completely dissociated into its ions in solution, so that a high concentration of these ions is produced when it dissolves. This effects the equilibrium between the undissociated molecules of the weak electrolyte and its component ions, causing some of the latter to combine so that their concentrations are restored to a level in keeping with the dissociation constant of the weak electrolyte. This effect is well illustrated by considering a saturated solution of hydrogen sulphide.

Ignoring the two-stage dissociation of hydrosulphuric acid, we can express its partial dissociation by the equation :

$$H_2S \rightleftharpoons 2H^+ + S^{2-}$$

so that at constant temperature $\qquad \dfrac{[H^+]^2 \cdot [S^{2-}]}{[H_2S]} = K_{H_2S}$

If a strong acid such as hydrochloric acid is now added to the solution, the concentration of H^+ ions is enormously increased thus:

$$HCl \rightarrow H^+ + Cl^-$$

As a consequence H^+ and S^{2-} ions unite to form undissociated H_2S molecules until the expression

$\dfrac{[H^+]^2 \cdot [S^{2-}]}{[H_2S]}$ is again equal to K_{H_2S} at that temperature. Thus the effect of adding the hydrochloric

acid is to suppress the dissociation of the hydrosulphuric acid and substantially lower the concentration of free S^{2-} ions in solution. Since H_2S is a weak electrolyte only slightly dissociated into ions in saturated solution, the value of $[H_2S]$ is relatively large and shows little percentage change, so the product $[H^+]^2 \cdot [S^{2-}]$ is approximately constant. Thus the sulphide ion concentration is approximately inversely proportional to the square of the hydrogen ion concentration and a tenfold increase in the latter reduces the former to about a hundredth of its original value.

Another example is provided by the addition of ammonium chloride (or ammonium nitrate) to a solution of ammonium hydroxide, which has the effect of partially suppressing its dissociation and lowering the concentration of OH^- ions in solution. The common ion here is the NH_4^+ ion, as the following equations indicate :

$$NH_4Cl \rightarrow NH_4^+ + Cl^-$$
$$NH_4OH \rightleftharpoons NH_4^+ + OH^-$$

Nor is the common ion effect limited to cases in which the dissociation of the weak electrolyte is reduced. For example, if ammonium hydroxide solution is added to a saturated solution of hydrosulphuric acid, H_2S, the removal of H^+ ions from the system by combining with OH^- ions from the ammonium hydroxide causes the acid to dissociate more extensively. Here OH^- ions are common to the dissociation of both the ammonium hydroxide and the water and the H^+ ions to the dissociations of the water and the hydrosulphuric acid.

The common ion effect is used to separate the metals into various groups in qualitative analysis. Thought will show that no new principle is involved in the common ion effect, however, and that the result it achieves can be successfully predicted and explained by applying Le Chatelier's principle to the various ionic equilibria concerned.

One example of the common ion effect which is frequently quoted is the addition of hydrogen chloride gas to a nearly saturated solution of sodium chloride, when precipitation of the salt occurs. The explanation which is usually given is that on dissolving in the water the hydrogen chloride dissociates strongly into H^+ and Cl^- ions, and the high concentration of the common chloride ion causes the solubility of the sodium chloride to be exceeded. There is undoubtedly some truth in this, but the common ion effect does not wholly account for the precipitation since a precipitate of sodium chloride is also obtained if hydrogen bromide gas is bubbled into the solution. It seems likely that when these very soluble gases dissolve and dissociate, substantial numbers of water molecules are used for hydrating the resulting ions and this increases the effective concentration of the sodium chloride causing its precipitation.

Solubility Product

We can determine the conditions for precipitation by applying the law of mass action to the equilibrium that exists between a sparingly soluble salt in the solid state and its ions in solution. Let us consider a saturated solution of the binary electrolyte remove. MA in contact with an excess of undissolved solid at constant temperature. The following equilibrium is set up, since such a salt is a strong electrolyte completely dissociated into its ions in solution:

$$MA \text{ solid} \rightleftharpoons M^+ + A^-$$

Applying the law of mass action to this and taking the concentrations of the individual ions as their active masses, we get:

$$\frac{[M^+][A^-]}{[MA \text{ solid}]} = \text{constant}$$

Now the active mass of any solid is a constant at constant temperature, so [MA solid] is constant.

$\therefore$ $[M^+][A^-]$ = constant = $S.P._{M.A.}$, the solubility product of the salt MA at that temperature.

For example, the solubility product of silver bromide, $S.P._{AgCr}$, is $[Ag^+][Br^-]$ and has the value 4×10^{-13} mol dm^{-3}. Similarly, for an electrolyte M_pA_q which dissociates thus:

$$M_pA_q \rightleftharpoons pM^{m+} + qA^{n-}$$

The solubility product is the product $[M^{m+}]^p [A^{n-}]^q$, in which the ionic concentrations are raised to the appropriate power. Thus the expression for the solubility product of calcium phosphate, which has the formula $Ca_3(PO_4)_2$, is $[Ca^{2+}]^3 . [PO_4^{3-}]^2$.

The importance of solubility product is that it indicates the maximum possible value of the ionic product in a solution when the ionic concentrations are expressed in mol dm^{-3}; in an unsaturated solution, of course, the ionic product will have a lower value. The concept of solubility product can only be applied to sparingly soluble electrolytes because only in those cases are the ionic concentrations sufficiently low to be taken as their active masses, and it is altogether wrong, therefore, to refer to the solubility product of soluble salts such as sodium chloride. A consequence of this limitation is that the numerical value of solubility products is always very small, rarely exceeding 10^{-2} and sometimes being as low as 10^{-50} for salts of extremely low solubility. Since solubility product varies with temperature, the latter should always be quoted if it is other than the arbitrarily selected standard of 25°C. Solubility products are usually determined experimentally in this way by first finding the solubility of the sparingly soluble salt either by volumetric analysis or from its molar conductance.

It is important to realise that although the solubility product of a salt is constant at constant temperature, the concentration of the individual ions in its saturated solution may vary over a very wide range. Naturally these ions are present in equivalent concentrations when the saturated solution is prepared by simply dissolving the salt in the solvent, but there may be a big difference in their concentrations when the solution is prepared by double decomposition or by mixing two solutions with a common ion. In such circumstances the solubility product can be used to determine when precipitation will begin and what concentration of residual ions will remain in solution.

Application of Solubility Product to Qualitative Analysis

Group I

The solubility products of silver chloride, mercurous chloride, and lead chloride at 15°C are 1.2×10^{-10} $mol^2 \ dm^{-6}$, $2.5 \times 10^{-18} \ mol^2 \ dm^{-6}$, and $2.0 \times 10^{-4} \ mol^3 \ dm^{-9}$ respectively. Thus if any of these three metals are present in the solution when the hydrochloric acid is added they will be precipitated as their chlorides because these solubility products will be exceeded. For example, if a silver salt is present and the concentration of Cl^- ions in the solution after precipitation is $0.5 \ mol \ dm^{-3}$ owing to the excess of hydrochloric acid, if follows that the concentration of Ag^+ ions remaining in the solution cannot exceed the very low figure of $2.4 \times 10^{-10} \ mol \ dm^{-3}$, since $[Ag^+][Cl^-] = 1.2 \times 10^{-10}$ at room temperature.

The solubility product of lead chloride increases markedly with temperature, which explains why it is always necessary to cool a group I solution thoroughly before proceeding to group II, lest the bulk of any lead present should escape precipitation in group I because the solubility product of its chloride is not exceeded. Even at room temperature precipitation of lead chloride is not complete and enough lead ions usually remain in solution to give a precipitate of lead sulphide in group II. Another interesting feature of lead is its ability to form a soluble complex ion with Cl^- ions when the concentration of the latter is very high :

$$Pb^{2+} + 4Cl^- \rightleftharpoons [PbCl_4]^{2-}$$

This explains why a group I solution which has been made by dissolving the solid to be analysed in concentrated hydrochloric acid should always be diluted, since the formation of the complex ion so reduces the concentration of the simple lead ion that again the solubility product of lead chloride is often not exceeded.

Group II

The precipitating reagent here is hydrogen sulphide in the presence of hydrochloric acid. As has already been explained, the hydrochloric acid suppresses the dissociation of hydrosulphuric acid by the common ion effect, lowering the concentration of S^{2-} ions from about $10^{-15} \ mol \ dm^{-3}$ (which is its values in a saturated solution at room temperature) to about $10^{-15} \ mol \ dm^{-3}$. Since the concentration of metal ions in the group solution will be of the order of $10^{-1} \ mol \ dm^{-3}$, this means that only the least soluble sulphides with solubility products below about 10^{-26} will be precipitated and the others will remain in solution. Thus sulphides such as bismuth (10^{-72}). mercuric mercury (10^{-54}), copper (10^{-45}), cadmium (10^{-29}), lead (10^{-29}), and tin appear in this group, whereas those of nickel (10^{-24}), zinc (10^{-23}), cobalt (10^{-22}), and manganese (10^{-15}) do not. This enables us to limit group II to some seven or eight metals and so facilitates their identification.

If the concentration of hydrochloric acid is made too high, then the dissociation of hydrosulphuric acid may be suppressed to such an extent that even lead, cadmium, and stannous sulphides, whose solubility products are near the cut-off value, may not be precipitated in this group. To guard against this danger, after treatment with hydrogen sulphide the group solution should always be diluted and tested again with hydrogen sulphide before proceeding to group III.

Group III

The precipitating reagent in this group is ammonium hydroxide in the presence of a high concentration of ammonium chloride. As already explained the dissociation of the ammonium hydroxide is suppressed by the common ion effect exerted by the ammonium chloride, so that the solubility products of some of the more soluble hydroxides are not exceeded and their precipitation does not occur in this group. The solubility products of the hydroxides of chromium, aluminium, and ferric irons are so low (10^{-30}, 10^{-33}, and 10^{-38} respectively), that these three are completely precipitated as their hydroxides. The following calculation shows that this is not the whole truth, however, and that some other factor must be at work as well to prevent the precipitation of nickel, cobalt, and zinc as their hydroxides. Let us assume that the ammonium hydroxide in the group solution is decinormal :

$$NH_4OH \rightleftharpoons NH_4^+ + OH^-$$

Then, since it is a weak electrolyte and its degree of dissociation α is small, when no ammonium chloride is added,

$$\alpha^2 / V = K_{NH_4OH} = 1.8 \times 10^{-5}, \text{ and } V = 10 \text{ dm}^3$$

$$\therefore \qquad a^2 = 1.8 \times 10^{-4} \quad \text{and } a = 1.3 \times 10^{-3}$$

$$\therefore \qquad [OH^-] = \frac{\alpha}{V} = 1.3 \times 10^{-3} \text{ mol dm}^{-3}$$

On adding the ammonium chloride the concentration of the NH_4^+ ion is raised to about 1.5 M, since this salt is a strong electrolyte completely dissociated into ions in solution :

$$NH_4Cl \rightarrow NH_4^+ + Cl^-$$

$\therefore$ $[NH_4^+] \simeq 1.5$ since the NH_4^+ ions from the ammonium hydroxide can be neglected compared with those from the strong electrolyte ammonium chloride.

But $\qquad \dfrac{[NH_4^+][OH^-]}{[NH_4OH]} = K_{NH_4OH} = 1.8 \times 10^{-5} \text{ mol dm}^{-3}$

$$\therefore \qquad \frac{1.5 \times [OH^-]}{0.1} = 1.8 \times 10^{-5}, \text{ and } [OH^-] = 1.2 \times 10^{-6} \text{ mol dm}^{-3}$$

Thus the effect on the OH^- ion concentration of adding the ammonium chloride is to reduce it about a thousandfold from about 10^{-3} to 10^{-4} mole per cubic decimetre.

Now for a trivalent metal the solubility product is given by

$$[M^{3+}][OH^-]^3 = \text{S.P.}_{M(OH)_4}.$$

Assuming that the concentration of the metal ion $[M^{3+}]$ is originally 10^{-1} mol dm^{-3}, the ionic product will be $10^{-1} \times (10^{-3}) = 10^{-10}$ without ammonium chloride and $10^{-1} \times (10^{-6})^3 = 10^{-9}$ in the presence of ammonium chloride, so in either case the hydroxides of aluminium, chromium, and ferric iron will be precipitated.

For a divalent metal, however, the solubility product is given by $[M^{2+}][OH^-]^2 = S.P._{M(OH)_2}$. Again assuming that $[M^{2+}]$ is originally 10^{-1} mol dm^{-3}, the ionic product will be about $10^{-1} \times (10^{-3})^2 = 10^{-7}$ in the absence of ammonium chloride and about $10^{-1} \times (10^{-6})^2 = 10^{-13}$ in its presence. Now the solubility products of nickel, cobalt, zinc, and manganese hydroxides are about 10^{-19}, 10^{-18}, 10^{-17}, and 10^{-14} respectively, so according to the above calculation, all these divalent metals would be expected to be precipitated as their hydroxides in this group, even in the presence of the ammonium chloride, since all their solubility products would be exceeded.

The main explanation of this conflict between experimental fact (that only the hydroxides of aluminium, chromium, and ferric iron are precipitated in this group) and theoretical prediction based upon the solubility product concept almost certainly lies in the formation of soluble complex ions with ammonia molecules, which abound in ammonium hydroxide solution. Three of the metals concerned, nickel, cobalt, and zinc, are known to form such complex ions very freely, e.g. $[Ni(NH_3)_6]^{2+}$, $[Co(NH_3)_6]^{3+}$, and $[Zn(NH_3)_2]^{2+}$. This would remove many of the simple metallic ions from the solution and prevent the solubility products of their hydroxides from being exceeded in the presence of ammonium chloride. Manganese, however, is not much affected in this way, and being near the border line of the solubility product cut-off it tends to be precipitated slowly as its hydroxide in group III if insufficient ammonium chloride is added. This is particularly liable to happen if one of the group III metals is present as well as manganese; the phenomenon, which is known as co-precipitation, is best avoided by using ample ammonium chloride and by filtering the group III solution promptly.

Group IV

In this group the precipitating reagent is either hydrogen sulphide in the presence of ammonium hydroxide solution or ammonium sulphide solution. In either case the concentration of S^{2-} ions in solution is very high and those metals with insoluble sulphides which escaped precipitation in group II (i.e. nickel, cobalt, zinc, and manganese) will be precipitated here.

Group V

The precipitating reagent is ammonium carbonate which gives a high concentration of CO_3^{2-} ions in solution and brings about the complete precipitation of calcium, strontium, and barium carbonates which have solubility products of 10^{-8}, 10^{-9}, and 8×10^{-9} respectively.

Application of Solubility Product to Volumetric analysis

The concept of solubility product can be applied to precipitation reactions in volumetric analysis enabling us to calculate the ionic concentrations at the end-point. A good example is afforded by the titration of a neutral chloride solution with silver nitrate solution by Mohr's method, using potassium chromate as the indicator:

e.g.
$$AgNO_3 + NaCl = AgCl \downarrow + NaNO_3$$

Now
$$AgCl \text{ solid} \rightleftharpoons Ag^+ + Cl^-$$

∴
$$[Ag^+][Cl^-] = S.P._{AgCl} = 1.2 \times 10^{-10} \text{ mol}^2 \text{ dm}^{-6}$$

The desired end-point of this titration is when neither silver nor chloride ions are in excess in the solution, i.e. when

$$[Ag^+] = [Cl^-] = \sqrt{1.2 \times 10^{-5}} = 1.1 \times 10^{-5} \text{ mol dm}^{-3}$$

We must aim to use, therefore, that concentration of chromate indicator which just gives a precipitate of silver chromate when the concentration of Ag^+ ions in solution is 1.1×10^{-5} dm^{-3}.

Now
$$Ag_2CrO_4 \rightleftharpoons 2Ag^+ + CrO_4^{2-}$$

$\therefore$
$$[Ag^+][CrO_4^{2-}] = S.P._{Ag_2CrO_4} = 2.4 \times 10^{-12} \text{ mol}^2 \text{ dm}^{-9}$$

$\therefore$
$$[CrO_4^{2-}] = \frac{2.4 \times 10^{-12}}{[Ag^+]^2} = \frac{2.4 \times 10^{-12}}{1.2 \times 10^{-10}} \text{ mol dm}^{-3}$$

$$= 2.0 \times 10^{-2} \text{ mol dm}^{-3}$$

Thus the concentration of chromate ion should be about M/100 at the end-point, which means that it ought to be about M/50 at the start of the titration. This can be achieved by adding about 0.5 cm^3 of bench potassium chromate solution to each 25 cm^3 of halide solution in the conical flask.

This case illustrates very well that one substance may be less soluble than another and yet have a higher solubility product. As the example in shows, the solubility of silver chloride is 1.1×10^{-5} mol dm^{-3} or 1.60×10^{-3} g dm^{-3} at 20°C. The solubility of silver chromate can be calculated from its solubility product as follows :

$$[Ag^+]^2[CrO_4^{2-}] = S.P._{Ag_2CrO_4} = 2.4 \times 10^{-12} \text{ mol}^3 \text{ dm}^{-9}$$

Now in its saturated solution $[Ag^+] = 2[CrO_4^{2-}]$

$\therefore$
$$4[CrO_4^{2-}]^3 = 2.4 \times 10^{-12} \text{ mol}^3 \text{ dm}^{-9}$$

and
$$[CrO_4^{2-}] = \sqrt[3]{\frac{2.4}{4}} \times 10^{-4} = \sqrt[3]{0.6} \times 10^{-4} = 0.84 \times 10^{-4} \text{ mol dm}^{-3}$$

Thus the solubility of silver chromate is 8.4×10^{-5} mole per cubic decimetre or $8.4 \times 10^{-5} \times 332 = 2.8 \times 10^{-2}$ g dm^{-3} at 20°C. The greater solubility of silver chromate is explained by the occurrence of the term $[Ag^+]^2$ in the expression for its solubility product.

Solubility of Salts of Weak Acids in Strong Acids

It is an experimental fact that salts of weak acids are much more soluble in a solution of strong acid than they are in water. For example, calcium orthophosphate is only sparingly soluble in water at room temperature, yet it dissolves freely in dilute hydrochloric acid. This phenomenon can be explained by considering the ionic equilibria existing in the solution.

When solid calcium phosphate is added to water and left to achieve equilibrium a saturated solution is formed in which the concentrations of calcium and phosphate ions are very low and are governed by the relationship :

$$Ca_3(PO_4)_2 \text{ solid} \rightleftharpoons 3\ Ca^{2+} + 2\ PO_4^{3-}$$

where
$$[Ca^{2+}]^3[PO_4^{3-}]^2 = S.P._{Ca_3(PO_4)_2} = 10^{-20}$$

The water is also dissociated, of course, but only to a very small extent, so that the concentration of H^+ and OH^- ions in solution is also very low:

$$H_2O \rightleftharpoons H^+ + OH^-$$

Under these circumstances very little combination of H^+ and PO_4^{3-} ions takes place, despite the fact

that orthophosphoric acid is a weak acid, because the expression $\dfrac{[H^+]^3[PO_4^{3-}]}{[H_3PO_4]}$ soon attains the value

corresponding to $K_{H_3PO_4}$:

$$H_3PO_4 \rightleftharpoons 3H^+ + PO_4^{3-}$$

Now when hydrochloric acid is added it dissociates completely into ions, so the concentration of H^+ ions in the solution is raised to a relatively very high level :

$$HCl \rightarrow H^+ + Cl^-$$

This causes hydrogen ions and phosphate ions to unite in large numbers to form undissociated

molecules of orthophosphoric acid, so that eventually the expression $\dfrac{[H^+]^3[PO_4^{3-}]}{[H_3PO_4]}$ is restored to its

original low value of $K_{H_3PO_4}$. In the process the PO_4^{3-} ion concentration is reduced to such a low level that the ionic product of calcium phosphate falls well below its solubility product causing more of the solid calcium phosphate to dissolve. The interaction between the various equilibria continues until a high concentration of orthophosphoric acid molecules exists in solution and much of the solid calcium phosphate has gone into solution. Thus the overall effect of the hydrochloric acid is to suppress the already small dissociation of the weak orthophosphoric acid by means of the common ion effect and thereby artificially to lower the concentration of phosphate ions in the solution so that more of the salt dissolves. A similar explanation can be provided for the solubility of salts of other weak acids, e.g. oxalates and tartrates.

Indicators

An indicator for acid-alkali titrations is a substance which shows different colours when the pH of its environment is altered. It is usually a weak organic acid or base capable of undergoing a tautomeric change within its molecule, which accounts for the colour change, but for our purposes it will suffice to regard it simply as a substance which dissociates into an ion of different colour thus :

$$HA \rightleftharpoons H^+ + A^-$$
Colour 1 Colour 2

Since the dissociation is only partial and is reversible, its extent will depend very much upon the hydrogen ion concentration of the solution to which it is added. In a strongly acidic solution, as Le Chatelier's principle predicts, the dissociation would be suppressed and colour 1 would predominate, whereas in a solution with a low concentration of H^+ ions the dissociation would be largely complete and the colour of the anion, colour 2, would be seen. In practice, although particles of both colours will always be present in the solution, only one colour will be noticeable, the predominant one, unless the ratio of the concentrations of the two types of particle is less than 10 : 1, when some intermediate colour will be observed. This means that as far as the human eye is concerned the colour change of an indicator takes place gradually over a band of about two units of pH, only being complete when there is a hundredfold change in H^+ ion concentration. It follows that if an indicator is to give a precise end-point by changing colour sharply on adding a single drop of acid or alkali, it is necessary that the pH of the solution should change suddenly by at least two units.

Now although indicators are weak electrolytes, they vary greatly in their dissociation constants, which means that they change colour over widely different bands of pH. This point is illustrated in Fig. 11.10, which shows the ranges of some well-known indicators. The H^+ ion concentration of the centre-point of each range is easily calculated from the value of the dissociation constant, since $\dfrac{[H^+][A^-]}{[HA]} = K_{Ind}$.

and so $[H^+] = \dfrac{K_{Ind}[HA]}{[A^-]}$. Now at the centre-point the two types of coloured particle will be present in equal concentrations, i.e. $[HA] = [A^-]$, and $[H^+]$ will therefore be equal to K_{Ind}. This explains why air-free distilled water is alkaline to methyl orange ($K_{Ind} = 2 \times 10^{-4}$ mol dm^{-3}, so central pH is 3.7) neutral to litmus ($K_{Ind} = 1 \times 10^{-7}$ mol dm^{-3}, so central pH is 7.0), and acidic to phenolphthalein ($K_{Ind} = 2 \times 10^{-10}$ mol dm^{-3}, so central pH is 9.7). At first sight the fact that various indicators change colour at different H^+ ion concentrations might seem to be a disadvantage in judging end-points since they would appear to disagree in their verdicts, but in practice, as we shall see in later sections, this difference between them is of great value and importance.

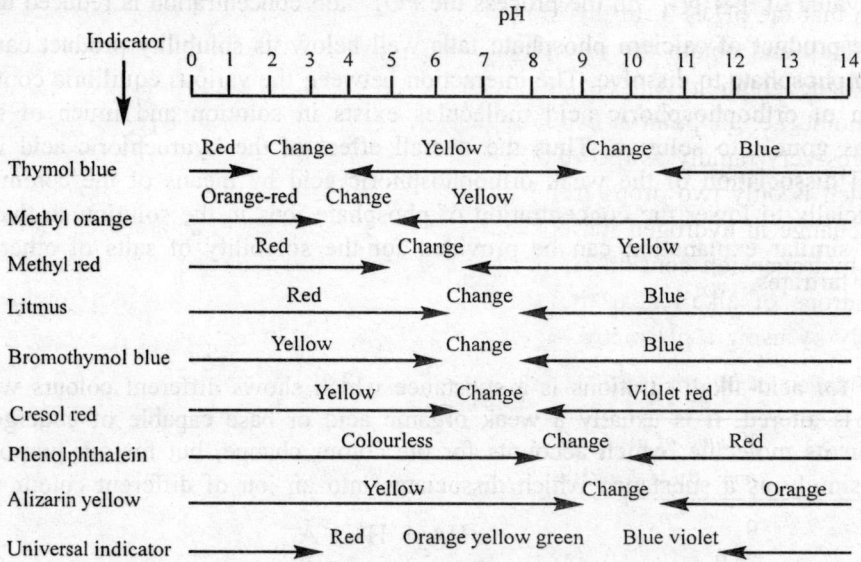

Fig. 11.10. Colours and ranges of common indicators.

When a carbonate is titrated with an acid, the solution becomes saturated with carbon dioxide and its pH changes to about 4.4. Although this is a big enough change to interfere when methyl red is the indicator, it does not affect methyl orange, which is why the latter is preferred to titrations involving carbonates. If methyl red is used for such titrations it is necessary to boil off the carbon dioxide from the solution before judging the end-point.

When water is stored in an open vessel an equilibrium is soon set up with the gases in the atmosphere, and the carbon dioxide which dissolves changes the pH to about 5.7. This is important when phenolphthalein is being used as the indicator and alkali is being added from the burette, since owing to the action of carbon dioxide upon pH the pink-purple colour which marks the end-point disappears again after a short period. If a further drop of alkali is added the pink colour reappears, since the pH is then temporarily restored to about 10, only to fade away again slowly on standing in air as the pH of the solution falls. The end-point in such a titration is clearly the moment when the indicator colour persists on shaking for, say, five seconds, since otherwise the alkali will be used for titrating the carbon dioxide in the atmosphere.

Titration Curves

These are of several distinct types.

Strong acid/strong alkali

The changes in hydrogen ion concentration which take place when a strong acid is titrated with a strong base can easily be calculated. For example, if we start with a conical flask containing 50 cm^3 of 0.1 M hydrochloric acid and run in 49.0 cm^3 of 0.1M sodium hydroxide solution from the burette, there remains in the flask 1 cm^3 of the decimolar acid diluted to 99 cm^3. Thus the concentration of the residual acid is approximately 1% of 0.1M, i.e. 10^{-3} M, and the H^+ ion concentration will be 10^{-3} mole per cubic decimetre, so that the pH is 3. In this way we can determine the pH of the solution after adding various volumes of alkali and obtain the results listed in Table 11.3. and plotted in Fig. 11.11 (curve I). From the table and the graph it will be seen that the pH changes only slowly at first, but that near the equivalence-point (i.e. the point at which acid and alkali have been added in exactly equivalent amounts) the pH changes very rapidly as the alkali is added. Indeed the figures show that he addition of 0.1 cm^3 of alkali, which is only two drops from the burette, changes the pH from 4.3 to 9.7, producing a half-million fold change in hydrogen ion concentration. If normal solutions are used the effect is even more striking, the hydrogen ion concentration being made over ten million times as small by the addition of only a few drops of alkali near the equivalence point. These calculated figures can be confirmed experimentally by using a pH metre.

Table 11.3. Titration of 50 cm^3 of 0.1M HCl with strong alkali.

Volume of alkali added (cm³ 0.1M NaOH)	[H⁺] mol dm⁻³	pH
0	10^{-1}	1.0
25.0	3.3×10^{-2}	1.5
45.0	5×10^{-2}	2.3
49.0	1×10^{-2}	3.0
49.5	5×10^{-4}	3.3
49.9	1×10^{-4}	4.0
49.95	5×10^{-5}	4.3
50.00	1×10^{-7}	7.0
50.05	2×10^{-10}	9.7
50.1	1×10^{-10}	10.7
50.5	2×10^{-11}	10.0
51.0	1×10^{-11}	11.0
55.0	2×10^{-12}	11.7

Now the aim of any titration is to determine the equivalence-point, i.e. to find what volumes of the two solutions are chemically equivalent to each other. Ideally the end-point, which is the point at which the titration is stopped, should coincide exactly with this equivalence-point, and to achieve this the indicator used must show a sharp colour change at the right pH. From titration curve I it should be clear that any indicator which changes colour between pH 4 and pH 10 will be satisfactory when titrating a strong acid with a strong base in view of the sudden and very wide swing in pH at the equivalence-point. Thus methyl orange and phenolphthalein will both change colour at about the same time in this type of titration and either may be used with confidence.

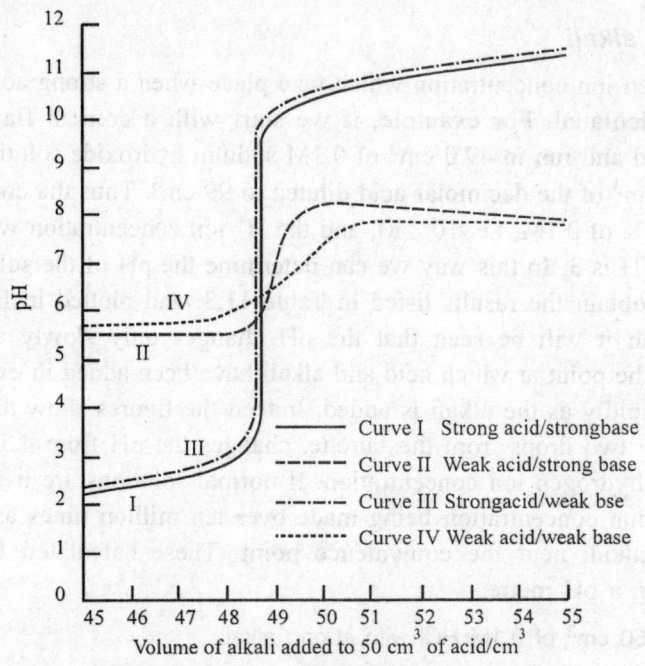

Fig. 11.11. Titration curves.

Weak acid/strong alkali

When a 0.1M solution of acetic acid (a typical weak acid) is titrated with a 0.1 M solution of sodium hydroxide, the change in pH is as shown in curve II of Fig. 11.11. There is very little variation at first as the alkali is added, the pH being considerably higher for a given volume of alkali than in curve I because the salt formed is extensively hydrolysed, but near the equivalence-point there is again a rapid swing in pH, although the change is not as great as before. These results are calculated by making certain assumptions which are only approximately true. For example, the presence of acetate ions in solution is regarded as being solely due to the complete dissociation of sodium acetate, so their concentration is taken as the concentration of alkali added. Again the concentration of undissociated acetic acid molecules present is calculated from the amount of residual acid, since it is a weak acid only slightly dissociated in solution.

Then since $\dfrac{[H^+].[CH_3COO^-]}{[CH_3COOH]} = K_{CH_3COOH} = 1.8 \times 10^{-5}\,\text{mol dm}^{-3}$

$$[H^+] = 1.8 \times 10^{-5} \times \frac{[CH_3.COOH]}{[CH_3.COO^-]} \text{ mol dm}^{-3}$$

$$= 1.8 \times 10^{-5} \times \frac{[\text{Residual acid}]}{[\text{Alkali added}]} \text{ mol dm}^{-3}$$

Thus when the amount of alkali added is enough to neutralise exactly half the acid originally taken, so that the concentrations of acetic acid molecules and acetate ions are the same, the hydrogen ion concentration is numerically equal to the dissociation constant of the acid, i.e. 1.8×10^{-5} mol dm^{-3}, and the pH then 4.74.

In this titration the best indicator is one which changes colour when the pH is about 9, since it is then that the equivalence-point is reached and the pH changes most rapidly with added alkali (see curve II in Fig. 11.11). Thus phenolphthalein, which changes colour in this region is a very suitable indicator, whereas the end-point indicated by methyl orange (at pH 3.7, say) would be seriously different from the equivalence-point and would not, in any case, be precise, because at that stage the addition of a few drops of alkali produces only a small change in pH.

Strong acid/weak alkali

Using a similar procedure it is possible to construct the titration curve, as it is called, for the neutralisation of a strong acid by a weak base, e.g. hydrochloric acid and ammonium hydroxide solution. This is shown as curve III in Fig. 11.11. From this curve it is clear that the equivalence-point is reached when the pH is about 4 or 5 and a suitable indicator must change colour in that region. Moreover, it is then that the pH is most responsive to addition of alkali, so a sharp and precise end-point is possible. Methyl orange and methyl red are both good indicators, therefore, for this type of titration, whereas phenolphthalein is useless and will give a large error.

Weak acid/weak alkali

Curve IV in Fig. 11.11 shows that the changes in pH which occur when a weak acid such as acetic acid is titrated against a weak base such as ammonium hydroxide. The extensive hydrolysis of the salt formed in this type of titration prevents any sudden swing in pH near the equivalence-point. No indicator will be satisfactory, therefore, since none will give a sharp and reliable endpoint. For this reason this type of titration must be avoided even if it means titrating the weak acid and the weak base against some strong electrolytes whose relative concentrations are known.

Polybasic acids

Another type of titration curve which is of particular interest is the one obtained when an alkali is added to a polybasic acid. Fig. 11.12 shows the changes in pH which occur when orthophosphoric acid, H_3PO_4, is titrated with sodium hydroxide. This titration curve shows two rapid swings in pH, neither of them very sharp, corresponding to the first and second stages of the acid's dissociation. A third inflection of the curve, this time very indistinct, occurs in strongly alkaline solution. This particular diagram explains why methyl orange is suitable for detecting the first end-point and phenolphthalein the second, and why no indicator is satisfactory for judging the position of the third dissociation.

A curve with two inflections is obtained when sodium carbonate is titrated with a strong acid. When the carbonate has been changed to bicarbonate the pH is about 9, so phenolphthalein responds, but methyl orange is needed to detect the second stage of the neutralisation at a pH of about 5.

Buffer Solutions

These are solutions which resist changes in pH on dilution or on addition of acid or alkali. They usually consist of a solution of a weak acid or a weak base in the presence of one of its salts, e.g. sodium acetate and acetic acid, or sodium citrate and citric acid, or borax and boric acid. Mixtures of two acid salts, such as disodium hydrogen orthophosphate and sodium dihydrogen orthophosphate, are also used.

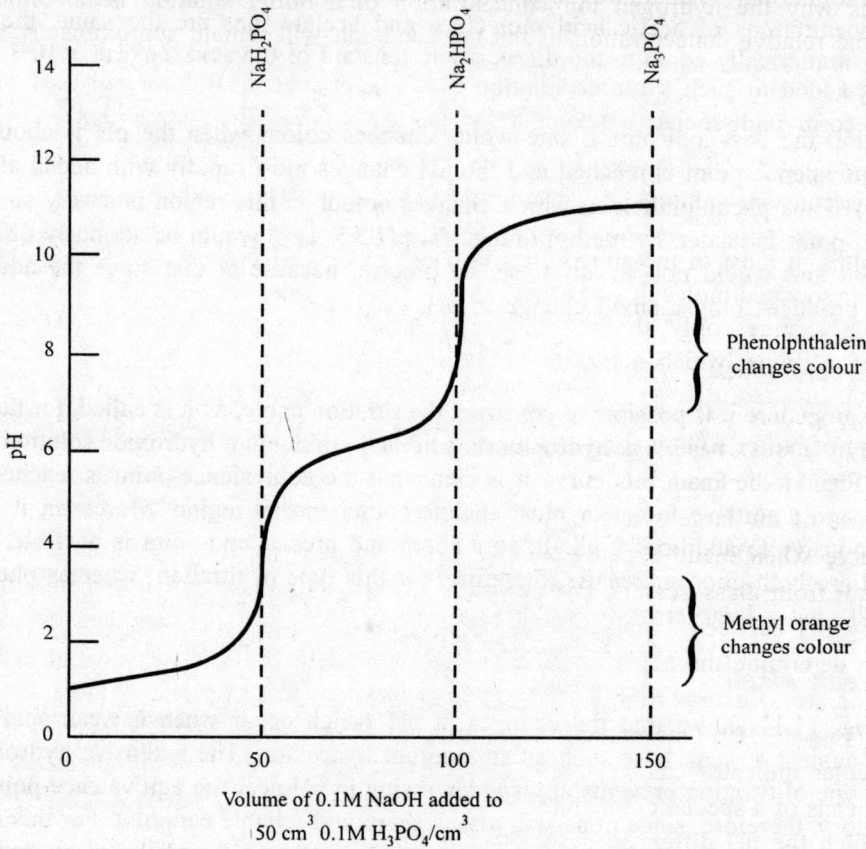

Fig. 11.12. Titration curve of a polybasic acid.

To understand how a buffer solution works let us consider a hypothetical but typical example of a weak acid HA and its salt MA, where H stands for hydrogen, M^+ for metal ion, and A^- for the common anion.

Then $$HA \rightleftharpoons H^+ + A^-$$

and $$MA \rightarrow M^+ + A^-$$

Now $$\frac{[H^+][A^-]}{[HA]} = K_{HA}$$

$$[H^+] = K_{HA} \times \frac{[HA]}{[A^-]}$$

where [HA] is effectively the concentration of weak acid taken, since it will be only very slightly dissociated in the presence of its salt, and where [A$^-$] is effectively the concentration of the salt because this is completely dissociated into its ions.

$$\therefore \quad [H^+] = K_{HA} \times \frac{[Acid]}{[Salt]}$$

This explains why the hydrogen ion concentration of a buffer solution is not much affected by dilution, since the relative concentration of the acid and salt will remain approximately the same.

If an acid is added to such a buffer solution most of the added H$^+$ ions will combine with the A$^-$ ions present to form undissociated molecules of the weak acid HA in order to keep the expression

$\dfrac{[H^+][A^-]}{[HA]}$ at its constant value, K_{HA}, and the pH will be little altered as a result. If an alkali is added

to the buffer solution, most of the added OH$^-$ ions are removed from solution by combination with H$^+$ ions to form undissociated molecules of water, further H$^+$ ions being furnished by the dissociation of HA, so again the pH changes only slightly. Thus the stability of the pH can be attributed to the high concentration of A$^-$ ions, which act as a trap for added H$^+$ ions, and the large concentration of HA molecules, which provide a reservoir of H$^+$ ions for combining with added OH$^-$ ions.

The chief use of buffer solutions in the laboratory is for making solutions of known and constant pH. Such solutions cannot be made accurately by adding acids and alkalies to each other in calculated quantities because of the large swing in pH which occurs near the equivalence-point. Nor would they be easily preserve when made because of the sensitivity of their pH to dissolved gases (e.g. carbon dioxide) or alkali from glass vessels. By a suitable choice of substances it is possible to prepare buffer solutions of any pH between 1 and 13. These solutions of known pH can then be used to calibrate indicators or to determine the pH of a solution colorimetrically. Special mixtures of indicators (known as universal indicators) are available, which show various characteristic colours for each pH over a wide range (Fig. 11.10). One of these can be used to find the pH of a solution approximately, and then one particular indicator can be applied to the solution and the colour obtained can be carefully compared by means of a special comparator with the colours shown by that indicator in a series of buffer solutions in which the pH differs by only 0.1.

It is important to be able to calculate the pH of any particular buffer solution and also the composition of the mixture needed to make a buffer solution with a given pH. We have shown that

$$[H^+] = K_{HA} \times \frac{[Acid]}{[Salt]}$$

Taking logarithms of both sides we get

$$\log_{10}[H^+] = \log_{10} K_{HA} + \log_{10} \frac{[acid]}{[salt]}$$

Multiplying each side by -1 gives

$$- \log_{10} [H^+] = - \log_{10} K_{HA} - \log_{10} \frac{[acid]}{[salt]}$$

Now
$$- \log_{10} [H^+] = pH$$

$\therefore$
$$pH = - \log_{10} K_{HA} - \log_{10} \frac{[acid]}{[salt]}$$

It follows from this equation that the pH of a given buffer solution depends upon the ratio of the concentrations of the acid and salt, and not upon their actual values. In the special case when acid and salt are present in equimolar concentrations, the hydrogen ion concentration is equal to K_{HA}. Clearly the magnitude of this dissociation constant is the predominant influence upon the pH of a buffer solution, although the relative concentrations of acid and salt taken determine the exact value within a narrow range.

Buffer solutions are also of great importance in medicine and biochemistry, where pH values are often critical and have to be maintained at a steady value. For example, the blood contains dissolved carbonic acid and bicarbonates and also phosphates which maintain its pH at about 7.4, and a change of $\pm$ 0.5 would probably be fatal. Injections into the bloodstream, therefore are usually buffered so as not to upset this delicate balance. Similarly, many fermentation processes and enzyme reactions depend critically upon the pH, which must vary only within narrow limits. Buffer solutions are widely used in electroplating and also in processed foods and drinks to prevent excessive acidity.

Electrometric Titrations

Electrometric titrations are generally of two types: (i) conductometric; and (ii) potentiometric titrations.

Conductometric titration

This involves measuring the conductivity of the solution at various stages of the titration and plotting a graph as in Fig. 11.13. As already explained hydrogen ions and hydroxyl ions have a much greater conductivity than other ions, so that during the titration of a strong acid with a strong base the conductivity of the solution in the conical flask will reach a minimum near the equivalence-point, when the total concentration of these highly conducting ions is least. This minimum in the conductivity is taken as the end-point, as in Fig. 11.13(a). A different graph, like that shown in Fig. 11.13(b) is obtained when alkali is added to a solution of a weak acid. The conductometric method has the advantage of being applicable to very dilute or coloured solutions for which the usual colorimetric methods are unsuitable.

Potentiometric titration

This uses a pH meter i.e. a system involving a calomel reference electrode and a special glass electrode which develops a potential directly proportional to the pH of the solution in which it is immersed. From the readings of potential difference as measured with a potentiometer a titration curve like those in Fig. 11.11 can be plotted. The end-point is taken as the point at which the pH changes most rapidly with added titrant. The method is accurate, even when the solutions are dilute, and can be applied to coloured or opaque solutions. It has the further advantages of needing only small quantities of titrant, of being applicable to many types of titration including redox and precipitation reactions, and of being adaptable for automatic use.

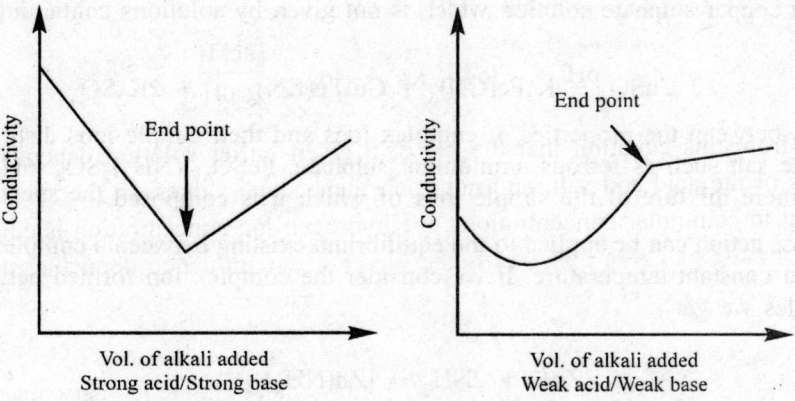

Fig. 11.13. Conductometric titrations.

Complex Ions

Many cases are known where an ion, particularly an ion of a transition metal joins with two or more ions of opposite charge or with two or more neutral molecules to form a larger ion which has completely different properties from the simple ion(s) it contains. For example, copper(II) forms the following ions:

$$Cu^{2+} + 4NH_3 \rightleftharpoons [Cu(NH_3)_4]^{2+}$$

$$Cu^{2+} + 4CN^- \rightleftharpoons [Cu(CN)_4]^{2-}$$

$$Cu^{2+} + 4H_2O \rightleftharpoons [Cu(H_2O)_4]^{2+}$$

$$Cu^{2+} + 4Cl^- \rightleftharpoons [CuCl_4]^{2-}$$

Larger ions such as these are known as complex ions; it is conventional to enclose them in square brackets as shown. The surrounding ions or molecules are linked to the central ion by co-ordinate linkages, which means that the complexes have definite shapes. When there are six ligands, for example, the arrangement is octahedral as in Fig. 11.14. A complex ion bears a net charge which is the algebraic sum of the separate charges of its constituent ions. Thus the ferricyanide ion carries a triple negative charge because the single ferric ion contributes three positive charges and each of the six cyanide ions one negative charge :

$$Fe^{3+} + 6CN^- \rightleftharpoons [Fe(CN)_6]^{3-}$$

All complex ions dissociate to some extent in solution into their simple ions, although some are so stable that the concentration of simple ions is insufficient to show the usual reactions. For example the hexacyanoferrate(II) ion, formed by adding an excess of potassium cyanide solution to a iron(II) salt, is very stable and when crystals of potassium hexacyanoferrate(II) are dissolved in water the solution gives no precipitate of iron(II) hydroxide with sodium hydroxide solution because the concentration of Fe^{2+} ions is too low to cause the solubility product of the hydroxide to be exceeded. Nor does it give a precipitate of silver cyanide with silver nitrate solution, showing that the concentration of simple cyanide ions must also be extremely low :

$$Fe^{2+} + 6CN^- \rightleftharpoons [Fe(CN)_6]^{4-}.$$

On the other hand, a solution of potassium ferrocyanide gives a gelatinous brown precipitate of cupric ferrocyanide with copper sulphate solution which is not given by solutions containing simple iron(II) or cyanide ions :

$$2\ CuSO_4 + K_4Fe(CN)_6 = Cu_2Fe(CN)_6 \downarrow + 2K_2SO_4$$

These differences between the properties of complex ions and their simple ions distinguishes a complex ion from a double salt such as ferrous ammonium sulphate, $FeSO_4 . (NH_4)_2SO_4 . 6H_2O$, which behaves in solution as a mere mixture of the simple ions of which it is composed.

The law of mass action can be applied to the equilibrium existing between a complex ion and its simple constituent ions at constant temperature. If we consider the complex ion formed between zinc ions and ammonia molecules we get :

$$Zn^{2+} + 2NH_3 \rightleftharpoons [Zn(NH_3)_2]^{2+}$$

$$\therefore \qquad \frac{[Zn(NH_3)_2]^{2+}}{[Zn^{2+}][NH_3]^2} = K_{stab}.$$

where K_{stab} is known as the stability constant of the iron.

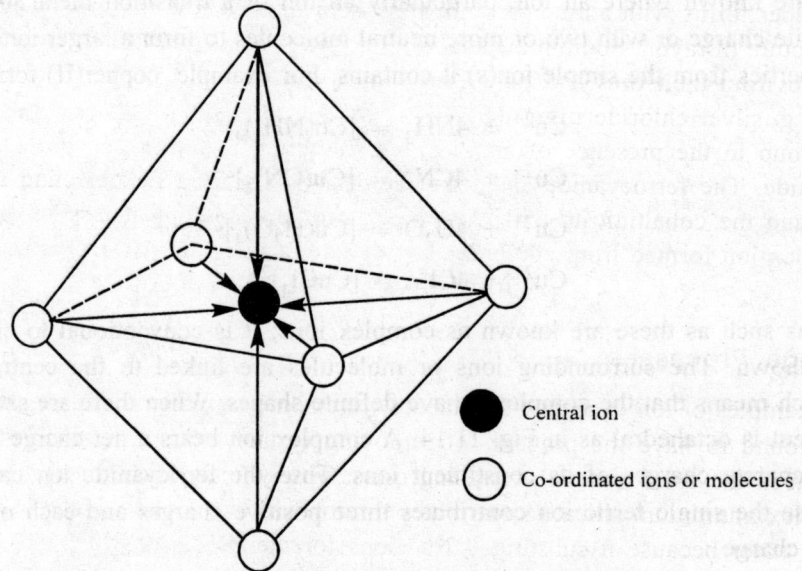

Fig. 11.14. Six co-ordinated octahedral complex ion.

- ● Central ion
- ○ Co-ordinated ions or molecules

Detection of complex ions

The presence of complex ions can be detected in various ways :

1. Metals which form complex anions, e.g. $[Ag(CN)_2]^-$, move towards the anode in electrolysis showing that the particle of which they are a part bears a negative charge overall.

2. When complex ions are formed their solutions show abnormally low colligative effects and lower conductivity than expected because they contain fewer particles of solute than would otherwise be present. For example, when a little mercuric iodide is added to potassium iodide solution the freezing point depression is smaller than before because the simple ions have united as follows :

$$2K^+ + 2I^- + HgI_2 \rightleftharpoons 2K^+ + [HgI_4]^{2-}$$

When the two salts are present in the molecular proportion of two of potassium iodide to one of mercuric iodide, the depression is only three-quarters of its original value since there are only three ions in solution for every four that existed before.

3. When a salt which is only sparingly soluble in water dissolves freely in an aqueous solution, it is usually a sign that a soluble complex ion is formed. For example, silver chloride dissolves readily in ammonium hydroxide solution, which contains a high concentration of ammonia molecules :

$$Ag^+ + Cl^- + 2NH_3 \rightleftharpoons [Ag(NH_3)_2]^+Cl^-$$
$$\text{Soluble complex salt}$$

4. Changes in colour often occur when a complex ion is formed. The familiar deep blue colour of the complex ion $[Cu(NH_3)_4]^{2+}$ is a good example.

5. X-ray analysis often reveals the presence of complex ions in the crystal.

Uses of complex ions

The soluble complex ion formed between silver ions and thiosulphate ions is used in photography as a means of 'fixing' the negative, i.e. removing the unchanged silver halide after exposure. The complex ion which copper forms with ammonia, known as Schweitzer's solution, dissolves cellulose and is used in the manufacture of rayon. Sodium or potassium argentocyanide and auricyanide are used for extracting silver and gold from their ores and for electroplating. In analysis complex ions are used for separating precipitates, e.g. silver chloride from mercurous chloride by adding ammonia solution, or detecting one metal of a group in the presence of another, e.g. cadmium in the presence of copper by forming the complex cyanide. The ferrocyanide and ferricyanide ions are used for detecting ferric and ferrous ions respectively, and the cobaltinitrite ion $[Co(NO_2)_6]^{3-}$ for precipitating potassium from solution in group 6. The complex ion formed from ethylene diamine tetra-acetic acid (EDTA) is used to remove hardness from water.

Ion Exchange Processes

The earlier examples of ion exchange substances were some naturally occurring silicates such as zeolites which were found to have the property of removing certain metallic ions from aqueous solution and retaining them in the solid material, releasing other loosely-held ions into solution in their place. From these a complex aluminium silicate known as permutit was developed which has been extensively used for softening water because it substituted Na^+ ions for the Ca^{2+} and Mg^{2+} ions present in hard water and could be regenerated with concentrated brine.

More recently whole ranges of synthetic resins have been made with very valuable ion exchange properties. These resins consist of solid cross-linked polymers of high molecular weight derived from certain organic substances such as phenols, formaldehyde, and styrene, to which are attached specific ionisable groups. If these groups are strongly acidic ones such as sulphonic acid, —SO_3H, or carboxyl, —COOH, then the resins are used for exchanging cations, but if basic groups such as amino or quaternary ammonium groups are present then they act as anion exchangers. The exchange of ions is a physical and completely reversible process, and like permutit these polymeric resins can be regenerated when exhausted and used over and over again.

Ion exchange processes are of great chemical and industrial importance. They are used for extracting and purifying uranium from its ores, for separating nuclear fission products from each other, for the purification of water, for separating the lanthanides or rare earth elements, for recovering precious metals from solutions, and for treating and disposing of very poisonous or dangerously radioactive waste. Many clays in soil also act as ion exchangers, retaining ammonium ions from added fertiliser or manure. Ion exchange chromatography has been very successfully applied to amino-acids by using anion exchange resins, and to the rare earths and the trans-uranic elements using cation exchange resins.

Example 11.6. The charge on a single electron is 1.6021×10^{-19} coulomb. Calculate Avogadro's number from the fact that 1 F = 96,487 coulombs.

Solution.

$$\text{Avogadro's Number} = 9.6487 \times 10^4 \text{ coulombs} \left(\frac{1 \text{ electron}}{1.6021 \times 10^{-19} \text{ coulomb}} \right)$$

$$= 6.0225 \times 10^{23} \text{ electrons}$$

Example 11.7. At 25°C the equivalent conductance of $0.100N$ $HC_2H_3O_2$ is 5.2 cm²/ohm equivalent: A_0 for acetic acid is 390.7 cm²/ohm equivalent. What is the degree of dissociation of $0.100N$ $HC_2H_3O_2$ at 25°C?

Solution.

$$\alpha = \frac{\Lambda}{\Lambda_0}$$

$$\alpha = \frac{5.2 \text{ cm}^2 / \text{ohm equivalent}}{390.7 \text{ cm}^2 / \text{ohm equivalent}}$$

$$\alpha = 0.013$$

At 25°C, 0,100N acetic acid is 1.3% ionised.

Chapter 12

Acids and Bases

INTRODUCTION

Acids (Latin *acidus* = sour) and alkalies (Arabic *al-quili* = ashes of a plant) were known as two distinct classes of substances even before chemistry was developed as a separate subject for systematic study. The general characteristics of acids are: (i) sour taste; (ii) ability to change the colour of certain vegetable dyes; (iii) high dissolving ability; and (iv) evolution of hydrogen with metals in most cases.

It was also observed that many non-metals (C, N, S, P) burn in oxygen and the product forms an acid with water. Oxygen got its name from these observations (Greek *Oxus* = sour; *gennae* = I produce) and it was considered to be an essential constituent of all acids. But that hydrochloric acid contained no oxygen.

Alkalies were known to produce a 'soapy feel' in water, affect certain vegetable dyes and "counteract" the action of acids. Certain oxides and hydroxides of metals were themselves insoluble in water but could neutralise acids forming salts. These were covered by the more general term base, introduced by Rouelle in mid-eighteenth century.

At present, the study of acids and bases has gained extreme importance in chemistry. The span of the terms has been extended and modified in several stages by different approaches to include a wide variety of chemical compounds.

A very large number of chemical reactions may now be treated as essentially acid-base reactions, including even redox reactions and formation of complex compounds of metals. We shall briefly discuss the main lines of approach to develop the sense associated with the terms "acids" and "bases". These so-called "theories" of acids and bases are practically definitions, each trying to provide a wider coverage for different aspects of acid-base chemistry; they are not theories in the sense in which we come across a theory in other fields of study, as, for example, in the theories of chemical bonding.

ARRHENIUS THEORY

In his theory of electrolytic dissociation, Arrhenius defined an acid as any hydrogen-containing compound which gives hydrogen ions in aqueous solution; a base was similarly defined as any compound which gives hydroxyl ions in aqueous solution.

$$HCl \xrightarrow{\text{Water}} H^+ + Cl^- \quad Acid$$

$$KOH \xrightarrow{\text{Water}} K^+ + OH^- \quad Base$$

The theory was highly successful in explaining different aspects of acid-base behaviour in water solution. A few points are mentioned:

1. The strength of an acid or base could be expressed quantitatively in terms of an ionisation equilibrium of the type :

$$HA \rightleftharpoons H^+ + A^- \; ; \; K = C_{H+} \times C_{A^-} / C_{HA}.$$

2. Neutralisation was explained as the combination between H^+ and OH^- ions forming H_2O. This justified the observed constancy of molar heat of neutralisation for any strong acid–strong base pair.

3. The observed catalytic property of acids in many reactions could also be correlated with the availability of hydrogen ion from an acid (i.e., its acid strength).

Limitations of Arrhenius Theory

In spite of its wide applicability in aqueous solutions, the Arrhenius theory for acids and bases had certain limitations :

1. The acidic or basic property of a substance was not inherent in it, but was dependent on the presence of water.

2. The existence of the bare hydrogen ion, H^+, was questioned. The free proton has the size of the nucleus, 10^{-15} m in radius, which is much smaller than the radii of other ions (around 10^{-10} m). Bronsted has shown that the free energy change of the process $H^+ + H_2O \, (l) = H_3O^+$ is highly negative. It is therefore reasonable to expect that the hydrogen ion would be hydrated strongly in presence of water. In fact, a hydrated species, H_3O^+, has been observed in electrical discharges through water vapour, and in some hydrated crystals (e.g., $HClO_4$, $H_2O \equiv H_3O^+ \, ClO_4^-$). The H_3O^+ ion is variously known as the hydronium ion, hydroxonium ion or oxonium ion.

Further hydration is not impossible; the species $H_9O_4^+$ is also indicated. For simplicity we shall use $H_3O^+ \, (aq)$ or H_3O^+ to express the aquated proton. Even H^+ may be used in some cases for simplicity, the hydration being always implied.

Accepting the hydrated proton, the acid property is not due to dissociation to furnish a proton, but due to transfer of proton. Thus, HCl is an acid not due to the reaction

$$HCl \longrightarrow H^+ + Cl^-$$

but because it transfers a proton to a molecule of water as:

$$HCl + H_2O \longrightarrow H_3O^+ + Cl^-$$

3. Acid-base behaviour in non-aqueous solvents, or in the gas phase were not covered by the Arhenius theory.

4. Basic properties not necessarily involve hydroxyl ions.

SOLVENT SYSTEM (DEFINITION)

Studies on certain non-aqueous systems, particularly on liquid ammonia, revealed that reactions in these solvents may be explained by assuming autoionisation of the solvent similar to water :

$$H_2O + H_2O \rightleftharpoons H_3O^+ + OH^-$$

$$NH_3 + NH_3 \rightleftharpoons NH_4^+ + NH_2^-$$

Just as substances furnishing H_3O^+ ions in water act as acids in aqueous medium, substances furnishing NH_4^+ ions in liquid ammonia may also be expected to behave as acids in that medium. Similarly, substances furnishing NH_2^- ions in liquid ammonia are basically similar to the compounds giving hydroxyl ions in water.

Acids in liquid ammonia: NH_4Cl, NH_4NO_3, $(NH_4)_2 SO_4$ etc.

Bases in liquid ammonia: KNH_2, $NaNH_2$ etc.

Neutralisation in liquid ammonia may also be compared to that in water :

In water:	HCl	+	NaOH $\longrightarrow$ NaCl	+	H_2O
	Acid		Base — Salt		Solvent
In liquid NH_3:	NH_4Cl +		$NaNH_2 \longrightarrow$ NaCl	+	$2NH_3$
	Acid		Base — Salt		Solvent

These observations may be generalised for other solvents as well and we may define acids and bases in general as follows :

An acid is a solute that gives the cation characteristic of the solvent (either by direct dissociation or by reaction with the solvent). A base is a solute that furnishes the anion characteristic of the solvent (in the same manner as mentioned above).

From these definitions, acid-base systems may be considered in aqueous as well as non-aqueous media. The case of liquid ammonia has already been mentioned. Liquid sulphur dioxide (SO_2) and phosgene ($COCl_2$) provide other typical examples for such systems:

1. *Liquid SO_2:*

Autoionisation:	$SO_2 + SO_2$	=	SO^{++}	+	$SO_3^=$
Acid:	$SOCl_2$	=	SO^{++}	+	$2Cl^-$
Base:	$CaSO_3$	=	Ca^{++}	+	$SO_3^=$
Neutralisation:	$SOCl_2 + CaSO_3$	=	$CaCl_2$	+	$2SO_2$

2. *$COCl_3$*

Autoionisation:	$COCl_2$	=	$COCl^+$	+	Cl^-
Acid:	$AlCl_3 + COCl_2$	=	$COCl^+$	+	$AlCl_4^-$
Base:	$CaCl_2$	=	Ca^{++}	+	$2Cl^-$
Neutralisation:	$2COCl^+ AlCl_4^- + CaCl_2$	=	$Ca [AlCl_4]_2$	+	$2COCl_2$

The chief merit of the solvent system definition lies in its ability to include many non-aqueous as well as non-protonic acid-base systems. Each solvent may be looked upon as giving rise to an acid-base system of its own. Separately, these may be treated along the same line of approach as that of Arrhenius for aqueous systems.

But the definition fails to explain the inherent acidic or basic character of a substance independent of a solvent. At the same time, it places too much emphasis on the autoionisation of the solvent. For example, when thionyl chloride is dissolved in liquid sulphur dioxide prepared from S^{35}, almost no exchange of S^{35} is observed. This shows that the ionisation steps written earlier for $SOCl_2$ and liquid SO_2 (either or both) may be erroneous.

PROTONIC THEORY

This was forwarded independently by Bronsted and Lowry. According to this concept, any species that tends to give up a proton is an acid, and any substance that tends to accept a proton is a base. Any proton donating species will be an acid according to this definition—molecules (HCl, HNO_3 etc.), cations (NH_4^+, H_3O^+ etc.) or anions (HSO_4^-, HCO_3^- etc). The donation of a proton in aqueous solution may be represented as

$$HCl + H_2O \rightleftharpoons H_3O^+ + Cl^-$$

$$NH_4^+ + H_2O \rightleftharpoons H_3O^+ + NH_3$$

$$HCO_3^- + H_2O \rightleftharpoons H_3O^+ + CO_3^=$$

Similarly, any proton-acceptor will be a base. These may be molecules (NH_3, Pyridine etc.) or ions (OH^-, $CO_3^=$ etc). The equilibrium representing transfer of a proton to such species may be represented as

$$NH_3 + H_2O \rightleftharpoons NH_4^+ + OH^-$$

$$CO_3^= + H_2O \rightleftharpoons HCO_3^- + OH^-$$

The definition can handle acid-base behaviour in protonic non-aqueous solvents like liquid ammonia, sulphuric acid or acetic acid (anhydrous).

In liquid NH_3: $\quad CO(NH_2)_2 + NH_3 \rightleftharpoons NH_4^+ + NH_2CONH^-$

In CH_3COOH: $\quad HCl + CH_3COOH \rightleftharpoons CH_3COOH_2^+ + Cl^-$

Reactions in gaseous phase may also be covered by this approach :

$$NH_3 (g) + HCl (g) \rightleftharpoons NH_4Cl (s)$$

Some other examples of acids and bases according to this definition are given in Table 19.1.

Table 19.1. Some typical conjugate acid-base systems.

Acid₁	+	*Base₂*	⇌	*Acid₂*	+	*Base₁*	
HNO_3	+	H_2O	⇌	H_3O^+	+	NO_3^-	
H_2SO_4	+	H_2O	⇌	H_3O^+	+	HSO_4^-	... Note 1
HSO_4^-	+	H_2O	⇌	H_3O^+	+	$SO_4^=$	... Note 1
$[Fe(H_2O)_6]^{3+}$	+	H_2O	⇌	H_3O^+	+	$[Fe (H_2O)_5 OH]^{2+}$	
NH_4^+	+	H_2O	⇌	H_3O^+	+	NH_3	
H_2O	+	NH_3	⇌	NH_4^+	+	OH^-	
H_2O	+	$CO_3^=$	⇌	HCO_3^-	+	OH^-	... Note 1
H_2O	+	HS^-	⇌	H_2S	+	OH^-	... Note 2

(Cont'd ...)

Acid$_1$	+	Base$_2$	⇌	Acid$_2$	+	Base$_1$	
HCl	+	CH_3COOH	⇌	$CH_3COOH_2^+$	+	Cl^-	... Note 3
$HClO_4$	+	H_2SO_4	⇌	$H_3SO_4^+$	+	ClO_4^-	... Note 4
$HClO_4$	+	HF	⇌	H_2F^+	+	ClO_4^-	... Note 4
HF	+	HNO_3	⇌	$H_2NO_3^+$	+	F^-	... Note 4
$CO(NH_2)_2$	+	NH_3	⇌	NH_4^+	+	NH_2CONH^-	... Note 5
HCOOH	+	$CO(NH_2)_2$	⇌	$H_2N.CO.NH_3^+$	+	$HCOO^-$	... Note 5
H_2	+	OD^-	⇌	DOH	+	H^-	... Note 6
$C_6H_5CH_3$	+	CH_3Na	⇌	CH_4	+	$C_6H_5CH_2^-Na^+$	... Note 7

Note 1: HSO_4^- may thus behave both as an acid and a base. The same remark applies to HCO_3^- and HS^-. The conjugate acid of HCO_3^- is H_2CO_3 (not shown above).

What is the conjugate base of HS^-?

Note 2: The reaction explains why an aqueous solution of NaHS will be basic and smell of H_2S.

Note 3: In pure acetic acid medium. Acetic acid acts as a base.

Note 4: The first reaction occurs when pure perchloric acid is dissolved in concentrated sulphuric acid. Sulphuric acid acts as a base!

The second and third reactions occur in anhydrous hydrogen fluoride medium. In the second reaction HF reacts as a base. In the third reaction HF is the acid and HNO_3 behaves as the base.

Note 5: In liquid ammonia, urea may behave both as an acid or a base.

Note 6: This reaction occurs when hydrogen is bubbled through molten potassium deuteroxide. The hydride ion cannot, however, be detected. But the hydrogen evolving from the melt is enriched with deuterium. Note that hydrogen gas acts here as an acid!

Note 7: This reaction occurs in hydrocarbon solvents. Note that both methane and toluene are weakly acidic, but toluene is more acidic than methane.

Conjugate Acid-Base Pair

Donation of a proton by a molecule of HCl in aqueous medium is represented by the equilibrium:

$$HCl \quad + \quad H_2O \quad ⇌ \quad H_3O^+ \quad + \quad Cl^-$$
$$\text{Acid} \qquad\qquad \text{Base} \qquad\qquad \text{Acid} \qquad\qquad \text{Base}$$

In the forward reaction, the molecule of water functions as a base by the Protonic definition, since it accepts a proton from the HCl molecule. In the reverse reaction, the chloride ion tends to receive a proton from the hydronium ion. Hence, the chloride ion may now be looked upon as a base and the hydronium ion as an acid according to the protonic definition. The overall reaction takes place mainly towards right, since water appears to be a better proton acceptor, i.e., a stronger base, than the chloride ion. The equilibrium thus involves two acid-base systems functioning simultaneously, i.e., (i) HCl (acid$_1$, say) and its corresponding base Cl^- (base$_1$, obtained by elimination of a proton from HCl); (ii) H_2O (base$_2$, say) and its corresponding acid H_3O^+ (acid$_2$; obtained by adding a proton to H_2O):

$$HCl \quad + \quad H_2O \quad ⇌ \quad H_3O^+ \quad + \quad Cl^-$$
$$\text{Acid}_1 \qquad\qquad \text{Base}_2 \qquad\qquad \text{Acid}_2 \qquad\qquad \text{Base}_1$$

The acid-base pairs HCl–Cl⁻ or H_3O^+ –H_2O may be called conjugate acid-base pairs. Chemical species which differ from one another only to the extent of a proton transfer are termed a conjugate acid-base pair. A few conjugate acid-base pairs are listed in Table 19.1.

The stronger the tendency of a species to donate a proton in a given medium, the stronger will be its acid character. The corresponding conjugate base of a strong acid must be weak, in as much as its tendency towards the reverse combination with the proton is less marked. So it may be concluded that conjugate bases of strong acids must be weak in nature. Similar consideration apply to bases and one may conclude that conjugate acids of strong bases will be weak in nature. The manifestation of the acidic or basic character of a substance is ultimately determined by the tendency of the medium to accept or release a proton. Interaction between the conjugate acid-base pair of the medium and the conjugate acid-base pair of a substance leads to combination between the stronger acid and stronger base of each conjugate pair to form the weaker acids and bases.

We note further that a molecule of water acts as a base towards a stronger acidic substance and forms the weak conjugate acid H_3O^+. Similarly, it can be used as an acid towards a more strongly basic substance when it gives the conjugate base OH⁻. Water is thus said to be an *amphiprotic* solvent. Any solvent capable of functioning as an acid as well as a base by releasing or accepting a proton called an amphiprotic solvent. Liquid ammonia, anhydrous hydrofluoric acid and acetic acid are other examples of amphiprotic solvents.

The protonic concept includes a large variety of substances in the realm of acids and bases, as exemplified earlier. Relative strengths of acids and bases can also be expressed according to this theory in the same manner as in the Arrhenius definition. For the dissociation of an acid HA in water, represented as

$$HA + H_2O = H_3O^+ + A^-$$

the equilibrium constant K in terms of concentrations is given as

$$K = \frac{C_{A^-}.C_{H_3O^+}}{C_{HA}.C_{H_2O}}; \quad \text{or, } K.C_{H_2O} = \frac{C_{A^-}.C_{H_3O^+}}{C_{HA}}$$

Since C_{H_2O} is comparatively very large in an aqueous solution, it may practically be taken to be a constant and we may replace $K.C_{H_2O}$ by some other constant K;. The strength of the acid HA may then be expressed in terms of K;, which has the same expression as the ionisation constant of an acid in the Arrhenius definition, allowing hydration of the proton. The procedure may also be applied to other media to determine relative acid-base strengths of substances. Hydrolysis of salts and catalytic activity of acids may also be explained in the light of this theory.

But the Protonic theory puts sole emphasis on the proton in determining acid-base criterion. This may be considered a drawback of the definition as it fails to include substances which do not contain any proton and yet exhibit acidic character in certain reactions (e.g., $AlCl_3$, BF_3, SO_3 etc.).

LEWIS DEFINITION

Lewis attributed acid-base character of a substance to its inherent electronic structure. According to him "An acid is any species which can accept a pair of electrons to form a covalent bond" and "a base is any species that can donate a pair of electrons to establish a covalent bond".

Lewis Acids

Lewis acids should, therefore, possess an empty orbital which may accommodate a pair of electrons. Three main types of such substances are:

1. *Molecules containing an atom with incomplete valence shell:* The boron trihalides and sulphur trioxide are typical examples for such molecules. The incomplete valence shell of boron or sulphur readily accept a pair of electrons from a base (Fig. 19.1).

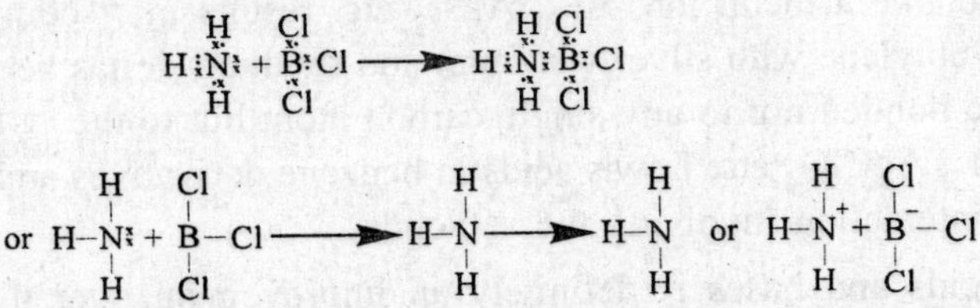

$$
\begin{array}{c} H \\ H\!:\!N\!:\! \\ H \end{array} + \begin{array}{c} Cl \\ B\!:\!Cl \\ Cl \end{array} \longrightarrow \begin{array}{c} H \; Cl \\ H\!:\!N\!:\!B\!:\!Cl \\ H \; Cl \end{array}
$$

$$
\text{or} \quad H\!-\!\underset{\underset{H}{|}}{\overset{\overset{H}{|}}{N}}\!: + B\!-\!Cl \longrightarrow H\!-\!\underset{\underset{H}{|}}{\overset{\overset{H}{|}}{N}} \longrightarrow H\!-\!\underset{\underset{H}{|}}{\overset{\overset{H}{|}}{N}} \quad \text{or} \quad H\!-\!\overset{+}{\underset{\underset{H}{|}}{\overset{\overset{H}{|}}{N}}}\!+\!B\!-\!Cl
$$

Fig. 19.1. Formation of a co-ordinate covalent bond between BCl, a typical lewis acid and NH_3, a typical lewis base.

2. *Cations with a stable, low-lying orbital which is empty or partially filled:* Ions of transition metals are good examples of this type of Lewis acid. Coordination of such ions by different donor ligands may thus be looked upon as an acid-base interaction, e.g.

$$Ag^+ + :NH_3 \longrightarrow (Ag:NH_3)^+ \ldots \qquad \text{(proceeds further)}$$

3. *Compounds with double bonds,* other than $C = C$ bonds. Thus CO_2 acts as a Lewis acid as

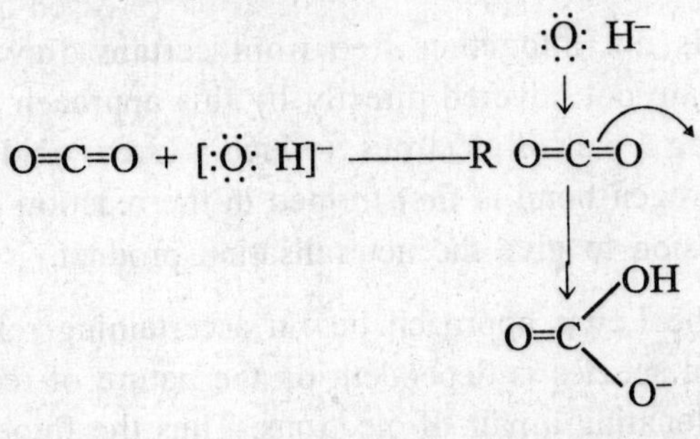

In addition, the central atom in certain molecules may expand their octet to accommodate a lone pair of electrons; these behave as Lewis acids. $SnCl_4$, $TiBr_4$, I_2 are typical examples.

Lewis Bases

Lewis bases are capable of donating a pair of electrons, may also be similarly classified as:

1. *Molecules having an atom with two or less unshared pairs of electrons:* H_2O and NH_3 are well known examples of such molecules.

   Surprisingly, the basic character decreases with increasing number of unshared pairs of electrons, as between ammonia and water. HF, with three unshared pairs, is very feebly basic, only in its reactions with strong acids like SO_2: $HF + SO_3 \longrightarrow HSO_3F$. The neon atom, with four unshared pairs of electrons, have no basic character. The reason for this trend is not well understood.

2. *(Almost) all negative ions:* Typical reactions of the F^-, Cl^- and I^- ions are shown:

$$F^- \quad + \quad BF_3 \quad \longrightarrow \quad BF_4^-$$

Base $\qquad$ Acid $\qquad$ Adduct

$$Cl^- \quad + \quad AlCl_3 \quad \longrightarrow \quad AlCl_4^-$$

Base $\qquad$ Acid $\qquad$ Adduct

$$I^- \quad + \quad I_2 \quad \longrightarrow \quad I_3^-$$

Base $\qquad$ Acid $\qquad$ Adduct

3. *Compounds containing carbon-carbon double bonds:* The donation of an electron pair from a π-bond to a Lewis acid like a metal ion, BF_3, $AlBr_3$ etc. results in π-complexes. In a number of complex compounds of ethylene with silver, platinum and rhodium, it has been definitely established that the metal atoms are bonded not to any single carbon atom but to the π-bond directly. The high solubilities of BF_3, $AlBr_3$, $AgClO_4$ etc. Lewis acids in benzene derivatives and in olefinic compounds also involve acid-base interaction involving the π-bonds.

The Lewis concept of acids and bases is definitely an improvement over the Bronsted concept. It covers the Bronsted acids and bases (mostly) and at the same time includes the non-protonic substances. These Lewis acids and bases, can be titrated to an end point using a coloured indicator. The inclusion of complex formation by metals is another interesting feature of the concept. Side by side, the Lewis concept introduces a parallelism between acid-base character and oxidising or reducing power of a substance. A strong base, being a good donor of an electron pair, would be a strong reducing agent. Similarly a strong acid should be a good oxidising agent. However, there are limitations to such generalisations.

However, the Lewis approach is not altogether free from certain drawbacks. The conventional protonic acids like H_2SO_4 and HCl are not covered directly by this approach as they do not apparently establish a covalent bond by accepting a pair of electrons, (which a Lewis acid ought to). A compromise can be done by assuming that a hydrogen bond is first formed in the reaction of an acid HX and a base B as X—H ... B; this undergoes fission to give the neutralisation product.

The second major difficulty in the Lewis approach lies in ascertaining relative acid-base strengths. Here the acidic or basic strength of a species is dependent on the nature of reaction studied and not on the absolute power of accepting or donating a pair of electrons. Thus the fluoride complex of beryllium (II) is more stable than the corresponding complex of copper (II). But the amine complex of copper (II) is far more stable than that of beryllium (II), suggesting copper (II) to be a stronger Lewis acid.

One of the starting points of the Lewis approach was that acid-base reactions should be fast. But this is not so in practice for many reactions due to kinetic factors although they definitely involve acids and bases. (e.g. the formation of $Cr(NH_3)_6^{3+}$). Lastly, the catalytic activities of acids and bases, one of the preliminary considerations by Lewis, are not in harmony between the protonic acids and Lewis acids. The reactions catalysed by Lewis acids are generally not catalysed by the protonic acids.

USANOVICH DEFINITION

M. Usanovich proposed a very wide definition of acids and bases. According to him, an acid is any chemical species which (i) reacts with a base, or (ii) accepts anions or electrons, or (iii) furnish cations; and a base is any chemical species that would (i) react with an acid, or (ii) combine with cations or (iii) furnish anions or electrons.

The scope of the definition is thus very wide, covering all the previous definitions. Some additional examples of acid-base reactions according to this would be

Acid		Base		Salt	
SO_2	+	Na_2O	=	Na_2SO_3	SO_2 accepts the anion $O^=$.
Cl_2	+	$2Na$	=	$2NaCl$	Chlorine accepts an electron
$Fe(CN)_2$	+	$4KCN$	=	$K_4Fe(CN)_6$	$Fe(CN)_2$ adds further anions (CN^-).

Usanovich considered (i) the general trend of acid-base character in the periodic table; (ii) the aspect of "coordination unsaturation".

1. In this context, his main observations may be summarised as follows :

    (a) Acidic character increases from left to right in the periodic table (Na_2O: strongly basic; Cl_2O_7: strongly acidic); acidic character decreases from top to bottom (As_2O_3 is more acidic than Sb_2O_3).

    (b) The higher valence of an element is associated with greater acidic character (SnS_2 is more acidic than SnS).

    (c) High charge of cations favours acidic property (Na_2O vs Al_2O_3), while high charge of anions promote basic property (Na_2O vs NaCl).

2. The term "coordination unsaturation" was used to mean the tendency of an atom to increase its covalence. Thus the sulphur atom in sulphur dioxide molecule is coordinatively unsaturated; as such, sulphur dioxide can accept an anion like O^{2-} and act as an acid.

The Usanovich definition is most general of the different acid base definitions. It includes all acids and bases defined by the Lewis definition; at the same time it states that the donation or acceptance of electrons need not take place as shared pairs. Accordingly, oxidation-reduction reactions may be classified as acid-base reactions.

The only point raised against the Usanovich definition is its extreme generality—its scope is so large that almost all chemical reactions may be considered as acid-base reactions. The convenience of treating a particular type of compound as acid or base is thereby lost.

Thus, several different approaches have been made to define acids and bases. Each definition is correct when applied to the particular range for which this was meant. So it would be meaningless to try to select the "right" or "wrong" definition.

STRENGTHS OF ACIDS AND BASES

For convenience, acids and bases may be broadly classified as protonic and non-protonic compounds. Where a proton transfer is involved, the strengths of acids and bases may be compared by the ease with which the proton is released (by an acid) or captured (by a base). Such strengths may be expressed by the dissociation constant as developed by the protonic definition; as shown earlier, this yields the same result as the Arrhenius definition for aqueous solutions. In this section we shall try to correlate strengths of acids and bases in terms of their dissociation constants. General trends of acid-base strengths in different compounds will also be considered.

Relative strengths of different acids in aqueous solution have also been compared from their efficiency in catalysing the inversion of sucrose.

Strengths of Lewis acids and bases have been correlated in terms of the enthalpy of formation of an acid-base adduct (ΔH) and acid or base parameters involving susceptibility of the species involved (i) to undergo electrostatic interaction (the E-parameter); and (ii) to form covalent bonds (the C-parameters).

Dissociation Constants of Acids and Bases

According to the Arrhenius definition, the strength of an acid HA may be expressed in terms of the equilibrium constant for its ionisation in aqueous solution furnishing a proton :

$$HA \rightleftharpoons H^+ + A^-; \quad K = (C_{H^+} \times C_{A^-})/C_{HA}$$

The strength of a base is similarly expressed by the equilibrium constant for ionisation to furnish a hydroxyl ion in aqueous solution.

As already shown, similar expressions may be obtained from the protonic definition taking water as a reference base (for acids) or acid (for bases) :

(i) $\quad HA + H_2O \rightleftharpoons H_3O^+ + A^-$
$\quad\quad\quad$ Acid$_1$ Base$_2$ $\quad$ Acid$_2$ $\quad$ Base$_1$

(ii) $\quad B + H_2O \rightleftharpoons BH^+ + OH^-$
$\quad\quad\quad$ Base$_1$ Acid$_2$ $\quad$ Acid$_1$ $\quad$ Base$_2$

From (i) :

$$K = \frac{C_{H_3O^+} \cdot C_{A^-}}{C_{HA} \cdot C_{H_2O}} \quad \text{or,} \quad K' = K \cdot C_{H_2O} = \frac{C_{H_3O^+} C_{A^-}}{C_{HA}}$$

Similarly we may express the dissociation constant for the base B.

It is often convenient to express the dissociation constant of an acid by K_a and that of a base by K_b. Polybasic acids and polyacidic bases have more than one dissociation constant corresponding to each successive stage of dissociation. Dissociation constants for a number of common acids and bases at 298°K are given in Table 19.2.

Table 19.2. Dissociation constants of acids and bases in aqueous solution at 298°K.

Acids	Equilibrium			K_a or K_b
Acetic acid	CH_3COOH	$+ H_2O = H_3O^+$	$+ CH_3COO^-$	1.8×10^{-5}
Hydrogen cyanide	HCN	$+ H_2O = H_3O^+$	$+ CN^-$	4.8×10^{-10}
Hydrogen fluoride	HF	$+ H_2O = H_3O^+$	$+ F^-$	6.8×10^{-4}
Nitrous acid	HNO_2	$+ H_2O = H_3O^+$	$+ NO_2^-$	4.5×10^{-4}
Monochloro acetic acid	$ClCH_2COOH$	$+ H_2O = H_3O^+$	$+ ClCH_2COO^-$	1.4×10^{-3}
Phenol	C_6H_5OH	$+ H_2O = H_3O^+$	$+ C_6H_5O^-$	1.3×10^{-10}
Hydrogen sulphide	H_2S	$+ H_2O = H_3O^+$	$+ HS^-$	1.1×10^{-7} ($= K_1$ for H_2S)
Bisulphide ion	HS^-	$+ H_2O = H_3O^+$	$+ S^=$	1×10^{-14} ($= K_2$ for H_2S)

(Cont'd ...)

Acids	Equilibrium			K_a or K_b
Carbonic acid	$H_2O + CO_2$	$+ H_2O = H_3O^+$	$+ HCO^-_3$	4.2×10^{-7} $(= K_1$ for $H_2CO_3)$
Bicarbonate ion	HCO_3^-	$+ H_2O = H_3O^+$	$+ CO_3^=$	4.8×10^{-11} $(= K_2$ for $H_2CO_3)$
Phosphoric acid	$H_3PO_4^-$	$+ H_2O = H_3O^+$	$+ H_2PO_4^-$	1×10^{-8} $(= K_1$ for $H_3PO_4)$
Dihydrogen phosphate ion	$H_2PO_4^-$	$+ H_2O = H_3O^+$	$+ HPO_4^=$	6.2×10^{-8} $(= K_2$ for $H_3PO_4)$
Hydrogen phosphate ion	$HPO_4^=$	$+ H_2O = H_3O^+$	$+ PO_4^{3-}$	1×10^{-12} $(= K_3$ for $H_3PO_4)$
Ammonia	NH_3	$+ H_2O = NH_4^+$	$+ OH^-$	1.8×10^{-5}
Methylamine	CH_3NH_2	$+ H_2O = CH_3NH_3^+$	$+ OH^-$	5.0×10^{-4}
Dimethylamine	$(CH_3)_2NH$	$+ H_2O = (CH_3)_2NH^+$	$+ OH^-$	7.4×10^{-3}
Trimethylamine	$(CH_3)_3 N$	$+ H_2O = (CH_3)_3NH^+$	$+ OH^-$	7.4×10^{-3}
Aniline	$C_6H_5NH_2$	$+ H_2O = C_6H_5NH_3^+$	$+ OH^-$	4.0×10^{-10}

Note 1: The law of mass action as applied above to obtain K_a or K_b is valid only for weak electrolytes which undergo reversible dissociation in water. For strong electrolytes, dissociation is almost complete and there is little or no equilibrium between the ions and the undissociated molecules in solution.

Note 2: The use of concentration terms in expressing K_a or K_b is only approximate. Strictly speaking, one should use the activities of the respective species; these are obtained by multiplying the concentrations with suitable moderating coefficients, known as activity coefficients. For dilute solutions, however, the activity coefficients may be roughly taken as unity and the activities may be fairly approximately replaced by concentration terms.

Note 3: The successive dissociation constants of a polybasic acid become gradually smaller, as may be observed from the data for hydrogen sulphide or phosphoric acid ($K_1 > K_2 > > K_3$). It appears that the release of a positively charged proton becomes more and more difficult as the negative charge on the anion increases.

Note 4: The values quoted are only approximate. They may differ from one book to another.

Dissociation constants may be determined from conductance measurements. More accurate results are obtained from e.m.f. study of a suitably devised cell containing the weak acid and its salt. Catalytic activities may also be used to compare dissociation constants.

Levelling Effect of Water

When the strengths of different acids or bases are interpreted in terms of their dissociation constants, the role of the solvent should always be remembered in the background. In aqueous solutions, the extent of dissociation of an acid depends upon the tendency of the water molecules to accept the proton, i.e. on the basic nature of water. Similarly, the extent of dissociation of a base in water is dependent on the tendency of the water molecules to donate a proton, i.e. on the acidic nature of water. Thus the basic or acidic nature of water plays a significant role in determining the strengths of acids or bases.

Referring back to the equilibria

(i) $\quad HA + H_2O \rightleftharpoons H_3O^+ + A^-\quad$ and $\quad$ (ii) $\quad B + H_2O \rightleftharpoons BH^+ + OH^-$
$\quad\quad$ Acid$_1$ $\quad$ Base$_2$ $\quad\quad$ Acid$_2$ $\quad$ Base$_1$ $\quad\quad\quad\quad$ Base$_1$ $\quad$ Acid$_2$ $\quad$ Acid$_1$ $\quad$ Base$_2$

we may infer that all acids stronger than the hydronium ion would displace the equilibrium (i) far to the right while all bases stronger than the hydroxyl ion would shift the equilibrium (ii) to the right.

In fact, water acts as an appreciably good base towards all the common "strong" acids like HCl, H_2SO_4 and HNO_3. They are almost completely dissociated in aqueous solutions; as such, all of them appear equally strong in aqueous solution. Similarly, water acts as an appreciably good acid for all common "strong" bases (NaOH, KOH etc.) and they appear equally strong in water. This tendency of water to equalise the strengths of many acids (or bases) is termed its levelling effect. Comparison of acid or base strengths in water is rendered difficult by its levelling effect.

Other solvents may have their own levelling effects, too. Thus, benzoic acid is a very weak acid in aqueous solution. But in liquid ammonia, it acts as a very strong acid. In general, it may be expected that acidic solvents (like acetic acid) will exert less levelling effect on acid strengths (and similarly basic solvents on base strengths). Relative strengths of the common mineral acids have been obtained by conductance measurement in anhydrous acetic acid. The order found is

$$HClO_4 > HBr > H_2SO_4 > HCl > HNO_3$$

Strength of Lewis Acids and Bases (Hard and Soft Acids and Bases)

The Lewis definition recognises acids and bases in terms of their ability to accept or donate electron pairs. As such, the strength of any acid or base is determined by the very nature of the reaction involved in any particular electron-transfer process. Accordingly, assignment of any single consistent criterion for acid-base strength becomes very difficult in the Lewis definition. Attempts have been made to correlate various strengths factors from enthalpy changes (ΔH) of acid-base combinations.

However, a qualitative correlation between the various Lewis acids and bases has been achieved. This is done by classifying the acids or bases into two classes : hard and soft. Acid-base reactions are then treated by the general principle that hard acids prefer to combine with hard bases and soft acids prefer to combine with soft bases (and vice versa).

The criterion of hardness (or softness) is ascribed to the "hardness" of the electron cloud associated with any species. A firmly held electron-cloud with low polarisability makes a species "hard", while an easily polarisable electron cloud characterises the species as "soft". A third category with intermediate character appears in the borderline. The principal distinguishing features of hard and soft acids and bases may then be summarised in Table 19.3.

Table 19.3. Principle distinguishing features of hard and soft acids and bases.

Hard acids	Soft acids
Acceptor atom marked by	Acceptor atom marked by
(i) small size	(i) large size
(ii) high positive oxidation state	(ii) zero or low positive oxidation state
(iii) absence of any outer electrons which are easily excited to higher states.	(iii) several easily excitable valence electrons
H^+, Li^+, Na^+, K^+	Cu^+, Ag^+, Hg^+
$Be^{2+}, Mg^{2+}, Ca^{2+}, Sr^{2+}$	$Cd^{2+}, Pt^{2+}, Hg^{2+}, Pt^{4+}$
$Al^{3+}, Fe^{3+}, Co^{3+}, Cr^{3+}$	I^+, I_2, Br_2, Br^+
$CO_2, SO_3,$	Cl, Br, I, N

(Cont'd ...)

Hard acids	Soft acids
CO_2, SO_3,	Cl, Br, I, N
BF_3, BCl_3, $AlCl_3$.	M^o (metal atoms) and bulk metals.
HX (hydrogen-bonding molecules).	
Fe^{2+}, Co^{2+}, Ni^{2+}, Cu^{2+}, Zn^{2+}, Pb^{2+}, Bi^{3+}, $SO2$	

Donor atom marked by	*Donor atom marked by*
(i) high electronegativity	(i) low electronegativity
(ii) low polarisability	(ii) high polarisability
(iii) presence of filled orbitals; empty orbitals may exist at high energy level	(iii) partially filled orbitals; empty orbitals are low-lying.

Examples	
H_2O, OH^-, F^-, Cl^-	H^-, I^-,
$CH_3CO_2^-$, PO_4^{3-}, SO_4^{2-}	SCN^-, CN^-, $S_2O_3^{2-}$
NO_3^-, CO_3^{2-}, ClO_4^-	CO, C_2H_4, C_6H_6
ROH, R_2O	R_2S, RSH, RS^-
NH_3, N_2H_4, RNH_2	R_3P, R_3As.
$C_6H_5NH_2$, C_5H_5N, N_3^-, N_2, NO_2^-, Br^-, SO_3^{2-}	

Usefulness: In the hard-soft terminology, a large number of chemical reactions may be understood. The guiding principle is that hard-hard and soft-soft combinations of acids and bases are always more stable than hard-soft combinations. A few examples will illustrate this point.

1. Ammonia, amines, water and fluoride ion prefer to combine with Be^{2+}, Ti^{4+}, Co^{3+} etc. giving stable complexes. On the other hand, phosphines (R_3P), thio ethers (R_2S) and other species with P and S as donor atoms prefer to combine with Pt^{2+}, Pd^{2+}, Hg^{2+} etc.

2. BF_3, a hard acid, combines readily with a further fluoride ion; but BH_3 (present as B_2H_6), a soft acid, prefer the softer base hydride ion.

$$BF_3 + F^- \longrightarrow BF_4^- \; ; \; B_2H_6 + 2H^- \longrightarrow 2BH_4^-$$

It is now easy to understand why the following reactions proceed as written

$$BF_3H^- + BH_3F^- \longrightarrow BF_4^- + BH_4^-$$
$$CF_3H + CH_3F \longrightarrow CF_4 + CH_4$$

3. Hard-hard and soft-soft combinations similarly determine the course of many other reactions; for example,

$$LiI + CsF \rightleftharpoons LiF + CsI$$
$$HgF_2 + BeI_2 \rightleftharpoons BeF_2 + HgI_2$$

These reactions illustrate the "Pearson Pauling paradox" of chemistry. In the Pauling concept of electronegativity, cesium (or mercury) and fluorine should form the more stable bond as their electronegativity difference is greater! In reality, however, LiF is more stable than CsF (LiF: 3; CsF: 502 kJ mol^{-1}) due to very large contribution of electrostatic interaction in the former.

In fact, the major driving force for the above reactions (and similar others) come from the stability of the hard-hard combination joining small atoms to each other by ionic and/or covalent bonding. The soft-soft combination contributes little or nothing to the driving force, except when other factors like pi-bonding are involved.

4. Numerous chemical aspects have been interpreted by the hard-soft concept. The catalytic power of metals may be understood from the fact that the soft metal atoms adsorb soft bases on their surface. Solubility may be justified by the tendency of hard solvents to dissolve hard solutes and soft solvents to dissolve soft solutes. Even speeds of electrophilic and nucleophilic substitution reactions may be related to hardness and softness of the species involved.

Limitations: The hard-soft classification is a useful concept, but it does not lead directly to a scale of acid-base strength. Inherent acid-base strengths are also not taken into consideration. Thus the OH⁻ and F⁻ ions are both hard bases.

Interpretation of many reactions by splitting the participants into acid-base fragments is also arbitrary to some extent. The very common reaction between ethanol and acetic acid may be interpreted in either of two ways :

$$CH_3COO^-H^+ + C_2H_5 + OH^- \quad \text{or} \quad CH_3CO^+OH^- + C_2H_5O^- + C_2HO^- H^+$$

Either interpretation is justifiable by the hard-hard combination of H⁺ with OH⁻. The actual involvement of the second mechanism has to be understood by other means. Sometimes the hard-soft principle fails to keep parity with inherent acid-base strengths. The reaction :

$$CH_3^+(g) + H_2(g) \longrightarrow CH_4(g) + H^+(g)$$

should be favourable in view of the soft-soft combination of CH_3^+ with H⁻. But actually the combination is endothermic by about 360 kJ per mole. (This has to be explained by the greater acidity of H⁺ relative to CH_3^+).

Similarly, hard-soft combinations take place in many cases like

$$SO_3^{2-} + HF \longrightarrow HSO_3^- + F^-$$

Here the soft base SO_3^{2-} replaces the hard base F⁻ and combines with the hard acid H⁺. The soft SO_3^{2-} is thus stronger than the hard F⁻.

However, we should restrict our expectations in view of the statement made by Pearson: "It should be stressed that the HSAB principle is not a theory but is a statement about experimental facts. Accordingly an explanation of some observation in terms of hard and soft behaviour does not invalidate some other, theoretical explanation".

General Trends in Acid Strength

We have critically discussed several definitions of acids and bases. However, the popular sense of acids and bases is conventionally associated only with a limited number of compounds mostly covered by the protonic definition. It is interesting to study the general trend in acidic or basic strength among such compounds.

The protonic acids may be broadly classified as (i) hydroacids; and (ii) oxoacids. In hydroacids, the acidic hydrogen atom is directly linked to some second element. Binary hydrides like HF, HCl, H₂S etc. are such examples. In oxoacids, the acidic hydrogen atom is linked to some central atom via an oxygen atom, as for example in H_2SO_4, HNO_3 or $HClO_4$. It is convenient to discuss the general trend in acid strength for these two classes separately.

Strengths of hydro acids

In the periodic table, acidity increases from left to right along a period. Thus, acid strength increases in the order :

$$NH_3 \ll H_2O \ll HF; \quad \text{or} \quad PH_3 \ll H_2S \ll HCl$$

Along a group, acidity increases as one passes downward. Thus :

$$H_2O < H_2S < H_2Se < H_2Te; \quad \text{or} \quad HF < HCl < HBr < HI$$

The variation of acidity along a period may be explained from the increasing electronegativity of the elements linked to hydrogen. Here size of the atom does not change appreciably from one element to the next; greater electronegativity of an element makes proton release easier due to greater shift of the bonding electron cloud towards it. So the acidity increases.

But consideration of electronegativity alone does not explain the increase of acidity down a group. (Rather this leads to opposite expectations). The size of the atom increases rapidly from one element to the next and upsets the electronegativity factor. As the atoms become larger, bonding with the small hydrogen atom becomes weaker (poorer overlap); the increase of acid character parallels this decrease in bond strength.

The strength of a hydro acid in aqueous solution may be treated thermodynamically as follows. The acidity is obviously controlled by the process

$$HX\ (aq) \longrightarrow H^+\ (aq) + X^-\ (aq)$$

that is, dissociation of an aquated acid molecule to furnish a proton and the corresponding anion, both aquated. This may be supposed to take place in a number of steps for which the energy changes are known :

Process	Energy term
1. $HX\ (aq) \longrightarrow HX\ (g)$	ΔH_1 (heat of dehydration)
2. $HX\ (g) \longrightarrow H\ (g) + X\ (g)$	D (heat of dissociation)
3. $H\ (g) + X\ (g) \longrightarrow H^-\ (g) + X^-\ (g)$	$I_H - E_X$ (Ionisation potential of hydrogen and electron affinity of X).
4. $H^+\ (g) \longrightarrow H^-\ (g)\ (aq) + X^-\ (aq)$	$-\Delta H_{hydration}$ (of the both H^+ and X^-).
$HX\ (aq) \longrightarrow H^+\ (aq) + X^-\ (aq)$	$\Delta H_1 + D + I_H - E_X - \Delta H$ hydration ($=\Delta H$ acid).

These steps may be arranged in a Born-Haber type cycle as shown in Fig. 19.2. If $\Delta H_{acid} = \Delta H_1 + D + I_H - E_X - \Delta H_{hydration}$. The magnitude of ΔH_{acid} is thus found to depend on these different energy terms. Some values are quoted below for HF, HCl, HBr and HI (in kJ mol^{-1}).

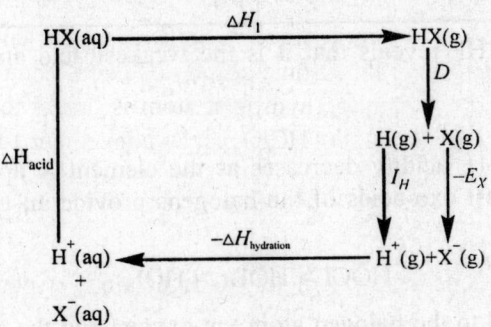

Fig. 19.2. Haber type cycle.

	ΔH_1	D	I_H	E_X	$\Delta H_{hydration}$		ΔH_{acid}
					H^+	X^-	
HF	48	566	1311	−333	−1091	−515	−14
HCl	18	431	1311	−348	−1091	−381	−60
HBr	21	366	1311	−324	−1091	−347	−64
HI	23	299	1311	−295	−1091	−305	−58

It is apparent that the total ΔH term for the acidic dissociation of HF (ΔH_{acid}) is largely affected by the high values of (i) heat of dissociation and (ii) heat of dehydration (ΔH_1). The high value of heat of dissociation is a reflection of the strong, short H–F bond (H–F bond length: 10 nm; whereas H–I bond length : 17 nm). The high heat of dehydration is in conformity with the formation of strong hydrogen bonds in aqueous solutions.

Thus the main causes of the weakness of HF as an acid in aqueous solution are (i) the great strength of the HF bond and (ii) the high heat of dehydration of the HF molecule (hydrogen bond with water). Side by side, the heat of hydration of the fluoride ion is large (favouring acidity), but this cannot overcome the former factors.

In concentrated aqueous solutions (5-15 M), however, acidity of HF increases owing to ionisation to H_3O^+, HF_2^- and more complex $(H_nF_{n+1})^-$ species.

The acid strength is, however, more fundamentally related to the Gibbs free energy term (ΔG) of the process : HX $(aq) \longrightarrow H^+ (aq) + X^- (aq)$, when $\Delta G = -RT \ln K$.

K, the equilibrium constant for the process is also called the dissociation constant of the acid. Thermodynamically, ΔG is related to ΔH as

$$\Delta G = \Delta H - T\Delta S$$

The values of ΔG, $T\Delta S$ and K for the halogen hydracids are given below (energy terms in kJ mol^{-1}).

	ΔH_{acid}	$T\Delta S$	ΔG	$K_{(app)}$
HF	– 14	– 29	+ 15	10^{-3}
HCl	– 60	– 13	– 47	10^{8}
HBr	– 64	– 4	– 60	10^{10}
HI	– 58	+ 4	– 62	10^{11}

The smallest value of K for HF reveals that it is the weakest acid among the series.

Strengths of Oxo-Acids

For oxo-acids of the type X–O–H, acidity decreases as the element X appears lower in a given group of the periodic table. The simplest oxo-acids of the halogens provide an example of this generalisation, acidity decreasing as

$$HOCl > HOBr > HOI$$

The acidic hydrogen is linked to the halogen atom via oxygen and the acid strength is determined by the ease of breaking of the O–H bond. The electronegativity of the atom X plays the most important role in the matter, by pulling the charge cloud of the X–O bond to itself; the effect is now relayed by the oxygen atom to the O–H bond (inductive effect), facilitating release of the acidic hydrogen. This is why acidity decreases down the group with decreasing electronegativity of the element X:

Acid	Order of K_a	Acid	Order of K_a
HOCl	10^{-8}	H_2SO_3	10^{-2}
HOBr	10^{-9}	H_2SeO_3	10^{-3}
HOI	10^{-13}	H_2TeO_3	10^{-6}

When the element X has additional oxygen atoms attached to it, inductive effect of these oxygen atoms makes the acid stronger—acidity increases with the number of oxygen atoms :

$$\text{H—O—Cl;} \quad \text{H—O—Cl;} \quad \text{H—O—Cl;} \quad \text{H—O—Cl} \longrightarrow \text{O}$$

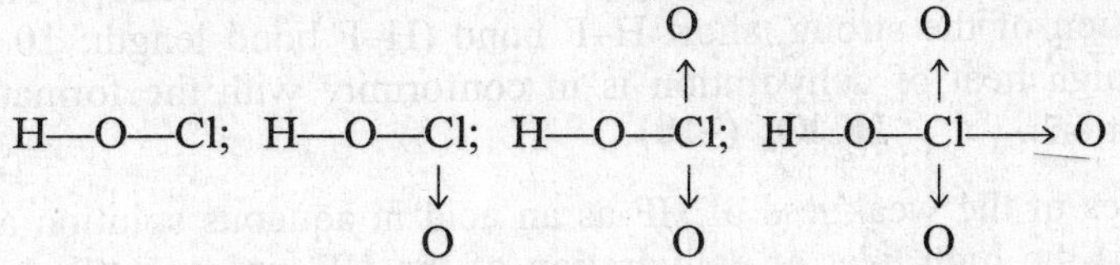

O.N. of chlorine	+1	+3	+5	+7
Order of K_a	10^{-8}	10^{-2}	Large	Very large

Similarly, sulphuric acid (H_2SO_4; $HO.SO_2.OH$) is a stronger acid than sulphurous acid (H_2SO_3; $HO.SO.OH$) since the sulphur atom in sulphuric acid has more coordinated oxygen atoms.

While considering such acid strengths, one should consider the distribution of the inductive effective of the coordinated oxygen atoms, over all the acidic sites (i.e. O–H bonds). Thus, among the three oxyacids of phosphorus, H_3PO_2, H_3PO_3 and H_3PO_4, acidity actually decreases as $H_3PO_2 > H_3PO_3 \geqslant H_3PO_4$. This is contrary to our expectation from the number of oxygen atoms alone. But a consideration of the structures of these acids explains the observed trend :

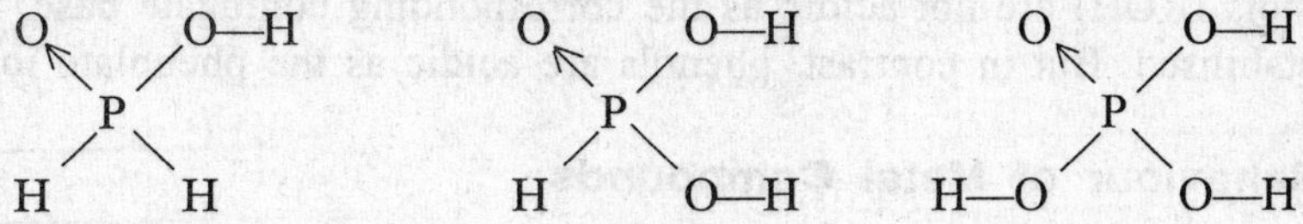

$$H_3PO_2 \qquad\qquad H_3PO_3 \qquad\qquad H_3PO_4$$

The hydrogen atoms bonded directly to phosphorus are not acidic as the P–H bond is not polar enough due to equal electronegativity of phosphorus and hydrogen. Acidity arises from the —O—H groups, one in H_3PO_2, two in H_3PO_3 and three in H_3PO_4. The inductive effect of the single oxygen atom acts on the single —O—H bond in H_3PO_2, but it is distributed over two and three —O—H bonds in H_3PO_3 and H_3PO_4 respectively. Accordingly, acidity decreases from H_3PO_2 to H_3PO_4.

Pauling's rules

Pauling proposed two empirical rules to predict the strengths of mono-nuclear oxo-acids :

1. For an oxo-acid of type $O_nX(OH)_m$, $pK_a \simeq 8 - 5_n$. The acid strength is thus related to the number of non-hydroxylic oxygen atoms.

2. For polyprotic acids ($m > 1$), the successive pK_a values increase by five units.

The success of these simple rules may be appreciated from the estimated pK_a values for different oxo-acids as shown below (experimental values in parentheses).

$n = 0$	$pK_a = 8$	HOCl (7.2); Si$(OH)_4$ (10)
$n = 1,$	$pK_a = 3$	H_2CO_3 (3.6); $HClO_2$ (2.0); H_3PO_4 (2.1; 7.4; 12.7)
	$pK_2 = 8$	H_3PO_3 (1.8; 6.6) H_3PO_2 (2.0).
	$pK_3 = 13$	

$n = 2.$ $pK_1 = -2$ HNO_3 (−1.4); H_2SO_4 (−2.0; 1.9)

$pK_2 = 3$ $HClO_3$ (−1.0)

$pK_3 = 8$

$n = 3.$ $pK = -7$ $HClO_4$ (−10)

Organic Acids

Inductive effect of electronegative elements or electron withdrawing groups gives rise to stronger acidic character in an organic compound. Thus, monochloroacetic acid, $ClCH_2COOH$ is a stronger acid than acetic acid, CH_3COOH, due to inductive effect of the chlorine atom. Similarly, one may explain why dichloro- or trichloro acetic acid is still stronger. Nitrophenols are stronger acids than phenols due to inductive effect of the nitro group which withdraws charge from the benzene ring and facilitates cleavage of the —O—H bond.

The strength of an acid is also increased when its conjugate base is stabilised by delocalisation of electrons; this is commonly represented by writing resonating structures for the conjugate base. The carboxylic acids, RCOOH, form typical examples. The conjugate base, the carboxylate anion, $RCOO^-$ is very stable which may be interpreted by writing a number of resonating structures for it. On the other hand, the alcohols (ROH) are not acidic as the corresponding conjugate base, the alkoxide ion, RO^-, is not similarly stabilised. But in contrast, phenols are acidic as the phenolate ion is stable.

Acid-Base Behaviour of Metal Compounds

Basicity of metal oxides

Metal oxides are in general basic in nature—only a few are amphoteric. In a given periodic group, the oxides become gradually more basic as one proceeds down the periodic table. Thus, BeO is amphoteric, while MgO is distinctly basic; and the basicity increases as MgO < CaO < SrO < BaO. This trend parallels the increase in size of the cation (charge remaining the same) and is related to the polarising ability of the metal ion (Fajan's rules).

When dissolved in water, metallic salts furnish metal ions which are hydrated in the solution :

$$Na^+ + nH_2O \longrightarrow [Na(H_2O)_n]^+$$

Hydration energy increases as the charge/size ratio of the cation increases.

Acidity caused by metal ions

When the hydration energy is sufficiently large, rupture of an O—H bond takes place and the solution becomes acidic :

$$Al^{3+} + 6H_2O \longrightarrow [Al(H_2O)_6]^{3+} \xrightarrow{+H_2O} [Al(H_2O)_5OH]^{+2} + H_3O^+$$

Clearly this is a case of hydrolysis. Tendency of a cation to hydrolyse increases with increasing charge and decreasing size of the cation—acidity of the hydrated cation increases in a parallel manner. Thus, Fe^{3+} ions, As explained in connection with Fajan's rules, cations of the transition elements possess greater polarising power than a cation of a non-transition metal of the same size (poorer shielding of the nuclear charge by the d-electrons). As such, Fe_3^+ ions hydrolyse to a greater extent than Al^{3+} ions (Fe^{+3} : $r = 6.7$ nm, $K_a \simeq 10^{-3}$; Al^{3+} : $r = 5$ nm, $K_a \simeq 10^{-5}$).

Structural and Steric Factors

Commonly, the formation of a Lewis acid-base adduct involves rearrangement of the spatial distribution of the orbitals around one or both of the participants. Thus the planar BF_3 molecule changes to tetrahedral hybridisation on forming an adduct with a base (like F^-); the tetrahedral SiF_4 molecule similarly changes to octahedral SiF_6^{2-} ion. Such changes are ultimately favoured by the energy of adduct formation.

The stability of the acid-base adduct may be largely influenced by steric effects. Thus 2-methylpyridine is a stronger base toward a proton, as expected from the inductive effect of the methyl group. In contrast, it is a much weaker base than pyridine toward the Lewis acid $B(CH_3)_3$. The ΔH values of adduct formation are illustrative:

Adduct of B $(CH_3)_3$ with	pyridine	2 methyl pyridine	3-methyl pyridine
ΔH (kJ mol^{-1})	-71	-42	-74

We find that shifting the methyl group to the meta position largely increases the acidity. Introduction of two methyl groups in both the ortho-positions does not allow adduct formation to take place at all, i.e., 2,6-dimethylpyridine does not react at all with $B(CH_3)_3$.

All these observations may be rationalised in terms of steric interactions between substituents on the acids and bases which appear "in front" during adduct formation. Such strain is termed F-strain. Such strain is lowered when the acid and base can form adduct from a larger distance. Thus $Al(C_2H_5)_3$ forms a weak adduct with 2,6-dimethylpyridine ($\Delta H = -13$ kJ mol^{-1}). The longer Al–N bond reduces steric hindrance between the substituents.

A similar effect operates due to the change in the geometry of the central atom on adduct formation. When a boron-atom changes from trigonal planar to pyramidal disposal of its valence orbitals, the substituents come closer. If they are branched, they will now introduce steric hindrance in the adduct; this is sometimes referred to as back-strain or B-strain (The crowding groups are on the "back-side" of the acceptor site). Thus $B(CH_3)_3$ forms a stronger adduct with NH_3 than $B(i-C_3H_7)_3$. Trimesitylboron is inert even to such strong bases as methoxide ion because crowding of the mesityl groups prevents change of geometry of the boron atom from planar to pyramidal.

Change of Bond Length in Acid-Base Interaction

The formation of an acid-base adduct changes the adjacent bond lengths remarkably, as may be seen from the following examples :

Species	B–N distance (pm)	B–F distance (pm)
BF_3	—	130
H_3NBF_3	160	138
Me_3NBF_3	158	139

The B—F distance increases when the boron atom forms an adduct with ammonia. The distance increases further when a stronger adduct is formed by Me_3N (B—N distance decreases). This may be stated as an empirical rule (Gutmann): the stronger and shorter the bond formed in an adduct between a Lewis acid and a Lewis base, the greater will be the lengthening of the adjacent bonds in both the donor and acceptor molecules. The bond formed in the adduct displaces electron density from the donor atom in the base to the acceptor atom in the acid. The donor atom acquires less partial negative charge than before and the acceptor atom becomes less positive. Both these effects reduce the polarity in the bonds in the acid and base molecules—ultimately leading to their lengthening. Proceeding in the same line of approach, one can also understand that bond lengths should increase with increase in coordination number of an atom. This is supported by the following bond lengths :

Species	Bond length (pm)	Species	Bond length (pm)
SiF_4	154	SO_2	143
SiF_6^{2-}	171	SO_3^{2-}	150
$SbCl_5$	231	I_2	266
$SbCl_6^-$	247	I_3^-	283

Acid-Base Strength and Hardness (or Softness)

As already explained, the hard-soft classification is based on the polarisability of the electron cloud associated with a species. Accordingly, the proton or H^+ ion is the hardest acid as it has no more electron to ionise and has the smallest size with highest positive field density. The methylmercury cation, CH_3Hg^+ is, on the other hand, a typical soft acid which is often convenient as a reference in studying acid-base equilibria. A large size and low positive field density imparts very feeble polarising power to the cation. The Hg–C bond is not involved in the competitive acid-base equilibrium under study, say,

$$BH^+ + CH_3Hg^+ \rightleftharpoons CH_3Hg B^+ + H^+$$

A hard base will move the equilibrium to the left while a soft base will drive it to the right.

The hardness or softness of acids and bases are quite different from their inherent acid-base strength. Let us consider the reactions

(i) $CHCl_3.NH_3 + BMe_3.C_6H_6 \rightleftharpoons CHCl_3.C_6H_6 + BMe_3.NH_3$

(ii) $SO_3^{-2} + HF \rightleftharpoons HSO_3^- + F^-$

(iii) $OH^- + CH_3Hg SO_3^- \rightleftharpoons CH_3Hg OH + SO_3^{2-}$

The equilibrium in each reaction lies largely toward right, contrary to our expectation from hard-hard and soft-soft combinations. In (i), $CHCl_3$ and NH_3 are both hard, BMe_3 is borderline, while C_6H_6 is soft. The forward reaction thus involves replacement of the hard base NH_3 from the hard acid $CHCl_3$ by the soft base C_6H_6. Similarly, reaction (ii) involves displacement of the F^- ion (hard base) from the H^+ ion (hard acid) by the softer base SO_3^{2-} (borderline). In (iii), the OH^- ion (hard base) displaces the SO_3^{2-} ion (borderline) from the soft acid CH_3Hg^+. Actually in all the forward reactions, the stronger base displaces the weaker one: $NH_3 > C_6H_6$; $SO_3^{2-} > F^-$; $OH^- > SO_3^{2-}$. thus acid-base interactions should be considered in the background of both hardness (softness) and strength. In a competitive situation involving both strength as well as hardness, the hard-soft interpretation works well :

$$CH_3HgF + HSO_3^- \rightleftharpoons CH_3HgSO_3^- + HF$$

$$CH_3HgOH + HSO_3^- \rightleftharpoons CH_3HgSO_3^- + HOH$$

In both cases the equilibrium lies well to the right.

LUX CONCEPT OF ACIDS AND BASES

According to this concept, an acid is an acceptor of oxide ion and a base is a donor of oxide ion :

$$Base \rightleftharpoons Acid + O^{2-}$$

This view is particularly useful in high temperature chemistry, as in the fields of ceramics and metallurgy :

$$CaO \rightleftharpoons Ca^{2+} + O^{2-}$$

$$SiO_2 + O^{2-} \rightleftharpoons SiO_3^{2-}$$

$$\underset{\text{Base}}{CaO} + \underset{\text{Acid}}{SiO_2} \rightleftharpoons \underset{\text{Salt}}{CaSiO_3}$$

Similarly, ores of Ti (as well as Ta, Nb) are dissolved in sodium pyrosulphate (an acid in the Lux sense) around 800°C :

$$TiO_2 + Na_2S_2O_7 = (TiO)SO_4 + Na_2SO_4$$

Similarly, SiO_2 displaces the more volatile bases P_4O_{10} or SO_3 :

$$6\,SiO_2 + 2\,Ca_3\,(PO_4)_2 \longrightarrow 6CaSiO_3 + P_4O_{10}$$

$$SiO_2 + CaSO_4 \longrightarrow CaSiO_3 + SO_3$$

Zinc oxide shows amphoterism as it can both donate or accept oxide ions :

Base $\qquad\qquad ZnO + S_2O_7 = Zn^{2+} + 2SO_4^{2-}$

Acid $\qquad\qquad Na_2O + ZnO = Na_2 ZnO_2$

The definition can be extended to cover acid-based reactions in terms of any negative ion transfer :

$$NaF + AlF_3 \longrightarrow Na^+ + AlF_6^{3-}$$

$$Na_2S + CS_2 \longrightarrow Na_2 CS_3,$$

$$EtNa + Et_2Zn \longrightarrow Na^+ + Zn\,Et_3^-.$$

Example 12.1: Predict how the acidity will change on adding SiO_2 to a molten mixture of Fe + FeO.

Solution: SiO_2 acts as an acceptor of oxide ion (forming $FeSiO_3$) from FeO: the acidity of the solution thus increases.

$$SiO_2 + FeO \longrightarrow FeSiO_3.$$

Complex Compounds

INTRODUCTION

Chemists of the late nineteenth century had difficulty in understanding how "molecular compounds" or "compounds of higher order" are bonded. The formation of a compound such as $CoCl_3.6NH_3$ was baffling–particularly in this case since simple $CoCl_3$ does not exist. In 1893, Alfred Werner proposed a theory to account for compounds as $[Co(NH_3)_6]Cl_3$. He assumed that the six ammonia molecules are symmetrically "coordinated" to the central cobalt atom by "subsidiary valencies" of cobalt while the "principal valencies" of cobalt are satisfied by the chloride ions.

Many practical applications have been derived from the study of complex compounds; advances have resulted in such fields as metallurgy, analytical chemistry, biochemistry, water purification, textile dyeing, electrochemistry, and bacteriology. In addition, the study of these compounds has enlarged our understanding of chemical bonding, certain physical properties (e.g., spectral and magnetic properties), minerals (many minerals are complex compounds), and metabolic processes (both heme of blood and chlorophyll of plants are complex compounds).

STRUCTURE

A complex ion or complex compound consists of a central metal cation to which several anions and/or molecules (called ligands) are bonded. With few exceptions, free ligands have at least one electron pair that is not engaged in bonding.

$$:\ddot{C}l:^- \qquad :C{\equiv}N:^- \qquad H-\ddot{O}: \quad H-\ddot{N}-H$$
$$\qquad\qquad\qquad\qquad\qquad\qquad | \qquad\quad |$$
$$\qquad\qquad\qquad\qquad\qquad\quad H \qquad\quad H$$

These electron pairs may be considered to be donated to the electron-deficient metal ions in the formation of complexes; ligands, therefore, are substances, that are capable of acting as Lewis bases. The bonding of complexes, however, shows a wide variation in character—from strongly covalent to predominantly ionic. The ligands are said to be coordinated around the central cation in a first coordination sphere. In the formulae of complex compounds, such as $K_3[Fe(CN_6)]$ and $[Cu(NH_3)_4]Cl_2$, the first coordination sphere is indicated by square brackets. The ligands are disposed about the central ion in a regular geometric manner (Fig. 13.1). The number of atoms directly bonded to the central metal ion, or the number of coordination positions, is called the coordination number of the central ion.

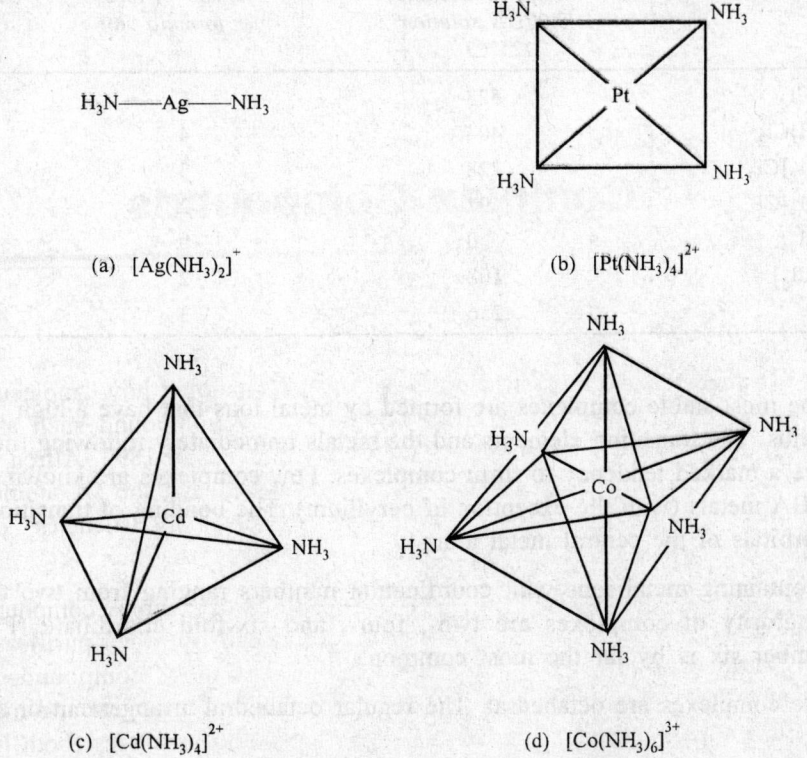

$$H_3N\!-\!\!-\!Ag\!-\!\!-\!NH_3$$

(a) $[Ag(NH_3)_2]^+$

(b) $[Pt(NH_3)_4]^{2+}$

(c) $[Cd(NH_3)_4]^{2+}$

(d) $[Co(NH_3)_6]^{3+}$

Fig. 13.1. Common configurations of complex ions: (a) Linear; (b) Square planar; (c) Tetrahedral; and (d) Octahedral.

The charge of a complex is the sum of the charges of the constituent parts; complexes may be cations, anions, or neutral molecules. Thus, in each of the complexes of platinum(IV)—$[Pt(NH_3)_5Cl]^{3+}$. $[Pt(NH_3)_2Cl_4]^0$, and $[PtCl_6]^{2-}$ – the platinum contributes 4^+, each chlorine contributes 1^-, and the coordinated ammonia molecules do not contribute to the charge of the complex.

An interesting series of platinum(IV) complexes appears in Table 13.1. The list is headed by the chloride of the $[Pt(NH_3)_6]^{4+}$ ion, and in each subsequent entry an ammonia molecule of the coordination sphere is replaced by a chloride ion. The coordinated ammonia molecules and chloride ions are tightly held and do not dissociate in water solution; however, those chloride ions of the compound that are not coordinated to the platinum are ionisable.

Hence, aqueous solutions of the last three compounds of the table do not precipitate silver chloride upon the addition of silver nitrate, whereas solutions of the first four precipitate AgCl in amounts proportional to 4/4, 3/4, 2/4, and 1/4, respectively, of their total chlorine content. In each case the total number of ions per formula unit derived from conductance data agrees with the formula listed in the table.

Table 13.1. Some platinum(IV) complex compounds.

	Molar conductance 0.001 N solution[a] (25°C)	Number of ions per formula unit	Number of Cl⁻ ions per formula unit
$[Pt(NH_3)_6]Cl_4$	523	5	4
$[Pt(NH_3)_5Cl]Cl_3$	404	4	3
$[Pt(NH_3)_4Cl_2]Cl_2$	228	3	2
$[Pt(NH_3)_3Cl_3]Cl$	97	2	1
$(Pt(NH_3)_2Cl_4]$	0	0	0
$K[Pt(NH_3)Cl_5]$	108	2	0
$K_2[PtCl_6]$	256	3	0

[a] cm^2/ohm mol.

In general, the most stable complexes are formed by metal ions that have a high positive charge and a small ionic radius. The transition elements and the metals immediately following (notably the IIIA and IVA metals) have a marked tendency to form complexes. Few complexes are known for the lanthanides and the IA and IIA metals (with the exception of beryllium). The bonding of transition-metal complexes involves the d orbitals of the central metal atom.

Complexes containing metal ions with coordination numbers ranging from two to nine are known. However, the majority of complexes are two-, four-, and six-fold coordinate (Fig. 13.1), and the coordination number six is by far the most common.

Six-coordinate complexes are octahedral. The regular octahedral arrangement of atoms is frequently represented as:

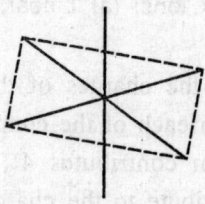

This representation is a convenient way to create a three-dimensional illusion. However, no difference between the bonds of the vertical axis and the other bonds should be inferred from this drawing; all the bonds are equivalent. Tetragonal geometry is a distorted form of the octahedral in which the bond distances along one of the axes are longer or shorter than the remaining bonds.

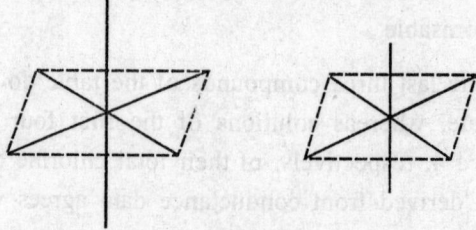

Four-coordinate complexes are known in tetrahedral and square-planar configurations. The square-planar configuration is the usual one for Pt^{II}, Pd^{II}, and Au^{III} and is assumed by many complexes of Ni^{II} and Cu^{II}. Examples include $[Pt(NH_3)_4]^{2+}$, $[PdCl_4]^{2-}$, $[AuCl_4]^{-}$, $[Ni(CN)_4]^{2-}$, and $[Cu(NH_3)_4]^{2+}$. In some instances there is evidence that square complexes may be tetragonal forms with two groups, along a vertical axis, located at greater distances from the central ion than the ligands of the plane. Thus, $[Cu(NH_3)_4]^{2+}$ in water solution may have two water molecules coordinated in the manner shown in Fig. 13.2. Hence, square-planar and octahedral geometries may be considered to merge.

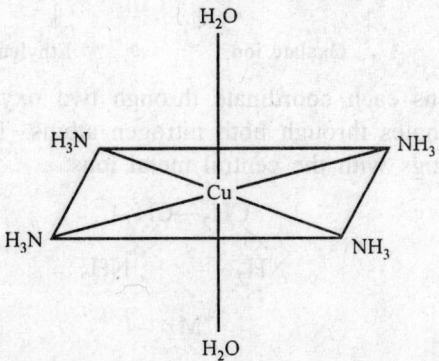

Fig. 13.2. Configuration of $[Cu(NH_3)_4(H_2O)_2]^{2+}$.

For four-coordinate complexes the tetrahedral configuration is encountered more frequently than the square-planar and is particularly common for complexes of the non-transition elements. Certain complexes of Cu^I, Ag^I, Au^I, Be^{II}, Zn^{II}, Cd^{II}, Hg^{II}, Al^{III}, Ga^{III}, In^{III}, Fe^{III}, Co^{II}, and Ni^O are tetrahedral; examples include $[Cu(CN)_4]^{3-}$, $[BeF_4]^{2-}$, $[AlF_4]^{-}$, $[FeCl_4]^{-}$, $[Cd(CN)_4]^{2-}$, $[ZnCl_4]^{2-}$, and $[Ni(CO)_4]^0$. The oxyanions of certain transition metals (such as VO_4^{3-}, CrO_4^{2-}, FeO_4^{2-}, and MnO_4^{-}) are tetrahedral, resembling the tetrahedral oxyanions of non-metals (such as SiO_4^{4-}, PO_4^{3-}, AsO_4^{3-}, SO_4^{2-} and ClO_4^{-}). Although a number of tetrahedral complexes of transition elements are known, the majority of complexes of these elements are octahedral.

Linear, two-coordinate complexes are not so common as the other forms previously mentioned. However, well-characterised complexes of this type are known for Cu^I, Ag^I, and Hg^{II}; examples are $[CuCl_2]^{-}$, $[Ag(NH_3)_2]^{+}$, $[Au(CN)_2]^{-}$, $[Hg(NH_3)_2]^{2+}$, and $[Hg(CN)_2]^0$.

In general, each metal exhibits more than one coordination number and geometry in its complexes. Although all known complexes of Co^{III} are octahedral, most cations form more than one type of complex. For example, Al^{III} forms tetrahedral and octahedral complexes, Cu^I forms linear and tetrahedral complexes, and Ni^{II} forms square-planar, tetrahedral, and octahedral complexes.

Complexes of the transition elements are frequently highly coloured. Examples of complexes that exhibit a wide range of colours are $[Co(NH_3)_6]^{3+}$ (yellow), $[Co(NH_3)_5(H_2O)]^{3+}$ (pink), $[Co(NH_3)_5Cl]^{2+}$ (violet, $[Co(H_2O)_6]^{3+}$ (purple), and $[Co(NH_3)_4Cl_2]^{+}$ (a violet form and a green form).

The ligands discussed above have been capable of forming only one bond with the central ion; they are referred to as unidentate (from Latin, meaning "one toothed") ligands. Certain ligands are capable of occupying more than one coordination position of a metal ion. Ligands that coordinate through two bonds from different parts of the molecule or anion are called bidentate.

Examples are:

$$\begin{array}{ccc} \overset{\displaystyle :\!O\!:}{\underset{\displaystyle \overset{\displaystyle \|}{C}}{}} & \overset{\displaystyle :\!O\!: \quad :\!O\!:}{\underset{\displaystyle \overset{\displaystyle \| \quad \|}{C\!-\!\!-\!C}}{}} & \overset{\displaystyle H \quad H}{\underset{\displaystyle H\!-\!\!C\!-\!\!C\!-\!H}{}} \end{array}$$

Carbonate ion   Oxalate ion   Ethylenediamine

The carbonate and oxalate ions each coordinate through two oxygen atoms; the ethylenediamine molecule (abbreviated *en*) coordinates through both nitrogen atoms. These positions are marked with arrows. Bidentate ligands form rings with the central metal ions.

$$\begin{array}{c} CH_2\!-\!\!CH_2 \\ \diagup \qquad\qquad \diagdown \\ NH_2 \qquad\qquad NH_2 \\ \cdot\cdot\qquad\qquad\cdot\cdot \\ M \end{array}$$

and the resulting metal complexes are called chelates (from Greek, meaning "claw"). The formation of five- or six-membered rings is generally favoured.

Multi-dentate ligands have been prepared that can coordinate at 2, 3, 4, 5, or 6 positions. In general, chelates are more stable than complexes containing only unidentate ligands. Thus, the sexadentate complexing agent ethylenediaminetetraacetate ion (EDTA)

$$\begin{array}{ccc} {}^-O_2C\!-\!\!CH_2 & & CH_2\!-\!\!CO_2^- \\ \diagdown & & \diagup \\ & N\!-\!\!CH_2\!-\!\!CH_2\!-\!\!N & \\ \diagup & & \diagdown \\ {}^-O_2C\!-\!\!CH_2 & & CH_2\!-\!\!CO_2^- \end{array}$$

is capable of forming a very stable complex with the calcium ion—an ion with one of the least tendencies toward the formation of complexes (Fig. 13.3).

Heme of haemoglobin is a chelate of Fe^{2+}, and chlorophyll is a chelate of Mg^{2+}. In both these substances the metal atom is coordinated to a quadridentate ligand, which may be considered to be derived from a porphin structure (Fig. 13.4) by the substitution of various groups for the H atoms of the porphin. The substituted porphins are called porphyrins.

Coordination around the iron atom of heme is octahedral. Four of the coordination positions are utilised in the formation of the heme (which is essentially planar), the fifth is used to bond the heme to a protein molecule (globin), and the sixth is used to coordinate either H_2O (Haemoglobin) or O_2 (Oxyhemoglobin). Coordination about this sixth position is reversible

$$\text{Haemoglobin} + O_2 \rightleftharpoons \text{Oxyhaemoglobin} + H_2O$$

and dependent upon the pressure of O_2. Hence, haemoglobin picks up O_2 in the lungs and releases it in the body tissues, where it is used for the oxidation of food. Haemoglobin reacts with carbon monoxide to form a complex that is more stable than oxyhaemoglobin, and therefore CO is toxic. Chlorophyll, the green pigment of plants, serves as a catalyst for the process of photosynthesis in which CO_2 and H_2O

are converted into a carbohydrate (glucose) and O_2. The energy for photosynthesis comes from sunlight, and the chlorophyll molecule initiates this process by absorbing a quantum of light.

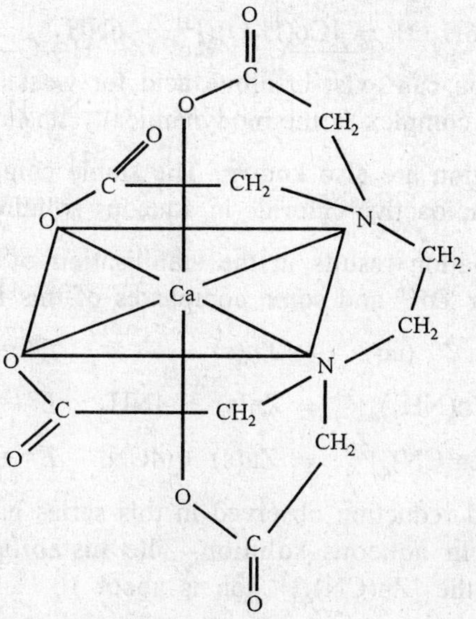

Fig. 13.3. Ethylenediaminetetraacetate complex of Ca^{2+} ($[Ca(EDTA)]^{2-}$).

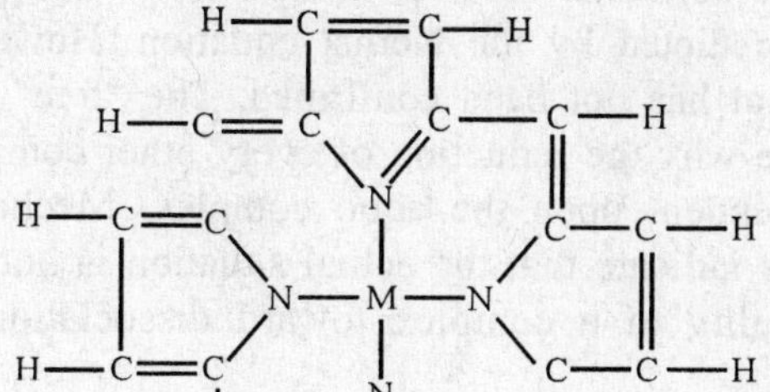

Fig. 13.4. Metal porphin.

LABILE AND INERT COMPLEXES

Certain complexes (which are called labile) rapidly undergo reactions in which ligands are replaced; other complexes (non-labile, or inert, complexes) do not undergo these substitution reactions or do so slowly. This distinction applies to the rate of attainment of equilibrium and has no bearing on the position of equilibrium. With the exception of the complexes of Cr^{III} and Co^{III}, most octahedral complexes of the fourth period transition elements are very labile; exchange reactions come to equilibrium almost as fast as the reagents are mixed. The reason that the complexes of Co^{III} have been studied more than any other group of complexes is that these substances undergo ligand exchange reactions at a slower, more convenient rate.

The inertness of a complex should not be confused with its thermodynamic stability. Although $[Co(NH_3)_6]^{3+}$ is stable in aqueous solution

$$[Co(NH_3)_6]^{3+} + 6H_2O \rightleftharpoons [Co(H_2O)_6]^{3+} + 6NH_3 \qquad K \cong 10^{-34}$$

the complex is unstable in aqueous acid

$$[Co(NH_3)_6]^{3+} + 6H_3O+ \rightleftharpoons [Co(H_2O)_6]^{3+} + 6NH_4^+ \qquad K \cong 10^{22}$$

Nevertheless, the $[Co(NH_3)_6]^{3+}$ ion can exist in dilute acid for weeks; the latter reaction must have a high activation energy. Thus, the complex is thermodynamically unstable and at the same time inert.

Examples of the reverse situation are also known. The stable complex $[FeCl_6]^{3-}$ is very labile and undergoes rapid exchange with radioactive chloride in aqueous solution.

In many cases complex formation results in the stabilisation of metal ions toward oxidation or reduction. Electrode potentials for Zn^{2+} and some complexes of this ion are

$$2e^- + Zn^{2+} \text{ (aq)} \rightleftharpoons Zn(s) \qquad E° = -0.763 \text{ V}$$

$$2e^- + [Zn(NH_3)_4]^{2+} \rightleftharpoons Zn(s) + 4NH_3 \qquad E° = -1.04 \text{ V}$$

$$2e^- + [Zn(CN)_4]^{2-} \rightleftharpoons Zn(s) + 4CN^- \qquad E° = -1.26 \text{ V}$$

The increasing stability toward reduction observed in this series parallels increasing stability of the complexes toward dissociation in aqueous solution. The instability constant of $[Zn(NH_3)_4]^{2+}$ is approximately 10^{-10} and that of the $[Zn(CN)_4]^{2-}$ ion is about 10^{-18}.

It is tempting to ascribe the lower electrode potentials of the complexes to the decreased concentrations of Zn^{2+}(aq) in the solutions of these complexes. Such an interpretation is in agreement with the concentration effects predicted by the Nernst equation. However, this explanation implies a mechanism for the reductions that has not been confirmed. The "free" zinc ion, Zn^{2+} (aq), is an aquo complex, and it is difficult to see why the reduction of every other complex of Zn^{2+} should be required to proceed through and be dependent upon the aquo complex. Mechanism studies of the oxidation-reduction reactions of complexes indicate that the actual situation is not this simple. There is, however, a relationship between the stability of a complex toward dissociation and its tendency to undergo oxidation or reduction.

In many instances complex formation results in the stabilisation of a metal in a rare or otherwise unknown oxidation state. A classic example is afforded by the complexes of Co^{III}. Simple compounds containing cobalt in an oxidation state of 3^+ are rare. The electrode potential

$$e^- + [Co(H_2O)_6]^{3+} \rightleftharpoons [Co(H_2O)_6]^{2+} \qquad E° = +1.81 \text{ V}$$

indicates that the hydrated Co^{3+} ion is a strong oxidising agent which is capable of oxidising water to oxygen and hence incapable of prolonged existence in water.

In the presence of many complexing agents, such as NH_3, the 3^+ oxidation state of cobalt is much more stable toward reduction.

$$e^- + [Co(NH_3)_6]^{3+} \rightleftharpoons [Co(NH_3)_6]^{2+} \qquad E° = +0.11 \text{ V}$$

The $[Co(NH_3)_6]^{3+}$ ion can exist in aqueous solution. It does not oxidise water to oxygen, and, in fact, the reverse reaction—the oxidation of Co^{II} complexes by a stream of air—is used to prepare Co^{III} complexes. The ammonia complex of Co^{III}, which has an instability constant of about 10^{-34}, is much more stable toward dissociation than the ammonia complex of Co^{II}, which has an instability constant of approximately 10^{-5}.

The electrode potentials for copper and its ions may be diagramed

$$Cu^{2+} \xrightarrow{\;+0.153\ V\;} Cu^{+} \xrightarrow{\;+0.521\ V\;} Cu$$

From this diagram we can see that the Cu^{+} ion is unstable in water solution and disproportionates

$$2Cu^{+}(aq) \rightleftharpoons Cu^{2+}(aq) + Cu(s) \quad E^{\circ} = +0.368\ V$$

The equilibrium constant for this disproprotionation may be derived from the cell potential.

$$K = \frac{[Cu^{2+}]}{[Cu^{+}]^{2}} \cong 10^{6}$$

With certain complexing agents the stability of the 1^{+} state is enhanced more than that of the 2^{+} state. For example, with ammonia $K \cong 10^{-2}$, and the $[Cu(NH_3)_2]^{-}$ ion is stable toward disproportionation. Other complexing agents, such as ethylenediamine, have a greater affinity for Cu^{II} than for Cu^{I}. For ethylenediamine, $K \cong 10^{5}$, and an ethylenediamine complex of Cu^{I} does not exist in water solution.

A similar situation is observed for the ions of gold; Au^{+} is theoretically unstable toward disproportionation into Au^{3+} and Au metal. In addition, both gold ions are strong oxidising agents and are capable of oxidising water. However, stable complexes of both Au^{I} and Au^{III} are known.

NOMENCLATURE

Since thousands of complexes are known and the number is constantly expanding, a system of nomenclature has been adopted for these compounds. The following paragraphs summarise the important rules of the system; they are adequate for naming the simple and frequently encountered complexes.

1. If the complex compound is a salt, the cation is named first whether or not it is the complex ion.

2. The constituents of the complex are named in the following order: anions, neutral molecules, central metal ion.

3. Anionic ligands are given -o endings; examples are: OH^{-}, hydroxo; O^{2-}; oxo; S^{2-}, thio; Cl^{-}, chloro; F^{-}, fluoro; CO_3^{2-}, carbonato; CN^{-}, cyano; CNO^{-}, cyanato; $C_2O_4^{2-}$, oxalato; NO_3^{-}, nitrato; NO_2^{-} nitro; SO_4^{2-}, sulphato; and $S_2O_3^{2-}$, thiosulphato.

4. The names of neutral ligands are not changed. Exceptions to this rule are: H_2O, aquo; NH_3, ammine; CO, carbonyl; and NO, nitrosyl.

5. The number of ligands of a particular type is indicated by a prefix: di-, tri-, tetra-, penta-, and hexa- (for two to six). For complicated ligands (such as ethylenediamine), the prefixes bis-, tris, and tetrakis- (two to four) are employed.

6. The oxidation number of the central ion is indicated by a Roman numeral, which is set off by parentheses and placed after the name of the complex.

7. If the complex is an anion, the ending -ate is employed. If the complex is a cation or a neutral molecule, the name is not changed. Examples follow:

$[Ag(NH_3)_2]Cl$	Diamminesilver(I) chloride
$[Co(NH_3)_3Cl_3]$	Trichlorotriamminecobalt(III)
$K_4[Fe(CN)_6]$	Potassium hexacyanoferrate(II)

[Ni(CO)$_4$]   Tetrcarbonylnickel (0)

[Cu(en)$_2$]SO$_4$ Bis(ethylenediamine)copper (II) sulphate

[Pt(NH$_3$)$_4$][PtCl$_6$]   Tetrammineplatinum (II) hexachloroplatinate (IV)

[Co(NH$_3$)$_4$(H$_2$O)Cl]Cl$_2$ Chloroaquotetramminecobalt (III) chloride

Common names are frequently employed when they are clearly more convenient than the systematic name (e.g., ferrocyanide rather than hexacyanoferrate (II) for [Fe(CN)$_6$]$^{4-}$) or when the structure of the complex is not certain (e.g., the aluminate ion).

ISOMERISM

Two compounds with the same molecular formula but different arrangements of atoms are called isomers; such compounds differ in their chemical and physical properties. Structural isomerism is displayed by compounds that have different ligands within their coordination spheres; several types of structural isomers may be identified.

The following pair of compounds of Co^{III} serve as an example of ionisation isomers.

$$\text{(a)} \quad [Co(NH_3)_5(SO_4)]Br$$
Red

$$\text{(b)} \quad [Co(NH_3)_5Br]SO_4$$
Violet

Conductance data show that both compounds dissociate into two ions in aqueous solution. In the first compound, the SO$_4$$^{2-}$ ion is a part of the coordination sphere, and the Br$^-$ ion is ionisable. Hence, an aqueous solution of compound (a) gives an immediate precipitate of AgBr upon the addition of AgNO$_3$, but since the SO$_4$$^{2-}$ ion is not free, no precipitate forms upon the addition of BaCl$_2$. For compound (b), the reverse is true. An aqueous solution of this compound gives a precipitate of BaSO$_4$ but not AgBr, since the SO$_4$$^{2-}$ is ionisable and the Br$^-$ is coordinated. Note that SO$_4$$^{2-}$ functions as a unidentate ligand and that the charge on the complex ion of compound (a) is 1+ and that of compound (b) is 2+. There are numerous additional examples of ionisation isomers, for example,

$$[Pt(NH_3)_4Cl_2]Br_2 \qquad\qquad [Pt(NH_3)_4Br_2]Cl_2$$

Hydrate Isomerism

Hydrate isomerism is analogous to ionisation isomerism and is probably best illustrated by the following series of compounds that have the formula CrCl$_3$.6H$_2$O.

(a) [Cr(H$_2$O)$_6$]Cl$_3$
Violet

(b) [Cr(H$_2$O)$_5$Cl]Cl$_2$·H$_2$O
Green

(c) [Cr(H$_2$O)$_4$Cl$_2$]Cl·2H$_2$O
Green

In a mole of these compounds there are six moles of water. However, in compound (a), six water molecules are coordinated; in compound (b), five; and in compound (c), four. The uncoordinated water, which probably occurs in the crystals as lattice water, is readily lost when compounds (b) and (c) are exposed to desiccants; however, the coordinated water is not so easily removed. Further evidence for the structures of the compounds is afforded by conductance data (the compounds are composed of four, three, and two ions, respectively) and the quantity of AgCl precipitated (the compounds have three, two, and one ionisable chloride ions, respectively). Another example of hydrate isomerism is given by the following pair of compounds.

$$[Co(NH_3)_4(H_2O)Cl]Cl_2 \qquad\qquad [Co(NH_3)_4Cl_2]Cl\cdot H_2O$$

Coordination Isomers

Coordination isomers may exist in compounds that have two or more centres of coordination. Isomers arise through the exchange of ligands between these coordination centres. In simple examples which involve only two complex ions per compound, the coordinated metal ions may be the same

$$[Cr(NH_3)_6][Cr(NCS)_6] \qquad\qquad [Cr(NH_3)_4(NCS)_2][Cr(NH_3)_2(NCS)_4]$$

or different

$$[Cu(NH_3)_4][PtCl_4] \qquad\qquad [Pt(NH_3)_4[CuCl_4]$$

and the oxidation state of the metals may vary

$$[Pt(NH_3)_4][PtCl_6] \qquad\qquad [Pt(NH_3)_4Cl_2][PtCl_4]$$

Tetraammineplatinum(II)-hexachloroplatinate(IV)   Dichlorotetraammineplatinum(IV)-tetrachloroplatinate(II)

Linkage Isomerism

Linkage isomerism is a rare but interesting type; it arises when ligands are capable of coordinating in two ways. For example, the nitrite ion, NO_2^-, can coordinate through an oxygen atom (—ONO, nitrito compounds) or through the nitrogen atom (—NO$_2$, nitro compounds).

$$[Co(NH_3)_5(NO_2)]Cl_2 \qquad\qquad [Co(NH_3)_5(ONO)]Cl_2$$

(Yellow)   (Red)
Nitropentaaminecobalt(III)-   Nitritopentaaminecobalt(III)-
chloride   chloride

Theoretically, many ligands might be capable of forming linkage isomers: CN$^-$ might coordinate either through the C atom or the N atom; SCN$^-$, through N or S; CO, through C or O. However, few authentic cases of linkage isomerism are known.

STEREO-ISOMERISM

Stereoisomerism is a second general classification of isomers. Compounds are stereoisomers when they both contain the same ligands in their coordination spheres but differ in the way that these ligands are arranged in space. There are generally two types of stereo-isomeism. 1. Geometric or *Cis-trans* isomerism. 2. Optical isomerism. An example of geometric isomerism is afforded by the *cis* and *trans* isomers of the square planar dichlorodiammineplatinum(II) (Fig. 13.5). In the *cis* isomer the chlorine atoms are situated on adjacent corners of the square (along an edge), whereas in the *trans* isomer they occupy opposite corners (along the diagonal). Since all the ligands of a tetrahedral complex have the same relationship to one another, *cis-trans* isomerism does not exist for this geometry. Many geometric isomers are known for octahedral complexes, however. There are two isomers of the dichlorotetraamminecobalt(III) ion: a violet *cis* form and a green *trans* form (Fig. 13.6).

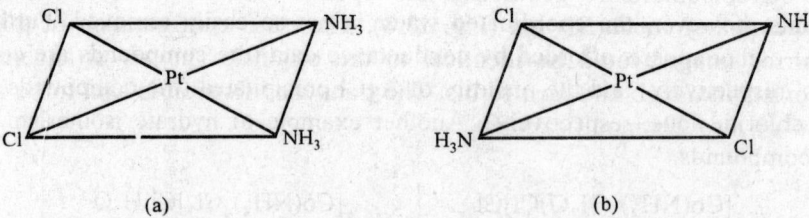

Fig. 13.5. (a) *Cis* and (b) *trans* isomers of dichlorodiammineplatinum(II).

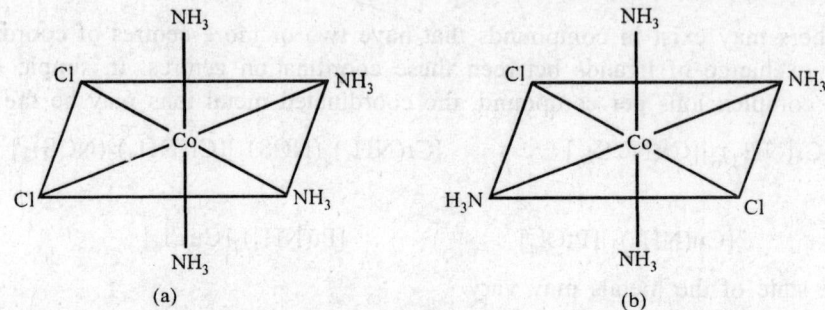

Fig. 13.6. (a) *Cis* and (b) *trans* isomers of the dichlorotetraamminecobalt(III) ion.

A second type of stereoisomerism is optical isomerism. Some molecules and ions can exist in two forms that are not superimposable and that bear the same relationship that a right hand bears to a left hand. That the hands are not superimposable is readily demonstrated by attempting to put a left-handed glove on a right hand. Such molecules and ions are spoken of as being dissymmetric, and the forms are called enantiomorphs (from Greek, meaning opposite forms) or mirror images (since the one may be considered to be a mirror reflection of the other).

Enantiomorphs have identical physical properties except for their effects on plane-polarised light. Light that has been passed through a polariser consists of waves that vibrate in a single plane. One enantiomorph (the dextro form), whether pure or in solution, will rotate the plane of the plane-polarised light to the right; the other (the levo form) will rotate the plane an equal extent to the left. For this reason enantiomorphs are called optical isomers or optical antipodes. An equimolar mixture of enantiomorphs, called a racemic modification, has no effect on plane-polarised light since it contains an equal number of dextrorotary and laevorotary forms.

The tris(ethylenediamine)cobalt(III) ion exists in enantiomorphic forms. Examination of the diagrams of Fig 13.7 confirms that the enantiomorphs are not superimosable and that the ions are dissymmetric. Note that bidentate chelating agents can span only *cis* positions.

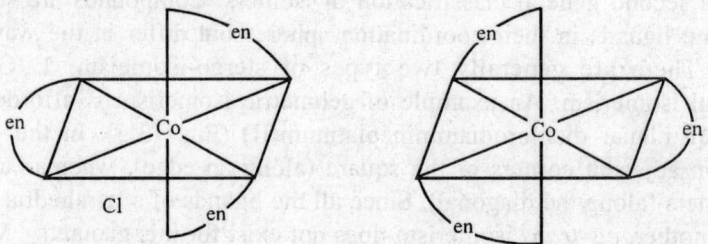

Fig. 13.7. Dextro and levo forms of the tris(ethylenediamine)cobalt(III) ion.

Both types of stereoisomerism—geometric and optical—are illustrated by the isomers of the dichlorobis(ethylenediamine) cobalt(III) ion (Fig. 13.8). The *trans* configuration of this ion is optically inactive, and a mirror image would be identical to the original. The *cis* arrangement, however, is dissymmetric and exists in *dextro* and *levo* forms. The trans modification is said to be a *diastereoisomer* of either the *dextro* or *levo cis* modification.

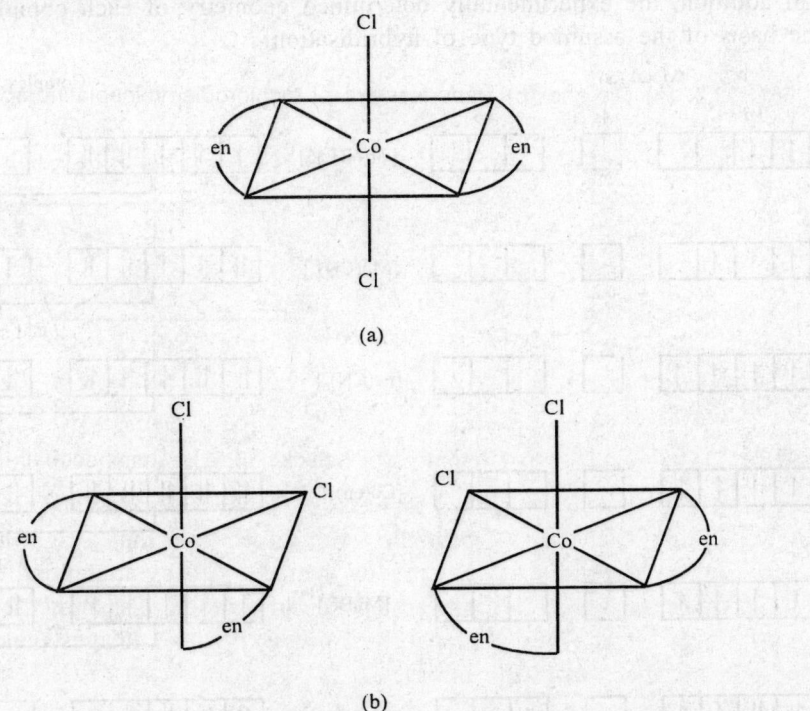

Fig. 13.8. Isomers of the dichlorobis(ethylenediamine)cobalt(III)ion: (a) Trans isomer, (b) Optical isomers of the *cis* form.

BONDING IN COMPLEXES

There are several approaches to the theoretical treatment of the bonding in complexes. The valence bond theory describes the bonding in terms of hybridised orbitals of the central metal ion. The metal ion is assumed to have a number of unoccupied orbitals available for complex formation equal to its coordination number. Each ligand donates a pair of electrons toward the formation of a coordinate covalent bond, and a bond arises from the overlap of a vacant hybrid orbital of the metal atom and a field orbital of a ligand. The bonds usually have an appreciable degree of polarity. The common types of hybridisation are:

sp	linear	dsp^2	square planar
sp^3	tetrahedral	d^2sp^3	octahedral

For octahedral coordination, d_{z^2} and $d_{x^2-y^2}$ orbitals, as well as p_x, p_y, and p_z orbitals, are directed toward ligands; these orbitals are combined with an s orbital into a d^2sp^3 set. Square-planar (dsp^2) hybridisation utilises the $d_{x^2-y^2}$, p_x, p_y, and s orbitals.

The valence bond diagrams of some simple complexes together with the electronic configurations of the central metal ions are shown in Fig. 13.9, Hund's rule applies to the electrons occupying the non-bonding d orbitals, and, accordingly, some of these electrons are unpaired. In each of the complexes shown, the number of unpaired electrons determined by magnetic susceptibility measurement agrees with the structure diagrammed. Thus, $[Co(en)_3]^{3+}$ and $[Ni(CN)_4]^{2-}$ are diamagnetic and the others are paramagnetic. In addition, the experimentally determined geometry of each complex agrees with that predicted on the basis of the assumed type of hybridisation.

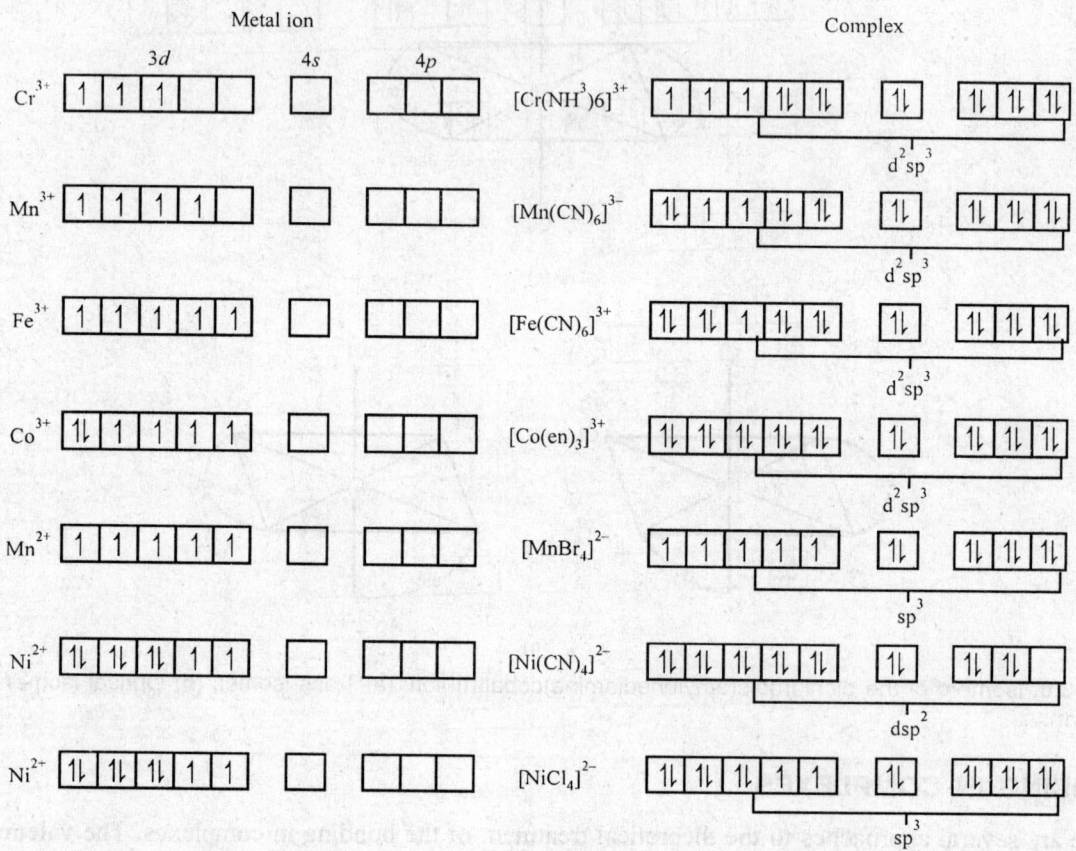

Fig. 13.9. Valence-bond diagrams for some simple complexes.

There are some defects in this attractively simple picture. From the diagrams of Fig. 13.9 one might suppose that all of the octahedral complexes of Fe^{III} would have one unpaired electron (as in $[Fe(CN)_6]^{3-}$ and that all of the octahedral complexes of Co^{III} would be diamagnetic (like $[Co(en)_3]^{3+}$). Such is not the case. The magnetic moments of $[FeF_6]^{3-}$ and $[CoF_6]^{3-}$ indicate the existence of five and four unpaired electrons, respectively, in these complexes; these numbers of unpaired electrons are the same as those found in the free Fe^{3+} and Co^{3+} ions. At one time, the bonding in this type of complex was thought to be ionic with the d orbitals of the central ion undisturbed by complex formation. According to a later explanation, the $4d$, rather than the $3d$, orbitals are assumed to be used for the octahedral hybridisation. Hence, inner (d^2sp^3) and outer (sp^3d^2) complexes are postulated (Fig. 13.10). Since it is impossible to clear two inner d orbitals of the Ni^{2+} ion for inner d^2sp^3 hybridisation, all octahedral Ni^{II} complexes must be assumed to be of the outer type.

A further complication arises in the case of $[Co(NO_2)_6]^{4-}$ —an octahedral complex with one unpaired electron. In order to fit this complex into the d^2sp^3 category, it is necessary to postulate the promotion of an electron into the $4d$ sub-level.

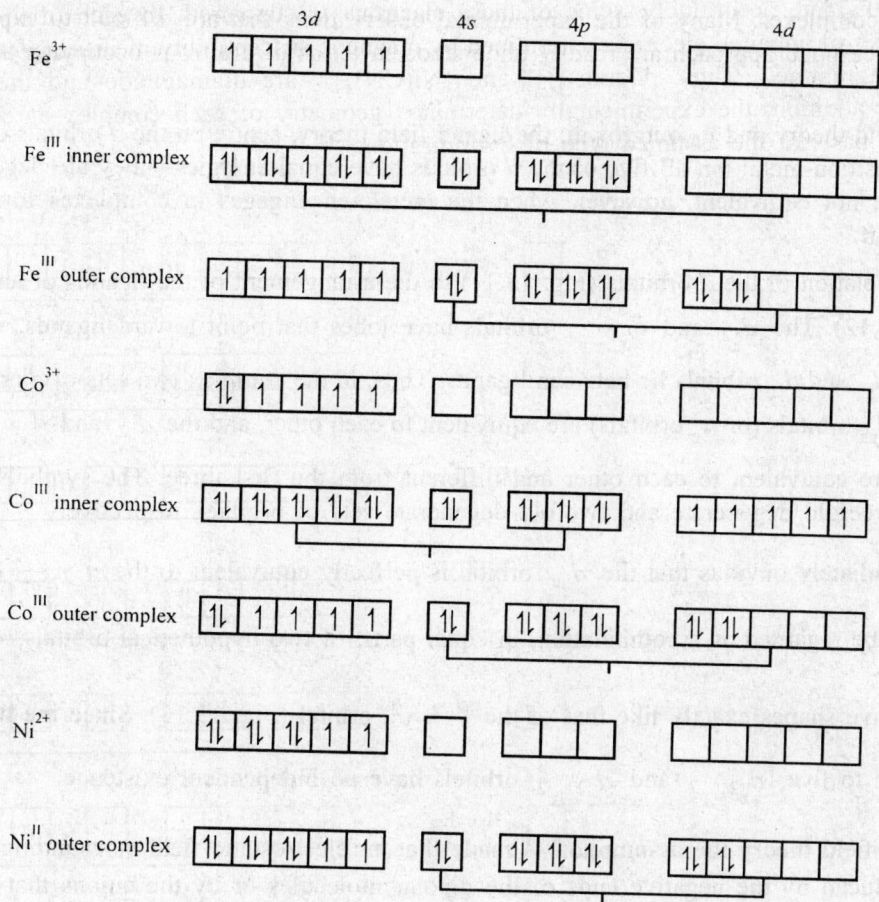

Fig. 13.10. Inner and outer complexes of Fe^{III}, Co^{III}, and Ni^{II}.

It is also necessary to assume such an unlikely promotion for all of the square complexes of Cu^{II}.

The fundamental defect of the valence bond theory is that it fails to take into account the existence of the anti-bonding molecular orbitals that are produced, along with bonding molecular orbitals, in the formation of the complexes. Many of the experimental observations that are difficult to explain on the basis of the valence bond approach are readily understood in terms of electrons occupying anti-bonding orbitals.

The crystal field theory and its outgrowth, the ligand field theory, centre on the d orbitals of the metal ion. In a free transition-metal ion all five of the d orbitals have equal energies—they are degenerate. All the d orbitals are not equivalent, however, when the metal ion engages in complexes formation; the degeneracy is split.

Consider the relation of the d orbitals (Fig. 13.11) to the arrangement of the ligands of an octahedral complex (Fig. 13.12). The d_{z^2} and $d_{x^2-y^2}$ orbitals have lobes that point toward ligands, whereas the lobes of the d_{xy}, d_{xz} and d_{yz} orbitals lie between ligands. Thus, in the complex two sets of d orbitals exist; the d_{xy}, d_{xz}, and d_{yz} orbitals (or t_{2g} orbitals) are equivalent to each other, and the d_{z^2} and $d_{x^2-y^2}$ orbitals (or e_g orbitals) are equivalent to each other and different from the first three. The symbols t_{2g} and e_g are applied to threefold degenerate and twofold degenerate sets of orbitals, respectively.

It is not immediately obvious that the d_{z^2} orbital is perfectly equivalent to the $d_{x^2-y^2}$ orbital. The d_{z^2} orbital may be regarded as a combination, in equal parts, of two hypothetical orbitals, $d_{z^2-y^2}$ and $d_{z^2-x^2}$ which have shapes exactly like that of the $d_{x^2-y^2}$ orbital (Fig. 13.13). Since the number of d orbitals is limited to five, $d_{z^2-y^2}$ and $d_{z^2-x^2}$ orbitals have no independent existence.

In the crystal field theory the assumption is made that an electrostatic field surrounding the central metal ion is produced by the negative ends of the dipolar molecules or by the anions that function as ligands. A metal-ion electron in a d orbital that has lobes directed toward ligands has a higher energy (owing to electrostatic repulsion) than an electron in an orbital with lobes that point between ligands. In octahedral complexes the orbitals of the e_g group, therefore, have higher energies than those of the t_{2g} group. The difference between the energies of the t_{2g} and e_g orbitals in an octahedral complex is given the symbol Δ_o.

In a square-planar complex (Fig. 13.12), the d orbitals exhibit four different relationships. The lobes of the $d_{x^2-y^2}$ orbital point toward ligands, and this orbital has the highest energy. The lobes of the d_{xy} orbital lie between orbitals but are coplanar with them, and hence this orbital is next highest in energy. The lobes of the d_{z^2} orbital point out of the plane of the complex, but the belt around the centre of the orbital (which contains about a third of the electron density) lies in the plane; therefore, the d_{z^2} orbital is next highest in energy. The d_{xz} and d_{yz} orbitals, which are degenerate, are least affected by the electrostatic field of the ligands since their lobes point out of the plane of the complex; these orbitals are lowest in energy.

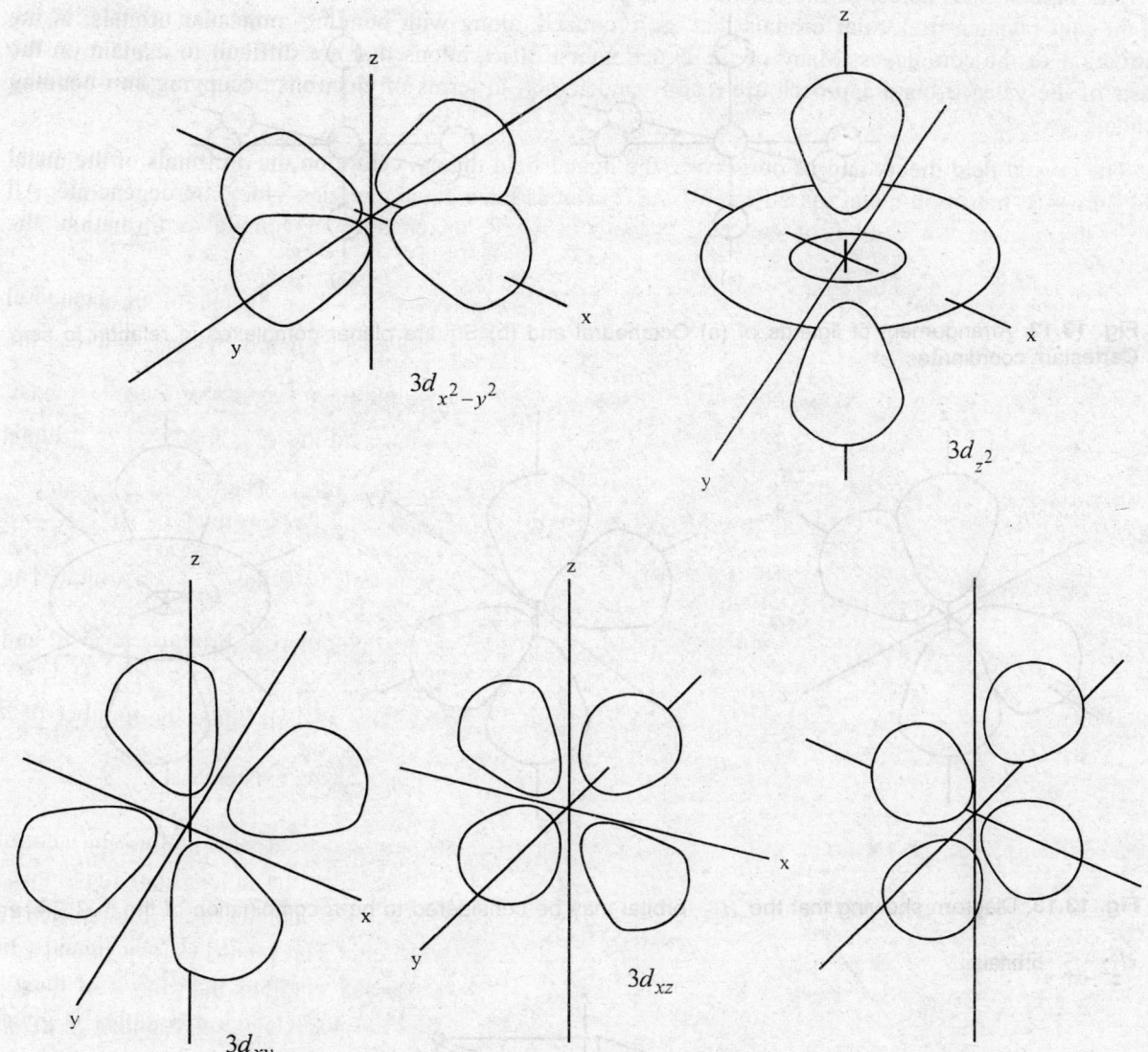

Fig. 13.11. Boundary surface diagrams for the *d* orbitals.

The order of splitting of the *d* orbitals in a tetrahedral complex may be derived from an examination of Fig. 13.14, which shows the relation of tetrahedrally arranged ligands to a system of Cartesian coordinates and a hypothetical cube. The lobes of the d_{xy}, d_{xz}, and d_{yz} orbitals point toward cube edges, and the lobes of the d_{z^2} and $d_{x^2-y^2}$ orbitals point toward the centres of cube faces. Notice that the distance from the centre of a cube face to a ligand is farther than the distance from the centre of a cube edge to a ligand. Therefore, the order of orbital energies is the reverse of that for octahedral coordination, and the threefold degenerate set, t_{2g}, is of higher energy than the two-fold degenerate set, e_g. The difference in energy is given the symbol Δ_t.

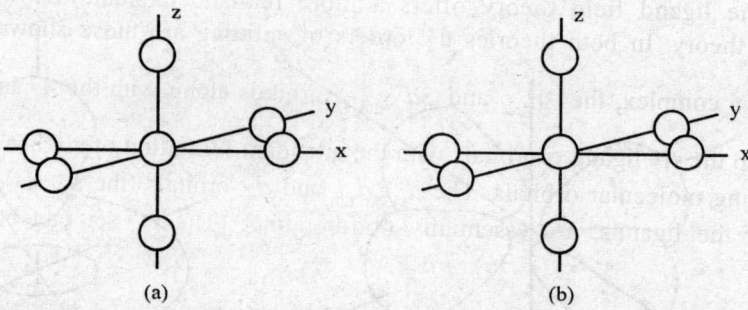

Fig. 13.12. Arrangement of ligands of (a) Octahedral and (b) Square planar complexes in relation to sets of Cartesian coordinates.

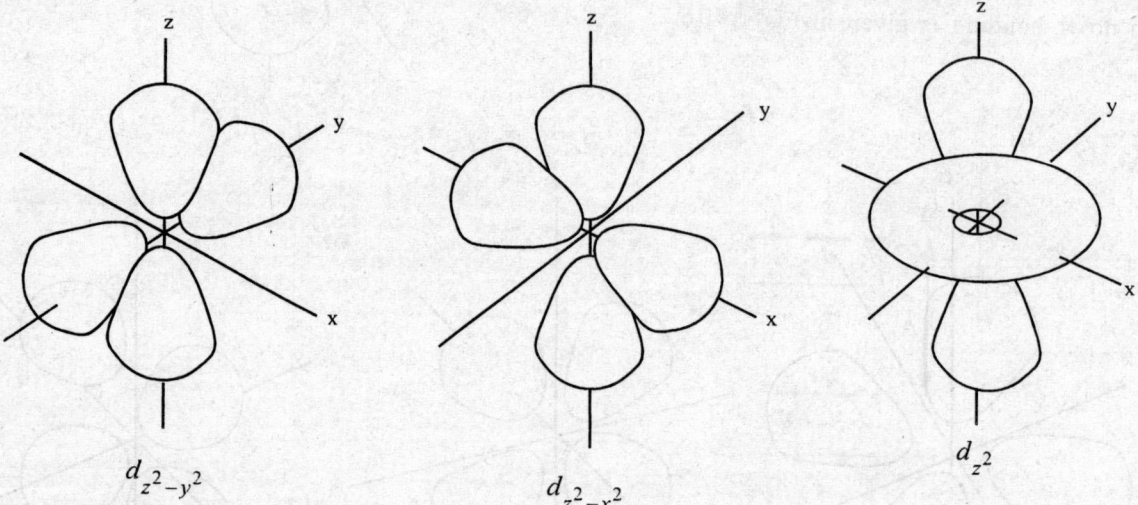

$$d_{z^2-y^2} \qquad d_{z^2-x^2} \qquad d_{z^2}$$

Fig. 13.13. Diagram showing that the d_{z^2} orbital may be considered to be a combination of the $d_{z^2-y^2}$ and $d_{z^2-x^2}$ orbitals.

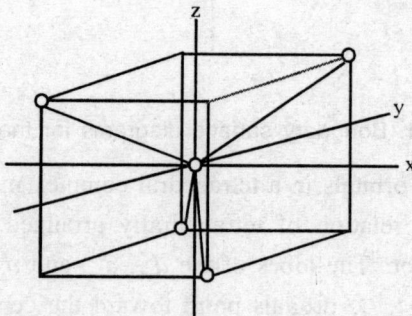

Fig. 13.14. The relation of tetrahedrally arranged ligands to a set of Cartesian coordinates.

The splitting of d-orbital energies in tetrahedral, octahedral, and square-planar complexes is summarised in Fig. 13.15. The crystal field theory fails to take into account the mixing of metal and ligand orbitals and the covalent character of the bonding in complexes, and hence it is ultimately unsatisfactory. The ligand field theory offers a more realistic explanation, which is derived from molecular orbital theory. In both theories tl ʊ orders of splitting are those shown in Fig. 13.15.

In an octahedral complex, the $3d_{z^2}$ and $3d_{x^2-y^2}$ orbitals along with the $4s$ and three $4p$ orbitals are assumed to overlap the six ligand σ orbitals with the attendant formation of six bonding molecular orbitals and six anti-bonding molecular orbitals. The d_{xy}, d_{xz}, and d_{yz} orbitals (the t_{2g} set), which do not overlap the Σ orbitals of the ligands, are essentially non-bonding. (The t_{2g} set can be used in π bonding, however).

A bonding molecular orbital concentrates electron density between the atoms and is of relatively low energy in comparison to an antibonding molecular orbital, which has a low electron density between the atoms and acts as a disruptive force. A molecular orbital energy level diagram for an octahedral complex with no π bonding is given in Fig. 13.16.

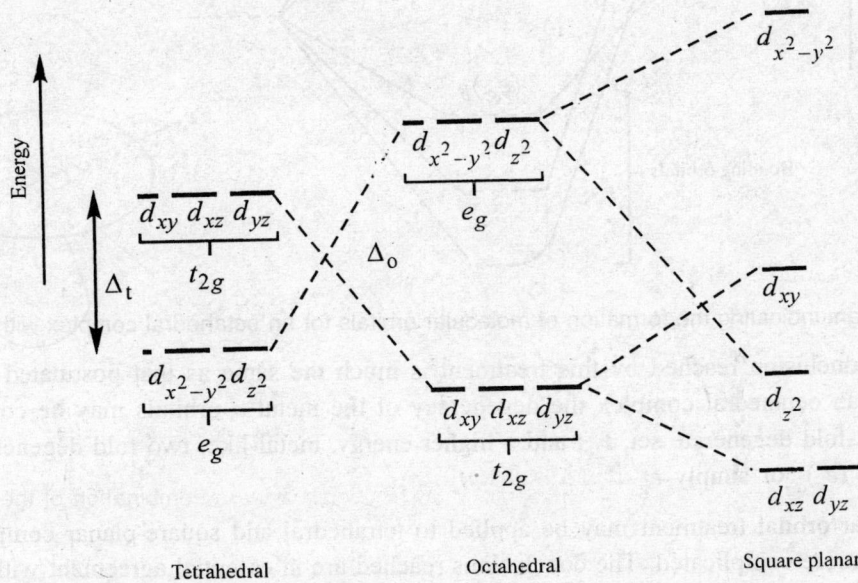

Fig. 13.15. The splitting of *d*-orbital energies by ligand fields of three different geometries.

Whenever two atomic orbitals of different energies combine, the character of the resulting bonding molecular orbital is predominantly that of the atomic orbital of lower energy, and the antibonding molecular orbital has mainly the character of the higher energy atomic orbital. In an octahedral complex the bonding orbitals have predominantly the character of ligand orbitals. The antibonding orbitals resemble metal orbitals more than ligand orbitals. The t_{2g} set, which are non-bonding, may be considered as purely metal orbitals.

In an octahedral complex the six electron pairs from the ligands completely occupy the bonding molecular orbitals. The d electrons of the central metal ion are accommodated in the t_{2g} non-bonding orbitals and the $(e_g)_a$ anti-bonding orbitals; the difference between the energies of these sets is Δ_o. The four remaining antibonding orbitals are never occupied in the ground states of any known complex.

Metal ion orbitals   Molecular orbitals   Ligand orbitals

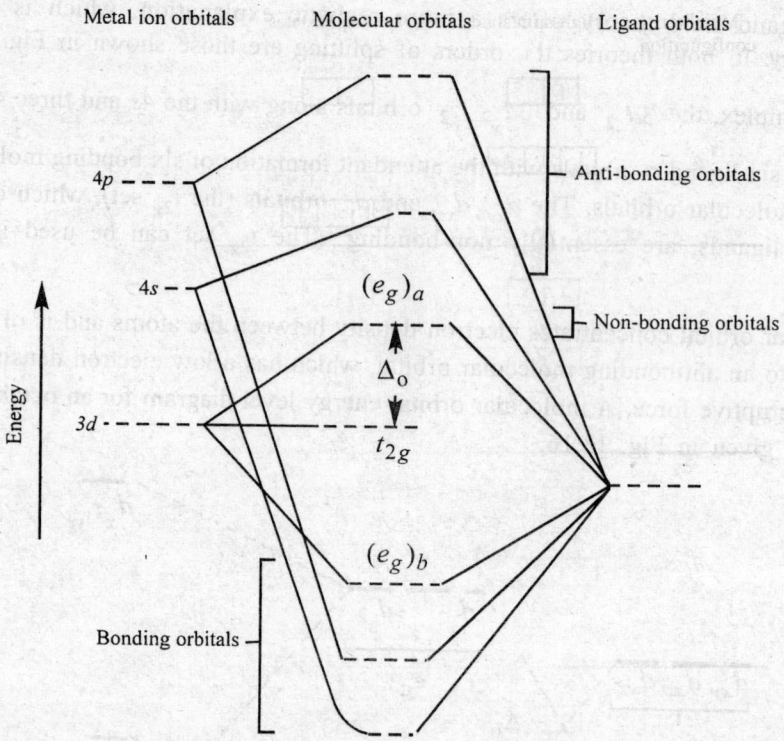

Fig. 13.16. Diagram indicating the formation of molecular orbitals for an octahedral complex with no π bonding.

Hence, the conclusion reached by this treatment is much the same as that postulated by the crystal field theory: in an octahedral complex the degeneracy of the metal d orbitals may be considered to be split into a three-fold degenerate set, t_{2g}, and a higher energy, metal-like, two-fold degenerate set, which may be labelled $(e_g)_a$ or simply e_g.

The molecular orbital treatment may be applied to tetrahedral and square-planar complexes, but the applications are more complicated. The conclusions reached are in essential agreement with the splittings diagrammed in Fig. 13.15.

For the octahedral complexes of a given metal ion, the magnitude of Δ_o is different for each set of ligands, and the electronic configurations of many complexes depend upon the size of Δ_o.

For complexes of transition-element ions with one, two, or three d electrons (which are referred to as d^1, d^2, or d^3 ions), the orbital occupancy is certain and is independent of the magnitude of Δ_o. The electrons enter the lower energy t_{2g} orbitals singly with their spins parallel. For d^4, d^5, d^6, and d^7 ions, a choice of two configurations is possible (Fig. 13.17).

In the case of an octahedral complex of a d^4 ion, the fourth electron can singly occupy a higher energy e_g orbital or it can enter a t_{2g} orbital and pair with an electron already present. The former configuration has four unpaired electrons and is called the high-spin state; the latter configuration, which has two unpaired electrons, is called the low-spin state. Which configuration is assumed depends upon which is energetically more favourable.

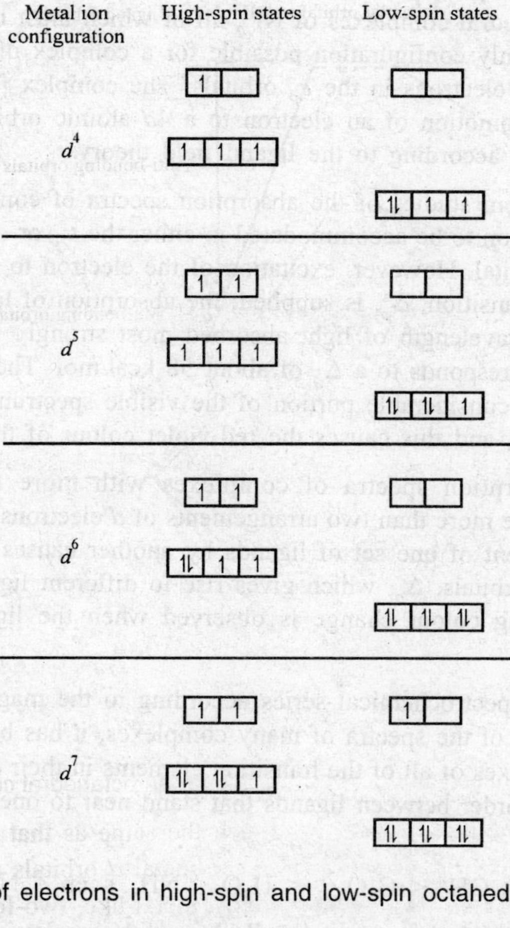

Metal ion configuration   High-spin states   Low-spin states

d^4

d^5

d^6

d^7

Fig. 13.17. Arrangements of electrons in high-spin and low-spin octahedral complexes of d^4, d^5, d^6, and d^7 ions.

If Δ_o is small, the electron may be promoted to the e_g level where it occupies an orbital singly. However, if Δ_o is large, the electron may be forced to pair with an electron in a t_{2g} orbital even though this pairing requires the expenditure of a pairing energy, P, to overcome the inter-electronic repulsion. Thus, the high-spin configuration results when

$$\Delta_o < P$$

and the low-spin configuration results when

$$\Delta_o > P$$

The values of P depends upon the metal ion; Δ_o is different for each complex.

This conclusion is also valid for complexes of d^5, d^6, and d^7 ions. For a complex of a d^8 ion there is only one possible configuration: six electrons paired in the t_{2g} orbitals and two unpaired electrons in the e_g orbitals. Likewise, complexes of d^9 and d^{10} ions exist in only one configuration.

The complexes that were fitted into the valence-bond scheme only with difficulty are readily classified by the ligand field theory. Hence, the inner complex $[Fe(CN)_6]^{3-}$ with one unpaired electron is a low-spin state of this d^5 ion, and the outer complex $[FeF_6]^{3-}$ with five unpaired electrons has the high-spin configuration of a d^5 ion. The octahedral complexes of Ni^{II}, all of which must be assumed to be of the outer valence-bond type, have the only configuration possible for a complex of a d^8 ion: six electrons in the t_{2g} orbitals and two unpaired electrons in the e_g orbitals. The complex $[Co(NO_2)_6]^{4-}$, for which it was necessary to postulate the promotion of an electron to a $4d$ atomic orbital in the valence-bond treatment, is a low-spin d^7 complex according to the ligand field theory.

Values of Δ_o can be obtained from studies of the absorption spectra of complexes. In the complex $[Ti(H_2O)_6]^{3+}$, there is only one electron to be accommodated in either the t_{2g} or e_g orbitals. In the ground state this electron occupies a t_{2g} orbital. However, excitation of the electron to an e_g orbital is possible when the energy required for this transition, Δ_o, is supplied; the absorption of light by the complex can bring about such excitations. The wavelength of light absorbed most strongly by the $[Ti(H_2O)_6]^{3+}$ ion is approximately 4900 Å, which corresponds to a Δ_o of about 58 kcal/mol. The single absorption band of this complex spreads out over a considerable portion of the visible spectrum; however, most of the red and violet light is not absorbed, and this causes the red-violet colour of the complex.

The interpretation of the absorption spectra of complexes with more than one d electron is considerably more complicated, since more than two arrangements of d electrons are possible. In general, for a given metal ion the replacement of one set of ligands by another causes a change in the energy difference between the t_{2g} and e_g orbitals, Δ_o, which gives rise to different light-absorption properties. Hence, in many instances a striking colour change is observed when the ligands of a complex are replaced by other ligands.

Ligands may be arranged in a spectrochemical series according to the magnitude of Δ_o they bring about. From the experimental study of the spectra of many complexes, it has been found that the order is generally the same for the complexes of all of the transition elements in their common oxidation states with only occasional inversions of order between ligands that stand near to one another on the list. The order of some common ligands is

$$I^- < Br^- < Cl^- < F^- < OH^- < C_2O_4^{2-} < H_2O < NH_3 < en < NO_2^- < CN^-$$

The value of Δ_o induced by the halide ions are generally low, and complexes of these ligands usually have high-spin configurations. The cyanide ion, which stands at the opposite end of the series from the halide ions, induces the largest d-orbital splittings of any ligand listed; cyano complexes generally have low-spin configurations.

A given ligand, however, does not always produce complexes of the same spin type. Thus, the hexaamine complex of Fe^{II} has a high-spin configuration, whereas the hexaamine complex of Co^{III} which is iso-electronic with Fe^{II} has a low-spin configuration.

For each metal ion there is a point in the series that corresponds to the change from ligands that produce high-spin complexes to ligands that form low-spin complexes. For example, Co^{II} forms high-spin complexes with NH_3 and ethylenediamine, but the NO_2^- and CN^- complexes of Co^{II} have low-spin configurations. The actual position in the series at which this change from high- to low-spin complex formation occurs depends upon the electron-pairing energy, P, for the metal ion, as well as the values of Δ_o for the complexes under consideration.

Example 13.1. Blood is said to be isotonic with 0.85% NaCl solution at 40°C. Assuming complete dissociation of NaCl; calculate total concentration of various solutes in blood. What is its approximate freezing point ? [$K_f = 1.86$].

Solution. As NaCl completely dissociates into two ions its osmotic pressure

$$\pi = 2 \times \frac{g}{v} \cdot \frac{RT}{M}.$$

Here g = 0.85 gm; v = 100 c.c. ≡ 0.1 litre; T = 313°A; M_{NaCl} = 58.5.

So
$$\pi = 2 \times \frac{0.85}{0.1} \times \frac{0.082 \times 313}{58.5} = 7.46 \text{ atm}.$$

As it is isotonic with blood at 40°C; blood also has 7.46 atm. osmotic pressure. So if C(M) is conc. of dissolved solutes then

$$C = \frac{\pi}{RT}$$

or
$$C = \frac{7.46}{0.082 \times 313} = 0.29(M).$$

For aqueous solution molar concn. can be approximately taken as molal conc. So molal conc. – 0.29 (Cm).

Now $\Delta T_f = K_f Cm = 1.86 \times 0.29 = 0.54$°C.

So freezing point of blood will be –0.54°C.

Example 13.2. A 0.2 molal aqueous solution of KCl freezes at –0.680°C. Calculate i and the osmotic pressure at 0°C. [K_f = 1.86].

Solution. As KCl is an electrolyte, so

$$\Delta T_f = i.K_f Cm$$

or
$$i = \frac{\Delta T_f}{K_f.Cm} = \frac{0.680}{1.86 \times 0.2} = 1.83.$$

For aqueous soln. molar conc. is approximately equal to molal conc.

So   Cm = C. Now
$$\pi = iCRT$$
$$\pi = 1.83 \times 0.2 \times 0.182 \times 273 \text{ atm}.$$

or
$$\pi = 8.19 \text{ atm}.$$

Chapter 14

Elements of Chemical Thermodynamics

INTRODUCTION

Thermodynamics is the study of the energy effects; in particular, it summarises the relations between heat, work, and other forms of energy that are involved in all types of changes. The laws of thermodynamics can be used to predict whether a particular chemical or physical transformation is theoretically possible under a given set of conditions. Thermodynamics, however, has nothing to say about the rate at which a predicted change will occur. This question is the concern of chemical kinetics. The stable form of carbon under ordinary conditions is graphite, not diamond according to thermodynamic principles. Therefore, the transformation of a diamond crystal into a graphite crystal is spontaneous. This change, however, is so extremely slow that it is not observed at ordinary temperatures and pressures.

FIRST LAW OF THERMODYNAMICS

Many scientists of the late eighteenth and early nineteenth centuries studied the relation between mechanical work and heat; thermodynamics had its origin in these studies. By the 1840's it became clear that heat and work are two manifestations of a larger classification—energy, that different forms of energy are interconvertible, and that energy is conserved. The first law of thermodynamics is the law of conservation of energy: energy can be converted from one form into another but cannot be created or destroyed.

The careful and convincing experimental work of James Joule did much to establish this law. Joule studied the conversion of mechanical and electrical work into heat. He used the work done by a falling weight to turn a puddle wheel that was immersed in a container of water and determined the heat produced by measuring the increase in the temperature of the water. In a series of such experiments, he used different weights and different quantities of water, as well as mercury in place of the water. Joule also studied the heating effects of electric currents and the conversion of work into heat by the compression of gases. Joule found that a given amount of work always produces the same quantity of heat. In modern terms the relation is

$$4.1840 \text{ J} = 1 \text{ cal}$$

One joule (J) is the amount of work done when a force of one newton (1 N = 1 kg m/sec^2) acts through a distance of one metre. The calorie, which was originally defined as the amount of heat required to raise the temperature of 1 g of water from 14.5°C to 15.5°C, is now defined by its joule equivalent. Since it is possible to make electrical measurements with greater precision than calorimetric measurements, the joule (which is a volt coulomb) is a better primary standard than the calorie.

In the International System of Units (SI) the joule is the unit for all energy measurements. However, thermodynamic values have been almost exclusively recorded in terms of calories and kilocalories in the past, and therefore one must be familiar with both units.

In applying thermodynamic concepts, we frequently confine our attention to the changes that occur within definite boundaries. That portion of the universe included within these boundaries is called a system; the remainder of the universe is called the surroundings. Work done on a system need not always result in an increase in the temperature of the system or in the conversion of work into heat. For example, the charging of a storage battery by an automobile engine results in an increase of the "chemical energy" of the battery; doing work on (or adding heat to) a sample of ice at 0°C could result in melting a part of the ice with no increase in the temperature of the system. A system is assumed to have an internal energy, E, which includes all possible forms of energy attributable to the system (some of which are the attractions and repulsions between the atoms, molecules, ions, and subatomic particles comprising the system and the kinetic energies of all of its parts). According to the first law of thermodynamics, the internal energy of an isolated system is constant. The actual value of E for any system is not known and cannot be determined. However, thermodynamics is concerned only with changes in internal energy, and such changes are measurable.

The internal energy of a system depends upon the state of the system and not upon how the system arrived at that state. Internal energy is therefore called a state function. Consider a sample of an ideal gas that occupies a volume of 1 litre at 100°K and 1 atm pressure (state A). At 200°K and 0.5 atm (state B), the sample occupies a volume of 4 litres. According to the first law, the internal energy of the system in state A, E_A, is a constant, as is the internal energy of the system in state B, E_B. It follows that the difference in the internal energies of the two states, ΔE, is also a constant and is independent of the path taken between state A and state B. It makes no difference whether the gas is heated before the pressure change, whether the heating is done after the pressure change, or indeed whether the total change is brought about in several steps

$$\Delta E = E_B - E_A$$

If ΔE were not independent of the manner in which the corresponding change was brought about, it would be possible to create or destroy energy, in violation of the first law, by taking a system from A to B by one route and then returning it to state A by a different route. Instead, in a cyclic process (in which a system is returned to its original state), the increase in E in one direction exactly equals the decrease in E in the opposite direction—there is no net gain or loss in energy for the cycle.

The change in internal energy of a system may be determined by measurement of the heat absorbed by the system from its surroundings, q, and the work done by the system on its surroundings, w. In these terms,

$$\Delta E = q - w$$

It is important to keep in mind the conventions regarding the signs of these quantities :

q, positive = heat *absorbed* by the system

q, negative = heat *evolved* by the system

w, positive = work done *by* the system

w, negative = work done *on* the system

The values of q and w involved in changing a system from state A to state B depend upon the way in which the change is carried out. However, the value of $(q-w)$ is a constant, equal to ΔE, for the change no matter how it is brought about. If a system undergoes a change in which the internal energy of the system remains constant, the work done by the system equals the heat absorbed by the system. If a system undergoes a change that reduces its internal energy, the energy released will be the sum of the heat evolved and the work done.

ENTHALPY

For ordinary chemical reactions the work term generally arises as a consequence of pressure–volume changes–for example, the work done against the atmosphere if the system expands in the course of the reaction. The term PV has the dimensions of work. Pressure, which is force per unit area, may be expressed in newtons per square metre; multiplying pressure by volume, which is expressed in cubic metres, gives a product in newton metres (or joules). A newton metre (or joule) is a dimension of work, since work is force times distance. In like manner, litre atmospheres are units of work. If the pressure is held constant, the work done in expansion from V_A to V_B is

$$w = P(V_B - V_A) = P\Delta V$$

No pressure-volume work can be done by a process carried out at constant volume, and $w = 0$. Thus, at constant volume the equation

$$\Delta E = q - w$$

becomes

$$\Delta E = q_v$$

where q_v is the heat absorbed by the system at constant volume.

Processes carried out at constant pressure are far more common in chemistry than those conducted at constant volume. If we restrict our attention to pressure-volume work, the work done in constant pressure processes is $P\Delta V$. Thus, at constant pressure the equation

$$\Delta E = q - w$$

becomes

$$\Delta E = q_P - P\Delta V$$

or, by rearranging,

$$q_P = \Delta E + P\Delta V$$

where q_P is the heat absorbed by the system at constant pressure. This relationship may be written

$$q_P = (E_B - E_A) + P(V_B - V_A)$$

or, by rearranging,

$$q_P = (E_B + PV_B) - (E_A + PV_A)$$

The thermodynamic function enthalpy, H, is defined by the equation

$$H = E + PV$$

Therefore,

$$q_P = H_B - H_A = \Delta H$$

Thus, the heat absorbed by a process conducted at constant pressure is equal to the change in enthalpy, and ΔH is related to the change in the internal energy of the system by the expression

$$\Delta H = \Delta E + P\Delta V$$

Enthalpy, like internal energy, is a function of the state of the system and is independent of the manner in which the state was achieved. The validity of the law of Hess rests on this fact.

The standard enthalpy of formation of a compound (given by the symbol $\Delta H°_f$) is the change in enthalpy for the reaction (at 25°C and 1 atm) in which one mole of the compound in its standard state is prepared from its constituent element in their standard states. The standard state of a substance is the stable form of the substance at 25°C and 1 atm. The use of $\Delta H°_f$ values are also discussed. The number of standard enthalpies of formation are listed in Table 14.1.

Table 14.1. Enthalpy of formation at 25°C and 1 atm.

Compound	$\Delta H°_f$ (kcal/mol)	(kJ/mol)[a]	Compound	$\Delta H°_f$ (kcal/mol)	(kJ/mol)[a]
$H_2O(g)$	−57.80	−241.8	$COCl_2(g)$	−53.3	−223.0
$H_2O(l)$	−68.32	−285.9	$SO_2(g)$	−70.96	−296.9
$HF(g)$	−64.2	−269.0	$CO(g)$	−26.42	−110.5
$HCl(g)$	−22.06	−92.30	$CO_2(g)$	−94.05	−393.5
$HBr(g$	−8.66	−36.2	$NO(g)$	+21.60	+90.37
$HI(g)$	+6.20	+25.9	$NO_2(g)$	+8.09	+33.8
$H_2S(g)$	−4.82	−20.2	$HNO_3(l)$	−41.40	−173.2
$HCN(g)$	+31.2	+130.5	$NH_4NO_3(s)$	−87.27	−365.1
$NH_3(g)$	−11.04	−46.19	$NaCl(s)$	−98.23	−411.0
$PH_3(g)$	+2.21	+9.25	$MgO(s)$	−143.84	−601.83
$CH_4(g)$	−17.89	−74.85	$CaO(s)$	−151.9	−635.5
$C_2H_6(g)$	−20.24	−84.68	$Ca(OH)_2(s)$	−235.80	−986.59
$C_2H_4(g)$	+12.50	+52.30	$CaCO_3(s)$	−288.45	−1206.9
$C_2H_2(g)$	+54.19	+226.7	$Ca_3P_2(s)$	−120.50	−504.17
$C_6H_6(l)$	+11.72	+49.04	$BaO(s)$	−133.4	−588.1
$CH_3OH(g)$	−48.08	−201.2	$BaCO_3(s)$	−291.3	−1218.0
$CH_3OH(l)$	−57.02	−238.6	$Al_2O_3(s)$	−399.09	−1669.8
$CH_3NH_2(g)$	−6.7	−28.0	$Fe_2O_3(s)$	−196.5	−822.2
$NF_3(g)$	−27.2	−113.0	$AgCl(s)$	−30.36	−127.0
$CF_4(g)$	−218.3	−913.4	$HgBr_2(s)$	−40.5	−169.0
$CHCl_3(l)$	−31.5	−132.0	$ZnO(s)$	−83.17	−348.0

[a] 1 kcal = 4.184 kJ (exactly).

When a bomb calorimeter is used to make a calorimetric determination, the heat effect is measured at constant volume; ordinarily, reactions are run at constant pressure. The relationship between change in enthalpy and change in internal energy is used to convert heats of reaction at constant volume ($q_v = \Delta E$) to heats of reaction at constant pressure ($q_p = \Delta H$). The appropriate conversion is made by considering the change in volume of the system—the total volume of the reactants subtracted from the total volume of the products. The changes in the volumes of liquids and solids are so small that they are neglected; if only liquids and solids are involved in the reaction, the equation $q_v = q_p$ is sufficiently accurate for most purposes.

For reactions involving gases, however, volume changes may be significant. Let us say that V_A is the total volume of gaseous reactants, V_B is the total volume of gaseous products, n_A is the number of moles of gaseous reactants, n_B is the number of moles of gaseous products, and the pressure and temperature are constant.

$$PV_A = n_A RT \quad\text{and}\quad PV_B = n_B RT$$

Thus

$$P\Delta V = PV_B - PV_A$$
$$= n_B RT - n_A RT$$
$$= (n_B - n_A)RT$$
$$= (\Delta n)RT$$

Therefore,

$$\Delta H = \Delta E + (\Delta n)RT$$

where Δn is the number of moles of gaseous products minus the number of moles of gaseous reactants.

In order to solve problems using this equation, we must express the value of R in appropriate units. As noted previously in this section, litre atmospheres are dimensions of energy; therefore, the value of R in litre atm/°K mol may be converted to cal/°K mol or to J/°K mol (Table 14.2).

Table 14.2. Value of the gas constant, R, in various units.

R	Units
0.082056	Litre atm/°K mol
8.3143	J/°K mol
1.9872	Cal/°K mol

Example 14.1. The heat of combustion at constant volume of $CH_4(g)$ is measured in a bomb calorimeter at 25°C and is found to be –211,613 cal/mol. What is the ΔH?

Solution. For the reaction

$$CH_4(g) + 2O_2(g) \rightarrow CO_2(g) + 2H_2O(l) \quad \Delta E = -211.613 \text{ cal}$$
$$\Delta n = 1 - (2 + 1) = -2$$

Therefore,

$$\Delta H = \Delta E + (\Delta n)RT$$
$$\Delta H = -211,613 \text{ cal} + (-2 \text{ mol})(1.9872 \text{ cal/°K mol})(298.2°K)$$
$$= -211,613 \text{ cal} - 1185 \text{ cal}$$

$$= -212,798 \text{ cal} = -212.798 \text{ kcal}$$

The change in internal energy could have been given in joules;

$$\Delta E = -885,389 \text{ J}$$

$$\Delta H = \Delta E + (\Delta n)RT$$

$$= -885,389 \text{ J} + (-2 \text{ mol})(8.3143 \text{ J/°K mol})(298.2°K)$$

$$= -885,389 \text{ J} - 4958 \text{ J}$$

$$= -890,347 \text{ J} = -890.347 \text{ kJ}$$

BOND ENERGIES

The value of ΔH for the dissociation of a gaseous diatomic molecule into gaseous atoms is a measure of the strength of the covalent bond of the molecule and is called the bond dissociation energy. For example, 118.3 kcal are absorbed when 1 mol of $O_2(g)$ is dissociated into oxygen atoms.

$$O_2(g) \rightarrow 2O(g) \qquad \Delta H = +118.3 \text{ kcal}$$

This value is also called the heat of atomisation of $O_2(g)$.

A bond dissociation energy of a polyatomic molecule refers to a process in which a specific bond of the molecule is broken. The dissociation of the molecule may be imagined to occur in steps; the ΔH values for the steps are not the same For example,

$$HOH(g) \rightarrow H(g) + OH(g) \qquad \Delta H = +119.7 \text{ kcal}$$

$$OH(g) \rightarrow H(g) + O(g) \qquad \Delta H = +101.5 \text{ kcal}$$

The sum of these bond dissociation energies is the heat of atomisation of gaseous H_2O.

$$H_2O(g) \rightarrow 2H(g) + O(g) \qquad \Delta H = +221.2 \text{ kcal}$$

In many instances the individual ΔH values pertaining to the successive dissociations of a polyatomic molecule are not known. However, the heat of atomisation can be calculated by use of the heat of formation of the gaseous compound together with the heats of atomisation of the elements that make up the compound (Example 14.2).

Example 14.2. The bond dissociation energies (or heats of atomisation) of $H_2(g)$ and $O_2(g)$ are +104.2 kcal/mol and +118.3 kcal/mol, respectively, and the heat of formation of $H_2O(g)$ is –57.8 kcal/mol. What is the heat of atomisation of $H_2O(g)$?

Solution. The following equations and corresponding ΔH values are derived from the thermochemical data given.

$$H_2O(g) \rightarrow H_2(g) + 1/2\ O_2(g) \quad \Delta H = +57.8 \text{ kcal}$$

$$H_2(g) \rightarrow 2H(g) \qquad\qquad \Delta H = +104.2 \text{ kcal}$$

$$1/2\ O_2(g) \rightarrow O(g) \qquad\qquad \Delta H = +59.2 \text{ kcal}$$

The sum of these enthalpy changes is the heat of atomisation of $H_2O(g)$.

$$H_2O(g) \rightarrow 2H(g) + O(g) \qquad \Delta H = +221.2 \text{ kcal}$$

An *average bond energy* can be derived from the heat of atomisation of a molecule in which all the bonds are alike. For the O—H bond of the water molecule, the average bond energy is one half the heat of atomisation of the H_2O molecule since there are two O—H bonds in the molecule (221.2/2 = 110.6 kcal/mol). The average bond energy of the N—H bond in the ammonia molecule is one-third the heat of atomisation of NH_3.

The bond energy of a given type of bond is not the same in all molecules containing that bond. However, the variation from compound to compound is small in most cases, and an average value is usually a satisfactory approximation. Hence, a heat of atomisation for a molecule together with established average bond energies can be used to derive a new average bond energy. For example, the hydrogen peroxide molecule (H—O—O—H) contains two H—O bonds and one O—O bond. The heat of atomisation of hydrogen peroxide, which is the energy required to break all three bonds, is 254.3 kcal/mol. Subtraction of 221.2 kcal for the two H—O bonds from 254.3 kcal gives 33.1 kcal/mol as the bond energy of the O—O bond. Average bond energies are given in Table 14.3. The values listed for diatomic molecules are, of course, also bond dissociation energies. Average bond energies may be used to obtain approximate heats of reaction.

Table 14.3. Average bond energies.

Bond	Average bond energy (kcal/mol)	(kJ/mol)	Bond	Average bond energy (kcal/mol)	(kJ/mol)
H—H	104	435	N—Cl	48	201
H—F	135	565	C—C	83	347
H—Cl	103	431	C=C	148	619
H—Br	87	364	C≡C	194	812
H—I	71	297	C—H	99	414
F—F	37	155	C—O	80	335
Cl—Cl	58	243	C=O	169	707
O—O	33	138	C—F	116	485
O_2^a	118	494	C—Cl	78	326
O—H	111	464	C—N	70	293
O—F	44	184	C≡N	210	879
O—Cl	49	205	S—H	81	339
N—N	38	159	S—Cl	66	276
N≡N	225	941	P—H	76	318
N—H	93	389	P—Cl	78	326

^{a} Double bond of molecular oxygen.

SECOND LAW OF THERMODYNAMICS

A chemical reaction or a physical change is said to be spontaneous if it has the potential to proceed of its own accord under the conditions specified. The first law of thermodynamics puts only one restriction on chemical or physical changes—energy must be conserved. However, the first law provides no basis for determining whether a proposed change will occur spontaneously. The second law of thermodynamics establishes criteria for making this important prediction.

The thermodynamic function entropy, S, is central to the second law. Entropy may be interpreted as a measure of the randomness, or disorder, of a system; a highly disordered system is said to have a high entropy. Since a disordered condition is more statistically probable than an ordered one, entropy may be regarded as a probability function. One statement of the second law of thermodynamics is: every spontaneous change is accompanied by an increase in entropy.

As an example of a spontaneous change, consider the mixing of two ideal gases. The two gases, which are under the same pressure, are placed in bulbs that are joined by a stopcock (Fig. 14.1). When the stopcock is opened, the gases spontaneously mix until each is evenly distributed throughout the entire apparatus. Why did this spontaneous change occur? The first law cannot help us answer this question. Throughout the mixing, the volume, total pressure, and temperature remain constant. Since the gases are ideal, no intermolecular forces exist, and neither the internal energy nor the enthalpy of the system is affected.

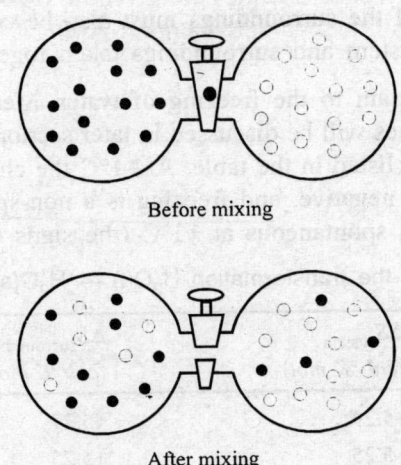

Before mixing

After mixing

Fig. 14.1. Spontaneous mixing of two gases.

This change represents an increase in entropy. The final state is more random and hence more probable than the initial state. The random motion of the gas molecules has produced a more disordered condition. The fact that the gases mix spontaneously is not surprising; one would have predicted it from experience. Indeed, it would be surprising if the reverse were to be observed–a gaseous mixture spontaneously separating into two pure component gases, each occupying one of the bulbs.

For a given substance the solid, crystalline state is the state of lowest entropy (most ordered); the gaseous state is the state of highest entropy (most random); and the liquid state is intermediate between the other two. Hence, when a substance either melts or vapourises, its entropy increases. The reverse changes, crystallisation and condensation, are changes in which the entropy of the substance decreases. Why, then, should a substance spontaneously freeze at temperatures below its melting point, since this change represents a decrease in the entropy of the substance?

If entropy is to be used as a criterion of spontaneity, all the entropy effects that result from the proposed change must be considered. When two ideal gases mix by the process previously described, there is no exchange of matter or energy between the isolated system in which the change occurs and its surroundings; the only entropy effect is an increase in the entropy of the isolated system itself. Usually, however, a chemical reaction or a physical change is conducted in such a way that the system is not isolated from its surroundings, and the total change in entropy is equal to the sum of the change in the entropy of the system (ΔS_{system}) and the change in entropy of the surroundings ($\Delta S_{surroundings}$).

$$\Delta S_{total} = \Delta S_{system} + \Delta S_{surroundings}$$

When a liquid freezes, the heat of fusion is evolved by the liquid and absorbed by the surroundings. This heat increase the random motion of the surrounding molecules and therefore increases the entropy of the surroundings. Hence, the spontaneous freezing of a liquid at a temperature below the melting point occurs because the decrease in entropy of the liquid (ΔS_{system}, negative) is more than offset by the increase in entropy of the surroundings ($\Delta S_{surroundings}$, positive) so that there is a net increase in entropy.

The total change in entropy should always be considered to determine spontaneity. When a substance melts, the entropy increases, but this effect alone does not determine whether or not the transformation is spontaneous. The entropy of the surroundings must also be considered, and spontaneity is indicated only if the total entropy of system and surroundings taken together increases.

The data of Table 14.4 pertain to the freezing of water. Measurement of entropy changes and the meaning of the units of ΔS values will be discussed in later sections. For the moment, let us be concerned only with the numerical values listed in the table. At $-1°C$ the change is spontaneous; ΔS_{total} is positive. At $+1°C$, however, a ΔS_{total} is negative, and freezing is a non-spontaneous change. On the other hand, the reverse change, melting, is spontaneous at $+1°C$ (the signs of all ΔS values would be reversed).

Table 14.4. Entropy changes for the transformation $H_2O(l) \rightarrow H_2O(s)$ at 1 atm.

Temperature (°C)	ΔS_{system} (cal/°K mol)	$\Delta S_{surroundings}$ (cal/°K mol)	ΔS_{total} (cal/°K mol)
+1	−5.29	+5.27	−0.02
0	−5.25	+5.25	0
−1	−5.22	+5.24	+0.02

At $0°C$, the melting point, ΔS_{total} is zero, which means that neither freezing nor melting is spontaneous. At this temperature a water-ice system would be in equilibrium, and no net change would be observed. Note, however, that freezing or melting could be made to occur at $0°C$ by removing or adding heat, but neither change will occur spontaneously.

Thus, the ΔS_{total} of a postulated change may be used as a criterion for whether the change will occur spontaneously. The entropy of the universe is steadily increasing as spontaneous changes occur. Rudolf Clausius summarised the first and second laws of thermodynamics as: "The energy of the universe is constant; the entropy of the universe tends toward a maximum."

Entropy, like internal energy and enthalpy, is a state function. The entropy, or randomness, of a system in a given state is a definite value, and hence, ΔS for a change from one state to another is a definite value depending only on the initial and final states and not on the path between them.

It must be emphasised that, whereas thermodynamic concepts can be used to determine what changes are possible, thermodynamics has nothing to say about the rapidity of change. Some thermodynamically favoured changes occur very slowly. Although reactions between carbon and oxygen, as well as between hydrogen and oxygen, at $25°C$ and 1 atm pressure are definitely predicted by theory, mixtures of carbon and oxygen and mixtures of hydrogen and oxygen can be kept for prolonged periods without significant reaction; such reactions are generally initiated by suitable means. Thermodynamics can authoritatively indicate postulated changes that will not occur and need not be attempted, and it can tell us how to alter the conditions of a presumably unfavoured reaction in such a manner that the reaction will be thermodynamically possible.

GIBBS FREE ENERGY

The type of change of primary interest to the chemist is, of course, the chemical reaction. Determination of ΔS_{system} for a given reaction will be discussed further in this chapter. The $\Delta S_{surroundings}$ for a reaction conducted at constant temperature and pressure may be calculated by means of the equation

$$\Delta S_{surroundings} = -\frac{\Delta H}{T}$$

where ΔH is the enthalpy change of the reaction and T is the absolute temperature. The change in the entropy of the surroundings is brought by the heat transferred into or out of the surroundings because of the enthalpy change of the reaction. Since heat evolved by the reaction is absorbed by the surroundings (and vice versa), the sign of ΔH must be reversed. Hence, the larger the value of $-\Delta H$, the more disorder created in the surroundings and the larger the value of $\Delta S_{surroundings}$.

On the other hand, the change in the entropy of the surroundings is inversely proportional to the absolute temperature at which the change takes place. A given quantity of heat added to the surroundings at a low temperature (where the randomness is relatively low initially) will create a larger difference in the disorder of the surroundings than the same quantity of heat added at a high temperature (where the randomness is relatively high to begin with). Entropy is therefore measured in unit of cal/°K or J/°K. In the last section we noted that

$$\Delta S_{total} = \Delta S_{system} + \Delta S_{surroundings}$$

If $-\Delta H/T$ is substituted for $\Delta S_{Surroundings}$ and if the symbol ΔS (without a subscript) is used to indicate the entropy change of the system, the following equation is obtained.

$$\Delta S_{total} = \Delta S - \frac{\Delta H}{T}$$

Multiplication by T gives

$$T\Delta S_{total} = T\Delta S - \Delta H$$

Since ΔS is measured in units of cal/°K or J/°K, the term $T\Delta S$ is expressed in cal or J; therefore, all of the terms of this equation ($T\Delta S$ as well as ΔH) are energy terms. By reversing the signs of the terms of this equation, we get

$$- T\Delta S_{total} = \Delta H - T\Delta S$$

The Gibbs free energy function, G, is defined by the equation

$$G = H - TS$$

For a change at constant temperature,

$$\Delta G = \Delta H - T\Delta S$$

Hence,

$$\Delta G = -T\Delta S_{total}$$

Since ΔS_{total} is greater than zero for a spontaneous change, $T\Delta S_{total}$ must also be greater than zero, and $- T\Delta S_{total}$ must be less than zero. Therefore, for a spontaneous change

$$\Delta G < 0$$

When a spontaneous change occurs at constant temperature and pressure, the free energy of the system decreases. Rather than use the positive value of $T\Delta S_{total}$ as a measure of spontaneity, the negative value of this quantity is employed so that values of ΔG conform to thermodynamic sign convention (negative values indicating energy liberated by the system). For a system in equilibrium.

$$\Delta G = 0$$

since $\Delta S_{total} = 0$ in such an instance. Notice that all of the terms used in defining ΔG pertain to changes in the properties of the *system*; therefore, the use of free energy values removes the necessity of considering changes in the surroundings.

What is the driving force of a chemical reaction? The earliest answer to the question was that reactions proceed because the reactants have a chemical affinity for each other: this "explanation" is still valid but unfortunately not very illuminating. In an effort to relate chemical affinity to some measurable quantity, Julius Thomsen and Marcellin Berthelot proposed that chemical reactions proceed spontaneously only if they evolve heat; presumably the heat evolved measures the chemical affinity of the reactants in going from a more energetic to a lesser energetic (more stable) state. Most chemical reactions are exothermic; however, the hypothesis of Thomsen and Berthelot is not tenable because spontaneous endothermic reactions are known and many spontaneous physical processes absorb heat (e.g., the melting of ice at room temperature).

The flaw of the hypothesis of Thomsen and Berthelot is that it ignores the role of entropy. The materials of a chemical reaction do indeed seek a minimum in energy, but they also seek a maximum in randomness (entropy). At times, these two factors work together; at other times, they oppose one another. The change in free energy takes into account both factors

$$\Delta G = \Delta H - T\Delta S$$

A system is more ordered at low temperatures than at high temperatures; as the temperature increases, the randomness increases. The factor $T\Delta S$ includes both temperature and change in entropy.

The most favourable circumstance for a negative value of ΔG, which indicates a spontaneous reaction, is a negative value of ΔH together with a positive value of ΔS (reaction 'a' in Table 14.5). However, a large negative value of ΔH can outweigh an unfavourable entropy change, resulting in a negative value of ΔG (reaction 'b' in Table 14.5). In addition, a large positive value of $T\Delta S$ can overshadow an unfavourable enthalpy change, giving rise to a negative value of ΔG (reaction 'c' in Table 14.5). For most chemical reactions at 25°C and 1 atm, the absolute value of ΔH is much larger than the value of $T\Delta S$; under these conditions exothermic reactions are usually spontaneous no matter how the entropy changes

Table 14.5. Thermodynamic values for some chemical reactions at 25°C and 1 atm (kcal).

	Reaction	ΔH	–	$(T\Delta S)$	=	ΔG
(a)	$H_2(g) + Br_2(l) \rightarrow 2HBr(g)$	−17.32	–	(+8.13)	=	−25.45
(b)	$2H_2(g) + O_2(g) \rightarrow 2H_2O(l)$	−136.64	–	(−23.25)	=	−113.39
(c)	$Br_2(l) + Cl_2(g) \rightarrow 2BrCl(g)$	+7.02	–	(+7.45)	=	−0.43
(d)	$2Ag_2O(s) \rightarrow 4Ag(s) + O_2(g)$	+14.62	–	(+9.44)	=	+5.18

However, with increasing temperature the value of $T\Delta S$ increases and the influence of the entropy effect on ΔG increases. Both ΔH and ΔS usually do not change greatly with increasing temperature. However, the term $T\Delta S$ includes the temperature itself so that at high temperatures $T\Delta S$ can be sufficiently large to be the dominant influence on ΔG. For example, consider reaction 'd' in Table 14.5 (a typical decomposition reaction) for which both ΔH and ΔS are positive. If we assume that ΔS does not change with increasing temperature (the actual change is small), at 300°C the value of $T\Delta S$ would be +18.94 kcal because of the increase in the value of T; therefore, $T\Delta S$ would be larger than ΔH (which also does not change greatly with increasing temperature), and ΔG would be negative.

The values in Table 14.5 pertain to the differences between the free energy of the products and the free energy of the reactants at 25°C and 1 atm. However, in some cases (notably reaction 'c') the reaction goes to an intermediate, equilibrium state rather than to completion because the free energy of the equilibrium state is lower than the free energy of the state at which the reaction is complete. Until equilibrium is discussed; we should refrain from drawing conclusions as to the degree of completion of a reaction from such data.

MEASUREMENT OF ΔG AND ΔS

Any spontaneous reaction has the capacity to do work. If work must be done on a system to bring about a chemical change, that change is not spontaneous. For a spontaneous reaction conducted at constant temperature and pressure, the decrease in Gibbs free energy is a measure of the maximum useful work that the system can do. The word useful is used to indicate that ΔG does not include pressure–volume work.

Some reactions are accompanied by an increase or decrease in volume (because gases are produced or used). If such reactions are run at constant pressure, work is done by the system to push back the atmosphere (in the case of an expanding volume) or work is done on the system by the surroundings (if the volume of the system contracts as the reaction proceeds). Such pressure-volume work must be done if the reaction is to take place at constant pressure, and therefore work of this type is not available for any other purpose. The change in free energy measures the maximum amount of work, other than pressure-volume work, that can be derived from a reaction under conditions of constant pressure. Since only a spontaneous reaction has the capacity to do work, ΔG is a measure of spontaneity. Consider the following spontaneous reaction

$$2Ag(s) + Cl_2(g) \rightarrow 2AgCl(s) \qquad \Delta H = -60.72 \text{ kcal}$$

When this reaction occurs at 25°C and 1 atm, 60.72 kcal of heat is evolved, but no useful work is done. One way to obtain useful work from the reaction is to make it produce electrical work by having it serve as the cell reaction of a voltaic cell. However, in the normal operation of a spontaneously discharging cell, less than the maximum amount of useful work is obtained because of the internal resistance of the cell and the concentration changes within the cell that occur when it delivers current.

We can, however, use the value of the reversible emf of the cell (which can be derived from standard electrode potentials) to calculate the maximum amount of useful work that the cell can theoretically produce. For the reaction under study,

anode:	$2Ag(s) + 2Cl^-(aq) \rightleftharpoons 2AgCl(s) + 2e^-$	$E_{ox} = -0.222$ V
cathode:	$2e^- + Cl_2(g) \rightleftharpoons 2Cl^-(aq)$	$E = +1.359$ V
cell reaction:	$2Ag(s) + Cl_2(g) \rightarrow 2AgCl(s)$	$E = +1.137$ V

The reversible emf of the cell, +1.137 V, is the maximum voltage that the cell can develop.

The maximum quantity of electrical work that can be obtained from the reaction, which is the decrease in Gibbs free energy for the reaction, can be calculated from

$$\Delta G = -nFE$$

where n is the number of moles of electrons transferred in the complete reaction (or the number of faradays), F is the value of the faraday, and E is the reversible emf of the cell in volts. If 96,487 coulombs is used as the value of the faraday, the answer is expressed in volt-coulombs, which are joules. On the other hand, the faraday also equals 23.061 kcal/V (since 1 J = 4.1840 cal), and if this value is used, the answer is expressed in kcal. For the complete reaction of silver and chlorine, 2 faradays of electricity is produced, and therefore

$$\Delta G = -2(23.06 \text{ kcal/V})(+1.137 \text{ V})$$

$$= -52.44 \text{ kcal}$$

Gibbs free energy, like the other thermodynamic functions we have discussed, is a state function; the value of ΔG depends only upon the final and initial states of the system and not upon the path taken between those states. When a spontaneous reaction occurs, the free energy of the system declines. Even though the reaction is conducted in such a way that no useful work is obtained from it, the possibility of getting the work defined by ΔG is lost. The value of ΔG for the reaction at 25°C and 1 atm is –52.44 kcal no matter how the reaction is conducted. Therefore, for the reaction

$$2Ag(s) + Cl_2(g) \rightarrow 2AgCl(s)$$

at 25°C and 1 atm, ΔH is –60.72 kcal, and ΔG is –52.44 kcal. From these two values we can derive a value for the entropy change of the reaction.

Since

$$\Delta G = \Delta H - T\Delta S$$

$$-52.44 \text{ kcal} = -60.72 \text{ kcal} - T\Delta S$$

$$T\Delta S = -8.28 \text{ kcal}$$

But

$$T = 298°K; \text{ therefore,}$$

$$\Delta S = \frac{T\Delta S}{T}$$

$$= \frac{-8.28 \text{ kcal}}{298°K}$$

$$= -0.0278 \text{ kcal/°K} = -27.8 \text{ cal/°K}$$

The fact that ΔS is negative indicates that the system becomes more ordered (less random) as the reaction proceeds. Notice that a mole of gas is consumed during the course of the reaction; the decrease in the entropy of the system is therefore not surprising.

The relations between these thermodynamic functions are summarised (Fig. 14.2). If the reaction is conducted outside the cell (a), heat equivalent to $-\Delta H$ is evolved. In the case of the ideal, reversible cell (b), the maximum useful work is obtained ($-\Delta G$) and heat equivalent to $-T\Delta S$ is evolved (which arises from the entropy change of the reaction). Since $\Delta G = \Delta H - T\Delta S$,

$$\Delta H = \Delta G + T\Delta S$$

as the figure illustrates. The ideal, reversible cell is an abstraction, not an operating device. When the reversible emf of a cell is measured, the emf of the cell is balanced against an external emf in such a way that no current flows; in this way the maximum voltage that the cell is capable of producing is measured. In an operating cell (c), less than the maximum amount of work is done (w) and an amount of heat greater than $-T\Delta S$ is evolved (q).

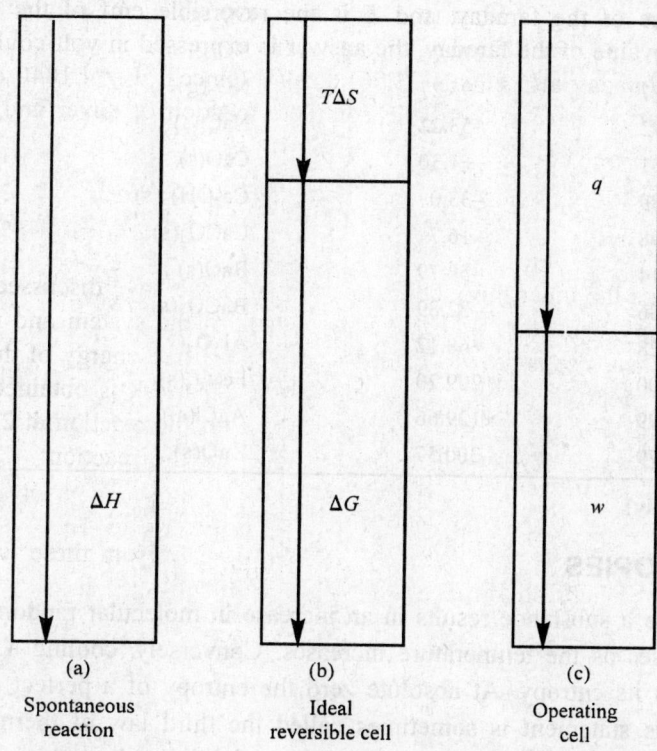

Fig. 14.2. Relations between thermodynamic functions for the reaction $2Ag(s) + Cl_2(g) \rightarrow 2AgCl(s)$ at $25°C$ and 1 atm.

STANDARD FREE ENERGIES

A standard free energy change, which is given the symbol $\Delta G°$, is the free energy change for a reaction at $25°C$ and 1 atm in which the reactants in their standard states are converted to the products in their standard states. The value of $\Delta G°$ for a reaction can be derived from standard free energies of formation in the same way that $\Delta H°$ values can be calculated from standard enthalpies of formation.

The standard free energy of formation of a compound, $\Delta G°_f$, is defined as the change in standard free energies when one mole of the compound is prepared from its constituent elements. According to this definition, the standard free energy of formation of any element is zero. The value of $\Delta G°$ for a reaction is equal to the sum of the standard free energies of formation of the products minus the sum of the standard free energies of formation of the reactants. Some standard free energies of formation are given in Table 14.6. A negative value means that the compound can be prepared from its constituent element under standard conditions.

Table 14.6. Gibbs free energy of formation at 25°C and 1 atm.

Compound	$\Delta G°_f$ (kcal/mol)	$(kJ/mol)^a$	Compound	$\Delta G°_f$ (kcal/mol)	$(kJ/mol)^a$
$H_2O(g)$	–54.64	–228.61	$CO(g)$	–32.81	–137.28
$H_2O(l)$	–56.69	–237.19	$CO_2(g)$	–94.26	–394.38
$HF(g)$	–64.7	–270.7	$NO(g)$	+20.72	+86.69
$HCl(g)$	–22.77	–95.27	$NO_2(g)$	+12.39	+51.84
$HBr(g)$	–12.72	–53.22	$NaCl(s)$	–97.79	–384.05
$HI(g)$	+0.31	+1.30	$CaO(s)$	–144.4	–604.2
$H_2S(g)$	–7.89	–33.0	$Ca(OH)_2(s)$	–214.33	–896.76
$NH_3(g)$	–3.98	–16.7	$CaCO_3(s)$	–269.78	–1128.76
$CH_4(g)$	–12.14	–50.79	$BaO(s)$	–126.3	–528.4
$C_2H_6(g)$	–7.86	–32.89	$BaCO_3(s)$	–272.2	–1138.9
$C_2H_4(g)$	+ 16.28	+68.12	$Al_2O_3(s)$	–376.77	–1576.41
$C_2H_2(g)$	+50.00	+209.20	$Fe_2O_3(s)$	–177.1	–741.0
$C_6H_6(l)$	+30.99	+129.66	$AgCl(s)$	–26.22	–109.70
$SO_2(g)$	–71.79	–300.37	$ZnO(s)$	–76.05	–318.19

^{a} 1 kcal = 4.184 kJ (exactly).

ABSOLUTE ENTROPIES

The addition of heat to a substance results in an increase in molecular randomness. Hence, the entropy of a substance increases as the temperature increases. Conversely, cooling a substance makes it more ordered and decreases its entropy. At absolute zero the entropy of a perfect crystalline substance may be taken as zero. This statement is sometimes called the third law of thermodynamics and was first formulated by Walther Nernst. The entropy of an imperfect crystal, a glass, or a solid solution is not zero at 0°K.

On the basis of the third law, absolute entropies can be calculated from heat capacity data by extrapolating to absolute zero. The standard absolute entropy of a substance, $S°$, is the entropy of the substance in its standard state at 25°C and 1 atm; some $S°$ values are given in Table 14.7. The $\Delta S°$ value for a reaction is equal to the sum of the absolute entropies of the products minus the sum of the absolute entropies of the reactants. Note that the absolute entropy of an element is not equal to zero and that the absolute entropy of a compound is not the entropy change when the compound is formed from its constituent elements.

Table 14.7. Absolute entropy at 25°C and 1 atm.

Substance	$S°$ (cal/°K mol)	$(J/°K mol)^a$	Substance	$S°$ (cal/°K mol)	$(J/°K mol)^a$
$H_2(g)$	31.21	130.6	$HBr(g)$	47.44	198.5
$F_2(g)$	48.6	203.3	$HI(g)$	49.31	206.3
$Cl_2(g)$	53.29	223.0	$H_2S(g)$	49.15	205.6

(Cont'd ...)

Substance	$S°$ (cal/°K mol)	(J/°K mol)^a	Substance	$S°$ (cal/°K mol)	(J/°K mol)^a
$Br_2(l)$	36.4	152.3	$NH_3(g)$	46.01	192.5
$I_2(s)$	27.9	116.7	$CH_4(g)$	44.50	186.2
$O_2(g)$	49.003	205.03	$C_2H_6(g)$	54.85	229.5
S(rhombic)	7.62	31.9	$C_2H_4(g)$	52.45	219.5
$N_2(g)$	45.77	191.5	$C_2H_2(g)$	48.00	200.8
C(graphite)	1.36	5.69	$SO_2(g)$	59.40	248.5
Li(s)	6.7	28.0	CO(g)	47.30	197.9
Na(s)	12.2	51.0	$CO_2(g)$	51.06	213.6
Ca(s)	9.95	41.6	NO(g)	50.34	210.6
Al(s)	6.77	28.3	$NO_2(g)$	57.47	240.5
Ag(s)	10.21	42.72	NaCl(s)	17.30	72.38
Fe(s)	6.49	27.2	CaO(s)	9.5	39.8
Zn(s)	9.95	41.6	$Ca(OH)_2(s)$	18.2	76.1
Hg(l)	18.5	77.4	$CaCO_3(s)$	22.2	92.9
La(s)	13.7	57.3	$Al_2O_3(s)$	12.19	51.00
$H_2O(g)$	45.11	188.7	$Fe_2O_3(s)$	21.5	90.0
$H_2O(l)$	16.72	69.96	HgO(s)	17.2	72.0
HF(g)	41.47	173.5	AgCl(s)	22.97	96.11
HCl(g)	44.62	186.7	ZnO(s)	10.5	43.9

^{a} 1 cal = 4.184 J (exactly).

FREE ENERGY AND CHEMICAL EQUILIBRIUM

The standard free energy change, $\Delta G°$, for the reaction

$$N_2O_4 \ (g) \rightleftharpoons 2NO_2(g)$$

is + 1.29 kcal. Since $\Delta G°$ is positive, one might predict that N_2O_4 in its standard state at 25°C and 1 atm would not dissociate into NO_2 at all and that the reverse reaction, the formation of N_2O_4 from NO_2, would go to completion. Both these predictions are incorrect. The reaction is reversible, and as we have seen, reversible reactions tend to go to equilibrium, not to completion. Furthermore, equilibrium can be approached from either direction; both the forward reaction (to the equilibrium state) and the reverse reaction (to the equilibrium state) should be spontaneous.

The free energy of a system in which this reaction occurs at 25°C and 1 atm is plotted against the fraction of N_2O_4 dissociated in Fig. 14.3. Point A represents the standard free energy of one mole of N_2O_4, point B represents the standard free energy of two moles of NO_2, and the intervening points on this curve represent the free energies of mixtures of N_2O_4 and NO_2 such that a material balance exists. Absolute values of free energies are not known, and no scale is indicated on the vertical axis of the diagram; however, differences in free energies can be calculated so that the shape of the curve is accurately represented.

The free energy curve exhibits a minimum at the equilibrium point, E, where 16.6% of the N_2O_4 is dissociated. The difference between the standard free energy of two moles of NO_2 (point B) and the standard free energy of one mole of N_2O_4 (point A) is $\Delta G°$ for the reaction (+ 1.29 kcal) and is indicated on the diagram. However, ΔG for the preparation of the equilibrium mixture (point E) from one mole of N_2O_4 (point A) is –0.20 kcal, which indicates that N_2O_4 will spontaneously dissociate until equilibrium is reached.

The graph shows that equilibrium can be approached from either direction. Thus, $\Delta G = -1.49$ kcal for the preparation of the equilibrium mixture (point E) from two moles of pure NO_2 (point B). The negative values of ΔG for both changes (from A to E and from B to E) indicate that both changes are spontaneous.

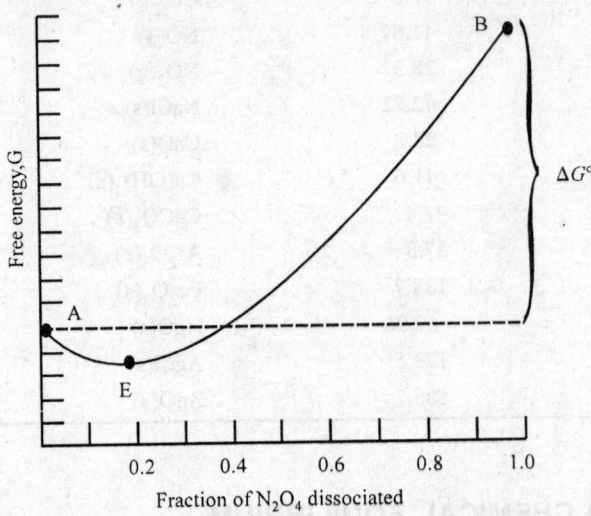

Fig. 14.3. Free energy of a system that contains the equivalent of one mole of N_2O_4 as the reaction N_2O_4 (g) $\rightleftharpoons$ 2 NO_2(g) occurs (25°C and 1 atm).

Why does the free energy curve exhibit a minimum? The value of ΔG for the change from pure N_2O_4 to the equilibrium state, in which 16.6% of the N_2O_4 has dissociated, might be expected to be 0.166 times the $\Delta G°$ value (+1.29 kcal), or +0.22 kcal; instead, ΔG for this change is –0.20 kcal. The value of ΔH for this change (N_2O_4 to equilibrium) is indeed 0.166 times $\Delta H°$, but ΔS for this change is larger than 0.166 times $\Delta S°$. The value 0.166 $\Delta S°$ is equal to the difference between (i) the weighted average of the entropies of pure N_2O_4 in its standard state and pure NO_2 in its standard state (weighted according to the relative amounts of each gas present in the equilibrium mixture) and (ii) the entropy of one mole of N_2O_4 in its standard state.

The entropy of the equilibrium mixture is larger than this weighted average because of the entropy of mixing—an "extra" increment of randomness brought about by the mixing of unlike gases. Notice that neither gas is in its standard state at equilibrium; the partial pressure of each gas is less than 1 atm (the total pressure is 1 atm). The entropy (randomness) of a gas increases when its volume expands and pressure decreases. Since $\Delta G = \Delta H - T\Delta S$, the term $T\Delta S$ is subtracted from ΔH in the calculation of ΔG, and the increase in ΔS over the "expected" produces a minimum in the free energy curve.

One criterion of chemical equilibrium is that $\Delta G = 0$ for a reaction in equilibrium. If we start with one mole of N_2O_4 and allow it to dissociate, the free energy of the system declines until equilibrium is attained. At that point the free energy of the system is a minimum, no further change in free energy is observed, and $\Delta G = 0$. At equilibrium there is no net change in the system that requires work or that can be harnessed to do work. The macroscopic properties of this system do not change with time (the concentrations of all substances present are constant) even though the reaction is proceeding in both directions. If we assign the Gibbs free energy of a mole of NO_2 in its standard state (unit activity) the symbol $G^o_{NO_2}$, the free energy of a mole of NO_2 at another activity, G_{NO_2} is given by the equation

$$G_{NO_2} = G^o_{NO_2} + RT \ln a_{NO_2}$$

where R is the constant, T is the absolute temperature and $\ln a_{NO_2}$ is the natural logarithm of the activity of NO_2. Notice that since the natural logarithm of 1 is zero, $G = G^o$ when the activity of NO_2 is 1, which means that the standard state is characterised by unit activity. For two moles of NO_2,

$$2 G_{NO_2} = 2 G^o_{NO_2} + 2RT \ln a_{NO_2}$$

or

$$2 G_{NO_2} = 2 G^o_{NO_2} + RT \ln (a_{NO_2})^2 \qquad ...(14.1)$$

Similarly, for one mole of N_2O_4,

$$G_{N_2O_4} = G^o_{N_2O_4} + RT \ln a_{N_2O_4} \qquad ...(14.2)$$

The free energy change for the reaction

$$N_2O_4(g) \rightleftharpoons 2NO_2(g)$$

is given by the equation

$$\Delta G = 2 G_{NO_2} - G_{N_2O_4} \qquad ...(14.3)$$

and the standard free energy change is

$$\Delta G^o = 2 G^o_{NO_2} - G^o_{N_2O_4} \qquad ...(14.4)$$

By substituting equations (14.1) and (14.2) into equation (14.3), we get

$$\Delta G = 2 G^o_{NO_2} - G^o_{N_2O_4} + RT \ln (a_{NO_2})^2 - RT \ln a_{N_2O_4}$$

Combining terms gives

$$\Delta G = 2 G^o_{NO_2} - G^o_{N_2O_4} + RT \ln \left(\frac{(a_{NO_2})^2}{a_{N_2O_4}} \right)$$

From equation (14.4),

$$\Delta G = \Delta G^o + RT \ln \left(\frac{(a_{NO_2})^2}{a_{N_2O_4}} \right)$$

In this equation the logarithmic term corrects the standard free energy change, $\Delta G°$, to the free energy change for a more general condition in which the activities of the substances involved are not unity. We shall use the approximation that the activities of gases are given by their partial pressures.

At equilibrium $\Delta G = 0$, and therefore

$$0 = \Delta G° + RT \ln \left(\frac{(p_{NO_2})^2}{p_{N_2O_4}} \right)$$

Since the system is at equilibrium, the fraction in this equation is the equilibrium constant K_p. Therefore,

$$\Delta G° = - RT \ln K_p$$

or

$$\Delta G° = -2.303 \; RT \log K_p$$

where R is the gas constant (1.987 cal/°K mol or 8.3143 J/°K mol) and T is the absolute temperature. For the dissociation of N_2O_4

$$\Delta G° = - 2.303 \; RT \log K_p$$

$$+ 1290 \text{ cal/mol} = - (2.303)(1.987 \text{ cal/°K mol})(298°K) \log K_p$$

$$\log K_p = - 0.946 = 0.054 - 1$$

$$K_p = 1.13 \times 10^{-1} = 0.113$$

The numerical value of K obtained by use of this equation depends upon the definition of the standard states used in the determining $\Delta G°$. Since the standard state of a gas is defined in terms of its pressure (in atm), K_p is obtained in this case. The equation relating $\Delta G°$ with the equilibrium constant is an important one. It may be used to derive values of equilibrium constants from thermodynamic data. In addition, since $\Delta G° = - nFE°$, electrochemical data can be used to calculate equilibrium constants for reactions that can be studied in voltaic cells. This equation also gives us the ability to interpret more fully the meaning of $\Delta G°$ in relation to reaction spontaneity.

Values of K corresponding to various $\Delta G°$ values are listed in Table 14.8. A large negative value of $\Delta G°$ means that K for the reaction is a large positive value, and therefore the reaction from left to right will go virtually to completion. On the other hand, if $\Delta G°$ is large positive value, K will be extremely small, thus indicating that the reverse reaction, from right to left, will go virtually to completion. Only if the value of $\Delta G°$ is neither very large nor very small (Table 14.8) will be value of K indicate a situation in which the reaction will not go essentially to completion in one direction or the other.

Table 14.8. Values of K corresponding to $\Delta G°$ values according to the equation $\Delta G° = -RT \ln K$.

$\Delta G°$		K
kcal	kJ	
−50	−209.2	4.5×10^{36}
−25	−104.6	2.1×10^{18}
−10	−41.84	2.1×10^{7}
−5	−20.92	4600

(Cont'd ...)

ΔG°		
kcal	*kJ*	*K*
−1	−4.184	5.4
0	0	1.0
+1	+4.184	0.18
+5	+20.92	0.00022
+10	+41.84	4.7×10^{-8}
+25	+104.6	4.7×10^{-19}
+50	+209.2	2.2×10^{-37}

Example 14.2. Calculate the standard entropy change of the reaction :

Solution. $Ag_2O\ (s) \longrightarrow 2Ag\ (s) + 1/2\ O_2\ (g)$

The standard entropy change $(\Delta S°)$ of the reaction is given by

$$\Delta S° = S°_{Products} - S°_{Reactants}$$
$$= (2 \times 42.67\ J\ °K^{-1} + 1/2 \times 205.01\ J\ °K^{-1}) - (121.75\ J\ °K^{-1})$$
$$= 66.09\ J\ °K^{-1}\ mole^{-1}$$

Example 14.3. Calculate the standard entropy change of the reaction :

Solution. $N_2\ (g) + O_2\ (g) \longrightarrow 2NO(g)$

The standard entropy change $(\Delta S°)$ for the reaction is given by

$$\Delta S° = S°_{Products} - S°_{Reactants}$$
$$= (2 \times 210.45\ J\ °K^{-1}) - (191.62\ J\ °K^{-1} + 205.01\ J\ °K^{-1})$$
$$= 24.27\ J\ °K^{-1}\ mole^{-1}$$

Chapter 15

Nuclear Chemistry

INTRODUCTION

In this chapter we shall consider certain special changes which directly involve the nucleus of the atom (e.g. radioactivity, atomic transmutations, and nuclear fission and fusion) and the products which are formed as a result of these changes.

NATURAL RADIOACTIVITY

Becquerel, whilst investigating the property of fluorescence, discovered that a crystal of uranium salt spontaneously emitted radiation which could penetrate matter opaque to light and affect a photographic plate. Experiment showed that metallic uranium itself and all its compounds possessed this remarkable property of radioactivity. Soon afterwards it was found that thorium also was radioactive. Pierre and Mme Curie discovered that a mineral of uranium called pitchblende was considerably more radioactive than could be explained from its uranium content. They concluded that another radioactive substance must be present as well and set out to isolate it. Eventually, as a result of prodigious effort under very trying circumstances they obtained specimens of two new and intensely radioactive elements, polonium and radium. The latter is several million times as a radioactive as uranium, mass for mass, but 10^3 kg of pitchblende contains less than 200 milligram of radium. Further another radioactive element, called actinium, was detected in pitchblende.

Investigation of these substances showed that they were invariably associated with other radioactive substances which were formed from them during radioactive changes and which could be separated from them by chemical means. On isolation these products it was found that for each of them the intensity of the radiation emitted decreased with time, but at widely different rates. About fifty naturally occurring radioactive substances are now known, although some of them are only weakly active (like the isotope potassium-40) and others are comparatively rare. Most of them belong to three distinct series known as the uranium-radium series, the thorium series, and the actinium series, which involve isotopes of the twelve elements from thallium to uranium. All isotopes with an atomic number of 84 or above are radioactive.

This phenomenon of radioactivity, although at first regarded as a mere curiosity, soon became a subject of tremendous importance and interest, for few studies have provided more convincing proof of the existence of atoms or thrown more light on their structure.

CHARACTERISTICS OF RADIOACTIVITY

Radioactive substances continually and spontaneously emit penetrating radiation at a rate which, unlike any chemical change, is completely independent of temperature and pressure. This radiation affects a photographic plate, causes gases through which it passes to ionise, and makes certain substances (e.g. crystalline zinc sulphide) fluoresce. It also has powerful physiological effects, some of them cumulative with time. For example, living cells suffer irreparable damage when exposed to intense radiation, and milder doses can cause anaemia or leukaemia in mammals or can induce cancerous growths. Even highly attenuated radiation may have serious genetic consequences. For these reasons great care should always be taken when working with radioactive substances to avoid exposure to dangerous amounts of radiation.

Radioactivity is always associated with the escape of a large amount of energy considering the mass of material taken—the energy released is, weight for weight, about a million times as great as that liberated during any chemical reaction. That it is a property of the atom is shown by the fact that if an element is radioactive, so are all its compounds to a degree which is proportional to the content of the element.

TYPES OF RADIATION

Naturally radioactive substances have been shown to emit three distinct kinds of radiation:

Alpha Rays (α-rays)

Experiment shows these to be fast-moving streams of positively-charged particles, each having a mass four times that of the hydrogen atom and bearing two units of positive charge. Thus each α-particle is, in fact, the nucleus of a helium atom. This has been demonstrated experimentally by collecting α-particles and identifying them as helium spectroscopically. Careful measurements have shown that one gram of radium, together with the decay products naturally associated with it, emits 1.4×10^{11} α-particles per second, providing 0.16 cm^3 of helium gas in a year.

α-rays have very little penetrating power and a sheet of paper or a layer of air about 7–8 cm thick is sufficient to absorb them completely. They do, however, have a very powerful ionising effect upon any gas through which they pass and it is this property which is most used for detecting them and measuring them intensity. Textbooks of physics should be consulted for details of the various electrical counting devices based upon this principle. It must suffice here to mention the important cloud chamber method developed by Wilson for detecting the actual paths followed by individual α-particles. In this apparatus the α-particle is allowed to pass through a gas which has just previously been supersaturated with water vapour. The ions formed in the track of the α-particle act like dust in serving as centres for the condensation of the water vapour, so that the path of the particle is revealed in much the same way that a vapour trail makes visible the course of a high-flying aircraft too small to be seen with the naked eye. The tracks persist long enough to be photographed so that a permanent record can be obtained of the movement of each α-particle through the gas.

Beta Rays (β-rays)

Investigation shows that these are very fast-moving streams of electrons. They are about a hundred times as penetrating as α-rays, but they are much less effective in ionising gases.

Gamma Rays (γ-rays)

These are a form of electromagnetic radiation and consequently they travel with the speed of light. They resemble X-rays, but have an even shorter wavelength. They possess very great powers of penetration. For example, to completely absorb the γ-radiation emitted by a sample of radium requires a layer of lead about 5 cm thick or correspondingly greater thicknesses of other substances, since absorption is roughly proportional to the density of the material.

The properties of these three forms of radiation are compared and contrasted in Table 15.1, and their separation illustrated in Fig. 15.1.

Table 15.1. Types of radiation.

	α-Rays	β-Rays	γ-Rays
Nature	Helium nuclei, He^{2+}	Electrons	Electromagnetic radiation
Velocity	About $\frac{1}{20}$ th speed of light	Varies widely—up to speed of light	Speed of light
Mass	4 units	$\frac{1}{1836}$ th of a unit	None
Charge	2 units (positive)	1 unit (negative)	None
Magnetic field	Deflected one way	Strongly deflected the opposite way	Not deflected
Electric field	Deflected one way	Strongly deflected the opposite way	Not deflected
Fluorescent screen	Scintillates	Scintillates	Scintillates
Photographic plate	Affected	Affected	Affected
Effects on gases	Rapid ionisation	Ionisation	Ionisation
Relative penetration	1	ca. 100	5,000–10,000

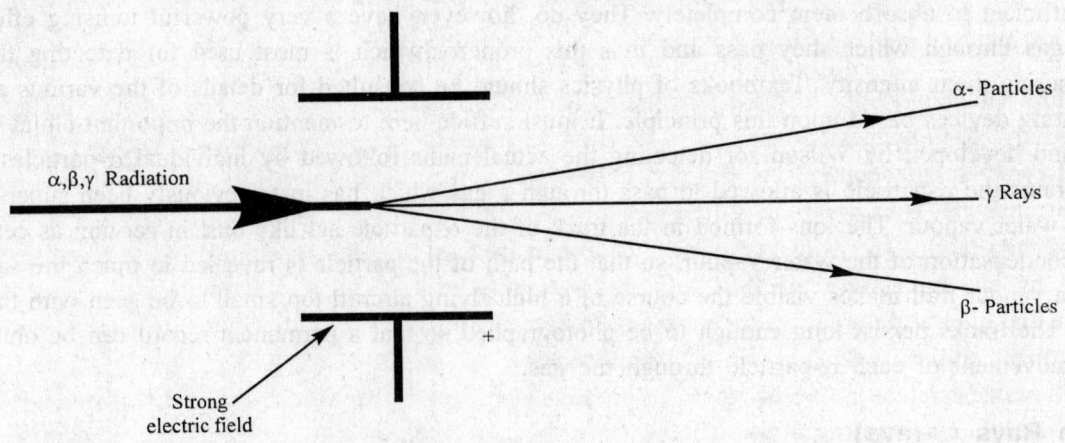

Fig. 15.1. Separation of the three types of atomic radiation.

SCATTERING OF α-PARTICLES

If thin sheets of metal foil are bombarded with α-particles, results of great importance are obtained which have been of tremendous help in arriving at our present theories of atomic structure. It is found, provided the foil is thin enough, that most of the α-particles pass straight through it, but that some are deflected through small angles and a very small proportion are 'reflected' back from the foil (Fig. 15.2). It was these few remarkable deflections through large angles which led Rutherford to conclude that atoms must largely consist of empty space with their mass concentrated into a very small positively-charged central particle which he called the nucleus. From his measurements on the deflected α-particles he calculated that in any case the radius of the nucleus was less than 10^{-14} m, i.e. less than 1/10,000th of the radius of the atom as a whole.

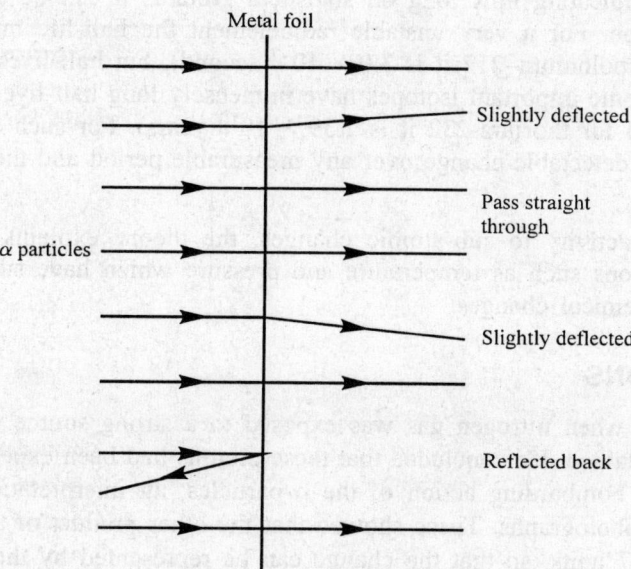

Fig. 15.2. The scattering of α-particles.

EXPLANATION OF RADIOACTIVITY

Rutherford and Soddy put forward the theory of nuclear disintegration to explain the phenomenon of radioactivity. They suggested that in a radioactive element the atomic nucleus was unstable and tended to disintegrate spontaneously (i.e. without the necessity of excitation of any kind) into the nucleus of a different element. Thus when an atom emits an α-particle it changes into another with atomic number two less than before and with atomic weight four units less; emission of a β-particle raises the atomic number by one but hardly affects the atomic weight. The new element formed in this way is itself unstable and after an interval which may vary from a few microseconds to millions of years it undergoes further disintegration with the release of more energy. A series of changes may occur, therefore, until finally a stable nucleus is produced. By reasoning along these lines it was predicted that the uranium thorium series of naturally radioactive substances would both end in the production of stable isotopes of lead of weight 206 and 208 respectively. The experimental confirmation of these predictions was one of the triumphs of the disintegration theory.

The theory also solved the problem of the source of the enormous amount of energy released during radioactivity as this was now seen to come from the nucleus of the unstable atom. Exactly when any given atom disintegrates releasing its excess or energy is purely a matter of chance and defies prediction, but clearly for any one element the number of nuclei breaking up per minute, say, is likely to be a fixed proportion of the number of unstable atoms present. This explains the experimental observation that the intensity of radiation from radioactive substances decreases exponentially with time, since as the nuclei disintegrate fewer and fewer unstable atoms remain. The rate at which radiation decays varies very widely from one radioactive substance to another and can therefore be used to characterise a substance. It is usually expressed in terms of its half-life, which is the time taken for the intensity of radiation to fall to half of its original value, or, what amounts to the same thing, the time taken for half of the total number of atoms in the sample to disintegrate. It may be regarded as an 'expectation of life' for any one atom of that substance, indicating how long on statistical grounds it can be expected to survive before undergoing disintegration. For a very unstable radioelement the half-life may be measured in fractions of a second (e.g. for polonium–212 it is 3.0×10^{-7} second), but half-lives of hours, days, or years are more common and some important isotopes have immensely long half-lives (e.g. for uranium-238 it is 4.51×10^9 years and for thorium-232 it is 1.39×10^{10} years). For such substances as these the rate of emission shows no detectable change over any measurable period and the half-life has to be determined by indirect means.

Lastly, in attributing radioactivity to sub-atomic changes, the theory explains why it is entirely unaffected by physical conditions such as temperature and pressure which have such a marked effect upon the rates of ordinary chemical changes.

ATOMIC TRANSMUTATIONS

Rutherford also observed that when nitrogen gas was exposed to a strong source of α-rays, a supply of fast-moving protons was obtained. He concluded that these protons had been expelled from the nuclei of the nitrogen atoms by the bombarding action of the α-particles, an interpretation which was later confirmed by cloud chamber photographs. These showed that the other product of the collision was an isotope of oxygen weighing 17 units, so that the change can be represented by the equation:

$$^{14}_{7}N + {}^{4}_{2}He = {}^{17}_{8}O + {}^{1}_{1}H$$

where the superscript numbers indicate mass numbers" and the subscript numbers refer to atomic number. This was the first time that an artificially induced atomic transmutation, as the change of the one element into another is called, had been observed, although spontaneous transmutations were continually occurring, of course, in radioactive substances. Artificial transmutations such as this were, however, very difficult to achieve with particles of such low energy and only about one in every quarter of a million α-particles brought about a nuclear change.

It was found that bombardment of certain light elements such as beryllium or boron with α-particles did not give protons but a new type of radiation instead. Chadwick showed that this consisted of a stream of uncharged particles called neutrons with a mass approximately that of a proton, so that the transmutation taking place was as follows:

$$^{9}_{4}Be + {}^{4}_{2}He = {}^{12}_{6}C + {}^{1}_{0}n$$

The neutrons, which are ejected with great speed and are consequently called fast neutrons, are extraordinarily penetrating and will pass through up to 50 cm of lead. They are not easily detected because being uncharged they have little or no ionising action upon gases through which they pass.

Many hundreds of different atomic transmutations can now be carried out by bombarding various elements with fast-moving atomic particles such as neutrons, protons, deuterons (nuclei of deuterium atoms), and α-particles. In general, neutrons, especially slow-moving ones, are the most useful projectiles because they are so readily available in the atomic pile and because, being uncharged, they are not repelled electrostatically by the nuclei of atoms with high atomic numbers. More advanced books should be consulted for details of these changes and it must suffice here to stress three important points. Firstly, in every case there is a change in either the mass or the charge (and often in both) of the nucleus of the atom undergoing bombardment. Secondly, many of the products formed are isotopes which do not exist naturally; some 700, most of them radioactive, have now been made artificially in this way. Here is the near-fulfilment of the alchemist's dream. Admittedly it is not economical to make gold in this way, but many of these radioactive isotopes have proved to greater value than gold in view of their industrial importance and uses, some of which are described. Lastly, by suitable transmutations we can make atoms with atomic numbers above 92, thereby synthesizing new elements not previously known to man. In the last 50 years with the help of powerful cyclotrons eleven such elements have been made, neptunium, plutonium, americium, curium, berkelium, californium, einsteinium, fermium, mendelevium, nobelium, and lastly, lawrencium, although as yet some of them exist only in exceptionally small quantities. These transuranic elements, as they are called, are members of the action on series and are therefore very similar chemically. They are all radioactive with half-lives which are short compared with the age of the earth, so that even if they did occur naturally at one time, they have long since disintegrated into lighter, stabler substances.

GROUP DISPLACEMENT LAW

If an atom of a radioactive element emits an α-particle, its mass decreases by four units and its nuclear charge by two units, since the escape of the α-particle removes two protons and two neutrons. If a β-particle is lost there is negligible change in mass, but the positive charge on the nucleus increases by one owing to the loss of a negatively-charged electron from the nucleus. Thus the loss of an α-particle causes an element to change into another with an atomic number two less than before and occupying a space two places before in the Periodic Table, whereas the loss of a β-particle causes an element to change into another with an atomic number one higher than before. This Group Displacement Law, as it is called, can be illustrated by considering the natural decay of uranium-235, the first three stages of which are:

$$\underset{\text{uranium-235}}{^{235}_{92}\text{U}} = \underset{\text{thorium-231}}{^{231}_{90}\text{Th}} + \underset{\alpha\text{-particle}}{^{4}_{2}\text{He}} \qquad \text{(half-life: } 7.1 \times 10^8 \text{ years)}$$

$$\underset{\text{thorium-231}}{^{231}_{90}\text{Th}} = \underset{\text{protactinium-231}}{^{231}_{91}\text{Pa}} + \underset{\beta\text{-particle}}{^{0}_{-1}\text{e}} \qquad \text{(half-life: } 25.6 \text{ hours)}$$

$$\underset{\text{protactinium-231}}{^{231}_{91}\text{Pa}} = \underset{\text{actinum-227}}{^{227}_{89}\text{Ac}} + \underset{\alpha\text{-particle}}{^{4}_{2}\text{He}} \qquad \text{(half-life: } 3.4 \times 10^4 \text{ years)}$$

These changes can be summarised in the following way :

$$^{235}_{92}\text{U} \xrightarrow[\text{(7.1} \times 108 \text{ yrs.)}]{-\alpha} {^{231}_{90}\text{Th}} \xrightarrow[\text{(25.6 hrs.)}]{-\beta} {^{231}_{91}\text{Pa}} \xrightarrow[\text{(3.4} \times 104 \text{ yrs.)}]{-\alpha} {^{227}_{89}\text{Ac}}$$

Similarly the 238 isotope of uranium undergoes radioactive decay with the loss of an α-particle and the production of thorium-234; this is an extremely slow change, the half-life being about 4.5×10^9 years

$$\underset{\text{uranium-238}}{^{238}_{92}\text{U}} = \underset{\text{thorium-234}}{^{234}_{90}\text{Th}} + \underset{\alpha\text{-particle}}{^{4}_{2}\text{He}}$$

It should be noted that for nuclear changes such as these there must be a balance with respect to mass number and atomic number on both sides of the equation. The mass number, which is written as a superscript, is the relative weight of the atom to the nearest whole number and is equal to the sum of the neutrons and protons in the nucleus.

NEUTRON-PROTON RATIO

For the lighter elements with atomic number below twenty, the numbers of neutrons and protons present in the nucleus of the stable isotopes is either identical or very similar and the neutron–proton ratio is near unity. In elements of higher atom number, however, the most stable nuclei contain more neutrons than protons and the neutron–proton ratio rises steadily with atomic number as shown in Fig. 15.3. From this Fig. it is clear that the ratio of neutrons to protons bears a definite relationship to the stability of the nucleus and for stable atoms varies between unity for the elements of low atomic number and a value of about 1.5 for those with atomic number around eighty. Atoms with a neutron-proton ratio lying outside the band of stability shown in the Fig. tend to be unstable and to undergo radioactive decay.

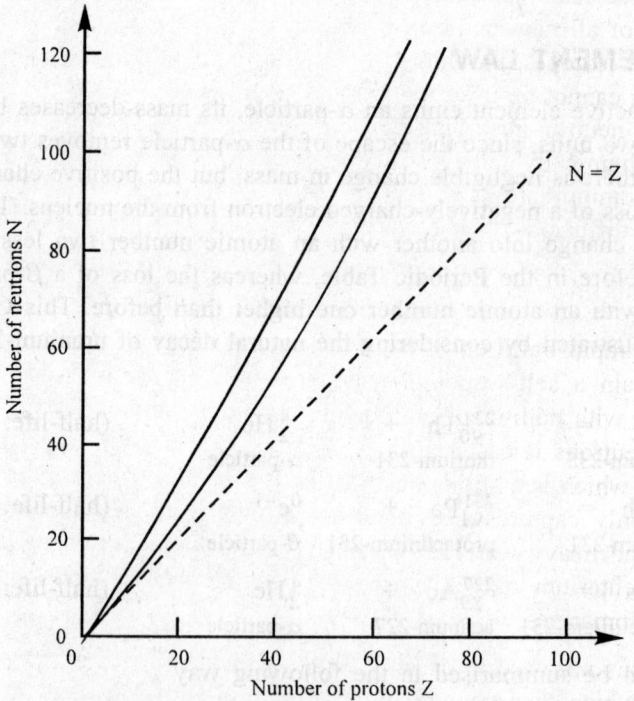

Fig. 15.3. The neutron-proton ratio in stable nuclei.

It is interesting to note that nearly all stable nuclei contain either an even number of protons or an even number of neutrons, or, as is true in the majority of cases, even numbers of both. In fact, only four stable nuclei are known in which the numbers of protons and neutrons are both odd, ^2_1H, ^6_3Li, $^{10}_5B$, and $^{14}_7N$.

NUCLEAR FISSION

When certain heavy atoms (e.g. uranium-235 or plutonium-239) are bombarded with neutrons of suitable energy, their nuclei break up into two smaller nuclei with the release of two or more new fast-moving neutrons and a large amount of energy. This process is known as nuclear fission and the fragments formed, which are intensely radioactive, as fission products. If things are so arranged that at least one of the neutrons expelled during each fission collides with another fissionable atom, then a chain reaction can be set up which will maintain itself indefinitely. This is the principle of the atomic pile or reactor, and also, when the chain reaction builds up explosively quickly, of the so-called atomic bomb. The same amount of energy may be released in both instanced, but in the reactor the rate of nuclear fission is carefully controlled so that the heat evolved can be dissipated as it is produced.

To maintain a chain reaction in some fissile material it is essential to have so much of it present in one lump that only a small proportion of the neutrons escape from the surface without causing further fission. This minimum quantity of fissile material is called the critical mass; for pure uranium-235 it is of the order of 30 kilogram. Thus an atomic bomb can be exploded by suddenly bringing together two portions of the material each weighing just over half the critical mass.

If every neutron released during nuclear fission is to bring about the fission of other atoms so that the chain reaction builds itself up at maximum speed, as it must in an atomic bomb, then the fissile material must be free of all neutron-absorbing impurities. Now ordinary uranium as extracted from its ores consisting of three isotopes, uranium-238 99.28%, uranium-235 0.715%, and uranium-234 0.005%. Thus ordinary uranium cannot be used for making a bomb and it is first necessary to separate the fissile uranium-235 from the neutron-absorbing isotopes. This presents a very difficult problem, since all the isotopes are identical chemically and even their physical differences are only slight. The problem has been solved by converting uranium into its volatile hexafluoride UF_6 and separating the fluorides of the isotopes by means of gaseous diffusion.

ATOMIC PILE

In a nuclear reactor a rapid build up of the chain reaction is not an essential or even desirable feature. It is possible to obtain a self-sustaining chain reaction by using ordinary uranium which has been considerably enriched with additional uranium-235 provided some material is present in the reactor to slow down the fast neutrons released during fission so that not too many of them are captured by the atoms of 238 isotope, which is a strong absorber of fast neutrons, before they can collide with the fissile uranium-235 which only captures slow neutrons readily. The best moderators, as such materials are called, are graphite and heavy water, D_2O, since they are both low absorbers of neutrons. Graphite is cheaper, but when deuterium oxide is used the reactor can be made small enough to be used for propelling ships or submarines.

Most of the neutrons not used in causing further fission are absorbed by the uranium-238 converting it into uranium-239, which then changes into neptunium and plutonium by radioactive decay thus:

$$^{238}_{92}U + ^1_0n \longrightarrow ^{239}_{92}U \xrightarrow{-\beta} ^{239}_{93}Np \xrightarrow{-\beta} ^{239}_{94}Pu$$

In this way plutonium, which is a valuable fissile material for use in bombs and reactors, can be made from uranium in large quantities. After a suitable period the uranium fuel is removed from the reactor and the plutonium recovered by chemical methods, but the intense radioactivity of the plutonium and the various fission products mixed with it makes the whole process hazardous and expensive.

The fissionable material in the reactor is usually contained in metal cans to facilitate loading and removal and to prevent the escape of the highly radioactive fission products. The metal chosen must be strong and easy to machine and weld and it must be a good conductor of heat. It should have as high a melting point as possible and, above all, it must be a low absorber of neutrons. An alloy of magnesium has been widely used, and zirconium alloys and stainless steel are also suitable. At one time beryllium promised to be the best metal for the purpose, but brittleness after cooling eventually led to its rejection.

A coolant is essential in a pile to remove the considerable heat energy produced as a result of nuclear fission; air, water, carbon dioxide, helium, and liquid sodium have all been employed for this purpose. Where the pile is being used for producing power a heat exchanger is fitted so that the heat can be used for generating steam to drive turbines. An important part of a nuclear reactor is the system for preventing the fission process from getting out of control. The usual method is to insert rods of boron steel or cadmium into the pile, since, these are strong absorbers of neutrons and will rapidly bring the chain reactions to a halt. Lastly, the inside of a pile is an intense source of neutrons and γ-rays, so it is necessary to protect the operators by surrounding it with a thick shield of steel and concrete. Unfortunately this rounding it with a thick shield of steel and concrete. Unfortunately this makes it a heavy and very bulky object, ill suited for propelling cars or aircraft.

Apart from its uses for producing plutonium or for generating electric power, a pile can also be used for effecting artificial transmutations. Elements inserted temporarily into the pile are subjected to intense irradiation which converts them into radioactive isotopes, many of which, as described in the next section, are of great industrial and medical importance.

The typical processes taking place in an atomic pile containing enriched uranium are shown diagrammatically in Fig. 15.4.

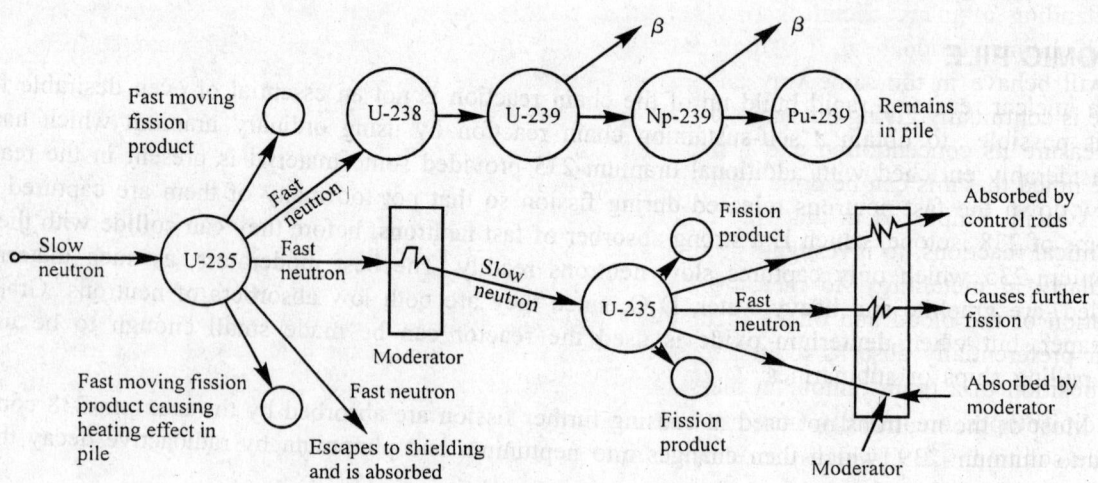

Fig. 15.4. Diagrammatic representation of fission processes in an atomic pile.

If the fissile material in the core of a reactor is surrounded with a layer of uranium-238, then many of the neutrons escaping from the core are captured by the uranium-238 producing uranium-239 which soon decays through neptunium into plutonium-239, itself a valuable fissile material. Similarly, if an outer layer of thorium-232 is used for surrounding the core of a reactor, it is converted into thorium-233 which then decays through protactinium into the highly fissile isotope uranium-233. In these ways it is possible to produce more fissile material than is consumed. Such a system is referred to as a breeder reactor and is of great interest because in effect it makes use of the more abundant isotopes such as uranium-238 and thorium-232 as sources of nuclear energy.

USES OF RADIOISOTOPES

Those isotopes which are intense sources of γ-rays have found extensive use in medicine for the radiation treatment of cancer. In particular cobalt-60 (half-life 5.25 years) has largely replaced the more expensive radium for this purpose. Small pellets of other radioisotopes, e.g. gold-198 (half-life 2.73 days), have been implanted in tumours where their radiation has served to destroy the cancerous cells around them. Iodine-131 (8 days) is especially useful for treating cancer of the thyroid gland because of its natural tendency to accumulate there after ingestion.

Cobalt-60 is also widely used for industrial radiography since it makes possible rapid inspection of welds and castings for cracks or flaws. Other isotopes which are less powerful forces are used for very accurately measuring the thickness of materials such as paper, plastic, or metal foil by measuring the attenuation of their radiation on passing through the material. Such sources are also used for ionising the air and thereby preventing the accumulation of static electricity during industrial processes where there is a risk of fire, and for detecting the position of leaks in buried pipelines, and for sterilising food, drugs, and bandages. Irradiation of certain plastics has been found to improve their physical properties, whilst gentle irradiation of seeds offers a chance of causing genetic mutations of value in agriculture.

Perhaps the most important use of all is as tracers, since this has led to a great advance in our understanding of many chemical and biological changes. Since isotopes of the same element are, in general, chemically identical, if a small proportion of a radioactive isotope is mixed with a stable isotope, both will behave in the same way and participate in exactly the same reactions. Now the radioactive isotope is continually giving out radiation it is possible by means of a sensitive counter to trace its position and measure its concentration and hence follow its course through a series of changes or reactions. It will be noted that this can be done without interfering in any way with the normal processes taking place in the system. Employed in this way radioisotopes can be used to study the mechanisms and kinetics of chemical reactions, to investigate complex changes like photosynthesis, to control industrial processes particularly in metallurgy, to measure the wear caused by friction, and to facilitate a study of the circulation of the blood and of the metabolism of plants and animals. For example, because a growing tumour preferentially absorbs iodine, injected radioactive iodine compounds can be used to locate the exact position of a brain tumour in the skull by means of its radiation. When administered to the body the radioisotopes are usually incorporated in ordinary compounds, e.g. $^{24}_{11}Na$ (15 hours) in sodium chloride solution, $^{33}_{15}P$ (14 days) in phosphates, $^{131}_{53}I$ (8 days) in sodium iodide, $^{14}_{6}C$ (5600 years) as a sugar or amino-acid, $^{3}_{1}H$ (12.3 years) as tritium oxide, and so on.

The $^{32}_{15}P$ isotope of phosphorus has proved particularly useful in studying the metabolism of plants. By using it as a tracer in phosphates, it has been possible to discover at what rate the growing plant absorbs phosphate fertilisers from the soil under various conditions and how the absorbed phosphates are distributed throughout the stems and leaves. Phosphorous-32 has also been used in the treatment of certain types of cancer because it tends to be concentrated by the body at the centres of malignant growth.

The use of carbon-14 for dating organic material and isotopes with even longer half-lives have been used by archaeologists and geologists for estimating the age of rocks and even the earth itself.

NUCLEAR FUSION

From the relative values of the mass defects it is clear that the synthesis of helium from atoms of hydrogen would result in the release of an enormous amount of energy. It is almost certain that transmutations of this kind are the main source of energy in the stars, many of which consist principally of hydrogen and helium. Direct fusion of four hydrogen atoms cannot yet be achieved on earth, but by exploding a fission bomb it has been found possible in the so-called hydrogen bomb to produce high enough temperatures (e.g. 10,000,000°C) locally to enable tritium to fuse with deuterium thus

$$^2_1D + ^3_1H = ^4_2He + ^1_0n$$

The tritium, a radioactive isotope of hydrogen, is made *in situ* by neutron bombardment of lithium deuteride:

$$^7_3Li + ^1_0n = 2^3_1H + ^2_1D$$

Devices of this kind are not subject to any critical size and can therefore be used to produce explosions of terrifying power; one tested in 1961 was equivalent to the simultaneous explosion of 5×10^{10} kg of TNT. Just as the energy provided by nuclear fission can be released in a controlled manner in an atomic pile and used for peaceful purposes, so it may eventually become possible to obtain from thermo-nuclear changes an abundance of energy for the benefit of all mankind.

Example 15.1. The isotope $^{60}_{27}Co$ has a half-life of 5.27 years. What amount of a 0.0100 g sample of $^{60}_{27}Co$ remains after 1.00 year?

Solution. The rate constant for this disinegration is

$$k = \frac{0.693}{t_{1/2}}$$

$$= \frac{0.693}{5.27 \text{ years}} = 0.132 / \text{year}$$

The fraction remaining undecomposed at the end of 1.00 year may be found in the following way

$$\log\left(\frac{N_0}{N}\right) = \frac{kt}{2.30}$$

$$= \frac{(0.132 / \text{year})(1.00 \text{ year})}{2.30} = 0.0573$$

$$\log\left(\frac{N}{N_0}\right) = -0.0573$$

$$\frac{N}{N_0} = 0.876$$

Therefore, the amount remaining after 1.00 years is

0.876 × 0.0100 g = 0.0088 g.

Example 15.2. A sample of carbon from a wooden artifact is found to give 7.00 $^{14}_{6}$C counts per minute per gram of carbon. What is the approximate age of the artifact?

Solution. The half-life of $^{14}_{6}$C is 5770 years. Therefore,

$$k = \frac{0.693}{t_{1/2}} = \frac{0.693}{5770 \text{ years}} = 1.20 \times 10^{-4} \text{ / year}$$

The $^{14}_{8}$C from wood recently cut down decays at the rate of 15.3 disintegrations per minute per gram of carbon. Therefore,

$$\log \left(\frac{N_0}{N} \right) = \frac{kt}{2.30}$$

$$\log \left(\frac{15.3 \text{ disintegrations / min}}{7.00 \text{ disintegrations / min}} \right) = \frac{(1.20 \times 10^{-4} \text{ / year})t}{2.30}$$

$$t = \frac{2.30 \log 2.19}{1.20 \times 10^{-4} \text{ / year}}$$

$$= 6520 \text{ years}$$

Chapter 16

Thermochemistry and Energetics

INTRODUCTION

Thermochemistry is concerned with the heat changes which take place during chemical reactions. Every chemical change is accompanied by the evolution or absorption of heat, although in many case this is not obvious because the amount is so small that no noticeable change in temperature occurs. Reactions which take in heat as they proceed are said to be endothermic, and those which give out heat exothermic.

These heat changes are given the general name of heats of reaction, but it is often found more convenient to classify them and refer to them as heats of formation, combustion, neutralisation, and solution, according to the type of change involved. In every case the convention is followed that heat evolved is negative in sign and heat absorbed is positive. This practice, which is now generally adopted throughout the world, has the advantage that an exothermic reaction is depicted as a negative heat change corresponding to the fall in the heat content of the system; heat is given out and the products contain that much less energy than the reactants did. Similarly, an endothermic reaction, which is regarded as involving a positive heat change, results in a gain in heat energy for the system.

Two other conventions generally adopted in thermochemistry also need explanation. In SI the unit of heat and energy is the joule (J), which is the energy used when a force of one newton acts over a distance of one metre. However, since the heat changes associated with chemical reactions are usually large, amounting to many thousands of joule, and are rarely known to the nearest joule, it is preferable to express them in kilojoule. This larger unit (which, as the name implies, is equal to 1000 joule) is distinguished form the latter by using the abbreviation 'kj'. Secondly, when it is desired to convey information about the heat change which accompanies a reaction, this is usually done by placing the symbol ΔH (which denotes the heat of reaction at constant pressure) at the end of the equation as in the examples quoted in this chapter. The Greek letter Δ (Pronounced 'delta') conventionally stands for the change in a quantity and H for the heat content or enthalpy.

In a reversible reaction the heat change of the forward (i.e. left-to-right) reaction is equal in magnitude but opposite in sign to the heat change of the backward or right-to-left reaction. This statement, which is known as the law of Lavoisier and Laplace after the scientists who discovered it can be deduced from a mush wider generalisation, the law of conservation of energy, since otherwise it would be possible to obtain an unlimited amount of energy merely by repeatedly reversing a reaction in which, more heat was given out than taken in. In reversible thermochemical equations the sign of the heat change is always understood to refer to the forward reaction.

HESS'S LAW

Hess summarised his experimental findings in his law of constant heat summation which states that the heat change in a given reaction depends only upon the initial and final states of the system and is independent of the path followed (provided that heat is the only form of energy to enter or leave the system). This may be illustrated by supposing that a substance A can be converted into another substance D either directly, when the heat of reaction is w joule, or indirectly by way of substances B and C when the heats of reaction of the three stages are x, y, and z joule respectively, thus:

$$A \rightarrow D; \Delta H = -w \text{ J}$$
$$A \rightarrow B; \Delta H = -x \text{ J}$$
$$B \rightarrow C; \Delta H = -y \text{ J}$$
$$C \rightarrow D; \Delta H = -z \text{ J}$$

Then according to Hess's law the heat of reaction in going from A to D is the same whether the change is achieved directly or through a series of separate stages, and $w = x + y + z$.

Hess's law is important because it enables us to add and subtract heat changes algebraically and so calculate heats of reaction which it is impossible to determine directly by experiment. This process is demonstrated in the worked examples given later in the chapter. Like the law of Lavoisier and Laplace, Hess's law can be deduced theoretically form the law of conservation of energy.

The definition of the various terms used in thermochemistry are considered and each of them is illustrated with an example.

Heat of Reaction

This is the amount of heat absorbed or evolved when the quantities of reactants stated in the equation, expressed in mole, react together. For example, in the synthesis of ammonia from its elements, which proceeds according to the equation:

$$N_2 + 3H_2 = 2NH_3; \Delta H = -92 \text{ kJ mol}^{-1}$$

the heat of reaction is 92 kilojoule because that amount of heat is liberated when 28 gramme of nitrogen combine with 6 gramme of hydrogen to form 34 gramme of ammonia. The experimental measurement of heat of reaction is described in and its significance in terms of chemical affinity and energetics is this chapter also discussed.

The heat of a reaction depends not only upon the weight of reagents used, but also upon the temperature and pressure of the reacting system and the physical states of the reactants and products, since latent heats of fusion and evaporation may be involved as well. To overcome this difficulty heats of reaction are usually quoted for 25°C and one atmosphere pressure and where any doubt or ambiguity exists the phases of the substances involved in the reaction are indicated by inserting into the thermochemical equation the symbols g, l,s, and aq, standing for gas, liquid, solid, or dilute aqueous solution respectively. For example, the following equations may be used to represent the reaction of hydrogen with iodine at 25°C and 375°C respectively:

$$H_2(g) + I_2 (s) = 2HI (g); \Delta H = + 11.3 \text{ kJ mol}^{-1}$$
$$H_2 (g) + I_2 (g) = 2HI (g); \Delta H = - 53.6 \text{ kJ mol}^{-1}$$

The heat of sublimation of iodine accounts for about 38 kJ^{-1} mol of the difference, the rest being due to the change in the heat of this reaction with temperature.

The value of a given heat of reaction also depends upon whether the measurement is made at constant volume or at constant pressure. For reactions involving only solids and liquids there is hardly any difference between the two, since the change in volume in such cases is usually negligible, but where the reactants or products are gases, the heat of reaction at constant pressure, ΔH, may differ appreciable from ΔU, the heat of reaction at constant volume. The relation between these two is like that between the two specific heat capacities of a gas, C_p and C_v, since $\Delta H = \Delta U + P\Delta V$, where P is the constant pressure of the system and ΔV is the change in volume that occurs at pressure P. Thus $\Delta H = \Delta U +$ external work done by the system at constant pressure. Since most experiments are carried out at atmospheric pressure, the heat change measured is usually ΔH, but when a bomb calorimeter is used the volume of the system is constant and the result obtained, ΔU, must then be corrected slightly if the heat of reaction at constant pressure is required.

Heat of Combustion

This is the amount of heat evolved when one mole of a given substance is burned in an excess of oxygen. For example, the heat of combustion of methane is 890 kilojoule because this amount of heat is liberated when 16 gram of it are burned completely in oxygen:

$$CH_4 \text{ (g)} + 2O_2 \text{ (g)} = CO_2 \text{ (g)} + 2H_2O \text{ (l)}; \Delta H = -890 \text{ kJ mol}^{-1}$$

The reference to combustion in excess of oxygen is necessary to exclude the possibility of only partial combustion to, for example, carbon monoxide, when the heat evolved would be appreciable less.

The heat of combustion of a substance is important for several reasons. Firstly, it is readily determined experimentally by using bomb calorimeter. This determination is applied industrially to suitable substances such as hydrocarbons in order to find their heating value as fuels; its use as a means of calculating heats of formation is referred to in the next section. Secondly, by comparing the heat of combustion of two different allotropes of an element it is possible to estimate the difference in their energy content. For example, when diamond is burned in excess of oxygen slightly more heat is evolved than form an equal mass of graphite, showing diamond to be the allotrope with the higher intrinsic energy.

The experimental figures are:

$$C \text{ (diamond)} + O_2 \text{ (g)} = CO_2 \text{ (g)}; \quad \Delta H = -395.4 \text{ kJ mol}^{-1}$$
$$C \text{ (diamond)} + O_2 \text{ (g)} = CO_2 \text{ (g)}; \quad \Delta H = -393.5 \text{ kJ mol}^{-1}$$

Applying Hess's law to these results, we conclude that 1900 joule would be absorbed if 12 gram of graphite were changed completely into diamond. Similar energy differences exist between other pairs of allotropes, the more stable form of an element always possessing less energy.

Heat of Formation

This is the amount of heat absorbed or evolved when one mole of any given compound is made from its constituent elements in their normal state. The reference to normal states in necessary because the heat of formation depends to some extent upon the physical conditions of the elements and their allotropic form. For example, the heat of formation of carbon dioxide is −406 kilojoule because that is the quantity of heat which is given out when 44 gramme of it are formed form oxygen and graphite (the stable allotrope of carbon) at 25°C and 101 325 N m^{-2} pressure:

$$C + O_2 = CO_2; \Delta H = -406 \text{ kJ mol}^{-1}$$

Since the heat of formation is defined in terms of one mole of a compound, thermochemical equations are often written for convenience with fractional coefficients before the molecules of reactants thus:

$$H_2 \text{ (g)} + 1/2O_2 \text{ (g)} = H_2O \text{ (l)}; \Delta H = -284 \text{ kJ mol}^{-1}$$

(This means that 284 kilojoule of heat are evolved when 18 gramme of water are formed form 2 gramme of hydrogen and 16 gramme of oxygen at 25°C and atmospheric pressure. Although the use of fractions in this way may seem rather odd at first sight, it is perfectly acceptable. Where, however, it is desired to eliminate fractions by multiplying both sides of the equation by an appropriate factor, the value of ΔH must be increased appropriately. Thus the above equation becomes:

$$2H_2 \text{ (g)} + O_2 \text{ (g)} = 2H_2O \text{ (l)}; \Delta H = -568 \text{ kJ mol}^{-1}$$

and the heat of reaction is twice as great in this case as the heat of formation.

The heat of formation of a compound is a useful guide to tis stability and reactivity. Most compounds are formed exothermically, but some such as acetylene, carbon disulphide, ozone, the oxides of nitrogen, and hydrogen peroxide, have positive heats of formation:

$$\text{e.g. } N_2 + O_2 = \rightleftharpoons 2NO; \Delta H = + 181 \text{ kJ mol}^{-1}$$

In general endothermic compounds (i.e. ones whose formation form their elements involves the absorption of heat) are unstable and highly reactive; when heated many of them decompose readily into simpler substances with the release of energy. The heat of formation of a binary compound is also closely related to the relative electronegativity of its constituent elements. Generally speaking, the greater the difference in electronegativity, the more exothermic the heat of formation and the more stable the compound formed. This explains why heats of formation have been used as a means of comparing and estimating the electronegativities of various elements.

Heats of formation are rarely measured directly, since direct synthesis of a compound form its elements is usually impracticable, but the heat of formation is generally calculated form the heat of combustion of the compound, as shown in the following examples:

Example 16.1. Calculate the heat of formation of carbon monoxide, given that its heat of combustion is –283 kilojoule per mole and that the heat of formation of carbon dioxide is –394 kilojoule per mole.

Solution. Let the heat of formation of carbon monoxide be –Q kJ mol^{-1}

Then
$$C + 1/2O_2 = CO; \Delta H = -Q \text{ kJ mol}^{-1} \quad\quad\quad (1)$$
$$CO + 1/2O_2 = CO_2; \Delta H = -283 \text{ kJ mol}^{-1} \quad\quad\quad (2)$$
$$C + O_2 = CO_2; \Delta H = -394 \text{ kJ mol}^{-1} \quad\quad\quad (3)$$

Now by Hess's law, (1) + (2) = (3)

∴ $Q + 283 = 394$

∴ $Q = 111$ kJ mol^{-1}, so that $\Delta H = -111$ kJ mol^{-1}

i.e. 111 kilojoule of heat would be evolved in forming one mole of carbon monoxide form its elements.

Examples 16.2. Calculate the heat of formation of methane, given that the heats of combustion of carbon, hydrogen, and methane are 393 kJ mol^{-1}, 285 kJ mol^{-1}, and 887 kJ mol^{-1} respectively.

Solution.

$$C + Oz_2 = CO_2; \qquad \Delta H = -393 \text{ kJ mol}^{-1} \tag{1}$$
$$H_2 + 1/2O_2 = H_2O; \qquad \Delta H = -285 \text{ kj mol}^{-1} \tag{2}$$
$$CH_4 + 2O_2 = CO_2 + 2H_2O; \qquad \Delta H = -887 \text{ kj mol}^{-1} \tag{3}$$

Multiplying (2) by 2:

$$\therefore\ 2H_2 + O_2 = 2H_2O; \qquad \Delta H = -570 \text{ kJ mol}^{-1} \tag{4}$$

Adding (1) and (4):

$$\therefore\ C + 2H_2 + 2O_2 = CO_2 + 2H_2O;\ \Delta H = -963 \text{ kJ mol}^{-1} \tag{5}$$

Substracting (3) form (5):

$$C + 2H_2 = CH_4; \qquad \Delta H = -76 \text{ kJ mol}^{-1}$$

Thus the heat of formation of methane is –76 kilojoule per mole.

Example 16.3. Given that the heats of combustion of acetylene, ethylene, and hydrogen are –1310 kJ mol–1, –1393 kJ mol–1, and –285 kJ mol–1 respectively, calculate the heat change for the reaction: $C_2H_2 + H_2 = C2H4$.

Solution.

$$C_2H_4 + 3O_2 = 2CO_2 + 2H_2); \qquad \Delta H = -1393 \text{ kJ mol}^{-1} \tag{1}$$
$$C_2H_2 + 21/2O_2 = 2CO_2 + H_2O; \qquad \Delta H = -1310 \text{ kJ mol}^{-1} \tag{2}$$
$$H_2 + 1/2O_2 = H_2O; \qquad \Delta H = -285 \text{ kJ mol}^{-1} \tag{3}$$

Adding (2) and (3):

$$CH_4 + H_2 + 3O_2 = 2CO_2 + 2H_2O; \qquad \Delta H = -1595 \text{ kJ mol}^{-1} \tag{4}$$

Substracting (1) from (4):

$$C_2H_2 + H_2 = C_2H_4; \qquad \Delta H = -202 \text{ kJ mol}^{-1}$$

Hence 202 kilojoule are given out when one mole of acetylene is converted into ethylene.

Heat of Neutralisation

This is the amount of heat evolved when one mole of H+ ions form an acid reacts with one mole of OH⁻ ions from an alkali. Thus the heat of neutralisation of sodium hydroxide by hydrochloric acid is 57.36 kilojoule because this amount of heat is given out when 40 gram of sodium hydroxide react with 36.5 gram of hydrochloric acid in dilute solution:

$$NaOH\ (aq) + HCl\ (aq) = NaCl\ (aq) + H_2O;\ *H = -57.36 \text{ kJ mol}^{-1}$$

Other examples are:

$$KOH\ (aq) + 1/2H_2SO_4\ (aq) = 1/2K_2SO_4\ (aq) + H_2O;\ \Delta H = -57.94 \text{ kJ mol}^{-1}$$

$$NaOH\ (aq) + HNO_3\ (aq) = NaNO_3\ (aq) + H_2O + H_2O;\ \Delta H = -57.32 \text{ kJ mol}^{-1}$$

$$1/2Ca(OH)_2\ (aq) + HCl\ (aq) = 1/2CaCl_2\ (aq) + H_2O;\ \Delta H = -57.40 \text{ kJ mol}^{-1}$$

$$1/2Ca(OH)_2\ (aq) + HNO_3\ (aq) = 1/2\ Ca(NO_3)_2\ (aq) = H_2O;\ \Delta H = -57.36 \text{ kJ mol}^{-1}$$

$$1/2Ba(OH) + HCl\ (aq) + HCl\ (aq) = 1/2Ba(Cl_3)_2\ (aq) + H_2O;\ \Delta H = -57.36 \text{ kJ mol}^{-1}$$

$$NaOH\ (aq) + 1/2H_2SO_4\ (aq) = 1/2\ Na_2SO_4\ (aq) + H_2O;\ \Delta H = -59.50 \text{ kJ mol}^{-1}$$

These figures demonstrate that in the neutralisation of a strong acid by a strong base the amount of heat liberated is approximately the same in every case. The evidence which this provides in support of the ionic theory where it is explained that the fundamental change common to all neutralisations is believed to be $H^+ + OH^- = H_2O$. On the other hand when weak acids or weak bases are involved the heat of neutralisation is less, as the following examples show:

$$NH_4OH \text{ (aq)} + HCl \text{ (aq)} = NH_4Cl \text{ (aq)} + H_2O; \Delta H = -51.5 \text{ kJ mol}^{-1}$$

$$NH_4OH \text{ (aq)} + HCN \text{ (aq)} = NH_4CN \text{ (aq)} + H_2O; \Delta H = -5.4 \text{ kJ mol}^{-1}$$

The explanation is that these weak electrolytes are only partially dissociated into ions in solution so that when neutralisation takes place some energy is used in drawing apart the undissociated ions before they can react. Thus the heat of neutralisation observed in such cases is only the net heat change and is less than 57.4 kJ mol^{-1} by the amount of energy needed to bring about complete dissociation of the electrolytes into ions.

The basicity of an acid (i.e. the number of hydrogen atoms in one molecule which are replaceable by a metal) can be determined from measurements of the heat evolved during the neutralisation of the acid, provided its molecular weight is known. For example, if 50 cm^3 portions of a molar solution of a dibasic acid are mixed in three calorimeters with 50 cm^3, 10 cm^3, and 150 cm^3 of M alkali solution respectively and the heat change is determined each time, it is found that altogether about twice as much heat is liberated in the second case as in the first, but that with 150 cm^3 of alkali the heat change does not increase further. Hence complete neutralisation of 50 cm^3 of a molar solution of an acid can be accomplished by using 100 cm^3 of a molar solution of alkali, showing the acid to be dibasic.

Heat of Solution

This is the amount of heat absorbed or evolved when one mole of a given substance is dissolved in so much water that further dilution results in no detectable heat change. For example, the heat of solution of anhydrous sodium carbonate is –23 kilojoule because that amount of heat is given out when 106 gram of it are dissolved in a large volume of water:

$$Na_2CO_3 \text{ (S)} + nH_2O = Na_2CO_3 \text{ (aq)}; \Delta H = -23 \text{ kJ mol}^{-1}$$

Some substances dissolve very exothermically (e.g. sodium hydroxide or sulphur trioxide), whilst others (e.g., sodium nitrite) absorb heat when they dissolve. The reason for these wide differences in behavior is that the heat of solution is only the resultant of a number of simultaneous and relatively large heat change. For example, when anhydrous calcium chloride is added to water a certain amount of energy is required to separate the ions from each other in solution (an endothermic change) but a much larger amount of heat is evolved as a result of the hydration of the ions (highly exothermic change). Again, some heat is absorbed during the dilution of the concentrated solution of calcium chloride, so that the heat of solution of the anhydrous salt is the net effect of these various changes and is, in fact, exothermic. On the other hand, crystals of the hexahydrate $CaCl_2.6H_2O$ dissolve endothermically because the ions in this form are already extensively hydrated and the heat liberated on this account does not predominate over the endothermic changes. This examples emphasizes how important it is to specify clearly the degree of hydration of a salt when quoting its heat of solution.

The heat of solution of a substance can be related to the way in which its solubility varies with temperature by applying Le Chatelier's principle to the physical equilibrium existing between a saturated solution and excess of undissolved solid. Where a solid dissolves endothermically in such circumstances one would expect an increase in temperature to cause more of it to dissolve, since in that way heat is taken in and the rise in the temperature of the system is opposed. Thus such a substance would, like most substances, be more soluble at higher temperatures and less soluble at lower temperatures. A word of warning is needed here, however, to guard against the common error when applying the principle in this way of using the heat change that occurs when a solid is dissolved in pure water rather than in its already nearly saturated solution as a guide to its solubility curve. Sodium hydroxide, for example, gives out a great deal of heat when added to pure water and dissolves so exothermically that the water sometimes boils, but it dissolves endothermically in its nearly saturated solution. It is this second heat change that matters when applying Le Chatelier's principle, and when it is used the increase in solubility of sodium hydroxide with rising temperature can be correctly predicted.

MEASUREMENT OF HEAT CHANGES

The method used depends upon the type of change. Where a large heat change such as a heat of combustion is being measured, or where the reactants or products include gases, bomb calorimetry is employed, but heats of solution or neutralisation are best determined by variations of the standard method of mixtures. An example of each method is described below.

The Bomb Calorimeter

This consists of a strong cylindrical steel vessel (Fig. 16.1) lined with enamel to prevent corrosion. If it is desired to determine the heat of combustion of carbon, for example, a known mass (about a gramme) of the element is placed in a platinum crucible inside the bomb, the lid is screwed on tightly, and oxygen is pumped in through a valve until the pressure inside is about 2×10^6 N m^{-2} (20 atm). The bomb is then immersed in a known mass of water in a calorimeter as in Fig. 16.2 and left to attain a steady temperature, which is carefully measured with a Beckmann thermometer to the nearest $\frac{1}{100}$ °C . This is necessary because the combustion may produce a rise in temperature of only a few degrees in the water. The carbon is ignited electrically by passing a current through the platinum coil and the temperature of the water, which is stirred continuously, is recorded at the end of each minute. Readings are continued until the temperature has shown a fall for five consecutive minutes, to enable a cooling correction to be made (Fig. 16.3. Allowance must be made for the thermal capacities of the two calorimeters together with the stirrer and thermometer, and for the heat generated by the ignition current. This gives the heat evolved from the combustion of a known mass of carbon, from which the heat of combustion can easily be calculated.

Method of Mixtures

The apparatus used is shown in Fig. 16.4. The first step is to determine the water equivalent of the calorimeter with stirrer and thermometer by finding the temperature rise which occurs when a known mass of hot water of known temperature is added. If, for example, the heat of neutralisation of a M solution of an acid by a M solution of an alkali is required, then equal volumes of each solution are taken and their initial temperatures are carefully measured. The two solutions are then mixed together in the calorimeter and stirred continuously, the temperature of the mixture being recorded at half-minute intervals until a definite decline is apparent. The readings may be plotted on a graph and a cooling correction applied in the usual way, although when a vacuum flask is used as the calorimeter this correction is hardly necessary because the heat losses are so small. From the results the heat evolved from molar quantities of acid and base can be calculated.

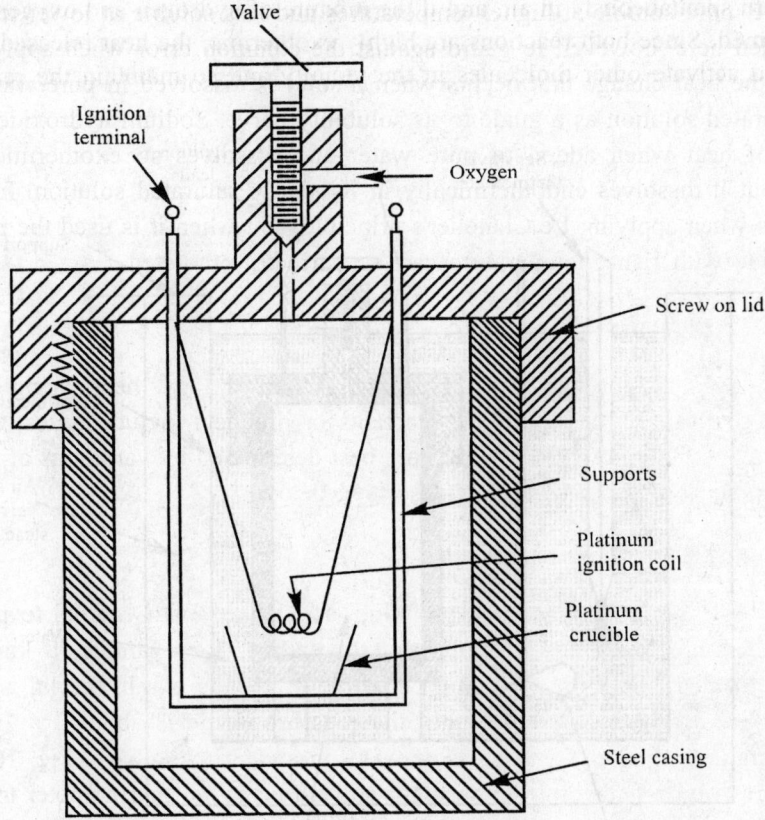

Fig. 16.1. Bomb calorimeter.

ENERGETICS

The natural tendency of any system when left to itself is to assume the lowest possible energy state; this is seen in all branches of science and forms the basis of one of its greatest generalisations, the second law of thermodynamics. The way in which water flows to the lowest possible level and heat flows from a hotter body to a cooler one, and a gas diffuses throughout a vessel, and an electric charge flows from one point to another of lower potential, are common examples of this universal tendency. It is not surprising, therefore, to find that in all chemical reactions which proceed spontaneously there is a loss of energy in the system. In most cases, but by no means in all, there is also a net evolution of heat, i.e. the reaction is exothermic overall.

This raises the problem of why all such reactions do not proceed with vigour and speed. Why, for example, should a lump of coal not burst into flames when exposed to the air, since its combustion to carbon dioxide is a highly exothermic change? Similarly, why is it possible for a mixture of hydrogen and oxygen to remain unchanged in a gas jar apparently indefinitely when the combination of the two gases to form water would result in a system of much lower energy? The accepted explanation is that before two substances can combine together their molecules must posses at the moment of impact at least a certain amount of energy, known as the energy of activation of that reaction. Without this energy the collision is unfruitful and no reaction occurs. Thus if the lump of coal is first heated until it is nearly red-hot it will burn spontaneously in air, and if the mixture of hydrogen and oxygen is ignited or sparked, water will be formed. Since both reactions are highly exothermic, the heat released by the initial changes will then serve to activate other molecules in the vicinity and so maintain the reaction.

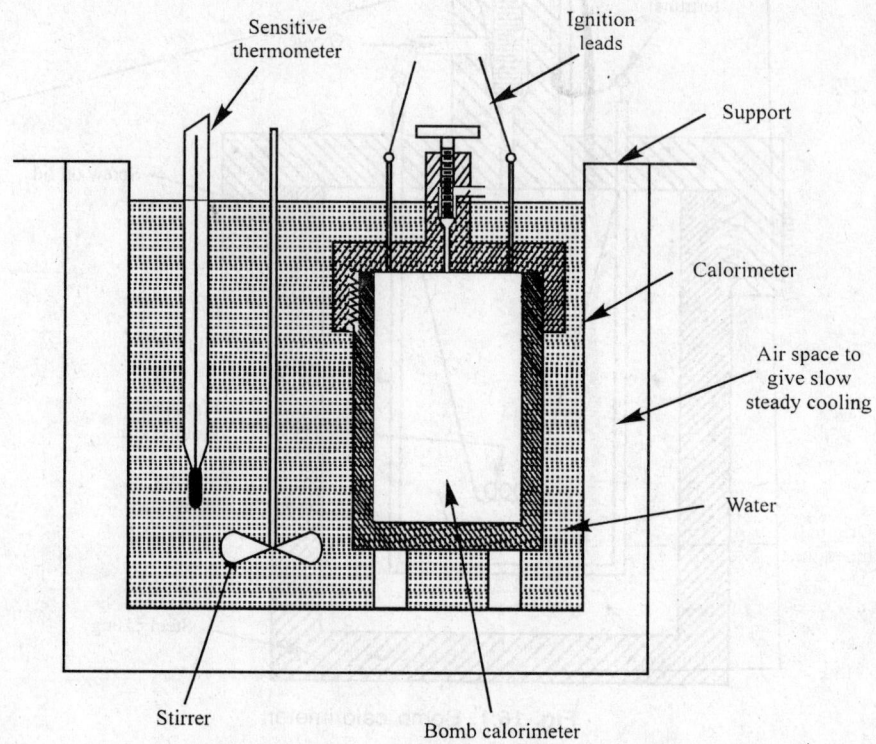

Fig 16.2. Apparatus for determining the heat of combustion.

The energy of activation probably represents the energy needed to pull apart some of the atoms in the original molecules so that they are in a suitable condition to react together. The very high speed of reactions between ions in solution, which indicates their low activation energies, provides evidence in favour of this suggestion, since in these ionic reactions many of the particles are already separated as a result of electrolytic dissociation. Similarly, investigation of certain very fast reactions such as those between methane and chlorine or between hydrogen and chlorine has revealed the presence of an abundance of free radicals in these systems during the reactions, which explains their low energies of activation and their extremely high speeds.

The energy changes occurring during an exothermic reaction are illustrated diagrammatically in Fig. 16.5. The level A represents the average energy of the reactants, B the energy needed before reaction occurs, and C the average energy of the products. The difference between A and B is the energy of activation of the forward reaction E_f, and the difference between B and C is the energy of activation of the backward reaction, E_b. From the diagram it will be seen that the heat of the reaction at constant volume, ΔU, is equal to the difference between these two energies of activation.

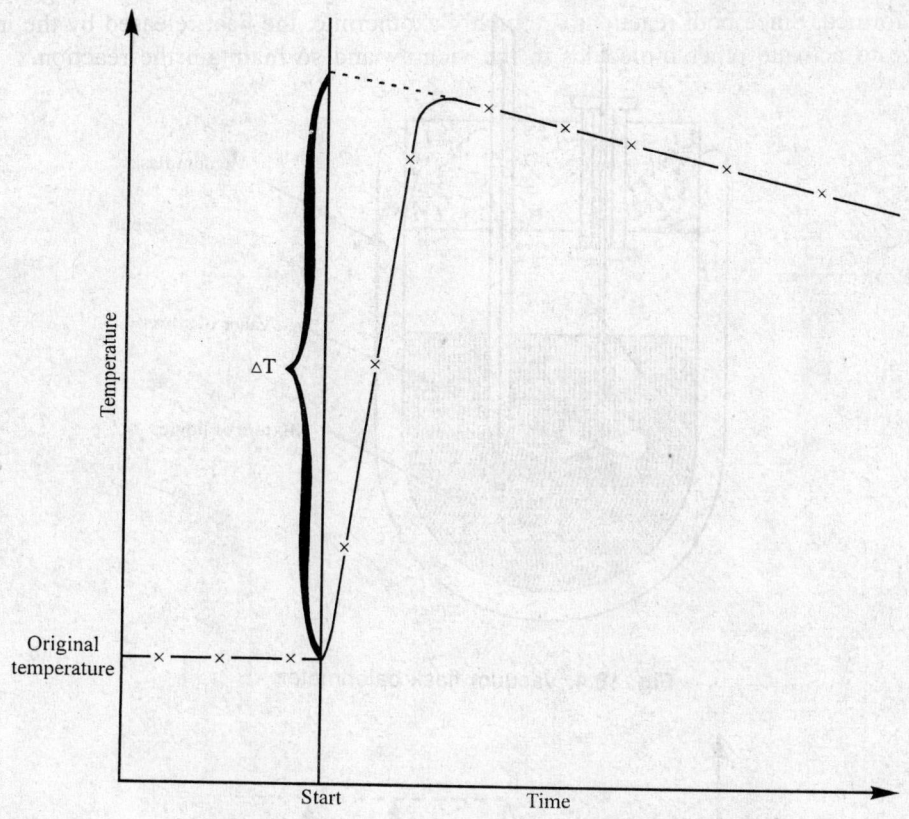

Fig. 16.3. Typical cooling curve correction.

The effect of using a catalyst to accelerate a reaction is shown diagrammatically in Fig. 16.6. The energies of activation of the intermediate stages are usually much smaller than that of the original reaction, so it proceeds comparatively rapidly although the overall energy change is the same.

ENTROPY AND FREE ENERGY

From the fact that the majority of spontaneous reactions (i.e. reactions which under suitable conditions are capable of proceeding of their own accord) are exothermic and result in a decrease in the heat content or enthalpy of the system, it might be concluded that this was the only factor which determined whether any particular reaction is spontaneous or not. In fact, some spontaneous reactions take place endothermically so that heat is absorbed as they proceed and there is a net gain in enthalpy, as the following examples illustrate:

$$N_2 + O_2 \rightleftharpoons 2NO; \Delta H = + 181 \text{ kJ mol}^{-1}$$

$$3O_2 \rightleftharpoons 2O_3; \Delta H = + 284 \text{ kJ mol}^{-1}$$

$$C_{(s)} + H_2O_{(g)} = CO_{(g)} + H_{2(g)}; \Delta H = + 131 \text{ kJ mol}^{-1}$$

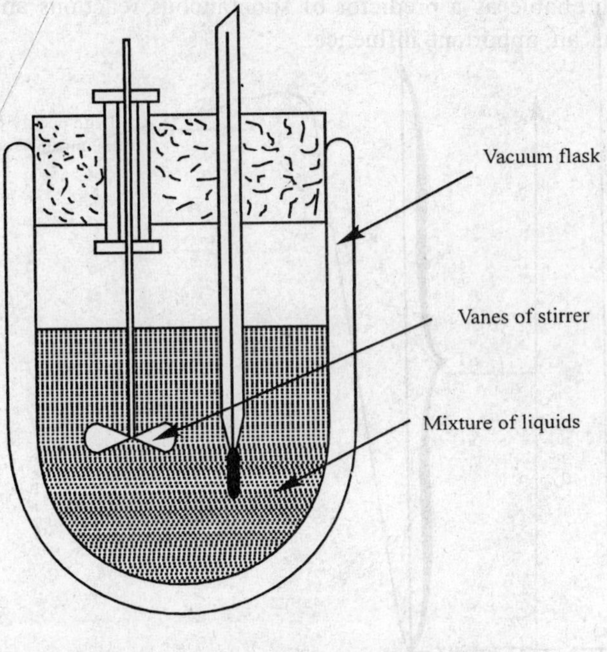

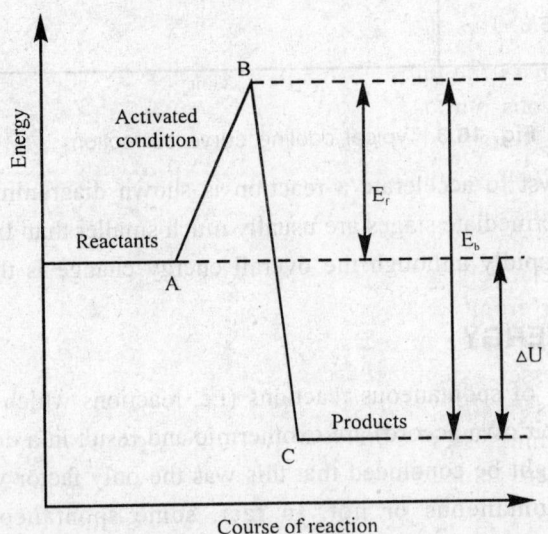

Fig. 16.4. Vacuum flask calorimeter.

Vacuum flask

Vanes of stirrer

Mixture of liquids

Fig. 16.5. Relationship between heat of reaction and energy of activation.

The readiness with which some salts dissolve in water with strong absorption of heat also demonstrates the limitation of exothermicity as a criterion for spontaneous changes. For example, when crystals of ammonium nitrate, ammonium chloride, or sodium nitrite are dissolved in water, so much heat is absorbed that their solutions are noticeably cooled. At first sight all these examples seem contrary to the general tendency to lose energy which was described in the previous section as a characteristic of all spontaneous changes. The explanation of this paradox lies in the realisation that the change in enthalpy , alone is completely unreliable as a predictor of spontaneous reactions and that some other factor must exist which also exerts an important influence.

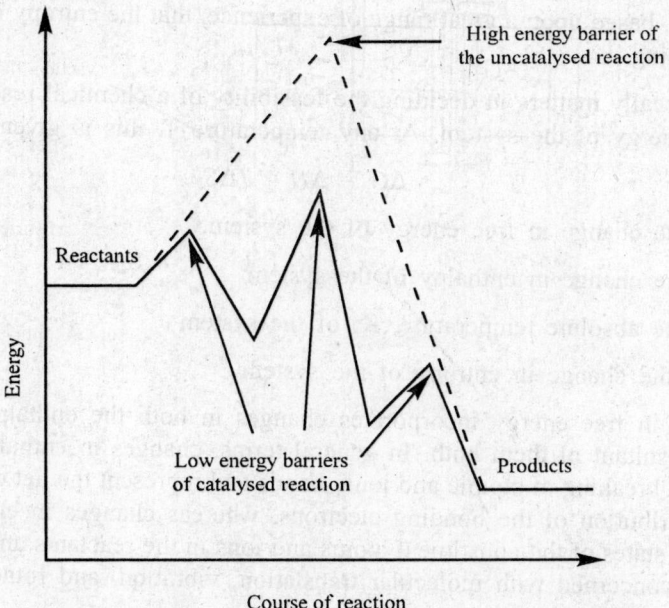

Fig. 16.6. The effect of a catalyst upon energy of activation.

This other factor is known as the entropy of the system. Although entropy is really a sophisticated thermodynamic concept, for our purposes it is sufficient to regard it as a measure of the degree of randomness or disorder of the system. A crystalline substance with a highly ordered and regular structure usually has low entropy, whereas a liquid in which the molecules are arranged in a much less orderly way tends to have a higher molecules are arranged in a much less orderly way tends to have a higher entropy value. In a gas, where the molecules are in continual and random motion and there is very little order at all, the entropy is relatively large. It follows that when any substance undergoes a change of state from solid to liquid or from liquid to vapour, there will be a substantial increase in entropy corresponding to the marked decrease in order and regularity resulting from melting and vapourisation. In the same way but to a lesser extent, the entropy of a substance increases with rising temperature because the constituent atoms or ions undergo greater vibration (in solids) and more rapid motion (in liquids and gases) and this diminishes their orderliness. It is convenient to regard a substance in the crystalline state at absolute zero temperature as having zero entropy because its atoms or ions are then in rigid order and perfect regularity. Its entropy value at any other temperature is known as its molar entropy, S_m, and is expressed in entropy units of joule per kelvin or JK^{-1}, the standard value, $S°_m$, being the entropy value at 25°C and 101 325 N m^{-2} pressure.

Many spontaneous physical and chemical changes involve an increase in entropy. For example, when two pure gases are mixed and allowed to diffuse into each other there is a gain in entropy corresponding to the decrease in order. The molecules become randomly mixed instead of being separated into two distinct samples. Similarly, when a solid dissolves in a liquid the resulting solution has less regularity than the original crystal because the particles become scattered throughout the solution and mix homogeneously with the solvent. Diffusion of liquids, evaporation of a liquid, osmosis, sublimation enantiotropic transition, thermal dissociation and thermal decomposition are other examples where spontaneous changes are accompanied by gains in entropy. Because of this natural tendency in spontaneous physical and chemical changes towards greater disorder and randomness, it is a fundamental belief of all scientists, based upon a great range of experience, that the entropy of the universe as a whole is continually increasing.

The thing which really matters in deciding the feasibility of a chemical reaction is the change which occurs in the free energy of the system. At any temperature T, this is given by the expression

$$\Delta G = \Delta H - T\Delta S$$

where ΔG = the change in free energy of the system.

ΔH = the change in enthalpy of the system

T = the absolute temperature, K, of the system

ΔS = the change in entropy of the system

Thus the change in free energy incorporates changes in both the enthalpy and the entropy of the system and is the resultant of them both. In general terms, changes in enthalpy during a reaction arise from the making and breaking of atomic and ionic bonds and represent the net changes in energy involved in changing the distribution of the bonding electrons, whereas changes in entropy are concerned with the organisation and states of the constituent atoms and ions in the reactants and products and the relative amounts of energy concerned with molecular translation, vibration, and rotation.

Only reactions which bring about a decrease in the free energy of the system can proceed spontaneously, since this fall in free energy is the essential driving force of the reaction. If a reaction is exothermic and also results in greater disorder, then the changes in enthalpy and entropy combine to ensure that there is a large decrease in free energy and such a reaction is fully capable of proceeding of its own accord. On the other hand, reactions which would result in absorption of heat and a loss in entropy are not spontaneous because in such reactions the free energy change ΔG would be positive. When a reaction is endothermic however, or when it involves a decrease in entropy, then it will only be spontaneous if the resultant of the two opposing influences leads to a net decrease in free energy. In such cases the sign of ΔG at any temperature will depend upon which term is greater and therefore represents a balance of opposing influences. At normal temperatures the term involving entropy is not usually large, amounting to only a few kilojoule per mole, but at higher temperatures the product $T\Delta S$ will tend to be much more significant and may even become predominant.

The reaction used industrially for making water gas provides a good example of the balance between enthalpy and entropy:

$$C(\text{graphite}) + H_2O(g) = CO(g) + H_2(g); \quad \Delta H = + 131 \text{ kJ mol}^{-1}$$

This reaction is strongly endothermic, and although there is a gain in entropy in forming the water gas from the reactants (ΔS is positive), the subtracted value of $T.\Delta S$ does not nearly compensate at room temperature for the large gain in enthalpy, so that at room temperature ΔG is positive and the reaction is not spontaneous. However, at temperatures in excess of 700°C the entropy term $T\Delta S$ becomes so substantial that it exceeds the positive value of ΔH. It follows that at these high temperatures ΔG is negative corresponding to a decrease in free energy and the reaction is therefore spontaneous.

Another example of industrial importance is the synthesis of ammonia:

$$N_2(g) + 3H_2(g) \rightleftharpoons 2NH_3(g); \quad \Delta H = -92 \text{ kJ mol}^{-1}$$

Here the formation of ammonia results in a decrease in entropy, so ΔS is negative. At room temperature the enthalpy term predominates over the value of $T\Delta S$ and there is a fall in free energy ΔG, so the reaction is spontaneous and a good yield of ammonia can theoretically be expected. It must be stressed here that considerations of free energy act as a guide to the position of equilibrium only and give no information at all on the rate of reaction, which in this case is infinitely slow at room temperature. Above 600°C the entropy term becomes so significant that it outweighs the change in enthalpy and ΔG becomes positive, so the reaction is not spontaneous above this temperature, a conclusion in keeping with experimental findings.

It should be noted that ΔG is given by the difference between ΔH, the total energy transferred to or from the environment during the reaction to keep the temperature constant, and $T\Delta S$, the energy associated with the internal changes within the system. Thus the change in free energy in a closed system is in effect a measure of the maximum amount of energy which is available to do work outside the system, although in practice in any particular reaction it may not be possible to harness all this free energy to do useful work.

The change in free energy, ΔG, for a particular reaction will vary with temperature because the components from which it is derived are dependent upon temperature. For purposes of comparison it is usual to consider the standard free energy change, ΔG^{ϕ}, which is the change in free energy which occurs with molar quantities of reactants and products at 25°C and at 101 325 Nm^{-2} pressure. For a reversible reaction at constant temperature the standard free energy change ΔG^{ϕ} is related to the equilibrium constant K_v at an absolute temperature T by the expression:

$$\Delta G^{\phi} = -RT \log_e K_v = -2.303\ RT \log_{10} K_v$$

which is known as the van't Hoff Isotherm. Its importance lies in its use for calculating the equilibrium constant at any given temperature from a knowledge of ΔG^{ϕ} or alternatively in determining ΔG^{ϕ} from experimentally measured values of K_v. Thus in a reversible reaction ΔG^{ϕ} and K_p both indicate how far reaction will proceed spontaneously and they define the position of equilibrium which will eventually be attained at that temperature. The greater the negative value of ΔG^{ϕ}, the farther the reaction will go towards completion until eventually a state is reached where no further fall in free energy is possible either by a decrease in enthalpy or by an increase in entropy. At this point ΔG^{ϕ} is zero and the system is in equilibrium under those conditions. It is important to realise, however, that ΔG^{ϕ} and the equilibrium constant give no information at all on the rate of the reaction which may be very slow indeed, depending upon the conditions.

BOND DISSOCIATION ENERGY

As explained in the previous section, the change in enthalpy ΔH represents the net energy change per mole which results from the breaking and making of chemical bonds during a reaction. By determining experimentally the value of ΔH in a series of carefully chosen reactions involving covalent bonds and by using spectroscopic measurements of heats of dissociation of diatomic molecules, it is possible to ascribe definite amounts of energy to the breaking or making of specific covalent bonds. The energy required to break a particular covalent bond under standard conditions (25°C, 101 325 Nm^{-2} pressure, for one mole of substance) and separate the two atoms from each other completely is known as its bond dissociation energy. A list of bond dissociation energies has been compiled, some of which are given in Table 16.1.

Table 16.1. Bond dissociation energies/kJ mol^{-1}.

C—C	343	C—O	339
C=C	611	O—H	460
C≡C	803	N≡N	941
C—H	414	H—H	435

The main value of these bond energies is that they can be added together to give an approximate value of an unknown heat of formation or to calculate a heat of reaction, as the following examples show:

Example 16.4. Calculate the heat of formation of ethane given that the bond dissociation energies of C—C and C—H bonds are 343 kJ mol^{-1} and 414 kJ mol^{-1} respectively, the heat of dissociation of hydrogen is 435 kJ mol^{-1}, and the heat of vapourisation of graphite is 711 kJ mol^{-1}.

Solution.

$$2C \text{ (graphite)} + 3H_2 \text{ (g)} = C_2H_6 \text{ (g)}; \Delta H = ?$$

Now $\qquad\qquad C \text{ (graphite)} \rightarrow C \text{ (g)}; \Delta H = +711 \text{ kJ mol}^{-1}$

and $\qquad\qquad H_2 \text{ (g)} \rightarrow 2H; \Delta H = +435 \text{ kJ mol}^{-1}$

$\therefore \qquad\qquad 2C \text{ (graphite)} \rightarrow 2C \text{ (g)}; \Delta H = +1422 \text{ kJ mol}^{-1}$ ...(16.1)

and $\qquad\qquad 3H_2 \text{ (g)} \rightarrow 6H; \Delta H = +1305 \text{ kJ mol}^{-1}$ ...(16.2)

Thus the total energy required to break bonds in forming a molecule of ethane from graphite and hydrogen is the sum of (16.1) and (16.2), i.e. + 2727 kJ mol^{-1}.

Now the bond energy of C—C is –343 kJ mol^{-1} ...(16.3)

and the bond energy of C—H is –414 kJ mol^{-1}

so the energy for (C—H) bonds is $6 \times -414 = -2484$ kJ mol^{-1}. ...(16.4)

Thus the total energy released by making bonds in a molecule of ethane is the sum of (16.1) and (16.2), i.e. –2827 kJ mol^{-1}.

$\therefore$ the net energy change is $+2727 - 2827 = -100$ kJ mol^{-1}

$\therefore$ the net energy change is $+2727 - 2827 = -100$ kJ mol^{-1}

Hence the heat of formation of ethane is –100 kJ mol^{-1}, i.e. 100 kilojoule of heat are evolved when one mole of ethane is formed from its elements in their standard states.

Example 16.5. From the bond dissociation energies given in Table 16.1, calculate the heat of the reaction:

Solution.

$$C_2H_2 + H_2 = C_2H_4$$

$$H-C{\equiv}C-H + H-H = \quad \begin{matrix} H & & H \\ & C{=}C & \\ H & & H \end{matrix} \quad ; \Delta H = ?$$

Breaking one C≡C bond requires + 803 kJ mol^{-1}

Breaking one C—H bond requires +828 kJ mol^{-1}

Breaking one H—H bond requires +435 kJ mol^{-1}

So the total energy required bonds is – 2267 kJ mol^{-1}.

Making one C=C bond releases –611 kJ mol–1

Making four C—H bond releases –1656 kJ mol^{-1}

Thus the net energy change, ΔH, is –200 kJ mol^{-1}, i.e. 200 kilojoule of heat would be evolved when one mole of ethylene was made by hydrogenation of acetylene.

In calculations of this kind it is assumed that the energy associated with any particular covalent bond is always the same irrespective of the molecule in which it occurs. In practice this is usually true, at least approximately, but sometimes there is a significant difference in bond energy values between different molecules because of a degree of polarity in the residual part of the molecule. For example, the bond dissociation energy in methane, CH_4, is 435 kilojoule for the first C—H bond broken, 433 kilojoule for the second and third, and 339 kilojoule for the fourth. Similar variations in the bond dissociation energy of O—H are found when the water molecule is disrupted. In such cases it is usual in theoretical calculations to accept the average value of bond dissociation energy as the energy needed to break or make the bond in question.

Occasionally heats of formation calculated from the generally accepted values for the bond dissociation energies show a marked discrepancy from the values found experimentally. Where the actual heat of formation of a compound is appreciably higher than the value expected from bond energy considerations, then the explanation is usually that the compound concerned has a resonance structure. Typical examples are the oxides of nitrogen and carbon and aromatic compounds such as benzene and toluene.

LATTICE ENERGY AND THE BORN–HABER CYCLE

The energy liberated when one mole of an ionic crystal lattice is formed from its constituent ions in the gaseous state is known as the lattice energy of that compound. For example, the lattice energy of sodium chloride is the energy involved in the change:

$$Na^+(g) + Cl^-(g) \rightarrow Na^+Cl^-(s); \quad \Delta H = \text{lattice energy}$$

Lattice energy is in effect potential energy stored in the ionic crystal lattice. It is the resultant of all the electrostatic attractions and repulsions between the ions in the three-dimensional structure. It cannot be determined directly by experiment, but it can be calculated theoretically by regarding a lattice simply as a collection of spherical ions, each with its charge evenly distributed, spaced apart at distances corresponding to the ionic radii as determined by X-ray analysis.

Alternatively, lattice energies can be obtained indirectly by applying Hess's law to a cycle of energy changes stage of which can be determined experimentally. Such a cycle, known as the Born–Haber cycle, traces the energy changes involved in converting the elements in their standard atomic states into an ionic lattice by two alternative routes. By adding the energy changes of the various individual stages and comparing the totals for the two parts of the cycle it is possible to deduce a value for the lattice energy. The process is illustrated for sodium chloride in Fig. 16.7. Equating the two overall energy changes for the reaction

$$Na(s) + 1/2 \, Cl_2(g) = Na^+ \, Cl^-(s)$$

gives

$$109 + 121 + 494 - 394 + \text{lattice energy} = -410$$

Hence the lattice energy of sodium chloride = 770 kJ mol^{-1}.

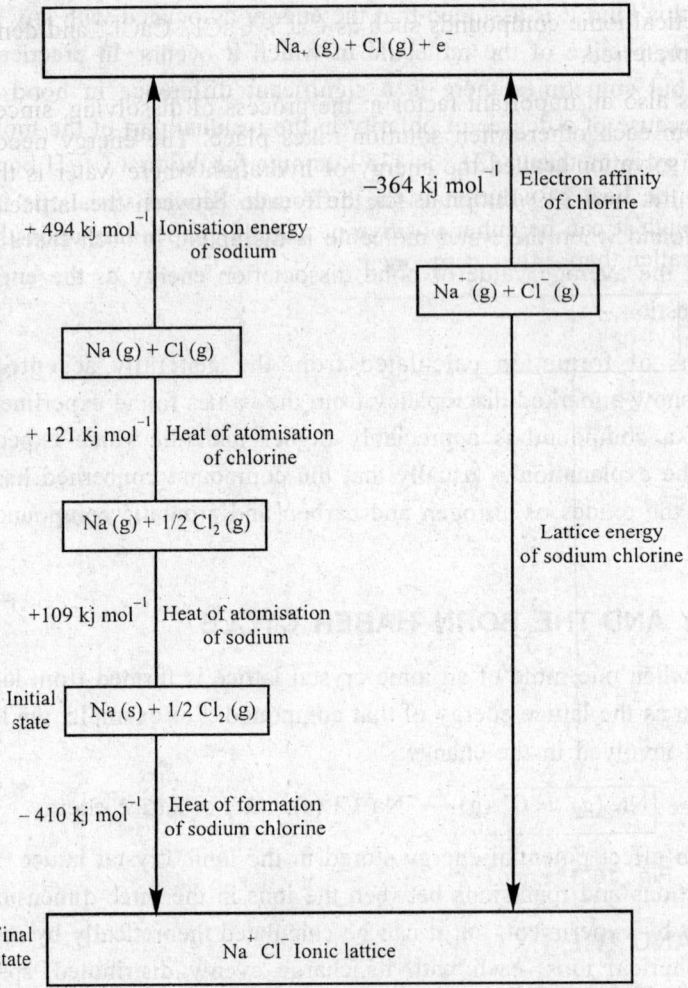

Fig. 16.7. The Born–Haber cycle for sodium chloride.

Now the value of the lattice energy of sodium chloride obtained by theoretical calculation is 766 kJ mol⁻¹. The very close agreement between this and the value determined by applying the Born–Haber cycle is typical of the halides of the alkali metals in general and lends powerful support to the concept of the ionic lattice for these compounds. Where there is a marked discrepancy between the two values, as in the silver halides and zinc sulphide, it suggests that in these cases the structure is not a purely ionic one and that the bonding between the elements is partially covalent in character.

It should be noted that the predominant influences upon the value of the heat of formation of an ionic lattice from its elements are the two largest items, the ionisation energy and the lattice energy. These are, of course, of opposite sign, since ionisation of a metal is always an endothermic change and ionic lattice formation is always an exothermic process. It follows that in general an ionic compound will only be formed where the lattice energy exceeds the ionisation energy of the metal, so that the energy released during lattice formation is enough to compensate for the energy required to bring about ionisation. By using the simple ionic model as a theoretical basis we can calculate the lattice energies we should expect for various hypothetical ionic compounds such as $CaCl$, $CaCl_2$, $CaCl_3$, and demonstrate why on energetic grounds $CaCl_2$ is preferred.

Lattice energy is also an important factor in the process of dissolving, since the ions in a crystal have to be separated from each other when solution takes place. The energy needed for this comes mainly from the energy of solvation (called the energy of hydration where water is the solvent). It will be seen from fig. 16.8 that the heat of solution is the difference between the lattice energy and the energy of solvation, and although it can be either positive or negative depending upon their relative magnitudes, it is always much smaller than either component.

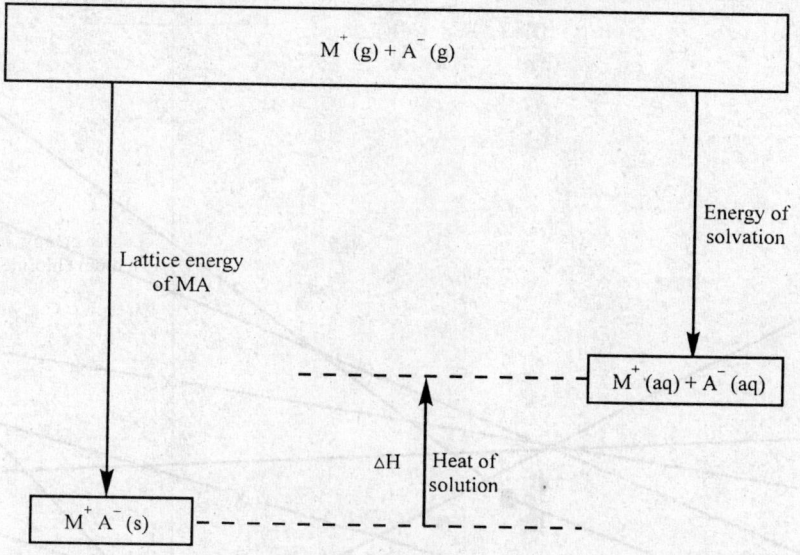

Fig. 16.8. Energy relationships during solution of an ionic solid.

FREE ENERGY AND THE EXTRACTION OF METALS

Many metals occur as minerals in the form of the oxide, because o the preponderance of oxygen in the earth's crust (about 50% by mass) and its high electronegativity. Other metals which occur naturally as sulphides are usually converted into oxides as the first step in their extraction. It follows that the common feature of many metal extraction processes is the heating of the oxide with some element or compound which has a greater affinity for oxygen than the metal has, so that reduction occurs. The choice of reducing agent is often complicated by the necessity of avoiding reaction between the metal and the reducing agent, so that, for example, carbon cannot be used to educe aluminium, uranium, or titanium oxides because these metals form carbides at the operating temperatures, but the essential thermodynamic requirement is that the particular reducing agent chosen will cause the metal to part from the oxygen with which it is combined.

The best guide to relative affinity for oxygen is to consider the free energy changes which occur when various elements combine with a fixed amount of gaseous oxygen to form oxides and to compare these values with each other. The oxide with the greatest negative free energy of formation will be the most stable and will be thermodynamically capable of reducing any oxide with a less negative value of $\Delta G\phi$. However such comparisons are not as straightforward as they might appear because the free energy changes involved in oxide formation can vary widely with temperature since the absolute temperature T is a term in the expression

$$\Delta G^{\phi} = \Delta H^{\phi} - T \, \Delta S^{\phi}$$

So unless the change in entropy ΔS^{ϕ} happens to be zero, the value of ΔG^{ϕ} for any one reaction will vary with temperature particularly at the very high temperatures commonly encountered in pyrometallurgical processes. For metals ΔS is normally negative because there is a decrease in entropy when a metal combines with oxygen gas to form a crystalline solid with an orderly structure. It follows that in such cases the negative value of ΔG^{ϕ} will diminish with rising temperature as shown in Fig. 16.9.

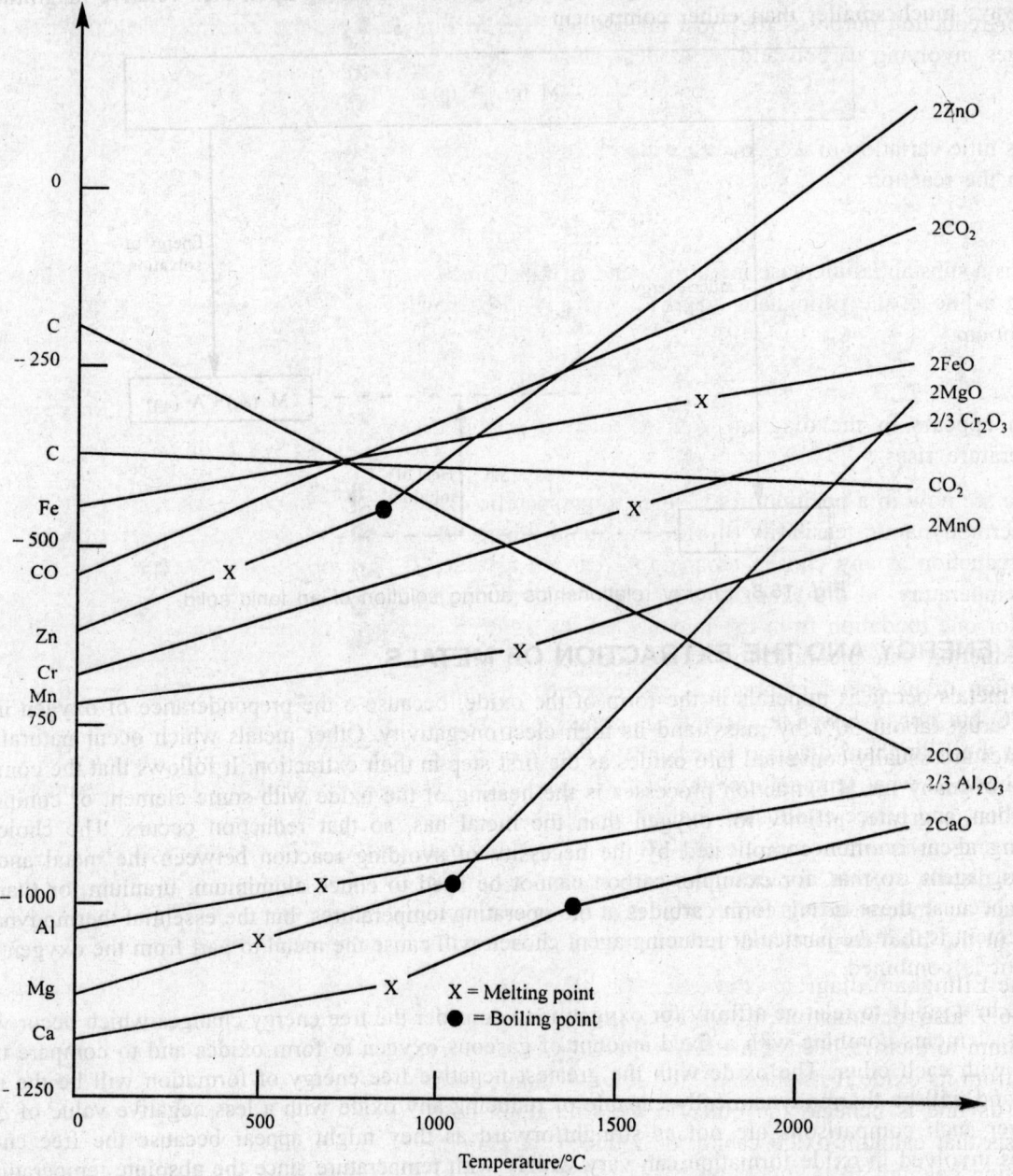

Fig. 16.9. Ellingham diagram for oxides.

This figure is known as an Ellingham diagram after this originator; it shows the variation in standard free energy changes with temperature in a number of oxidation reactions for one mole of oxygen at a pressure of 101 325 N m^{-2}. It will be seen from the diagram that iGϕ tends to vary linearly with temperature, although there are changes in gradient at the melting point and, more obviously, the boiling point of each metal. These abrupt changes in slope result from the increased randomness of the reactants, particularly at the boiling point of the metal, where its entropy increases by the molar latent heat of evaporation divided by the absolute temperature.

For reduction purposes the most interesting lines in Fig. 16.9 are those representing the free energy changes involving carbon and its oxides. The reaction

$$C + O_2 \; CO_2$$

shows little variation of ΔG^ϕ over a wide range of temperature and its line runs approximately horizontal, but in the reaction

$$2C + O_2 = 2CO$$

there is a substantial increase in entropy and ΔG^ϕ becomes steadily more negative with rising temperature giving a line in the Ellingham diagram with a marked negative slope. The contrary is the case for the equilibrium

$$2CO + O_2 \rightleftharpoons 2CO_2$$

which appears in the diagram with a positive gradient because ΔG^ϕ decreases numerically as the temperature rises.

We are now in a position to consider some specific examples of oxidation reactions and to decide on the thermodynamic feasibility of bringing about a particular reduction. The standard free energy change of a reduction at any chosen temperature can be assessed by drawing a vertical line corresponding to that temperature on the Ellingham diagram and subtracting the intercept corresponding to the value of ΔG^ϕ for one oxidation from the intercept of the other; a decrease in standard free energy indicates that the reduction can theoretically take place, although like all thermodynamic judgements it gives no indication of its speed in practice. For example, magnesium oxide can be reduced by carbon above 1650°C but not at lower temperatures because only above this temperature does the $2C + O_2 = 2CO$ line in the Ellingham diagram line beneath the line for the oxidation $2Mg + O_2 = 2MgO$. The diagram also shows at what temperatures the zinc oxidation line rises above the lines representing the oxidation of carbon, and hence establishes that the minimum temperatures at which zinc oxide can be reduced by carbon and carbon monoxide are above the boiling point of the metal. This explains why zinc (unlike iron) always distills over as a vapour from the retort or furnace during the extraction process. Similarly we can gain an understanding of the vital role of carbon monoxide in bringing about the reduction of iron oxide in the blast-furnace and the importance of using the right operating temperature.

The Ellingham diagram is not limited, however, to explaining reductions by carbon and it oxide. Fig. 16.9 also demonstrates why aluminium can be used to reduce the oxides of manganese and chromium to their respective metals in the Goldschmidt process; in general any element can replace any other from its oxide if its standard free energy change during oxidation ΔG^ϕ has a higher negative value, i.e. if its line is beneath the other in the Ellingham diagram at that temperature. Although Fig. 16.9 suggests that calcium oxide cannot be reduced by aluminium, this is in fact the method now used for extracting calcium from quicklime. This reduction is possible thermodynamically in vacuo because at very low pressure the calcium oxidation line has a much steeper slope and rises at high temperatures above the line for the oxidation of aluminium giving a net fall in the free energy of the system when the reduction occurs.

Example 16.6. Heat of neutralisation between HCl and NaOH is 13.7 K.cal. and between HCN and NaOH is 3,000 cal. at 25°C. What is the heat of ionisation of HCN?

Solution. Neutralisation between HCl and NaOH occurs in one step as

$$H^+ + OH^- = H_2O \quad \Delta H = -13.7 \text{ K.Cal.}$$

But neutralisation between HCN and NaOH occurs in two steps, which are :

$HCN = H^+ + CN^- \quad \Delta H = \text{let x kcal.}$

$H^+ + OH^- = H_2O \quad \Delta H = -13.7 \text{ kcal.}$

x kcal is heat of ionisation. So here

$x - 13.7 = -3 \qquad$ [given heat of neutralisation]

or $x = 10.7$ kcal $(= -3000 \text{ cal} = -3 \text{ K. Cal.})$

Example 16.7. Heat of neutralisation of (i) NH_4OH and HCl is -51.46 KJ/mole (ii) CH_3COOH and NaOH is -50.63 KJ/mole. Calculate heat of neutralisation of NH_4OH and CH_3COOH. Given heat of neutralisation for NaOH and HCl is -57.54 KJ/mole.

Solution.

Given

1.	$NH_4OH + HCl = NH_4Cl + H_2O$	$\Delta H = -51.46$ KJ/mole
2.	$CH_3COOH + NaOH = CH_3COONa + H_2O$	$\Delta H = -50.63$ KJ/mole
3.	$NaOH + HCl = NaCl + H_2O$	$\Delta H = -57.54$ KJ/mole

Adding (1) and (2) and subtracting (3) we get

$NH_4OH + CH_3COOH = CH_3COONH_4 + H_2O \qquad \Delta H = -44.55$ KJ/mole

Chapter 17

The Colloidal State

INTRODUCTION

Colloidal chemistry is a subdivision of physical chemistry comprising the study of phenomena characteristic of matter when one or more of its dimensions lie in the range between 1 millimicron (nanometer) and 1 micron (micrometer). The science thus includes not only finely divided particles but also films, fibres, foams, pores, and surface irregularities. It is the dimension that is critical, rather than the nature of the material. Colloidal particles may be gaseous, liquid, or solid and occur in various types of suspensions (imprecisely called solutions), e.g., solid/gas (aerosol), solid/solid, liquid/liquid (emulsion), gas/liquid (foam). In this size range, the surface area of the particle is so much greater than its volume that unusual phenomena occur; for example, the particles do not settle out of the suspension by gravity, and are small enough to pass through filter membranes. Macromolecules (proteins and other high polymers) are at the lower limit of this range; the upper limit is usually taken to be the point at which the particles can be resolved in an optical microscope. Though the term is often used synonymously with surface chemistry, in a strict sense it is limited to the size range mentioned in at least one dimension, whereas surface chemistry is not. Natural colloidal systems include rubber latex, milk, blood, egg-white, etc.

A colloid is not a particular type of substance, as was first thought, but a substance in a particular state of sub-division. In fact most substances can be obtained in a colloidal condition sufficient trouble is taken, although some substances assume it more readily than others. A substance is colloidal when it is dispersed or scattered as very fine particles, droplets, or bubbles in another substance of different phase. It is the presence of two distinct phases in a colloidal mixture which distinguishes it from a true solution. Finely-divided material is known as the disperse phase and the continuous medium as the dispersion medium.

Protective colloid is a hydrophilic high polymer whose particles (molecules) are of colloidal size, such as protein or gum; it may be either naturally present in such systems as milk and rubber latex, or intentionally added to mixtures to prevent coagulation of coalescence of the particles of fat or other dispersed material. Protective colloids are also called stabilising, suspending, or thickening agents; they also act as emulsifiers. Examples are: (i) hydrocarbon particles of latex that are covered with a layer of protein which keeps them from cohering as a result of the impact due to their Brownian motion; (ii) gelatin, sodium alginate, or gum arabic which are added to ice cream to inhibit formation of ice particles, and to confectionery, and other food products to obtain a smooth, creamy texture. They are readily adsorbed by the suspended particles and reinforce the protective effect of proteins that may be naturally present.

TYPES OF COLLOIDAL SYSTEMS

Finely divided particles of any substance with diameters lying within 10–2000Å range dispersed in any medium constitute what we term a colloidal system. A colloidal system is thus a two-phase system consisting of a continuous phase or dispersion medium in which extremely minute particles, lying within the colloidal range, of a second substance termed as discontinuous phase or dispersion phase, are suspended.

The dispersed phase may not necessarily be a solid always. It may be a liquid or even a gas as well. Similarly, the dispersion medium may be as gas or a liquid or even a solid. Thus, several different types of colloidal systems, depending upon the states of aggregation of the dispersed phase and the dispersion medium, are possible, as shown below in Table 17.1.

Table 17.1. Different colloidal system.

Dispersion medium	Dispersed phase	Examples
Gas	Liquid	Clouds, mists, fogs
Gas	Solid	Smoke, volcanic dust
Liquid	Gas	Foams, whipped cream
Liquid	Liquid	Emulsions (milk, cod-liver oil)
Liquid	Solid	Starch, proteins, arsenic sulphide, gold
Solid	Gas	Adsorbed or occluded gases
Solid	Liquid	Jellies, gels, cheese
Solid	Solid	Coloured precious stones, rock-salt

A gas dispersed in another gas does not exist since the gases are completely miscible with one another. When dispersion medium is a gas, the colloidal system is called aerosol. The systems with solids as dispersed phase and a liquid as dispersion medium are known as sols. When the liquid medium is water, the system is called hydrosol or aquasol. When this is alcohol or benzene or any other organic liquid, the system is referred to as alcosol, benzosol or organosol, respectively.

CLASSIFICATION OF COLLOIDS

Substances such as proteins, starch and rubber, whose molecule are large enough to be close to the lower limit of colloidal range, pass readily into colloidal state whenever mixed with a suitable solvent. Such colloids are known as lyophilic colloids (lyophilic implies 'love' for liquid). These colloids have strong interactions with the dispersion medium. On the other hand, substances like arsenic sulphide, ferric hydroxide, gold and other metals, which are sparingly soluble and whose molecules are much smaller than the lower colloidal limit, change into colloidal state by aggregation of many individual molecules. These substances, therefore, do not pass into colloidal state readily and are called lyophobic colloids (implying 'hatred' for liquid). Lyophilic systems are much stable and are also known as reversible colloids since on evaporating the dispersion medium (say, water), the residue can again pass into colloidal state on the mere addition of the liquid. Lyophobic colloids are much less stable and the residue left on evaporation cannot readily be reconverted into by ordinary means. Such systems are said to be irreversible colloids. The main differences between the two types of sols are summed up in Table. 17.2.

Table 17.2. Essential differences between lyophobic and lyophilic sols.

Property	Lyophobic sols	Lyophilic sols
Surface tension	Surface tension is usually the same as that of the medium (i.e., the liquid in which the particles are dispersed).	Surface tension is generally lower than that of the medium (i.e., the liquid in which the particles are dispersed).
Viscosity	Viscosity is about the same as that of the medium.	Viscosity is much higher than that of the medium.
Visibility	The particles, though invisible, can be readily detected under ultra-microscope.	The particles may migrate in either direction or not not at all in an electric field.
Migration in an electric	The particles migrate either towards anode or towards cathode in an electric field.	The particles may migrate in either direction or not at all in an electric field.
Action of electrolytes	The addition of small quantities of electrolytes can cause precipitation (coagulation).	The addition of small quantities of electrolytes has little effect. Much larger quantities are needed to cause precipitation.
Reversibility	These are irreversible.	These are reversible.
Hydration	The particles are not hydrated to any large extent.	These particles are extensively hydrated. This is due to the presence of a number of polar groups in the molecules of lyophilic colloids as, for example, in polysaccharides, proteins, etc.

Polymer Solutions

It is possible by suitable polymerisation reactions, to join together by primary valancies, a large number of small molecules to get giant molecules called macromolecules. Each macromolecule may consist of hundreds or even thousands of simple molecules. The materials containing such macromolecules called polymers, have high molecular weights, sometimes of the order of a few millions. Solutions of these polymers are called macromolecular solutions.

Since the size of a macromolecule lies within the colloidal range, macromolecular solutions, behave in many ways like lyophilic sols. The examples are synthetic rubber, polystyrene, polyethylene, nylon, etc. The naturally occurring polymers such as proteins, polysaccharides, gums and resins, which have been mentioned under lyophillic colloids also belong to the category of macromolecular solutions.

True Solution, Colloidal Solution and Suspension

When a few crystals of a water-soluble substance, like sugar or sodium chloride, are put into water, they dissolve to give a homogeneous solution. The particles of a solute are now of molecular size. They are invisible and do not settle on standing. Such mixtures are termed as true solutions or molecular solutions. On the other hands, if an insoluble substance, like barium sulphate or clay or sand, is put into water, the particles are large enough to be visible, if not to a naked eye, at least under a microscope. They settle down on standing for some time. Such mixtures are called suspensions. In between these two extremes, lie particles which are bigger than molecules but are too small to be visible even under a most powerful microscope. When suspended in a liquid or even when present in air, they have a tendency to settle down, through extremely slowly. These particles are said to belong to the colloidal state and when suspended in a liquid, they are referred to as colloidal solutions.

Size Range

The upper limit of colloidal range is fixed as the lower limit of microscopic visibility. It is well-known that it is not possible to see any particle whose diameter is less than half the wave-length of the light used. The lowest wavelength of visible light is about 0.4μ, ie,. 4.0×10^{-50} cm. Thus, it is not possible to see a particle if its diameter is less than 0.2μ. Therefore, 0.2μ or $200 m\mu$ or 2000Å is regarded as the upper limit of the colloidal range. The lower limit is generally taken as $1 m\mu$ or 10Å which is of the order of diameters of molecules of common substances.

The size range and other characteristics of true solutions, colloidal solutions and suspensions are summed up in Table 17.3.

Table 17.3. Characteristics of true solutions, colloidal solutions and suspensions.

True solutions (Molecular solutions)	Colloidal solutions (Colloidal dispersions)	Suspensions (Coarse dispersions)
Particles have the sizes of the same order as the molecules, i.e., of the order of $1 m\mu$ or 10Å.	Particles generally lie in the size range $1-200 m\mu$, i.e., $10-2000\text{Å}$.	Particles have size greater than $200 m\mu$ or 2000Å.
Particles are invisible under all circumstances.	Particles themselves are invisible even under a most powerful microscope but the scattering effect can be viewed using a special device known as ultra-microscope.	Particles are visible even to a naked eye or at least under a microscope.
Diffuse readily through parchment membrane	Diffuse slowly through parchment membrane.	Generally do not diffuse.
Pass readily through filter paper as well as parchment membrane.	Pass readily through filter paper but very slowly through parchment membrane.	Do not pass through filter paper or parchment membrane
Do not scatter light.	The particles scatter light. The phenomenon is known as Tyndall effect.	Do not show tyndall effect.

PREPARATION OF COLLOIDAL SOLUTIONS

The primary consideration in the preparation of sols is that the dispersed particles should be within the size range of $1-200 m\mu$. The lyophilic sols can be readily prepared, as already mentioned. The lyophobic sols require special treatments. The methods consist either in :

1. Breaking down the coarser aggregates into particles of colloidal size; or
2. Grouping molecules into larger aggregates of colloidal size.

The methods belonging to these categories are known as dispersion and condensation methods, respectively.

Dispersion Methods

1. *By mechanical grinding:* The most obvious method of dispersion consists in breaking down the coarser solid by mechanical grinding. This is done in the so-called 'colloid mill' which generally consists of two metal discs, held at a small distance apart from one another and capable of revolving at a high speed in opposite directions. The material to be ground is fed in between the two discs in the form of a wet slurry. However, it is doubtful if this method produces particles uniformly of colloidal dimensions.

2. *Peptisation:* Certain precipitates such as silver chloride, ferric hydroxide, aluminium hydroxide, can be converted into colloidal state by the addition of small amount of a suitable electrolyte. This process is known as peptisation. For example, if to a freshly formed precipitate of ferric hydroxide, a small amount of a dilute solution of ferric chloride is added, a reddish brown sol of the substance is obtained. The excess of ferric chloride, if any, can be removed by dialysis. Similarly, if to a precipitate of stannic oxide, a small amount of dilute hydrochloric acid is added a stable sol of stannic oxide is obtained. The conversion of a freshly formed precipitate into a sol by the addition of a small amount of a suitable electrolyte is called peptisation.

3. *By Bredig's arc method:* If an arc is struck between two electrodes of a metal, like platinum, gold, silver or copper, in water containing traces of an alkali, the metal passes into colloidal solution of a reasonable, though not high, concentration (Fig. 17.1). It is believed that the metal first changes into vapours (molecular state) on account of the heat of the spark and the vapours then condense in water to give aggregates of colloidal range.

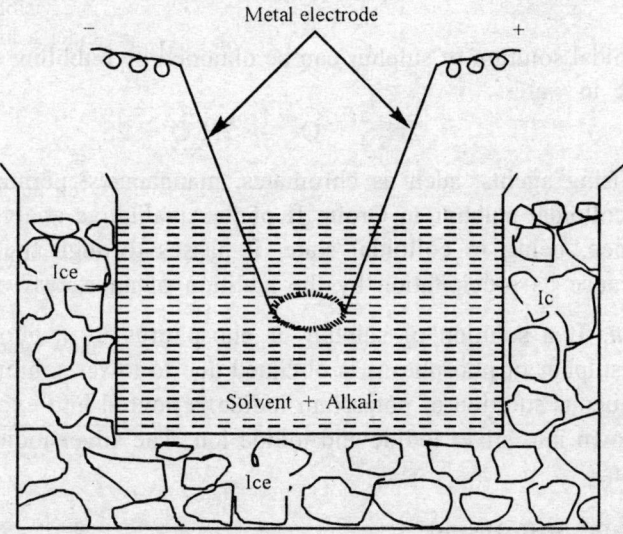

Fig. 17.1. Bredig's arc method.

Condensation Methods

The condensation methods usually involve chemical reactions in a medium in which the dispersed phase is sparingly soluble. A condition of super-saturation is produced but the actual precipitation is avoided. A few common methods are given below:

1. *By double decomposition:*

    (a) A sol of arsenic sulphide is obtained by passing hydrogen sulphide gas through a cold solution of arsenious oxide in water.

    $$As_2O_3 + 3H_2S \rightarrow As_2S_3 + 3H_2O$$

The excess of hydrogen sulphide is removed by boiling. If the solution of arsenious oxide is hot when hydrogen sulphide is passed, arsenic sulphide is obtained as a precipitate and not as a sol.

    (b) When a concentrated solution of ferric chloride is added, drop by drop, to a large excess of hot distilled water, a red sol of hydrated ferric oxide, commonly known as ferric hydroxide sol, is obtained.

    $$2FeCl_3 + 6H_2O \rightarrow Fe_2O_3.3H_2O$$

The excess of ferric chloride and the hydrochloric acid formed are removed by dialysis.

2. *By reduction:* A number of metals, such as silver, gold and platinum, have been obtained in colloidal state by reducing aqueous solutions of their salts by reagents, such as formaldehyde, phenyl hydrozine, hydrogen peroxide, carbon monoxide, phosphorus, tannic acid, etc. Thus a solution of gold chloride, when treated with formaldehyde or hydrogen peroxide, gives a collidal solution of gold. Faraday prepared gold sol by reducing a gold chloride soluton with white phosphorus dissolved in ether.

Platinum, mercury, silver, bismuth and copper sols have been produced by such methods. The presence of a small amount of a lyophilic sol (e.g., gelatin) is helpful in increasing the stability of such systems.

3. *By oxidation:* A colloidal solution of sulphur can be obtained by bubbling oxygen through a solution of hydrogen sulphide in water.

    $$2H_2S + O_2 \rightarrow 2H_2O + 2S$$

The presence of oxidising agents, such as chromates, manganates, permanganates and nitric acid, causes the formation of colloidal sulphur in Group II of the qualitative analysis. This causes difficulty in removing sulphur, since, being in colloidal state, it passes through the filter paper. It becomes necessary, therefore, to cause its precipitation by the addition of an excess of an electrolyte.

4. *By change of solvent:* If a solution of sulphur or phosphorus in alcohol is poured into water, a colloidal solution of sulphur or phosphorus is obtained due to lower solubility in water. If a solution of silver iodide in aqueous solution of potassium iodide is poured into excess of water, the complex ion $[AgI_2]^-$ breaks down into silver iodide and iodide ion. The silver iodide is obtained in the form of a colloidal solution.

Purification of Colloidal Solutions

The presence of impurities, particularly the electrolytes, renders the sols unstable. These must be eliminated by suitable means. Two methods are generally employed.

1. *Dialysis:* It has already been stated that while particles in true solution, that is, the molecules, can easily diffuse through parchment and other fine membranes, the colloidal particles, being much larger, cannot do so readily. If a mixture, containing colloidal particles as well as particles in true solution, is placed in a parchment bag which is then held in a wider vessel containing pure water

(Fig. 17.2), the substances in true solution pass out while the colloids remain in the bag. The distilled water in the wider vessel is renewed frequently.

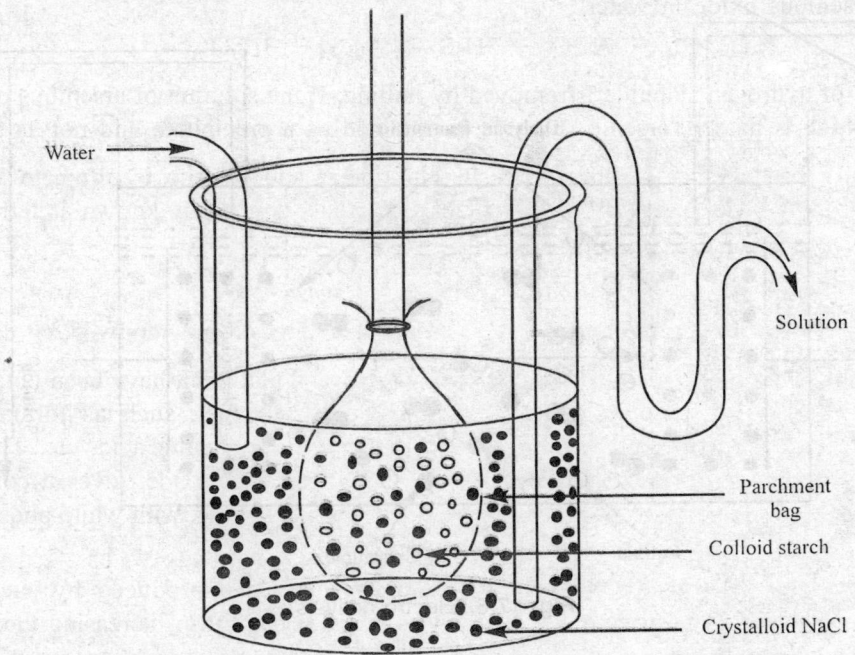

Water

Solution

Parchment bag

Colloid starch

Crystalloid NaCl

Fig. 17.2. Purification of colloidal solutions by dialysis.

The process of separating substances in colloidal state from those present in true solution with the help of fine membranes, is known as dialysis and the membrane used for the purpose is known as dialyser. Dialysis is regarded as a process of fractional diffusion of solutes and colloids. Ordinarily, the process is quite slow but it can be quickened by applying an electric field if the substance in true solution is an electrolyte. The process is then called electrodialysis. The mixture is placed between two dialysing membranes, while pure water is contained in a compartment on each side (Fig. 17.3). There is one electrode in each compartment by means of which the required emf (electro motive force) is applied. The ions of the electrolyte migrate out to the oppositely charged electrodes while the colloidal particles are held back.

2. *Ultra-filtration:* The separation of solutes from colloidal system can also be carried out by the process known as ultra-filtration. Ordinarily, filter papers have pores larger than 1μ (i.e., 1000 mμ) so that the colloidal particles (which are less than 200 mμ) can readily pass through along with the ions or molecules in solution. But the pores can be made smaller by soaking the filter paper in a soaking in formaldehyde. The pores thus become very small and the colloidal particles may be retained on the treated filter paper. The treated filters are known as ultra-filters. This process of separating colloids from solutes is known as ultra-filtration. A series of graded ultra-filters can be prepared by soaking filer papers in solutions of collodion of different concentrations. The pores even in the finest ultra-filters will be large enough to permit the passage of ions or molecules in true solution but these will be small enough to withold the colloidal particles. By using a series of graded ultra-filters, it may be possible to separate colloidal particles of different sizes from one another. The

process is very slow and sometimes a little pressure is needed to drive the particles through the filters.

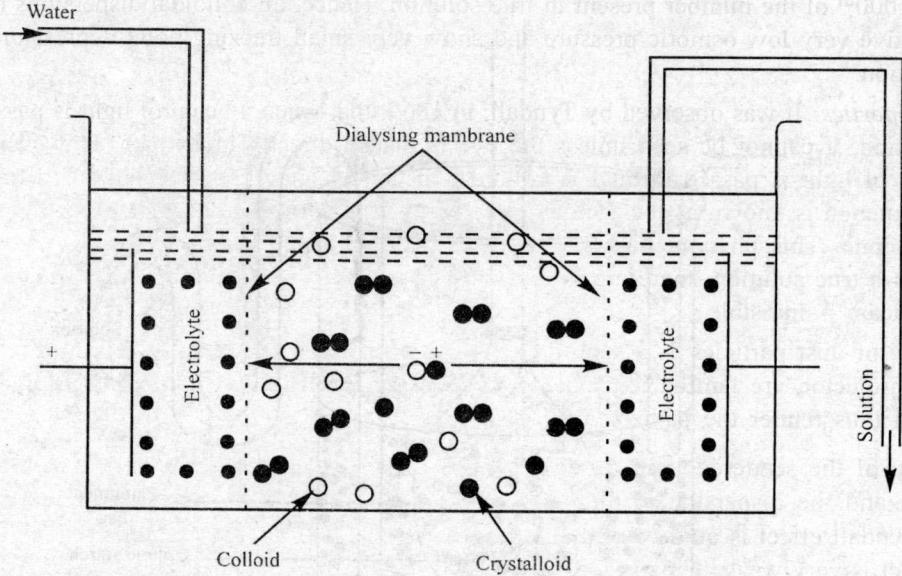

Fig. 17.3. Electro-dialysis.

PROPERTIES OF COLLOIDAL SYSTEMS

1. *Heterogeneous character:* As already stated, colloidal systems unlike true solutions, are heterogeneous in character. They consist of two phases: (i) the dispersed phase; and (ii) the dispersion medium.

2. *Diffusibility:* The colloidal particles constituting the dispersed phase do not readily diffuse through parchment or other fine membranes.

3. *Filtrability:* The colloidal particles readily pass through ordinary filter-papers along with any dissolved material. This is because even an extremely fine filter-paper has pores bigger than the colloidal dimensions.

4. *Visibility:* It is not possible to see colloidal particles even with the help of a most powerful microscope. A gold sol, for instance, appears to be as clear as a true solution of gold chloride in water. The reason of the 'insibility' of colloidal particles is not far to seek. It is impossible to see a particle whose diameter is less than half the wavelength of the light used. The shortest wavelength of the visible light (extreme visible violet end of spectrum) is about 4000Å, that is 400 mμ. Hence, no particle of diameter less than 200 mμ can be seen.

Attempts have been made in recent times to use ultra-violet rays or cathode rays. But, these rays make no impression on the retina of the eye. However, the images formed by them can be photographed. The electron microscope, for instance, makes use of a beam of cathode rays and by a combination of special types of the lenses, images of colloidal particles can be obtained on photographic plates.

5. *Colligative properties:* The magnitude of osmotic pressure, lowering of vapour pressure, depression in freezing point and elevation of boiling point, depends upon the number of solute particles present

in a given weight of the solvent. Now, colloidal particles are not simple molecules. These are physical aggregations of molecules. In arsenic sulphide sol, for instance, each particle is composed of about 1000 molecules. Thus for a given mass of arsenic sulphide, the number of particles in the sol will be only $1/1000^{th}$ of the number present in true solution. Hence, all colloidal dispersions (unlike true solutions) give very low osmotic pressure and show very small freezing point depression or boiling point elevation.

6. *Optical properties:* It was observed by Tyndall, in 1869, that when a beam of light is passed through a true solution, it cannot be seen unless the eye is placed directly in its path. However, when the same beam of light is passed through a colloidal dispersion, it becomes visible as a bright streak. This phenomenon is known as the Tyndall effect and the illuminated path (streak of light) is known as Tyndall cone. This phenomenon is due to the scattering of light from the surface of colloidal particles. In a true solution, there are no particles of sufficiently large diameter to scatter light and hence the beam is invisible.

The visibility of dust particles in a semi-darkened room when a sun beam enters or when a light is thrown from a projector, are familiar examples of Tyndall effect. The dust particles are large enough to scatter light and thus render the path of light visible.

The intensity of the scattered light depends on the difference between the refractive indices of the dispersed phase and the dispersion medium. In lyophobic colloids, this difference is appreciable and, therefore, the Tyndall effect is quite well-defined. But in lyophilic sols, the difference is very small and the Tyndall effect is very weak. Thus in the case of freshly prepared silicic acid, blood serum, albumins, etc., there is little or no Tyndall effect.

The Tyndall effect has been used by Zsigmondy and Siedent of in devising the ultra-microscope. A strong beam of light from an arc lamp or any other source is condensed by a system of lenses and passed through the colloidal solution and the scattered beam (Tyndall beam) is viewed through a microscope, placed at right angles to the beam (Fig. 17.4). In this way, the colloidal particles, which are too small to be seen under an ordinary microscope, can be detected as spots of light moving irregularly. It may be emphasised that we do not see the actual particles they are too small to be visible) but only the light scattered by them. Our eye pictures various spots of light as round or spherical particles.

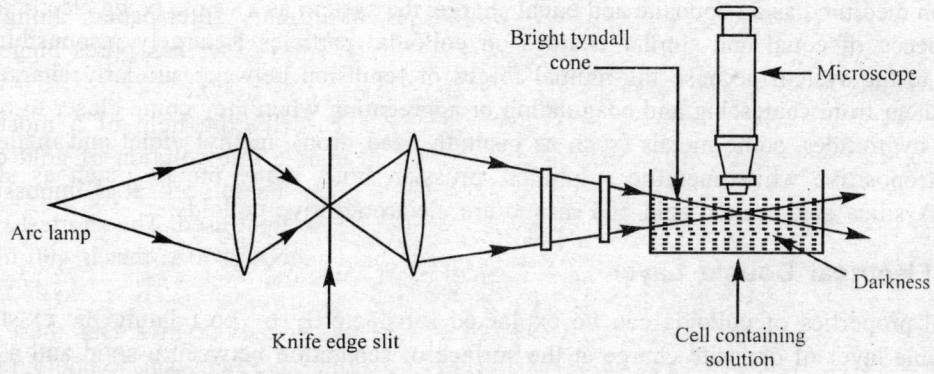

Fig. 17.4. Principle of ultra-microscope.

7. *Brownian movement:* Robert Brown, an English botanist, in 1827, observed that pollen-grains in aqueous suspensions were in constant motion. Later on when ultra microscope was invented, it was

found that particles of lyophobic sols were also in a state of ceaseless erratic and random motion (Fig. 17.5), similar to pollengrains. This kinetic activity of particles suspended in a liquid is called Brownian movement. The Brownian movement is due to the bombardment of colloidal particles by molecules of dispersion medium which are in constant motion like molecules in a gas. As a result, the colloidal particles acquire almost the same energy as possessed by the molecules of the dispersion medium. But, since the colloidal particles are considerably heavier than the molecules of the dispersion medium, their movement is considerably slower than that of the molecules of the medium.

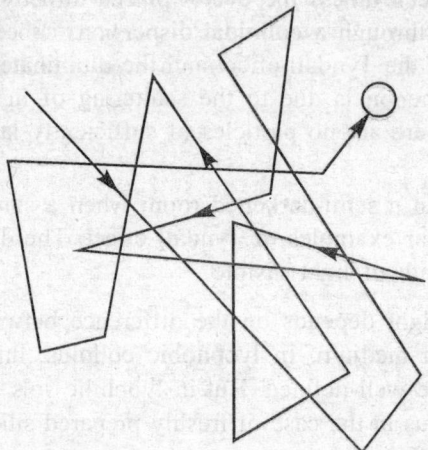

Fig. 17.5. Brownian movement.

The Brownian movement is not observed in ordinary suspensions because the mass of each particle in this case is so large that the bombardment of molecules of the dispersion medium produces little effect on them.

Brownian movement offers a visible proof of the random kinetic motion of molecules in a liquid.

8. *Electrical properties:* The most important property of collidal dispersions is that the particles carry an electric charge. All the particles in a given colloidal system carry the same charge and the dispersion medium has an opposite and equal charge, the system as a whole being electrically neutral. The presence of equal and similar charges on colloidal particles is largely responsible in giving stability to the system because the mutual forces of repulsion between similarly charged particles prevent them from coalescing and coagulating or aggregating when they come closer to one another. Metallic hydroxides, some metals (such as bismuth, lead, iron), methyl violet and methylene blue, are electropositive while metallic sulphides, prussian blue, many metals (such as silver, gold, platinum), silica acid, tannic acid and mastic are electronegative colloids.

Theory of Electrical Double Layer

The electrical properties of colloids can be explained satisfactorily by postulating the existence of an electrical double layer of opposite charge at the surface of separation between a solid and a liquid, i.e., at a solid-liquid interface.

According to the modern views, when a solid is in contact with a liquid, a double layer on ions appears at the surface of separation, as shown in Fig 17.6. One part of the double layer is fixed of the surface of the solid. This is known as the fixed part of the double layer. It consists of either positive ions or negative ions. The second part consists of a diffuse or a mobile layer of ions which extends into the liquid phase. This layer consists of ions of both the signs but its net charge is equal and opposite to that on the fixed part of the double layer. In one case (Fig. 17.6(a)), the fixed part of the double layer carries positive charge while in the second case (17.6(b), the fixed part carries negative charge. It will be seen that in each case, the net charge on the diffuse layer is equal and opposite to that on the fixed layer.

This theory is applicable to colloidal systems also. The ions which are preferentially adsorbed by the colloidal particles, are held in the fixed part of the double layer. These ions give the characteristic charge to the colloidal particles. The ions carrying the opposite charge, i.e., the counter ions, are present more in the diffuse potion of the double layer, giving a net opposite charge to this layer.

The existence of charge of opposite signs on the fixed and diffuse parts of the double layer leads to the appearance of a difference of potential between the two layers, as shown in Fig. 17.6. This difference of potential between the two layers is known as zeta potential or electrokinetic potential.

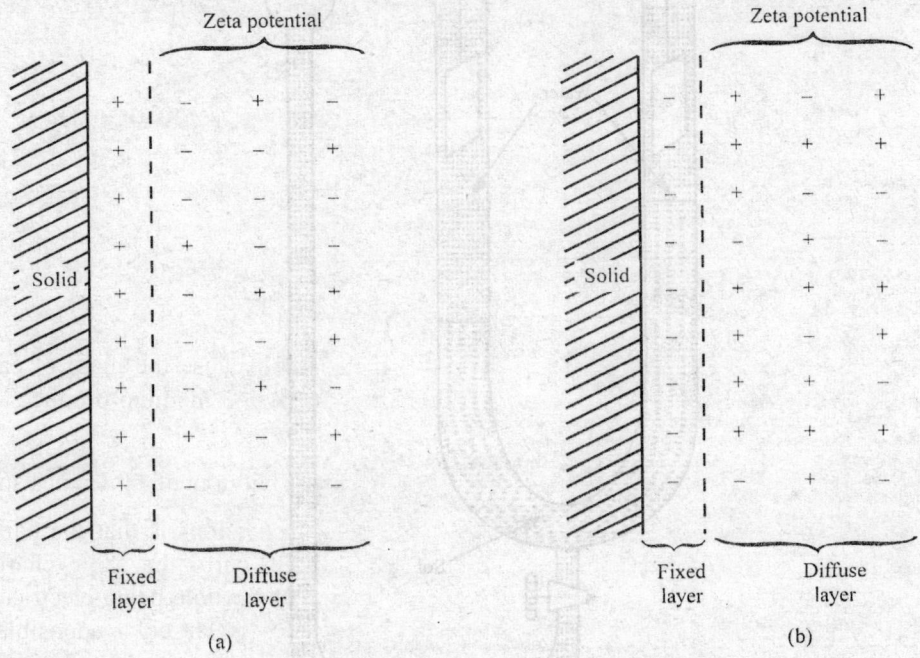

Fig. 17.6. Electrical double layer.

Since the solid particles and the liquid medium carry opposite charges, it is obvious that when an electric field is applied, the particles and the liquid will migrate in opposite directions. When experiments are so arranged that the particles can move, but not the medium, we have the phenomenon of electrophoresis. If, on the other hand, the experiments are designed in such a way that the medium can move, but not the particles, we come across the phenomenon of electro-osmosis. Both these phenomena are discussed below.

Electrophoresis

The electrophoretic effect can be studied by the simple apparatus shown in Fig. 17.7. This consists of a U-tube provided with a stop-cock through which it is connected to a funnel-shaped reservoir. A small amount of water is first placed in the U-tube and a reasonable quantity of the sol is taken in the reservoir.

The stop-cock is then slightly opened and the reservoir gradually raised so as to introduce the sol into the U-tube gently. The water is displaced upwards producing a sharp boundary in each arm. A voltage of 50 to 200 volts is then applied by means of the platinum electrodes which are immersed in water layer only. The movement of the particles can readily be followed by observing the position of the boundary by means of a naked eye or a lens or a cathetometer. When the particles are negatively charged (as in the case of arsenic sulphide sol), the boundary on the negative electrode side is seen to move down and that on the positive side to move up showing that the particles move towards the positive electrode. Thus by noting the direction of motion of the particles in the electric field, it is possible to determine the sign of the charge carried by the particles. It is also possible by this technique is determine the rate at which colloidal particles migrate in an electric field.

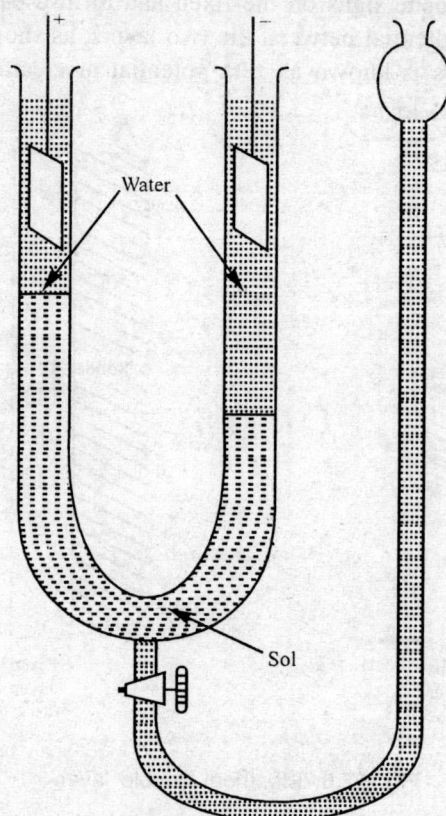

Fig. 17.7. Electrophoresis.

The electrophoretic mobility of colloidal particles is defined as the distance travelled by them in one second under a potential gradient of one volt per centimetre. It has been found that the migration velocities of colloidal particles are of the same order as those of ions, under similar conditions, that is, of the order of $10–60 \times 10^{-5}$ cm./sec./volt/cm.

By considering the electrical double layer as a parallel plate condenser, it has been shown that zeta potential (z) is given by the expression :

$$z = \frac{4\pi\eta u}{D} \qquad \qquad \qquad ...(17.1)$$

where u is the electrophoretic mobility of the particles, η is the coefficient of viscosity and D is the dielectric constant of the dispersion medium.

Since different colloidal materials have different colloidal materials have different mobilities, it is possible to separate them from one another from their mixtures. This method has been used for the fractionation of proteins, polysaccharides, nucleic acids and other complex substances.

Electro-osmosis

When electrophoresis of dispersed particles in a colloidal system is prevented by some suitable means, it is observed that the dispersion medium itself begins to move in an electric field. This phenomenon is known as electro-osmosis. A simple apparatus for studying electro-osmosis is shown in Fig. 17.8.

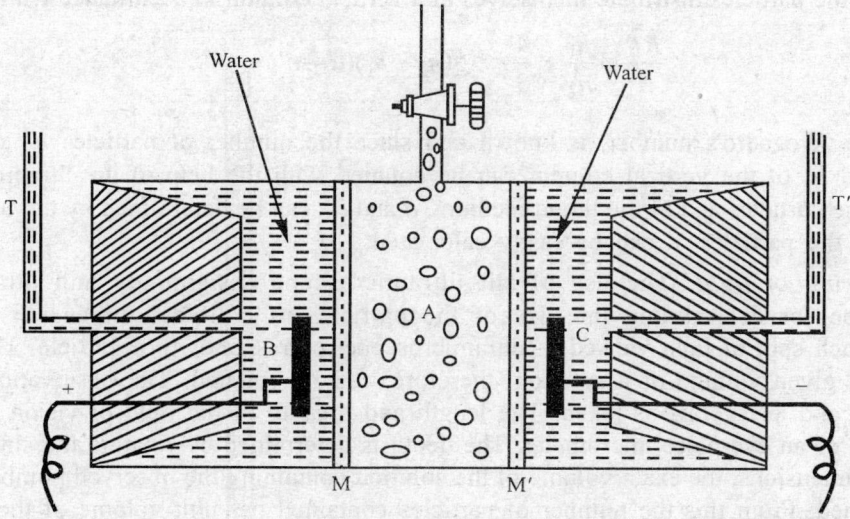

Fig. 17.8. Electro-osmosis.

The colloidal system is placed in the central compartment A which is separated from the compartments B and C filled with water, by the dialysing membranes M and M'. The water in the compartments B and C also extend to the side tubes T and T', as shown. The membranes prevent the movement of the colloidal particles. Therefore, when a potential difference is applied across the electrodes held close to the membranes in the compartments B and C, as shown, the water begins to move. If the particles carry positive charge, the water will carry negative charge. Therefore, it would start moving towards the anode and hence the level of water in the side tube T would be seen to rise. If, on the other hand, the particles carry negative charge, the water, which now carries positive charge, will start moving towards the cathode and the level of water in the side tube T' would start rising.

DETERMINATION OF SIZE OF COLLOIDAL PARTICLES

There are a number of methods for the determination of size of colloidal particles. Some of these are given below.

1. *By using ultra-filters:* An approximate idea about the size of particles in a colloidal system can be use of ultra-filters. These are prepared by impregnating filter papers with collodion or gelatin which

are subsequently hardened by immersing in formaldehyde. The pores can be made small enough to retain particles of colloidal dimensions. The size of the pores depends on the particular filter paper employed and the concentration of the collodion or gelatin solution used for impregnating it. It is thus possible to obtain a series of graded ultra-filters by means of which a colloidal solution may be separated into fractions containing particles of different sizes. An approximate estimate of the size of particles can be obtained from a knowledge of the dimensions of the pores of the ultra-filters. The latter parameter is determined from the pressure required to force air or water through the pores. The results obtained by this method are very approximate because pore size is by no means the only factor which determines whether a given particle will pass through an ultrafilter or not.

2. *From Brownian movement:* Colloidal particles suspended in a liquid medium are subjected to Brownian movement. They also tend to settle down due to gravitation. Under the influence of both these effects, the particles distribute themselves in a vertical column in accordance with the equation:

$$\frac{RT}{N} \ln \frac{n_1}{n_2} = \frac{4}{3} \pi r^3 g (h_2 - h_1)(d - d') \qquad \qquad ...(17.2)$$

Since N_0, the Avogadro's number, is known and since the number of particles n_1 and n_2 at two depths h_1 and h_2 of the vertical column can be counted with the help of an ultramicroscope and densities of the particles and of the liquid medium, d and d', can be determined by the usual methods, the radius of the particles, r, can be easily calculated.

3. *From scattering of light:* The use of slit ultramicroscope. Zsigmondy and others used the ultramicrscope for determining the size of the particles of colloidal dimensions. As already mentioned, each spot of light viewed in ultramicroscope, corresponds to a particle. The number of particles in a given volume of a solution, therefore, can be counted. The observation is repeated several times and an average is taken. The length and breadth of the field of vision are measured with the help of an eye-piece micrometer. The depth is determined by rotating the slit through 90°. From these dimensions, the exact volume of the solution containing the observed number of particles can be obtained. From this the number of particles contained per unit volume of the solution can be obtained. Let this number be *n*.

The next step is to evaporate a known volume of the colloidal solution to dryness. From the weight of the residue, the mass of colloidal particles per unit volume can easily be obtained. Let this be m gram. Now two assumptions are made. Firstly, the particles are supposed to be spherical. Secondly, the density (d) of the colloidal particles is supposed to be the same as that in the bulkstate. Evidently, the volume of colloidal phase is m/d and, therefore,

$$\frac{m}{d} = \frac{4}{3} \pi r^3 \times n \qquad \qquad ...(17.3)$$

or
$$r = \left(\frac{3m}{4 \pi d n} \right)^{\frac{1}{3}} \qquad \qquad ...(17.4)$$

4. By ultracentrifuge.

Origin of the Charge on Colloidal Particles

The origin of charge on colloidal particles has not been completely understood. However, it has been observed that sols are invariably associated with minute quantities of electrolytes and that if the latter are invariably associated with minute quantities of electrolytes and that if the latter are completely removed by persistent dialysis, the sols become unstable. It is believed, therefore, that the charge on the colloidal particles is due to preferential adsorption of either positive or negative ions on their surface. If the particles have a preference to adsorb positive ions, they acquire positive charge and vice versa. According to this view, the positive charge on ferric hydroxide sol prepared by the hydrolysis of ferric chloride is due to preferential adsorption of Fe^{3+} ions on the surface of the particles. The ferric ions come from the ionisation of ferric chloride which is always present in traces in the sol.

Similarly, the negative charge on arsenic sulphide sol is due to the preferential adsorption of sulphide ions on the surface of the particles. The sulphide ions are furnished by the ionisation of hydrogen sulphide which is present in traces. Likewise, the negative charge on metal sols prepared by the Bredig's arc method is due to the adsorption of hydroxyl ions furnished by the traces of the alkali added. It should be remembered that the ion which is more nearly related chemically to the colloidal particles is preferentially adsorbed by it. Thus, in ferric hydroxide sol, ferric and not chloride ion is preferred. Similarly, in arsenic sulphide sol, sulphide and not hydrogen ion is preferred.

An interesting case is furnished by stannic oxide sol. If a freshly formed precipitate of stannic oxide is peptised by a small amount of hydrochloric acid, the sol carries a positive charge, but if peptised by a small amount of sodium hydroxide, the sol carries a negative charge. In the former case a small amount of stannic chloride, $SnCl^4$, is formed and Sn^{4+} ion is preferred and the sol is positively charged. In the latter case, a small amount of sodium stannate, Na_2Sn_3, is formed and now the negatively charged $SnO_3{}^{2-}$ ion is preferred and accordingly the sol is negatively charged. The structure of the colloidal particles in the two cases is represented as $[SnO_2]Sn^{4+} : 4Cl^-$ and $[SnO_2]SnO_3{}^{2-} : 2Na^+$, respectively. The chloride and sodium ions, known as counter ions, are directed towards the liquid phase and constitute the diffused path of the electrical double layer.

Another interesting case is furnished by the formation of positively as well as negatively charged sols of silver iodide. If a dilute solution of silver nitrate is added to a slight excess of a solution of potassium iodide, a negatively charged sol of silver iodide is formed. This is due to the adsorption of iodide ions. The structure of the particle is represented as $[AgI]I^- : K^+$. But, if dilute solution of potassium iodide is added to a slight excess of silver nitrate solution, a positively charged sol of silver iodide is formed due to the adsorption of silver ions. The structure of the particles is now represented as $[AgI]Ag^+ : NO_3{}^-$. However, if silver nitrate and potassium iodide are mixed in equivalent amounts, there is precipitation of silver iodide and no sol is formed. It may be mentioned once again that the ions preferred by colloidal particles are those which are common to them. For example, in the first case, silver iodide prefers iodide ion and not potassium ion. In the second case, it prefers silver ion and not nitrate ion.

The counter ion, in every case, is directed towards the liquid phase. Thus, on the surface of colloidal particles there is an electrical double layer of opposite signs, as discussed above.

Another possible way in which colloidal particles acquire charge sometimes is by direct ionisation of the material constituting the particles. This phenomenon is observed mostly in the case of acidic and basic dyestuffs. An acid dyestuff, for example, ionises, yielding hydrogen ions in solution and thereby leaving an equivalent amount of negative charge on the particles. The structure of the colloidal particles of the dye may be represented as shown in Fig. 17.9(a).

A basic dyestuff, on the other hand, ionises yielding hydroxyl ions in solution and thereby leaving an equivalent amount of positive charge on the particles. The structure of the colloidal dye particles in this case may be represented as shown in Fig. 17.9(b).

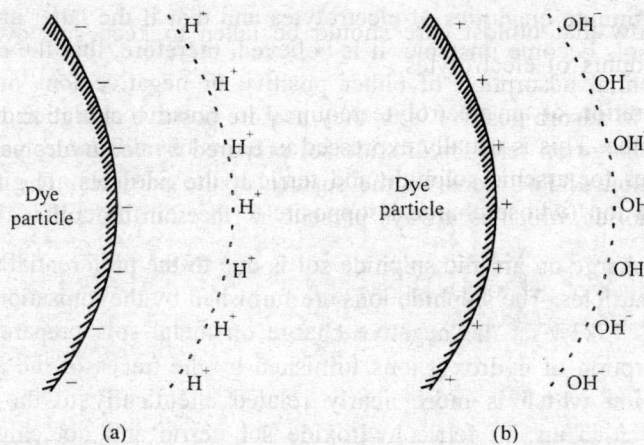

Fig. 17.9. Origin of charge in dyestuffs from ionisation of surface groups.

Coagulation of Colloidal Solutions

The stability of colloidal state rests on the existence of electrical charge on the particles. This charge, on account of electrical repulsion, prevents the particles from approaching sufficiently close to one another and to coalesce. However, when the charge is neutralised or lowered to a certain critical value, the particles approach close enough to coalesce and form bigger aggregates which lie outside the colloidal range. This phenomenon of change of colloidal state to suspended state is known as coagulation or flocculation of colloidal solutions. Coagulation of colloids is generally brought about by any one of the following treatments:

By the action of electrolytes

If traces of electrolytes are essential for stability of sols, the presence of larger amounts causes their coagulation. The reason is that colloidal particles take up ions carrying charge opposite to that present on themselves. This results in partial or complete neutralisation of the charge on the particles leading to their coagulation or flocculation. The ion carrying the opposite charge is called the flocculating ion. For example, if an electrolyte like barium chloride or aluminium sulphate is added to the negatively charged arsenic sulphide sol, barium or aluminium ions are taken up on the surface of the particles. This causes neutralisation of the charge and hence precipitation of the colloidal solution. Similarly, if the same electrolytes are added to positively charged ferric hydroxide sol, the chloride or sulphate ions are taken up on the surface of the particles, neutralising the charge and causing coagulation of the colloidal solution.

It has been found that generally the greater the valence of the flocculating ion added, the greater is its power to cause precipitation. This is known as Hardy-Schulze rule. Thus, in the coagulation of negatively charged arsenic sulphide sol, the flocculating power decreases in the order $Al^{3+} > Ba^{2+} > Na^+$ and in the coagulation of the positively charged ferric hydroxide sol, the flocculating power decreases in the order $[Fe(CN)_6]^{4-} > [PO_4]^{3-} > [SO_4]^{2-} > Cl^-$. However the rule is only approximate and several departures are known.

Generally, the precipitating power of a trivalent ion is nearly 500–1000 times as high as that of a monovalent ion while that of a bivalent ion is about 100–500 times as high as that of a univalent ion.

These observations show that utmost care should be taken to keep sols away from environments containing even small amounts of electrolytes.

The minimum concentration of an electrolyte required to cause coagulation or flocculation of a sol is called its flocculation value. This is usually expressed in terms of milli-moles per litre. The flocculation values of a few electrolytes for arsenic sulphide and ferric hydroxide sols are given in Table 17.4. The valency of the flocculating ion (whose charge is opposite to that on the colloidal particle) is also given in an appropriate column.

Table 17.4. Flocculation values of different electrolytes

Electrolyte	Arsenic sulphide sol		Electrolyte	Ferric hydroxide sol	
	Cation valency	Flocc. value (millimoles per litre)		Anion valency	Flocc. value (millimoles per litre)
NaCl	1	52	HCl	1	132
KCl	1	52	KBr	1	138
HCi	1	30	KNO_3	1	132
1/2 K_2SO_4	1	64	$KBrO_3$	1	131
$MgSO_4$	2	0.72	K_2CrO_4	2	0.315
$CaCl_2$	2	0.69	K_2SO_4	2	0.210
$ZnCl_2$	2	0.68	$K_2C_2O_4$	2	0.238
$AlCl_3$	3	0.093	$K_3[Fe(CN)_6]$	3	0.096

It is evident that polyvalent ions are more effective in causing coagulation since much smaller concentrations are needed for the purpose.

Certain departures from the Hardy-Schulze rule are quite evident. For example, the flocculation value of hydrochloric acid for arsenic sulphide sol and that of potassium bromate for ferric oxide sol are relatively low. The reason is that H^+ ions amongst the monovalent cations in the first case and BrO_3^- ions amongst the monovalent anions in the second case are strongly adsorbed.

Even more striking departures are known. For instance, the flocculation values of potassium citrate, potassium acetate and potassium formate for arsenic sulphide sol are found to be 270, 115 and 85 millimoles per litre. Since potassium ion which is effective in the present case is common, one would have expected the flocculation value of each electrolyte to be about the same. The marked discrepancy in the flocculation values shows that the ion carrying the same charge as the colloidal particles is also effective in some way in determining the flocculation value of an electrolyte. It has been established by experiment that citrate, acetate and formate ions in the above electrolytes are also adsorbed to different extents on the surface of arsenic sulphide particles thus raising the negative zeta potential of particles to different extents. Hence different concentrations of potassium ions are needed to bring about neutralisation of the negative charge on the arsenic sulphide sol particles in order to cause their flocculation.

By the mutual action of sols

When two sols carrying opposite charges are mixed together in suitable proportions, mutual precipitation occurs. For example, when negatively charged arsenic sulphide sol is added to positively charged ferric hydroxide sol, in suitable proportions, precipitation of both the sols takes place simultaneously.

By persistent dialysis

It has been reported earlier that traces of electrolytes are invariably associated with colloidal systems and that this is essential for their stability. If the sols are subjected to prolonged dialysis, these traces of electrolytes also pass out through the dialyser and the colloids become unstable.

PROTECTIVE COLLOIDS

Lyophilic colloids, as has been discussed earlier, are much more stable than lyophobic colloids. The reason is that they are extensively hydrated (or solvated, using a general term). It is believed that the colloidal particles are covered by a sheath of the liquid in which they are dispersed. This sheath acts as a barrier and prevents them from coming together and forming larger aggregates. A strong evidence in favour of this view is that lyophilic sols have high viscosity.

Lyophilic colloids also possess the property of protecting lyophobic colloids from precipitation by the action of electrolytes. When a lyophilic sol is added to a lyophobic sol, it is believed that lyophilic particles form envelopes around the lyophobic particles and thus protect the latter from the action of electrolytes. Lyophilic colloids used for such purposes are called protective colloids. Thus, if a little gelatin (lyophilic colloid) is added to a gold sol (lyophobic sol), the latter is 'protected' and a much higher concentration of sodium chloride solution is required now to cause its coagulation.

The lyophilic colloids differ in their protective powers. The term 'gold number' is introduced to measure the protective powers of different colloids. Is defined gold number as the weight in milligrams of a protective colloid which prevents the coagulation of 10 ml of a given gold sol on adding 1 ml of a 10 per cent solution of sodium chloride. Evidently, the smaller the gold number of a lyophilic colloid, the greater is its protective power. The gold numbers and their reciprocals in the case of a few lyophilic colloids are given in Table 17.5. Thus, gelatin has the maximum and starch has the minimum coagulating power.

Table 17.5. Gold numbers and their reciprocals.

Lyophilic sols	Gold number	Reciprocal
Gelatin	0.005–0.01	200–100
Caseinate	0.01	100
Haemoglobin	0.03	33
Gum arabic	0.15	6.6
Sodium oleate	0.4	2.5
Gum tragacanth	2	0.5
Potato starch	25	0.04

EMULSIONS

Emulsions are colloidal systems in which dispersed phase as well as dispersion medium are normally liquids. The examples are: (i) milk which consists of particles of liquid fat dispersed in water; and (ii) cod-liver oil emulsion in which particles of water are dispersed in oil. The particles of dispersed phase in emulsions are generally bigger than those in sols and are sometimes visible under the microscope. These particles usually carry negative charge and are sensitive to the addition of electrolytes. They show Tyndall effect and Brownian movement. Emulsions are of considerable technical importance.

Types of Emulsions

In most cases, one of the liquid phases is an oil and the other is water. Accordingly, there are two types of emulsions: oil-in-water type in which oil is the dispersed phase and water is the dispersion medium and water-in-oil type in which water is the dispersed phase and oil is the dispersion medium. The two types may be identified in a number of ways.

One method consists in adding a small amount of an oil-soluble dye to the emulsion under examination. If it is of water-in-oil type, i.e., if oil is the continuous medium it will take up the colour of the dye readily. But if the emulsion is of oil-in-water type, the dye will not mix and the emulsion will remain colourless.

Another method involves measurement of electrical conductivity. The conductivity is appreciable if water is the dispersion medium but negligible if oil is the dispersion medium. The type of emulsion formed in a given case depends largely, though not entirely, upon the relative proportions of the two constituents present. The constituent that is in excess, generally, forms the dispersion medium.

Emulsifiers

When two immiscible liquids, such as a hydrocarbon oil and water are shaken together vigorously, a milky looking solution results. This is an emulsion consisting of small globules of oil suspended in water. However, the globules remain suspended in water for a short time only. On standing, the two liquids soon separate the oil globules rising to the top and forming a separate oil phase. In other words, the emulsion formed by merely shaking the two immiscible liquids together, is highly unstable. In order to get stable emulsions of fairly high concentrations, it is necessary to add another substance, known as an emulsifier or emulsifying agent, in a small quantity.

The emulsifying agents are generally long chain compounds with polar groups, such as soaps of various kinds, long chain sulphonic acids and alkyl sulphates. Their function is to lower the interfacial tension between water and oil so as to facilitate the mixing of the two liquids. The soap molecules get concentrated at the interface between water and oil in such a way that their polar end (—COONa) is dipping in water and the hydrocarbon chain (R-) in the oil, as shown in Fig. 17.10. This helps in bringing the two liquid phases in more intimate contact with each other.

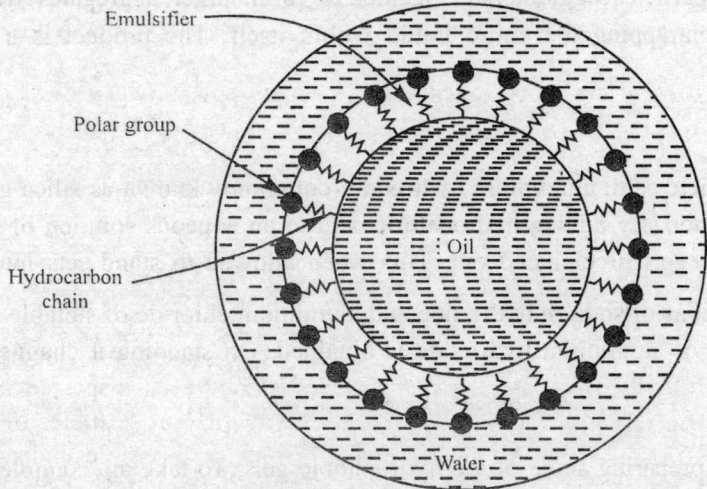

Fig. 17.10. A cross-section of an emulsion.

It may be mentioned that the detergent action of soaps and compounds of long chain sulphonic acids is also due to their ability to emulsify the grease in water. The sodium or potassium soaps yield oil-in-water type emulsions, while calcium or aluminium soaps yield water-in-oil type emulsions, provided the two liquids are present in proper proportions. There is no entirely satisfactory explanation for this behaviour.

Some lyophilic colloids, such as proteins, gums and agar-agar, also act as emulsifiers. It is believed that they act as protective agents by forming a protective layer round the globules of dispersed phase. Thus, in milk, casein (a protein) is known to form a protective layer round fat globules dispersed in water. Milk is, therefore, a fairly stable emulsion.

GELS

Several lyophilic sols and a few lyophobic sols as well, when coagulated under certain conditions, change into a semi-rigid mass, enclosing the entire amount of the liquid within itself. Such a product is called a gel. The process of transformation of a sol into a gel is known as gelation. Gel represents a liquid-solid system, i.e., a liquid dispersed in a solid. Amongst lyophilic sols, the examples are: gelatin, agar-agar, gum arabic, mastic and gamboge sols, etc. Amongst lyophobic sols, the examples are: silicic acid, ferric hydroxide, ferric phosphate sols, etc. The sols should be of sufficiently high concentration to facilitate the gelation process.

Preparation of Gels

Gels may be prepared by any one of the following methods:

Cooling of sols of moderate concentrations

Gels of gelatin and agar-agar, etc., are obtained by cooling their sols of moderate concentrations prepared in hot water. As has been mentioned earlier, the particles of hydrophilic sols are extensively hydrated. When cooled, the hydrated particles agglomerate together to form larger aggregates which ultimately form a semi-solid network entrapping the entire liquid within itself. The product is a semi-rigid gel structure.

Double decomposition

The hydrophobic gels like silicic acid, aluminium hydroxide (commonly known as silica gel and alumina gel) are prepared by this method. By adding hydrochloric acid to an aqueous solution of sodium silicate a highly hydrated silicic acid gets precipitated out. This when allowed to stand sets into a gel.

Similarly, by mixing solutions of sodium hydroxide and aluminium chloride of suitable concentrations, a highly hydrated precipitate of aluminium hydroxide is obtained. On standing it changes into a gel.

Change of solvents

This method is also used for preparing some of the hydrophobic gels. To take an example, when ethanol is added rapidly to a solution of calcium acetate of fair concentration, the salt separates out to give a colloidal solution. When allowed to stand, it undergoes gelation. The ultimate product is a semi-rigid gel of calcium acetate. The entire liquid is entrapped within.

Elastic and Non-elastic Gels

Gels are divided into two categories depending upon their properties. These are: elastic gels and non-elastic gels. The two varieties are distinguished chiefly by their behaviour on dehydration and rehydration. Elastic gels are reversible. When partially dehydrated they change into a solid mass which, however, changes back into the original form on simple addition of water followed by slight warming, if necessary. Non-elastic gels, on the contrary, are irreversible. When dehydrated they become glassy or change into a powder which on addition of water and followed by warming does not change back into the original gel.

Gelatin, agar-agar and starch are examples of elastic gels. In these cases, dehydration and rehydration on exposure to water vapour are almost reversible even when the cycle is carried out more than once.

Silica, alumina and ferric oxide gels are examples of non-elastic gels. Thus if silica gel is dehydrated, addition of water will not reset it into the form of a gel.

There is another point of difference between elastic and non-elastic gels. While elastic gels can imbibe water when placed in it and undergo swelling, non-elastic gels are incapable of doing so. This phenomenon is known as imbibition e.q. or swelling.

Another characteristic property possessed both by elastic and non-elastic gels is to undergo shrinkage in volume when allowed to stand. This phenomenon is called syneresis.

Some of the gels, particularly gelatin (reversible gel) and silica (irreversible gel) liquefy on shaking changing into the corresponding sol. The sol on standing reverts back to the gel. This phenomenon of reversible sol-gel transformation is generally referred to as thixotropy.

Colloids and Surface Area

It can easily be shown that surface area of a given mass of any material can be progressively increased when it is divided and sub-divided. Consider, for instance, a cube of edge = 1 cm. It has surface area of 6 sq. cm. If it is divided into 8 cubes of 1/2 cm length, the total surface area of the eight cubes will be $1/2 \times 1/2 \times 6 \times 8 = 12$ sq. cm. This is illustrated in Fig. 17.11. If it is divided into 1000 cubes, each with 1 mm edge, the total surface area of the small cubes will be $1000 \times 1/100 \times 6 = 60$ sq. cm. Similarly, If it is divided to 10^6 cubes, each cube with 0.1 mm. edge, the total surface area will be 600 sq. cm. It can be shown that if this process of division and sub-division is continued so that ultimately 10^{21} cubes with edges equal into 1 mμ (10^{-7} cm). are obtained, the total surface presented by the particles will amount to 6×10^7 sq. cm. or roughly equal to 105 acres. This is an amazing surface indeed! It may be noted that the same mass of material initially had a surface area of only 6 sq cm. Since colloids are very minute particles, lying within the range mμ–200 mμ, they have enormous specific surface (i.e., surface area per unit mass or per unit volume). On the contrary, the particles of suspensions (such as sand or precipitates) are so large that their specific surface is almost negligible in comparison. A number of properties of colloids are due to their large surface, the foremost of which is adsorption.

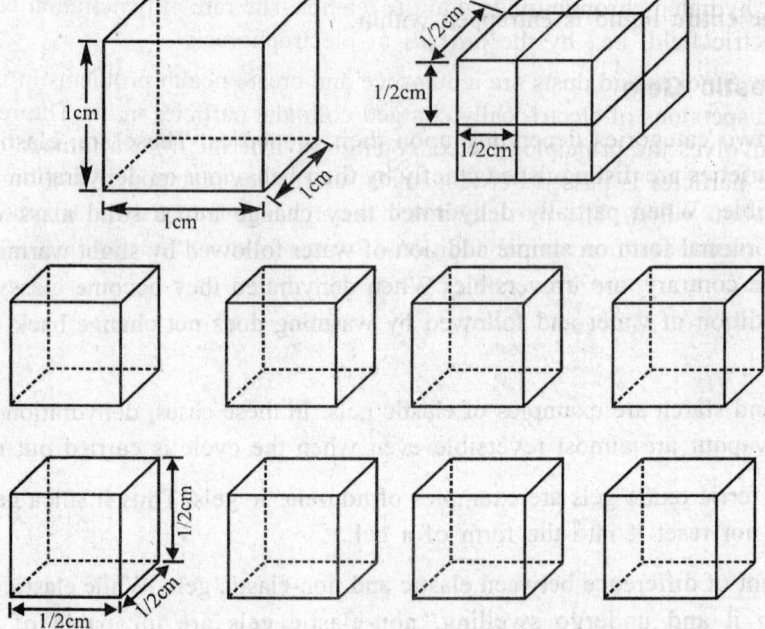

Fig. 17.11. Increase in surface area of a cubic body on sub-division.

APPLICATIONS OF COLLOIDS

Colloids play a very important role in everyday life as well as in industry, agriculture, medicine and biology. Some of the important applications of colloids are given below :

1. *Foods:* Many of our foods are colloidal in nature. For example, milk is an emulsion of fat dispersed in water. It is stabilised by casein which itself is a lyophilic colloid and, being a protein, is a nutrient of great value. Gelatin is added to ice cream as a protective agent so as to preserve its 'smoothness'. Whispered cream, fruit jellies, salad dressings, eggs and a host of other materials used as foods are colloidal in nature.

2. *Medicines:* A number of medicinal and pharmaceutical preparations are emulsions, i.e., colloidal in nature. It is believed that in this form they can be more effective and are easily assimilated. Colloidal calcium and gold, for instance, are administered by injections to raise the vitality of the human system.

3. *Industrial goods:* Soaps, the index of modern civilisation, is a colloidal electrolyte. The same is true of a series of newer detergents and wetting agents that have been produced in recent years. Paints varnishes, enamels, celluloses, resins, gums, glues and other adhesives rayon, nylon, terylene, textiles, leather, paper, etc., are all colloidal in nature. Latex, from which rubber is obtained, is a suspension of negatively charged colloidal particles of rubber. Industrial processes such as tanning, dyeing, lubrication, etc., are of colloidal in nature. This list is by no means exhaustive and can be continued further.

4. *Rubber plating:* The negatively charged particles of rubber (latex) are made to deposit on to wires or handles of various tools (in order to insulate them) by electrophoresis. The article to be rubber-plated is made the anode. The rubber particles migrate in an electric field towards the anode and get deposited on it.

5. *Chrome tanning:* The chrome tanning of leather is brought about by the penetration of positively charged particles of hydrated chromic oxide into the leather. The rate of penetration can be increased by applying an electric field, i.e., by the process of electrophoresis.

6. *Cottrell precipitator:* Smokes and dusts are a nuisance and create health problems in industrial areas. Actually these are dispersions of electri- cally charged colloidal particles in air. The removal of these particles from air involves the principle of elect rophoresis. The air from a furnace or an industrial plant carrying these particles is passed between metal electrodes maintained at a high difference of potential (about 50,000 volts). The particles are discharged and deposited as precipitates on the oppositely charged electrodes from which they can be scrapped mechanically (Fig. 17.12).

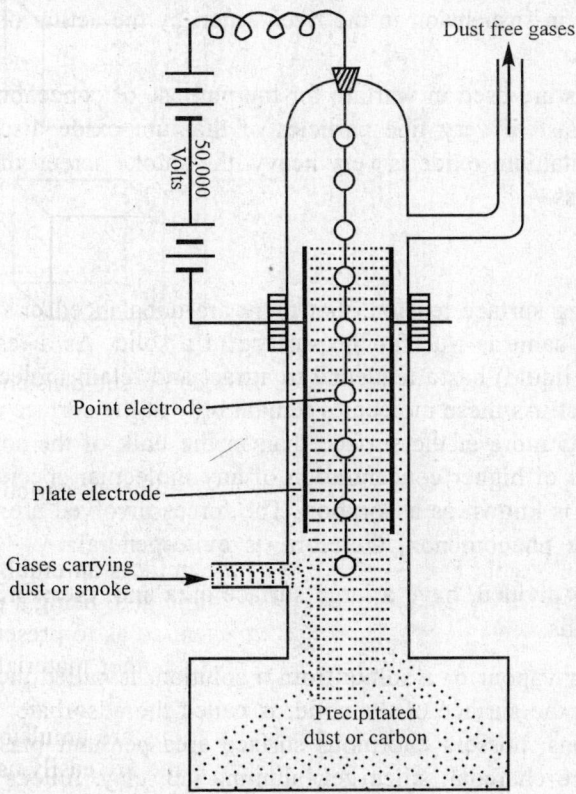

Fig. 17.12. Cottrell precipitator.

7. *Sewage disposal:* Sewage water consists of particles of dirt, rubbish, mud, etc., which are of colloidal dimensions and carry electric charge and, therefore, do not settle down easily. On creating an electric field in the sewage tank, these particles migrate to the oppositely charged electrodes, get neutralised and settle down at the bottom. It will be seen that here, too, the electrophoretic property of colloids has been made use of.

8. *Clarification of water:* Sometimes slight turbidity is noticed in water. This is due to the presence of negatively charged particles of very fine clay. The addition of potash alum or aluminium sulphate furnishes the trivalent aluminium ions (Al^{3+}) which cause the coagulation of the clay particles, which, therefore, settle down leaving water in clear state.

9. *Detergent action of soap:* Most of the dirt or dust sticks on to grease or oily materials which somehow gather on cloth. As grease is not readily wetted by water, it is difficult to clean the garment by water alone. The addition of soap lowers the interfacial tension between water and grease and this causes the emulsification of grease in water. The mechanical action, such as rubbing, etc., releases the dirt.

10. *Artificial rain:* Clouds consist of charged particles of water dispersed in air. Rain is caused by the aggregation of these minute particles. Some workers have succeeded in causing such aggregation by artificial means such as by throwing electrified sand from aeroplanes.

11. *Formation of deltas:* The deltas at the mouths of great rivers are formed by the precipitation of the charged clay particles, carried in suspension in the river-water, by the action of salts present in sea-water.

12. *Smoke screens:* Smoke screens are used in warfare for the purpose of concealment and camouflage. Smoke screens generally consist of very fine particles of titanium oxide dispersed in air and are ejected from aeroplanes. As titanium oxide is very heavy, the smoke screen drops down rapidly as a curtain of dazzling whiteness.

ADSORPTION

It was shown earlier while studying surface tension, that there are unbalanced or residual forces acting along the surface of a liquid. The same is true for the surface of a solid. As a result of these residual forces, the surface of a solid (or a liquid) has a tendency to attract and retain molecules of other species with which it is brought into contact. As these molecules remain only at the surface and do not go deeper into the bulk, their concentration is more at the surface than in the bulk of the solid (or the liquid) as the case may be. The phenomenon of higher concentration of any molecular species at the surface than in the bulk of a solid (or a liquid) is known as adsorption. The forces involved are, evidently, molecular forces (vander Waals' forces). The phenomenon, therefore, is quite general.

Solids, particularly when finely divided, have a large surface area and, therefore, show this property to a much larger extent than liquids.

The solid that takes up a gas or vapour or a solute from a solution, is called the adsorbent while the gas or the solute, which is held to the surface of the solid, is called the adsorbate. Colloids, on account of their extremely small dimensions, possess enormous surface area per unit mass and are, therefore, good adsorbents. The examples are charcoal, silica gel, alumina gel, clay, fuller's earth, etc.

Adsorption is to be distinguished carefully from absorption. The latter term implies that a substance is uniformly distributed throughout the body of a solid or a liquid. Thus, water vapour is absorbed by anhydrous calcium chloride while it is adsorbed by silica gel. Similarly, while ammonia is absorbed in water, it is adsorbed by charcoal. When a hot crucible is cooled in atmosphere, a film of moisture collects at the surface. This is a case of adsorption of water vapour on the material of the crucible. Charcoal when mixed with a coloured solution of sugar adsorbs the colouring matter and is used as a decolouriser. The term sorption should be used to describe a process in which both absorption and adsorption take place simultaneously.

Adsorption of Gases by Solids

Several methods for determining adsorption of gases on solid adsorbents, under given conditions, have been devised. In one such method, the gas is contained in a vessel of known volume at a given temperature. The pressure of the gas is measured on a manometer attached to the vessel. The adsorbent is then introduced into the vessel by a suitable device. Adsorption takes place fairly quickly and the pressure of the gas falls. This is noted on the manometer. Knowing the fall of pressure, the quantity of the gas adsorbed by the slid can be calculated, assuming Boyle's law to hold good. It is necessary to apply correction due to the volume of the adsorbent added.

Example 17.1. A sample of charcoal weighing 6.00 gm was brought into contact with a gas contained in a vessel of one litre capacity at 27°C. The pressure of the gas was found to fall from 700 to 400 mm of mercury. Calculate the volume of the gas (reduced to N.T.P.) that is adsorbed per gm. of the adsorbent under the conditions of the experiment. The density of the charcoal sample was 1.5 gm per c.c.

Solution.

Initial pressure of the gas = 700 mm

Initial volume of the gas = 1 litre

Final pressure of the gas = 400 mm

Let the final volume of the gas be V ml

∴ $700 \times 1000 = 400 \times V$

∴ $V = \dfrac{700 \times 1000}{400} = 1750$ ml

But, the volume of the gas in the flask = 1000– the volume of the solid.

$$= 100 - \frac{6.00}{1.50} = 996 \text{ ml}$$

∴   Volume of the gas adsorbed by charcoal

   = 1750 – 996 = 754 ml

∴   Volume of the gas adsorbed per gm of charcoal

   $= \dfrac{754}{6} = 125.6$ ml

∴   Volume of the gas V_1 at *N.T.P.* adsorbed per gm. of charcoal will be given by

∴ $V_1 = \dfrac{VPT_1}{TP_1} = \dfrac{125.6 \times 400 \times 273}{300 \times 760}$

   = 60.15 ml

Heat of Adsorption

As a result of adsorption, there is decrease in residual forces acting along the surface of the adsorbent. Consequently, there is decrease of surface energy which appears as heat. Adsorption, therefore, is invariably accompanied by evolution of heat. The amount of heat evolved when one mole of a gas or vapour is adsorbed on a solid is known as molar heat of adsorption. This value depends largely upon the nature of the gas. It is of the same order but always greater than the molar heat of condensation of the gas.

Adsorption is invariably accompanied by decrease in enthalpy of the system, i.e., ΔH of the process is invariably negative. Further, since absorbate changes from the more random gaseous (or solution) state to the less random adsorbed state on the surface of a solid, adsorption is also accompanied by decrease in entropy of the system, i.e., $T\Delta S$ of the process is also negative. Adsorption is thus accompanied by decrease in enthalpy as well as decrease in entropy of the system. The free energy change, ΔG, for any feasible process, as given by the well known thermodynamic equation

$$\Delta G = \Delta H - T\Delta S \qquad \qquad ...(17.5)$$

should be negative. Since ΔH and ΔS are both negative, it is evident that ΔH must have a sufficiently high negative value so that the net result of the expression on the right hand side of equation (17.5) should be negative, i.e., the value of $\Delta H - T\Delta S$ should be negative. This is actually the case. However, we know from experiment that the heat of adsorption per mole of the adsorbate goes on decreasing, i.e., ΔH becomes less and less negative, as adsorption proceeds further and further. Ultimately, ΔH becomes equal to $T\Delta S$ and ΔG becomes zero. This is the stage at which equilibrium is attained.

Factors Influencing Adsorption

The magnitude of gaseous adsorption depends upon the following factors: (i) temperature; (ii) pressure; (iii) nature of the gas and; (iv) nature of the adsorbent.

Effect of temperature and pressure

Since adsorption is invariably accompanied by evolution of heat, therefore in accordance with Le Chatelier's principle, the magnitude of adsorption should increase with fall in temperature. This actually happens. Further, since adsorption of a gas leads to decrease of pressure, the magnitude of adsorption increases with increase in pressure.

Thus, decrease of temperature and increase of pressure both lend to cause increase in the magnitude of adsorption of a gas on a solid.

The variation of adsorption with pressure at a constant temperature is generally expressed graphically as in Fig. 17.13. Each curve is known as adsorption isotherm for the particular temperature. The relationship between the magnitude of adsorption and pressure can be expressed mathematically by an empirical equation commonly known as Freundlich adsorption isotherm, viz.,

$$a = k\,p^n \qquad \qquad ...(17.6)$$

where a is the amount of gas adsorbed per gram of adsorbent at the pressure p, and k and n are constants depending upon the nature of the gas and adsorbent. The value of n is less than 1 and therefore, a does not increase as rapidly as p, as is evident from the isotherms (Fig. 17.13).

Nature of the gas and nature of the adsorbent

It has been found that the more readily soluble and easily liquefiable gases such as ammonia, hydrochloric acid, chlorine and sulphur dioxide are adsorbed more than so called 'permanent' gases such as hydrogen, nitrogen and oxygen. The reason is that van der Waals' or molecular forces which are involved in adsorption are more predominant in the former category than in the latter category of gases.

Since adsorption is a surface phenomenon, it is evident that the greater the surface area per unit mass of the adsorbent, the greater is its capacity for adsorption under given conditions of temperature and pressure.

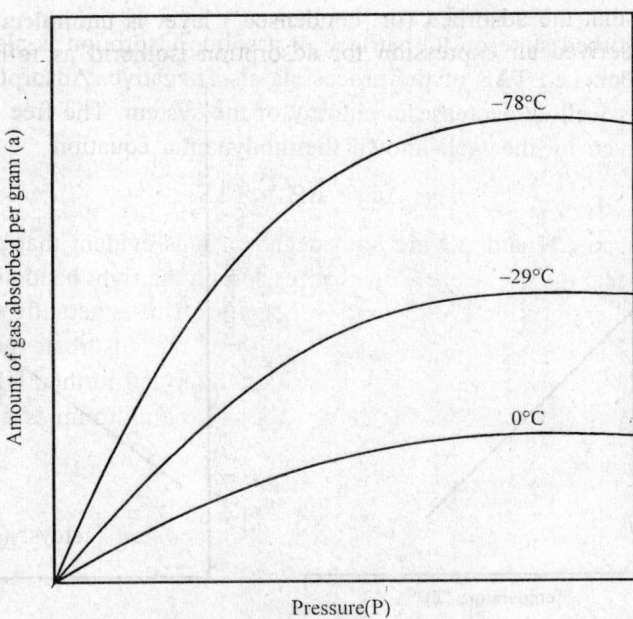

Fig. 17.13. Adsorption isotherms.

Adsorption Isobar and Adsorption Isostere

The effect of temperature (T) on the extent of adsorption (a), at a given pressure of the adsorbate, is also expressed graphically, as shown in Fig. 17.14.

The curve showing the effect of temperature on the extent of adsorption at a given pressure is called an adsorption isobar. The amount of adsorption, evidently, decreases with rise in temperature. This is in accordance with Le Chatelier's principle, the process of adsorption being exothermic. The curve showing the variation of pressure (P) with temperature (T), for a given amount of adsorption, is called an isostere (Fig. 17.15).

Since rise of temperature tends to lower the extent of adsorption, it is evident that in order to get the same amount of adsorption at the higher temperature, we have to raise the pressure of the system as well. Hence a straight line relationship between temperature and pressure is generally obtained.

LANGMUIR'S UNIMOLECULAR ADSORPTION

Langmuir, in 1916, put forth the view that gases adsorbed on a solid surface cannot form a layer more than one molecule in thickness. According to him, the magnitude of adsorption of gas on a solid at a given temperature, goes on increasing with increase in pressure until the entire surface of the solid gets completely covered by a unimolecular layer of the gas. Since adsorption is unimolecular, no further adsorption is possible. The adsorption isotherm, therefore, take the form as shown in Fig. 17.13. The thickness e.q. the adsorbed layer, therefore, is the same as the thickness of a molecule of the adsorbate.

Langmuir visualised the adsorption process to consist of two opposing tendencies of gaseous molecules: (i) the tendency of molecules to condense from gaseous phase on to the surface of a solid; and (ii) the tendency of these molecules to evaporate back into the gaseous phase.

He assumed further that the adsorbed (or 'condensed') layer is unimolecular in thickness. **Based on** these assumptions, he derived an expression for adsorption isotherm as follows.

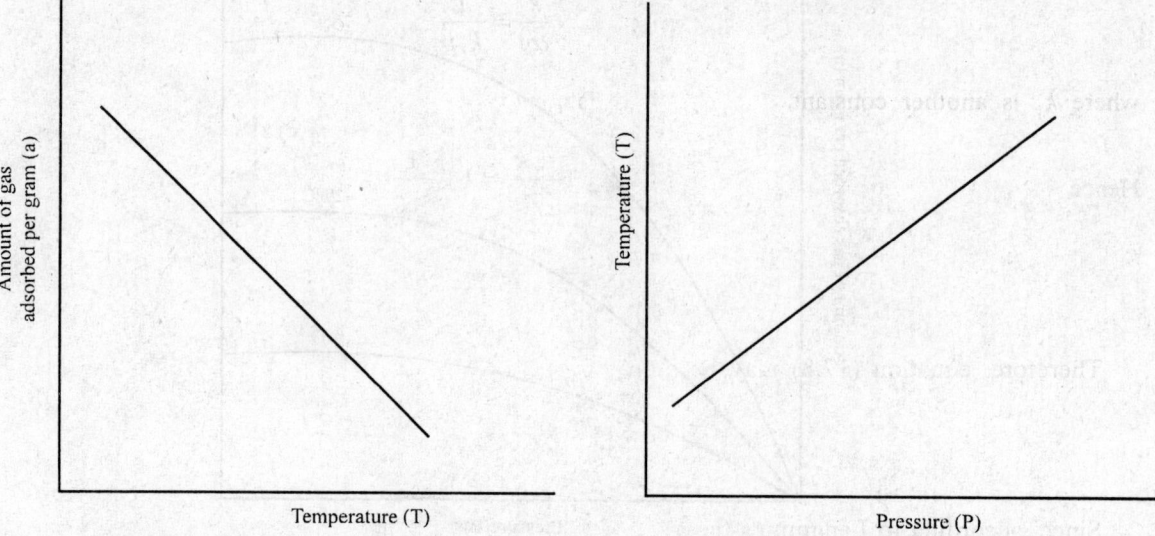

Fig. 17.14. Adsorption isobar.　　　　　　**Fig. 17.15.** Adsorption isostere.

Imagine 1 sq. cm of a surface exposed to a gas maintained at the pressure p. Let n be the number of molecules of the gas (proportional to the pressure of the gas) which strike one square centimetre of the surface per second and let α (a constant depending upon the nature of the gas) be the fraction of the molecules that adhere or 'condense' on the surface. This, an is the number of molecules which condense per sq. cm of the available surface per second. Now, suppose θ is the fraction of the surface already covered by the gaseous molecules. Then, $(1-\theta)$ sq. cm. of the surface will be bare and available for condensation of the molecules. The number of molecules condensing per second on the surface will, therefore, be $\alpha n (1-\theta)$. In other words:

Rate of 'condensation' of gaseous molecules $= \alpha n (1 - \theta)$

The rate of 'evaporation' of the gaseous molecules will be proportional to the area of the surface covered θ.

$\therefore$ Rate of evaporation of gaseous molecules $= k\theta$

Since, at equilibrium, the rate of condensation must be equal to the rate of evaporation, hence

$$\alpha n (1-\theta) = k\theta \qquad \qquad ...(17.7)$$

or

$$\theta = \frac{\alpha n}{\alpha n + k} \qquad \qquad ...(17.8)$$

Now

$$\frac{\alpha n + k}{\alpha n} = 1 + \frac{k}{\alpha n}$$

Since, k and α are constants and n is proportional to the pressure of the gas p, it follows that

$$\frac{k}{\alpha n} = \frac{1}{k_1 p} \qquad \qquad ...(17.9)$$

where k_1 is another constant.

Hence

$$1 + \frac{k}{\alpha n} = 1 + \frac{1}{k_1 p}$$

$$= \frac{k_1 p + 1}{k_1 p} \qquad \qquad ...(17.10)$$

Therefore, equation (17.8) may be put as :

$$\theta = \frac{k_1 p}{1 + k_1 p} \qquad \qquad ...(17.11)$$

Since, according to Langmuir's theory, only a single layer of molecules can be adsorbed, the fraction θ is proportional to the amount of gas adsorbed per unit mass of the adsorbent. Representing this quantity by α, we may express the statement as:

$$a \propto \theta$$

or

$$a = k_2 \theta \qquad \qquad ...(17.12)$$

where k_2 is another proportionality constant.

Substituting θ from equation (12.11), we have

$$a = k_2 \theta = \frac{k_1 k_2 p}{1 + k_1 p} \qquad \qquad ...(17.13)$$

The amount of adsorption per gram of adsorbent (a) is generally represented by x/m where x is the total amount of gas adsorbed on m gram of adsorbent. The equation (17.13), therefore, is written more frequently as :

$$a = \frac{x}{m} = \frac{k_1 k_2 p}{1 + k_1 p} \qquad \qquad ...(17.14)$$

Now, two extreme cases of adsorption may be considered :

1. When adsorption is very little, i.e., when the pressure of the gas is very low, θ may be negligibly small in comparison to unity and, therefore, equation (17.7) may be put as :

$$\alpha n = k\theta$$

or

$$\theta = \frac{\alpha n}{k}$$

$$= k_1 p \qquad \qquad ...(\text{from equation } 17.9)$$

Substituting in equation (17.12), we have:

$$a = \frac{x}{m} = k_1 k_2 p \qquad ...(17.15)$$

or

$$\frac{x}{m} \infty \, p \qquad ...(17.16)$$

Thus, at low pressures, the magnitude of adsorption is directly proportional to the pressure.

2. On the other extreme, when adsorption is considerable, i.e., when the pressure of the gas is high, θ may be taken as close to unity and the equation (17.7) may be put as:

$$\alpha n \, (1-\theta) = k$$

$$1 - \theta = \frac{k}{\alpha n}$$

$$\theta = 1 - \frac{k}{\alpha n}$$

$$= k_2 - \frac{k_2}{k_1 p} \qquad ...(\text{from equation } 17.9)$$

Substituting in equation (17.12), we have:

$$a = \frac{x}{m} = k_2 \left(1 - \frac{1}{k_1 p} \right)$$

$$= k_2 - \frac{k_2}{k_1 p} \qquad ...(17.17)$$

Evidently, as p increases, $\dfrac{k_2}{k_1 p}$ becomes smaller and smaller and, therefore, the magnitude of adsorption, x/m, tends to approach a limiting value close to k_2. Thus at high pressures,

$$a = k_2 \qquad ...(17.18)$$

In other words, when the pressure is high, adsorption approaches a constant value and becomes independent of further rise in pressure.

Combining the results of equations (17.15) and (17.18), it is evident that according to this theory, the magnitude of adsorption, at a given temperature should first increase in proportion to increase of pressure and later on should tend to attain a certain limiting value. This is exactly as indicated by the adsorption isotherms plotted in Fig 17.13.

The results can also be looked at in another way. Since at low pressures, adsorption is proportional to pressure, i.e.,

$$\frac{x}{m} = kp^1 \qquad \text{...(from equation 17.15)}$$

and, at high pressures,

$$\frac{x}{m} = k = kp^0 \qquad \text{...(from of equation 17.18)}$$

it follows that, at intermediate pressure, the following expression should hold good :

$$\frac{x}{m} = kp^n \qquad \text{...(17.19)}$$

where n lies between 0 and 1.

This equation is exactly the same as the Freundlich experimental equation (17.6). This agreement was taken as a proof for the theory of unimolecular adsorption advanced by Langmuir. But it is now known that the same equation can be deduced from other considerations as well.

Limitation of Langmuir's Theory

There is now little doubt that the Langmuir theory of unimolecular adsorption is valid only at comparatively low pressures and high temperatures. As the pressure is increased or temperature is lowered, additional layers are formed. This has led to the modern concept of multimolecular adsorption.

ADSORPTION FROM SOLUTION PHASE

Solids can adsorb substances from solution phase as well. The use of activated carbon in decolourising sugar solutions by adsorbing the coloured impurities on its surface is probably a very early example of this phenomenon. Similarly, when a solution of acetic acid in water is shaken with charcoal, a part of the acid is adsorbed by the charcoal and the concentration of the solution decreases. This type of adsorption is also influenced appreciably by temperature, the extent of adsorption decreases with rise in temperature. The increase in concentration of the solution also causes an increase in the magnitude of adsorption at a given temperature. The effect of concentration can be represented by Freundlich's equation, which may be written as:

$$a = k\, c^n \qquad \text{...(17.20)}$$

where a is the amount of solute adsorbed by a unit mass of adsorbent from solution of concentration c, at the given temperature and k and n are constants for the given adsorbent and solute. The value of n is less than unity.

Therefore, on plotting a against c, a parabolic curve is obtained, as shown in Fig. 17.16.

Thus, on plotting log a against log c, a straight line (Fig. 17.17) should be obtained. This is generally the case provided the concentration varies within a small range only. The slope of the line gives the value of n and the intercept on the y-axis gives the value of log k from where the value of the constant k can be calculated.

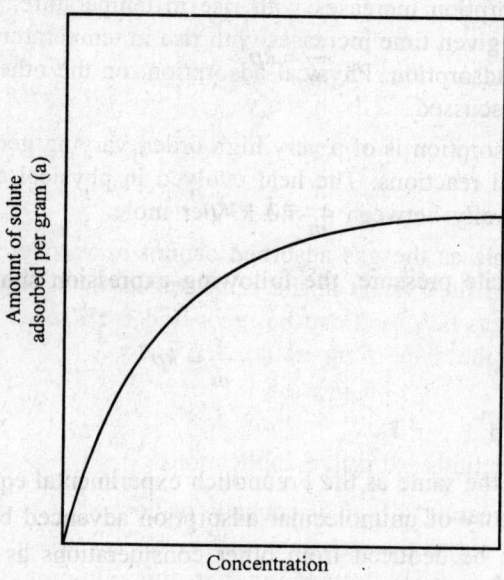

Fig. 17.16. Adsorption from solution phase.

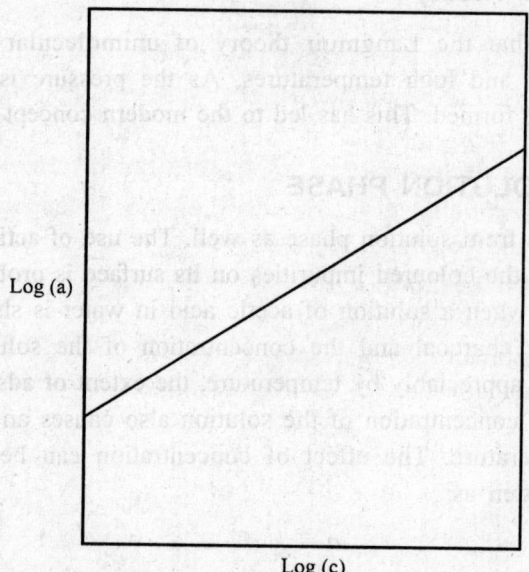

Fig. 17.17. Adsorption from solution phase.

Chemisorption

Adsorption is not necessarily a physical phenomenon always. It may as well be a chemical process involving chemical interaction between the surface-atoms of the adsorbent and the atoms of the adsorbate. This type of adsorption is known as chemisorption. For example, oxygen is chemisorbed by tungsten as well as carbon and hydrogen is chemisorbed by nickel under suitable conditions. In each case, a stable surface compound, frequently referred to as surface complex, results. Chemisorption differs from physical adsorption in three important respects:

1. The magnitude of chemisorption increases with rise in temperature. This is just as the magnitude of a chemical reaction in a given time increases with rise in temperature. Chemisorption is, therefore, often termed as activated adsorption. Physical adsorption, on the other hand, decreases with rise in temperature, as already discussed.

2. The heat evolved in chemisorption is of a very high order, varying generally between 40–400 kJ per mole, as in many chemical reactions. The heat evolved in physical adsorption, on the other hand, is quite low, varying generally between 4–40 kJ per mole.

3. Chemisorption is irreversible as the gas adsorbed cannot be recovered from the adsorbent as such on lowering the pressure of the system at the same temperature. Physical adsorption, on the other hand, is reversible as the gas adsorbed can be recovered from he adsorbent easily on lowering the pressure of the system at the same temperature.

Applications of Adsorption

The phenomenon of adsorption finds extensive applications. A few examples are given below:

1. A very good method of creating a high vacuum is to connect a bulb of charcoal cooled in liquid air to a vessel which has already been exhausted as far as possible by vacuum pump. Since the magnitude of adsorption at such low temperatures is quite high, the remaining traces of air in spite of the low pressure, are adsorbed by the charcoal almost completely.

2. Activated charcoal is used in gas masks in which all toxic gases and vapours are adsorbed by the charcoal while pure air passes through its pores practically unchanged.

3. Silica and alumina gels are used as adsorbents for removing moisture and for controlling humidities of rooms.

4. Animal charcoal is used as a decolouriser in the manufacture of cane-sugar.

5. Soil contains small amounts of colloidal fractions in the form of very fine particles of clay. It can, therefore, always adsorb and retain certain amount of moisture in which nutrients, such as compounds of nitrogen, phosphorus and potassium, can dissolve and pass up to the plant through the roots.

6. Adsorption also plays an important role in heterogeneous catalysis, e.g., the use of finely divided iron in the manufacture of ammonia and that of finely divided nickel in the hydrogenation of oils.

Chapter 18

Stereoisomerism

INTRODUCTION

The name isomerism is given to the phenomenon whereby two or more different compounds can exist having the same molecular formula but different arrangements of their atoms in the molecule. Isomerism is of two main kinds, structural isomerism, in which the same atoms are linked together in different ways so that the isomers have different structural formulae, and stereoisomerism, in which the same atoms are linked together in the same ways but are arranged differently in space. It follows that stereoisomers have identical structural formulae, since the different spatial configurations of the atoms cannot be shown in such formulae. Table 18.1 shows the various types of isomerism and refers to examples of each type. Since many instances of structural isomerism have already been encountered, this kind should now be thoroughly understood. Consequently this chapter will be devoted to the various types of stereoisomerism.

OPTICAL ISOMERISM

Light may be regarded as a transverse wave motion. When it passes through certain solids it becomes plane-polarised, i.e. it vibrates at right angles to the direction of propagation in only one plane instead of in an infinite number of such planes. Fig. 18.1, which represents a cross-sectional diagram of an oncoming ray of light, should make this clear; ordinary light is shown as vibrating in all directions, whereas light that has passed through a Nicol prism made from pieces of calcite or through a sheet of polaroid material vibrates only in one direction, say vertically up and down. If this polarised light is passed through certain substances it is found that the plane of polarisation is rotated. If the rotation is in a clockwise direction as observed in the eyepiece the substance is said to be dextro-rotatory, and if in an anti-clockwise direction, levo-rotatory. Substances which cause rotation of the plane of polarised light in this way are said to be optically active.

They are of two kinds, those which are optically active only in the solid state, such as quartz, and those compounds which are optically active when molten or in solution as well as when crystalline. In this chapter we are concerned only with the latter kind.

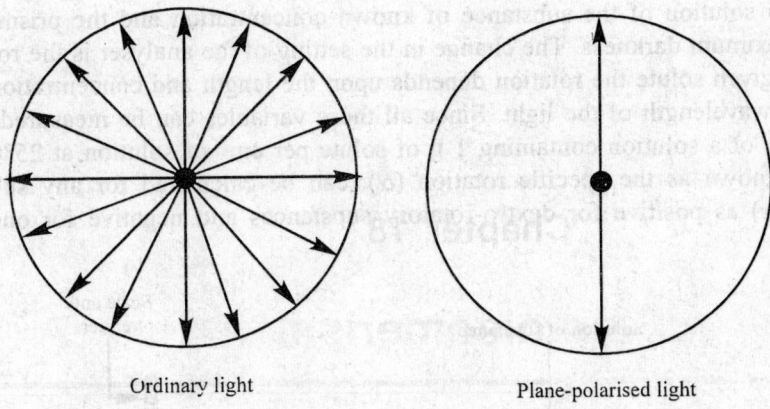

Ordinary light   Plane-polarised light

Fig. 18.1. Polarisation of light.

Table 18.1. Types of isomerism.

Structural
Isomerism
(atoms linked in
different ways)

Chain Isomerism — Different carbon chains or skeletons, e.g., the paraffins

Metamerism — Different radicals linked to the same functional group, e.g., the ethers

Position Isomerism — Substituent groups in different, e.g., the propyl alcohols
(35.4) and o,m, + p. aromatic nuclear derivatives.

Series Isomerism — Involves compounds in different homologous series,
e.g., { ethyl alcohol and dimethyl ether / acetic acid and methyl formate / acetone and propionaldehyde }

Isomerism

Tautomerism — A dynamic equilibrium exists between the isomers,
e.g., { Glycine / Urea }

Optical Isomerism — Occurs in compounds with asymmetric carbon atoms.
e.g., { Lactic acid / Tartaric acid }

Stereos Isomerism
(atoms arranged
differently in space)

Geometrical Isomerism — Occurs in certain compounds containing double bonds,
e.g., maleic and fumaric acids.

Cyclic Isomerism — Occurs in certain ring molecules,
e.g., { Glucose / Benzene hexachloride }

The apparatus used for studying optical activity is called a polarimeter. It consists of a glass tank placed between two Nicol prisms as shown diagrammatically in Fig. 18.2. To measure the optical activity of a given substance the prism B is first adjusted to a position where minimum light is observed in the eyepiece and its setting is noted. In this position the axes of the two prisms are exactly at right angles so that none of the polarised light produced by the polariser A is transmitted through the analyser B. The tank is then filled with a solution of the substance of known concentration and the prism B is rotated to the new position of maximum darkness. The change in the setting of the analyser is the rotation caused by the solution. For any given solute the rotation depends upon the length and concentration of solution, the temperature, and the wavelength of the light. Since all these variables can be measured or specified, the rotation per decimetre of a solution containing 1 g of solute per cm^3 of solution at 25°C for sodium light of D wavelength, known as the specific rotation (α), can be calculated for any substance. It is conventional to regard (α) as positive for dextro-rotatory substances and negative for ones which are laevo-rotatory.

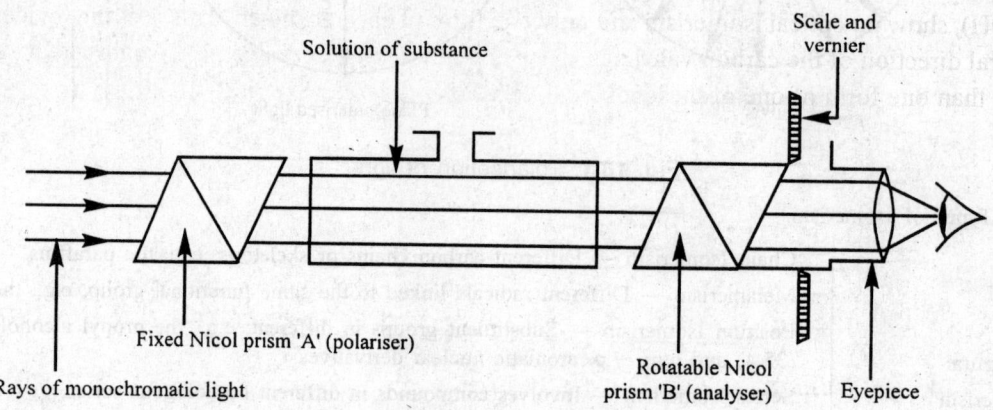

Fig. 18.2. The polarimeter.

Only a small proportion of organic substances are optically active. When these substances are thoroughly investigated each is found to exist in two or more forms which differ in the direction in which they rotate polarised light. These forms, called optical isomers, are very similar in their physical and chemical properties, the main difference between them being their opposite effect upon polarised light. The two isomers are distinguished by prefixing their names by d- and l-. These prefixes stand for dextro and laevo respectively, and indicate the direction in which the plane of polarised light is rotated by a solution of the substance. They are not intended to denote the absolute configurations of the atoms in the molecule, which should be indicated by the prefixed D- and L-, and which are not considered in this book.

When the structures of these optically active substances are determined it is found that they all have one feature in common—the presence of one or more asymmetric carbon atoms, i.e. carbon atoms linked to four different atoms or groups. Fig. 18.3 shows a theoretical example, Cabcd, projected on to a plane, where a, b, c, and d are all different. The carbon atom itself is not asymmetrical, but the arrangement of atoms around it is, giving an asymmetry to the molecule as a whole. All molecules containing asymmetric carbon atoms give rise to optical isomers, the maximum number being 2^n, where n is the number of asymmetric carbon atoms in the molecule.

The explanation of the existence of optical isomers lies in the tetrahedrally-directed valencies of the carbon atom, (as suggested by van't Hof and Le Bel). The molecule Cabcd can be regarded as a tetrahedron with the groups a, b, c, and d at the corners and the carbon atom at the centre, so that the angle between each of the bonds is 109°28'. This view is confirmed by X-ray analysis, electron diffraction, and other physical evidence. On consideration it will be seen that two forms of such a molecule are possible, as shown in Fig. 18.4. These two forms are mirror images of each other and are known as enantiomorphs. They cannot be superimposed upon each other, their relationship being similar to that of a right-hand glove and a left-hand glove. These enantiomorphs are identical in their physical properties except in their effect upon polarised light, which they rotate by equal amounts in opposite directions. Thus optical isomerism is accounted for by the possible alternative spatial configurations of the atoms when four different atoms or groups are arranged tetrahedrally around a carbon atom.

Molecules of the types Ca_4 (e.g. CH_4), Ca_3b (e.g. CH_3Cl), Ca_2b_2 (e.g. CH_2Cl_2), and Ca_2Bc (e.g. $CH_2Cl.COOH$), show no optical isomerism and only one form of each is known. This is strong evidence for the tetrahedral direction of the carbon valencies, since any other spatial arrangement would inevitably give rise to more than one form of one of these molecules and some stereoisomerism would be expected.

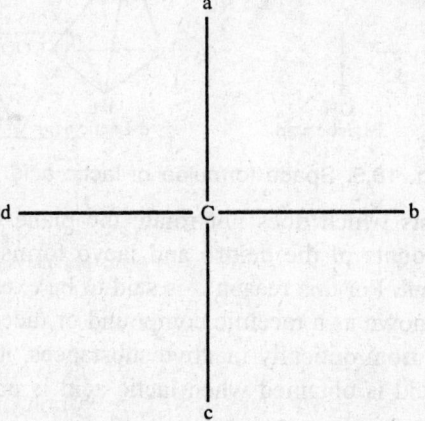

Fig. 18.3. Asymmetric carbon atom.

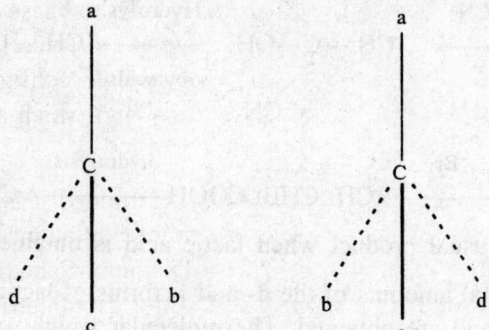

Fig. 18.4. Enantiomorphs.

Lactic Acid

$CH_3CH(OH).COOH$, this compound is a colourless hygroscopic solid with a sour taste and smell, but it is so difficult to crystallise that it is usually encountered as a syrupy liquid. It is very soluble in water.

Lactic acid is one of the simplest compounds to show optical isomerism, existing in dextro and laevo forms. The molecule contains an asymmetric carbon atom and the two isomers correspond to the two enantiomorphous arrangements of the atoms shown in Fig. 18.5. In the space formulae on the left-hand side, bonds in the plane of the paper are represented by ordinary lines, bonds above the plane of the paper by think heavy lines, and bonds below the plane of the paper by broken lines. In the alternative space formulae the molecule is represented by a solid tetrahedron with the asymmetric carbon atom (not shown) at the centre and the various atoms or groups at the corners. Both types of formula show the mirror image relationship between the two isomers which accounts or their opposite effects upon polarised light, but the student should construct some three-dimensional models to satisfy himself that the difference is clearly understood.

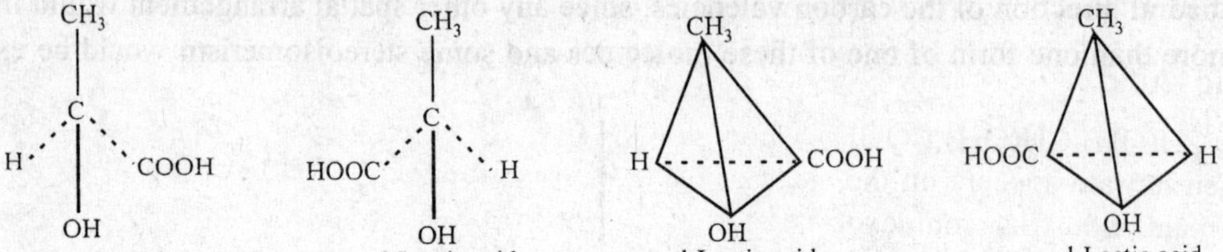

Fig. 18.5. Space formulae of lactic acid.

A third form of lactic acid exists which does not rotate the plane of polarised light. This optically inactive form consists of equal amounts of the dextro and laevo forms combined together so that their effects upon polarised light cancel out. For this reason it is said to be externally compensated and is called a dl compound. Alternatively it is known as a racemic compound or racemate. When a compound capable of optical isomerism is synthesised from optically inactive substances, it is the racemic compound which is formed. For example, dl-lactic acid is obtained when lactic acid is prepared in either of the following ways :

1.

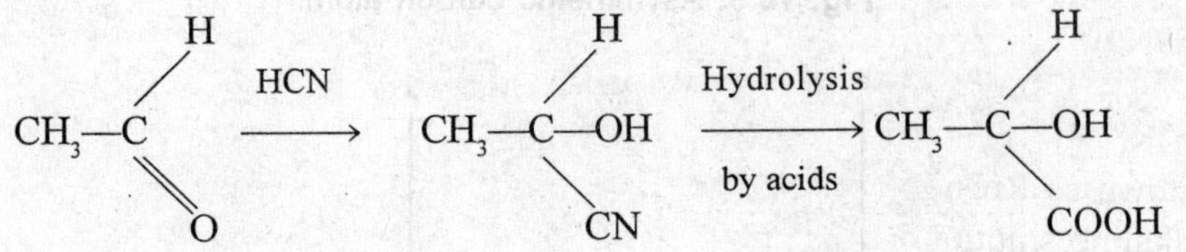

2. $CH_3.CH_2.COOH \xrightarrow{\quad Br_2 \quad} CH_3.CHBr.COOH \xrightarrow{\quad hydrolysis \quad} CH_3.CH(OH).COOH.$

The racemic acid is also the usual product when lactic acid is obtained from sour milk.

When solutions containing equal amounts of the d- and l- forms of lactic acid are mixed and allowed to crystallise, crystals of the dl acid are obtained. The molecular weight of these crystals is twice that of the individual d- and l- forms, suggesting that the racemic compound resembles a double salt. It melts at 18°C and dissolves in water giving a solution containing equal amounts of each optically active isomer :

$$dl\text{-lactic acid} \rightleftharpoons d\text{-lactic acid} + l\text{-lactic acid}$$

For a few substances, but not for lactic acid, the position of this equilibrium is so far over to the right-hand side that separate crystals of the two optically active isomers are deposited when the solution is concentrated giving a racemic mixture. Such a mixture is optically inactive when dissolved in water because it contains equal amounts of each isomer. In rare cases the crystals in a racemic mixture can be distinguished from each other by their enantiomorphous shapes and can be separated by hand picking.

d-Lactic acid, m.p. 26°C, occurs naturally in muscle and is usually prepared from meat extract. The laevo form, m.p. 26°C, does not occur naturally, but it can be prepared from the racemic compound by adding an optically active base such as d-cinchonine (which occurs naturally in Peruvian bark), or l-brucine, or l-strychnine. If d-cinchonine is used, the two salts d-lactic acid d-cinchonine and l-lactic acid d-cinchonine are formed. These salts differ in their solubility and can therefore be separated from each other by fractional crystallisation. They are then reconverted into the separate d and l forms of lactic acid by acidification. This is the best general method of separating or resolving racemic forms into their optically active constituents. Other methods, such as the use of selective fermentation of one form by moulds, or the physical separation of enantiomorphous crystals in a racemic mixture by handpicking, are of only very limited application and consequently of little practical importance.

Tartaric Acid

COOH.CH(OH).CH(OH).COOH, when grape juice is fermented to make wine, impure potassium hydrogen tartrate appears on the sides of the vessel as a brown crystalline deposit known as argol. This is the main industrial source of tartaric acid, which is used, mixed with sodium bicarbonate, as an ingredient of baking powders and health salts because the mixture effervesces when water is added.

Tartaric acid is a colourless crystalline solid, very soluble in water. It decomposes when heated, leaving a black residue of carbon. Being a dibasic acid, it gives rise to two series of salts, e.g. normal

potassium tartrate,
$$\begin{array}{l} CH(OH.COOK \\ | \\ CH(OH).COOK, \end{array}$$
and the acid salt potassium hydrogen tartrate
$$\begin{array}{l} CH(OH.COOK, \\ | \\ CH(OH).COOK \end{array}$$

which is commonly known as cream of tartar. This acid salt is important because it is only slightly soluble in water and can therefore be used for detecting potassium in analysis. When its solution is neutralised with sodium carbonate and evaporated, crystals of sodium potassium tartrate,

$$\begin{array}{l} CH(OH).COONa \\ | \qquad\qquad .4H_2O \\ CH(OH).COOK \end{array}$$

are obtained, known as Rochelle salt. The latter is used in the laboratory for making Fehling's solution by adding its alkaline solution to a solution of copper sulphate.

Tartaric acid and its salts are strong reducing agents. They decolourise potassium permanganate solution and give a silver mirror with an ammoniacal solution of silver oxide. Tartrates can be distinguished from oxalates by adding calcium chloride solution because the white precipitate of calcium tartrate, unlike calcium oxalate, dissolves in acetic acid.

Two optically active isomers of tartaric acid are known which are identical in all their properties except in their effect upon the plane of polarised light, which they rotate in opposite directions. They are referred to as the d- and l- forms accordingly. In addition, two optically inactive isomers exist, the racemic compound or dl acid, and mesotartaric acid. These two forms differ from each other and from the d- and l- forms in their physical properties and in their crystalline shape. Isomers of this kind which are not enantiomorphous and differ physically are called diastereoisomers.

When tartaric acid is prepared from argol in the usual way, the d-form is obtained. If this acid is heated with water under pressure to about 165°C it is converted into a mixture of racemic and mesotartaric acids which can be separated by fractional crystallisation. This process whereby optically active isomers change into the racemic form when heated is called racemisation. The laevo acid is prepared from the racemic acid by resolving it with an optically active base, as for lactic acid.

The stereoisomerism of tartaric acid can be readily accounted for in terms of the three possible configurations of the atoms in a molecule which contains two asymmetric carbon atoms. These configurations are shown by means of space formulae and tetrahedral formulae in Fig. 18.6. In explaining the difference between these isomers it is helpful to regard the molecule of tartaric acid as composed of two similar groups linked together, each containing one asymmetric carbon atom. If the configurations of these two groups are such that they are both dextro-rotatory, for example, then the two halves of the molecule reinforce one another in their effect and the molecule as a whole will be dextro-rotatory too, and we shall have the d-form of the acid. Conversely, if the effect of both groups is to rotate the plane of polarised light in an anti-clockwise direction, then the molecule will be that of l-tartaric acid. As Fig. 18.6 makes clear, these two forms will be enantiomorphous with each other.

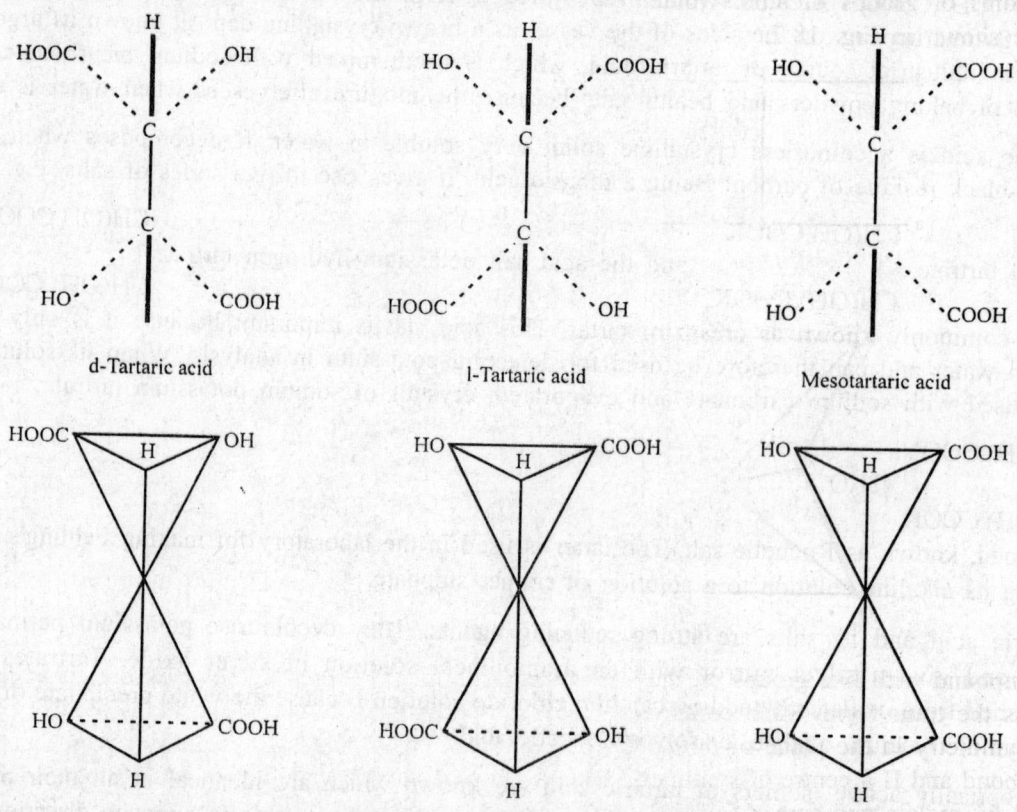

Fig. 18.6. Space formulae of tartaric acid.

A third type of molecule will exist in which a group with dextro-rotatory configuration will be linked to one with laevo-rotatory properties. In this case the two parts of the molecule will have equal but opposite effects upon polarised light and the molecule as a whole will be optically inactive. Mesotartaric acid is produced in this way and is said to be internally compensated. It cannot be separated into the d- and l-forms because it does not contain either of them; its molecular weight is the same as that of the optically active forms.

The fourth isomer of tartaric acid, the racemic acid, is composed of equal amounts of the d- and l-forms. It is optically inactive by external compensation, therefore, and like the racemic form of lactic acid it can be separated into the d- and l-forms by the usual methods. Its melting point (206°C) is higher than that of the d- and l-forms (170°C) showing that it is a distinct compound and not merely a mixture of the two isomers.

GEOMETRICAL ISOMERISM

Two atoms linked by a single covalent bond may be regarded as generally free to rotate independently about the axis of the bond, for if this were not so a compound such as ethylene dichloride, $CH_2Cl.CH_2Cl$, would show stereoisomerism. The same is not true of two atoms joined by a double bond, for this confers upon a molecule a certain rigidity giving rise to a form of stereoisomerism known as geometrical isomerism. The point is best illustrated by the hypothetical compound Cab : Cab, where a and b are different atoms or groups of atoms joined to the two carbon atoms. The configurations of the two isomers are shown in Fig. 18.7.

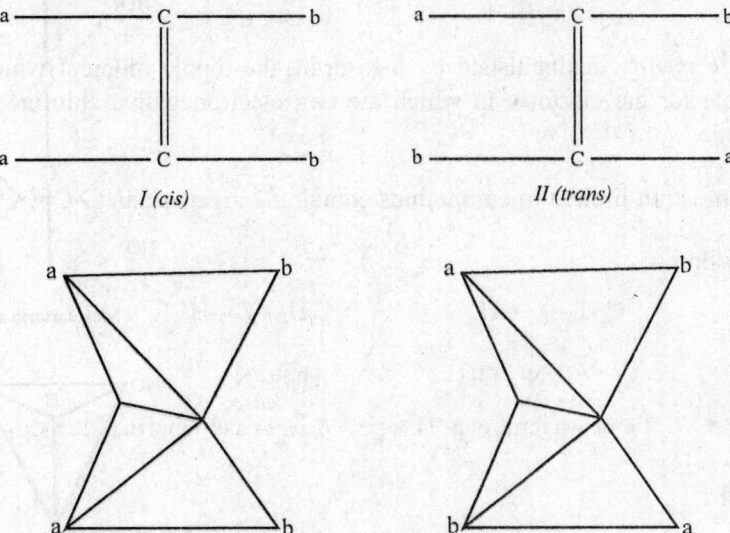

Fig. 18.7. Geometrical isomers.

The compound with like groups on the same side of the double bond is known as the cis form, and the other as the trans form. Such compounds are not optical isomers because both molecules contain planes of symmetry in the plane of the paper. In addition, I has a plane of symmetry at right angles to the C = C bond and II a centre of symmetry. One of the best-known examples of this type of isomerism is provided by maleic and fumaric acids, which may be represented by the formulae :

H—C—COOH
‖
H—C—COOH

amaleic acid
m.p. 130°Cp 1.59 g cm⁻³
solubility: 79 g in 100 g water
at 25°C

H—C—COOH
‖
HOOC—C—H

fumaric acid
m.p. 287°Cp 1.64 g cm⁻³
solubility: 0.7 g in 100 g water
at 25°C

These two forms also differ chemically, maleic acid readily forming an anhydride when heated to 160°C owing to the proximity of the two carboxyl groups, whereas fumaric acid gives the anhydride only above 230°C :

$$
\begin{array}{ccc}
H\!-\!C\!-\!COOH & & H\!-\!C\!-\!C\overset{\displaystyle O}{} \\
\parallel & = & \parallel \quad\quad >\!O \; + \; H_2O \\
H\!-\!C\!-\!COOH & & H\!-\!C\!-\!C\underset{\displaystyle O}{}
\end{array}
$$

Another example is afforded by 1, 2-dichloroethylene :

$$
\begin{array}{cc}
H\!-\!C\!-\!Cl & \qquad H\!-\!C\!-\!Cl \\
\parallel & \qquad \parallel \\
H\!-\!C\!-\!Cl & \qquad Cl\!-\!C\!-\!H
\end{array}
$$

Cis form   *Trans* form
b.p. 60°C   b.p. 48°C

These two isomers are readily distinguished by measuring the dipole moment which is zero for the trans form but appreciable for the cis form in which the two electronegative chlorine atoms are on the same side of the molecule.

Geometrical isomerism is not limited to compounds containing an ethylenic, $\gg\!C\!=\!C\!\ll$, bond and is also shown by benzaldoxime thus :

$$
\begin{array}{cc}
C_6H_5\!-\!C\!-\!H & \qquad C_6H_5\!-\!C\!-\!H \\
\parallel & \qquad \parallel \\
N\!-\!OH & \qquad HO\!-\!N
\end{array}
$$

Cis or syn form, m.p. 34°C   *Trans* or anti-form, m.p. 129°C

CYCLIC ISOMERISM

This is really a special case of geometrical isomerism in which two carbon atoms are not free to rotate because they form part of a cyclic or ring compound. The simplest example is shown in Fig. 18.8, where one form is shown with all three X groups on one side of the ring and the other form with two X groups on one side and one on the other side.

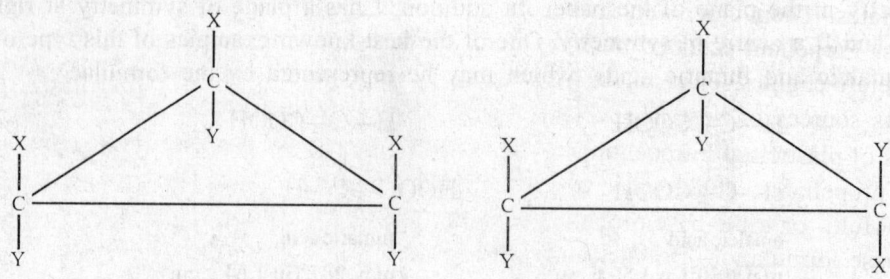

Fig. 18.8. Cyclic isomers.

Chapter 19

Explosives and Propellants

INTRODUCTION

An explosive may be defined as a material, which under the influence of thermal or mechanical shock, decomposes rapidly and spontaneously with the evolution of a great amount of heat and large volume of gases. At the temperature of the explosion, the volume of the evolved gases is perhaps, 15000–20000 times as that of the explosive itself. The first real explosive, called gun powder or black powder was discovered by Roger Bacon. It is an explosive mixture of sulphur, charcoal and saltpetre and its action as explosive is at the expense of the oxygen presence in the saltpetre. Generally gun powder consists of about 15% charcoal, 10% sulphur and 75% saltpetre. Various physical and mechanical improvements have now been made in gun powder into modern times. When cellulose is allowed to react with a mixture of conc. H_2SO_4 and conc. HNO_3, various compounds of cellulose with nitric acid are formed, the composition and properties of which depend on the strength of acid mixture and the temperature, as well as on the duration of the reaction. Such compounds are called nitrocellulose.

Explosives have all been derived from the products of distillation of coal. Picric acid and trinitro toluene are well known examples of explosives derived from the products of distillation of coal. A wide variety of other chemicals which have military or commercial use have been abandoned or replaced by more effective explosives, although explosives are most frequently used for the destruction purposes, their importance to help a man in creative purposes is also not less important. Explosives are also used for the creation of dams, tunnels, roads in hills, mining of coal, metals and non-metals etc. A comparison of the characteristics associated with propellant burning, explosive detonation, and the performance of conventional fuels is shown in Table 19.1. The greatest difference is the rate at which energy is evolved.

Propellants are mixtures of chemical compounds that produce large volumes of gas at controlled, predetermined rates. Their chief applications are in launching projectiles from guns, rockets, and for missile systems. Propellant-actuated devices are used to drive turbines, move pistons, operate rocket vanes, start aircraft engines, eject pilots, jettison stores from jet aircraft, pump fluids, shear bolts and wire, and act as sources of heat in special devices. They are employed in guns in the form of dense grains or sheets of plasticised nitrocellulose. Those that contain only plasticised nitrocellulose are known as single-base propellants; double-base propellants also contain a liquid explosive plasticiser such as nitroglycerine. Multi- or triple-base propellants incorporate a crystalline explosive such as nitroguanidine in the double-base formulation. Double- and triple-based nitrocellulose propellants are used in rockets as well as guns. Typical components of nitrocellulose propellants are listed in Table 19.2.

Table. 19.1. Characteristics of burning and detonation.

Characteristics	Burning		Explosive detonation
	Fuel	Propellant	
Typical material	Coal-air	Propellants	Explosives
Linear reaction rate, m/s	10^{-6}	10^{-2}	2–9×10^{-3}
Type of reactions	Oxidation–reduction	Oxidation–reduction	Oxidation–reduction
Time for reaction completion, s	10^{-1}	10^{-3}	10^{-6}
Factor-controlling reaction rate	Heat transfer	Heat transfer	Shock transfer
Energy output, J/g[a]	10^4	10^3	10^3
Power output, W/cm^2	10	10^3	10^{9b}
Most common initiation mode	Heat	Hot particles and gases	High temperature-high pressure shock waves
Pressure developed, MPa[c]	0.07–0.7	0.7–7×10^2	7×10^3 –7×10^4
Uses	Source of heat and electricity	Controlled gas pressure, guns, and rockets	Brisance, blast, munitions, civil engineering

[a] To convert J to cal, divide by 4.184.

[b] This may be compared with the electric-generating capacity of about 30×10^6 kW.

[c] To convert MPa to psi, multiply by 145.

EXPLOSIVE DETONATION

Detonations proceed as a result of a reaction front moving in a direction normal to the surface of the explosive. However, detonation is a hydrodynamic phenomenon that differs in a fundamental sense from burning. Upon initiation, burning first occurs at an increasing rate for period of time up to several microseconds. A high pressure shock wave is formed which passes through the explosive at high velocity. As it does so, it causes exothermal decomposition of the explosive. The continued passage of the wave is supported by transfer of energy from the spent explosives to the unreacted explosive by shock compression. The rate of reaction depends on the rapid rate of transmission of a shock wave rather than on the relatively slow rate of heat transfer associated with propellant burning. The detonation rate is stable and constant and is primarily governed by the physical and chemical properties of the explosive, its geometry, degree of confinement, and particularly its density ρ, with which the detonation rate varies in a linear fashion for most explosives:

$$D_i = D_o + M(\rho_i - \rho_o)$$

where D_i = linear detonation rate dx/dt at density ρ_i, D_o = linear detonation rate at density ρ_o, and M = a constant characteristics of the explosive composition. Typical values of D_o at ρ_o = 1.0 g/cm^3 are about 5000–6000 m/s. Values of M are about 3000–4000 m/s.

Table 19.2. Typical components of nitrocellulose propellants and their function.

Component	Application
Nitrocellulose	Energetic polymeric binder
Nitroglycerine, metriol trinitrate, diethylene glycol dinitrate,	Plasticisers
Triethylene glycol dinitrate, dinitrotoluene	Energetic
Dimethyl, diethyl or dibutyl phthalates, triacetin	Plasticisers
	Fuels
Diphenylamine, diethyl centralite, 2-nitrodiphenylamine	Stabilisers
Organic and inorganic salts of lead; e.g., lead stannate, lead stearate, lead salicylate	Ballistic modifiers
Carbon black	Opacifiers
Lead stearate, graphite, wax	Lubricants
Potassium sulphate, potassium nitrate, cryolite (potassium aluminium fluoride)	Flash reducers
Ammonium perchlorate, ammonium nitrate	Oxidisers inorganic
RDX, HMX, nitroguanidine and other nitramines	organic
aluminium	Metallic fuels cross-linking catalysts
Lead carbonate tin	Defouling agents

Primary Explosives

Explosives are commonly categorised as primary, secondary or high explosives and propellants in order of decreasing sensitivity to energy output.

Primary, or initiator, explosives are the most sensitive to heat, friction, impact, shock, and electrostatic energy. They have been studied in considerable detail because of their almost unique capability, even when present in small quantities, to transform rapidly a low energy stimulus into a high intensity shock wave. Most recent evidence indicates that there is a minimum thickness or run-up distance before steady state detonation occurs and a gradual transition from an unreacted shock to a stable detonation.

Primary explosives are used to initiate the next element in a series which consists of explosives of increasing mass and decreasing sensitivity. They are arranged in sequence to amplify the input stimulus to an output level of sufficient intensity to maximise the probability of initiating the main charge. Overall energy intensification is about 10^7:1. Primary explosives are used in military detonators, commercial blasting caps, and in stab and percussion primers.

Mercury fulminate

Mercuric cyanate, [Hg(ONC)$_2$], is a gray-white powder, which explodes at 160°C, has a detonation rate 5.4 km/s at 4.2/ g/cm^3; is soluble to the extent of 0.1 g/100 H$_2$O at 20°C. It is prepared by the Chandelen process, essentially a large-scale laboratory process.

Lead azide

Lead azide $(Pb(N_3)_2)$, is the primary explosive used in military detonators. Detonation rate 5.1 km/s at 4.8 g/cm^3. The azides are among the very few useful explosive compounds that do not contain oxygen. Lead azide is made in small batches buffered by the reaction solutions of lead nitrate or lead acetate with sodium azide:

$$Pb(NO_3)_2 + 2NaN_3 \rightarrow 2NaNO_3 + Pb(N_3)_2$$

Silver azide

Silver azide, AgN_3, has received attention as a potential replacement for lead azide because it may be used in smaller quantities as an initiator and, therefore, offers the possibility of miniaturisation of fuse components. Silver azide is made in the same manner as lead azide, except that silver nitrate is used in the reaction with sodium azide.

Lead styphnate

Lead styphnate, lead 2,4,6-trinitroresorcinate, is one of a number of compounds used in priming compositions to start the ignition-to-detonation process in the explosive sequence. Detonation rate 5.2 km/s at 2.9 g/cm^3. Lead styphnate monohydrate is precipitated as the basic salt from a mixture of solutions of magnesium styphnate and lead acetate followed by conversion to the normal form by acidification with dilute nitric acid.

Diazodinitrophenol (DDNP)

DDNP is an orange-yellow compound made by diazotising picramic acid with sodium nitrite and hydrochloric acid, washing the product with ice water, and recrystallising it from hot acetone. It is more effective for some purposes than lead azide and is used as an initiator in commercial blasting caps.

Tetrazene

Tetrazene is a pale-yellow crystalline explosive (explodes at 140–160°C) made by adding sodium nitrite to a solution of l-aminoguanidine hydrogen carbonate in dilute acetic acid at 30°C. It is used as a component in priming compositions.

Secondary Explosives

Aliphatic nitrate esters

Aliphatic nitrate esters are among the most powerful explosives available. They are generally less stable than aromatic nitro compounds or nitramines because they tend to hydrolyse autocatalytically to form nitric and nitrous acids which further accelerate decomposition.

Nitroglycerine

Nitroglycerine, NG, glyceryl trinitrate, is primarily used as an explosive in dynamites and as a plasticiser for nitrocellulose in double- and multi-base propellants. It is very sensitive to shock, impact, and friction and is employed only when desensitised with other liquids or absorbent solids or when compounded with nitrocellulose. It is readily soluble in many organic solvents, acting as a solvent for many explosives ingredients. Unconfined nitroglycerine burns without exploding if present in thin layers and in small quantities but detonates if confined. Aerated nitroglycerine and other liquid explosives containing microbubbles are especially sensitive to shock.

Nitroglycerine is made from very pure glycerol to ensure stability of the product. Mixed acid (90% nitric acid and 25–30% oleum) is used in both batch and continuous processes.

Nitrocellulose

Nitrocellulose provides mechanical strength as well as readily available energy to gun and rocket propellants. It is manufactured by nitration of cellulose with mixed acid.

Dry nitrocellulose burns rapidly and furiously. It may detonate if present in large quantities or if confined and is dangerous to handle in the dry state since it is sensitive to friction, static electricity, impact and heat. Nitrocellulose is always shipped wet with water or alcohol. The higher the nitrogen content, the more sensitive it tends to be. Even nitrocellulose with 40% water detonates if confined and sufficiently activated.

The batch nitrogen process have included the pot process, the centrifugal process, the Thompson displacement process, and the Mechanical Dipper process. In the batch process, the raw materials proceed by gravity through the processing operations. The nitration of cellulose occurs very rapidly at first and the nitrocellulose is separated from spent nitrating acid. Semi-continuous nitration processes have been developed for military and industrial grades. Semi-continuous nitration uses a multiple-cascade system and a continuous wringing operation. The controlling factors in the nitration process are the rates of diffusion of the acid into the fibres, the composition of the mixed acid, and the temperature.

Nitramines

The four most important nitramines are: cyclotrimethylenetrinitramine (RDX); cyclotetramethylenetetranitramine (HMX); nitroguanidine (NQ); and 2,4,6-trinitrophenylmethylnitramine (tetryl). Tetryl has been increasingly replaced by RDX; both RDX and HMX are used as high energy explosives and may also be incorporated in high performance rocket propellants. Nitroguanidine is employed almost exclusively in gun propellants.

RDX and HMX

The properties of RDX and HMX are quite similar; HMX has a higher density and a higher detonation rate, yields more energy per unit volume and has a higher melting point and higher explosion and cook-off temperatures.

Both are white, stable crystalline solids, less toxic than TNT and capable of being handled with no physiological effect if appropriate precautions are taken to assure cleanliness of operations. Both RDX and HMX detonate to form mostly gaseous, low molecular weight products with little intermediate formation of solids.

Both RDX and HMX are substantially desensitised by mixing with TNT to form cyclotols with (RDX) and octols (with HMX) or by coating with waxes, synthetic polymers, and elastomeric binders. The two most common processes use hexamethylene tetramine (hexamine) as starting material. But the Bachmann process is now used exclusively. In the Bachmann process an 80–84% yield is obtained, ca 5–10% of which is cyclotetramethylenetetranitramine (HMX). The Woolwich process gives a 70–75% yield containing only a trace of HMX.

Nitroguanidine (NQ)

Nitroguanidine has been used to some extent as an industrial explosive but not as a military explosive because of its relatively low energy content and difficulty of initiation. It is stable and non-hygroscopic. Produced in the alpha crystalline form with a bulk density of about 0.2 g/cm^3, the crystals are needle like and often hollow and are about 5μm in diameter and 15 μm long.

Nitroguanidine can be made by several methods. In the typical Welland process, calcium cyanamide is first made from calcium carbonate and converted to cyanamide by acidification of a water slurry. The dimer dicyandiamide, which is then formed by filtration and evaporation of the filtrate, is fused with ammonium nitrate to form guanidine nitrate. Dehydration with 96% sulphuric acid gives nitroguanidine. In these processes, guanidine nitrate is the intermediate which is then dehydrated with sulphuric acid.

Tetryl

Tetryl has been used in pressed form mostly as a booster explosive and as a base charge in detonators and blasting caps because of its sensitivity to initiation by primary explosives and its relatively high energy content. It is highly stable, losing virtually no weight on prolonged storage at 80°C. It is manufactured by a batch process in which dimethylanisidine dissolved in concentrated sulphuric acid is nitrated with mixed acid.

Nitroaromatics

The commonly used nitroaromatic explosives contain three NO_2 groups, generally in the 1, 3, and 5 positions. Aromatics are most often nitrated to the trinitro stage with mixed acid. Further nitration is difficult, and aromatics with four or more nitro groups attached to the ring tend to be relatively unstable. The most extensively used explosive is trinitrotoluene (TNT); however, hexanitrostilbene (HNS), hexanitroazobenzene (HNAB), and di- and triaminotrinitrobenzene (DATB and TATB) have found increasing application because of their low sensitivity to impact, shock, and friction and their excellent stability at elevated temperatures. Ammonium picrate (AP) has been used in armor-piercing gun projectiles because of its insensitivity to impact and shock.

Trinitrotoluene

2,4,6-trinitrotoluene (TNT) is very stable and may be stored indefinitely at temperate conditions without deterioration. TNT is non-hygroscopic and relatively insensitive to impact, friction, shock and electrostatic energy. It has been fired in high acceleration gun projectiles with reported prematured rates of less than one in a million (10^6). Bombs and projectiles are filled with steam-melted TNT by the casting process. Melted TNT also serves as the liquid carrier for RDX, HMX, aluminium, ammonium nitrate, and other high melting ingredients to form a wide range of castable slurries.

Trinitrotoluene is made by batch or continuous process. Toluene is nitrated in a three-stage operation by using increasing temperatures and mixed-acid concentrations to successively introduce nitro groups to form mononitrotoluene (MNT), dinitrotoluene (DNT), and trinitrotoluene (TNT). The steps used in each stage are similar and include acid mixing, addition of the oil, digesting the reaction to completion, cooling and settling the mix, and separating the oil from the acid. The TNT is purified by treatment with sodium sulphite solution.

Picric acid and ammonium picrate

Picric acid (2,4,6-trinitrophenol) (PA) is of historic interest as the first modern high explosive to be used extensively as a burster in gun projectiles. It was first obtained by nitration of indigo, and used primarily as a fast dye for silk and wool. Picric acid has an energy content somewhat greater than that of TNT and a higher detonation rate. A large disadvantage is its tendency to form sensitive salts with calcium, lead, zinc, and other metals. Picric acid is no longer used as a military explosive. It can be made by gradually adding a mixture of phenol and sulphuric acid at 90–100°C to a nitration acid containing a small excess of nitric acid. The picric acid crystals are separated by centrifuging, washed, and dried.

Ammonium picrate (explosive D) (AP) is now used only where a high explosive is required that is particularly insensitive to shock. It has been used in a pressed form primarily as a burster in naval projectiles or armor penetration.

Uses

Military

TNT, RDX or HMX, nitrocellulose, and nitroglycerine are the single-component explosives most commonly used for military compositions. Nitrocellulose and nitroglycerine are used exclusively to make propellants. Other binary mixtures include the octols (HMX + TNT), cyclotols (RDX + TNT), pentolites (PETN + TNT), tetrytols (tetryl + TNT), amatols (ammonium nitrate + TNT), and picratols (ammonium picrate + TNT).

Industrial

The quantity of industrial explosives used in India continues to increase, but the type used has changed radically in the last twenty years. The use of black powder and liquid oxygen disappeared completely and dynamites declined significantly, whereas the consumption of dry and water-based ammonium nitrate consumption increased greatly.

PROPELLANTS

Propellants as already discussed are mixtures of chemical compounds that produce large volumes of gas at controlled, predetermined rates. Their chief applications are in launching projectiles from guns, rockets, and for missile systems.

Polymer-based rocket propellants, referred to as composite propellants, contain a cross-linking polymer which acts as a viscoelastic matrix for holding a crystalline inorganic oxidiser such as ammonium perchlorate and for providing mechanical strength. Many other substances may be added including metallic fuels, plasticisers, extenders, and catalysts. Polymer-based composite propellants are too erosive to be used in guns because of the residues formed after repeated firings. Typical components of composite propellants are listed in Table 19.3.

Table 19.3. Typical components of composite rocket propellants.

Typical components	Characteristics
Binders	
Polysulphides	Reactive group, mercaptyl (—SH), is cured by oxidation reactions; low solids loading capacity and relatively low performance; now mostly replaced by other binders
Polyurethanes Polythers, Polyesters	Reactive group, hydroxyl (—OH), is cured with isocyanates; intermediate solids loading capacity and performance
Polybutadienes copolymer, butadiene and acrylic acid	Reactive group, carboxyl (—COOH) or hydroxyl (—OH), is copolymer, butadiene cured with difunctional epoxides or aziridines; intermediate solids loading capacity and better performance than polyurethanes; less than adequate cure stability and mechanical characteristics
Terpolymers of butadiene, acrylic acid acrylonitrile	Superior physical properties and storage stability
Carboxy-terminated polybutadiene	Cured with difunctional epoxides or aziridines; have very good solids loading capacity, high performance and good physical properties.
Hydroxy-terminated polybutadiene	Cured with diisocyanates; have very good solids loading and performance characteristics and good physical properties and storage stability.
Oxidisers	
Ammonium perchlorate	Most commonly used oxidiser; it has a high density, permits a range of burning rates, but produces smoke in cold or humid atmosphere
Ammonium nitrate	Used in special cases only, it is hygroscopic and undergoes phase changes, has a low burning rate and forms smokeless combustion products.
High energy explosives (RDX-HMX)	Have high energy and density; produce smokeless products; have a limited range of a low burning rates.
Fuels	
Aluminium	Almost commonly used; has a high density; produces an increase in specific impulse and smoky and erosive products of combustion.
Metal hydrides	Provide very high impulse, but generally inadequate stability, give smoky products, and have a low density
Ballistic modifiers	
Metal oxides	Iron oxide most commonly used.
Ferrocene derivatives other	Permit a significant increase in burning rate coolants for low other burning rate and various special types of ballistic modifiers
Modifiers for physical characteristics	
Plasticisers	Improves physical properties at low temperatures, and processability; may vapourise or migrate; can increase energy if nitrated.
Bonding agents	Improve adhesion of binder to solids.

Principle of the Rocket

Fig. 19.1. is a simplified illustration of how a rocket motor or engine works. Fig. 19.1a shows a chamber with a hole in one end in which gas is being generated rapidly enough to maintain pressure, P_c. Since the pressure is exerted equally in all directions, there will be a pressure imbalance forcing the chamber to the left because pressure is not exerted on the area where there is a hole on the right. The force, F, that will propel the rocket to the left is equal to the chamber pressure, P_c, times the area of the hole of throat, A_t.

In an actual rocket this accounts for only part of the thrust. As shown schematically in Fig. 19.1b, an increase in thrust is obtained by acclerating the gas through the nozzle and expanding it to P_e, the pressure at the nozzle-exit plane. The new equation for force which is called thrust is:

$$F = A_t P_c C_f$$

where C_f is the thrust coefficient whose value depends on the ratio of P_c to P_e and the area of the throat to the area at the exit plane. Thus a nozzle-expansion cone is needed for high performance and is used on almost all rockets.

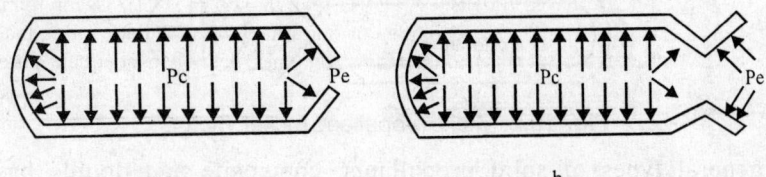

a

b

Fig. 19.1. Principle of rocket propulsion.

Since thrust is a function of motor design, it is not useful for comparing the performance of propellants. A much better criteria of rocket-propellant performance is specific impulse which can be defined as the thrust delivered per unit flow weight of propellant, which is the same as weight burned in unit time.

$$I_{sp} = \frac{1}{w} \int_{t_1}^{t_2} F dt$$

and the units are lb-sec (force)/lb of propellant. This is often abbreviated as sec. Measured I_{sp} values are determined by firing motors on elaborately instrumented thrust stands and converting the measured thrust according to the above equation.

It is time consuming and expensive to optimise propellant I_{sp} by firing numerous motors. Computer programes have been devised to calculate the I_{sp} for various propellant compositions. With a computer the calculations can be made rapidly, thus allowing many compositions to be investigated and the ratio of ingredients to be optimised. The results of the calculations are only as good as the thermodynamic data input to the computer program, but in most cases these data have been refined to the point that theoretical I_{sp}'s are dependable.

Types of Propellants

There are two general types of propellants in widespread use; these are solid propellants and liquid propellants. Both of these require an oxidiser and a fuel. If a single compound is used then it is called a monopropellant. There is also a combination of these two types known as a hybrid propellant in which a solid propellant fuel is used with a liquid oxidiser.

Solid propellants

A solid-propellant rocket motor is relatively simple in that it has very few moving parts (Fig. 19.2). The propellant in the motor is referred to as the grain. This is a carry-over from cannon powder terminology where each individual piece is called a grain. The grain is usually bonded to the metal- or fibreglass-reinforced plastic case and burns only on the interior surface called the perforation. Since the unburned propellant protects the case from heat this allows lighter weight cases to be used. The geometry of the perforation controls the shape of the pressure-time curve after the motor is ignited. Since the propellant burns perpendicularly to the surface, the pressure-time curve can be neutral (level), progressive, or regressive. A progressive curve leads to increasing acceleration, while a regressive curve can yield constant acceleration because of weight loss as the propellant burns. Various perforation shapes can be used including cylindrical, star, and cruciform. In some cases, end-burning grains are used without a perforation. The latter produce rather low thrust but long burning times.

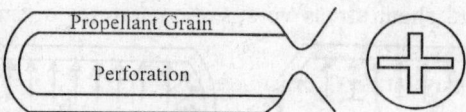

Fig. 19.2. Solid propellant rocket motor.

There are two general types of solid propellants, composite and double base. The composite propellants consist of a solid oxidiser and a metallic fuel intimately mixed in an elastomeric material which serves as both a binder and a fuel. Double base propellants use nitrocellulose plasticised with an energetic nitrato compound, and they may also contain a solid oxidiser and a metallic fuel. The double base propellants which contain a solid oxidiser and a metallic fuel are called composite-modified double base propellants.

Composite propellants

Binders for composite propellants are based on liquid polymers that contain functional groups which allow the polymer to be chain extended and cross-linked (cured) after they have been mixed with the solid oxidiser and fuel. When cured, the propellant exhibits rubbery properties.

The common oxidisers used in composite propellants are ammonium perchlorate, ammonium nitrate, and potassium perchlorate. Ammonium perchlorate is used in almost all composite propellants; ammonium nitrate and potassium perchlorate are used for specialised applications.

Metallic fuels are added to composite propellants primarily to increase the temperature of the gas which increases the specific impulse. While the metal oxides produced are usually liquids or solids which reduce specific impulse, the amount of heat released more than offsets this loss. The common metallic fuels are aluminum and beryllium. The addition of metals also increases propellant density which allows more propellant to be put in a given volume. Since many missile or space rocket motors have only a certain volume available, the denser propellants greatly improve performance.

Double-base propellants

Double-base propellants differ from composite propellants by having the oxidising and reducing functions contained in the same molecule. Nitrocellulose is swollen by a nitrate plasticiser such as nitroglycerine (glyceryl trinitrate) to give a rubbery gelled structure which, although it is not cross-linked, does have viscoelastic properties. The nitrocellulose desensitises the shock-sensitive nitroglycerine so that it can be used.

The addition of NH_4ClO_4 or HMX (cyclotetramethylene tetranitramine) or both, and Al increases both I_{sp} and density which can raise the theoretical I_{sp} to the 265 to 270 sec, and the density to 1.80 to 1.85 g/cc. The latter type of propellants are referred to as composite modified-double-base propellants.

Characterisation of solid propellants

Solid propellants must be characterised so that a process can be designed which is able to handle the uncured and cured propellants. In addition, physical and mechanical properties must be defined so that a satisfactory motor can be designed, and the consequent changes in the propellant needed to meet design requirements must be determined. Safety tests such as impact, friction, static-spark sensitivity, and thermal stability are conducted to determine whether the propellant can be handled safely. Rheological properties, including viscosity and shear stress versus shear rate as a function of time, assist in deciding whether the propellant will process satisfactorily. To ascertain whether the propellant grain in the motor can withstand the thermal and gravitational loads, stress, strain, and modulus are evaluated at various temperatures and strain rates under uniaxial, biaxial, and even triaxial conditions; stress-relaxation and strain-endurance values are also obtained. The ballistic properties evaluated include I_{sp}, burning rate, and ratio of burning surface to nozzle throat area (K_n) as a function of pressure, temperature sensitivity of pressure (π_k) and burning rate (σ_p); these are used to design the propellant grain and nozzle throat size.

Manufacture of Propellants

Composite solid propellants are relatively easy to manufacture. The first step is to grind part of the oxidiser (usually ammonium perchlorate) to a particle size of 1 to 30μ, depending on the burning rate required. The finer the oxidiser the higher the burning rate. A premix containing the liquid polymer, the metallic fuel, and the curing agent is often made. Either this premix is added slowly to the vertical or horizontal Baker-Perkins mixer containing the oxidiser, or the oxidiser is added slowly to the mixer containing the premix. After mixing a sufficient time to blend the ingredients, and raising the temperature to 60°C or so to increase fluidity, the propellant is cast (poured) through a slit plate into the evacuated motor. The vacuum removes entrapped air. A mandrel installed in the motor forms the perforation in the grain. After heating for several days at a temperature above room temperature (usually between 45 and 60°C) the propellant is a tough viscoelastic solid which is bonded directly to the walls of the motor case. A liner which is usually made from the same polymer as the propellant binder covers the case wall for better adhesion of the propellant.

There are two main methods of manufacturing double-base-propellant rocket motors, the casting-powder process and the slurry casting process. In the casting-powder process, the solid ingredients, including the fibrous nitrocellulose, are mixed with a volatile solvent which yields a plastic mass. The plastic is consolidated in a blocking press, extruded, cut in small pieces and dried. The casting powder is placed in the motor and the casting solvent which includes the nitroglycerine or other nitroglycerine or other nitrato plasticiser is sucked into the evacuated motor to fill the voids. The motor is then heated to 40 to 60°C which causes the casting solvent to swell the casting powder and form a strong grain.

In the slurry casting process, a granular form of nitrocellulose is used. The plasticiser and solid ingredients including nitrocellulose, ammonium perchlorate, HMX, and aluminium can be added directly to the mixer. The mixture is stirred until it is homogeneous and then poured into the motor and cured. Since the plasticiser viscosity is low, it is sometimes desirable to predissolve some nitrocellulose in the plasticiser to increase its viscosity so that the solids do not settle.

Solid propellants have found wide application in rockets for both military and space applications because of their high reliability, low cost, and instant readiness. It is expected that they will continue to be used in many applications.

Liquid propellants

The liquid bipropellant system is used today for the largest rockets. In this system the liquid fuel and the liquid oxidiser are stored in separate tanks and are fed separately to the combustion chamber. The propellants are fed either by means of pumps or by pressurisation with an inert gas. In the case of the pum-fed rocket, the pumps are driven by gas turbines. The gas for these turbines is supplied by a generator which is actually a small rocket motor operating off a side stream of the main rocket propellants (Fig. 19.3) or, in some cases, a separate propellant system. In the largest rockets, thousands of pounds per second of propellants must be pumped to the engine, and the pump drives must develop thousands of horsepower. The propellants enter the thrust chamber through an injector which has a function similar to that of a carburretor in an internal combustion engine. The injector has to atomise and mix the propellant so that the oxidiser and fuel will be in the right proportions to vapourise and burn smoothly.

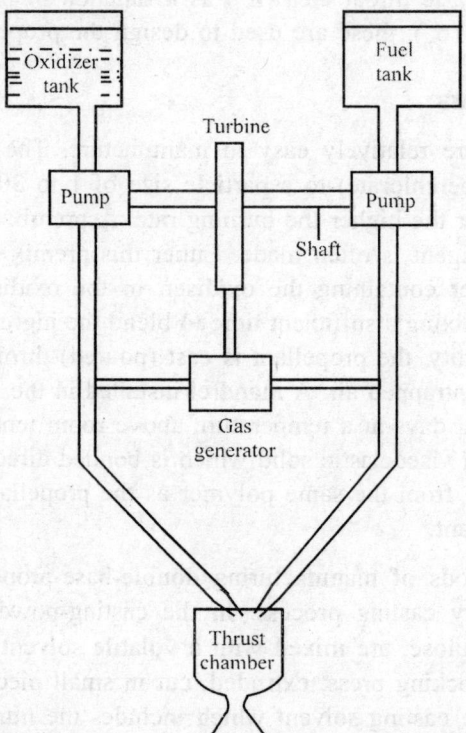

Fig. 19.3. Liquid bipropellant rocket engine.

Cryogenic propellants

The ability to design and fabricate the large, high-performance rockets needed for space applications depends in large measure on advances in the engineering technology used in handling such intractable materials as liquid oxygen (b.p.−183°C), liquid fluorine (b.p.−188°C), and liquid hydrogen (b.p.−253°C).

Storable propellants

Typical examples of storable liquid propellants are nitrogen tetroxide (N_2O_4), chlorine trifluoride (CIF_3). hydrazine (N_2H_4), and unsymmetrical dimethylhydrazine (UDMH).

Nitrogen tetroxide is very stable at room temperature, but at 150°C it begins to dissociate into nitric oxide and free oxygen. Upon cooling, nitrogen tetroxide is reformed. It is not corrosive to most common metals if the oxidiser is dry (< 0.1% H_2O); carbon steels, aluminium, stainless steel, nickel, and inconel can be used. Under wet conditions 300-series stainless steel should be used. As a result of its good properties, nitrogen tetroxide is currently the most important operational earth-storable liquid oxidiser.

The catalyst oxidation of ammonia is the most important commercial method of preparing, N_2O_4. A mixture containing about 9 per cent ammonia in air is oxidised at pressures between 25 and 100 psig. The gaseous products containing nitric oxide (NO) and water are cooled and dried by scrubbing with cold nitric acid. The dry nitric oxide containing unreacted nitrogen from the air is further oxidised by the addition of air to form NO_2. Nitrogen dioxide forms an equilibrium mixture with its dimerised form, dinitrogen tetroxide, N_2O_4, and on cooling the equilibrium shifts to favour the formation of N_2O_4. The reaction sequence is illustrated by the following equations:

$$4NH_3 + 5O_2 \rightarrow 4NO + 6H_2O$$
$$2NO + O_2 \rightarrow 2NO_2$$
$$2NO_2 \rightarrow N_2O_4$$

Chlorine trifluoride is highly corrosive but is stable to shock, heat and electrical spark. Because it forms an inert fluoride film on some metals, it can be used with copper, brass, steel, aluminium and nickel. Chlorine trifluoride has been prepared by the reaction of chlorine and fluorine at 280°C.

Hydrazine is stable to shock, heat, and cold. It begins to decompose at 160°C if no catalysts are present. It freezes at 1.4°C, but it contracts and does not decompose. It is compatible with stainless steel, nickel, aluminium, Teflon, and Kel-F.

Hydrazine is produced by the controlled oxidation of ammonia in the well-known Raschig synthesis. Sodium hypochlorite, formed by the chlorination of caustic soda, reacts with excess ammonia in a two-step process to form, successively, chloramine and aqueous hydrazine :

$$NH_3 + NaOCl \rightarrow NH_2Cl + NaOH$$

$$NH_2Cl + NaOH + NH_3 \rightarrow N_2H_4 + NaCl + H_2O$$

Hydrazine and water form an azeotrope which boils at 120.1°C. Anhydrous hydrazine is obtained by dehydration of the azeotrope with caustic, followed by a vacuum distillation. In another method, the azeotrope is broken by the addition of aniline as a third component, followed by a distillation in which the water-aniline azeotrope is taken overhead and anhydrous hydrazine is obtained as a bottoms product. Using this latter method hydrazine may be produced at 98 + per cent purity.

Unsymmetrical dimethyl hydrazine (UDMH) is neither shock- nor heat-sensitive, being stable to 249°C, its critical temperature. Furthermore, it is compatible with nickel, Monel, and stainless steel, but it does attack aluminum if water is present.

UDMH has been prepared by a modification of the Raschig process by reacting the chlorimine intermediate which dimethylamine instead of ammonia. It can also be produced commercially by nitrosation of dimethylamine to N-nitrosodimethylamine followed by reduction and purification. UDMH is usually used as a 50/50 mixture with hydrazine; in fact, this is the most common earth-storable fuel.

Many applications have been found for liquid propellants because of their generally high performance. They have been widely used in space pr ogrammes but have also found many military applications.

Selection Of Propellants

In addition to energy and burning-rate considerations, a propellant must meet other criteria including mechanical characteristics, stability, sensitivity, cost of manufacture, and uniformity of performance.

Burning Process

The mass rate of propellant burning at a given pressure and temperature depends on the amount of heat evolved during decomposition and the amount of heat transferred to the burning surfaces of the propellant from the hot gases above it. It is also influenced by the tangential velocity of the propellant gases and the radiation from the surrounding gases. Propellants burn in parallel layers so that the surface recedes in all directions normal to the original surface. The burning rate at gun pressures usually varies somewhat less than the first power of the pressure. It changes more slowly at rocket pressures of 3.45–10.34 MPa (500–1500 psi), often to less than the square root of the pressure.

The operating pressure of the system has the predominant effect on the burning rate of propellants. Photographic evidence shows that increasing the pressure decreases the distance between the flame zone in the gas phase and the propellant surface. The rate of heat transfer to the propellant surface increases accordingly.

Typical characteristics and uses of important explosives and propellants are given in Table 19.4.

Table 19.4. Characteristics and uses of the more important explosives.

Primary explosives

Name	Composition or chemical formula	Density (g/cc)	Detonation velocity[a] (km/sec)	Detonation pressure (kilobars)	Detonation temperature[a] (°K)
Mercury fulminate	$Hg(ONC)_2$	3.6	4.7	220	6900
Lead azide	PbN_6	4.0	5.1	250	5600
Lead styphnate	$C_6H(NO_2)_3$ O_2Pb	2.5	4.8	150	–
Nitromannite (Mannitol hexanitrate)	$C_6H_{58}(ONO_2)_6$	1.73	8.3	300	6000
Dinitrodiazophenol (DDNP)	$C_6H_2N_4O_5$	1.5	6.6	160	–

Secondary high explosives

Name	Composition or chemical formula	Density (g/cc)	Detonation velocity[a] (km/sec)	Detonation pressure (kilobars)	Detonation temperature[a] (°K)
Ammonia gelatin dynamites	30–90% grades same as straight gelatins except for some NG and $NaNO_3$ replacement by NH_4NO_3	1.2–1.5		4–6.5	0.75–1.15
Semigelatin dynamite	15–20% NG 1–2% DNT oil, AN-SN dope	1.2	(depends on diameter)	3.5–5	0.9

(Cont'd ...)

Name	Primary explosives Composition or chemical formula	Density (g/cc)	Detonation velocity[a] (km/sec)	Detonation pressure (kilobars)	Detonation temperature[a] (°K)
Prilled AN-oil	94/6 NH_4NO_3/oil	0.8–0.9	1.5–4	0.81–0.83	
Slurry explosives	TNT	17–40	1.4–2.0	5–8	0.7–1.8
TNT-SE	Oxidiser*	30–65			
	H_2O	12–25			
	Al	0–20			
	Other	0.3–1.5			
Smokeless powder SE	SP	20–40	1.35–1.9	4–7	0.65–1.7
	Oxidiser*	30–60			
	H_2O	3–25			
	Al	0.20			
	Other	0.3–10			

[a] Most important properties of detonators.

* AN, SN, perchlorates, etc.

Sensitivity	Major characteristics	Uses
Very high	Best primary explosive for single-component (fuse) detonators; easily detonated by flame, spark, heat, or friction; easily dead-pressed.	In fuse caps (mixed with $KClO_3$); propellant primer; in fuses for shells; small arms cartridges caps.
Very high (higher than NG; less than mercury fulminate)	Powerful detonator but requires strong igniters, e.g., lead styphnate.	Primary explosive in composition (EB) caps; military fuses.
Exceedingly high	Extremely sensitive to sparks, static electricity; explodes rapidly on ignition; good thermal stability.	Igniter in composition caps, military fuses; very satisfactory detonator explosive for fast ignition.
Very high (greater than NG; less than lead azide)	Stronger and more brisant than NG, RDX, PETN.	In composition caps and fuses.
Very high (less than lead azide)	Does not dead-press. About 3/4 as strong as TNT.	In composition caps and fuses.
High	More economical; only slightly less brisant than straight gelatin; exhibits low-order detonation with threshold priming and high pressures.	General small and large diameter blasting in hard rock and under water.
High	Stringy, plastic; easily loaded in "uppers;" economical; high strength; moderate brisance.	Popular small diameter metal-mining explosive.

(Cont'd ...)

Sensitivity	Major characteristics	Uses
Low (requires booster)	One of the cheapest sources of explosive energy available today; flammable and will explode when ignited under strong confinement; no water resistance; adaptable to do-it-yourself operations.	Open-pit and underground blasting where dry conditions prevail; most adaptable to soft, easy shooting.
Low (requires booster)	Gel or thick pea-soup consistency; capable of detonation at high pressures, excellent water resistance.	Large diameter, open-pit, small diameter underground, oil well, submarine, water-filled boreholes, deep-water bombs.
Low (requires boosters)	Generally similar to TNT slurry.	Large diameters, open-pit blasting.

Name	Composition or chemical formula		Density (g/cc)	Detonation velocity (km/sec)	Available energy (kcal/g)
Al-SE	Al	1.0–10	1.1–1.5	2–5	0.7–1.3
Slurry blasting agents	Al	0–35	1.1–1.6	2–6	0.7–2.0
	Oxidisers*	50–80			
	H_2O	4–18			
	Other	0.2–10			
Nitrostarch powders	Nitrostarch in place of NG		1.2	4–5	0.8–1.0
Composition B	40/59/1 TNT/RDX/wax		1.7	7.8	1.1
Composition B-3	40/60 TNT/RDX		1.73	7.9	1.15
Haleite or EDNA	$(CH_2NHNO_2)_2$		1.6 (pressed)	7.9	1.2
Ammonium picrate (Explosive D)	$(ONH_4)C_6H_2(NO_2)_3$		1.56 (pressed)	6.6	0.7
Nitrostarch	Mixtures of various nitro esters of starch		1.4 (pressed)	6.4	0.95
Tetryl	$(NO_2)_3C_6H_2CH_3N_2NO_2$		1.45 (pressed)	7.0	0.95
PETN (pentaerythritol tetranitrate)	$C(CH_2ONO_2)_4$		1.6 (pressed)	7.92	1.31
Pentolite	50/50 TNT/PETN		1.63 (cast)	7.7	1.1
Trinitrotoluene (TNT)	$CH_3C_6H_2(NO_2)_3$		1.59 (cast)	6.9	0.9
			1.45 (pressed)	6.9	–
			1.03 (Pelletol)	5.1	–
			0.8 (grained)	4.2	0.8

(Cont'd ...)

Name	Composition or chemical formula	Density (g/cc)	Detonation velocity (km/sec)	Available energy (kcal/g)
Amatols	50/50 AN/TNT	1.55 Cast	5–6.5 (depending on diameter)	0.95
	80/20 AN/TNT	1.0 (loose) 1.45 (pressed)	4 (large diameter) 5.6 (large diameter)	0.93

Sensitivity	Major characteristics	Uses
From cap sensitive to very low–depends on alumnium fineness.	Gelatin; no explosive ingredients.	Small diameter, underground and general blasting.
Low (requires boosters)	Gelatin to thick or thin pea-soup consistency.	Large diameter, underwater, wet- and dry-hole blasting, large bombs.
Moderately high, but less than dynamites	Good "fumes"; fair water resistance; powerful; economical.	Small diameter blasting.
Average	Very high brisance.	Bursting charge and special weapons.
High	Very high brisance.	Experimental standard.
High	High brisance; less sensitive RDX and PETN.	In Ednatols for bursting charges.
Very low	Insensitive to shock and friction; melts with decomposition; shells filled with high-pressure pressing.	Armor-piercing shells.
High	Highly inflammable white powder	Demolition blocks and Trojan blasting explosives.
High	Very sensitive; rapidly reacting; easily pressed with 1–2% graphite; high brisance.	Booster; base charge in caps; in tetrytols for bursting charges.
High	Very powerful and sensitive (more sensitive than RDX, less than NG).	In Primacord fuse; base charge in caps.
Moderate	High pressure or brisance; primacord sensitive	Booster and special weapons; commercial booster for prilled AN-fuel oil and slurry explosives.
Low	Easily melted and cast; suitable liquid for slurrying with other explosives; easily pressed into blocks; completely waterproof.	Military; "Nitropel" TNT used in slurry explosives and in filling annulus between charge and bore-hole in water filled holes; in amatols.
Low	Insensitive; hygroscopic, not waterproof; less brisant but stronger than TNT; 50/50 can be cast; 80/20 eitheter pressed or granulated.	Military; oil well shooting; quarrying; dry-hole booster for very low-sensitive either types.

Name	Composition or chemical formula	Density (g/cc)	Detonation velocity (km/sec)	Available Energy (kcal/g)
Dinitrotoluene (DNT)	$CH_3C_6H_3(NO_2)_3$	1.28 (liquid)	5	0.7
		0.8 (granular solid)	2–3.5 (depending on diameter)	
Nitromethane (NM)	CH_3NO_2	1.12	6.2	–
Cyclonite (RDX)	$C_3H_6N_6O_6$	1.2 (loose)	6.8	1.32
		1.6 (pressed)	8.0	
HMX	$C_4H_8N_8O_8$	1.89	9.1	1.35
HBX	Mixtures of RDX, TNT, aluminum, and wax	1.78	7.5	1.5
Plastic explosives (compositions A,C, C-2, C-3, C-4)	Waxed RDX	1.45–1.6	8.0	1.1–1.3
PBX 9404	94/3/HMX/binder/ nitrocellulose	1.84	8.8	1.3
Nitroglycerine (NG)	$C_3H_5(ONO_2)_3$	1.59	7.8	1.41
Ethylene glycol dinitrate (EGDN)	$C_2H_4(ONO_2)_2$	1.48	7.4	1.43
Straight dynamites[b]	20–60% NG, in balanced SN dope 20% grade ≡ 20% NG, etc.	1.3	4–6	0.55–0.85[b]
Ammonia dynamites (and permissibles)	As above except NH_4 NO_3 replaces part of NG and $NaNO_3$	0.8–1.2[b]	1.5–5.5 Depends on AN particle size, NG content.	0.7–0.9[b]
Blasting gelatin	92/8/NG/nitrocotton ("Solidified" NG contains some wood pump to minimise low-order detonation).	1.55 (1.45)	7.5 (7.2)	1.45 (1.4)

[b] Depends on grade.

Sensitivity	Major Characteristics	Uses
Very low	Reddish brown or yellow liquid.	Sensitiser in "Nitramons"; 60/40 NG/DNT in oil well shooting; up to 20% in TNT bursting charges; in FNH (flashless) propellant; 6% in small-arms ammunition (with guncotton).
Moderate	Clear, watery liquid	Special demolition, experimental studies of liquid explosives.

Sensitivity	Major Characteristics	Uses
High	High thermal stability in solid state expressively sensitive in pure state; 1.65 times as strong as low density TNT; 1.45 times as strong as cast TNT.	Major ingredients in plastic explosives: one of most brisant explosives in cast explosives; one of most base charge in caps.
High	Better than RDX in all respects.	Same as RDX.
Average	Very powerful.	Underwater explosive.
Moderate	Plastic, easily moulded or pressed.	Specialised military demolition.
Moderate	Plastic bonded.	Specialised military demolition.
Very high (almost a primary explosive)	Oily, toxic liquid; volatile above 50°C; gelatinised by nitrocotton; exhibits low-order detonation with threshold priming.	Shooting oils wells; main explosive in dynamites; used in double-base powders.
Very high	Closely resembles NG; more volatile, toxic, slightly stronger but less brisant (owing to lower density).	Used in solution with NG as freezing point depressant.
High	Cheesy, plastic substance; packed in paper cartridges; may be slit and tamped in borehole for greatest blasting effect; fired by detonator as are all dynamites; heat, friction, shock, and flame sensitive.	Ditching, stumping, other uses where high propagation-by-influence "sensitiveness" is required.
High	Cheaper than comparable grade straight dynamites; must be waterproofed by special additives.	General small and large dynamite blasting; permissible (some grades).
High	Strongest, most brisant dynamite; completely waterproof; exhibits low-order detonation with threshold priming and under high pressures.	Oil well and submarine blasting, tunnel drilling, demolition.

Propellants

Name	Composition or chemical formula	Sensitivity
Colloidal nitrocellulose (NC) Powders	Pyrocotton: cellulose nitrate with 12.6%N. Guncotton: cellulose nitrate with 13.3% N.	Low
Double-base powders	60–80% Nitrocellulose. 20–40% Nitroglycerine.	Moderate
Cordite	65% N.C., 30% NG, 5% vaseline	Low
FNH (Flashless non-hygroscopic powders)	Either straight N.C. or double-base powders with addition of coolants, etc., to prevent muzzle flash, and decrease water adsorption.	Low
Albanite; DINA powder	Di(2-nitrooxyethyl)nitramine.	Low
Rocket powder (solventless powder)	Nitrocellulose plasticised with about 50% NG, plus stabilisers and potassium salts.	Low

(Cont'd ...)

Propellants

Name	Composition or chemical formula	Sensitivity
Chemical propellants[c]	Hydrogen perioxide, 80–90% H_2O_2 plus Ca, Na, or K permanganate (solid or aqueous solution). Hydrazine hydrate plus methyl alcohol. Fuming nitric acid-aniline. Mixed acid-monoethylaniline. Liquid oxygen-kerosene.	
Black powder	75% KNO_3 (or $NaNO_3$), 15% charcoal, 10% sulphur.	High

[c] Many new rocket propellants have been described; the best ones are under security classifications.

Sensitivity	Major Characteristics	Uses
High	Jelly-like substance; powerful, waterproof; exhibits low-order detonation under threshold primering and high pressure.	In hard rock; mudcapping demolition; submarine blasting.

Major characteristics	Important uses
Burning rate controlled by graining, hygroscopic, smokeless flame, with intense flash; gelatinised with alcohol-ethel.	Combined with stabilisers and modifiers to make smokeless powders for artillery, small arms, and sporting ammunition.
Pyrocotton and guncotton are usually blended to secure an average of 13.15% N.	Dry guncotton in fibre form is used in primers fired by an electric current.
Very rapid burning rate, controllable by surface area; more powerful and more readily ignitable than straight N.C. powders; causes erosion of gun bores; can be detonated and is subject to DDT.[d]	Propellant for mortars and sporting ammunition; not used by U.S. armed forced as cannon powder because of bore erosion.
Gelatinised with acetone.	Propellant for large caliber naval guns (English).
Like other smokeless powders, but can be rolled into sheets; flash reduced by DNT,potassium salts, etc	Propellant for small armor-piercing rockets such as the "Bazooka" (NG base); for naval ammunition (NG base).
Better flashless powder than FNH powders. Very rapid, uniform burning rate; can be made with thick section since no solvent need be removed.	Naval ammunition. For rockets up to 4.5 inches.
Catalytic decomposition into water and O_2 releases about 1000 Btu per pound. Supplies oxygen to burn petroleum fuel.	For driving turbines on submarines: V-2 rocket-fuel pumps; jet motors; launching device for ram-jets.
Rapid combustion; fuel and oxidiser are both liquid.	For torpedo turbine drives.
Rapid rate of reaction generates heat and gases.	For launching device for ram-jets; jet motors.
As above.	As above. Rocket motors.
Cheap; excellent "heaving action," persistent smoky flame; very sensitive to friction, spark, and heat; hygroscopic.	Time (delay) fuses for blasting and shell; in igniter and primer assemblies for propellants; pyrotechnics; $NaNO_3$ powder, in commercial black powder and for practice bombs and saluting charges; (it is being discontinued as blasting charge).

Example 19.1. Which will be most basic toward BMe_3 in each row :

(i) Me_3N Et_3N.

(ii)   2-MePy   4-MePy.

(iii) $2-MeC_6H_4CN$ C_6H_5CN.

Solution.

(i) Me_3N (less F-Strain in adduct).

(ii)   4-MePy (less F-strain in adduct).

(iii) $2-MeC_6H_4CN$. (Inductive effect of the methyl group wins; steric effect cannot interfere since the donor nitrogen atom is situated at a larger distance (not directly in the ring as in Py).

Example 19.2. Arrange according to increasing Lewis acid character :

(a) BF_3, BCl_3, BBr_3, BI_3

(b) $B (n - Bu)_3$, $B (t - Bu)_3$

(c) $\cdot SiF_4$, $SiCl_4$, $SiBr_4$, SiI_4

Solution. (a) On the basis of the relative electronegativities of the halogens, one might expect the B atom in BF_3 to be most electron deficient and hence most acidic. However, the compounds involve extensive π-interaction from a filled p orbital on the halogen and an empty p-orbital on B. The small F-atom forms the most stable $p - p$ π-bond by efficient match of energy and size of the orbitals. Formation of an adduct with a base converts the geometry of bonds around boron from planar to pyramidal, thus rupturing the π-bonds. The Lewis acidity therefore increases with decrease in the extent of π-conjugation, as the halogen atom increases in size giving poorer overlap, i.e., from BF_3 to BI_3. Hence the order is

$$BF_3 < BCl_3 < BBr_3 < BI_3.$$

(b) The highly branched tertiary butyl group involve appreciable back-strain (B-strain) when the boron atom changes to pyramidal environment on adduct formation. This destabilises the adduct. Hence the order is

$$B (t - Bu)_3 < B (n - Bu)_3.$$

(c) The order in this case is the reverse of that for BX_3. π-conjugation from the halogen p-orbital to the Si–d orbital is not as intense as in the case of BX_3 and the order of acidity follows the increase in electron withdrawing power of the halogen from I to F. Hence the order is

$$SiI_4 < SiBr_4 < SiCl_4 < SiF_4.$$

Chapter 20

Organic Reaction Mechanisms

INTRODUCTION

This chapter is devoted to a consideration of the mechanisms of organic reactions, i.e. the actual courses we believe are followed by the reactants in combining together and the various stages involved in reaching the final products. From this study we hope to arrive at an understanding of why particular compounds and functional groups react in the way they do and obtain explanations of their reactivity and their acidic or basic character. Before delving into organic mechanisms, however, we must first consider the covalent bond in greater detail, because organic reactions primarily consist of breaking and making covalent bonds and the nature of that bond has a profound and far-reaching influence upon the course of these reactions.

POLARISATION EFFECTS IN THE COVALENT BOND

There are a number of ways of describing the covalent bond; it can be portrayed as a pair of electrons positioned between two atoms, or as an overlap of their atomic orbitals, or as a region of electronic charge under the joint control and influence of two adjacent nuclei. The essential feature of all these representations is that the two electrons involved in the bond are shared between the atoms in some way to give a rigid and directional linkage. With two identical atoms the covalent bond is regarded as completely symmetrical with the electrons shared exactly equally between them. When the two atoms joined by the covalent bond are different, however, this no longer holds true and the electrons are then shared unequally between them, with the more electronegative atom having the greater affinity for electrons possessing the greater share. This effect explains that such an electron shift can cause a pronounced polarity in the bond and, when the molecular shape is appropriate, a high dipole moment in the molecule as a whole.

This asymmetrical distribution of electrons in bonds between atoms of different electronegativity, which is known as the inductive effect, can be depicted in several ways. For example, a covalent bond between carbon and chlorine can be shown with an arrow on it thus $\diagup$C$\rightarrow$Cl to convey that the electron pair is attracted towards the electronegative chlorine atom and displaced towards it. Alternatively the notation $\overset{\delta+}{\diagup}$C—$\overset{\delta-}{Cl}$ may be used to indicate the polarity which arises, δ^+ and δ^- standing of small ionic charges on the carbon and chlorine atoms respectively. Such conventional representations aim to

466

make clear that as a result of the inductive effect of covalent bond in question has a partially ionic character and that there is a significant deficiency of electrons at the carbon atom and an excess of negative charge at the chlorine atom.

Similar displacement effects occur in unsaturated covalent bonds between atoms of widely different electronegativity. For example in the carbonyl group, which is conventionally portrayed as $>C=O$, there is a marked polarity because of the electron-attracting influence of the strongly electronegative oxygen atom. Such a bond is sometimes written $>C=O$ with a curved arrow to indicate electron shift or as $>C^+=O^-$ to emphasise the very uneven electron density in the bond.

It should be noted that when a carbon atom is linked to a more electropositive atom or group, the resulting electron shift is then towards the carbon atom. For example, alkyl groups are weakly electron-donating and cause a slight displacement towards the carbon atom thus $R \rightarrow\!\!-C<$.

These inductive effects can be passed on to some extent to adjacent bonds in the molecule, although in saturated compounds the repercussions of a particular electron shift do not extend very far. We shall see that this relay effect is of considerable importance in influencing the extent to which organic compounds containing an —OH group undergo dissociation and display acidic character.

Another notable feature of covalent bonds is their polarisability i.e. their readiness to adopt a temporary polarity when it is induced in them by the close proximity of a charged or highly polarised group. For example, a bromine molecule which consists of two atoms linked covalently together by a symmetrical sharing of an electron pair will develop a temporary dipole moment thus $Br^{\delta+}\!\!-\!\!Br^{\delta-}$ When one end of it approaches closely to a centre of high electron density such as a double or triple bond. As we shall see later, this induced polarisation effect has an important bearing on the mechanism of certain organic reactions.

Yet another major factor influencing organic reactivity is the ability of certain substances, molecules or ions, to adopt a resonance structure in which any displaced electrons are delocalised i.e. spread over the whole structure and not concentrated entirely on one particular carbon atom. Delocalisation of this kind, known as the mesomeric effect, leads to the formation of ions such as the carboxylate ion $RCOO^-$ and the phenoxide ion $C_6H_5O^-$ which are more stable than the alternative structural forms.

TYPES OF ORGANIC REACTION

When a single covalent bond between two atoms breaks it can do so in three different ways depending upon the nature of the particular atoms and the prevailing conditions. For example, the compound AB can split up in the following ways :

$$A : B \rightarrow A\cdot + B\cdot \qquad\qquad ...(20.1)$$

$$A : B \rightarrow A: + B \qquad\qquad ...(20.2)$$

$$A : B \rightarrow A + B: \qquad\qquad ...(20.3)$$

In the first case, which is known as homolysis or homolytic fission, the bond divides symmetrically so that each atom retains one of the electrons which constituted the original bond. The products A· and B· are free radicals; they possess an unpaired electron and are very reactive.

In cases (20.2) and (20.3), which are known as heterolysis or heterolytic fission, the bond divides unequally so that either A or B captures both electrons and becomes a negatively charged ion whilst the other atom acquires a positive charge owing to the loss of its share in the pair of electrons :

$$A : B \rightarrow A^- + B^+$$

or

$$A : B \rightarrow A^+ + B^-$$

Thus in homolytic fission the products are neutral particles, whereas in heterolysis charged ions are formed as intermediates.

HOMOLYTIC REACTIONS

As explained in the previous section, in homolytic reactions the covalent bond is broken symmetrically yielding two radicals each having an unpaired electron which is depicted by a single dot thus :

$$A : B \rightarrow A \cdot + B \cdot$$

These free radicals are extremely reactive; their energy of activation is so small, often only a few kilojoule, that they tend to combine readily with the first molecules they encounter and so have a very brief existence.

Homolytic reactions usually occur in the gaseous phase or in non-polar solvents and give rise to very rapid chain reactions. They are often initiated photochemically by ultra-violet light or by means of high temperatures or the use of a catalyst which encourages the formation of free radicals.

The halogenation of alkanes (paraffins) provided numerous examples of homolytic organic reactions. Above 250°C or in diffused daylight chlorine reacts with methane, for example, giving various substitution products. The first step involves the fission of a chlorine molecule into two chlorine radicals thus

$$Cl_2 \rightarrow Cl \cdot + Cl \cdot$$

These single chlorine atoms then rapidly combine with methane molecules setting up a chain reaction in the mixture.

Many addition polymerisations are homolytic reactions. For example, the polymerisation of vinyl chloride to polyvinyl chloride is initiated by traces of some substance such as a peroxide which decomposes thermally producing free radicals, X·. These then react with molecules of the vinyl chloride monomer to produce a succession of new radicals in the following way :

As each stage of the chain reaction is completed the size of the polymer radical grows until eventually the process is terminated by the combination of two of the free radicals to form a complete molecule. A similar mechanism applies to the polymerisation of vinyl cyanide (acrylonitrile) to polyacrylonitrile and ethylene to polyethylene by the high pressure method.

The pyrolysis (i.e. thermal decomposition) of hydrocarbons is normally a homolytic reaction. For example, in the cracking of petroleum the larger molecules undergo homolytic fragmentation producing a mixture of free radicals which then combine with each other to give molecules of simpler hydrocarbons. Similarly when natural gas is raised to a high temperature it decomposes homolytically.

These examples illustrate three common features of homolytic organic reactions, (i) their great speed, (ii) their sensitivity to ultra-violet light, high temperatures, catalysts, and inhibitors (substances which react with free radicals and remove them from the reaction mixture), and (iii) their indiscriminate nature, since a mixture of different products is usually obtained.

HETEROLYTIC REACTIONS

This kind of reaction in which the covalent bond divides asymmetrically with the creation of two ions, can be written as:

$$A : B \rightleftharpoons A : + B$$

or

$$A : B \rightleftharpoons A^- + B^+$$

These ionic intermediates are often resonance structures which have an enhanced stability because of the delocalisation of their ionic charge; benzene and its derivatives are good examples of this. A carbon atom or radical bearing a positive charge is called a carbonium ion, whilst one with a negative charge is known as a carbanion.

The tendency to heterolysis is greatest in a covalent bond between two atoms of widely different electronegativity, because in such a bond the electron pair is already displaced towards the more electronegative atom. Most heterolytic reactions take place in solution in polar solvents because these encourage the separation of charges and the formation of ions.

It is found that reagents participating in organic reactions are of two kinds, electrophiles and nucleophiles. The electrophiles or electrophilic reagents are electron pair acceptors. They are acidic in character and oxidising in nature. Most of them are positively charged or are capable of assuming positive charges when polarised by electron-rich centres. Their one common feature is their readiness to form a covalent bond by accepting a pair of electrons from another atom or ion. Examples of electrophiles are the hydrogen ion H^+, nitric acid HNO_3, sulphuric acid H_2SO_4, nitrous acid HNO_2, chlorine Cl_2, and bromine Br_2.

Nucleophiles or nucleophilic reagents are electron pair donors. They are basic in character and reducing in nature. They are usually negatively charged. Their essential characteristic is the possession of a pair of unshared electrons which they can use to form a covalent bond with another atom. Examples are the hydroxyl ion OH^-, the halide ions Cl^-, Br^-, and I^-, the bisulphite ion HSO_3^-, the cyanide ion CN^-, and ammonia NH_3.

Just as reducing agents react with oxidising agents in redox reactions and acids with bases in neutralisations, so electrophiles and nucleophiles combine with each other in heterolytic reactions, for one compound is providing the electron pair which is to constitute the new covalent bond and the other is accepting it. To this extent it can be misleading to refer to any particular reaction as specifically electrophilic or nucleophilic, but it is conventional in discussing organic mechanisms to regard one reactant as the reagent and the other as the substrate or 'molecule under attack' and to classify a reaction in terms of the electrophilic or nucleophilic nature of the reagent, and this is the practice we shall follow in succeeding sections.

ELECTROPHILIC REACTIONS

As will be clear from the previous section, these are reactions in which the reagent is an electrophile i.e. it is capable of forming a new covalent bond with another atom or ion by accepting a pair of electrons donated by it. Amongst the examples we have encountered the reactions involving substitution in the benzene ring such as nitration, halogenation, sulphonation, and alkylation (Friedel–Craft reaction) and also the important processes of acetylation and benzoylation. Other examples include the numerous addition reactions across double or triple bonds such as occur in alkenes (olefines) and alkynes (acetylenes). For the sake of brevity we shall consider only four of these in detail, chosen because they are so important and typical:

1. *Nitration of benzene:* There is overwhelming evidence available from cryoscopic and spectroscopic measurements that to the nitrating agent is the nitronium ion, NO_2^+, which is produced in high concentration in a mixture of concentrated nitric and sulphuric acids :

$$2H_2SO_4 + HNO_3 \rightleftharpoons 2HSO_4^- + H_3O^+ + NO_2^+$$

A study of the kinetics of the nitration reaction shows it to be a two-stage process. In the first step the electrophilic nitronium ion is attracted to the electron-rich benzene molecule and attaches itself covalently to one of the carbon atoms by accepting an electron pair. This confers upon the intermediate product a positive charge which is spread by delocalisation over the resonance structure of the benzene ring to give a stable carbonium ion:

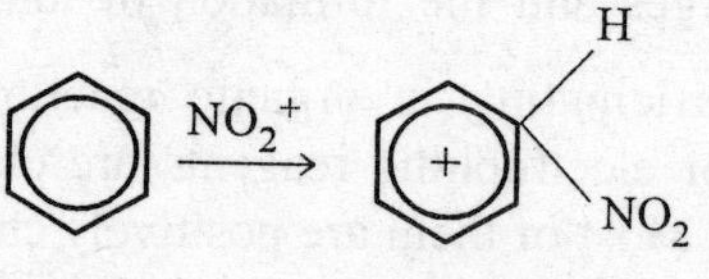

The second stage involves the loss of a proton H^+ to the base HSO_4^-, leaving a neutral molecule of nitrobenzene :

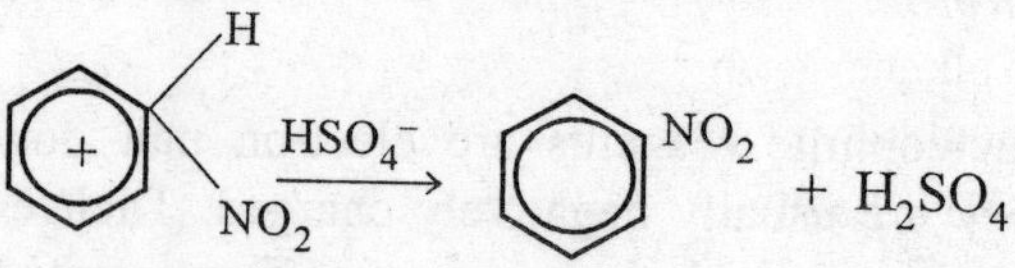

2. *Addition of bromine to ethylene:* This reaction is believed to occur in three stages. In the first a bromine molecule approaching the ethylene becomes strongly polarised under the influence of the high electron density of the double bond and splits heterolytically into two ions Br^+ and Br^-.

$$\underset{\underset{H}{\overset{|}{C}}=\underset{\overset{|}{H}}{\overset{H}{C}}}{} \quad \rightarrow \quad \underset{\underset{H}{\overset{|}{C}}=\underset{\overset{|}{H}}{\overset{H}{C}}}{}$$

$$Br^{\delta+} \qquad\qquad Br^{+}$$
$$|$$
$$Br^{\delta-} \qquad\qquad Br^{-}$$

The positively-charged bromine ion is then attracted to one of the carbon atoms and combines electrophilically forming a carbonium ion as follows:

$$\underset{H}{\overset{H}{C}}\!\cdots\!\underset{Br^{+}}{\overset{H}{C}} \quad \rightarrow \quad \underset{H}{\overset{H}{C}}\!-\!\underset{Br}{\overset{H}{C}}\!-\!H$$
$$Br^{-} \qquad\qquad Br^{-}$$

In the final stage a residual Br^- ion is attracted to the positively-charged carbon atom of the carbonium ion and combines with it to give the dibromide of ethylene, 1,2-dibromoethane :

$$\underset{H}{\overset{H\;\;Br^{-}\;H}{C^{+}}}\!-\!\underset{Br}{\overset{|}{C}}\!-\!H \;\rightarrow\; H\!-\!\underset{H}{\overset{Br\;\;H}{C}}\!-\!\underset{Br}{\overset{|}{C}}\!-\!H$$

3. *Addition of hydrogen bromide to ethylene:* The mechanism resembles reaction (2) above, but in this case the reagent hydrogen bromide is already strongly polarised thus $H^{\delta+}$—$Br^{\delta-}$ before the reaction begins owing to the difference in electronegativity of its constituent atoms. On approach to the double bond, the HBr molecule splits heterolytically into a hydrogen ion H^+ and Br^- ion; the former, attracted by the electron-rich double bond, adds electrophilically to the ethylene to give a carbonium ion to which a residual Br– ion is attracted as in reaction (2) :

$$\underset{H}{\overset{H}{C}}\!\cdots\!\underset{Br^{+}}{\overset{H}{C}} \;\rightarrow\; \underset{H}{\overset{H\;\;H}{C}}\!-\!\underset{H}{\overset{|}{C}}\!-\!H \;\rightarrow\; H\!-\!\underset{Br}{\overset{H\;\;H}{C}}\!-\!\underset{H}{\overset{|}{C}}\!-\!H$$
$$Br^{-}$$

4. *Addition of hydrogen bromide to propylene (propene):* This reaction differs from (3) in that in this case the molecule of alkene is unsymmetrical about the double bond and is therefore preferentially polarised in one particular way before the reaction begins as a result of the inductive effect of the electron-donating methyl group.

$$CH_{3}\!\rightarrow\!\underset{\overset{|}{H}}{\overset{\delta+}{C}}=\underset{\overset{|}{H}}{\overset{\overset{H}{|}\;\;\delta-}{C}}$$

Thus when the polar hydrogen bromide molecule approaches the propylene and splits heterolytically into H^+ and Br^- ions, the positively-charged hydrogen ion is naturally attracted to the carbon atom bearing the fractional negative charge:

$$
\underset{\overset{|}{H}}{CH_3-C} \overset{H}{\underset{\overset{|}{H}}{=C}} \quad \xrightarrow{} \quad CH_3-\overset{\overset{H}{|}}{\underset{\overset{|}{H}}{C}}-\overset{+}{\underset{\overset{|}{H}}{C}}-H
$$

$$H^+$$

This results in a secondary carbonium ion (i.e. a carbonium ion in which the carbon bearing the positive charge has two alkyl groups attached to it). Not only is the formation of a secondary carbonium ion favoured by the inductive effect but it is also preferred energetically in that it is more stable than a primary carbonium ion because the two electron-donating alkyl groups partially compensate for the electron deficiency of the positively charged carbon atom. The final stage of the addition consists of the approach of a Br^- ion to the carbonium ion and its attachment to the carbon atom bearing the positive charge :

$$
\underset{\overset{|}{H}}{H}-\overset{\overset{H}{|}}{\underset{\overset{|}{H}}{C}}-\overset{\overset{Br^-}{+}}{\underset{\overset{|}{H}}{C}}-\overset{\overset{H}{|}}{\underset{\overset{|}{H}}{C}}-H \rightarrow \underset{\overset{|}{H}}{H}-\overset{\overset{H}{|}}{\underset{\overset{|}{H}}{C}}-\overset{\overset{Br}{|}}{\underset{\overset{|}{H}}{C}}-\overset{\overset{H}{|}}{\underset{\overset{|}{H}}{C}}-H
$$

Thus the eventual product of the addition is isopropyl bromide (2-bromopropane) and not n-propyl bromide in accordance with Markownikoff's rule.

A similar mechanism explains the addition of hydrogen halides to isobutylene, $(CH_3)_2.C{=}CH_2$, the homologue of propylene. In this case there are two methyl groups exerting an inductive effect and the tertiary carbonium ion $(CH_3)_3C^+$ is formed in preference to the primary one because of its greater stability, giving tertiary butyl bromide $(CH_3)_3.CBr$ in accordance with Markownikoff's rule.

NUCLEOPHILIC REACTIONS

These are reactions in which the main reagent is a nucleophile i.e. a substance possessing a pair of unshared electrons available for bond formation. The most important examples are the reactions of alkyl halides which lead to the replacement of the halogen atom by various functional groups, the characteristic addition reactions of the carbonyl group, and the reactions of alcohols and it is these three types which will be considered in detail.

1. *Replacement reactions of halides:* Let us consider a primary alkyl halide RCH_2X where X is an atom of chlorine, bromine, or iodine. As explained earlier, the bond between the carbon atom and X will be polarised because of the pronounced electron shift towards the more electronegative halogen atom. This inductive effect gives the carbon atom an electron deficiency (or partial positive charge) which strongly attracts any nucleophilic reagents such as Y^- that approach the halide molecule. We believe that the reaction mechanism probably passes through a brief transition state when Y links itself loosely to the carbon atom by means of its pair of unshared electrons whilst the halogen atom still remains joined to the carbon by a loose and highly polar bond.

$$Y^- + H\!-\!\overset{\displaystyle H}{\underset{\displaystyle R}{\overset{|}{\underset{|}{C}}}}\!\rightarrow X \;\rightarrow\; \left(Y \quad \overset{\displaystyle H\;\; H}{\underset{\displaystyle R}{\overset{|}{C}\!\rightarrow\!\!\rightarrow X}} \right) \;\rightarrow\; Y\!-\!\overset{\displaystyle H\;\; H}{\underset{\displaystyle R}{\overset{|}{C}}} + X^-$$

transition state

This transition state is an unstable and short-lived one and corresponds to the energy peak B in, it is not an intermediate compound, which always corresponds to a trough or minimum in the energy diagram of the reacting system.

In the final stage the halogen atom breaks away from the carbon atom completely leaving a neutral molecule RCH_2Y and an X^- ion.

e.g.
$$C_2H_5I + K^+CN^- \rightarrow C_2H_5CN + K^+I^-$$
$$C_2H_5Br + K^+OH^- \rightarrow C_2H_5OH + K^+Br^-$$

Whilst the nucleophilic reagent Y is usually a negatively-charged ion such as OH^- or CN^-, it may also be an uncharged molecule such as ammonia which contains a lone pair of electrons with which to attach itself to the carbon atom :

$$C_2H_5Br + NH_3 \rightarrow C_2H_5NH_3^+ + Br^-$$

A study of kinetics makes it clear that replacement reactions of tertiary alkyl halides are unimolecular and have a different mechanism involving the formation of a carbonium ion.

2. *Addition reactions of the carbonyl group:* The strong polarisation of the carbonyl group $>\!C\!=\!O$ has

already been explained above where the conventional notations of $>\!C\!=\!O$ or $>\!C^{\delta+}\!=\!O^{\delta-}$ were

introduced. As a result of this effect a nucleophilic reagent is readily attracted to the carbon atom bearing the positive charge and attaches itself by means of its unshared electron pair:

$$\underset{\underset{\displaystyle Y^-}{\displaystyle R'}}{\overset{\displaystyle R}{\underset{\displaystyle \overset{\delta-\;\delta-}{C\!=\!C}}{\quad}}} \longrightarrow \underset{\displaystyle R'\quad Y}{\overset{\displaystyle R\quad O^-}{C}}$$

In the second stage the negative charge on the oxygen atom attracts a proton to form a hydroxyl group and so complete the course of the addition :

$$\underset{\displaystyle R'\quad Y}{\overset{\displaystyle R\quad O^-}{C}} + H^+ \rightarrow \underset{\displaystyle R'\quad Y}{\overset{\displaystyle R\quad OH}{C}}$$

Hydrogen cyanide gives a cyanohydrin, for example:

$$\underset{\underset{R'}{\overset{R}{\diagdown}}}{\overset{\delta^+ \ \delta^-}{C=O}} \quad \xrightarrow{\ \ CN^-\ \ } \quad \underset{\underset{R'\quad CN}{}}{\overset{R\quad O^-}{C}} \quad \xrightarrow{\ H^+\ } \quad \underset{\underset{R'\quad CN}{}}{\overset{R\quad OH}{C}}$$

Similarly, sodium hydrogen sulphite (sodium bisulphite) and ammonia add nucleophilically as follows:

$$\underset{\underset{R'}{\overset{R}{\diagdown}}}{\overset{\delta^+ \ \delta^-}{C=O}} \ \underset{HSO_3^-}{Na^+} \quad \rightarrow \quad \underset{\underset{R'\ \ SO_3H}{Na^+}}{\overset{R\quad O^-}{C}} \quad \rightarrow \quad \underset{\underset{R\ \ SO_3^-}{Na^+}}{\overset{R\quad OH}{C}}$$

$$\underset{\underset{R'}{\overset{R}{\diagdown}}}{\overset{\delta^+ \ \delta^-}{C=O}} \ {NH_3} \quad \rightarrow \quad \underset{\underset{R'\ \ NH_3^+}{}}{\overset{R\quad O-}{C}} \quad \rightarrow \quad \underset{\underset{R'\ \ NH_2}{}}{\overset{R\quad OH}{C}}$$

A study of the kinetics of these reactions reveals that the first step is the slow rate-determining stage, the acquisition of the proton being relatively fast.

With lithium aluminium hydride and sodium borohydride the first step in the reaction is the transfer of a nucleophilic hydride ion H^- from the reducing agent to the carbon atom of the carbonyl group :

$$\underset{\underset{H^-\ AlH_3\ Li^+}{R'}}{\overset{\delta^+ \ \delta^-}{\underset{R}{C=O}}} \quad \rightarrow \quad \underset{\underset{R'\quad H}{}}{\overset{R\quad OAlH_3^-Li^+}{C}}$$

On adding water or acid the intermediate compound is rapidly hydrolysed giving a secondary alcohol.

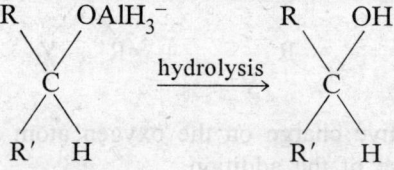

3. *Reactions of alcohols:* Alcohols, like water, can act as weak bases when brought into contact with strong acids. In these circumstances they accept a proton, which links to the alcohol molecule electrophilically by making use of a lone pair of electrons belonging to the oxygen atom, to form an alkyloxonium ion thus:

$$R\text{—}\overset{\cdot\cdot}{\underset{\cdot\cdot}{O}}\text{—}H + H^+ \rightleftharpoons \left(R\text{—}\overset{\overset{H}{|}}{\underset{\cdot\cdot}{O}}\text{—}H\right)^+$$

This can be compared to the analogous tendency of a water molecule to accept a proton and form a hydroxonium ion thus :

$$R\text{—}\overset{\cdot\cdot}{\underset{\cdot\cdot}{O}}\text{—}H + H^+ \rightleftharpoons \left(R\text{—}\overset{\overset{H}{|}}{\underset{\cdot\cdot}{O}}\text{—}H\right)^+$$

This ability to act as a base and form an alkyloxonium ion constitutes the first step in the reaction of alcohols with hydrogen halides. With hydrogen bromide, for example, the Br^- ion is attracted to the alkyloxonium ion and reacts nucleophilically with it leading to a transition state which rapidly decomposes with the loss of a water molecule leaving an alkyl bromide :

$$Br + R.\overset{+}{O}H_2 \rightarrow (Br\frown R \rightarrow OH_2) \rightarrow Br\text{—}R + H_2O$$
$$\text{Transition state}$$

In some secondary and most tertiary alcohols the water molecule tends to be discarded by the alkyloxonium ion first and the Br^- ion then combines nucleophilically thus :

$$R.\overset{+}{O}H_2 \rightarrow R^+ + H_2O$$

$$R^+ + Br^- \rightarrow R\text{–}Br$$

This tendency of alcohols to behave as weak bases and form alkyloxonium ions also plays an important part in the preparation of alkenes. For example, propan-2-ol or isopropyl alcohol reacts thus:

$$\underset{CH_3.CH.CH_3}{\overset{OH}{|}} + H^+ \rightleftharpoons \underset{CH_3.CH.CH_3}{\overset{\overset{+}{O}H_2}{|}}$$

The alkyloxonium ion then dissociates readily into a carbonium ion and water :

$$\underset{CH_3.CH.CH_3}{\overset{\overset{+}{O}H_2}{|}} \rightleftharpoons CH_3.\overset{+}{C}H.CH_3 + H_2O$$

Now under certain circumstances (e.g. high temperature or the presence of a base) this carbonium ion tends to lose a proton by discarding a hydrogen ion from one of the carbon atoms adjacent to the one bearing the positive charge :

$$CH_3.\overset{+}{C}H.CH_3 \rightleftharpoons CH_3.CH{=}CH_2 + H^+$$
$$\text{(propylene)}$$

The net result of the sequence of changes is to eliminate a molecule of water from the alcohol and create a double bond, so the reaction is essentially a dehydration and a means of preparing alkenes.

ACIDITY IN ORGANIC COMPOUNDS

An acid is best defined in Bronsted Lowry terms as a proton donor and a base as a proton acceptor. The acidity of carboxylic acids arises from their ability to dissociate in aqueous solution yielding hydroxonium ions thus :

$$R.COOH + H_2O \rightleftharpoons R.COO^- + H_3O^+$$

In this dissociation the hydroxyl group undergoes heterolytic fission into a proton and a negatively charged ion. The readiness with which this dissociation occurs depends upon the degree of polarity in the O—H bond. This polarity is much greater in carboxylic acids than in alcohols because of the influence of the adjacent carbonyl group, itself highly polarised by the inductive effect as described above which leaves the carbon atom with a small positive charge. This attracts electronic charge from the oxygen of the hydroxyl group and encourages dissociation :

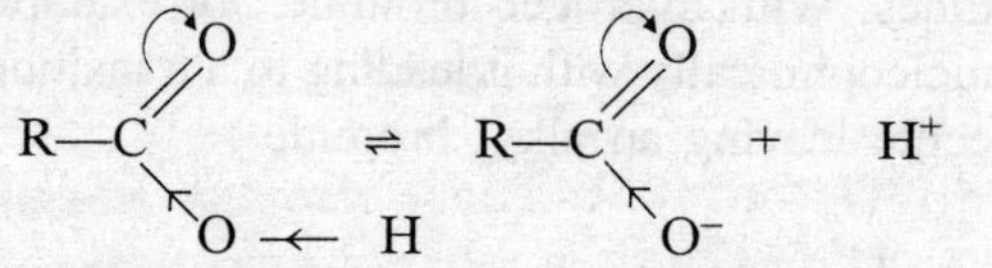

Another important factor here is the ability of the carboxylate ion R.COO$^-$ to adopt a resonance structure which is a hybrid of the two canonical forms 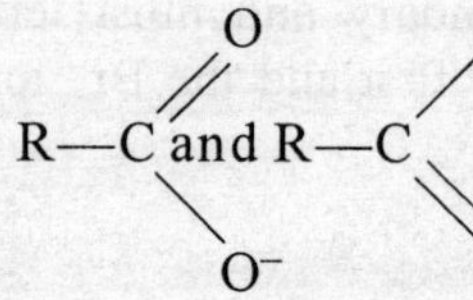 By facilitating delocalisation of the negative charge this gives the resulting ion lower energy and an increased stability and hence promotes the dissociation.

As Table 20.1. demonstrates, the strength of a carboxylic acid depends upon the nature of the radical R. There is not much difference between acetic acid and its higher homologues, but formic acid is significantly stronger because it is free of the weak electron-donating inductive effect of alkyl groups. When we consider the three chloro-substituted acetic acids the marked influence of substituent groups becomes very clear. Each substituted chlorine atom causes an electron shift away from the carbon atom to which it is attached, thereby decreasing the electron density and enhancing the dissociation of the carboxyl group. As a consequence the monochloro acid is about 80 times stronger than acetic acid and the di- and trichloro acids about 2700 and 11,000 times as strong respectively.

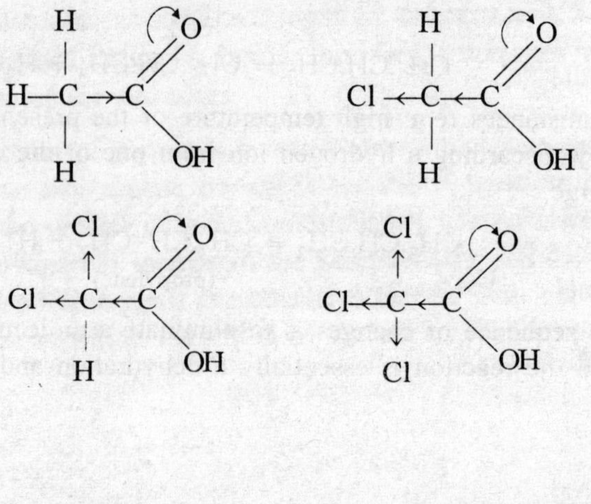

These effects are much less pronounced if the substituent groups are attached to the β and γ carbon atoms, since they are then too remote to have much inductive influence upon the dissociation of the carboxyl group, as is clear from the figures in the Table for the three chlorobutyric (chlorobutanoic) acids. When another electron-attracting carboxyl group is attached to the α carbon atom, however, as in oxalic acid $(COOH)_2$, then there is a very marked increase in the dissociation constant. Benzoic acid, in which R is a phenyl group C_6H_5—, is only slightly stronger than acetic, but the presence of substituted nitro groups in the benzene ring markedly increases the degree of dissociation.

Table 20.1. Strengths of organic acids.

Name	Acid		Dissociation Constant K $\times 10^{-5}$ mol dm^{-3}	pKa
		Formula		
Formic		H.COOH	18	3.75
Acetic		CH$_3$.COOH	1.8	4.75
Propionic		C$_2$H$_5$.COOH	1.3	4.89
α-Butyric		CH$_3$.CH$_2$.CH$_2$.COOH	1.5	4.82
Monochloroacetic		CH$_2$Cl.COOH	150	2.82
Dichloroacetic		CHCl$_2$.COOH	5000	1.31
Trichloroacetic		CCl$_3$.COOH	20000	0.70
α-chlorobutyric		C$_2$H$_5$.CHCl.COOH	140	2.85
β-chlorobutyric		CH$_3$.CHCl.CH$_2$.COOH	10	4.00
γ-chlorobutyric		CH$_2$Cl.(CH$_2$)$_2$.COOH	3	4.52
Oxalic		HOOC.COOH	6000	1.22
Benzoic		C$_6$H$_5$.COOH	6.4	4.19
o-Nitrobenzoic		o–O$_2$N. C6H4.COOH	680	2.17
Phenol		C$_6$H$_5$OH	0.00001	10.0
g-Nitrophenol		o-O$_2$N.C$_6$H$_4$.OH	0.006	7.22
Trinitrophenol		2, 4, 6-(NO$_2$)$_3$.C$_6$H$_2$.OH	9500	1.02

Phenol undergoes slight dissociation of its hydroxyl group thus

$$C_6H_5OH + H_2O \rightleftharpoons C_6H_5O^- + H_3O^+$$

The mild electron-attracting influence of the phenyl group facilitates this dissociation but the main reason why phenol is acidic and alcohols are not lies in the stabilising effect on the phenoxide ion, $C_6H_5O^-$, of its resonance structure in which the negative charge is delocalised over the benzene ring. Phenol itself is a very weak acid $(K = 1.3 \times 10^{-10}$ mol dm$^{-3})$ but the presence of substituted groups, particularly electrophilic nitro groups, has a notable effect upon its dissociation, as the Table shows. Thus trinitrophenol, sometimes known as picric acid, is a strong acid comparable in strength to trichloroacetic and the mineral acids. The effect of substitutents in phenol is not only to increase greatly its strength as an acid but also to make the benzene ring more reactive, especially in its response to further substitution by electrophilic groups. Thus once a nitro group or bromine atom has been introduced its effect on the distribution of electronic charge is relayed around the benzene ring by the mesomeric effect and further substitution is facilitated.

The inductive effect can also be used to explain why secondary amines are stronger bases than primary amines or ammonia, since the alkyl groups are more electron-repelling than hydrogen and encourage the nitrogen atom to take up a proton. When an electron-attracting carbonyl group is introduced, however, its inductive effect acts strongly in the opposite direction and so amides are only very weakly basic substances.

CONCLUSION

Although substitution reactions in alkanes and arenes have a superficial resemblance, we have been in this chapter that the mechanisms of these two reactions are entirely different. Alkane substitution consists of homolytic free-radical chain reactions, whereas aromatic substitution entails electrophilic attack by positively charged ions and highly polarised groups. Similarly, addition reactions can proceed by completely contrasting mechanisms e.g. additions to alkenes and alkynes across the double or triple bond are initiated by the approach of an electron-seeking electrophilic atom or group whereas additions across the carbonyl bond are restricted to nucleophilic reagents, most of them negatively charged which themselves provide the electron pair required to form the bond. A study of mechanism gives an appreciation of these differences and a deeper understanding of the various types of organic reaction.

Example 20.1. 0.30 g. of an organic substance on combustion gave 0.66 g. of CO_2 and 0.36 g. of water. Its V.D. was 30. Calculate its molecular formula. Write the structures of the three substances having this formula.

Solution.

(a) Empirical formula : C_3H_8O (worked out as usual)

E.F. wt. = 36 + 8 + 16 = 60 and mol. wt. = 2 × V.D. = 2 × 30 = 60.

Since, the molecular weight is the same as empirical formula weight, the molecular formula of the substance is also

C_3H_8O. $(C_nH_{2n+2} O)$

The compound may be an alcohol or an ether and amongst alcohols also, a primary or a secondary alcohol because a tertiary alcohol should have at least four carbon atoms. The possible structural formulae are :

(i) $CH_3CH_2CH_2OH$ (normal propyl alcohol)—Propanol-1

(ii) $CH_3 CH_3$ CHOH (isopropyl alcchol)—Propanol-2

(iii) $CH_3.O.CH_2CH_3$ (methyl-ethyl ether)—Methoxy ethane.

Note—Taking into account the valencies of carbon (4), hydrogen (1) and oxygen (2), various possible arrangements have been suggested for the compound having one oxygen atom.

Example 20.2. A monobasic organic acid gave the following results on analysis.

(a) 0.1890 g. in ultimate analysis gave 0.1761 g. of CO_2 and 0.0540 g. of water.

(b) 0.2079 g. in Carius determination gave 0.3157 g. silver chloride.

(c) 0.3150 g. required 33.33 ml. of N/10-NaOH for complete neutralisation.

Deduce the molecular formula of the substance and suggest its possible structural formula.

Solution.

(a) Empirical formula : $C_2H_3ClO_2$ (worked out as usual)

(b) Molecular formula :

        (i) E.F. weight = 24 + 3 + 35.5 + 32 = 94.5

        (ii) Molecular weight : 33.33 ml. of N/10-NaOH ≡ 0.3150 g. acid.

∴ 1000 ml. of normal-NaOH (its equivalent quantity)

$$\equiv \frac{0.3150 \times 1000 \times 10}{33.33} = 94.51 \text{ g. of acid}$$

Hence, Eq. wt of the acid = 94.51

and Mol. wt. = Eq. wt. × basicity = 94.51 × 1 = 94.51.

Since, the molecular weight is the same as the empirical formula weight, the molecular formula is also $C_2H_3ClO_2$.

(c) Structural formula : Since the compound is a monobasic acid, therefore, by setting aside one —COOH group, the remaining part CH_2Cl group may be attached to it. Hence, the given compound is

$CH_2ClCOOH$—Monochloro-acetic acid.

Chapter 21

Noble Gases

INTRODUCTION

In chemical terminology the term *noble gas* describes an element which either is completely unreactive or reacts only to a limited extent with other elements. Six noble gases constitute a group in the periodic table which is variously called the zero group (as the first three of its members have a valence of zero), the inert gas group, and the noble gas group. The last is preferable, as three of the gases, though unreactive, are not inert. The gases of this group are helium, neon, argon, krypton, xenon, and radon. The noble metals are generally considered to be gold, silver, platinum, palladium, iridium, rhodium, mercury, ruthenium, and osmium. The term has no reference to the commercial value of these metals.

Helium, neon, argon, krypton, xenon and radon constitute Group 0 of the periodic table. Elements of this group have been called by several names, such as gases of the atmosphere, inert gases, rare gases, and noble gases. They have been called gases of the atmosphere because all of them are gases and occur in the atmosphere; inert gases, because they are chemically inert and till recently were not known to form any chemical compound; rare gases, because they occur in nature in small amounts and were obtained in the beginning only in small quantities; noble gases, because like noble metals, they occur in the free state and are characterised by lack of chemical reactivity. The above names, except the last one are now considered to be inappropriate for these elements. The first name is considered to be unsuitable, because these are not the only elements which occur as gases in the atmosphere; the second because a number of authentic chemical compounds of krypton, xenon and radon are now known; the third because they are now being produced on a large scale owing to the increased industrial demand for them.

POSITION IN THE PERIODIC TABLE

Elements of zero group were not known till 1869—the year in which Mendeleeff compiled his periodic table. He did not leave any vacant spaces for them as he did for a number of elements which he expected would be discovered later. When argon and helium were discovered (1892-1894), difficulty was experienced in assigning position to these elements in the periodic table. Juliot Thomsen, from a critical examination of the periodic table, observed that there cannot be an abrupt change in going from the halogens (Gr. VIIB) to the alkali metals (Gr. IA), He, therefore, predicted the existence of a group of elements with atomic weights, 4, 20, 36, 84, 132 and 212, possessing zero chemical character between the halogens and the alkali metals.

Ramsay and Travers soon discovered the remaining four elements of this group and placed all the six elements between the halogens and the alkali metals as suggested by Thomsen. This arrangement was vehemently criticised by some contemporary chemists and warmly supported by others. The chief points of criticism were :

1.   Argon cannot be placed between potassium and calcium as required by its atomic weight.

2.   It has upset Newlands' law of octaves. With the adoption of atomic number as the basis of classification of elements, the first criticism disappeared. The second criticism is purely sentimental and is therefore not serious. Had these elements been known at the time of Newlands, he could have suggested some other suitable analogy for the recurrence of similarities then between the first and the ninth elements. The more important points in favour of the classification are :

(a) These elements because of their lack of chemical reactivity act as a bridge between the highly electronegative halogens and the highly electropositive alkali metals.

Gr. VIIB	Gr. 0	Gr. IA
H	He	Li
Cl	Ne	Na
F	Ar	K
Br	Kr	Rb
I	Xe	Cs
At	Rn	Fr

(b) In Lothar Meyer Curve, it is seen that noble gases fall between the halogens and the alkali metals. The question whether the noble gases should be placed in Group VIII as a B sub-group after the halogens or in Group 0 before the alkali metals, has been the subject matter of much discussion in the past.

When these elements are placed in Group VIII, they occupy the last column in each period of the p-block elements and therefore, act as terminal elements closing a period—a position which is in conformity with their atomic structures.

However, the horizontal relationships which exist between the elements in any period, alter abruptly after the halogens and they show no similarities to the noble gases, except that the halogens and noble gases both are non-metallic elements. Furthermore, following the natural order of atomic numbers, each noble gas precedes an alkali metal and there is an abrupt change in chemical behaviour from a non-reactive non-metal to a highly reactive metal. Noble gases, therefore, bear no relationship to its neighbouring groups viz., Groups VIIB and IA. These elements however, form a very homogeneous group of the periodic table as is evident from the regular trends observed in the physical properties in going from helium to radon (Table 21.1) in the 0 group.

From many points of view, the trends in the magnitude of the physical properties for this group may be regarded as almost ideal. It may be noted that there is a negligible departure in the ratios of specific heats (C_p/C_v) in each case from the ideal value of 1.667 for a monoatomic gas. The fact that these gases can be liquefied and solidified shows the existence of van der Waals forces, because all other forces which might bring the atoms closer together are absent.

Table 21.1. Some physical properties of noble gases.

	He	Ne	Ar	Kr	Xe	Rn
Atomic number	2	8	18	36	54	86
Electron configuration	2	2,8	2,8,8	2,8,18,8	2,8,18,18,8	2,8,18,32, 18,8
Atomic radius (A°)	1.2	1.6	1.9	2.0	2.2	–
Ionisation potential (eV)	24.5	21.5	15.7	13.9	12.1	10.7
Melting point (°K)	0.9	24	84	116	161	202
Boiling point (°K)	4	27	87	121	164	211
Atomic volume (liq) c.c.	31.8	16.8	28.5	32.2	42.9	50.5
Density (g/c.c.)	0.126	1.204	1.65	2.6	3.06	4.4
Heat of fusion kcal/mole	0.033	0.080	0.280	0.341	0.549	–
Heat of vapourisation kcal/mole	0.025	0.405	1.600	2.240	3.100	3.600
Ratio of specific heats (C_p/C_v)	1.65	1.64	1.65	1.69	1.67	–
Critical temp. (°K)	13.8	14.7	37.9	73	110.9	–
Critical pressure (atm.)	2.26	26.86	47.97	54.3	58.2	62.4
Thermal conductivity at °C $(K \times 10^6)$	343	111.2	38.2	21.2	12.4	–

However, the very low melting and boiling points of the noble gases indicate that these interatomic forces are very weak. The interaction between the noble gas atoms probably results from the slight mutual polarisation of the atoms so that in each atom the centres of positive and negative charges no longer coincide. The polarity is thus induced in each atom making it an induced dipole. The polarised atoms are attracted to each other by induced dipole-induced dipole interaction. Such interaction is the basis of van der Waals forces existing between atoms of the noble gases. The magnitude of van der Waals forces is proportional to the polarisability of the noble gas atoms. The polarisability of these atoms increases with increasing atomic radius thus leading to the increased van der Waals forces between the atoms. This is in agreement with the increase in melting and boiling points on going from helium to radon in the noble gas group. Helium (low van der Waals forces) is unique amongst the elements in that it cannot be solidified by cooling alone, no matter how low the temperature is. It forms a solid when subjected to low temperature and a pressure of about 25 atm. Helium exists in two liquid phases, viz., He^{-I}, a normal liquid, and He^{-II} a superfluid. Helium-II is a liquid with the properties of a gas. It has such a low energy that the thermal motion of the atoms stops for all practical purposes but van der Waals forces are not strong enough to form a solid. When helium-I changes to helium-II, many physical properties such as density, specific heat, viscosity, thermal conductivity and surface tension change abruptly. Liquid helium-II is characterised by a very low viscosity, very high thermal conductivity and peculiar flow properties which enable it to flow uphill. The reasons for the strange behaviour of helium-II are not fully understood. Helium-II, because of its peculiar properties, is sometimes classed as the fourth state of matter.

In 1785, Cavendish passed an electric discharge through air mixed with excess of oxygen and absorbed the resulting oxides of nitrogen in caustic soda solution and the unreacted oxygen by an absorbent such as potassium pentasulphide, K_2S_5. He found that a small residue of the gas was always left which would not combine with oxygen, and refused to undergo any chemical combination. He estimated its amount to be not more than 1/120th part of the original air. Unfortunately, this important observation escaped notice for more than a hundred years until in 1892 Lord Rayleigh found that the density of nitrogen obtained from air was about half a per cent more than the density of nitrogen prepared chemically, the actual values being 1.2572 and 1.2505 gm/litre respectively. The discrepancy was not due to an experimental error which was estimated to be not more than 0.13 per cent. Various suggestions were put forward such as the presence of heavy allotrope, e.g., N_3 in atmospheric nitrogen, but they were soon shown to be untenable. Lord Rayleigh, in collaboration with William Ramsay, repeated Cavendish's experiment and confirmed the presence of small unabsorbed residue which was examined spectroscopically and shown to be a hitherto unknown gas. The above investigators named the newly discovered element as argon from a Greek word *argos* meaning inert or idle.

Janssen (1868) spectroscopically examined the chromosphere of the sun during a total eclipse and observed a new yellow line differing in position from D_1 and D_2 lines of sodium. Frankland and Lockyer suggested that the new line was given by an element not known before. The new element thus discovered was named helium from a Greek word *helios* meaning sun. In 1889, Hillebrand obtained a sample of this gas by heating the mineral cleveite but wrongly identified in with nitrogen. Ramsay (1895) from a spectroscopic examination of the gas showed it to be identical with helium discovered previously in the chromosphere of the sun.

Ramsay had now begun to suspect that argon prepared from the atmosphere was not pure and so he with Travers took the supposedly crude argon, liquefied it and submitted it to low-temperature fractionation. The first fraction when examined spectroscopically showed the presence of a new element which was named neon from a Greek word *neos* (=*new*). The same investigators discovered two further new elements from the spectroscopic examination of the less volatile fraction and named them krypton and xenon from the Greek words *kryptos* (=*hidden*) and *xenos* (=*stranger*) respectively.

The family became complete with the discovery of radon by Dorn in 1901 as a result of the spectroscopic examination of an emanation from radium. In the same year, Kayser detected trace amounts of the new element in the atmosphere. Ramsay and Rayleigh were awarded the Nobel prize (separately) in chemistry and physics respectively for their discovery of noble gases.

It may be noted that the whole family of noble gases was discovered spectroscopically. Further, analytical balance played a vital role in the discovery of argon. The discovery of argon is rightly called the triumph of the fourth decimal place.

SOURCES OF NOBLE GASES

The noble gases occur to the extent of about 1 per cent in the atmosphere. Most of it is made up of argon; the remaining noble gases accounting for about 0.003 per cent by volume of the total. The atmospheric content of noble gases (in per cent volume) is given below :

He	0.0005
Ne	0.0018
Ar	0.94
Kr	0.0001
Xe	0.00001
Rn	6×10^{-10}

The most important source of helium at present is the natural hydrocarbon gases found in certain wells in North America and Canada. These gases contain sometimes up to 7 per cent of helium which is undoubtedly formed by α-decay of radioactive minerals. Certain radioactive minerals, such as monazite, cleveite and thorianite, contain notable amounts of helium.

ISOLATION

The isolation of noble gases from the atmosphere involves two steps; viz., (i) the preparation of the noble gas residue by the removal of nitrogen, oxygen, moisture and carbon dioxide from the atmospheric air by chemical methods; and (ii) the individual separation of the elements from the resulting noble gas residue by physical methods.

Preparation of Noble Gas Residue

Rayleigh and Ramsay considerably improved Cavendish's method of removing nitrogen and oxygen from air by sparking for the preparation of the noble gas residue. Their apparatus is shown diagrammatically in Fig. 21.1.

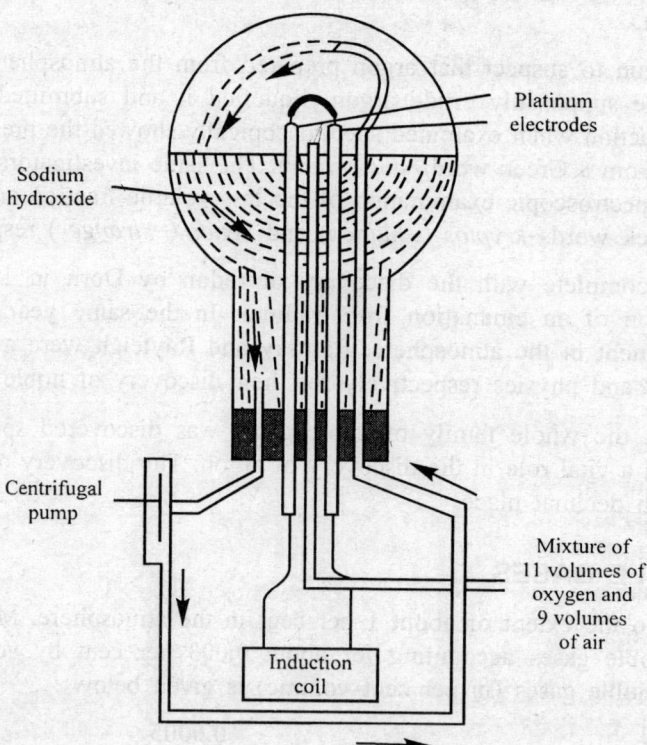

Fig. 21.1. Rayleigh and Ramsay's method of the preparation of noble gas residue.

An electric discharge of 6000–8000 volts is passed from an induction coil through a mixture of 11 volumes of oxygen and 9 volumes of air taken into a glass globe of 50 litre capacity provided with heavy platinum electrodes. A solution of sodium hydroxide is circulated into the glass bulb to dissolve the resulting oxides of nitrogen. The reactions which occur are given below:

$$N_2 + O_2 \rightarrow 2NO$$

$$2NO + O_2 \rightarrow 2NO_2$$

$$2NO_2 + 2NaOH \rightarrow NaNO_2 + NaNO_3 + H_2O$$

Noble gases together with a little unreacted oxygen accumulate in the flask. The oxygen is removed from the residual gas by absorption through an alkaline solution of pyrogallol. The second method of the preparation of noble gas residue is due to Ramsay.

In this method, air is passed over red hot copper to remove oxygen as CuO and then through soda lime and concentrated sulphuric acid to remove carbon dioxide and moisture respectively. The nitrogen is finally removed as Mg_3N_2 by passing the residual gas repeatedly over strongly heated magnesium. A more efficient chemical method of removing nitrogen and oxygen from air is due to Fischer and Ringe which consists in circulating air over calcium carbide mixed with about 10 per cent of calcium chloride at 800°C. The nitrogen and oxygen are absorbed as calcium cyanamide and calcium carbonate respectively according to the equations :

$$CaC_2 + N_2 \rightarrow CaCN_2 + C$$

$$C + O_2 \rightarrow CO_2$$

$$2CaC_2 + 5O_2 \rightarrow 2CaCO_3 + 2CO_2$$

$$CO_2 + C \rightarrow 2CO$$

Some carbon monoxide produced during the process is removed by passing the gaseous mixture over heated copper oxide; carbon dioxide is finally removed by absorption through a potash tower.

$$CO + CuO \rightarrow CO_2 + Cu$$

$$CO_2 + 2KOH \rightarrow K_2CO_3 + H_2O$$

Water vapour, which may affect the carbide, is removed in the beginning by concentrated sulphuric acid.

Individual Separation of Noble Gases

From the noble gas residue, the individual gases are separated by physical methods such as fractional distillation and selective adsorption by charcoal. The boiling points of the noble gases increase from helium to xenon. Hence during fractional distillation, the gases escape from the mixture in the order of their volatility. The separation by selective adsorption method is based on the following two principles:

1. The most easily liquefied gases are most readily adsorbed by porous bodies like charcoal.

2. Desorption is most rapid with those gases which are adsorbed with the greatest difficulty.

In this method, the inert gas residue is cooled to −180°C in contact with charcoal in a glass bulb when Ar, Kr and Xe get adsorbed, the unadsorbed helium and neon are kept in contact with a fresh charcoal contained in another glass bulb at −225°C when neon gets adsorbed and helium is left behind. On warming the first charcoal to −80°C, almost pure argon can be evolved. The charcoal containing Kr and Xe is next placed in contact with a fresh charcoal at −180°C when Kr passes into the second charcoal, leaving Xe in the old charcoal. A schematic representation of the separation of the noble gases by this method is shown in the flow diagram in Fig. 21.2.

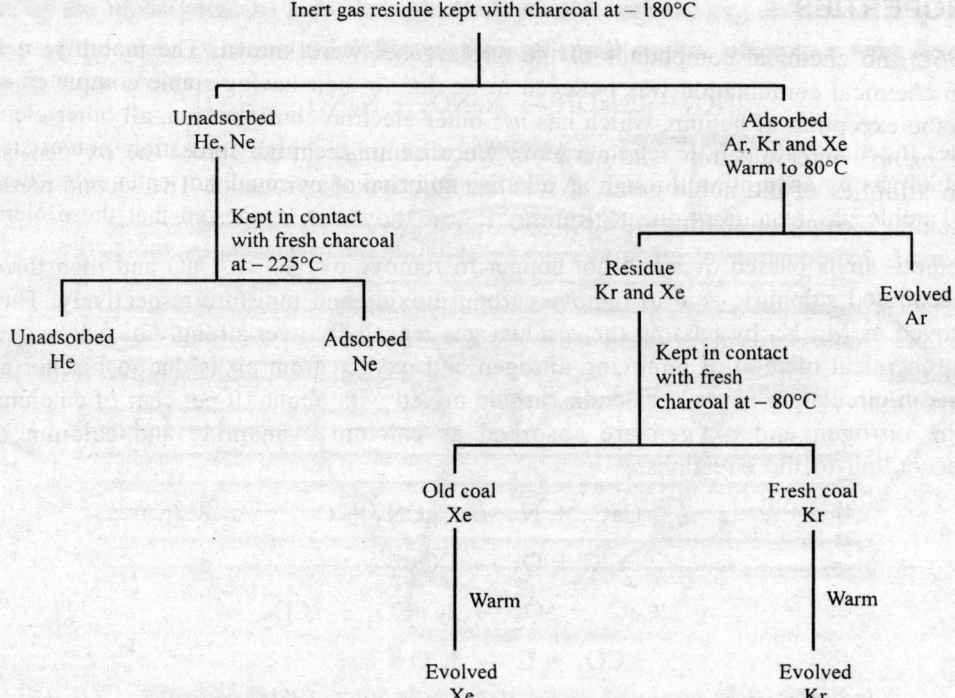

Fig. 21.2. Separation of noble gas by charcoal.

Isolation of Noble Gases by Fractional Distillation of Liquid Air

All noble gases, with the exception of radon, are now obtained industrially by the fractional distillation of liquid air. During fractional distillation of liquid air, neon, nitrogen and helium come in the more volatile fractions, whereas oxygen, argon, krypton and xenon are concentrated in the residual liquid. A second fractionation of the less volatile portion gives argon of about 50 per cent purity, the removal of the residual oxygen from this (by burning with hydrogen and passing over heated copper) yields a krypton-xenon mixture. The separation of krypton from xenon can be effected by further fractionation or by selective adsorption with charcoal at low temperature. The mixture of helium, neon and nitrogen in the volatile fractions is led through a spiral column placed in the stream of evaporating liquid nitrogen when most of the nitrogen is liquefied. The remaining nitrogen is removed by chemical means, e.g., with calcium carbide. The separation of helium from neon is then affected either by cooling in liquid hydrogen when neon solidifies and helium passes on, or by selective adsorption by active charcoal at low temperature. Previously, helium was obtained largely from radioactive minerals, such as monazite, by heating them directly to about 1000–1200°C or by decomposing them with sulphuric acid. This process is inconvenient and very expensive. Helium is now mostly obtained from natural gases. It is isolated from these gases simply by intense cooling when the accompanying gases condense, leaving helium in the gaseous state. For the preparation of radon, radium chloride is dissolved in water, the resulting oxygen, nitrogen, hydrogen, helium and radon mixture is pumped off from the solution and exploded to remove oxygen and hydrogen; water formed being removed by phosphorus pentoxide. The radon is finally separated from helium and nitrogen by passing the gaseous mixture through a tube cooled in liquid air when radon alone gets condensed as a liquid.

CHEMICAL PROPERTIES

Until recently (1962), no chemical compounds of the noble gases were known. The inability of these gases to enter into chemical combination was believed to be due to their having stable completed shells of electrons. With the exception of helium, which has ns^2 outer electron configuration, all other elements of Group 0 have ns^2np^6 outer electron configuration. Because of the high ionisation potentials and negligible electron affinities of the noble gases, it was thought that they could not enter into chemical combination by gaining, losing or sharing of electrons. It was, however, suggested that these elements could form compounds by donation of electron pairs to suitable electron acceptor molecules.

Attempts were accordingly made to prepare coordination compounds of the elements with strong electron acceptors, like boron trifluoride. Thus, Booth and Wilson (1935) from a thermal analysis of the argon-boron trifluoride system at low temperatures ($-127°$ to $-133°C$) reported the formation of compounds, $Ar.xBF_3$ (where x = 1,2,3,6,8 and 16) and represented them as shown :

$$Ar \rightarrow (BF_3)_x$$

Parallel studies on krypton-boron trifluoride system by Wiberg and Karbe (1948) gave no evidence of compound formation, even though the formation of compound could be expected to be better in this system. A reinvestigation of argon-boron trifluoride system by the same investigators failed to confirm the formation of any of the compounds reported earlier by Booth and Wilson. It is noteworthy that Wiberg and Karbe were also unable to obtain evidence for compound formation between xenon and a number of electron acceptor molecules, such as diboron, trimethyl boron, and dimethyl ether.

Excepting helium and neon, the gases form crystalline hydrates and deuterates of variable compositions on freezing water (light or heavy) in the presence of the noble gas under pressure. The number of H_2O or D_2O molecules per gas atom approach six with increasing atomic number. The formation of these unstable hydrates and deuterates probably result from the dipole-induced dipole interaction. In the presence of H_2O or D_2O, which are strong dipoles, the atoms of noble gases get polarised to such an extent that they become induced dipoles which interact with dipoles of H_2O or D_2O.

CLATHRATES

Compounds of krypton, xenon and radon with phenol of 1:2 stoichiometry have been prepared. These compounds are also believed to result from dipole-induced dipole interaction. Dipole-induced dipole interactions are expected to be very weak and it is probably more correct to regard these compounds, particularly the noble gas hydrates and deuterates, as clathrate or enclosure compounds.

Powell (1950) prepared cage-like compounds of argon, krypton and xenon with quinol of the type $NG.3C_6H_4(OH)_2$ (NG = noble gas) by crystallising quinol from aqueous or alcoholic solutions under considerable pressure of the inert gas. The crystals formed contain noble gas atoms trapped in the quinol lattice. These compounds are usually called clathrate or enclosure compounds. The term clathrate or enclosure compound refers to the physical trapping of one component of a compound within cavities in the crystal lattice of the other component and there is no chemical bond formation between the two components.

Helium and neon fail to form quinol clathrates, presumably due to the smaller size of their atoms which enables them to escape from the quinol lattice. Helium in the discharge tube forms a short-lived helium molecule ion, He_2^+ which is believed to contain a three electron bond $(He ... He)^+$. Under these excited conditions a $1s$ electron in the helium atom is promoted to a higher energy level, probably $2s$. On the valence bond theory, the formation of He_2^+ may occur by the overlapping of the singly occupied $1s$ orbitals of the excited helium atom and helium ion.

Despite the theoretical evidence for the inertness of noble gases, a few chemists did predict the existence of compounds of these gases. Thus, Pauling in 1933 predicted the possibility of the formation of KrF_6, XeF_6, XeF_2, xenic acid (H_4XeO_6) and salts of this acid of the types Ag_4XeO_6 and AgH_3XeO_6. These predictions were based on the considerations of ratios of ionic radii and the coordination number which is allowed by the packing of the anions round the central cation. Yost and Kaye following Pauling's predictions attempted to react krypton and xenon with fluorine and chlorine both. The mixtures were subjected to electrical discharges and ultra-violet radiation. Although they were unable to isolate any compound, the nature of their results was such that the existence of noble gas halides could not be ruled out.

Bartlett and Lohman (1962) prepared dioxygenyl hexafluoroplatinate (V), $O_2^+[PtF_6]^-$ by reaction of molecular oxygen with PtF_6 vapour. Since the ionisation potentials of molecular oxygen (12.2 eV) and xenon (12.13 eV) are closely similar, it occurred to them that an analogous compound could be prepared with xenon. They were able to confirm this prediction readily and prepared xenon hexafluoroplatinate (V), $Xe^+[PtF_6]^-$ by reacting xenon and platinum hexafluoride at room temperature.

$$Xe + PtF_6 \rightarrow Xe[PtF_6]$$

It is an orange yellow solid insoluble in non-polar solvents, such as carbon tetrachloride. The preparation of the above compound was rapidly confirmed by Argonne National Laboratory, USA.

The compounds of ruthenium ($XeRuF_6$) and rhodium ($XeRhF_6$), are prepared in a manner similar to the preparation $XePtF_6$. Clifford and Zeilenga (1964) prepared Xe_2SiF_6, $XePF_6$, and $XeSbF_6$ by the reaction of xenon with fluorine in the presence of non-metal fluoride.

$$2Xe\ (g) + F_2\ (g) + SiF_4\ (g) \rightarrow Xe_2SiF_6\ (s)$$

$$2Xe\ (g) + F_2\ (g) + 2PF_5\ (g) \rightarrow 2XePF_6\ (s)$$

$$2Xe\ (g) + F_2\ (g) + 2SbF_5\ (vap) \rightarrow 2XeSbF_6\ (s)$$

The silicon and phosphorus compounds decompose around room temperature, but the antimony compound decomposes much above room temperature. These compounds are believed to contain Xe^+ ion. This simple but dramatic experiment forced a revision of the commonly held view that noble gases are incapable of forming normal chemical compounds.

Thermal decompositions of PtF_6 and RuF_6 shown below :

$$PtF_6 \rightarrow PtF_4 + F_2$$

$$2RuF_6 \rightarrow 2RuF_5 + F_2$$

suggest that these metal fluorides actually behave as fluorinating agents rather than as simple electron acceptors as the original comparison between Xe and O_2 suggested. The investigators at Argonne, therefore, tried direct reactions between xenon and fluorine and succeeded in preparing three different colourless solid fluorides of xenon; viz., XeF_2, XeF_4 and XeF_6.

XeF_4 (m.p. 100°C) was obtained by heating xenon and fluorine (excess) in a nickel tube at 400°C; XeF_2 (m.p. 120–140°C) was obtained by irradiating a mixture of xenon and fluorine with ultra-violet light; XeF_6 (m.p. 46°C) was prepared by increasing fluorine-Xe ratio and heating the gases under pressure. There are unconfirmed reports of the preparation of XeF_8. The xenon fluorides react quantitatively with hydrogen.

$$XeF_2 + H_2 \rightarrow 2HF + Xe$$

$$XeF_4 + 2H_2 \rightarrow 4HF + Xe$$

$$XeF_6 + 3H_2 \rightarrow 6HF + Xe$$

In water, xenon difluoride is hydrolysed completely yielding, xenon, oxygen and hydrogen fluoride.

$$2XeF_2 + 2H_2O \rightarrow 2Xe + O_2 + 4HF$$

Partial hydrolysis of the hexafluoride yields xenon oxytetrafluoride, $(XeOF_4)$

$$XeF_6 + H_2O \rightarrow XeOF_4 + 2HF$$

whereas more complete hydrolysis yields an explosive trioxide.

$$XeF_6 + 3H_2O \rightarrow XeO_3 + 6HF$$

Hydrolysis of xenon tetrafluoride is less simple; disproportionation of the Xe(IV) to xenon gas and Xe(VI) species occurs predominantly in acidic and neutral media.

$$3XeF_4 + 6H_2O \rightarrow 2Xe + XeO_3 + \tfrac{3}{2}O_2 + 12HF$$

Hydrolysis with basic solution yields salts of perxenic acid. Thus hydrolysis of XeF_6 with sodium hydroxide yields sodium perxenate, $Na_4[XeO_6].H_2O$, in which the xenon is in the octavalent state. Xenates and perxenates are powerful and potentially useful oxidising agents.

Hydrolysis of the hexafluoride with excess of an acidic or neutral solution produces a solution of XeO_3; xenic acid, H_4XeO_4 is believed to be present in such solutions. XeO_3 has been prepared in quantity and is quite well characterised. It is violently explosive when dry but behaves well in aqueous solution. This is a powerful oxidising agent and may be used for this purpose extensively in the future.

Two fluorides of krypton, KrF_2 and KrF_4, have also been prepared, but these compounds are more difficult to form and decompose than the xenon compounds. KrF_2 is prepared by passing an electric discharge through a mixture of krypton and fluorine. It is an unstable white crystalline solid which sublimes below 0°C. It is highly reactive fluorinating agent and decomposes spontaneously at room temperature. It hydrolyses at −30°C to −60°C yielding kryptic acid. It reacts with SbF_5, yielding $KrF_2.2SbF_5$ which is more stable and less volatile than KrF_2. The radon tetrafluoride, RnF_4, due to the extremely small amounts of radium available and the difficulty of working with it (because of the intense radiations emitted) is still less well characterised.

BONDING AND STEREOCHEMISTRY

A full range of modern techniques for structural determination including infra-red and Raman spectroscopy, NMR spectroscopy, Mössbauer spectroscopy, and X-ray, electron and neutron diffractions have been employed in studying the xenon fluorides and their reaction products. Both valence bond theory and molecular orbital theory have been applied to suggest the type of bonding in the above compounds.

In XeF_2, there are 10 electrons (Fig. 21.3(a)) around the central xenon atom, out of which two are bonding and three non-bonding pairs. Valence bond theory predicts a linear structure for the molecule with sp^3d hybridisation around xenon. In the trigonal pyramid, the three equatorial positions are occupied by lone pairs.

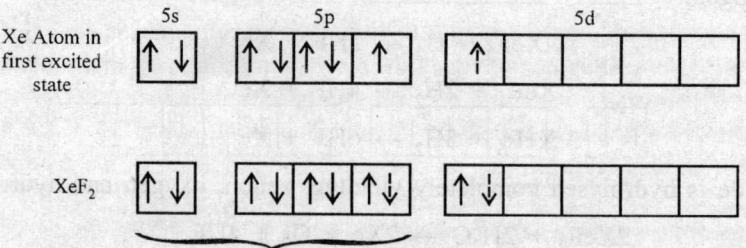

sp^3d-hybridisation with three positions of trigonal bipyramid occupied by lone pairs.

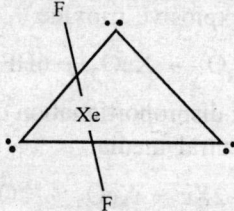

Fig. 21.3(a). Showing 10 electrons around the central xenon atom.

The symmetrical linear structure for XeF_2 is confirmed by X-ray diffraction.

Similarly in XeF_4, sp^3d^2 hybridisation would give an octahedral structure with two octahedral positions occupied by lone pairs, the resulting shape is square planar. The square planar structure for XeF_4 has been established by X-ray diffraction studies (Fig. 21.3(b)).

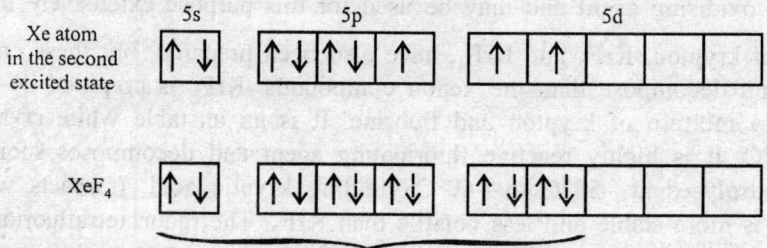

sp_3d_2-hybridisation with two positions of octahedron occupied by lone pairs.

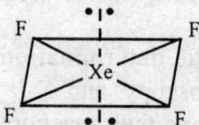

Fig. 21.3(b). Showing square planar structure of XeF_4.

In $XeOF_4$ one lone pair of electrons is replaced by a doubly bonded oxygen atom (Fig. 21.3(c)).

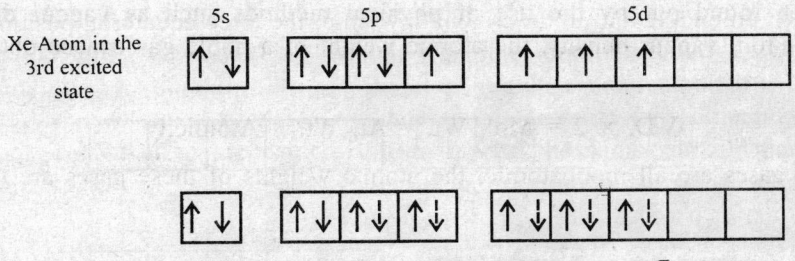

sp^{3}d^3-hybridisation with one position of the pentagonal bipyramid occupied by the lone pair.

Fig. 21.3(c). Showing pentagonal bipyramid occupied by lone pair.

sp^3d^3 hybridisation in XeF_6 would produce a pentagonal bipyramid with one lone pair, i.e., a distorted octahedron (Fig. 21.3(d)).

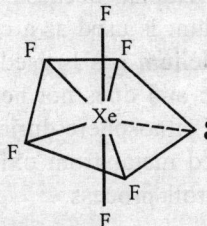

Fig. 21.3 (d). Showing a trigonal bipyramid witha non-bonding electron pair at the apene.

The trioxide would be a trigonal bipyramid with a non-bonding electron pair at the apex. (Fig. 21.3(e))

Fig. 21.3(e). Showing a square planar XeF_4 and an octanedral XeF_6.

According to valence-bond theory, 5s-,5p- and 5d-orbitals of xenon are involved in bonding. The molecular orbital approach favours the use of only the $5p^6$-electrons of the xenon and forms molecular orbitals from combination of the 5p orbitals of the xenon and the 2p-orbitals of fluorines. The proposal predicts a linear XeF_2, a square planar XeF_4 and an octahedral XeF_6. The two theories predict the correct structures for XeF_2, and XeF_4, but differ over the shapes of XeF_6. Unfortunately, the experimental evidence so far available does not allow a distinction to be made between the two theories.

ATOMIC WEIGHT DETERMINATION

Since noble gases, particularly the lighter ones, are for all practical purposes devoid of chemical properties, their atomic weights cannot be determined by chemical methods. The atomic weights of these elements have been found out by the use of physical methods such as vapour density and mass spectrophotometry. From vapour density, the atomic weight of a noble gas can be obtained by making use of the relation :

$$\text{V.D.} \times 2 = \text{Mol. Wt.} = \text{At. Wt.} \times \text{Atomicity}$$

Since the noble gases are all monoatomic, the atomic weights of these gases are the same as their molecular weights.

USES

The chief application of helium is in the filling of balloons and dirigibles where it has replaced hydrogen, because of its non-inflammability. It has a lifting power only about 8 per cent less than that of hydrogen. An oxygen-helium mixture is used for artificial breathing in deep-sea diving because helium being less soluble in blood than nitrogen does not cause caisson sickness or 'bends' (nitrogen bubbles out of the blood as the diver comes back to the surface and the pressure is suddenly released). An oxygen-helium mixture is also used in the treatment of asthma, because it diffuses more rapidly than oxygen through the constricted lung passages. Liquid helium is used as a coolant and heat transfer medium in cryogenic studies in the neighbourhood of 0.5°A. Helium gas is used as a heat transfer agent in gas-cooled atomic reactors, because it is chemically inactive and does not become radioactive. Both helium and argon are used in large amounts to provide an inert atmosphere during the welding of certain metals, e.g., Mg and Al. The noble gas protects the hot-welded metal from oxidation by the air. An atmosphere of argon is used in the production of titanium by Kroll process.

When an electric discharge is passed through neon gas at low pressures, a bright orange-red glow is produced. This glow is the basis of the principal use of neon today in the familiar luminous neon sign tubes. When mixed with other gases, it gives variety of colours. Besides its use in arc-welding, argon is used in filling incandescent and fluorescent lamps. The use of argon in incandescent lamps retards the sublimation of the filament and thus increases the life of the lamp and minimises its blackening. In fluorescent lamps, argon serves as an ionisable gas for starting the lamp and for carrying the current during its operation. Krypton and xenon are even more efficient than argon in gas-filled lamps because of their lower thermal conductivities, but are used for this purpose to a much smaller extent because of their comparative scarcity. Argon is often used in chemical laboratories to provide an inert atmosphere for handling of substances that are reactive towards moisture, oxygen or nitrogen of the air. Very pure argon is employed as a carrier gas in gas-chromatography. A krypton-xenon mixture has been used in some flash tubes for high speed photography. Radon is used in the radium-therapy for the treatment of cancer and other malignant growths. Noble gas lasers have recently been developed.

Example 21.1. A reaction mixture was prepared at 25°C by filling a 1.0 litre nickel container with F_2 gas at 8 atm. and Xe gas at 1.7 atm. The reaction mixture was maintained at 400°C for 1 hr. Then it was cooled to 25°C and the contents of the nickel container were analysed. All the xenon gas had reacted to form a solid xenon fluorine compound, but some of the fluorine gas had not reacted. The pressure of the F_2 gas was 4.6 atm. What formula would you propose for the xenon fluorine compound?

Solution. Since all pressures were measued at the same temperature while the volume of the container was unchanged, i.e., T and

V are constant in the equation $PV = n\text{RT}$ or $\left(P = \dfrac{n\text{RT}}{V} \right)$.

Therefore, we can say that $P \propto n$ i.e., pressure can be taken as a measure of the number of moles of reactants present.

Pressure of F_2 gas before reaction = 8 atm.

Pressure of F_2 gas after reaction = 4.6 atm.

∴ Pressure difference = 8–4.6 = 3.4 atm.

The pressure difference is proportional to the number of moles of F_2 consumed in the reaction. Similarly pressure difference of Xe gas = 1.7–0 = 1.7 atm.

So that ratio of moles used $= \dfrac{\text{moles of } F_2 \text{ used}}{\text{moles of Xe used}} = \dfrac{3.4}{1.7} = 2.$

One mole of fluorine gas contains 2 g atoms, whereas one mole of Xenon gas contains 1 g. atom (∵ it is monatomic). The formula which satisfies these conditions is XeF_4.

Chapter 22

Corrosion and its Control

INTRODUCTION

The phenomenon of metal corrosion is manifested by a heterogeneous chemical or electrochemical surface reaction, which results in a change of the metal to its oxidised state. The most important foundations of corrosion science are, therefore, based on two allied disciplines, namely those of metallurgy and physical chemistry. In the matter of corrosion prevention, knowledge of the properties of various protective coatings, organic and inorganic, also plays an important role. The mechanical and design aspects of corrosion prevention are also of no lesser importance. In addition, because of its tremendous economic significance the corrosion and its control at the present time, exceeds the limits of physical chemistry and physical metallurgy, and has become a particularly fascinating independent branch of science. Besides, it is one of the major problems of many industries, and the consequences of corrosion losses have been well recognised. Hence, corrosion control considerations are no longer subordinate to aspects like good operation and mechanical maintenance of plant and equipment.

The great practical significance of the science of corrosion and corrosion inhibition, and the necessity to use it as a basis for the solution of the multitude of complex problems of modern technology naturally presupposes all out scientific studies of the basic corrosion phenomena. Of late, the scope of the term 'corrosion' has been extended to: (i) corrosion science; and (ii) corrosion engineering in consideration of the fundamental and the applied aspects of this subject. The former aims at the establishment of the general laws that govern: (i) the rate and extent of destruction of metal under the influence of physico-chemical action of the external environment; and (ii) the processes that control the resulting damage; while the latter is the practical application of these laws economically and safely for abatement of corrosion damage under the most varied conditions of service to objects, such as machines, apparatus, buildings, and means of transportation etc. Corrosion engineering can only grow in effectiveness as its science develops. For example, the corrosion scientist develops better criteria of cathodic protection, while the corrosion engineer, in turn, employs this method on a large scale to prevent corrosion of buried pipelines thereby making large savings in repair costs. Thus, for an accurate diagnosis of corrosion failures and for the selection of proper remedial measures, both the scientific and engineering viewpoints supplement each other. Hence services of specialised corrosion engineers have now assumed an important role in industry. Scientific facts and principles are the basic tools for the corrosion engineer. Just as cures for the ills of man depend on developments in medical research which then are applied by the physician, so cures for ills of the equipment depend on corrosion research developments applied by the corrosion engineer. Hence, for a successful battle against corrosion of metals, a conjoint action of a corrosion scientist and a corrosion engineer is essential, since relatively few engineers receive formal educational training in corrosion. The frontiers of corrosion science and corrosion engineering have expanded so rapidly, especially within the past decade, that now-a-days it is impossible to conceive of a metallurgical engineer, a chemical engineer, or a general technologist who is net conversant with the basic principles of corrosion and its inhibition.

Corrosion is deterioration or decay occurring when a material reacts with its surroundings or the fluid being transported or contained. It may be either uniform, where the material corrodes at the same rate over the entire surface, or localised with only small portions affected.

TYPES OF CORROSION

Generally, there are twelve types of corrosion.

Uniform corrosion

Uniform corrosion occurs over the entire metal surface at the same rate. It can be controlled by proper selection of material and using protection methods such as coatings.

Galvanic corrosion

Galvanic corrosion occurs when two different metals are in contact in a conductive solution. An electrical potential exists between the different metals which serves as a driving force to pass current through the corrodent. The result is more corrosion of one of the metals in the couple. The more active metal becomes the anode, and corrodes at a faster rate than the cathode. When, for example, joints and valves of two different piping materials come in contact, the possibility of galvanic corrosion exists.

Erosion corrosion

Erosion corrosion occurs from movement of a corrosive over a surface increasing the rate of attack due to mechanical wear and corrosion. Erosion is attributed to the stripping or removal of protective surface film or adherent corrosion products. Erosion appears as smooth-bottomed shallow pits. Attack may also have a directional pattern which is attributed to the corrodent path moving over the pipe surface. Rate of corrosion is increased under high velocity conditions, especially during turbulence or impingement. Fast moving slurries containing hard or abrasive particles are likely to generate such corrosion. There are several methods which can be used to prevent this type of corrosion. First, of course, is the selection of a more resistant material. Erosion can also be reduced in transport applications by increasing pipe diameters which decreases velocity and turbulence. Also, flared tubing can reduce problems at the inlet of tube bundles. Generally, erosion occurs sporadically.

Cavitation corrosion

Cavitation corrosion is a special form of erosion corrosion. Cavitation is produced by the rapid formation and collapse of vapour bubbles at a metal surface. High pressures that are produced can deform the underlying metal and remove protective films. Smooth pipes reduce sites of bubble formation and lessen cavitation corrosion.

Fretting corrosion

Fretting corrosion is another form of erosion corrosion. Fretting occurs when metals slide over each other, producing mechanical damage to their surfaces. Vibration generally causes sliding. Corrosion products cause continued exposure of fresh surface that actively corrodes. The use of harder materials reduces friction.

Crevice corrosion

Crevice corrosion is common at gaskets and lap joints, and comes from dirt deposits and corrosive products. This type of corrosion can be attributed to one of the three things: (i) changes in acidity in the crevice; (ii) lack of oxygen in the crevice; or (iii) the build-up of a harmful iron species or the depletion of an inhibitor. Alloys are generally less susceptible to crevice corrosion than pure metals.

Pitting corrosion

Pitting corrosion is the formation of holes on an unattached surface with the shape of the pit responsible for continued growth. This is generally a slow process, taking months or sometimes even years before first traces are apparent.

Exfoliation corrosion

Exfoliation corrosion begins on a clean surface, but spreads below it. The attack has a laminated appearance, and entire areas can be eaten away. Exfoliation is marked by a blistered or flaky surface with aluminium alloys most commonly attacked. It is combated by heat treatment and alloying.

Selective leaching or parting corrosion

Selective leaching or parting corrosion is the removal of one element in an alloy. An example is the leaching of zinc in copper-zinc alloys (dezincification). Leaching is detrimental since it adds a porous metal to the effluent, and is combated by utilising non-susceptible alloys.

Intergranular corrosion

Intergranular corrosion involves an attack upon grain boundaries. When molten metal is cast, it solidifies at randomly distributed nuclei and grows in a regular atomic array to form grains. The planes of atoms in neighbouring grains do not match up because of the random nucleation. This area of mismatch between the grains is known as the grain boundary. The atomic mismatch offers an ideal place for segregation and precipitation. Corrosion takes place because the corrodent attacks the grain-boundary phase.

Under severe conditions, entire grains are dislodged due to complete deterioration of their boundaries. A surface that has undergone intergranular corrosion appears rough and feels sugary. The grain-boundary phenomenon that produces intergranular corrosion is heat sensitive. Such an attack is generally a by-product of a heat treatment, welding, or a stress relieving operation. The cure is another heat treatment, or selection of a modified alloy.

Stress-corrosion cracking

Stress-corrosion cracking is due to the combined action of tensile strength and a corrodent. It is the most serious of all corrosion problems because many alloys will undergo stress-corrosion cracking. Stresses that cause cracking are due to residual cold work, welding and thermal treatment, or may be externally applied by mechanical injury.

The cracks are generally in intergranular or transgranular paths, and such corrosion usually takes a long time. Preventative measures include: stress relieving; removing the critical environment species; or proper selection of a more resistant material.

Corrosion fatigue

Corrosion fatigue is a special form of stress-corrosion cracking. It is caused by repeated cyclic stressing, and occurs in the absence of corrodents. It is common in structures which are subject to continued vibration. The presence of a corrodent increases susceptibility to fatigue.

Most construction materials are expected to undergo some type of corrosion. It, therefore, becomes important to determine what effects chemicals in an environmental system will have on materials. Careful analysis must be made of effluents, and existing piping and construction materials should also be examined and compared. The following factors influencing the extent of corrosion should be considered: (i) concentration of major constituents being handled; (ii) pH of effluent; (ii) temperature of effluent; (iv) degree of aeration (limited aeration may enhance certain types of corrosion); (v) velocity of the fluid stream in the transport system; (vi) inhibitors; and (iv) startup and downtime procedures.

Construction materials for pipe and tanks are available in numerous materials, metallic and non-metallic. Physical properties should be thoughtfully examined before any final selection is made.

CORROSION IN INDUSTRIES

Corrosion in Boiler Plant

An increasing awareness of the need for the more efficient use of fuel has in the past two decades led to the development of steam plants of greatly increased size, operating at increased steam pressure. In water-tube-boilers steam is generated on the surface of the tubes by the hot gases and flames produced by the combustion of the fuel. The steam travels through the tubes and is delivered to the boiler drums.

High-pressure boilers which drive turbines operate on a recirculatory system, and only make use of so-called 'make-up' quantities of water, which are very carefully treated by means of modern ion-exchange methods. Low-pressure boilers, particularly those that provide either hot water or process steam, use less well treated water. The major boiler troubles, due to the use of unsuitable water, may be classified as: (i) corrosion; (ii) scale formation; and (iii) caustic embrittlement.

Corrosion

Corrosion is one of the most serious problems created by the use of untreated water. Boiler tubes, economisers, superheaters and condensers are the most affected parts. The corrosion problem extends even to parts which are not directly in contact with boiler water because gases like O_2 and CO_2 are released during heating of water which have corrosive effects. Hence in studying the corrosion phenomena in a boiler installation, it is necessary to distinguish between two entirely separate aspects, viz.: (i) the corrosion on the 'water'side of the boiler, i.e., inside the boiler tubes, the superheater, economiser and condensing unit; and (ii) the corrosion on the outer surfaces, i.e. 'fire' side of the boiler.

1. *Corrosion on the water-side of the boiler:* Water used for steam raising often contains dissolved solid

and gaseous impurities. These can cause scaling and corrosion in the boiler plant. Corrosion can be attributed to the following:

(a) Dissolved oxygen;

(b) Mineral acids;

(c) Dissolved carbon dioxide; and

(d) Galvanic cell formation.

2. *Dissolved oxygen:* Oxygen dissolved in water is mainly responsible for corrosion in boilers. Accordingly, the higher the pressure, the greater will be the dissolved oxygen content, and the higher the temperature the lesser the oxygen content. The amount of oxygen dissolved is also influenced by the other dissolved matters present, solubility of oxygen becoming less in water containing dissolved matters. In a boiler, oxygen is introduced through the raw make-up water supplied through it and also through the infiltration of air into the condensate system. As the water is heated in the boiler the dissolved oxygen is liberated and iron is corroded.

Corrosion inside a boiler can only be restricted by keeping oxygen concentration very low; this stops the cathodic reaction from taking place. The dissolved oxygen can be removed by :

1. *Mechanical deaeration methods* using (i) distillation; (ii) steam scrubbing; (iii) desorption; and (iv) flash-type deaeration techniques. These techniques permit oxygen concentration to be reduced to about 0.01 ppm.

2. *Chemical treatment* often follows the above methods for the final reduction of the oxygen concentration to virtually zero. This is done by adding certain reducing agents such as hydrazine, sodium sulphite (oxygen scavengers).

3. *Ion exchange techniques* are also able to reduce the oxygen concentration in boiler water to very low values.

4. *Mineral acids:* Most of the natural waters are alkaline. Waters in the mining areas are often acidic. Sometimes, in industrial areas the water may become acidic because of the discharge of acidic industrial wastes into the surface water. Some of the inorganic salts may hydrolyse to produce acidity causing corrosion of the boiler tubes. This attack is normally accompanied by hydrogen embrittlement caused by hydrogen formed in the acid attack, penetrating the steel so that failure occurs very quickly.

5. *Carbon dioxide:* Water contains some dissolved carbon dioxide. This gas will also be produced by the decomposition of some bicarbonates if present in the boiler water. Carbon dioxide coming in contact with water produces carbonic acid, H_2CO_3, which causes local corrosion called pitting.

$$Fe + CO_2 + H_2O \rightarrow FeCO_3 + H_2$$
$$4FeCO_3 + O_2 + 10H_2O \rightarrow 4Fe(OH)_3 + 4H_2O + 4CO_2$$
$$\text{(dissolved)}$$
$$4Fe(OH)_3 \rightarrow 2Fe_2O_3 + 6H_2O$$

Carbon dioxide in water can be removed by the addition of lime or by heating the water.

6. *Galvanic cell formation:* Corrosion (pitting) can also be due to simple galvanic cells which iron forms with some of the boiler fittings made of other materials or with impurities. This can be avoided by suspending zinc plates when zinc is anodic to iron and gets dissolved first and iron is saved.

Corrosion on furnace side of a boiler plant: It can be prevented by two general methods, viz.: (i) purification (mainly sulphur removal) of fuel; and (ii) addition of chemicals.

1. *Purification of the fuel:* Methods of removing sulphur from heavy fuel oil are as follows:

   (a) In hydro-desulphurisation process sulphur present in the oil is converted into hydrogen sulphide by reaction with hydrogen using cobalt molybdenite as a catalyst.

$$H_2 + S \rightarrow H_2S$$

   (b) The Cat-Ox system, developed in the USA utilises a miniature sulphuric acid contact plant for removal of sulphur.

2. *Addition of chemicals:* Dolomite has been used with success, as it reacts with free sulphuric acid to give high fusion point sulphates with light friable deposits instead of heavy bonded and corrosive scales. Pure magnesia or a higher proportion of magnesium carbonate (% > 45 per cent), though more expensive, can give better results.

Scale formation

From the standpoint of the chemical constituents present in water, a boiler may be considered to be a huge concentration of impurities. Water containing impurities is fed into the boiler and pure water, in the form of steam, is removed, leaving behind the deposits of impurities inside the boiler tubes. Some of the deposits stick to the metal surface and are known as scales. If they are in the form of soft muddy deposits, which can be flushed out easily, or in the form of suspensions, they are known as sludges.

Scale formation may be prevented by the following methods :

1. *External treatment:* This treatment is given outside the boiler before the feed water enters it. Attempt is made to remove or reduce the amounts of those substances, mainly calcium and magnesium salts and silica, which form deposits.

   For the removal of Ca, Mg and other ions the following processes viz. (i) lime-soda process (hot and cold); (ii) zeolite process; (iii) demineralisation process; and (iv) sequestration are used.

   Silica can be removed by special ion exchange techniques.

2. *Internal treatment:* In spite of previous treatment some salt often remains in the boiler. Attempt is made, by adding chemicals in the boiler, to convert the remaining salts into more soluble salts or such salts that the deposit will be in the form of non-sticky sludge which can be easily removed.

   (a) Inorganic and organic materials such as sodium silicate, soda ash, kerosene, glycerine, tannin, etc. are added to boilers. These substances form a coating over the scale forming particles which prevents their coalescence; it results in the formation of scales which are not very adherent or are suspended in water and can be easily removed by blowing off.

   (b) Calcium is removed from water of the boiler as calcium carbonate or calcium phosphate which settles down as a sludge. This prevents the formation of calcium sulphate scales.

$$CaSO_4 + Na_2CO_3 \rightarrow CaCO_3 \downarrow + Na_2SO_4$$
$$3CaSO_4 + 2Na_3PO_4 \rightarrow Ca_3(PO_4)_2 \downarrow + 2Na_2SO_4$$

   The best results are obtained when phosphate treatment is done at pH 10.5.

   (c) Sodium hexametaphosphate (calgon) designated by the formula $Na_2(Na_4P_6O_{18})$ may also be added to the boiler as a sequestering agent. This compound forms a complex with calcium having the formula $Na_2(Ca_2P_6O_{18})$ due to which calcium is not present in the form of free calcium ion and cannot be precipitated as calcium sulphate which forms adherent scales.

Caustic embrittlement

Embrittlement is the name that has been given to boiler failures due to development of certain types of crack resulting from excessive stress and chemical attack. These cracks are intergranular which distinguishes them from cracks resulting from strain without chemical attack which are predominantly transgranular (corrosion fatigue). In steam boiler operation, the chemicals that are believed to be responsible are caustic soda and silica. Caustic soda as such as not found in natural water but often originates from sodium carbonate, generally added for water treatment purpose.

$$Na_2CO_3 + H_2O \longrightarrow 2NaOH + CO_2$$

a reaction which takes place at elevated temperature. Any minute cracks or grain boundaries in a stressed section of a boiler may be affected.

Steam boilers, which were made by riveting construction, rather than being welded as they usually are today, are particularly liable to damage by caustic alkali solutions (caustic cracking). This cracking is not due to corrosion and the cracks have the appearance of a brittle fracture due to which it has been given the name 'caustic embrittlement'. Following are the conditions under which caustic embrittlement takes place: (i) the metal of the boiler is highly stressed due to internal stresses or due to inaccurate fitting of riveted parts; and (ii) the stressed metal, when it comes in contact with a solution capable of causing the embrittlement.

Principal methods of preventing caustic embrittlement

The principal methods of preventing caustic embrittlement are as follows :

1. *Addition of sodium sulphate:* This chemical is added to the boiler water in a sufficient quantity to ensure that the weight ratio, $Na_2SO_4/NaOH$, always exceeds 2.5. Sometime other sulphuric acid or magnesium sulphate is added to boiler water already containing free NaOH when sodium sulphate is produced *in situ*.

$$2NaOH + H_2SO_4 \longrightarrow Na_2SO_4 + 2H_2O$$
$$2NaOH + MgSO_4 \longrightarrow Na_2SO_4 + Mg(OH)_2$$

Sodium sulphite is often added to water to reduce the oxygen concentration and this also produces sodium sulphate,

$$2Na_2SO_3 + O_2 \longrightarrow 2Na_2SO_4$$

The mechanism by which sodium sulphate prevents caustic cracking is that, instead of sodium hydroxide alone penetrating the crack, it is now a mixture of sodium sulphate and sodium hydroxide which enters. Long before the sodium hydroxide solution reaches its critical concentration of above 10 per cent the sodium sulphate crystallises thus barring the entry of any further quantities of solution into the crack or intergranular space. Thus it becomes impossible for the solution to reach its critical concentration of 10 per cent NaOH. The disadvantage of sodium sulphate treatment is that calcium sulphate may be formed which is one of the main scaling agents of boiler plant.

2. *Addition of sodium phosphates:* Polyphosphate $(NaPO_3)_6$ is added to boiler water and can react as follows :

$$(NaPO_3)_6 + 6H_2O \longrightarrow 6NaH_2PO_4$$

The sodium dihydrogen phosphate formed then neutralises the calcium and sodium hydroxides introduced in the water softening process. Free calcium ions in solution form sparingly soluble calcium phosphate which is precipitated without scale formation. Polyphosphate treatment is carried out in such a way that there is always an excess of about 10 per cent of NaH_2PO_4 in solution in excess of any hydroxide present. Too much of NaH_2PO_4 is, however, to be avoided otherwise the pH of the boiler water may fall below the recommended level (8.5 to 9.0).

3. *Addition of sodium nitrate:* For boilers operating at pressures between 7 and 40 bar, sodium nitrate is added to the boiling water so that the weight ratio, $NaNO_3/NaOH$, is maintained above unity. Sodium nitrate appears to inhibit caustic cracking extremely effectively, as it makes portions liable to be affected passive.

4. *Addition of organic agents:* For low pressure boilers (< 20 bars) caustic cracking can be avoided by the addition of certain organic reagents such as tannin, lignin, quebracho, etc. The function of these is same as that of sodium sulphate mentioned in above.

5. *Use of crack-resisting steels:* Certain steels, such as those which have had a reasonable quantity of aluminium added during manufacture to eliminate traces of intergranular iron oxide formation, appear to be resistant to caustic cracking.

Corrosion in Chemical Industries

Selection of suitable materials of construction which can resist corrosion under severe corrosive conditions is of considerable importance for economic operation of chemical plants. Any material used has to satisfy certain basic requirements of physical, mechanical and corrosion resistance properties under the operating conditions. We shall mainly confine in this chapter, without reference to any specific chemical industry, to the (i) reactions dependent on environmental conditions; (ii) different types of corrosive media and their behaviour; and (iii) engineering materials and their corrosion resistance.

Environmental factors

It is important that the equipment should have proper corrosion resistivity at the operating environmental conditions frequently met in the process industries. One of the fundamental properties of the environment is its ability to form or destroy a protective film on the metal. Change in the hydrogen ion concentration of the solution is one of the most critical factors affecting the rate of corrosion. As an example of the influence of pH on the corrosion velocity, we can cite the rapid increase in the dissolution rate of Fe, Zn, Mg and a number of other metals by a transition from neutral media to solutions of non-oxidising acids or by an increase in the concentration of the acid (e.g. HCl). Indirect influence of pH on corrosion consists of change in the solubility of the corrosion products and the possibility of protective film formation with change in pH. In this respect the metals can be divided into three groups :

1. Metals (noble) which are usually quite stable in acid as well as in alkaline solution.

2. Metals, e.g. Zn, Al, Pb an Sn—the oxides (amphoteric) of which are soluble in acids as well as in alkalies—are not resistant to acid or alkaline solutions.

3. Metals whose oxides are easily soluble in acids but are insoluble in alkalies, such as Ni, Cu, Co, Cr, Mn, Cd, Mg and Fe.

Oxidising acids

Oxidising acids (HNO_3, conc. H_2SO_4) passivate some metals, viz. Fe, Co, Ni, Al, W, Ti, etc. under certain conditions, and corrosion practically ceases. Besides oxidising acids, a number of other reagents, viz. $HClO_3$, $K_2Cr_2O_7$, $KMnO_4$, etc. as well as oxygen can serve as passivating agents. Among activating salts are salts of halogen acids, giving F^-, Cl^-, Br^-, and I^- ions on dissociation. These ions (anodic accelerators) often prevent the formation of a passive film on metals.

Corrosives

For this purpose a chemical may be said to either: (i) dissolve a material uniformly, the rate depending on pH (uniform corrosion) or (ii) non-uniformly leading to pitting corrosion. Various corrosives may, in general, be classified as: (i) mineral acids; (ii) acids; (iii) alkalies; (iv) corrosive vapours.

1. *Mineral acids:* With respect to processes of corrosion, mineral acids may be grouped into two types, viz. oxidising acids and non-oxidising acids. Non-oxidising acids may be defined as those in which the cathodic process occurs only by hydrogen depolarisation and oxidising acids as those in which the predominant cathodic process is oxidation depolarisation, i.e., reduction of the acid anions. In practice, however, there are possible intermediate class of gradual transition from one type of action of the acid on the metal to another type dependent on concentration, corrosion potential, temperature, passivity.

2. *Acids:*

   (a) Sulphuric acid is used directly or indirectly in nearly all industries and is a vital commodity in our national economy. Corrosion problems occur in plants, viz., HCl, fertilisers, dyes, drugs, pigments, explosives, rayons, textiles, petroleum refining, rubber etc., where it is utilised under variety of conditions. Sulphuric acid is a strong acid and is normally non-oxidising. Its maximum corrosiveness occurs at concentrations of 60 to 70 per cent. Concentrated sulphuric acid appears to be less corrosive, particularly to iron and steel, than other concentrations at room temperature. Cast iron and steel can safely handle concentrated sulphuric acid for this reason.

   (b) Hydrochloric acid is more active than sulphuric acid and its handling is a problem of great difficulty in industrial plant operations. Factors which help to explain its higher corrosivity are the greater solubility of chlorides, less tendency to form insoluble basic salts, and higher mobility of he chloride ion. Most concentrations of the acid are severely corrosive to most of the common metals and alloys. Increase in temperature, concentration and velocity all tend to accelerate the attack. Many unexpected failures in service often take place due to the presence of minor impurities like oxygen (aeration), oxidising agents like HNO_3, ferric or cupric chlorides, etc.

In consideration of the above, the limits for so-called acceptable corrosion rates are usually raised when considering materials of construction for handling hydrochloric acid. Materials that show very low rates of corrosion are often not economically feasible. Good judgement is required to obtain a balance between service life and cost of equipment. Where contamination is a problem, e.g. in steam tubes for heating chemically pure acid, expensive materials such as tantalum are the only ones that can be utilised. Molybdenum is an important constituent of the alloys generally used as corrosion resistance materials to hydrochloric acid. Durichlor is a high-silicon iron containing Mo and is excellent corrosion resistant to hydrochloric acid. This alloy is used in industry for all concentrations of hydrochloric and muriatic acids at moderate temperatures. Aeration does not affect corrosion resistance. Durichlor pumps give satisfactory service in 30 per cent HCl and also sludges containing 10 per cent acid at ambient temperatures. Nickel-base alloys with large Mo contents, viz. Chlorimet-2 and Hastelloy B show good corrosion resistance to all concentrations of hydrochloric acid up to boiling temperatures. These alloys, however, are attacked if aeration or oxidising ions are present. Chlorimet-3 and Hastelloy C are good in dilute acids at moderate temperatures and exhibit better resistance to oxidising environments because of their high chromium contents. Copper, the various bronzes, cupronickels, Monel, Inconel, Ni-Resist, Hastelloy D and stainless high alloys are susceptible to influences other than the acid itself and must be used with caution with a definite knowledge of the specific conditions. Ordinary carbon steels, cast irons, aluminium and its alloys, lead and its alloys are never used for hydrochloric acid service.

Non-metallics have found widespread use in hydrochloric applications because of their good resistance and immunity to attack by oxidising ions. Subject to their relative temperature limitations most of the plastics and rubbers are suitable for all concentrations of hydrochloric acid. Rubber-lined steel can be used for many years for vessels and piping for hydrochloric acid service. Wood also finds application for dilute acid.

(c) *Hydrofluoric acid* is similar to hydrochloric acid in many respects. Aeration and the presence of other oxidising agents increases the corrosivity of the acid. Temperature has similar effect.

Magnesium resists attack by hydrofluoric acid and some shipping containers for this acid are made of Mg. Steel is suitable for handling concentrations from 60 to 100 per cent and aqueous mixtures below 60 per cent evolve hydrogen. Wrought Monel can resist all concentrations of HF at all temperatures up to boiling point.

(d) Phosphoric acid is more like sulphuric acid than hydrochloric acid in terms of corrosivity. Aeration and other oxidising agents increase the corrosiveness. The impure acid frequently contains fluoride salts and oxidising compounds which also tend to increase the rate of attack. Thus, the high-silicon irons, ceramics, and tantalum are appreciably affected by the presence of hydrofluoric acid.

Two of the most widely used alloys as materials of construction up to 85 per cent concentration of H_3PO_4 and at boiling point are type 316 stainless steel and Durimet-20. Lead and its alloys are also used at temperatures up to 475°K at concentration up to 80 per cent for pure and 85 per cent for impure acid. High-silicon irons, glass, and stoneware show good resistance to pure acids. High nickel-molybdenum alloys exhibit good resistance to pure acid in absence of oxidising impurities and aeration.

(e) Nitric acid is a strong oxidising agent and influences the corrosion rate by oxidising the polarising film of hydrogen. Only metals which form protective oxide films, such as al, Ti, the stainless steels, etc. are resistant.

The choice of metals and alloys for nitric acid services is quite limited. High silicon (14.5 per cent) alloys (duriron) find useful application for handling nitric acids. Chromium (17 per cent) iron alloys are also known for their considerable resistance to oxidising acids, 18–8 S.S. is the most widely used of all the materials for handing nitric acids. Aluminium has been used for transporting nitric acid as its corrosion rate is considerably less at low temperatures in very low and very high concentrations of nitric acid including fuming acid. Aluminium equipment is used in the manufacture and handling of strong nitric acids including the fuming varieties. Teflon shows outstanding corrosion resistance to nitric acid and glass-filled Teflon finds wide application for rotating rings in mechanical seals for nitric acid pumps.

Impurities like sulphuric acid enhance the corrosion rate. Ferrous metals and alloys may be used for handling concentrated nitric acid when mixed with sulphuric acid.

3. Alkalies:

Alkalies such as caustic soda and caustic potash are not particularly corrosive and can be handled in pure iron and steel. Corrosion problems, however, arise during high temperature evaporation of concentrated alkali solutions. Pb, Zn, Al and Sn are readily corroded in sodium or potassium hydroxide solutions and form plumbates, zincates, aluminates and stannates. Nickel, Monel metal and nichrome are relatively unattacked with caustic solutions and have extensive applications in caustic solutions.

Corrosion by alkalies is often characterised by pitting and localised attack. Another undesirable phenomenon often connected with alkali corrosion is a form of stress corrosion cracking at certain temperature and concentrations.

A considerable number of gases, in addition to those already discussed, are encountered frequently in the chemical and process industries. These include O_2, H_2, N_2, CO_2 and CO. All of these gases are corrosive in one form or another at elevated temperatures. O_2 causes oxidation of the metal while H_2 penetrates the metal at elevated temperatures and can cause embrittlement N_2 forms nitrides on the surface of the metal, and CO_2 and CO may carburise the metal.

4. *Corrosive vapours:* In addition to above, in chemical plants which produce acids, the vapours of hydrochloric, nitric, sulphurous and other acids influence severely the corrosion of metals. Steel and iron pipes can be employed for handling dry chlorine and hydrochloric acid gas. Cast iron pipes are used for handling hot SO_2 gases and lead may be employed when the gases become sufficiently cooled. For the construction of apparatus for the synthesis of ammonia it has been reported that low-carbon steel containing 2 to 2.5 per cent Cr can be used with sufficient corrosion resistance.

In conclusion, it may be said that it is a problem of great difficulty to select a required material suitable for the construction of equipment for chemical plants which can withstand corrosion, as the corrosion rate is generally dependent on various factors like the temperature, concentration of corrodant, aeration and velocity, etc.

Materials of construction

A brief outline for construction of materials for chemical plants is given in Table 22.1.

Table 22.1. Materials of construction in chemical industries.

Material of construction	Applicable	Not applicable
1. *Ferrous metals*	Nitric acid	
a. Chromium steels (17 to 27% Cr)	Sulphur gases	
b. Chromium-nickel steels (18% Cr, 8%, Ni)	Atmospheric exposure. High temperature stresses	
c. High silicon-iron alloys (14 to 16% Si)	Most acids	Halogen acids Fuming sulphuric acid, alkalies
d. Steel	Concentrated sulphuric acid	
2. *Non-ferrous metals*		
a. Copper	Acetic acid	Oxidising acids, non-oxidising acid with air
b. Copper-nickel alloys (Monel metal)	Non-oxidising acid (except HCl) Hydrofluoric acid	Oxidising acids
c. Nickel and nickel alloys	Alkalies	
d. Aluminium	Nitric acid, C > 85% formaldehyde, synthetic resin. Distilled and deionised water storage tanks,	Alkalies

(Cont'd ...)

Material of construction		Applicable	Not applicable
		piping and condenser.	
e.	Titanium	Hot, concentrated oxidisers.	
f.	Lead	Dilute sulphuric acid.	HCl, HNO_3; caustic alkalies.
g.	Alloys of Co, Cr, W or Mo	High resistance to many chemicals, to high temperatures, to abrasion.	(high temperature)
h.	Ta	Extreme resistance to chemicals	Expensive
3.	*Non-metallic materials*		
a.	Porcelain enamel	Distillation of citric, maleic, lactic acids.	
b.	Cement lined steel pipe	Salt water, sulphur bearing oils.	
c.	Chemical stoneware	Acids, alkalies	
d.	Glass	Acids	HF
e.	Fused silica	Acids (hot and cold). Condensing, cooling and concentrating apparatus.	HF, H_3PO_1 above 700°K Alkalies fragile expensive.
f.	Wood	Weak acids	Strong oxidising agents. Alkalies, crystallising salts.
g.	Natural rubber	Lining material for steel tank for chemical equipment.	
h.	Synthetic rubber	Many chemicals, acids, HF	

Corrosion in Petroleum Industry

The petroleum industry contains a wide variety of corrosive environments. For example, oil fields are situated in tropical areas where high humidity, salt bearing winds and air borne sand take the toll of structures and equipment. Costly pipelines convey the crude oil—often itself actively corrosive towards iron and steel—to long distances, either to refineries or to coastal installations where ocean-going tankers may be loaded via a submarine pipeline. In the refineries, the vast quantities of cooling water required for their operation, often necessitate the use of sea water, so that intake lines, condensers and coolers all require special protection against corrosive attack. Finally, the refined products must be distributed giving rise to special corrosion problems in oceangoing tankers and underground pipelines. It is convenient to group all these environments in relation to corrosion problems, mainly into three broad areas, viz. (i) production; (ii) transportation and storage; (iii) refinery units. The corrosion experienced in these areas may be divided into two classes; viz.: (i) that due to the fluid being produced and hence usually internal; and (ii) that due to environment in which the equipment is placed and hence usually external.

Production

The internal corrosion experienced in typical oil and gas wells is normally associated with hydrogen sulphide, carbon dioxide and organic acids present in the oil, brine or gas. Internal corrosion is normally referred to as being sour (from 'Sour Oil Wells') or sweet (from 'Sweet Oil Wells') according to the higher or lower sulphur content (mainly H_2S) of the oil. The gas may be present in the well or may result from the activity of sulphate reducing bacteria. Sulphide corrosion results in large deep pits and heavy iron sulphide scale. This attack is not restricted to the well equipment only, but continues on into the pipelines and tankers also. Corrosion in the absence of hydrogen sulphide is most frequently associated with carbon dioxide as the chief corrosive agent, with organic acids contributing to the attack. Three methods are used to mitigate this corrosion, viz. coated tubing, inhibitors, and special alloy steels. Coated tubing has found the most favour, and air-dried and baked epoxy resins are now being used in increasing amounts for almost all coating installations.

The external corrosion of well casings is now recognised as a major problem, due to the huge repair costs involved. The most common causes of casing corrosion are due to (i) sulphate reducing bacteria; and (ii) local concentration cells. A variety of corrosion prevention methods are used to mitigate casing corrosion.

1. Adding inhibitors so that these are uniformly dispersed over the entire casing.
2. Cathodic protection, with sacrificial anodes or impressed currents, of the casing.
3. Using protective barriers like cementing the casing.
4. Insulating the well from its flow line.

Transportation and storage

Petroleum products are transported by tankers, pipelines, railway tank cars, and tank trucks. The most severe internal corrosion problem occurs in tankage. If the crude is sour, early perforation of the fixed roof sheets is likely. The use of floating or aluminium roofs, and coatings are the most common preventive methods used. The coatings in most common use are coal tar based. The tank may also be subjected to external corrosion attack. This can be prevented with coatings or by using cathodic protection. Whether the attack is external or internal or both, providing the tank with a concrete bottom prevents further corrosion.

Ordinary carbon steel which is used for the construction of tanker is exposed to an aggressive natural environment of salt water and marine atmosphere. Most serious exposure of steel to sea water occurs during the return (ballast) voyage when the tanks are void of cargo. Although corrosion during cargo voyage is probably relatively small, the nature of different types of cargo has a profound effect on the overall corrosion. Gasoline-carrying tankers present a more severe internal corrosion problem than oil tanks because the gasoline keeps the metal too clean. Oil leaves a film which serves an effective barrier against general corrosion. For ballast tanks a galvanic system (Mg anode) is always used.

Internal corrosion of storage tanks is due chiefly to saline water which settle sand remains on the bottom. Coatings based on vinyl or epoxy resins and cathodic protection are mainly used. For domestic fuel oil tanks alkaline sodium chromate (or sodium nitrate) is an effective inhibitor.

Paint is the traditional material for protecting hulls and super structures of steel ships. In more recent years chemical resistance coatings have been developed which neither affect cargoes nor are affected by cargoes.

Rust formation on the internal walls of the pipelines caused by water precipitated from products may reduce the line throughout and may give rise to contamination of the product. Formation of the rust may be prevented by lining the pipe or rust formation may be inhibited by the injection of inhibitors (a few parts per million) such as amines and nitrites into the product stream. Corrosion of the external walls of the pipelines varies enormously according to the nature of the soil or water, the temperature, access of oxygen and other factors. To combat soil corrosion, both coatings and cathodic protection are used.

The three basic types of coating and wrapping now finding use in oil industry are: (i) hot-applied coaltar enamel and asphalt coatings; (ii) plastic tapes; and (iii) coal-tar/epoxy coating. Two principal tapes at present offered for buried pipe-line protection are those of polyvinyl chloride, and polyethylene. In the past few years, the application of cathodic protection to products transmission lines have become normal practice. It is now generally recognised that the combinations of good coating and cathodic protection is the best method of ensuring a long leak-free life to buried pipes.

Refinery units

Corrosion problems in the refinery tubular heat exchange equipment can be solved by mainly relying on the selection of materials. Crude oil always contains impurities which frequently lead to severe corrosion problems in processing. Condensed products from the distillation processes are frequently contaminated with such substances as sulphuric acid, naphthenic acid, hydrogen sulphide and hydrogen chloride. Considerable corrosion is, therefore, liable to occur on the product side of the condenser and cooler tubes. Corrosion generally takes the form of uneven general wastage and insoluble corrosion products such as copper sulphide are frequently formed. Pitting may proceed rapidly beneath these non-protective deposits. The most widely used tube materials are brasses with a high zinc content. When sea water or brackish water is used for cooling, there is a possibility of corrosion on the water-side in the heat exchange equipment. Carbon steel, stainless steel or monel tubes cannot be used, owing to their susceptibility to pitting in sea water. Brass, arsenical Admiralty metal, red brass, and cupronickels are used satisfactorily. On the project-side of heat exchange tube, corrosion can be minimised by injection of alkali or ammonia to keep the pH at a controlled figure in conjunction with the addition of one or other of film forming amine inhibitors. On the cooling water-side, benefits may be obtained by the application of proper systems of cathodic protection in condenser and cooler channels and heaters. Zn or Mg can be used as sacrificial anodes.

Sea water cooling with circulating pumps in refineries provides a major problem in corrosion control. Protective measures employed involve the use of devices to overcome galvanic couples which encourage local attack and pitting. When cooling water is through steel pipes, a combination of thick coatings based on coal tar pitch plus cathodic protection is effective. Cooling water circulating pumps are made of bronze and cast iron, which are fitted with cathodic protection from an external source of impressed current.

For tubing in stills and gas-cracking tubes austenitic stainless steels are used. In some cases, a single tower is lined with two or three different materials to take care of the changing corrosiveness from the top to the bottom of the tower. Corrosion by sour crudes increases with temperature and with increasing sulphur content. Chromium is the most beneficial alloying element in steel for resistance to sulphur compounds. Thus, the Cr content of steel is increased with increasing sulphur and temperature starting as low as 1 per cent Cr. Experience indicates that 2.5 per cent Cr, 1 per cent, Mo steel is generally adequate for less than 0.2 per cent H_2S in the gas stream while higher amounts of H_2S require 5 per cent or higher amount of Cr in the steel. Four to 6 per cent Cr, 0.5 per cent Mo steels are widely used in refineries.

Non-metallic materials are resistant to chemical corrosion and are free from contamination of the product. These render them very attractive to refinery and petro-chemical industries. Natural rubber has been used as a structural material and as a lining for vessels to prevent contamination and corrosion. Flexible pipes are widely used as temporary connections in the storage and transport of materials. Graphite has a very good thermal conductivity, and this property combined with its resistance to steam and highly corrosive fluids opens up the field of heat-exchange applications, in addition to its use in vessels, valves, pumps and piping.

Petro-chemical corrosives

These may be classified into two general categories, viz. (i) those present in crude oil; and (ii) those associated with these processes.

Table 22.2. gives an outline of the general nature of these corrosives. In conclusion, it may be said that petroleum industry has much in common with the chemical industry since most of the corrosion problems in refineries are due to inorganics (water, H_2S, CO_2, H_2SO_4, NaCl) rather than organics.

Table 22.2. Petro-chemical corrosives and their actions.

Name	Nature
1. Water (in crude oils)	Acts as an electrolyte. Hydrolyses salts (e.g. chlorides producing acids).
2. Carbon dioxide (in gas wells)	Chief corrosive agent with organic acids.
3. Salt water (in oil wells and in crude oil)	Salts of $CaCl_2$, $MgCl_2$ and NaCl. Forms acid due to hydrolysis which is removed by ammonia addition. Removed by desalting methods.
4. Sulphide compounds (viz. H_2S, RSH etc.) (in crude and gases)	Reaction leading to severe attack (Sulphide corrosion) in refining process. Removed by caustic addition.
5. Nitrogen (in some crude) (from air in burning operations)	Ammonia. Damage heat exchanger made of copper-bearing alloy. Cyanide interferes with diffusion of hydrogen into steel. Controls pH of water.
6. Oxygen (air) (in refinery operation)	Responsible to oxidation.
7 Sulphuric acid (in refinery operation)	Attack stainless steel in presence of sludges containing carbonaceous materials. Copper-base alloys can be used.
8. Hydrochloric acid (in refinery operation)	Formed through hydrolysis. Present in distillation columns and in condensed petroleum fractions.
9. Caustic and lime (for H_2S removal and for neutralisation of HCl)	Causes deposits, clogging, and stress corrosion.
10. Naphthenic acid (in oil)	Corrosive above 800°K. Stainless, steel can be used.

Corrosion in Fertiliser Industries

Like other chemical and process industries, the fertiliser industry also requires special materials of construction to suit the corrosive nature of the media handled and the various process parameters like temperature, pressure, pH, velocity as well as prevention of product contamination. Various types of stainless steels find wide application on account of their excellent corrosion resistance, strength and toughness, good fabrication properties and fine surface finish.

The various chemical processes adopted for production of different types of fertilisers, process parameters involved, types of corrosion and corrosion problems encountered in these process industries are discussed briefly.

Nitrogeneous fertilisers

Nearly 70 per cent of India's nutrient consumption is in the form of nitrogen. Major fertilisers in this category are ammonium sulphate, nitrate, urea etc.

1. *Ammonium sulphate:* The major part of ammonium sulphate is produced as a by-product from the high temperature carbonisation of coal. The process involves handling of weak sulphuric acid of 1 to 20 per cent concentration, ammonium sulphate crystal slurry, coke oven gas laden with sulphuric acid mist, ammonia vapours etc. The corrosion problems encountered are due to handling of highly corrosive low concentration sulphuric acid, ammonia vapour and liquor, presence of H_2S and HCN in coke oven gas and abrasion due to ammonium sulphate crystal slurry. The materials of construction used are lead and acid proof brick-lined mild steel, S.S. conforming to AISI-316, cast iron alloy-20 etc., for various equipment, viz. saturator, evaporator, ammonia column, tanks, centrifuges, driers, etc.

2. *Urea:* Urea is now the most common nitrogeneous fertiliser. This is primarily due to its low production cost. Modern urea plants are both energy efficient and make maximum use of the raw materials ammonia and carbon dioxide. In addition, advanced designs have resulted in lower maintenance requirements and longer service life. A significant factor in the design has been development of economical corrosion resistant material.

The principal steps in the manufacture of urea are: (i) the synthesis of ammonia from synthesis gas mixture; and (ii) processing ammonia into finished fertiliser.

In a nitrogeneous fertiliser factory there would be two main plants, viz. ammonia plant and finished fertiliser plant, e.g. urea plant where urea is the finished product. Another major plant that must exist alone with the above two plants is the steam generation plant. Here all discussion are centred on a modern Naptha Based Nitrogenous Fertiliser Plant with urea as the finished product since most of the information on corrosion problems in fertiliser industry have been regarding ammonia and urea plant.

Synthesis of ammonia

Major steps involved in a typical naphtha based ammonia plant are: (i) desulphurisation of naphtha; (ii) naphtha steam reforming; (iii) purification of synthesis gas; and (iv) ammonia synthesis. Besides these there are ammonia cracking section and water cooled heat exchanger. The ammonia process is considered to be moderately corrosive. The corrosive environments in different sections are given in Table 22.3.

Table 22.3. Corrosive environments in various sections.

Sections	Corrosives	Materials of construction
1. Hydrodesulphuriser	S, Organic sulphur compounds like mercaptan, mono and poly sulphides, thiophenes.	Carbon steel, 13% Cr and 18-8 Cr-Ni austenitic steel for sulphur compounds.
2. Catalyst steam reforming (Primary and secondary)	High temperature CO_2, H_2 and O_2	HK 40-25% Cr, 20% Ni, 0.35–0.45% C and rest Fe (more popular); IN-519-24 Cr, 24Ni, 1 Si, 1 Mn, 1.5 Nb, 0.25–0.35 C, rest Fe IN-657-49 Cr, 49 Ni, 1.5 Nb, 0.5 Si, 0.1 C Paralloy CR 32 N-20 Cr, 32 Ni, 1.3 Nb, 1 Mn, 0.8 Se, 0.1 C, 0.03 S and 0.03 P, rest Fe

(Contd ...)

Sections	Corrosives	Materials of construction
3. Carbon monoxide conversion (high temperature and low temperature)	Hot CO-CO_2-H_2 gas mixture	Carbon and low alloy steels
4. Decarbonation	Hot aqueous CO_2 solution	Austenitic stainless steels, primarily types 304 and 316 in both wrought and cast form
5. Methanation	Nil	Plain carbon and low alloy steels
6. Compression	Erosion-corrosion	Type 410, Type 430
7. Ammonia synthesis	Hot compressed H_2 and condensed ammonia	Type 304 for catalyst cartridge and integral heat exchanger. Low alloy high strength stainless steels for piping and equipment around the converter.

Corrosion problems

Numerous types of corrosion problems have been identified in the ammonia process. These are stress corrosion, pitting corrosion, crevice corrosion, galvanic corrosion, intergranular corrosion, end grain corrosion, high temperature corrosion like carburisation, oxidation, decarburisation, sulphidation and nitriding.

The overall problem in the reformer, where catalytic reforming of the feed occurs at elevated temperatures and pressures is, in particular, the materials used for the tubes inside this reformer and the support and ancillary pipe work that may be exposed to high temperatures and pressures. The pipe work system expands at this severe operating conditions aiding crepe occurrence and loss of metal elasticity. The most commonly used tube alloy in the modern ammonia plants is similar in composition to type 310—nominally 25 per cent Cr and 20 per cent Ni, balance iron—but with a carbon content in the range of 0.35 to 0.45 per cent for improved high temperature crepe and stress rupture strength (HK–40).

Corrosion failure

The following corrosion failures in reformation section may be mentioned as typical examples :

1. Leak caused by intergranular microcracks in the header assemblies in the steam/naphtha reformation section due to formation of ferrite and carbide, the material having been DIN/1.4863. Presence of ferrite and carbide caused microcrack which propagated and caused leak due to high heating rate during start up, viz. 550°K per hour as against the standard practice of 320°K per hour and high frequency of shut down and start up.

2. Failures with deformation and bulging of reformer tubes made of centrifugally cast HK-40 alloy and operating at 12 Kg/cm^2 pressure and 1100°K temperature. The failure was due to carburisation, sigma phase formation which caused microcracks, and overheating led to propagation of microcracks, deformation and bulging.

3. Failure in the form of longitudinal crack on the outer side of the outlet pig tail of reformer tube (21 Cr, 31 Ni, 0.07C, 1.5 Mn, 0.25 Si, 0.35 S) due to carburisation followed by oxidation, sulphidation and overheating.

4. Chloride/caustic stress corrosion failures of inlet piping materials (HK-40) of reformer section.

   Potash corrosion in H.T. shift converter is a problem is the carbon monoxide conversion unit. Change of direction of the flow of fluids often causes deposition in dead ends. At times, potash gets leached out from the catalyst and gets deposited in dead legs. There have been instances of failure of the snuffing steam line by potash corrosion attack and corrective action of potash corrosion is to radiograph the line during shut down, to detect potash deposits and/or crack caused by the deposit and to take pre-emptive repair action.

Failures have been reported to occur, by erosion in the lean liquor draw off line from the reboilers in the decarbonation section. Incorporating stainless steel sheath around the zone of attack considerably reduces the rate of attack.

Preheater coils made up of AISl-321 type stainless steel have been found to fail due to nitriding caused by overheating of the tubes much above the maximum allowable temperature ($550°–650°K$) in the Ammonia Cracking Section during the start up of the fertiliser plant.

Corrosion in ammonia synthesis section offers the greatest problem. The ammonia synthesis converters are pressure vessels containing a removable cartridge filled with catalyst. The unreacted gas at high pressure and temperature can be destructive to carbon and low alloy steels and other metals due to hydrogen embrittlement, hydrogen attack and nitriding. Condensed ammonia is also a corrodent and in the anhydrous state can cause stress corrosion cracking of stressed carbon and low alloy steels. The remedial measures for these types of corrosion are proper selection of materials of construction, proper design and strict adherence to set control process parameters. Type 304 is specified for catalyst cartridge and integral heat exchanger. Austenitic stainless steels are resistant to nitriding, hydrogen embrittlement and hydrogen attack. Increased Ni and Ni-base alloys offer greater resistance to nitriding. Corrosion problems in high pressure steam generators are the same as those with high pressure boiler systems.

Corrosion in water-cooled heat exchanger systems also pose many problems. Achieving desired ammonia production depends to a very large extent on the efficient heat transfer/heat dissipation from hot process fluids through metallic surface of heat exchangers into cold cooling water (recirculating (open) cooling water system). Corrosion encountered in ammonia plant cooling system are: (i) electrochemical corrosion; (ii) fouling and fouling-induced corrosion. Although these phenomena are very much similar to other cooling systems, the problems in ammonia cooling systems have got some unique features. These are :

1. The coolers and condensers are mostly made from carbon steel, for economic reasons, which are most susceptible to aqueous corrosion.

2. There are huge number of coolers and condensers in the system and due to their widely scattered locations, uniform water distribution and uniform maintenance of water velocity through all the equipment in the circuit are rare phenomena; as a result coolers deprived of proper water flow repeatedly suffer from corrosion failures.

3. Some of the coolers have water in the shell-side which create inherent stagnant zones, e.g. near baffle corner, shell wall, longitudinal baffle wall etc.

4. Ingress of ammonia, oil/naphtha from the system cause fouling due to precipitation or oil sludge, supply abundant nutrients to micro-organisms leading to microbiological problems, cause wide pH fluctuations, e.g. pH rises when ammonia enters and falls when nitrification occurs; nitrification also produces corrosive nitrate ions; oil also makes corrosion inhibitors ineffective by film formation.

Electrochemical corrosion in the cooling system is brought about by dissolved oxygen and the product of corrosion is FeO or Fe_2O_3. Since cooling water is maintained near neutral or alkaline, corrosion products (Fe_2O_3) accumulate on the metal surface unless water velocity is very high or efficient surface active agents-cum-antifoulant are employed; and it is found that electrochemical corrosion ultimately leads to deposition in practice unless corrosion rate is too low. Corrosive mineral ions, temperature, and other common factors increase the rate of electrochemical corrosion.

In cooling system, failure due to uniform or general corrosion is rare. Problems occur due to fouling-cum-deposit and underdeposit corrosion which makes the control of cooling system failures extremely difficult. Under the deposit catastropic rate of corrosion and unimaginable rapid failure of equipments due to the combined effect of the following phenomena are found to occur.

1. Hydrolysis of Fe_2O_3 produced by corrosion or iron bacteria, to give H^+ ions, the concentration of which develop very high, giving rise to acid corrosion sometimes again giving ferric oxide or salt as the corrosion product; and the cycle of hydrolysis and acid corrosion continues.

2. The area beneath the deposit shelters anaerobic bacteria which produces weak acids like H_2CO_3 and H_2S which corrodes iron surface rapidly by different mechanisms and in multiple ways.

3. Overheating of the metal surface below the deposit associated with differential oxygen concentration make the area below the deposit strongly anodic with respect to surrounding surface thus leading to underdeposit electrochemical corrosion. Also overheating weakens the metal structure by high temperature intergranular corrosion.

4. Deposit also creates stress on the metal surface leading to stress corrosion.

5. Deposits inhibit corrosion inhibitors from reacting on the metal surface to impart protection and electrochemical corrosion between inhibited and non-inhibited zones occurs.

Fouling met in ammonia cooling system are of the following types :

1. Crystallisation fouling caused by precipitation of scale forming salts like $CaCO_3$, $Ca_3(PO_4)_2$ etc.

2. Particulate fouling caused by deposits of water-borne suspended particulate matters like suspended matters, dust etc.

3. Microbiological fouling caused by slime forming bacteria, fungi and algae.

4. Oil/Hydrocarbon fouling caused by oil, naphtha etc.

5. Corrosion fouling caused by in-site deposits of corrosion products.

6. Foreign material fouling caused by ingress of foreign materials, e.g. wood piece, pebbles, aquatic insects, etc. into the system.

Keeping metal surfaces in contact with water clean and free from deposits is the ultimate solution to get rid of the problem. This can be achieved by :

1. Proper microbiological control and monitoring.

2. Proper maintenance of water velocity through equipments and monitoring.

3. Avoiding stagnant zones in 'water in shell soda coolers' by reverting to latest design modification.

4. Incorporating suitable antiscalant (e.g. organic phosphonates) and antifoulant (e.g. polyacrylates) along with chemical inhibitory treatment.

5. Applying on steam desludging and chemical cleaning whenever deposits are felt to form.

6. Arresting ingress of foreign debris into the system, e.g. by incorporating modification of strainer system in the cooling tower.

Only when the heat transfer surfaces are kept clean and free from any fouling and deposit, the corrosion in the cooling system may follow predictable electrochemical mechanisms and the conventional corrosion inhibition vocabulary become applicable.

In conclusion, it may be remarked that proper selection of materials of construction, proper design of equipments and strict process control are the primary requirements and must be given due consideration before starting fabrication of plant equipments. In most of the sections, corrosion is due to process-side environments and once materials of construction and design are fixed nothing could be done excepting controlling recommended process parameters which then also become inefficient to resist corrosion of faulty equipments. In other areas such as boilers and heat exchangers (water cooled) where corrosion occurs due to externally controllable aqueous environments, faulty selection of materials and faulty design would give rise to permanent source of corrosion failures and it can not be arrested even with the most efficient treatment programme of the corrosive aqueous environment.

Production of urea

The rapid growth of the urea industry has been associated with a strong demand for a cheap fertiliser with a high nitrogen content. There are two basic urea process types on the market today. One is the 'total recycle process', while the other is the 'stripping process', offered by Snam Progetti (ammonia stripping) and Stamicarbon (carbon dioxide stripping). In all these processes the basic chemical reactions are the same; ammonia and carbon dioxide are reacted at high temperature and pressure to form ammonium carbamate.

$$2NH_3 + CO_2 \rightarrow NH_2COONH_4 + heat$$

The ammonium carbamate is dehydrated upon subsequent heating to form urea and water.

$$NH_2COONH_4 + heat \rightarrow NH_2CONH_2 + H_2O$$

In practice it is not possible to obtain complete conversion to urea. The processes essentially differ in the methods by which urea is separated from the ammonium carbamate and the by-products are recycled. An important contributing factor in the growth of urea industry is the introduction of total recycle process, where separation of urea from carbamate and dissolved gases is performed by stepwise reduction of pressure in combination of heating. The residual gases are condensed and feed back to the reactor at high pressure.

Generally speaking urea is not corrosive to stainless steels (the usual material of construction for urea reactor), but ammonium carbamate solutions are. For this reason, most of the corrosion problems in urea plants are encountered where carbamate solutions at high temperature and concentrations are handled, e.g. reactors, strippers, condensers and decomposers. Probably the most severe condition in an entire urea plant can occur in the reactor.

As to the selection of materials of construction, besides corrosion resistance, consideration is to be given to factors like mechanical properties, workability, weldability and economies. Selecting the proper materials of construction, is therefore, a highly specialised job. Stainless steels used extensively for this duty are the austenitic grade AISI 316L, 317L, Sandvik grade / 2RE69 and UHB 725 LN etc. Roughly 2 per cent Mo content (as in 316L, 317L and UHB 725 LN) imparts rapid passivation in the carbamate solution. Stainless steel owes its corrosion resistance to the presence of a protective oxide layer on the metal. Corrosion of stainless steel occurs during contact with liquid phase. Under equilibrium conditions the oxygen content of the liquid is determined by the partial pressure of the oxygen in the corresponding gas phases and the Huey Coefficient for the system O_2—process liquid. It is stated that in the Stami carbon process 0. to 3.0 per cent volume of oxygen reduces corrosion to a negligible level.

Stainless steels with high Mo content are prone to intermetallic phase formation in the heat affected zone. Formation of ferrite strings in welds are to be avoided by the use of proper welding technique as it may give rise to selective attack in carbamate containing media. In order to minimise the chances of carbide precipitation and reduce changes of intercrystalline corrosion due to ammonium carbamate, the steels used must be either stabilised or should have carbon content of less than 0.03 per cent. In the Toyo recycle process, the urea reactor is a titanium lined multilayer carbon steel vessel. Injection of air in titanium clad reaction vessels helps in protection through passivation. It is, however, susceptible to certain problems, for example erosion-corrosion. All equipment in the finishing sections such as centrifuges, driers, cyclones, melters, prill heads, coolers etc. are made up of type 304 S.S. that are resistant to corrosion by urea up to its solution. Most of the corrosion problems are usually carefully monitored by regular NDT test methods, more so during turn around and thus costly replacements are avoided.

Phosphatic fertilisers

Phosphoric acid is a intermediate for fertilisers like ammonium phosphate, triple super-phosphate, nitro-phosphates, high analysis NPK fertilisers, and liquid fertilisers. The wet process, mostly followed today for the production of phosphoric acid either by dihydrate or hemidrate or anhydrite process, consists of reactors, filters, heat exchangers, evaporators and pumps. Phosphate rocks are normally not corrosive, though mildly abrasive, but the impurities, fluorine, chloride, and silica (reactive) are the constituents which influence corrosive characteristics of phosphate rocks directly or indirectly. A low silica content allows free hydrofluoric acid to remain in the reaction slurry and thereby making it more corrosive. Thus the unpredictable nature of phosphate rock and its impurities, make the selection of materials of construction much more difficult in phosphoric acid and phosphatic fertiliser services.

In the reaction vessel, large percentage of solid is kept in suspension by circling enough liquid from filtration section. This ensures efficient reaction of sulphuric acid with phosphate rock. The equipments involved are reaction tanks, pumps, piping, valves, fittings etc., and require careful selection for their constructional materials. Reasonably suited to this purpose are stainless steels confirming to AISI 316L, 317 and 317L. In case if the chloride content of phosphate rock exceeds 0.5 per cent, use of AISL317L is preferred to AISI 316 and 317. But these materials are not suitable for use in concentration section.

By and large, reaction tanks are constructed in RCC or mild steel with natural or synthetic rubber lining, though concrete tanks require less maintenance cost. Trouble free service is obtained by using rubber lined pumps for handling slurry medium. The filter cloth is polypropylene and polyesters. For most of the pipe lines and instruments carbon steel with rubber lining is commonly used.

Other special alloys which have been found suitable for most of the plants in particular application are Aloyco-20, Carpenter 20, Cooper FA-20, Durimet-20, HV-7, HV-9, HV-90 and other equivalent alloys developed by different producers. Though it is difficult to generalise the suitability of a particular material for all the purposes, materials like HV-9, Narloy, and CAA series have been found satisfactory under severe corrosive conditions existing in phosphoric acid and phosphatic fertiliser plants.

Potash fertilisers

The most commonly used potash fertilisers are potassium chloride, sulphate nitrate, hydroxide and carbonate. The entire demand of potash fertilisers of this country is met from import till this date since suitable raw materials for the production of these fertilisers are not available indigenously.

Corrosion of Other Metals

The steel-zinc data are valid only for two years exposure. The corrosion of zinc is approximately linear with time, but the corrosion rate of steel is not. If other steel data are compared to this, it should be calculated from the corrosion for two years.

If a corrosion problem exists with other metals such as aluminium or magnesium, it should not be a assumed that the steel-zinc data estimate the severity of the environments for alloys of these metals. However, if outdoor testing is done with a specific metal or alloy, it will be useful to test zinc and steel at the same time, or to test the metal in question at a site that has been calibrated in order to take advantage of a possible comparison with the existing data.

Solving Corrosion Problems

Outdoor corrosion can be stopped by isolating the metal surface from the moisture. Or, it can at least be inhibited to an acceptable degree by restricting the rate at which it reaches the surface. This is done with impermeable corrosion-resistant metals (nickel/chromium), slightly permeable coatings (paints) or sacrificial coatings (zinc or cadmium). Specifications should be developed by the pollution engineer for acceptable coating thicknesses and quality testing assurance to protect abatement systems.

Unfortunately, most system designs cannot wait for securing long-time data. When such is true, then existing data as close to expected service as possible should be examined. In the meantime it is advisable to start an outdoor testing programme.

MODERN PROTECTIVE COATING TECHNOLOGY

Paints and coatings are usually of all concern both in specification and actual job site practice. They are usually the first area from which the dwindling overall capital budget is cut. As a result, they are often the most obvious area of failure in a waste or water treatment or air pollution control facility.

A small pump, which might control the operation of a facility, can fail and not be noticed until too late. Moreover, when paint begins to chip and peel and rust appears, poor housekeeping is readily evident to everyone. Also, unless frequent recoating is done, equipment, tanks, and other steel must be replaced prematurely.

Paint and coating technology has advanced rapidly in recent years. Proper selection of paints and coatings, carefully worded specifications, skilled application, and close field inspection greatly enhance plant performance. At the same time, they reduce both initial capital expenditures and ultimate maintenance costs.

Selection of Paints and Coatings

There are many conditions which must be considered when making a specific end use recommendation for a protective coating system. It is important to remember that the conventional system is not necessarily optimal either in price or performance.

The first area of consideration should be the type of service involved. Will it be immersion, splash, spillage, fumes, weathering, or an environment of salt, air, or industrial chemical fallout graded as to severity? The substrate (steel, aluminium galvanising, concrete, masonry or wood) over which the coating will be applied must be considered. Then, a decision must be made on the type of surface preparation. It must be decided whether the work will be done in the fabricating shop or at the job site.

The physical and mechanical abuse to which the coating will be subjected must be known. For floor coatings the severity of foot traffic, steel-wheeled traffic and rubber-wheeled carts affects the coating choice. The temperature to which the coating will be exposed as well as thermal shock and extremes of both heat and cold are important.

Another area of consideration is the wear factor. This is becoming especially important in pipe and tank linings. There will be abrasion resistance requirements for a coating inside a pipe, slurry tanks, precipitator scrubber, or clarifier mechanism. This relates to the size and nature of the particles, chemical and physical makeup, and the flow rates. In general, it relates to what will be the wear phenomenon which would reduce the life of the coating.

Finally, after all of these performance characteristics have been defined, the cost of the protection enters into the equation. This includes both the initial cost in terms of capital dollars and the ultimate yearly maintenance costs. Maintenance costs are becoming more important.

The selection of protective coating for steel surfaces, especially in new construction, is governed by the degree of surface preparation allowable or tolerable. It is desirable in new construction to specify the form of surface preparation, making certain directions are followed.

The life of a coating system is proportionate to the quality of the surface preparation over which it is applied. It is unwise to apply paints or coatings over mill scale-bearing steel. The steel under the mill scale will be in contact with water and oxygen. Technically with conventional paint systems, mill scale absorbs water which has penetrated the coating. The mill scale is "popped" from the steel when iron oxide forms. The protective coating also breaks loose. Corrosion undercutting then proceeds at a fast rate.

The Steel Structures Painting Council estimates that a coating applied over a blast-cleaned surface will last approximately three and one-half times longer than the same coating applied over power tool-cleaned steel. Therefore, in order to minimise future maintenance costs, it is always recommended that blast cleaning be used wherever possible.

Note that hand blast-cleaning figures do not include costs of scaffolding and removal of sand necessary if this were done in the field. From this, it is apparent that automatic blast cleaning or modified acid pickling is the most economical method of preparing steel.

When shop blasting is specified, priming of the steel is essential. The primer must be capable of protecting the steel during fabrication, and through the final erection and topcoating phases of the project. It must withstand physical abuse during shipment and erection of the steel. Today, primers are readily available which permit the primed steel to be welded or bolted with high-tension bolts. They have excellent corrosion resistance and require negligible touchup in the field. These materials can save up to 50 per cent in painting costs when properly specified and applied.

Selection of Shop Primers for Steel

Steel corrodes through an electrolytic cell reaction. In order for this to happen, there must be an anode, a cathode, and an electrolyte which become a small electrical cell. Anodes and a cathodes may themselves be found on the surface of the steel. Variations in amounts of iron, carbon, or other materials in an alloy, placement of iron oxides as mill scale, or surface contaminants cause this. When water as an electrolyte is present, an electric cell is completed and corrosion begins.

One common way of protecting steel is to remove the water by forming a barrier over it. Protective coatings are a barrier. Another way is to protect it cathodically which is similar to galvanising. Zinc, the most commonly used metal for this type of protection, will sacrifice itself to protect the steel underneath. Inorganic zinc primers, which are used widely, rely on sacrificial protection.

Protective coatings generally are of two basic types, organic and inorganic. Organic coatings include alkyd (often called red lead, zinc chromate, or iron oxide), epoxy polyamide, vinyl, chlorinated rubber and organic zincrich. Inorganic zinc primers, which protect cathodically, fall into the second category. Inorganic zinc primers are classed as alkali silicate inorganic zincs, or ethyl silicate-based inorganic zincs. Of these, the alkali silicates, which are often called water-based, are less desirable. The ethyl silicate-based inorganic zinc primers have a reduced tolerance for surface preparation. Most alkali silicate-based inorganic zincs require a minimum of a near white metal blast cleaning for adequate adhesion over steel. Ethyl silicate-based inorganic zinc can be applied over mill scale, brush-off blasted steel, or commercially blasted steel.

As mentioned earlier, shop applied primer must protect the steel during shipment and construction phases prior to topcoating. A properly chosen primer will require minimum touchup prior to topcoating and have a long protective life.

Selection of Field Topcoats

The major physical and chemical properties of inorganic zinc primers have been covered up to this point. It should be stated that inorganic zinc primers may be used in virtually all areas of a water or waste treatment plant or an air pollution control facility. In order to complete the coating systems, a decision must be made on topcoats to be used over the inorganic zinc primer. These general recommendations serve as a guide for topcoat selection.

Non-immersion service

There is a variety of topcoats which may be used over inorganic zinc primers. However, for enclosed steel surfaces, topcoats are not required since aesthetics and direct exposure to chemicals are not considerations.

General weathering

Coastal or inland or areas where architectural colours are desired. Where resistance to corrosive fumes is not required or in a mild fume environment, an acrylic latex is an excellent topcoat over inorganic zinc. This type has excellent weathering and chalking resistance, and retains both colour and gloss. It is water-based and available in a wide variety of architectural colours.

Acid or caustic fumes

In areas of acid and caustic fumes, a high-build vinyl or chlorinated rubber topcoat is recommended over the inorganic zinc primer. These have been successfully used for a number of years in waste-water treatment facilities. However, they have normally been used over their own generic type primers. This has proven to be costly due to premature failure of the primer and high cost of replacement.

Solvent spillage

In solvent spillage areas, it is recommended that the inorganic zinc primer be topcoated with an epoxy polyamide topcoat. These topcoat are resistant to mild acids and alkalies to heavy solvent conditions.

However, these catalysed materials are more inconvenient to use than single component vinyl and chlorinated rubber topcoats commonly used in municipal water and waste treatment facilities. The epoxy polyamide high build materials have greater acceptance in industrial waste treatment facilities.

Severely corrosive environments

These exposures exist in advanced waste-water treatment facilities of the tertiary type or in industrial waste treatment facilities. For recommendations, it is wise to consult with the protective coating suppliers. Generally, modified phenolic and modified epoxy phenolic coatings, applied over inorganic zinc primers, perform well when exposed to acids, alkalies and solvents. However, conventional painters have difficulty applying them, and they are more costly than vinyls, chlorinated rubbers, or epoxy polyamide materials.

High temperature steel coatings

Steel, when operating at temperatures above 250°F normally does not require coating since water will not condense at this temperature. Unfortunately, continuous operations above this temperature cannot be guaranteed. Most high-temperature stacks, incinerators, and furnaces cycle up and down in temperature. Weekend shutdowns of incinerators are common, causing a cool-down of the entire plant. During thermal cycling and periodic plant turnarounds, the steel cools down, moisture condenses on the steel, and rust forms. For this reason, the steel must be protected.

Surface preparation consists of blast cleaning the steel to a near white metal finish. Next, the steel should be shop primed with 3 mills of an ethyl silicate-based, partially hydrolised inorganic zinc primer. A silicone topcoat is applied in the field. Pure silicones are used for temperature ranges between 1000° and 1200°F. Silicones may be modified, but the degree and type will determine the temperature range, performance, and cost. Some silicones are modifies for use between 750° and 1000°F, others for use between 500 and 750°F, and still others for use between 250° and 500°F. Since costs increase in proportion to the quantity of silicone resin in the formulation, the system used should be recommended for the temperature range specified.

Immersion service

In immersion service, a white metal blast-cleaned surface conforming to National Association of Corrosion Engineers Standard NACE #1, or pickling to a white finish in accordance with SSPC SP 8-63 with manufacturer's modifications should be specified. This is followed with a 3-mil partially hydrolysed, ethyl silicate-based inorganic zinc primer. Topcoats may then be selected for a given service.

Potable water service

The lining of potable water tanks is commonly done with four or five coats of low solids, vinyl materials. While these materials hold up very well in service, they have poor abrasion resistance. During winter months, ice chunks fall down from the tops of tanks and damage the bottoms. They are very costly because of the labour of applying multiple coats.

A better recommendation is high film build, high abrasion resistant, hard durable coatings for the interior of potable water tanks and process equipment. These are catalysed epoxy materials, either epoxy polyamide or epoxy amine. Normally, they are applied in two coats at dry film thicknesses between 4 and 6 mills per coat. They are applied over inorganic zinc primers to take advantage of shop surface preparation and shop priming. Epoxy polyamides are preferred to epoxy amines as topcoat over inorganic zinc primers.

Waste-water and process water

Here, the topcoat recommendation is a coal tar epoxy material. As a rule, coal tar epoxies are applied in two coats, 7 to 9 mills each, over the inorganic zinc primer.

Other environments and corrosives

It is recommended that a quality coating manufacturer be contracted for recommendations on specific services other than those outlined.

Topcoats for scrubbers

Flyash from both fossil-fired electric generating stations and incinerator waste service systems are acidic and abrasive which must be considered when selecting systems. Most organic tank lining materials used for scrubber installations are not recommended above 250°F. If higher temperatures will be encountered, materials such as acid-proof brick or mortar linings should be used. Such conditions are generally encountered prior to the quenching of scrubber gases. For gases that have been quenched, it is generally possible to line scrubbers with organic coatings.

If organic zinc primers are to be used in scrubber installations, the primed surface should be brush blasted prior to the application of lining material. This will leave a very thin coat of inorganic zinc over the steel, or none at all. Preferably, the scrubber lining material should be applied over practically bright white metal steel. A thin film (approximately 1/2 mill of inorganic zinc) would be allowable as a primer in order to hold the blasted surface. The surface should be cleaned prior to application of the protective lining system.

Materials which have found their greatest application in scrubber lining and demister lining work have been bisphenol A fumerate, flakeglass polyester lining materials. These are normally applied either by spray or trowel methods. The trowel-applied material must be rolled to align the glass flakes after they are trowel applied. Spray-applied materials do not require this extra step, and are normally less expensive to install. Application generally consists of two coats, 20 mills per coat. Some manufacturers have insisted upon a third coat to yield a dry film thickness of 60 mills for additional safety in scrubber linings and demister lining application areas. The same cautions and design parameters as taken in tank linings should be observed in this application.

Scrubber ductwork

The ductwork of a scrubber system and the stacks coming off it present a different problem from the normal scrubber. The duct stacks are generally made of very light gauge steel. This steel is subject to rapid thermal shocks and severe flexing. Because of this, it is necessary to use a chemically resistant, abrasion resistant, elastomeric material rather than a rigid flakeglass polyester coating. The selection of this elastomeric materials is very critical, since there are many on the market which have experienced failures. Fortunately, there are some which are specially formulated for this service and have proven successful.

The most successful used elastomeric material for steel ductwork has been an abrasion-resistant, chemically-resistant, polyurethane lining. This lining is applied in two coats (20 mils each) over an epoxy polyamide tie coat and inorganic zinc primer. This system has now been accepted and is performing satisfactorily on scrubber installations. Earlier installations of rigid materials, such as flakeglass polyester, had failed in both stacks and ductwork lining areas.

Galvanised steel surfaces

The procedure has always been to passivate the galvanised steel by pretreatment with a material such as two-package wash primer, phosphoric acid, or a vinegar wipe. Then, it is primed with an inhibitive primer, followed by a tie coat and a finish coat. This four-step operation is costly and, in fact, has only marginal performance. The galvanising itself protects the steel very well because of the cathodic protection of the zinc. It is painted, essentially, to improve the appearance. There are available on the market today self-priming vinyl copolymers with high film build characteristics. They are applied at a dry film thickness of 3 mills. and only one coat is necessary. Surface preparation consists of solvent wiping the galvanised surface to remove grease and contaminants. These one-coat systems perform far better than the traditional multiple-coat systems. Over well-aged galvanising, acrylic latex coatings show excellent performance characteristics. Here, the state of the coating art has progressed from an expensive four-step to an easy one-coat system. The vinyl system may also be used over brass, bronze, and aluminium with the same reduction in cost as experienced with galvanised surfaces.

CORROSION-RESISTANT LININGS FOR STACKS AND CHIMNEYS

Power plants and industries discharging gases from sulphur-bearing fuels have always had a need for corrosion-resistant linings in chimneys and stacks. With lower operating temperatures changes in fuels, and growing use of wet scrubbers to met environmental regulations, corrosion resistance is even more critical. In the past, independent liners constructed with acid-proof brick or steel have proven very successful. With the decrease in flue gas temperature and an increase in moisture content, condensation is higher inside the chimney and corrosion is accelerated. Unprotected steel liners are no longer considered suitable. Independent brick linings will still meet these new requirements. A new generation of corrosion-resistant mortars has been developed for use in power plants, refineries, steel mills, incinerators, and chemical plants.

Mortars based on soluble silicate comprise some of the original corrosion-resistant cements used. The first silicate mortars were simply mixtures of fillers, such as silica, quartz, gannister, clays, or barytes, and sodium silicate solution. Mortars of this type harden by loss of water and require exposure to air or heat to set. Construction with such a mortar is extremely low. Although thin joints are used, the fluid mixture squeezes out if more than three or four courses of brick are laid at one time. In most cases, not over 6 ft. of brickwork per day can be installed. Very careful drying is also necessary. A 30-day period is usually recommended before putting the structure in service. Air-drying mortars are no longer used for brick linings due to these drawbacks and the development of improved mortars.

Chemical Setting Mortars

Chemical setting sodium silicate mortars developed utilise a setting agent which reacts with the soluble sodium silicate to cause the mixture to harden. The setting agent may be either an acid or a compound which will decompose and liberate acid to accelerate the cure. Typical setting agents are: ethyl acetate, zinc oxide, sodium fluorosilicate, glyceral diacetate, hexamethylene tetramine, formamide, and other amides and amines.

Chemical setting mortars are supplied as two-component systems. They consist of liquid sodium silicate solution and filler powder incorporating selected aggregates or as one-part systems in powder form to be mixed with water. Chemical setting mortars take initial sets in 15 to 45 min., and final sets in 24 to 96 hr. or longer, depending on the temperature. Continuous bricklaying is possible because of chemical reaction, and does not require exposure to air or heat. Large quantities of chemical setting sodium silicate mortars have been successfully used in industry for the past 40 years. Mortars of this type are still employed today for many types of acid service.

Potassium Silicate Mortars

New chemical setting mortars have been developed using potassium silicates instead of sodium silicates. Several fundamental properties of potassium silicates combine to make them preferable to sodium silicates. Potassium silicate mortars have better workability due to their smoothness and lack on tack. They do not stick to the trowel nor run or flow from the joints of the brickwork. They possess greater resistance to strong acid solutions as well as to sulphation, and have greater refractoriness. Moreover, they do not effloresce or bloom and have less tendency to form hydrated crystals in the hardened mortar.

Potassium silicate mortars are supplied in two parts—the silicate solution and the filler powder. Mortars are available which utilise organic or inorganic setting agents or a combination of the two. The properties of the mortar are determined by the setting agent used. Such properties as absorption, porosity, strength, and water resistance are affected by the choice of setting agent. For example, organic setting agents will burn out at low temperatures, increasing porosity and absorption. The organic setting agents are water-soluble and can be leached out if the mortar is exposed to steam or moisture. Due to crystal structure formation, the mortars take a longer time to gain strength, remaining in a plastic state for 96 hour or more.

Modified Silicate Mortars

As previously mentioned, the one-part powder form chemical setting silicate mortars have been commercially available for some time. They have not been used extensively because of their higher cost. These one-part silicate mortars are similar to the two-component mortars except they have somewhat lower physical properties. Research and field testing have recently produced new modified silicate mortars with characteristics not previously available in either one- or two-part systems. These new modified silicate mortars utilise different classes of setting agents. The properties of the mortar are dependent upon the setting agent selected.

Monolithic Linings

Many chimneys and stacks have linings of calcium aluminate cements or refractory concrete applied by Guniting, cast-in-place, or troweling. Calcium aluminate cement linings are unsatisfactory if the pH of acid condensate is below 3.5. The alumina gel common to these cements dissolves rapidly in such an environment. Refractory concrete is not acid-resistant, and its additional cost is not warranted when operating temperatures are below 500°F.

Monolithic corrosion-resistant lining materials have been used successfully to restore and repair brick linings in concrete chimneys and steel stacks. These monolithic linings may be applied to both new and existing steel liners by guniting. The original acid-proof concrete lining is an inorganic silicate composition which resists all acids except hydrofluoric, water, oil, most solvents, and temperatures to 1750°F. It is recommended for use over a pH range of 0.0 to 7.0. Supplied in two parts, the powder and liquid are mixed together when used. The liquid is substituted for the water normally used in Guniting.

Modified silicate-base cements designed for use as monolithic linings, to be applied by the cement gun process, are now also being supplied. These are single-component systems which produce a high-strength acid-proof lining when mixed with water. They have extremely good adhesion to concrete, brick and steel, and require only a minimum of surface preparation. These linings are not affected by acids, except hydrofluoric and acid fluoride salts from the lowest pH to 9.0.

Chapter 23

Photochemistry

INTRODUCTION

Photochemistry is a science which deals with chemical reactions caused by exposure of reactants to light radiations. The light radiations of the visible and ultraviolet regions lying between 8000Å to 2000Å wavelengths are chiefly concerned in bringing about such reactions. Such reactions are known as photochemical reactions.

Photochemical reactions have been defined, therefore, as those reactions which occur with the adsorption of light radiations (photons). The photons supply the necessary energy to the reactants enabling them to react to yield products.

Photochemical reactions are of many types including oxidation, reduction, decomposition, hydrolysis and polymerisation. These are caused by exposure to light of suitable wavelengths. Some of the photochemical reactions such as conversion of oxygen into ozone and decomposition of ammonia into nitrogen and hydrogen are accompanied by increase of free energy unlike ordinary (or 'dark') reactions which are invariably accompanied by decrease of free energy. Photosynthesis of carbohydrates from carbon dioxide and water taking place in nature is another reaction which is accompanied by increase of energy. The reason in all such cases is that some of the light energy absorbed by the reactants is converted into free energy of the products.

It may be recalled that a photon is a unit (or quantum) of electromagnetic radiation. If the radiation has frequency v, the energy (E) of its photon will be given by the expression:

$$E = hv \qquad \qquad ...(23.1)$$

where h is Planck's constant.

Further, knowing that frequency (v) of a radiation is related to the wavelength (λ) as:

$$v = \frac{c}{\lambda} \qquad \qquad ...(23.2)$$

where c is the velocity of light, a quantum of energy (or a photon of energy) may also be put as:

$$E = h\frac{c}{\lambda} \qquad \qquad ...(23.3)$$

The energy value of a photon is expressed sometimes in terms of wave number (v') which is defined as the number of wavelengths per centimetre (or per metre as in SI system). Wave number, evidently, s the reciprocal of wavelength (λ) expressed in centimetres, i.e.,

$$v' = \frac{1}{\lambda}$$

...(23.4)

Hence energy associated with a photon may also be written as :

$$E = hcv'$$

...(23.5)

It may be emphasised that a photochemical reaction requires absorption of photons of a definite energy, by the reactants.

LAWS OF PHOTOCHEMISTRY

Lambert's Law

Lambert's law states that when a beam of monochromatic radiation passes through a homogeneous absorbing medium, the rate of decrease of intensity of the radiation with thickness of absorbing medium is proportional to the intensity of the incident radiation. Mathematically, this statement may be expressed as :

$$-\frac{dI}{dx} = k\,I$$

...(23.6)

where I is the intensity of radiation after passing through a thickness x of the medium, dI is the infinitesimally small decrease in the intensity of radiation on passing through infinitesimally small thickness dx of the medium. In other words, $-dI/dx$ gives the rate of decrease of intensity of radiation with thickness of the absorbing medium. The proportionality constant k is called the absorption coefficient. Its value depends upon the nature of the absorbing medium.

If I_0 is the intensity of radiation before entering the absorbing medium (i.e., when $x = 0$), then the intensity of radiation I after passing through any finite thickness, x, of the medium, can be obtained by integrating equation (23.6) between these limits.

$$\int_{I_0}^{I} \frac{dI}{I} = -\int_{x=0}^{x=x} k\,dx$$

...(23.7)

or

$$\ln \frac{I}{I_0} = -kx$$

...(23.8)

or

$$I/I_0 = e^{-kx}$$

...(23.9)

or

$$I = I_0 e^{-kx}$$

...(23.10)

The Lambert's law is usually expressed in the form of equation (23.10). The intensity of the radiation absorbed, I_{abs}, is evidently given by

$$I_{abs} = I_0 - I = I_0\,(1 - e^{-kx})$$

...(23.11)

If desired, the natural logarithm can be changed to base 10 in which case Lambert's law equation (23.10) is written as :

$$I = I_0 \; 10^{-ax} \qquad \qquad ...(23.12)$$

where a is called extinction coefficient of the absorbing medium. It is related to absorption coefficient k, evidently, by the expression :

$$a = \frac{k}{2.303} \qquad \qquad ...(23.13)$$

Beer's Law

If the absorbing substance is present in solution, the decrease in intensity of radiation with thickness of the solution is given by Beer's law.

According to this law, when a beam of monochromatic radiation is passed through a solution of an absorbing substance, the rate of decrease of intensity of radiation with thickness of the absorbing solution is proportional to the intensity of incident radiation as well as to the concentration of the solution.

Mathematically, the law may be stated as:

$$-\frac{dI}{dx} = k' I c \qquad \qquad ...(23.14)$$

where c is the concentration of the solution in moles per litre and k', called the molar absorption coefficient, is proportionality constant the value of which depends upon the nature of the absorbing substance. Let I_0 be the intensity of radiation before entering the absorbing solution (i.e., when $x = 0$). Then, integrating equation (23.14) between the same limits as before and assuming the concentration c to remain constant, we have

$$\int_{I_0}^{I} \frac{dI}{I} = -\int_{x=0}^{x=x} k' \; c \; dx \qquad \qquad ...(23.15)$$

or

$$\ln \frac{I}{I_0} = -k' c \, x \qquad \qquad ...(23.16)$$

or

$$\frac{I}{I_0} = e^{-k'cx} \qquad \qquad ...(23.17)$$

or

$$I = I_0 e^{-k'cx} \qquad \qquad ...(23.18)$$

Changing the natural logarithm to the base 10, the equation (23.18) may be written as :

$$I = I_0 \; 10^{-a'cx} \qquad \qquad ...(23.19)$$

where a' is called the molar extinction coefficient of the absorbing solution. It is related to molar absorption coefficient of the absorbing solution, k', by the expression:

$$a' = \frac{k'}{2.303} \qquad \qquad ...(23.20)$$

Thus Beer's law may be expressed mathematically by either of the equation (23.18) or (23.19).

The law may also be stated as: "When a monochromatic radiation is passed through a solution of an absorbing substance, its absorption remains constant when the concentration (c) and the thickness of the absorption layer (x) are changed in an inverse ratio".

Example 23.1. A monochromatic radiation is incident on a solution of 0.05 molar concentration of an absorbing substance. The intensity of the radiation is reduced to one-fourth of the initial value after passing through 10 cm. length of the solution. Calculate the value of molar extinction coefficient as well as molar absorption coefficient of the substance.

Solution :

According to Beer's Law :

$$I = I_o\, 10^{-a'cx}$$

or
$$I/I_o = 10^{-a'cx}$$

$\therefore$
$$0.25 = 10^{-a'} \times 0.05 \times 10$$

$\therefore$ Molar extinction coefficient, $\qquad a' = 1.204$

Further,
$$a' = \frac{k'}{2.303}$$

or
$$k' = 2.303\, a'$$
$$= 2.303 \times 1.204$$

Thus, molar absorption coefficient, $\quad k' = 2.773$

Grotthus-Draper Law

This law, first discovered by Grotthus in 1818, and rediscovered by Draper in 1839, states that only those radiations which are absorbed by a reacting system are effective in producing chemical changes.

The significance of this law is that photochemical reactions result only from the absorption of radiations. This does not mean, however, that all the light radiation that is absorbed can be effective in bringing about a chemical reaction. This also does not mean that just any light radiation that is absorbed, necessarily brings about a chemical reaction. In several cases, the light radiation that is absorbed is merely converted into kinetic energy of translation of the molecules. In several other cases, energy of light radiation that is absorbed, is re-emitted as light radiation of the same or of some other lower frequency. This phenomenon of absorption of light radiation of one frequency and re-emission of radiation of the same or lower frequency is known as fluorescence.

Stark-Einsteins' Law of Photochemical Equivalence

Towards the beginning of the present century, several scientists attempted to correlate the extent of a photochemical reaction with the amount of light energy absorbed. It was Einstein and Stark, who made an important advance in this regard. The significance of this work was realised when Einstein deduced the law of photochemical equivalence on the basis of thermodynamics. This law, sometimes referred to as Stark-Einstein law, states :

In a photochemical reaction, each molecule of a reacting substance absorbs one quantum of the radiation causing that particular reaction.

This law provided a great stimulus to research in photochemistry.

Suppose, v is the frequency of the absorbed light. Then the corresponding quantum of energy absorbed per molecule will be hv where h is Planck's constant. The quantum of energy (E) absorbed per mole of the reacting substance is, therefore, given by

$$E = N_0hv \text{ ergs} \qquad \qquad ...(23.21)$$

$$= \frac{N_0hv}{4.184 \times 10^7} \text{ calories} \qquad \qquad ...(23.22)$$

where N_0 is Avogadro's number.

Since $\qquad \qquad N_0 = 6.022 \times 10^{23}$ and $h = 6.626 \times 10^{-27}$ erg-sec

$$\therefore \qquad E = \frac{6.022 \times 10^{23} \times 6.626 \times 10^{-27} \times v}{4.184 \times 10^7}$$

$$= 9.536 \times 10^{-11} \, v \text{ calories}$$

Now, $v = \dfrac{c}{\lambda}$, where c is the velocity of light (= 3×10^{10} cm. per sec.) and λ is the wavelength of the radiation absorbed. Substituting these values in the above equation, we get:

$$E = \frac{9.536 \times 10^{-11} \times 3 \times 10^{10}}{\lambda} \qquad \qquad ...(23.23)$$

$$= \frac{2.86}{\lambda} \text{ calories} \qquad \qquad ...(23.24)$$

In the above equation, λ is expressed in centimetres. But, generally, λ is expressed in Angstrom units (Å = 10^{-8} cm.). Then

$$E = \frac{2.86}{\lambda} \times 10^8 \text{ calories}$$

$$= \frac{2.86}{\lambda} \times 10^5 \text{ kcal.} \qquad \qquad ...(23.25)$$

TREATMENT IN SI UNITS

Quantum of energy absorbed per mole = N_0hv

Substituting $N_0 = 6.022 \times 10^{23}$ and $h = 6.626 \times 10^{-34}$ J-s, we have:

$$E = 6.022 \times 10^{23} \times 6.626 \times 10^{-34} \, v = 39.90 \times 10^{-11} \, v \text{ joules}$$

$$= \frac{39.90 \times 10^{-11} \times c}{\lambda} = \frac{39.90 \times 10^{-11} \times 3 \times 10^8}{\lambda} \text{ joules}$$

In this equation, λ is in metres. Expressing it in Angstrom units (1Å = 10^{-10}m), we have:

$$E = \frac{0.1197}{\lambda} \times 10^{10} \text{ joules} = \frac{11.97 \times 10^5}{\lambda} \text{ kilojoules}$$

The quantity E, i.e., the energy absorbed per mole of the reacting substance is called one Einstein. It is evident that its numerical value varies inversely as the wavelength of the light absorbed; the shorter the wavelength the greater is the energy absorbed. For example, for violet light of wavelength 3750Å, the value of Einstein is 319.2 kJ, while for red light of wavelength 7500Å, the value of Einstein is 159.6 kJ. The values of Einstein for a few other wavelengths are given in Table 23.1.

Table 23.1. Values of Einstein for light of different wavelengths.

Wavelength (Å)	Colour range	Values of Einstein (kilo joules)
2000	Ultra-violet	598.5
3000	Ultra-violet	399.0
4000	Violet	299.3
5000	Blue-green	239.4
6000	Yellow-green	199.5
7000	Red	171.0
8000	Near infra-red	149.6

If the law of photochemical equivalence is correct, one mole of a reactant should decompose for every $\frac{11.97}{\lambda} \times 10^5$ kJ of light radiation absorbed. In other words, $\frac{\lambda}{11.97 \times 10^5}$ mole of the absorbing substance should decompose per kJ of radiation energy absorbed by it. The law, therefore, can be verified experimentally.

Quantum Efficiency

The experimental results are expressed in terms of quantum efficiency or quantum yield, ϕ, defined by the equation :

$$\phi = \frac{\text{Number of moles reacting in a given time}}{\text{Number of Einsteins absorbed in the same time}}$$

In other words, quantum yield may be defined as the number of moles of the light absorbing substance which react per Einstein of light absorbed by it. Evidently, the quantum yield may also be expressed as :

$$\phi = \frac{\text{Number of molecules reacting in a given time}}{\text{Number of quanta absorbed in the same time}}$$

If the law is correct, the quantum efficiency should be unity.

Experimental Verification

In order to find out quantum yields of photochemical reactions, two types of determinations are needed. These are:

1.   Determination of the number of moles of the light absorbing substance that react in a given time.

2.   Determination of number of Einsteins of light of required wavelength that is absorbed by the same substance in the same time.

Determination of number of moles reacting

The number of moles reacting in a given time can be determined by the usual analytical techniques used in chemical kinetics. The reaction cell is usually made of glass unless violet or ultra-violet radiation is to be used. In that case the cell is made of quartz. The incident radiation is made to fall at right angles to it in the form of a parallel beam. The design of the cell varies with the nature of the reaction depending upon the fact whether it involves gases or liquids or both. The extent of photochemical reaction depends upon the intensity of the light used irrespective of its wavelength. It has been calculated that 1 candle power of light falling on one square centimetre of surface per second corresponds to 2×10^{14} quanta. This shows that if the law is strictly valid, 2×10^{14} molecules or 3.32×10^{-10} moles of the light absorbing substance would react per square centimetre of surface in one second. This is a very small quantity indeed. It is necessary to use light of high intensity. It may be remembered that energy of light radiation falling per square centimetre of surface is given by the product of intensity of light and the time of exposure of the surface.

Monochromatic light

In all photochemical experiments, it is desirable to work with light of a single wavelength, i.e., a monochromatic light, as far as possible. If polychromatic light is used, some other reactions may also take place simultaneously.

This is usually done by employing discharge tubes which give atomic line spectra. A common source of light used for this purpose is the mercury-vapour lamp. Iron and carbon arcs and, in some cases, metal filament lamps (e.g., tungsten lamp) are also used as light sources. The spectrum given out by these sources consists only of a few sharp lines out of which the line of a desired wavelength can be isolated by means of suitable filters. For very accurate work, a special device called monochromator is used. It acts like a spectrometer in which the wavelength of light within a narrow range can be determined and isolated. It consists, essentially, of a suitable source of light and a prism which is ordinarily of glass or of quartz if the monochromatic light to be used is in the violet or ultra-violet region. The light of the required wavelength is isolated and made to fall on the reaction cell.

Determination of the number of Einsteins of light absorbed

The energy of monochromatic radiation (in terms of the number of Einsteins absorbed) is determined accurately by employing a thermopile which consists of a number of thermocouples joined in series. One set of junctions of the thermocouples is blackened so as to absorb all the radiations. The other set is protected from radiation and maintained at a constant temperature. The heat radiations associated with light when falling on one set of junctions of the thermocouple generate thermoelectric current in the circuit. The e.m.f. set up due to thermoelectric effect is measured, from which, the energy of the incident radiation can be easily calculated.

The measurements are made before and after passing the light through the cell. The difference gives the energy of the radiation absorbed by the reacting substance.

The energy of radiation can also be measured, though not so accurately, by employing an actinometer. In this device, a standard photochemical reaction is used to estimate the energy of the radiation absorbed.

One of the actinometers, which is in common use, is the uranyl-oxalate actinometer. It consists of a dilute solution of oxalic acid mixed with uranyl sulphate. The latter serves to sensitise the decomposition of oxalic acid on exposure to violet or ultra-violet light. The wavelength of the required light falls in the range of 2540Å to 4350Å. The extent of decomposition of oxalic acid is determined at the conclusion of the experiment by titration against potassium permanganate. The assumption made is that the amount of decomposition is proportional to the product of the intensity of light of a given wavelength and the time of exposure. The apparatus is first standardised with respect to radiations of different wavelengths. It is possible, therefore, to evaluate the amount of energy of light radiation of a given wavelength absorbed in the case of a photochemical reaction. The measurements are made before and after passing the light through the reacting cell.

Example 23.2. Photobromination of cinnamic acid to dibromo cinnamic acid was carried out in blue light of wavelength 4400Å at 35°C using light intensity of 1.5×10^{-3} J per second. An exposure of 20 minutes produced a decrease of 0.075 millimole of bromine. The solution absorbed 80 per cent of the light passing through it. Calculate the quantum yield of the reaction.

Solution.

$$\left.\begin{array}{l}\text{Energy associated with a quantum}\\ \text{of light of wavelength 4400Å}\end{array}\right\} = \frac{hc}{\lambda} = \frac{6.626 \times 10^{-34} \times 3 \times 10^8}{4400 \times 10^{-10}}$$

$$= 4.51 \times 10^{-19} \text{ J}$$

Intensity of light
$$= 1.5 \times 10^{-3} \text{ J per sec.}$$

$$\therefore \left.\begin{array}{l}\text{Radiation energy absorbed}\\ \text{in 20 minutes}\end{array}\right\} = 1.5 \times 10^{-3} \times 20 \times 60 \times \frac{80}{100} \text{ J} = 1.44 \text{ J}$$

$$\text{Number of quanta absorbed} = \frac{1.44}{4.51 \times 10^{-19}} = 3.19 \times 10^{18}$$

$$\left.\begin{array}{l}\text{Number of molecules of}\\ \text{bromine reacting}\end{array}\right\} = \frac{0.075}{1000} \times 6.022 \times 10^{23}$$

$$= 45.15 \times 10^{18}$$

$$\therefore \text{ Quantum yield} = \frac{45.15 \times 10^{18}}{3.19 \times 10^{18}} = 14.15$$

Some Photochemical Reactions and their Quantum Yields

Bodenstein studied quantum yields of a large number of photochemical reactions and found that while some reactions followed the law of photochemical equivalence, the others did not. The results obtained in the case of some important photochemical reactions together with the effective wavelengths are given in Table 23.2.

Table 23.2. Effective wavelengths and quantum yields of photochemical reactions.

Reaction	Effective wavelength ($\mathring{A}$)	Quantum yield
$2NH_3 \rightarrow N_2 + 3H_2$	2100	0.2
$2HI \rightarrow H_2 + I_2$	2070–2820	2
$2HBr \rightarrow H_2 + Br_2$	2070–2530	2
$H_2 + Cl_2 \rightarrow 2HCl$	4000	10^5
$CO + Cl_2 \rightarrow COCl_2$	4000–4360	10^3
$SO_2 + Cl_2 \rightarrow SO_2Cl_2$	4200	1
$2NO_2 \rightarrow 2NO + O_2$	4050	0.7
$H_2S \rightarrow H_2 + S$	2080	1
$3O_2 \rightarrow 2O_3$	1700–1900	3
$CH_3COCH_3 \rightarrow CO + C_2H_6$ (vapour)	3000	0.1
Maleic acid $\rightarrow$ fumaric acid	2000–2800	0.04
$2Fe^{2+} + I_2 \rightarrow 2Fe^{3+} + 2I^-$	5790	1
$2H_2O_2 \rightarrow 2H_2O + O_2$	3100	> 7
$H_2 + Br_2 \rightarrow 2HBr$	5100	0.01

It is evident that the law of photochemical equivalence is strictly valid for a very few reactions only. The various reactions can be divided into following three categories :

1. Those in which the quantum yield is a small integer, such as 1, 2, or 3; e.g., combination of sulphur dioxide and chlorine, dissociation of hydrogen iodide or hydrogen bromide, ozonisation of oxygen, etc.

2. Those in which the quantum yield is less than 1; e.g., dissociation of ammonia, nitrogen dioxide or acetone vapour and transformation of maleic acid into fumaric acid, etc.

3. Those in which the quantum yield is extremely high; e.g., combination of carbon monoxide and chlorine and of hydrogen and chlorine.

In order to explain the variations, Bodenstein pointed out that photochemical reactions involve two distinct processes :

Primary processes

Primary processes in which light radiation is absorbed by an atom or a molecule giving rise to the formation of an excited atom or an excited molecule, as the case may be. Thus :

$$A \quad + \quad h\nu \quad \longrightarrow \quad A$$

A	$h\nu$	A
Atom or molecule	One quantum of light	Excited atom or molecule

Another possibility is that the molecule which absorbs light may get dissociated yielding atoms (some in excited state) or free radicals.

Secondary processes

Secondary processes which involve the excited atoms or molecules or free radicals produced in the primary stage. The action of light is merely confined to primary processes. The secondary processes are such as can take place in the 'dark' as well.

Bodenstein and others emphasised that the law of photochemical equivalence can be applied only to primary processes in which each molecule capable of entering into chemical reaction absorbs one quantum of radiation. The secondary processes take place of themselves quite independent of the light radiation. As a result, the quantum efficiency of the reaction, as whole, gives distorted or misleading picture with regard to the applicability of the law of photochemical equivalence. In the light of this discussion, we may consider a few typical photochemical reactions in some details.

Photochemical Decomposition of Hydrogen Iodide

As it has been shown in Table 23.2, the photochemical decomposition of hydrogen iodide takes place in the radiations of wavelengths between 2070Å and 2820Å and the quantum yield of the reaction is 2. It has been established from spectroscopic data that the primary stage of the reaction involves absorption of one quantum of light of an appropriate wavelength per molecule which then dissociates as represented by the equation:

(i) $$HI + h\nu \longrightarrow H + I$$

This is in accordance with the law of photochemical equivalence. The possible secondary reactions which may follow are :

(ii) $$H + HI \longrightarrow H_2 + I; \qquad \Delta E = -133.9 \text{ kJ}$$

(iii) $$I + HI \longrightarrow I_2 + H; \qquad \Delta E = +146.4 \text{ kJ}$$

Reaction (iii), being highly endothermic, cannot take place at ordinary temperatures. The final step in the reaction is :

(iv) $$I + I \longrightarrow I_2$$

The mechanism of the reaction is obtained by adding steps (i), (ii) and (iv) which gives :

$$2HI + h\nu \longrightarrow H_2 + I_2$$

This accounts for the observed quantum yield of 2.0.

The quantum yield of the reaction falls below 2.0 after some time. This is due to the fact that as sufficient molecules of iodine produced in step (iv) accumulate, the following reaction, regenerating HI, is also set up :

$$H + I_2 \longrightarrow HI + I$$

Photochemical Decomposition of Hydrogen Bromide

The quantum yield of the reaction involving the photochemical decomposition of hydrogen bromide in radiations of wavelengths between 2070Å and 2530 Å is about 2 initially. The mechanism of this reaction, therefore, is considered to be very similar to the one discussed above for the decomposition of hydrogen iodide. The quantum yield, however, falls below 2 more readily. It appears, therefore, that the reaction involving the regeneration of HBr takes place more readily than the corresponding reaction involving the regeneration of HI.

Hydrogen-bromine reaction

The quantum yield of the reaction :

$$H_2 (g) + Br_2 (g) \longrightarrow 2HBr(g)$$

is extremely small, being only 0.1 mole per Einstein of energy absorbed.

Bromine absorbs light in the green region of the spectrum (4500–5500Å). The primary process is dissociation of the molecule into atoms :

(i) $\quad\quad\quad\quad\quad\quad Br_2 + h\nu \longrightarrow 2Br;$ $\quad\quad\quad\quad\quad\quad$ Rate constant = k_1

The secondary reactions should be the same as are known to occur in the thermal process as well. These are given below :

(ii) $\quad\quad\quad\quad\quad\quad Br + H_2 \longrightarrow HBr + H;$ $\quad\quad\quad\quad\quad$ Rate constant = k_2

(iii) $\quad\quad\quad\quad\quad H + Br_2 \longrightarrow HBr + Br;$ $\quad\quad\quad\quad\quad$ Rate constant = k_3

(iv) $\quad\quad\quad\quad\quad H + HBr \longrightarrow H_2 + Br;$ $\quad\quad\quad\quad\quad$ Rate constant = k_4

(v) $\quad\quad\quad\quad\quad\quad Br + Br \longrightarrow Br_2;$ $\quad\quad\quad\quad\quad\quad\quad$ Rate constant = k_5

The bromine atoms formed in step (i) attack hydrogen yielding hydrogen bromide and a hydrogen atom. The latter then attacks bromine forming another molecule of hydrogen bromide and another bromine atom. In this way if steps (ii) and (iii) take place one after the other, a very high quantum yield will be obtained. But the reaction (ii) is highly endothermic and requires high energy of activation. This step is, therefore, very slow at ordinary temperatures. If it proceeds for a while, the reverse of it which, in fact, is step (iv), becomes increasingly important and the rate of formation of HBr decreases. As bromine atoms accumulate, since these are not used up in step (ii), the step (v) involving recombination of bromine atoms to form molecules takes place readily. Hence the quantum yield of the reaction is extremely low.

If, however, the temperature is raised, the rate of reaction (ii) increases. This enhances the quantum yield of the overall reaction to some extent.

Kinetics of the Hydrogen-Bromine Reaction

The kinetics of the overall reaction may now be examined. The rate of formation of bromine atoms in step (i) is photochemical and, therefore, depends upon the intensity of the light used. The rate of the formation of bromine atoms is thus given by the expression:

$$\frac{d[Br]}{dt} = k_1 I_{abs} \quad\quad\quad\quad\quad ...(23.26)$$

where I_{abs} is the rate of absorption of light and k_1 is a constant.

The product of the reaction, namely hydrogen bromide, is formed in steps (ii) and (iii) and is consumed in step (iv). Hence the net rate of formation of HBr may be represented as :

$$\frac{d[HBr]}{dt} = k_2[Br][H_2] + k_3[H][Br_2] - k_4[H][HBr] \qquad \qquad ...(23.27)$$

Bromine and hydrogen atoms are merely transitory intermediates as these are used up in subsequent steps. They are present in too small amounts to be measured directly. We have, therefore, to express their concentrations in terms of measurable quantities. To simplify, it is assumed that the rate at which they are formed is equal to the rate at which they disappear. Thus hydrogen atoms are produced by reaction (ii) and removed by reactions (iii) and (iv). Hence

$$k_2[H_2][Br] = k_3[H][Br_2] + k_4[H][HBr] \qquad \qquad ...(23.28)$$

Similarly, bromine atoms are produced by reactions (i), (iii) and (iv) and removed by reactions (ii) and (v). Hence

$$k_1 I_{abs} + k_3[H][Br_2] + k_4[H][HBr] = k_2[H_2][Br] + k_5[Br]^2 \qquad \qquad ...(23.29)$$

Subtracting equation (23.28) from equation (23.29), we have :

$$k_1 I_{abs} = k_5[Br]^2 \qquad \qquad ...(23.30)$$

$$\therefore \qquad [Br] = \sqrt{\frac{k_1 I_{abs}}{k_5}} \qquad \qquad ...(23.31)$$

Substituting this value of [Br] in equation (23.28), it is possible to express the concentration of hydrogen atoms as :

$$[H] = \frac{k_2[H_2]\sqrt{\dfrac{k_1 I_{abs}}{k_5}}}{k_3[Br_2] + k_4[HBr]} \qquad \qquad ...(23.32)$$

Inserting the values of [Br] and [H] from equations (23.31) and (23.32) into the rate equation (23.27) for the formation of HBr, we have :

$$\frac{d[HBr]}{dt} = \frac{2k_2\sqrt{\dfrac{k_1}{k_5}} \times [H_2]\sqrt{I_{abs}}}{1 + \dfrac{k_4[HBr]}{k_3[Br_2]}} \qquad \qquad (23.33)$$

$$= \frac{k\sqrt{I_{abs}}[H_2][Br_2]}{k'[Br_2] + [HBr]} \qquad \qquad ...(23.34)$$

where k and k' are new constants. This equation was found to agree with the experimental observations of Bodenstein. This also explained the surprising observation made by several workers that the rate of the reaction varies as the square root of the intensity of light.

Hydrogen-Chlorine Reaction

The photochemical combination of hydrogen and chlorine is highly interesting as, in contrast to the reactions considered above, its quantum yield is extremely high, varying between 10^4 to 10^6, in the absence of oxygen. The presence of oxygen slows down the rate of the reaction and lowers the quantum yield also.

The extremely high quantum yield has been attributed to the setting up of a chain. According to the mechanism proposed by Nernst, and largely accepted at present, the primary process is the dissociation of chlorine molecules into atoms as a result of absorption of radiation of energy. Thus

(i) $\qquad\qquad\qquad Cl_2 + hv \longrightarrow 2Cl;$ $\qquad$ Rate constant $= k_1$

The following secondary processes then follow :

(ii) $\qquad\qquad\qquad Cl + H_2 \longrightarrow HCl + H;$ $\qquad$ Rate constant $= k_2$

(iii) $\qquad\qquad\qquad H + Cl_2 \longrightarrow HCl + Cl;$ $\qquad$ Rate constant $= k_3$

The regenerated chlorine atom, as in equation (iii), reacts with another molecule of hydrogen to give another molecule of hydrogen chloride and the hydrogen atom produced thereby reacts with another molecule of chlorine to give still another molecule of hydrogen chloride, and so on and so forth. Thus, the chain set up by the dissociation of one molecule of chlorine into atoms, on the absorption of one quantum of energy, is propagated.

It will appear curious at first sight that although a similar mechanism was suggested for the combination of hydrogen and bromine to form gaseous hydrogen bromide, the quantum efficiency of that process was found to be extremely low. The reason is that while the reaction between bromine atom and hydrogen molecule immediately following the primary process, is highly endothermic and, therefore, extremely slow at ordinary temperatures, the corresponding reaction between chlorine atom and hydrogen molecule, represented by equation (ii) above, is exothermic and takes place almost instantaneously.

The quantum yield of the reaction depends upon the number of times the reactions in steps (ii) and (iii) are repeated before the termination of the chain. The chain-terminating reaction appears to be the recombination of chlorine atoms to form molecules on the walls of the reaction vessel represented as:

(iv) $\qquad\qquad\qquad\qquad Cl + Cl \longrightarrow Cl_2;$ $\quad$ Rate constant $= k_4$

In the presence of oxygen the chain may also be terminated by the reaction :

$$H + O_2 \longrightarrow HO_2$$
$$\text{(radical)}$$

In this way some of the hydrogen atoms are converted into HO_2 radicals which do not propagate the chain further. The quantum yield of the reaction is consequently lower in the presence of oxygen. The chain-mechanism put forward above receives supports from the following observations :

1. The combination of hydrogen and chlorine in the form of a chain reaction can take place even in the absence of light by introducing either hydrogen or chlorine atoms into the reaction vessel.

2. The quantum yield of the reaction is decreased very largely if the reaction is allowed to take place in capillary tubes. This supports the view that the chain is terminated at the walls of the containing vessel.

Kinetics of the hydrogen-chlorine reaction

Since chlorine atoms are formed in reactions (i) and (iii) and they disappear by reactions (ii) and (iv), hence, at the stationary state :

$$k_1 I_{abs} + k_3 [H][Cl_2] = k_2 [Cl][H_2] + k_4 [Cl^2] \qquad \text{...(23.35)}$$

where I_{abs} is intensity of the absorbed radiation.

Similarly, since hydrogen atoms are formed in step (ii) and removed in step (iii), hence, at the stationary state :

$$k_2 [Cl][H_2] = k_3 [H][Cl_2] \qquad \text{...(23.36)}$$

Combining equations (23.35) and (23.36) we have :

$$k_1 I_{abs} = k_4 [Cl]^2 \qquad \text{...(23.37)}$$

or

$$[Cl] = \sqrt{\frac{k_1}{k_4} I_{abs}} \qquad \text{...(23.38)}$$

Hence the overall rate of formation of HCl is given by

$$\frac{d[HCl]}{dt} = k_2 [Cl][H_2] + k_3 [H][Cl_2]$$

$$= 2k_2 [Cl][H_2] \qquad \text{...(23.39)}$$

Substituting the value of [Cl] from equation (23.38) in equation (23.39), we get :

$$\frac{d[HCl]}{dt} = 2k_2 \sqrt{\frac{k_1}{k_4} I_{abs}} [H_2] \qquad \text{...(23.40)}$$

$$= k \sqrt{I_{abs}} [H_2] \qquad \text{...(23.41)}$$

where k is a new constant.

According to equation (23.41), the rate of the reaction should vary directly as the square root of intensity of light as well as pressure of hydrogen. Both the conclusions have been borne out by experiment.

Decomposition of Ammonia

The photochemical reaction involving decomposition of ammonia has been found to take place by radiations of wavelength 2100Å. Its quantum yield is 0.15 mole per Einstein at 20°C and 0.50 mole per Einstein at 200°C. The primary stage in this case involves dissociation of ammonia as :

$$NH_3 + hv \longrightarrow NH_2 + H$$

Since the final products are known to be nitrogen, hydrogen and hydrazine, the secondary stage is represented by the following equations :

$$NH_2 + NH_2 \longrightarrow N_2 + 2H_2$$

$$NH_2 + NH_2 \longrightarrow N_2H_4$$

$$H + H \longrightarrow H_2$$

As the quantum yield is less than 1, it appears that NH_2 and H produced in the primary stage recombine to form ammonia.

Photochemical Equilibrium

There are some cases in which a photochemical reaction is followed by a thermal reaction in the opposite direction. The rate of the photochemical reaction is proportional to the intensity of the light radiation while that of the thermal reaction is proportional to the concentration of the photochemical product. The state of affairs is represented as :

$$A \underset{k_2}{\overset{k_1}{\rightleftharpoons}} B$$

where k_1 is the rate constant of the photochemical reaction and k_2 is the rate constant of the thermal reaction. The net rate of reaction is evidently given by the expression :

$$\frac{dx}{dt} = k_1 I_{abs} - k_2 [B] \qquad ...(23.42)$$

where I_{abs} is the intensity of absorbed radiation. Ultimately, a state of equilibrium is approached when

$$k_1 I_{abs} = k_2 [B] \qquad ...(23.43)$$

This is a case of photochemical equilibrium or photostationary state.

From equation (23.43)

$$[B] = \frac{k_1}{k_2} \times I_{abs} \qquad ...(23.44)$$

Evidently, the concentration of the product formed at the equilibrium state is directly proportional to the intensity of the light absorbed.

Some of the photochemical reactions occurring in solution provide examples of photochemical equilibrium.

Photochemical Reactions in Solutions

Two examples of photochemical reactions taking place in solution phase now be considered.

1. *Isomeric transformation of maleic acid into fumaric acid:* Warburg found that the conversion of maleic acid into fumaric acid, as well as the reverse transformation, could take place by exposing their aqueous solutions to light of wavelength 2070Å. The quantum yield for the conversion of maleic acid into fumaric acid was found to be about 0.04. The reason, obviously, is that the direct as well as the back reactions are both photochemical in nature and, therefore, ultimately, a state of equilibrium is reached.

$$\text{Maleic acid} \underset{\text{u.v. light}}{\overset{\text{u.v. light}}{\rightleftharpoons}} \text{Fumaric acid} \qquad ...(23.45)$$

2. *Polymerisation of anthracene:* Anthracene polymerises to dianthracene when its solution in benzene or toluene is exposed to ultra-violet light. In the dark, the reverse reaction takes place, i.e., the dianthracene breaks down into simple anthracene molecules.

$$\underset{\text{Anthracene}}{2C_{14}H_{10}} \underset{\text{Dark}}{\overset{\text{u.v. light}}{\rightleftharpoons}} \underset{\text{Dianthracene}}{C_{28}H_{20}} \qquad\qquad ...(23.46)$$

Therefore, ultimately, a stationary state is attained since as much of dianthracene is formed by the action of light as is decomposed in the thermal (or dark) reaction.

The primary process involves the formation of an excited molecule of anthracene for each quantum of light energy absorbed.

$$\underset{\text{Anthracene}}{C_{14}H_{10}} + hv \longrightarrow \underset{\text{Excited molecule}}{C_{14}H_{10}}$$

The excited molecule then on collision with an ordinary molecule gives dianthracene.

$$C_{14}H_{10} + C_{14}H_{10} \longrightarrow C_{28}H_{20}$$

The quantum yield, according to the above mechanism, should be 2, but actually it is only about 0.5. This is due to the simultaneous occurrence of the back thermal reaction, as mentioned above. The concentration of dianthracene at the steady state is in accordance with equation (23.46).

Photosensitisation

Some photochemical reactions take place by the absorption of light by one of the non-reactants which may be present in the system. This was first discovered by Vogel when he found that silver halides were affected even in yellow or red light in photographic processes if blended with suitable dyestuffs. This is know as photosensitisation.

A striking example of photosensitisation is the action of chlorophyll in causing combination of carbon dioxide and water in sunlight to form carbohydrates :

$$CO_2 + H_2O + hv \longrightarrow \underset{\text{Glucose}}{1/6\ (C_6H_{12}O_6)} + O_2$$

Carbon dioxide and water themselves being colourless do not absorb any portion of visible light, but chlorophyll, the green colouring matter of the plants, can do so. Although the exact mechanism of this important life process is not understood, it appears that the chlorophyll passes on the radiation energy absorbed by it to the carbon dioxide and water molecules which then combine, as represented above, even in the absence of ultra-violet light.

Another interesting example of photosensitisation is provided by the decomposition of hydrogen molecules into hydrogen atoms when hydrogen gas mixed with mercury vapour is exposed to ultra-violet light of wavelength 2537Å coming from a mercury vapour lamp. The energy associated with this radiation is 46.8 kJ (per mole of photons). The energy required to dissociate hydrogen molecules into atoms is 435.1 kJ. per mole. In this case, ultra-violet light is absorbed by mercury vapour which is then made available to hydrogen for its decomposition into atoms. The reaction may be represented as:

$$Hg + hv \longrightarrow Hg$$

$$Hg + H_2 \longrightarrow Hg + H$$

Evidently, in this reaction mercury acts as a photosensitiser. Photo-decomposition of oxalic acid by uranyl ions used in actinometer has already been referred to.

Photo-inhibitors

It has been observed that even traces of certain substances are able to reduce the rate of a photochemical reaction. For example, mere traces of nitric oxide and propylene lower the quantum yield of the photochemical combination of hydrogen and chlorine. Such substances are known as photo-inhibitors. It is believed that these substances interrupt the chain reactions by removing the atoms or radicals which normally act as chain carriers.

Chemiluminescence

The phenomenon of emission of light in a chemical reaction is known as chemiluminescence. It is, therefore, the reverse of photochemical reaction. The glow of phosphorus and of its trioxide is a familiar example of chemiluminescence. The oxidation of Grignard reagents by air or oxygen emits greenish blue luminescence. On mixing a solution of strontium chloride with dilute sulphuric acid in a dark room, feeble glow is emitted along with the precipitation of strontium sulphate. The emission of the so called 'cold light' by glow-worm is another well-known case of chemiluminescence. The chemical reaction involved is the oxidation of luciferon—a protein, by atmospheric oxygen in the presence of an enzyme called *luciferase*.

Example 23.3. The extinction co-efficient of $Coen_2Br_2{}^+$ is 40 M cm^{-1} at 650 mμ. Calculate the per cent transmission for a 5 cm cell filled with 0.01 M solution. Neglect solvent.

Solution.

Optical density $= \log \dfrac{I_O}{I} = \varepsilon.C.l.$

Here $\varepsilon = 40$ M cm^{-1}, C = 0.01 molar; 1 = 5 cm.

So $\qquad \log \dfrac{I_O}{I} = 40 \times 0.01 \times 5 = 2$

or $\qquad \dfrac{I_O}{I} = 100$

or $\qquad \dfrac{I}{I_O} = 0.01$

So % of transmitted light $= 0.01 \times 100 = 1\%$.

Example 23.4. A 0.003 M solution of $Co(NH_3)_6{}^{3+}$ transmits 75% of incident light of 500 mμ if the path length is 1 cm. Calculate the extinction coefficient and the per cent absorption for a 0.01 M solution.

Solution.

Given $\qquad$ I = 75 when I_0 = 100.

So $\qquad$ O.D. $= \log \dfrac{100}{75} = \varepsilon \times 0.003 \times 1$

or Extinction coefficient $\varepsilon = 41.65$ M. cm^{-1}.

For a 0.01 M solution in the same cell

O.D. $= 41.65 \times 0.01 \times 1 = 0.4165$

So here $\log \dfrac{100}{I'} = 0.4165$

or % of transmitted light $I' = 38.33$

So % of absorbed light $= (100 - 38.33) = 61.67\%$.

Chapter 24

Chemical Methods of Separation

INTRODUCTION

The sub-division of chemistry concerned with identification of materials (qualitative analysis) and with determination of the percentage composition of mixtures or the constituents of a pure compound (quantitative analysis). The gravimetric and volumetric (or "wet") methods (precipitation, titration, and solvent extraction) are still used for routine work; indeed, new titration methods have been introduced, e.g., cryoscopic, pressure-metric (for reactions that produce a gaseous product), redox methods, and use of a fluoride-sensitive electrode. However, faster and more accurate techniques (collectively called instrumental) have been developed in the last few decades. Among these are infra-red, ultra-violet, and x-ray spectroscopy, where the presence and amount of a metallic element is indicated by lines in its emission or absorption spectrum; colorimetry, by which the percentage of a substance in solution is determined by the intensity of its colour; chromatography of various types by which the components of a liquid or gaseous mixture are determined by passing it through a column of porous material, or on thin layers of finely divided solids; separation of mixtures in ion-exchange columns; and radioactive tracer analysis. Optical and electron microscopy, mass spectrometry, microanalysis, nuclear magnetic resonance (NMR), and nuclear quadrupole resonance (NQR) spectroscopy all fall within the area of analytical chemistry.

VOLUMETRIC METHODS

Titrimetry

Qualitative analysis deals with the determination of the nature of the constituents of a given material. On the other hand, quantitative analysis deals with the determination of the proportions of the various constituents of a given material. One of the most important branches of quantitative analysis is volumetric or titrimetric analysis which depends on the accurate measurements of volumes of solutions undergoing a chemical change. In short, a volumetric experiment consists in dissolving a known weight of a substance in water and making up the solution to an exact volume. A measured volume of this solution is completely reacted with another solution in presence of a substance called indicator. The volume of the latter solution is determined. The whole of this process is known as titration. Knowing the volume of the reacting solutions, the results are calculated. Titrimetric analysis is a quick process and thus has the major advantage over the other types of analysis such as gravimetric analysis.

It may be pointed out that titrimetric analysis can be carried out only if a suitable reaction is possible between the reactants, e.g., an acid and a base. A volumetric experiment between the reactants must satisfy the following conditions:

1. A suitable reaction should be possible between the substance to be taken in the burette and the titration flask, e.g., a reaction between an acid and a base; an oxidising agent and a reducing agent.
2. There should be only one reaction occurring between the solutions taken in the burette and the titration flask. There should be no side reaction which may complicate the process of calculations.
3. The reaction between the solution taken in the burette and the titration flask should be complete within a reasonable time, i.e., reaction should be quite rapid.
4. The reaction between the solution taken in the burette and the titration flask should be simple so that a definite weight relationship exists between the reactants.
5. The reaction between the solution taken in the burette and the titration flask should be possible at about room temperature, i.e., at not very high temperature; in dilute solutions and not any very special conditions.
6. A suitable indicator must be available which can help to locate the exact end point.

Keeping the above points in view, chemists have developed the following types of volumetric analysis:

1. Acid base titrations (acidimetry and alkalimetry).
2. Oxidation-reduction titrations (redox titrations).
3. Precipitation titrations.
4. Complexometric titrations.

Let us discuss the theory of the above types of titrations.

Acid-Base Titrations

It is a well known fact that an acidic solution has a pH value of less than 7 while an alkaline solution has a pH value greater than 7. Therefore, when an acidic solution is reacted with an alkaline solution, pH value changes. These changes in pH value can be determined by any one of the following methods:

1. *Visual method:* Organic compounds such as methyl orange, phenolphthalein, methyl red and thymol blue are sensitive to pH value and have different colours at different pH values. Using any one of these compounds as an indicator, it is possible to determine the changes in pH value and thus the end point.
2. *Electrical conductivity measurements:* We know that electrical conductivity depends upon the number and mobility of the ions. As the acid and base undergo neutralisation, electrical conductivity changes which helps to locate the end points. Titrations based on electrical conductivity measurements are called conductometric titrations.
3. *E.M.F. measurements:* Since the electrical potential of hydrogen electrode varies with the concentration of H^+ ions (i.e., pH value) with which it is in contact, it is possible to determine the

changes in pH when an acid is neutralised by a base by measuring e.m.f. Such type of titrations which are based on the measurement of electrical potential are called potentiometric titrations.

Quite commonly acid-base reactions are studied in the laboratory using organic compounds known as indicators. Examples of commonly used indicators in acid-base titrations are methyl orange, phenolphthalein or methyl red.

Oxidation-Reduction (or Redox) Titrations

Oxidation involves loss of electrons while reduction involves gain of electrons. Therefore, it is possible to titrate an oxidising agent (which will accept electrons) against a reducing agent (which will lose electrons). Such type of reactions in which oxidation and reduction occur at the same time are known as redox reactions and titrations involving redox reactions are known as redox titrations. Some of the commonly used oxidising and reducing agents in the redox titrations are as follows:

1. Oxidising agents
   (a) $KMnO_4$ and dil. H_2SO_4
   (b) $K_2Cr_2O_7$ and dil. H_2SO_4
   (c) Ceric sulphate $[Ce(SO_4)_2]$ and dil. H_2SO_4
   (d) Iodine solution
   (e) Potassium iodate (KIO_3) in HCl solution

2. Reducing agents
   (a) Ferrous sulphate, $FeSO_4.7H_2O$
   (b) Ferrous ammonium sulphate (or .Mohr's salt) $FeSO_4.(NH_4)_2SO_4.6H_2O$
   (c) Oxalic acid $\begin{matrix} COOH \\ | \\ COOH \end{matrix}$ .2H_2O
   (d) Oxalates such as potassium oxalate
   (e) Sodium thiosulphate
   (f) Potassium iodide
   (g) Arsenious oxide, As_2O_3
   (h) Antimonius compounds such as tartar emetic.

There is no universal oxidising agent which can be titrated against every reducing agent and vice-versa. Hence, the choice of an oxidising agent to be used against a particular reducing agent depends upon the reaction conditions and standard reduction potential of the oxidising agent. Therefore, it will be useful to give a summary of the various type of redox titrations.

Permanganometry

Permanganate ion in acidic solution is a strong oxidising agent and its action can be represented by the following reaction:

$$MnO_4^- + 8H^+ + 5e^- \rightleftharpoons Mn^{2+} + 4H_2O; \quad E° = 1.15 \text{ V}$$

[*Note:* In permanganometry, potassium permanganate is used as an oxidising agent as it is easily available in good purity].

It is evident from the above equation that the equivalent weight of $KMnO_4$ is one-fifth of its formula weight, i.e.

$$\frac{39 + 55 + 64}{5} = 31.6$$

Permanganate titrations are carried out in presence of dil. H_2SO_4 as it has no action on permanganate ion. Hydrochloric acid is not used in permanganate titrations particularly because HCl is liable to be oxidised to chlorine by $KMnO_4$. Then more of $KMnO_4$ will be consumed in the redox titration. It is also not advisable to use dil. HNO_3 in these titrations because HNO_3 is itself an oxidising agent and will interfere with the oxidation action of $KMnO_4$.

Since potassium permanganate has dark pink colour, it itself acts as an indicator and the end point is the appearance of a permanent pink colour in the solution.

The reducing agents which are generally titrated against acidified $KMnO_4$ are ferrous salt such as ferrous sulphate, ferrous ammonium sulphate known as Mohr's salt, oxalic acid, oxalates such as potassium oxalate and ferrous oxalates.

It may be mentioned that ferrous salts, are titrated against acidified $KMnO_4$ at room temperature. However, oxalic acid and oxalates are titrated while keeping the oxalic acid, oxalate solutions quite warm (around 60–70°C). This is due to the fact that oxalic acid and oxalates on oxidation with acidified $KMnO_4$ produce carbon dioxide according to the equation:

$$\begin{array}{l} COOH \\ | \qquad\qquad + [O] \longrightarrow 2CO_2 + H_2O \\ COOH \end{array}$$

In order to expel the CO_2 gas produced in the above reaction, the reaction mixture has to be kept hot so that the reaction proceeds in the forward direction.

It is also possible to estimate ferric salts by carrying out titration with acidified $KMnO_4$ after reduction of ferric salts with zinc and sulphuric acid.

Moreover, non-reducing cations like Ca^{2+}, Zn^{2+} which form precipitated oxalates can also be estimated by dissolving the washed precipitate in dilute sulphuric acid and then titrating the resulting solution against acidified $KMnO_4$.

Dichrometry

Acidified potassium dichromate is also a powerful oxidising agent though not as powerful as a potassium permanganate. However, potassium dichromate has the following advantages over potassium permanganate as an oxidising agent.

1. It is obtainable in very pure state so that it serves as an excellent primary standard.
2. Its aqueous solutions are stable for a long time as its solution is not decomposed by light.
3. Its solution has no action on organic matter such as rubber so that its solution can be used even in burettes having rubber taps.

4. Its titrations can be carried out even in presence of hydrochloric acid provided the concentration of the acid is not more than 20%.

Acidified dichromate acts as an oxidising agent according to the following equation:

$$Cr_2O_7^{2-} + 14H^+ + 6e^- \rightleftharpoons 2Cr^{3+} + 7H_2O$$

It is evident from the above equation that the equivalent weight of $K_2Cr_2O_7$ is one-sixth of its formula weight, i.e.,

$$\frac{2 \times 39 + 2 \times 52 + 7 \times 16}{6} = 49$$

The end point in potassium dichromate titrations is determined by any one of the following methods:

By the use of external indicator

Potassium dichromate cannot act as its own indicator because it forms green coloured chromic salts during redox reaction. Therefore, potassium ferricyanide is used as external indicator during potassium dichromate titrations. On a white tile, drops of the potassium ferricyanide solutions are placed with a glass rod. During titration a drop of the solution of ferrous ions taken in the titration flask is touched with the drop of the indicator on the tile. The development of blue colour indicates the presence of ferrous ions according to the reaction.

$$2K_3[Fe(CN)_6] + 3Fe^{2+} \longrightarrow Fe_3[Fe(CN)_6]_2 + 6K^+$$
$$\text{Blue}$$

At the end point, there will be no more ferrous ions so that no blue colour is produced rather a brownish yellow colour is formed.

By the use of internal indicator

Since the use of external indicator is quite cumbersome, analytical chemists have developed internal indicators for dichromate titrations. These are:

1. 1% solution of diphenylamine in conc. H_2SO_4; or
2. 1% solution of diphenylbenzidine in conc. H_2SO_4; or
3. 0.2% aqueous solution of sodium diphenylamine sulphonate.

With the above indicators, phosphoric acid is always used which lowers the oxidation potential of ferrous-ferric system by forming a complex $[Fe(HPO_4)]^+$. The above indicators impart green colour to the ferrous ions which changes to intense-purple or-violet at the end point.

However, now-a-days N-phenylanthranilic acid is used as indicator for dichromate titrations as with this indicator there is no need to add phosphoric acid. In this case at the end point, the colour changes from green to violet red.

Redox reactions involving iodine

An aqueous solution of iodine is a mild oxidising agent according to the following equation.

$$I_2 + 2e^- \rightleftharpoons 2I^-$$

Therefore, an aqueous solution of iodine is used to estimate reducing agents such as sodium thiosulphate, sodium sulphite, arsenites and antimonites according to the reaction:

$$2S_2O_3^- + I_2 \rightleftharpoons S_4O_6^- + 2I^-$$

$$SO_3^{2-} + I_2 + H_2O \rightleftharpoons SO_4^{2-} + 2H^+ + 2I^-$$

$$AsO_3^{2-} + I_2 + H_2O \rightleftharpoons AsO_4^{3-} + 2H^+ + 2I^-$$

$$SbO_2^- + I_2 + H_2O \rightleftharpoons SbO_3^- + 2H^+ + 2I^-$$

All these titrations which involve the use of standard iodine solution for the estimation of a reducing agent are called iodiometric titrations.

On the other hand, it is possible to estimate active oxidising agents like permanganates, dichromates, hydrogen peroxide and cupric ions by treating with excess of potassium iodide to liberate iodine. The liberated iodine is then estimated by titration against standard sodium thiosulphate solution. The following reactions are involved in these titrations:

With acidified $KMnO_4$ solution

$$2KMnO_4 + 3H_2SO_4 \longrightarrow K_2SO_4 + 2MnSO_4 + 3H_2O + 5(O)$$

$$2KI + Of + H_2SO_4 \longrightarrow K_2SO_4 + H_2O + I_2$$

With acidified $K_2Cr_2O_7$ solution

$$K_2Cr_2O_7 + 4H_2SO_4 \longrightarrow K_2SO_4 + Cr_2(SO_4)_3 + 4H_2O + 3(O)$$

$$2KI + Of + H_2SO_4 \longrightarrow K_2SO_4 + H_2O + I_2$$

With copper sulphate solution

$$CuSO_4 + 2KI \longrightarrow CuI_2 + K_2SO_4$$

$$2CuI_2 \longrightarrow Cu_2I_2 + I_2$$

All those titrations in which on oxidising agent is estimated by first reacting with potassium iodide and estimating the liberated iodine with standard sodium thiosulphate solution are known as iodiometric titrations.

In the titrations involving iodine solution, freshly prepared starch solution is used as an indicator which forms blue coloured complex with iodine. It may be pointed out that in iodometry, starch solution indicators may be added at any time while in iodiometric titrations, starch solution is added only when the solution is faintly yellow indicating the close approach of the end point.

Precipitation Titrations

There is another class of reactions which involve the formation of a precipitate when a solution taken in the burette reacts with a solution taken in the titration flask. Such type of titrations are known as precipitation titrations. The best example of such type of titrations is the reaction between silver nitrate solution and a halide solution such as sodium chloride solution.

$$AgNO_3 + NaCl \longrightarrow AgCl \downarrow + NaNO_3$$

Titrations involving the use of silver nitrate to estimate chloride, bromide, iodide or thiocyanate content in a solution are called argentometric titrations.

Complexometric Titrations

There are many reactions which involve the formation of a soluble undissociated stoichiometric complex at the end point when a substance taken in burette reacts with a substance taken in the titration flask. Such type of titrations are called complexometric titrations. For example, a reaction between silver nitrate solution and sodium cyanide solution results in the formation of a very stable and soluble complex $Na[Ag(CN)_2]$ at the end point.

$$AgNO_3 + 2NaCN \longrightarrow \underset{\text{Complex}}{Na[Ag(CN)_2]} + NaNO_3$$

In the complexion $[Ag(CN)_2]^-$, silver is the central atom while cyanide ions are attached to it are called ligands. Since two cyanide ions are attached to silver, it means co-ordination number of silver in this complex is two. In general, ligands contain highly electronegative atoms such as nitrogen, oxygen or halogens. Such ligands contain lone pairs of electrons which act as donor sites. Some metal ions such as aluminium, bismuth and lead easily form complexes with ligands having oxygen donor sites. However, metal ions like iron, cobalt, nickel, copper, zinc, cadmium and mercury form stable complexes with ligands containing nitrogen donor atoms. In some cases, the central metal ion links with donor atoms of the ligand in such a way that a ring is formed. Such a ligand is called chelating agent. Complexes formed by chelating agents are more stable than those formed by non-chelating agents. Commonly used chelating agents in analytical chemistry are EDTA (Ethylenediamenetetracetic acid), DMG (Dimethylglyoxime) and 8-hydroxyquinoline (known as oxine). In volumetric analysis, EDTA finds numerous applications as it forms very stable complexes with metal ions such as magnesium and calcium. Titrations involving chelating agents are called chelometric titrations. However, titrations involving the use of EDTA are known as EDTA titrations.

EDTA is commonly known as complexone III, sequesterene, versene, nullapon etc. It is a polyprotic acid and so its abbreviated formula is H_4Y. It dissolves readily in water and is obtainable in high degree of purity. It can co-ordinate with metal ions through its two nitrogen atoms and four oxygen atoms of the carboxyl groups.

Indicator in EDTA titrations

In these titrations, Eriochroma Black T (Erio T) is used as indicator. Erio-T has the interesting property that it can also form complexes with metal ions through its oxygen and nitrogen donor sites but the complexes thus formed are less stable than metal-EDTA complexes. Consequently, the indicator releases metal ions at the end point and then undergoes change in the colour. Most of the EDTA titrations are performed in presence of buffers having pH 8 to 10 wherein the indicator has blue colour. Therefore, the titration of a solution containing magnesium ions in presence of a buffer (such as NH_4Cl in NH_4OH having pH 10) against EDTA using Erio Black T changes its colour from red to blue at the end point.

Advantages of EDTA titrations

1. EDTA forms stable and soluble complexes which have definite composition as compared to other titrants.
2. EDTA is commonly used in the form of its disodium salt having the abbreviated formula Na_2H_2Y which is itself a primary standard.
3. Since EDTA forms soluble complexes, the end point is reached quite readily.

Indicator Theories

Indicator is a substance which indicates by its colour change the completion of a reaction. For example, in the neutralisation reaction between sodium hydroxide and hydrochloric acid (taken in the burette), phenolphthalein turns pink to colourless. Now let us try to find out as to why do the indicators change colour at the end point. This can be best explained with reference to acid-base titrations. It has been found out that the indicators used in neutralisation titrations are complex organic compounds which are themselves weak acids or bases. Therefore, like other weak acids and bases, they associate or dissociate according to pH of the solution. Hence, they change their colour with change in pH as shown in Table 24.1.

Table 24.1. Colour changes of indicators with pH.

Indicator	pH range	Colour	
		Acid solution	Alkaline solution
Thymol blue	1.2–2.8	Red	Yellow
Methyl orange	3.1–4.4	Red	Yellow
Methyl red	4.2–6.3	Red	Yellow
Litmus	5.5–7.4	Red	Blue
Phenolphthalein	8.3–10.0	Colourless	Red

The following two theories have been put forward to explain the colour change of indicators at the point in neutralisation reactions.

Ostwald theory

The main points of this theory are as follows:

1. An indicator is a weak acid or a weak base.

2. Ionisation of the indicator causes change in its colour. In other words, the unionised molecules of the indicator possess different colour than the ions.

3. Being weak electrolytes, indicators are feebly ionised in solution. However, the addition of a strong acid or strong base increases their ionisation considerably which causes change in the number of coloured ions.

4. An acidic indicator such as phenolphthalein must possess a coloured anion while a basic indicator such as methyl orange must possess a coloured cation.

In the light of this theory, let us explain the behaviour of phenolphthalein as well as methyl orange in neutralisation reactions.

Phenolphthalein as an indicator

Phenolphthalein is a weak organic acid having the formula HPh. Being a weak acid, it exists mostly as unionised molecules. These unionised molecules are colourless while on ionisation it produces H^+ ions (colourless) and Ph^- ions (pink) according to the reaction:

$$HPh \rightleftharpoons H^+ + Ph^- \qquad \qquad ...(24.1)$$

$$\text{(Colourless)} \qquad \text{(Colourless)} \qquad \text{(Pink)}$$

Since the reaction (24.1) is in equilibrium, the addition of an acid (or H^+ ions) increases the H^+ ion concentration and shifts the equilibrium in the backward direction making the solution colourless. Thus in acidic solution phenolphthalein remains colourless.

On the other hand, addition of a base like NaOH solution, produces OH^- ions which react with H^+ ions to produce feebly ionised water molecules as shown below:

$$HPh \rightleftharpoons H^+ + Ph^-$$
$$NaOH \rightleftharpoons OH^- + Na^+$$

$$\Updownarrow$$

$$H_2O \ (Freebly \ ionised)$$

Consequently, the equilibrium of reaction (24.1) is shifted in the forward direction resulting in the formation of greater number of Ph^- ions which are pink coloured. Therefore, in basic solution phenolphthalein gives pink colour.

This theory could explain as to why a weak base like NH_4OH cannot be titrated against a strong acid like HCl using phenolphthalein as indicator. This is because a weak base like NH_4OH is feebly ionised and hence produces very small number of OH^- which are insufficient to shift the equilibrium of reaction (24.1) towards the right until a large excess of NH_4OH is added. Therefore, solution remains colourless, i.e., phenolphthalein fails to work as an indicator for such titrations which involve weak bases.

Methyl orange as an indicator

Methyl orange is a weak organic base having the formula MeOH. Being a weak base, it exists mostly as unionised molecules. These unionised molecules have yellow colour while on ionisation. It produces Me^+ ions (red) and OH^- ions (colourless) according to the reaction :

$$MeOH \rightleftharpoons Me^+ + OH^- \qquad \qquad ...(24.2)$$
$$\text{(Yellow)} \qquad \text{(Red)} \qquad \text{(Colourless)}$$

Since the reaction (24.2) is an equilibrium, the addition of a base like NaOH increases the concentration of OH^- ions and thus pushes the equilibrium in the backward direction making the solution yellow in colour. Therefore, in basic solution methyl orange gives yellow colour.

On the other hand, addition of an acid like HCl solution, produces H^+ ions which react with OH^+ (already present in solution due to reaction 24.2) to produce feebly ionised water molecules as shown below:

$$MeOH \rightleftharpoons Me^+ + OH^-$$
$$HCl \rightleftharpoons Cl^- + H^+ \qquad \qquad ...(24.3)$$

$$\Updownarrow$$

$$H_2O \ (freebly \ ionised)$$

Consequently, the equilibrium of the reaction (24.2) is shifted in the forward direction resulting in the formation of greater number of Me^+ ions which are red coloured. Therefore, in acidic solution, methyl orange gives red colour.

This theory could explain as to why methyl orange cannot be used as an indicator in neutralisation reactions involving a weak acid like acetic acid (CH_3COOH). This is because a weak acid like acetic acid produces a very small number of H^+ ions which are insufficient to shift the equilibrium of reaction (24.2) to the forward direction until a very large excess of the acetic acid is added. Therefore, methyl orange fails to work as an indicator in such titrations which involve weak acids.

Quinonoid Theory

The main points of this theory are as follows:

1. The acid-base indicators are organic aromatic compounds which can exist in at least two tautomeric forms.

2. The tautomeric forms of the indicator exist in equilibrium with one another.

3. One of the tautomeric form of the indicator can exist in an acid solution while the other form can exist in an alkaline solution.

4. The two tautomeric forms possess different colours depending upon the pH value of the solution. Therefore, as the pH of the solution changes, the colour of the solution changes due to conversion of one form into the other.

5. The two tautomeric forms possess different structures known as benzenoid (benzene-like) form and quinonoid (quinone-like) form.

6. The quinonoid form is usually deeper in colour than the benzenoid form.

On the basis of this theory, let us explain the colour changes with change in pH using phenolphthalein or methyl orange as indicator.

Phenolphthalein as indicator

Phenolphthalein is a diprotic acid which is colourless in acidic medium as it possesses benzenoid structure. It gives pink to deep red colour in basic medium while it is colourless in acidic medium.

Methyl orange as indicator

Methyl orange is an organic base which has a yellow colour in the benzenoid form. On the addition of an acid, it forms a cation which exists in quinonoid form wherein it has pink colour.

Advantage of quinonoid theory

This theory has the advantage that it could explain not only the behaviour of indicators in acid-base titrations but even in redox titrations. For example, diphenyl amine is used as an indicator in dichromate titrations. In presence of reducing agent diphenylamine is colourless having benzenoid structure. On oxidation, it changes into diphenyl-benzidine.

Errors in Titrimetry and their Rectifications

In titrimetrical analysis, the following types of errors can crepe in:

1. *Operational errors:* They arise due to the individual analyst who may not be following proper analytical techniques. Examples are:

   (a) Improper washing of the apparatus such as burette, pipette and titration flask.

(b) Use of improper quantity of the indicator in each titration.

(c) Improper weighing of the reagents.

2. *Personal errors:* They arise from constitutional inability of an individual analyst to make correct observations. Examples are:

(a) Due to colour blindness, a person may not be able to judge the correct end point in a visual titration.

(b) An analyst having small height may not be able to correctly read the burette.

3. *Instrumental errors:* They arise from faulty construction of the balances, uncalibrated weights, uncalibrated burettes and pipettes.

4. *Reagent errors:* They arise from use of reagents which are substandard, i.e., impure acids, bases and indicators.

5. *Errors of method:* These errors arise due to faulty choice of method of analysis. Examples are:

(a) Incompleteness of a reaction. For example, incomplete neutralisation between an acid and a base.

(b) Occurrence of side reactions. For example, in titration between liberated iodine from $CuSO_4$ solution and sodium thiosulphate solution, atmospheric oxygen may take part in the reaction and cause a side reaction.

(c) Reaction of substances other than the constituent being determined may take place. For example, water containing oxygen may oxidise ferrous sulphate solution rather than being oxidised by acidified $KMnO_4$.

Rectification of errors in titrimetry

1. Operational errors can be minimised by carefully washing the apparatus; weighing the reagents correctly; and using the same quantity of the indicator for each titration.

2. Personal errors can be minimised by taking the help of a person who does not suffer from constitutional inability.

3. Instrumental errors can be minimised by using good quality balance, calibrated weights and properly calibrated glass apparatus.

4. Reagent errors can be rectified by using reagents of analar grade chemicals of E. Merck or B.D.H.

5. Errors of methods can be rectified by carefully carrying out the experiment in such a way that external factors do not harm the main reaction.

CHEMICAL METHODS OF SEPARATION

Recently, new methods have been developed to separate inorganic compounds at the micro-level. Some of these methods are ion exchange, solvent extraction and chromatography. They are quite simple, rapid and do not require any sophisticated apparatus. In this unit, we shall explain briefly the underlying principles of these methods.

Ion Exchange Method

Originally, this method was used for softening (or deionising) water. But now this technique has been modified to such an extent that it can be used for separation of anions, cations (particularly lanthanides) which are otherwise difficult to separate. It is possible to separate a mixture of amino acids by ionexchange method.

The early ion exchangers were complex inorganic compounds such as silicates, fuller's earth, and synthetic aluminosilicates (zeolites). But now organic synthetic resins have replaced inorganic ionexchangers. Organic ion-exchange resins should have the following special properties:

1. They should have reactive —OH, —COOH or —SO$_3$H group as exchange sites which are reversible.
2. They should be almost insoluble in water as well as organic solvents.
3. They should have an open, permeable molecular structure so that ions and solvent molecules can move freely in and out of the molecular structure.

Types of ion exchangers

There are two types of ionexchange resins as:

Cation exchangers

They are acidic resins as they contain sulphonic acid groups (RSO$_3^-$H$^+$) or carboxylic acid groups (RCOOH). The resins containing sulphonic acid groups are known as strong acid resins and find numerous applications in chemistry. The resins containing carboxylic acid groups are known as weak acid resins.

These type of resins are made by first copolymerising styrene and divinyl benzene, the polymer is then sulphonated to get the requisite type of cation exchange resin.

Anion exchangers

They are basic resins as they are either quarternary or tertiary amines or contain hydroxyl groups. The resins which are quarternary or tertiary amines are known as strongly basic anion exchangers. On the other hand, resins which contain —OH groups are known as weakly basic anion exchangers. Structurally in anion exchangers, —SO$_3$H group of the cation resin is replaced with —CH$_2$—NH$_3^+$ or —OH$^-$ group.

Selectivity of ionexchange process depends upon the following factors:

1. At the same concentration, the greater the charge on the ions, the strongly they are held by the resin.
2. At the same concentration and ions having the same charge the larger the ion (less hydrated), the more strongly it is held by the resin.
3. Selectivities widen with increase in cross-linkage of the resin.

Mechanism of ion exchange

Different views have been put forth to explain the mechanism of ionexchange. These are: (i) crystal lattice exchange; (ii) double layer theory; and (iii) donnan membrane theory.

Crystal lattice exchange

According to this view, the ion exchange in resins is quite similar to exchange of crystal lattice ions. The exchange of ions on the resins occurs throughout the entire gel structure of the resin and not on the exposed surface only.

Double layer theory

According to this theory, as in colloids, a double layer exists with inner fixed layer and outer mobile layer of charges. The charged layers are due to absorbed ions and the concentrations of the ions on the outer layer depends upon the concentration and pH of the external solution. As soon as the foreign ions present in the external solution come in contact with the outer layer new ions enter and replace the old ions in the outer layer and consequently, new equilibrium is established.

Donnan membrane theory

First as Donnan membrane theory, there is unequal distribution of ions on two sides of a membrane, in the same way, in ionexchange equilibria, the interface between solid and liquid phases is considered as membrane. There is greater activity of ions on one side of the membrane (which is free of non-diffusible ions, i.e., colloidal micelle to which exchangeable ions are attached) as compared to the other. The exchange of ions occurs so that the activity ratio become equal on both sides.

Applications of ion exchange method

In recent years, ion exchange methods have become quite popular and have numerous applications in analytical chemistry as well as in industry. Some of their applications are listed below :

1. This method was first of all used for softening or deionising water and is still used quite extensively.

2. Using ammonium citrate-citric acid, buffered at pH–5.5 cation exchange resins are used to separate lanthanide ions.

   Similarly, zirconium hafnium, niobium and tantalum can be separated by cation exchange resins.

3. Using EDTA solutions, it has been possible to separate quantitatively a mixture of alkali and alkaline earth metals.

Solvent Extraction Method

This method has become quite popular in recent years as it has the following advantages: (i) this method is quite simple and clean; (ii) it is quite rapid and convenient; (iii) it does not require any sophisticated apparatus or instrumentation; (iv) it can be carried out at macro-level as well as micro-level; and (v) it can also be used to purify an inorganic compound.

Solvent extraction is defined as the process by which a substance may be extracted from dilute solution (usually aqueous or inorganic solutions) into an immiscible solvent (usually organic solvent) with or without the use of a complexing agent. The organic solvents used for extraction may be polar like chloroform, nitrobenzene, tetrahydrofuran, butyl acetate, ethers, methyl isobutyl ketone etc., or non-polar like benzene, carbontetrachloride, petroleum ether, kerosene, xylene, *n*-hexane, etc.

Nerst distribution law governs the partition of a solute or substance say X between two immiscible solvents expressed by the equilibrium:

$$X_{aq} \rightleftharpoons H_{org}$$

The ratio of activities of X in organic (*org*) or aqueous (*aq*) phases will be constant and independent of the total concentration of X at a given temperature.

$$K = \frac{[X_{org}]}{[X_{aq}]}$$

K = partition coefficient which is independent of the total solute concentration in either of the phases

$[X_{org}]$ = activity of X in organic phase

$[X_{aq}]$ = activity of X in aqueous phase.

At constant temperature, the value of X is constant. This is applicable theoretically to ideal solutions only but in practice many substances follow the expression and activities may be replaced by molar concentration. Quite often, K is approximately equal to the ratio of the solubility of X in each solvent.

Generally, ionic substances are insoluble in non-polar solvents but soluble in polar solvents with high dielectric constant due to the solvation of ions. On the other hand, covalent compounds are mostly soluble in non-polar solvents. Therefore, the extraction of an ion from aqueous solution into organic phase can take place only if the ion forms a species showing preference for organic phase either by: (i) chelate formation; or (ii) solvate formation; or (iii) ion-pair formation.

Chelate formation

Chelating agents form an important class of organic bases which can interact with a metal atom/ion from more than one position in the ligand molecule in such a way that the metal atom/ion is bound in a stable 4 or 5 or 6 membered ring. Such complexes are called chelates and organic bases forming these complexes are called chelating agents. A widely used chelating agent is 8-hydroxyquinoline (C_9H_7NO). Most of its chelates are soluble in organic solvents.

A wide variety of chelating agents are in use for the separation of metal ions at micro as well as macro levels. For example, extraction of uranium with 8-hydroxyquinoline is done in $CHCl_2$ and that of Fe with cupferron is done in CCl_4 solvent.

Solvate formation

Extraction by this method is very clean and can be performed at microgram concentration of the metal ions as well as macrogram or pilot plant level. It involves solvation of the extracted species into the organic phase. There are several kinds of solvating solvents, for example, ether, ethylacetate, methylisobutyl ketone (MIBK), tributylphosphate ($C_4H_9O)_3PO$(TBP), tributylphosphine oxide $(C_4H_9)_3PO$ trioctylphosphine oxide $(C_8H_{17})_3PO$, etc. This method is frequently used for the extraction of transition metals and inner transition metals. For example, uranium can be conveniently separated from lead and thorium if the nitrate in water is extracted with etherial phase as a solvate $[UO_2(NO_3)_2.2H_2O.2Et_2O]$. Lead and thorium are left in aqueous phase. Uranium can be extracted on industrial scale using TBP as an extractant (for solvate formation) in kerosene phase. This method is based on the principle of solvation of the metallic species into the organic phase.

A number of covalent compounds like $ZnCl_2$, $HgCl_2$, $CeCl4$, OsO_4, RuO_4, etc., undergo easy extraction into organic phase of hydrocarbons, halohydrocarbons or ethers. The mechanism involves solvation of these substances into the organic phase.

Ion-pair formation

Another category of extractants involve extraction by non-pair formation. The extractants are mostly high molecular weight amines. As for example, tertiary amines like trioctylamine TOA ($C_6H_{17})_3N$, triiso-octylamine [OA (i-$C_8H_{17})_3N$ trioctylamine oxide $(C_8H_{17})_3NO$, etc. are best extractants for anionic complexes with mineral acids. A typical example showing mechanism of extraction of a solution of $FeCl_3$ in concentrated hydrochloric acid, $HFeCl_{14}$ is as follows :

$$Fe_{aq}^{3+} + 4Cl^-_{aq} \rightleftharpoons FeCl_4^-{}_{aq}$$

$$R_3NO_{org} + H_{aq}^+ + Cl^-_{aq} \rightleftharpoons R_3NH_{org}^+ Cl_{org}^-$$

$$R_3NH_{org}^+Cl_{org}^- + FeCl_4^-{}_{aq} \rightleftharpoons [R_3NH^+FeCl_4^-]_{org} + Cl_{org}^-$$

Ion-pairs involving big metallic cations or anions with ions of the opposite charges, such as, tetraphenylarsonium perrhenate $[C_6H_5)_4As]^+$ ReO_4^- or tetraphenylarsonium permanganate $[(C_6H_5)_4As]^+$ MnO_4^- undergo easy extraction with organic phase of chloroform.

Choice of a solvent

The following aspects must be looked into while considering the choice of a solvent or an extractant:

1. The solvent should be practically insoluble in the aqueous phase.
2. There should be significant difference in their densities and organic phase should preferably form the upper layer.
3. The organic solvent should have low toxicity and if possible low inflammability.
4. The dissolved metal ion in the organic phase should be easily recoverable.
5. The distribution ratio of the extractable metal ions should be high and that of impurities low.

Applications of solvent extraction method

Some examples involving the use of solvent extraction procedures for the separation of metal ions are given below:

1. Uranium can be conveniently separated from impurities of lead and thorium by the etherial extraction of aqueous solution of uranium (VI) (uranyl UO_2^{2+}) in saturated solution of NH_4NO_3 with 1.5 M concentration of HNO_3.
2. Extraction of plutonium and uranium and their separation from a solution of uranium fuel elements in a nuclear reactor is based on the process of extraction.

    Using oxidation-reduction procedures and extraction or precipitation, bulk of unwanted fission products are removed. The aqueous solution containing UO_2^{2+} and PuO_2^{2+} is then treated with an organic phase containing methylisobutyl ketone (MIBK). Both UO_2^{2+} and PuO_2^{2+} get extracted into organic phase which on washing with aqueous SO_2 affects their separation. Plutonium (VI) is reduced by SO_2 to Pu^{4+} which goes into aqueous phase, whereas UO_2^{2+} remains unaffected in the organic phase.

    In another method, UO_2^{2+} and Pu^{4+} are extracted into an organic phase of kerosene containing 30% TBP (tributylphosphate) as complexing agent from an aqueous solution of $6NHNO_3$. The organic phase containing UO_2^{2+} and PU^{4+} is then reduced with aqueous SO_2, when PU^{4+} gets changed to Pu^{3+} and goes into aqueous phase and UO_2^{2+} remain in organic phase.
3. Extraction and separation of a mixture of Zr (IV) and Hf (IV) or a mixture of Nb (V) and Ta (V) is based on the extraction by solvation of their halide from the acidic aqueous medium into an organic phase using MIBK as extractant.
4. Absolute alcohol is used for the separation of calcium from a mixture of Ca, Ba and Sr nitrates. Calcium nitrate, $Ca(NO_3)_2$ forms a solvated complex with alcohol and goes into solution whereas others (Sr and Ba nitrates) remain insoluble.

ADSORPTION

Just as the particles on the surface of a liquid experience inward pull resulting in surface tension, in the same way, particles on the surface of a solid experience inward pull because of unbalanced attractive interaction with other particles which surround them only on one side and not all sides. Moreover, when a solid is broken, some inter-atomic bonds break. As a result, valencies of the atoms at the surface remain unsatisfied. Consequently, the surface of a solid has a tendency to attract and hold molecules of a gas, a liquid or a dissolved solute. In other words, there is greater concentration of a substance (which comes in its contact) on the surface of solid than in the bulk. This is known as adsorption. Therefore, adsorption is defined as the phenomenon of higher concentration of any molecular species at the surface than in the bulk of a solid (or liquid). For example, if a piece of coconut charcoal is introduced into a gas jar containing ammonia, after a few minutes it is found that, there is no smell of ammonia in the jar indicating that the whole of ammonia has been taken up by charcoal.

The substance on whose surface absorption takes place is called adsorbent. In the above example, charcoal is the adsorbent. The substance which gets adsorbed on the adsorbent is called adsorbate. In the above example, ammonia is the adsorbate.

It has been established that non-polar substances like nitrogen get adsorbed on the surface of a solid through van der Waals' forces (known as physical adsorption) while polar substances like carbon monoxide may get adsorbed on the surface of a solid through chemical interaction known as chemical absorption or chemisorption. The phenomenon of adsorption differs from absorption in the following respects:

1. In adsortion, there is increased concentration of the particles of a substance on the surface while in adsorption the particles actually penetrate into the entire bulk of the absorbing substance.

2. Adsorption is rapid in the beginning but slows down with the passage of time while absorption occurs at a uniform rate.

Some of the important characteristics of adsorption are as follows :

1. The process of adsorption is selective and specific depending upon the nature of adsorbate and adsorbent.

2. Adsorption of gases on the surface of solids is a rapid and reversible process.

3. Adsorption of gases on solid increases with increase in the pressure of the gas. Similarly, in case of solutions, the adsorption on a solid increases with increase in the concentration of the solute in the solution.

4. Since adsorption is an exothermic process, according to Le Chateliers' Principle, adsorption decreases with increase in temperature.

Mechanism of Adsorption

Langmuir put forth certain views about the mechanism of adsorption. On the basis of his views, he also derived a mathematical equation known as Langmuir's adsorption isotherm which could explain the phenomenon of adsorption of gases on solids in a fairly satisfactory manner. To derive Langmuir adsorption isotherm, Langmuir made the following assumptions:

1. A gas molecule on striking the surface of a solid gets condensed there known as inelastic collision. After some time, the molecule evaporates off; the time lag between condensation and evaporation causes adsorption.

2. An equilibrium always exists between the adsorbed and the unadsorbed gas.

3. The adsorbed layer of the gaseous molecules on the surface of the adsorbent is normally one molecule thick.

Langmuir derived the well-known equation known as Langmuir adsorption isotherm which is as follows:

$$\frac{x}{m} = \frac{k_3 p}{1 + k_1 p}$$

where x is amount of the gas adsorbed, m is the mass of the adsorbent, k_1 and k_3 are constants, characteristics of the system, and p is the pressure of the gas. Langmuir's Adsorption Isotherm has been verified at various pressures and has been found to hold good.

Langmuir's adsortion equation for solution is obtained by substituting concentration C for pressure. Then we have,

$$\frac{x}{m} = \frac{k_3 C}{1 + k_1 C}$$

Applications of Adsorption

The phenomenon of adsorption is extensively employed in the industrial and laboratory processes. Some important applications of adsorption are as follows:

Ion exchange resins

These are either inorganic polymers containing silicate groups or organic polymers containing —COOH, —SO$_3$H, —NH$_2$ or >NH groups which have the property of selective adsorption of ions from a solution. They are used in (i) softening of water on an industrial scale, and (ii) separating a mixture of rare-earths into individual components.

Adsorption indicators

In titrations involving precipitation reactions, dyes such as eosin, fluorescein are used as indicators and are known as adsorption indicators. The working of these indicators is based on their preferential adsorption by the precipitates.

Chromatography

All chromatographic techniques are based on the selective adsorption of different substances by an adsorbent.

Miscellaneous applications

The phenomenon of adsorption finds applications in decolourisation of oils; sugar juice; gas masks; dyeing of cloth; in detergents and dehumidisers.

Chromatography Method

Tswettt described a new technique known as chromatography to separate a mixture of coloured substances into individual components. The original technique has been modified and extended so that it has been used to separate a mixture of coloured as well as colourless substances into individual components. Moreover, this technique can also be used to test the purity of a substance.

Principle of chromatography

Chromatography is based upon the selective removal of the components of one phase from it as it flows through another phase which remains stationary. The selective removal itself may be due to either of the following two processes:

1. *Adsorption:* In this process, the components of a mixture in the liquid or gas phase undergo selective adsorption to the surface solid phase.
2. *Partitioning:* In this process, the components of a mixture undergo selective dissolution in immiscible solvents in which they have different relative solubilities.

Types of chromatography

The main types of chromatography are as follows: (i) column (or adsorption) chromatography; (ii) paper chromatography; (iii) thin layer chromatography; (iv) ion exchange chromatography; (v) gas chromatography; and (vi) high performance liquid chromatography (HPLC). Let us discuss briefly the various types of chromatographic techniques.

Column (or adsorption) chromatography

This technique is used to separate non-polar substances and constituents of low volatility.

Principle of adsorption chromatography

The various constituents of the mixture get adsorbed to different extents on the fixed solid phase such as alumina or silica gel which is packed as a long column. The constituents get separated into distinct bands which are removed with the help of a solvent. The extracted component along with the solvent is known as elute. From the elute, the solvent is removed to get the individual component.

It is believed that there is always a competition between solute and solvent molecules for the adsorption sites. Consequently, a dynamic equilibrium gets established at the interface where the solute and solvent molecules get attracted and settled at the solid surface for a short while and then leave the solid surface to re-enter the mobile liquid phase. However, the desorbed molecules which have greater affinity for the solid surface flow into the mobile layer slowly while those desorbed molecules which have lesser affinity for the solid surface flow into the mobile layer quickly.

Characteristics of good adsorbent

1. It should possess high and selective adsorption power.
2. It should be uniform in size.
3. It should be finely divided to offer greater surface area for adsorption.
4. It should be pure as impurities cause irreversible adsorption.
5. It should be chemically inert.

Commonly used adsorbents are activated alumina, activated charcoal, magnesium oxide, silica gel, starch and fuller's earth.

As a result of experimental investigations, it has been found out that greater the number of double bonds and hydroxyl groups present in a molecule, greater is the adsorption. Thus, we have adsorption series as:

Acids, bases > alcohols > ketones, aldehydes > unsaturated hydrocarbons > saturated hydrocarbons.

Adsorption chromatography is mainly used to separate a mixture of organic compounds such as a mixture of ortho and paranitroanilines; a mixture of blue and red dyes.

Paper chromatography

Mechanism of paper chromatography

This technique is based upon a mechanism which is partly adsorption and partly partition. A mixture of the solute is placed on a strip of chromatographic paper such as Whatmann No. 1 or 3 and a solvent is allowed to move along the paper strip. The solvent extracts the solute because of distribution of the solute between the two solvents.

Types of paper chromatography

There are three main types of paper chromatgraphic techniques as

1. Ascending paper chromatography
2. Descending paper chromatography
3. Radical paper chromatography

Advantages of paper chromatography

Its main advantages are as follows:

1. The procedure is very simple and precise.
2. It is a reasonably rapid process.
3. Only small quantities of the material are required.
4. No costly apparatus as well as reagents are required.

Thin layer chromatography (TLC)

Although paper chromatography is a good technique for separation of a mixture yet it suffers from the following defects:

1. Quite often, the sample spots tend to spread out on the filter paper as the fibres of the filter-paper are quite coarse.
2. It is not possible to remove the spot from the filter-paper either for identification or quantitative analysis.
3. It works only for those substances which are separated on cellulose.

To overcome the above defects, a new technique known as thin layer chromatography has been developed. In this technique, a solid adsorbent such as alumina or silica gel is mixed with a little binder such as hydrated calcium sulphate or starch. It is then spread as a thin layer on a glass plate or plastic sheet. The plate thus formed is known as chromatoplate. By means of a fine capillary, solution of the sample to be analysed is applied on chromatoplate and then dried. Then it is allowed to stand in jar containing solvent covered with lid. After some time, spots are located either through naked eye or by applying suitable reagent.

Mechanism of thin layer chromatography

Its mechanism is the same as that of paper chromatography.

Advantages of thin layer chromatography

1. It is a very rapid process as compared to paper chromatography.
2. It is used for large number of sample separations.
3. In this method sharp spots are obtained while in paper chromatography diffused spots are formed.
4. In this method, even a solution containing acid or alkali can be separated whereas paper chromatograph cannot be used for such solutions.

Applications of thin layer chromatography

Some important applications of thin layer chromatography are as follows:

1. Quantitative and qualitative analysis of organic and inorganic compounds.
2. Checking impurities in a solvent.
3. Separation of plasticisers, inks, dyes and anti-oxidants.

Ion Exchange Chromatography

Essentially, this technique of chromatography is similar to column or thin layer chromatography. However, the adsorbent is either a cellulose derivative or a synthetic ion-exchange resin. For example, if a sample solution containing electrolyte and non-electrolyte is allowed to pass through a column of ion-exchange resin, the electrolyte flows down the column while non-electrolyte particles diffuse into the resin particles. Therefore, electrolyte will appear first in the effluent. Thereafter, the water is made to flow down the column of resin when non-electrolyte also gets detected in the effluent. In this way, it is possible to separate ionic from non-ionic substances. For example, $NaCl$ and ethanol; $NaCl$ and ethylene glycol; HCl and CH_3COOH can be separated by this technique.

Gas Chromatography

Essentially, this technique is similar to liquid-liquid chromatography except that a mobile liquid phase is replaced by a moving gas phase.

Types of gas chromatography

There are two types of gas chromatographic techniques as:

1. *Gas-liquid chromatography:* In this type, the separation is effected by partitioning the sample between a mobile gas phase and a thin layer of non-volatile liquid coated on a solid support.
2. *Gas-solid chromatography:* In this type, the separation is effected by partitioning the sample between a mobile gas phase and solid of large surface area. For example, separation of CO_3 from H_2, O_2 N_2, CO or C_2H_2 using silica gel column.

In this method, the vapours of the sample are allowed to enter the column inlet. The solutes get adsorbed at the head of the column. The separated components are detected by special detectors such as thermal conductivity, flame ionisation or electron capture detectors.

High Performance Liquid Chromatography (HPLC)

So far we have described adsorption, partition and ion exchange chromatography which are examples of liquid column chromatography. The classical liquid column chromatography involved the use of a long (0.5 to 5 m) glass tube with diameter $12–50 \times 10^{-3}$ m containing large sized particles of the solid having diameter $150–200 \times 10^{-6}$ m which acts as a stationary phase. The mobile liquid phase moves down the column on account of its own pressure leading to separation of components. But the rate of flow was very low and so the process of separation took a very long time. The use of vacuum and by pumping did not help to increase significantly the rate of flow. Later experimental studies revealed that the rate of flow did increase by: (i) increasing the height of the column; and (ii) decreasing particle size of the packings.

In view of the above facts, an improved chromatographic method has been developed which is known as high performance liquid chromatography (HPLC). The diagram indicating the layout of the equipment employed in HPLC is shown in Fig. 24.1. Let us briefly discuss each component of the apparatus.

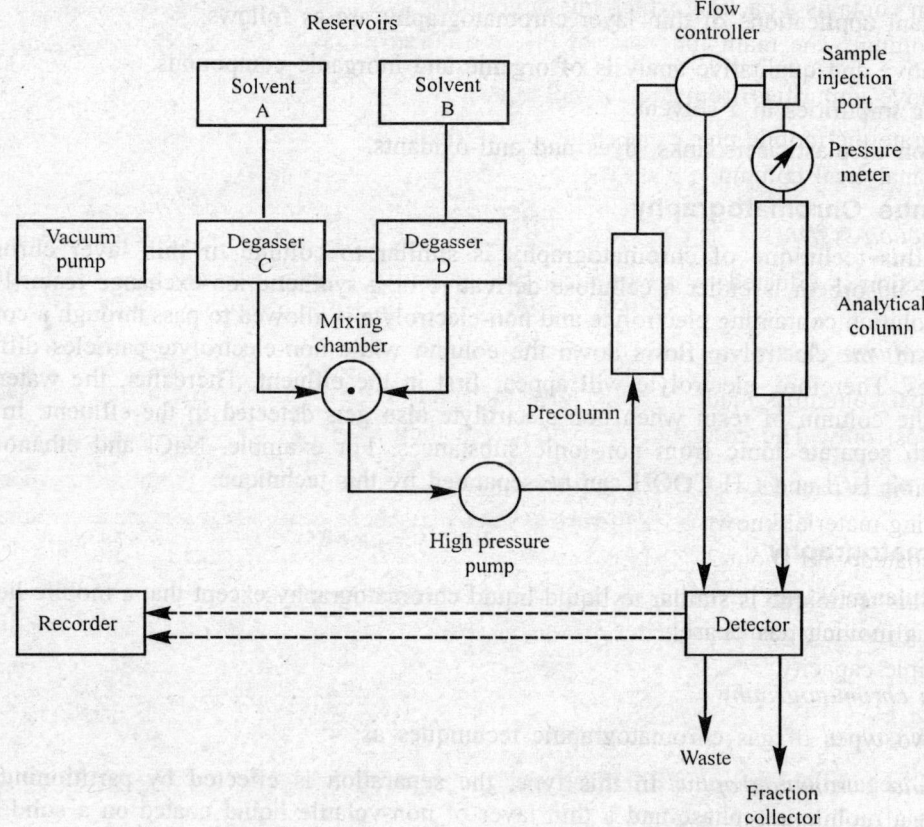

Fig. 24.1. A typical high performance liquid chromatography unit.

Solvent reservoir and degassing system

There are one or more reservoirs (A and B) which can hold about 2 litres of a solvent. A separation which is carried out using a simple solvent is called isocratic elution and is not very efficient. Therefore, two or more solvents having different polarities are used instead of a simple solvent for more efficient separation. The pressure of dissolved gases such as nitrogen or oxygen in the solvents form bubbles in the column and the detector. These bubbles cause band spreading and interfere with the working of the detector. Therefore, the solvents have to be degassed (i.e., gases have to be removed). Degassers (C and D) are either vacuum pumping system or a distillation system. The degassed solvents from reservoirs are led into a mixing chamber at varying rates.

Pumps

In HPLC, screw-driven, reciprocating or pneumatic pumps are used which have output of 4000 to 6000 psi with a flow delivery rate of at least 3 ml min^{-1}. Although high pressures are generated by these pumping devices, yet they are not dangerous because liquids are not very easily compressible.

Precolumns

A precolumn contains a packing which has larger particle size but is chemically identical with that in the analytical column. The main functions of the precolumn are:

1. To remove impurities from the solvent and thus prevent contamination of the analytical column.
2. To saturate the mobile phase and thus prevent the stripping of the stationary phase from the packing of the analytical column.

Sample injection system

Sample injection is effected by means of a syringe and self-sealing septum of silicone or teflon.

Analytical column

The analytical column is made of either glass or stainless steel. Glass column can be used for pressure below 600 psi only. The column is 0.15 to 1.5 m long with internal diameter of 2 to 3×10^{-3} m.

The column is packed with finely divided silica gel, alumina or celite (diatomaceous earth). Recently, a new packing material known as *pellicular particles* has been developed. It consists of small glass beads which are coated with about 2×10^{-6} layer of porous material such as silica gel, alumina or ion exchange resin. Pellicular packings have the advantage that the rate at which equilibrium is established between the phases is high which results in better efficiency. However, pellicular packings have the disadvantage of limited sample capacity.

Temperature control

Most HPLC separations are carried out at room temperature. However, if temperature control is desired, water-jacketed columns are used.

Detectors

Depending upon the nature of the sample, a detector is used. Most commonly used detectors are ultraviolet, visible, infra-red absorption detectors. Even mass spectrometry has been employed as a sensitive detector.

Recorders

The chromatogram is recorded with potentiometric recorders in conjunction with the particular detector.

Applications of HPLC

1. It is used in separation of components in pharmaceutical and pesticide industries.
2. In conjunction with ion exchange, HPLC has been used to separate nucleic acids and vitamins.
3. It has been used to determine the molecular masses of polymers and biochemicals.
4. HPLC is particularly suitable for non-volatile substances (including inorganic ions) and thermally unstable materials.

ELECTROPHORESIS

It is a well established fact that the particles of colloids bear either positive or negative charge due to the presence of adsorbed ions. Therefore, on passing electric current through a colloidal solution, the charged colloidal particles migrate to the oppositely charged electrodes which is known as electrophoresis. Hence, electrophoresis was defined as the migration of colloidal particles through the solution under influence of an electrical field. Later on, it was found that the process of electrophoresis can be applied to colloidal aggregates as well as monodispersed ions. Methods based on electrophoresis provide powerful tools in the hands of analytical chemists to separate components of a wide variety of biological materials such as proteins, gastric juices, nucleic acids and vitamins.

Types of Electrophoretic Methods

Methods employed to separate a mixture into its components based on the property electrophoresis are called electrophoretic methods. These methods are of two types as described below:

Free Solution Method

In this method, the sample solution is introduced as a band at the bottom of a U-tube having electrodes near the ends of the tube which is filled with a buffered solution. On passing electric current, the charged particles start moving towards one or the other electrode. However, the rates of migration of different particles of the mixture are different:

1. Different particles have different charge to mass ratios.
2. Different particles have different inherent mobilities in the medium.

Consequently, different species get deposited at the electrodes at different intervals of time and at different places leading to separation of the mixtures.

Drawbacks of free solution method

Although this method played an important role in the development of biochemistry yet it suffers from the following defects:

1. The separated components have a tendency to get mixed by convection currents.
2. The bands of separated species can be detected only by using elaborate optical systems.

Stabilising Medium Method

The various drawbacks associated with free solution method get removed if the process of electrophoresis is carried out in a stabilising medium such as paper, a layer of finely divided solid or a column packed with a suitable solid. Consequently, components of a mixture get separated by electrophoresis in combination with adsorption or ion exchange method. Depending upon the type of stabilising medium, different types of electrophoretic methods are known such as electrochromatography, zone electrophoresis, electromigration and ionophoresis. Of all these methods, the most common method is known as electrophoresis or electrochromatography which involves the migration of changed solute particles through the influence of electrical field. It has been found that the migration of charged particles in paper depends on the following factors:

1. Nature and magnitude of charge on the solute particles.
2. Concentration of the electrolyte.
3. pH value of the medium.
4. Temperature of the medium.
5. Viscosity of the medium.
6. Adsorption capacity of solute particles.
7. Voltage applied.
8. Mobility of the ions in the opposite directions.

The distance which a charged particle travels under the influence of electric field is given by the relationship:

$$D = \frac{Ute}{qk}$$

where D is the distance travelled;

U is the mobility of ion at time t;

e is current applied;

q is the cross-sectional area; and

k is the conductivity.

Experimental Methods of Electrochromatography

There are numerous methods for carrying electrochromatography but we shall discuss here only three simple and common methods of electrochromatography. These are as follows:

Horizontal Strip Method

A strip of filter paper is stretched horizontally between two containers. A and B which are filled with a buffer solution. The filter paper is well soaked with the buffer solution. To check evaporation of the buffer solution, the apparatus is housed in an air-tight container (Fig. 24.2).

The sample is placed at the centre of the strip and a direct current of 100 to 300 volts is applied across the two electrodes. After some time, the filter paper is removed, dried and bands of components are detected by colorimetric reagents.

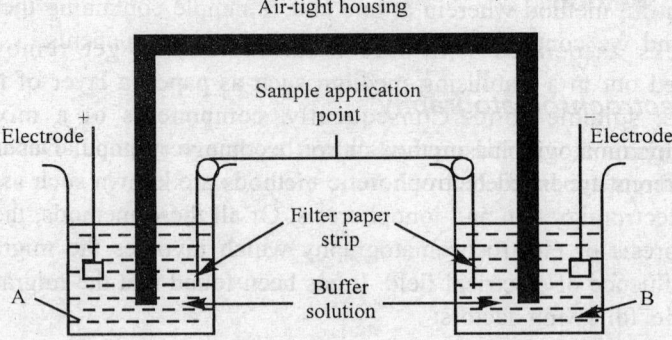

Fig. 24.2. Horizontal strip method.

Inverted V-Strip Method

This method is quite similar to the horizontal strip method. However, the filter paper strip is in the form of an inverted V (Fig. 24.3).

The sample under test is placed at the apex of the V. After electrolysis, the cationic species move down one arm of V while anionic species move down the other arm of V. Thereafter, each species is separately detected.

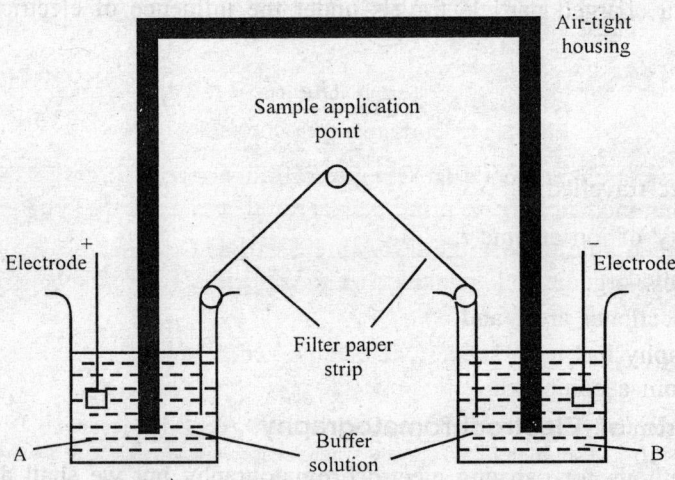

Fig. 24.3. Inverted V-strip method.

Curtain electrochromatography

In this method, a curtain of filter paper immersed in a buffer solution is held vertically (Fig. 24.4). The sample is continuously applied on the filter-paper from the sample reservoir. One end of the filter paper is in fluted form.

On passing electric current, separation of components occurs and the separated components can be detected by special reagents. Alternatively the separated components are allowed to fall in different sample collection tubes.

Thus, this is a beautiful method wherein at one end, a sample containing the mixture is continuously fed and at the other end we continue to collect the individual components.

Applications of electrochromatography

In recent years, electrochromatographic methods have become very popular as analytical tools. Some of their important applications are listed below:

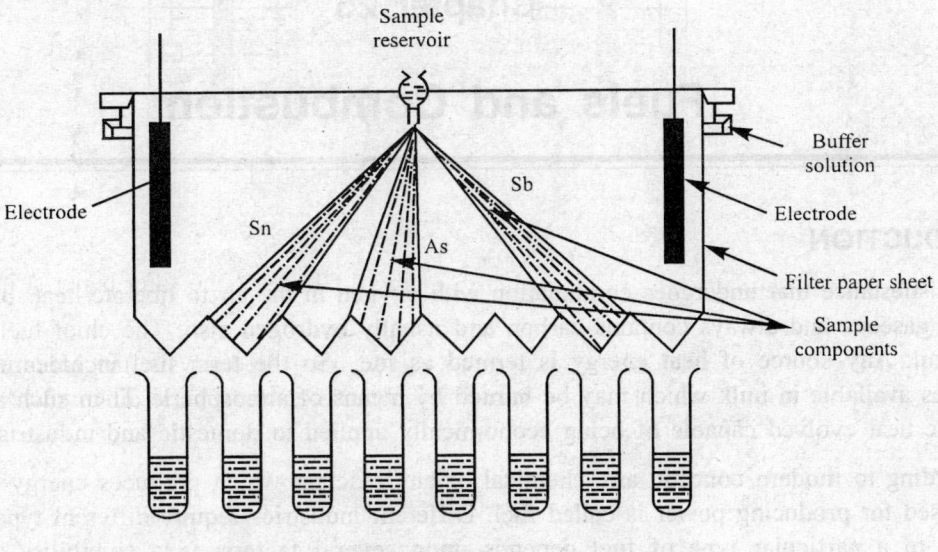

Fig. 24.4. Curtain electrochromatography.

1. For clinical diagnosis, a clinical chemist separates through electrochromatography proteins and other large molecules contained in serum, urine, spinal fluid, gastric juices and other body fluids.
2. Biochemists have been able to separate through electrochromatography alkaloids, antibiotics, amino acids, carbohydrates, organic acids, natural pigments, nucleic acids, steroids and vitamins from natural sources.
3. Electrochromatography has been successfully employed to separate a mixture containing a number of metallic ions from a complexing medium.

Example 24.1. The molar absorptivity or molar extension coefficient of a coloured compound is 3.20×10^3 at 240 nm. Calculate the absorbance 'A' of a 5.0×10^{-5} M solution in 5.0 cm cell when measured at wave length 240 nm.

Solution.

$A = \varepsilon.c.l.$

$A = 3.20 \times 10^3 \times 5.0 \times 10^{-5} \text{ M} \times 5.0 \text{ cm.} = 0.800$

When concentration is moles per litre and path length is in centimetres, the constant 'ε' is called molar absorptivity.

Example 24.2. Calculate the optical path length of a coloured complex which has the molar absorptivity of 1480 and shows an absorbance of 0.750 when its concentration is 1.4×10^{-3} M.

Solution.

$A = \varepsilon.c.l.$

$$\therefore \quad l = \frac{A}{\varepsilon.c} = \frac{0.750}{1480 \times 1.4 \times 10^{-3}} = 0.36 \text{ cm.}$$

Chapter 25

Fuels and Combustion

INTRODUCTION

Fuel is a substance that undergoes combination with oxygen in the air to liberate heat. It may be solid, liquid or gaseous and always contains carbon and usually hydrogen also. The chief fuels are coal and mineral oil. Any source of heat energy is termed as fuel. So the term fuel includes all combustible substances available in bulk which may be burned by means of atmospheric air in such a manner as to render the heat evolved capable of being economically applied to domestic and industrial purposes.

According to modern concept, any chemical or any reactant which produces energy in a form that can be used for producing power is called fuel. Different industries require different types of fuel. The selection of a particular type of fuel depends upon several factors, viz., suitability to the process concerned, supply position, cost and cleanliness of the fuel.

MODERN CONCEPT OF FUELS

It is now well established that combustion is not necessary for a fuel to give out energy. For example, electrical energy when used as a source of heat is also called fuel. Nuclear energy may also be used as a source of heat. Chemical compounds, such as power alcohol, hydrazine, ammonia etc. are used as fuels.

According to modern definition, a fuel is any fissionable material, chemical or reactant which produces energy in a form that can be used for producing power. For example, fuel cell is a battery which uses reactants or chemicals that can be very easily replaced and some times continuously. These reactants are called fuels which are generally methane, propane and kerosene which are oxidised by air in the cells.

CLASSIFICATION OF FUELS

Fuels are classified into: (i) natural or raw fuels; and (ii) manufactured or processed fuels.

Natural fuels

1. Solid e.g., wood, peat, lignite, bituminous and anthracite coal.
2. Liquid, e.g., petroleum.

3.   Gaseous, e.g., Natural gas (from petroleum wells).

Manufactured or processed fuels

1.   Solid, e.g., charcoal, high and low temperature coke.
2.   Liquid, i.e., petrol or gasoline, kerosene, diesel oil, benzol, methanol, ethanol, etc.
3.   Gaseous, i.e., coal gas, cokeoven gas, producer gas, water gas, blast furnace gas, oil gas, etc.

Fuels are further classified as: (i) primary fuels; and (ii) secondary fuels.

Primary fuel

Primary fuel is a fuel that is directly used for the function of heat and its technical utilisation as such. It may be solid, liquid, or gaseous e.g., coal, wood, petroleum.

Secondary fuel

A secondary fuel is one from which is manufactured other fuels which is then utilised as a fuel. Hence, a secondary fuel is obtained from a primary fuel. For example, coal gas, water gas and producer gas are secondary fuels obtained from the primary fuel coal.

Calorific Value

Calorific value may be defined as the amount of heat liberated in calories by the complete combustion of a combustible material with oxygen and the condensation of the products to the desired temperature. Consider the combustion of carbon and hydrogen, which may be represented by the following equations.

$$C \quad + \quad O_2 \quad \rightarrow \quad CO \quad + \quad 97{,}644 \text{ cals}$$
$$\text{12 gms} \qquad \text{32 gms} \qquad \text{44 gms}$$
$$H_2 \quad + \quad O \quad \rightarrow \quad H_2O \quad + \quad 69{,}000 \text{ cals}$$

Since 12 gms of carbon give 97644 cals of heat, 1 gm. of carbon would give 97644/12 = 8137 cals of heat. Similarly 2 gms of hydrogen give 69000 cals of heat 1 gm of hydrogen would give 69000/2 cals = 34500 cals of heat, when steam is condensed to water.

A British Thermal Unit (B.T.U.) represents the amount of heat required to raise the temperature of one pound of water through 1°F (from 60°F to 61°F). Heat energy is frequently measured in calories, a calorie being the amount of heat required to raise the temperature of 1 gm of water through 1°C (from 15°C to 16°C). A B.T.U. is equal to 252 calories.

The mean calorific value of any fuel containing carbon and hydrogen can be calculated as :

$$\text{Total calories} = \frac{(\% \text{ carbon} \times 8137) + (\% \text{ hydrogen} \times 34500)}{100}$$

If a fuel contains oxygen in combination, a small amount of this gas will be utilised for their combustion. Hence less heat will be produced in proportion.

Criterion of Selection of Fuel

The following factors are of prime importance while selecting a fuel for a particular type of purpose.

1.   *Suitability to process:* For example, for foundry fuel, coke made from bituminous coal is the best, although charcoal or anthracite may also be used. Solid, liquid, gaseous fuels or even electricity, may be used in boilers for generating steam.

2. *Supply position:* Supply position with regard to availability in sufficient quantity should be considered.

3. *Cost of fuel:* The fuel should have a low cost and should not change its properties substantially on long storage. The cost of fuel is governed by various factors such as cost of fuel per unit of heat value, efficiency of utilisation, cost of equipment, labour and convenience, refuse handling, auxiliary power. When the use of a particular fuel does not involve any dirt or dust nuisance, the fuel is considered to be clean.

4. *Calorific value:* The fuel must give on combustion an appreciable amount of heat per unit mass or volume and produce no harmful gases which might pollute the atmosphere and affect the structural materials of the furnace or drier.

The most important requirements of a fuel are thus: (i) reasonably high velocity of combustion; (ii) release of thermal energy of reasonably high value; (iii) proper ignition temperature, (iv) low percentage of non-combustibles; and (v) low cost.

Characteristics of a Good Fuel

1. A fuel should have high calorific value.
2. It should have low ignition temperature.
3. It should have low moisture content.
4. During combustion it should not produce much smoke or poisonous gases.
5. After combustion it should not leave much ash.
6. It should be easy for handling and transportation.

Properties of Fuels

1. *High calorific value:* The calorific value of a fuel is the number of units of heat evolved during complete combustion of unit weight of the fuel. In case of liquids and solids it is usually expressed in BTU per pound or calories per gram. 1 calorie per gram = 1.8 BTU per lb. For gaseous fuels the heating value is expressed as BTU per cubic feet at 60°F and 30 inch - Hg.

The grade of solid fuel is expressed in BTU. The various factors that affect BTU and hence the quality of the solid fuel are:

(a) Fixed percentage of carbon which indicates the organic matter that does not undergo volatilisation when a known weight (1 gm) of the powdered solid fuel is combusted at a fixed temperature (950° ± 20°C) for a fixed period (generally 7 minutes).

(b) Moisture content of the fuel, which is determined from the loss in weight during heating 1 gm of the powdered fuel at 100°C for 60 minutes.

(c) Percentage of volatile matter, which expresses the organic and inorganic matters (e.g. NH_3) obtainable from the thermal decomposition of the fuel.

2. *Composition of fuel:* Another important characteristic of the fuel is the composition of a dry fuel i.e., of a fuel dried up to a constant mass. It is used in all thermal calculations.

3. All fuels are characterised by the specific heat of combustion or heating (calorific) value. The heating value of a particular kind of fuel may vary within a wide range, depending on the moisture and ash

content and other properties. It can be related to the organic, combustible, dry or working mass of a fuel. The moisture that forms on combustion can be present in the combustion products as liquid or vapour.

The specific heat of combustion of a fuel can be determined by burning a small amount of fuel, measuring the heat liberated on combustion, and recalculating it per kilogram in case of solid and liquid fuels and per cubic metre in case of gaseous fuels. It is also possible to calculate it on the basis of chemical composition of the fuel.

4. *Ignition temperature:* The temperature at which a solid, liquid or gaseous fuel catches fire and continues to burn without further heat is called ignition temperature which has a definite value for a given fuel.

5. *Flame temperature:* The maximum temperature to which an object can be heated by a flame is called flame temperature. This is very important property of a gaseous fuel. However, it can not be determined with accuracy. So a quantity known as theoretical flame temperature has been proposed which is regarded of the same order as the actual flame temperature. The flame temperature generally increases with the increase in the number of combustibles and increase in the proportion of heavier hydrocarbons.

6. *Explosive range:* The limiting composition of a gas-air mixture beyond which the mixture will not ignite and continues to burn is called explosion limit or limit of inflammability of a gaseous fuel or an inflammable vapour. For every gaseous fuel there exists an upper as well as a lower limit of flammability, up to atmospheric pressure, the explosion range increases with pressure, but above atmospheric pressure, the explosion range has been found to decrease with increase in pressure, in case of a number of gaseous fuels. The explosion range generally has been found to increase with increase in temperature.

7. *Flash point:* This is the property of liquid fuel and defined as the minimum temperature at which a sample of a fuel oil gives off enough vapour that catches fire but does not continue to burn, in presence of air, by a naked flame or electric discharge. The flash point depends upon vapour pressure of the oil and the manner in which it is determined. The higher the volatility the smaller the flash point. The flash points of kerosene and gasoline are 70°F and 45°F respectively.

8. *Fire point:* The temperature at which the oil vapours will catch fire and continue to burn is called fire point. Both flash point and fire point are used to detect dissolved impurities, much more volatile than the bulk of the oil.

9. *Aniline point:* The temperature at which equal volumes of a given sample of diesel fuel and aniline cease to be miscible in each other so as to give homogeneous solution is called aniline point. Aniline point to close to 70°C for high speed fuels. In case of diesel, the ignition quality index and aniline point is related according to the expression,

$$\text{Aniline point} = (F°) = \frac{\text{Diesel index} \times 100}{\text{A.P.I. Sp. gravity}}$$

10. *Knocking:* It is the property of the liquid fuel.

11. *Specific gravity:* It is also an important property of liquid fuel and used for comparing different grades of different samples of oils used for a specific purpose. In petroleum industry, the specific gravities are usually expressed in A.P.I degrees or in Baume scale. Degree of A.P.I is expressed as,

$$\text{Degree of A.P.I.} = \frac{141.5}{\text{Sp. gravity} / 60°\text{F}} - 131.5$$

As the specific gravity decreases, API gravity increases. Baume scale is generally used for those liquids which are heavier than water and expressed as,

$$(^0\text{Be}') = 145 - \frac{145}{\text{Sp. gravity} / 4°\text{C}}$$

For liquids lighter than water, it is expressed as

$$(^0\text{Be}') = \frac{140}{\text{Sp. gravity} / 4°\text{C}}$$

12. *Pour point:* This is an important property of lubricating oils and defined as the temperature at which oil ceases to flow when cooled under standard conditions.

13. *Cloud point:* It is an index of dissolved wax concentration and may be defined as the temperature at which the cloudiness appears in a sample of oil due to separation of wax under standard conditions.

14. *Coke number:* Different lubricating oils used in the moving parts of an internal combustion engine produce free carbon on thermal decomposition in varying amounts. The amount of carbon deposited by a weighed sample of oil in a standard apparatus under a standard set of experimental conditions is called coke number. Better grade lubricating oils are those which have lower coke numbers.

15. *Viscosity:* This is the property of lubricating oils. The unit is centipoise. In USA, Saybolt Universal Viscosity (SUV) is generally employed in petroleum industry and is expressed as the number of seconds required for 60 cc of the oil to run through a standard orifice from a cylinder filled to a definite level at a standard temperature.

SOLID FUELS

The main solid fuels include wood, peat, lignite, coal, charcoal and briquetted fuels. In addition to these, certain agricultural and industrial wastes such as bagasse, spent tan, rice husk, coconut and nut shells are also employed as fuels.

Wood

Wood has been the main source of fuel until recent times on account of its relatively rapid growth and production and ease of obtaining the supplies. However, large scale deforestation and the increasingly large demands of energy by the industries led to the more extensive use of other types of fuels. Freshly cut wood possesses greater water content (25 to 50%) than dry wood (15%) and its heating value is directly proportional to the water content. Wood mainly consists of cellulose, ligno-cellulose as well as some cell sap associated with traces of mineral ash. On the dry and ash-free basis, the average composition of wood is 50% C, 43% O, 6% H and 1% nitrogenous and resinous material. As a fuel, dry wood is very combustible, easily handled and burns with a long non-smoky flame. It gives maximum heat intensity very quickly. However, the calorific value of dry wood is only 19.7 to 21.3 MJ/Kg (4710 to 5085 Kcal/kg). The ash content of all wood is very low and lies in the range 0.3 and 0.6%. Wood is largely used as domestic fuel. It is rarely used in industry except for special purposes where dirt and smoke are undesirable. Due to its high flame emissivity, it is preferentially used for space heating.

Wood charcoal is obtained by the destructive distillation of wood. The carbonisation is performed usually in closed retorts. The charcoal is not pure carbon because even when the carbonisation is conducted at high temperatures, it rapidly absorbs some gas and moisture. It also contains some inorganic residues derived from the wood. Charcoal was widely used for metallurgical operations formerly, but it has now been replaced by coke excepting for some special applications. The major use of wood charcoal today is for producing activated carbon which finds extensive application for decolourisation (e.g., in sugar industry), adsorption of gases and vapours and recovery of solvents from gases and air. Charcoal is also used in the production of CaC_2, ferro-alloys, and special quality pig iron in small furnaces.

Advantages of wood

1. It burns readily with a bright cheerful flame, due to which it can be used in starting the burning of other fuels.
2. The amount of ash and soot produced after burning the wood is very small. Hence wood can also be used with great advantage in some industries such as melting of glass, firing of porcelain etc.
3. Easily available.
4. The ash being rich in potassium derivatives, has been used in agriculture. The ash from some selected varieties of wood have good detergent properties.
5. It has reasonably high calorific value (6500–9200 BTU per lb) suitable for meeting the domestic requirement.

Disadvantages of wood

1. It has high percentage of moisture present in it. Presence of about 15–20% moisture even in properly dried wood decreases its calorific value considerably and so it is very rarely used in Industry. When wood is burned, water is converted into steam and evaporation process absorbs large amount of heat from the fuel giving it a low calorific value.
2. The ash being very light accompany the fuel gas generated from wood.

Peat

In the peat deposits of the world there exists an enormous potential supply of fuel. Originally it was used as a domestic fuel but in recent years it has been used on a large scale for power production. Peat is a compact fibrous substance, either brown or black and contains large quantities of water and inorganic matter. Since it contains a large proportion of water, even when dry, it burns only very slowly and the heating value or calorific value of this fuel is very low. Important deposits of peat occur in England, Scotland, Ireland, Northern Europe, Russia, Sweden and Germany. In India peat deposits occur mainly in the west coast of Tamil Nadu, in Palani and Nilgiri Hills, and in Kashmir state.

Peat is product of anaerobic decomposition of various vegetable remains under moist conditions. Peat is the first stage of coal formation from cellulose. It is neither wood nor coal in the proper sense, but something intermediate between the two, which is formed by gradual decaying of vegetable matter in moist places through an incomplete transformation of plant to coal. Peat has a brown or black look and compared to coal, has large amount of water. In dry conditions, it has the composition,

$$C = 57\%, \; H_2 = 6.1\% \text{ and } O = 34.9\%$$

General composition of peat is: moisture = about 85% which is reduced on drying, volatile matter = 10.4%, fixed carbon = 4.6%. The composition of organic matter present in peat depends on the degree of decomposition. Greater the decomposition, greater the carbon and lesser the oxygen.

Lignite or Brown Coal

Lignite or brown coal is an immature coal and has composition intermediate between peat and bituminous coals. Lignite may be amorphous, fibrous or woody in texture. They contain a high proportion of water as moisture and burn with a long brown flame and have a low calorific value (4000–6000 cals/kg). It is extensively used for heating boilers and evaporating pans in industry and for domestic purpose.

Brown coals are characterised by a high content of non-combustible matter (ballast) and therefore, a large difference between the heating value of the combustible mass and that of working mass. They are usually brown in colour but on exposure to air the colour darkens. These coals contain a small quantity of resins and contain 25–40% of moisture which dries up on exposure to air and thus they undergo shrinking. Lignite on exposure to air absorbs oxygen readily and gets ignited spontaneously. Hence it should not be stored in open. Lignite burns with a long smoky flame and has a calorific value of 6000–7500 BTU per lb. Like peat, it has high percentage of volatile matter which under conditions of carbonisation forms tar, convertible to synthetic petrol by high pressure hydrogenation, combustible gases and NH_3 etc. Taking into consideration the content of carbon, lignite is inferior to coals but superior to peat.

Coal

Coal is one of the two principal fossil fuels, the other being petroleum. It is the product of the partial decomposition of the vegetable matter under anaerobic conditions and under heat and pressure. Coal is a combustible solid. It usually stratified and formed by the burial of partially decomposed vegetation in past geological ages. After that pressure and temperature converted them to coal. Chemically, coal is highly complex organic matter with varying amount of water, trapped as well as combined, together with nitrogen (0.75 to 1.75%), and sulphur in the form of organic and inorganic matter. The contents of impurities in coal make different grades of coal. Different grades of coal are :

1. Anthracite (86–88% fixed carbon).
2. Bituminous coals–low volatile type (78–86% fixed carbon), medium volatile type (69–78% fixed carbon), and high volatile type (< 69% fixed carbon).
3. Sub-bituminous coals or black lignite.

Bituminous Coals

This is the penultimate stage in the road to maturity of the coal. Containing upto 92% free carbon in dry coal and about 10% moisture these coals are sub-divided into three types on the basis of carbon content. Sub-bituminous, bituminous and semi-bituminous coals contain 75–80, 80–90 and 90–92% carbon respectively. Volatile matter decreases with increasing carbon content. Calorific value ranges from 30 to 36 KJ/Kg. All of them are black in colour and have banded structure. Their main uses are in the manufacture of metallurgical coke, coal gas and in steam raising.

Anthracite

This highest rank coal contains 92–98% carbon, 7% volatile matter and less than 2% moisture. Calorific value is as high as 37 KJ/Kg. Since volatile matter is low, it does not ignite readily, but it burns without a flame with high calorific intensity. Found in the Himalayan ranges, it is used for making metallurgical coke and in steam raising.

Sub-bituminous coal

It is a dull black coal which is rather difficult to distinguish from ordinary bituminous coal. Its heating value or calorific value is less than that of bituminous coal. It is denser and harder than lignite. It also has much lesser moisture content than lignite. The moisture content in it varies from about 10%–25%. It ignites very easily and has a calorific value of 7000–15000 BTU per lb.

Analysis of Coal

For effecting proper utilisation and for comparing different coal samples, analysis has to be done. Two types of analysis with different purposes are carried out: (i) proximate analysis; and (ii) ultimate analysis.

Proximate analysis

This involves the determination of moisture, volatile matter, ash and fixed carbon. This is an empirical or functional analysis which indicates the efficiency of the coal as a fuel.

Significance of this analysis lies in the identification of the factors affecting the calorific value of the coal and its ignitability. Moisture needs latent heat for vapourisation. This heat is used at the expense of the calorific value of the fuel. However, a little moisture makes the coal bed uniform and reduces the risk of the ash flying off. Free moisture can be removed by air-drying. Inherent moisture is retained after air-drying.

High volatile matter in coal means easy ignitability. This advantage is offset by the formation of a long, smoky flame. Therefore, in the final reckoning high volatile matter is not desirable. Volatile matter includes moisture.

Ash in coal originates from its mineral matter content. The non-combustible inorganic matter left as the residue after combustion of the fuel is called the ash. The mineral matter may be sand, clay, mud or ore particles in which case the ash is called free ash. Ash may arise from original mineral matter in the plant debris which became the coal substance. This ash is called inherent ash. If the mineral matter with the decaying vegetation during coal formation gets sandwiched between layers of coal, the resulting ash is called adventitious ash.

Since ash is not combustible, it does not add to the calorific value of the fuel. For a coal containing much of this non-fuel component transportation, storage and handling costs are wastefully high. As estimate says that for each 1% ash from coal the heat loss is 1.5% (being the sensible heat of the ash).

Some types of ash such as those containing oxides of sodium, potassium, calcium and magnesium, melt easily. The fused ash forms clinkers or big lumps which block the grates. The clinkers may also trap some coal leading to loss of fuel. As coal and coke are used as reducing agents in metallurgy, the ash may enter the metal or the slag and hamper metallurgical operations and affect the purity of the metal being won.

Fixed carbon is the most welcome constituent of coal. It is a measure of the non-volatile carbonaceous matter. % Fixed Carbon = 100 – (% moisture + % volatile matter + % ash). A high percentage of fixed carbon means a high calorific value and a high calorific intensity (because the flame is short or absent). The calorific value of a high-ranking coal is obtained by the Goutel formula :

Calorific value = 82 × FC + a × VM; here FC = % of fixed carbon, VM = % of volatile matter and a = a factor which depends on the nature of the volatile matter and inversely related to its amount.

Proximate analysis is carried out as follows

Moisture. An accurately weighed amount of coal (1 gm) is taken in a weighed capsule and dried at 105–110°C for one hour. The capsule with the coal is cooled in a desiccator and again weighed. The percentage loss in weight suffered by the coal sample is the moisture content of the coal. For one gm of coal % moisture = loss in weight of coal × 100.

Volatile matter. An accurately weighed amount of coal (1 gm) is taken in a weighed crucible (with lid) and placed in a muffle furnace. The crucible is heated to 950°C ± 20°C and maintained at this temperature for exactly 7 minutes. The crucible with the coal is rapidly cooled and weighed. The per cent loss in weight suffered by the sample gives the % of VM. For 1 gm of sample, % of VM. = loss in weight of the sample × 100. This includes moisture. In this experiment it is desirable to use platinum crucible.

Ash. An accurately weighed amount of the coal (1 gm) is taken in a weighed crucible and the sample is burnt on a bunsen burner. When no more black particles are seen, the ash sample is cooled and weighed. Heating, cooling and weighing are repeated till the weight becomes constant. For 1 gm coal, % ash = weight of residue (ash) × 100.

Fixed carbon. Fixed carbon is estimated by difference. % Fixed carbon = 100 – (% moisture + % VM. + % ash).

Ultimate analysis

This is essentially an elemental analysis where in carbon, hydrogen, oxygen, nitrogen, sulphur and metals (ash) are estimated.

Carbon and hydrogen in coal are estimated through a single experiment based on combustion. On burning C becomes CO_2 and H becomes H_2O. A known weight of coal (1 gm) is completely burnt in a stream of air in a combustion tube packed with CuO. The CO_2 and H_2O formed are absorbed in weighed KOH and anhydrous $CaCl_2$ bulbs respectively. Increases in weights of the two bulbs give the weights of CO_2 and H_2O formed. 44 gm of CO_2 contains 12 gm of C and 18 gm of water contains 2 gm of hydrogen. On this basis for 1 gm of coal:

$$\% \text{ Carbon} = \frac{\text{Wt. of } CO_2 \text{ formed} \times 12 \times 100}{44}$$

$$\% \text{ Hydrogen} = \frac{\text{Wt. of } H_2O \text{ formed} \times 2 \times 100}{18}$$

Nitrogen in coal is estimated by Kjeldahl method. A weighed amount of coal is digested with conc. H_2SO_4. All the nitrogen is converted into ammonium sulphate. When the solution becomes clear, it is distilled with excess NaOH. The evolved ammonia is quantitatively absorbed in a known excess of standard acid. The unreacted acid is titrated against a standard base. From this back titration the number of milli-equivalents of acid which has reacted with ammonia evolved can be calculated. This is the same as the number of milli-equivalents of ammonia, which, in turn, is the same as the number of milli-equivalents of nitrogen present in the weighed amount of coal. For one gm of coal.

$$\% \text{ Nitrogen} = (V_1 - V_2) \times N_{acid} \times 14 \times 100$$

Where

V_1 = Volume equivalent of base required to neutralise the volume of acid taken for the absorption of ammonia.

V_2 = Volume of base required to neutralise the unreacted acid, i.e., the acid left after the absorption of NH_3.

14 = Equivalent weight of nitrogen.

Nitrogen can also be estimated by the Dumas method.

Sulphur in coal is estimated by converting the sulphur into SO_2 by fusing with eschka mixture (MgO + Na_2CO_3). The sulphites so formed are oxidised by bromine to sulphates; the sulphate is then treated with excess of $BaCl_2$ solution. The ppt of $BaSO_4$ formed is weighed after filtration and drying.

For 1 gm of coal,

$$\% \text{ Sulphur} = \text{Wt. of } BaSO_4 \left(\frac{32}{233}\right) \times 100$$

Significance of ultimate analysis of coal lies in the calculation of the calorific value which is done on the basis of the Dulong's formula :

G.C.V. in Kcal/Kg = (1/100) × [(8080 × % C) + 34460 × (% H – % 0/8) + 2250 × % S)]

G.C.V. calculated on this basis agrees to within 2% with the value determined by the bomb calorimeter.

The higher the % of C and H, the larger will be the calorific value. Though sulphur also contributes to the calorific value, the combustion product, viz., SO_2, is likely to corrode the combustion equipment. Nitrogen does not burn and hence does not increase the calorific value. Similarly oxygen and ash also cannot burn and act as diluents in the coal. While arriving at the amount of oxygen needed for the combustion of a given coal the amount of oxygen already present in the coal must be taken into account. Oxygen in coal is always estimated by difference.

$$\% \text{ Oxygen} = 100 - (\% \text{ C} + \% \text{ H} + \% \text{ S} + \% \text{ ash})$$

Experimental determination of the calorific value of a solid or liquid fuel using bomb calorimeter :

The bomb calorimeter is shown in the Fig. 25.1. A known mass of the fuel is taken in a nickel crucible. The crucible is placed in a stainless steel cylinder called a bomb which is filled with oxygen at a pressure of 25 atm. The bomb is placed in water taken in a copper calorimeter. A known mass of water (M_1 gm) is taken. After noting the initial temperature of the water (T_1 K) using a sensitive thermometer, the fuel is ignited electrically. The heat evolved during combustion is absorbed by the water and the copper calorimeter. The final temperature is also noted (T_2K) using a time-temperature graph. Heat lost = heat gained. The heat absorbed by the water and copper calorimeter is given by $Q = (M_1 + M_2 \times s) \times (T_2 - T_1)$. Here M_2 is the mass of the calorimeter and 's' is the specific heat of the material (copper) of the calorimeter. Q is the heat evolved by the combustion of a given amount of the fuel. Hence the heat evolved for the combustion of a unit weight of the fuel can be calculated. In practice the heat capacity of the copper calorimeter is determined using a substance of known calorific value in the bomb and measuring the temperature rise. The calorific value is in cal/gm, because the specific heat of water is assumed to be 1 cal/gm in the above formula.

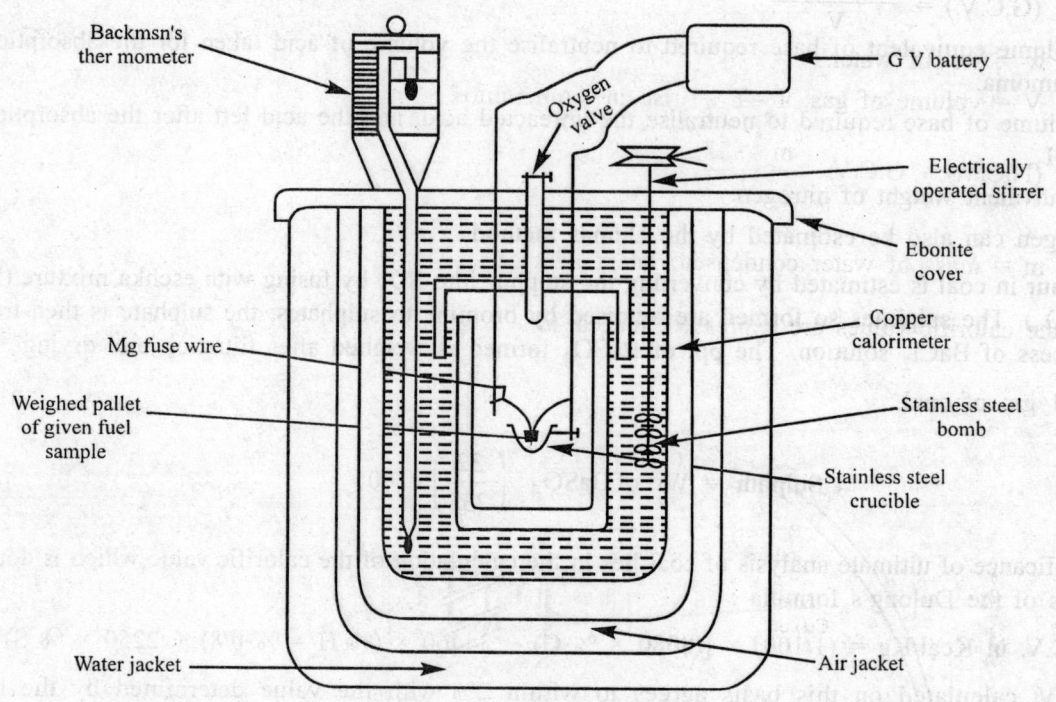

Fig. 25.1. Bomb calorimeter.

For ensuring the accuracy of the results the following corrections have to be made. The fuse wire used in the electrical ignition may also contribute to the heat evolved; some of the oxidations result in the formation of acids which are exothermic processes. These are in addition to the cooling correction usual with a calorimetric experiment. When 0.935 g of a fuel underwent complete combustion in excess of oxygen, the increase in temperature of water in a calorimeter containing 1.365 g of water was 2.40°C. Calculate the higher calorific value of the fuel, if the water equivalent of the calorimeter is 135 g.

Heat evolved = (Water equivalent of the calorimeter + mass of water) × rise in temperature.

$$= (135 + 1.365) \times 2.4 = 327.3 \text{ cal}$$

Calorific value in cal/g = 327.3/0.935 = 350

Determination of Calorific Value of a Gaseous Fuel (Boy's Calorimeter)

The calorimeter is essentially a gas burner in which a known volume of a gas is burnt at a uniform pressure at a uniform rate. Around the burner is a chimney which is coiled inside and outside with copper. Water is passed through the coil. Water enters through the top of the outer coil and reaches the bottom of the chimney and again ascends the inner coil. Two thermometers record the temperatures of the entering and outgoing water. The flowing water takes up the heat from the burner. The steam formed is condensed into water and collected. The following parameters are measured: volume of gas at given temperature and pressure, amount of water passing through the coil at the same time interval, rise in temperature and mass of water condensed during the time (Fig. 25.2).

$$\text{H.C.V. (G.C.V.)} = \frac{w(T_2 - T_1)}{V}$$

where w = wt. of water.

V = volume of gas; $T_2 - T_1$ = rise in temperature.

$$\text{L.C.V. (N.C.V.)} = \text{G.C.V.} - \frac{m \times 587}{V}$$

where m = mass of water condensed; latent heat of steam/$m^3 = \dfrac{587 \times m}{V}$

Here the calorific values are expressed in Kcal/m^3.

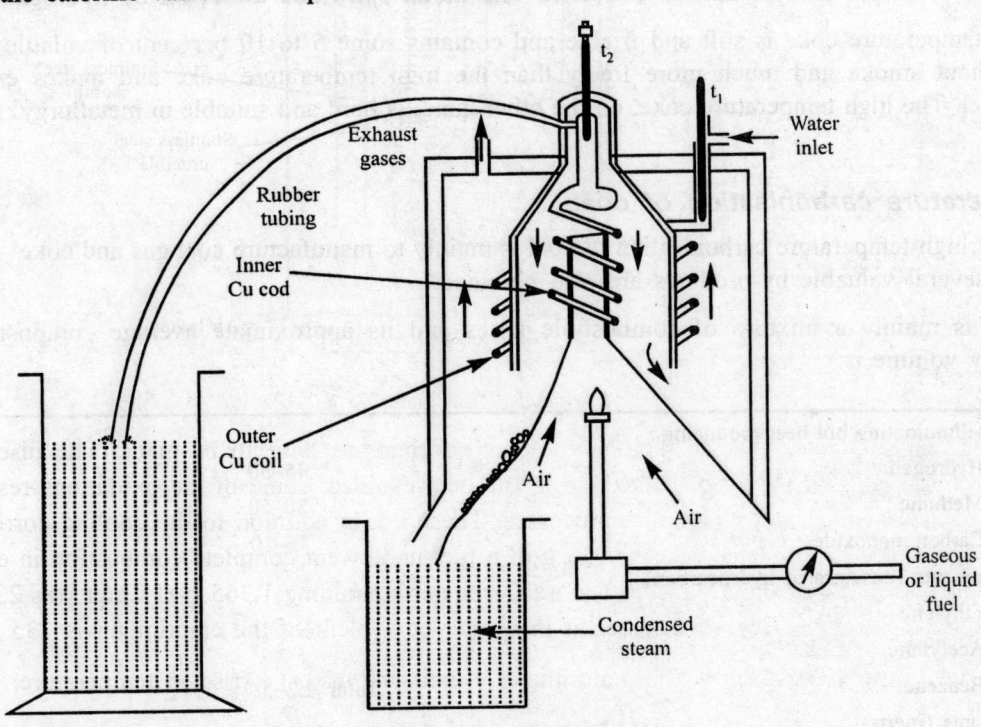

Fig. 25.2. Boy's calorimeter.

Carbonisation of Coal

Besides being used directly as a fuel, coal may be converted into: (i) coal gas and coke by destructive distillation; (ii) gaseous fuels such as water gas, producer gas, etc; and. (iii) liquid fuels by hydrogenation.

The destructive distillation of coal is known as the carbonisation of coal. Coal may be carbonised at high temperature (1200–1400°C) or at low temperature (600–650°C). In either case, the main products are coal gas, coal-tar and coke. But the products of the two processes differ in quality.

Product	Low temperature carbonisation	High temperature carbonisation
Coke		
lbs/100 lb coal	75–80	65–75
Per cent volatile matter	7–15	1–3
Gas cu ft per ton	4000–6000	10000–13000
Btu cu ft	700–900	530–600
Light oil and tar gallons per ton	20–40	10–17
Ammonium sulphate lb per ton	12–15	24–26

In the low temperature process, much less gas is obtained though it is of higher calorific value. The low temperature gives a high yield of a very fluid tar of low viscosity with a high percentage of phenol and nitrogen compounds. Its most important feature is the high percentage of tar acids it contains. It gives motor spirit and fuel oil on distillation. The ordinary or high temperature tar, on the other hand, is the principal source of aromatic compounds, such as benzene, toluene, naphthalene, anthracene and phenol. The low temperature tar can be converted into motor spirit and diesel oil by hydrogenation.

The low temperature coke is soft and friable and contains some 5 to 10 per cent of volatile matter. It burns without smoke and much more freely than the high temperature coke and makes excellent smokeless fuel. The high temperature coke, on the other hand, is hard and suitable in metallurgy; it burns with smoke.

High temperature carbonisation of coal

The object of high temperature carbonisation of coal is mainly to manufacture coal gas and coke—during the process several valuable by-products are also obtained.

Coal gas is mainly a mixture of combustible gases and its approximate average composition in percentage by volume is:

Non-illuminating but heat producing	
Hydrogen	45–50
Methane	30–50
Carbon monoxide	5–10
Illuminants (unsaturated hydrocarbon)	
Ethylene	
Acetylene	
Benzene	Total 2.2–5
Diluants (inerts)	
Nitrogen	2–10
Carbon dioxide	0.2
Oxygen	0.1
Impurities, H_2S, CS_2, etc., if any.	—

Besides these, H_2S, NH_3, CS_2, cyanogen, etc., are also present in coal gas as impurities.

Apart from, coal gas, the other by-products of distillation of coal are coal-tar, ammoniacal liquor, gas carbon and coke.

On account of the very heavy demand of coke for metallurgical processes, specially in iron and steel manufacturing, coke oven batteries are always set up along the iron and steel works.

In modern coke ovens, e.g., Otto-Hoffmann, Simon Carves or Koppers, coal is carbonised at high temperature in narrow fire-clay or silica brick retorts, 40 feet long, 15–18 feet wide and 12 feet high, which are heated by flues passing between them in which a part of the coke oven gas, mixed with air, preheated in regenerators, is burnt. The volatile products, leaving at the top, are cooked and separated into gas and tar in the usual way. The coke is pushed out by rams and quenched with water. The coke oven gas is similar to coal gas in composition.

The Bee-hive Coke Oven in Otto-Hoffmann Coke Oven–Otto-Hoffmann oven (Fig. 25.3) consists of number of chambers, each chamber is constructed 14–22 inches wide, 12 to 14 feet high and about 30–40 feet long.

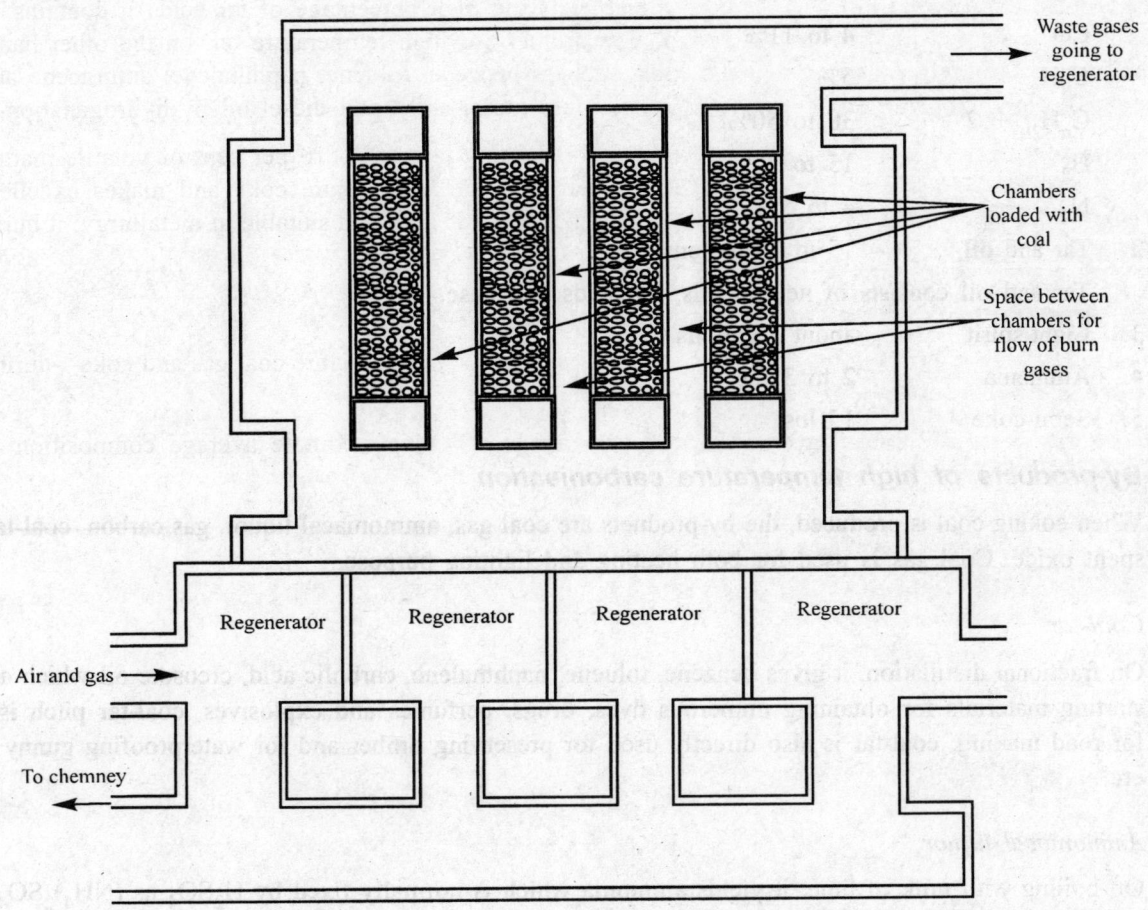

Fig. 25.3. Otto-Hoffmann oven for manufacture of coal.

At the top of each chamber, there are three holes for charging. In the space between two coking chambers, the ovens are heated by burning producer gas, the heating and coking chambers alternate each other. Every chamber at the top is connected with pipe which carries out volatile matter. Coal is charged out into chamber and the chambers are closed. Fuel gas and preheated air from the regenerator burn in the space between the chambers.

The coking takes place in about 24 hours and when coke has been formed, the red hot coke is pushed outside by means of a ram. The yield of coke is 75 per cent of the charge of coal.

Heat is circulated by regeneration of the hot exhaust gases, which are passed through brickwork when the latter is heated up. Now, if cold air is passed through this hot brickwork, it is heated up. This process of alternately passing of hot exhaust gases and cold air and gas through regenerators, is called regeneration.

In high temperature carbonisation, the yield from one ton of coal having 30–35 per cent ash content at 900°C is as follows :

1. Gas 3500 to 6000 cu ft of gas of calorific value 600–800 Btu/cu ft gas composition.

CO_2	4 to 5%
CO	4 to 11%
C_nH_n	5%
C_nH_{2n} + 2	30 to 50%
H_2	15 to 30%
N_2	4 to 5%

2. Tar and oil   15 to 20 gallons

   Tar and oil consists of neutral oils, tar acids and base.

3. Light spirit   about 3 gallons

4. Ammonia   2 to 3 lbs

5. Semi-coke   15 lbs

By-products of high temperature carbonisation

When coking coal is produced, the by-products are coal gas, ammoniacal liquor, gas carbon, coal-tar and spent oxide. Coal gas is used for both heating and lighting purpose.

Coal-tar

On fractional distillation, it gives benzene, toluene, naphthalene, carbolic acid, creosote oil which are the starting materials for obtaining numerous dyes, drugs, perfumes and explosives, coal-tar pitch is used for road making; coal-tar is also directly used for preserving timber and for waterproofing gunny bags, etc.

Ammoniacal liquor

On boiling with milk of lime, it yields ammonia which is normally fixed by H_2SO_4 as $(NH_4)_2SO_4$ used as fertiliser.

Gas carbon

It is obtained as a hard, dense deposit inside the retort due to thermal decomposition of gaseous hydrocarbons in contact with the hot walls of the retort. It is a good conductor of heat and electricity and is used in making electrodes. Spent oxides of iron and lime are formed as a result of absorption of gases like H_2S, HCN, NH_3, etc., present in coal gas. Spent iron oxide is used for preparation of potassium ferrocyanide and for the recovery of sulphur. Spent lime is used as a fertiliser.

Low temperature carbonisation

The objects of low temperature carbonisation are :

1. It is resorted to when low temperature semi-coke as an improved domestic fuel is required.
2. Low temperature carbonisation is done as a preliminary step when complete gasification of the coal is required.
3. A smokeless, free burning coke is required.
4. To obtain coal-tar suitable for conversion into liquid fuels for use as motor spirit, etc.

Difficulties of low temperature carbonisation

1. Heat cannot penetrate through the whole charge which is put in a retort heated externally.
2. After carbonisation, discharging is difficult, because the coal after carbonisation swells. Coal used for low temperature carbonisation should possess suitable coking properties. So generally the coal is treated in such a way that its swelling is reduced but the rate of heating is improved.

Yield of byproducts in low temperature carbonisation

1. In low temperature carbonisation, the yield of gas is low, but the heating value of the gas is high.
2. Quantity of tar is more and it contains higher percentage of paraffinic substances.
3. Ammonia yield is low.
4. The aromatic content of the tar is low.
5. The tar acid content is high and consists mainly of acids of higher molecular weights.

ARTIFICIAL SOLID FUELS

Charcoal

Charcoal is produced from wood either by burning wood in a limited supply of air in pits or kilns or by destructive distillation of wood.

In India, charcoal is generally produced by the partial combustion of wood which is generally covered with earth etc. when combustion is carried out, while in other countries wood is carbonised in closed retorts to produce charcoal. Generally hard woods are used for the manufacture of charcoal. As a result of carbonisation in retorts, gas, tars and an aqueous liquor are also obtained, in addition to charcoal. The yields are charcoal = 30–32%, gas = 20–23%, tars and oils = 5–6% of the wood on dry basis. The composition of the gas is approximately, H_2 = 10–11%, CH_4 = 10–12%, CO_2 = 30–31%, CO = 35%, N_2 = 2% and higher hydrocarbons = 8%. Further distillation of oils and tars yields turpentine and creosote oils.

Coke

Coke is prepared by destructive distillation of coal in a coke oven. When coking is carried out at about 500–750°C, the process is known as low temperature carbonisation and it produces a coke that is soft and contains 5–15% volatile matter. It is called semi-coke and is used for domestic purposes either directly or as briquette and as a furnace fuel too in the pulverised form. Coking of coal at 900–1400°C is known as high temperature carbonisation which produces hard metallurgical coke with higher mechanical strength but lower calorific value.

The best coke generally contains 90% carbon and less than 8% of ash. Coke is slightly hygroscopic, but does not absorb more than 1–2% moisture. The heating or calorific value of coke containing 90% of carbon and 0.5% of hydrogen is about 13400 Btu per lb. The best coke to be used in metallurgy is heavy, compact, homogeneous with bright light grey lustre. In addition to carbon and ash, coke generally contains small amounts of H_2 (0.2–1.2%), oxygen (1.2%), organic sulphur and total sulphur (0.8–1.8%). Coke is almost free from volatile matter and sulphur, but contains most of the ash forming constituents of coal. Bituminous coal is used for coking purposes. Coke made from highly volatile coal forming swelling coke in retorts is called soft coke, while coke obtained by heating a mixture of high volatile and low volatile coking bituminous coal is beehive ovens or coke ovens is called hard coke. Both hard as well as soft coke are obtained by high temperature carbonisation. The coke obtained by low temperature carbonisation is called semi-coke or smokeless fuel.

Briquettes

It is a kind of a compressed fuel, which can be prepared from powdered coal, peat, lignite or coke in the presence of some binding material such as tar pitch, asphalt pitch or molasses (5–8%) under an applied pressure of 1800–3000 psi.

INDUSTRIAL SOLID FUELS

Fossil coals, oil shales, furnace slags and peat are the various industrial fuels. The most important are the fossil coals, which are classified as brown coals, coals proper and anthracite. We have already discussed brown coals and anthracite. Coals proper are diverse in their composition and properties.

Oil shales are formed by rolling of clays, marl, limestone and some other rocks between decomposition products of some kind of organic substances. Oil shales are high ash fuels (Ash contents 40–60%). The percentage of volatile matter in oil shales may go upto 70% or even more. The heating value of combustible mass of oil shales is 27230–33520 kJ/kg. The heating value of the working mass is, however, 6285–8380 kJ/kg due to high percentage of ash. Oil shales are commonly used as local fuels.

Local fuels also include boiler slags and other fuel wastes containing much of unburnt carbon. Boiler slag contains upto 40% of combustible matter and have a heating value of 8380–16750 kJ/kg. Peat has also been included among local fuels.

LIQUID FUELS FROM COAL

The production of synthetic liquid fuels resembling petroleum is important in countries which do not possess large petroleum deposits. This is achieved by hydrogenation of coal by Bergius process.

When coal is subjected to pyrolysis, in the presence of a sulphur-resisting catalyst such as molybdenum, in an atmosphere of hydrogen, at a temperature of 300–500°C and at a pressure of 200–250 atmospheres hydrogen combines with nitrogen of the coal to produce ammonia and with oxygen of coal to produce water.

Coal with not more than 2.5 per cent ash content is cleansed and powdered and is mixed with equal quantity of heavy oil to form a paste. Hydrogen gas is prepared by mixing steam and water gas.

$$CO + H_2O = CO_2 + H_2$$

$$CO + H_2 + H_2O \xrightarrow{\hspace{2cm}} Fe_2O_3 + Cr_2O_3 \xrightarrow{\hspace{2cm}} CO_2 + 2H_2$$

Water gas   Steam

The mixture of carbon dioxide and hydrogen is passed through carbon dioxide absorber containing Triethanolamine which absorbs CO_2. The paste containing coal and heavy oil is sent to the converter. Hydrogen gas is bubbled through the bottom of the converter in which a pressure of 250 atmospheres and a temperature from 350 to 500°C are maintained for one and a half hours. After this, the pressure is released. The gases obtained are taken into a fractionating column and the heavy residue is withdrawn from the bottom of the converter. Three fractions are obtained :

1. Crude spirit which goes out of top.

2. Middle oil which is the middle fraction.

3. The heavy oil which is the residue.

The middle oil fraction is again sent to another convertor where it is subjected to heating under pressure in presence of hydrogen and catalyst. Thus the middle oil is converted into crude spirit. The crude spirit from the first and second converters are mixed together and fractionated and then used as motor spirit (Fig. 25.4).

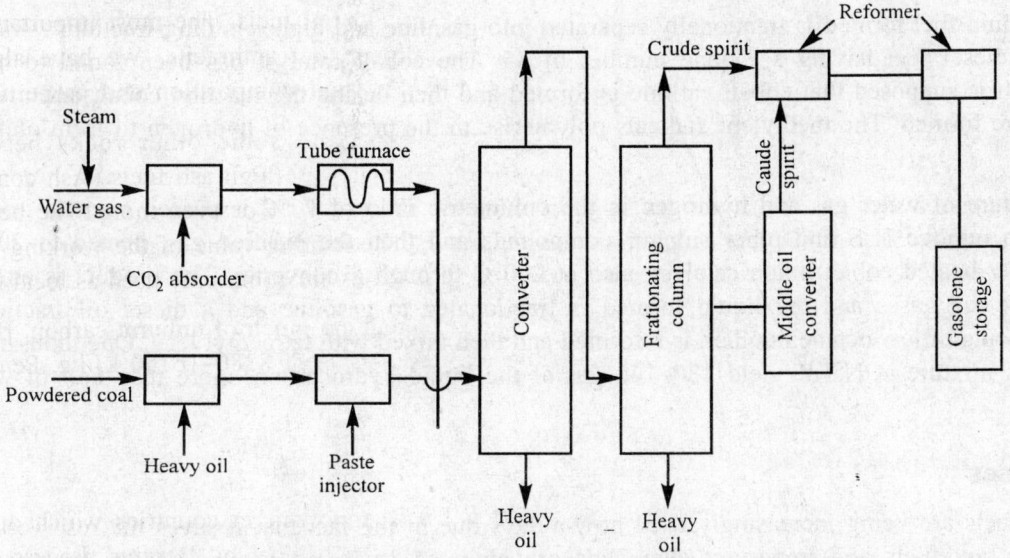

Fig. 25.4. Flow sheet for liquifaction of coal.

The catalysts used are compounds of lead, tin, titanium, zinc, iodine or nickel. By using suitable temperature, pressure, time, hydrogen concentration and the catalyst, various products may be obtained.

Fisher-Tropsch Process

Fisher and Tropsch found that if water gas containing excess of hydrogen is passed over iron and cobalt catalysts at a temperature between 200 and 300°C, there is formation of a mixture of saturated and unsaturated hydrocarbons. From this mixture, a lighter paraffin fraction can be obtained which is used as motor spirit.

$$CO + 2H_2 \longrightarrow (CH_2)'' + H_2O$$
$$\text{Methylene}$$

$$n(CH_2)'' \xrightarrow{\text{Polymerisation}} C_nH_{2n}$$

The excess of hydrogen present further reacts with olefins to form paraffins.

$$CH_2 + H_2 \longrightarrow CH_4$$
$$C_nH_{2n} + H_2 \longrightarrow C_nH_{2n+2}$$

Catalysts used are iron, cobalt, nickel, etc. The catalyst is converted into carbides which further reacts as shown in the following reactions.

$$CO + 2X \longrightarrow XC + XO$$
$$\qquad\qquad \text{Metal} \quad \text{Metal}$$
$$\qquad\qquad \text{carbide} \quad \text{oxide}$$

$$XC + XO + 2H_2 \longrightarrow 2X + (CH_2)'' + H_2O$$

By this process, water gas obtained from soft coke, semi-coke, and natural gas are hydrogenated.

The liquid fuel formed is fractionally separated into gasoline and higher boiling fractions which is an excellent diesel fuel having a Cetane number of 85. The cobalt catalyst has been found to be more efficient. It is supposed that cobalt carbide is formed and then during the reaction cobalt and methylene radicals are formed. The methylene radicals polymerise in the presence of hydrogen to form olefins and paraffins.

A mixture of water gas and hydrogen in the volumetric ratio of 1 : 2 is passed over an iron oxide catalyst to remove H_2S and other sulphur compounds and then the purified gas, heated to 200°C, is passed over heated cobalt oxide catalyst, also at 200°C through a converter. The product is condensed to separate the gases and the liquid formed is fractionated to gasoline and a diesel oil fraction. The gasoline being of low octane number, is reformed and then mixed with tetraethyl lead. One thousand litres of the gas mixture at N.T.P. yield 130–140 gm of the liquid hydrocarbon, more than half of which is gasoline.

Fuel Gases

Gaseous fuels are being increasingly used now-a-days due to the fact that it saves the reserves of coal and other raw fuels and important chemicals are obtained as by-products. Gaseous fuel has many advantages: (i) distribution of gas may easily be made over a wide area by pipes from source of production; (ii) smoke can be avoided and ash eliminated thereby keeping the place of work clean; (iii) the temperature can be easily controlled; (iv) the rate of combustion is high, and therefore higher temperature attained; (v) as the gas can be led around all the parts of the vessel which is being heated, more uniform heating is possible; and (vi) many of the gases have higher calorific values.

The following gaseous fuels are generally used: (i) producer gas; (ii) blue-water gas; (iii) semi-water gas; (iv) carburetted water gas; (v) coal gas; (vi) coke oven gas; (vii) natural gas, (viii) acetylene; and (ix) ethylene.

Producer gas

When air is passed over a bed of white hot coke or a solid carbonaceous fuel, a gas consisting mainly of carbon monoxide and nitrogen is produced. This is known as 'producer gas'.

$$C + 1/2O_2 + 2N_2 \text{ (air)} = CO + 2N_2 \quad \Delta H = -27.06 \text{ Kcal}$$

It is suggested that initially carbon dioxide is produced which reacts further with coke to give carbon monoxide.

$$2C + O_2 = 2CO - 57000 \text{ cal or } 104,000 \text{ Btu}$$

$$C + O_2 = CO_2 - 96500 \text{ cal or } 173,700 \text{ Btu}$$

$$C + O_2 = 2CO + 38700 \text{ cal or } 69.700 \text{ Btu}$$

$$C + H_2O = CO + H_2O + 38,900 \text{ cal or } 70,000 \text{ Btu}$$

Producer gas is produced by passing air through a burning bed of carbonaceous fuel at a temperature of 1000–1400°C. Although peat, wood waste, spent tan bark and other types of carbonaceous fuels may be used, the fuels mostly used as raw materials are coal and coke. Coke, anthracite or charcoal are used in smaller, but in the plants with larger capacities coal is used. Producer gas used in gas engines for power production is obtained from coke or anthracite or charcoal.

There is still divergence of opinion concerning the exact mechanism of the reactions and their relative velocities. There is, however, evidence to show that the oxidation zone, immediately above the ash zone, gives the predominating reaction $C + O_2 = CO_2$. As the partial pressure of free oxygen reaches a low value, or is practically nil, carbon monoxide appears and rapidly increases. Shortly above this point, steam decomposition begins and free hydrogen appears rapidly at first and then increases only very slowly. This thin zone is called the primary reduction zone. On the higher or secondary reduction zone, no reduction of steam by carbon occurs. This secondary reduction zone serves mainly as a heat interchanger, the hot gas serving to heat the incoming fuel. Above the secondary reduction zone is the distillation zone which is relatively unimportant with low volatile fuels, such as coke or anthracite, but which may contribute a considerable quantity of thermal energy in the form of gaseous hydrocarbons from high volatile fuels. It may represent as much as 40 per cent of the heating value of the gas. Above the fuel-bed, the heating value of the gas may drop somewhat because of the reversion $2CO \rightarrow CO_2 + C$ and to the leakage of CO_2 and steam around the edges of the fuel bed and through blow holes.

The capacity of a producer is a variable quantity, depending chiefly upon the quality of fuel supplied, the method of operation, the design of the producer and the character of the demand for gas with respect to quality and quantity.

Producer gas is manufactured in plants known as *gas producer*. The fuel is stacked on a grate in the cylindrical furnace, made of fireclay refractory bricks. Blast of air is blown through the grate. The fuel is fed from the top by a ball and hopper arrangement. The gas produced leaves the furnace through the gas outlet.

$$2C + O_2 = 2CO + 58000 \text{ calories} \qquad ...(25.1)$$

$$C + O_2 = CO_2 + 97000 \text{ calories} \qquad ...(25.2)$$

$$CO_2 + C = 2CO - 39000 \text{ calories} \qquad ...(25.3)$$

As the carbon dioxide formed is reduced to carbon monoxide by the heated coke and as the reaction is endothermic, the temperature of the bed is kept high at above 1000°C. The ratio $CO : CO_2$ increases with increase in temperature. The equilibrium volumes percentages of CO and CO_2 at atmosphere are shown as follows :

Temp°C	850°	900°	950°	1000°	1050°	1100°	1200°
CO	93.77	97.78	98.68	99.41	99.63	99.85	99.94
CO_2	6.23	2.22	1.32	0.59	0.37	0.15	0.06

The proportion of carbon monoxide increases with increase in temperature. The producer gas is mainly carbon monoxide and nitrogen; its average composition in percentage is H_2 10, CH_4 8, CO 20, CO_2 4, N_2 64. Its calorific value is about 150 Btu. per cu ft.

It is used as fuel for heating coal gas retorts, in furnaces and gas engines.

Blue-water gas

Water gas is often called 'blue gas' because of the colour of the flame when it is burned. Blue-water gas is produced by passing steam through a hot bed of coke at 1400–1000°C. The reaction is endothermic so the temperature falls from 1400 to 1000°C, some carbon dioxide is also formed during the reaction and the amount of the gas becomes considerable at 1000°C. Hence, as soon as the temperature falls to about 1000°C the blast of steam is stopped and an air blast is passed through the hot bed of coke for sometime to raise the temperature of the coke to 1400°C and then the air blast is stopped and the steam blast is resumed. As the mixture of air and hot water gas will explode, the water gas remaining in the generator is removed by passing steam for one minute which sweeps off the water gas (at the lower temperature, much water gas is not produced) this steam is known as purge steam.

$$C + H_2O \longrightarrow CO + H_2$$
$$\Delta H = + 53,850 \text{ Btu}$$

Another reaction also occurs, apparently at several hundred degrees lower temperature.

$$C + 2H_2O \longrightarrow CO_2 + 2H_2 \qquad \Delta H = + 39,350 \text{ Btu}$$

These reactions are endothermic and therefore tend to cool the coke bed rather rapidly, thus necessitating alternate 'run' and 'blow' period. During the 'run' period, the above gas reactions take place, and $(CO + H_2)$ at higher temperature and $(CO_2 + 2H_2)$ at a lower temperature results. During the 'blow' period air is introduced and ordinary combustion ensues, thus reheating the coke to incandescence and supplying thermal units required by the endothermic gas making reactions and making up the various heat losses of the system. During the 'run' and 'blow' periods, the reactions are :

$$C + O_2 \longrightarrow CO_2 \qquad \Delta H = - 173,930 \text{ Btu}$$
$$CO_2 + C \longrightarrow 2CO \qquad \Delta H = + 68,400 \text{ Btu}$$

The water gas always contains some carbon dioxide. Its average composition (in percentage) is $H_2 = 48$, $CH_4 = 1$, $CO = 42$, $CO_2 = 3$ and $N_2 = 6$. Its calorific value is about 300 Btu/cu ft.

Water gas is used as a source of hydrogen and as a reducing agent in metallurgy. It is also added to coal gas for street lighting. It is being increasingly used nowadays for the synthesis of chemicals such as alcohols and oxygenated compounds. The mixture $(CO + H_2)$ is also a step in low cost hydrogen for ammonia manufacture.

Carburetted water gas

To enhance the heating value of blue water gas, oil is atomised into hot blue gas. A plant producing this carburetted blue gas has, in addition to the ordinary generator, a carburettor and a superheater.

In making carburetted water gas, two towers packed with chequor brick, called carburettor and superheater, are placed after the water gas generator. During the 'blow' period (two minutes) air is blown through a hot bed of coke in the generator, which is thereby heated to incandescence, and the producer gas formed is mixed with air and burnt in the carburettor and the superheater to heat the chequor bricks. During the 'run' period (four minutes) steam is blown through the heated coke, and the water gas is passed through the hot carburettor down which is injected a spray of mineral oil which vapourises. The mixture of water gas and oil vapour then passes through the red hot bricks in the superheater, where the oil vapour is cracked. The gas is freed from tar and sent through a purifier in which H_2S is removed by hydrated Fe_2O_3 and finally stored in gas holders.

Semi-water gas

Semi-water gas is obtained by passing a mixture of air and steam continuously through a bed of red hot coke. The amount of steam mixed with air is so adjusted that the endothermic water gas reaction is just balanced by the exothermic producer gas reaction, and hence the temperature of the coke is kept constant. This is corresponding to a vapour pressure of 123 lbs of water vapour at a saturation temperature of 56°C. The average composition (in percentages) of semi-water gas is $H = 12$, $CH_4 = 3$, $CO = 30$, $CO_2 = 2$ and $N_2 = 53$.

Its calorific value is about 180 Btu per cu ft. It is used as a fuel and as a source of nitrogen and hydrogen in the manufacture of ammonia.

Coke oven gas

Coke oven gas is obtained as a co-product during the destructive distillation of coal for the manufacture of coal gas and coke. The co-product in the hydraulic main are passed first through a foul main and then through a tar extractor fitted with baffles where the coke oven gas is separated from the ammonia liquor and the coal-tar. The ammonia present in the gas is removed as ammonium sulphate by bubbling through sulphuric acid. The coal-tar products eg., benzene, toluene, naphthalene, are scrubbed in a packed light oil tower. Coke oven gas is used mainly as fuel gas in many steel plants, and also for firing the coke ovens themselves. It is also purified and sold for local consumption as gaseous fuel.

Mond gas

It is made with a mixture of air and a large excess of steam which keep the temperature low (650°C) and allows the recovery of nitrogen of coal as ammonia.

Natural gas

Natural gas is a mixture of gases suitable for use as gaseous fuel. Most important of the natural gas is the natural gas associated with crude petroleum which is obtained from oil wells. It accumulates in underground reservoirs either with or without petroleum oil. Natural gas is found and exploited extensively in the oil wells of the U.S.A. In India too with the finding of petroleum, natural gas is being obtained and arrangements for its utilisation as a fuel gas and also for obtaining petrochemicals are being made.

Natural gas derived from oil wells are of two types : (i) dry; and (ii) wet. The dry variety is obtained from wells which yields no crude petroleum. Such a variety does not contain any gasoline vapour. The dry gas mainly contains methane. The approximate composition of dry gas (in percentage) is CH_4–96, C_2H_4–0.8, N_2–3.2, another sample of dry gas contains CH_4=67.6 per cent, Ethane 31.3 per cent and N_2 = 1.1 per cent.

The gas obtained from oil producing wells is called wet natural gas. It contains, besides high percentage of methane, higher boiling hydrocarbons such as n-propane, n-butane, iso-butane, iso-pentane and a little percentage of gasoline.

Usually smaller quantities of propane, hydrogen, hydrogen sulphide and nitrogen are found along with methane and ethane—the percentage by volume of methane is 80–95 per cent; ethane and other hydrocarbons 5–20 per cent. Helium is also sometimes found in natural gas—this is recovered. Besides the USA, natural gas is also found in other European countries and Canada.

The gas obtained by the fermentation of organic matter in sewage is also called natural gas. The composition of such a gas is mainly methane (about 70 per cent) and carbon dioxide (30 per cent) with an average calorific value of 62.5 Btu/cu ft.

At present in India much activity is going on for utilising cowdung as source of natural gas for use as a fuel. Cowdung on fermentation gives gas with composition as in sewage. The practice will solve the fuel problem to some extent, without damaging the fertilising property of cowdung. In India, the farm manure is burnt everywhere as a fuel on account of great fuel shortage in rural areas. If the farm manure is fermented to obtain fuel gas, the residue can be used as a manure–hence both the needs are fulfilled by the fermentation.

Natural gas burns with a smoky flame in ordinary burners. If a non-luminous flame is required, then, special appliance are to be used. It can be liquefied under pressure and cooling to 121°C and absorbed at that temperature in fuller's earth.

Natural gas is generally used as a domestic fuel. It is also mixed with other fuel gas to increase the calorific values. Natural gas is used for the manufacture of carbon black. The methane of the natural gas is utilised for the manufacture of formaldehyde and methyl alcohol by controlled oxidation. The liquid hydrocarbons are recovered from wet natural gas. Helium is also recovered from natural gas.

LIQUID FUELS

Petroleum, also called rock oil, is liquid which is obtained from the ground either by natural seepage or by drilling wells to various depths under the ground. The oil may flow out by itself due to upward thrust of underground gas or hydrostatic pressure or the oil is mechanically pumped out.

Petroleum consists of liquid oils, hydrocarbons, substances such as natural gas, asphalts and waxes. It also contains organic compounds of sulphur, nitrogen and oxygen in small quantities. Some inorganic matter is also present in very small quantities.

Origin of Petroleum

Several theories have been proposed in order to explain the formation of petroleum :

1. The inorganic theories are based upon observations in the laboratory that certain heavy metal carbides when treated with water or steam yield hydrocarbons. It is supposed, according to this theory that petroleum has resulted from the action of steam or water on heavy metal carbides existing in the interior of the earth. This theory is not however favoured by the geologists. They prefer the organic theory of petroleum formation.

2. The organic theory suggests the formation of petroleum by the decomposition of marine organic matter deposit by bacterial action in the absence of oxygen. The presence of salt water suggests the marine origin. If the organic matter is not purely marine, it is certain that the organic matter deposit from which the oil was formed accumulated near the seacoast. Fats, waxes and oils are the most probable compounds from which petroleum was produced. A large proportion of the fatty matter must have come from whales and seals and other marine animals. Another theory suggests the decomposition of organic matter by the catalytic action of inorganic matter.

Composition of Petroleum

Crude petroleum is a mixture of mostly hydrocarbons, minor amount of naphthenic acids and also sulphur, nitrogen and oxygen derivatives of hydrocarbon. The average ultimate composition of petroleum is C–83 to 87 per cent H–11 to 14 per cent and sulphur, nitrogen and oxygen—0 to 4 per cent. The reason for the consistency in composition is that petroleum consists of only eight or nine homologous series of compounds and, therefore, though the proportions of individual compounds present in crudes of different sources may vary within wide limits, the effect on ultimate analysis is rather small.

The hydrocarbons present in petroleum are :

n-paraffins (straight chain open chain hydrocarbons) $C_n H_{2n + 2}$ iso paraffins (branched chain open chain saturated hydrocarbons) $C_2 H_{2n + 2}$.

Olefins (open chain unsaturated hydrocarbon) $C_n H_{2n}$.

Naphthenes (cycloparaffins-saturated ring hydrocarbons) $C_n H_{2n}$ Aromatics (benzene, naphthalene, anthracene and their alkyl and naphthenic derivatives).

Paraffins or saturated hydrocarbons are the most abundant of the hydrocarbons. The composition varies from CH_4 to $C_{20} H_{62}$. The distribution approximately is as follows :

Product	Boiling range	Composition
Light gasoline	20–100°C	C_5H_{12}—C_7H_{16}
Benzene	70–90°C	C_6H_{14}—C_7H_{16}
Ligroin	80–120°C	C_6H_{14}—C_8H_{18}
Petrol or gasoline	90–200°C	C_6H_{14}—$C_{11}H_{24}$
Solvent naphtha	200–300°C	$C_{12}H_{26}$—$C_{16}H_{24}$

Product	Boiling range	Composition
Kerosene	200–300°C	$C_{12}H_{16}$
Heavy oil or diesel oil	above 300°C	$C_{12}H_{18}$
Lubricating oil	above 300°C	$C_{16}H_{20}$
Grease, vaseline, petrolatum	above 300°C	$C_{18}H_{22}$
Paraffinic wax	above 300°C	C_{20}—C_{30}
Asphaltic bitumen	above 300°C	C_{30}—C_{40}

Next to paraffins, the naphthenes are the most abundant hydrocarbons. Aromatics are present in small percentages in most petroleum. Olefins are present in very small amount.

There are small percentages of naphthenic acids in petroleum. They include cyclopentane, carboxylic acid, cyclopentanyl acetic acid, 3-methyl cyclopentyl acetic acid, camphonaic acid, 4-methyl-cyclohexane carboxylic acid.

The sulphur compounds present include isobutyl mercaptan, methyl ethyl sulphide, pentamethylene sulphide and thiophene. The sulphur compounds are objectionable ingredients in the gasoline and other petroleum products. They are not possess disagreeable odour but the sulphur dioxide gas formed during combination of gasoline, kerosene and fuel oil produces sulphuric acid with moisture which would corrode the metal parts of the engine. Hence, the law does not allow the presence of more than 0.1 per cent of sulphur. Very small amounts of nitrogen compounds are present in petroleum oil. The nitrogen compounds are isolated 2-3-8-trimethyl quinoline, 3-4-8 trimethyl-quinoline and 2-4-dimethyl 8-sec-butyl quinidine.

Besides the above, phenols, free sulphur and hydrogen sulphide are also present in petroleum oil. In determining the composition of petroleum the points studied are: (i) aniline point (the temperature at which a given gasoline is miscible with equal volume of aniline. This determines the hydrocarbon composition approximately); (ii) refractive index; and (iii) specific gravity.

Classification of Petroleum

Petroleum is classified into three categories according to its compositions :

1. *Paraffin base petroleum:* It contains hydrocarbon of the paraffin series. The paraffins present may be straight or branched.

2. *Asphaltic base petroleum:* It contains hydrocarbons of naphthene ring or naphthene products with side chains.

3. Mixed base petroleum which contain both paraffin and asphaltic base petroleum.

Distillation of Petroleum

Crude petroleum contains sand and water in suspension and also some gas and, as such cannot be used directly. It is brought to the surface and pumped under pressure to cylindrical tanks.

On releasing the pressure and allowing to settle, the gas comes out first. The oil floats on the top layer and sand and water form the lower layer. These two layers are withdrawn separately.

The gas is drawn out as a wet gas directly from the well when the amount is large; or by separating from the crude by release of pressure and heating for a short time by steam coil in the storage cylinder.

The gas which is dissolved in the crude is also separated during distillation. If the temperature of the crude oil rises above 400°C, cracking takes places, and gases are formed. Hence, distillation is carried out below 400°C.

Crude petroleum is allowed to settle and then centrifuged to remove wet sand. Regaining of petroleum consists of three processes :

1. *Fractional distillation:* In this process, lighter and heavier portions separate out and gas, gasoline, fuel oils, lubricants, wax distillates and asphalts are produced. The primary products obtained are modified by decolourising, deodorising and by separating out hydrocarbon groups and impurities.

2. *Cracking:* By this process crude petroleum or its fractions are decomposed by heat to produce products which have lower drilling points.

3. *Treatings:* The process is of three kinds.

   (a) Process by which a product is improved by removing undesired components.

   (b) Process by which the quality of a product can be improved by adding some outside matter which either combines chemically or remains in solution.

   (c) Process through which the chemical structure of hydrocarbons constituting the oil fraction is changed.

Fractional distillation

The fractional distillation of crude petroleum is carried out in a still called, 'topping still'. The Figs. 25.5(a) and (b) give the essential parts of petroleum distillation methods. The topping still, in essence, consists of a pipe still and a bubble tower. In pipe still the crude petroleum is heated and in the bubble tower it is fractionated.

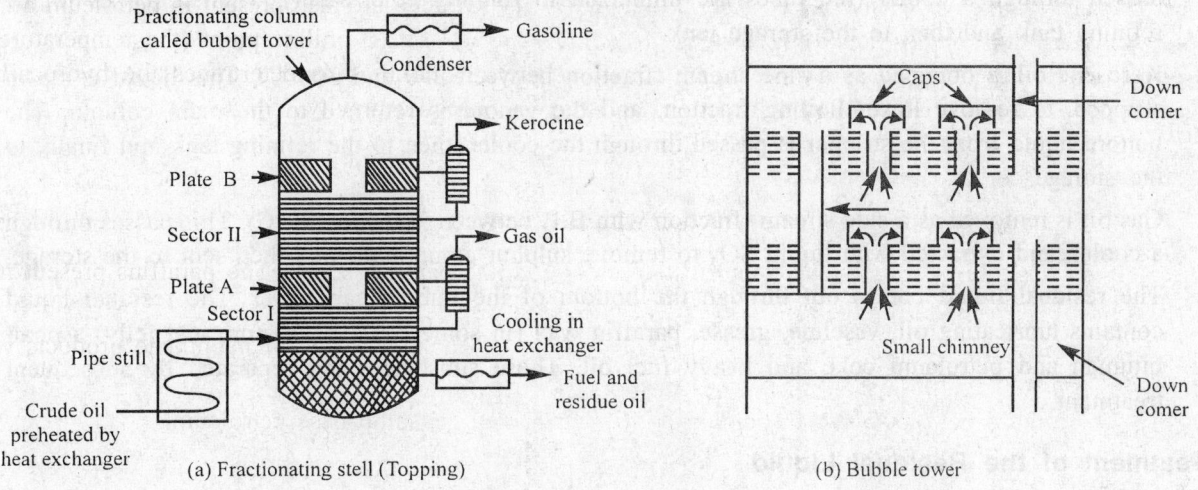

(a) Fractionating stell (Topping)   (b) Bubble tower

Fig. 25.5. Petroleum distillation.

The bubble tower consists of a number of plates. Each plate has a number of small chimneys for rising vapours. These small chimneys are covered with caps so that the vapours bubble through the cool refluxing oil flowing down the bubble tower over the plate. When hot gases bubble through the refluxing oil, these are cooled down and transfer their heat to the refluxing oil. A downcomer carries the liquid to the next lower plate. Baffles are placed in the plate so that the liquid has to pass a long distance before it reaches the next downcomer.

Crude oil is piped through a pipe still where it is heated to a temperature 316–371°C. All components, except the residue, are vapourised. All the products obtained are introduced in the first section of the bubbling tower. The residue is taken from the bottom. The vapours go up in the bubble tower. As the vapours go up, these become cooler. Due to this gradual cooling, vapours condense over these plates, has higher boiling range than the oil from plates at higher level.

The gasoline vapours go out from the top, while gas oil is collected from plate A and kerosene oil from plate B.

Gas oil, kerosene and gasoline are further treated in separate fractionating columns to further purify them and the higher portion from these are returned to the bubble tower for redistillation.

The distillation of crude petroleum is a continuous process. The following fractions are obtained simultaneously: (i) straight-run gasoline; (ii) solvent naphtha; (iii) kerosene oil; (iv) gas oil or diesel oil; and (v) residual.

1. Gasoline is obtained as overhead in the boiling fraction up to 200°C. The vapour is condensed in a condenser and passed to the refining tank to remove sulphur compounds and diolefins and then to the storage. After refractionation: (i) petroleum ether boiling between 40 and 60°C; (ii) benzene boiling between 70 and 90°C; and (iii) gasoline boiling between 90 and 200°C are obtained.

2. Solvent naphtha is obtained as a side stream in the condensed form, which is tapped from a plate on which a liquid accumulates above 200°C but below 250°C this contains some gasoline, hence the liquid is stripped by steam in a small tower containing a few plates, the gasoline vapour as an overhead is passed back into the main fractionating column and the bottom from the stripper is passed through a cooler (the tubes are immersed in running cold water). Then it passes to the refining tank and then to the storage tank.

3. Kerosene oil is obtained as a side stream (fraction between 250 and 300°C). This liquid is steam stripped, to remove lower boiling fraction, and the vapour is returned to the main column. The bottom liquid from the stripper is passed through the cooler, then to the refining tank and finally to the storage.

4. Gas oil is removed as a side stream (fraction with B.P. between 300 and 350°C). This passes through a cooler, and extracted with liquid SO_2 to remove sulphur compounds and then sent to the storage.

5. The residual liquid comes out through the bottom of the bubble cap tower. The residual liquid contains lubricating oil, vaseline, grease, paraffin wax (in some cases petrolatum and jelly) asphalt bituman and petroleum coke and heavy fuel oil. These substances are separated by subsequent treatment.

Treatment of the Residual Liquid

A part of the residual liquid is vapourised in a pipe still and fractionated under vacuum. The overhead is condensed to obtain diesel fuel. A part of the vapour is subjected to vapour phase cracking to obtain cracked gasoline and other gases. The side stream from the middle contains lubricating oil and paraffin wax and lower boiling asphalt. This is passed through a cooler and then is passed through asphalt separator (the recovered asphalt is sent to the asphalt storage). The lubricating oil, with dissolved paraffin wax, is chilled to separate the wax as a solid. The liquid and solid are filter-pressed and the liquid is fractionated and the overhead passed again through a condenser and chiller (the process is repeated to separate completely the wax from the lubricating oil). The liquid is further solvent-extracted to remove the wax, and the lubricating oil is distilled in vacuum, condensed, treated for decolourisation with absorber clay, filter-pressed and sent to storage. The residual product of the vacuum distillation is asphalt bitumen.

The other part of the liquid residue is subjected to cracking operation when gases, gasoline, a heavy liquid and petroleum coke are the products. Gases come out from the top of the cracking tower. The liquid is subjected to fractionation in bubble cap tower. Gasoline is obtained as overhead, is condensed and sent to cracked gasoline storage. The residual liquid from the last bubble cap tower is subjected to destructive distillation, when petroleum coke is obtained as a residue and heavy fuel oil from the condensation of volatile products. Casinghead gasoline, or natural gasoline, is obtained from wet natural gas obtained from crude oil. It has a low boiling point.

Gasoline

Refining of gasoline

The gasoline obtained from the bubble cap tower contains impurities, especially sulphur compounds, viz., hydrogen sulphide and mercaptans, etc. These sulphur impurities impart a bad odour, cause corrosion of the metal parts of internal combustion engines and also interfere with the efficiency of anti-knock compounds like tetraethyl lead.

To remove hydrogen sulphide, the gasoline, is treated with NaOH. Freed from hydrogen sulphide the oil is treated with a little 94 per cent sulphuric acid at ordinary temperature to remove any excess of NaOH. A heavy acid sludge is formed which separates as a lower layer and is easily removed. The mercaptans are converted into sulphides and the diolefins are polymerised. After removal of the acid sludge, the gasoline is washed with water and redistilled.

The cracked gasoline is treated with 80–85 per cent sulphuric acid at low temperature for a short time, because cracked gasoline contains higher proportions of unsaturated hydrocarbons which have high octane value. The cracked gasoline is usually refined by vapour phase treatment. The vapour is passed through a tower containing hydrated aluminium silicate containing nickel and cobalt oxides acting as catalyst. The vapour phase is passed over the catalyst at 310–340°C. The sulphur compounds are preferentially absorbed on the catalyst surface, which are removed as hydrogen sulphide and sulphur dioxide during regeneration of the catalyst. The sulphur compounds may also be removed by treatment with sodium plumbite or sodium hypochlorite solution.

Natural gasoline or casinghead gasoline

Natural gasoline is also called casinghead gasoline. It is present in the wet natural gas in the form of vapour. This shows that a portion of the most volatile petroleum oil passes into the gas. Natural gasoline has thus a lower boiling point and is highly volatile. It is utilised to make motor fuel by blending with refinery petroleum (both straight run and cracked) required for the easy starting of internal combustion engines in the cold weather.

Butane obtained from natural gasoline is converted into butadiene which is used for making synthetic rubber.

Aviation gasoline

Aviation gasoline is extremely light and has a high octane number. It is prepared artificially by the blending of alkylates (obtained by the reaction between isobutane and butylenes, etc. which are obtained from natural gasoline and alkylated by reaction with gaseous olefine such as ethylene, propylene and butylene and isobutylene) with catalytically cracked gasoline and straight run gasoline and mixed with requisite amount of tetraethyl lead. Unsaturated open-chain hydrocarbons and naphthenes have also higher octane number than straight-chain paraffin. Hence gasoline of high octane number has to be prepared by mixing alkylates (branched-paraffins) with cracked gasoline (unsaturated open chain hydrocarbon). The anti-knock agent, tetraethyl lead, increases the octane number indirectly.

Straight-run gasoline

The gasoline obtained from the bubble cap tower has a low octane value and is used as a cheap fuel for internal combustion engines. It is always mixed with tetracitly lead to increase the octane number.

Cracked gasoline

Cracked gasoline is produced by reforming straight-run gasoline and solvent naphtha. In the reforming furnace, the straight-chain paraffins partially undergo cracking resulting in the decomposition of straight chain paraffins into olefines of low molecular weight but high octane number and straight-chain paraffins The latter undergo isomerisation in the reforming furnace, the branched chain paraffins have high octane number. The residue from the bubble cap tower (in which crude petroleum is fractionated) is partly fractionated to obtain cracked gasoline. The overhead is cracked in the vapour phase to obtain cracked gasoline. The other part of the residue is subjected to mix phase cracking to obtain cracked gasoline. It is used directly or in mixture as motor fuel (The processes for cracking of gasoline are given later).

Uses of the products obtained from petroleum distillation (other than gasoline) :

1. Petroleum ether (B.P. 20–60°C)—used as a solvent.
2. Benzene (B.P. 70–90°C)—used in laundry for dry cleaning.
3. Ligroin (B.P. 80–120°C)—used as a solvent.
4. White spirit (B.P. 150–200°C)—used as solvent.
5. Solvent naphtha (B.P. above 200°C and below as that of kerosene)—used as solvent.
7. Kerosene (B.P. as that of solvent naphtha)—used as fuel and for illumination.
8. Lubricating oil (B.P. above 300°C)—used as lubricant for machines, internal combustion and other engines).
9. Vaseline and grease—used as lubricants for wheels. Highly refined vaseline is used in medicine and in preparing cosmetics.
10. Diesel oil (also known as gas oil)—used in diesel engines. It has a high calorific value.
11. Heavy fuel oil—used as a fuel for generating heat.
12. Paraffin wax—used for making candles, waxed paper and polishes. It is crystalline in structure.
13. Petrolatum wax—obtained from waxy residues from the distillate of certain types of crude oils. It is free from oil and is a plastic solid. Used to make waterproof coating of metals and other surfaces.
14. Petrolatum jelly—This is petrolatum wax containing some proportion of lubricating oil. It is used for coating metals. The highly refined variety is used in cosmetics.

15. Asphalt bitumen—Amorphous solid. Mainly used for road making, roofing felts, bituminised paper and paints and for coating pipes.

16. Liquid paraffin or white oil—It is made by treatment of a heavy distillate with fuming sulphuric acid. It is used in medicine.

Cracking of Petroleum

Cracking is the process by virtue of which crude petroleum or its fractions are decomposed by heat to produce products which have lower boiling points. The main object of cracking is not only to increase the amount of gasoline, but to yield also alkylates which essentially constitutes very high octane aviation gasoline. Cracking is of two kinds: (i) thermal cracking; and (ii) catalytic cracking.

Thermal cracking

Here the charge is heated to a temperature of 520°C and pressure of 600 psi. The pressure is kept at the outlet of the heating coil so that a dense phase and longer time are provided during the reaction.

$$C_{10}H_{22} \xrightarrow{\text{Cracking}} \underset{\text{Paraffin}}{C_5H_{12}} + \underset{\text{Olefin}}{C_5H_{10}}$$

The lower molecules may be further decomposed at a certain high temperature to lower hydrocarbons. The products from the cracking tower are passed to a chamber where the pressure is reduced. On reduction of pressure, the lower boiling fractions are vapourised. The residue is removed from the bottom and is used as a fuel oil residue. The vapourised materials are passed to the bottom of the fractionating column and fractionated. The overhead portion from the fraction column is condensed to form gasoline. A higher boiling fraction is obtained as a side stream somewhat below the top of the column, this is made free of gasoline by steam and then collected.

The ultimate products of thermal cracking are: (i) cracked gasoline; (ii) furnace distillate; (iii) gases containing CH_4, C_2H_6, C_3H_8, C_4H_{10}, C_2H_4, C_3H_6, C_4H_8 and hydrogen, (iv) fuel oil; and (v) petroleum coke as residue.

Catalytic cracking

Gasoline obtained by thermal cracking has an octane number up to 72. The use of catalyst in cracking increases the rates of decomposition of different types of hydrocarbons present in crude oil. Gasoline obtained by catalytic cracking is low in olefins and high in iso-paraffins and aromatic hydrocarbons.

The octane number of catalytically cracked gasoline is 80 and it is used as aviation fuel. This has several advantages over thermally cracked gasoline :

1. No fuel is required from outside for catalytic cracking. All the heat required is obtained from the burning coke placed on the fuel.

2. The pressure required is low.

3. The gasoline obtained has a high octane number.

4. The yield of gasoline is high.

5. Sulphur contents are low and there are less gum-like substances present.

The catalysts used are generally silica, aluminium oxide or bauxite itself. There are two types of catalytic cracking: (i) catalytic cracking where a fixed bed of catalyst is used; and (ii) moving bed catalytic process. The following catalytic processes are in general use :

1. *Hondry process:* In this process the catalyst is aluminium silicate with one per cent manganese oxide. The catalysts are fixed in towers. The temperature employed is 500°C, pressure 30 psi. The oil heated in a furnace is fed to the top of the cracking tower containing the catalyst, and passes through it in ten minutes.

   The products of cracking (vapour) are fractionated to obtain gas and gasoline as overhead. The gas is separated during the condensation of gasoline. A middle side stream distillate is collected, which may be used as domestic fuel. 40 to 50 per cent of gasoline by volume is obtained. The bottom liquid is reboiled to recycle the vapour in the fractionating column. Ultimately gas oil is obtained from the bottom. The octane value of the oil is high (about 80) and it contains only 0.2 per cent sulphur.

   These reactor towers are used alternately. After 10 minutes of passage of charge stock through the reactor, the catalyst is regenerated by burning away the carbonaceous matter by compressed air. While one reactor is working, the second one is purged with steam jet to remove the condensed oil—it requires about five minutes; then the carbonaceous matter is burnt by admitting hot compressed air for 10 minutes and then the waste gas is purged by steam jet ejectors in about five minutes.

2. *Thermofor process:* In this process also the catalyst is aluminium silicate in the form of beads. The temperature employed is 450 to 510°C and the pressure 10–15 psi. The hot catalyst beads flow down a tower in which it is intimately contacted with the hot vapour of the oil to be cracked. The beads of catalyst flow down through a series of baffles placed on the clay. In the tower the feed oil in the vapour phase undergoes cracking and carbonaceous matter collects on the catalyst beads. The vapours from the top are fractionated into gas, gasoline and bottom oil (which is recycled).

   The beads withdrawn from the bottom of the tower are freed from oil by superheated steam, and then transferred by bucket elevators to the top of the regenerating tower. Here they flow down through a series of burning and cooling chambers alternately. In the burning chamber, carbonaceous matter is burnt with air admitted into it, and in the cooling chamber the beads are cooled by coils in which steam is generated. Thus at each stage the temperature of the beads is kept at 620°C, until all the carbonaceous matter is burnt away. Then the beads are withdrawn from the bottom of the regenerator, and hot beads are transferred by means of a bucket elevator to the top of the cracking tower.

The fluid catalyst process

In this process, the catalyst is finely divided powder of alumina silica gel. The temperature employed is 454–537°C and pressure 20 psi. The catalyst continually flows through the reactor, the regenerator and the cyclone separators being suspended in gas or oil vapour or air (in the regenerator). The suspension remains in the flowing fluid condition, being in a violent state of agitation.

The crude oil is preheated in a tube furnace and then vapourised by injection into the hot fluid catalyst (coming from the regenerator).

The velocity of the particles diminishes in the reactor, and a relatively dense bubbling bed of catalyst particles is formed. The vapour passing through this bed is cracked rapidly, at a temperature of 480°C.

The product (vapour) carrying some catalyst with it passes through a settling zone in the reactor into a cyclone, where the vapour is made substantially free from the catalyst particles. The separated catalyst particles are introduced into the catalyst bed of the reactor, from the bottom of which the catalyst is withdrawn into an air stream. The air stream carries it into the regenerator. Here again, a denser bed, a violently agitated bed—is formed and the carbonaceous matter is burnt away in the air at a temperature of 590°C. The carbon free particles are separated from spent air, first in a settling zone above the catalyst bed and then in a cyclone. The catalyst is then returned to the reactor on the stream of hot oil vapour.

The cracked product vapour, substantially freed from catalyst particles, is fractionated to obtain gas, gasoline and gas oil. The gas oil containing some catalyst particles is recycled. A high octane gasoline is obtained.

Octane number

The octane number of a fuel is the percentage of iso-octane which must be added to n-heptane in order to obtain a mixture which matches the knocking characteristics of the fuel under examination, in a one cylinder motor engine under standard conditions. Thus octane number is an index of knocking of a gasoline in the internal combustion engine. The octane number of iso-octane (or 2,2,4-trimethyl pentane) is 100, and that of N-heptane is zero (these values, of course, are arbitrary). Iso-octanes and other branched paraffins, under high compression, have much less knocking tendency under ordinary conditions, hence the gasoline which contains larger amounts of branched paraffins has high octane value, i.e. less knocking tendency and is the better motor fuel for modern motor cars and aeroplanes.

Ordinary gasoline containing large amounts of straight-chain paraffins, and also small amounts of olefins, naphthenes and aromatics (all of which have a high octane number) has a low octane number varying from 73 to 20. The cracked gasoline containing larger amounts of olefins and other high octane constituents has a high octane number (approximately 80).

The petrol used in an aeroplane (called aviation gasoline or petrol) has an octane number 100. The octane number of cracked petrol is higher than those obtained by straight distillation. The latter is therefore subjected to many processes of refining, such as: (i) alkylation, (ii) isomerisation; (iii) cyclisation; (iv) aromatisation, etc. in order to improve its octane number. The 100-octane fuel for aircraft contains (i) 40 to 95 per cent iso-octane; (ii) 45 of 74 per cent octane aviation gasoline; (iii) 15 per cent of iso pentane; and (iv) 3 cc of tetraethyl lead (anti-knock compound).

The octane rating is determined by two methods which generally give different results. One is the CRF-M method (the motor method) which usually gives lower figures, the other is the CFR-R method (the research method). For high values, such as 100, both methods frequently agree.

Cetane number

In diesel engines, the air alone is compressed, the temperature rising to about 300°C. At the end of the upward (compression) stroke, the liquid fuel is sprayed in, which automatically ignites, no sparking being necessary.

For diesel engines, fuels of lower octane number are suitable. Normal hexadecane, known as cetane, is found to be most suitable (its value is taken to be 100) and methyl naphthalene the least (value taken as zero) efficient for use in diesel engines.

The suitability of a fuel for a diesel engine is expressed in terms of cetane number.

The cetane number of a fuel is defined as the percentage of cetane which is to be mixed with methyl naphthalene in order to obtain a fuel which matches the performances of the fuel under examination.

For heavy fuels, a cetane number of 30 gives a good performance, while for the lighter diesel engines, a lighter fuel testing 40–60 cetane is chosen.

Flash point

The flash point of a fuel is defined as the minimum temperature at which a fuel produces enough vapours to cause a momentary flash when a small flame is applied to it.

The flash point of gasoline is 45°C and that of motor lubricating oil about 230°C. The value largely depends on the manner of determination and the apparatus used. The flash point of kerosene should be sufficiently high to prevent accidents and is fixed by the governments of different countries. In India, the minimum flash point permitted is 111°F, but in England it is 73°F. The determination of flash point in Abel's apparatus is carried out as follows:

The sample of fuel contained in a metal cup is heated on a water bath and by means of an automatic device is brought in contact with a small flame from time to time by opening momentarily a slit in the cover. When the flash point is reached, the vapours ignite with a slight report and the flame extinguished for the first time.

The dangerous character of petrol is due to its having too low a flash point, which in a warm country like ours, may be below the hot summer temperature.

Knocking

Knocking is a sharp metallic sound produced in the internal combustion engine. It results in huge loss of energy. Gasoline or petrol is used as a fuel in internal combustion engines. The vapours of petrol mixed with air are compressed and allowed to burn in the cylinder of the engine. This produces large expansion resulting in the movement of piston and consequently of the automobile. But if the petrol used is of such a quality that its vapour mixed with air can ignite prior to high compression on passing of the firing spark, the energy produced by burning of the fuel is not utilised in moving the piston. It is, on the other hand, lost in creating a metallic sound and we say that knocking has occurred. Knocking is maximum in petrol containing straight chain paraffins, but is not produced to any extent in the presence of branched-chain paraffins.

Anti-knock compounds

Some compounds are added to gasoline or petrol to prevent or minimise its knocking tendency. These compounds are known as the anti-knock compounds. Tetraethyl lead ($Pb(C_2H_5)_4$) and iron carbonyl ($Fe(CO)_5$) are known as anti-knock compounds—of these, tetraethyl lead is largely employed to improve the anti-knock properties of gasoline or petrol. The anti-knock compounds retard the rapid combustion of gasoline vapour in the cylinder. Tetraethyl lead is mixed with 20 per cent ethylene dibromide and 9 per cent ethylene dichloride (formed during combustion of leaded gasoline) to form volatile lead bromide which is swept out in the exhaust gases.

Tetraethyl lead is prepared on a large-scale by the action of lead sodium alloy on ethyl chloride. The finely-divided alloy is mixed with ethyl chloride in a reactor, jacketted and fitted with a stirrer, and heated to 70°C to start the exothermic reaction. The reactor is then cooled to maintain the temperature between 40 and 60°C. After the reaction is over, the lead derivative is separated by steam distillation and the residue is melted down again to lead for re-use.

$$4C_2H_5Cl + 4NaPb = Pb(C_2H_5)_4 + 4NaCl + 3Pb$$

Utilisation of the gases obtained from natural gas

Besides large quantities of methane, natural gas yields propane, butane, pentane which have gained considerable importance now-a-days as fuels, as also raw materials for plastic and synthetic rubber industries. All these gases are liquefied and transported in cylinders as gaseous fuels.

Butane is converted into isobutane which is alkylated with butylene. Butane and propane are cracked to produce ethylene which is the essential ingredient for polythene, polyvinyl chloride plastic, ethylene glycol, from which terylene is produced. Butadiene obtained from butane is used for the manufacture of synthetic rubber.

Kerosene oil

Kerosene oil contains paraffins, naphthenes and aromatic hydrocarbons and so chemical and physical properties of kerosene depends upon the amount, structures and boiling range of these hydrocarbons. In fact kerosene oil is a straight run distillate whose boiling range lies between gas oil and gasoline. The initial and final boiling range of kerosene oil lie between 128°C–180°C and 235°C–330°C respectively. The specific gravity of kerosene oil varies from 0.775–0.850. Its flash point varies from 24°C–66°C. In India, the flash point of kerosene oil is 44°C. A paraffinic kerosene gives high flame without smoke, but the kerosene containing high percentage of aromatic and naphthenic hydrocarbons gives a lower flame with smoke. The kerosene oil is mostly used for domestic lighting and burning as fuel.

Gas oil

The gas oil is a petroleum distillate which distils between kerosene oil and lubricating oils. The boiling range of gas oils lies between 180°C and 450°C. They vary in colour from yellow to red. Some variety of gas oil also furnish bluish or greenish fluorescence. Gas oils undergo cracking at 725–775°C to form a gas which can be blended with water gas ($CO + H_2$) to give heating as well as illuminating properties.

Diesel oil

Diesel oil or gas oil is the fraction with the boiling range 250–350°C. It is used as fuel in compression ignition engines such as buses, lorries and trucks. A good diesel oil should have very little ash content, sulphur and aromatics. An ideal diesel fuel does not knock in the engine. Diesel knock is a minor explosion occurring in the engine due to an ignition delay, i.e., a time lag between injection of the fuel and ignition. Among other factors the chemical composition of the diesel hydrocarbons is an essential factor in determining the knocking tendency. Long chain hydrocarbons ignite readily and hence cause no ignition delay. Highly branched chain hydrocarbons and aromatics suffer ignition lag. Hence the former group knock less and the latter group knock most. Diesel knock is measured in terms of cetane number or cetane rating which is defined as the percentage by volume of cetane in a mixture of cetane and α-methylnaphthalene which mixture has the same knocking tendency as the given diesel fuel. In this scale cetane, which has the formula $C_{16}H_{34}$ is assigned a cetane number of 100, while α-methylnaphthalene is assigned a cetane number of zero.

In place of α-methylnapthalene another hydrocarbon called heptamethylnonane can also be used. Conditions for diesel knock are diametrically opposite to those for petrol knock. This is because the former is due to ignition delay, while the latter is due to premature ignition. It may be safely said that a good petrol is bad diesel and a good diesel is a bad petrol. In other words, a liquid fuel with a high octane number has a low cetane number and vice-versa.

The cetane number of a diesel fuel can be improved by adding some chemicals in very small amounts. Examples for these chemicals called dopants include ethyl nitrite and amyl nitrate.

A second important parameter for a diesel fuel is the aniline point which is the temperature at which equal volumes of aniline and the diesel fuel just cease to be completely miscible. The product of the aniline point and the specific gravity in proper units is called the diesel index. Usually the diesel index of an oil is three units higher than its cetane number.

Other characteristics of the diesel oil which have to be studied are the smoke point, flash point, cloud point and pour point which are important for lubricating oils as well.

PETRO-CHEMICALS

Some of the important petrochemicals are discussed below :

Acetaldehyde

It is colourless liquid; pungent, fruity odour. Sp. gr. 0.783 (18/4°C); b.p. 20.2°C; m.p. –123.5°C; vapour pressure 740.0 mm (20°C); flash point –40°F (–40°C) (open cup); specific heat 0.650; refractive index 1.3316 (20°C); wt. 6.50 lb/gal (20°C); miscible with water, alcohol, ether, benzene, gasoline, solvent naphtha, toluene, xylene, turpentine and acetone.

It is prepared by: (i) oxidation of ethylene; (ii) vapour-phase oxidation of ethanol; (iii) vapour-phase oxidation of propane and butane; (iv) catalytic reaction of acetylene and water (chiefly in Germany). It is highly flammable; toxic (narcotic). Dangerous fire and explosion risk. Explosive limits in air 4 to 57%. Tolerance, 100 ppm in air.

It is manufactured by acetic acid and acetic anhydride, n-butanol, 2-ethylhexanol, peracetic acid, aldol pentaerythritol, pyridines, chloral, 1,3-butylene glycol, and trimethylolpropane; synthetic flavours.

Acetic Anhydride

It is colourless mobile, strongly refractive liquid; strong odour; sp. gr. 1.0830 (20/20°C) b.p. 139.9° C; f.p. –73.1°C; flash point 121°F (49.4°C) (C.C.). Autoignition temp. 732°F (385°C); wt/gal (20°C) 9.01 lbs. Miscible with alcohol, ether, and acetic acid; soluble in cold water; decomposes in hot water to form acetic acid. Combustible.

It is prepared by: (i) oxidation of acetaldehyde with air or oxygen with catalyst; (ii) by catalysed thermal decomposition of acetic acid to ketone; (iii) reaction of methyl acetate and carbon monoxide; (iv) from carbon monoxide and methanol. It is strongly irritating and corrosive may cause burns and eye damage, tolerance 5 ppm in air. Moderate fire risk. It is used in cellulose acetate fibres and plastics, vinyl acetate, dehydrating and acetylating agent in production of pharmaceuticals dyes, perfumes explosives, aspirin, esterifying agent for food starch (5% wax).

Acetone

It is colourless volatile liquid; sweetish odour. M.p. –94.3°C; b.p. 56.2°; refractive index (20°C) 1.3591; sp. gr. (20/20°C) 0.792; wt/gal (15°C) 6.64 lb; flash point (open cup) 15°F (–9.4°C). Autoignition temp. 1000°F (537°C). Miscible with water, alcohol, ether, chloroform and most oils.

It is prepared by: (i) oxidation of cumene; (ii) dehydrogenation or oxidation of isopropyl alcohol with metallic catalyst; (iii) vapour-phase oxidation of butane; (iv) by-product of synthetic glycerol production. It is flammable; dangerous fire risk. Explosive limits in air 2.6 to 12.8%. Tolerance, 750 ppm in air. Narcotic in high concentrations. Low to moderate toxicity by ingestion and inhalation.

It is used in chemicals (methyl isobutyl ketone, methyl isobutyl carbinol; methyl methacrylate; bisphenol (A); paint, varnish and lacquer solvent; cellulose acetate, especially as spinning solvent; to clean and dry parts of precision equipment; solvent for potassium iodide and permanganate; delusterant for cellulose acetate fibres; specification testing of vulcanised rubber products.

Acrolein

It is colourless or yellowish liquid; disagreeable choking odour. Soluble in water, alcohol and ether. Polymerises readily unless inhibitor (hydroquinone) is added. Very reactive, B.p. 52.7° C; m.p. –87.0°C; sp. gr. (20/20°C) 0.8427; wt/gal (20°C) 7.03 lb; flash point (COC) below 0°F (–17°C). Autoignition temp. 532°F (277°C).

It is prepared by: (i) oxidation of allyl alcohol or propylene; (ii) by heating glycerol with magnesium sulphate; (iii) from propylene with bismuth-phosphorus-molybdenum catalyst. It is toxic by inhalation and ingestion; strong irritant to eyes and skin. Tolerance, 0.1 ppm in air. Flammable, dangerous fire risk. Explosive limits in air 2.8 to 31%.

It is used in intermediate for synthetic glycerol, polyurethane, and polyester resins, methionine, pharmaceuticals; herbicide; warning agent in gases.

Acrylonitrile

It is colourless, mobile liquid; mild odour; f.p. –83°C; b.p. 77.3–77.4°C; sp. gr. 0.8004 (25°C); flash point (TOC) 32°F (0°C). Soluble in all common organic solvents; partially miscible with water.

It is prepared from: (i) propylene, oxygen, and ammonia with either bismuth phosphomolybdate or a uranium-based compound as a catalyst; (ii) addition of hydrogen cyanide to acetylene with cuprous chloride catalyst; (iii) dehydration of ethylene cyanohydrin. It is toxic by inhalation and skin absorption. A known carcinogen. Flammable, dangerous fire risk. Explosive limits in air 3 to 17%. Tolerance 2 ppm (OSHA); no exposure permitted (ACGTH).

It is used as monomer for acrylic and modacrylic fibres and high-strength whiskers; ABS and acrylonitrile-styrene copolymers; nitrile rubber; cyanoethylation of cotton; synthetic soil blocks (acrylonitrile polymerised in wood pulp); organic synthesis grain fumigant; monomer for a semi-conductive polymer that can be used like inorganic oxide catalysts in dehydrogenation of *tert*-butyl alcohol to isobutylene and water.

Adipic Acid

It is white, crystalline solid. M.p. 152°C; b.p. (100 mm) 265°C; sp. gr. (20/4°C) 1.360; flash point (closed cup) 385°F (196°C). Slightly soluble in water; soluble in alcohol and acetone. Relatively stable. Combustible; low toxicity.

It is prepared by oxidation of cyclohexane, cyclohexanol, or cyclohexanone with air or nitric acid.

It is used in manufacture of nylon and of polyurethane foams; preparation of esters for use as plasticisers and lubricants; food additive (acidulant); baking powders; adhesives.

Allyl Alcohol

It is colourless liquid with pungent mustard like odour. B.p. 96.9°C; m.p. −129°C; sp. gr. (20/4°C) 0.8520; wt/gal (20°C) 7.11 lb; refractive index 1.4131; flash point (TOC) 70°F (21°C); autoignition point 713°F (375°C). Miscible with water, alcohol, chloroform, ether.

It is prepared by: (i) hydrolysis of allyl chloride (from propylene) with dilute caustic; (ii) isomerisation of propylene oxide over lithium phosphate catalyst at 230–270°C; (iii) dehydration of propylene glycol. It is toxic by ingestion and inhalation; strong irritant to eyes and skin. Flammable, moderate fire risk. Tolerance, 2 ppm in air.

It is used in esters for use in resins and plasticisers; intermediate for pharmaceuticals and other organic chemicals; manufacture of glycerol and acrolein; military poison gas; herbicide.

Allyl Chloride

It is a colourless liquid with unpleasant pungent odour; b.p. 45.0°C; f.p. −134.5°C; sp. gr. (20/4°C) 0.9382; wt/gal (20°C) 7.83 lb; refractive index 1.416; flash point (TOC) −20°F (−29°C). Autoignition temp. 737°F (390°C). Insoluble in water; miscible with alcohol, chloroform, ether, and naphtha.

It is prepared by gas-phase direct chlorination of propylene at 15 psi and 400–500°C. Method of purification - distillation. It is strong irritant to tissue. Toxic by ingestion, inhalation and skin absorption. Highly flammable, dangerous fire risk. Explosive limits in air 3.3 to 11.2% Tolerance, 1 ppm in air.

It is used in the preparation of allyl alcohol and other allyl derivatives; thermosetting resins for varnishes, plastics, adhesives; synthesis of pharmaceuticals, glycerol, and insecticides; precursor of epichlorohydrin.

Carbon Tetrachloride

It is colourless liquid; vapour 5.3 times heavier than air; sweetish, distinctive odour. sp. gr. 1.585 (25/4°C); b.p. 76.74°C; freezing point −23.0°C; refractive index 1.4607 at 20°C; vapour pressure 91.3 mm (20°C); wt/gal 13.22 lb (25°C); flash point. none. Miscible with alcohol, ether, chloroform, benzene, solvent naphtha, and most of the fixed and volatile oils; insoluble in water. Non-combustible.

It is prepared by: (i) interaction of carbon disulphide and chlorine in presence of iron; (ii) chlorination of methane or higher hydrocarbons at 250–400°C. It is toxic by ingestion, inhalation, and skin absorption. Narcotic. Tolerance, 5 ppm in air. Decomposes to phosgene at high temperature. Do not use to extinguish fire. A known carcinogen (OSHA).

It is used as refrigerants; metal degreasing; agricultural fumigant; chlorinating organic compounds; production of semi-conductors; solvent (fats, oils, rubber, etc.).

Chloroform

It is a colourless, highly refractive, heavy, volatile liquid; characteristic odour; sweet taste. Keep away from light. Miscible with alcohol, ether, benzene, carbon disulphide, carbon tetrachloride, fixed and volatile oils; slightly soluble in water. Sp. gr. 1.485 (20/20°C); b.p. 61.2°C; freezing point −63.5°C; wt/gal 12.29 lb (25°C); refractive index 1.4422 (25°C); non-flammable, but will burn on prolonged exposure to flame or high temperature.

It is prepared by: (i) reaction of chlorinated lime with acetone, acetaldehyde, or ethyl alcohol; (ii) by-product from the chlorination of methane. It is toxic by inhalation; narcotic. Prolonged inhalation or ingestion may be fatal. Tolerance, 10 ppm in air; 50 mg per cubic metre of air. A known carcinogen (OSHA). It has been prohibited by FDA from use in drugs, cosmetics and food packaging, including cough medicines, toothpastes, etc.

It is used as fluorocarbon refrigerants; fluorocarbon plastics; solvent; analytical chemistry; fumigant; insecticides.

Cresol

It is colourless, yellowish, or pinkish liquid; phenolic odour; sp. gr. 1.030–1.047; wt/gal 8.66–8.68 lb; flash point approx. 180°F (82°C); m.p. 11–35°C; b.p. 191–203°C. Soluble in alcohol, glycol, dilute alkalies and water.

It is prepared by coal-tar (from coke and gas works); also from toluene by sulphonation or oxidation. It is toxic and irritant; corrosive to skin and mucous membranes; absorbed through skin. Tolerance, 5 ppm in air.

It is used as disinfectant; phenolic resins; tricresyl phosphate; ore flotation; textile scouring agent; organic intermediate, mfg. of salicyladehyde, coumarin, and herbicides; surfactant; synthetic food flavours (para isomer only).

Dodecylbenzene

It is prepared by alkylation of benzene with isomeric dodecenes, obtained usually by polymerisation of propylene. It is moderately toxic by ingestion.

Dodecylphenol

It is a straw-coloured liquid; sp. gr. 0.94 (20/20°C); flash point 325°F (162.7°C); boiling range 310–335°F (154–168°C); phenolic odour; soluble in water, combustible.

It is moderately toxic and used as solvent, intermediate for surface-active agents; oil additives; resins; fungicides; bactericides; dyes; pharmaceuticals; adhesives; rubber chemicals.

Epichlorohydrin

It is highly volatile, unstable liquid. Chloroform-like odour; miscible with most organic solvents; slightly soluble in water. Sp. gr. 1.1761 (20/20°C); b.p. 115.2°C; wt/gal 9.78 lb; vapour pressure 12.5 mm (20°C); f.p. –25°C; viscosity 1.12 cp (20°C); refractive index (n 25/D) 1.4358; flash point 93°F (33.9°C) (TOC).

It is prepared by removing hydrogen chloride from dichlorohydrin. It is toxic by inhalation, ingestion, and skin absorption, strong irritant. A suspected carcinogen. Tolerance, 2 ppm in air. Flammable, moderate fire risk.

It is used as major raw material for epoxy and phenoxy resins; manufacture of glycerol; curing propylene-based rubbers; solvent for cellulose esters and ethers; high wet-strength resins for paper industry.

Epoxy Resin

A thermosetting resin based on the reactivity of the epoxide group. One type is made from epichlorohydrin and bisphenol A. Aliphatic polyols such as glycerol may be used instead of the aromatic bisphenol A. Molecules of this type have glycidyl ether structures, in the terminal positions, have many hydroxyl groups, and cure readily with amines.

Another type is made from polyolefins oxidised with peracetic acid. These have more epoxide groups, within the molecule as well as in terminal positions, and can be cured with anhydrides, but require high temperatures. Many modifications of both types are made commercially. Halogenated bisphenols can be used to add flame-retardant properties. The reactive epoxies form a tight cross-linked polymer network, and are characterised by toughens, good adhesion, corrosion, chemical resistance, and good dielectric properties.

Most epoxy resins are the two-part type, which harden when blended. A one-component liquid type for filament winding and a pelletised type for injection moulding are available under the trademark "Arnox".

It is strong irritant to skin in uncured state and is used in surface coatings, as on household appliances and gas storage vessels; adhesives for composites and for metals, glass, and ceramics; casting metal-forming tools and dies; encapsulation of electrical parts; filament-wound pipe and pressure vessels; floor surfacing and wall panels; neutron-shielding materials; cements and mortars; non-skid road surfacing; rigid foams; oil wells (to solidify sandy formations); matrix for stained glass windows.

Ethanolamine

It is colourless, hygroscopic, viscous liquid. Ammoniacal odour. Strong base. Miscible with water methanol, acetone. Sp. gr. 1.0179 (20/20°C); b.p. 170.5°C, melting point 10.5°C; vapour pressure 0.48 mm (20°C); flash point (open cup) 200°F (93.3°C); wt/gal 8.5 lb (20°C), combustible.

It is prepared by reaction of ethylene oxide and ammonia gives a mixture of mono-, di-, and triethanolamines. It is moderately toxic, skin irritant, tolerance 3 ppm in air.

It is used as scrubbing acid gases (H_2S, CO_2), especially in synthesis of ammonia, from gas streams; non-ionic detergents used in drycleaning, wool treatment, cal intermediates; pharmaceuticals; corrosion inhibitor, rubber accelerator.

Ethyl Alcohol

It is a pure 100% absolute alcohol (dehydrated) colourless, limpid volatile liquid; b.p. 78.3°c; f.p. – 117.3°C; ethereal, vinous odour; pungent taste. Miscible with water, methyl alcohol, ether, chloroform, and acetone.

(95%) Refractive index 1.3651 (15°C); surface tension 22.3 dynes/cm (20°C); viscosity 0.0141 poise (20°C); vapour pressure 43 mm (20°C); specific heat 0.618 cal/g (23°C); flash point 55°F (12.7°C); sp. gr. (15.56°C) 0.816; b.p. 78°C; f.p. –114°C. Autoignition temp. 793°F (422°C).

It is prepared by: (i) ethylene by direct catalytic hydration of with ethyl sulphate as intermediate; (ii) fermentation of sugar-rich organic wastes (black-strap molasses, whey, degraded corn, etc.); (iii) enzymatic hydrolysis of cellulose. It is flammable, dangerous fire risk; flammable limits in air, 3.3 to 19%. Tolerance, 1000 ppm in air. Ethyl alcohol is classified as a depressant drug. Though it is rapidly oxidised in the body and is therefore non-cumulative, ingestion of even moderate amounts causes lowering of inhibitions, often succeeded by dizziness, headache, or nausea. Larger intake causes loss of motor nerve control, shallow respiration, and in extreme cases unconsciousness and even death. Degree of intoxication is determined by concentration of alcohol in the brain. Of primary importance is the fact that intake of even moderate amounts of together with barbiturates or similar drugs is extremely dangerous, and may even be fatal.

It is used as solvent of resins, fats, fatty acids, oils, hydrocarbons; extraction medium; manufacture of acetaldehyde, acetic acid, ethylene, butadiene, 2-ethylhexanol, dyes, pharmaceuticals, elastomers, detergents, cleaning preparations, surface coatings, cosmetics, explosives; anti-freeze; beverages; anti-sepsis; gasohol yeast-growth medium.

Note: Ethyl alcohol from fermentation of biomass and hydrolysis of cellulose is a significant alternate energy source, especially as an automotive fuel. Its use in gasohol will continue to increase.

Ethylbenzene

It is colourless liquid; aromatic odour; vapour heavier than air; boiling point 136.187°C; refractive index 1.49594 (20°C) sp. gr. 0.867 (20°C); f.p. –95°C; wt/gal 7.21 lb (25°C); flash point 59°F (15°C); autoignition temp. 810°F (432°C); specific heat 0.41/cal/gal/°C; viscosity 0.64 centipoise (25°C). Soluble in alcohol, benzene, carbon tetrachloride, and ether; almost insoluble in water.

It is prepared by: (i) heating benzene and ethylene in presence of aluminium chloride; with subsequent distillation; (ii) by fractionation directly from the mixed xylene stream in petroleum refining. It is flammable, dangerous fire risk. Moderately toxic by ingestion, inhalation, and skin absorption. Irritant to skin and eyes. Tolerance, 10 ppm in air. It is used as intermediate in production of styrene; solvent.

Ethyl Chloride

It is prepared by gas at room temperature; when compressed, a colourless, volatile liquid. Ether-like-odour burning taste. Stable and non-corrosive when dry but will hydrolyse in the presence of water or alkalies. Miscible with most of the commonly used solvents; slightly soluble in water. Sp. gr. 0.9214; f.p. –140.85°C; b.p. 12.5°C; critical point 187.2°C (52 atm; sp. gr. 0.33); vapour pressure 1000 mm (20°C); flash point (closed cup) –58°F (–50°C). Autoignition temp. 966°F (518°C).

It is prepared from: (i) ethylene and hydrogen chloride; (ii) by passing hydrogen chloride into a solution of zinc chloride and ethyl alcohol. It is highly flammable, severe fire and explosion risk. Flammable limits in air 3.8–15.4%. Moderately toxic; irritant to eyes. Tolerance, 1000 ppm in air.

It is used in manufacture of tetraethyl lead and ethylcellulose; anaesthetic; organic synthesis; alkylating agent; refrigeration; analytical reagent; solvent for phosphorus; sulphur, fats, oils, resins and waxes; insecticides.

Ethylene

It is colourless gas with sweet odour and taste; freezing point –169°C; boiling point –103.9°C; flash point –213°F (–135°C); sp. gr. of liquid at 0°C 0.610; vapour density 0.975 (air = 1.29); critical temperature 9.5°C; autoignition temp. 1009°F (543°C); critical pressure (absolute) 744 psi. Purity not less than 96% ethylene by gas volume, not more than 0.5% acetylene, not more than 4% methane and ethane; 13.4 cu ft/lb (15.6°C), slightly soluble in water, alcohol, and ethyl ether. An asphyxiant gas.

It is prepared by: (i) thermal cracking of hydrocarbon gases: (ii) dehydration of ethyl alcohol. It is highly flammable; dangerous fire and explosion risk. Explosive limits in air, 3 to 36% by vol.

It is used in polyethylene, polypropylene, ethylene oxide, ethylene dichloride, ethylene glycols, aluminium alkyls, vinyl chloride, vinyl acetate ethyl chloride, ethylene chlorohydrin, acetaldehyde, linear alcohols, polystyrene, styrene, polyvinyl chloride, SBR, polyester resins, trichloroethylene, etc.; refrigerant; welding and cutting and metals; anaesthetic; in orchard sprays to accelerate fruit ripening.

Note: Under development is a method of producing ethylene directly from petroleum by so-called flame-cracking. The advanced cracking reactor (ACR) process involves combustion at about 2000°C of a mixture of crude oil and high-temperature gases. Combustion in supported by pure oxygen. If the prototype unit is successful, large-scale application of this method will be possible within the decade.

Ethylene Dibromide

It is a colourless, non-flammable liquid. Sweetish odour, emulsifiable, miscible with most solvents and thinners; slightly soluble in water Sp. gr. 2.17–2.18 (20°C); wt/gal 18.1 lb; b.p. 131°C; vapour pressure 17.4 mm (30°C); f.p. 9°C; refractive index 1.5337 (25°C); flash point, none.

It is prepared by bromine on ethylene and is toxic by inhalation, ingestion, and skin absorption. Carcinogen in test animals. Strong irritant to eyes and skin.

It is used as Scanvenger for lead in gasoline; grain fumigant; general solvent; waterproofing preparations; organic synthesis; fumigant for free crops.

Ethylene Dichloride

It is colourless, oily liquid, chloroform-like odour; sweet taste. Stable to water, alkalies, acids, or active chemicals. Resistant to oxidation will not corrode metals. Miscible with most common solvents; slightly soluble in water. B.p. 83.5°C; f.p. –35.5°C; sp. gr. 1.2554 (20/4°C); wt/gal 10.4 lb; refractive index 1.444; flash point 56°F (13.3°C).

It is in action of chlorine on ethylene with subsequent distillation, with metallic catalyst; also by reaction of acetylene and HCl. It is toxic by ingestion, inhalation, skin absorption. Strong irritant to eyes and skin. Tolerance, 10 ppm in air Flammable, dangerous fire risk; explosive limits in air 6 to 16%. May be carcinogenic.

It is used as vinyl chloride solvent; lead scavenger in anti-knock gasoline; paint, varnish and finish removers; metal degreasing; soaps and scouring compounds; wetting and penetrating agents; organic synthesis; ore flotation.

Ethylene Glycol

It is clear, colourless, syrupy liquid; sweet taste; hygroscopic; lowers freezing point of water. Relatively non-volatile. Odourless Soluble in water, alcohol and acetone. Sp. gr. .1155 (20°C); b.p. 197.2°C; f.p. –13.5°C; wt/gal 9.31 lb (15/15°C); refractive index 1.430 (25°C); flash point 240.8°F (116°C); combustible; autoignition temp. 775°F (412°C).

It is prepared by: (i) air oxidation of ethylene followed by hydration of the ethylene oxide formed; (ii) acetoxylation; (iii) from carbon monoxide and hydrogen (synthesis gas) from coal gasification; (iv) oxirane process. It is toxic by ingestion and inhalation.

Ethylene Oxide

It is colourless gas at room temperature; liquid below 12°C; soluble in organic solvents; miscible with water in all proportions; f.p. – 111.3°C; b.p. 10.73°C; sp. gr. (20/20°C) 0.8711; wt/gal (20°C) 7.25 lb; viscosity (0°C) 0.32 cp; flash point (TOC) below 0°F (–17.7°C) autoignition temp. 804°F (429°C).

It is prepared by: (i) oxidation of ethylene in air or oxygen with silver catalyst; (ii) action of an alkali on ethylene chlorohydrin. It is highly flammable; dangerous fire and explosion risk. Flammable limits in air 3 to 100%. Toxic; irritating to eyes and skin. Tolerance, 10 ppm in air.

It is used in manufacture of ethylene glycol and higher glycols; surfactants; acrylonitrile; ethanolamines; as petroleum demulsifier; fumigant; rocket propellant; industrial sterilant; e.g., medical plastic tubing; fungicide.

Formaldehyde

A readily polymerisable gas. It is commercially offered as a 37 to 50% aqueous solution, which may contain up to 15% methanol to inhibit polymerisation. These commercial grades are called formalin. It is one of the few organic compounds known to exist in outer space.

It is strong, pungent, odour; vapour density 1.08; b.p. –19°C; f.p. –118°C; autoignition point 806°F (430°C); soluble in water and alcohol. (Acqueous 37% solution with 15% methanol): B.p. 96°C; flash point 122°F (50°C); (methyanol-free): b.p. 101°C; flash point 185°F (85°C). Sp. gr. 1.08. It is prepared by oxidation of synthetic methanol or low-boiling petroleum gases such as propane and butane. Silver, copper, or an iron-molybdenum oxide are the most common catalysts.

Grades

Aqueous solutions: 37%, 44%, 50% inhibited (with varying percentage of methanol) or stabilised or unstabilised (methanol-free); also available in solution in n-butanol, methanol, or urea.

It is toxic by inhalation; strong irritant. Moderate fire risk. Tolerance, 2 ppm in air. Explosive limits in air 7 to 73% (Solution): Avoid breathing vapour and skin contact.

It is used in urea and melamine resins; polyacetal resins; phenolic resins; ethylene glycol; pentacrythritol; hexamethylenetetramine; fertiliser; dyes, medicine (disinfectant, germicide); embalming fluids; preservative; hardening agent; reducing agent, as in recovery of gold and silver; corrosion inhibitor in oil wells; durable-press treatment of textile fabrics; industrial sterilant; treatment of grain smut.

Isobutylene

It is colourless volatile liquid or easily liquefied gas; coal gas odour; b.p. –6.9°C; f.p. –139°C; flash point –105°F (–76°C); sp. gr. 0.6 (20°C); soluble in organic solvents. Polymerises easily and also reacts easily with numerous materials. Non-toxic. Autoignition point 869°F (465°C).

It is prepared by fractionation of refinery gases; catalytic cracking of MTBE. It is highly flammable; dangerous fire and explosion risk. Explosive limits in air 1.8 to 8 8%.

It can be used in production of isooctane, high-octane aviation gasoline; butyl rubber, polyisobutene resins, tert-butyl chloride, tert-butanol methacrylates; copolymer resins with butadiene, acrylonitrile, etc.; methyl tert-butyl ether.

Isobutyl Alcohol

It is colourless liquid. Partially soluble in water. Soluble in alcohol and ether. Sp. gr. 0.806 (15°C); b.p. 107°C; flash point 100°F (37.7°C) (open cup); f.p. –108°C; refractive index 1.397 (15°C). Autoignition point 800°F (426°C).

It is prepared by by-product of synthetic methanol production, purified by rectification. It is flammable, moderate fire risk. Moderately toxic, strong irritant. Tolerance, 50 ppm in air.

It is used in organic synthesis, latent solvent in paint and lacquers; intermediate for amino coating resins; substitute for n-butyl alcohol. Paint removers; fluorometric determinations; liquid chromatography; fruit flavour concentrates.

Isophthalic Acid

It is colourless crystals; m.p. 345–348°C; sublimes. Slightly soluble in water; soluble in alcohol and acetic acid; insoluble in benzene and petroleum ether. Combustible; low toxicity.

It is prepared by: (i) oxidation of meta-xylene; (ii) liquid phase oxidation of mixed xylenes; (iii) direct oxidation of mixed alkyl aromatics with heavy metal salts and bromine as catalysts.

It is used in polyester, alkyd, polyurethane, and other high polymers; plasticisers.

Isopropyl Alcohol

It is a colourless liquid. Pleasant odour; b.p. 82.4°C; sp. gr. 0.7863 (20/20°C); refractive index 1.3756 (20°C); sp. ht. 0.65/cal/g. f.p. –86°C; critical temperature 235°C; critical pressure 53 atmospheres; vapour pressure 33 mm at 20°C; flash point 53°F (11.7°C) (TOC); heat of combustion 14,346 Btu/lb; heat of vapourisation 288 Btu/lb; viscosity 2.1 cp (25°C). Autoignition temp. 850°F (453°C). Soluble in water, alcohol and ether.

It is prepared by treatment of propylene with sulphuric acid and hydrolysing. It is purified by rectification. It is flammable, dangerous fire risk. Toxic by ingestion and inhalation. Tolerance, 400 ppm in air. Explosive limits in air 2 to 12%. Ingestion of 100 cc can be fatal.

It is used in manufacture of acetone and its derivatives; manufacture of glycerol and isopropyl acetate; solvent for essential and other oils, alkaloids, gums, resins, etc.; latent solvent for cellulose derivatives; coatings solvent; deicing agent for liquid fuels; lacquers; extraction processes; dehydrating agent; preservative; lotions; denaturant.

Malathion

It is yellow, high-boiling liquid; (b.p. 156–157°C, under 0.7 mm with slight decomposition); m.p. 3.0°C; refractive index (n 25/D) 1.4985; sp. gr. 1.2315 (25°C); vapour pressure (20°C) approximately 0.00004 mm. Miscible with most polar organic solvents. Slightly soluble in water. Combustible. It is prepared from diethyl maleate and dimethyldithiophosphoric acid. It is toxic by ingestion and inhalation. Absorbed by skin. Tolerance, 10 mg per cubic metre of air. Cholinesterase inhibitor.

Maleic hydrazide

It is a crystal; m.p. 297°C; slightly soluble in hot alcohol; more soluble in hot water. It is prepared by treating maleic anhydride with hydrezine hydrate and is toxic by ingestion.

It is used as systematic herbicide; treatment of tobacco plants; post-harvest sprouting inhibitor; weed control; sugar content stabiliser in beets.

Methyl Ethyl Ketone

It is a colourless liquid; acetone-like odour; b.p. 79.6°C; sp. gr. 0.8255 (0/4°C), 0.805 (20/4°C), and 0.7997 (25/4°C); refractive index 1.379 (20°C); sp. heat 0.549 cal/g; f.p. −86.4°C; viscosity 0.40 cp (25°C); solubility in water 22.6 wt %; solubility of water 9.9 wt %; flash point (TOC) 24°F (−4.4°C); 6.71 lb/gal (20°C). Autoignition temp. 960°F (515°C). Soluble in benzene, alcohol, and ether; miscible with oils.

It is prepared from: (i) mixed n-butylenes and sulphuric acid to cause hydrolysis, followed by distillation to separate sec-butyl alcohol, which is dehydrogenated; (ii) by controlled oxidation of butane; and (iii) by fermentation.

It can be used as solvent in nitrocellulose coatings and vinyl films; "Glyptal" resins; paint removers; cements and adhesives; organic synthesis; manufacture of smokeless powder; cleaning fluids; printing; catalyst carrier; acrylic coatings. It does not dissolve cellulose acetate and most waxes.

It is flammable, dangerous fire risk. Explosive limits in air 2 to 10%. Tolerance, 200 ppm in air. Narcotic by inhalation.

Nonylphenol

It is a mixture of isomeric monoalkyl phenols, predominantly para-substituted. It is a pale yellow, viscous liquid with a slight phenolic odour. Insoluble in water; soluble in most organic solvents. Sp. gr. (20/20°C) 0.950; b.p. 293°C; f.p. −10°C (set to glass below this temperature); viscosity 563 cp (20°C); flash point 285°F (140.5°C). It is combustible.

It is used as non-ionic surfactant (nonbiodegradable); lube oil additives; stabilisers, petroleum demulsifiers, fungicides, anti-oxidants for plastics and rubber.

Perchloroethylene

It is colourless liquid; ether-like odour. Extremely stable. Resists hydrolysis. Sp. gr. (20/20°C) 1.625; b.p. 121°C; f.p. −22.4°C; weight 13.46 lb/gal (26°C); refractive index 1.5029 (25°C); flash point, none. Miscible with alcohol, ether, and oils, in all proportions. Insoluble in water. Non-flammable.

It is prepared by: (i) chlorination of hydrocarbons, and pyrolysis of the carbon tetrachloride also formed; (ii) from acetylene and chlorine via trichloroethylene. It is purified by distillation. It is moderately toxic. Irritant to eyes and skin. Tolerance, 100 ppm in air.

It is used as dry-cleaning solvent; vapour-degreasing solvent; drying agent for metals and certain other solids; vermifuge; heat-transfer medium; manufacturing of fluorocarbons.

Phenol

(i) A class of aromatic organic compounds in which one or more hydroxy groups are attached directly to the benzene ring. Examples are phenol itself (benzophenol), the cresols, xylenols, resorcinol, naphthols. Though technically alcohols, their properties are quite different; (ii) phenol (carbolic acid; phenylic acid; benzophenol; hydroxybenzene).

It is a white, crystalline mass which turns pink or red if not perfectly pure or if under influence of light; absorbs water from the air and liquefies; distinctive odour; sharp burning taste. When in very weak solution it has a sweetish taste; sp. gr. 1.97; m.p. 42.5–43°C; b.p. 182°C; flash point 172.4°F (78°C) (C.C.). Soluble in alcohol, water, ether, chloroform, glycerol, carbon disulphide, petrolatum, fixed or volatile oils and alkalies. Combustible. Autoignition temp. 1319°F (715°C).

It is prepared by the oxidation of cumene, yielding acetone as a by-product. The first step in the reaction yields cumene hydroperoxide, which decomposes with dilute sulphuric acid to the primary products, plus acetophenone and phenyl dimethyl carbinol. Several other benzene-based processes have been used in the past; derivation from benzoic acid is also possible. It can be purified by rectification.

It is toxic by ingestion, inhalation, and skin absorption. Strong irritant to tissue. Tolerance, 5 ppm in air.

It is used in phenolic resins; epoxy resins (bisphenol-A); nylon-6 (caprolactam); 2,4-D, selective solvent for refining lubricating oils; adipic acid; salicyclic acid; phenolphthalein; penta-chlorophenol; acetophenetidine; picric acid; germicidal paints; pharmaceuticals; laboratory reagent; dyes and indicators; slimicide; general disinfectant.

High-boiling phenols are mixtures containing predominantly meta-substituted alkyl phenols. Their boiling point ranges from 238 to 288°C; they set to a glass below −30°C. They are used in phenolic resins, as fuel-oil sludge inhibitors, as solvents and as rubber chemicals.

Methyl Alcohol

It is clear, colourless, mobile, highly polar liquid. Miscible with water, alcohols and ether; sp. gr. 0.7924; f.p. −97.8°C; b.p. 64.5°C; wt/gal 6.59 lb (20°C); refractive index 1.329 (20°C); surface tension 22.6 dynes/dm (20°C); viscosity 0.00593 poise (20°C); vapour pressure 92 mm (20°C); flash point (open cup) 54°F (12.2°C); autoignition temp. 867°F (464°C).

It is prepared by: (i) high-pressure catalytic synthesis from carbon monoxide and hydrogen; (ii) partial oxidation of natural gas hydrocarbons; (iii) destructive distillation of wood. It can be purified by rectification.

It is flammable, dangerous fire risk. Toxic by ingestion (causes blindness). Tolerance, 200 ppm in air. Explosive limits in air 6–36.5% by volume.

It is used in the manufacture of formaldehyde and dimethyl terephthalate; chemical synthesis (methyl amines, methyl chloride, methyl methacrylate, automotive fuels); antifreeze; solvent for nitrocellulose, ethylcellulose, polyvinyl butyral, shellac, rosin, manila resin, dyes; denaturant for ethyl alcohol; dehydrator for natural gas; fuel for utility plants (methyl fuel); feedstock for manufacture of synthetic proteins by continuous fermentation; source of hydrogen for fuel cells; home heating oil extender.

Polyethylene

Thermosetting white solid; high temperature-resistant; excellent resistance to chemicals and to crepe; high impact and tensile strength; high electrical resistivity; insoluble in organic solvents; does not stress-crack. Combustible. Non-toxic.

It is prepared by: (i) irradiating linear polyethylene with electron beam or gamma radiation; cross-linking taking place through a primary valence bond; (ii) by chemical cross-linking agent such as an organic peroxide (e.g., benzoyl peroxide). All grades of polyethylene and most copolymers can be chemically cross-linked.

It can be used wire and cable coatings and insulation (low-density grades); pipe and moulded fittings (high-density grades). Special types having low electrical resistivity can be made; these can be regarded as semiconductors.

In moulding cross-linked polyethylene, the desired part must be formed before cross-linking is initiated, as material will not change its shape after cross-linking. The variations in composition and wide range of properties approach the ideal of a universal material more closely than most polymers.

The density of polyethylene and other thermoplastic polymers is affected by the shape and spacing of the molecular chains; low-density materials have highly branched and widely spaced chains, whereas high-density materials have comparatively straight and closely aligned chains. Polymers of the latter type are called linear. The physical properties are markedly affected by increasing density.

Low-density (branched chain)

Sp. gr. 0.915; crystallinity 50–60%; m.p. 240°F; tensile strength 1500 psi; impact strength over 10 ft-lb/in./notch; thermal expansion 17×10^{-5} in./°C; soluble in organic solvents above 200°F; insoluble at room temperature.

It is prepared by: (i) ethylene is polymerised in a free-radical-initiated liquid phase reaction at 1500 atm (22,000 psi) and 375°F, with oxygen as catalyst (usually from peroxides); (ii) a much more effective and cheaper process uses pressures of only 100 to 300 psi at less than 212°F; the catalyst is undisclosed and reaction is vapour-phase.

It is used in packaging film, especially for food products; paper coating; liners for drums and other shipping containers; wire and cable coating; toys; cordage; refuse and waste bags; chewing-gum base; squeeze bottles; electrical insulation.

High-density (linear)

Sp. gr. 0.95; crystallinity 90%; m.p. 275°F; tensile strength 4000 psi; impact strength 8 ft-lb/in. notch; high electrical resistivity; film is gas-permeable, hydrophobic, does not resist nitric acid.

It is prepared by ethylene polymerised by Ziegler catalysts at from 1 to 100 atm (15 to 1500 psi) at from room temp. to 200°F. Catalyst is a metal alkyl, e.g., triethylaluminum, plus a metallic salt ($TiCl_4$) dissolved in a hydrocarbon solvent. A vapour-phase modification of this process was developed in 1965. Another method uses such metallic catalysts as Cr_2O_3 at 100–500 psi with solvents such as cyclohexane or xylene.

It can be used as blow-moulded products; injection-moulded items; film and sheet; piping; fibres; gasoline and oil containers.

Ethylene may be copolymerised with varying percentages of other materials, e.g., 2-butene or acrylic acid; a crystalline product results from copolymerisation of ethylene and propylene. When butadiene is added to the copolymer blend, a vulcanisable elastomer is obtained.

Low molecular weight

Molecular weight from 2000 to 5000. Translucent white solids; excellent electrical resistance; abrasion-resistant; resistant to water and most chemicals; sp. gr. 0.92; slightly soluble in turpentine, petroleum naphtha, xylene, and toluene at room temperature; soluble in xylene, toluene, trichloroethylene, turpentine, and mineral oils at 82.2°C; practically insoluble in water; slightly soluble in methyl acetate, acetone, and ethanol up to the boiling point of these solvents. Available as emulsified and non-emulsified forms. Combustible.

It can be used in mould-release agent for rubber and plastics; paper and container coatings; liquid polishes and textile finishing agents.

Polyethylene Glycol

It is clear, colourless, odourless, viscous liquids to waxy solids; soluble or miscible with water and for the most part with alcohol and other organic solvents; heat-stable; inert to many chemical agents; do not hydrolyse or deteriorate; have low vapour pressure. Combustible; non-toxic.

It is prepared by condensation of ethylene glycol, or of ethylene oxide and water. It is used in chemical intermediates (lower molecular weight varieties); plasticisers; softeners and humectants; ointments; polishes; paper coating; mould lubricants.

Propionic Acid

It is colourless oily liquid; rancid odour; sp. gr. 0.9942 (20/4°C); f.p. −20.8°C; refractive index 1.3862 (20°C); b.p. 140.7°C; soluble in water, alcohol, chloroform and ether. Flash point 130°F (54.4°C). Combustible. Autoignition temp. 955°F (512°C).

It is prepared by reaction of ethyl alcohol with carbon monoxide, using a boron trifluoride catalyst; also by the reaction of carbon monoxide with hydrogen and olefins or alcohols. It can be purified by rectification.

It is moderate in fire risk; strong irritant. Tolerance, 10 ppm in air.

It is used in propionates, some of which are used as mould inhibitors in bread and fungicides in general; herbicides; preservative for grains and wood chips; emulsifying agents; solutions for electroplating nickel; perfume esters; artificial fruit flavours; pharmaceuticals; cellulose propionate plastics.

Polypropylene Glycol

It is one of a group of compounds comparable to polyethylene glycols, but more oil-soluble and substantially less water-soluble. Classified by approximate molecular weight, as 425, 1025, and 2025. Non-volatile, non-corrosive liquids; lower molecular weight members are soluble in water. Solvents for vegetable oils, waxes, resins. Combustible.

It is used as hydraulic fluids; rubber lubricants; antifoam agents; intermediates in urethane foams, adhesives, coatings, elastomers; plasticisers; paint formulations; laboratory reagent.

Polystyrene

Polymerised styrene, a thermoplastic synthetic resin of variable molecular weight depending on degree of polymerisation.

It is transparent, hard solid; high strength and impact resistance; excellent electrical and thermal insulator. Attacked by hydrocarbon solvents but resists organic acids, alkalies, and alcohols. Not recommended for outdoor use; unmodified polymer yellows when exposed to light, but light-stable modified grades are available. Easily coloured, moulded and fabricated. Copolymerisation with butadiene and acrylonitrile and blending with rubber or glass fibre increase impact strength and heat resistance. Nontoxic.

It is prepared by polymerisation of styrene by free radicals with peroxide initiator.

Forms: Sheet, plates, rods, rigid foam, expandable beads or spheres. It is combustible; autoignition temperature about 800°F.

It is used as packaging, refrigerator doors; air conditioner cases, containers and moulded household wares; machine housings; electrical equipment; toys; clock and radio cabinets. (As foam): thermal insulation; light construction as in boats, etc.; ice buckets, water coolers; fillers in shipping containers; furniture construction. (As spheres): radiator leak stopper.

Propylene Dichloride

It is colourless, stable liquid. Chloroform-like odour. B.p. 96.3°C; sp. gr. 1.1583 at 20/20°C; wt/gal 9.6 lbs (20°C); refractive index 1.068 (20°C); flash point 61°F (16.1°C); solubility in water 0.26% by wt. (20°C); freezing point −80°C. Miscible with most common solvents; insoluble in water. Autoignition temp. 1035°F (557°C).

It is prepared by the action of chlorine on propylene.

It is flammable, dangerous fire risk. Explosive limits in air 3.4 to 14.5%. Toxic by ingestion and inhalation. Tolerance, 75 ppm in air.

It is used as an intermediate for perchloroethylene and carbon tetrachloride; lead scavenger for anti-knock fluids; solvents for fats, oils, waxes, gums, and resins; solvent mixtures for cellulose esters and ethers; scouring compounds; spotting agents; metal degreasing agents; soil fumigant for nematodes.

Propylene Oxide

It is colourless liquid, ethereal odour. Sp. gr. 0.8304 at 20/20°C; b.p. 33.9°C; vapour pressure 445 mm (20°C); flash point −35°F (−37.2°C); wt 6.9 lb/gal (20°C); freezing point −104.4°C; partially soluble in water; soluble in alcohol and ether.

It is prepared by: (i) chlorohydration of propylene, followed by saponification with lime; (ii) peroxidation of propylene; (iii) epoxidation of propylene by a hydroperoxide complex with molybdenum catalyst.

It is highly flammable, dangerous fire risk; explosive limits in air 2 to 22%. Tolerance, 20 ppm in air. Moderately toxic and irritant.

It is used in polyols for urethane foams; propylene glycols; surfactants and detergents; isopropanol amines; fumigant; synthetic lubricants; synthetic elastomer (homopolymer); solvent.

Meta-xylene

It is clear, colourless liquid, soluble in alcohol and ether; insoluble in water. sp. gr. (15°C) 0.8684; f.p. -47.4°C; b.p. 138.8°C; refractive index (20°C) 1.4973. Flash point 85°F (29.4°C). Autoignition temp. 982°F (527.7°C).

It is prepared by selective crystallisation or solvent extraction of meta-para mixtures. It is flammable, moderate fire risk. Tolerance, 100 ppm in air. It is used as a solvent; intermediate for dyes and organic synthesis, especially isophthalic acid; insecticides; aviation fuel.

Ortho-xylene

It is clear, colourless liquid; soluble in alcohol and ether; insoluble in water. Sp. gr. (20/4°C) 0.880; f.p. –25°C; b.p. 144°C; refractive index (20°C) 1.505. Flash point 115°F (46.1°C) (TOC). Autoignition temp. 867°F (463.8°C). It is combustible. It is moderate fire risk. Tolerance, 100 ppm in air.

It is used in the manufacture of phthalic anhydride; vitamin and pharmaceutical syntheses; dyes; insecticides; motor fuels.

Example 25.1. The following data were obtained in a gas calorimeter experiment : Volume of gas used 0.1 m^3 at NTP; weight of water heated = 25 kg; Temperature of inlet and outlet water are 20°C and 33°C; and weight of steam condensed is 0.025 kg. Calculate the higher and lower calorific value per m^3 at NTP. Heat liberated in condensing water vapour and cooling the condensate is 580 k. cals/kg.

Solution. We know H.C.V. $= \dfrac{W(T_2 - T_1)}{V}$

In the present problem, $V = 0.1$ m^3, $W = 25$kg, $T_1 = 20$°C, $T_2 = 33$°C and $m = 0.025$ kg.

Thus H.C.V. $= \dfrac{25(33 - 20)}{0.1} = 3.250$ k.cals/m^3.

Now, L.C.V. = H.C.V. – (m/V) × 580 = [3.250 – (0.025 × 580)/0.1] = 3.250 – 0.145 = 3.105 k.cal/m^3.

Example 25.2. A producer gas has the following percentage composition by volume : CO_2 = 8%, CO = 27.6%, H_2 = 10%, CH_4 = 1.2%, N_2 = 52.6% and O_2 = 0.6%. Calculate the gross calorific value of the gas. Calorific values of H_2, CO and CH_4 are 3100, 2970 and 9260 k.cals/m^3 respectively.

Solution. In the above producer gas the combustible constituents are H_2, CO and CH_4. Their volumes per m^3 of the producer gas are 0.10 m^3, 0.276 m^3 and 0.012 m^3 respectively. Thus gross calorific value of the producer gas,

= 0.10 × 3100 + 0.276 × 2970 + 0.012 × 9260 = 310.0 + 819.7 + 111.1 = 1240.8 k.cal/m^3.

Chapter 26

High Polymers

INTRODUCTION

Cellulose have extremely large molecules, the molecules are made up of recurring units. For instance, it may be recalled that the formula for cellulose may be written $(C_6H_{10}O_5)_n$. There is a unit, $C_6H_{10}O_5$, that is repeated a large number of times in the molecule. Cellulose is an example of a naturally occurring high polymer. It will be recalled that celluose is in some contrast to a substance such as glucose that has a rather modest molecular weight. Similarly cellulose nitrate and cellulose acetate may be regarded as plastics, or high polymers. These are examples of synthetic polymers, at least to the extent that the original molecule has been considerably modified.

The formation of a polymer may be regarded in the following way. Suppose we have a molecule X, and that this molecule is able to unite with itself to form X_2. (Such a case actually occurs in the equilibrium $2NO \; 1 \; N_2O_4$). But suppose that the molecule X is able to continue uniting with itself to form very long chains which might be indicated thus.

$$X–X–X–X–X–X–X–X$$

The resulting large molecule is called a polymer of X. The word "polymer" is derived from the Greek meaning "many times". In speaking of polymers it is the custom to refer to the single unit (X in the above case) as a monomer. If the molecular weight of the polymer is very large, the substance is said to be a high polymer.

Plastics are high polymer, usually synthetic, combined with other ingredients, such as curatives, fillers, reinforcing agents, colourants, plasticisers, etc.; the mixture can be formed or moulded under heat and pressure in its raw state, and machined to high dimensional accuracy, trimmed and finished in its hardened state. The thermoplastic type can be resoftened to its original condition by heat; the thermosetting type cannot.

Plastics in general (including all forms) are sensitive to high temnperatures, among the more resistant being fluorocarbon resins, nylon, phenolics, though even these soften or melt above 260°C. Other types are combustible when exposed to flame for a short time (polyethylene, acrylic polymers, polystyrene), and still others burn with evolution of toxic fumes (polyurethane).

Engineering plastics are those to which standard metal engineering equations can be applied; they are capable of sustaining high loads and stresses, and are machinable and dimensionally stable. They are used in construction, as machine parts, automobile components, etc. Among the more important are nylon, acetals, polycarbonates, ABS resins, PPO/styrene, and polybutylene terephthalate.

Fibres, films, and bristles are examples of extruded forms. Plastics may be shaped by either compression moulding (direct pressure on solid material in a hydraulic press) or injection moulding (ejection of a measured amount of material into a mould in liquid form). The latter is most generally used, and articles of considerable size can be produced. Because of their dielectric and non-toxic properties, plastics are essential components of electrical and electronic equipment (especially for use within the human body).

Plastics can be made into flexible and rigid foams by use of a blowing agent; they are light and strong, and the rigid type is machinable. These forms are collectively called cellular plastics. Plastics can also be reinforced, usually with glass or metallic fibres for added strength. They are laminated to paper, cloth, wood, etc. for many uses in the packaging, electrical and furniture industries; they also can be metal-plated. Plastic pipe is widely used for underground transportation of gases and liquids over long distances as well as intraplant.

Several natural materials (waxes, clays, and asphalts) have rheological properties similar to synthetic products, but as they are not polymeric, they are not considered true plastics. Certain proteins (casein, zein) are natural high polymers from which plastics are made (buttons, and other small items), but they are of decreasing importance.

Plastics have permeated industrial technology. Not only have they replaced and improved upon many materials formerly used, but also have made possible industrial and medical applications that would have been impracticable with older technologies.

Synthetic resins. A man-made higher polymer resulting from a chemical reaction between two (or more) substances, usually with heat or a catalyst. This definition includes synthetic rubbers and silicones (elastomers), but excludes modified, watersoluble polymers (often called resins). Distinction should be made between a synthetic resin and a plastic. The former is the polymer itself, whereas the latter is the polymer plus such additives as fillers, colourant, plasticisers, etc.

The first truly synthetic resin was developed by Backeland (phenol-formaldehyde). This was soon followed by a petroleum-derived product called coumarine-indene, which did indeed have the properties of a resin. The first synthetic elastomer was polychloroprene, and later called neoprene. Since then many new types of synthetic polymers have been synthesised, perhaps the most sophisticated of which are nylon and its congeners (polymamides, and the inorganic silicone group). Other important types are alkyds, acrylics, aminoplasts, polyvinyl halides, polyesters, epoxides and polyolefins.

In addition to their many applications in plastics, textiles, and paints, special types of synthetic resins; are useful as ion-exchange media.

Rubber is the name given to elastic, high-molecular compounds capable of changing their shape (deforming) when acted upon by an external force and of rapidly resuming their original shape when the force is removed. The elsatic properties of rubber are due to the fact that the linear macromolecules of the material which normally are of a bent, curled or curved shape, easily stretch out of contract under the effect of an applied force; when the force is removed, the molecules return to their initial shape and length.

If sulphur is added to natural rubber which is then heated, the sulphur will form links between the linear macromolecules of the rubber, i.e., form bridges between them; by this larger macromolecules of a web structure will be produced. This process is called vulcanisation of the rubber, and the product of reaction between the rubber and sulphur is vulcanised rubber. Vulcanisation improves the mechanical properties of the rubber and increases the range of temperatures within which the elasticity of the material is retained; it also increases the resistance of the rubber to organic solvents, etc.

Natural and synthetic rubber is the basic material in the fabrication of rubber-textile and rubber-metal articles which are so extensively used in various branches of the economy. These are motor car and aircraft tyres, drive belts and conveyer belts, flexible hoses and sleeves, machine parts and various gaskets, electric insulating materials, rubberized fabrics and rubber footwear, articles of personal hygiene and many other rubber products. The consumption of rubber has increased year by year. This explains the attention devoted to developing the rubber industry in India and in many other countries. Rubbers are classified as natural rubber and synthetic rubber. Until the nineteen thirties only natural rubber, obtained from the latex of rubber plants which grow in countries with a damp tropical climate (mainly in Malaya, Indonesia, etc.), was used.

MECHANISM OF POLYMERISATION

There are several useful materials which have high molecular weight. Examples include alkaloids, dyes, etc., and a group of materials known as polymer. The difference between the two types lies in the fact that the former has a very complex structure whereas the polymer has rather a very simple chemical formula. A polymer is a large molecule built up by successive repetition of small chemical unit (mer) which, in turn, is derived from simple compound(s) called monomer(s). The number of mers present in a polymer molecule is called *degree of polymerisation (DP)*. Hence,

$$\text{Molecular weight of a polymer} = \text{DP} \times \text{mer weight}$$

The molecular weight of polymers may vary between 10^2 and 10^6 range. However, as long as the molecular weight is low and within few thousands, the polymer is known as oligomer. When the molecular weight is 5000 or above they are called high polymer, super-molecules, giant molecules, macromolecules.

$$n\mathrm{CH_2} = \mathrm{CH} \longrightarrow n\mathrm{CH_2}\!-\!\mathrm{CH}\!-\!\longrightarrow (-\mathrm{CH_2}\!-\!\mathrm{CH})_n\!-$$

CN	CN	CN
Acrylonitrile monomer	Acrylonitrile-mer	Polyacrylonitrile

$n = $ Degree of polymerisation

Though derived from simple chemical compound (monomer), polymers exhibit properties that are very much different from those of the monomer of any other low molecular weight compound. The differences are a matter of degree rather than of kind, i.e., they are mainly physical and mechanical rather than chemical. The unique properties of high polymers are due to unusual shape and large size of the molecules. The large size and surface result in a large magnitude of total force of intermolecular attraction. Also due to great length of the molecules, polymers are capable of undergoing enormous changes in dimension in passing from a straightened to a coiled or folded condition. In simple molecules, · both these possibilities are absent.

CLASSIFICATION OF POLYMERS

Polymers may be classified as natural (such as cellulose, wool, protein, natural rubber, etc.) and synthetic (e.g., nylon, bakelites, styrene, butadiene rubber etc.). They may also be of organic origin like wool or nucleic acid and inorganic, e.g., diamond, ceramics, etc. More scientifically, high polymers are usually classified on the basis of the outline of the principal chains of atoms. When the polymers contain isomers joined together in the form of a rope or chain, it is called linear.

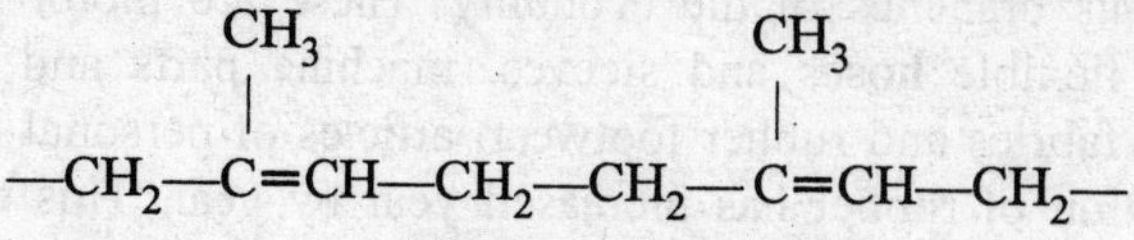

$$-CH_2-\underset{\underset{CH_3}{|}}{C}=CH-CH_2-CH_2-\underset{\underset{CH_3}{|}}{C}=CH-CH_2-$$

Linear polyisoprene

When the linear structure bears side chains composed of the same linear structure as the main chain, it is called branched polymer.

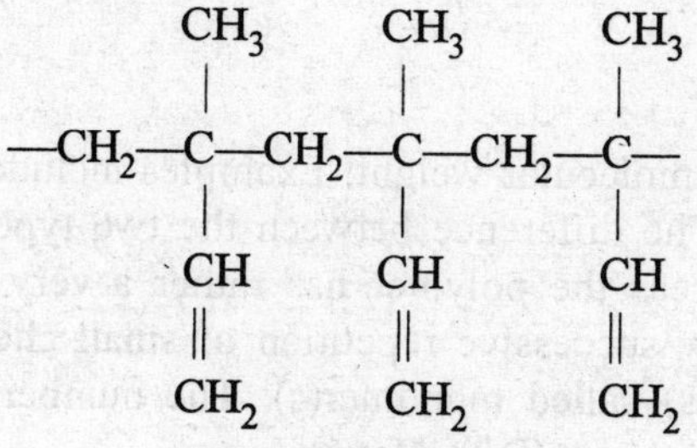

$$-CH_2-\underset{\underset{CH}{\overset{CH_3}{|}}}{C}-CH_2-\underset{\underset{CH}{\overset{CH_3}{|}}}{C}-CH_2-\underset{\underset{CH}{\overset{CH_3}{|}}}{C}-$$

Branched Polyisoprene

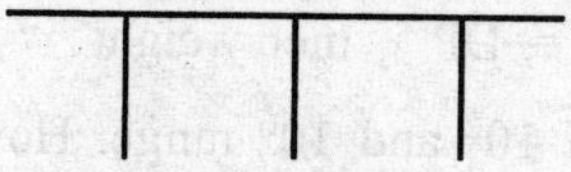

Branched polymer

If the branches are of a different structure than the linear polymer chain (backbone polymer) it is called a graft polymer.

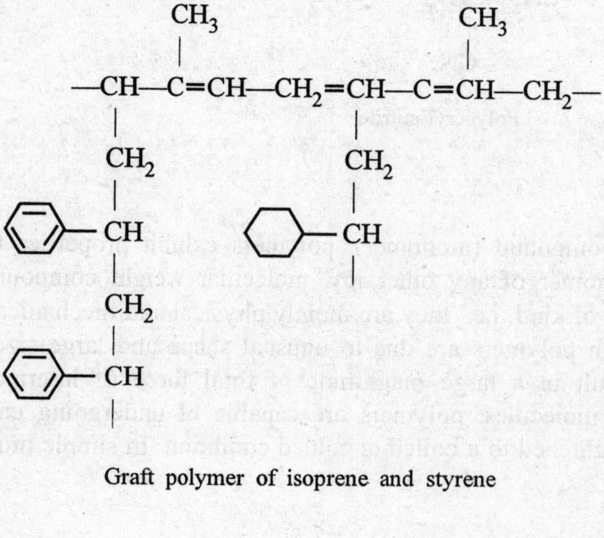

Graft polymer of isoprene and styrene

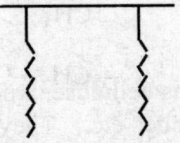

Graft polymer

There are several other types of branched polymers which are shown below :

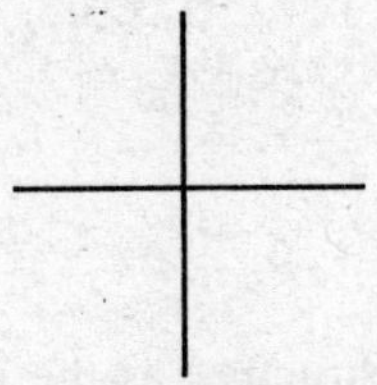

Cruci form

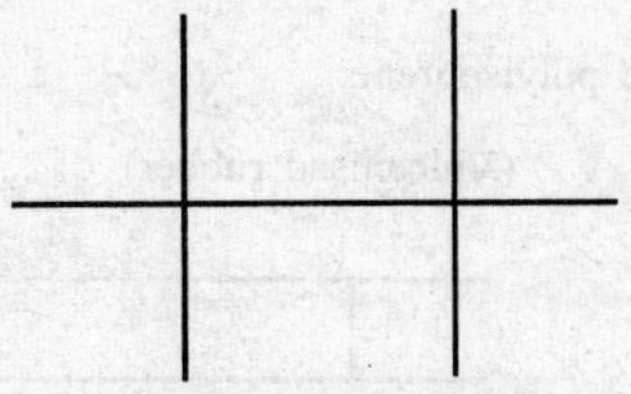

Double cruci form

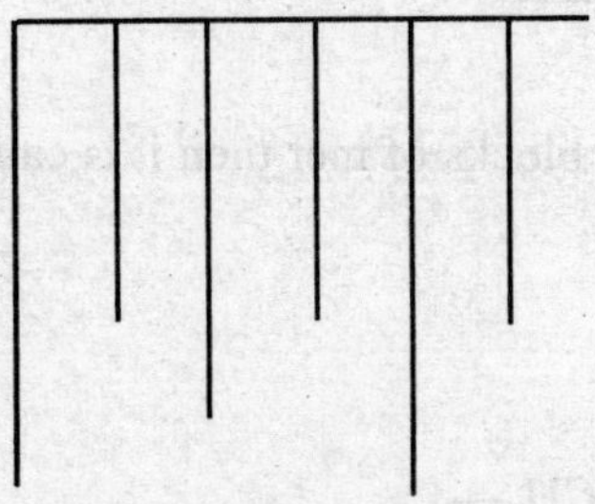

Random comb

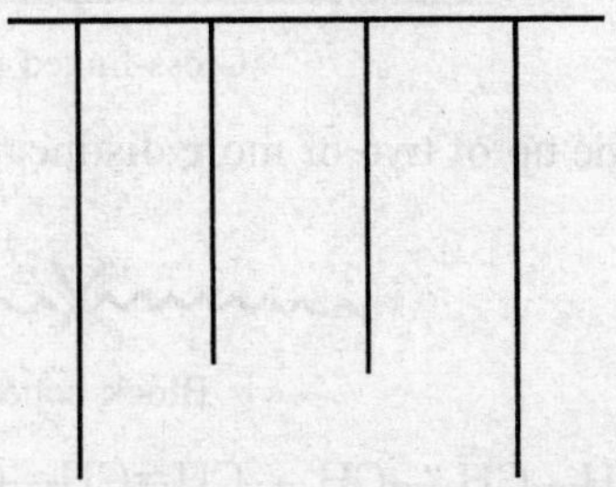

Regular comb

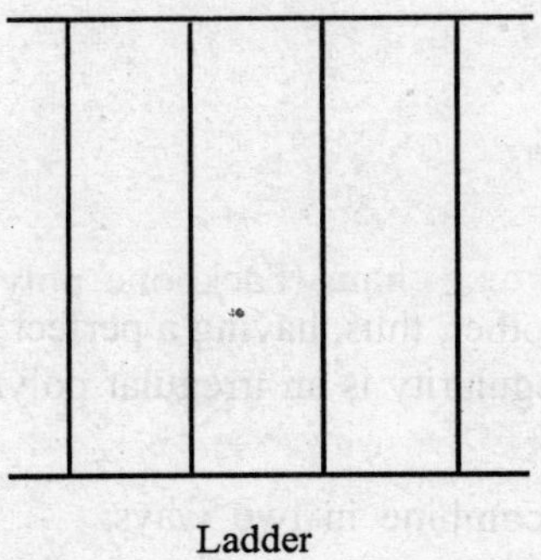

Ladder

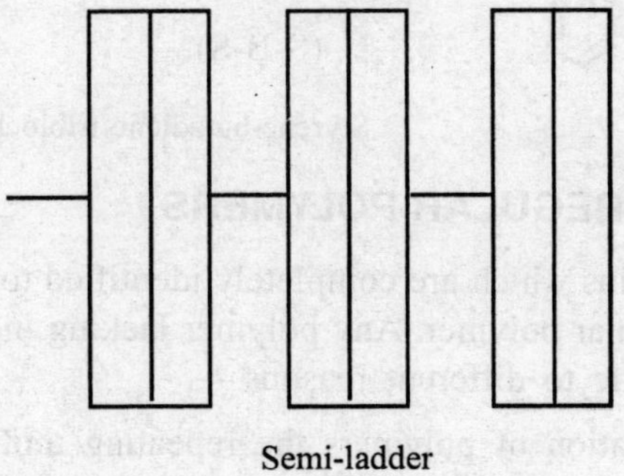

Semi-ladder

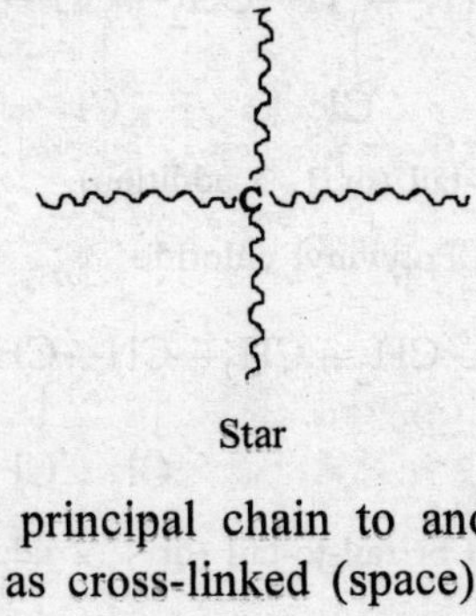

Star

When the branches connect one principal chain to another so that a three dimensional network develops then the polymer is called as cross-linked (space) polymer.

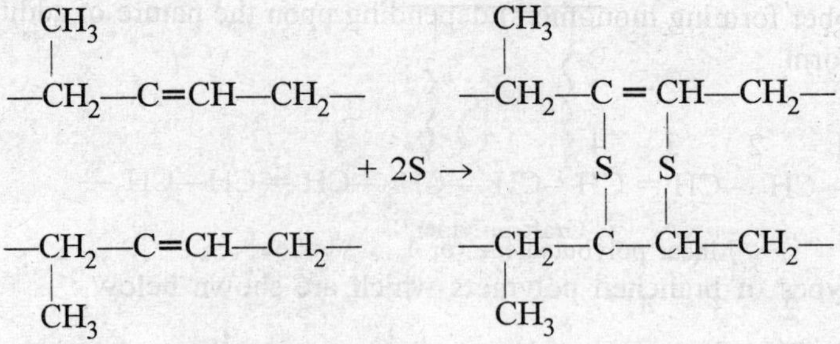

$$
\begin{array}{ccc}
& CH_3 & & CH_3 \\
& | & & | \\
-CH_2-C=CH-CH_2- & & -CH_2-C=CH-CH_2- \\
& & +2S \rightarrow & \quad S \quad S \\
& & & \quad | \quad | \\
-CH_2-C=CH-CH_2- & & -CH_2-C-CH-CH_2- \\
& | & & | \\
& CH_3 & & CH_3
\end{array}
$$

Cross-linked polyisoprene

(Vulcanised rubber)

Cross-linked polymer

When a polymer is made up of two or more distinct but different blocks of mer then it is called a block copolymer.

Block copolymer

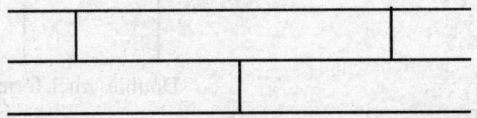

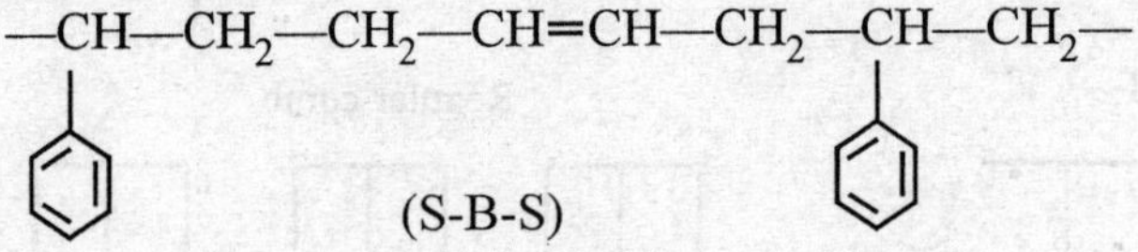

$$
-CH-CH_2-CH_2-CH=CH-CH_2-CH-CH_2-
$$

(S-B-S)

Styrene-butadiene triblock polymer

REGULAR AND IRREGULAR POLYMERS

A polymer having chains which are completely identified to one another, thus, having a perfect long-range order of units is a regular polymer. Any polymer lacking in such regularity is an irregular polymer. Chain irregularity may be due to different reasons.

Thus, during formation of polymers the repeating units may combine in two ways:

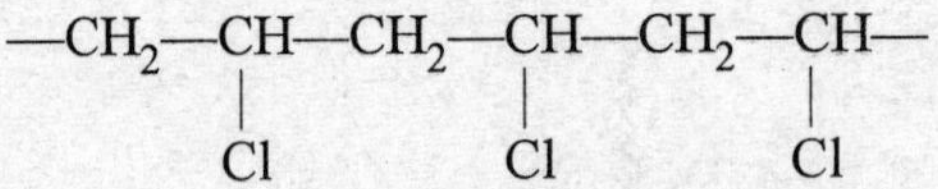

$$
\begin{array}{cccc}
-CH_2-CH-CH_2-CH-CH_2-CH- \\
\quad | \qquad\quad | \qquad\quad | \\
\quad Cl \qquad\quad Cl \qquad\quad Cl
\end{array}
$$

Head-to-tail (or 1, 3 addition)

Polyvinyl chloride

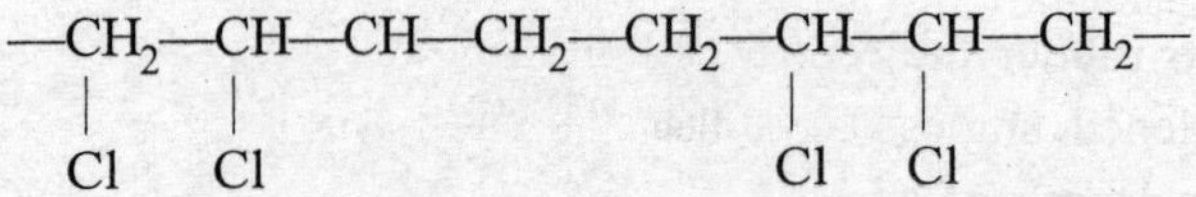

$$
\begin{array}{cccc}
-CH_2-CH-CH-CH_2-CH_2-CH-CH-CH_2- \\
\quad | \quad | \qquad\qquad\quad | \quad | \\
\quad Cl \quad Cl \qquad\qquad\quad Cl \quad Cl
\end{array}
$$

Head-to-head or tail-to-tail (or 1, 2 addition)

In the case of dienes (rubber forming monomers) depending upon the nature of addition either a linear or branched polymer may form.

$$\underset{1}{-CH_2}-\underset{2}{CH}=\underset{3}{CH}-\underset{4}{CH_2}-\underset{1}{CH_2}-\underset{2}{CH}=\underset{1}{CH}-\underset{2}{CH_2}-$$

Linear polybutadiene (or 1, 4 addition)

$$\underset{1}{-CH_2}-\underset{2}{CH}=\underset{3}{CH}-\underset{4}{CH_2}-\underset{1}{CH_2}-\underset{2}{CH}=\underset{3}{CH}-\underset{4}{CH_2}-\underset{1}{CH_2}-\underset{2}{CH}-$$

$$\overset{|}{CH}$$
$$\overset{||}{4CH_2}$$

Branched polybutadiene (or 1, 2 addition)

In the isolated state the macromolecule may change its shape, and each polymer may turn differently relative to one another. When polymers crystallise, the units of certain sections of the polymer becomes fixed. Thus the molecular polymer chain is made up of repeating sections of identical spatial structure called repeat units. The distance from the beginning of one repeat unit to beginning of the next is called repeat distance or identity period. Polymers having same chemical composition but differing in spatial arrangement (and thus in identity period) are called *cis*- and *trans*-isomer. Thus:

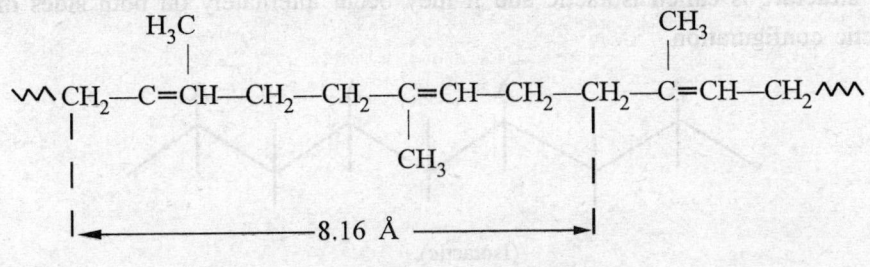

cis-isomer (Natural rubber)

trans-isomer (Gutta percha)

The repeat distance of crystalline rubber is 8.16 Å, while that of gutta-percha (a plastic) is 4.8 Å. Going from gutta-percha to rubber the repeat distance should have been 9.6 Å (i.e. 4.8 × 2), but it is less due to changes in valency angles. Crystalline polyethylene is a plane zig-zag carbon chain with identity period equal to the length of one zig-zag:

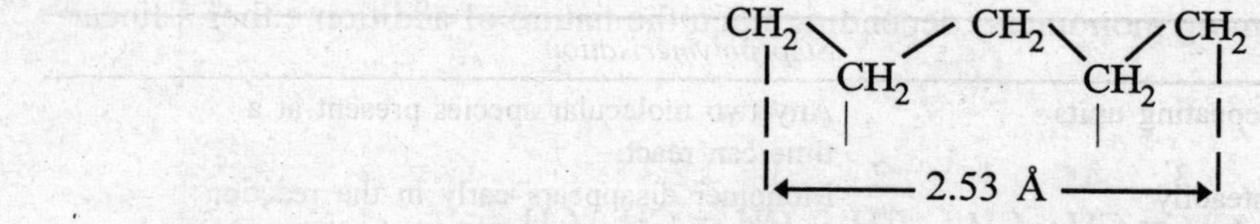

$$\longleftarrow 2.53 \text{ Å} \longrightarrow$$

Another type of configurational isomerism is D-isomerism, which is due to the presence of an asymmetric carbon atom in the polymer chain.

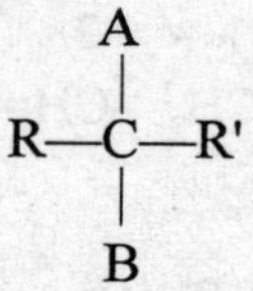

The asymmetry is determined by unequal length and possible differences in space configuration of the two parts of the molecular chains connected to each carbon atom (R≠R'), in the presence of two different substituents A and B. Polymers in which all adjacent asymmetric carbon atoms over the length of at least one macromolecular chain have the same space configuration are called *isotactic polymers*. Polymers whose macromolecules are made up of units with opposite space configuration of each consequent carbon atom (asymmetric) in the chain are called syndisotactic polymers. A simpler (though crude) way of looking at these is that if the substituent (say A) appears on the same side of the chain in space then the structure is called isotactic and if they occur alternately on both sides of chain then it is in syndisotactic configuration.

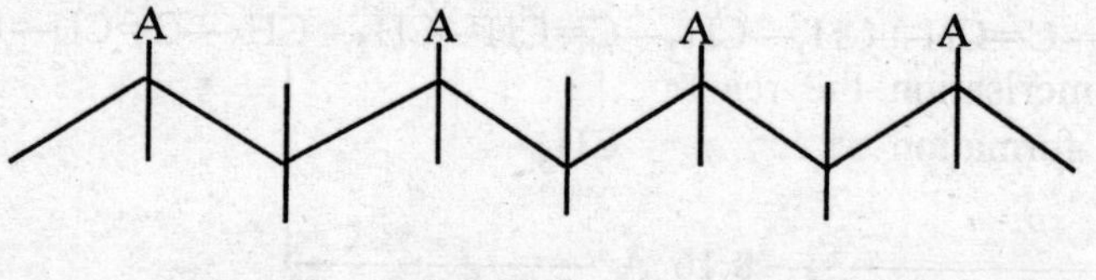

(Isotactic)

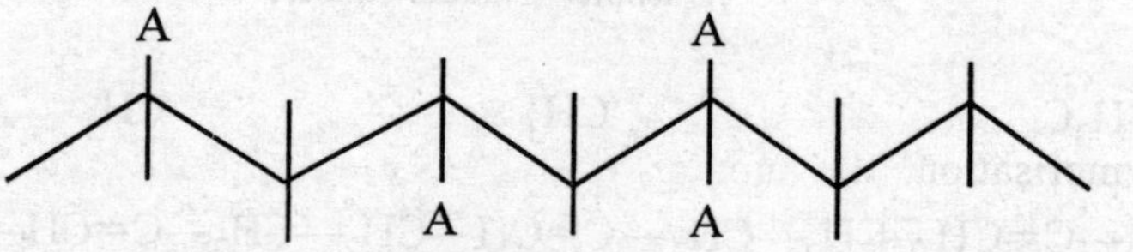

(Syndisotactic)

Polymers with an irregular arrangement in space are called *atactic*.

POLYMERISATION

It is a process by which simple chemical compounds are reacted to form a macromolecule. Essentially the reaction takes place intermolecularly and is capable of proceeding indefinitely (to grow to an infinite length) functionally. Basically, they are of two types—condensation polymerisation and addition polymerisation. According to the mechanism of the reaction they are called as step polymerisation and chain polymerisation. The essential distinguishing features of the two are listed below:

	Chain polymerisation	Step polymerisation
1.	Only growth reaction adds repeating units one at a time to the chain.	Any two molecular species present at a time can react.
2.	Number of units decreases steadily throughout the reaction.	Monomer disappears early in the reaction at DP 10, 1% monomer remains.
3.	High polymer is formed at once; polymer molecular weight changes little throughout the reaction.	Polymer molecular weight rises steadily throughout the reaction.
4.	Long reaction time gives high yields but affect molecular weight little.	Long reaction times are essential to obtain high molecular weight.
5.	Reaction mixture contains only monomer, polymer, any other added agent and about species are present	At any stage all molecular species are present in a calculable distribution.

STEP POLYMERISATION

This type of polymerisation can take place in two ways. In the step growth system, monomers react intermolecularly (condensation) with the formation of the polymer accompanied by a small, volatile molecule.

$$nA \longrightarrow P + X$$

$$n\text{Cl}-\text{CH}_2-\text{CH}_2-\text{Cl} + \text{Na}_2\text{S}_x \longrightarrow \left[-\text{CH}_2-\text{CH}_2-\overset{\displaystyle |}{\underset{\displaystyle S}{S}}-\overset{\displaystyle |}{\underset{\displaystyle S}{S}}- \right]_n + 2\text{NaCl} + (n-x)\,\text{S}$$

Ethylene dichloride   Sodium polysulphide   Thiokol rubber

In the step polymerisation the reaction takes place by addition of mers but the product is not accompanied by the formation of any small molecule.

$$nB \longrightarrow (B)n$$

$$n\text{OC}-(\text{CH}_2)_5-\text{NH} \longrightarrow \left[-\text{OC}-(\text{CH}_2)_5-\text{NH} \right]_n$$

Caprolactam   Polycaprolactam
                               (Nylon-6)

For a step polymerisation, the monomer(s) must have certain characteristics which include the following:

(1) The monomer structure should be such that it is incapable of forming 5- or 6- membered ring.

(2) The average functionality of the monomer or monomers should be at least two or more, if it is two a linear polymer will be formed and if it is more than two a cross-linked polymer will be formed. The functionality (f), DP (n) and extent of polymerisation (P) are related by Carother's equation:

$$P = \frac{2}{f} - \frac{2}{nf}$$

ADDITION POLYMERISATION

Addition (chain) polymerisation is the polymerisation of unsaturated (ethylenic) monomers typically involving in a chain reaction. A chain reaction is a series of consecutive reactions in which an activated species converts reactants to products, another activated molecule being formed in the process. The most common chain reaction proceeds via a free radical mechanism, though it may also take place through a cationic, anionic or coordination complex mechanism.

The monomers which are most susceptible to addition polymerisation are derivative of ethylene (vinyl compounds, diene, etc.) producing an unsymmetrical double bond.

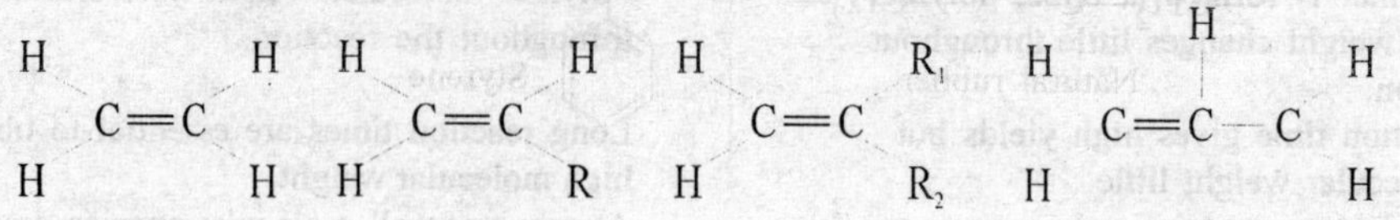

Ethylene   Vinyl   Vinylidene   Diene
compound   compound

Depending upon the nature of attacking agent (free radical or ionic) the double bond may undergo either homolytic (free radical) or heterolytic fission.

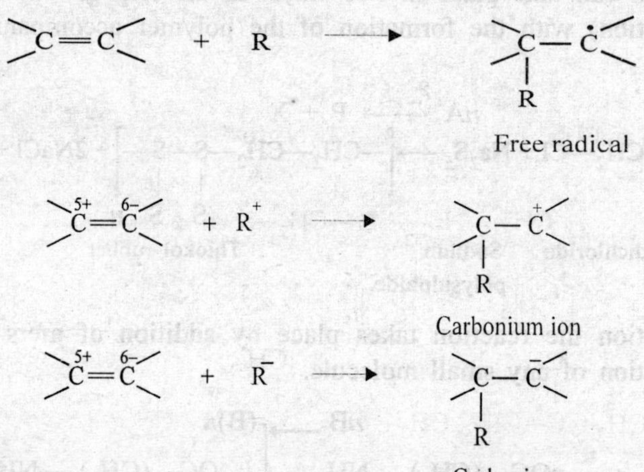

Free radical

Carbonium ion

Carbonium

COPOLYMERISATION

This is a process by which two or more monomers are polymerised together to produce a polymer having a mer derived from contributions by each of the monomers:

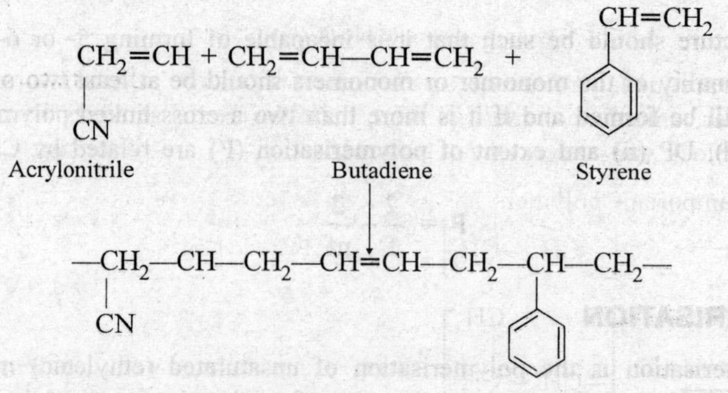

$CH_2=CH + CH_2=CH—CH=CH_2 +$

CN
Acrylonitrile   Butadiene   Styrene

$—CH_2—CH—CH_2—CH=CH—CH_2—CH—CH_2—$

CN

Copolymer of acrylonitrile-butadiene-styrene or ABS

Another type of copolymerisation is done by using a polymer substrate on which a reactive site is developed followed by reaction with a monomer activated by the reactive site. This process is known as *graft copolymerisation.*

Natural rubber   Styrene

Styrene grafted natural rubber

Block copolymerisation may be prepared when two growing polymer chains react only through end groups.

Styrene - isoprene block copolymer

Copolymerisations are usually carried out by the same type of initiators used for making homopolymers. However, anionic polymerisation and Ziegler Natta catalysts have been found most suitable for producing block and stereoregular polymers.

Structures of some important polymers are given as under:

$$n CH_2 = CH_2 \longrightarrow \left[-CH_2 - CH_2 \right]_n \quad \text{Polyethylene}$$

$$n CH_2 = \underset{\underset{CH_3}{|}}{CH} \longrightarrow \left[-CH_2 - \underset{\underset{CH_3}{|}}{CH} \right]_n \quad \text{Polypropylene}$$

$$n\text{CH}_2 = \underset{\underset{\text{CH}_3}{|}}{\overset{\overset{\text{CH}_3}{|}}{\text{C}}} \longrightarrow \left[-\text{CH}_2 - \underset{\underset{\text{CH}_3}{|}}{\overset{\overset{\text{CH}_3}{|}}{\text{C}}} - \right]_n \qquad \text{Polyisobutylene}$$

$$n\text{CH}_2 = \underset{\underset{\text{Cl}}{|}}{\text{CH}} \longrightarrow \left[-\text{CH}_2 - \underset{\underset{\text{Cl}}{|}}{\text{CH}} - \right]_n \qquad \text{Polyvinyl chloride}$$

$$n\text{CH}_2 = \underset{\underset{\text{OR}}{|}}{\text{CH}} \longrightarrow \left[\text{CH}_2 - \underset{\underset{\text{OR}}{|}}{\text{CH}} \right]_n \qquad \text{Polyvinyl ether}$$

$$n\text{CH}_2 = \underset{\underset{\text{CN}}{|}}{\text{CH}} \longrightarrow \left[\text{CH}_2 - \underset{\underset{\text{CN}}{|}}{\text{CH}} \right]_n \qquad \text{Polyacrylonitrile}$$

$$n\text{CH}_2 = \text{CH} - \text{CH} = \text{CH}_2 \longrightarrow \left[-\text{CH}_2 - \text{CH} = \text{CH} - \text{CH}_2 - \right]_n \quad \text{Polybutadiene}$$

$$n\text{CH}_2 = \underset{\underset{\text{CH}_3}{|}}{\text{C}} - \text{CH} = \text{CH}_2 \longrightarrow \left[-\text{CH}_2 - \underset{\underset{\text{CH}_3}{|}}{\text{C}} = \text{CH} - \text{CH}_2 - \right]_n \quad \text{Polyisoprene}$$

$$n\text{CH}_2 = \underset{\underset{\text{Cl}}{|}}{\text{C}} - \text{CH} = \text{CH}_2 \longrightarrow \left[-\text{CH}_2 - \underset{\underset{\text{Cl}}{|}}{\text{C}} = \text{CH} - \text{CH}_2 - \right]_n \quad \text{Polychloroprene}$$

$$n\text{NH}_2 - (\text{CH}_2)_6 - \text{NH}_2 \rightarrow \left[\text{H} - \text{NH} - (\text{CH}_2)_6 - \text{NH} - \text{OC} - (\text{CH}_2)_4 - \text{CO} \right]_{n-1} \text{OH}$$

Hexamethylene diamine $\qquad \rightarrow \qquad$ Nylon 66

$+$ $\qquad\qquad\qquad\qquad\qquad\qquad +$

$n\text{HOOC} - (\text{CH}_2)_4 - \text{COOH}$ $\qquad\qquad\qquad$ H_2O

Adipic acid

$$n\text{NH}_2 - (\text{CH}_2)_5 - \text{CO} \longrightarrow \left[\text{NH} - (\text{CH}_2)_5 - \text{CO} \right]_n$$

Nylon 6

$n\text{HO } (\text{CH}_2)_4 - \text{OH}$
Butanediol

$$+ \longrightarrow \text{H} \left[\text{O}(\text{CH}_2)_4 - \text{O} - \overset{\overset{\text{O}}{\|}}{\text{C}} - \text{NH} - (\text{CH}_2)_6 - \text{NH} - \text{CO} \right] \text{OH}$$

$n\text{OCN} - (\text{CH}_2)_6 - \text{NCO}$ $\qquad\qquad\qquad$ **Perlon**
Hexamethylene diisocyanate

$$nCl\ (CH_2)_x\ Cl + nNa_2S \longrightarrow \left[-(CH_2)_x-S-\right]_n + nNaCl$$

Polysulphide
(Thiokol A)

$$nHO-\underset{\underset{CH_3}{|}}{\overset{\overset{CH_3}{|}}{Si}}-OH \longrightarrow HO\left[\underset{\underset{CH_3}{|}}{\overset{\overset{CH_3}{|}}{Si}}-O\right]_{n-1}H + (n-1)\ H_2O$$

Polydimethylsiloxane
(Silicone)

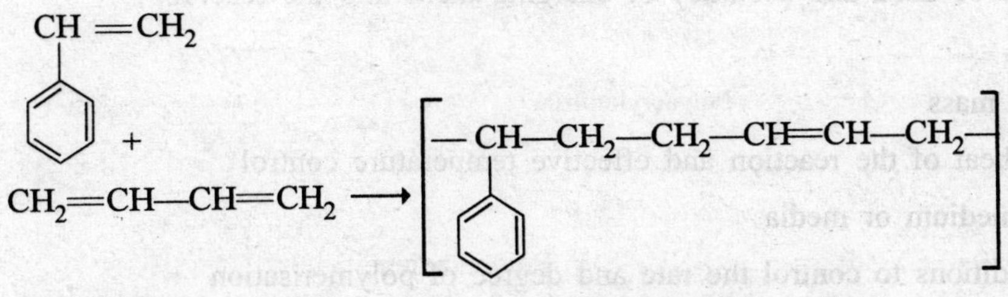

Styrene butadiene rubber

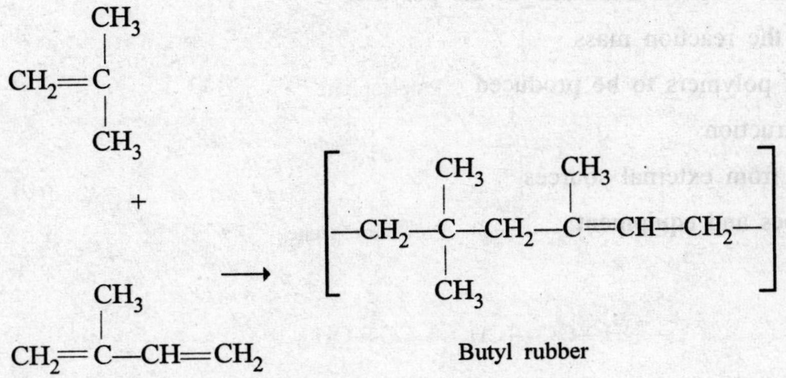

Butyl rubber

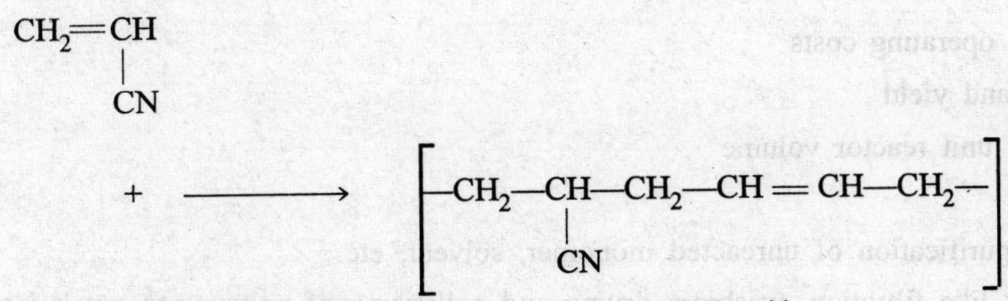

Nitrile rubber

METHODS OF POLYMERISATION

There are several methods of conducting a vinyl polymerisation. They only vary in the physical state of dispersions. But numerous theoretical and practical considerations dictate the use of one method in preference to the other for a particular monomer. The most important of these procedures are only four

(1) Bulk polymerisation

(2) Solution polymerisation

(3) Emulsion polymerisation

(4) Suspension polymerisation

Apart from the nature of chemical reaction involved and the type of polymer to be produced, the selection of a polymerisation technique for production of polymers on an industrial scale is based on several technical and economic factors.

Technical Factors

(1) Type of raw materials to be used and accuracy of charging them into the reactor

(2) Reactor design

(3) Agitation of the reaction mass

(4) Removal of exothermic heat of the reaction and effective temperature control

(5) Choice of heat transfer medium or media

(6) Control of operating conditions to control the rate and degree of polymerisation

(7) Movement and handling of the reaction mass and of the polymer

(8) Separation of polymer from the reaction mass

(9) Type, number and quality of polymers to be produced

(10) Choice of materials of construction

(11) Prevention of contamination from external sources

(12) Provision for cleaning of pipes and equipment

(13) Flexibility of operation

(15) Safety hazards

Economic Factors

(1) Capital investment and operating costs

(2) Cost of raw materials and yield

(3) Rate of production per unit reactor volume

(4) Scale of operation

(5) Cost of recovery and purification of unreacted monomer, solvent, etc.

(6) Cost of unit operations like filtration, washing, drying and pelletising of polymer to get it into a processable form

BULK POLYMERISATION

This is the simplest method and involves the use of monomer and initiator only as the major components. So the polymer could be obtained in a high state of purity and the maximum utilisation of the production capacity can be made. However, the process is not without disadvantages, some of the more important ones being the problems connected with heat transfer, agitation, control of molecular weight, removal of residual monomer, initiator, etc. The molecular weight distribution is also broad.

High temperature polycondensation may also be considered as a type of bulk polymerisation. Most of the points discussed below are equally applicable to high temperature polycondensation also (Fig. 26.1).

The heats of polymerisation of vinyl monomers are high and fall in the range of 20 kcal/g mole. Heat dissipation becomes a problem with increasing viscosity. This may lead to local superheating leading to large variations in size distribution, lower molecular weight, discolouration and added danger of cross-linking leading to useless product formation which may stick to the walls of the reaction kettle and can only be removed with great difficulty. Also, there are chances of explosion and fire hazards. Sometimes, at low conversion percentages auto acceleration effect sets in and the reaction be better stopped. Due to these reasons, this method was not previously recommended for general industrial practice as a continuous method. However, recent advances in mitigating the heat of polymerisation has made it possible to use this method for continuous operation, and the advantages of this method are becoming more and more evident. Due to these advantages, the recent tendency is to polymerise vinyl polymers by this method.

The advantages of this method as compared to other processes are as follows:

(1) Simplicity of equipment required
(2) Near theoretical yields of polymer
(3) High purity of products
(4) Utilisation of the maximum capacity of the reactor.

Also if the heat dissipation problem could be solved then the polymer properties would also improve.

The feasibility of employing bulk polymerisation depends upon, among several considerations, the chemical properties of the monomer and polymers, the equipment necessary and the use of the polymer. It is most obvious method for polymerisation of liquid vinyl monomers. Bulk polymerisation is homogeneous when polymer dissolves in the monomer and is heterogeneous when the polymer precipitates out from the mixture during reaction.

In case of vinyl polymerisation, large amount of heat (10-30 kcal/g. mole) is given out during the process of conversion of double bonds into single bond. This heat of polymerisation must be carefully and quickly dissipated in order to minimise explosion hazard, uncontrolled polymerisation, charring, broad molecular weight distribution of the product, etc. The overall heat transfer coefficients in bulk polymerisation systems are in the range of 5-30 BTU/hr. -sq. ft.-°F, depending upon various factors, such as polymer concentration, viscosity, amount of turbulence, etc. In case of large reaction vessels, where surface to volume ratio is very small, the heat dissipation becomes really difficult.

Free radical initiators are better because of their safe decomposition temperature ranges and wide choice of initiators over a wide range of temperatures. With ionic initiators the activation energy is low (4-10 kcal/g. mole) and rate of polymerisation is very fast. So control of the reaction is difficult and it is always advisable to avoid their use if a suitable free radical initiator is available.

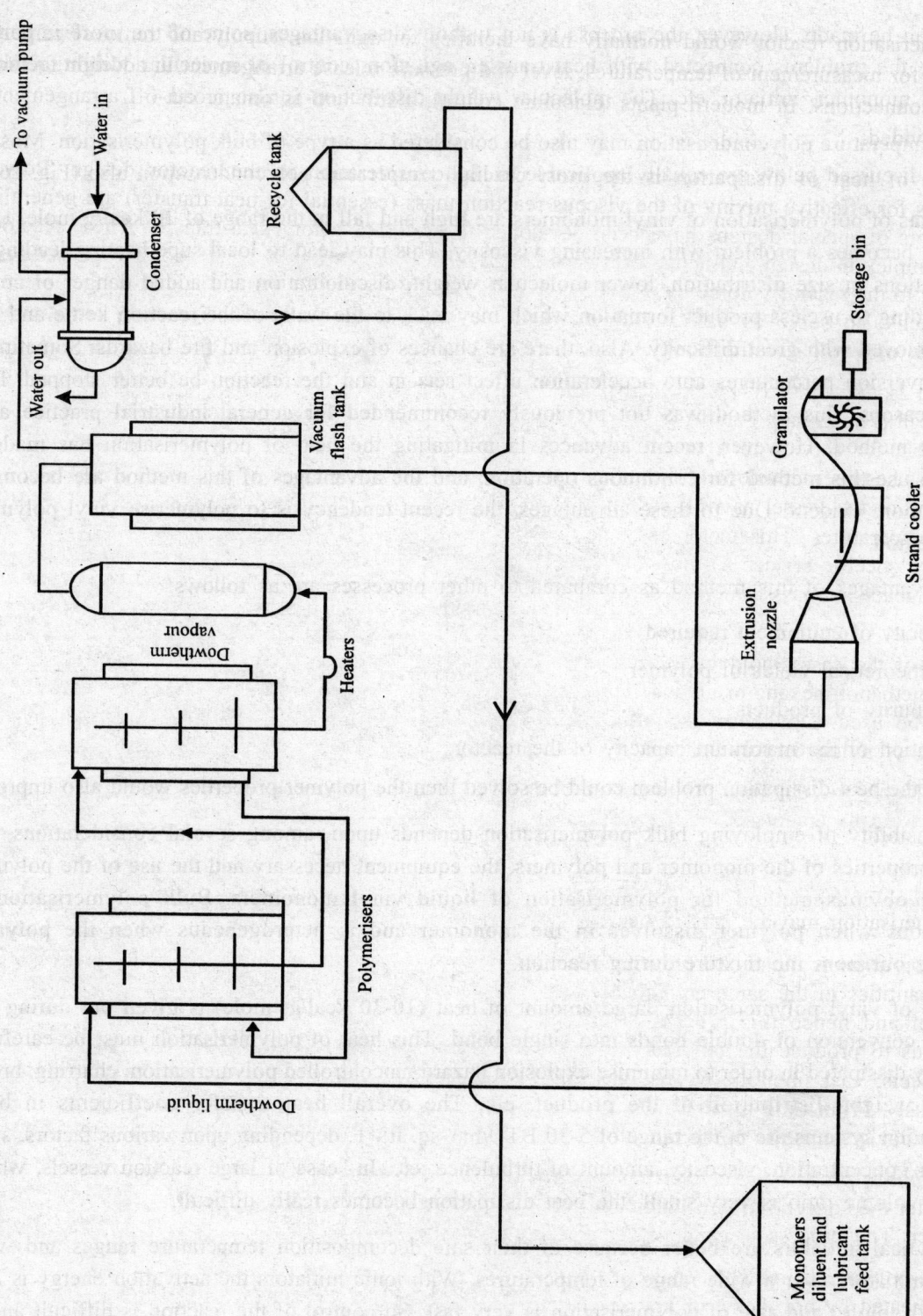

Fig. 26.1. Schematic flowsheet–mass polymerisation process.

A polymerisation reactor would normally have facilities for agitation, supply and removal of heat, connections for measurement of temperatures, level and pressure release arrangement in addition to inlet and outlet connections. In modern plants, elaborate control, alarm and automatic cut-off arrangements are also provided.

The rate of heat of dissipation is the most crucial consideration in the reactor design. Power requirements for effective mixing of the viscous reaction mass (essential for heat transfer) are generally high. Reactors of special designs have been developed to give satisfactory heat transfer. These reactors are quite complex in design and hence expensive. Most of the reactor designs employ high ratios of heat transfer area to the reactor volume by providing jacket and large number of internal cooling coils or plate coolers which also serve as baffles. Some processes use evaporative cooling and condensation of vapours with or without compression.

Agitation is done mostly by mechanical agitators. In view of high viscosities encountered special agitator design is necessary. Paddle, anchor, screw or ribbon agitators are generally used. A double-ribbon mixer suitable for service in a vertical, cylindrical reactor at 100 r.p.m. for a 200,000 cp viscosity reaction mass has been developed. The power requirement for agitation is usually high and depends upon the geometry of the vessel as well as the agitator and physical characteristics of the reaction mass at operating temperatures. This factor has to be taken into account before selecting the type and size of agitator for a specific service as most of the power input to the agitator drive is converted into heat in the viscous reaction mass which is sometimes as high as 30-40% of the total heat generated in the process.

In most of the cases, cold water (35°–65°F) is normally used as the *heat-transfer medium*. Other agents are methanol, hexane or glycols. Direct expansion of the refrigerant in the reactor jacket may also sometimes be used. Dowtherm is normally used for high temperature processes like those for the manufacture of nylon 6, polyester, etc.

For the highly viscous polymerisation mix, *movement of the reaction mass* is preferably carried out with positive displacement screw or jacketed gear, pumps equipped with mechanical seals.

Methods

Bulk polymerisation may be carried out either by batch process or by continuous process.

Batch operation is mostly employed when it is desired to produce a number of small formulations in small quantities in the same equipment. Only about 70% of the cycle time is used for carrying out the reaction and hence this process is uneconomic for large-scale preparation. But this process is advantageous to produce the polymer in the desired form and shape from the monomer directly like perspex sheets, cast phenolic resins, etc. In a batch reactor the monomer and catalyst are added simultaneously. From time to time little amount of fresh monomer is to be added. The polymer which forms a particular shape is taken out after proper residence time in the reactor.

In *continuous process* the conversion of monomer to polymer may be carried out to certain per cent and thereafter the concentration of monomer-polymer is maintained and fed into another reactor for better heat-transfer. The process is generally carried out in two stages.

The pre-polymerisation (first stage) is carried out in nitrogen atmosphere in the reaction vessels, which are heated by hot jacket or hot water coils and the monomer is added continuously with stirring at certain rate into the vessel. The monomer is polymerised to about 30%.

The polymerisation mix, which is highly viscous and in a clear state, is fed into the upper end of a vertical reactor, which in turn, is further subdivided into a number of zones, for post-polymerisation (second stage) operation. Reactor zones are kept at desired temperatures by means of superheated steam, except the lowest one, which is electrically heated to get maximum heat necessary for complete conversion of monomer. The average residence time into the respective zones is the deciding factor for the success of the process because viscosity increases with polymerisation and agitation problems appear. The highest possible conversion of monomer is carried out at the lowest zone and the resulting molten polymer is discharged into the extruder unit. Polymer is conveyed through conveyor belts over an air or water cooling systems, followed by chopping and grinding of the solid polymer.

SOLUTION POLYMERISATION

In this method, the monomer is dissolved in a suitable solvent and initiator is then added to start the polymerisation reaction. The heat of polymerisation is mostly mitigated by evaporation (followed by condensation and recycling) of the solvent. After the required extent of polymerisation, the polymer is precipitated by adding a non-solvent to the mix. Sometimes the polymer itself is insoluble in the polymerisation. The solid polymer is then filtered off, washed free of any adhered monomer or solvent and dried before use. The solvent-unreacted monomer-non-solvent mixture is separated by distillation or any other suitable methods and reused.

The disadvantages of this polymerisation technique are bulk handling, tiresome mechanical separation processes, purification of the solvents for reuse and difficulty to remove the solvent from the monomer completely. However, the most serious problem is due to chain transfer (Table 26.1) to solvent restricting the polymer molecular weight value, and/or reducing the rate of polymerisation. Solvents also may determine the stereoregularity of the polymer. Sometimes the induction period of the polymerisation may be quite long increasing uselessly the time of production. For these reasons, this method is not normally used for industrial production of polymer.

There are certain systems in which water can be used as solvent for polymerisation (N-vinylpyrolidone, acrylonitrile, acrylic acid, etc.) thus providing a cheap solvent and also permitting the use of inexpensive inorganic redox initiator.

Table 26.1. Recipe for emulsion polymer styrene-butadiene rubber (SBR).

Ingredients	Parts by wt.
Butadiene	50-85
Styrene	50-15
Emulsifier	1-5
Buffer	1-5
Catalyst	0.1-1
Orientation catalyst	01-1
Water	100-400

(Temperature 35-60°C, pressure 3-5 atmosphere, time 5-20 hrs.)

EMULSION POLYMERISATION

This method combines the advantages of bulk and solution methods, namely, ease of temperature control, rapid production of polymer and high average molecular weight. This method is applicable to almost all vinyl compounds and dienes except those which have a good water solubility like acrylonitrile and vinyl acetate (recent techniques had made it possible to polymerise these monomers). Capital investment is low. A number of products like elastomers, plastics and latices are prepared industrially by this method.

Bulk polymerisation differs from suspension polymerisation in initiation, which in this case occurs in the aqueous phase.

In this method, the monomers are dispersed as fine droplets in a large amount of water and then emulsified with common soap or special emulsifying agent. To stabilise the emulsion, protective colloids like casein, gum, gelatin or dextrine are used. The initiator is usually a redox type which may be either water-soluble or organo-soluble. The reaction mass remains in the form of an emulsion after the polymerisation is over. It may be used as such in a number of applications like adhesive, surface-coating and textile finishing. For other uses, the emulsion is broken by adding an electrolyte or a solvent, followed by washing and drying of the coagulated polymer (Fig. 26.2).

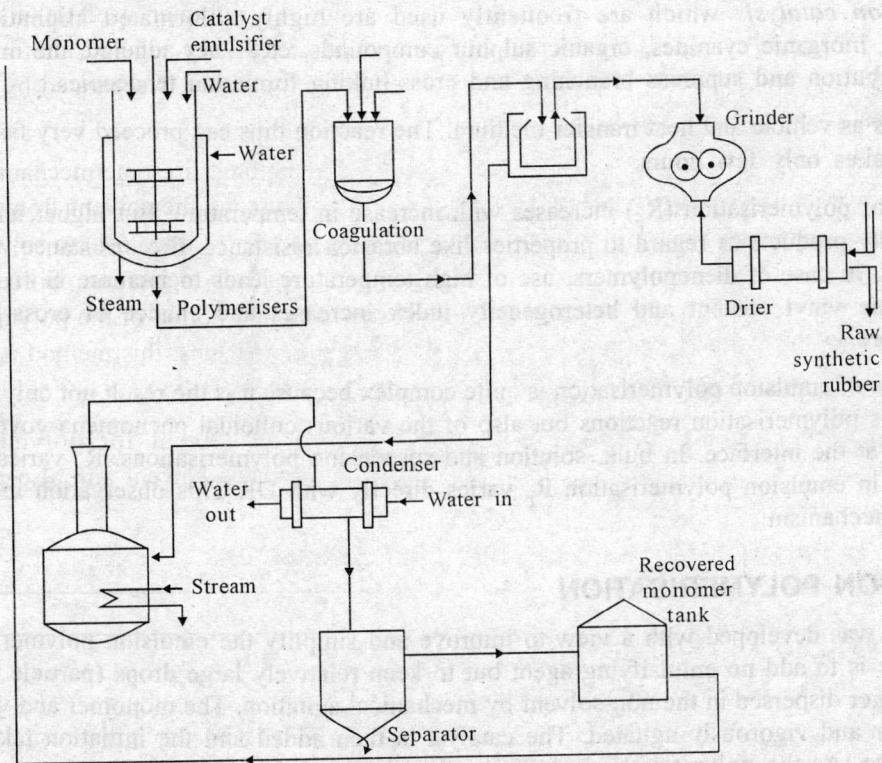

Fig. 26.2. Schematic flowsheet—emulsion polymerisation process.

A *disadvantage* of this method is that impurities, e.g., emulsifier, stabiliser, etc. left in the product may damage electrical or other properties.

Emulsifiers derived from natural products, such as, fatty acids and resins dominate the field. As unsaturated compounds hinder polymerisation so saturated compounds should be used as emulsifier.

The rate of polymerisation is dependent on number of carbon atoms in the soap upto C_{12} and then becomes independent of it. Industry normally uses a soap containing 16–18 carbon atoms. Rosin acid soap is better to form higher tacky products. But they impart an yellowish tinge in the polymer. Synthetic detergents, which are available now economically from petrochemical sources, are finding increasing use as emulsifier. The advantages are:

(1) They are stable at neutral and slightly acidic media.

(2) Monomers soluble in water and affected at high pH like acrylonitrile and acrylic acid can be polymerised.

(3) Emulsifier can be easily washed off the polymer after coagulation yielding purer product.

To conduct the polymerisation, the pH of the medium must be controlled and in normal case should be slightly alkaline. To achieve this, *buffers* like sodium acetate or sodium phosphate are often used.

One of the major advantages of emulsion polymerisation is the use of inexpensive inorganic redox free-radical catalysts. Peroxides, peracids, perborates and persulphates both of inorganic and organic origin may be used.

Orientation catalysts which are frequently used are highly chlorinated aliphatics like carbon tetrachloride, inorganic cyanides, organic sulphur compounds, etc. They regulate the molecular weight and its distribution and suppress branching and cross-linking formation tendencies.

Water acts as vehicle and heat transfer medium. The reaction thus can proceed very fast and monomer conversion takes only few hours.

The rate of polymerisation (R_p) increases with increase in temperature. But higher temperature leads to poor quality products as regard to properties like abrasion resistance, flux resistance, tensile strength, modulus, etc. In case of dienepolymers, use of high temperature leads to increase in *cis*-content. But at the same time vinyl content and heterogeneity index increases and chance of cross-links formation becomes more.

The nature of emulsion polymerisation is quite complex because it is the result not only of the interplay of the various polymerisation reactions but also of the various colloidal phenomena governing reactions which occur at the interface. In bulk, solution and suspension polymerisations, R_p varies inversely with DP whereas in emulsion polymerisation R_p varies directly with DP. This observation indicates towards a different mechanism.

SUSPENSION POLYMERISATION

This method was developed with a view to improve and simplify the emulsion polymerisation method. The principle is to add no emulsifying agent but to keep relatively large drops (particle size 0.1-1 mm) of the monomer dispersed in the non-solvent by mechanical agitation. The monomer and water are placed in the reactor and vigorously agitated. The catalyst is then added and the initiation takes place in the aqueous phase. As the polymerisation proceeds, the dispersed drop increase in viscosity and becomes sticky. At this point if the suspension is not stabilised by any agent then upon collision the droplets agglomerate and drop away, thus, breaking the suspension. This is prevented by the addition of a small amount of suspension stabiliser. If the stabilising agent works properly, polymerisation continues and the drops pass beyond the sticky phase and become hard without agglomerating upon collision. When the reaction is complete, the reaction product is in the form of pearls (hence the name pearl or bead polymerisation) and does not have to be recovered by coagulation. The stabiliser is removed from the surface of the polymer simply by washing. The material is dried and used (Fig. 26.3).

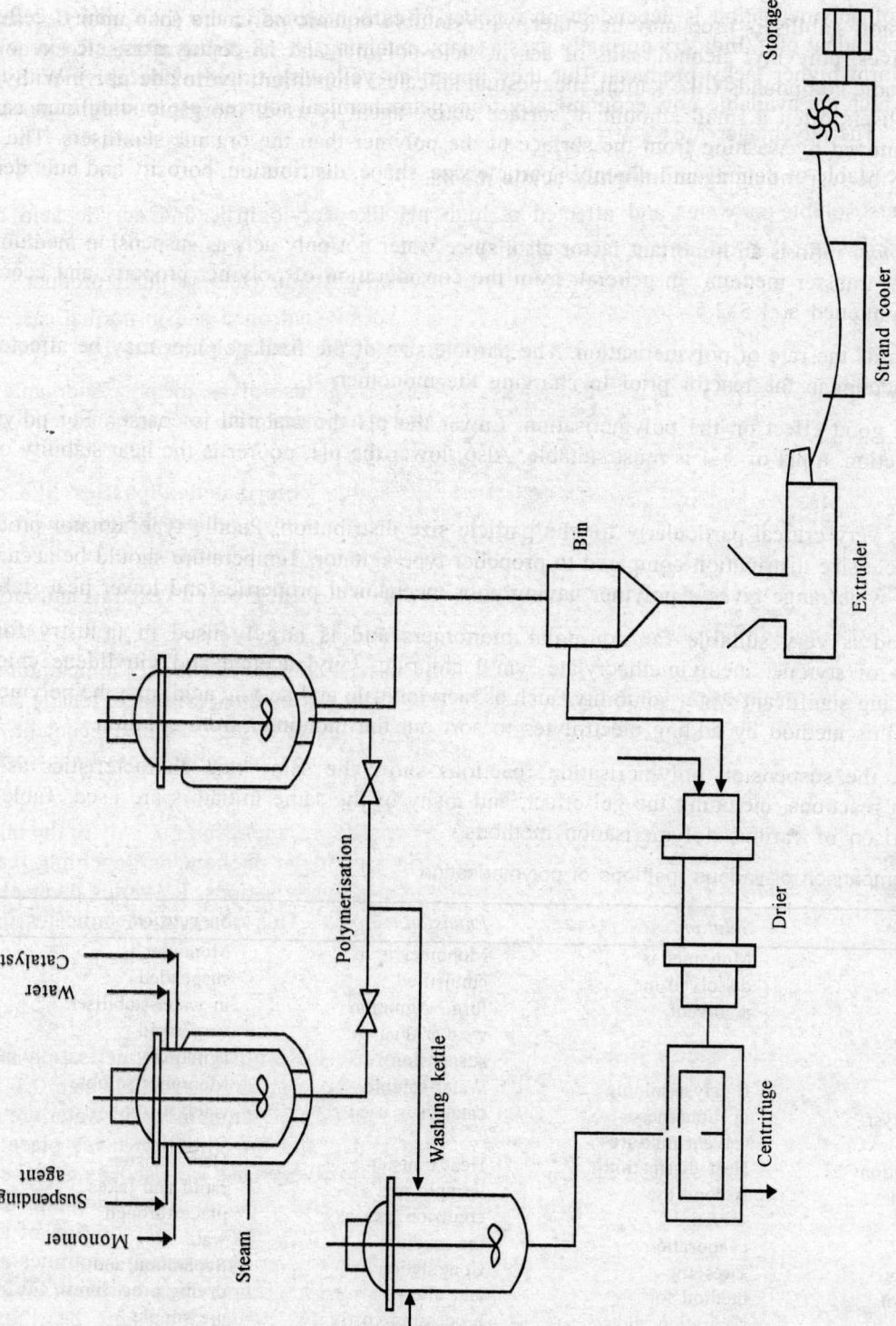

Fig. 26.3. Schematic flowsheet—suspension polymerisation.

The suspension stabilisers used may be either water-soluble organic polymers (like methyl cellulose and its derivatives, polyvinyl alcohol, salts of acrylic acid polymer and its copolymers, etc.) or water-insoluble inorganic compounds (like kaolin, magnesium silicates, aluminium hydroxide, etc.). With these inorganic stabilisers often a small amount of surface active agent is used. Inorganic stabilisers can be more easily removed by washing from the surface of the polymer than the organic stabilisers. The type and quantity of the suspending agent influences particle size, shape, distribution, porosity and bulk density of the polymer.

Water-monomer ratio is an important factor also, since water not only acts as suspension medium but also as the heat transfer medium. In general, from the consideration of polymer property and economy the ratio is maintained at 1.5–2.5.

Oxygen retards the rate of polymerisation. The particle size of the final polymer may be affected by varying the vacuum in the reactor prior to charging the monomer.

pH has got good effect on the polymerisation. Lower the pH the material is coarser. For polyvinyl chloride production, a pH of 3–4 is most suitable. Also, lower the pH, poorer is the heat stability of the polymer.

Agitation is very critical particularly for the particle size distribution. Paddle type agitator produces a sharper particle size distribution compared to propeller type agitator. Temperature should be accurately maintained. A wide range gives a polymer having poor mechanical properties and lower heat stability.

This method is very suitable for non-polar monomers and is largely used in industry for the polymerisation of styrene, methylmethacrylate, vinyl chloride, vinyl acetate and vinylidene chloride. Monomers having significant water solubility, such as, acrylonitrile and acrylic acid, may be polymerised efficiently by this method by adding electrolytes to sort out the monomer from solution.

In general, the suspension polymerisation reactions show the same rate characteristics as bulk polymerisation reactions, including the gel effect, and many of the same initiators are used. Table 26.2 shows comparison of various polymerisation methods.

Table 26.2. Comparison of various methods of polymerisation.

Bulk	*Solution*	*Emulsion*	*Suspension*
1. Monomer is polymerised in its own medium only	Monomer is dissolved in a solvent	Monomer is in emulsified form. Agitation weaker than suspension	Monomer is suspended in water-stabiliser, very rapid agitation
2. Monomer soluble catalyst is used	Catalyst soluble in monomer-solvent mixture	Water-soluble catalyst is used	Monomer soluble catalyst used
3. Heat dissipation is a problem	Heat dissipation is done by solvent evaporation	Heat transfer is rapid and emulsion acts as the medium	Heat transfer rapid and takes place through water
4. No separation and drying of the polymer	Tiresome method for separation and drying	Coagulation and after-processing costly	Separation and drying procedures are simple

(Cont'd ...)

Bulk	Solution	Emulsion	Suspension
Polymer very pure	Polymer contains solvent molecule	Purity of polymer is poorest	Polymer pure but has lower stability and clarity than in bulk method
High Rp and DP	Rp depends upon solvents and DP limited	Rp is fastest and DP also very high	High Rp and DP
7. Equipment simple	Equipment complex	Equipment simplest	Equipment simple
8. Safety hazards are more	Safety hazards are there	No safety hazards	No safety hazards.
9. Capital investment low	Capital investment high	Capital investment low	Capital investment low
10. Conversion cost high.	Conversion cost highest.	Conversion cost lowest.	Conversion cost higher than emulsion but lower than bulk.

PLASTICS

Plastics are an arbitrary group of artificial materials, generally of synthetic organic origin which at some stage in manufacture are in a plastic condition during which they are shaped, often with the aid of heat and pressure and often in a mould.

Difference Between Synthetic Resins and Plastics

All plastics are based on synthetic resins. A synthetic resin may be used as such for making plastics or it may be mixed with other ingredients such as fillers, plasticisers, dyes and pigments, catalysts and lubricants to form plastics. For example, polyvinyl chloride and polyethylene may be used as such for shaping articles out of them. Phenol formaldehyde resin is mixed with other ingredients to form plastics. So a synthetic resin is an essential ingredient of a plastic and it may or may not be mixed with other ingredients to form plastics.

Classification of Polymers

Polymers are classified into three groups, viz. thermoplastics, thermosets and elastomers. In addition, thermoplastics are divided into amorphous, crystalline and liquid crystalline polymers.

Thermoplastics are linear or branched polymers which can be melted upon the application of heat. They can be moulded and remoulded using conventional techniques. Thermosetting plastics or thermosets are heavily cross-linked polymers which are normally rigid and intractable. They consist of a dense three-dimensional molecular network and like rubbers, degrade rather than melt on the application of heat.

Elastomers or rubbers are polymeric materials which display elastomeric properties, i.e., they can be stretched easily to high extensions and will spring back rapidly when the stress is released.

Crystalline polymers are thermoplastic materials having regularly arranged molecular structure. In these polymers the chemical structure allows the polymer chains to fold on themselves and pack together in an organised manner. The resulting organised regions show the behaviour characteristics of crystals.

Amorphous polymers are thermoplastic materials having irregularly arranged molecular structure. When the molecules solidify they acquire random coily structure as in the moten state. Molecular packing will be very poor in these materials. All the crystalline polymers have amorphous regions between and connecting the crystalline regions. For this reason, crystalline polymers are often called semicrystalline polymers.

Liquid crystalline polymers are best thought of as being a separate and unique class of plastics. The molecules are stiff, rodlike structures that are organised in large parallel arrays or domains in both the melt and solid states. These large, ordered domains provide liquid crystalline polymers with unique characteristics compared to those of crystalline and amorphous polymers.

Organic and inorganic polymers

Further, polymers are divided into organic polymers and inorganic polymers. Organic polymers are formed by reacting organic monomers. The main chain will predominently contain carbon atoms. Inorganic polymers are formed by reacting inorganic monomers. The main chain will contain atoms like Si and Ge.

Commodity, Transition and Engineering Plastics

Based on applications, plastics are further classified as commodity plastics, transition plastics and engineering plastics. Commodity plastics are low strength materials, predominently used in noncritical consumer applications. Engineering plastics are hgh strength materials mainly used in engineering applications. Transition plastics are having strength in between that of commodity and engineering plastics. Some plastics have exceptional properties like high temperature resistance, chemical resistance and very high strength. Those plastics are known as speciality thermoplastics. There are some thermoplastic materials which are having elastomeric properties and they are classified as thermoplastic elastomers.

Polymer and monomer

A polymer is a substance which consists of molecules which are, at least approximately, multiples of low molecular weight units, (In Greek *polys* means many and *meros* means part of unit). The low molecular weight unit is monomer. The linking of small molecules (monomers) to form large molecules (polymers or macromolecules) is called polymerisation.

Homochain polymers and heterochain polymers

Also polymers are divided into homochain polymers and heterochain polymers. In homochain polymers, only one type of atoms combine together to form the main chain. They are formed by the addition polymerisation technique. In heterochain polymers, more than one type of atoms combine together to form the main chain. They are formed by the condensation polymerisation technique.

Homopolymer

If only one monomer is polymerised the product is called a homopolymer.

Copolymer

The polymerisation of a mixture of two monomers of about equal reactivity leads to the formation of a copolymer and the two types of monomer units enter the chain in a random fashion, as shown below:

A—B—A—A—B—B—B.

A and B are two types of monomer units.

Polyblend

A mixture of macromolecules containing only one type of monomeric units with another macromolecules containing essentially only another type of monomeric unit is called a polyblend or a mixture of homopolymers.

Block copolymer

It is a copolymer which contains longer stretches (or chain) of two or more monomeric units linked together by chemical valencies in one single chain.

A—A—A—A—B—B—B—B—A—A—A—A.

where A and B are two different monomes.

Graft copolymer

A graft polymer contains branches of varying length made up of different monomeric units on a common backbone or trunk chain.

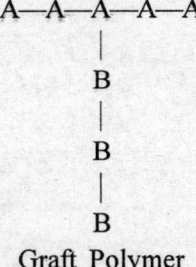

Graft Polymer

The detailed structure of block and graft copolymers is determined by the relative reactivity ratios of the various building units.

Isotactic polymer

This is a formation of a polymer in which the monomers have entered the chain in a regular fashion.

Atactic (random) polymer

This is a formation of a polymer in which the monomers have entered the chain in a random fashion.

Syndyotactic polymer

In this formation of a polymer the monomers have entered the chain in alternating fashion. It may be noted here that isotatic, atactic or syndyotactic polymers are formed depending upon the presence of complex catalysts.

Classification of Plastics

The plastics are divided into two groups: (i) thermosetting materials and (ii) thermoplastic materials.

(1) *Thermosetting materials:* Thermosetting materials are those plastics which require heat and pressure to mould them into shape. When heat is applied, they first become soft and plastic and on further heating they undergo chemical change and set hard. The process is called thermosetting or thermohardening. When a material is thermoset, it is permanently set and and does not soften to any appreciable extent when again heated. However, intense heating will bring about the breakdown of the material by burning. The following are some of the thermosetting materials :

    (a) Alkyds and Polyesters

    (b) Aminos (urea and formaldehyde resins and plastics)

    (c) Casein

    (d) Epoxides

    (e) Phenolics

    (f) Silicones.

(2) *Thermoplastic materials:* Thermoplastic materials are those plastics which soften on the application of heat, with or without pressure but they require cooling to set them to shape. As the hardening in thermoplastic materials is not due to any chemical action, so the shaped articles from thermoplastic materials will resoften on heating. The following are some of thermoplastic materials :

    (a) Cellulose Acetate

    (b) Cellulose Acetate Butyrate

    (c) Cellulose Nitrate

    (d) Ethyl Cellulose

    (e) Nylon 6 : 6

    (f) Nylon 6 : 10

    (g) Nylon 6

    (h) Nylon 10

    (i) Polyurethanes

    (j) Polyethylene

    (k) Polypropylene

    (l) P.T.F.E. or Polytetra-fluoroethylene

    (m) Polystyrenes

    (n) Styrene Acrylonitrile

    (o) Polymethyl Methacrylate

    (p) Polyvinyl chloride.

Thermoplastics

The general-purpose thermoplastic resins such as acrylonitrile-butadiene-styrene (ABS), vinyl acetate and acrylonitrile-styrene and the engineering thermoplastics such as polycarbonates, acetals, and nylons thermoplastics. Some of the important ones are described below.

Polyethylene

Industrially polyethylene is manufactured by three distinct processes, namely high, medium and low pressure. The products of these processes exhibit different properties as shown in Table 26.3. It may be appended that polymerisation can be conducted in gas phase or liquid phase (solution/emulsion/slurry).

Table 26.3. Properties of polyethylene produced by different methods.

	High pressure	Low pressure	Medium pressure
Crystallinity %	54–68	80–90	95
Density Kg/M^3	915–925	940–965	950–965
Melting point °C	108–110	120–140	140–143
Brittle point °C	–70 & lower	–70 (lower)	
Tensile strength kg/cm^2	120–16	220–450	400–450
Elongation % (cross-wise)	400–700	100–1000	20
Dielectric strength			
Kv/mm	45–60	45–60	
Organic solvents not affecting	Paraffins	Paraffins	
	Aromatics	Aromatics	
	dissolve	Serviceable	
	Serviceable	upto 80°C	
	upto (50°C)		
Branching & unsaturation per 1,000 C-atoms (as methyl groups)	20–30	7–10	2
Melt index	0.2 to 0.3	0.1 to 4.0	
Resistant to heat	Upto 98°C	Upto 110°C	

High pressure polyethylene (low density polyethylene)

The polymerisation is carried out under a very high pressure (pressures ranging upto 350 MPa) in a temperature range of 200°C to 240°C. The reaction is credited to free radical mechanism. This being a chain reaction, the exothermic heat of 100 kJ/mole has to be successfully dissipated. Otherwise the reaction mixture goes uncontrollable at a rate of temperature rise of 10° to 12°C per 1% ethylene converted. Chain reaction to be perpetual, ethylene of oncrous purity is essential. Such purity of immense nature obviously craved for the solvent/emulsion polymerisation, which can go with less purity stocks.

Process description of high pressure polyethyene is represented in Fig. 26.4. The main steps involved in the process are :

(1) Purification and compression of ethylene in two stage (50 bars and 300 bars).
(2) Introduction of free radical initiators.

(3) Raising the reaction mixture to a temperature of 200°C and maintaining around 250°–270°C.

(4) Release of pressure in a number of stages.

(5) Separation of polymer from unreacted ethylene which is recycled.

Polymer grade ethylene (99.9%) mixed with 600–800 ppm oxygen is compressed mainly in two stages c_2 and c_1 upto 35 MPa. The compressed gas is passed into a tubular reactor : (i) (the reactor is tube-in-tube type), where heat dissipation is conductively carried out by circulating dowtherm or hot water. Another method of heat removal is to heat the incoming gas by out going mixture. The temperature of the reactants in the entering zone is around 150°C, but at the end of the coil, the temperature reaches 240°C. The temperature in the reactor outlet should never drop below 200°C, as the crystallisation of polyethylene begins, giving obstruction to flow. The reaction products are directed to pass into high pressure separator; (ii) The pressure of the gases, obtained from this stage is still above 35 MPa, hence, this after purification can go directly to second stage compression (c_1). From the high pressure separator, the melt goes into a second stage separator (low pressure); (iii) where the remaining ethylene is expelled and routed to 1st stage (c_2) compressor, through purification system. In fact, after first high pressure separator, there may be a series of separators to catch polymer dust. The gases are rich with ethylene and oxygen but contain some carbondioxide and formaldehyde. Hence purification is essential before the gas enters 2nd stage compression. A safety system is desirable while handling such high pressure gas, to actuate exit automatically, if disturbance in heat balance or flow obstruction appears in the reactor. It is always desirable to blow the whole system with nitrogen before the polymer gas enters the reactor. As high as 80% ethylene is recycled due to inklingly low conversions per pass. LDPE formed is collected from low pressure and high pressure separators. The gases leaving out of the system pass through a separator where some liquid polymers gets condensed.

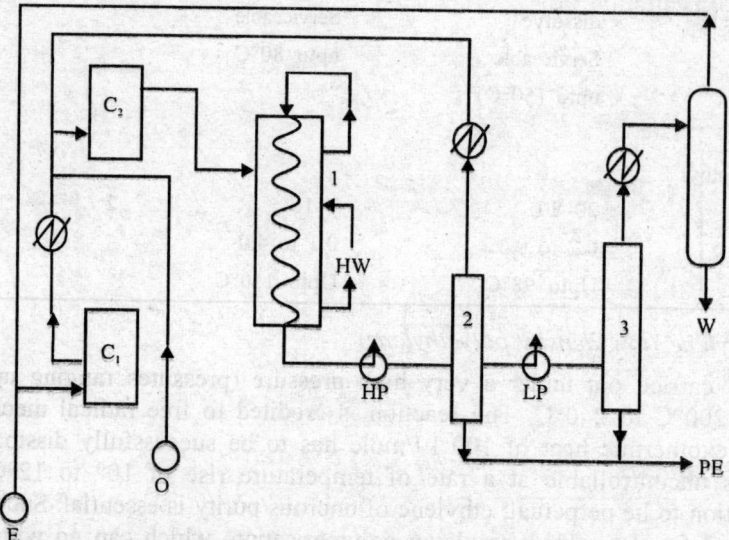

Fig. 26.14. Manufacture of polyethylene (low density). (E) Ethylene; (O) Oxygen; (W) Waxes and liquid polymers; (PE) Polyethylene; (HW) Hot water circulation; (HP) High pressure reliever; (LP) Low pressure reliever; (C₁ & C₂) compressors; (1) Reactor (tubular); (2) Separator & cyclone (high pressure); (3) Separator; (low pressure).

LDPE can also be produced by solution polymerisation, where the pressure and temperature are aminably low order, 100 MPa, and temperature 190°C. The solvent constitutes a equivolume mixture of ·

benzene and water. Oxygen as in the former case acts as initiator which is introduced into ethylene stream. After reaction, the products are separated, and unconverted ethylene goes back. This process gives a high conversion rate upto 20% per pass.

Medium pressure polyethylene

Flow sheet of medium pressure process based on chromium oxide catalyst is shown in Fig. 26.5. The polymerisation is conducted at a pressure of 5 MPa in solvent medium; either in a slurry fashion or in a solution. Phillips solution catalysts is prepared by muffling chromium-impregnated silica-alumina supports at selected conditions. Usually hexavalent chromium oxide on any inert base (silica-alumina) is a poorer solvent for polyethylene; will serve as polymer catalyst. Some active ingredients like calcium, barium, tungsten oxides are also used. Some times halogenated chromium oxide is prevalent in the industry to increase the molecular mass of the polymer.

The polyethylene obtained from this system is linear and highly crystalline. There are other catalysts like nickel, cobalt, molybdenum oxides. With different catalysts, the conditions of the reactor are likely to vary.

Medium pressure process employs a pressure of 35 MPa and a temperature of 130–170°C. 0.1% Cr_2O_3 treated with HF gives a molecular mass of 45,000 otherwise the polymer mass does not exceed 20,000. In this process cyclohexane is used as solvent and the concentration of ethylene in solution is kept around 5%. The solvent serves the dual role as a solvent to ethylene and polymer, and permit the chain length to grow irrespective of the presence of chain breakers. Viscosity, heat transfer problems and control of ethylene consumption are also attended successfully by solvents.

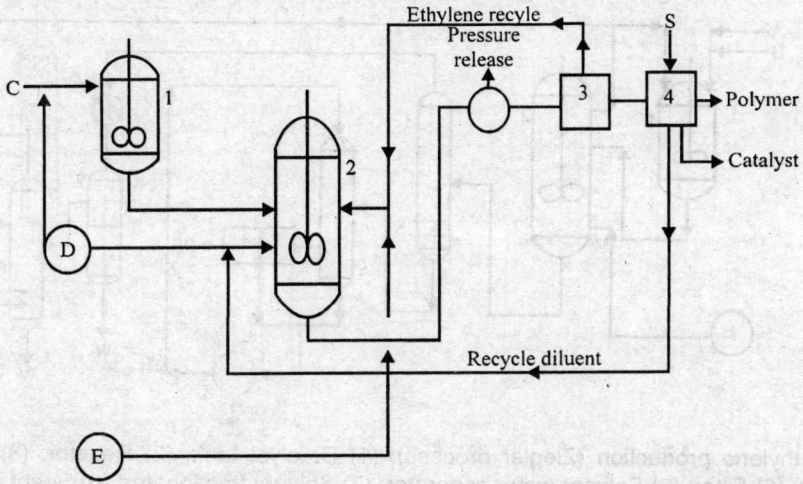

Fig. 26.5. Manufacture of polyethylene (medium pressure process). (1) Catalyst tank; (2) Reactor; (3) Separator; (4) Washer-evaporator; (C) Catalyst; (D) Diluent; (E) Ethylene; (S) Solvent.

In recent versions, solution process gives high solids (40%) content with no de-ashing problem. In the slurry process higher yield of product per given reactor is an advantage. Further, because of poor solvency nature, the polymer or copolymer separates from the diluent in fine particles. Thus the viscosity of solution does not increase as seen in solution process. Of course slurry process takes a longer time (2–4 hours). Further all slurry processes are credited with yielding very high mole weights, perhaps of no commercial value, hence chain stoppers are inevitably sent into the system.

Low pressure polyethylene (Ziegler's process)

Karl Ziegler developed a composite catalyst system for polymerising ethylene at low pressure and low temperature. The conditions required are almost ambient conditions, viz. atmospheric pressure and a temperature of 60°C though it was a profounding discovery the manufacturers were reluctant to replace the older technology. Providentially, Giulio Natta extended the complex catalyst to get polypropylene of high molecular weight, a linear polymer, of course stereo-regular. This prompted others to work on these catalysts and ultimately all types of polymers have become a reality. The discovery thus paved the way to polymerise different monomers, where free radical polymerisation failed.

Production of HDPE (high density polyethylene)

Fig. 26.6 shows the outlines of the process. In the catalyst vessel a fresh and dilute solution of composite catalyst (max. 0.1 gm/litre) in pentane or hexane is made and this acts as a reservoir too. The solution is fed continuously into reactor where solvent and ethylene streams enter. A stirrer in the reactor is helpful for homogenising the mixture. The polymer mixture after reaction, is allowed to enter a flasher, where unreacted monomer and inerts are vented out. The solid material is given a thorough wash with water to decompose the catalyst. The strippers separate the organic solvent from water-hydrocarbon azeotrope and solvent is recycled. The polymer material is washed many times, either with water or with methanol, after which it is filtered and dried. Catalyst is obtained in a slurry and is recovered for usage. Additives for preserving the quality of polymer are added at the stage of granulation. It is found that by alcoholic wash, the ash content of the polymer decreases very much. Before start of the operation all units are submitted to nitrogen breathing, and is essentially important to see nowhere water or air seeps in.

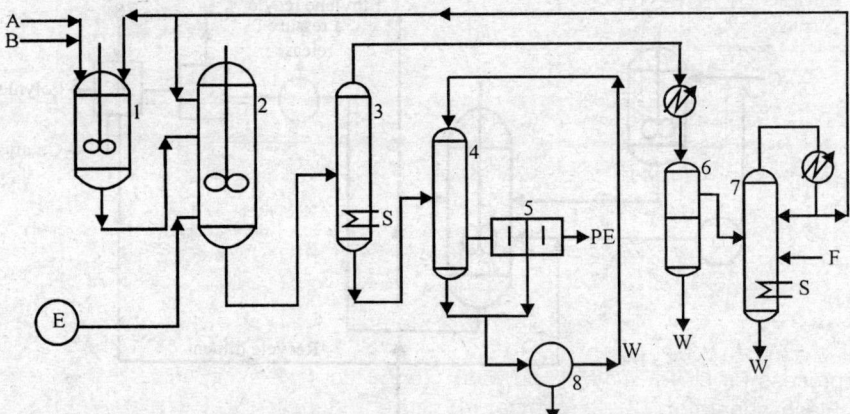

Fig. 26.6. Polyethylene production (Ziegler process). (1) Catalyst tank; (2) Reactor; (3) Solvent stripper; (4) Washing column; (5) Filter; (6) Solvent-water separator; (7) Solvent fractionater; (8) Spent catalyst recovery unit; (A) Alkyl aluminium; (B) Titanium halide; (E) Ethylene; (S) Steam; (W) Water; (P) Solvent recycle; (PE) Polythylene (HD); (F) Fresh solvent.

DU Pont process. This is a solution process. Du Pont uses soluble Ziegler catalysts, although it differs from Phillips process in a true solution way, i.e. all things remain in solution. These catalysts are comprised of $TiCl_4$, VCl_5, $VOCl_3$ and the residence time for polymerisation is 3 to 4 minutes. Molecular weight can be easily controlled by hydrogen and temperature. All solution processes are generally run at moderately high pressures and high temperatures. This gives rise to increased efficiency of catalyst (500 kg/gm). A polymer concentration of 35–40% is common in such solution process. Du Pont process is also suitable for co-polymer production. These polymers from this process are found to have less ash content because the catalyst can be dislodged very easily and effectively.

Solvay process. The catalysts are formed by reacting transition metal compound ($TiCl_4$) with one of three different supports at a time, followed by activation with organo-aluminium compounds. Polymerisation is carried out in a loop type reactor using n-hexane as diluent. Reactor conditions are moderate; 25–35 bars and 50–90°C. The resultant slurry is centrifuged. Wide range of molecular weights are possible with this process. Fig. 26.7. shows flow sheet of the process. The reactor is tubular provided with a jacket for circulating cold water. Hydrogen is incorporated in the reacting mixture, so that copolymerisation is smoothly controlled. Further, this is a closed loop reactor, provided with a surge drum. Polymer fluff is available in hexane slurry. A depressuriser flash vessel is provided to separate the unreacted reactants for circulation.

Polypropylene

Polypropylene is commercially produced by only one method that is anionic polymerisation. The process is accomplished in hydrocarbon solvents with the same Ziegler type composite catalyst, but with a difference, titanium trichloride is opted over tetrachloride.

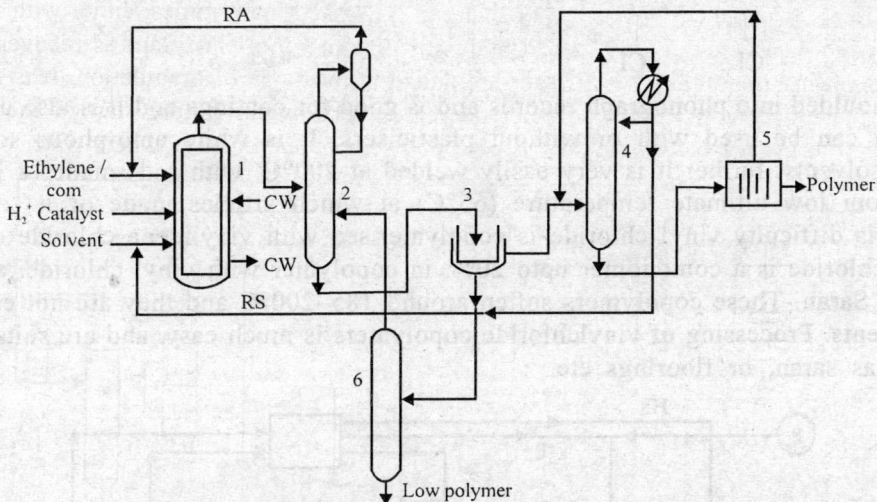

Fig. 26.7. Solvay process for HDPE. (CW) Cold water; (RS) Recycle solvent; (RA) Recycle olefins; (1) Reactor water cooled; (2) Flash separator; (3) Centrifuge; (4) Solvent stripper; (5) Filter; (6) solvent purifier.

The reaction is carried out at 10–20 bars in the temperature range of 65–70°C. Polymerisation is done in medium for which a solvent is chosen to act either as solvent or medium for slurry. Though conventionally slurry process is more pronouncive, solvents like heptane or gasoline with no unsaturates is preferred. Propylene of very high grade purity (99.9%) is essential, propane-propylene (35%) mixture is also used, at times. Polymerisation is carried out in batch or continuous wise.

Fig. 26.8 illustrates the process. A 10% solution composite catalyst is made in a solvent usually heptane and kept in a tank: (i) The solution is proportionally charged into reactor; (ii) where olefin and some more solvent are admitted. The reactor is kept under required pressure and temperature for about 3 to 4 hours to complete the reaction. Feed gas and catalyst thoroughly mix due to agitation. After the reaction, products are transferred into a stabiliser; (iii) where pressure is decreased, to let off unreacted gas. The catalyst is decomposed by means of alcohol. All the polymer product is centrifuged; (iv) and the solvent-alcohol mixer is separated by steam stripping in a stripper; (v) recovered solvent is distilled for purification; to begin circulation. A separate stripper-purifier section is essential for this. The heavy polymer with adhering solvent is stripped in another stripper; (vi) by means of steam, and the recovered solvent mixture is processed for alcohol and solvent in section; (vii) Polypropylene collected from the bottom of the stripper is further centrifuged, mixed with stabilisers and sent to storage. Polypropylene requires about 1.1 tons of monomer per each ton of polymer.

Polyvinyl Chloride (PVC)

This is one of the most widely used big tonnage thermoplastic credited with diverse applications. Two types of homopolymers are produced one being flexible and the other a rigid type. Flexible polymers have a good elasticity possessing a density of 1150 to 1600 Kg/m^3. Rigid type polymers exhibit a density of 1150 to 1800 kg/m^3 with a good tensile strength 42–60 MPa. Rigid polymers are suitable for making articles of hard type sheets, papers, gears etc. Both the types show low crystallinity due to random orientation of chlorine atom. Polyvinyl chloride is obtained by radical polymerisation leading to head to tail configuration as shown here.

$$-CH_2-CH-CH_2-CH-CH_2-$$
$$\qquad\quad | \qquad\qquad |$$
$$\qquad\quad Cl \qquad\quad Cl$$

PVC is moulded into phonograph records and is good for coatings and it is also used in the form of sheets. It can be used with or without plasticisers. It is white amorphous solid, soluble in chlorinated solvents, further it is very easily welded at 200°C, with rods made of PVC. However, it suffers from low ultimate temperature (65°C) at which articles made of it can be used. To overcome this difficulty vinyl chloride is copolymerised with vinylidene chloride or acrylonitrile. Vinylidene chloride is a comonomer upto 20%, in copolymer with vinyl chloride, and the product is known as Saran. These copolymers soften around 185–200°C and they are not easily soluble in organic solvents. Processing of vinylchloride copolymers is much easy, and are suitable for making fibres such as saran, or floorings etc.

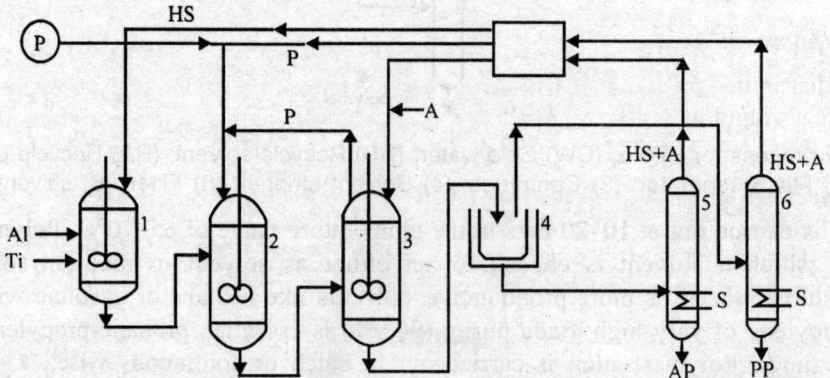

Fig. 26.8. Production of polypropylene. (Al) Aluminium (Triethyl); (Ti) Titanium halides; (P) Propylene (A) Alcohol; (HS) Hydrocarbon solvent; (S) Steam; (AP) Attactic polymer; (PP) Polypropylene.

Polystyrene

Next to polyethylene and polypropylene, polystyrene is the widely used plastic. Polystyrene can be made in a number of ways to suit the end use. As the molecular weight increases, it slowly turns from liquid to solid state. High strength polystyrene is usually having a molecular weight more than 100,000; these are difficult to mould, the intermediate.range ploystyrene 70,000–100,000 is easy to mould. As the molecular weight drops below 70,000; the tensile strength, impact strength and elongation fall of rapidly: however, these are classified as general purpose polystyrenes. Polystyrene has got many favourable properties, like ease of fabrication high thermal stability, low specific gravity (900 kg/cu. m) etc. Homopolymers of polystyrene are extremely brittle and rigid hence copolymers are opted.

Polystyrene production is carried out by free radical initiation or by co-ordinated catalyst. Bulk, suspension and emulsion polymerisations are in use. The polymer produced by initiators is atactic, while Ziegler's polymers are isotactic. Polystyrene is a linear polymer, though the phenyl groups often prevent ordered arrangement of macromolecules and crystalline structure. At 80°C and above amorphous polystyrene passes into rubbery state till it reaches 150°C, whereby it becomes plastic. Further heating above 300°C results in decomposition. It dissolves in most of the aromatic solvents. Isotactic polymer melts at 220°C. Its brittleness decreases with the addition of plasticisers, thus lowering its glass transition temperature.

Cellulose

Cellulose is the name given to a long chain of atoms consisting of carbon, hydrogen and oxygen arranged in a of plant life, especially in cotton and wood.

Cellulose and cellulose derivatives

Cellulose, as it occurs in plant life, cotton or wood cannot be shaped, moulded, cast or extruded. To carry out these processes cellulose is modified in the following two ways :

(1) *Regenerated cellulose.* Cellulose is brought into solution form, which is a temporary modified form of cellulose. From the modified form fibres or films are shaped by spinning or casting and then regenerated to the original cellulose chain, forming rayon or cellophane.

(2) *Cellulose derivatives.* Cellulose is brought into a solution which is a permanently modified form, such as cellulose acetate, cellulose nitrate or cellulose acetate butyrate. From this are produced moulded or extruded article, films or fibres.

Manufacture of cellulose in its pure form

Cellulose is wide distributed in nature. It is the essential constituent of trees, corn stalks and grasses. In its purest from it is found as cotton fibres. The commercial sources of cellulose are wood and cotton fibres.

(1) *From wood.* Trees are debarked, chipped and treated chemically to remove the unwanted lignin and other structural constituents of wood which are mixed with cellulose in its natural form. About 60% of cellulose is recovered from the wood in the form of chemical wood pulp. It is supplied to the manufacturers of cellulosics in the form of large rolls or sheets of blotting paper or as compressed bales.

(2) *From cotton linters.* The small fibres left on the cotton seed after the long fibres are removed are known as cotton linters. About 82% of purified cellulose is recovered from raw linters. Cellulose from cotton linters is generally supplied in compressed bales.

Regenerated cellulose. If cellulose is brought into a solution by a chemical reaction and it is reprecipitated out of the solution to give essentially the same chemical structure but in different form (film or fibre), it is said to be regenerated cellulose. In this transformation of shape chemical and physical properties do not change much. The following are the two forms of the regenerated cellulose :

(1) *Cellophane:* Cellulose obtained from wood pulp or cotton linters is taken and is treated with caustic forming soda cellulose. Soda cellulose is shredded into pieces and is aged under controlled conditions of temperature and humidity. During ageing, the original high polymer molecule breaks down a little. Breakdown is indicated by better solubility in solvents.

Aged and shredded pieces of soap cellulose are dissolved by treating with carbon disulphide followed by an aqueous solution of caustic soda. A viscous orange solution which is known as viscose is formed.

Viscose solution is filtered to remove foreign bodies, and then vacuum debubblising is done to remove bubbles etc. Finally ageing is done to adjust the final viscosity.

Viscose solution is then extruded through a narrow slot into an acid solution (sulphuric acid). Acid bath acts as a coagulating bath and cellulose is regenerated from the viscose solution.

(2) *Rayon:* Rayon is the generic name given to fibres (filaments cut into small lengths) and continuous filament yarns made of regenerated cellulose. The method of manufacture is the same as in the manufacture of cellophane. The only difference is that the viscose solution is extruded through spinnerets (small holes) into the acid bath instead of extruding it through a narrow slot into the acid bath. Fibres are formed. The fibres of regenerated cellulose are similarly washed, desulphurised, bleached and re-washed to remove traces of all the chemicals and finally dried.

If fibres are dried and sold in the form of continuous filament yarn, it is called viscose rayon yarn. If fibres are cut into short lengths before drying, they are called staple fibre. These short fibres called staple fibres are very useful for blending with other fibres and spinning into threads.

Cellulose acetate

Cellulose acetate is manufactured in the following operations :

(1) *Pretreatment of Cellulose.* Pure cellulose is dried at a temperature of about 50°C to a moisture content of less than 4%. After this it is pretreated with small amounts of mineral acids in order to increase the reactivity of cellulose to break it down to make it more soluble.

(2) *Esterification with Acetic Anhydrides.* To 100 parts (dry) pretreated cellulose is added to a mixture of 200 parts of glacial acetic acid, 100 parts acetic anhydride and 100 parts concentrated sulphuric acid. The mixture is kept below 10°C. After two hours the mixture first becomes fluid and then starts thickening up. The process is finished when no cotton fibres remain. It takes generally about 5 hours.

(3) *Hydrolysis or Ripening.* The material formed after acetylation is cellulose triacetate which is soluble only in chloroform, trichloro ethylene etc. It is also known as primary acetate. Water is carefully added so that there is no undue rise in temperature while residual acetic anhydride is converted into dilute acetic acid. Ripening is carried out by allowing the mixture to stand in large, open copper pans at about 25°C for 4 to 5 days in presence of dilute acetic acid. During ripening acetyl groups are split off and the acetyl value of the cellulose acetate decreases. Ripening is allowed until the acetyl

value has reached the corresponding value of cellulose diacetate. As the ripening continues the material becomes golden syrup. Degree to which primary acetate is ripened depends upon the purpose for which it is required. Cellulose triacetate has acetyl value of 62.5%, cellulose diacetate has 48.8%, and cellulose monoacetate 29.4%.

(4) *Precipitation.* Water is added to the syrup very carefully. Cellulose acetate is precipitated as a finely divided white material. It is centrifuged and washed.

(5) *Stabilisation and Drying.* It is stabilised by boiling with very dilute mineral acid for about 2 hours. Sulphuric esters are removed by so doing. Cellulose acetate is then dried at low temperature until moisture content is below 3%. Then it is ground to a coarse powder.

Properties of cellulose acetate

Cellulose acetate is a white, odourless, tasteless, non-toxic solid. It is manufactured having substitutions ranging from 2.3 to 3.0 or from 53 to 62.5% combined acetic acid. The triacetate having 60.0 to 62.5% acetic acid is not used much on account of its poor solubility in ordinary solvents. Cellulose acetate which finds commercial application has 53 to 59% combined acetic acid. Cellulose acetate used for plastics has acetic acid value between 52 and 54% and for lacquers it has acetic acid value from 54 to 55%.

Polyvinyl acetate

It is a thermoplastic resin and is a vinyl acetate polymer. Polyvinyl acetate is prepared from vinyl acetate which is a liquid boiling at 72°C and is obtained by the interaction of acetylene and acetic acid in the presence of a complex mercury catalyst.

Polyamide resins

Polyamide resins are synthetic polymers which have recurring amide groups. Ther term nylon refers specifically to those synthetic polyamides which are capable of forming fibres.

Methods of preparation

The following are the brief general methods of their preparation.

(1) *By the reaction of dicarboxylic acids with diamines.* For example nylon 6, 6 and nylon 6, 10 are products of the condensation reaction of hexa-methylene diamine with adipic acid and with sebacic acid.

(2) *By the condensation of amino acids.* For example, nylon 6 and nylon 11 are obtained by the self-condensation of ε-amino caproic acid and ω-amino undecanoic acid.

(3) *By the reaction of so-called polymerised vegetable oil acids with polyamines.* The unsaturated fatty acids in vegetable oils such as linoleic acid may dimerise or polymerise to low polymers through their unsaturated groups. The di or polycarboxylic acids obtained yield tough polyamides by condensation with di or polyamines.

Polyamide resins are horny, whitish, translucent, high melting polymers. Polyamide resin can be essentially transparent and amorphous when their melts are quenched. On annealing or cold drawing, they become highly crystalline and translucent. These polymers are used for fibres, bristles, bearings, gears, moulded objects, coatings and adhesives. The term nylon refers specially to those synthetic polyamides which can be spun into fibres.

Thermosetting

Thermosettings are long chain molecules that give the polymer cross links in three dimensional infusible structures. These are produced by condensation reaction. Setting is caused by unidirectional polymerisation under heat and pressure resulting in hard rigid masses; the way they differ from thermoplastics. Economically they are not regarded as important as there is no scrap value, hence they are not produced in such large tonnage as thermoplastics.

Cross linking occurs, during curing or polymerisation or with cross linking agents. The demands for strong and stiffest polymers for special uses cannot be over looked, hence they have been produced on a limited scale in the category of engineering plastics. Reinforcement to improve the quality of thermosets · is a common practice. The most widely used material for reinforcement is glass fibre; others include synthetic fibres, cotton, paper.

Phenol-formaldehyde, Urea-formaldehyde, Polycarbonates are among the older engineering plastics. These plastics are exceptionally strong, dimensionally stable, and resist abrasion, wear and impact; and work well over a wide temperature range.

Sheet moulding compounds (SMC) are made by sandwiching resin, reinforcement fillers etc. with the aid of thickners between two sheets of plastics. Thermosetting (plastics) resins are used in several ways, like moulding powders casting resins, impregnating resins or manufactured goods, sheets, rods and tubes (extruded shapes). Polysulphones, polyacetals because of the transparent nature, toughness, high heat distortion temperature have penetrated into medical appliances and devices. New combinations of resins like Styrene-maleic anhydride (SMA), impact modified Nylons-6; ABS/Nylons, ABS/PCs are also classified as new engineering polymers because of the high performance and rare properties they exhibit. And new developments to make them fire-proof to replace metals are in progress round the world. Plastics are divided into the following categories depending upon the prominence of use.

Resins for moulding purpose

(i) Phenol-formaldehyde; (ii) Phenol-furfural; (iii) Urea-formaldehyde; and (iv) Melamine formations.

Resins for adhesives

(i) Urea-formaldehyde; (ii) Melamine-formaldehyde; and (iii) Phenol-formaldehyde.

Resins for coatings, varnishes etc

(i) Phthalic anhydride-glycerine; (ii) Maleic anhydride-glycerine; (iii) Phenol-formaldehyde; (iv) Urea-formaldehyde; and (v) Melamine-formaldehyde.

Phenol-Formaldehyde Resins

These are widely used in electrical and automative industry. Phenol-formaldehyde polymers can be produced in a wide variety of forms because of bi and tri functional groups. The (condensation) reaction between phenol-formaldehyde results in two types of resins known as Novolacs and Resols. The condensation reaction proceeds in alkaline as well as in acid medium producing two distinct types of resins. When phenol and formaldehyde react in acidic medium with a slight excess of phenol or equimolal quantities of phenol and formaldehyde o and p-hydroxy benzyl alcohols results.

These resins are soluble in many organic solvents and have a good demand in paint and varnish industry. To produce thermoset from this resin, it is heated and cured by mixing with a small amount of hexamethylene tetramine to effect cross-linking. Instead of acidic medium if alkaline medium with excess of formaldehyde is used, the condensation reaction leads to di; or trimethylol phenols in stages. In the first stage phenoxide is formed.

In the manufacture of phenolic resins, it is possible to produce a resin fit for many different applications, like resins for moulding, heat resistance duties, or coatings, enamels, or adhesives, specially for plywood industry.

Urea-formaldehyde resins

These fall under aminoplastics. Urea and foamaldehyde condensation products are as old as phenol formaldehydes and condensation reaction follows a number of steps. As in the above case the ratio of the reactants and the conditions of reaction alter the course of reaction. Nucleophilic addition of urea to formaldehyde results in monomethylol urea or dimethylol urea.

Condensation (Polymerisation) of methylol urea takes place when heated in presence of acid or hexamethylene tetramine. The reaction proceeds slowly under carefully regulated conditions, temperature is kept at 40°C and pH 7–8. Heating is continued to distil the water. Usually large amount of formaldehyde; 3 to 4 times of urea is used. After water removal the resinous mass becomes viscous. In case of phenol-formaldehyde resins water separates out, but in this case it does not. Hence water repellent materials like thiourea or cellulose are added to aid the separation of water from the product. Highly viscous resin is allowed to set gradually after fillers are added. Acid catalysts like bromine, tartaric acid are added before the resin is kneaded. Under vacuum the acid catalysts are driven off, subsequently heat promotes the polymerisation. Thermosetting resin is mixed with ammonium chloride, which works as a hardner, before moulding. While it is only urea that can give a polymer with aldehyde, other substituted ureas cannot give, as methylene bridges with nitrogen are absent ($-N=CH_2$) i.e. NH_2 group acts as a primary amine and an amide.

Urea-furfural resins

These resins have been known only for chemical interest and confined to laboratory only.

Melamine-formaldehyde resin

Melamine is produced from urea under high pressure in an atmosphere of ammonia. Amino compounds, primarily the melamines, can be modified chemically or by use of fillers. Melamines are moulded at 150–175°C under a pressure of 150–350 bars. Thermosetting moulding compounds are produced with α-cellulose as the fuller.

Articles made from amine resins appear water-clear, hard, strong but can be broken. They are used like urea formaldehyde resins in laminations. These are good adhesives and can be safely used for moulds too. Textile treatment is recommended because of the less water absorption capacity. The single greatest use of melamine moulding compounds is in the production of dinnerware. Like other amino resins these can be used for binding purposes, such as abrasive paper, wood boards, flooring and furniture assembly.

Epoxy resins

Epoxy resins are termed as general purpose polymers and contain at least two epoxide groups made by reacting diphenol with epichlorohydrin. These can be produced in forms ranging from low viscosity liquids to high melting solids just by varying the repeating units in polymer molecule. Epoxy resins are cross linked by reaction with hardners, leaving excellent wetting adhesive properties and can give composites with fillers like glass fibres, asbestos, paper fibres, cotton etc. Infact most important attribute of epoxies which they share with polyester resins is their ability to convert from liquids to thermosets without pressure and without any volatile by-products. Resins having melting point range of 64–76°C or 97–103°C are used for coatings. Commercially these resins are made by reacting bisphenol-A and epichllorohydrin. Condensation takes place in basic medium at 60–70°C. It is a ring opening reaction; a neucleophilic substitution. The acid liberated is neutralised by base present in the medium. 4% caustic acts as catalyst. Toluene is added to separate the salt from the crude product. Toluene can be later removed by distillation.

Most of these resins are soluble in organic solvents like acetone, toluene, MEK. They are having good mechanical properties, exhibit good electrical insulation properties even in presence of water. High adhesive strength, good chemical and heat resistance have made them useful in composites with different fillers. Further delicate electrical circuits and instruments are encapsulated in epoxy castings. Obviously properties of these can be modified by chlorine blends, curing agents, modifiers etc. Most of the glycidyl type epoxy resins are made of mono, bi or poly nuclear phenols. Epoxy resins are extensively used to bond concrete blocks and in other structural works; matrix materials in high performance composites, such materials are used in aggresive environments. Moisture affects badly the strength of resin. These resins can be made with aliphatic polyols or aromatic polyols.

Aliphatic polyols of ethylene, propylene and pentaerithrytol have been used in aliphatic resins. Ephchlorohydrin reacts with polyols in presence of Lewis acids to form chlorohydrin ether which on hydrodechlorination in presence of sodium aluminate gives epoxy resins. Epoxy-phenolics, Epoxy modified silicones are also in service for better performance.

ABS plastics

These are two phase systems consisting of acrylonitrile; butadiene-styrene copolymers and truly thermoplastic type. There are two basic types of ABS polymers. One type is just a mechanical blend of styrene-acrylonitrile copolymer and butadiene-acrylonitrile rubber. In the other type styrene acrylonitrile copolymer grafted on butadiene-acrylonitrile copolymer of varying amounts. Acylonitrile gives the blend, the desired properties of chemical resistance and tensile and flexural strength styrene gives good thermoplastic characteristics. The heart of the process is grafting and follows (Table 26.4).

Table 26.4. ABS resins properties.

Melt flow (G/Min)	2–45
Tensile strength (kg/cm^2)	350–540
Flexural strength (kg/cm^2)	520–800
Rockwel hardness (kg/cm^2)	94–110
Flexural modulus (kg/cm^2)	18,000–27,000
Impact strength (kg-cm/cm)	10–40
Specific gravity	1.03–1.05

Styrene and acrylonitrile monomers are added to a latex of polybutadiene and the mixture is agitated at 50°C. This allows the monomer to be absorbed by the polymer latex. Potassium persulphate is used as a catalyst in this system. The emulsion is coagulated after copolymerisation is over. The mix is washed and dried and compounded.

Production of copolymers can be done by usual methods like mass, suspension, emulsion and solution polymerisation. Free radical producing catalysts are common initiators. Emulsion polymerisation is used, where potassium persulphate acts as catalyst and sodium stearate acts as emulsifying agent. Modifiers are usually mercaptans. The reaction is carried out in absence of oxygen at around 0°C. ABS resins may be fabricated by most methods used for thermoplastics, injection moulding gives housings for radios, telephones etc. extrusion gives pipes and fittings and also sheets. Decorative pieces are added attractions; can be metallised, plated. Although these are flammable; are serviceable in a temperature range of −40°C to 107°C.

Fig. 26.9. shows the production flow sheet of ABS plastics. It contains two sections. First section gives butadiene-acrylonitrile polymer, while the second gives acrylonitrile styrene polymer. These two polymers are then blended to yield ABS polymer.

Reactor (1) contains the water-emulsion, modifiers, short stop and monomers. The polymerisation is carried out and the polymer is stripped off (2) to separate monomers for recirculation. From mixer (3) the acrylonitrile and styrene monomers in emulsion-water is carried to polymerisation reactor.

The polymer material is separated in the form of latex. This latex and latex from the other section are sent proportionally into another graft-polymeriser (5) reactor where antioxidants and modifiers are added. The crumb is washed with water in a tank (4) to remove catalyst and chemicals and sent to drier and mill section.

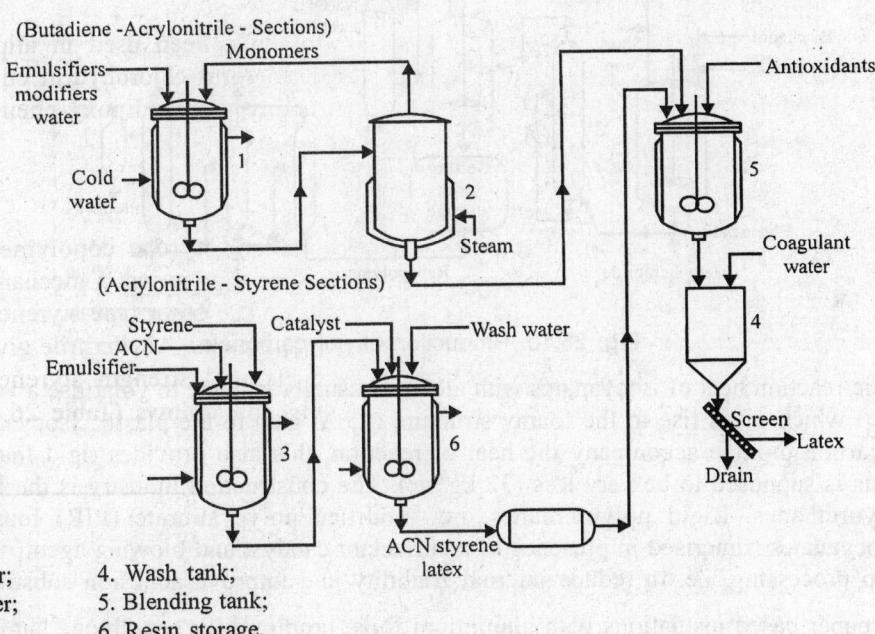

1. Reactor;
2. Stripper;
3. Mixer;
4. Wash tank;
5. Blending tank;
6. Resin storage.

Fig. 26.9. Styrene-butadiene-acrylonitrile plastics.

Poly carbonates (PC)

These are relatively transparent and expensive thermoplastics. These are esters of carbonic acid HOCOOH, and are produced by condensation polymerisation of sodium salt of bisphenol A and phosgene. They have excellent impact resistance with high heat distortion temperature. The credit of transparency and chemical resistance, are widely used in specialised applications.

The reaction is similar to epoxy resin condensation. An acqueous solution of disodium salt of bisphenol A is brought into contact with phosgene in a solvent like methyl chloride. A phase transfer catalyst such as tetrabutyl ammonium chloride has to be added to initiate the reaction between the two immiscible phases. Sodium chloride is precipitated and the solvent assists this. Later the solvent is separated by distillation. Fig. 26.10 shows the outlines of the process.

Polyurethanes

Polyurethanes are generally used for load bearing, abrasion resistance works. Their low conductivity and lightness with strength has been widely accepted in refrigeration and packaging industry respectively. Diisocyanates react with any compound having an active hydrogen, such as hydroxyl terminated epoxies, polyesters, furans and cellulose, rubber and groups like SH, NH, NHR, $CONH_2$, substituted carbamide CONHR, SO_2, NH_2, substituted sulphoamide SO_2NHR, thionides, substituted thioamides etc. Only not very active hydrogen containing groups require the assistance of alkaline catalysts rather than metallic or acidic catalysts. Isocyanates of all types are extremely reactive, with water they produce carbon dioxide, when a large amount of heat. The produced urea again reacts with isocyanates to form substituted ureas to form prepolymers. High molecular weight polyethenes, polypropylene glycols, polyesters are used in the soft foams. Rigid foams do not require high purification of reactants. Polymethylene poly phenyl isocyanate is the most widely used isocyanate in rigid foams.

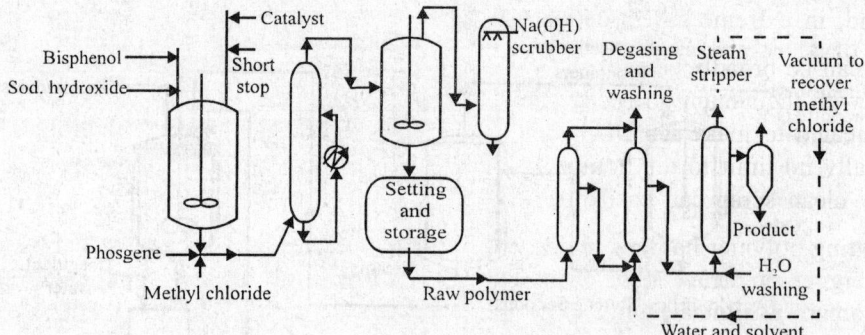

Fig. 26.10. Production of polycarbonate.

Exothermic reaction heat of isocyanates with alcohols usually is made to volatilise a volatile liquid like Freon ($CFCl_3$) which gives rise to the foamy structure (95 Y. gas) to the plastic. Isocyanate when reacts with water, carbon dioxide accompany the heat of reaction, this also provides rigid foams. The density of these foams is supposed to be very less (32 kg/m^3). The construction industry is the largest consumer of rigid polyurethanes. Rigid polyurethanes, i.e. modified polycyanurate (PIR) foams comprise of polymeric isocyanates trimerised in presence of a surfactant catalyst and blowing agent; while polyols are added to help processing, i.e. to reduce internal friability and improve adhesion substrates.

Urethane paper based insulations with aluminium foils, produced by continuous laminations are used in buildings for insulation purpose. Glass fibre reinforced laminates yield fire-proof insulation material.

Polyimides

Polyimides are heat resisting plastics and stable up to 500°C. Polyimides are produced by two stage condensation process. The initial condensation of pyromellitic anhydride with diaminodiphenyl ether in anhydrous dimethyl acetamide yields a soluble acid. This latter is heat cured to an insoluble infusible poly aromatic imide. These can be used for filled polymers, and other industrial applications like piston rings, valve seats, bushings and glass sheets.

Poly Acetals

Poly vinyl alcohol films and fibres are made water resistant by immersion in aqueous formaldehyde. The acetal reaction takes place between hydroxyl groups and aldehyde group resulting vinyl formal form. This form is soluble in many organic solvents. These are suitable for coatings, and laminating purpose. Some polyacetals derived from butyraldehyde can be plasticised and can be utilised in coating textiles.

Equipments used for Polymer Processing

The equipment used for polymer processing tends to be more complex, mainly because molten plastic is a very viscous liquid whereas molten metal is far more free flowing. In simple compression-moulding processes this is of little consequence as the raw plastic is in the form of granules which are introduced directly into the mould cavity. In injection or transfer moulding, however, the granules are plasticised (changed to a molten state) before moulding takes place and viscosity then becomes an important consideration.

All polymer processing techniques require the input of heat to soften the polymer and enable the forming process to produce the required shape. The temperature and period of heating are critical, and are the main source of processing defects. If the polymer is exposed to excessive or prolonged heating, thermal degradation will take place. This will show itself in a reduction of viscosity, oxidation or hydrolysis and, in extreme cases, depolymerisation.

Polymers can be broadly categorised as thermoplastic, thermosetting and elastomers, the names being synonymous with their main characteristic. Thermoplastic polymer softens under the application of heat and so it is heated to make it mouldable, then cooled to hold the shape. The process is reversible and there is virtually no limit to the number of times that this softening/hardening cycle can be carried out. Consequently clean scrap can easily be recycled.

Thermosetting polymer hardens under the application of heat and so it is heated to hold it shape. The polymer undergoes an irreversible chemical change during the heating cycle and cannot therefore be reused. Elastomers exhibit a highly elastic, rubberlike behaviour. They are really a subgroup of either thermoplastic or thermosetting polymers, but behave differently in that they take up their original shape upon removal of a deforming force.

Polymerisation is the reaction in which the raw materials (the monomers) are converted into the products (the polymers). As the reaction takes place, the material is constrained by a mould or die so that the desired shape or form is achieved. The material used in all the moulding processes is supplies in the form of liquids, sheets, granules or powder. This makes the handling of them makes the handling of them very convenient and problems do not arise until the material is being manipulated in the mould.

Both viscosity and elasticity can be reduced by increasing the working temperature, but there is an upper limit at which the polymer will begin to degenerate in some way. In addition different types of polymers have different working temperature ranges. PVC for example has a very narrow range in comparison to polythene, whose viscosity can be varied by a factor of ten. Viscosity can also be reduced by the addition of plasticisers. These are liquids which enter the structure of the polymer and enable the molecular chains to slide more easily over one another. The visco-elastic characteristic can also be modified by compounding two or more different polymers.

Apart from altering the visco-elastic nature of the polymer, a number of additives can be introduced either at the mixing stage or during processing, to achieve other desired effects. The type of additive and its effect are summarised in Table 26.5.

Table 26.5. Types of additives and their effects.

Type of Additive	Purpose
Gases	Expanding and foaming.
Liquids	Plasticisers, lubricants, stabilisers.
Pigments and dyes	Protection of material, protection of packaged contents, decoration.
Fillers (wood, paper pulp, asbestos, chalk, talc, graphite)	Modification of properties, e.g. flammability. Reduce brittleness, Increase stiffness. Reduce material cost by acting as extenders.
Metallic	Modification of electrical and magnetic properties and heat transfer characteristics. Protection, e.g. lead-filled PVC radiation screen.
Flexible cotton and synthetic fibres	Improvement in mechanical strength, stiffness and impact resistance.
Rigid asbestos and glass fibre	Considerable increase in Young's modulus and tensile strenght.
Metallic fibres	Extends use in high temperature applications.

Compression moulding

Compression moulding is normally used for thermosetting plastics, although certain thermoplastic articles are made by this process. A measured amount of polymer is charged to the mould, which is then closed and heated. The polymer can be in the form of a loose material charge (i.e. powder of granules) or be a preformed shape, as shown in Fig 26.11. The preform is a low density moulding that has been lightly compacted in a cold mould.

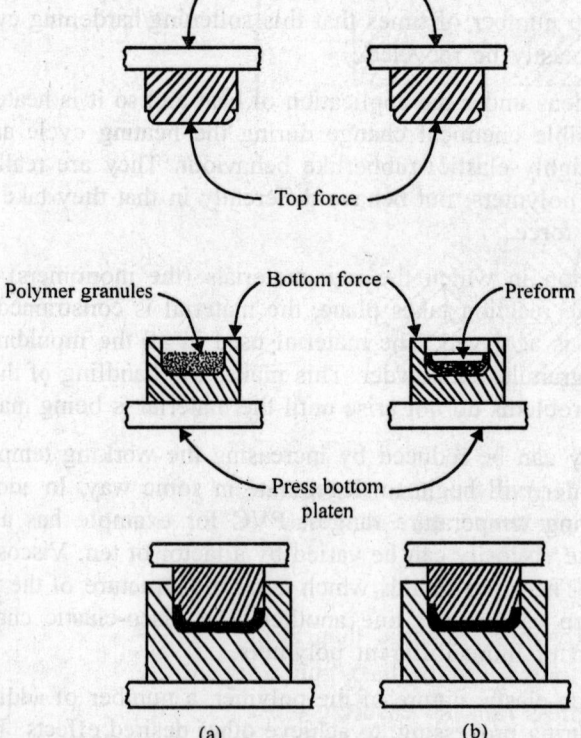

Fig. 26.11. Compression moulding using (a) A loose material charge and (b) A low density preform.

The pressure exerted by the plunger is usually pre-determined to give the required density of the finished article. The heat and pressure cause the polymer to plasticise and flow throughout the cavity. The pressure required may be governed initially by certain additives in the polymer. Heavy, coarse fillings will cause poor flow characteristics which may be impossible to counteract by temperature alone. The shape of the component may also affect the polymer flow. This is particularly true of intricate shapes and deep sections. For this reason, compression moulding is better suited to relatively shallow components having no intricate shapes or inserts. Preforming helps to overcome the flow problem as the uniform, low density preform will not need to flow as much as loose material charged directly into the heated mould. The design of compression moulds varies quite markedly as shown in Fig 26.12. All these would be used in upstroking or downstroking presses incorporating heated platens. They each have various advantageous and limitations as outlined below.

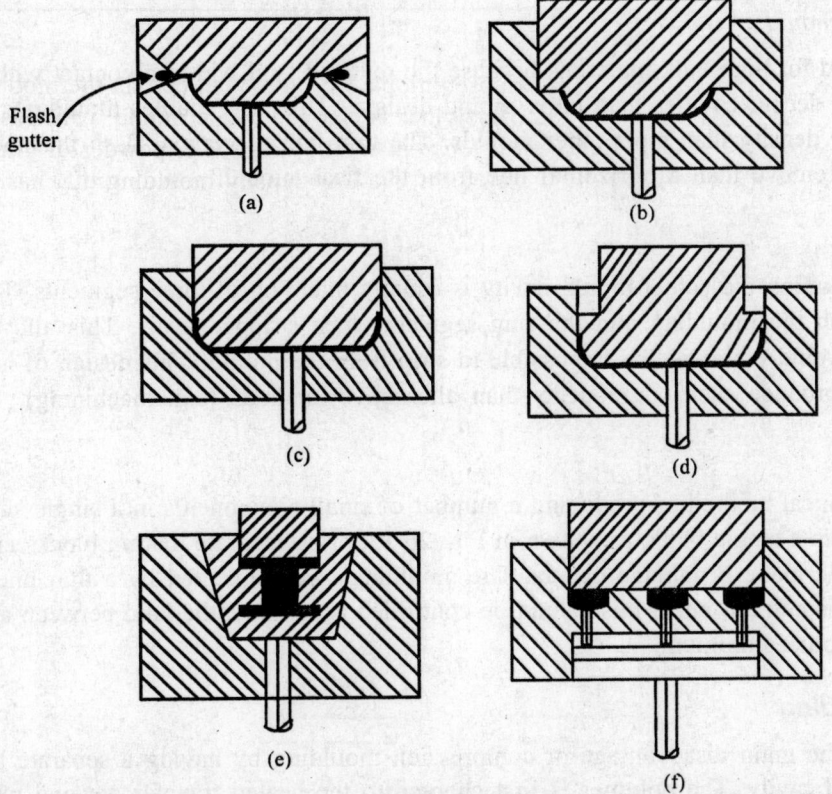

Fig. 26.12. Types of compression mould all with a bottom ejector rod (a) Flash mould; (b) Semi-positive mould; (c) Positive mould; (d) Landed plunger mould; (e) Split mould; and (f) Gang mould.

Flash mould

The cavity is charged with slightly more material than is needed to produce the finished component, which results in the excess being forced into the flash gutter. A pressure will thus be set up in the cavity balancing the pressure from the press ram, so ensuring the mouldings are all of uniform density. This is the simplest type of mould and therefore the cheapest to manufacture.

Semi-positive mould

Also known as positive flash moulds, because excess material will escape vertically until the plunger engages with the cavity walls. Consequently the mould can be overcharged to ensure the mouldings are all of uniform density. This type of mould is particularly suitable for high bulk factor materials or deep mouldings, and the vertical flash is usually easier to trim then horizontally produced flash.

Positive mould

This will overcome the problem of a moulding whose density varies, which can happen when the leak-off in a semi-positive mould varies around the cavity. However, any variation in the amount of material charged into the cavity will result in density variations between individual moulding. The charge in usually preformed to help overcome the problem.

Landed plunger mould

These can be used for high bulk materials because the reduced land width in contact with the inside face will permit much deeper cavities than other mould designs. This also enables mouldings to be produced of a much higher density than most other moulds. The thin line of vertical flash that inevitable ensures is less costly to remove than a horizontal line from the flash mould.

Split mould

The inner ring forming the actual mould cavity is usually made up of three segments. The inner ring is ejected along with the moulding and the ring segments then stripped away. This allows shapes to be moulded which would otherwise be impossible to strip from the mould. This design of mould is a much cheaper way to produce such components than alternative methods (e.g. machining).

Gang mould

This is an economical method of producing a number of small components in a single shot. The cavities can be machined in a single block as shown in Fig. 26.12, or as separate mating blocks enabling cheaper replacements to be made if damage occurs. The mouldings are connected by a thin line of flash if the measured charge is correct, but if this cannot be controlled accurately, the land between adjacent cavities is kept to an absolute minimum.

Transfer moulding

This overcomes the main disadvantage of compression moulding by having a separate heated chamber outside the mould cavity. The polymer is first charged to the heated transfer pot and when sufficiently molten, forced by the plunger into the mould below. After cooling the moulding has to be broken from the sprue, the feed pipe from the transfer gate Fig 26.13. Unfortunately this creates an increase in the downtime of the process, as the material left in the transfer gate (the cull) has to be removed after each moulding cycle and discarded. This adds to the curing time of between one and five minutes for which the polymer has to remain in the mould.

The main criterion in the design of the mould is the positioning of the transfer gate. For articles of consistent shape the gate should be positioned so that the polymer is forced to travel the shortest possible distance in filling the mould cavity. However, due to the polymer being plasticised before it enters the mould cavity, thick sections do not suffer from density variations. As this leads to transfer moulds being used for articles having both thick and thin sections, it is better to place the transfer gate above a thick rather than thin section. Just like the multiple or gang compression mould, a multi-cavity transfer mould can be used. This has the cavities arranged radially around the central transfer plot.

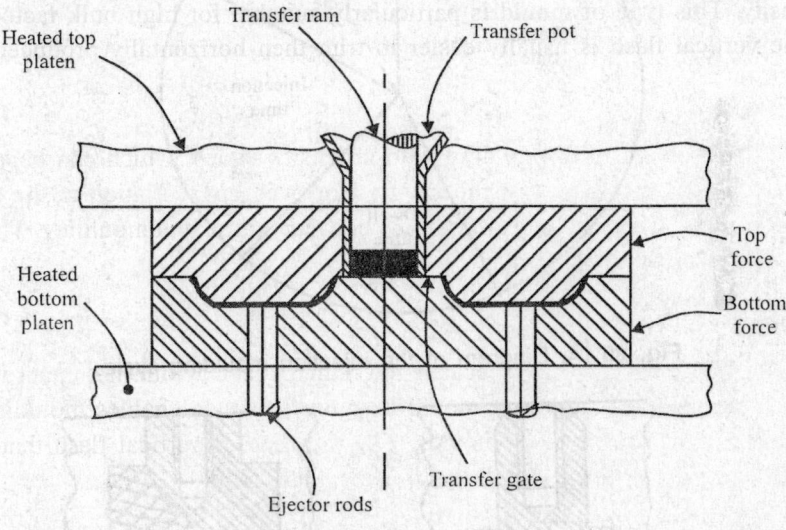

Fig. 26.13. Transfer mould.

The closure of the transfer mould can be done by operating the transfer pot with a separate press cylinder to the ram, or allowing the ram pressure to hold the mould cavity closed. In the latter case the area of the transfer chamber should be at least 25 per cent greater than the area of the contact face of the mould, otherwise there is a danger of the internal pressure forcing open the cavity.

Inserts in plastics

A fairly common feature of polymer components is some form of insert which has been incorporated during the forming process. Most inserts are metallic and are usually of the form of an external pin or internal bush. The most common type of insert is the threaded insert for although threads can be easily formed in the polymer, the material is often too weak to accommodate the forces exerted on the thread form.

Three different designs for a threaded insert are shown in Fig 26.14. Each illustrates a different method of preventing the insert from being pulled out or turning. Feathered edges alone are to be avoided as the thickness of polymer preventing the insert from being pulled out can easily be sheared. Similarly square or hexagonal inserts can cause the stress, which is developed as a result of the tendency to rotation, to be concentrated at the corners. This can cause the polymer to crack or split.

Consideration has to be given to the possibility of the polymer encroaching on the face of the insert if this is required to be flush with the surface. One answer is to arrange for the insert to be the same diameter as the carrier pin and to be screwed on to it. In this situation, the carrier pin would need flats on the shank to enable it to be unscrewed after ejection. Where possible, ejection pins are arranged to push on the insert carrier, or are designed as a combined insert carrier/ejector pin Fig 26.15.

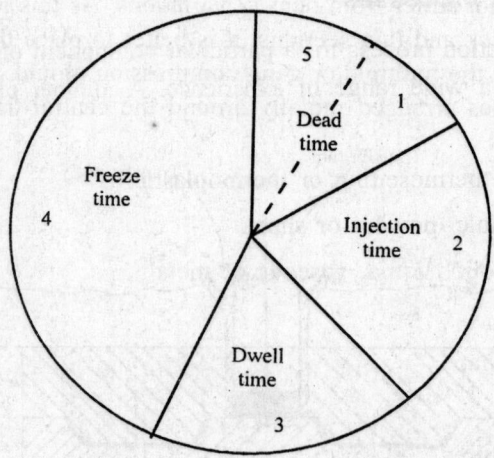

Fig. 26.14. Diagram of the injection moulding cycle.

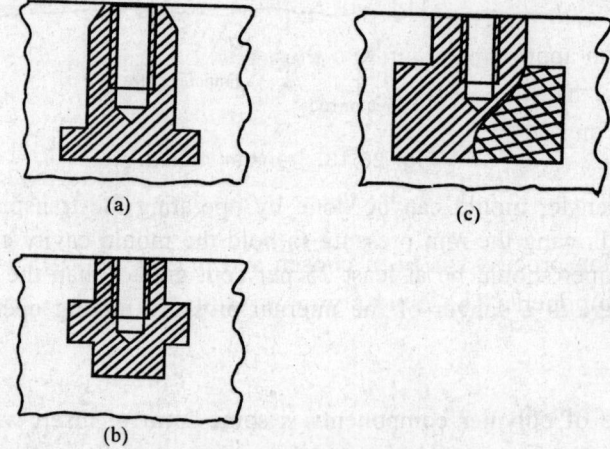

Fig. 26.15. Types of threaded insert (a) Feathered edge; (b) Knurled periphery; and (c) Squared ends.

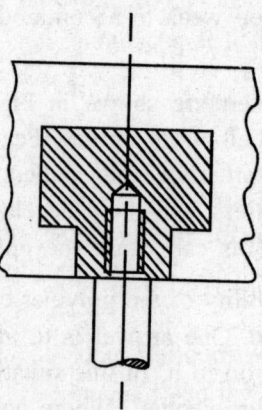

Fig. 26.16. Insert located on a carrier or ejector.

Selection of production processes

The correct choice of production process for a particular component often requires a deep understanding of polymer processing and a wide range of experience. A number of factors and variables have to be considered, such as:

(1) The type of polymer – thermosetting or thermoplastic.

(2) The form – liquid, granule, powder or sheet.

(3) The type of additives–solid, liquid, gaseous or metallic.

(4) The size of the article.

(5) The complexity of its shape.

(6) The quantity required.

(7) Whether inserts are required.

(8) The design of the tooling.

(9) The tool cost.

Table 26.6 gives an approximate guide to the relative merits of the three basic processes. It should be appreciated, however, that no process is strictly limited, as certain factors can sometimes be modified to give greater freedom and adaptability. Injection moulding, for example, appears more suited to very high output of small articles. Despite this the process has been successfully applied to the production of car bodies and boat hulls.

After the production process has been chosen and put into operation, there are numerous faults that can occur in the component. The most common faults and their probable causes ($\star$) are shown in Table 26.7.

Other Processing Techniques

Compression, transfer and injection moulding are the three most widely used methods of polymer processing. There are a number of other basic techniques used by industry for producing the vast array of consumer products, not the least of which is packaging, which accounts for so much of the polymer that is manufactured.

Calendering

Three or four heated rolls in a single vertical stand are used for producing continous thermoplastic sheet. The sheet can be used as the raw material for further processing or as a finished product.

Extrusion

This uses a machine similar to the screw injection machine with an extrusion die in place of the split mould. As well as solid sections, tubes of various diameters can be extruded, including tubular film for making bags and sacks.

Table 26.6. A guide to the three basic moulding processes.

Process	Polymer type	Output rate	Complex shapes	Factor suitable for inserts	Waste material	Component size	Machine capital	Mould cost cost
Compression moulding	Mainly thermosetting	Low or medium	No	No	Little or none	Medium to large	Medium	Low
Transfer moulding	Mainly thermosetting	Low	Yes	Yes	Considerable	Medium to large	Medium	Medium
Injection moulding	Thermo Plastic	High	Yes	Yes	None	Small to medium	High	High

Table 27.7. Common injection-moulding faults.

Fault	Operating temperature	Injection pressure	Injection time	Cooling time	Mould locking force	Gate design
Brittleness	★	★				
Bubbles		★				★
Crazing	★	★	★			
Distortion	★	★	★	★		
Flashing	★	★		★	★	
Gloss	★					
Hollows	★					
Sticking	★					

(Probable cause)

Blow moulding

Tube is extruded intermittently form an extrusion machine mounted above a split mould as shown Fig. 26.17. The closing of the mould traps the bottom of the tube which can then be blown to the shape of the mould be means of an air nozzle incorporated with the extruding die. The technique is widely used for bottles and containers, and a threaded neck can be formed by the mould if required.

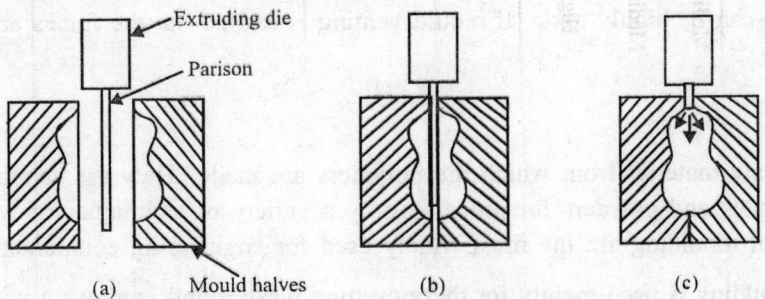

Fig. 26.17. Blow moulding a bottle (a) Tube of plastic (a parison) is extruded between the halves of a split mould; (b) The mould halves close, pinching and sealing the end of the parison; and (c) Air is blown into the parison, stretching and forming it to the shape of the mould cavity.

Thermoforming

This describes a number of techniques in which a heated thermoplastic sheet is clamped around its circumference above a male or female die. Air pressure is invariably used to force the sheet to the die shape, either by increasing the air pressure above the sheet or by creating a vacuum inside the die cavity. The process is used for items such as buckets, bowls and wide-necked containers.

Slush casting

A female mould is heated and charged with the slush, a fine polymer powder suspended in a liquid. After allowing time for the slush to coat the inside of the mould, the excess is poured out. Upon solidification, the polymer can be stripped from the inside of the mould. The outside surface of the article reproduces the mould detail to a high degree and the technique is used extensively for manufacturing toys such as dolls and balls.

Dip coating

This is very similar to the slush casting process but is the inverse, in that a male mould is dripped into the slush. Items such as gloves are made in this way, and metal components can be heated and plastic coated in a similar manner.

Safety in plastics moulding

The most dangerous aspect associated with polymer moulding is in the use of power presses with compression and transfer moulds. In addition, when presses are used for polymer moulding the platens are heated, usually by steam but sometimes electrically. In both cases the platen's surfaces are extremely hot to ensure that sufficient heat is conducted to the moulds. Heat-resistant gloves should, therefore, be worn at all times.

Considerable pressures are exerted during transfer and injection moulding and although locking devices are employed on the mould halves, these have been known to fail for various reasons.

Face visors of safety glasses must be worn during the injection cycle unless the machine is fitted with a shield or screen. In the case of the latter, this should activate an interlock switch which prevents operation until it is in position.

Suitable ventilation or extraction equipment is necessary with all the moulding processes. When the polymer is curing in the moulds, fumes are given off which at the least are very unpleasant, but in certain polymer compounds can be highly toxic. If mould venting is carried out the fumes are released in great quantities.

Consolidation

Monomers are the raw material from which the polymers are made. They are supplied in the form of liquids, sheets, granules and powders for processing by a variety of techniques of which compression, transfer and injection moulding are the most widely used for engineering components.

Compression moulding is used mainly for thermosetting plastics and employs a wide range of mould designs to counteract some of the many problems inherent in the process. The main problem is the poor flow of the polymer in the cavity and although this is lessened to some extent by preforming, it is still the main disadvantage.

Transfer moulding overcomes this disadvantage by plasticising the polymer outside the mould cavity and then forcing it through the transfer gate in a much more fluid Condition. This in itself creates a problem, however, as the thermosetting plastic cures in the gate along with moulding. This cull has to be removed and discarded, lengthening the production time and adding to the costs.

Injection moulding is a very similar process which is used for thermoplastics. By keeping the injection nozzle short and recycling the sprue, however, it does not suffer form the same problems. The screw extruder is more widely used than the ram injector and is more suited to continuous production techniques.

Moulded inserts overcome many of the weaknesses of plastic components, by allowing parts such as bushes and threads to the incorporated in more suitable materials.

It should be borne in mind that there are many other polymer-processing methods which, although having limited use for engineering components, are widely used for consumer products and packing. The most commonly used processes are these: *Calendering*—the use of heated rolls for producing continous plastic sheet; *extrusion*—using a screw extruding machine, similar to the screw fed injection moulder, to produce continuous lengths of material; *blow moulding*—extruding a tube into a closed, split mould and then inflating the tube to the shape of the cavity; *thermoforming*—forcing a softened sheet of plastic on to a male form tool by air pressure, or into a female mould by creating a vacuum; *slush casting*—coating the inside of a heated mould with fine polymer powder suspended in a liquid (a slush); *dip coating*——dipping a heated former into a slush (the inverse of the slush casting process).

RUBBER

An elastomer is a linear polymer which has two unique properties. It is capable of undergoing long deformations upon application of the stress and the deformed elastomer can recover spontaneously and almost completely upon removal of the stress.

Rubber is almost a synonym to elastomer. Natural rubber is a linear high polymer of isoprene. Hence, it may also be called Polyisopropene. The long chain molecules in the natural rubber are arranged int he coiled form. During elongation, the polymetric chains partially uncoil. The moment the stress is taken off, the rubber chains coil again.

NATURAL RUBBER

Natural rubber can be produced from many plants, however actually the latex is obtained from a small number of so-called rubber plants. The most important of these is a tree-like plant *Hevea brasilliensis*, whose latex contains 20 to 35% of a rubber constituent, formed by biochemical synthesis. The latex is drawn from notches in the tree bark in the same way as turpentine. This is called tapping.

Latex contains the rubber in the form of tiny pear-like particles or globules, which are suspended in water and bear a negative electric charge; this prevents them from coagulating. Latex also contains a small amount of protein substances (2%), resins (up to 2%), sugary substances (0.4%), and ash (0.5%). To remove the charge from the rubber particles formic or acetic acid diluted with water is added to the latex; this causes the latex to coagulate and the rubber is separated in the form of doughy, bulky mass. The rubber is filtered off, thoroughly washed with water and passed through corrugated rollers. The rubber sheets produced are dried and smoked in a smoke chamber so that they will keep without spoiling. The sheets are packed in bales. Natural rubber is a polymer of isoprene, $CH_2=C-CH=CH_2$. The macromolecules of natural rubber are made up of

$$\left(\begin{array}{c} H \quad CH_3 \\ | \quad\quad | \\ CH_2{=}C{-}C{=}CH_2 \end{array} \right)_x$$

The degree of polymerisation exceeds 2,000 and the molecular weight is from 136,000 to 340,000. Natural rubber possesses high strength and elasticity within a wide range of temperatures. Natural latex is also used directly, without separating the rubber from it. It is used as a glue, for impregnating textiles and making microporous rubbers.

SYNTHETIC RUBBER

Mainly synthetic rubber is used in the rubber industry. At the present time the chemical industry turns out a wide range of synthetic rubbers, both made from one individual monomer by addition polymerisation or from a blend of two or three monomers by copolymerisation. The synthetic rubbers can be divided into the two groups of general-purpose and special-purpose rubbers. This division is purely nominal. General-purpose rubbers are used for making tyres and other rubber articles of mass consumption. Special rubbers have certain improved properties. Such rubbers can operate under severe conditions, for instance at low or at high temperatures, resist highly corrosive media (acids, alkalies, oxidants, solvents, etc.)

Styrene-Butadiene Rubber

Styrene-butadiene rubber (SBR) comprises more than half of world's production of synthetic rubber and exceeds natural rubber supply. In fact, tyre industry depends more on SBR than on natural rubber because it is available in plenty and serves the purpose better after the incorporation of fillers.

Raw materials

The principal raw materials for SBR manufacture are monomers styrene and butadiene. Other raw materials are oils used for extending, soaps antioxidants, short stopping agents, coagulating agents, catalysts, modifiers, water and hydrocarbon solvents.

Styrene

Styrene can be obtained as follows:

$$\text{Benzene} + CH_2=CH_2 \xrightarrow[95°C]{AlCl_3} \text{Ethyl benzene (}CH_2CH_3\text{)} \xrightarrow[\text{bauxite } 600° -H_2]{\text{Metallic oxides or}} \text{Styrene (}CH=CH_2\text{)}$$

Butadiene

Butadiene is chiefly obtained by thermal cracking of petroleum. It can also be obtained by oxidation of alcohol.

$$C_2H_5OH + Air \xrightarrow[\text{Catalyst}]{\text{Catalyst}} CH_3CHO + H_2O$$

$$CH_3CHO + C_2H_5OH \xrightarrow{\text{Catalyst}} CH_2 = CH - CH = CH_2 + 2H_2O$$

Manufacture

Different ratios of styrene and butadiene can be used but the ratio mostly used is styrene 25% and butadiene 75%. If the percentage of styrene is increased to 50% or above, a tough product-a reinforcing resin is obtained. SBR is obtained mostly by emulsion polymerisation using free radical catalyst. Fig 26.18 Thus:

$$\text{Styrene (25\%)} (CH=CH_2) + CH_2 = CH - CH = CH_2 \text{ Butadiene (75\%)} \rightarrow$$

$$\left[CH_2-CH = CH - CH_2 - CH_2 - CH \right]_n$$

Polymerisation at 5°–10°C is called cold process and the product is called cold SBR while the one made at 50°C is called hot SBR. The cold initiators are redox combinations such as *p*-methane hydroperoxide/ferrous sulphate which result in higher molecular weight at lower temperatures. Hot process uses initiators such as potassium persulphate. There is higher degree of branching. Cold SBR gives better abrasion resistance and dynamic properties. Higher molecular weights obtained also allow higher levels of oil extension.

Structure

SBR is a random copolymer. The butadiene units are present about 20% in 1,2 configuration, 20% in *cis* 1.4 and 60% in *trans* 1.4 if polymerisation is brought about at 50°C which increases if made at low temps. Because of irregular structure, it does not crystalline.

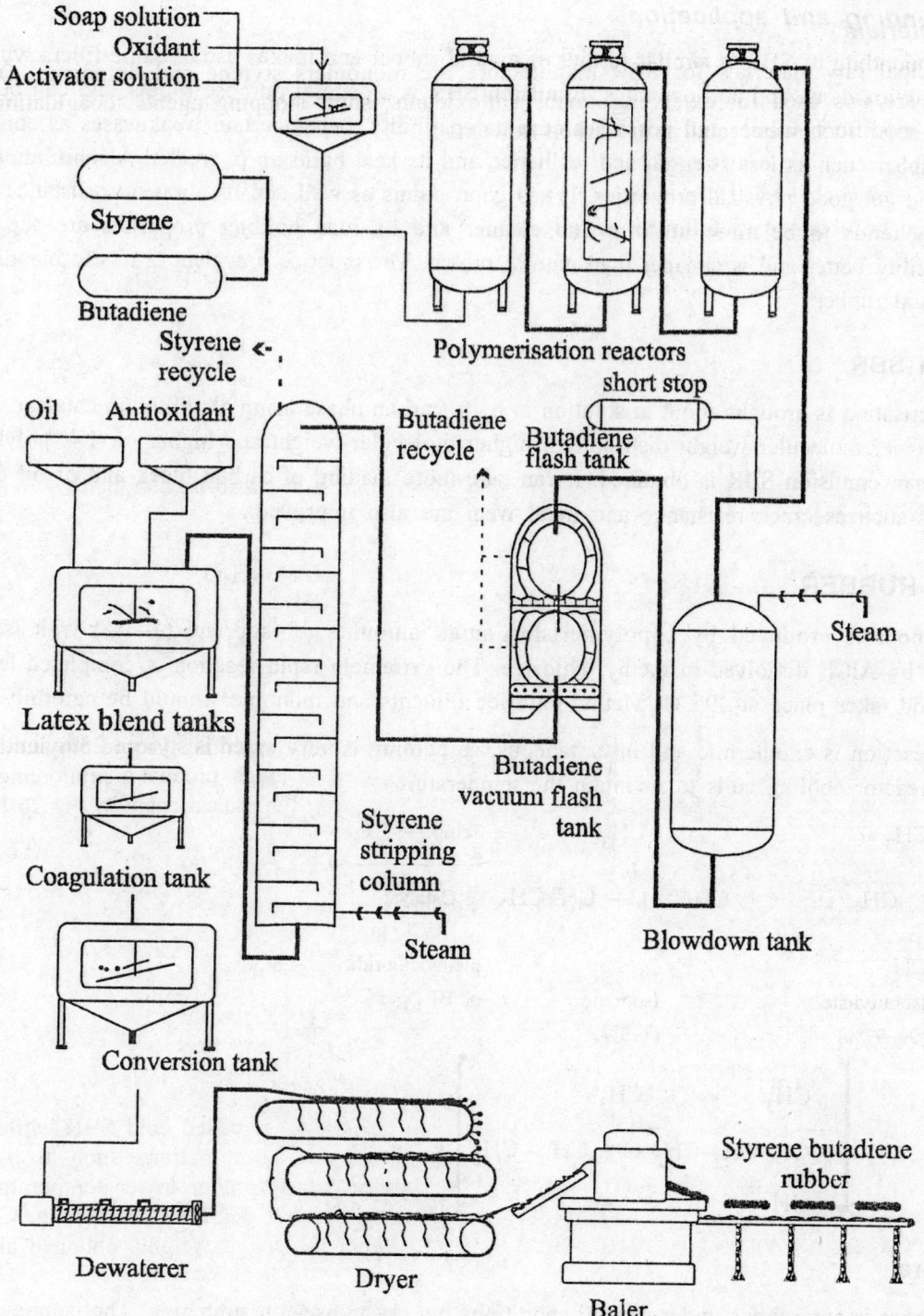

Fig. 26.18. Manufacture of emulsion styrene-butadiene rubber (SBR).

Compounding and applications

The compounding of SBR is similar to that of natural rubber and makes use of same fillers, vulcanising systems, antioxidants and anti-ozonates. Emulsion SBR is a general purpose rubber and can be blended with any other such rubber and possesses cure compatibility. It has certain weaknesses as compared to natural rubber such as less strength and resilience and its heat build up is greater. A reinforcing filler is essential to get good physical properties. It has good points as well and its abrasion resistance is better, its quality tends to be more uniform and cleaner and its road holding properties are superior and weatherability better and is cheaper than natural rubber. The practice prevalent is to use blends of SBR with natural rubber.

Solution SBR

If polymerisation is brought about in solution in hydrocarbon phase using alkyllithium catalyst, a product with narrower molecular weight distribution, higher molecular weight and higher *cis*-1,4, polybutadiene content than emulsion SBR is obtained. It can take more loading of carbon black and of oil. Desirable properties such as crack resistance and tread wear are also improved.

BUTYL RUBBER

Butyl rubber is produced by copolymerising small amounts of isoprene (1–3%) with isobutylene catalysed by $AlCl_3$ dissolved in methyl chloride. The extremely rapid reaction is completed less than a second and takes place at –95°C. Methyl chloride diluents and monomer should be carefully dried.

The reaction is exothermic and instantaneous, so cooling is very essential. Liquid ethylene is boiled through reactor cooling coils to maintain the temperature.

$$
\underset{\substack{\text{Isobutylene} \\ (99\text{–}97\%)}}{\overset{\displaystyle CH_3}{\underset{\displaystyle CH_3}{\overset{|}{\underset{|}{C=CH_2}}}}} \quad + \quad \underset{\substack{\text{Isoprene} \\ (1\text{–}3\%)}}{\overset{\displaystyle CH_3}{\overset{|}{CH_2=C-CH=CH_2}}} \quad \xrightarrow[\substack{\text{Catalyst} \\ AlCl_3 \text{ in} \\ \text{methylchloride} \\ \text{or } BF_3 \text{ gas}}]{\text{Temp. } -95°C}
$$

$$
\left[\overset{\displaystyle CH_3 \qquad\quad CH_3}{\underset{\displaystyle CH_3}{\overset{|}{\underset{|}{-C}}} - CH_2 - CH_2 - \overset{|}{C} = CH - CH_2} \right]_n
$$

Structure

Butyl rubber is amorphous under normal conditions but crystallises on stretching. The isoprene units are mostly present in *trans* 1.4 structure and the polymer is linear. The molecular weight is controlled at 2,00,000–3,00,000 level by addition of chain transfer agents during its manufacture.

Properties

Butyl polymers are among the most widely used synthetic elastomers in the world and rank third among synthetic rubber production. Because of very low residual unsaturation, they posses unique age-resistant properties.

The special propety of the rubber is its being impermeable to air and because of its low unsaturation being very resistant to attack by oxygen, ozone chemicals and moisture.

Applications

About three fourth of butyl rubber produced is used for inner tubes for tyres. Other uses are mainly in the field of mechanical goods. The ozone resistance of butyl rubber together with moisture resistance of its saturated structure finds use as a high quality electrical insulation.

Compounding

Butyl rubber can be vulcanised by three methods (i) sulphur vulcanisation (ii) crosslinking with dioxime or denitiose compounds and (iii) polymethylol-phenol resin.

Sulphur vulcanisation

$$
\text{Thiuram} \quad
\begin{matrix} S \\ \| \\ (R_2\text{–N–C–S})_2 \end{matrix}
\quad \text{or dithiocarbamate} \quad
\begin{matrix} S \\ \| \\ (R_2\text{–N–C–S})_x \end{matrix}
$$

metal, where N is accelerators which are very potent are used along with sulphur and zinc oxide. Thiozoles such as mercaptobenzothiazole, if used reduce scorching during processing. Vulcanisation temperature range from 135–190°C.

Dioxime cure

The cure can be brought about with *p*-quinone dioxime in the presence of PbO_2 which acts as an oxidising agent.

Polymethylol-phenol resin cure

Phenolmethylol groups of phenol-formaldehyde resin bring about curing by reacting with the isoprenoid units. Thus

If hydroxyl group of the methylol group is replaced by bromine, a more reactive cure is obtained.

Synthetic Polyisoprene Rubber

Manufacture

Isoprene is derived from petroleum. It is mixed with a hydrocarbon solvent such as *n*-pentane. Two types of catalysts can be used. One is based on a coordination catalyst of titanium tetrachloride and an aluminium alkyl such as triisobutyl aluminium. (Al–Ti) while the other anionic catalyst is butyl lithium. Reaction takes place at 50°C and moderate pressure. The process makes use of solution polymerisation.

Properties and applications

The properties are nearly identical with those of natural rubber and this polymer is preferred because of greater cleaniness and uniformity. It is virtually colourless due to high purity and freedom from contamination and is used for medicinal purposes. It has high elongation break which leads to its use in rubber threads for golf balls. Tensile strength and tear resistance are some what low as compared to natural rubber, other properties viz., ageing, hysterisis are similar to natural rubber.

Polybutadiene Rubber

Manufacture

It is manufactured similar to polyisoprene employing solution polymerisation using pure dry monomer and hydrocarbon solvents. Organic lithium compounds or coordination catalysts containing metals in reduced valence states are employed. Care has to be taken to eliminate completely air, moisture and impurities which can have detrimental effect on the properties of the resultant polymer.

Properties

The high *cis*-1,4-polybutadiene exhibits excellent dynamic properties, low hysteresis and good abrasion resistance. A glass-transition temp. of 102°C is reported for this polymer which accounts for its excellent low temperature, performance, high resilience and abrasion resistance. It also processes well in blends with other polymers e.g., natural rubber and SBR. It should not however be used more than 40-50%, above which it imparts poor skid resistance.

Ethylene-Propylene Rubber

There are two types of ethylene-propylene rubbers, EPM and EPDM. EPM stands for simple copolymer of ethylene and propylene ['E' stands for ethylene and 'P' for propylene and 'M' for polymethylene– $(CH_2)_x$–) type backbone]. In case of EPDM, 'D' denotes a third comonomer, a diene which is introduced to bring about unsaturation in the molecule.

They are the fastest growing elastomer, because of its outstanding properties such as good weather resistant in ozone environment.

Manufacture

The polymerisation is brought about by solution or slurry process, the solvents being hexane, heptane or octane. The catalyst system used for polymerising consists of vanadium halide along with a hologenated aluminium alkyl.

The ethylene content in EP elastomer varies between 75–455. The monomers are randomly distributed resulting in amorphous type copolymer. Higher ethylene containing polymers possess crystallinity and consequently higher green strength but poorer mill processing behaviour.

Applications

The outstanding properties of ethylene propylene rubber is due to its complete saturation which leads to air-ageing properties and ozone resistance.

It is used in type sidewalls, tank linings, wire and cables, mechanical goods, agricultural equipments and automotive applications. It was considered as a special purpose rubber. However, it has found use in wide variety of applications and has replaced natural rubber, styrene-butadiene rubber and butyl rubber where resistance to ozone and weathering is most desired.

Processing

EPDM rubber requires reinforcement to be of practical value since the mechanical properties of the unfilled rubber are poor. Carbon black is the most useful filler, although silica clay and talc may also be used. Naphthenic oils provide the best compatibility. The rubber is inherently not tacky. If tack is desired, a tackifier has to be added. Sulphur vulcanisation is possible. Common accelerators may be used.

Modified Polyethylene or Chlorosulphonated Polyethylene

Manufacture

Chlorosulphonated polyethylene (Hypalon) is unique among elastomers as it is derived from a plastic. It is prepared by dissolving polyethylene in a suitable solvent and passing simultaneously SO_2 and Cl_2 in the presence of a free radical catalyst to bring about chlorination and chlorosulphonation.

Properties

The rubber is completely saturated and so has excellent resistance to chemicals, weather and ozone. It has flame resistance as well due to the presence of chlorine.

Applications

It is used for making white tyre sidewalls after blending with Neoprene and N.R. and gives a beautiful white colour and a surface free from cracking and ozone attack. It finds use of hose, electrical applications and in reservoir- liner construction.

Polychloroprene (Neoprene)

Neoprenes are among the earliest synthetic rubbers produced from chloroprene (2-chloro-1, 3-butadiene) by emulsion polymerisation using free radical catalyst.

$$CH_2 = \overset{\overset{\displaystyle Cl}{\displaystyle |}}{C} - CH = CH_2 \longrightarrow \left[CH_2 - \overset{\overset{\displaystyle Cl}{\displaystyle |}}{C} = CH - CH_2 \right]_n$$

Vulcanisation

Vulcanisation of neoprene is different from that of other rubbers and it can be vulcanised by heat alone. Zn oxide and magnesium oxide are good vulcanising agents. Sulphur vulcanises very slowly and accelerators are not effective.

Properties

The rubber possesses very high tensile strenght (3500–4000 psi) even without the use of carbon black. Crystallising is an inherent property of Neoprener. It stands well to ageing, resists aliphatic hydrocarbons and is flame proof.

Neoprene is better than natural rubber, butyl or SBR as regards oil resistance but inferior to nitrile rubber. The dynamic properties of this rubber are superior to those of most other synthetic rubbers though slightly less than those of natural rubber.

Applications

It is used in the manufacture of industrial hoses, coal mine belts, wire and cable coatings, and solid tyres. Because of a combination of polarity and crystallinity, neoprene occupies an important place among rubbers for use in adhesives. It finds application in aircraft, automobiles, furniture shoes and industrial components. Neoprene latex is used in the manufacture of gloves and coated fabrics.

Nitrile Rubbers

Manufacture

This rubber is obtained by copolymerisation of butadiene and acrylonitrile by emulsion polymerisation using free radical catalyst. Acrylonitrile can be prepared by passing acetylene and hydrogen cyanide over heated carbon.

Compounding

Basically nitrile rubbers are compounded in much the same way as N.R. or SBR. Reinforcing fillers are necessary to obtain optimum properties. Vulcanisation can be achieved with sulphur and accelerator or peroxides when transparent compounds are obtained.

Properties

Oil resistance is the most important property of nitrile rubber. It is available in several grades of oil resistance based on the acrylonitrile content of the polymer. As the acrylonitrile content increases oil resistance, tensile strength, hardness, abrasion resistance, heat resistance and gas impermeability increase while low temp. flexibility, resilience and compression set decrease. It resists both aliphatic and aromatic hydrocarbons and is second only to sulphide rubbers in this respect.

Applications

Because of oil and fuel resistance characteristics, nitrile rubber finds uses in applications where these properties are required such as gasoline hoses for automotive, marine, aircraft, industrial air hose, curburettor parts, oil-drilling industry, adhesives, cements and brakes linings.

Acrylic Rubbers

Manufacture

Acrylic elastomers are obtained by emulsion polymerisation of methyl or ethyl acrylate and chloro ethyl acrylate or chloro ethyl vinyl ether. Chlorine is introduced to ease vulcanisation which can be carried out by: (i) polyamine + sulphur; (ii) peroxides; and (iii) bases (sodium metasilicate).

Properties

The elastomer has excellent combination of heat and oil resistance. It resists heat up to 200°C and is second to silicones only. It also possesses resistance to oxidation and ozone. Its resistance to sulphur bearing oils is also good. Its shortcomings are that it is attacked by acid, alkali, water and steam. Its low temperature performance is also not good.

Compounding

Acrylic rubbers need reinforcing agents for useful properties as they do not have high gum strength. They have a saturated backbone so need incorporation of copolymerised reactive curve sites.

Applications

Finds use in hose, tubing, belting and tank linings.

Polysulphide Rubbers (Thikols)

Manufacture

Polysulphide rubbers are obtained by condensation polymerisation of aliphatic dihalides and sodium polysulphide.

$$n\text{Cl}(CH_2)_2\text{Cl} + n\,Na_2\,Sx \rightarrow [(CH_2)_2Sx]_n + 2_n\,NaCl$$

Variation in properties is obtained by using different dihalides or their mixture and Na_2S_x of different ranks where x is known as rank.

Properties

The outstanding properties of these type of rubbers are their excellent resistance to both aliphatic and aromatic solvents (depending on sulphur content) very good low temperature properties and because of lack of unsaturation, high ozone resistance and low weathering. Some of the drawbacks are difficulty in processing, bad odour, poor compression set and moderate physical properties.

As sulphur content increases in the rubber, oil resistance increases but odour gets bad.

Compounding

Curing can be brought about with a metal oxide such as zinc oxide or lead oxide or with peroxides. Reinforcement is best obtained with various grades of carbon black.

Applications

The main use of these rubbers is in lining of fuel storage tanks, petrol and paint spray hoses, as a diaphragm of gas meters and in rollers for printing and lacquer coating industries.

Polyurethane Elastomers (Vulkollan)

Polyurethane elastomers contain repeating Urethane groups. Other groups such as urea, ester also may be included. Such a diversity of structure makes a very wide range of polymer properties possible.

Manufacture

There are three stages in their preparation:

(1) Production of a polyester by condensing adipic acid with a glycol until a molecular weight approx. 5000 is reached. Several other glycols can be used.

(2) To prepare a prepolymer by reaction with diisocyanate, excess isocyanate over that calculated is used to ensure that end groups are isocyanate.

(3) Further linking together the molecules by the action of chain extender a glycol, diamine or water on the terminal iscocynate groups.

Properties

Outstanding properties of this class of rubber are: (i) high abrasion resistance and tear strength even without carbon black; (ii) resistance to aliphatic solvents; and (iii) non-inflammable.

Applications

They are widely used in foams, for making thread for elastic garments and sports goods, top lift for ladies high heel soles and also engineering components of resilient type.

Silicone Rubbers (Silastomers)

Silicones are inert because of lack of unsaturation, absence of plasticisers and their elemented silicon constituent. These rubbers are extensively used in air craft applications and about 250 kg of these rubbers find use in a single bomber. They had spectacular growth during the past 50 years since the produce was introduced commercially. They are also used in wire and cable insulation and are valuable in surgical and prosthetic devices. These polymers may be straight-chain, or cross-linked with benzoyl peroxide or other free radical initiator, with or without catalyst. Stable over temperature range from −50 to +250°C. Very low surface tension; extreme water repellency; high lubricity; excellent dielectric properties; resistant to oxidation, weathering, and high temperature; permeable to gases. Soluble in most organic solvents; unhalogenated types are combustible. Non-toxic.

Properties

These are special purpose costly rubbers. The outstanding properties of these rubbers are:

(1) Thermal and oxidation stability upto as high as 260°C.

(2) Retention of flexibility even up to as low as-100°C.

(3) Inertness to weathering, chemicals, oils and solvents.

(4) Desirable electrical properties.

(5) Non-toxicity.

(6) Absence of taste and odour.

Manufacture

Their manufacture involves following steps :

(1) Preparation of chlorosilane

$$CH_3Cl + Si \xrightarrow[230°C]{Cu\ powder} \text{Mixture of chlorosilanes obtained}$$

Most useful is Dimethyl dichlorcsilane

(2) Hydrolysis

$$R_2\ SiCl_2 + 2H_2O \longrightarrow \underset{\underset{R}{|}}{\overset{\overset{R}{|}}{OH - Si - OH}} + 2HCl$$

Silicol

(3) Polycondensation with loss of water

$$- \underset{\underset{R}{|}}{\overset{\overset{R}{|}}{Si}} - O - \underset{\underset{R}{|}}{\overset{\overset{R}{|}}{Si}} - O - \underset{\underset{R}{|}}{\overset{\overset{R}{|}}{Si}} - O \quad + nH_2O$$

Molecular weight

The polymer varies from 3,00,000–800,000 in molecular weight and has 4,000–11,000 silicone atoms per molecule.

Vulcanisation

Vulcanisation is carried out with the help of peroxides. R can be CH_3 or a mixture of CH_3 and C_6H_5 and depending on that different set of properties are obtained.

If R = CH_3 rubber obtained is best as regards heat resistance and can resist temp. upto 260°C.

If R= mixture of CH_3 + C_6H_5–5%, rubber obtained can resist extremely low temp. i.e., upto –100°C.

If R = $-CH_2CH_3CF_3$ (trifluoropropyl), oil resistance improved.

Fluoro Carbon Rubbers (Fluorel/viton)

These are exciting new group of elastomers and serve needs for fuel and chemical resistant rubbers for service under extremely low or high temperatures.

Properties and Applications

Properties with introduction of fluorine imparts are: (i) heat and chemical resistance; (ii) flame retardance and; (iii) electrical characteristics. Some of these rubber retain their flexibility even up to 500°C and resist even fuming nitric acid. They are the costliest among all the rubbers and have best resistance of all the rubbers to heat, chemicals and solvents.

In fluorocarbons, stability of the molecules is due to primary forces between carbon and fluorine. Fluorine forms a much stronger bond to carbon than either hydrogen, chlorine, silicon or carbon to carbon and hence imparts the highest order of stability. Also small fluorine atoms fit nicely and uniformly around carbon chain and form a protective shield which effectively prevents chemical attack on weaker C–C linkage.

Vulcanisation

Curing or vulcanisation is carried out with the help of peroxides. Examples of a few fluoro rubbers are:

(1) Polyfluoreprene from 2 fluoro butadiene

$$(CH_2 - \overset{\overset{\displaystyle F}{\displaystyle |}}{C} = CH - CH_2)_n$$

(2) Fluorinated olefins $(CH_2 - CF_2)_n$

(3) Fluorinated silicones which are known for best temp flexibility i.e., from –50°C to 500°C.

These elastomers are prepared in emulsion using peroxy compounds are initiators.

Monomers must be pure. They range in molecular weight from 5,000 to 200,000. There are many types of fluorocarbon elastomers available with special properties to fit particular requirement

Thermoplastic Elastomers

The term thermoplastic elastomers is used for materials which have elastomeric properties, at room temperature but process like thermoplastics thus removing the need of vulcanisation. They can thus be processed on plastic equipments such as injection moulder, blow moulder sheet extruder etc.

The advantages offered are significant and have resulted in a variety of applications. The commercially important thermoplastic elastomers are:

(1) Polystyrene/elastomer block copolymers such as with butadiene or isoprene.

(2) Polyurethane block copolymers.

(3) Polyester block copolymers.

(4) Polypropylene/Ethylene propylene copolymers.

(5) Polyamide block copolymers.

There are two phases present in the thermoplastic elastomers. One phase is a hard polymer at room temperature but becomes fluid when heated. The second phase is a soft rubbery polymer. In the block copolymers, the two phases are formed from segments of the same chain. If X represents hard plastic and Y rubbery polymer, the structure X–Y–X is the simplest arrgangement of a three block system. Both ends of the rubbery Y segment must end in the hard X segment.

The polyurethane, polyester and polyamide block copolymers have the structure X–Y–X–Y–X. The hard segments are polyamide while soft rubbery segments are polyethers and polyesters.

The polypropylene/ethylene propylene copolymers are multi-phase systems and are cross-linked.

Manufacture

An alkyl lithium initiator is made use of for initiating the polymerisation of a monomer X. The resulting product can be used to initiate polymerisation of a second monomer Y thus giving a block copolymer X_n–Y_n. The polystyrene/elastomer block copolymers are made by this technique.

Anionic polymerisation is employed to prepare block copolymers of styrene, butadiene and isoprene.

The polyester thermoplastic elastomers are prepared by condensation polymerisation of three monomers. A long-chain glycol such as polytetramethylene ether glycol is made to react with dimethyl terephlhalate to give the elastomeric segments. Butanediol is made to react with dimethyl terephthalate to give hard segments. The polyurethane and polyamide block copolymers are manufactured following the same technique.

Polypropylene/ethylene propylene (EPDM) are mixed by intensive shear-mixing, resulting in grafting to produce polypropylene/ethylene propylene copolymer.

Relationship between molecular structure and properties

Molecular weight

Low molecular weight polymers are easily processed but do not give so good properties.

Ratio of hard to soft phase

The resulting product becomes stiffer with the increase in hard phase e.g., styrene content. When styrene content is about three fourth, the product is glass-like hard and possesses high impact resistance.

Application

The products based on polystyrene/elastomer block copolymers can be processed by conventional thermoplastic machines such as injection and blow moulders which work fast. The major use of polystyrene/elastomer block is in footwear. When mixed with resins and oils, tack increases and so the products find use in adhesives, coatings and sealants.

Polyester, Polyurethane and Polyamide block copolymers can replace conventional vulcanised rubber. They can be extruded or moulded just like thermoplastics at temperatures 175–225°C. Scrap can be used again. They are flexible, tough and solvent resistant. Some of them can also be used as hot-metl adhesives. Their hardness varies between 60 Shore A to 70 Shore D.

Polypropylene/Ethylene-Propylene Copolymer can also replace conventional vulcanised rubber. They find use in wire insulation. Their hardness varies between 60 Shore A to 50 Shore D.

Compounding

The term compounding in rubber technology is used for selection and incorporation of various additives into rubbers so as to give a homogeneous mix for processing. Every recipe contains a number of components, each having a particular function to perform either during processing, reinforcement or vulcanisation and ultimately in the end use of the product.

Ingredients

The materials used in compounding can be categorised under the following heads.

Elastomers

These form the backbone and are the basic components. They can be natural or any synthetic rubber or thermoplastic elastomer and are selected to give specific physical properties to the final product e.g., oil, heat or aging resistance.

Fillers

These are used for reinforcing or reducing the cost.

Processing acids

As the name suggests, these materials help during processing which may include mixing, extrusion, calendering or moulding.

Vulcanising agents

These substance bring about cross-linking resulting in a marked improvement in the physical properties of the rubber products.

Accelerators

These ingredients reduce the time of vulcanisation by accelerating the rate of vulcanisation.

Activators

This addition is essential for deriving the maximum benefit from vulcanisation. They form complexes with accelerates.

Anti-oxidants

They slow down the process of deterioration in service of the rubber compounds by checking oxidation.

Anti-ozonants

These ingredients slow down the deterioration due to attack by ozone.

Miscellaneous additives

These substances are not normally incorporated in the rubber compounds but may be used for specific purposes. Examples are blowing agents, colours, odorants, retarders etc.

Elastomers or rubbers-classification

Rubbers are classified into following three major classes:

General purpose

They are for general purpose where specific resistance to the action of petroleum or heat is not required. Examples are natural, SBR, polyisoprene polybutadiene, EPDM, butyl rubbers.

Solvent resistance type

They are used where specific resistance to the action of petroleum solvents is required. Examples are polysulphide, nitrile, neoprene, polyurethene rubbers.

Heat resistance type

They are used where effect of exposure to high temperatures can be tolerated. Examples are silicon and fluoro rubbers.

Fillers and other additives

Fillers are used to reinforce or reduce the cost of the product. There are three main types of fillers: (i) carbon black; (ii) non-black fillers; and (iii) organic fillers.

(1) Carbon blacks are the most important fillers used for rubbers. Their incorporation brings about marked improvement in properties such as hardness, tensile, strength, tear and abrasion resistance.

(2) The non-black fillers include clays (semireinforcing), precipitated silica (reinforcing), calcium carbonate (extending fillers).

(3) Organic fillers include high styrene resin, phenolic resin, cyclised rubber, lignin etc.

Vulcanising agents

Sulphur is the most common vulcanising agent. General purpose rubbers which form the bulk of rubbers used contain unsaturation and vulcanisation with sulphur is possible. Sulphur brings about cross-linking and converts the rubber to form three-dimensional structure. Thus a weak material is transformed into a strong elastic product having many desirable properties. Usually 0.5–5.0 phr. of sulphur is used.

The saturated rubbers cannot be cross-linked with sulphur and an organic peroxide is required for vulcanisation. Free radicals are generated when a peroxide decomposes and these make polymer chains to combine. Usually 2.5 phr of a peroxide is used. Dicumnyl peroxide is most prominent.

Accelerators

The vulcanisation with sulphur is slow and can be speeded up many times by the addition of small amounts of organic or inorganic compounds known as accelerators. Most common accelerators are: amines; aldehydes-amine; guanidines; thiazoles; thiouruams; sulphenamides; dithiocarbonates; and xanthates.

Activators

These compounds activate the accelerators to work more effectively and increase the cure rate. Most common example is of zinc oxide which is used (3–5 phr) along with a fatty acid such as stearic acid (1–3 phr).

Antioxidants

Rubbers are affected by aging process when in service resulting in lowering in physical properties. Antioxidants are added to slow down and improve the aging behaviour and to enhance the service life.

Antiozonates

These compounds are effective to check the action of ozone. They are mostly derivatives of paraphenyldiamine. Waxes are also used for this purpose.

Processing aids

These compounds are used usually 2–10 phr and help proper mixing. The processing aid must be completely compatible with the rubber and other compounding ingredients. Incompatibility will result in improper processing or bleeding in the final product.

The aids can be petroleum or coal tar products such as mineral oils, coal-tar pitch, cumar reins, vegetable oils, pine products such as resin, pine tar or resins such as coumarone-indene, phenol-formaldehyde, shellac or waxes, fatty acids, esters etc.

Miscellaneous ingredients

These substances are used for some specific effect in a vulcanisate.

Blowing agents

For manufacture of microprocess rubber or blown sponge, these chemicals are added to generate gases during vulcanisation. Examples are carbonates and ozo compounds.

Flame retardants

To reduce flammability, chlorinated hydrocarbons, phosphates and antimony compounds may be added.

Colourants

For increasing customer appeal, these ingredients are added. They should be stable and fast in colour.

Odorants

For masking odour of rubber compounds, these compounds (essential oils) are added generally in wearing apparel and drug sundries.

Retarders

These chemicals reduce the effect of accelerators and prevent scorching during processing and storage. These are generally organic acids e.g., salicyclic acid.

Mixing Methods

For incorporation of compounding ingredients into rubber, either a mill or a Banbury (internal) mixer is used.

Reinforcement

Reinforcement implies incorporation into rubber of substances which give to the vulcanisates high abrasion resistance, tensile strength, tear resistance and enhancement in stiffness and hardness. These properties can be upgraded by the addition of certain fillers into the rubbers before vulcanisation.

Classes of fillers

Fillers can be divided into two classes viz., inert fillers such as clay and whiting which have little effect on the physical properties and reinforceing fillers which improve these properties substantially.

Black filler

Carbon black. Carbon black is the most outstanding reinforcing filler for both natural and synthetic rubbers and is inexpensive. For obtaining maximum benefit of carbon black reinforcement, it should be thoroughly dispersed into the polymer. Small particle size (high surface area) carbon blacks are difficult to incorporate. High surface area carbon blacks give more reinforcement in a rubber compound. The particle size affects the abrasion resistance, resilience, tensile strength and tear resistance of the rubber compounds.

Carbon black increases tensile strength of natural rubber, SBR nitrile rubber and even the polysulphide rubber but has no effect in enhancing the tensile strength of neoprene or butyl rubber.

Non-black fillers. The non-black fillers used in rubber industry are clay, calcium carbonate and silica. Clays are used to the largest extent followed by calcium carbonates and silicas.

Non-black fillers include non-enforcing to highly reinforcing. The particle size plays the most important role in deciding the nature of the filler. While silicas (highly reinforcing) may have particle size of 0.01 micron, calcium carbonates (inert) have particle size of 10 microns i.e. 1000 fold.

Organic reinforcement

If fillers above 80 phr are used, processing is difficult, there is tendency for scorch and heat build up. Rubbers can take up much higher loadings of organic than of inorganic fillers. There are various types of organic fillers which can be used :

(1) *High styrene resin:* It is used in shoe compounds.
(2) *Phenolic resins:* Find use in high quality battery, boxes, V belting, roller compounds and hoses.
(3) *Cyclised rubber:* It behaves like high styrene resin.
(4) *Lignin:* It behaves like phenolic resins.
(5) *Graft polymers:* Monomers such as methylmethacrylate and styrene can be grafted to natural rubber.

Vulcanisation

The chemistry of vulcanisation is very complex and has not been well understood as yet. The process transforms an elastomer with weak thermoplastic properties into a strong, elastic and tough substance with useful properties and introduces a network of cross-links into the elastomer.

Sulphur is used 0.5–5 phr for bringing about vulcanisation. If higher amounts (30–40 phr) of sulphur are used ebonite is obtained which is very hard. It is thus clear cross-linking should not be too much.

Tensile strength is maximum at optimum cure. Beyond this point, the stock is said to be overcured. It gets stiffer and harder but weaker and less extensible.

Sulphur vulcanisation

Vulcanisation takes place in the presence of sulphur by heating but the process is quite slow. When small amounts of substances called accelerators are used, the process can be made faster many times. Also the presence of activators is essential to get full benefit. Thus the sulphur vulcanisation system should consist of :

(1) Sulphur (0.5–5 phr) rhombic form normally used.
(2) An organic accelerator (0.5–2 phr).
(3) Metallic oxide such as zinc oxide (3–5 phr).
(4) Fatty acid such as stearic acid (1–3 phr).

It is believed during vulcanisation rubber-soluble metallic soap is formed by the reaction of metallic oxide and fatty acid. Sulphur vulcanisation is used with rubbers which are general-purpose type and contain unsaturation. With sulphur cross-links and cyclic structures are created as shown in Fig. 26.19

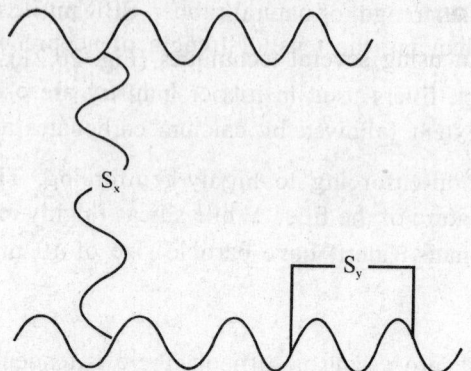

Fig. 26.19. Sulphur cross-links cyclic structure.

In efficient accelerated curing system, x is 1–2 with no cyclic groups formed while in inefficient systems, x can go up to 8 and many cyclic structures are formed. For most rubbers, one cross-link for 200 monomer units in the chain is sufficient to produce a suitable vulcanised product (molecular weight between cross-links equals ca 8000 to 12,000). Selenium or tellurium can be used in place of sulphur where excellent heat resistance is needed.

Non-sulphur vulcanisation peroxides

The saturated rubbers such as silicone or fluoro cannot be cross-linked by sulphur and accelerators. Organic peroxides are used for their vulcanisation. The peroxides decompose forming free radicals on the polymer chains. These chains may then combine to form cross-links (Fig. 26.20).

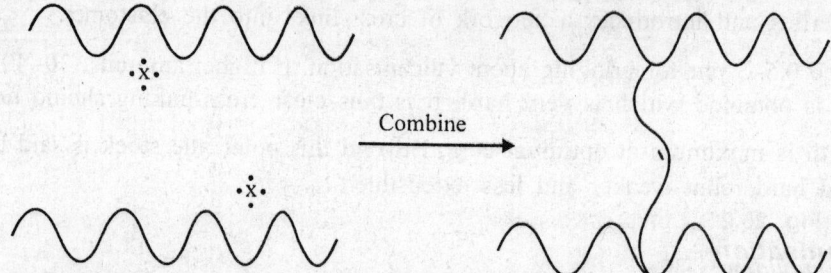

Fig. 26.20. Peroxides decompose forming free radicals on the polymer chains.

Carbon to carbon bonds are thus created which are very stable. Peroxides are used generally 2–5 phr. Dicumyeperoxide is most common.

Metal oxides

Polychloroprene (neoprene) is vulcanised using zinc oxide. Chlorosulpharated polyethylene (Hypolan) is cross-linked using litharge (PbO) or magnesia (MgO). Polysulphide can also be vulcanised using zinc oxide or lead oxide.

Difunctional oxides

Some difunctional compounds can also be used for crosslinking rubbers. Quinone dioxime is used for butyl rubber, epoxy resins are used for nitrile and diamines for fluoro rubbers.

Techniques of vulcanisation

Vulcanisation can be carried out using several techniques (Fig. 26.21).

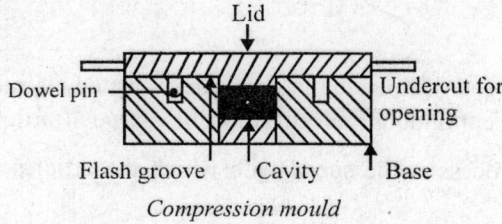

Compression mould

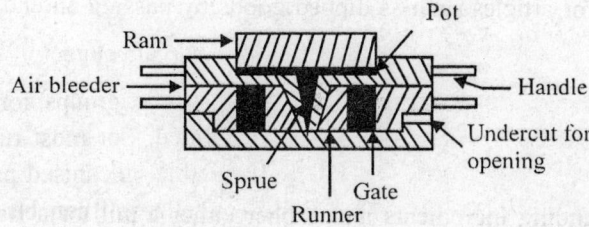

Transfer mould

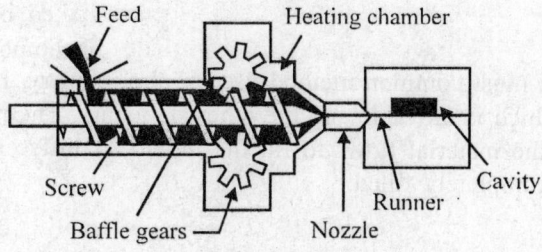

Injection moulding

Fig. 26.21. Compression, transfer and injection moulding techniques.

(1) *Compression moulding:* This is the most common method used in the industry. Here the blank is placed in a 2 piece mould, one part of which is movable and the other stationary. The mould is closed and heat and pressure are applied so that the material flows to fill the mould. A little excess of the material is used to ensure that the mould is completely filled.

(2) *Transfer moulding:* This technique is a variation of compression moulding. The blank is placed in a separate chamber called the pot, preheated to a temperature below cure temperature and transferred into the heated mould where cure takes place. Although in this technique, moulds are more expensive, the process takes shorter cure time due to use of higher temperature and there is better heat transfer.

(3) *Injection moulding:* Injection moulding is used mostly for plastics. However, rubber articles can also be moulded using this technique. The polymer is preheated in a cylindrical chamber to a temperature at which it will flow and is then forced into a comparatively cold, closed mould cavity by means of high pressure.

(4) *Continuous vulcanisation process:* Continuous vulcanisation involves some form of heating so that vulcanisation occurs immediately after the rubber article is formed. The process is used for extruded articles, conveyor belts, flooring and coated wiring.

Fluidised beds

Fluidised beds consist of glass beads suspended in a stream of heated air and are very efficient vulcanising system. The beds are used for continuous vulcanisation of extruded articles.

A great advantage of the process is the speed with which the articles can be produced (in seconds).

Hot-air tunnels

Hot-air tunnels can be used for articles such as dipped goods by passing an endless conveyor carying the articles.

Processing Methods

Mixing

For incorporation of compounding ingredients into rubber either a mill or a Banbury (internal) mixer is used. These are available in many sizes e.g., mills having size from 6" × 8" provided with motor of 7.5 H.P. capable of taking 1 kg material to 28" × 84" provided with motor of 250 H.P. capable of taking 150 kg and Banbury capable of taking batch of 500 g (H.P. of motor 7.5) to 500 kgs. (H.P. of motor 500).

Moulding

Compression moulding is the most common method used in the industry. Here the blank is placed in a 2 piece mould, one part of which is movable and the other stationary. The mould is closed and heat and pressure are applied so that the material flows to fill the mould. A little excess of the material is used to ensure that the mould is completely filled.

Extrusion

In the extrusion process, polymer or rubber is propelled continuously along a screw through regions of high temperature and pressure where it is melted and compacted and finally forced through a die shaped to give the final product. Several shapes can be given by extrusion such as rods, tubing, hose, sheeting and film.

The screw of an extruder can be divided into three zones. The feed zone picks up the polymer from a hopper and propels it into the main part of the extruder. In the compression zone, the loose feed is compressed when air escapes and material melts homogeneously. Some external heat is applied but much heat is generated by friction. The metering zone gives uniform flow rate. It serves as metering pump from where molten mass goes to the die at constant volume and pressure.

Modern trend in extruder design include twin-screw or mutipple screw in which screws turn side by side in opposite directions (Fig. 26.22).

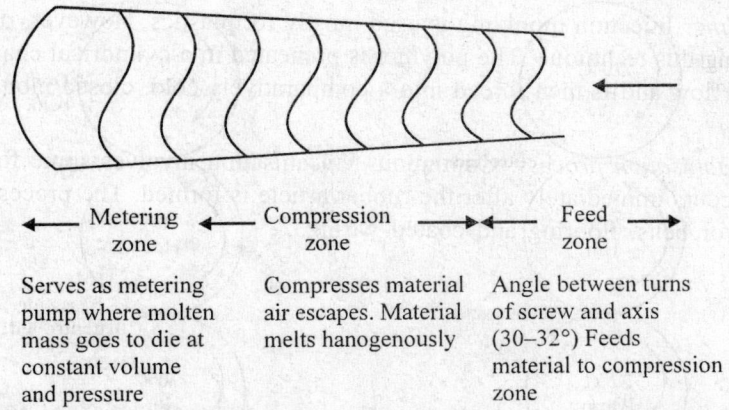

← Metering zone ←	Compression zone →	← Feed zone →
Serves as metering pump where molten mass goes to die at constant volume and pressure	Compresses material air escapes. Material melts hanogenously	Angle between turns of screw and axis (30–32°) Feeds material to compression zone

Fig. 26.22. Modern trend in extruder design include twin-screw or multiple screw.

Spider is the device that supports pin in extrusion. Breaker plate is a perforated steel plate. Its function is to delivr polymer at even pressure and correct temperature to the die. Supports fine screen (30–40 mash) on the side where screw ends. The screen filters out unmelted solid particles. Bullet shape given to upstream of spider so that volume of polymer diminishes gradually and progressively. For good extrusion:

(1) Particles having flat plate like structure give good surface and easier flow e.g., china clay and mica.

(2) Rubber content should be between 10–70%. Above 70% rough surface obtained.

(3) Superior rubber or high filler loading or high styrene resin gives better performance and speeds extrusion.

(4) Factice increases rate of extrusion.

(5) Mixing in internal mixer is recommended. Tough stock is good.

(6) There should be rest period between mixing and extrusion.

Calendering

Calendering is a process used for: (i) the manufacture of smooth sheet of uniform controlled thickness (ii) to apply rubber to textiles or; (iii) embossing. Rubber in the process is passed between pairs of highly polished heated rolls under pressure (Fig. 26.23(a–d)).

Proper calendering requires precise control of roll temperature, pressure and speed of rotation. By maintaining a slight speed differentiation between a roll pair, it is possible to impart an exceedingly high gloss to the film or sheet surface. It is also possible to produce an embossed design on the surface by means of a calender roll which is engraved.

The rolls are of chilled cast iron. Channels are built up inside so that steam or water can reach nearest to surface. The temperature control is accurate and it is nearly same inside and outside of rolls. The rolls can be considered as rotating beams under stress with resulting deflection. The rolls are cambred. The middle roll is grind cylindrically while the lower one is slightly concave. The upper roll has a tendency to go up while middle and bottom rolls have a tendency to deflect down so bottom roll is given concave shape. Top and bottom rolls can be moved independently to adjust thickness. High rubber mixes are difficult to calender and require higher temperature. Incorporation of fillers ease calendering. Calendering can can be done bank or double bank.

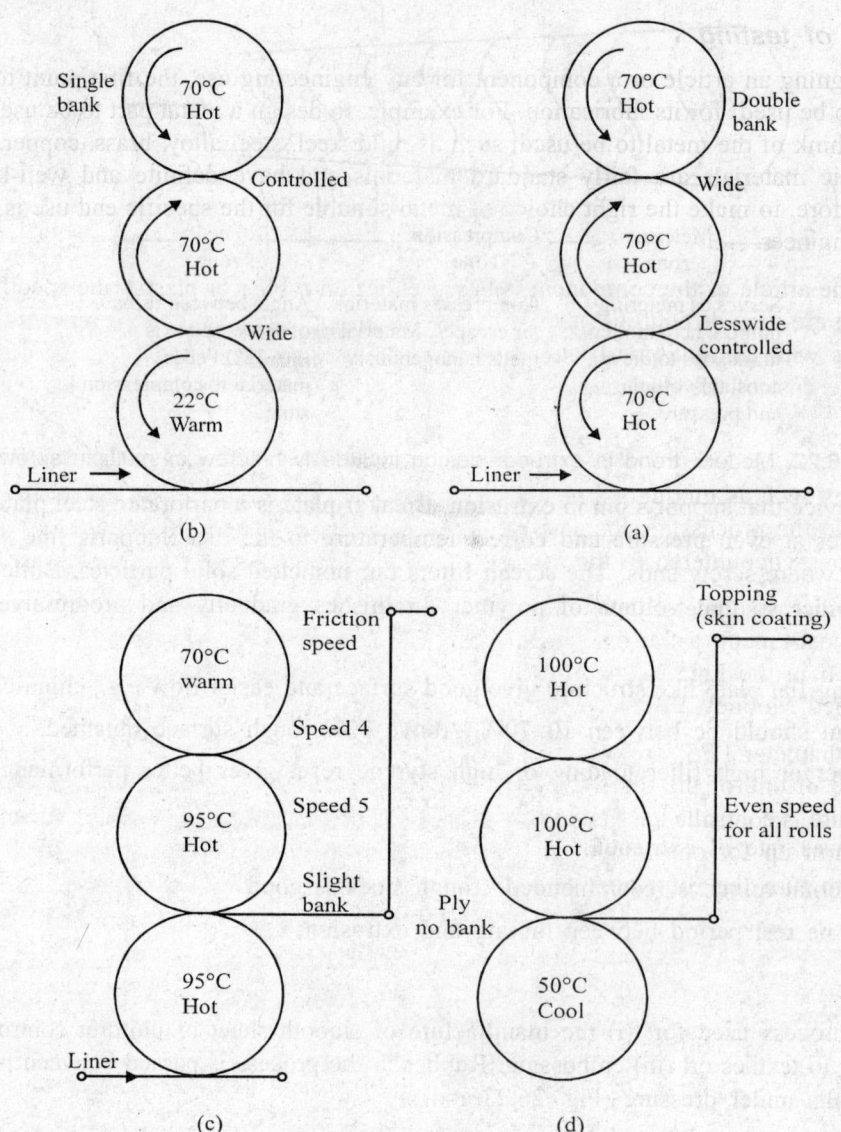

Fig. 26.23(a–d). Calendering process used for manufacture of smooth sheet of uniform controlled thickness.

Testing of Rubber

Rubbers are complex materials and the primary consideration before designing any article or component based on them, is to test the properties of vulcanised rubber.

To properly consider which properties should be measured and how the test should be carried out, it is necessary to first clearly identify why we are testing, because the requirements for each of the reasons will be different. The need of the testing is to prove performance—that is material has to pass through various specifications in a process, where it is mutually agreed for the particular properties of material. For example, oil seal, it has to be oil resistant, dimension stability, of all these are effected to the product.

Need/purpose of testing

Thus, while designing an article or a component for any engineering use, the first point to be considered is 'the material to be used' for its fabrication. For example, to design a metal part to be used in a machine, one would first think of the metal to be used, such as mild steel, steel alloy, brass, copper, stainless steel, etc. Such metallic materials are fairly standard materials and have definite and well-known physical properties. Therefore, to make the right choice of metal suitable for the specific end use is not all difficult for the design engineer.

However, if the article or the component is based either on rubber or plastic, the specific requirements of the component are very difficult to specify because one can design a rubber or a plastic product having desired properties with continuous variations over a very wide range in the composition of the product, durability and the cost of the product.

Plasticity test

Various operations such as mixing, extrusion, calendering and moulding require different levels of shear. Tests to determine the processing characteristics of a compound should expose the test sample to shear level similar to one encountered in the actual operation. However, it is not possible to do so.

The shearing disc viscometer developed by Melvin Mooney has become very popular in the rubber industry. In this instrument, a flat disc rotates in a mass of rubber contained in a grooved cavity under pressure as shown in the Fig. 26.24. The torque required to rotate the disc at 2 rpm at a temperature of 120°C is termed Mooney Viscosity.

Two discs of diameter 1.75" are cut from rubber sheet not less than 0.4" thick. A hole 0.5" in diameter is made in centre of one of the discs which is kept below the rotor. The platens are electrically heated and the temperature is controlled at the desired level. The top platen is pneumatically lowered. The sample is allowed to warm up for one minute after the platens are closed. The motor is then started and the torque required to turn the rotor at 2 rpm is recorded after the prescribed time.

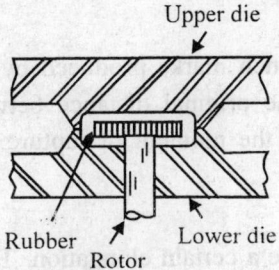

Fig. 26.24. Mooney chamber and rotor.

There are two rotors for the Mooney viscosity test a large one (1–5" dia) for general use and a smaller one (1.2" dia) for tough rubbers. Reading taken after 4 minutes is called Mooney Plasticity. The result is expressed as follows:

$$40 - ML \; 1 + 4 \; (120°C)$$

where 40-M is the Mooney Plasticity number, L indicates that large rotor has been used, 1 is the time in minute for which the specimen has been preheated, 4 is the time in minutes after starting the rotor at which the reading has been taken and 120°C is the temperature of the test. Scorch time is taken as the time when plasticity is 5 units higher than minimum value.

Physical testing of vulcanisates

No tests should be performed until at least 16 hours after vulcanising as significant post vulcanisation changes in structure of the meterial may occur during this period.

Tensile tests

These tests are the most widely used test in the rubber industry. They are used to determine rate of cure and optimum cure, quality check on the compound, effect of immersion in liquids etc. Two shapes of test specimen are most popular-dumb bells and rings.

The dumb bell so called because of its shape has tabbed ends for gripping in the test machine tapers to a central constricted section of uniform width (Fig. 26.25). Bench mark 1" apart is stamped on the constructed section to facilitate observation of the elongation during test. The dumb bell specimen is clicked out with the help of a die.

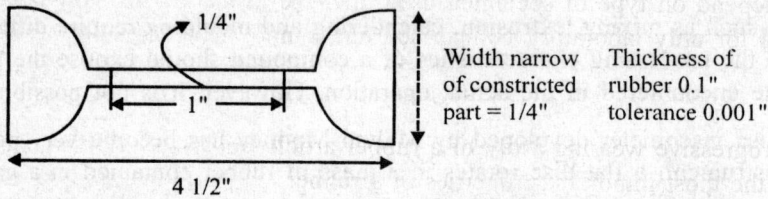

Width narrow of constricted part = 1/4"

Thickness of rubber 0.1" tolerance 0.001"

Fig. 26.25. Dumb bell specimen.

Tensile strength

Tensile strength is the maximum tensile stress reached in stretching a test piece to its breaking point. The force required is expressed as force per unit area of the original cross-section of the test length.

Elongation

Elongation is the extension between bench marks produced by a tensile force applied to the test piece and is expressed as a percentage of the original distance between the marks. Elongation at break or ultimate elongation is the elongation at the moment of rupture.

Modulus

Modulus is the stress required to produce a certain elongation. The standard temperature for conditioning and testing the samples is $23 \pm 2°C$. The tests should be made along the grain in the rubber i.e., parallel to direction of passage through the mill.

Hardness

Hardness, as applied to rubber, may be defined as the resistance to indentation under conditions that the rubber is not punctured.

Hardness is one of the most useful and often quoted properties of rubber. However, the values mentioned can be misleading due to mechanical limitations of the instrument and personal error. The spring-loaded pocket durameter is the most popular instrument for measuring hardness of rubbers.

Rebound tests

Resilience is the ability of a rubber vulcanisate to return the energy used to deform it during its recovery from deformation. It is expressed in percent. The simplest test for resilience is the falling ball rebound 3. If the drop height of the steel ball is divided into 100 equal parts, the rebound height is equal to the resilience.

In a pendulum rebound test, the specimen is held at the rest position of the pendulum (zero degree) and is impacted by the centre of percussion of the arm. The angle of rebound is noted on a scale and resilience calculated by the formula:

$$R = \frac{1 - \cos \text{ angle of rebound}}{1 - \cos \text{ angle of fall}} \times 100$$

Tear test

Tear test results depend on type of specimen used, the rate of tearing and the temperature. The method is therefore useful for only laboratory comparison and cannot be applied for service evaluation.

Abrasion tests

Abrasion is the progressive wearing away of a rubber article in service. In many applications, resistance to wear is one of the most important properties of a rubber article although it is one of the most difficult to measure. The type of wear undergone by an automobile tyre on a road is very different from the wear on a hose used for sand-blasting.

Abrasion resistance is a complicated phenomenon and is dependent on many properties such as resilience, stiffness, resistance to cutting and tearing etc. and different applications require these properties in varying proportions.

Ageing tests

Heat ageing

Tests for heat ageing are carried out for two purposes. Firstly to establish changes in physical properties at elevated service temperatures. Secondly they serve as accelerated ageing tests at high temperatures which try to predict the long-term life at low temperature. ISO 188 lays down methods using an air oven or an oxygen pressure chamber.

The test consists of ageing test pieces for a given period at a specified temperature and then measuring the physical properties such as tensile strength, modulus, elongation at break and hardness. Comparison is thereafter made with non-aged test pieces.

The ageing period should not be too long to prevent determination of ultimate physical property. After vulcanisation of specimen, a minimum period of 24 hours and maximum of 14 days prior to commencement of ageing should be allowed.

Resistance to liquids

Elastomer when in contact with a liquid may absorb the liquid or the liquid may extract some soluble constituents from the elastomer or there may be chemical reaction between the two. Absorption is normally greater than extraction resulting in increase in volume which is called swelling. The

International Standard for resistance to liquids in ISO 1817 which specifies choice of liquid, conditioning, temperature and duration of test.

The size of the test pieces when measuring change in volume should be 1–3 cm with a uniform thickness of 2 mm and should not be longer or broader than 50 mm.

The test consists of completely immersing the test piece in the liquid in a stoppered glass bottle at the required temperature. The volume of the liquid must be at least 15 times the volume of the test piece.

Compression set

Rubbers deform under load and the difference between the original and final dimensions is known as permanent set and can be measured in compression.

Fatigue-flex cracking

After prolonged action of stress or strain on rubber, there is a change in its properties which is termed as 'Fatigue'. Flex cracking test is a fatigue test which strains the test piece in flexure and this type of deformation is experienced in service by footwear, tyre and belting. The De Mattia machine is the most widely used apparatus for this test.

Example 26.1. In the polymerisation of ω-hydroxycaproic acid, $HO(CH_2)_5 COOH$, a 2% impurity is present. Calculate the degree of polymerisation of the polymer formed.

Solution.

$$\overline{DP} = \frac{1 + N_1/N_0}{1 - p + N_1/N_0} = \frac{1 + \frac{2}{98}}{1 - 0.98 + \frac{2}{98}} = 25$$

Example 26.2. In the polymerisation of hexamethylenediamine and adipic acid, a 2% excess of adipic acid is present. Calculate the degree of polymerisation of the polymer formed for 98% conversion.

Solution.

$$\overline{DP} = \frac{1 + r}{2r(1 - p) + (1 - r)} = \frac{1 + 0.98}{2 \times 0.98(1 - 0.98) + (1 - 0.98)}$$
$$= 33.44 \approx 33$$

Chapter 27

Surface Coating Industries

INTRODUCTION

Products of the surface-coating industries are essential for the preservation of all types of architectural structures, including factories, from ordinary attacks of weather. Uncoated wood and metal are particularly susceptible to deterioration, especially in cities where soot and sulphur dioxide accelerate such action. Besides from their purely protective action, paints, varnishes, and lacquers increase the attractiveness of manufactured goods, as well as the aesthetic appeal of a community of homes and their interiors.

PIGMENTS

Pigments may be classifed as inorganic and organic.

Inorganic Pigments

Inorganic pigments are an integral part of decorative, protective, and functional coatings. They provide mass colouration for fibres, plastics, paper, rubber, elastomers, glass, cement, glazes, and porcelain enamels, and they are colourants in inks, cosmetics, and markers.

Pigments can retard corrosion and act as fungistats and antistatic agents. It can be coloured, colourless, black, white, or metallic. Most pigments are insoluble solids. When dispersed in a binder or medium, most pigments remain insoluble.

Properties

Colour results from the pigment's selective absorption of visible light. Large pigment particles also may scatter light and thereby influence the opacity of the binder. Because inorganic pigments involve both transition and non-transition elements, several different theories explain how colour is produced by these compounds.

The intense yellow colour of lead chromate, $PbCrO_4$, is attributed to a charge-transfer process between oxygen and chromium. The presence of two different oxidation states for iron in iron blue, $Fe(III)NH_4- [Fe(II)(CN_6))]$, results in charge-transfer and the associated intense blue colour. The colour of cadmium sulphide is based on its semi-conductor properties and the charge transfer transitions between the valance band and the conductance band. The colour of ultra-marine blue results from electronic transitions within S_3.

Most inorganic pigments are crystalline. Polymorphism is displayed by pigments such as titanium dioxide (rutile, anatase, and brookite), lead chromate (light and primrose type) and calcium carbonate (calcite and aragonite).

The average ultimate particle size of most commercial pigments, excluding extenders, is 0.01–1.0 μm dia. (Table 27.1). The hiding power of a pigment depends primarily upon the ability of the dispersed particles to scatter light. Factors that influence hiding power are refractive index and particle size. The effective particle size for most pigments should be ca one-half the wavelength of visible light.

Microscopy, gas absorption, and sedimentation techniques are used to determine particle size. Particle shape can be noted by transmission electron microscopy.

Another property of pigments is surface area. Carbon blacks have surface areas of 6-1100 m^2/g, whereas most other pigments have areas of 1-100 m^2/g. To achieve complete dispersion, the surfaces of the pigment particles must be properly wetted. Surface-area determination can be made by the BET method, which is based on the absorption of nitrogen at extremely low temperatures and low pressures.

Table 27.1. Particle size of some inorganic pigments, μm.

Pigment	Particle size, μm
Iron blue	0.01-0.2
Titanium dioxide	0.2-0.3
Red iron oxide	0..3-4.0
Natural crystalline silica	1.5-9.0
Strontium chromate	0.3-20.0
Hydrated aluminium oxide	0.4-60.0
Micaceous iron oxide	5.0-100.0

Manufacture

Many pigments are manufactured by precipitation, i.e., lead chromate; or by precipitation and calcination, i.e. cadium sulphide; or by calcination, i.e. chromium oxide. Titanium dioxide is prepared by a vapour-phase reaction.

Care is taken in selection of raw materials and control of manufacturing parameters, i.e., temperature, concentration, and pH excesses, and agitation. Other process steps in manufacture include chemical end treatment, filtration, washing, drying, grinding, and packaging.

White Pigments

The principal white hiding pigment is titanium dioxide; others are zinc oxide, zinc sulphide, lithopone, lead pigments, and antimony oxide (Table 27.2).

Titanium dioxide. The stable crystal form of TiO_2 is rutile; anatase can be converted to rutile at ca 700–950°C (Table 27.3). Titanium dioxide, mp > 1800°C, is insoluble in water, organic solvents, alkalies, and most acids. It is attacked by sulphuric and hydrofluoric acids after long contact and at high temperature. Titanium dioxides are not ideal whites because they do not reflect all wavelengths of visible light equally. They absorb below 430 nm in the blue or violet end of the spectrum.

Table 27.2. Characteristics of some white hiding pigments.

Pigment	Chemical composition	Refractive index[a]	Average particle size, μm
Titanium dioxide, anatase	TiO_2	2.55	0.2
Titanium dioxide, rutile	TiO_2	2.70	0.2-0.3
Zinc oxide	ZnO	2.01	0.2-0.35
Zinc sulphide	ZnS	2.37	0.2-0.3
Lithopone	28% ZnS, 72% $BaSO_4$	1.84	0.2-0.3
Lead carbonate, basic white lead	$2PbCO_3.Pb(OH)_2$	2.0	1.0
Lead sulphate, basic	$PbSO_4.PbO$	1.93-2.02	0.8
Antimony oxide	Sb_2O_3	2.1	1.0

[a] A common refractive index for a paint vehicle is 1.6.

Zinc oxide. Zinc oxide is manufactured from sphalerite. In the indirect process, zinc metal is vapourised, followed by oxidation and collection of the zinc oxide powder. In the direct process, zinc sinter and coal are burned and the resulting zinc vapour is oxidised. Zinc oxide, together with sulphur and organic accelerators, is used in the vulcanisation of elastomers. In addition, zinc oxide absorbs uv radiation and thereby protects organic binders from photodegradation and reduces chalking. Leaded zinc oxide is prepared by cofuming lead and zinc ores or by blending lead-free zinc oxide and basic lead sulphate or lead silicate. It commonly contains 12–55 wt% basic lead sulphate.

Table 27.3. Typical pigment properties of anatase and rutile TiO_2.

	Anatase	Rutile
Density, g/cm^3	3.8-4.1	3.9-4.2
Refractive index	2.55	2.76
Oil absorption, g oil/100 g pigment	18-30	16.48
Tinting strength, Reynolds method	1200-1300	1650-1900
Particle size (av), μm	0.3	0.2-0.3

Zinc sulphide. It is soluble in acids but insoluble in water. It is softer than TiO_2 and does not have the yellowish undertone associated with TiO_2. It is produced from $ZnSO_4$ and Na_2S. It has low toxicity.

Lithopone. Lithopone is a mixed pigment which originally consisted of 28–30 wt% ZnS and 70–72% $BaSO_4$; grades with higher zinc sulphide contents (ca 60 wt%) have been developed. It is prepared by coprecipitation of ZnS and $BaSO_4$. Lithopones are used in water-based paints and in rubber and plastics.

Lead pigments. White lead is the oldest white hiding pigment. Today's usage of lead pigments, e.g., basic lead sulphate, basic lead silicate, basic lead silicosulphate, and dibasic lead phosphite is limited because of restrictions on lead compounds.

Antimony oxide. Antimony oxide is an effective fire retardant controls chalking, and improves tint retention in exterior enamels. It is prepared by roasting stibnite, Sb_2S_3.

Zirconium oxide and zircon. These are used in porcelain or vitreous enamels. Vanadium-doped zircon is a blue pigment used in glazes and enameling.

Extender pigments

Extender pigments are inexpensive colourless or white pigments with refractive indexes < 1.7. Kaolin is a hydrous aluminosilicate mineral. It is used as a filler in the paper and paper-board industry and in paper coatings. Calcined kaolins are white and hard; they are used in water-based and traffic paints.

Clays

Clays are graded according to particle size, i.e., fine, intermediate, and coarse. Mined clay is slurried at the open-pit mine site and the slurry is pumped to a degritting station and then stored in tanks. Slurries are blended, and particle-size distribution is controlled through processing. Attapulgus clay is a crystalline, hydrated magnesium aluminium silicate (Fuller's earth) used in paints as a flattening agent and in sealants.

Calcium carbonates

Natural calcium carbonate, a widely used pigment, is quarried or mined. Limestone is the raw material for precipitated grades. Synthetic calcium carbonates are available in several grades of calcite and aragonite. High calcium types are used in paints, plastics, rubber, paper, adhesives, and joint fillers.

Barium sulphate

Barium sulphate can be prepared synthetically or from the mineral barite. It is used mainly as a weighting material in drilling muds. It is an excellent extender pigment because of its inertness and high refractive index.

Silicates

Magnesium silicate (talc) is derived from natural sources such as soapstone, steatite, and asbestine. Talcs reduce chalking, blistering, and cracking of paint film.

Calcium silicate is derived from wollastonite. The synthetic material is prepared hydrothermally from diatomaceous silicas and lime. It is used in coatings for water-thinned emulsion paints where hiding is achieved by air voids. Natural silicas are obtained from tripolitic ores. They are used as abrasives, reinforcing fillers for silicone rubber, and in caulking compounds and porcelain insulators.

Diatomaceous silica pigments are hydrous silicas obtained by open or strip mining. Diatomaceous silica controls gloss and sheen of flattening paints and is used in water-thinned emulsion flat wall paints, flat varnishes, primers, concrete, and stucco finishes. Precipitated synthetic silicas have large surface areas (100–300 m^2/g) and high oil absorption. Pyrogenic silicas are extremely pure SiO_2 prepared by high temperature reaction. They are used in rubber and plastics. Aerogel and hydrogel silicas have surface areas of 100–400 m^2/g and high porosity; they are used as flattening agents in coatings.

Micas

Micas are complex aluminium potassium silicates; muscovite is the most important. They are used in wallpaper and coated paper, primer and corrosion coatings, and in joint cements and lubricants.

Other extender pigments

Hydrated aluminium oxides, manufactured by seed nucleation of a saturated alumina solution, are fine white pigments. Pumice is a glossy volcanic lava used to impart a rough non-slip surface. Asbestos, a fibrous magnesium silicate, functions as a viscosity-control agent. Calcium sulphate is used in composite pigments and as a filler in plastics and primer paints. Sodium silicoaluminate is an ultrafine synthetic pigment used as a filler in paper and as an extender of TiO_2. Sodium, potassium, aluminium silicate (nepheline syenite) is used in the glass and ceramic industries.

Coloured pigments

Iron oxides are characterised by low chroma and excellent lightfastness; they are non-toxic and inexpensive. They are processed from haematite, limonite, siderite, and magnetite, and provide a range of reds, yellows, purples, browns, and blacks. Raw sienna is a dark yellow that is converted to burnt sienna, a deep brown, by calcination. Ochres are yellow iron oxides, lighter in colour than siennas. Cyprus is the source of fine raw umber, commonly called Turkey umber. Burnt umber is produced by calcinating raw umber. The flake like crystalline structure of gray micaceous iron oxide resembles that of mica. It is used in metal-protective coatings.

Synthetic iron oxides are red, yellow, brown, and black. Advantages over the natural oxides include chemical purity, uniform particle size and size distribution, and the possibility of use in pre-dispersed vehicle systems by flushing techniques. Ferrous sulphate heptahydrate, $FeSO_4.7H_2O$ (copperas) is the main source of iron for these pigments. For example, calcination gives copperas red.

$$FeSO_4.7H_2O \xrightarrow{\Delta} FeSO_4.H_2O + 6H_2O$$

$$6FeSO_4.H_2O \xrightarrow{\Delta} 2Fe_2O_3 + Fe_2(SO_4)_3 + 6H_2O + 3SO_2$$

$$Fe_2(SO_4)_3 \xrightarrow{\Delta} Fe_2O_3 + 3SO_3$$

Direct precipitation of red iron oxides involves growth of iron oxide particles on specially prepared nucleating particles or seeds of Fe_2O_3.

Synthetic red iron oxides are prepared in various grades from light to dark reds and are sold as Indian red, Turkey red, and Venetian red. The Penniman-Zoph process for the manufacture of yellow iron oxides involves the preparation of a seed or nucleating particle by alkali precipitation of ferrous sulphate. Magnetic gamma iron oxide, used in recording tape, is prepared from an acicular yellow iron oxide precursor. Zinc and magnesium ferrites are tan pigments formed by the interaction of iron oxides with metallic oxides. The principal use of transparent iron oxides is in metallic automotive finishes. Chemically they are the same as their opaque counterparts but are much smaller in particle size.

Lead chromate and molybdate oranges

These pigments provide a gamut of hues from greenish yellow through orange to light red. Medium chrome yellows are essentially pure $PbCrO_4$ and provide the reddest yellow hue. Primrose chrome yellow and light chrome yellows are solid solutions to lead chromate and lead sulphate. Chrome orange are $PbCrO_4.PbO$ and the variation from light to dark orange is a consequence of change in particle size. Molybdate oranges are solid solutions of lead chromate, lead molybdate, and lead sulphate. They display hues from orange to red. They are used in industrial finishes; where the risk of health hazards is minimal. Medium-yellow shades are used in traffic-paint formulations. Blends with organic reds and violets produce inexpensive, durable automotive finishes. Heat-resistant types, are suitable as colourants in plastic materials.

Lead chromates are prepared by precipitation from soluble salts in aqueous media. Primrose and light chrome yellows require coprecipitation of lead chromate and lead sulphate. The light chrome yellows are precipitated hot and with excess lead. Chrome oranges are precipitated from alkaline solutions. Chrome yellows and molybdate oranges are soft textured. Their dispersibility results in proper pigment-grind development and high gloss suitable for topcoat finishes. Normal lead silica chromate is manufactured by coating a core of silica with a medium-yellow lead chromate. The pigment is used primarily for traffic paint.

Cadmiums

Cadmium pigments provide brilliant colours ranging from the greenish primrose yellow through orange, red, and maroon. They display clean, bright hues; heat stability up to and in some cases above 400°C; excellent bleed and alkali resistance; fair acid resistance; and good tint light fastness when protected from moisture. They are used in artists' paints, heat-resistant coatings, printing inks, and latex paints. Mercadmium pigments are based on cadmium sulphide-mercury sulphide solid solutions; variation in chemical composition yields a hue range similar to those of the selenide oranges, reds, and maroons (Table 27.4).

Table 27.4. Typical chemical compositions and related hues for cadmium pigments.

| | *Compositions,* | | *wt %* | |
	CdS	*ZnS*	*CdSe*	*HgS*
Cadmium yellow, concentrated				
primrose	79.5	20.5		
lemon	90.9	9.1		
golden	93.4	6.6		
deep golden	98.1	1.9		
Cadmium sulphoselenide orange-reds, concentrated				
light orange	85.0		15.0	
light red	67.5		32.5	
medium light red	58.8		41.5	
medium red	51.5		48.5	
dark red	44.8		55.2	
maroon	35.0		65.0	
Mercadium orange-reds				
deep orange	89.0			10.9
light red	83.4			16.6
medium light red	81.0			19.0
medium red	78.5			21.5
dark red	71.6			23.9
maroon	73.5			26.5

Synthetic mixed metal oxides

These pigments are classified by crystal structure, e.g., mixed-phase rutile pigments or mixed-phase spinel pigments. They are produced by solid-state chemical reactions at high temperature. The dry or wet raw materials are mixed and blended and then calcined. These pigments show excellent heat stability as well as weather resistance and lightfastness. They are used in the glass and ceramics industries. Nickel titanates are colourants for PVC siding materials.

Iron blue

Iron blue, as first developed, was hard-textured and difficult to grind. The modern iron blues are ferric ammonium cyanides used for colouring paper, for bluing, and in printing inks. Their poor alkali resistance can be improved by treatment with nickel compounds. Preparation is based on the oxidation of Berlin white, $Fe(NH_4)_2[Fe(CN)_6]$, which is produced from sodium ferrocyanide, ferrous sulphate, and ammonium sulphate.

Ultramarine (Lapis lazuli blue) provides good brilliance, heat resistance, lightfastness, and alkaline stability. It is prepared from intimate mixtures of china clay, sodium carbonate, sulphur, silica, sodium sulphate, and charcoal or pitch. The intensity of the colour depends on reactant concentrations, residence time, and temperature. Manganese violet and cobalt violet are prepared by precipitation. The former is used in cosmetics and is also used to tone white pigments.

Chrome oxide green is a calcined pigment prepared by reduction of sodium bichromate with sulphur or carbonaceous materials. Because it reflects irradiation and is similar in reflective properties to chlorophyll, it is used extensively in camouflage coatings. Transparency permits formulations of polychromatic finishes. These pigments are manufactured by hydrolysing a complex chromium borate obtained by heating sodium bichromate with boric acid.

Chrome greens are blends of variable composition of primrose chrome yellow and iron blue. They are used in paints, printing inks, flooring materials, plastics, and paper goods because of their high chroma and excellent hiding power. Natural mercuric sulphide, HgS, can be derived from the mineral cinnabar. As the pigment vermilion, it occurs in a red crystalline or black amorphous form. Antimony vermilion, Sb_2S_3, varies in hues from orange to red. Van Dyke brown is a natural pigment, consisting of as much as 92 wt% organic matter, water, and traces of iron oxide, alumina, and alkali oxides. It is found in peat beds in the FRG.

Black pigments

Carbon black, one of the oldest pigments, is used in plastics, paints, and printing inks.

Speciality pigments

Speciality pigments include zinc yellow (corrosion-inhibitive primers), basic zinc chromate (a wash primer), strontium chromate (metal protection and corrosion resistance), red lead (protection of iron and steel surfaces), cuprous oxide (an antifoulant agent for marine paints and colourant for ceramic glazes and glasses), calcium plumbate (anticorrosion pigment for steel), basic lead silicochromate (electro-deposition of water-based automotive coatings), white molybdates (corrosion inhibitors), modified barium metaborate (mold and enzyme inhibition, corrosion and chalk resistance), zinc phosphate (corrosion-resistant steel undercoatings), and nacreous pigments (pearl essence for decorative effects).

Metallic pigments

Metallic pigments are prepared from metallic elements and their alloys, e.g., aluminium, copper, bronze, and zinc. Aluminium flake pigment is sold in a paste form, which typically contains 65 wt% flake metal and 35 wt% volatile hydrocarbon. Bronze pigments or powders are manufactured from granular alloys of copper and zinc; they are used for coating paper and cardboard. Gray metallic zinc powder or dust is widely used on iron and steel because of its cathodic effect. Luminescent pigments are based on zinc sulphide and zinc-cadmium sulphide that have been doped with an activating material; they are used for indoor decoration, safety signs, television and electronics, and military coatings.

Organic Pigments

Pigments are coloured, colourless, or fluorescent particulate solids which are usually insoluble in and essentially physically and chemically unaffected by the vehicle or medium in which they are incorporated. They alter appearance either by selective absorption or by scattering of light. Pigments may be organic or inorganic chemicals. For application, they are usually dispersed in vehicles, e.g., inks or paints. In some cases, the substrate may serve as a vehicle, e.g., in the mass colouration of polymeric materials.

Colour and structure

In organic compounds, colour is associated with double bonds. The unsaturated, conjugated double bonds in chromophores contribute to the selective absorption of light. Chromophores include nitroso, nitro, carbonyl, thiocarbonyl, azo, azoxy, azomethine, and ethenyl groups, whereas auxochrome groups include amino, alkylamino, dialkylamino, methoxy, and hydroxy groups.

Properties

The characteristics that control the performance of a pigment include its chemical composition; chemical and physical stability; solubility; particle size and shape; degree of dispersion or aggregation; crystal geometry; refractive index; specific gravity, absorption; extinction coefficients; surface area and character; and the presence of impurities, extenders, or modifying agents. The inherent strength of a pigment is controlled by its light-absorbing properties, which are related to molecular and crystalline structure. The intensity is a measure of brightness or cleanness, as opposed to dullness, or shade. Generally, a pigment having a molecular structure containing two or more chromophores is less intense than one containing a single chromophore.

Fastness defines the inherent ability of a pigment to withstand the chemical and physical factors to which it is exposed in its application. Durability defines the ability to withstand weather, light, water, gases, and industrial effluents. Dispersibility is measured by the effort required to develop the full tinctorial potential of a pigment in a vehicle; it affects the maximum surface area and homogeneity of the pigment system. Texture describes the ease of dispersion. The working properties include compatibility, oil absorption, contribution to rheology, ease of grinding, wettability, gloss, bronzing, hiding power, flocculation, etc.

Classification

Organic pigments are classified as azo and non-azo pigments. The former contain chromophore (—N=N—) groups and are subdivided into pigment dyes and precipitated azos. Pigment dyes are insoluble on formation, whereas precipitated azos include products containing salt-forming groups, principally sulphonic or carboxylic acids.

Azo pigments

The Hansa yellows are semi-opaque, intense, monoazo pigments which are used in emulsion paints and paper coatings. The intense colour of diarylide yellows provides greater tinctorial strength and resistance to bleeding and heat than is provided by the Hansa yellows. Nickel azo yellow is a greenish yellow, chelated nickel azo pigment with high transparency and excellent durability.

Nickel Azo Yellow

Hansa orange, Dinitraniline orange, and pyrazolone orange all exhibit intense, high strength colour with varying degrees of lightfastness and chemical resistance.

Azo reds and maroons

Toluidine red is one of the most popular red pigments for industrial enamels. It exhibits good bake and chemical resistance but poor lightfastness. Para red is similar but darker and bluer. Chlorinated para reds are intense, very light yellowish red. Parachlor Red is one of the more lightfast azo-pigment dyes. Lithol reds are intense reds of dark shades. They are formed by coupling diazotised Tobias acid to 2-naphthol and precipitated as the Ba, Ca, and Sr salts. Lithol reds are inexpensive and are used widely where extreme durability is not required.

The BON reds and maroons derive their name from β-oxynaphthoic acid as the second component in the coupling of various diazotised amines containing salt-forming groups. Lithol Rubine exhibits a wide range of intense shades suitable for industrial enamels. Permanent red 2B defines the calcium, barium, and manganese precipitations of the coupling from diazotised 2-chloro-4-aminotoluene-5-sulphonic acid with 3-hydroxy-2-naphthoic acid. They are used in printing inks, plastics, and automotive finishes. Yellow BON maroon is a lightfast, yellow maroon with superior solvent bleed resistance suitable for outdoor finishes. The manganese toner of Lithol Red 2G is intense in colour and is used mainly in blends with inorganic molybdate orange for durable reds; the calcium toner is used in printing inks.

The pyrazolone reds are disazo pigments which provide high colour intensity, excellent masstone lightfastness, high transparency and good bleed, bake, and chemical resistance. The search for pigments with improved properties has resulted in a group of higher molecular weight disazo products.

The monoazo naphthol reds and maroon pigments provide a wide range of colours from light reds to dark maroons.

Lakes are either dry toner pigments extended or reduced with a solid diluent or an organic pigment prepared by precipitation of a water-soluble dye on an adsorptive surface. Alizarine Red B or madder lake is a co-ordination complex of alizarin with alumina and calcium plus sulphated castor oil and a phosphate. It has an intense, deep bluish-red of fair lightfastness. Helio Fast Rubine 4BL is a precipitated quinizarinsulphonic acid derivative used to shade pigments for organic coatings.

A basic-dye pigment is the precipitated product of the reaction between a basic dye and either complex inorganic heteropoly acids or precipitants, e.g., tannic acid, tartar emetic, clay, or rosin soaps. Despite their high costs, these pigments are widely used for their high strength and intensity.

Phthalocyanines

Phthalocyanines are characterised by excellent lightfastness, intensity, bleed and chemical resistance, and heat stability. Copper Phthalocyanine Blue exists in a red-shade blue alpha form and the more stable green-shade beta form. Chlorination gives Copper Polychlorophthalocyanine Green which has excellent pigmentary properties. Copper polybromochlorophthalo cyanines provide yellower shades of green than the polychloro compound.

Quinacridone pigments offer outstanding fastness properties in the orange, maroon, scarlet, red, magenta, and violet ranges. An example of a violet with excellent fastness properties is Pigment Violet 23, a dioxazine pigment derived from 3-amino-N-ethylcarbazole.

Vat pigments

A number of vat dyes based on anthraquinone and substituted and/or condensed anthraquinones have found use as pigments because of fastness and durability characteristics. In general, vat dyes must be modified chemically and physically to develop pigmentary strength and colour intensity.

The thioindigos provide a broad range of hues from a yellow shade of red to violet. The perinone pigments are diimides of naphthalene-1,4,5,8-tetracarboxylic acid formed by condensation of o-diamines with the acid or acid anhydride. The perylene pigments are diimides of perylene-3,4,9,10-tetracarboxylic acid and generally are stronger and more resistant to chemicals, heat, and solvent bleeding than the thioindigos.

Among the anthraquinone vat dyes, indanthrone is redder but less intense than the phthalocyanine blue pigments, but as durable as Copper Phthalocyanine Blue. It is used in automotive and other high quality finishes. Partially chlorinated indanthrone blues are greener than the unchlorinated but not as bake resistant. Other anthraquinone pigments include Isodibenzanthrone Violet, Dibromoanthranthrone Orange, Flavanthrone Yellow, and Anthrapyrimidine Yellow.

Pigments can be standardised only in terms of performance, colour, durability, and working properties, and often only for the specific application or vehicle system for which they are intended. In general, organic pigments have no toxic effects.

Organic pigments are used for decorative and functional effects; they provide colour, hiding power, and high visibility and contribute to durability. They contribute to countless applications involving colouration.

Dispersed

A dispersed pigment concentrate is an extremely fine distribution of colour pigment in any medium; it is suitable for supplying colour to surface printing and coating or a complete mass colouring.

Dispersion

Reduction of the pigment to primary pigment particle size is necessary in order to develop the optimum visual performance and economic properties. The extent of the reduction is determined by the nature of the pigment, the dispersion system and processing equipment, product-use requirements, and economics.

The maximum aggregate size that adversely affects physical properties is a small fraction of the thickness of the film, coating, or thin mass-coloured material. The optimum size to which a pigment should be reduced is the primary pigment particle. In dispersed pigments, the term primary pigment particle refers to individual crystals and tightly held aggregates. Agglomerates are larger associations of the primary pigment particles, i.e., crystals and aggregates. Ideally, a dispersion consists mainly of primary pigment particles and a minimum of loosely held aggregates. The system of wetted primary particles must be stabilised in order to avoid reversal of the dispersion process.

Flushing is the direct transfer of pigments in an aqueous phase to an oil or non-aqueous phase without intermediate drying. Because of the difficulties involved in flushing inorganic pigments, it is often more practical to disperse them in dry form. Flushing is therefore confined mostly to organic pigments, e.g., Diarylide Yellow. It is usually carried out in a heavy-duty dough mixer and produces dispersions of better gloss, transparency, and strength than methods using dry pigments.

Diarylide Yellow

Equipment

Kneader dispersers process systems with extremely high plastic viscosity, i.e., up to 10^7 Pa.s (10^8 P). The Banbury mixer is the most efficient mixer for preparing large amounts of concentrated pigment dispersions in rubber and thermoplastic carriers. The flusher or heavy-duty mixer is used primarily in the ink and paint industries. The two-roll mill can be considered a kneader or internal mixer; clearance between the rolls is large (up to 15 mm) and the mass to be processed is subjected to kneading just before it enters the nip itself. These mills are used to prepare dispersions in elastomers and thermoplastic resins.

Close-tolerance mills can be classified as cylindrical roll mills, rotating-disk mills, and cone mills. Cylindrical rolls mills of several types are used for processing pigment dispersions in paste form for inks, paints, and other uses. Rotating-disk mills, cone mills, stone mills, and colloid mills essentially are rotors in the shape of disks or truncated cones which rotate at high speed (3000–5000 rpm) and at a small distance (10–100 μm) from a stationary surface.

Impeller and impingement high speed, fluid-energy mills are used for the preparation of fairly low viscosity mill bases for inks and paints.

Dispersion-process equipment employing a large number of dispersing surfaces, whose movement is not precisely controlled, are categorised into ball or pebble mills (media size > 3.0 mm) and bead, shot, and sand mills (media size 0.2–3.0 mm). The most common design of a ball or pebble mill is a horizontal cylinder, partially filled with balls or pebbles, which rotates on the horizontal axis at a rate that cascades the balls or pebbles rather than centrifuges them or allows them to slide. These mills operate more efficiently with small particle size media. They have been largely replaced by bead, shot, and sand mills, which are vertical or horizontal cylindrical containers filled with grinding medium under intense agitation.

Uses

A dispersed pigment concentrate must be suitably formulated for manufacturing the concentrate and converting it into an application product. High pigment concentrations provide maximum flexibility and economy. For heavy-duty mixers and impingement mills, the formulation used for the dispersion process must be modified. Ink films 0.002–0.01 mm thick require relatively high pigment concentration of ca 10–20 wt%. Manufacturers offer a wide variety of pigment concentrates in liquid, paste, and solid form.

In various countries, most letterpress and lithographic oil inks are manufactured from dispersed pigment concentrates, mainly in flushed form. Most gravure and flexographic liquid inks are made from dry pigment or presscake (Table 27.5). Gravure inks are characterised by very fluid viscosities (ca 0.1 Pa.s 1P) when press ready; these inks dry by solvent evaporation, leaving a film ca 10 μm thick. Flexographic inks, like gravure inks, are pigment-resin-solvent systems of low viscosity (ca 0.3 Pa.s = 3P) and film thickness of ca 8 μm; they dry by solvent evaporation.

Environmental and safety considerations will affect future development in ink manufacture. Concern for organic-solvent vapour emission favours aqueous or water-based inks. Today, flexographic ink is the only one based on water as a solvent.

Coatings are generally heavier films, a fraction of to a few millimetres thick; they require a lower pigment content than inks. Inorganic pigments are hydrophilic and are easily dispersed in water-based systems. Organic pigments, on the other hand, are hydrophobic and require wetting agents for dispersion.

Table 27.5. Distribution of usage of paste ink colourants by pigment form, %.

Type of Ink	Flush	Dry	Other
Letterpress news	95	5	
Heat-set letterpress	95	5	
Glycol letterpress	0	70	30
Other letterpress	90	10	
Lithographic news	95	5	
Heat-set web offset	95	5	
Quick-set sheet offset	90	5	ca 5
uv offset	0	100	
Conventional metal decorating	0	100	
Dry offset metal and plastic	50	50	

Dispersed-pigment or colour concentrates for use in plastics vary widely in pigment concentration and physical form. Some are manufactured as liquids and high viscosity gels which are usually metered directly into the processing equipment. Pellet concentrates are the most popular form. The effect on the plastic is a consideration in pigment selection. Acetals, ABS, fluoropolymers, and nylon present special problems.

Mass colouration, i.e., spin-colouration or dope-dyeing of synthetic fibres, particularly polyolefins, can be accomplished with pigment dispersions. In this process the colourant is incorporated before the fibre is formed.

Pulp colourants are water pastes derived from pigments, extended pigments, or lakes. The principal applications are in paper and textile printing, leather and paint.

PAINTS

Paint is a substance composed of solid colouring matter suspended in a liquid medium and applied as a coating to various types of surfaces.

The purpose of the coating may be decorative, protective, or functional. Decorative effects may be produced by colour, gloss, or texture. The protective coating may be the paint on a wooden boat which serves as a barrier against moisture and prevents rotting; the interior lining of metal cans or drums which prevents corrosion from foods or chemicals; the coating on electrical parts to exclude moisture; the fire-retardant paint which protects combustible surfaces; the coating on plaster or concrete which makes for ease of cleaning, etc. An example of the functional use of paint would be as a traffic paint which marks the centre or edge of a road for safer driving.

The paint industry serves two distinct types of markets, trade sales (shelf goods) and industrial sales (chemical coatings).

Trade sales is the large consumer-oriented portion of the business. Trade-sale paints are house paints and other products marketed through wholesale and retail channels to the general public and professional painters for use on new construction or for the maintenance of old buildings. Also included in this category are the paints sold to garages and repair shops for automobile refinishing, marine finishes for boats, paints for graphic arts such as signs, paints for refinishing machinery and equipment, and paints sold to government agencies, especially the traffic paints used for road-marking.

Industrial sales comprise the coatings sold directly to the manufacturer for factory application. The products on which paints are applied include durable goods such as automobiles, appliances, and house sidings, and non-durable goods such as cans for food and beverages.

All coatings contain a resinous or resin-forming constituent called the *binder*. This can be a liquid such as a drying oil, or a resin syrup that can be converted to a solid gel by chemical reaction. In some instances, where the binder is either a solid or is too viscous to be applied as a fluid film, a volatile solvent or *thinner* is also added. This evaporates after a film is deposited, causing solidification of the film. The binder plus solvent is known as the *vehicle*. Most paints also contain *pigments* which are described in the first part of this chapter.

In addition to pigment, binders, and thinners, a paint may contain many additives, such as defoamers, thickeners, flow agents, and driers, to improve specific properties.

Paints can be classified according to the binder used, e.g., alkyds, vinyls, and epoxies. Paints are also classified according to their properties or end use. Alkyd enamels, for instance, are gloss paints with good abrasion resistance and good cleanability, while alkyd flat wall paints are characterised by very low sheen and good film build.

Paints applied directly to the surface are called undercoaters or primers. Primers are used to aid the adhesion of the top coat to a surface and to prevent absorption of the top coat into a porous surface. Primers can also be used to prevent corrosion of a metal surface. Fillers or surfacers are types of primers that fill scratches and surface imperfections to give a smooth surface. Paints used as the final coat are referred to as finish coats or top coats. Some top coats are self-priming.

Binders

The protective properties of a coating are determined primarily by the binder. In the early days of paint technology, binders were limited to materials of natural origin, such as drying oils, congo resins, and asphalts. These still find some usage in the protective-coating industry.

Coatings in liquid form can be divided into two types, *solutions* and *dispersions*. In solution systems the resin binder is dissolved in a solvent. In dispersion systems the resin is in the form of tiny spheres (usually 10 microns or less in size) suspended in a volatile liquid carrier. If the liquid in a dispersion system is water, the system is called an emulsion; if it is an organic material, it is called an organosol. When the liquid evaporates, a mixture of soft resin and pigment is left behind and fuses into a continuous film. The general types of resins used are listed below. Mixtures of two or more types may sometimes be used to improve certain properties.

(1) *Oils:* easy application; soluble in aliphatic solvents.

(2) *Alkyds:* all purpose; combined with other resins; most are soluble in aliphatic solvents.

(3) *Cellulosics:* (nitrate and acetate)–used in lacquers; fast dry.

(4) *Acrylics*: good colour and durability.

(5) *Vinyls:* good durability; abrasion resistant.

(6) *Phenolics*: good chemical resistance; yellow colour.

(7) *Epoxies:* good chemical resistance.

(8) *Polyurethanes:* good flexibility; abrasion resistance.

(9) *Silicones:* good heat resistance.

(10) *Amino resins:* (ureas and melamines)–blended with alkyds for baking finishes; tough, good colour

(11) *Styrene-butadiene*: low cost; alkali resistant.

(12) *Polyvinyl acetates*: low cost; good colour retention.

(13) *Acrylics*: good colour and durability.

Solvents

The type of binder determines the type of solvent or thinner used in a paint formulation. The preferred type of thinner is an odourless aliphatic hydrocarbon which can be used in all areas including the home. Unfortunately, aliphatic thinners do not dissolve all resins. In such cases, strong solvents such as aromatic hydrocarbons, esters, and ketones are used.

Solvents are generally classified as low boilers, medium boilers, and high boilers depending upon the evaporation rate. Various ranges are required depending on the specific application. Examples of solvents classified by their chemical composition are as follows:

(1) *Hydrocarbons*: naphtha and mineral spirits; and aromatics–benzene, toluene, and xylene.

(2) *Alcohols:* methyl alcohol, ethyl alcohol, and butyl alcohol.

(3) *Ethers*: dimethyl ether and ethylene glycol monoethyl ether.

(4) *Ketones:* acetone, methyl ethyl ketone, and methyl isobutyl ketone.

(5) *Esters:* ethyl acetate, butyl acetate, and butyl lactate.

(6) *Chlorinated:* tetrachloroethane.

(7) *Nitrated*: nitromethane, nitroethane, and l-nitropropane.

Latex or emulsion paints use water as the volatile component so that they may be used in the home.

General Properties of the Various Types of Paints

Acrylics

Acrylic resins are used in protective and decorative lacquers for paper, fabrics, leather, plastics, wood, and metal. White baking enamels having excellent resistance to chemicals and chemical fumes can be made from acrylic resins. Acrylic vehicles are used in luminescent paints. The present automobile finishes are based on acrylic resins.

Thermosetting types of acrylic resins have been developed which cross-link upon heating to form hard, insoluble, infusible coatings for appliance finishes that will withstand severe service on washing machines, stoves, and dishwashers.

Alkyds

Alkyds are oil-modified phthalic resins that dry by reacting with oxygen from the surrounding air. Alkyd finishes are usually of the general-purpose type and are available as clear or pigmented coatings. Paints are available in flat, semi-gloss, and high-gloss finishes in a wide range of colours. They are easy to apply and may be used on most surfaces with the exception of fresh concrete, masonry, and plaster which are alkaline. Alkyd finishes have good colour and gloss, and retain these characteristics in normal interior and exterior environments except under corrosive conditions.

Alkyd finishes are available in odourless formulations for use in hospitals, kitchens, sleeping quarters, and other areas where odour during painting might be objectional.

In trade sales applications, alkyds are used for interior walls and woodwork in both flat and gloss, and on the exterior as trim paints. In industrial finishes, alkyds are combined with amino resins to produce hard baking finishes for use on appliances, metal furniture, etc.

Cellulosics

The most widely used cellulosic derivatives are the esters, particularly cellulose nitrate and cellulose acetate. Cellulose nitrate has been combined with many different resins to produce useful coatings. Perhaps the most outstanding combinations have been those with alkyd and amino resins to produce tough, hard, durable lacquers capable of withstanding the severe service requirements of automotive, aircraft, and other industrial finishes.

Successful coatings have also been based on blends on nitrocellulose with natural resins and drying oils, resin derivatives, phenolic resins, acrylic resins, certain vinyl resins, and other materials. Applications include coatings for metal, wood, paper, fabrics, leather, and cellophane.

Epoxies

Epoxy binders are of two types: (i) the oil-modified compositions, which dry by oxidation; and (ii) the two-component materials, which comprise epoxies and amine or polyamide hardeners and are mixed just prior to use. In the latter type, when the two ingredients are mixed they react to form the cured coating. These coatings have a limited pot-life, usually a working day. Anything left at the end of the day must be discarded.

Epoxy paints can be used on any surface and can be applied with high-solids content, thus producing high film build per coat. The cured film has outstanding hardness, adhesion, flexibility, and resistance to abrasion, alkali, and solvents, as well as being highly corrosion resistant. Their major uses are as tile-like glaze coatings for concrete and masonry and for the protection of structural steel in corrosive environments. Their cost per gallon is high, but this is offset by the higher solids content and the reduced number of coats required to provide adequate film thickness. When used on exterior surfaces, epoxy paints tend to chalk to low-gloss levels and fade. Apart from this their durability is excellent.

Epoxy-coal tar coatings are made by adding coal tar as an ingredient to epoxy paints, thereby reducing the cost. They have outstanding corrosion resistance and are used for interior and submerged surfaces. Colour choice is limited to black.

Oils

Linseed oil is the major binder used in oil house paints. These paints are the oldest type of coatings in use. They are used primarily on exterior wood and metal since they dry too slowly for interior use. They are sensitive to alkaline masonry. Oil paints are easy to use and give high film build per coat. They also wet the surface very well so that surface preparation is less critical than with other types of paints for metal. Oil paints are not particularly hard or resistant to abrasion, chemicals, or strong solvents, but they are durable in normal environments.

Oleoresinous

These binders are made by processing drying oils with hard resins, such as natural resins, rosin esters, and hydrocarbon resins. They are generally used as spar varnishes or as mixing vehicles for aluminium paint.

Phenolics

Phenolic binders are made by processing a drying oil with a phenolic resin and, thus, are a type of oleoresinous binder. They may be used as flat (lustreless) or high gloss finishes which are clear or pigmented in a range of colours. The clear finishes may be used on exterior wood and as mixing vehicles for producing aluminium paints. The durability of the clear finish is very good for this class of material, i.e., one to two years; the durability of the aluminium paints is excellent.

Phenolic paints are used as top-coats on metal in extremely humid environments and as primers for fresh-water immersion. These paints require the same degree of surface preparation as alkyds, but they are slightly higher in cost than alkyds. Phenolic coatings have excellent resistance to abrasion, water, and mild chemical environments. They are not available in white or light tints because of the relatively dark colour of the binder.

Phenolic and alkyd binders are often blended to combine the hardness and resistance of phenolics with the colour and colour-retention of the alkyds. This may be done either by blending phenolic varnish with the alkyd vehicle or by the addition of phenolic resins during the processing of the alkyd resin.

Silicones

Silicones resins are used for heat-resistant finishes. They have good water-resistance and outstanding gloss-retention. When pigmented with aluminium, heat-resistant organic finishes containing a high concentration of silicone resins have the ability to withstand temperatures up to 1200°F. A combination of silicone and alkyd resins provides some heat resistance at a lower cost.

Urethanes

Urethane binders are of three types: (i) oil-modified, which are cured by oxygen from the air; (ii) moisture-curing, which are cured by moisture in the air; and (iii) two-part systems which are mixed just prior to use.

Oil-modified urethanes are similar to phenolic varnishes. Although somewhat more expensive, they have better initial colour and colour-retention, dry more rapidly, are harder, and have better abrasion-resistance. They can be used as exterior spar varnishes or as tough floor finishes. Oil-modified urethanes can be used on all surfaces. In common with all clear finishes, they have limited durability when used on exterior surfaces.

Moisture-curing urethanes are used in a manner similar to other one-package coatings except that the containers must be full to exclude moisture during storage. They have outstanding abrasion resistance and chemical resistance.

Vinyls

Lacquers based on modified polyvinyl chloride resins are used on steel where the ultimate in durability under abnormal environments is desired. They are moderate in cost, but they have low solids and require the most extensive degree of surface preparation to secure a firm bond. Because of their low solids, vinyl finishes require numerous coats to achieve adequate dry film thickness; thus the total cost of painting is higher than with most other paints. Since vinyl coatings are lacquers, they are best applied by spray, and they dry quickly, even at low temperatures. Recoating must be done with care to avoid lifting by the strong solvents which are present. In addition, these solvents present on odour problem. Vinyls can be used on metal or masonry, but they are not recommended for use on wood. They have exceptional resistance to water, chemicals, and corrosive environments, but they are not resistant to solvents.

Vinyl-alkyd combinations offer a compromise between the excellent durability and resistance of vinyls with the lower cost, higher film build, ease of handling, and adhesion of alkyds. They can be applied by brush or spray, and they are widely used on structural steel in marine and moderately severe corrosive environments.

Rubber-base

So-called rubber-base binders are solvent-thinned and should not be confused with latex binders which are often called rubber-base emulsions. Four types are available: chlorinated rubber, styrene-butadiene, vinyl-toluene-butadiene, and styrene-acrylic. They are lacquer-type products which dry rapidly to form finishes which are highly resistant to water and mild chemicals. Recoating must be done carefully to avoid lifting by the strong solvents used. Rubber-base paints are available in a wide range of colours and levels of gloss. They are used for exterior masonry, also for areas which are wet, humid, or subject to frequent washing, e.g., swimming pools, wash rooms, shower rooms, kitchens, and laundry rooms.

Latices

Latex paints are based on aqueous emulsions of three basic types of polymers: polyvinyl acetate, polyacrylics, and polystyrene-butadiene. They dry by evaporation of the water, followed by coalescense of the polymer particles to form tough insoluble films. They have little odour, are easy to apply, and dry very rapidly. Interior latex paints are generally used either as a primer or finish coat on interior walls and ceilings made of plaster or wall board. Exterior latex paints are used directly on exterior masonry or on primed wood. They are non-inflammable, economical, and have excellent colour and colour-retention. Latex paint films are somewhat porous so that blistering due to moisture vapour is less of a problem than with solvent-thinned paints. They do not adhere readily to chalked or dirty surfaces nor to glossy surfaces under eaves. Therefore, careful surface preparation is required for their use.

Latex paints are very durable in normal environments, at least as durable as oil paints. The popularity of latex paints is due mainly to easy cleaning of brushes, etc., with water.

Inorganics

The major inorganic binders used in paints are sodium, potassium, lithium, and ethyl silicates. These binders are used in zinc-dust-pigmented primers in which they react with the fine zinc metal to form very hard films. These films are extremely resistant to corrosion in humid or marine environments. Many of these primers also contain substantial concentrations of lead oxides which react with the silicates in conjunction with the zinc to form an even more corrosion-resistant coating.

Another type of inorganic binder is Portland cement. The paint is supplied as a powder to which water is added before use. Cement paints are used on rough surfaces such as concrete, masonry. They dry to form hard, flat, porous films. When properly cured, cement paints of good quality are quite durable; when improperly aired, they chalk excessively on exposure, and then they may present problems on repainting. A typical paint formula for an outside house paint is shown in Table 27.6.

Table 27.6. Analysis of a white outside house paint.

	Pounds	Gallons	Per cent (by wt)
Titanium dioxide (rutile)	105	3.05	7.4
White lead—basic carbonate	160	1.88	11.3
Zinc oxide (35 per cent leaded)	370	7.55	26.2
Talc	240	14.70	16.9
Linseed oil—bodied 3	85	10.54	6.0
Linseed oil alkali refined	350	45.40	24.8
Drier–manganese naphthenate 6 per cent	2	00.26	—
Drier–lead naphthenate 24 per cent	7	00.72	—
Mineral spirits	105	15.90	7.4
Total	1419	100.00	100.0
Constants			
Viscosity	90 KU		
PVC	30 per cent		
Vehicle solids	80 per cent		
Total solids	92.5 per cent		
Wt/gallon	13.1 lbs		

Manufacture of Paints

The manufacture of paint involves the following operations: mixing, grinding, thinning, adjusting, and filling (Fig. 27.1).

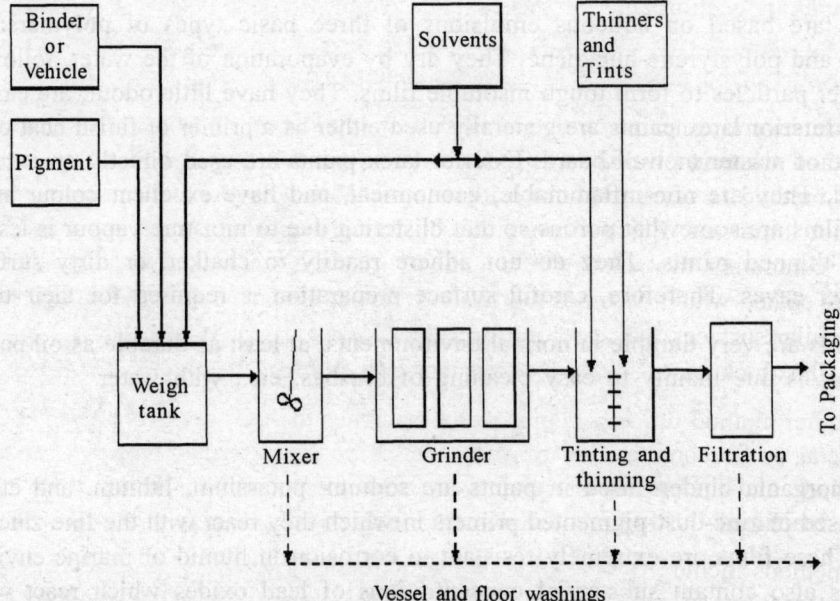

Binder or Vehicle — Solvents — Thinners and Tints — Pigment — Weigh tank — Mixer — Grinder — Tinting and thinning — Filtration — To Packaging

Vessel and floor washings

Fig. 27.1. Typical flow chart for paint production.

One of the older methods consisted of mixing all the pigment and part of the vehicle to make a paste of suitable consistency in a tub with rotating blades. This paste is then fed by means of a trough into a slow-speed stone mill which has two circular stones, one stationary and the other rotating above it. The pressure developed between the two stones is sufficient to disperse the pigments and liquids initimately, eliminating unwetted agglomerates and air pockets. This paste is fed in a continuous stream into a third piece of equipment, the thinning tank, which has rotating blades. At this point the remainder of the liquids is added, and the completed product was tested for viscosity, colour, and other physical properties pertinent to the formulation being prepared. The batch, having been approved, is then strained and filled into appropriate size containers. The term "grinding" is a misnomer, as little or no breakdown of pigment particles takes place.

Dispersion by old-fashioned low-speed stone mills is extremely slow and costly and is obsolete. Various types of more efficient equipment are now in use. The general principle, however, is the same that is, to wet each individual particle thoroughly with the vehicle and to eliminate flocculated aggregates. Mills in use today include steel roller, ball, pebble, sand, high-speed impeller, Morehouse (a high speed stone type), and Cowles disperser.

Application of paint

Paints can be applied in many different ways. Although most architectural paints are applied with a brush or roller, much paint is now applied by professional painters with air or airless spray equipment. With airless spray equipment, the paint is atomised by forcing it through a very small orifice under very high pressure. Other types of spray application are electrostatic spraying, hot spraying, steam spraying, two-component spraying, and aerosol spraying.

In electrostatic spraying the atomised paint is attracted to the conductive object to be painted by an electrostatic potential between the two. Advantages of this process include efficient use of coating material, rapid application, and relative ease of coating irregular shapes uniformly. Two-component spray equipment consists of two lines leading to the spray gun so that two materials, for example, an epoxy and a catalyst, can be mixed in the gun just before application.

There are several other methods for industrial application of paints. Dip application of coatings is a simple method wherein objects to be coated are suspended and dipped into a large tank containing the paint. This method is often used for undercoating objects where the uniformity and appearance of the paint are not important. Electrodeposition consists of depositing paint on a conductive surface from a water bath containing the paint. The negatively charged paint-component particles are attracted to the object being coated which becomes the anode when an electrical potential is applied. Paint can be applied to very irregular surfaces with very uniform thicknesses and little loss. The system is limited to one coat of limited film thicknesses and the equipment cost is high.

Yet, another method utilises roller-coating machines to apply paint to one or both sides of flat objects, such as metal or fibreboard. The thickness of the coating can be controlled by the clearance between a doctor blade and the applicator rolls. Decorative effects such as wood-grain patterns can be applied with these machines.

Other methods include flow coating, where the paint is allowed to flow over the object being coated, and powder coating, in which paint in dry powder form is fused on the surface of the object being coated.

VARNISHES

The term "varnish" is applied to clear, transparent coating materials that dry by a process comprising evaporation of the solvent, followed by oxidation and polymerisation of the drying oils and resins. Varnishes are a homogeneous mixture of resins, drying oils, driers, and solvents, and they contain no pigment. Varnishes fall into three general classes, *spar varnishes* (exterior), *floor varnishes*, and *furniture finishes*. Typically flow chart for varnish manufacture is shown in Fig. 27.2.

The types of oils and resins and the ratio of oil to resin are the principal factors which determine the properties of a varnish. The selection depends on the compatibility of different oils and resins and on the intended use of the varnish. It is generally accepted that the oils in the finished coatings contribute to its elasticity, and the resin, to its hardness. In oleoresinous varnishes the ratio of oil to resin is expressed as the number of gallons of oil that are combined with 100 pounds of resin, and it is commonly referred to as the "length" of the varnish. Thus, where 50 gallons of oil are used with 100 pounds of resin, the varnish has a 50-gallon length. Varnishes containing less than 20 gallons of oil per 100 pounds of resin are usually classed as *short-oil* varnishes. A *medium-oil* varnish contains from 20 to 30 gallons of oil, and a *long-oil* varnish is one in which 30 gallons or more are used. Short-oil varnishes dry more rapidly than long-oil varnishes. They are used primarily where hardness, a high degree of impermeability, or resistance to alcohols, alkalies, or acids is desirable and where elasticity in the film is relatively unimportant. Short-oil varnishes are especially suitable where a "rubbed" finish is desired, e.g., on furniture, but they are too brittle for floors for which a medium-oil varnish is best suited.

Varnishes for exterior exposure are usually of a long-oil formulation because of the beneficial effect of the drying oil in providing elasticity and resistance to weathering. Spar varnish, a high-grade exterior varnish formulated originally for use on the wooden spars of ships, is a varnish of this type.

Interior varnishes are often based on linseed oil in combination with ester gum or a maleic-modified-rosin resin. Rosin-modified phenolic resins form varnishes with fairly good water-and alkali-resistance, but for exterior durability a long-oil varnish is required. The most durable spar (wood) varnishes for exterior exposure are usually made from tung oil combined with 100 per cent phenolic resin. The phenolics as a group yield varnishes of relatively dark colour. Terpene-phenolic resins are lighter in colour and lower in cost than the comparable phenolic resins and have very good water- and alkali-resistance but only fair exterior durability. The terpene resins themselves are low in cost and make good light-coloured interior varnishes with good colour retention and good resistance to water and alkali.

Coumarone-indene resins are thermoplastic resins available in a wide range of hardness. Although poor in colour and weathering, they contribute excellent water- and alkali-resistance and good dielectric properties to varnishes.

Petroleum hydrocarbon resins are soluble in drying oils and form varnishes with good water-, alkali-, and alcohol-resistance, but they have only fair colour.

Natural resins, such as, congo, kauri, pontlianak, and batu are often used in varnishes because of their low cost and their ability to impart specific properties such as hardness, gloss, and moisture-resistance.

Alkyd varnishes differ from the conventional type in that the resin is formed in the kettle and co-reacted with the oil.

The principal uses of varnishes are for interior woodwork, floors, and furniture, and outdoors for buildings, furniture, and boats.

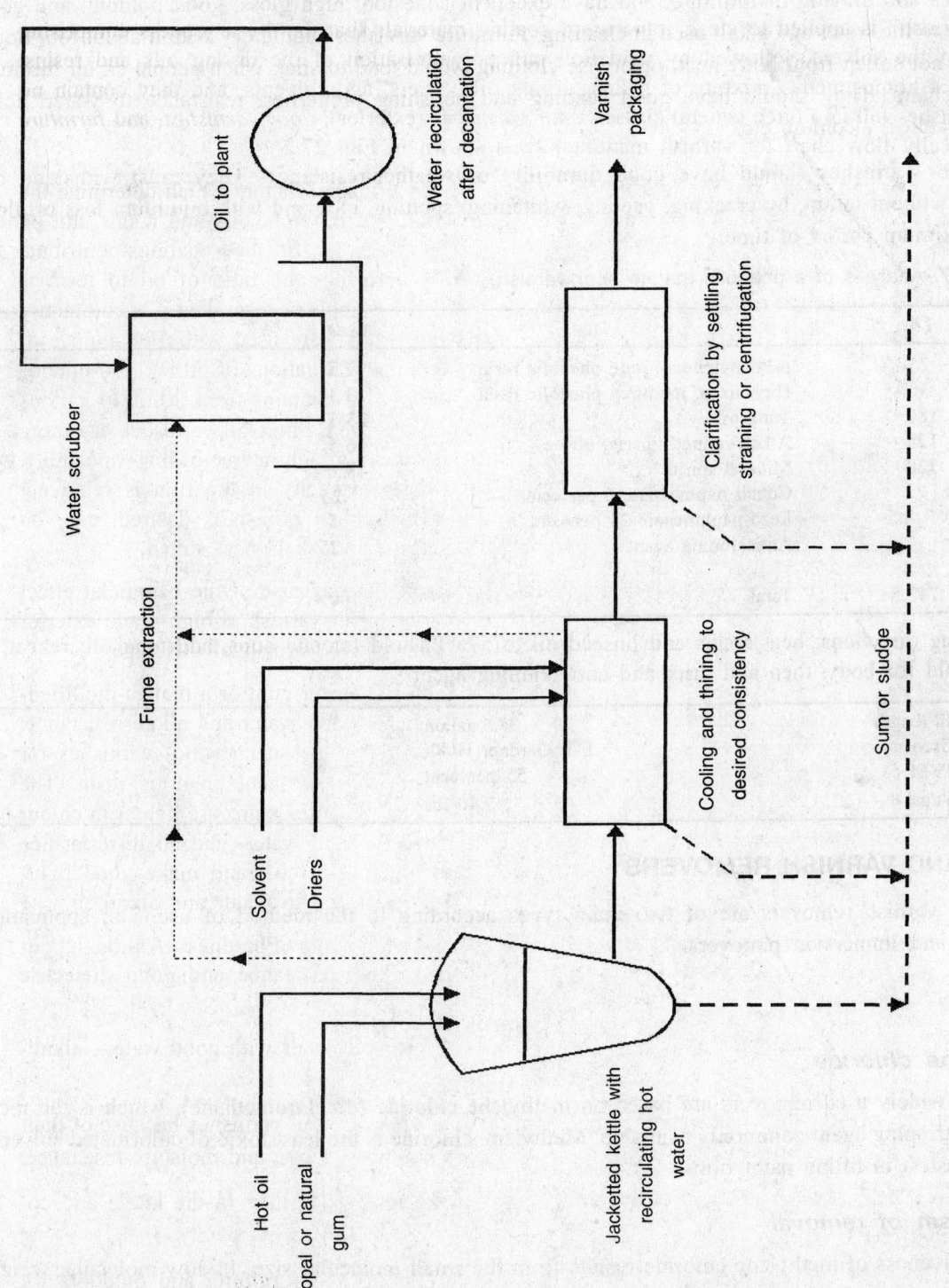

Fig. 27.2. Typical flow chart for varnish manufacture.

A floor varnish should dry to a tack-free finish within four hours. It must be able to stand scuffing from shoes and moving of furniture, and have excellent adhesion, high gloss, good holdout, and good resistance to the water and alkali used in cleaning. Furniture varnishes should dry within about four hours and must not soften from body heat, otherwise clothing would tend to stick when people sit on furniture such as chairs. They should have good sanding and polishing properties, resistance to water, acids (food), alkalies, alcohhol, etc.

Exterior varnishes should have good durability or weather-resistance. They must withstand the elements without failure by cracking, peeling, whitening, spotting, etc., and with minimum loss of gloss for a maximum period of time.

Table 27.7. Analysis of a phenolic marine spar varnish.

Lbs		Gals
75	p-Phenylphenol pure phenolic resin	7.5
30	High m. p. modified phenolic resin	3.3
180	Tung oil	23.1
125	Alkali-refined linseed oil	16.5
328	Mineral spirits	48.0
2	Cobalt naphthenate 6 per cent	.25
4.5	Lead naphthenate 24 per cent	.5
2.0	Antiskinning agent	.25
743.5	Total	100.4

Cooking directions: heat resins and linseed oil to 575°F; hold for one hour; add tung oil; reheat to 450°F; hold for body, then add drier and anti-skinning agent.

Oil length	37.5 gallon
Viscosity	E-F Gardner Holdt
NVM	55 per cent
Wt/gal	7.40 lbs

PAINT AND VARNISH REMOVERS

Paint and varnish removers are of two main types according to the method of use, i.e., application removers and immersion removers.

Organic

Methylene chloride

The most widely used removers are based on methylene chloride (dichloromethane), which is the most versatile stripping agent commonly available. Methylene chloride is the least toxic of chlorinated solvents and the fastest in lifting paint films.

Mechanism of removal

The effectiveness of methylene chloride results from the small molecular size. Its low molecular weight enables it to penetrate rapidly into a coating, and its intermediate solvency enables the coating not to be dissolved so that redeposition on the substrate is avoided. When the methylene chloride has reached the substrate, it swells the film to several times its original volume.

Functions of components

An application remover that is based on methylene chloride has several components, including solvents, e.g., methylene chloride for other chlorinated hydrocarbons; cosolvents; activators and corrosion inhibitors; evaporation retarders; thickeners; emulsifiers; and wetting agents.

Modifications

Methylene chloride removers are modified to increase stripping power for special purposes. The modifying chemicals include amines, alkalies, and organic acids.

Manufacture

Most formulae are made by simple mixing of components. Order of addition is important and care must be taken to incorporate the thickener properly.

Other organic paint and varnish removers

Some finishes are relatively easy to strip and can be dissolved by inexpensive solvents and blends of solvents which may be formulated with thickeners and waxes. Shellac is stripped by alcohols, and lacquers are removed by blends of alcohol and acetates, e.g., butyl acetate. One remover for exterior paints is a slowly evaporating oil-in-water (o/w) emulsion that is based on xylene and dimethylformamide.

Inorganic

Caustic soda

Caustic soda is one of the most common industrial strippers. It is used to clean paint-making and paint-application equipment, including jigs, hangers, and conveyors. Caustic soda baths also are used to salvage ferrous metal parts with defective finishes. Some alkali-resistant coatings are not removed in caustic systems. Caustic baths usually must be heated to ca 93°C to be effective.

Others

Paint can be removed by other alkali systems, inorganic acids, fused alkali, and molten-salt baths.

Choosing a Paint and Varnish Remover

Selection of the best compound or method for industrial or commercial paint and varnish removal can be complicated by many factors. Often the problems are best solved by companies who specialise in paint-stripper supplies and services. Selection of a paint stripper depends on the substrate, type of coating to the stripped, how much attack, if any, is allowed on the substrate, number of coats and primer to be stripped, available equipment, time and temperature limitations, odour and flammability restrictions, and disposal of spent stripper.

Health and safety factors

Many potent removers used in the past are no longer used because of carcinogenic and other toxicological effects. These include benzene, benzene derivatives, phenol, cresols, and some chlorinated and fluorinated hydrocarbons.

Handling and safety

Any chemical or formulation that blisters an organic coating film is likely to blister human skin. When working with ordinary methylene chloride removers, one should wear protective clothing and safety glasses. Good ventilation is necessary; established inhalation standards are associated with many of the chemicals and solvents.

Environment

Fume emission must be considered in industrial stripping operations. Emission standards have been established, for many of the solvents used in paint removers, by state and local governments. Disposal of liquid waste and sludge is a serious problem. Minimising quantities by reusing filtered or decanted stripper solutions is a partial solution.

LACQUERS

A lacquer is a protective coating which dries by evaporation of volatile components. The film-forming constituent is usually a cellulosic ester, i.e., nitrate, acetate, acetate-butyrate, or other high molecular weight polymer. Other types of lacquers are based on acrylics, polyurethanes, vinyls, etc. There are several differences between varnishes and lacquers. As stated previously, lacquers dry essentially by evaporation, while varnishes dry by a combination of oxidation and polymerisation. Lacquers are characterised by very rapid drying and distinctive odour. They are usually based on high molecular weight polymers which require low-boiling solvents of high solvency power such as alcohols, ketones, and esters. The film will redissolve in the original solvent. Lacquers are usually available in a solids range of 20–30 per cent, while varnishes have a solids range of 45–55 per cent. Fig. 27.3 highlights the manufacture of lacquers.

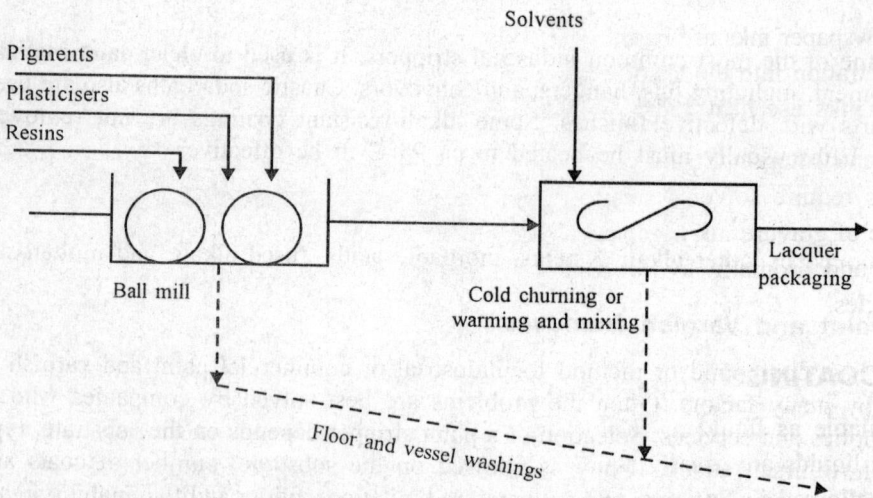

Fig. 27.3. Typical flow chart for lacquer manufacture.

Lacquers dry "tack-free" in 5 to 15 minutes and to a firm film in 30 minutes to 4 hours. They are usually applied by spraying, brush application often being impractical because of the very rapid drying. Pigmented lacquers predominate in use; clear lacquers are used where colourless, tough films are desired. Clear lacquers are not generally considered as durable as high grade varnishes on exposure to sunlight and moisture.

Nitrocellulose, which is one of the principal film-formers, will be considered in detail.

In addition to nitrated cotton, a nitrocellulose lacquer contains, for example: (i) a solvent mixture which usually contains a ketone, an alcohol, an ester, and frequently an ether-alcohol; (ii) a resin such as an alkyd, a phenolic, or ester gum to increase solids and improve adhesion; (iii) a plasticiser for flexibilising the film; (iv) an inexpensive volatile diluent such as toluene; and (v) a dye or pigment which is omitted if a clear lacquer is desired.

The main outlets for lacquers are automobile finishes, furniture finishes, metal finishes, and plastic, rubber, paper, and textile finishes.

PRINTING INKS

Printing inks are a mixture of colouring matter dispersed or dissolved in a vehicle or carrier, which forms a fluid or paste which can be printed on a substrate and then dried. The colourants used are generally pigments, toners, and dyes or a combination of these materials. The vehicle acts as a carrier for the colourant during the printing operation and binds the colourant to the substrate. Printing inks are applied to many different substrates such as, paper, paperboard, metal sheets, metallic foil, plastic film, moulded plastic parts, textiles, and glass.

There are four major classes of printing inks, which vary considerably in physical appearance, composition, method of application, and drying mechanism. They are letterpress and lithographic (litho) inks, commonly called oil inks or paste inks, and flexographic (flexo) and rotogravure (gravure) inks, which are referred to as solvent inks.

Letterpress newspaper inks are based mainly on mineral oil which is sometimes combined with rosin. Drying is by penetration into the absorbent paper stock. Lamp black is used as the pigment for the black ink. Lithographic inks use heat-bodied oils and alkyds as the vehicles. They are heat-dried in ovens.

Flexographic and rotogravure inks are low viscosity materials and contain highly volatile solvents. Flexographic inks require solvents such as alcohols that do not attack the rubber rollers and printing plates. In the case of gravure ink no contact with rubber is involved, and it is permissable to use solvents such as ketones and aromatic hydrocarbons. Flexographic inks use nitrocellulose, polyamides, and acrylates as vehicles.

INDUSTRIAL COATINGS

Coatings are available as liquid or fusible compositions. Table 10.8 gives a classification of coatings formulations. The liquids are usually aqueous or organic solutions. The coatings are applied by the user to the substrates, allowed to flow out smoothly, chiefly by forces of surface tension, and then cured to the final solid form. An industrial organic coating usually comprises an organic binder, pigments, a carrier liquid (sometimes omitted), and various additives.

The binder exists in the final film as a polymer of high molecular weight that may or may not be cross-linked. It is primarily responsible for the plastic quality of the film. Binders are grouped into certain overlapping classes such as acrylic, vinyl, alkyd, polyester, etc. The molecular structure of the binder and the forces operating between the molecules largely determine the mechanical properties.

The pigments, which may be organic or inorganic, contribute primarily to opacity and colour, in addition to durability, hardness, adhesion, and particular rheological properties of the coating in fluid form. Coatings are classified as primers that are applied directly to a substrate or as topcoats. The latter are applied over a primer and are usually the last coat. Intermediate coats are called sealers. In some uses, one coat satisfies all requirements.

Table 27.8. Formulation possibilities of synthetic resins in coatings formulation.

Vehicle system	Uses
Class 1: Vehicles containing oil-modified alkyds or other polymers containing drying oil	
(1) Oxidising alkyd resins (sometimes mixed with oleoresinous varnishes)	Architectural enamels, house paints, interior paints, flat wall paints, baking and airdrying undercoats and enamels for machinery, prefab housing structural units, and other factory products.
(2) Alkyd and phenoplast Alkyd and nitrocellulose Alkyd and chlorinated rubber Alkyd and polystyrene Alkyd and diisocyanate Alkyd and vinyl and epoxy	Air-drying or low-temperature baking under coats and enamels for metals products) that have more plasticlike film properties than to possible to attain with alkyds alone
(3) Alkyd and aminoplast Alkyd and aminoplast and epoxy Alkyd and silicone	Similar uses as above, but where a high premium is placed on colour retention, and superior chemical and heat resistance
(4) Oil-modified epoxy resins and aminoplast.	Air-drying or baking-type undercoats or enamels; improved baking enamels and undercoats
Class 2: Vehicle systems containing no alkyd or drying oil	
(5) Vinyl acetals and/or phenolic allylaminoplast and 2,4,6-trimethylolphenyl alkyd ether epoxy	Chemically resistant baking under- coats and enamels
(6) Phenoplasts (with or without epoxy, or vinylacetal, or aminoplast)	Thermosetting undercoats and/or enamels for high corrosion protection, especially in thin films (not resistant to discoloration); room-temperatu--re setting mastics for corrosion and abrasion protection
(7) Polyester and triazine resin allyl polyester silicone thermosetting acrylics complex amino resins some other polyesters	Chemical and discoloration resistant glossy, clear films and pigmented baking enamels for metallic and non-metallic production goods (thermosetting filmformers)
(8) Vinyl acetate–chloride, copolymers vinylidene or vinyl chloride-acrylonitrile copolymers butadiene copolymers acrylic copolymers poly(vinyl acetate)	Thermoplastic lacquers, baking or air-drying emulsion paints for production finishes on non-metallic goods such as acoustical board or molded plastics, fire-retardant and corrosion-protective mastics; exterior house paints and interior decorators' paints of emulsion type
(9) Nylons some cellulosic esters and ethers electrical insulation poly(vinyl acetals) saturated polyesters unsaturated polyesters and styrene epoxy and polyamide copolymers of ethylene or propylene	Special type of coatings, potting compounds mastics, etc. for polyurethanes polytetrafluoroethylene and corrosion production.

Industrial coatings are generally applied with specialised machinery representing a large investment. Coating properties must be controlled carefully to ensure continuous and satisfactory application. Properties to be controlled include rheological characteristics, stability, colour, smoothness or gloss, metallic lustre, durability, adhesion, permeability, hardness, flexibility, and protection of the substrate against deterioration.

MARINE COATINGS

Ships, offshore working platforms, and onshore waterfront structures are damaged by contact with the harsh marine environment. Control of this destructive action is best achieved through a programme of: (i) selection of the materials most resistant to deterioration; (ii) design to minimise conditions favourable to corrosion; and (iii) effective utilisation of protective coatings and/or cathodic protection (an electrical method of preventing metal corrosion in a conductive medium by placing a charge on the item to be protected) to deter corrosion. Protective (anti-corrosive) coatings impart protection to the substrate by forming a barrier to the water, salt, and oxygen which accelerate corrosion. Antifouling paints control the attachment and growth of marine fouling organisms on immersed areas.

The modern synthetic coatings have three common ingredients: solvent, binder, and pigment. Coatings that cure by chemical reaction of two component parts are the most widely used in submerged marine applications. For example, epoxies, coal-tar epoxies, urethanes, and polyesters are durable and resistant to water, solvents, and chemicals.

Modern synthetic marine coatings provide much longer protection than earlier ones, but require both complete cleanliness and surface profile (tooth) in order to obtain adequate bonding of prime coats. High speed abrasive (sand, grit, or shot) blasting by conventional air-pressured equipment or by newer equipment that utilises centrifugal force to propel the abrasive helps provide an appropriate surface. Actions taken by personnel to eliminate particulate emissions during abrasive blasting include the use of hard, sharp, and properly sized abrasives that produce minimum emission of particulates, blasting inside a building or under a temporary shroud, and using equipment that automatically moves across a regular surface and picks up and recycles the spent abrasive.

Manufacturers of marine coatings always have printed information available on the use of their products. This information includes recommendations on the equipment to be used, mixing of components, time and temperature requirements, coverage rate at a recommended dry-film thickness, and good application practices.

Protective (anticorrosive) coatings include vinyls, chlorinated rubbers, epoxides, coal-tar epoxies, urethanes, polyesters, inorganic zincs, zinc-rich organics, and specialised coatings (e.g., solvent-free epoxies, powder coatings, plastic-coated steel electrical conduits and fittings, and petrolatum-coated tapes).

Currently, the only two biocidal materials that are used extensively in antifouling paints are cuprous oxide and organotin compounds (usually tributyltin oxide or tributyltin fluoride). Most commercial antifouling paints use a vinyl binder, although products with other binders are also marketed. Rosin or some other leaching agent must be added to cuprous oxide formulations to permit its controlled release into sea-water where it is lethal to fouling larva forms. Organotins usually do not require leaching agents to dissolve slowly in seawater.

Recent research has lead to the development of organometallic polymers for use in antifouling paints. A sheet material of black neoprene rubber impregnated with tributyltin is currently marketed. Because the sheet is so much thicker than an antifouling coating system (usually about 100 μm) it has a larger reservoir of biocide that can result in longer-lasting fouling control.

RESISTANT COATINGS

Resistant or high performance coatings or linings are specialty products used to give long-term protection under difficult corrosive conditions to industrial structures. This contrasts with paint, which is used for general appearance and shorter-term protection against milder atmospheric conditions, and industrial coatings, i.e., coatings that are applied to manufactured products.

There are two basic methods by which coatings or linings protect the surface. The first is based on the principle of impermeability. The second method uses anodically active or inhibitive pigments in the primer or in the coating to regulate corrosion. The inert impervious system performs best as a lining where it is subject to continual moist, wet, or immersion conditions and where it is subject to little or no physical abrasion. Inhibitive coatings perform best in areas where the coating is subject to weathering, atmospheric conditions, high humidities, or chemical fumes.

Components of Resistant Coatings and Linings

The similarity to paint extends to the essential ingredients of the coating or linings which include the binder, colour-carrying pigments, inert and reinforcing pigments, inhibitive pigments for primer only, and solvents. In general, the same equipment and manufacturing procedures are used for paint or industrial coatings.

Application of Resistant Coating Systems

The characteristics provided by each part of a coating system are as follows :

(1) *Primer:* Good adhesion to surface, satisfactory bonding surface for next coat, ability to retard the spread of corrosion, enough chemical and weather resistance to protect the surface until application of next coat, chemical resistance equivalent to the remainder of the system.

(2) *Intermediate coats:* Adequate film thickness for the coating or lining system, uniform bond between the primer and the topcoats, a superior barrier to moisture vapour and aggressive chemicals.

(3) *Finish coats:* Must be pleasing in appearance as well as resistant, may serve to provide a non-skid surface, a matrix for antifouling agents or other specialised purposes, must have sufficient weather, chemical, and abrasion resistance in any environment to ensure its remaining intact and providing protection to the substrate. Proper surface preparation and coating or lining application are essential to the performance of the resistant coating.

Resistant coatings or linings represented about 9% of the overall paint or coatings market in 1983. Three health and safety factors are important to consider both during manufacture and application of protective coatings: air pollution, health hazards to the individual and fire and explosion hazards.

New Coating Developments

As a consequence of the new and increasing governmental environmental standards as well as health and safety requirements, there are three avenues of development that will have a strong influence on the future of protective coatings: the use of water-based coatings; the use of solvent-free coatings or 100% solids coatings; and the use of inorganic materials as coatings, e.g. organic zinc coatings.

Chapter 28

Silicate Industry: Ceramic, Refractories, Cement and Glass

INTRODUCTION

The technology of silicate processes is concerned with the production of minerals or their mixtures, wares made from them, and also of different kinds of glass and glass ware. The materials and wares manufactured in the silicate industry possess diverse and valuable properties. The properties typical of most silicates result from the specific structure of their molecules, the basic structural element being the tetrahedric group SiO_4^{4-}. A characteristic feature of this structure is the high strength of the bond between the Si^{4+} and O^{2-} ions, as a result of which most silicates are extremely hard and have a high melting point. Other properties common to most silicates are chemical stability, resistance to high temperatures, and also comparatively low cost, due to the availability of the raw materials.

The raw materials for the silicate industry are very widespread in nature, where they are found in the form of deposits of such common minerals as clays, marls, limestone, chalk, dolomite, quartz sand, tuff, tripoli, quartzite, feldspars, nepheline, etc. Besides naturally occurring raw materials, various synthetic and artificial raw materials are used in manufacturing silicates—soda ash, borax, sodium sulphate, oxides and other compounds of various metals, etc. Waste products of ferrous and non-ferrous metallurgy and of a number of chemical industries–blast-furnace slag, shale cinder, nepheline sludge of alumina-manufacturing processes, etc.—can be utilised as raw materials for the silicate industry. At present, silicate technology covers the production of many minerals and products which do not contain silicon dioxide and its compounds, for instance, the manufacture of high-refractory oxides and wares from them—cermets, magnesia-, chrome-magnesia-and graphite-refractories, air-binding materials (gypsum, lime). Production of these materials and products is nominally classed as belonging to silicate technology due to the use of manufacturing methods similar to those employed in making true silicates.

The silicate industry produces extremely important products, and the variety of the different products is tremendous. The silicate industry has several independent branches, of which the most important are the manufacture of ceramics and refractories, cements and plasters, glass and pyrocerams.

Ceramics is the name given to ware made from ceramic bodies (mixtures with various amounts of moisture in them which, as a rule, contain clay) by forming of casting, followed by drying and firing

719

to the sintering point. According to their main fields of use ceramic materials and ware can be divided into the following basic groups:

1. *Structural ceramics:* Articles used mainly in constructing buildings and various other structures. Building brick–common brick or hollow tile, brick blocks, roof tile, drain tiles, rock goods (clinker brick, ceramic slabs for floors, sewer pipe etc.) belong to this group.

2. *Facing material:* Articles used for internal and external facing of building and structures–facing bricks and slabs and oven tiles.

3. *Refractories:* Materials which retain their mechanical properties at high temperatures (above 1000°C) and which are used in making various parts of industrial furnaces, ovens and apparatus for operation at high temperatures.

4. *Fine ceramics:* A group which includes wares, mainly porcelain and glazed pottery, used in various fields–domestically (dishes, wash basins, sinks, decorative articles), in the electrical industry (electrotechnical porcelain), in the laboratory (chemical ware and apparatus), etc.

5. *Special types of ceramics:* A group of articles with specific properties utilised in the radio industry, aviation, instrument manufactures, etc.

Commercially ceramics are classed as rough wares, which include building materials and refractories, made from coarse-grained ceramic dough and which have a porous biscuit of a non-uniform structure, and fine wares, which include sintered or fine-pore items with a uniform structure of the biscuit–porcelain, glazed pottery and special ceramics.

REFRACTORIES

The word refractory implies resistant to melting or fusion. In technology, refractory refers to those materials which are used to withstand the effect of thermal, chemical and physical effects that are not met with in furnace procedures. Whenever very high temperatures are involved, for example, in furnaces, kilns and electrical heating apparatus, including resistances, the refractories provide the linings, supports and other filaments. Thus refractories are those materials which are used for the construction of furnaces, kilns, ovens, crucibles and retorts etc., on account of their resistance to heat, when they are subjected to the cutting action of flue gases, sudden changes in temperature and to influence of slages. Refractories are sold in the form of the bricks, silica magnesite, chromite, magnesite, chromite bricks, silicon carbide and zirconia refractories etc. The fluxes required for binding together the particles of the refractories are kept at a minimum in order to reduce vitrification.

Cement is the general term given to the powdered materials which initially have plastic flow when mixed with water, or other liquid, but have the property of setting to a hard solid structure in several hours with varying degree of strength and bonding properties. The latter continue to improve with age.

Portland cement is a greenish-grey impalpable powder. Its essential constituents are lime, silica and alumina which are combined to form tricalcic silicate, $3CaO.SiO_2$; tricalcium aluminate, $3CaO.Al_2O_3$ and dicalcium silicate, $2CaO.SiO_2$; these are unstable compounds, which on being wetted rearrange with different speeds. Tricalcium silicate acts rapidly, forming gelatinous calcium hydrate and gelatinous silica, and to this change is due the initial set which occurs in three hours; the hydration continues, the gelatinous material binding the grains of sand which are always added, and the crushed stone filler, to a hard mass. Tricalcium aluminate acts with the same rapidity as tricalcium silicate, but does not produce a strong bond. Dicalcium silicate acts only after months have elapsed. The hardening of Portland Cement continues for years and the concrete made form it increases in strength. As time passes, the gelatinous calcium hydrate crystallises adding further strength.

In mixing water with Portland Cement to give concrete, 4 to 8 gallons of water are used per bag of cement; not more then 2½ gallons will chemically combine with each bag of cement, to become part of its structures. The rest evaporates. Too rapid evaporation is undesirable; it is retarded by keeping the surface covered with straw and by wetting it (which is termed 'curing'). The purpose of wetting is to prevent evaporation until after hydration has proceeded well along to completion.

The setting of Portland Cement is so rapid that if uncontrolled it would be useless. The addition of a small percentage of $CaSO_4$ (as gypsum) gives the desired retardation. The initial set of pure cement occurs in 6 minutes. Three per cent of gypsum added lengthens the period to 3 hours.

In the manufacture of Portland Cement the proportions of lime (as limestone), alumina and silica are carefully adjusted and the mixture is sintered. Considerable cement is now-a-days made from iron Blast Furnace slag. Portland Cement sets under fresh water as well as in air. It is, therefore, also a 'hydraulic Cement'. For sea-water, the iron content should be raised to several parts per hundred.

GLASS

Any substance which has solidified from the liquid state without crystallisation is known as glass. Generally, however, the term glass refers to the product that is obtained by the fusion of mixture of silica, basic oxides and a few other compounds. Sometimes glass is called a 'super-cooled' liquid as its ingredients can not be identified.

CERAMIC

Basic Raw Materials for the Ceramic Industries

The three main raw materials for the ceramic industry are clay, felspar and sand. Clays are more or less impure hydrated aluminium silicate that have resulted from the weathering of igneous rocks in which felspar was the principal original material. Clays include kaolinite (Al_2O_3, $2SiO_2$, $2H_2O$); montmorillonite (MgCa)O. $Al_2O_3.5SiO_2$. nH_2O and illite (K_2O. MgO. Al_2O_3. H_2O). Clays are plastic and mouldable when finely pulverised and wet, rigid when dry and vitreous when fired at a suitable high temperature. Upon these properties depend the manufacturing procedures in the ceramic industries.

Along with clay minerals are varying amounts of felspar, quartz and other impurities such as iron oxide. But the basic clay used is kaolin or bentonite. The property of plasticity and workability of clay is influenced mostly by physical conditions of the clay and varies greatly among the different types. Clays are chosen according to the requirements of particular products and are often blended.

There are three types of felspar—potash $K_2O.Al_2O_3.6SiO_2$; soda $Na_2O.Al_2O_3.6SiO_2$ and lime $CaO.Al_2O_3.6SiO_2$—all of which are used in the ceramic industries. The third main constituent of ceramics is sand or flint. For light coloured ceramic products, sand with a low iron content is used.

Besides the above, a large variety of inorganic materials are used as fluxing agents and special refractory ingredients. Common fluxing agents which lower the temperature are: borax, boric acid, soda ash, sodium nitrate, pearl ash (potassium carbonate), calcined bones, apatite, fluorspar, cryolite, iron, antimony and lead oxides, lithium and barium minerals.

The common special refractory ingredients used are: alumina, olivine, chromite, aluminium silicates, magnesite, lime, zirconium and titanium oxides, carborundum, dolomite, thorium oxide.

An actual ceramic body contains many more ingredients than clay itself. Hence the chemical reactions are more involved and there will be other chemical species besides mullite and christobalite present in the final product. For example, various silicates and aluminates of Ca, Mg and possibly the alkali metals may be present. However, the alkali portion of the felspar and most of the fluxing agents become a part of glassy or vitreous phase of ceramic body. All ceramic bodies undergo a certain amount of vitrefication or glass formation during heating and the degree of vitrefication depends upon the relative amounts of refractory and fluxing oxides in the composition and the final temperature of heating. The vitreous phase gives desirable properties to some ceramic body, e.g., acting as a bond and imparting translucent properties in chinaware. Even in refractories some vitrefication is desirable to act as a bond but extensive vitrefication destroys the refractory property.

The degree of vitrefication or the progressive reduction in porosity, provides the basis for a useful classification of ceramic products as follows:

1. *Whitewares:* Varying amounts of fluxes; heat at moderately high temperature; varying degree of vitrification.
2. *Heavy-clay products:* Abundant fluxes; heat at low temperatures; little vitrefication.
3. *Refractories:* Little fluxes; heat at high temperatures; little vitrification.
4. *Enamels:* Very abundant fluxes; heat at moderate temperature; complete vitrefication.
5. *Glass:* Moderate fluxes; heat at high temperatures; complete vitrification.

Different Classes of Wares

Whitewares

Whiteware is a generic term for ceramic products which are usually white and of fine texture. These are based on selected grades of clay together with varying amounts of fluxes, and heated to a moderate temperature (1200–1500°C) in a kiln. Because of the differing amounts and kinds of fluxes, there is a corresponding variation in the degree of vitrefication among the whitewares. These may be broadly defined as:

1. Earthenware, sometimes called semi-vitreous dinnerware, is porous, non-translucent with a soft glaze.
2. Chinaware is a vitrefied translucent ware with a medium glaze which resists abrasion to a degree. It is used for non-technical purposes.
3. Porcelain is a vitrefied translucent ware with a hard glaze which resists abrasion to the maximum degree. It includes chemical, insulating and dental porcelain.
4. Stonewares, one of the oldest ceramic wares, were used long before porcelain was developed. In fact, it may be regarded as crude porcelain, the raw material being of poorer grade and not so carefully fabricated.
5. Sanitaryware, formerly made from clay, was usually porous. Hence a vitreous composition is presently used. Prefired and sized vitreous grog is sometimes included with the triaxial composition.
6. Whiteware tiles available in a number of special types are generally classified as floor tiles, resistant to abrasion and impervious to stain penetration, glazed or unglazed and like wall tiles (in variety of colours and textures) which also have a hard permanent surface.

General Properties of Ceramics

1. *Chemical and physical properties:* The components present in ceramics, such as oxides, carbides etc. give high chemical stability to ceramics. Most of the constituent oxides are usually resistant to highly oxidising and reducing atmospheres and also to fluctuations in temperatures. Compactness of the crystal structure, high directional character of chemical binding and high field strengths of small cations of high charge are also responsible for the stability of the ceramics.

2. *Optical properties:* Ceramics are opaque as well as transparent. However, most of the ceramics are transparent over a wide range of wave length regions of the electromagnetic radiations. Optical transmission or absorbance is due to the interaction of the electromagnetic field of the incident beam with the polarisable electron fields of the constituent atoms or ions present in the crystal lattice of the ceramics. The optical properties of ceramics depend upon its composition and crystal structure, because extent of polarisation mainly depends upon or controlled by ionic size, bonding energy and crystallographic direction. It should be noted that isotropic crystals exhibit identical properties in all the directions, but anisotropic crystals, which have higher refractive index, exhibit properties only along the closed packed crystallographic directions. The presence of vacant electronic energy band impurities, with donor and acceptor sites also contribute to the absorption in the visible region. These impurities cause colour centres in a normally colourless material because of non-stoichiometry in the cation anion ratio.

   The translucency or opacity in a ceramic body is due to scattering of light resulting due to difference in the refraction at the boundaries of a polyphase polycrystalline material.

3. *Mechanical properties:* Ceramics are brittle solids which are very resistant to compression. The strength of ceramics is mainly controlled by following important factors: (i) temperature (ii) size and shape (iii) composition (iv) surface conditions, and (v) micro structure.

4. *Electrical and magnetic properties:* Oxide ceramics are generally bad conductors or insulators in their normal oxidation states. The non-oxide ceramics, however, act as semiconductors. Ceramics containing transition metal ions have also been found to show magnetic properties, because of spins associated with unpaired electrons. Magnetic ceramics are those which contain molecules with odd number of electrons and incompletely filled 3d, 4f and 5f ions. Unlike magnetic metals, magnetic ceramics are bad conductor of electricity and hence do not conduct electricity. Magnetic ceramics respond to magnetic field, known as magnetic susceptibility.

Permeable (porous) and impermeable (non-porous) wares

It is well evident from the above classification that a more technical classification is to separate ceramic products into permeable, (porous) and impermeable (dense, non-porous, vitrefied) ware. Since glazes are always impermeable, this classification is only applicable to the body. Permeable (porous) pottery absorbs water and shows a rough fracture and sticks to the tongue when tasted in this fashion. Heavy clay ware, refractory ware, teracotta and earthen ware are the examples of permeable (porous) ware. Impermeable (non-porous) ceramic materials absorb very little or no water. Stone ware, vitreous china (similar to stone ware but its colour is white), china and porcelain are the examples of impermeable ceramic materials.

Manufacturing Process

Both glazed and unglazed products are manufactured. The glazed product is obtained by applying glaze on the unglazed body either before or after firing. It is, therefore, clear that in all cases the unglazed product has to be manufactured. Following steps are usually involved in the manufacture of potteries.

Grinding of raw material

Raw materials other than clay are ground finely before mixing to get uniform mixture of clay and other raw materials. There are various methods of grinding or disintegrating the raw materials. Weathering is an old process and mainly applied to clays. They are left in the open air and are exposed to sun, rain and froast. They expand, contract and break up. Calcination (heat treatment) is used to make flint more brittle and more easily crushable. Mechanical crushing machines, such as stone crushers or jaw crushers are used for breaking up large pieces of hard materials. These are subsequently reduced to finer sizes by treatment in pan mills, edge roller mills, dust mills and alsing cylinders (ball or pebble mills). Edge runner mills are frequently used for crushing calcined flint, sand glass and frits. Edge runners are of two types. In the first types, the pan is stationary and the rolls rotate around a vertical shaft. This type is known as chaser mill. In the second type, the pan rotate and the shaft and rolls are stationary. Crushing rolls are used in clay working for breaking up lumps of clay and for disintegrating hard material embedded in clays. All such machines (described above) are suitable for grinding non-plastic dry and wet materials and for mixing and grinding clays and clay mixtures in plastic and semiplastic states. These mills are therefore extensively used in the manufacture of bricks, roofing tiles and refractory materials. Alsing cylinders or ball mills are used for uniform and thorough reduction of unplastic materials grinding.

Mixing or preparation of bodies

After the raw materials have been broken up into small pieces and then purified (if necessary) the next stage is mixing. The raw materials are mixed in definite proportions so as to give the fired body the properties required, and also to give the clay mixture the right consistency for the particular shaping process, which is to be used. After grinding, hard material is mixed with the required amount clay and water in a mixing tank called blunger.

Mixing can, therefore, be achieved by the following methods: (i) the clay is manipulated in a plastic state, either soft or stiff; (ii) the clay is mixed in a dry state and necessary amount of water added after mixing; and (iii) dissolving the clay in water, then mixing the clay slip with dry or wet (dissolved in water) pulverised non-plastic material and removing the water in filter presses. The use of plastic clay (method i) is least costly, because the clay used in the mixing process has the rigid consistency for the shaping process. (Method ii) is useful where a fine texture and uniformity of the body is essential, e.g., in manufacturing of high grade stone ware and refractory materials. (Method iii) is suitable where a fine texture and an intimate mixture is essential, e.g., in the manufacture of earthen ware, porcelain and china.

Body preparation using clay in plastic state

If the clays has the rigid consistency, they can be placed directly in the mixing machine in the required proportions. If clay is too dry, water is added and thoroughly mixed with the clay. The process is called tempering. If the material is too moist, it is mixed with dry powdered clay or non-plastic material by making use of edge runner mills. Further quantities of water are added either in trough mixers or pug mills, in order to mix the materials in plastic or semi-plastic state.

Body preparation using dry clay

Thorough mixing can, however, be achieved if the plastic materials are first dried and then crushed. The drying of the clay may be carried out in various types of clay dryers, such as rotary dryers, tunnel dryers etc. Grinding of the dried clay is carried out by various machines already described. The ground material is shifted and coarse particles which do not pass the sieve are returned to the mill for further grinding.

The mixing of various materials is done either in alsing cylinders, dust mills or in rotating propeller machines contained in cylinders or arranged on pans. After complete mixing, the mixture is usually moistened to give a sufficient plasticity for subsequent pressing.

Body preparation using clay slip

Filtering

The mixed mass, called the slip is pumped into filter presses in order to remove excess of water. The filter press is used for turning the water slip into a plastic body. Continuous filters, such as vacuum filters or rotary vacuum filters are used. The rotary vacuum filter consists of a cloth covered drum suspended in a tank containing the material to be filtered in slip f rm. The level of the slip is maintained at a certain height in the tank. The periphery of the drum is divided into a number of compartments. Each compartment is joined through a common valve to a vacuum pump. As the drum rotates, each compartment passes through the slip in the tank and removes water through the cloth, forming a cake. The cake is separated from the drum by a string. It should be noted that vacuum in each compartment can be maintained separately.

Kneading

The filter cake is taken from the filter press or vacuum filters and then put into a pug mill, which is used to mix the materials in plastic or semi-plastic state.

Articles other than round shape are formed in plaster moulds by different methods. The simplest mould is that used for ordinary bricks and consists of a wooden frame. For articles of irregular form plaster of Paris moulds are used. Plaster of Paris absorbs water from the clay and does not adhere to the soft clay body, so that the clay can be easily removed from the body.

Jollying

Jollying is a mechanical process of making a large number of similar articles economically. In this process, the body is formed partly by a plaster of Paris mould and partly by means of a profile tool having a partial outline of the articles to be made. The body in the plastic state is placed into or on a mould of plaster of Paris fixed on a rotating wheel (Jolly) and pressed to shape by profile. Jolly is a device for carrying a tool termed as profile which gives shape to one side of the ware, the other being formed by plaster mould. In the case of a cup. e.g., the outside is formed by the mould and the inside by the tool. Similarly, the face of a plate is shaped by the plaster of Paris mould and the back of the plate by the profile tool. Automatic jollying machines have been developed for cup and plate making.

Slit casting

Slit casting is a process of giving shapes to clay wares from clayey liquid. The slip used for slip casting is prepared from plastic body by adding a small amount of sodium carbonate, sodium silicate or caustic soda. When any of the chemical is mixed with plastic clay, it becomes thin enough to be poured out of the vessel. In other words, clay particles lose their plasticity and turn into a liquid. Usually 0.2–.3 per cent of these chemicals are added enabling the thin clay liquid to get filled in hollow casting mould (made of plaster of Paris) of the article to be made. The inside of the mould will now be found to be covered with a layer of clay, some water having been absorbed by the plaster. In this manner a uniform layer of clayey body is deposited round the inside of the mould. When the inner layer acquires some thickness, extra clayey liquid is poured and if this layer is left for some time in the mould, it will become hard. After drying, whole casting is taken out of the mould.

Pressing

Pressing is the process used for making plastic, semi-plastic and powder bodies. Plastic bodies are used in the manufacture of tiles, roofings, drain pipes etc. and they can be prepared by placing clay inside the moulds made in two halves and then pressing either manually or mechanically. In making semi-plastic body, the plastic body is dried, powdered and mixed with small amounts of water and oil. Oil avoids sticking of the body to dies when pressed hard. For making powdered body, very powerful presses are employed as they do not possess the property of flowing like plastic or semi-plastic bodies. Powdered body is used in the manufacture of walls and floor tiles. Pressing in steel dies may be wet pressing, dry pressing or semi-dry pressing.

Extrusion

This method consists in forcing the plastic paste through a die or mouth piece of the required shape. Vertical and horizontal presses have widely been used for this purpose, but de-airing pug mills are also not uncommon. The shape of the extrusion die corresponds to the cross-section of the rod, tube or bar to be extruded. After extrusion the material is cut into pieces of suitable length. The process of extrusion has been used in the manufacture of bricks, rods, pipes, bars etc. and in the manufacture of porcelain, stone ware and refractory material.

Turning

Turning is carried out on lathes and before turning a ceramic body it is essential that the body should be dry to a reasonably hard state. In case of medium and large articles, the body may be in semi hard state. Small articles, such as sparking plugs are turned in hard state. In pottery manufacture, turning is frequently used after jollying or casting. Thus grooves or under cut profiles can be turned in cups, pots and vases produced by either jollying or casting. In the manufacture of insulators and other technical porcelain, turning is used for producing grooves, cylindrical tubes, bushings etc.

Drying

Articles made by any of the above mentioned methods are dried well before they are found fit for firing. Drying is most essential, because imperfect drying may cause cracking of the article during firing. Apart from the nature of the body used, the speed of drying depends upon the following factors :

1. Shape of the articles
2. Temperature of the surrounding air
3. Rate of the air circulation
4. Humidity content of the surrounding air.

Firing

The action of heat on clay and mixture of clay with other materials is really the basis of the ceramic industry. These substances acquire durability and many other favourable qualities not encountered in other substances, when they are heated at higher temperatures. Finally dried articles are known as blanks. In such cases glaze is put on the blanks before firing, but in most cases blanks are fired to produce bisques or biscuits. In such cases glazed products are obtained by putting glaze on cold biscuits and then firing. The vitrification of ceramic products and the prior chemical conversion of dehydration, oxidation and calcination are carried out in kilns that may be operated either in a continuous or in a periodic manner. Muffle kilns, in which the flames heat the outside of a refractory chamber continuing the article to be fired, are also in use. The flames in these kilns do not enter the chamber. Gas, coal and oil are the most common fuels for firing. In some cases electricity has also been used. Various chemical and physical changes taking place during firing are :

1. Dehydration of adsorbed and chemically combined water. The rate of dehydration is maximum at about 500°C.

2. Oxidation of oxidisable organic matter, sulphur compounds and ferrous compounds between 300°C–900°C.

3. Decomposition of carbonates of calcium and magnesium to form oxides and CO_2.

4. Vitrification or the melting of some of the constituents to a glassy mass so as to cover up and bond the unmelted mass or particles on cooling. During vitrefication, volume shrinkage takes place and extent of this shrinkage is a measure of progress of firing. The degree of shrinkage mainly depends upon temperature (which is between 950–1050°C for common clays) and time required for the process.

5. *Reduction:* It involves colour dilution of the processed article. During reduction, the air supply to the kilns is cut off. As a result, transition metal ions are reduced from higher to lower oxidation states.

6. *Annealing:* It is the slow cooling of the fired article.

7. *Decoration:* Under glaze decoration developed on the ware below a layer of transparent glaze and over glaze is applied over the glaze which may be opaque.

Glazing

Glazing is important in white wares and table wares. A glaze is a thin coating of glass that is melted onto the surface of the more or less porous ceramic material. In other words, glaze is applied on the surface to cover, the pores which are present almost in all classes of ceramics, except hard porcelain. In the case of common clay products obtained at low temperature, the number of pores is large, but they are also present in small number in white wares, obtained at comparatively higher temperature. In the case of hard porcelain, these pores are closed up as a result of virtification to about 1400°C (at which porcelain is produced). A glaze is a fine powder, consisting of a mixture of glass forming materials of proper composition, e.g., lead silicates, borosilicates etc. A glaze contains refractory materials, such as feldspar, silica and china clay and fluxes, such as soda, potash, fluorspar and borax in different proportions. Different combinations of these materials and different temperatures at which they are fired give a wide range of texture and quantity. The glaze may be put on the ware by dipping, spraying, pouring or brushing. In actual practice, glazing may be performed as follows: The glazing ingredients are finely powdered and mixed in definite proportions. The mixture is then mixed with requisite amount of water. In order to get a slip of necessary consistency. The slip is then applied either to the dry blanks or more frequently to the biscuits by dipping, pouring, spraying or brushing with a great care.

The Purpose of glazing is many fold

(i) to produce decorative effect; (ii) to improve the appearance of the article; (iii) to improve the durability of the ceramic article; (iv) to provide a smooth and glossy surface to the glazed material; (v) to protect the article from environmental and atmospheric effects; (vi) to make the surface impervious to liquids.

In fact, glazes are useful, both in the case of porous as well as in dense bodies. In porous bodies the glazes act as seals and prevent penetration of water, while in dense bodies they make the surface smoother and brilliant. A glaze is formed on the surface of the shaped ceramic articles during the firing. Before firing, the glaze covers the surface of the ceramic articles in the form of dry powder. A glaze on the surface of a ceramic article must not get too fluid. If it were fired to the degree of fluidity required for flowing a glass bottle, it would run off the ceramic article. Glazes are the products of reaction between acidic and basic oxides. The glazing mixture free from iron and other colouring pigments forms colourless glaze.

Glazes can be divides into the following three classes, according to the way in which they are prepared: (i) raw glazes, prepared from materials which are insoluble in water; (ii) frit glazes, in these some materials, which are soluble in water are first melted with silica and then mixed with other materials insoluble in water; and (iii) salt glazes, are produced by throwing salt into the fire mouth in the later stages of firing.

Salt glazing and liquid glazing

Glazing can be accomplished either by salt glazing or by liquid glazing. In salt glazing, common salt is used for getting glossy film over the earthen ware. In this process common salt is thrown into the furnace in which the articles to be salt glazed are in red hot condition. The common salt undergoes volatilisation under intensive heat and reacts with silica of the article to form glossy and impervious film of sodium silicate. In liquid glazing, which is much superior then salt glazing, fine powder of glaze mixture and requisite amount of colouring pigments are mixed with water to form a colloidal solution known as slip glaze. The articles to be glazed are then burned in a kiln at low temperature and then taken out and dipped momentarily in the slip glaze.

Frits

The preparation of frit is very important for the composition of glaze suitable for temperatures lower than 1200°C. We have seen that glazes are applied to the surface of ceramic articles in the form of finely powdered particles suspended in water. Various fluxes, such as soda, potash, boric acid and borax are water soluble and important for low temperature glazes. These are converted into a non-water soluble substance by mixing them with silica and heating the mixture at or above its melting point.

Lead oxide is poisonous and hence it should not be used directly in the glaze slip. It is first added to the frit where it forms a water insoluble silicate. Most lead silicates are insoluble in water and dilute acids (such as those formed in the stomach of a human being). Lead should never be used in frits which contain borax or boric acid. Lead should be used as lead bisilicate in cases where the frits contain boric acids, as lead bisilicate is insoluble in water. It is so necessary to add it with mill materials (water insoluble materials, which together with the ground frit, form the glaze) to the ground frit particles.

Decoration

There are two types of decorations under glaze and over glaze. In the case of under glaze decoration the colours and the glaze fuse directly together, while in over glaze decoration, they fuse together with the help of a soft flux. In case of under glaze the decorations are first prepared by first calcining colouring metal oxides with such material as china clay, feldspar, alumina etc. The calcined mixture is then disintegrated to extremely fine powder. The powder is mixed with gum or oils and applied to the biscuit either by a brush or by spraying. In case of china or porcelain, the under glaze decoration is dried at room temperature. The ware, after drying is covered with the glaze as usual and then fired. It should be noted that glaze must react with the colour, but not dissolve the colour. In case of over glaze, the decoration is put on the glaze and then fired. In other words, over glaze colours are applied on top of the glaze and they are prepared by mixing the colouring oxides with the frits of low softening temperature. Coloured glaze can be prepared by the addition of colouring oxides.

Applications of colours to pottery

Colours to the pottery are usually applied by the following important methods: (i) painting: (a) hand painting, (b) lining banding; (ii) spraying; (iii) stencilling; (iv) stamping; (v) printing lithography; (vi) silk screen printing; (vii) ground laying; (viii) sgraffito decoration; and (ix) gilding.

Manufacture of porcelain

Porcelain may be prepared in three ways. The wet process porcelain used for production of fine-grained highly-glazed insulators for high voltage service; the dry process porcelain employed for rapid production of more open-textured quality; low voltage process and the cast porcelain necessary for making of pieces too large and too intricate for the earlier two processes. These three processes are based on same raw materials, the difference in the manufacturing being largely in the drying and forming steps.

The flow chart (Fig. 28.1) given here may be broken into the following sequences. The raw materials of proper proportions and properties to furnish porcelain of desired quality are weighed from overload hoppers into the weighing car.

The felspar, clay and flint are mixed with water in the clay-water mixer and then passed over a magnetic separator, screened and stored.

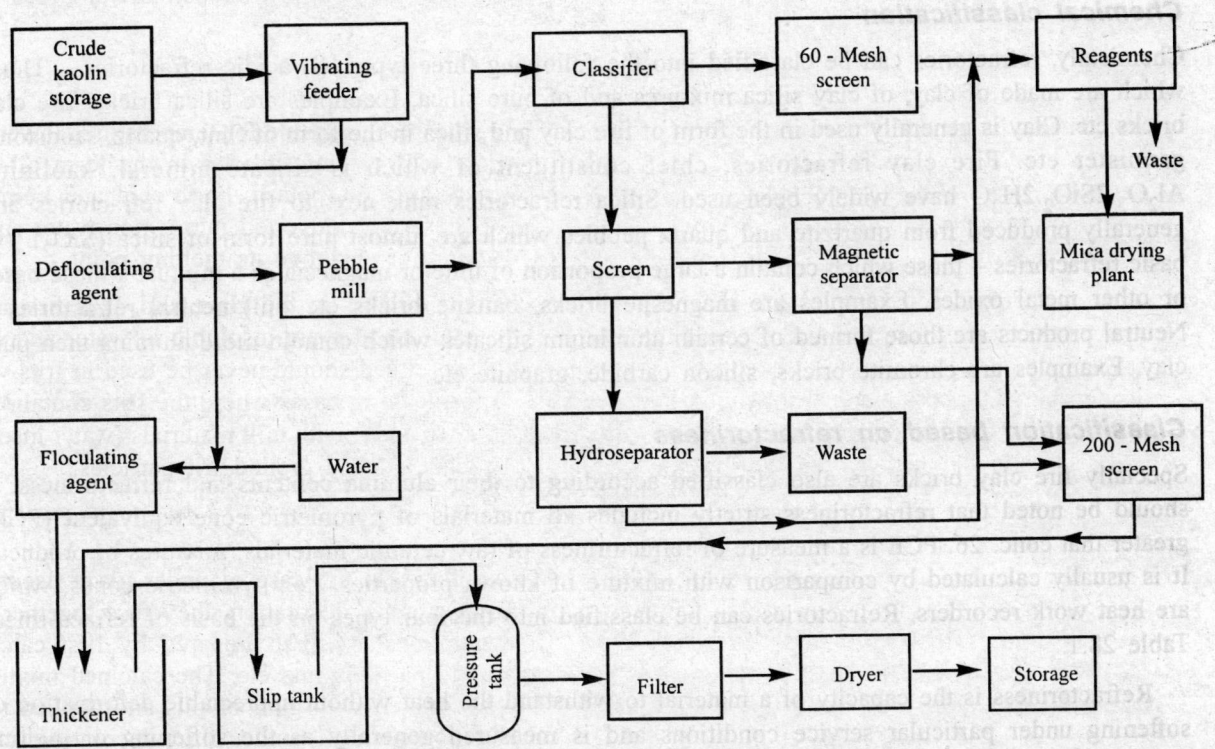

Fig. 28.1. China clay beneficiation

Most of the water is removed in the filter press. All the air is taken out in the pug mill, assisted by vacuum and slicing knives. This results in a dense and stronger porcelain. The prepared clay is formed into block in a hydraulic press or by hot pressing in suitable moulds. The blocks are preliminarily dried, trimmed and finally completely dried, all under carefully controlled conditions. A high surface lustre is secured by glazing with selected materials. The vitrification of the body and the glaze is carried out in tunnel kilns with exact control of temperature movement.

The porcelain articles are protected by being placed in saggers fitted one on top of the other in the cars. This represents a one fire process wherein body and glaze are fired simultaneously. The porcelain pieces are rigidly tested electrically and inspected before storage for sale.

Glazing is important in whitewares and particularly in table wares. A glaze is a thin coating of glass that is melted on to the surface of the more or less porous ceramics ware. It contains ingredients of two distinct types in different proportions: refractory materials, such as felspar, silica and china-clay; and fluxes such as soda, potash, fluorspar and borax. Different combinations of these materials at the temperatures at which they are fired, give a wide range of texture and quality. The glaze must bond with the ware and its co-efficient of expansion must be sufficiently close to that of the ware to avoid defects, such as 'crazing' and 'shivering'. The glaze may be put on by dipping, spraying, pouring and brushing. Decoration of such ware may be 'under glaze' or 'over glaze'. Earthenware should be glazed between 1050°C and 1100°C, stoneware between 1250°C and 1300°C.

Classification of Refractories

Chemical classification

Chemically, refractories can be classified into the following three types. (i) acidic refractories – Those which are made of clay, of clay silica mixtures and of pure silica. Examples are silica bricks, fire clay bricks etc. Clay is generally used in the form of fire clay and silica in the form of flint, quartz, sandstone, gannister etc. Fire clay refractories, chief constituent of which is silicate mineral, kaolinite, $Al_2O_3.2SiO_2.2H_2O$ have widely been used. Silica refractories rank next to fire clay refractories and generally produced from quartzite and quartz pebbles which are almost pure form of silica (SiO_2). (ii) basic refractories – those which contain a large proportion of lime or magnesia or a mixture f these bases or other metal oxides. Examples are magnesite bricks, bauxite bricks etc. (iii) neutral refractories – Neutral products are those formed of certain aluminium silicates which contain more alumina then pure clay. Examples are chromite bricks, silicon carbide, graphite etc.

Classification based on refractoriness

Specially fire clay bricks are also classified according to their alumina contents and refractoriness. It should be noted that refractoriness strictly includes all materials of pyrometric cone equivalent (PCE) greater that conc. 26. PCE is a measure of refractoriness of raw ceramic materials, mixtures or products. It is usually calculated by comparison with mixture of known properties, i.e., pyrometric cones, which are heat work recorders. Refractories can be classified into the four types on the basis of refractoriness Table 28.1.

Refractoriness is the capacity of a material to withstand the heat without appreciable deformation or softening under particular service conditions and is measured generally as the softening or melting temperature of the material. It should be noted that refractory materials generally do not have sharp fusion temperatures, because of the fact that they are mixtures of various metallic oxides and other substances. It is, therefore, common practice to determine softening temperature, rather than fusion temperature. The softening temperature of refractory materials is generally determined by making use of pyrometric cones or seger cones test. It is important to note that a material can only be used as a refractory if its softening temperature is much higher than the operating temperature of the furnace where it has to be used. The inner refractory lining in a furnace is at a much higher temperature than the outer lining. Hence, unless the refractory melts away completely it can usually be regarded to withstand a temperature higher than its softening temperature, since the outer end of refractory lies at a lower temperature and remains in solid state and provides the necessary strength.

Table 28.1. Classification of refractories on the basis of refractoriness.

Cones of P.C.E. 26–28	Moderate heat duty.
Cones of P.C.E. 28–31	Intermediate heat duty.
Cones of P.C.E. 31–33	High duty.
Cones of P.C.E. 33–34	Super duty.

Properties of Refractories

1. *Fusion point:* Since a large number of metallurgical and other operations are performed in industries at very high temperatures using refractories, the refractories must be able to withstand high temperatures and must remain unaffected at the temperature of the chemical process being carried out. Fusion point, which is the temperature at which the refractory starts to soften, is an index of its suitability for a particular process at a given temperature, and determines the temperature upto which a particular refractory can be used. The fusion temperature of the usual refractories in use ranges from 1600–2700°C. Fusion points of refractories are usually found by use of pyrometric cones. Most commercial refractories do not have sharp melting points, because of the fact that they are composed of several different crystalline and amorphous materials.

2. *Chemical properties:* Since refractories are acidic, basic and neutral, it is extremely important to select appropriate refractories to withstand the chemical action of slags, fuel ashes, furnace gases as well as products such as glass or steel. An acid brick should not be used in contact with an alkaline product or vice versa. Chemical reactions of the product for which the refractory is to be used, must first be taken into consideration, before its use. Chemical interaction of a refractory with its environment is very important factor contributing to its slow withering away on continuous use.

3. *Porosity:* Porosity is directly related to many other physical properties of brick, including resistance to chemical attack. The greater the porosity of the brick, the more easily it is penetrated by molten fluxes and gases. For a given class of brick, the one with the lowest porosity has the greatest strength, thermal conductivity and heat capacity. Hence porosity of a refractory is the deciding factor of the degree of penetration by molten fluxes and gases and brings about disintegration. So, as seen above, the greater the porosity, the greater is the susceptibility of the refractory to chemical attack by molten fluxes and gases. In general, a decrease in porosity has the general effect of increase in strength, heat capacity, thermal conductivity and chemical resistance.

4. *Spalling:* Refractories should be able to withstand spalling, e.g., cracking and flaking of the bricks due to uneven expansion on contraction. Spalling is thus the fracture or flaking off a refractory because of uneven expansion of a refractory due to heat. Refractories which expand unevenly by heating usually undergo flaking.

5. *Strength:* They must also be able to withstand abrasion or erosion of the furnace charge and also pressure of the load. Strength is the resistance of the refractory to compressive loads, tension and shear stresses in cold as well as hot working conditions. The strengths of refractories are different for different refractories.

6. *Thermal conductivity:* The thermal conductivity of densest and least porous brick is highest. This is probably due to the absence of air in voids. Insulation is also desirable in some special refractories.

7. *Resistance to rapid temperature changes:* They should also be able to withstand sudden changes in temperature during the introduction of cold charge or sudden rush or cold air in empty furnace.

8. *Heat capacity:* Furnace heat capacity depends upon thermal conductivity, the specific heat and the specific gravity of the refractory. The light weight bricks, which absorb low quantity of heat are used in furnaces operated intermittently. This is due to the fact that working temperature of the furnace can be achieved in less time with fuel.

Manufacture of Refractories

The manufacture of refractories consists of the following important steps. (i) crushing; (ii) grinding; (iii) screening; (iv) mineral dressing; (v) storage; (vi) mixing; (vii) drying; and (viii) firing.

1. *Crushing:* The clays in the form of big lumps are crushed to suitable size in single or double roll crushers, jaw crushers or roll crushers. For crushing harder clays e.g., grog, jaw crushers are most suitable.

2. *Grinding:* After crushing the lumps down to 25 mm in size, the materials are ground in suitable grinding machines. Mechanical crushing machines such as stone crushers or jaw crushers and Alsing cylinders or ball mills are used for this purpose. Hardinge conical mills have also been used.

3. *Screening:* Screening is carried out in order to separate fine particles from coarse material. After screening, the desired size material is passed on to the brick making machine and over size material is recycled to grinding machine, in order to get again the particles of desired size. Shaking screen, closed type screen and air separators are used for this purpose.

4. *Mineral dressing:* In order to produce good refractories, it is most essential that the raw materials should be as pure as possible. So in order to purify the raw materials, materials dressing (which is also known as concentration method), is used.

5. *Storage:* After mineral dressing, the pure raw materials are kept in storage bins.

6. *Mixing:* The function of mixing is the distribution of plastic material so as to coat throughtly the non-plastic constituents. Mixing provides a lubricant during the moulding operation and permits the bonding of the mass with minimum voids.

7. *Moulding:* Since refractory bricks of greater density, strength volume and uniformity are used in large number of factories, the dry pressure method of moulding (in which a high pressure is applied either manually or mechanically) is used. When moulding is done by hand the density and strength of the refractory is not as high as in the case of mechanical moulding. The density as well as the strength of the refractory can be increased by mechanical moulding by de-airing of the refractory. The dry press method is particularly applicable to batches that consist primarily of non-plastic materials.

8. *Drying:* Drying is carried out to remove the moisture from refractories. Drying is performed very slowly under specified conditions of humidity and temperature depending upon the refractory. Rate

of drying should be so maintained that neither voids are left in the refractory nor shrinkage causing in the production of internal stresses takes place.

9. *Firing:* The refractories are fired in order to stabilise and strengthen the structure. Firing or burning is usually carried out in down draught kilns or continuous tunnel kilns. The function of burning or firing is vitrification and development of stable mineral forms. During firing the following changes usually occur: (i) water of hydration no longer exists; (ii) calcination of carbonates takes place; (iii) ferrous ion undergoes oxidation; (iv) shinkage in volume may be as high as 30%; and (v) due to shrinkage great stresses may occur in the refractory. The shrinkage can be avoided by prestabilisation of the materials used and consists of appropriate sizing and pressing.

Some special refractories

Fire clay bricks

Fire clays are the most widely used refractory materials since they are suitable for a variety of applications. Fire clays are hydrated aluminium silicates having the general composition, $Al_2O_3.2SiO_2.2H_2O$. with impurities of sand and gravel, alkalies, oxide, sulphate and sulphide or iron, silicates and carbonates of Ca and Mg and small quantity of TiO_2. Fire clays form acidic refractories and may be of three different types: These are flint or hard clay, plastic soft clays and clays of intermediate character between hard clays and soft clays. In the manufacture of fire clay bricks both flint and plastic clay have been used.

Fire clay goods are composed of fire clays or china clays, with the addition in case of plastic clays, of grog of free silica. Grog is nothing but broken granulated fired refractory clay and is made from rejected fire clay works, broken saggers and crucibles etc. For very plastic clay, the proportion of opening material may be as great as two parts by weight of grog to one part of clay. The use of silica as an opening agent changes the chemical composition and reduces the refractoriness. For best fire clay goods, it is desirable to use only the grog made of fire clay.

Manufacture

Fire clay is dug and then allowed to weather for a long time in order to increase its plasticity. Weathering may also be replaced by de-airing for the same purpose. Clay and grog are now mixed in a pug mill and requisite amount of water is added. In modern factories, clay powder and grogs of different grain sizes are collected in hoppers placed in an upper story so that they come down under gravity into the pug mill or other mixing machine. A more uniform and intimate mixture is obtained, if dry clay is mixed with the grog and desired amount of water is added after mixing. If the grain size of the grog used is large, then the resulting goods are more resistant to sudden temperature changes, but if grog has a very small grain size, the fire clay material is less porous and more resistant to chemical attacks. A compromise is, however, necessary, if refractory with resistance both to temperature changes as well as to the chemical action of materials (with which it may come in contact) is required. Fire clay bricks are then shaped by any of the following methods: (i) hand moulding from plastic mixture; (ii) machine pressing from a coarse damp dust; (iii) extrusion through dies of a plastic material; (iv) tamping of damp dust or plastic clay by rammers into strongly made moulds; (v) casting.

These shaping processes give fire clay refractories for boiler furnaces, blast furnaces, lime kilns, cement kilns and metallurgical furnaces for melting, reheating and heat treatment of iron, steel and non-ferrous metals.

Various types of kilns such as down draught intermittent ovens, continuous chamber kilns, tunnel kilns have been employed for firing of the clay refractories. Usually time required for firing the fire clay refractories varies form six to ten days, according to the kiln used and the type of materials to be fired. The temperature of firing also varies from 1200°C–1400°C and in the case of fire clay goods rich in alumina, the temperature of firing is 1400°C or even more. Cooling process also takes about seven to ten days, according to the type of kiln employed. Extreme care should be taken during cooling in order to avoid cracking of the ware as a result of sudden decrease in temperature.

Properties

The colour of fire clay bricks may be white, cream, light buff, pale buff, slightly brown etc, depending upon the composition of clay, duration and temperature of firing. The composition of fire clay refractories also depends upon the raw materials used. The composition, e.g., of a high grade variety is: SiO_2 = 53.5%, Fe_2O_3 = 2.2%, Al_2O_3 = 41%, CaO = 0.3%, Na_2O = 0.9%, MgO = 0.3%, TiO_2 = 1.6%. Hardness varies according to its firing temperature. Properly fired fire brick is as hard as steel. The fusion point is 1500°C–1750°C. Porosity of fire clay refractories lies between 8% to 24%. At high temperature, fire clay refractories, being acidic refractories, react with alkalies and other bases (such as soda, potash, lime, iron oxide, magnesia etc.) and also with other salts such as carbonates, sulphates and chlorides to form fusible compounds such as aluminates, silicates and alumino silicates. Refractories fired at very high temperatures have been found to have greater resistance to slags, particularly those which contain sodium or potassium carbonate or nitrate. Fire clay refractories are also capable of decomposing CO gas, if present in flue gases, with the deposition of carbon at high temperature usually between 400°–900°C. There are four important changes that take place, when a fire clay refractory is heated. These are baking, vitrefying, softening and fusing.

Uses

The steel industries are the largest consumers of these refractories for the linings of blast furnaces, stoves, open hearths and other furnaces. Some other industries which are using these refractories are lime kilns, pottery kilns, brass and copper furnaces, continuous ceramic and metallurgical kilns, boilers, gas generating sets and glass furnaces. For example, these refractories are used in the construction of reverberatory furnaces for annealing, roasting etc., checker work of regenerative furnace, laddles of different designs and gas producers, lining of flues and stacks, crucible furnace for the melting of brass and silver etc.

High alumina refractories

Alumina occurs in nature as crystalline alumina in the form of corundum and emery or as hydrated alumina such as bauxite, laterite, diaspore ($Al_2O_3.H_2O$) etc. Pure bauxite is the best refractory material, but is rarely used for the manufacture of refractories because it is too costly. Naturally occurring bauxite, having the composition, alumina (Al_2O_3) = 50–90%, silica (SiO_2) 3–25%, iron oxide (Fe_3O_4) = 1–12% and water (H_2O) = 10–30%, si generally used for the manufacture of alumina refractories.

Manufacture

High alumina bricks are made from clay rich in bauxite and diaspore and usually embrace those which contain more than 45% alumina. The alumina content increases the refractoriness and the temperature of incipient vitrefection. Strictly speaking, only those goods which contain more alumina than pure clay can be called high alumina or aluminous refractories and they are practically inert to carbon monoxide and are not disintegrated by natural gas atmosphere upto 1000°C. Aluminous refractory bodies are prepared by adding alumina in some form or the other to a pure fire clay. Bauxite, which is hydrated alumina ($Al_2O_3.H_2O$) is frequently used for this purpose and it increase the refractoriness of the fire clay alongwith increasing the aluminium content. It has to be calcined and ground before being added to the body. The alumina content in bauxite bricks varies from 45–80%. The firing temperature for bricks containing more than 55% alumina is 1400°C or even more. The various binding materials used for uniting the bauxite particles are fire clay, bentonite, silica, magnesite, lime, plaster of Paris, sodium silicate, aluminium nitrate etc. Tar, heavy mineral oils, molten paraffin wax etc. have also been used as binding materials temporarily. There are four important varieties of the bauxite refractories. These are bauxite refractories which do not have any binding material, those which have 10% clay as binding material, those containing about 3% lime and those containing about 50% clay. The latter type of refractories are also known as semi-bauxite refractories.

Properties

Bauxite refractories show a great tendency of shrinkage even at elevated and higher temperatures. Fused bauxite refractories, however, do not show shrinkage and so shrinkage effect can be eliminated by manufacturing the refractory from fused bauxite. Bauxite refractories are basic in nature and so remain unaffected by basic slags, but they are attacked by lime. The heat resisting capacity of bauxite is quite high. They are heat resistant from 1790°C to 1880°C. They can be heated upto 1700–1750°C. When heated under a pressure of 50 lbs per sq. inch, bauxite refractories soften at about 1500°C. They are highly porous refractories and so their durability decreases when they are used in the furnace. The mechanical strength of bauxite refractories is quite high. For example, a bauxite refractory having 25% fire clay softens only at about 1500°C under a pressure load of 50 lbs per. sq. inch. The bauxite refractory has a great resistance to abrasion when bauxite is sufficiently heated before converting it into a refractory.

Uses

Bauxite bricks are more costly to produce than fire clay bricks and they are, therefore, used in special circumstances. They are usually used where very high temperatures are expected or where resistance to basic slag is required. High alumina bricks are, however, employed in the cement industry and paper mill factories. They are also used in the linings of glass furnaces, oil fired furnaces, high pressure oil stills and in generator checkers of blast furnaces.

Silica bricks

Silica bricks contain about 95–96% SiO_2 and about 2% of lime added during grinding to furnish the bond. Silica and a bonding material are thus the two important raw material used for the manufacture of silica bricks. Gannister, a hard, dense, fine grained variety of 90–96% pure silica is a good refractory material. However, silica occurs in nature as amorphous silica (precipitate chalk, flint, opal and hydrated silica, known as geyserite), crystalline silica (quartz, quartzite, sand, sandstone, rock crystal etc) and as cellular silica (kieselguhr (diatomite), tripole, randanite etc. In fact, silica bricks are not used as a general refractory, as the fall into pieces, if they are heated or cooled rapidly. They do not contain clay, but the physical strength of silica bricks, when heated is much higher that those made from clay. As a result, they are very suitable for arches in large furnaces.

Manufacture

The raw material most widely used for the manufacture of silica bricks is quartzite, which is a rock composed of crystals of almost pure silica, bound together by a cement or similar material in such a manner as to produce an almost smooth surface. The binding materials are lime, clay, sodium silicate, alum, aluminium sulphate, magnesia, magnesium silicate plaster of Paris, colloidal silica, calcium phosphate, barium sulphate and even waste products such as tar, molasses, pitch, heavy mineral oils etc. The rocks are reduced to a convenient size by making use of rotary crushers. The weighed amount of the material is then fed into heavy grinding mills in order to reduce quartzite to a proper size for moulding. The power is charged into the mill and requisite amounts of water and lime are added. Water makes the material slightly plastic and lime acts as a cement in the firing process. The mixed charge is then shaped as desired and then placed on plates and fixed in racks to dry. Down draught intermittent kilns and tunnel kilns are usually employed for firing or burning of silica bricks.

Properties

The colour of silica refractory various from white to buff, depending upon the raw materials used. The common variety has the composition, SiO_2 = 96.7%, Al_2O_3 = 0.75%, Fe_2O_3 = 0.5%, CaO = 2% and MgO = 0.5%. Specific gravity varies from 2.3–2.6, depending upon the firing temperature. As silica bricks are acidic refractories, they are not much affected by acidic slags or acids, except hydrofluoric acid and phosphoric acids. Silica bricks are attacked by hydrofluoric acid in cold but by phosphoric and at about 400°C. Thermal conductivity is greater than fire clay bricks, being 0.0013 g. calories per cm per second at 200°C. Per cent porosity is 20–30%. Slightly affected by reducing agents above 1050°C. Fusion point is 1700–1750°C.

The most important characteristics of a silica refractory are: (i) high tensile strength when hot; (ii) high crushing strength; (iii) refractoriness as high as 1690°C; (iv) high percentage of silica, but low concentration of other oxides; (v) no volume change as far as possible when used; (vi) little or no spalling when subjected to sudden fluctuations in temperature.

Uses

Open hearth furnaces have silica bricks in their main arch side walls, port arches and bulkheads. Because of their high thermal conductivity, silica bricks have been utilised in co-product coke oven and gas retorts. In general, silica bricks can be used where no shrinkage in refractory and high resistance to heat is required, because silica bricks have a tendency to expand on heating.

Sillimanite refractories

If refractory material of great refractoriness than silica bricks are required, then sillimanite or kyanite (both have same chemical composition $Al_2O_3.SiO_2$) are frequently used. Sillimanite and kyanite are not plastic and they are made workable by mixing with clay.

Manufacture

The sillimanite ware is manufactured by first calcining it at 1500°C for about two days and then grounding the calcined sillimanite in suitable crushers. The water suspension of ground sillimanite, called the slip is then passed through sieves into a mixer or blunger, where it is mixed with clay. The amount of clay added is dependent on the refractoriness required. Clay is much cheaper than sillimanite and ordinary ware is made of mixture containing 40–60% sillimanite and 60–40% fire clay. Ordinary sillimanite ware is fired between 1500–1600°C, and firing is done both in intermittent ovens and tunnel kilns. Sallimanite ware is shaped in the same way as other refractories. In the manufacture of large articles, such as bricks, sillimanite having a large grain size is used, while for the manufacture of smaller articles, such as pyrometer tubes, sillimanite having small grain size is suitable. Sillimanite, because of no volume shrinkage during heat treatment, requires no prior calcination, but kyanite requires a preliminary baking.

Properties

They are almost neutral refractories, fusion point is about 1800°C. Highly resistant to the corrosive action of slags. Excellent resistance to abrasion. Can resist oxidising and reducing atmospheres at high temperature. Specific gravity is 3.23. No volume change when heated at high temperatures. Coefficient of thermal expansion is very low, being 0.45×10^{-5}. As a result the refractory is capable of with standing sudden changes in temperature. It is perfectly stable upto softening point. Resistance to spalling. Very high. Refractoriness. Greater than silica bricks. Electrical conductivity. Very low. Hardness. 6 to 7 mhos. Colour; Pure sillimanite is colourless, but generally it is brown because of the presence of iron in it.

Uses

Sillimanite refractories are used in the construction of industrial furnaces of all kind such as glass furnace, pottery furnace etc. and also in making bricks, blocks, crucibles, saggers etc. The refractory properties of sillimanite has also been utilised in the manufacture of spark plugs, pyrometer tubes and similar wares. It is also used in the fillings for electrical goods and tubes for surface combustion.

Magnesite refractories

Magnestie is naturally occurring magnesium carbonate ($MgCO_3$) which constitutes the raw material for magnesit refractories and in nature it occurs as crystalline magnesium spar or crypto crystalline (or microscopically crystalline) magnesite. Magnesite also occurs as hydrated carbonate of magnesia or as breunnerite [$MgFeCO_3$] as crystalline magnesite continuing 5–30% of ferrous carbonate.

Manufacture

The magnesite refractories are made from domestic magnesites, dead burnt magnesia or magnesia extracted from brines. The magnesite is calcined at about 1600°C and then crushed to fine powder in suitable crushers. Now a bonding material, such as iron ore is added. The prepared powder is pressed is pressed into bricks in hydraulic presses. Drying is done very slowly and carefully in a way similar to that described in case of silica bricks. After about a week, the bricks are loaded into kilns for firing. Magnesite bricks do not stand much load at elevated temperatures (this difficulty has now been overcome by blending with chrome ores). Hence they have to be loaded in sections supported by silica bricks, in order to reduce the load to which they would be subjected. A chrome ore is usually used between silica and magnesia, in order to avoid chemical reaction between them during firing. The kiln temperature is slowly increased to 1500°C and then the kiln is slowly cooled. The whole cycle completes in about 4 weeks, magnesia bricks have a tendency to spall or split when subjected to temperature changes. Their resistance to temperature changes may be increased by mixing magnesite with chrome ores, without decreasing the refractoriness. Suitable graded mixtures of chrome ore and magnesite give bricks of high refractoriness and good spalling resistance. Spalling resistance of magnesite bricks may also be improved by adding 2–4% of alumina.

Properties of magnesite refractories

These are generally grey or brown in colour, composition - MgO = 87.5%, SiO_2 = 4.0%, Al_2O_3 = 2.0%, Fe_2O_3 = 6.0%, CaO = 0.5%. Specific gravity varies form 3.05 to 3.58 depending upon the temperature of firing and amount of impurities present. Fusion point is 2800°C. Hardness – 4 to 5 mhos. Resistance to slags – Basic slags have little or no corrosive effect, because these are basic refractories. Readily attacked by acids slags such as slags from open hearth steel furnaces. Specific heat higher than that of fire clay and the value is 0.291 between 29°C and 1300°C. Heat conductivity is very low. Slightly attacked by reducing agents and carbon at about 1450°C.

Iron oxide has corrosive action and steam reduces it to powder. Carbon also corrodes it and reduces, it to Mg and O_2. Magnesium thus volatilises. Low resistance to abrasion. Their mechanical strength is more than that of ordinary fire clay brick. Porosity is low if manufactured without any binding material. Porosity is more in presence of binding material such as cellulose, starch etc. Electrical resistance is very high and so very useful in electric furnaces. Thermal conductivity. 0.015 cal. per cm per degree when cold and 0.008 cal. per.cm. per degree at 1300°C.

Uses

These refractories are used in open hearth and electric furnace walls, in the burning zones of cement kilns, and in roofs of non-ferrous reverberatory furnaces. Hard burned chrome magnesite bricks are used in the basic open hearth furnaces. On a commercial scale, magnesite refractories have been used in the construction of soaking pits, basic open hearth furnace, are furnaces, hot metal mixers, converters for Cu, Pb, Ni etc., and furnaces for refining noble metals.

Silicon carbide or carborundum refractories

Silicon carbide is manufactured by the fusion of sand and coke in an electric furnace. These bricks are extremely refractory and possess and high thermal conductivity, low expansion and high resistance to abrasion and spalling. They are strong mechanically and withstand loads in furnaces to temperatures upto 1650°C. Carborundum and graphite are the materials which are intermediate between metals and ceramic materials as far as heat conductivity is concerned. They are intermediate electrically between conductors and insulators and hence used for the manufacture of electrical resistors. Carborundum refractories may be manufactured by mixing carborundum (SiC) with clay and shaping by the various shaping methods used in the ceramic industry. The firing temperature varies from 1400–1600°C, according to the amount of carborundum present in the body. Firing is done in a neutral or reducing atmosphere.

Silicon carbide (SiC) is manufactured by heating together a mixture of sand (SiO_2) = 52–54%, Coke (C) = 35%, saw dust = 7–11% and salt = 1.5–4% in an electric furnace at 1300–2200°C. SiO_2 + 3C → SiC + 2CO.

The carborundum so formed has the composition, Silicon = 65%, Carbon = 30%, and Impurities = 5%. Ground carborundum and binding material (such as lime, clay, dolomite, feldspar, mineral oils, tar, molasses, resin, glycerine, Plaster of Paris etc.) are mixed with water in a *pan mill* and the mixture is moulded into bricks by hand or by presses. The moulded bricks are fired in an electric furnace upto 1400–2000°C, depending upon the property needed. The wares fabricated are generally of three types: Lime bonded, clay bonded and tar bonded.

Properties

Dark grey to blue-black. Fusion point about 2500°C. Thus refractoriness is also high. It however, starts decomposing at 2250°C. Thermal conductivity very high. Coefficient of thermal expansion is very low. Specific gravity: 3.17 to 3.21. Resistance to abrasion, high, resistance to spalling, high, flux resistance, not readily attacked by acidic fluxes and attacked by alkali sulphates and sulphides at high temperature. Action of oxidising and reducing atmosphere; Oxidised above 1750°C. Reducing agents do not have any action. Action of molten salts, not attacked by molten salts, mechanical strength, very high.

Uses

Owing to its high thermal conductivity, it is used to a great extent in muffles, in the manufacture of saggers, bais and other kiln furniture, for supporting the ware in tunnel ovens, etc.

Carbon or graphite refractories

Carbon refractories are very resistant to sudden temperature changes. The carbon used for the manufacture of these refractories is either mineral graphite or retort graphite and the refractoriness depends upon the type of carbon used. The shaping of graphite refractories is done by the usual methods of shaping and firing is done in the reducing atmosphere.

Manufacture

Carbon refractories are either manufactured from coke or from graphite. Manufacture of carbon refractory from coke can be carried out by crushing the blast furnce coke (containing about 8% ash) and then grinding in edge runner mill. The crushed and ground coke is then mixed with required quantity of pitch or tar as binding material and mixture is passed through pug mill, heated with a steam jacket. The mixture is moulded into bricks in moulds by hand or by presses. The bricks are dried and then fired in a kiln from 1100°C to 1300°C. Manufacture of carbon refractories from graphite is made by first removing impurities from graphite by grinding or by floatation. Crushed and finely ground graphite is then mixed with clay and fired at about 1300–1400°C. During firing, contact of the refractories with air is minimised by filling the space in between the refractories with a mixture of sand and powdered coke.

Properties

Light grey in colour. Sp. garavity, 1.5–1.9. Crushing strength, About 4500 lbs. per. sq. inch. they are highly refractory at any attainable temperature, if not attacked by air. When exposed to air and heated, they burn away.

In addition, they are resistant to slag action, thermal shock and chemical treatment, spalling, abrasion, molten salt etc. They undergo low shrikage on thermal treatment. They are capable of resisting oxidation of metal and they are oxidised to some extent by fused metallic oxides and fluxes only at high temperatures. Their softening point is above 2000°C.

Uses

Carbon bricks are used as bottoms and sides of iron and steel furnaces. They are also used as materials of construction of electrodes, lining of highly chemically resistant equipments, atomic reactors, electric furnaces, heat transfer systems, and for smelting of Cu, Pb and Al etc. Fire clay bonded graphite crucibles are in extensive use for metal melting. Graphite is used for large scale production of phosphorus, calcium carbide, ferrosilicon, ferrovanadium in electrothermal furnaces as refractory material for furnace linings. Use of carbon block cupolas carbon tap-holes, slagging troughs and dams, illustrates the metallurgical importance of carbon refractories.

Pure oxide refractories

The refractory industry is constantly faced with increased demands for products which will withstand higher temperatures and more difficult operating conditions. The refractories described so far are used only below 1700°C. In order to meet these requirements and for temperature above 1700°C, a group of special pure oxide refractories has been developed. Refractory oxides of importance are alumina, zircons, beryllia and thoria. Refractory oxides do not possess plastic properties.

Clays or bentonites are used to make the bodies plastic, but they reduce the refractoriness of pure oxides to a great extant. Clays and bentonite are, therefore, added only in very small amount, but even then they decrease the melting point. Organic plasticisers, such as dextrin and resins have also been used, but they also do not produce a plastic paste in reality and make dense bodies slightly porous. Alumina and the other oxides may be made plastic by treating it with hot hydrochloric acid for several days. Such treatment forms colloidal hydrates, which make the metal oxides plastic.

Magnesia is a basic refractory and is easily reduced at high temperatures. The magnesite bricks contain about 87–88% magnesia and their applications are limited to oxidising atmospheres at temperature not exceeding 4000°F.

Zircon ($ZrSiO_4$ zirconium silicate) refractories are used for electric furnaces, for high temperature insulating bricks, small high temperature saggers and laboratory wares. Pure zirconia undergoes a crystalline change from monoclinic to tetragonal form at about 1000°C, accompanied by a drastic volume change on inversion; stabilisation of the crystal structure to the cubic which is produced by evaporating zirconium nitrate to dryness and then dissolving the residue in water and again drying. This process is repeated at least for four time. Zirconia is calcined and powdered. The powdered zirconia is kept in water for some time and wet zirconia is mixed with binding material (colloidal zirconia or alumina) in an edge runner mill. The product is moulded into bricks by hand moulding or pressing. The moulded bricks are air dried and fired at about 1750°C. Since mineral zirconite undergoes volume changes on heating and cooling, it is stabilised by adding MgO or CaO.

Beryllia is superior to alumina in refractoriness, thermal conductivity and electric characteristics at temperatures higher than 1900°C. It is, however, not used commercially in heavy ware because of its high cost and volatilisation above 3000°F in the presence of water vapour. Beryllia bricks are made by firing moulded articles from powdered pure mineral at 1900–2000°C. Beryllia bricks process high fusion point (2550°C), low electrical conductivity, high thermal conductivity, good resistance to thermal shocks and inertness to CO and CO_2 upto 2000°C. They have considerable hot strength, due to which they are used in jet population fields. They are also used in making crucibles for melting of uranium and thorium. Due to its low neutron absorption capacity, it is also used as moderator in nuclear reactors and also in radiation shields in carbon resistance furnaces. Berylllia, however, undergoes volatilisation even at 1000°C, especially in the presence of water vapour. The beryllia vapour or dust, if inhaled, may cause serious health hazard.

Insulating refractories

Refractory materials used to insulate the furnace walls for avoiding excessive heat losses due to conduction are known as insulating refractories. Refractories of low thermal conductivities are used as heat insulators. These are of two types:

1. Bricks made from naturally occurring porous diatomaceous earth such as kieselguhr, asbestos, vermiculite etc. These heat insulators are generally employed for low operating temperatures upto 900°C. These heat insulators are not suitable for operations at high temperatures because kieselguhar undergoes large contraction. Vermiculite expands to a large extent on heating. Hence porous refractories are very suitable under these circumstances.

2. Synthetic, light weight bricks are made by mixing saw dust, coke powder and clay, moulding and firing in a continuous type kiln. Carbon burns during firing leaving a highly porous structure.

These bricks have low thermal conductivity and high porosity and can be used for high temperature operations because air inside the voids acts as a poor conductor of heat. The manufacture of porous refractories can best be carried out by incorporating combustibles such as saw dust, cork, husks etc to a usual refractory material before firing. Alternatively a slip made of clay, grog and water is mixed with chemically prepared foam (containing 45% rosin, 7% NaOH and rest water) and stabilised with glue and alum. This mixture is poured into moulds and fired.

CEMENT

Portland cement is one of the most important building materials at the present time. Portland cement is chemically defined as the finely ground mixture of calcium aluminates and silicates of varying compositions, which hydrate when mixed with water to form a rigid solid structure with good compressive strength. Portland cement is a mixture of the following compounds.

Types of Cements

There are various types of cements other than Portland cement. Some of them are briefly described below.

Sulphate resistant cement

This is a cement with higher C_2S/C_3S ($2CaO. SiO_2/3CaO.SiO_2$) ratio and can resist sulphate attack and corrosion by sea water.

High alumina cement

It is essentially a calcium aluminate cement and can be prepared by heating a mixture of limestone and bauxite (containing iron oxide, magnesia, silica and other impurities) at 1550–1600°C. High alumina cement has a very rapid rate of development of strength and superior resistance to sea and sulphate waters. The composition of another type of high alumina cement, which is prepared by heating ferruginous bauxite with limestone at 1500–1600°C. In addition, some minor constituents, such as C_3A_5 ($3CaO.5Al_2O_3$), C_5A_3 ($5CaO.3Al_2O_3$), C_2AS ($2CaO.Al_2O_3.SiO_2$), C_2S ($2CaO.SiO_2$) and C_4AF ($4CaO.Al_2O_3.Fe_2O_3$) are also present. This cement has tensile strength comparable to that of Portland cement, but its rate of hardening and compression strength are much higher.

Water-proof cement

The moisture loving property of Portland cement can be decreased by mixing it with varying amounts of calcium and aluminium stearates. The resulting cement is then called water-proof cement.

Hydraulic hydrated lime

The cement is low priced and had good strength. In this cement, $Ca(OH)_2$ is a major constituent, while C_2S and C_3A are minor. It is used for brick mortar composition.

Slag cement

It is a mixture of portland cement clinkers and blast furnace slags ground together with small quantity of gypsum which is added to control the setting. This cement is highly suitable for erecting structures exposed to sea water corrosion.

Acid resisting cement

This is a special type of cement which is manufactured taking into considereation its capability of resisting corrosion by acids in the post hardening stage. Very specialised combination of ingredients are used in its manufacture.

Super sulphate cement

This cement is manufactured by fusing together blast furnace slag, small amount of lime but large quantities of gypsum in a kiln. This is exceptionally resistant to chemical corrosion and is used specially in making highways.

White cement

The colour of the ordinary portland cement is greysh-black due to the presence of iron oxide present as one of the constituents. If iron oxide can be avoided by suitable selection of raw materials it is even possible to get white cement. Hence, if low iron feldpar, limestone and gypsum are fused together at 900°C and then leached with water, the fusion of the leached product at 1400°C gives white cement. This is mainly used for decorative constructions.

Coloured cement

Coloured cement can be obtained by mixing white or gery Portland cement with suitable coloured pigments, which must have fine state of subdivision, suitable colour and chemical composition not destroyed by the components of the cement. For getting blue, green, black and red cements, the pigments that are respectively added are cobalt blue (or ultramarine), chromium oxide, carbon black (or MnO_2 or iron oxide Fe_3O_4) and iron oxide (Fe_2O_3).

Pozzolan cement

This cement is made by grindng 2–4 parts of pozzolan (volcanic ash), wit I part of hydrated lime in presence of some burnt clay. It is mixed with portland cement as a cheap extender. In fact, it is a mixture of Portaland cement, volcanic rock, surki (brick dust), burnt clay and hydrated lime. It has certain advantages over portland cement, because it evolves less heat, more resistant to sulphate shock and has capacity to prevent leaching away of lime during hydration of cement, which increases the strength and decreases the porosity of the bonded structure.

Raw Materials of Cement

The essential raw materials for the manufacture of cement are limestone and clay, which supply all the four principal ingredients, viz, CaO, Al_2O_3, SiO_2 and Fe_2O_3. Calcium oxide and iron oxide are obtined from limestone, while silica and alumina are obtained from the clay. Thus raw materials include (i) calcareous materials such as limestone (containing 65–80% $CaCO_3$), calcium carbonate sludge, marl, calk and alkali waste which contains precipitated $CaCO_3$. It is obtained form the manufacture of NaOH. Calcareous materials supply lime and should be such that they contain 3–4% SiO_2, Fe_2O_3 and Al_2O_3 (combined) and less than 3.3% MgO. Siliceous or argillaceous materials, which supply silica, iron oxide and alumina. These include clay (e.g., laterite rich in Al_2O_3), blast furnace slag, siliceous stones, shale, slate etc. These materials should have 2.5–4 times more of silica than alumina.

The relative proportions of raw materials must be such that the analysis of the final cement falls within the following limits (Table 28.2).

Table 28.2. Proportion of raw materials.

Ingredient	limits	Average
Lime CaO	58–65%	61.5%
Silica SiO_2	20–25%	22.5%
Alumina Al_2O_3	4.0–11%	7.5%
Iron oxide Fe_2O_3	0–4%	2.0%
Magnesia MgO	0–4%	2.0%
Sulphur trioxide SO_3	0–1.75%	1.0%
Alkali Na_2O, K_2O	0–3%	1.5%

Selection of raw materials

The ratio of silica to alumina lies between 4 and 2.5, while that of calcium oxide to silica + alumina + ferric oxide should be as close to 2 as possible. If the lime is above the maximum limit then it is left uncombined and thus free lime gives an unsound cement, which cracks during setting. If lime is less than required by the above ratio, the resulting cement is weak in strength. The percentage of lime also depends upon thorough mixing, burning and grinding. If all these operation are upto the mark, higher percentage of lime can be tolerated without affecting the quality of the cemnt. Since tricalcium silicate is the most basic ingredient present in the cement, the molecualr ratio of CaO/SiO_2 must not be greater than 3. An excess of dicalcium silicate disintegrates the cement on setting and the cement containing an excess of $2CaO.SiO_2$ is not hydraulic. The compounds formed in presence of excess of lime are tricalcium silicate ($3CaO.SiO_2$) and tricalcium aluminate ($3CaO.Al_2O_3$). Hence upper limit for CaO can be expressed as $CaO + MgO/SiO_2 + Al_2O_3 = 3$. If CaO is below a certain limit, dicalcium silicate ($2CaO.SiO_2$) is formed and the lower limit for CaO is expressed as, $CaO + MgO/SiO_2 - (Al_2O_3 + Fe_2O_3)$ =3 or more. If the ratio is less than three, then an undesirable compound $2CaO.SiO_2$ is formed. According to some other standards, the lower limit for CaO should be $CaO/(SiO_2 + Al_2O_3) = 2.85$.

When SiO_2 is in excess, the amount of $(Al_2O_3 + Fe_2O_3)$ is decreased and the temperatue of burning is also increased. If $(Al_2O_3 + Fe_2O_3)$ is in excess, the amount of CaO is decreased. The resulting solution loses it cementing value. The amount of MgO should not exceed above 5%, because higher percentage is dangerous to the soundness of cement at later stages. There is very little free lime in the cement and it is present almost entirely in the combined state as calcium silicates and calcium aluminates. Tricalcium silicate and tricalcium aluminate are considered to be the best and dicalcium silicate is condidered to be poor and a danger to the cement. According to some scientists, clinker is a compound intermediate between the di-and tricalcium silicates. The functions of various ingredients are:

Lime (CaO) is the principal constituent of cement. Excess of lime reduces the strength of cement and presence of lime in amount lesser than needed also reduces the strength of cement and makes it quick setting. Silica (SiO_2) imparts strenght to cement. Alumina (Al_2O_3) increse the rate of setting but excess of Al_2O_3 weakens the strength of cement. Calcium sulphate or gypsum retards the rate of setting of cement and actually enhances the initial setting time of cement. Iron oxide (Fe_2O_3) gives colour, strength and hardness to cement. SO_3 is desirable in small proportion, because it imparts soundness to cement SO_3 in excess reduces the soundness of cement. Alkalies should be present in small amounts. In excess, they cause the cement to become efflorescent.

Manufacture

The chief raw materials for the manufacture of portland cement are limestone and clay and these are generally available in large amounts in the vicinity of cement factories. There are two methods of manufacturing portland cement. The wet process and the dry process. The wet process is older and used mostly in India and Europe. The dry process was invented in America and still used there. It should be noted that the two processes differ only in the treatment of the raw materials. (In dry process no water is added to the material in grinding and thus no slurry is made), otherwise very much same equipment is used in both. For example, the treatment of burned clinker is the same in both processes. The choice between two process is usually governed by the following factors: (i) physical conditions of the available raw materials; (ii) climate surrounding the place of manufacture and; (iii) cost of fuel. If limestone and clay are soft, climate is fairly moist and the fuel is cheap, the wet process is preferred. The dry process is employed, if the limestone and clay are hard. The cost, is however low in dry process, because in this process much less fuel is needed in burning the material subsequently in a rotary kiln. In the wet process, the limestone is crushed in a suitable mill to particles of suitable size and clay is washed with water in wash mills to remove foreign material such as flint and a slurry containing about 60% water is obtained. Crushed limestone and clay slurry are mixed in requisite proportions and then pulverised in a special type of ball mill. The resulting slurry, called, raw slurry, containing about 40% water is further ground in tube mills and thoroughly homogenised by making use of compressed air and then stored in the correcting tanks, where after analysis additions are made to adjust the proportion.

In the dry process, the raw materials are separately crushed and ground in suitable machine and dried. These are then mixed in proper proportions, pulverised in ball mills and finally homogenised , i.e. made slightly uniform by means of a compressed air mixing arrangement. The resulting mass is known as raw meal, which may, if necessary, be briquetted by making a stiff paste in a pug mill. The paste is then cut into briquettes, which are dried by hot air. In dry process crushing is done is gyratory crushers and drying is done by means of rotary driers. In wet process raw materials are crushed by gyratory crushers. Grinding of the raw material is carried out in two stages, first by ball mill and then by tube mill. The material from ball mill to the tube mill is converged by making use of screw driver. The material is mixed with water (30–40%) before grinding. Wet process is most common and is almost universally employed for the manufacture of cement (Fig. 28.2).

Dry process is slow, costly and cement produced is of low quality. In this process the fuel consumption is low, cost of production of cement is comparatively less but it can only be used when the raw materials are quite hard. In wet process, the fuel consumption is higher, process is comparatively faster, the cement produced is of superior quality and the process can be used for hard as well as soft raw materials. In general, dry process is costly and not used much and wet process is comparatively cheaper and widely been used. Conversion of raw material into cement is done in rotary kilns which are

ractory lined steel drums resting on rollers at an angle of 3 to 4° with the horizontal. The fuel burnt
the kiln may be coal dust, gaseous fuel or a fuel oil producing hot gases. Depending on the
mperature, different zones in the furnace are: (i) water evaporation; (ii) dehydration zone; (iii) limestone
sociation zone and; (iv) cement forming zone.

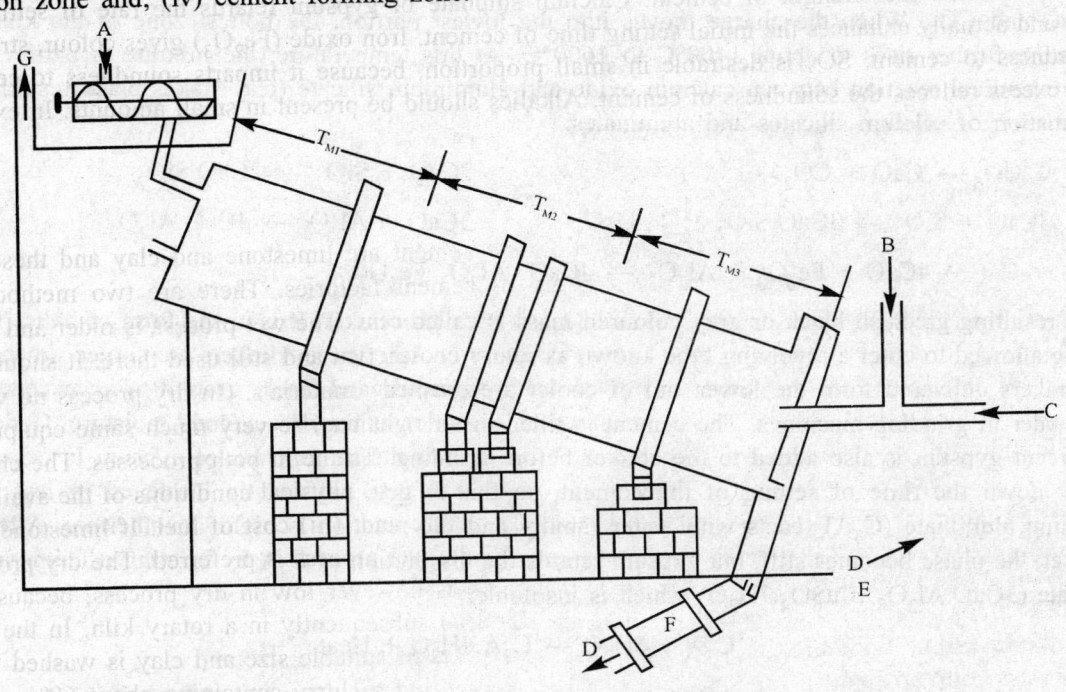

T_{M1} = Moderate temperature zone; T_{M2} = 950°C - Avg. temp. zone; T_{M3} = 1540°C. Max temp. zone
          (Moisture eliminated)   (Lime stone decomposed)

A = Raw material feed; B = Fuel inlet; C = Rimary air inlet for combustion; D = Secondary air inlet for cooling of clinker;
E = Discharge of clinker; F = Rotary cooler; G = Outlet for flue.

Fig. 28.2. Rotary kiln for manufacture of portland cement.

The raw slurry from the wet process or the raw meal from the dry process is now introduced into
opper provided on the upper part of a rotary kiln (for making the clinker), which consists of an inclined
ylinder made of sheet steel. The cylinder is about 200–350 ft long and 7 ft to 12 ft in diameter. The
wer end of the cylinder is provided with a fire proof hood, to which is attached a short rotating cylinder
clined in the opposite direction for the passing out of the hot clinker and for cooling it.

The charge slowly moves forward due to the rotary motion (30–60 turns per hour) given to the kiln
y means of girth gear situated near its middle and a train of reducing gears. The upper portion of the
iln is usually bare or lined with ordinary brick, while the middle portion is lined with thick fire brick
ining and the lower firing zone is lined with fire clay bricks. The lower firing zone may also be lined
vith a lining made of a concrete made of cement clinker pebbles and cement. A blast of burning coal
lust and air is blown from the lower end (firing end, which takes a sufficiently long time for the charge
o pass from one end to another). The hot air obtained by cooling the hot clinker is also introduced in
he kiln from the lower end. During its passage, the slurry first loses water (in the upper part of the kiln

by means of hot gases and the dry material gravities down the kiln and meets a powdered coal gas flame which is at a very high temperature. In the upper portion, whole of the moisture present in the materials is eliminated as in this region the temperature remains at about 750°C. When the charge enters the middle portion of the kiln, the temperature rises to about 1000°C. At this temperature, limestone is decomposed into CaO and CO_2. When the charge moves into the lowest portion, the hottest zone of the kiln, the temperature further rises to about 1400°C to 1600°C. At this temperature, the mixture is partly fused and the chemical reaction between calcium oxide and aluminium silicate (clay) takes place resulting in the formation of calcium silicates and aluminates.

$$CaCO_3 \rightarrow CaO + CO_2 \qquad\qquad 2CaO + SiO_2 \rightarrow 2CaO.SiO_2$$

$$3CaO + SiO_2 \rightarrow 3CaO.SiO_2 \qquad\qquad 3CaO + Al_2O_3 \rightarrow 3CaO.Al_2O_3$$

$$4CaO + Fe_2O_3 + Al_2O_3 \rightarrow 4CaO.\ Al_2O_3\ Fe_2O_3$$

The resulting greenish black or grey coloured mass is called clinker. It is in the form of balls. These balls are allowed to enter a revolving tube known as rotary cooler, where, clinkers are cooled down. The cold clinkers delivered form the lower end of cooler are crushed and finely ground to an exceedingly fine powder in grinding machines. The cement is filled in air tight bags to exclude moisture. Generally 2–3 percent gypsum is also added to the clinker before grinding. The function of gypsum in cement is to slow down the time of setting of the cement, so that it gets sufficiently hardened in less time. Tricalcium aluminate (C_3A) reacts with water rapidly and the reaction is highly exothermic. After the initial set, the phase becomes stiff, but gypsum retards the dissolution of C_3A by forming calcium sulpho aluminate ($3CaO.Al_2O_3.xCaSO_4.7H_2O$) which is insoluble.

$$C_3A + 6H_2O \rightarrow C_3A.6H_2O + Heat$$

Theories of the setting of cement

The setting of cement on the addition of water is explained by two theories.

1. *The crystalline theory:* According to this the hardening or setting of cement is due to the interlocking of the crystals during hydration.

2. *The colloidal theory:* According to this theory gels of hydrated silicates are formed and when these gels harden the set cement gains strength.

The second theory, i.e., the hydration theory, explains the setting of cement more satisfactorily. At ordinary temperature the hydrated calcium silicate is non-crystalline and the physical properties of set cement are also better explained by this theory.

The hydration of cement consists of two different chemical reactions: (i) hydrolysis- when water is added to cement hydrolysis of calcium silicate takes place. Calcium silicates decompose into calcium silicates of lower basicity and the released CaO forms $Ca(OH)_2$ with water. Tricalcium silicate undergoes hydrolysis with less or greater quantities of water to form $2CaO.SiO_2$ (aq) or $3CaO.SiO_2$ (aq) along with $Ca(OH)_2$. The dicalcium silicate present forms on hydrolysis $CaO.2SiO_2$ along with $Ca(OH)_2$; (ii) hydration-silicates and aluminates take up water on hydration as follows: (a) dicalcium silicate on hydration forms $2CaO.SiO_2.4H_2O$. (b) tricalcium aluminate on hydration $3CaO.Al_2O_3.6H_2O$.

Hydrated tricalcium aluminate then combines with the gypsum ($CaSO_4.2H_2O$) to form $3CaO.Al_2O_3.3CaSO_4.31H_2O$ and also reacts with $Ca(OH)_2$ to form tetracalcium aluminate 4CaO.

$Al_2O_3.13.5H_2O$. Dicalcium aluminate takes up water to form $2CaO.Al_2O_3.nH_2O$. The tetra calcium aluminium ferrite also forms $3CaO.Al_2O_3.6H_2O$ along with $CaO.Fe_2O_3$.

On hydration heat is evolved. This evolution of heat continues during the first week of hardening. The functions of the different constituents in regard to setting are: tricalcium aluminate is responsible for the initial set; tricalcium silicate for the first strength. Dicalcium silicate and tricalcium silicate are responsible for the final strength which occurs in about a year.

Properties of Cement

1. *Quality:* The quality of cement is expressed in terms of silica and alumina modules, expressed as.

$$\text{Silica Module (n)} = \frac{\%\,SiO_2}{\%\,Al_2O_3 + Fe_2O_3} \qquad\qquad \text{Alumina Module (p)} = \frac{\%\,Al_2O_3}{\%\,Fe_2O_3}$$

2. *Setting time:* When cement is brought in contact with water it starts setting or hardening. Involving stepwise hydration, followed by gradual crystallisation of various components. The initial setting or early strength is gained within 24 hours through the hydration and crystallisation of tricalcium illuminate and sulpho aluminate. This is followed by setting of tricalcium silicate (C_3S) which is complete in 7 days, Final setting of other components like dicalcium silicate is over within 28 days, and dicalcium silicate and tricalcium silicate provide full strength at one year.

3. *Shrinkage:* During setting of cement, hydration of various components causes an initial increase in volume, but it decreases due to possible crystallisation of these components. The volume shrinkage ranges between 5–10% and it may cause cracks in many cases because of improper setting.

   The volume shrinkage depends upon water/cement ratio, drying period and temperature fluctuations of the surrounding. The longer the drying period, the smaller the shrinkage.

4. *Soundness:* The capacity of the set or hardened cement to resist disintegration either on fixation in marshy area or alternative exposure to wet and dry conditions is called soundness. If the cement is sound, it will not undergo appreciable change in volume or indicate any disintegration on drastic weather conditions.

5. *Colour:* The greenish grey colour of ordinary portland cement is due to the presence of iron in it. If this iron is prevented by some suitable choice of raw materials, even white cement can be obtained.

6. *Heat of setting or hardening:* In addition to gel formation, cement is believed to be formed by hydrolysis and hydration of the components present in cement. The process of hydration is exothermic and the heat evolved is the resultant of the heats evolved during hydrolysis of various cement constituents. This is known as heat of setting or hardening. The compounds responsible for the evolution of heat in the decreasing order are: Tricalcium aluminate (C_3A) > Tricalcium silicate (C_3S) > Tetracalcium alumino ferrite (C_4AF) > Dicalcium silicate. (880 > 500 > 420 > 250 kJ/kg).

   The heat of hydration of cement can be decreased by lowering C_3A and C_3S by using higher percentage of Fe_2O_3 in the raw slurry. Fe_2O_3 combines with Al_2O_3 to form C_4AF and thus decreases C_3A in the cement. The cement having low C_3A is very useful in the construction of dams, because of their extraordinary crack resistant capacity.

7. *Strength:* The quality of hardening of cement is best judged in tensile strength and compressive strength of the cement through setting for a particular period. The increase in strength through

progressive setting is due to varying individual strength contributions of various cement componen

Example,

8. *Acid corrosion:* Cement constructions are attacked by all types of acids.

9. *Dissolved CO_2 corrosion:* The strength of cement structures is greatly affected by water containi dissolved CO_2, because it dissolves the free lime present in the cement to calcium bicarbon ($CaHCO_3$).

10. *Sulphate corrosion:* The hardened cement surfaces are attacked by water containing dissolv sulphates such as $MgSO_4$ and Na_2SO_4 and thus reduce the strength of the set cement. Th sulphates disturb the balance of free lime, silicate and aluminate of the set cement. For example, water affects the cement surface of the floating vessels.

Specifications and additional tests

Portland cement must conform to the specifications and additional tests which are official. The n unmixed cement must have a tensile strength of 500 lbs/sq in after 7 days; the cement mixed wit parts of sand must have a strength of 200 lbs after the same period.

The strength of the concrete which is composed of cement 1000 lbs, sand 1500 lbs and crushed st 3000 lbs is greatly increased by imbedding in it twisted square steel bars or wire-netting; it is then ca 'reinforced concrete'.

One reason for the almost universal use of Portland cement is the comparative ease of working Another is its strength which increases with age. A third is its uniformity, which permits calculation strength as reliable as those made for structural steel.

ISI Specification for cement

The ISI specifications for Portland cement are:

(a) $$\dfrac{CaO}{2.8\,SiO_2 + 1.2\,Al_2O_3 + 0.65\,Fe_2O_3}$$

Not greater than 1.02 and not less then 0.66.

$CaO.SiO_2\ Al_2O_3$ and Fe_2O_3 indicate their percentages in cement.

(b) $\dfrac{Al_2O_3}{Fe_2O_3}$ ratio not less than 0.66—Al_2O_3 and Fe_2O_3 indicate their percentages in cement.

(c) weight of insoluble residue not more than 1.5%.

(d) Weight of Magnesia (MgO) not more than 5%.

(e) Total Sulphur content calculated as SO_3 not more than 2.75%.

(f) Total loss on ignition not more than 4%.

Testing of cement

The quality of a sample of cement is determined form a number of measurements. For exampl Tensile Strength: It should not be less than 300 lbs/sq inch after 72 hours and not less than 2500 sq. inch after 7 days. 2. Compressive Strength: It should not be less than 1600 lbs/sq. inch after 3 days and not less than 2500 lbs/sq. inch after 7 days. 3. Soundness: It is estimated by the Le-Chatelier technique and expresses expansivity of cement set for 24 hours, between 80°F and the boiling point of water. According to I.S.I. specification, the value for ordinary cement being within 10 mm. 4. Finess: According to Turbidimetic method, the fineness of ordinary cement should be 1600 sq.cm/g 5. Specific Gravity: It should be 3.1 to 3.2.

Uses

Cement is one of the most important building materials now a days and used in the construction of roads, buildings, dams, bridges etc. For construction purposes, it is used in the form of a paste with sand and water. Concrete is usually mixture of cement, sand, gravel or small pieces of stones and water. The concrete sets to a very hard mass and hence used in foundations, floors, roads and walls of buildings, concrete has also been used in the buildings and roofs, dams and bridges. If the cement concrete is filled in and around wire netting or a skeleton of iron rods and allowed to set, the resulting structure is very hard and rigid and is known as reinforced concrete. Reinforced concrete being very hard and strong is used in the construction of bridges and roofs.

Mortars and Concrete

For construction work, cement cannot be used alone, because of its sensitivity to moisture and development of various internal stresses. These stresses are capable of developing cracks and also reduce the strength. In order to avoid these defects in cement, it is generally mixed with sand and stone. This process is generally called dilution and stabilisation of cement. This diluted and stabilised cement continue to gain strength with ageing when used for construction work and no cracks are developed in this manner.

Mortars

A mixture of slaked lime, sand and water is known as lime mortar. The function of sand in the mortar is to make the mass more porous and hard. It also prevents the cracks cause by excessive shrinkage. Mortar is used as building material. When sand is mixed with slaked lime, it slowly converts into a hard mass of calcium silicate. Lime also reacts with CO_2 to form $CaCO_3$.

$$Ca(OH)_2 + SiO_2 \rightarrow CaSiO_3 + H_2O \qquad\qquad Ca(OH)_2 + CO_2 \rightarrow CaCO_3 + H_2O$$

Setting of mortar is mainly due to the loss of water: When cement is added to mortar it becomes hard and water proof. It is called cement mortar. These mortars are mixture of cement and sand and sometimes other fine aggregates below 3/16 inch mesh size. They are used for bonding in masonry and in surface covering. When limestone containing more than 10% aluminium silica is burnt, hydraulic mortar is obtained.

Concrete

Concrete is a mixture of cement, sand (below 3/16 inch mesh size) with calculated the amount of water. The size of gravel or coarse aggregates varies with the purpose for which the concrete is required. Common maximum sizes of coarse aggregates are 0.75 inch or 1.5 inch, but coarse aggregate of even 0.5 inch have also been used for some purposes. In case of heavy mass concrete the size may be even 6 inch or more. The common proportions of cement, sand and coarse gravel may be in ratios: (i) 1 : 15 : 3; (ii) 1 : 2 : 4; (iii) 1 : 3 : 6.

When the cement concrete is filled in around a wire netting of iron rods and allowed to set, the resulting structure is called reinforced concrete. Concrete has high compressive strength and relatively low tensile strength. So in order to impart high strength as high tensile strength so that it can resist loads which tend to crush concrete, another form of concrete, known as reinforced concrete is used. The reinforced concrete can withstand not only high tensile strength but also the compressive stresses. The combination of steel and concrete produces a structure known as reinforced concrete construction (RCC), which is capable of bearing all types of loads RCC possesses greater rigidity, moisture and fire resistance than plane concrete, RCC is easier to make and cast into any desired shape, which can withstand all types of loads.

Curing of Concrete

The hardening of concrete is due to hydration of the cement constituents. The process continues indefinitely but maximum amount of strength as well as hardness is developed during few early days after placement. It is, therefore, necessary to keep the concrete damp for about 7 days in order to enable hydration reactions to go to completion. The chemical reaction taking palce between cement and water occur only under favourable conditions of temperature. At low temperature, the rate of reactions in concrete is slow but completely stops, when water in concrete is frozen. The process of dampening concrete by spraying is known as curing of concrete. Hence curing may be regarded as the process of maintaining a satisfactory moisture content and favourable temperature in concrete during the period immediately following placement, in order to allow the process of hydration to continue, until the desired properties such as strength are developed to a sufficient extent. In general, concrete can be protected by giving a coating of bituminous material which is capable of preventing direct contact between concrete and water. Decay of concrete can also be prevented by contacting the surface with silicon fluoride (soluble) together with ZnO, MgO or Al_2O_3. The CaF_2 so formed in the capillaries prevents the dissolution of lime in concrete.

Gypsum

Gypsum is an important non-metallic natural mineral of great economic importance. It is anhydrous sulphate of calcium, $CaSO_4.2H_2O$. The anhydrous variety of gypsum is called anhydride. For a number of years gypsum is used chiefly in the manufacture of cement, where it is used to prevent the rapid setting of the cement. In recent years it is being used increasingly in the manufacture of important fertiliser, ammonium sulphate. It is also used in the manufacture of sulphuric acid.

Pure gypsum is colourless to white, but due to mixing of impurities, it may be grey, brown, red or pink in colour It is a soft mineral, with hardness 1.5–2.0 and specific gravity of 2.31–2.33. Sometimes it may be fibrous and it may also be massive. When crystallised, it produces transparent or translucent tabular crystals known as selenite.

Gypsum is usually found in beds or bands in the sedimentary rocks such as limestone, sandstone and shales. Sometimes they are also found as scattered crystals and grains in clays. Gypsum is also obtained as a by product during the manufacture of salt from brines and sea water by direct solar evaporation method. Gypsum is found in various forms. For examples, albester is massive, dense and crystalline variety of gypsum. It is generally translucent and pure variety of gypsum. Being soft dense and fine textured, it is used for carving image and other art pieces. Selentite is another pure form of gypsum. It has monoclinic crystal structure and occurs in the form of sheets or plates. This is used in the laboratory equipment for the polarisation of light, because thin sheets of selentite polarise light. Anhydrite is the anhydrous variety of gypsum. Its hardness is 3–3.5 and sp. gravity is 2.8–2.9. It occurs sometime as dense mass and sometimes show tints of blue and pink. Finely crystalline gypsum mixed with loams clays, sands and humus is called gypsite in which gypsum content varies form 60–90%. Satin spar or gypsum spar is pure form of gypsum having monoclinic crystals and occurs in the form of parallel threads. It is translucent, when densely formed. Deposits of pure gypsum in the form of white sands are known as gypsum sands. On heating gypsum to 120°C, it gets converted into plaster of Paris $(CaSO_4.1/2H_2O)$.

$$2(CaSO_4.2H_2O) \xrightarrow{120°C} (CaSO_4)^2H_2O + 3H_2O \xrightarrow{200°C} CaSO_4$$

The temperature should not exceed above 120°C, otherwise at 200°C, gypsum is fully dehydrated. It is then called dead burnt gypsum which does not quickly set with water. In addition to the uses of gypsum already given above, gypsum is used in the preparation of plaster of Paris, chalk pencils or crayons etc. It is also used in the manufacture of cement to prevent rapid setting. It is used in the preparation of tiles, plasters, baking powder and for the treatment of the soils. Gypsum is used in paints, pharmaceuticals, paper filling, insecticide, water treatment etc. because it is tasteless, non-abrasive and chemically inert. It is used to lower the pH of water brewery.

Plaster of Paris

Chemically Plaster of Paris is $CaSO_4.1/2H_2O$. It is a white powder. It quickly sets to a hard mass when made into thin paste with water. A slight expansion takes place in the process and heat is evolved (This process is exothermic). The setting takes place in two stages.

$$\underset{\text{Plaster of Paris}}{CaSO_4.1/2H_2O} \quad \underset{H_2O}{\xrightarrow{\text{Setting}}} \quad \underset{\text{Orthorhombic}}{CaSO_4.2H_2O} \xrightarrow{\text{Hardening}} \underset{\text{Monorhombic gypsum}}{CaSO_4}$$

The setting is due to reformation of the dehydrate which forms interlacing needles. When mixed with water, the Plaster of Paris forms a plastic mass, which quickly hardness and a slight expansion takes. As a result, closely packed crystalline gypsum $CaSO_4.2H_2O$. The process of setting of Plaster of Paris can be accelerated by adding alkali sulphates such as Na_2SO_4, K_2SO_4. The latter initiate as well as accelerate the process of crystallisation.

Plaster of Paris is made from gypsum by heating gypsum in powdered form in a vertical kettle of retort provided with four flues for passing out gases. The capacity of the kettle varies from 10–15 tonnes. The kettle having steel bottom (which is convex upwards) is fired by gas, fuel oil or by coal. A vertical shaft carrying agitating arms is also provided with the kettle. A screw conveyor kettle transfers the powdered gypsum into the vertical kettle and as it strikes the hot bottom of the cattle., dehydration starts within 15–20 minutes and whole charge is placed into the hot kettle and the mass is heated to 120°C, by raising the temperature gradually. The temperature is not allowed to increase beyond 120°C. As seen above. The reaction is exothermic, the heat given out is used up in bringing about the dehydration (During the heating 1.5 molecules of water are expelled out of gypsum) and the product formed is hemihydrate $(CaSO_4.1/2H_2O)$. During heating the mass boils vigorously (turbulence is caused by going out of steam). After the conversion of whole mass into hemihydrate, the boiling stops and turbulence caused by going out steam is also stopped. Now temperature increases rapidly to 160°C. At this temperature the charge is thrown out and the product is cooled and stored as plaster of Paris or calcined gypsum $(CaSO_4.1/2H_2O)$.

If instead of dumping down the charge at 160°C, the temperature is further increased to about 190°C, the charge again starts boiling and it will continue till the 0.5–(1/2 molecule) of water present in hemihydrate is also expelled out and the charge settles down. Heating is continued upto 220°C. This charge is known as seconds settle Plaster of Paris which shows greater density and greater strength when made into castings and structural products. However, with storage ageing, it absorbs moisture and gets converted into ordinary hemihydrate, $CaSO_4.1/2H_2O$ known as Plaster of Paris. As seen above, the setting of Plaster of Paris takes place by hydration due to the formation of a solid crystalline hydrate and hardening of Plaster of Paris is hydration reaction, which is reverse of the dehydration of gypsum. Plaster of Paris is used in surgery for plastering fractured parts of the body. In laboratory it is used for making apparatus air tight. It is also uses in making moulds, statues and for making black board chalks. It also finds used in plastering the walls interior ceilings and decorations etc.

Lime

Lime is generally obtained by heating a natural mineral, known as limestone ($CaCO_3$), and consists either of calcium oxide (CaO) or calcium oxide (CaO) with some magnesium oxide (MgO). There are in fact many types of limes. For example.

1. *Fat limes:* These limes are also known as high calcium limes and are generally obtained by burning or heating limestone ($CaCO_3$) which is almost in pure form. Fat limes contain about 95–98% CaO. The rest are silica (SiO_2), alumina (Al_2O_3) and magnesia (MgO). These limes can be rapidly slaked when brought in contact with water and as a result, large amount of heat is liberated (Slaking is an exothermic process) and large volumes (2.0–2.5 times) of powdered lime is formed. Fat limes are non-hydraulic and their setting takes place only though drying. They never undergo setting in wet conditions and so they are not suitable to be used in damp conditions. Fat limes have been used for making mortars, in joining brick works, in white washing, in glass and metallurgical industries and also in water softening.

2. *Lean limes:* These limes are also known as poor limes, because they contain about 70–80% CaO and other impurities such as MgO (about 5%), silica (SiO_2) and iron oxide (Fe_2O_3) etc. These limes undergo slaking very slowly and exhibit very low shrinkage, better plasticity and workability than the high calcium or fat limes. These limes are mainly used for mortar purposes and interior work because they are very suitable for plaster finishing. It should be noted that these limes should be completely slaked before using for plaster finish, otherwise they are expected to expand slowly in the finished work and hence decreases the soundness.

3. *Hydraulic limes:* These limes generally contain 5–30% silica and alumina and less than 2% MgO, alongwith 70–80% CaO. These limes undergo setting to a hard mass when immersed in water and hence exhibit hydraulic properties. These limes do not undergo slaking easily and so they are generally used in the form of finely ground state. These limes do not shrink or crack and set very slowly. Hence they are used as cement substitutes in marine work, foundations, shafts of tall chimneys and thick walls where there is a limited supply or entry of the air.

Manufacture of Lime

The manufacture of lime is carried out by the calcination of limestone ($CaCO_3$), which occur in nature in the form of rocks and contains impurities of silica (SiO_2), aluminate (Al_2O_3), iron oxide (Fe_2O_3) and magnesium oxide (MgO) in trace amounts. Hydraulic lime is generally manufactured by the calcination of either limestone containing a good quantity of silica (SiO_2) and alumina (Al_2O_3) or by heating a mixture of pure limestone and clay in requisite proportions. Dolomitic lime can be prepared by heating dolomite ($CaCO_3 + MgCO_3$) in a limited supply of air, that is by calcination. Production of lime from limestone involved calcination (heating in limited supply of air) and hydration. Limestone has been found to decompose when heated at 880°C, but the reaction

$$CaCO_3 \rightleftarrows CaO + CO_2 + 42.5 \text{ k.cals}$$

is exothermic as well as reversible. The reaction is expected to proceed only in the forward direction (formation of lime, CaO), if CO_2 formed according to above reaction is quickly removed and temperature is not allowed to decrease below 900°C. Magnesium carbonate present in limestone also decomposes much earlier the $CaCO_3$ into MgO and CO_2. Thus dolomite, containing $MgCO_3$ and $CaCO_3$ is much easier to calcine than limestone ($CaCO_3$), because dolomite limestone decomposes at lower temperature.

In usual practice, calcination is generrly done between 1000–1200°C, so that complete dissociation of limestones, $CaCO_3$ into lime (CaO) may take place. At temperatures greater than 1200°C, the lime formed may combine with silica, alumina and iron oxide present as impurities and lime will be completely vitrified on the outside of the lumps. As a result, lime formed will be very slow slaking and will escape hydration. As a consequence, it expands or blows in the walls and causes blisters in the finished plaster works. In order to obtain hydraulic lime, limestone is heated at about 1500–1700°C. Under these conditions, decomposition of carbonate is followed by combination of lime produced with clayey matter silica and alumina. As result, calcium silicate and calcium aluminates are formed. Calcination of limestone is done in kilns, which are of two types-rotary kilns and vertical kilns Fig. 28.3.

The diameter of rotary kiln varies from 6.0 to 11.5 ft and length varies from 60–400 ft. The commonly used length is 150 ft. Rotary kilns are provided with preheaters and coolers so that fuel can be saved. The kiln rotates at a speed of one revolution in 95 seconds and the maximum temperature that can be attained in rotary kiln in about 1370°C. The lime formed as a result of dissociation of limestone is taken out of the kiln at a temperature of about 900°C. Size of limestone calcined in rotary kiln is 0.25 to 2.5 inches. Cooling air is led in by the draft on kiln into the discharge end. This air cools the lime and gets itself heated up. This hot air is used in the kiln as a second combustion air. A separate fan takes air from the outlet and this air is used as primary combustion air for drying coal used for calcination of limestone. The most important disadvantages of rotary kiln is high cost of fuel per unit of lime produced and thus high cost of investment.

Calcination of limestone is now almost carried out in continuous vertical kiln, which is a large chimney like tower, about 10–24 ft in diameter and 35–75 ft in height and lined inside with refractory bricks. The heating or firing of these kilns are carried out either by burning coal inside the fire boxes which are built in the side around lower part of the kiln or by the use of gaseous fuel such as producer gas. Vertical kiln burns limestone of 6–8 inches in size and limestone is fed at the top. Most modern kilns are gas fired or oil fired. The flames of the burning fuel sweeps in from the sides and decompose limestone into CaO and CO_2. A current of air is blown in at the base at the bottom of the furnace. This air cools the hot lime at the base of the furnace, furnishes heated air for combustion of fuel and removes CO_2 from the kiln in order to prevent reverse reaction. The lime is finally dumped into a trolley, run under the furnace. There are four important zones in the kiln. In cooling zone air and lime pass each other in opposite directions and exchange of heat takes place. Lime is cooled and air is heated up. In finishing zone, fuel is burnt with excess of air to complete the calcination of limestone in a zone of moderate temperature. In calcining zone calcination takes place, and in preheating zone gases are cooled and

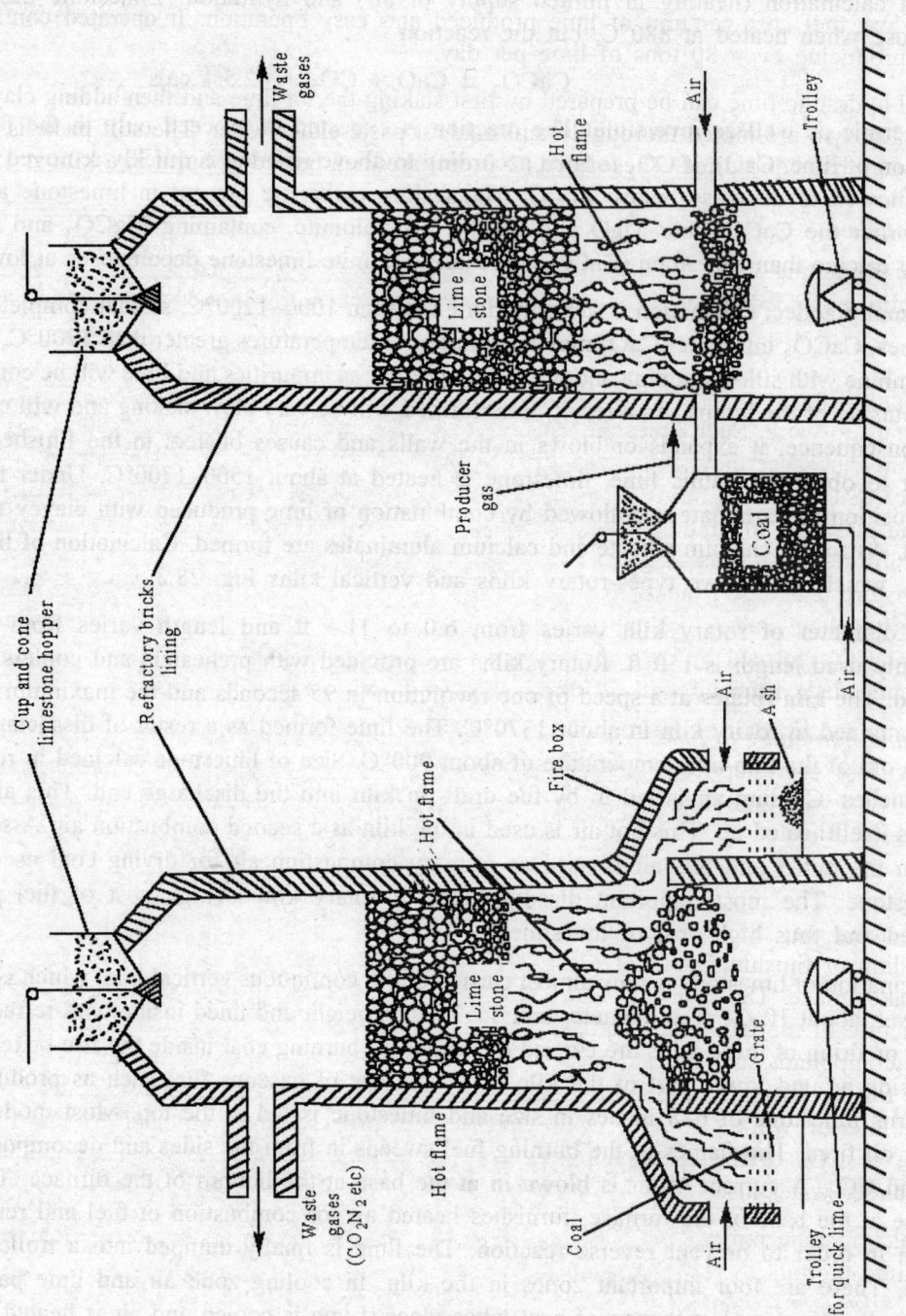

Fig. 28.3. Vertical section of a coal fired and gas fired continuous shaft lime kilns.

limestone is preheated. Arrangements are made in the kiln for the control of temperature, duration of calcination and feed of the fuel. The most important advantage of vertical kiln is small cost investment because of low fuel cost per unit of lime produced and easy operation. It operated continuously and capable of producing even 80 tons of lime per day.

Artificial hydraulic lime can be prepared by first slaking the fat lime and then adding clay in requisite proportion. The two are mixed thoroughly and made a paste either water. The stiff paste is kneaded and made into small balls and dried. These balls (2–3 cm) are then heated or burnt in a kiln at 1500–1700°C.

Properties of lime

Lime is a high melting white solid. When pure lime, called quick lime (CaO) is treated with water, the latter is readily absorbed through the pores of quicklime lumps and slaked lime, $Ca(OH)_2$ is formed

$$CaO + H_2O \rightarrow Ca(OH)_2 + 15.9 k.cals$$

Pure calcium limes are poorly plastic (plasticity is the ability of lime to spread during application), sticky and hard to work. Such limes are also called short limes. Limes containing some MgO are known as plastic limes, because they spread easily and smoothly. When lime is used as mortar or plaster, some sand or silica (SiO_2) is also mixed with it because silica decreases the shrinkage of lime during setting, increases the plasticity of lime and makes the mixture more economical. The sand carrying capacity of high calcium limes or fat limes is greater than that of dolomite limes. When lime is placed in air, it slowly absorbs moisture and CO_2 from at the atmosphere and swells as well as disintegrates into fine powder known as air soaked lime. This lime is not suitable of the manufacture of mortars and plaster because it contains $CaCO_3$ (formed by the combination of CO_2 and CaO and very little slaked lime ($Ca(OH)_2$). Hence mortars made from air soaked lime will not harden.

Setting and hardening of limes The setting and consequent hardening of lime or lime mortar involves dehydration (loss of water from slaked lime as a result of evaporation), carbonation (combination of atmospheric CO_2 with lime) and colloidal gel formation of part of $Ca(OH)_2$ in the mortar, which gradually hardens.

$$CaO + CO_2 \rightarrow CaCO_3 \quad or \quad Ca(OH)_2 + CO_2 \rightarrow CaCO_3 + H_2O$$

Hence setting may be regarded as the transformation of lime into a hard substance, $CaCO_3$ which acts as binding or finishing material. Because setting involves loss of water, a shrinkage in the volume of mortar takes place. The shrinkage in volumes of mortar can be avoided by mixing sand with lime in the manufacture of mortar. The decomposition of complex silicates of Ca and Al in presence of water into simple compounds such as calcium silicate, calcium aluminate and calcium hydroxide etc. is mainly responsible for the setting of hydraulic lime. The simple compounds so formed crystallise to form a hard mass in the interior, while $Ca(OH)_2$, which is soluble in water, comes to the surface and forms, $CaCO_3$, in presence of CO_2 of the atmosphere. This indicates that hydraulic lime is capable of setting and hardening under water also.

High calcium lime mortars set least hard in comparison to high magnesium or dolomite limes. The ability of lime mortars to resist erosion and impact of setting is called hardness.

Qualities of Glass

Requisite qualities of glass are: (i) the material must liquefy at a conveniently feasible temperature; (ii) on cooling the molten mass should not crystallise; (iii) the molten glass should be such that it can be easily given the desired shape and form and; (iv) the glass should be resistant to the conditions for which it is manufactured.

The are variety of glasses in use. The composition of each class of glass varies according to the use to which it is put to. The Table 28.3 give the average composition of some quality glass.

Besides the above Table 28.3 glass contains 2.90 per cent of manganese, plate glass contains 12.23 per cent of barium oxide and 3.20 per cent of arsenious oxide, bottle glass contains 0.40 per cent arsenious oxide and laboratory glass contains 0.9 per cent antimony oxide.

The main ingredients for the manufacture of glass, as will be evident from the above Table 28.3 are: silica, sodium carbonate (soda ash), borax, alumina and waste glass pieces or 'cullets'.

Raw Materials of Glass

The specification of the raw materials used are as follows:

Silica

Sand is used as the source of silica. The iron content of the sand must be as low as possible (0.5 per cent for table glass of 0.0753 for optical glass), as the presence of iron imparts colour to the glass.

Table 28.3. Average composition of various types of glasses.

	SiO_2	B_2O_3	Na_2O	K_2O	CaO	MgO	Al_2O_3	Fe_2O_3
Table glass	77.70	–	10.02	–	10.02	–	0.43	0.21
Plate glass	71.80	–	11.10	–	15.70	–	1.26	0.41
Bottle glass	63.51	–	9.50	–	14.32	3.87	2.54	2.36
Jena lab. glass	65.30	15.06	–	–	–	–	3.80	0.41
Sheet glass	72.42	–	13.13	–	11.97	–	2.13	0.35
Corning pyrex lab. glass	80.75	12.00	4.10	0.10	0.30	–	1.50	0.70
Spectacle glass	69.04	0.25	5.95	11.75	12.07	–	–	–

Soda

Sodium sulphate is usually used as the sources, but as sodium sulphate requires a higher temperature than sodium carbonate for reacting with silica, and also the amount of sodium sulphate required is more than sodium carbonate, the latter is sometimes used. The salt cake is said to remove scum from the glass tanks.

Borax

Borax is generally used as it (i) has high fluxing power; (ii) lowers the expansion coefficient of glass, and; (iii) increases chemical durability. Where both alkali and boron trioxide are required, borax is used. But where alkali is not required, borax acid is employed as the source of boron trioxide.

Alumina

Alumina is added to the charge to increase (i) the viscosity of the glass and also; (ii) resistance to sudden changes of temperature. Alumina is added in the form of felspar, $K_2O,Al_2O_3,6SiO_2$ or Cryolite, $AlF_3, 3NaF$.

Waste glass pieces or 'Cullets'

Addition of waste glass pieces to the charge facilitates melting and also helps to utilise the wastes.

Minor ingredients such as arsenic trioxide is added to facilitate the removal of the bubbles formed in the glass and nitrates of potassium or sodium help to oxidise iron and thereby decrease the chance attack of air on glass.

Coloured glasses: Coloured glasses are made by the addition of different colouring materials such as copper and cobalt oxides for pink glasses, green oxide of chromium for green glasses and manganese dioxide for violet glasses, etc.

Manufacturing Process

The manufacture of process consists of the following steps: (i) melting of the charge; (ii) shaping the article desired; (iii) annealing the article formed; (iv) finishing.

Melting

The raw materials are mixed in requisite proportions and a 'batch' is thus prepared. The batch is now melted either in a (i) pot furnace or; (ii) a tank furnace. The pot furnace is of batch type while the tank furnace works continuously.

Pot furnace

Two types of pots, the open crucible types or the closed covered type are in use. Those vessels are used where the melting batch is to be protected from the products of combustion. The closed covered types are specially used for the manufacture of optical glass. The pots are heated by supplying heat form external sources.

Tank furnace

In the furnace the batch materials are fed at one end and the finished glass in taken out at the other end. These furnaces are used where large quantities of glass are required. The furnace is of the Reverberatory type and the molten glass forms a pool over the hearth of the furnace. As it is difficult to stir the charge under process, the products from this type of furnace may not be of uniform composition. The undecomposed or unused materials, along with other impurities rise to the surface and is called 'glassgall' or 'sandiver'. The heating is done by combustion of gases as that is most economical. Only where special glasses are to be made, wood is still used as a fuel. The heat economy is effected either by the recuperative or regenerative system Fig. 28.4.

Chemical reactions in glass furnace

Glass is considered to be a mixture of sodium and calcium silicate. The main reactions between the basic compounds (that is, alkali) and acidic compounds (that is, silica) may be stated as.

$$Na_2CO_3 + xSiO_2 = Na_2O.xSiO_2 + CO_2$$

$$Na_2SO_4 + C = Na_2O + CO + SO_2$$

$$Na_2O + y\ SiO_2 = Na_2O.y\ SiO_2$$

$$CaCO_3 + z\ SiO_2 = CaO.\ zSiO_2 + CO_2$$

The mixture of CaO. $zSiO_2$.Na_2O. SiO_2 and Na_2O.$ySiO_2$ with some other compounds obtained is commercial glass.

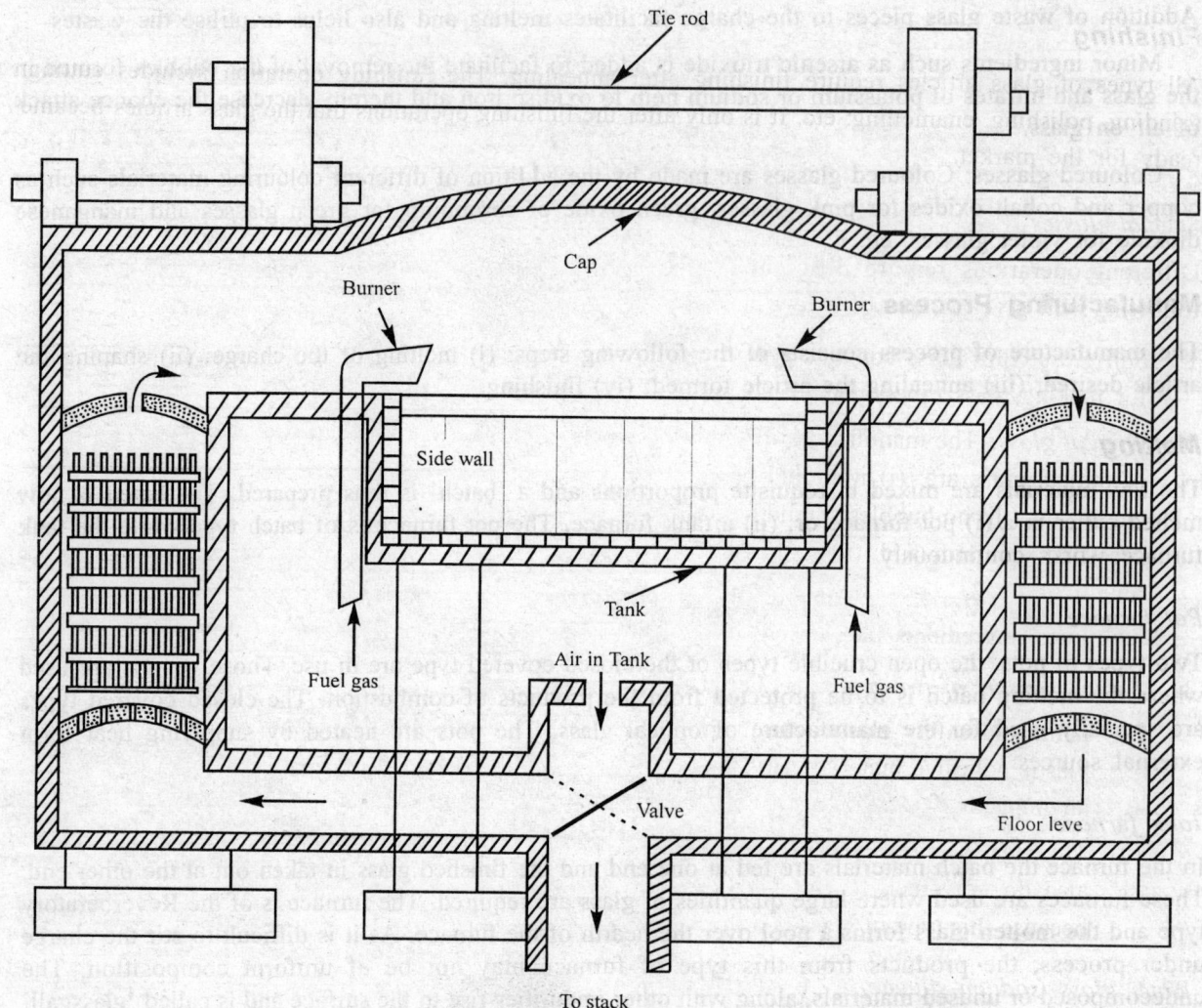

Fig. 28.4. Tank furnace for glass manufacture

Shaping or forming

The desired shape or form is given by blowing or, by moulds while the glass is still molten. In advanced countries, the blowing is still done by the mouth by skilled workers for manufacturing intricate glass articles.

Annealing

As soon as the glass is given the desired shape, it is passed through a series of cooling chambers, known as 'annealing chambers' to bring its temperature down gradually to the normal. If semi-solid, hot glass is exposed to atmosphere suddenly, unequal cooling will result. This imparts strain to the cooled glass, and it becomes brittle. This cooling operation is known as 'annealing'. Sometimes the articles are placed at the end of a tunnel or kiln as it passes out through the other end, cooled to about room temperature.

Finishing

All types of glass articles require finishing after annealing. The finishing operation includes cutting, grinding, polishing, enamelling, etc. It is only after the finishing operations that the glass articles became ready for the market.

Special glasses

Different operations require different types of glasses whose compositions necessarily vary. The following will give an idea of a few classes of special glasses :

1. *Soft glass:* Soft glass contains a high percentage of sodium and calcium silicates. It is widely used for making glass tubes, plate glass and ocmmion articles as this glass in easily fusible.

2. *Optical glass:* The manufacture of this kind of glass requires very special care. The special features of optical glass are: (i) it should be colourless; (ii) it should not contain any extraneous common chemicals; (iii) no bubbles should be present in its; (iv) the optical properties of the glass must follow specified standards and; (v) it should be able to withstand atmospheric conditions for a pretty long time.

    Sometimes, ingredients like cerium oxide are added to the optical glass to cut off ultraviolet rays.

3. *Safety glass is of two types:* (i) laminated; (ii) heat strengthened or tempered safety type.

    (i) laminated safety glasses are made by joining two thin sheets of plate glass with a non-brittle plastic adhesive material. This prevents the splinter from flying off when the glass is broken by an impact.

    (ii) The Heat Strengthened or Tempered Safety Type is actually resistant to bending; other forces also do not crack it. But they can not bear heavy impacts and fall to many pieces. This is because this type of glass is a thin plate, the outer layers of which are in a state of compression.

Water glass (sodium silicate)

Sodium silicate is manufactured by fusing soda ash with clean sand in furnaces especially suitable for silicate manufacture. These are generally of the Tank Furnace type meant for continuous production Fig. 28.5.

The heating and the operation are also similar to those of glass furnaces.

$$SiO_2 + Na_2CO_3 = Na_2O.SiO_2 + CO_2$$

The fused mass is taken out from time to time out of the furnace, cooled to solid blocks, and then broken into pieces. Sometime the liquid molten mass is broken to pieces by a stream of cold water. The pieces are then digested with steam and water under pressure in boilers. The solution thus obtained is then evaporated to required consistency and is then known as water glass.

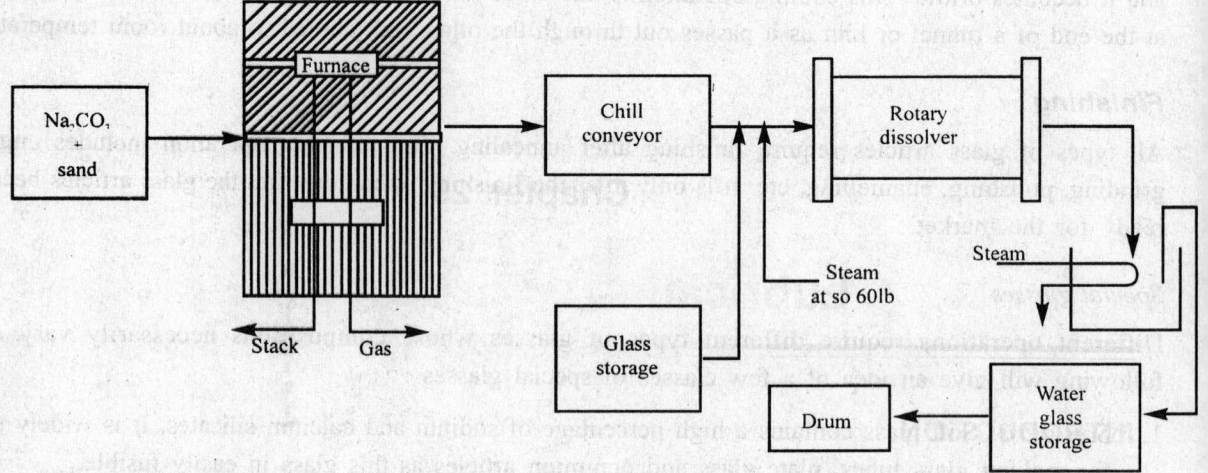

Fig. 28.5. Flowsheet for manufacture of sodium silicate (Water Glass).

Water glass is an important commercial article and used in firproofing, weigheing of silk, in preparation of adhesive. It is also used as a filler in cheap type soaps.

Chapter 29

Lubrication and Lubricants

INTRODUCTION

The process of reducing friction between two surfaces moving tangentially with respect to one another by interposing a substance of low shear strength between them is called lubrication and the interposed substance is known as a lubricant. Thus primary purpose of lubrication is suspension of moving surfaces to minimise friction and wear. The loss of energy due to friction is considerably reduced in this manner. The lubricant, in fact, acts in a number of manners, for example, it acts as a coolant or heat transfer medium by removing the heat of friction generated as a result of rubbing of surfaces. In lubrication, a thin film of lubricants is formed, between two rubbing surfaces, which does not allow a direct contact between the bubbling surfaces. In internal combustion engines, it acts as a seal by sealing the piston and cylinder wall at the compression rings. Thus there is no leakage of gases at high pressure in the combustion chamber.

FUNCTIONS OF LUBRICANTS

The main functions of the lubricants are as follows :

1. Reduces wear and tear by avoiding direct contact between the rubbing surfaces.
2. It reduces loss of energy in the form of heat by acting as coolant.
3. It increases the efficiency of the machine.
4. It minimises the liberation of heat.
5. It reduces the maintenance as well as running cost of machine.
6. It also acts as a seal to internal combustion engines.

Lubrication Principles

Several distinct regimes are commonly described to systematise the fundamental principles of lubricants. These range from complete separation of moving surfaces by a fluid lubricant, through partial separation in boundary lubrication to dry sliding where solid material properties and surface chemistry dominate. Presently, a change from sliding motion to rolling contact with elasto hydrodynamic lubrication also is possible.

761

TESTS OF LUBRICANTS

The most important tests of lubricants which are of prime importance in lubrication are copper strip corrosion, oxidation stability, inorganic sulphates, steam emission number, chlorides, viscosity, viscosity-temperature relation, viscosity index, volatility etc. In order to get efficient lubrication, there should be no change in these properties during lubrication. Moreover, there should be no effect of any chemical change such as decomposition, oxidation, reduction at higher temperature and emulsification during lubrication.

Viscosity

The behaviour of a lubricating oil or lubricant is determined by its viscosity. Viscosity is defined as the force in dynes necessary for the movement of 1 sq. cm. layer of a fluid with a velocity of 1cm. per sec., past another parallel layer 1cm away. The unit usually employed is centipoise. The measurement of viscosity is usually done in C.G.S. units and the unit is stoke. In case of thin oil which have very low viscosity, it is measured in centi-stokes. The viscosity measured by this method is called kinematic viscosity. In case of lubricants, viscosity depends upon the temperature. The viscosity has been found to decrease with increase in temperature.

Specific Gravity

The specific gravity of an oil is defined as the ratio of the weight of a given volume of oil to the weight of the same volume of water. The test can be conducted either by using a hydrometer or a density bottle if greater accuracy is desired. The temperature of the test is important and always specified by the standard specification. It has been calculated that for insulating oils specific gravity varies by 0.0005 per every degree centigrade, the value decreasing with increasing temperature. The specific gravity of the oil is considered to have no important bearing on the utility of the oil. However, an upper limit of 0.89 to 0.90 is specified so that any water present in the oil may remain at the bottom only and not tend to float on the oil. A value lower than 0.85 adversely affects other properties, particularly viscosity so that normal insulating oils have specific gravity ranging from 0.85 to 0.90. The specific gravity is used extensively to facilitate volume to weight conversions.

Copper Strip Corrosion

Crude petroleum usually contains sulphur compounds, most of which are removed during the refining processes (sweetening). This test is designed to detect any traces of free or combined sulphur that may be present in an oil. A clean polished strip of pure copper metal of the specified dimensions is heated in a fixed quantity of the oil at a fixed temperature for the specified period of time. A black, brown, or grey tarnish indicates the presence of corrosive sulphur. The presence of sulphur promotes oxidation of oil in service and causes excessive formation of sludge. The sulphur also corrodes the copper and silver metal parts of the equipment.

Oxidation Stability

The oxidation test is perhaps the most important test an oil must pass before it is accepted for service. An oxidation test submits the oil sample to accelerated oxidising conditions in a relatively short time than it is likely to meet in practice. The aim is to assess the behaviour of the oil during its long service from the results of the severe laboratory oxidation.

Though the conditions laid down in various oxidation tests of the standards are different, the accelerated conditions are brought by one or more of the following methods :

1   Increase in temperature to the order of 100°–170°C.

2   Inclusion of metals or soluble catalysts

3   Bubbling of air or oxygen in the oil during oxidation.

The oxidation stability test as laid down by the Indian Standard consists in ageing 100 grams of the oil at 150°C for a period of 45 hours during which purified air at the rate of two litres per hour is bubbled into the oil. A copper foil 51 × 32 × 0.1 mm after being cleaned and rolled into a cylinder is placed in the oil to act as a catalyst. The oxidation is carried out in a round bottomed flask of fixed dimensions fitted with a condenser through which cold water is circulated. This ensures that the volatile products of oxidation do not escape. At the end of the oxidation period the oil is diluted to four times its volume with normal-heptane and the amount of sludge and acids formed determined by analysis. The maximum limits fixed for new oils are acidity – 2.5 mg/KOH/g and sludges 1.2% by weight. Formation of deterioration products above this level has been correlated by experience, to a faster rate of deterioration in service. Following are some of the other variation of the oxidation test :

1.   The oil is heated in a metal bomb for a fixed period under specified conditions of temperature and pressure, in an atmosphere of oxygen. The sludge and acids formed are subsequently determined (American method).

2.   The oil is aged in the presence of an electrical field and oxygen and the deterioration products determined after a fixed period of time (Swedish method).

3.   The tensile strength of cotton fibres immersed in the oil, which is being aged continuously, is tested periodically. The sludge formed is also determined (Swiss method).

The repeatability of the oxidation tests is poor and standards often fix the limits of difference between duplicate tests. The rate at which an oil will deteriorate during service can be assessed by the oxidation absorption tests. Such tests though not specified by any standard are widely used to assess the quality of oils.

Inorganic Sulphates and Chlorides

The inorganic salts such as sulphates and chlorides may be present in the oil as left-overs from the refining processes. These salts and to corrosivity of the oil. Their presence is determined by making an extract of the oil with hot water. The aqueous extract is treated with Barium chloride solution for sulphates or silver nitrate solution for chlorides. Any presence of a white haziness or precipitate indicates sulphates or chlorides respectively. A new oil must not contain any quantity of these salts.

Steam Emulsion Number

Water normally cannot remain in suspension in the oil. However the presence of soaps, polar contaminants etc., tends to increase its water retention capacity. The steam emulsion number is a measure of this capacity. In this test a known amount of steam is passed into a fixed quantity of the oil under prescribed conditions. The mixture is allowed to stand and the time taken for the condensed steam to separate out is noted. This time in seconds is the steam emulsion number. This test confirms the absence of soluble impurities if the oil has a low steam emulsion number.

Viscosity index

It may also be regarded as the average decrease in viscosity per degree rise of temperature between 100°F and 210°F. In the case of most lubricants, the viscosity index (V.I.) ranges from –50 to +110.

Pour Point

It is an important property of a sample of lubricating oil and is defined as the temperature at which the oil ceases to flow when cooled under standard conditions. This property indicates the dissolved wax concentration in a given sample of lubricating oil.

Cloud Point

It is another index of dissolved wax concentration, which may be defined as the temperature at which the cloudiness appears in a sample of oil, under standard conditions, because of wax separation. Lubricants also undergo oxidation, hydrolysis and pyrolysis. The property of lubricants to undergo oxidation is different for different lubricants and mainly depends upon the composition and purity of the lubricant. As a result of oxidation, the lubricant undergoes decomposition and its lubricating property is also affected. Hence certain anti-oxidants are employed in the lubricants to avoid oxidation.

It is very important to test the *flash point* of the oil to used as lubricant. It is not advisable to use an oil as lubricant, which flashes or reaches combustion temperature under the conditions of the lubrication. This is due to the fact that spontaneous combustion of oil vapours occurs if the oxidation is so rapid that heat of oxidation is not dissipated so quickly as to avoid the reaching of flash point or combustion temperature. When lubricants such as esters are exposed to moisture at high temperatures and in the presence of acids and alkalies, they undergo decomposition as a result of hydrolysis. Polyethers and halides are hydrolysed only under drastic conditions of hydrolysis and poly siloxanes are hydrolysed easily by alkaline solutions. It is, therefore, necessary to use those lubricants, which are not capable of hydrolysing under lubricating conditions.

Interfacial Tension

The interfacial tension between the oil and water is measure of the molecular attractive force between their unlike molecules at the interface, It is expressed in dynescentimeter.

When an oil sample is poured on water the hydrocarbons comprising the oil arrange themselves at the interface and form a bond with the water molecules. This interface possesses a high tension. When the oil contains dissolved impurities which are polar in nature these tend to diffuse into the water through the interface consequently lowering the interfacial tension.

The interfacial tension is determined by means of suitable tensiometers (Torsion balance) in which the force required to lift a planar ring of platinum or a thin glass plate from the oil/water interface into the oil is measured. This value varies with temperature and ageing of the interface, and consequently the prescribed test conditions must be rigidly adhered to. A minimum interfacial tension value of 40 dynes/cm specified in some standards ensures freedom from all types of dissolved impurities.

Neutralisation Number

Neutralisation number or the acid content of an oil is defined as the number of milligrams of potassium hydroxide required to neutralise completely the acids present in one gram of the oil. Neutralisation number is the sum of the organic and inorganic acids in the oil. To determine this a weighed quantity of the oil sample is boiled with a previously neutralised mixture of alcohol and water (60:40). The resulting mixture is titrated while hot against a standard solution of potassium hydroxide using a suitable indicator. Organic acidity is the difference between the total and inorganic acidity. The maximum prescribed limit for total acidity in new oils is 0.05 mg/KOH/g.

Saponification Value

The saponification number is defined as the number of milligrams of potassium hydroxide required to saponify completely the combined and free acids present in one gram of the oil.

It is determined by refluxing a weighed quantity of the oil dissolved in ethyl methyl ketone with an alcoholic solution of potassium hydroxide. The unconsumed alkali is measured by titration with standard hydrochloric acid. The amount of potassium hydroxide thus consumed is calculated. Under these conditions many of the soaps, metals, sulphur and its compounds, halogens and nitrogen compounds consume the alkali, in addition to the substances indicated in the acidity test. Also these are potential contaminates and if present in large quantities lead to a faster rate of oil deterioration.

MECHANISM OF LUBRICANTS

Depending on the operating conditions three mechanisms may occur as:

1. Complete fluid or hydrodynamic lubrication.
2. Boundary lubrication.
3. Extreme pressure lubrication.

Complete Fluid or Hydrodynamic Lubrication

In such type of lubrication, one of the rubbing surfaces is slightly displaced in relation to the other and a wedge shaped film of the lubricant is formed between the two surfaces rubbing against each other. The pressure developed in the wedge shaped film supports the load and prevents the direct contact between the rubbing surfaces. Since a solid-solid interface has been replaced by two solid-liquid interfaces (Fig 29.1), viscosity of the lubricant becomes a crucial factor. An example of this type of lubrication is the rotation of a shaft with respect to a stationary bearing. In Fig 29.1, the lubricant occupies the annular space between the shaft and the bearing and forms a hydrodynamic wedge. So long the shaft rotates, the wedge will remain and prevent contact between two solids. When the load becomes very high, the lubricant is squeezed out of the wedge and friction occurs.

Complete fluid lubrication is provided to in delicate instruments such as watches, clocks, guns, sewing machines, scientific instruments etc. The most satisfactory and widely used lubricants for this type of lubrication are hydrocarbon oils, which are blended with selected long chain polymers in order to maintain suitable viscosity of the oil. Some anti-oxidants such as aminophenols are also added to hydrocarbon petroleum fraction which are susceptible to oxidation under operating conditions.

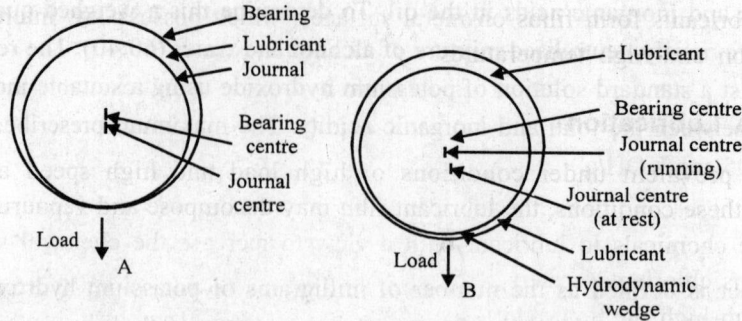

Fig 29.1. Fluid film formation in bearing "A" at rest. "B' rubbing.

Boundary Lubrication

Boundary lubrication takes place under any of the following conditions :

1. When the relative speed of the lubricated surface is insufficient to produce pressure in the lubricant film to carry the load. Such conditions are reached when a heavily loaded bearing is stopped or started.

2. When the load on bearing exceeds the pressure generated in the lubricant film.

3. When the surfaces are not perfect and the load is carried on high spots.

Boundary lubrication is therefore suitable when a continuous film of lubricant cannot persist and direct metal to metal contact is possible because of various reasons, for example, when a shaft starts moving from rest of speed is very low or load is very high or the viscosity of the oil is extremely low. Under all these circumstances friction between the lubricated surfaces lies on absorbed or boundary layer of lubricant. The load is carried by the layer of the absorbed lubricant or both the metal surfaces. The coefficient of friction in such a case is generally 0.05 to 0.15. This friction is generally dependent upon the nature of the material of the surfaces and also on some characteristics that can be determined by the chemical properties of the lubricant.

The effectiveness of boundary lubrication primarily depends upon the structure and chemical properties of the oil to be used. The lubricant used in such type of lubrication should have long chain and active atoms or groups capable of forming linkages with the metals or other surfaces. The other properties of boundary lubricants are good oilness, low pour point, high resistance to oxidation and heat and high viscosity index. It should be noted that in fluid lubrication there is no friction between the surfaces and the wear is almost negligible. So in fluid lubrication wear decreases as the friction decreases. In boundary lubrication, decrease in friction does not decrease the wear to the same extent as the decrease in friction and so minimum friction is not achieved by minimum wear.

Wear is always present in boundary lubrication and it can be reduced if metal surfaces in contact are protected from chemical attack by oxide films etc., provided the oxide films are adherent and quite hard. Certain additives such as sulphur and phosphorus are also good in preventing the surfaces from chemical attack. Compounds which contain strongly polar groups and capable of absorbing on the metal surfaces are very effective and wearing agents. Graphite and molybdenum disulphide are the two most usual solid lubricants, which either alone or as stable suspension in oil are also capable of producing boundary lubrication. These lubricants form films on metal surfaces, which possess low internal friction and can withstand compression and high temperatures.

Extreme Pressure Lubrication

This mechanism is prevalent under conditions of high load and high speed and occurs at high temperatures. Under these conditions, the lubricant film may decompose and vapourise. This necessitates the addition of some chemicals to lubricant with a view to increase the chemical stability of the film. The additives contains chlorinated esters, sulphated oils and aryl(tricresyl) phosphates. These react with metal surfaces to form high melting chlorides, sulphides and phosphides. Their action is chemical, they do not reduce friction but prevent formation of welded joints.

Substances used as lubricants

Vegetable oils and fatty acids are most important substances which are used as lubricants. Tallow, lard, whale oils, neat foot oil, rosin oil, castor oil, olive oil, coconut oil, palm oil, oleic acid etc. are the various substances used as lubricants. Mineral oils compounded with fatty acids are known as compounded oils and these are also suitable as lubricants.

Lubricating oil not only reduce wear and friction between two moving or sliding metallic surfaces by providing a continuous fluid film in between them, but they also serve as coolant sealing agent, corrosion preventors etc.

Characteristics of good lubricating oil

The characteristics of good lubricating oil are :

1. High boiling point or low vapour pressure
2. Suitable viscosity for particular service condition
3. Low freezing point
4. High thermal stability
5. High oxidation resistance
6. Non-corrosive nature
7. Resistance to decomposition at higher or elevated temperatures.

ADDITIVES FOR LUBRICATING OILS

Although the bulk of additives is ended in automotive lubricants, they also are used in oils for turbines, diesel and aircraft engines, two-cycle engines, hydraulic equipment, gears, and metal working. The common types of additives in the approximate order of their frequency of application are oxidation inhibitors, rust inhibitors, anti-wear agents, detergents-dispersants, pour-point depressants, viscosity index improvers and foam inhibitors.

Additives containing polar groupings such as fatty acids and their derivatives (e.g. multivalent metal soaps, known as metallic soaps), long chain alcohols and organic phosphides are used to ease boundary lubrication.

Certain additives such as long chain alkylated ring compounds, alkyl naphthalenes, polymers of long chain alkyl acrylated, metallic soaps of fatty acids etc. are used to reduce the temperature at which a liquid lubricant may become solid. This type of additives are known as pour depressants. Their function is colloidal and they prevent the wax crystals to grow.

Phenol and certain condensation products of chlorinated wax with naphthalene are also used as pour point depressing additives. They also prevent the separation of wax from the oil.

Additives such as phenols, their derivatives, amines, organic phosphides etc. are added to the lubricants to prevent the oxidation of lubricant and known as anti-oxidants.

The anti-oxidants or inhibitors retard the oxidation of oil by getting themselves preferentially oxidised. They are generally added to lubricant used in internal combustion engines, turbines etc., where oxidation causes a serious problem.

Additives containing strongly polar groups are added to increase the oiliness and film strength of the lubricants. Examples of such additives are amyl phenyl phosphate, methyl di-chloro stearate, dibenzyl disulphide, tricresyl phosphate, sulphurised sperm oil etc.

Oiliness of a lubricant can also be increased by adding oiliness carriers such as vegetable oils (coconut oil, castor oil etc.) and fatty acids (palmitic acid, oleic acid, stearic acid etc.)

Additives such as high molecular weight compounds like hexanol are added as viscosity index improvers. Additives are also added as detergents of deflocculents to the lubricants. Examples of such additives are salts of phenol, carboxylic acids, sulphonic acids and alkyl phosphoric acids etc. Additives added as wear inhibitors prevent wearing of the lubricating surface. Alkaline earth phenolates are generally used as anti-corrosive wear agent and tricresyl phosphate is generally used as anti-abrasive wear agent.

Corrosion preventatives are organic compounds of phosphorus or antimony. Their function is to protect the metal from corrosion by preventing contact between the metal surfaces and the corrosive substances.

The gummy and acidic constituents of lubricants can be removed by a suitable refining processes. The tendency of emulsification of the lubrication oil is checked by certain additives to the lubricant. The additives are also added or used for number of other purposes. For example, polymeric thickeners are used as additives for improving the viscosity index of the lubricant.

Additives containing polar grouping such as fatty acids and their derivatives, (for example metallic soaps), long chain alcohols and organic phosphides are used to ease boundary lubrication. Certain additives, such as long chain alkylated ring compounds, alkyl naphthalenes, polymers of long chain alkyl acrylated, metallic soaps of fatty acids are used to reduce the temperature at which a liquid lubricant may become solid. This type of additives are known as pour depressants and their function is colloidal and they prevent the wax crystal to grow. Phenol and certain condensate products chlorinated wax with naphthalene are also used as pour point depressing additives. They also prevent the separation of wax from the oil.

The anti-oxidants or inhibitors retard the oxidation of oil by getting themselves preferentially oxidised. These are generally added to lubricant used in internal combustion engines, turbines etc., where oxidation causes a serious problems.

Oiliness of lubricant can be increased by adding oiliness carriers such as vegetable oils and fatty acids.

Extreme pressure additives such as compounds containing phosphorus, lead sulphur and chlorine (e.g. fatty acids, esters ketones, chlorinated wax, sulphurised wax etc.) are added to make lubricating oil to bear extreme pressure. Under extreme pressure a thick film of oil is not capable of maintaining the film and it needs to have high oiliness high pressure additives are absorbed on the metal surface or chemically react with metal producing a surface layer of low shear strength on the metal surface. Hence tearing up of metal is prevented. High pressure additives also react on metal surface at high temperatures forming surface alloys and prevent welding to gather of the rubbing parts under severe operating conditions.

Additives such as fatty acids, oil soluble alkaline earth salts of fatty acids etc. are added to the lubricant as rust preventives. Additives such as alkali metal salts of carboxylic and sulphonic acids, non-ionic emulsifiers such as mono-esters of polyhydric alcohols, ether alcohols etc. are added to the lubricant which help the formation of emulsion with water. These additives are known as emulsifiers. It should be noted that additives are added to the lubricant taking all the factors including properties, application and effects of individual additives into consideration.

No single oil serves as the most satisfactory lubricant for various modern machineries. The properties of petroleum oils are greatly improved by adding various additives given above but the exact adjustment of the additives to a lubrication oil is one of the most complex problem. The so- called blended oils, are capable of giving desired lubricating properties, required for a particular machinery.

LUBRICANTS OF MINERAL ORIGIN

Lubricants are either of mineral origin or vegetable origin but lubricants of mineral origin are generally used as lubricants. Lubricants of vegetable origin such as vegetable oils and fatty acids are rarely used as lubricants, but they are generally used as additives to the mineral lubricants to improve their properties and qualities.

Large number of lubricants or lubricating oils are obtained from crude petroleum distillation as residue. These residues are either paraffins or asphaltic. Paraffinic residues are steam distilled to get an oil distillate and steam refined cylinder oil as the main fractions. Asphaltic residues are distilled to get various fractions with different viscosities. The lubricating oils obtained from mineral residues have a large number of impurities such as paraffin waxes, resins, asphalts etc. For a good lubricant these properties are undesirable, so it is necessary to remove them. For example, waxes increase the pour point and render the lubricant unfit for use at low temperatures. Asphaltic compounds undergo decomposition at higher temperature, causing carbon deposition and sludge formation. Resinous and asphaltic compounds are easily oxidisable and form sludge when the lubricant is used on the surfaces.

Methods for Removing Impurities

Sharpley's method

Sharpley's method is generally used for removing wax from lubricating oil. This method is based on the difference of density between precipitated wax and the solvent lubricating oil solution from which wax is separated by cooling or chilling. In this method, the lubricating oil containing wax and the solvent for dilution are mixed together and then heated to get a complete solution. The clear solution so formed is chilled down slowly to $-20°F$ and then charged into a centrifugal rotating machine. The wax is separated from the solution of the solvent and lubricating oil. It should be noted that the lubricating oil should be completely soluble in the solvent at moderately high temperature and wax should be so precipitated that it can be easily removed.

Acetone benzene solvent or propane dewaxing method

Wax can also be removed by using acetone–benzene solvent method or propane dewaxing method. In acetone–benzene solvent method acetone and benzene (1:20) are mixed and then added to the oil at 120°F. The oil solvent mixture is then chilled to –30°F. The precipitated wax is removed by using continuous filtration drum type filter. In propane dewaxing method, lubricating oil is mixed with liquid propane solvent (propane remains liquid) at a pressure of about 200 lbs. per square inch in the ration of 1:2. Asphalt is thus precipitated which is removed by cooling. It is then brought about by releasing the pressure and obtaining autogenous refrigeration. As a result, wax is thrown out.

Solvent Extraction

Aromatic compounds are removed by solvent extraction using different methods such as sulphur dioxide benzene process, nitrobenzene process, furfural process, phenol extraction process, chlorex process etc.

In sulphur dioxide benzene process, aromatic compounds are removed by using a mixture of benzene (30:70) and sulphur dioxide.

In nitrobenzene method, the solvent and lubricating oil are mixed and extraction temperature is brought to 40°F. Two layers are separated when the mixture is allowed to stand. The lower layer contains naphthenic hydrocarbons in naphthalene and upper layer contains paraffin compounds with some dissolved nitrogen. The two layers are continuously separated and the solvent is removed by evaporation in a vacuum evaporator.

In furfural method, furfural is used as a solvent for the extraction of lubricating oil. It removes olefinic and aromatic compounds but does not remove paraffinic compounds. The lubricating oil is then treated with 200% by volume of furfural for extraction.

In phenol extraction method, phenol is used as an intermediate solvent between selective solvents like nitrobenzene and furfural and the high potency solvents. Phenol is cheap, stable and can be removed easily from the oil. It also reduces the viscosity sulphur content, oxidation number and carbon content as residue.

In chlorex process, chlorex (b-b' dichloroethyl ether) is used for the extraction in the chlorex; oil ratio of 0.7:1.15. In this process viscosity and carbon residue of the lubricating oil are reduced and pour point is improved. Moreover, extraction also causes improvement in colour and oxidation stability.

Advantages

1. It is less expensive
2. It provides satisfactory removal of asphaltic impurities and hence minimises the chances of carbon deposit formation during the use of refined lubricating oil.
3. It gives high yield of refined oil and provides a very refined product which does not show much variation in viscosity due to increase in temperature.

Disadvantages

1. It yields oil which is less resistant to oxidation

2. The refined oil also posses lower oiliness and hence its oiliness is improved by adding suitable additives.

Acid treatment

Unsaturated hydrocarbons from the oil are removed by acid treatment. The lubricating oil is treated with 10% by volume of the oil of concentrated sulphuric acid. The principle of counter-current is used in the treatment. In order to avoid side reactions, generally the temperature is kept low. The oil is then separated from sulphuric acid sludge and is treated with dilute alkali and then with clay.

The acid refining can be followed after dewaxing process. If the oil still contains undesirable constituents like naphthenic and asphaltic impurities. These are eliminated by treating the oil with concentrated sulphuric acid and then agitated. Some of the impurities dissolved in the acid while others are converted into sludge. The later is removed by filtration and the filtrate is neutralised with requisite amount of NaOH so as to remove excess acid. The oil is finally decolourised by passing through fullers' earth at 100º–140ºC. It should be noted that during above treatment of lubricating oils, not only the objectionable material and oxidisable impurities are removed, but polar bodies in lubricants are also removed. These polar bodies attach themselves to lubricated surfaces and provide better lubricant because of their chemical affinity. Hence the additives are now-a-days added to ultra purified and non-reactive oily substances and these additives are capable of giving polar properties to lubricating oils.

Classification of Lubricants

Lubricants are classified on the basis of their states of aggregation or consistency. They are :

1. Liquid lubricants
2. Semi-solid lubricants
3. Solid lubricants.

Liquid lubricants (lubricating oils)

Most of the lubricating oils are of petroleum origin. When petroleum is distilled, a residual fraction collects at the bottom of the fractionating column. This oil on refractionation under reduce pressure gives three fractions—light, medium and heavy oils. These have to be purified before use as lubricants. The oil contains wax, asphalt (tar), coloured substances and oxidisable impurities. Wax prevents fluidity of the lubricant at low temperatures; asphalt deposits on the engine parts during use of the lubricant; and oxidisable impurities cause sludge formation.

Semi-solid lubricants or greases

A grease is a semi-solid lubricant obtained by thickening a lubricating oil through the addition of a metallic soap. The thickener also called a gelling agent, is usually a sodium, calcium or lithium soap. The long chain fatty acids forming the soap may be stearic acid, oleic acid or palmitic acid. If sodium soap is added to the lubricating oil, the grease is known as soda base grease; if calcium soap is added, lime base grease or cup grease results. The functions of the soaps are optimising the operating temperature, improving the resistance to oxidation and hydrolysis as well as adjusting the consistency. The different types of greases, their properties and uses are listed in Table. 29.1.

Table 29.1. Properties and uses of various types of greases.

Greases	Properties	Uses
Soda Base Grease	Low water resistance; hence can be used only in dry conditions	Used in high temperature lubrication upto 175°C in ball bearings
Lime Base or Cup Grease	General purpose greases. Can be used only upto 70°C. More resistant to water than soda base grease	Lubrication of pumps and tractors
Lithium-Soap Grease	Resistant to both heat and water.	For aircraft lubrication; from –55 to 150°C

Though soaps are the usual thickening agents in the greases, graphite, asphalt (tar) or bentonite (clay) may also be used. Among synthetic greases, the silicones are the foremost.

Conditions in which greases can be used

1. Oil is squeezed out due to heavier load or low speed, or the oil does not remain in place due to vertical flow or jerks.
2. The design of the machine does not allow the lubricant to be retained.
3. Lubricant should not contaminate the manufactured product, such as food, paper, textile etc.
4. Lubricant has to act as a screen to prevent entry of dirt or moisture.
5. Electrical motor bearings are to be fabricated.
6. Frequent application of lubricant is inconvenient, as in automobile wheel bearings.

The important properties of greases are consistency, penetration index and drop point, of which, dropping point determination helps in finding the maximum temperature of use for grease. The dropping point of grease is the lowest temperature at which the grease, when heated under specified conditions, becomes sufficiently fluid to drop from a cup through a hole of standard dimensions.

Solid Lubricants

The reasons for which greases are preferred over lubricating oils hold preference of solid lubricant as well. Only those solids which have laminar micro structure can be used as lubricants. Graphite and molybdenum disulphide are ideal lubricants on this basis.

Graphite may be used as powder or as colloidal dispersion in water or in oil. Colloidal dispersion of graphite is denoted by the dag (Deflocculated Acheason's Graphite). Dispersions in water is known as aquadag and oily dispersion is called oildag. Graphite is not effective as a lubricant in vacuum and gets oxidised above 370°C. Graphite is used as a lubricant in I.C. engines, air compressors, lathes etc.

SYNTHETIC LUBRICANTS

Many lubricants even with additives can be used under various abnormal conditions, for example when the temperature is extremely high or the atmosphere is chemically reactive. Under such conditions synthetic lubricants having appropriate properties to be used under abnormal conditions are generally employed. Such synthetic lubricants are prepared by hydrocarbons, polyglycols, esters, polycarbonates, amines, silicones and fluorine derivatives etc.

In fact synthetic lubricants are developed to meet the most sever and drastic operations such as those existing in aircraft engines in which oil may have to be pumped at –50°C. During take off or landing the same oil may get heated to as high a temperature as 250°C. Hence it is necessary for the oil to maintain satisfactory properties. The oil to be used at high temperatures and high operating pressures is expected to have high resistance to oxidation and high viscosity, and at the same time it must also possess sufficiently low freezing point and it should also be non-inflammable.

Modern Synthetic Lubricants

Modern synthetic lubricants are prepared by taking all these factors into consideration and have number of distinguishing characteristic features. They are inflammable and have high flash points, high viscosity index and high thermal stability at higher temperatures. For most common purposes blended petroleum oils are good and synthetic lubricants are needed only in special circumstances, e.g. in certain specialised military and industrial uses.

Polymerised Hydrocarbons

Polymerised hydrocarbons such as polymerised ethylene, propylene, butylene etc. having molecular weight ranging from 250 to 50,000 are used as synthetic lubricants. Although these hydrocarbons have chemical and physical properties similar to the lubricants obtained from petroleum fractions, but they have so distinct disadvantages over equivalent petroleum products. For example they can be used at high temperatures as residue free high temperature lubricants, they are light in colour and free from non-hydrocarbons, having better electrical properties.

Polyglycols and Related Compounds

Polyglycols and related compounds such as polyethylene glycol, polypropylene glycol, their ethers and esters, higher polyalkylene oxides, polythioglycols and polyglycidyl ethers can be used as water soluble and water insoluble lubricants.

Water Diluted Lubricants

Water diluted lubricants can be prepared by polyethylene glycol or polypropylene glycol to be used in rubber bearings and joints. After applying to metal surfaces they can be easily removed by water flushing. Higher polyalkylene oxides and polyglycidyl ether absorb a considerable amount of water, although they are not soluble in waters. Temperature has no effect on their viscosity. Polyalkylene oxides undergo decomposition into the volatile parts under high temperature and these volatile parts undergo oxidation. Hence these lubricants are very useful at high temperatures and used in rolling bearings of sheet glass machinery.

Natural Fats and Oils as Esters of Glycerides

They have high freezing point and can be easily oxidised. On account of this reason they cannot be used as lubricants. Diesters of glycol and aliphatic dibasic acid as well as that of aromatic dibasic acids and triesters of phosphoric acid etc. have been found to be good synthetic lubricants because their viscosity is not much affected by increase in temperature. Their volatility is very low, they can be easily oxidised by this property, this can be reduced by adding additives. They also act as good lubricant even at low temperatures. Esters as lubricants can however be decomposed by steam, acids and alkalies.

Organic Amines and Amides

Organic amines and amides are good synthetic lubricants having high viscosity index and low pour point. Silicones and the products obtained by hydrolysis and decomposition of organic chloro silicones are also employed as synthetic lubricants, because they have low viscosity temperature coefficients and are not readily oxidised upto about 200°C. They however oxidise quickly above 225°C. Silicones are very useful for low temperature lubrication of small parts.

Fluorocarbon and Other Compounds

Fluorocarbon and other compounds, e.g. tertiary amines, ethers, esters which can be completely fluorinated are also used as synthetic lubricants. They are chemically inert and resist to all chemicals except sodium. They are not decomposed easily be heat. They are also not easily oxidisable.

Preparation of Greases

By saponification of fats

Greases are generally prepared by saponification of fat such as tallow or fatty acid with alkali such as lime caustic soda, followed by adding hot lubricating oil with constant agitation soaps are also added which act as gelling agents and give an interconnected structure. At high temperature, the interconnected structure which is held by intermolecular forces ceases to exist, because soap (gelling agent) dissolves in the oil and hence the grease undergoes liquefaction.

From fatty acids and lubricating oil

Greases can also be prepared by dissolving fatty acid in part of lubricating oil. Metal hydroxide is then added to it in form of dry oil slurry or as aqueous solution. The mixture of soap and oil is dehydrated by stirring and the balance of lubricating oil is added and homogenised. Since the gelling agents in soap based grease are crystallities of fibrous shape. Their gelling property is due to their ability to form a tangle in the intersticies as a result of which the oil is held, the soap dissolved in oil. As a result, the fibrous structure no longer exists and liquefaction of grease takes place.

Greases can be made stable to oxidation and to provide lubrication under extreme conditions of load and sliding speed by adding additives to lubrication oils. The consistency of the greases considerably changes when the temperature is suddenly decrease, because the intersection of gelling agent and viscosity of oil changes with increase in temperature. The temperature at which the consistency of greases changes, determines the upper temperature limit of the applicability of the greases. Under certain conditions it is not possible to use liquid lubricant, so in such cases solid lubricants such as talcum, graphite, boron nitride, molybdenum disulphide are used.

Example 29.1. A lubricating oil has Saybolt universal viscosity of 58 seconds at 210°F and 560 seconds at 100°F. The low V.I. standard Gulf oil has Saybolt universal viscosity of 58 seconds at 210°F and 758 seconds at 100°F. The high V.I. standard Pennysylvanian oil gave the Saybolt universal viscosity value of 58 seconds at 210°F and 420 seconds at 100°F. Calculate the V.I of the lubricating oil.

Solution. We know

$$V.I.X = \frac{V_L - V_X}{V_L - V_H} \times 100$$

In the present problem V_L = 758 seconds, V_H = 420 seconds, V_X = 564 seconds. Thus,

$$V.I. = \frac{758 - 564}{758 - 420} \times 100 = \frac{194}{338} \times 100 = 57.396$$

Example 29.2. An oil of unknown V.I. has a Saybolt universal viscosity of 58 seconds at 210°F and 560 seconds at 100°F. The high viscosity index standard (i.e. pennsylvannian) oil has Saybolt viscosity of 58 seconds at 210°F and 420 seconds at 100°F. The low viscosity index standard (i.e. Gulf oil) has a Saybolt universal viscosity of 560 seconds at 210°F and 770 seconds at 100°F. Calculate the viscosity index of unknown oil.

Solution.

$$V.I. = \frac{V_L - V_X}{V_L - V_H} \times 100$$

In the present problem, V_L = 770 seconds, V_H = 420 seconds and V_X = 560 seconds. Thus

$$V.I. = \frac{770 - 560}{770 - 420} \times 100 = \frac{210}{350} \times 100 = 60$$

Chapter 30

Adhesives

INTRODUCTION

Adhesives are substances capable of holding materials together in a useful manner by surface attachment. An adhesive may be regarded as a substance or material capable of uniting or bonding two other materials together by adhering strongly to the surface of both. The bond formed, should, however, be strong internally, otherwise there would be no effective bonding. In recent years, adhesive is the general term used for bonding agents and also includes paste, glue, cement etc.

Glues, which are mostly used in wood work industry, are prepared from animal proteins such as hoofs, hides, mucilage and tendons. In recent years, all the adhesives used in wood work industry are known as glues.

Mucilage is an adhesive that can be prepared by vegetable gums and water. It is generally used for bonding paper. Pastes may be regarded as adhesive compositions having plastic consistency. They can be prepared by mixing starch and water and then cooling.

Cement means adhesive based on rubbers or thermoplastic resins dispersed in organic solvents. These adhesives undergo setting by dehydration or loss of solvent. The various important factors that are known to influence the adhesive action of an adhesive are surface tension, porosity of surface and relative smoothness, physical properties of adhesive film, thickness of adhesive film, viscosity of adhesive solution and methods of application of adhesives. A liquid adhesive is suitable for a particular surface, if it is capable of wetting that particular surface. The wettability of a particular adhesive depends upon viscosity, surface tension and a number of various other factors of the liquid adhesive.

The behaviour of adhesives applied to surfaces is greatly influenced by the porosity of the surface. In case of porous surfaces, the capillaries present on the surface conduct away the more mobile portions of the adhesive as a result of which equilibrium between solute and solvent is disturbed. In some cases, this is advantageous, e.g., in the case of silicate adhesives, the removal of water through capillary action in porous surface causes a rapid development of tack and quick drying properties. In other cases, as in the manufacture of plywood, quick disappearance of the mobile portions of the adhesive leaves the joint weak.

The adhesion of two surfaces is greatly influenced by physical properties such as compression strength, shear strength, tensile strength, creep rate, modulus of elasticity etc. The greater the tensile

776

strength, shear strength and compressive strength of a film of adhesive in between the two surfaces, the greater would be the adhesive strength. If the modulus of elasticity (the property which indicates the ability of adhesive to absorb and distribute loads from one surface to another) of adhesive film is almost equal to that of the materials bonded together, the bond is the strongest.

Adhesion of the adhesive film to bonded surfaces is independent of the thickness of the adhesive film. More viscous solution tends to form the thickest film with too many voids and with highly viscous adhesives the strength of the joint is low. The viscosity factor, however, has no effect on the strength of the joint when bonding or adhesion of surfaces is carried out under pressure and thickness of adhesive film is controlled by pressure. The strength of the joint will be independent of the thickness of adhesive film when totally liquid adhesive products are free of volatile matter. Solvent adhesives have been found to show marked decrease in strength with the increase in thickness of adhesive film.

The method of application of adhesives to bonded surfaces also depends upon various factors such as physical form of the adhesive (i.e., whether it is a prepared film, liquid, powder or combination of all these). The strength of the joint, in general, is influenced by temperature, pressure and time. Temperature and time determine whether the film of the adhesive is fully cured or dried. If surfaces to be joined are non-porous and flat, there is no need to apply pressure, because the adhering film may squeeze out leaving a poor adhesion on applying the pressure. Pressure is necessary for porous surfaces for developing sufficient cohesive strength to keep the joints together. Wood adhesives are applied by pressure.

CLASSIFICATION OF ADHESIVES

A broad scheme of classification is given in Table 30.1.

Table 30.1. Classification of various adhesives.

Origin and basic type		Adhesive material
Natural	Animal	Albumen, animal glue (including fish), casein, shellac, beeswax.
	Vegetable	Natural resins (gum arabic, tragacanth, colophony, Canada balsam, etc.); oils and waxes (carnauba wax, linseed oils); proteins (soyabean); carbohydrates (starch, dextrines)
	Mineral	Inorganic materials (silicate, magnesia, phosphates, litharge, sulphur, etc.); mineral waxes (paraffin); mineral resins (copal, amber); bitumen (including asphalt)
Synthetic	Elastomers	Natural rubber (and derivatives, chlorinated rubber, cyclised rubber, rubber hydrochloride)
		Synthetic rubbers and derivatives (butyl, polyisobutylene, polybutadiene blends (including styrene and acrylonitrile), polyisoprenes, polychloroprene, polyurethane, silicone, polysulphide, polyolefins (ethylene vinyl chloride, ethylene polypropylene)
		Reclaim rubbers
	Thermoplastic	Cellulose derivatives (acetate, acetate-butyrate, caprate, nitrate, methyl cellulose, hydroxy ethyl cellulose, ethyl cellulose, carboxy methyl cellulose)
		Vinyl polymers and copolymers polyvinyl acetate, alcohol, acetal, chloride, polyvinylidene chloride, polyvinyl alkyl ethers).
		Polyesters (saturated) polystyrene, polyamides (nylons and modifications)
		Polyacrylates (methacrylate and acrylate polymers, cyanoacrylates, acrylamide)
		Polyethers (polyhydroxy ether, polyphenolic ethers)
		Polysulphones

(Cont'd ...)

Origin and basic type	Adhesive material
Thermosetting	Amino plastics (urea and melamine formaldehydes and modifications)
	Epoxides and modifications (epoxy polyamide, epoxy bitumen, epoxy polysulphide, epoxy nylon)
	Phenolic resins and modifications (phenol and resorcinol formaldehydes, phenolic-nitrile, phenolic-neoprene, phenolic-epoxy)
	Polyesters (unsaturated)
	Polyaromatics (polyimide, polybenzimidazole, polybenzothiazole, polyphenylene)
	Furanes (phenol furfural)

Thermosetting and Thermoplastic Adhesives

Thermosetting adhesives are a general class of adhesives based upon cross-linked polymeric structure, and are incapable of being softened once solidified. Thermoplastic and thermosetting adhesives are cured (set, polymerised, solidified) by heat, catalysis, chemical reaction, free-radical activity, radiation, loss of solvent, etc., as governed by the particular adhesive's chemical nature.

Elastomers

Elastomers are a special class of thermoplastic adhesive possessing the common quality of substantial flexibility or elasticity.

Anaerobic Adhesives

Anaerobic adhesives are a special class of thermoplastic adhesive (polyacrylates) that set only in the absence of air (oxygen). The two basic types are: (i) *machinery*—possessing shear strength only; and (ii) *structural*—possessing both tensile and shear strength.

Pressure-Sensitive Adhesives

Pressure-sensitive adhesives are permanently (and aggressively) tacky (sticky) solids which form immediate bonds when two parts are brought together under pressure. They are available as films and tapes as well as hot-melt solids.

PROCESS OF BONDING

Due to the complex nature of adhesives and adherends and complexity of service requirements any exact bonding process can not be used in case of all the adhesives. However, there are various common factors, which are almost common to most of the adhesive applications, without consideration of type of adhesive and the nature of adherends involved in the bonded joints. The general factors in bonding process are briefly given below :

1. For the application of certain types of adhesives such as pressure sensitive cellophane tapes for wrapping packages, the adherents are bonded without any surface treatment. The adhesive should, however, have necessary specific adhesion to the adherent. When plastic surfaces which are moulded or pressed under high pressure and temperature are bonded together with an adhesive, they are lightly sand blasted. As a result, free surface is exposed and better specific adhesion is achieved. Certain adherent surfaces, such as metals are necessary to be treated chemically to remove dust, grease, oil and other contamination before adhesive bonding. Chemical treatment may involve solvent washing, vapour dressing, detergent washing, acid or alkali treatment or electrolytic methods etc.

2. For the use of certain adhesives, rigid adherends like wood or metal are machined to provide flat and true surfaces for proper joining of these films by using only a thin adhesive film. Mechanical surfaces also remove surface oxides and other contaminations over the surfaces and these may also interfere with the proper adhesion.

3. The preparation of adhesives can be carried out in the following steps:
   (a) Dissolving in some solvent
   (b) Melting solid adhesives
   (c) Thinning of liquid adhesives
   (d) Addition of additives such as catalyst, hardeners, fillers, extenders, plasticisers, fortifiers etc.

4. Uniform application of adhesives over the adherent surface can be carried out by any of the following processes:
   (a) By roller coating
   (b) By spraying of liquid adhesives
   (c) Roller or knife coating of molten adhesives
   (d) By brushing of adhesive over the surface
   (e) In the form of pressure sensitive tapes
   (f) By simple laying on a film of adhesive or sheet of dry adhesive.

The application of the adhesive in case of special chemically reactive adhesive system, is generally carried out in a different manner, in which one component (e.g thermosetting resin) is applied to one surface and other component (e.g. catalyst or reactivating solvent) is applied to other surface and then the two surfaces are assembled for the mixing of the component. Various reactive adhesives get hardened immediately as soon as the components are allowed to mix. Highly exothermic epoxy resin formulations belong to this type of adhesives.

5. Adhesives which are thermosetting or which are applied in molten state develop strength immediately or rapidly. When such adhesives are applied, the adherends should be immediately assembled. Adhesives which are thermoplastic or which are rubber solutions in organic solvents, are first applied on the surface of adherends and the surfaces are exposed to air for long time for drying and volatilisation of solvent, before the surfaces are assembled together. As a result of air drying of adhesive coated surfaces required tackiness is developed to form adhesion, and blisters or voids are avoided by evaporation of the volatile matter before assembling. Wood working adhesives are generally applied on surfaces and surfaces are assembled by laying them together in close contact and the assembled joint is given certain period of time before pressure is allowed to the joint. As a result, there is partial loss of solvent and completion of the reaction, specially in case of thermosetting resin adhesives. In the case of adhesives which are used for bonding metals, the coated adherends are heated in an oven either before or after the assembly of the joint and then pressure is applied. This process is known as pre-curing.

6. When the adhesive applied on adherends has reached the desired consistency, bonding pressure is applied to the assembled joint to cause the two surfaces to come in uniform contact and to allow the liquid adhesive to flow out uniformly over the whole surface, leaving no air bubbles or voids in the adhesive film. In case of pressure-sensitive or fast-setting adhesives, the applied pressure is immediately released. In case of slow setting adhesives (e.g. thermosetting resin adhesives), the pressure is applied for some time until chemical reactions and other required processes are complete

and full bond strength is developed. Temperatures of joints are maintained according to the type of mechanism used in adhesive film for the development of full strength. For example, cooling is necessary in case of molten adhesives, and heating (under pressure) is necessary in case of chemically reactive or drying adhesives.

7. Joints are conditioned after the bonding. Adhesives based on thermoplastic resins when applied, the joint is hot pressed for bonding. Such joints are cooled below a certain temperature before the pressure is released in order to develop full strength of the joint. In case of adhesives based on thermosetting resins, the joint may not have developed the full strength after the pressure is released. So the joint is further conditioned by keeping it for some more time after the release of pressure so that full strength may develop. In some joints, internal stresses may also develop after bonding. This may happen due to different temperatures and moisture contents of adherends and adhesive films. In such cases, conditioning is done of such bonded joints. The internal stresses in adhesive bonds can be reduced by a number of processes such as:

(a) Coefficient of thermal expansion of adherents and the adhesive film are brought close enough to each other by modifying the adhesive composition or by using special filters.

(b) The adherent surfaces are so prepared that there is no need for variable and excessively thick adhesive film.

(c) Mechanical properties of final dry film of adhesive are modified to such an extent that they approach those of adherent. The modification can be done by plasticising a hard brittle material to make it more flexible and tends to flow and hence relieve the internal stress which is expected to develop.

(d) The internal stress can also be avoided by avoiding the solvent in the adhesive film by any means. This can be achieved by using high concentration of adhesives in the solvents or by drying techniques.

Advantages and Disadvantages of Adhesive Bonding

Some of the advantages and disadvantages of adhesive bonding are as follows :

Advantages

Ability to bond similar or dissimilar materials of different thicknesses; fabrication of complex shapes not feasible by other fastening means; smooth external joint surface; economic and rapid assembly; uniform distribution of stresses; weight reduction; vibration damping; prevention or reduction of galvanic corrosion; insulating properties.

Disadvantages

Surface preparation; long cure times; optimum bond strength not realised instantaneously, service-temperature limitations; service deterioration; assembly fire or toxicity; tendency to creep under sustained load.

PREPARATION OF ADHESIVES

Preparation of some adhesives is given below.

Animal Glue

Animal glue is the oldest type of adhesive. It is derived from boiling animal hides, tendons or bonds which are high in collagen. These glues are of great importance because of their adhesive and gelling properties. Various adhesives can be prepared from animal glues by incorporating various additives, such as starch, plasticisers, wetting agents, fillers and preservatives, depending upon the manner of application or requirement. Animal glue obtained from hides is known as hide glue and that obtained by bones is known as bone glue.

Animal glue is widely used for glueing furniture, musical instruments, pencils, labels, stamps roll coating etc. It is also used in paper and paper board, straw boards, sizing etc.

Hide glue

Hide glue can be prepared by soaking hide trimmings in lime to dissolve and remove albuminous materials and to render the fat inactive by saponification. The glue obtained by first extraction is of best quality. The liquor is filtered and concentrated (a small amount of alum may be added as a clarifying agent) by evaporating it in small shallow pan under vacuum at low temperature. As a result, it sets firmly to a jelly, which is taken out and laid on drying nets to dry. The glue thus obtained contains about 65% moisture and may thus be pulverised.

Bone glue

Bone glue is prepared by washing the blood and dirt etc. from the crushed bones. These bones are cured or acidulated by soaking in 2–8% HCl. As a result, bone protein is converted into gelatin. The bones are degreased using benzene as a grease solvent. About 1 Ton of bones may be degreased in 4–6 hours by using 600 litres of benzene. The degreased bones are then extracted with hot water (cooking) as in case of hide glue in order to extract bone glue. The glue is filtered, concentrated, dried and pulverised as already described. It is well evident from the above discussion that preparation of animal glue includes a number of steps which include grinding of bones, cutting hides and scraps into small pieces, degreasing the material by percolating a grease solvent through it, liming and plumping, washing, making several extractions by hot water, filtering liquor, evaporating, chilling and drying etc. When dry, the slabs of glue are, flaked or ground or pulverised, blended, graded and barreled or bagged.

Other Protein Adhesives

Liquid glues made from waste materials of cod, haddock, cusk, hake and pollock are called as fish glues and have practically the same applications as other animal glues. Casein adhesives are prepared from milk derived protein, called casein and these adhesives have widely been used in wood work industry and in the manufacture of drinking cups, straws and ice cream containers. Soya protein adhesives are cheaper than casein adhesives, have similar chemical properties, but are not as good as casein adhesives. The latest protein based adhesives are peanut protein hydrates which have been found to be very suitable for making gummed tape and flexible glues for boxes and books.

Starch Adhesives

Industrial adhesives based on starch and dextrin find widespread applications in paper conversion, book binding industries etc. Adhesives based on different raw materials have different properties and are thus used for different applications. Starch adhesives or glues are used for sealing of wrappers, certain sealings, wood bonding, wall paper, office paste, laminating paper board, postage stamps, industrial paper tape etc. Starch based adhesives are usually manufactured from cornstarch, tapioca flour, potato starch, wheat flour and tamarind starch. Starch adhesives may be applied in cold and do not have the undesirable odour of some animal glues. These adhesives, however, have the disadvantage of less strength and water resistance than animal glues. Starch adhesives are less costly than synthetic resin adhesives. Native starch has widely been used as an adhesive for veneer, plywood, and corrugated and laminated boards, where water resistance is not important. In combination with 5–15% resins, such as urea formaldehyde, it is used on cartons, where water resistance is important.

Native starch

Native starch is the basic raw material for hydration to dextrin and British gums. These are modified starches used for envelope gums, gummed paper, labelling glues for glass, metal and wood, cartons, laminated boards, and padding glues.

Dextrins

Dextrins are modified starches made by heating a dry starch with dilute acid causing partial hydration. British gums are prepared by heating native starch with small amounts of catalysts. British gums are more gummier and more adhesive than dextrins.

In the manufacture of starch adhesives, dextrins, British gums or starches are mixed with a number of other chemicals in order to get a product of good quality. For example, borax is added to increase viscosity, rate of tack, gumminess and rate of production. A 50% dextrin solution is equal to 25% dextrin solution plus 10–15% borax, NaOH improves penetration for rosin-sized materials. Na_2CO_3, KOH and K_2CO_3 also function in the same manner, but to lesser degree. A number of other materials, such as plasticisers, deforming agents, preservatives, fluidifying agents, (they delay viscosity increase with ageing), colouring agents, flavouring agents and emulsifying agents are also added in the manufacture of starch adhesives, in order to get a product of good quality and required properties.

The starch is stirred constantly with water and caustic lye is added slowly. After about one and a half hour borax as well as acid is added slowly. Now the paste is placed in slightly alkaline urea to lower the viscosity of the paste. Dicyanidamide may also be used instead of urea for the same purpose. The pH of the liquor is adjusted by adding monosodium phosphate. There are other adhesives which are prepared by solution of gum arabic or acacia in water. Simple solution of gum arabic in cold water gives a pale yellow paste, which is quick drying and stable. Gum is soaked in water for a day and then stirred. The ratio of gum to water varies from 1 : 2 to 1 : 3. The solution is filtered through cloth bags and chemicals like borax, gelatin perborate and glycerine etc., along with preservatives like sodium benzoate are added to get the product of good and stable quality.

Tapioca dextrin in the adhesive has widely been used for postage stamps. Another quick setting machine adhesive, for kraft paper, envelope cover is based on dextrin. First heat dextrin with hot water maintaining the temperature at 85–90°C. Now add glucose and cool and solution to about 70–75°C. Add phosphoric acid and formaldehyde slowly.

Synthetic Resin Adhesives

Synthetic resin adhesives are of great practical importance, especially when water resistance is required and special conditions are to be met. Urea resin adhesives consist of urea-formaldehyde alone or combined with starch based adhesives, when water resistance is desirable. These adhesives are mostly used in wood work industry, where high water resistance is not essential.

Phenolic resin adhesives

Are based on phenol or its derivatives such as resorcinol, which are condensed with aldehydes and ketones. They can be completely polymerised in presence of proper curing and suitable catalyst. They are thermosetting adhesives and used in wood gluing, whether water resistance is most important.

Epoxy resins

Are obtained by the condensation of phenol, acetone and epichlorohydrin. They are particularly useful in metal bonding.

Alkyds, acrylates, methacylates, allyls, and hydrocarbon polymer

Silicone resins and natural and synthetic rubbers, all find applications as adhesives or bases. Vinyls are large and important class of adhesives which contain water soluble and insoluble members, polymers.

Rubber Based Adhesives

These adhesives are used for joining cloth and metals, rubber or rubber sponge with metals, rubber and metals with paper. These adhesives are heat resistant as well as water and weather resistant. They have very good cementing properties. Rubber based adhesives are available in a number of types. For example, one such type contains 40–45 per cent reclaimed rubber, resin, drier and solvent. Another type contains, gum, rubber resin and about 15% solids. A third type of such adhesive or cement contains compounded latex. Fourth type may contain reclaimed rubber, natural or synthetic rubber, resin and asphalt dispersed in water. Other type of such cement is also available which only contains reclaimed rubber, asphalt and solvent.

In addition to a number of solvents (such as petroleum ether, solvent naphtha, spirit, carbon tetrachloride, benzene and toluene etc.), various other types of ingredients are also used in the formulation and these materials impart different quality properties to the product.

Cellulose and Silicate Adhesives

Cellulose acetate, nitrate, butyrate and ethyl cellulose are most important cellulose adhesives for solvent soluble adhesives. Methyl and carboxy methyl cellulose are important for aqueous adhesives. There are a number of more products in the market. Examples are sodium silicate solutions, mucilage (a solution of gum arabic or acacia in water), asphalts, waxes, rosin, rubber and shellacs. All, except sodium silicate, are more or less additives and specially adhesives.

Silicate based adhesives are colloidal silicates as 37–47% solutions. These solutions are used alone or in combination with many other materials as adhesives for plywood, wall board, flooring and metal foils. Silicate adhesives have high viscosity and contain about 32–47% solids. They are insect-proof and thus can be used for corrugated boxes. The most important properties of such adhesives are fast setting, good contact and spreading, and formation of a stable, permanent, rigid water and heat resistant bond. Silicate adhesives are quite cheap and can be prepared by using starch, water and sodium silicate.

Starch is first stirred with water and then the paste is added to the silicate and the mixture is again stirred. It is now heated on a steam bath until the solution is clear. The wet track of the adhesives may be increased by adding some dextrin to it.

APPLICATIONS OF VARIOUS ADHESIVES

The applications of various adhesives are given as under:

1. *Thermoplastic adhesives:* Construction, textile and packaging.
2. *Thermosetting adhesives:* Wood-bonding, construction, foundry, and automobiles.
3. *Elastomeric adhesives:* Construction and rubber.
4. *Acrylic adhesives:* Construction, textile, packaging, automobiles, electrical, abrasives, and furniture.
5. *Cyanoacrylate adhesives:* Electronics, automobiles, and toys.
6. *Anaerobic adhesives:* Automobile and structural bonding.
7. *Polyvinyl acetate adhesives:* Construction, packaging, textiles, furniture and book-binding.
8. *Ethylene vinyl acetate adhesives:* Packaging, textile, book-binding, furniture, automobiles, construction, and footwear.
9. *Polyurethane adhesives:* Textiles, forest products, packaging, construction, automobiles, furniture, footwear.
10. *Construction adhesives:* Gypsum board, concrete, glass-fibre, rockwool insulation, ceramic tiles, vinyl flooring, wall covering, pipe joint cements, doors, safety glass, plywood floor, asphalt roofing, paper and film lamination.
11. *Furniture adhesives:* Lamination, epoxy, rubber, and wood.
12. *Automotive adhesives:* Autobody exterior (hood and deck lids), FRP bonding, autobody interior (carpet, package trays, fabric foam seats), safety glass, and tyre adhesives.
13. *Aircraft adhesives:* Metal bonding, composite bonding, and laminated glass.
14. *Electrical/Electronic adhesives:* Batteries, TV tubes, build-up mica, magnet bonding, flexible printed circuits.
15. *Appliance adhesives:* Refrigerators and freezers, washers and dryers, air conditioners, dishwashers, microwave ovens.
16. *Packaging adhesives:* Corrugated board, solid fibreboard, can assembly, folding carbon, bags, gummed tape/paper, envelopes, paperboard lamination, labelling, cigarette making.
17. *Pressure sensitive adhesives:* Tapes, natural rubber, butyl rubber, SBR plastics, electrical duct pipe wrap, medical.
18. *Textile adhesives:* Automobile, medical, sanitary towels, furniture, carpets, footwear, handbags, polyester, PVC.
19. *Consumer adhesives:* White glue, neoprene and natural rubber, epoxy, starch, silicone.
20. *Shoe adhesives:* Sole attaching, neoprene and natural rubber, polyester.

Chapter 31

Metals and Alloys

INTRODUCTION

Metals occur in nature in free as well as combined states. Metals having low reactivity or those at the bottom of the reactivity series have very low or little affinity for air, moisture, carbon dioxide or other non-metals under reaction conditions prevalent in nature. Such metals remain in elemental or native or free state in nature. Gold, platinum, silver, and copper may be found in free state in nature and are called 'native metals'.

On the other hand, most of the metals are active and combine with moisture, air, carbon dioxide and non-metals like carbon, nitrogen, oxygen, phosphorus, arsenic, sulphur, halogens, etc. to form their compounds. Such metals occur in nature in the form of their compounds or in the combined state. A natural material in which a metal or its compound occurs is called a mineral. A mineral from which a metal can be extracted economically is called an ore. Thus, all ores are minerals but all minerals may not be ores. Ores are invariably found in nature in contact with rocky materials. These rocky, clayey or earthy impurities accompanying the ores are termed gangue or matrix.

COMMON ORES

Most common ores of metals are their oxides, carbonates, silicates, phosphates, arsenics, sulphides, sulphates and halides. Some metals occur in native or free state. Some of the important ores are mentioned below.

Native Ores

These include the ores of metals (with very low reactivity) such as, silver, gold, platinum, copper, mercury, etc. These metals are found in contact with rocky materials, clay or sand. Lumps of pure metal may also be found sometimes as Nuggets. Metallic nodules are also said to be found at the seabed.

Oxide Ores

Important oxide ores of metals like iron, aluminium, zinc, manganese, tin, copper, etc., are haematite (Fe_2O_3), bauxite ($AlO_3.2H_2O$); zincite (ZnO); pyrolusite (Mr, O_2), cassiterite (SnO_2) and cuprite (Cu_2O), respectively.

Carbonate Ores

These are mostly formed by the combination of atmospheric carbon dioxide with metal oxides or hydroxides. Some of the common carbonate ores are: siderite ($FeCO_3$); calamine ($ZnCO_3$); dolomite ($MgCO_3$ $CaCO_3$); lime stone ($CaCO_3$); malachite [$CuCO_3.Cu(OH)_2$]; and cerrusite ($PbCO_3$).

Sulphide Ores

Many metals (especially the heavier ores) combine with sulphur in the soil to form metal sulphides. Some of the important sulphide ores are: iron pyrites (FeS_2); copper glance (Cu_2S); galena (PbS); zinc blende or blackjack (ZnS); cinnabar (HgS); nickel blende or millerite (NiS); silver glance or argentite (Ag_2S).

Sulphate Ores

Sulphate ores may undergo oxidation by atmospheric air to form sulphates. Some of the important sulphate ores are: barytes ($BaSO_4$); anglesite ($PbSO_4$); epsomiete ($MgSO_4.7H_2O$); anhydrite ($CaSO_4$) etc.

Arsenide Ores

The arsenical ores are formed by the combination of naturally occurring arsenic with the metal. Some of the important arsenical ores are: kupfernickel ($NiAs$); nickel glance ($NiAsS$), etc.

Phosphate Ores

There are not many phosphatic ores. Lead occurs as pyromorphite [$3Pb_3(PO_4)_2PbCl_2$]. Calcium occurs as rock phosphate $Ca_5(PO_4)_3(OH)$.

Silicate Ores

Most of the silicate ores require an expensive and difficult treatment for the extraction of metals. Some of the common silicate ores are: hemimorphite ($2ZnO.SiO_2.H_2O$), asbestos ($CaMg_3Si_4O_{12}$); mica ($KAl_3H_2Si_3O_{12}$), felspar ($K_2O.Al_2O_3.6SiO_2$).

Halide Ores

Metallic halide ores are not many but some of these are quite important. For example, horn silver ($AgCl$); carnallite ($KCl.MgCl_2 6H_2O$); common salt or rock salt ($NaCl$); fluorspar (CaF_2); cryolite (Na_3AlF_6); etc. Besides that chlorides of sodium, potassium and magnesium and bromides and iodides of magnesium and potassium are present in sea-waters as well as salt beds.

METALLURGICAL PRINCIPLES AND PROCESSES

The process of extraction of metals from their ores economically is called *metallurgy*. The process of extraction for each metal depends upon its nature and the nature of the ore. The ores are mostly oxides, carbonates, sulphides, sulphates, halides and silicates. Before taking up the stepwise metallurgical processes employed during the recovery of metal from an ore, we describe here a few broad principles used for the isolation of metals depending upon the chemical composition of an ore.

Metallurgical Principles

Oxide ores

Metals like iron, aluminium, zinc, tin, bismuth, chromium and manganese occur as their oxides. Such ores are normally reduced to metallic state by heating with charcoal. Some of the examples are given below:

$$Fe_3O_3 + C \rightarrow 2Fe + 3CO$$

$$ZnO + C \rightarrow Zn + CO$$

$$SnO_2 + 2C \rightarrow Sn + 2CO$$

$$Bi_2O_3 + 3C \rightarrow 2Bi + 3CO$$

However, in cases where the metal has a tendency to react with carbon at the reduction temperature, the oxide ore is generally reduced with aluminium powder (Goldschmidt-thermite process).

$$Cr_2O_3 + 2Al \rightarrow 2Cr + Al_2O_3$$

$$3Mn_3O_4 + 8Al \rightarrow 9Mn + 4Al_2O_3$$

Electrolytic reduction is done in cases where none of the above methods is applicable. As for example, metallic aluminium as obtained by the electrolytic reduction of alumina Al_2O_3. Since the melting point of Al_2O_3 is very high (2288K), therefore, it is not electrolysed directly at such high temperatures but as a solution in molten cryolite Na_3AlF_6 (melting point = 1270 K). The following reactions occur at electrodes:

$$Al_2O_3 \rightleftharpoons 2Al^{3+} + 3O^{2-} \qquad \text{(solution)}$$

$$2Al^{3+} + 6e^- \rightarrow 2Al \qquad \text{(cathodic process)}$$

$$3O^{2-} \rightarrow 3/2O_2 + 6e^- \qquad \text{(anodic process)}$$

Carbonate ores

Carbonates are generally heated to form oxides which are then reduced to metallic state as discussed below:

$$ZnCO_3 \rightarrow ZnO + CO_2$$

$$PbCO_3 \rightarrow PbO + CO_2$$

$$FeCO_3 \rightarrow FeO + CO_2$$

$$4FeO + O_2 \rightarrow 2Fe_2O_3$$

Sulphide ores

Sulphide ores are generally contaminated with sulphur and sometimes even with arsenic. Such ores are heated in air to convert sulphide into oxide with is then reduced with C or Al to get the metal.

1. Zinc blende ZnS is heated in a free supply of air

$$2ZnS + 3O_2 \rightarrow 2ZnO + 2SO_2$$

$$ZnO + C \rightarrow Zn + CO$$

2. Copper glance Cu_2S or cuprous sulphide Cu_2S, obtained by the treatment of copper pyrites $CuFeS_2$, is heated in air to form Cu_2O which combines with Cu_2S in the furnace itself to form metallic copper.

$$2CuS + 3O_2 \rightarrow 2Cu_2O + 2SO_2$$

$$\underline{Cu_2s + 2Cu_2O \rightarrow 6Cu + SO_2}$$

$$3Cu_2S + 3O_2 \rightarrow 6Cu + 3SO_2 + heat$$

3. Galena PbS is heated in air at 770 to 870 K to cause partial oxidation of the ore as follows:

$$2PbS + 3O_2 \rightarrow 2PbO + 2SO_2$$

$$PbS + 2O_2 \rightarrow PbSO_4$$

At this stage, air supply is cut off and heating is done at high temperatures when sulphide combines with oxide and sulphate to form Pb as follows:

$$PbS + 2PbO \rightarrow 3Pb + SO_2$$

$$PbS + PbSO_4 \rightarrow 2Pb + 2SO_2$$

4. Cinnabar, HgS is heated in a current of air to form mercury:

$$HgS + O_2 \rightarrow Hg + SO_2$$

Halide ores

Water soluble ionic chloride (such as those of alkali metals Li, Na, K, Rb, Cs and alkaline earths Mg, Ca, Sr, Ba are very stable and do not decompose easily to form metals. Such metal chlorides are electrolysed in their fused state.

$$NaCl \rightleftharpoons Na^+ + Cl^-$$

$$Na^+ + e^- \rightarrow Na \quad \text{(at the cathode)}$$

$$2Cl^- \rightarrow Cl_2 + 2e^- \quad \text{(at the anode)}$$

Insoluble chloride like that of silver AgCl is brought into solution by treatment with dilute solution of NaCN.

$$AgCl + 2NaCN \rightarrow Na[Ag(CN)_2]\ NaCl$$

The argentocyanide solution is made alkaline with NaOH and then treated with Zn or Al to obtain metallic silver.

$$2NaAg(CN)_2 + Zn \rightarrow Na_2[Zn(CN)_4] + 2Ag$$

$$Na_2[Zn(CN)_4] + 4NaOH \rightarrow Na_2ZnO_2 + 4NaCN.$$

Sulphate ores

The chief mineral of barium in nature is barytes $BaSO_4$. The powdered ore is heated with charcoal at 370–1070 K to form the sulphide as:

$$BaSO_4 + 2C \rightarrow BaS + 2CO_2$$

$$BaSO_4 + 4C \rightarrow BaS + 4CO$$

METALLURGICAL PROCESSES

Metal ores are invariably found mixed with rocky and earthy impurities. The metal content in the ore varies depending upon the impurities present and the chemical composition of the ore. Although the economical extraction of metal on industrial scale is greatly dependent on the proper choice of ore, yet there are some processes which are common in metallurgy: (i) crushing and pulverisation; (ii) concentration or ore-dressing; (iii) calcination and roasting; (iv) reduction to metallic state; (v) bessemerisation; (iv) purification and refining; and (vii) solvent extraction.

Crushing and Pulverisation

The ore is generally available in big lumps or pieces. In order to make metallurgical operation easier and efficient, the lumps or big stones of ore are broken into smaller pieces by crushing with the help of mechanical crushers. The lumps or ore stones are brought in between the plates of a crusher forming a jaw. One of the plates of the crusher is stationary while the other moves to and fro and the crushed pieces are collected.

The crushed pieces of the ore are then pulverised in a stamp mill. The heavy stamp rises and falls on a hard die to powder the ore which is then carried away through a screen by a stream of water.

Pulverisation can also be done in a ball mill. A ball mill can be visualised as a steel cylinder containing iron balls. The steel barrel is mounted in such a way, so that it can be rotated freely. After adding the crushed ore pieces into the barrel, it is set on revolving motion so that the steel balls are continuously striking against each other and the walls of the barrel. This pulverises the ore pieces when these come in between the balls or balls and barrel.

Concentration or Ore-dressing

The ore are usually found mixed with large proportions of non-metallic matter, such as quartz, felspar, mica, sand, lime stone, earthy and rocky impurities. These unwanted impurities in the ore are called gangue or matrix. The process of removal of gangue from the powdered or is called concentration or ore-dressing. It is necessary to concentrate an ore before proceeding to extract the metal from the ore. The concentration operation may utilise one or more of the following processes.

Hand-picking

It is sometimes possible to have an ore that may be isolable in a sufficient degree of purity by means of simple picking it up by hand and breaking away the adhering rocky materials with a hammer.

Gravity-separation

It is based on the difference in the specific gravity of metallic ore and the gangue particle. Therefore, it is used for the concentration of heavier oxide ores like haematite, tin-stone and native gold. In this process the powdered ore is agitated with water or washed with a strong current of water. The heavier ore settles down rapidly and the light sandy and earthy materials and washed away. The two common methods employed for the hydraulic washing of the ore are:

1. Hydraulic classifier
2. Wilfley's washing table.

Hydraulic classifier

It consists of a large conical reservoir (Fig. 31.1) fitted with an ore inlet at the top and water inlet at the bottom. There is a provision to remove concentrated ore from the bottom and light gangue particles from the top as shown in Fig. 31.1. Finely powdered ore is slipped into the reservoir while a strong stream of water is pushed up form the bottom. Lighter gangue particles are taken away by the current of water and the heavier ore particles settle down at the apex of the cone. The conical shape with increasing upward diameter helps in reducing the velocity of water thereby preventing the heavier ore particles from being carried away along with the gangue.

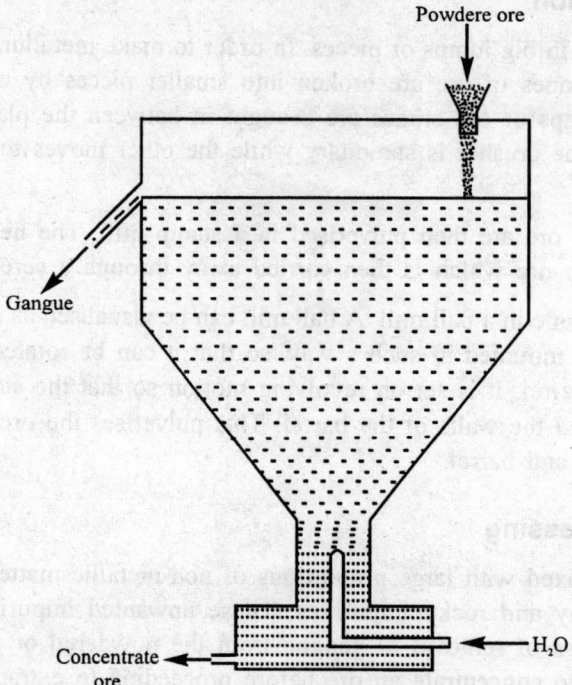

Fig. 31.1. Hydraulic classifier

Wilfley's washing table

It is particularly useful for the concentration of native gold ore and cassiterite (SnO_2). Wilfley's table consists of a wooden surface fixed with an inclination towards one side. Long wooden strips called closets are fixed to the table and the powdered ore is introduced at the top and a stream of water is made to flow across the table. The table is given a regular jerking motion. With each jerk the lighter gangue particles are thrown upwards and are carried across the table by the current of water. The heavier metal bearing particles (collecting at the cleats) are carried towards one side of the table by the jerking motion of the table. In this way, a reasonably good concentration takes place before the ore reached the bottom.

Magnetic Concentration

It is used specially for the separation of magnetic materials from the non-magnetic ones. As for example, tin-stone (SnO_2) is non-magnetic but contains magnetic impurities such as, iron and manganese tungstates ($FeWO_4$, $MnWO_4$). Similarly, magnetic titanium or (Ilmenite, $FeTiO_3$) can be separated form non-magnetic impurities of chlorapatite 3 $Ca_3(PO_4)_2CaCl_2$ by magnetic separation. Chromite $FeCr_2O_3$ is magnetic and can be separated form non-magnetic silicious gangue by this method.

The powdered ore is made to fall from a hopper on to a leather or rubber belt moving over two rollers a one of which is flitted with an electromagnet (M) as shown in Fig. 31.2. The ore travels slowly towards the right hand side magnetic roller and is influenced by the magnetic field. As a consequence, the magnetic material gets attracted and falls from the belt in a heap nearer the magnet and non-magnetic materials fall in heap away from the magnet due to the influence of the centrifugal force Fig. 31.2.

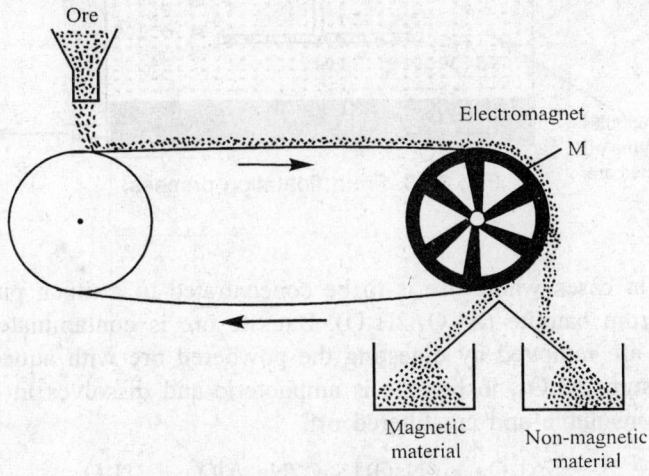

Fig. 31.2. Magnetic separator

Electrostatic concentration

It is used for the separation of metallic portions of the ore from that non-metallic or for the separation of better conductors of electricity from the poorer ones. It is based on the principle that an electrostatic field would charge the good conducting material immediately which when submitted to an electrode carrying a like charge is repelled and thrown away subjected to the influence of an electrostatic field. This method is employed for the separation of PbS and ZnS occurring together in an ore. The PbS being a good conductor is charged at once and thrown away from the roller, whereas, ZnS being a poor conductor drops vertically from the roller.

Froth flotation

This method is suitable especially for the concentration of low grade sulphide ores, such as, galena (PbS), zinc blende (ZnS), copper pyrites ($CuFeS_2$), etc. It is based on the different wetting properties of the surfaces of ore and gangue particles. The sulphide ore particles are wetted preferentially by oils and gangue particles by water. In the process, the powdered ore is mixed with water and proper wetting agent (usually pine oil) and fed into a vat. The slime is agitated vigorously by blowing in compressed air as shown in Fig. 31.3. The oil forms a foam or froth with air. The sulphide ore particles stick to the froth which rises to the top of the vessel. The gangue particles are left in water and sink to the bottom from where these are drawn off. The froth is skimmed off, collected and allowed to subside and dry.

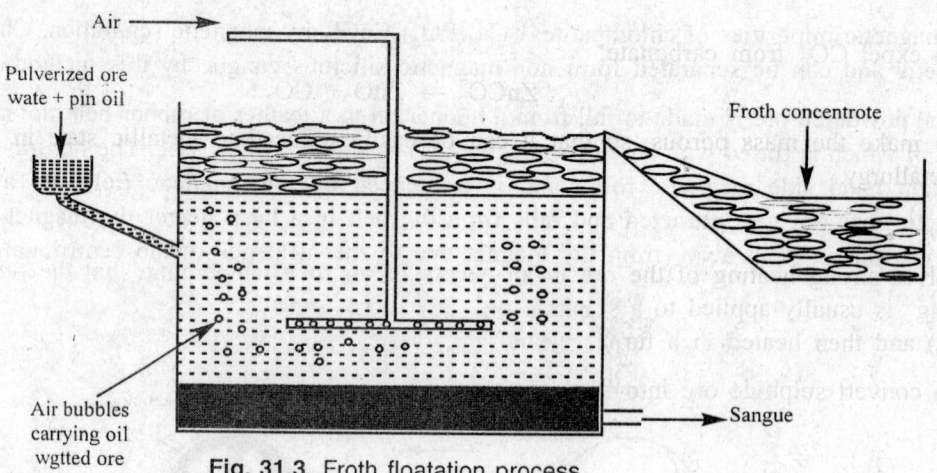

Fig. 31.3. Froth floatation process.

Chemical method

It is employed mostly in cases where ore is to be concentrated to a much purer form, this is done in aluminium extraction from bauxite ($Al_2O_3.2H_2O$). Bauxite ore is contaminated with Fe_2O_3, TiO_2 and SiO_2. These impurities are removed by digesting the powdered ore with aqueous solution of NaOH at 420–440 K under pressure. Al_2O_3, in bauxite is amphoteric and dissolves in NaOH, whereas, Fe_2O_3, SiO_2 and TiO_2 remain insoluble and are filtered off.

$$Al_2O_3 + 6NaOH \rightarrow 2Na_3AlO_3 + 3H_2O$$

The filtrate containing sodium aluminate Na_3AlO_3 is diluted with water and a little freshly precipitated $Al(OH)_3$ is added to cause precipitation of $Al(OH)_3$ from the bulk. It is filtered and ignited to get pure Al_2O_3.

$$Na_3AlO_3 + 3H_2O \rightarrow Al(OH)_3 + 3NaOH$$
$$2Al(OH)_3 \rightarrow Al_2O_3 + 3H_2O$$

Calcination and roasting

The concentrated ore is converted into a metal oxide by calcination or roasting in specially designed shallow furnaces.

Calcination

It involves strong heating of the ore in the absence of limited supply of air to a temperature that the ore does not melt.

The purposes of calcination are :

1. To remove water of hydration.

$$Al_2O_3.2H_2O \rightarrow Al_2O_3 + 2H_2O\uparrow$$

2. To expel water form a hydroxide.

$$2Fe(OH)_3 \rightarrow FeO_3 + 3H2O\uparrow$$

3. To expel CO_2 from carbonate.

$$ZnCO_2 \rightarrow ZnO + CO_2\uparrow$$

4. To make the mass porous, so that it can be easily reduced to metallic state in the next step of metallurgy.

Roasting

It involves strong heating of the ore in an excess of air to a temperature that the ore does not melt. Roasting is usually applied to a sulphide ore. The ore is mostly mixed with coal or coke (which acts as fuel) and then heated in a furnace. The purposes of roasting are:

1. To convert sulphide ore into oxide or sulphate.

$$2ZnS + 3O_2 \rightarrow 2ZnO + 2SO_2$$

$$PbS + 2O_2 \rightarrow +PbSO_4$$

2. To remove water of hydration.
3. To remove arsenic and sulphur impurities which escape as volatile oxides.

$$S + O_2 \rightarrow SO_2$$

$$4As + 3O_2 \rightarrow 2As_2O_3$$

(iv) To make the ore porous so that it undergoes easier reduction in the next step of metallurgy.

Calcination and roasting processes are generally done in a reverberatory furnace Fig. 31.4. Reverberatory furnace consists of a fire box and a big hearth lined inside with sand or any other **suitable** flux Fig. 31.4.

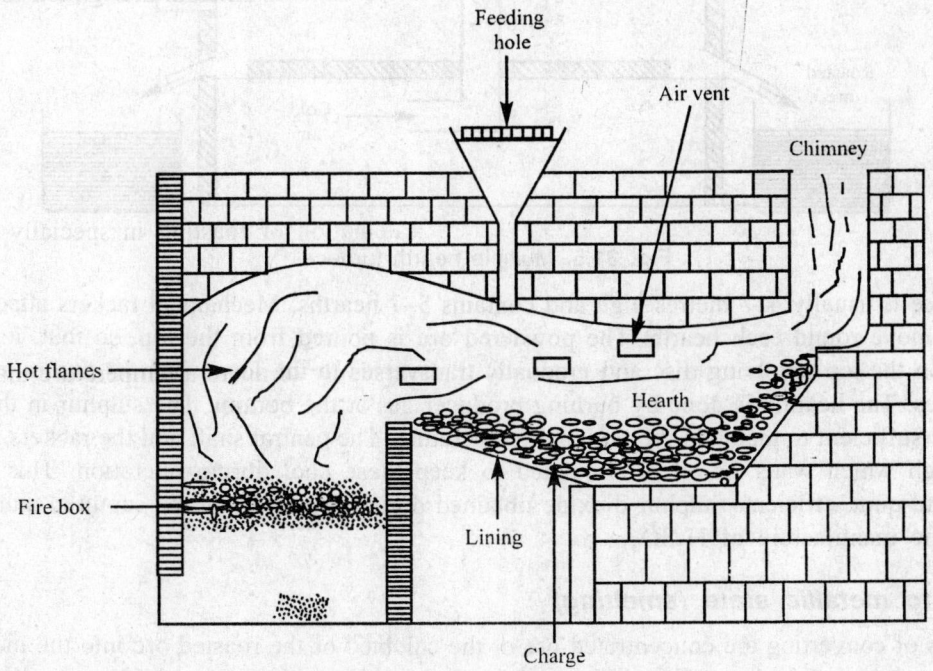

Fig. 31.4. Reverberatory furnace.

Air supply is manipulated by vents and direct blasts. The charge is placed on the hearth and heated by directing the flames the fire place. The concave roof of the furnace reflects or reverberates the rediant heat or hot flames on to the charge in the hearth. Stirring of the charge is done mechanically through locking doors.

Calcination and roasting are also carried out in an improved multiple hearth furnace as shown in Fig. 31.5.

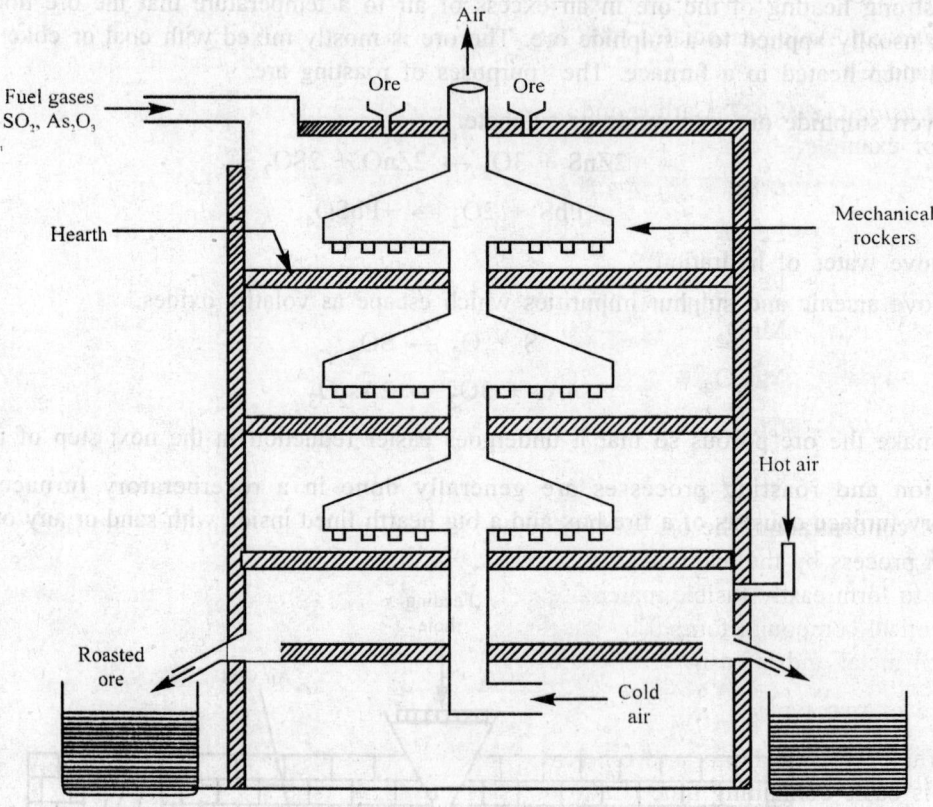

Fig. 31.5. Multiple hearth furnace.

The furnace is usually 4–7 metres high and contains 5–7 hearths. Mechanical rackers attached to the central shaft move round each hearth. The powdered ore is poured from the top, so that, it spreads in a thin layer on the top revolving disc and gradually transverses to the lower hearths due to a system of revolving discs. The heating is done by burning producer gas at the bottom. The sulphur in the or burns and is usually sufficient to maintain the required temperature. The central shaft and the rackers are usually hollow through which water or oil is circulated to keep these cool during operation. This process is continuous and quite efficient. Sulphur dioxide obtained during roasting may be captured and converted to SO_2 for the manufacture of H_2SO_4.

Reduction to metallic state (smelting)

It is a process of converting the concentrated ore or the calcined or the roasted ore into the metallic state by reduction using chemical or electrolytic method. Some commonly use methods are given below.

Smelting

The extraction of metal in the fused (molten) state is termed smelting. The roasted or calcined ore containing metal oxide is mixed with a reducing agent and heated. Depending upon its nature, the extraction of the metal can be brought about by one of the following reductants:

1. *Carbon and carbon monoxide:* In the metallurgy of Fe, Cu, Pb, Sn, Mg, Co and Zn.
2. *Aluminium, sodium, magnesium or hydrogen:* In the metallurgy of Cr, Mn, Ti, Mo and W.
3. *Water gas (CO + H$_2$):* In the metallurgy of Ni.
4. *Auto-reduction:* Anion associated with metal in the ore brings about the change in the metallurgy of Pb, Hg and Cu.

1. *Smelting using C and CO:* Carbon and carbon monoxide reduce the oxide to the free metal in molten state. For example,

Ore		Reductant		Metal		Volatile
Fe_2O_3	+	$3C$	→	$2Fe$	+	$3CO$
Fe_2O_3	+	$3CO$	→	$2Fe$	+	$3CO_2$
MnO_2	+	$2C$	→	Mn	+	$2CO$
Mn_2O_3	+	$3C$	→	$2Mn$	+	$3CO$
Co_3O_4	+	$4C$	→	$3Co$	+	$4CO$
Co_2O_4	+	$4CO$	→	$3Co$	+	$4CO_2$

In spite of concentration, the ore may still carry some gangue material which are finally removed in the reduction process by the addition of flux during smelting. The flux combines with gangue at higher temperatures to form easily fusible material called slag which is not soluble in the molten metal. Slag is a fusible chemical compound formed by the reduction of flux with gangue. The slag, being lighter, floats on the molten metal and is easily removed.

Fluxes are of two types:

1. Acidic fluxes—silica, borax and other non-metallic oxides are acidic fluxes. These are used when gangue is basic containing lime (CaO) or other metallic oxides.

$$\underset{\text{(Acidic flux)}}{SiO_2} \quad + \quad \underset{\text{(Basic gangue)}}{CaO} \quad → \quad \underset{\text{(Slag)}}{CaSiO_3}$$

2. Basic fluexes—lime stone (CaCO$_3$) magnestite (MgCO$_3$), etc., are basic fluxes. These are employed to remove acidic gangue, such as silica, silicates, etc.

$$\underset{\text{(Basic flux)}}{CaCO_3} \quad + \quad \underset{\text{(Acidic gangue)}}{SiO_2} \quad → \quad \underset{\text{(Slag)}}{CaSiO_3} \quad + \quad CO_2 \uparrow$$

This shows that the choice of flux depends upon the nature of ore and the metal to be extracted.

Smelting, using carbon as reductant, is generally done in blast furnace for the isolation of copper, lead or iron metals from their respective oxides. The working of blast furnace is detailed below:

Blast furnace consists of a tall cylindrical furnace made up of mild steel plates. It is lined inside with fire clay refractory bricks. The mouth of the furnace has a cup and cone feeder and two tapping holes are provided near the bottom; one for the removal of molten metal and the other for slag. Hot blasts of air are supplied to the furnace by circular pipes through a number of tuyers (Fig. 31.6). A water cooling system is provided in the brick work all around the hearth to take care of high temperatures.

A charge consisting of calcined ore, hard coke (fuel and reductant) and lime stone (flux) in appropriate ratios are introduced into the furnace and heated with hot air blasts rising upwards. The charge melts and the molten metal, slag are removed separately near the bottom (Fig. 31.6). The furnace can run non-stop for years.

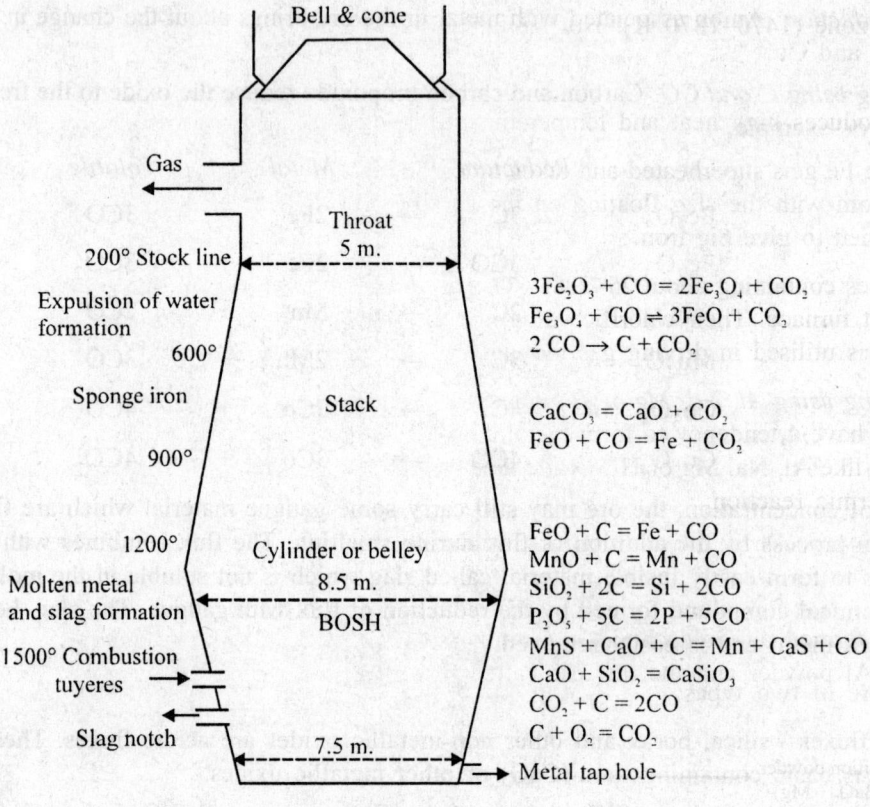

Bell & cone

Gas

Throat
5 m.

200° Stock line

Expulsion of water
formation
600°

Sponge iron

900°

1200°

Molten metal
and slag formation

1500° Combustion
tuyeres

Slag notch

Stack

Cylinder or belley
8.5 m.
BOSH

7.5 m.

Metal tap hole

$3Fe_2O_3 + CO = 2Fe_3O_4 + CO_2$
$Fe_3O_4 + CO \rightleftharpoons 3FeO + CO_2$
$2\,CO \rightarrow C + CO_2$

$CaCO_3 = CaO + CO_2$
$FeO + CO = Fe + CO_2$

$FeO + C = Fe + CO$
$MnO + C = Mn + CO$
$SiO_2 + 2C = Si + 2CO$
$P_2O_5 + 5C = 2P + 5CO$
$MnS + CaO + C = Mn + CaS + CO$
$CaO + SiO_2 = CaSiO_3$
$CO_2 + C = 2CO$
$C + O_2 = CO_2$

Fig. 31.6. Blast furnace

Different chemical reactions take place in different zones of the furnace depending upon the temperatures. Zones with their temperature ranges and reactions are mentioned below:

Preheating zone (470–620 K): It provides preheating effects and helps in the evaporation of moisture.

Reduction zone (620–1470 K): In the 620–1070 K range, the oxide ore is reduced to spongy metallic Fe and lime stone decomposes.

$$Fe_2O_3 + 3CO \rightarrow 2Fe + 3CO_2$$

$$CaCO_3 \rightarrow CaO + CO_2$$

In the range 1070–1470 K, the following reactions occur:

$$CO_2 + C \rightarrow 2CO$$

$$Fe_2O_3 + 3C \rightarrow 2Fe + 3CO + heat$$
$$CaO + SiO_2 \rightarrow CaSiO_3 \text{ (slag)}$$

At about 1470 K, side reactions involving other oxides occur and the elements, thus formed, are absorbed by metallic Fe.

$$SiO_2 + 2C \rightarrow Si + 2CO$$
$$P_2O_5 + 5C \rightarrow 2P + 5CO$$

Fusion zone (1470–1870 K). Hot air burns coke at the base to form CO.

$$2CO + O_2 \rightarrow 2CO + heat$$

This produces high heat and temperature in the vicinity of tuyers touches a maximum of 1870 K.

Metallic Fe gets superheated and takes up carbon at this temperature. The molten metal trickles down to the bottom with the slag floating on the surface. The molten metal is so poured into moulds where it is solidified to give pig iron.

Hot gases containing about 25–28% CO, 12–15% CO_2, 58–60% N_2 and small amount of H_2 pass out of the blast furnace. Their calorific value is quite high because of large amount of CO in the mixture. Their heat is utilised in driving gas blowing engines and for preheating the air blast.

2. *Smelting using Al, Na, Mg or H_2:* Ores which can not be reduced by carbon, CO or produce metals which have a tendency to form metal carbides in the process of reduction, are reduced by reducing agents like Al, Na, Mg or H_2. Oxide ores like Cr_2O_3 or Mn_3O_4 are reduced by Al powder in a highly exothermic reaction

$$Cr_2O_3 + 2Al \rightarrow 2Cr + Al_2O_3 + heat$$
$$3MnsO_4 + 8Al \rightarrow 9Mn + 4Al_2O_3 + heat$$

This is termed as Goldschmidt's Alumino thermite reductions. The oxide-ore is mixed with equal weight of Al powder and the mixture called thermite is placed in the cavity of fluorspar (Fig. 31.7).

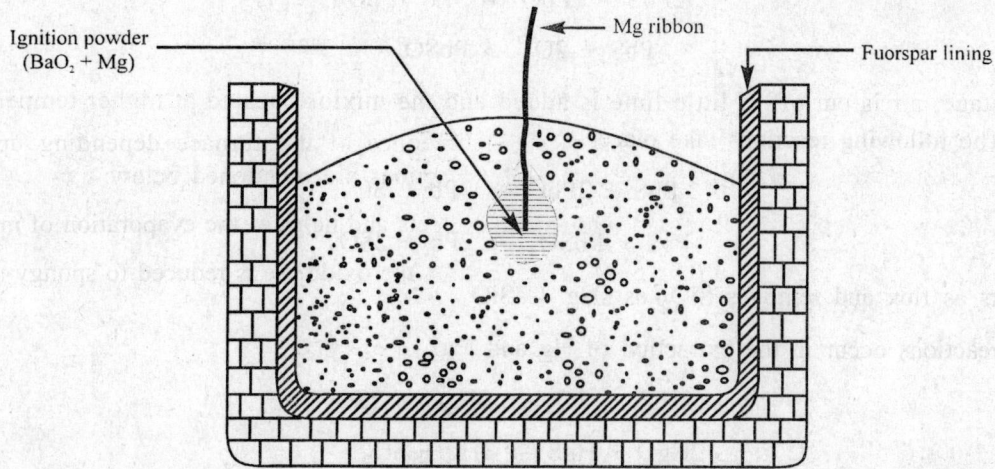

Fig. 31.7. Goldschmidt's alumino-thermite reduction.

A small cartridge of BaO_2 and Mg powder is placed under thermite and connected to a long Mg ribbon. The reaction is started by lighting the Mg ribbon with ignites the cartridge and the thermite with a big flash resulting in pure molten metal at the bottom with Al_2O_3 floating on it.

Strontium and barium metals can be obtained by the reduction of their oxides with Al powder. The reduction is done at higher temperatures in vacuum.

$$3BaO + 2Al \rightarrow 3Ba + Al_2O_3$$

Pure titanium, zirconium and vanadium metals are obtained by the reduction at higher temperatures (1070–1170 K) of their tetrachlorides with sodium or magnesium.

$$TiCl_4 + 4Na \rightarrow Ti\ 1 + 4NaCl$$

$$VCl_4 + 2Mg \rightarrow V + 2MgCl_2$$

Molybdenum and tungsten oxide ores are reduced in a current of hydrogen at elevated temperature (1270–1470 K).

$$WO_3 + 3H_2 \rightarrow W + 3H_2O$$

3. *Smelting using water gas:* Nickel oxide NiO, obtained from the roasting of the sulphide ore (NiS), is heated carefully in a tower at 600 K in which a current of water gas ($CO + H_2$) is passed.

$$NiO + CO \rightarrow Ni + CO_2$$

$$NiO + H_2 \rightarrow Ni + H_2O$$

4. Auto-reduction. Smelting can also be brought about by auto-reduction, where in, the anions associated with the metal bring about the change. This is possible in the sulphide ores of Pb, Hg and Cu.

The ore is heated in a supply of air at moderate temperatures (770–970 K) to effect partial conversion as :

$$2PbS + 2PbO \rightarrow 3Pb + SO_2$$

$$PbS + 2O_2 \rightarrow PbSO_4$$

At this stage, air is cut off, a little lime is added and the mixture heated at higher temperatures (1070 K). The following reactions take place:

$$PbS + 2PbO \rightarrow 3Pb + SO_2$$

$$PbS + PbSO_4 \rightarrow 2Pb + 2SO_2$$

Lime acts as flux and removes SiO_2 as slag, $CaSiO_3$.

Similar reactions occur in the extraction of Hg and Cu.

$$2HgS + 3O_2 \rightarrow 2HgO + 2SO_2$$

$$2HgO + HgS \rightarrow 3Hg + SO_2$$

$$Cu_2S + 2Cu_2O \rightarrow 6Cu + SO_2$$

Reduction by precipitation

Noble metals like gold and silver are extracted from their concentrated ores by dissolving metal ions in the form of their soluble complexes. The metal ion is then precipitated by adding Zn dust. For example, concentrated argentite Ag_2S is first treated with a diluted solution of NaCN to form the soluble complex argentocyanide.

$$Ag_2S + 4NaCN \rightleftharpoons 2Na\,[Ag(CN)_2] + Na_2S$$

This solution is decanted off and made alkaline by adding NaOH. It is then treated with Zn or Al to precipitate out Ag.

$$2Na[Ag(CN)_2] + Zn \rightarrow 2Ag + Na_2[Zn(CN)_4]$$
$$\text{(ppt.)}$$

$$Na_2[Zn(CN)_4] + 4NaOH \rightarrow Na_2ZnO_2 + 4NaCN$$

Electrolytic reduction

Very active metals like Na, K, Mg, Ca, Al, etc., are execrated by the electrolysis of their fused anhydrous extracted salts like halides, oxides or hydroxides. For example, sodium is obtained by the electrolysis of fused sodium chloride.

$$NaCl \rightleftharpoons Na^+ + Cl^-$$

At cathode, $Na^+ + e^- \rightarrow Na$ (metal)

At anode
$$Cl^- \rightarrow Cl + e^-$$
$$Cl + Cl \rightarrow Cl_2$$

Electrolyte	At cathode	At anode
$MgCl_2$	Mg	Cl_2
Al_2O_3	Al	O_2
NaOH	Na, H_2	O_2

Purification and refining

Except electrolysis, metal produced by any of the above mentioned methods is generally impure or in crude form. The impurities may be in the form of: (i) foreign metals; (ii) unreduced oxides of the metal; (iii) non-metals like C, Si, P, As, S, etc.; and (iv) slag or flux. Crude metal is refined by one or more of the following methods:

Liquation

Easily fusible metals like tin, lead and bismuth are refined by this process. Here the impure metal is poured on the sloping hearth of a reverbratory furnace (Fig. 31.8) and heated gently to a temperature little above the melting point of the metal. The pure molten metal drains out leaving behind the infusible matter (called dross) on the hearth.

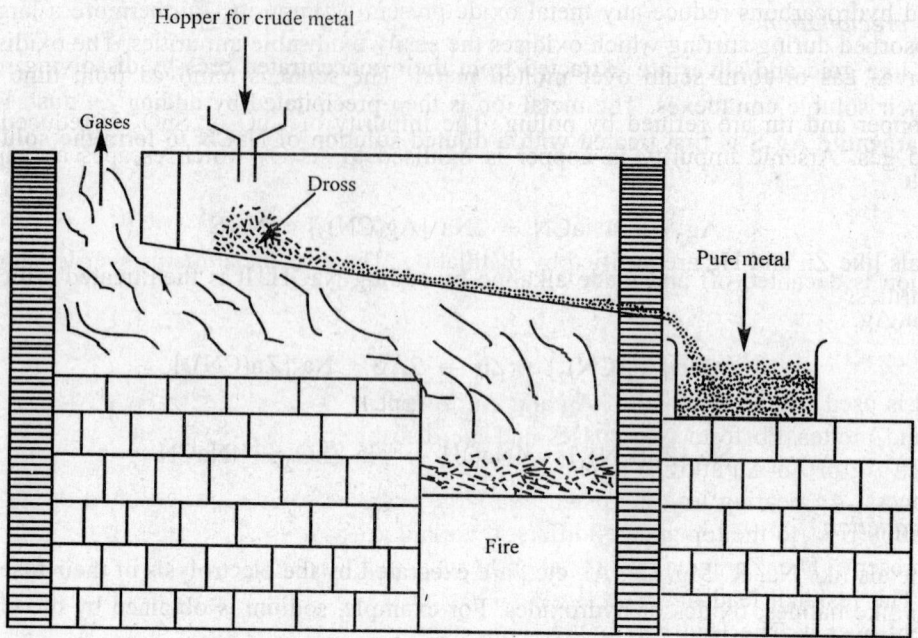

Fig. 31.8. Liquation in reverberatory furnace.

Poling

Poling consists in stirring the molten impure metal with green logs of wood or green bamboo sticks as shown in Fig. 31.9.

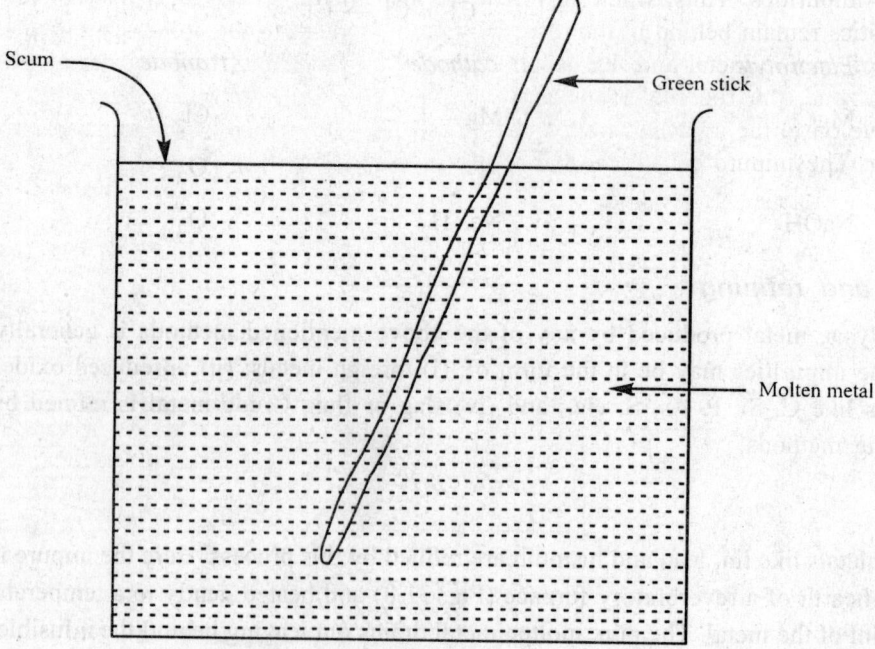

Fig. 31.9. Poling

The wood hydrocarbons reduce any metal oxide present as impurity. Furthermore a large quantity of air is also absorbed during stirring which oxidises the easily oxidisable impurities. The oxidised impurities escape either as gas or form scum over molten metal. The scum is removed from time to time.

Blister copper and tin are refined by poling. The impurity of CuO or SnO_2 is reduced to Cu or Sn by the wood gas. Arsenic impurity in copper is oxidised to As_2O_3 which escapes as vapour.

Distillation

Volatile metals like Zn and Hg are purified by distillation. The pure metal distils over leaving behind non-volatile impurities.

Distribution

This method is used for the removal of Ag and Au from lead. This method is based on the principle that molten Zn and molten Pb form two phases and the distribution ratio of Ag and Au is in favour of Zn phase. The Zn-Ag or Zn-Au alloy is lighter than Pb and freezes first and can be skimmed off as it cools. In actual process, Ag bearing lead is melted in a large vessel and 1–2% Zn is added to the molten mass. The Zn-Ag alloy rises to the top and solidifiers. It is removed and distilled with a little charcoal to reduce any oxide present. The Zn metal distils over leaving behind Ag. The crude Ag is then purified by cupellation. The residual lead in crude Ag is melted in small bone-ash dishes called cupels and a blast of air is swept over it. The lead gets oxidised to litharge PbO which forms scum over molten Ag or is carried away by the blast of air. The scum is removed and molten Ag is received in ingots.

Zone refining

This method is used for the preparation of high quality materials, such as silicon, germanium, gallium and tellurium (used as semi-conductors). It is based on the principle that the melting point of a substance is lowered by impurities. Thus, when an impure molten metal is cooled, crystals of pure metal appear and the impurities remain behind in the remaining molten metal. In the process, a circular heater is fitted around a rod of impure metal and the heater is slowly moved along the rod from one end to the other. At the heated zone, the rod melts but as the heater moves on the pure metal crystallises out while impurities move on to the adjacent molten part (Fig. 31.10). In this way, the impurities are swept to one end of the bar. The impure end of the bar is removed. This process may be repeated to get very pure metal.

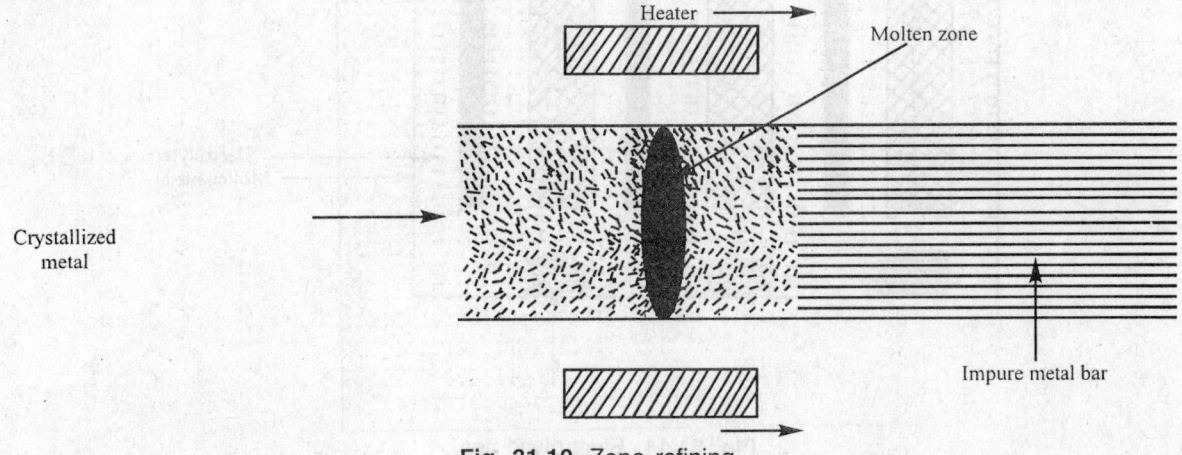

Fig. 31.10. Zone refining.

Van Arkel-de Boar method

Extremely pure Ti, Zr or V can be made on the laboratory scale by this method in which pure metal iodide is vapourised and decomposed on a hot wire in vacuum.

$$\underset{\text{(Crude)}}{Ti} + 2I_2 \rightarrow \underset{\text{(Pure)}}{TiI_4}$$

$$TiI_4 \xrightarrow[\text{Vacuum}]{\text{Hot wire}} \underset{\text{(very pure)}}{Ti + 2I_4}$$

Electrolytic refining

A large number of metals like Cu, Ag, Au, Cr, Zn, Sn, Pb, Ni, etc. are refined electrolytically. A block of impure metal is made the anode and a thin sheet of pure metal forms the cathode of the electrolytic cell containing suitable metal salt solution. On electrolysis, pure metal deposits at the cathode sheet while more electropositive impurities are left in solution and noble metallic impurities settle below the anode as "anode mud".

As for example, in the electrolytic refining of crude copper, a series of slabs of crude copper behaving as anode are suspended into a large tank containing the electrolyte made up, of 15% solution of $CuSO_4$ mixed with 5% H_2SO_4. Thin plates of pure copper, suspended in between, behave as cathode (Fig. 31.11). On passing the electric current, the metal from anode dissolves into the electrolyte as Cu^{2+} ions.

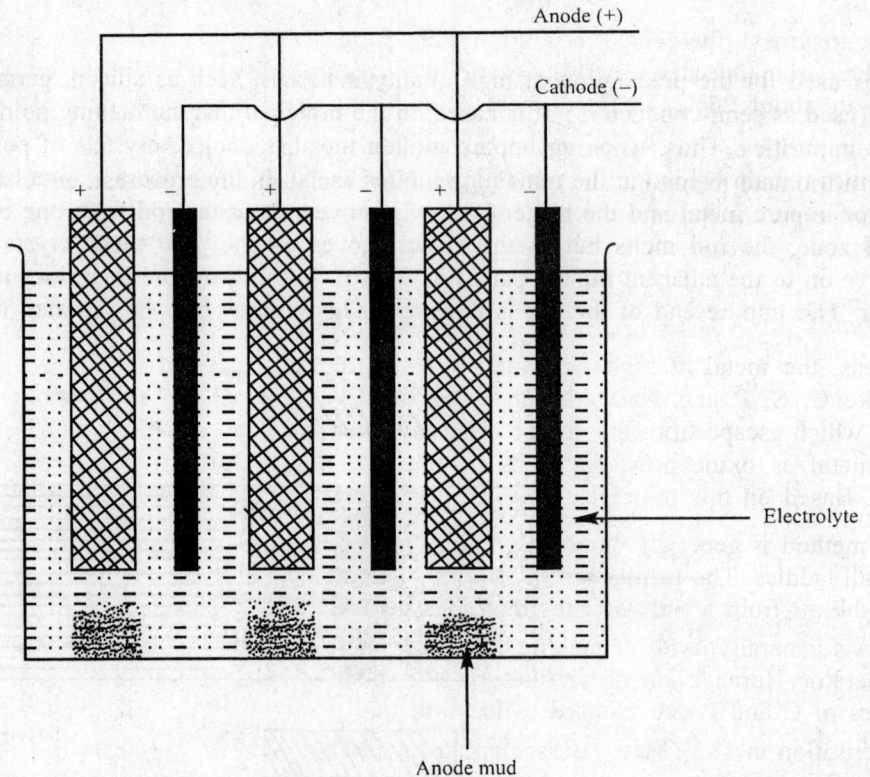

Fig. 31.11. Electrolytic cell.

These Cu^{2+} ions at the anode go into solution and equivalent amount of Cu^{2+} ions from the electrolyte get reduced to Cu at the cathode and are deposited there.

$$Cu^{2+} + 2e^- \quad \rightarrow \quad Cu$$

(In solution)   (At cathode)

The impurities of metals above Cu in the activity series (that is, more basic than Cu), such as, Zn, Pb, Ni, etc. also get ionised but remain in solution. On the other hand, metals below Cu in the activity series (that is, noble metals) like Ag, Au, etc. do not get ionised and fall below as anode mud. Anode mud is then treated suitably to isolate Ag and Au.

Parting

Refining of gold contaminated with silver is termed as parting of gold. Parting may be brought about by any one of the following methods:

1. *Chlorine parting:* The crude metal is melted in a crucible and dry chlorine is bubbled through the molten mass. Silver is converted into AgCl whereas gold remains unaffected. Silver chloride is removed as scum form the surface of the molten Au.

2. *Acid parting:* The crude metal is treated with hot concentrated H_2SO_4 or boiling HNO_3 when silver dissolves and gold remains unaffected.

$$2Ag + 2H_2SO_4 \rightarrow Ag_2SO_4 + 2H_2O + SO_2$$

$$Ag + 2HNO_3 \rightarrow AgNO_3 + H_2O + NO_2$$

For H_2SO_4 treatment, the content of gold in the crude metal should not be more than 25%, otherwise it protects silver from the attack of H_2SO_4. In such cases, silver is added from outside to reduce the gold content to about 25%. This treatment of reducing gold content to 25% is sometimes called quartation.

In case of nitric acid treatment, the crude gold is alloyed with silver, so that, gold content is reduced to 40–50%. Silver is recovered from the $AgNO_3$ solution by treatment with Cu turnings.

$$2AgNO_3 + Cu \rightarrow Cu(NO_3)_2 + 2Ag\downarrow$$

Oxidative refining

In this process, the metal at high temperatures or in molten state is exposed to air oxidation. The impurities like C, S, P and As are oxidised to their volatile oxides (CO_2, SO_2, P_2O_5 and As_2O_3 respectively) which escape from the charge. The unwanted metallic impurities get oxidised and float on the molten metal as oxide powders or as fusible slag after combination with flux SiO_2 at higher temperatures. Based on this principle, different methods are employed for different metals :

1. Tossing method is generally employed for the purification of lead. Here the molten-metal is tossed with small laddles. The laddles are filled with molten metal and then the molten metal is allowed to drop in the air from a little height. Impurities get oxidised and float on the molten metal.

2. Puddling is generally used for making wrought iron. The molten pig iron is placed on the hearth of a reverberatory furnace and the molten mass is puddled (stirred) with long bars of wrought iron. Impurities of C and P get oxidised to leave behind purer iron.

3. Bessemerisation in 1855 Henry Bessemer, an English steel maker, introduced his process (popularly known as Bessemerisation) in the steel making industry. Bessemerisation is the fastest and simplest of all the steel making processes and is carried out in a special furnace called Bessermer Converter.

Pig iron, obtained in the blast furnace, contains 92–94% iron, 2–5% carbon, some impurities like S, P, Mn, Si etc. Bessemerisation involves the making of steel by the chemical oxidation (using oxygen of the air) of impurities of pig iron into volatile gases and slag. This is followed by addition of requisite amounts of alloying elements like C, Mn, Ni etc. This process is carried out in Bessemer converter which is a larger pear-shaped steel vessel looking like a concrete mixer lined inside with refractory bricks. Fig. 31.12. It is mounted on axles, so that, it can be tilted for charging and discharging. The bottom of the convertor is perforated for blowing in the air. The nature of refractory lining depends upon the composition of pig iron used in the converter.

(a) If the pig iron does not contain much P (is 0.1%) refractory lining of silica bricks is used. This is called Acidic Bessemer Process.

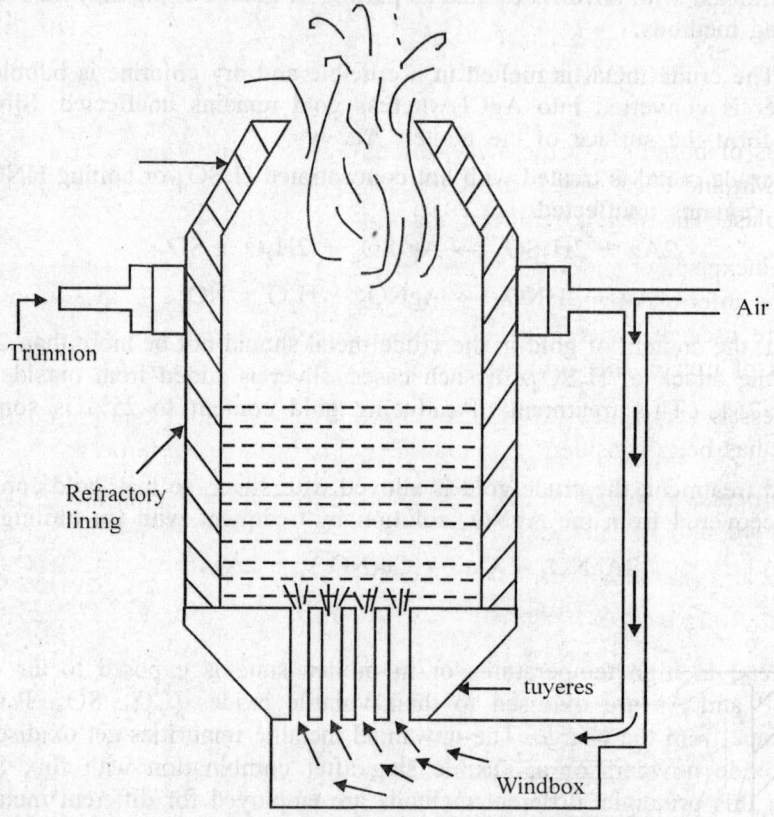

Fig. 31.12. Bessemer converter.

(b) If the pig iron contains large amounts of P (0.6–1.5%) refractory lining of calcined dolomite (CaO.MgO) bricks is used. This is called Basic Bessemer Process.

Working

The converter is tilted and a charge of about 10 tons of pig iron is charged into it directly form the blast furnace. The converter is brought back into vertical position and a hot blast of air is forced through the molten material. During the first few minutes, all impurities like C, Si, P, Mn, etc. get oxidised to their respective oxides which either escape or combine with flux to form slag.

Carbon is oxidised to carbon monoxide which burns with a blue flame at the mouth of the converter. The other reactions taking

$$2C + O_2 \rightarrow 2CO\uparrow$$

place in the converter are :

$$Si + O_2 \rightarrow SiO_2$$

$$2Mn + O_2 \rightarrow 2MnO$$

$$MnO + SiO_3 \rightarrow MnSiO_2 \text{ (Slag)}$$

In the basic process, P is oxidised to P_2O_5 which combines with basic dolomite to from calcium and magnesium phosphate slag.

$$4P + 5O_2 \rightarrow 2P_2O_5$$

$$3CaO.MgO + 2P_2O_5 \rightarrow Ca_3(PO_4)_2 + Mg_3(PO_2)_2$$
$$\text{(Thomas Slag)}$$

As soon as the flames of burning CO die away, the air blast is cut off and a calculated amount of Spiegel (an alloy of Fe, Mn and C) is added to get steel of required quality. The charge is remixed with the help of the hot air blast. The slag is removed and molten steel is poured out.

Bessemer process is inexpensive and rapid but the steel made by this process is not considered to be very fine. The steel is inferior due to "blow holes" (due to sudden escape of gases) and traces of slag in it. Such steels are used in concrete reinforcement for structural work. Bessemer steel is more liable to corrosion. Loss of iron in this process is quite significant (~15%).

Bessemerisation process is very quick (20 minutes) but difficult to conrol and the quality of product is variable. Therefore, it has been considerably replaced by the *open hearth process*.

In the *open hearth process*, a charge consisting of pig iron, scrap iron, iron ore (haematite Fe_2O_3) and lime stone (flux) is fed into the shallow hearth furnace lined with calcined dolomite (CaO.MgO) bricks (Fig. 31.13).

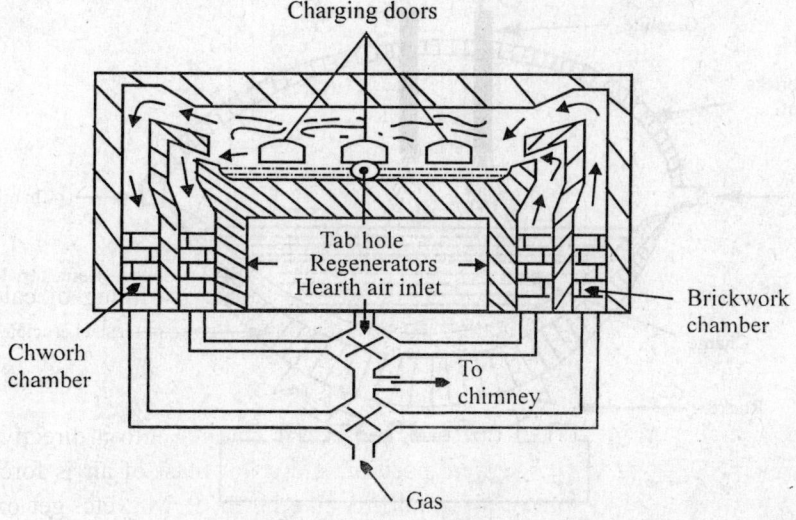

Fig. 31.13. Open-hearth process.

Preheated air and producer gas from checker work on one side of the furnace are fed into the hearth and burnt over it to produce high temperatures (~2000 K). The hot gases pass through the checker brick work on the opposite side and recycled.

During heating Fe_2O_3 oxidises impurities to their respective oxides which react with SiO_2 and P_2O_5 to form slag.

$$3Si + 2Fe_2O_3 \rightarrow 4Fe + 3SiO_2$$
$$6P + 5Fe_2O_3 \rightarrow 10Fe + 3P_2O_5$$
$$CaO + SiO_2 \rightarrow CaSiO_3 \text{ (slag)}$$
$$3CaO + P_2O_5 \rightarrow Ca_3(PO_4)_2 \text{ (slag)}$$

This process is long and takes 10–12 hours. The molten metal and slag are separated as usual. A calculated quantity of spiegel (C, Mn and Fe alloy) is added to get steel of required type. This process has some advantages over Bessemer's process.

1. It utilises waste scrap steel.
2. Loss of Fe in the process is low (~4%).
3. It is easy to control and gives steel of required quality.
4. Heating is economical and based on regenerative system of heat economy.
5. Steel of good quality is produced. It does not have blow-holes as in Bessemer process.

Since 1940, both Bessemer and open-hearth processes have been superseded by the Electric furnace. Earlier it was used for the production of steel from pig iron. Now it is generally used for the making of quality steel and alloy steels out of steel manufactured by Bessemer and open-hearth furnaces. Various modifications of electric furnaces are available but the most modern one is Heroult electric arc furnace.

Heroult electric arc furnace consists of a huge crucible made up of welded steel plates (Fig. 31.14). It is mounted on rocker which allow the furnace to tilt when required. The refractory lining of the furnace is acidic or basic. The furnace has a charging door and a tap hole. The roof of the furnace is made up of silica bricks from which two graphite electrodes extend into the charge.

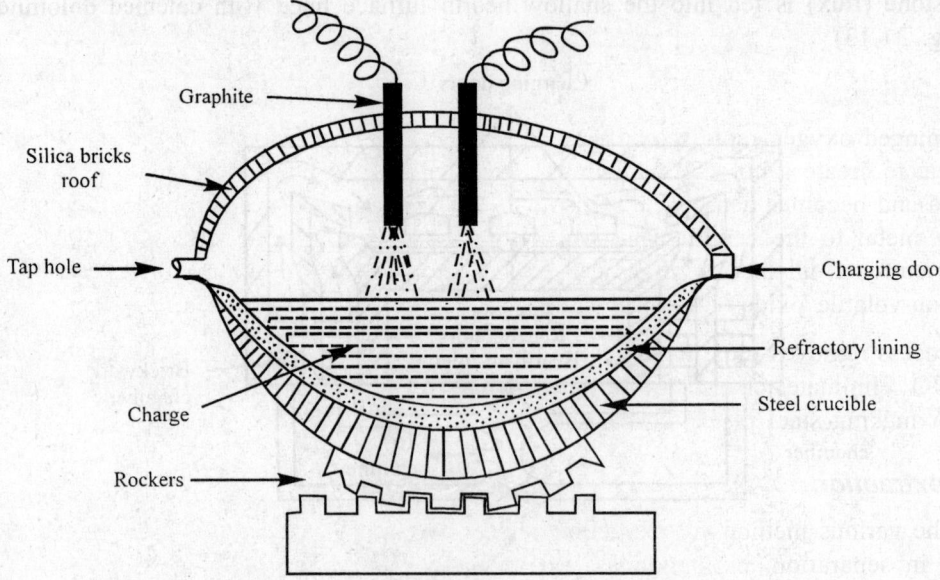

Fig. 31.14. Heroult Electric Arc Furnace.

Molten steel; from Bessemer/open-hearth and some scrap iron are placed in the crucible and electrodes are introduced into the charge. An electric arc is struck between the electrodes and the charge in the crucible. The temperature rises above 2000 K which melts and refines iron. During this process, lime and ferro-silicon are added as fluxes to combine with unwanted matter to form slag. Then required amount of Cr, Ni or other alloying elements are added to get proper alloy steel. It has advantage of producing very high quality steel because of its easier control. The process takes 4–12 hours.

The latest modification of Bessemer process as adopted at Rourkela steel plant in India is the Linz and Donauwitz process commonly called the L.D. process. It was developed in Austria in 1953. The converter is very similar to Bessemer converter but has a solid bottom. It is lined with a basic lining of calcined dolomite. A 300–400 kg charge consisting of molten pig iron containing about 20% scrap and a little lime is poured into the converter. A high pressure jet of very pure oxygen (>99.5%) is directed on the surface of the charge with the help of a Cu lance as shown in Fig. 31.15.

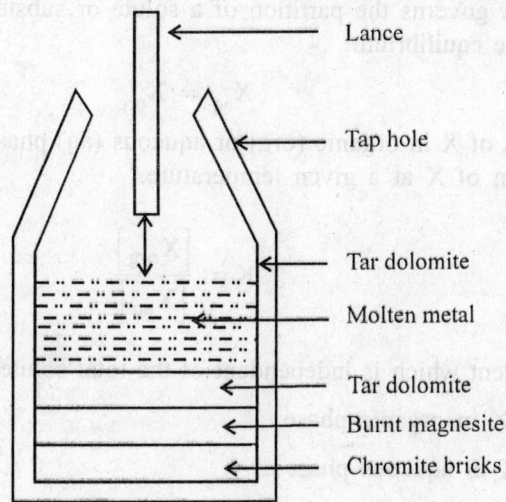

Fig. 31.15. L. D. furnace.

The impinged oxygen rapidly oxidises Si, Mn, P and C in the vicinity of the central zone liberating a lot of heat to create a hot spot at the position indicated in Fig. 31.15. The metal in the central zone gets refined and becomes denser and settles to the bottom. This creates convection currents which bring the impure metal to the central zone where it interacts with oxygen and gets refined. Due to these convection currents, all the impurities get oxidised in about 30–45 minutes. The lime acts as flux and converts non-volatile oxides into slag which is removed from the surface.

L.D. process removes impurities quickly and effectively. The plant runs on low energy requirements. The use of O_2 eliminates chances of nitrogen absorption, thereby, producing better quality steel. It is very suitable for making steel plates, sheets and wires.

Solvent extraction

Amongst the various methods of extraction, solvent extraction has acquired great importance due to its usefulness in separation, extraction as well as purification of metal ions with special reference to purification of nuclear fuels, nuclear wastes, separation of Lanthanides as well as Actinides which, otherwise, were separated by very tedious processes involving fractional crystallisation and fractional precipitation.

The extraction method has many advantages over fractional crystallisation or precipitation:

1. The method has high selectivity
2. The process is very quick
3. The operations are very neat and simple
4. The technique can be used for preparation, purification, enrichment and separation on all scales (micro as well as macro levels).

Solvent extraction is defined as the process by which a substance may be extracted from dilute solutions (usually aqueous or inorganic solutions) into an immiscible solvent (usually organic solvent) with or without the use of a complexing agent. The organic solvents used for extraction may be polar like chloroform, nitrobenzene, tetrahydrofuran, butyl acetate, ethers, methyl isobutyl ketone, etc. or non-polar like benzene, carbon tetrachloride, petroleum ether, kerosene, xylene, *n*-hexane, etc.

Nerst distribution law governs the partition of a solute or substance say X between two immiscible solvents expressed by the equilibrium:

$$X_{aq} \rightleftharpoons X_{org}$$

The ratio of activities of X in organic (org) or aqueous (aq) phases will be constant and independent of the total concentration of X at a given temperature.

$$K = \frac{\left[X_{org} \right]}{\left[X_{aq} \right]}$$

K = partition coefficient which is independent of the total solute concentration in either of the phase

$[X_{org}]$ = activity of X in organic phase

$[X_{aq}]$ = activity of X in aqueous phase

At constant temperature the value of K is constant. This is applicable theoretically to ideal solutions only but in practice many substances follow the expression and activities may be replaced by molar concentration. Quite often, K is approximately equal to the ratio of the solubility of X in each solvent.

Generally ionic substances are insoluble in non-polar solvents but soluble in polar solvents with high dielectric constant due to the solvation of ions. On the other hand, covalent compound are mostly soluble in non-polar solvents. Therefore, the extraction of an ion from aqueous solution into organic phase can take place only if the ion forms a species showing preference for organic phase either by:

1. chelate formation; or
2. solvate formation; or
3. ion-pair formation.

Chelate formation

Chelate agents form an important class of organic bases which can interact with metal atom/ion from more than one position in the ligand molecule in such a way that the metal atom/ion is bound in a stable 4 or 2 or 6 numbered ring. Such complexes are called chelate and organic bases forming these complexes are called chelating agents. A widely used chelating agent is 8-hydroxyquinoline (C_9H_7NO). Most of its chelates are soluble in organic solvents. The complex is soluble in organic solvents and the equilibrium is pH dependent. As a consequences, metals having different formation constants with the ligand can be separated through pH control of the acqueous phase. This way even trace amounts of metal can be separated.

Similarly acetylacetone $CH_3COCH_2COCH_3$ (acacH), diphenyl-thiocarbazone $C_6H_5NHCSN = NC_6H_5$, diamines, etc., are bidentate ligands and form complexes mostly soluble in organic slovens. The equilibrium between metal ion and the reagent is pH dependent and by controlling the pH of the aqueous phase, separation of metallic ions is done.

$$2(acacH)_{org} + (M^{2+})_{aq} \rightleftharpoons [M(acac)_2]_{org} + 2H^+_{aq}$$

A wide variety of chelating agents are in use for the separation of metal ions at micro as well as macro levels. For example, extraction of uranium with 8-hydroxyquinoline is done in $CHCl_3$ and that of Fe with cupferron is done in CCl_4 solvent.

Extraction by solvation

Extraction by this method is very clean and can be performed at microgram concentration of the metal ions as well as macrogram or pilot plant level. It involves solvation of the extracted species into the organic phase. There are several kinds of solvating solvents, for example, ether, ethyl-acetate, methylisobutyl ketone (MIBK), tributylphosphate $(C_4H_9O)_3PO$(TBP), tributylphosphine oxide $(C_4H_9)_3PO$, trioctylphosphine oxide $(C_8H_{17})_3PO$, etc. This method is frequently used for the extraction of transition metals and inner transition metals. For example, uranium can be conveniently separated from lead and thorium if the nitrate in water is extracted with etherial phase as a solvate $[UO_2(NO_2)_2 \cdot 2H_2O \cdot 2Et_2O]$. Lead and thorium are left in aqueous phase. Uranium can be extracted on industrial scale using TBP as an extractant (for solvate formation) in kerosene phase. This method is based on the principle of solvation of the metallic species into the organic phase.

A number of covalent compounds like ZnCl2, HgCl2, CeCl4, OsO4, RuO4, etc., undergo easy extraction into organic phase of hydrocarbons, halohydrocarbons or ethers. The mechanism involves solvation of these substances into the organic phase.

Ion-pair formation

Another category of extractants involve extraction by ion-pair formation. The extractants are mostly high molecular weight amines. As, for example, tertiary amines like trioctylamine TOA $(C_8H_{17})_3N$, triiso-octylamine TIOA (i-$C_8H_{17})_3N$, N, trioctylamne oxide $(C_8H_{17})_3NO$, etc., are base extractants for anionic complexes with mineral acids. A typical example showing mechanism of extraction of a solution of $FeCl_3$ in concentrated hydrochloric acid $HFeCl_4$ is as follows :

$$Fe^{3+}_{aq} + 4Cl^-_{aq} \rightleftharpoons FeCl_4^-{}_{aq}$$

$$R_3NH^+_{org} + H^+_{aq} \rightleftharpoons R_3NH^+_{org} + Cl^-_{org}$$

$$R_3NH^+_{org} + Cl^-_{org} + FeCl_4^-{}_{aq} \rightleftharpoons [R_3NH^+FeCl_4^-]_{org} + Cl^-_{aq}$$

Ion-pairs involving big metallic cations or anions with ions of the opposite charge, such as, tetraphenylarsonium perrhenate $[(C_6H_5)^4As]^+ReO^{4-}$ or tetraphenylarsonium permanganate $[(C_6H_5)_4As]^+MnO^{4-}$ undergo easy extraction with organic phase of chloroform.

Choice of a solvent

The following aspects must be looked into while considering the choice of a solvent or an extractant:

1. The solvent should be practically insoluble in the aqueous phase
2. There should be significant difference in their densities and organic phase should preferably form the upper layer
3. The organic solvent should have low toxicity and if possible low inflammability
4. The dissolved metal ion in the organic phase should be easily recoverable.
5. The distribution ratio of the extractable metal ions should be high and that of impurities low.

Uses of solvent extraction

Some examples involving the use of solvent extraction procedures for the separation of metal ions are given below :

1. Uranium can be conveniently separated from impurities of lead and thorium by the etherial extraction of aqueous solution of uranium (VI) (uranyl UO_2^{2+}) in saturated solution of NH_4NO_3 with 1.5 M concentration of HNO_3.
2. Extraction of plutonium and uranium and their separation form a solution of uranium fuel elements in a nuclear reactor is based on process of extraction. Using oxidation-reduction procedures an extraction or precipitation, bulk of unwanted fission products are removed. The aqueous solution containing UO_2^{2+} and PuO_2^{2+} is then treated with an organic phase containing methylisobutyl ketone (MIBK). Both UO_2^{2+} and PuO_2^{2+} get extracted into organic phase which on washing with aqueous SO_2 affects their separation. Plutonium(VI) is reduced by SO_2 to Pu^{4+} which goes into aqueous phase, whereas UO_2^{2+} remain in organic phase.

   In another method, UO_2^{2+} and Pu^{4+} are extracted into an organic phase of kerosene containg 30% TBP (tributylphosphate) as complexing agent from an aqueous solution of $6NHNO_3$. The organic phase containing UO_2^{2+} and PU_{4+} is then reduced with aqueous SO_2 when Pu^{4+} and goes into aqueous phase and UO_2^{2+} remain in organic phase.
3. Extraction and separation of a mixture of Zr(IV) and Hf(IV) or a mixture of Nb(V) and Ta(V) is based on the extraction by solvation of their halides from acidic aqueous medium into an organic phase using MIBK as extractant.
4. Absolute alcohol is used for the separation of calcium form a mixture of Ca, Ba and Sr nitrates. Calcium nitrate $Ca(NO_3)_2$ forms a solvated complex with alcohol and goes into solution whereas others (Sr and Ca nitrates) remain insoluble.

METALLURGY OF GOLD

Gold occurs mostly in native state. Native gold is occasionally found in the form of small and big nuggets as heavy as 190 lbs. Generally there are four forms in which the native gold occurs :

1. *Alluvial gold or placer gold:* It is present in sand, mud or gravel in the beds of rivers which pass over auriferrous rocks. Usually it contains one part gold for 1 to 15 million parts of gravel.
2. *Reef gold or vein gold:* It is found in hard rocks or in quartz veins mixed or alloyed with silver. Reef gold occurs in the ratio of 1 part gold to 50,000 parts of quartz.
3. *Auriferrous pyrites:* Some sulphide ores of Cu, Ag and Pb contain gold in small amounts (1 part gold: 50,000 parts pyrites)

4. *Sea water:* Gold is also present in sea water but the quantity of gold is too small (0.001–0.01mg per m^3) to be extracted. The total gold content in all the sea waters is estimated at many million tons but it is too dilute to be extracted profitably.

Gold ores are usually contaminated with silver, copper, bismuth, mercury, zinc, iron, lead, palladium and platinum. Very little gold occurs in the combined state in minerals like: (i) calaverite Au Te$_2$; and (ii) sylvanite Au Ag Te$_4$.

These ores are roasted in air to remove tellurinum as TeO$_2$ leaving behind gold or an alloy of gold or an alloy of gold and silver. Another rare mineral of gold is bismuth aurite AuBi. Natural gold consists of 100% isotope $^{197}_{79}$Au.

Treatment of Ore

Different methods are employed for the extraction of gold depending upon the nature of ore. Physical methods are adopted for the extraction of gold from rich sources of alluvial or reef gold. It is termed as mining. Chemical methods have to be employed for extraction from poorer sources of gold. Extraction of gold from different sources is done by the following methods.

Extraction of Alluvial Gold

Extraction of gold from alluvial deposits involves the washing of finely divided ore with water. Gangue particles, being lighter, are washed away by water stream while heavier gold particles are left behind. Any one or more of the following methods are employed for the extraction of gold from alluvial deposits:

1. *Panning:* It is simple and a primitive method workable on small-scale for alluvial deposits containing good concentration of free gold. In circular pans (about 1 foot diameter) of iron on zinc, gold bearing a sand mud is agitated with water by giving a rotary motion to the pan. As a consequence of rotary motion, heavier gold particles settle down and accumulate at the bottom while the sand is washed away over the edges. The pan has to be refilled with water several times before the extraction is complete. This method has now been replaced by placer mining.

2. *Placer mining:* In this process the gold bearing sand is introduced in a sluice way system and a powerful stream of water is passed over it. The sluice way system, shown in Fig. 31.16, consists of an inclined channel provided with a series of sluices, that is, troughs having cross-wise strips called riffles or cleats. The gold particles are retained in the sluices due to obstruction by the riffles while the lighter earthy waste is carried away by the water. Sometimes a little mercury is added into the sluices. It catches the gold better and forms amalgam. Amalgam is heated to distil away mercury and gold is left behind.

3. *Hydraulic mining:* Wherever gold deposits occur near an edge of a river, hydraulic mining is preferred. A powerful jet of water is forced through nozzles and focused on to the deposits of auriferrous sands. The resulting mixture of gravel and water is washed down a system of sluices containing riffles set at right angles to the length of the channels as shown in Fig. 31.16 in the previous method. Heavier particles of gold are retained in sluices and lighter sand particles are washed away. The sluices are cleaned once a week.

4. *Wilfley's Table:* In this the gold bearing sand or mud is mixed with water and allowed to fall on to the table. As the table jiggles, the lighter particles are washed away while the heavier gold particles accumulate behind the cleats. Sometimes a little mercury is added to make it more effective as gold amalgam is heavier. Gold is then separated from amalgam by distillation of mercury. Gold is left behind.

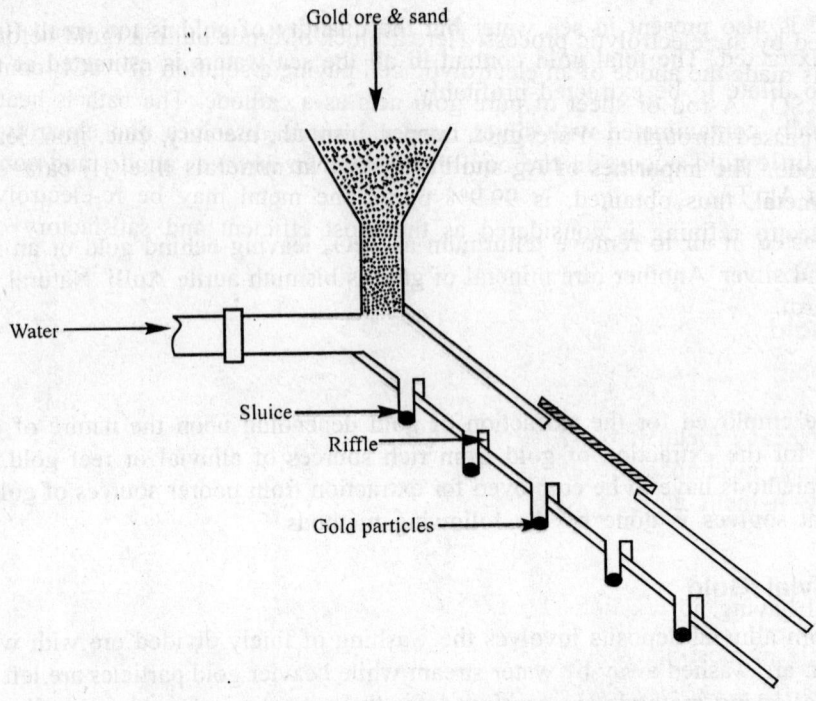

Fig. 31.16. The sluice-way system.

Refining or Purification of Gold

Fusion

Gold usually contains traces of iron, copper, zinc, silver, lead, etc. The metal is refined by fusion with borax (flux) and nitre (oxidising agent) in a crucible. The impurities, other than Ag, get oxidised and form scum at the molten surface and are removed.

Parting with H_2SO_4 or HNO_3

The refined metal obtained after fusion contains Ag as impurity. Purification of Au containing Ag is done by treating the crude metal with hot conc. H_2SO_4 or boiling HNO_3 when Ag dissolves and gold remains unaffected.

$$2Ag + 2H_2SO_4 \rightarrow Ag_2SO_4 + 2H_2O + SO_2$$
$$Ag + 2HNO_3 \rightarrow AgNO_3 + H_2O + NO_2$$

It has been found that if the percentage of gold in the impure sample is more than 25%, then the sample is not attacked by conc. H_2SO_4. In such case, silver is added from outside to the sample, so that the gold content in the sample is reduced to 25% gold. The two are fused together and dropped in water. The alloy, thus formed, is then treated with conc. H_2SO_4. The treatment of decreasing gold content to 25% is sometime called quartation.

In case of nitric acid treatment, the impure gold is alloyed with Ag to reduce its percentage to 40–45% gold. Concentrated HNO_3 dissolves Ag as $AgNO_3$. Silver is recovered from $AgNO_3$ solution by treatment with Cu turnings.

$$2AgNO_3 + Cu \rightarrow Cu(NO_3)_2 + 2Ag \downarrow$$

Electrorefining

Gold may be refined by the electrolytic process. Here a block of crude bullion (gold before making coins is termed bullion) is made the anode of an electrolytic cell having a solution of $AuCl_3$ containing 10–12% HCl and a little H_2SO_4. A rod or sheet of pure gold acts as a cathode. The bath is heated and then an electric current is passed through it. Pure gold from the anode dissolves into electrolyte and deposits slowly on the cathode. The impurities of Ag and Pb are thrown down as anode mud consisting of AgCl and $PbSO_4$. The metal, thus obtained, is 99.9% pure. The metal may be re-electrolysed to acquire ultimate purity. Electro refining is considered as the most efficient and satisfactory of all the other methods of refining.

Properties of Gold

Physical

Gold is a soft and heavy metal (sp. gr. 19.3). It has a beautiful yellow colour and bright lustre which do not fade or tarnish in atmosphere. It is highly malleable and ductile. It is a very good conductor of heat and electricity. It melts at 1336 K and appears green in molten state. It boils at 2870 K.

Chemical

It is a noble metal having no reaction with air, moisture, caustic alkalies and common acids.

Gold is insoluble in most acids but is attacked by certain acids in the presence of oxidising agents. For example: Aqua regia (3:1 mixture of con. HCl and HNO_3) dissolves gold to form $AuCl_3$ which is further converted with chloroauric acid $HAuCl_4$.

$$3HCl + HNO_3 \rightarrow 2H_2O + Cl_2 + NOCl$$
$$Au + Cl_2 + NOCl \rightarrow AuCl_3 + NO$$
$$AuCl_3 + HCl \rightarrow HAuCl_4$$

Gold combines with chlorine (or bromine) at higher temperatures to form gold chloride (or bromide).

$$2Au + 3Cl_3 \rightarrow 2AuCl_3$$

Gold is soluble in alkali cyanide solution in the presence of air.

$$2Au + H_2O + 1/2O_3 + 4KCN \rightarrow 2KAu(CN)_2 + 2KOH$$

Gold is precipitated from solutions of its chloride by reducing agents such as $FeSO_4$, $SnCl_2$, SO_2 oxalic acid, tartaric acid, sugar, etc.

Uses of Gold

Gold is precious noble metal. Since it retains its brightness, it is used in making ornaments and jewellery. Since gold is too soft to be used in jewellery, therefore, it is mixed with other metals like Ag, Cu or Al to make it hard. Pure gold is 24 carats. Gold coins are 22 carats that is 22 parts gold and 2 parts impurity of Ag or Cu. Jewellery is mostly 18 carats gold. 14 carats jewellery contains 14 parts gold and 10 parts Cu.

Gold is used for gold plating and gilding. Here $KAu(CN)_2$ is used as the electrolyte and the article is first Ag plated and then made the cathode. Gold leaves are used in Ayurvedic medicines. Colloidal gold in the form of "Purple of Cassius" is used for colouring pottery and glass.

METALLURGY OF ZINC

Occurrence

Zinc has been known to ancients as an alloy of Zn and Cu (very similar to brass) without knowing its actual composition. Zinc has been known in our country since long as *Yashada* commonly called *Jast*. Extraction of zinc in India was done as early as 1374 A.D. and exported to neighbouring countries. Zinc does not occur native. The chief ores of zinc are:

Zinc blende (Black jack)	ZnS
Calamine (Zinc spar)	$ZnCO_3$
Zincite	ZnO
Zinc Spinel	ZnO-Al_2O_3
Willemite	Zn_2SiO_4

Treatment of Ore

The principal source of zinc is zinc blende ZnS. The carbonate calamine $ZnCO_3$ and oxide Zincite ZnO are also used to some extent. There are two main methods or metallurgy of zinc: (i) distillation of Reduction method; and (ii) electrolytic method.

Distillation of reduction method

The following are the metallurgical steps :

1. *Concentration:* The ore is first dressed and unwanted portions of the rock are removed. It is then finely powdered and the pulverized ore is concentrated either by (a) froth flotation of (b) gravity separation or (c) electromagnetic separation.

   (a) *Froth flotation process:* It is especially undertaken for sulphide or Zinc blende. The pulverized ore is brought into flotation vessel in which water and frothing agent (pine oil or eucalyptus oil) are agitate by passing in compressed air. The sulphide ore is "wetted" by the froth and rises to the surface alongwith froth. The gangue particles remain behind and sink to the bottom of the flotation vessel.

   (b) *Gravity separation:* The powdered or (especially calamine) is washed with a strong current of water, whereby lighter gangue particles are carried over by water current heavier ore particles are left behind.

   (c) *Electromagnetic separation:* It is useful when zinc ore contains iron oxide as impurity. The powdered ore is passed over an electromagnet. Iron oxide is attracted by the magnet and falls in a separate heap away form the non-magnetic zinc ore.

2. *Roasting:* The concentrated sulphide ore is heated in a free supply of air in a multiple hearth roaster to obtain zinc oxide.

$$2ZnS + 3O \rightarrow 2ZnO + 2SO_3 \uparrow$$

Roasting is generally done at 1170 K, so that, ZnSO4, if formed decomposes into ZnO.

$$ZnS + 2O_2 \rightarrow ZnSO_4 \rightarrow ZnO + SO_2 + 1/2O_2$$

In case of calamine, the ore is calcined to eliminate carbon dioxide and moisture.

$$ZnCO_3 \rightarrow ZnO + CO_2 \uparrow$$

3. *Reduction:* The zinc oxide ZnO is mixed with carbon in closed retorts and reduced by strong heating at 1700–1900 K.

$$ZnO + C \rightarrow Zn + CO \uparrow$$

The vapours of zinc are cooled and collected form the condenser. This process of smelting is brought about by the following three methods.

(a) *Belgian method:* The roasted ore is mixed with about half its weight of powdered coal or coke and taken in circular bottle shaped fire-clay retorts closed at one and as shown in Fig. 31.17. retort is attached to air-cooled fire-clay receivers (to collect molten zinc) which in turn are connected to flexible tin bags to collect zinc dust Fig. 31.17. A retort can hold a charge of about 20 kg. A large number of retorts are fixed in slanting positions in a muffle furnace as shown in Fig. 31.17. The furnace is heated by burning producer gas on the regenerative principle of heat economy. The temperature inside the furnace is maintained in the range 1170–1370 K. As heating progresses, reduction takes place to give elementary Zn.

$$ZnO + C \xrightarrow{\text{heat}} Zn + CO \uparrow$$

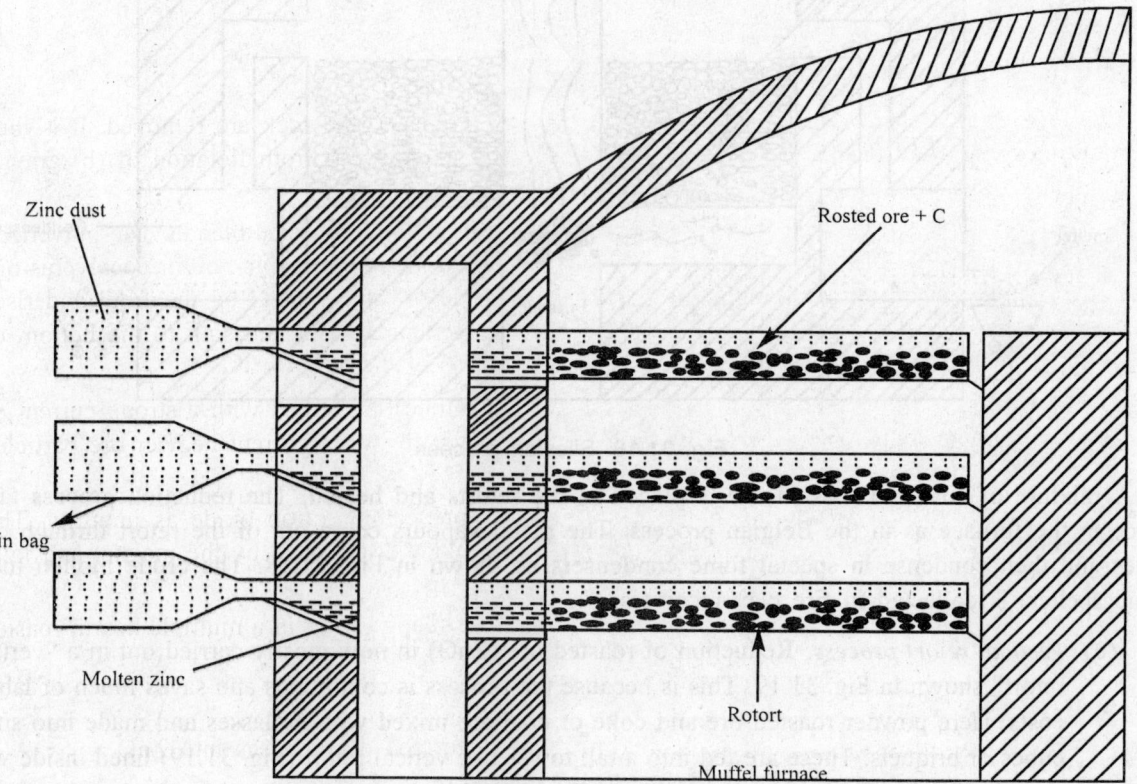

Fig. 31.17. Belgian retort method.

Carbon monoxide is evolved and burns with a blue flame at the end of fir-clay receiver. The size of

CO flame gives an indication of the reduction process. Reduction is complete when the size of the flame is the smallest. Zinc (b.p. = 1181.5 K) volatilises out and most of it condenses and collects into fire-clay receivers as crude molten metal. The rest of zinc vapours (5.13%) precipitate out as zinc dust in the tine bags attached to fire-clay receivers.

The molten crude zinc is approximately 97–98% and contains about 1–2% Pb and small amounts of Fe, Cd and As. The crude liquid metals called zinc spelter. The retorts are taken out, recharged and replaced. The process has a disadvantage that it is not continuous.

(b) *Silesian Process:* In this process, D-shaped fire-clay retorts are arranged in the bed of a reverberatory furnace (Fig. 31.18).

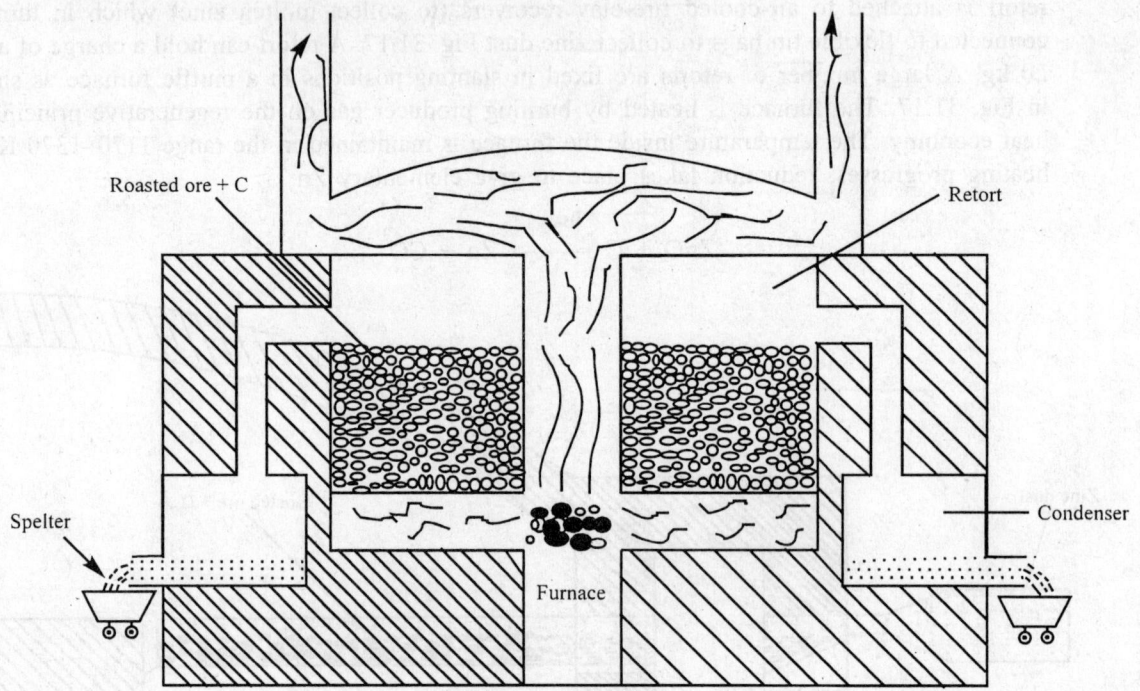

Fig. 31.18. Silesian process.

A mixture of coal and roasted ore is placed in the retorts and heated. The reduction process takes place in the furnace as in the Belgian process. The metal vapours come out of the retort through clay pipes and then condense in special fume condensers as shown in Fig. 31.18. The crude molten metal (zinc spelter) is collected in fire clay receivers.

(c) *Vertical retort process:* Reduction of roasted ore (ZnO) in now mostly carried out in a "vertical retort" shown in Fig. 31.19. This is because the process is continuous and saves much of labour costs. Here powder roasted ore and coke or coal are mixed with molasses and made into small cubes or briquets. These are fed into a tall tower like vetical retort (Fig. 31.19) lined inside with fire-clay bricks. The retort is heated to 1470–1570 K by burning producer gas in the outer jacket as shown in Fig. 31.19.

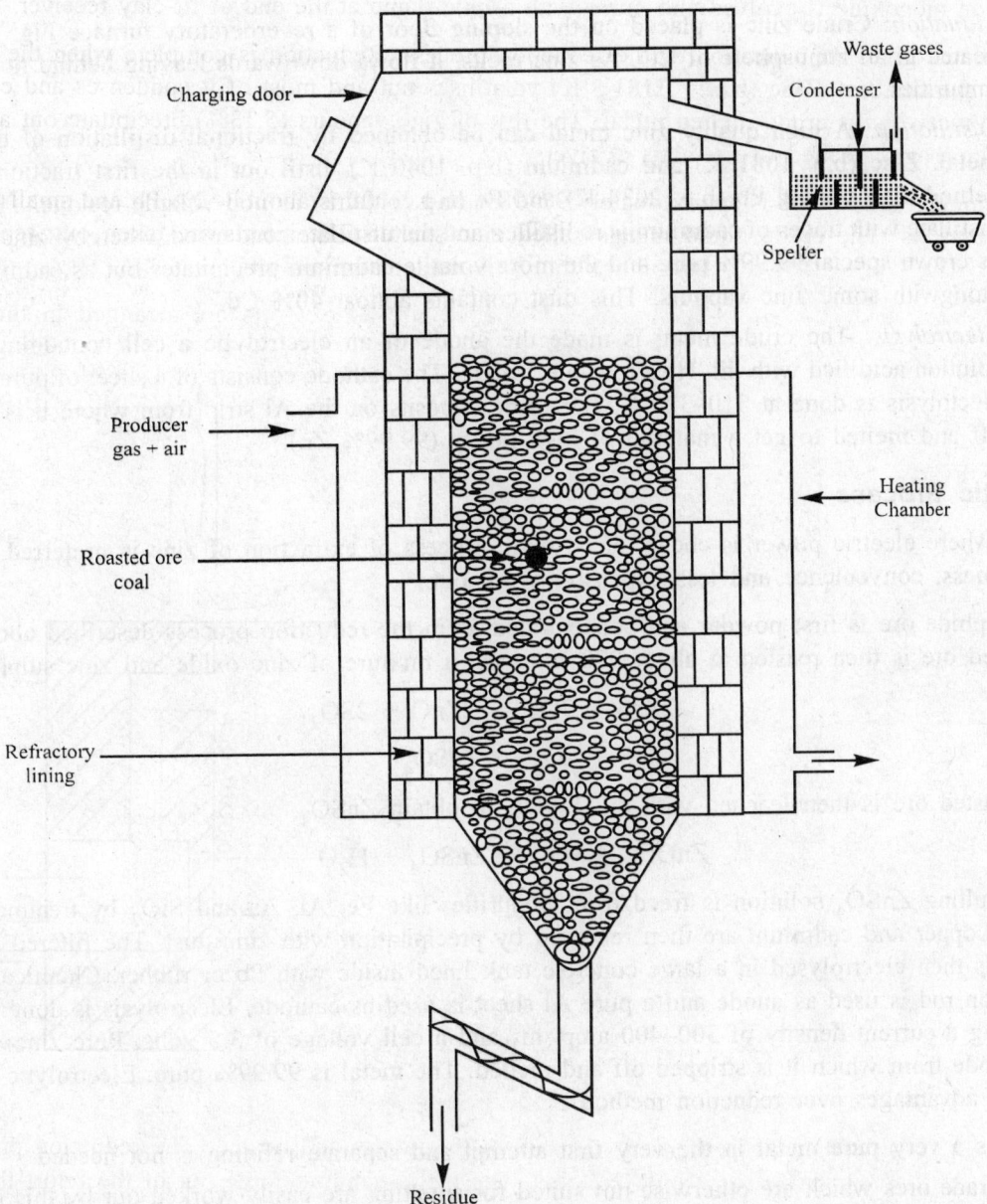

Fig. 31.19. Vertical retort method.

Zinc vapours, formed as a result of reduction of zinc oxide and carbon monoxide are fed into air-cooled condenser. Carbon monoxide escapes away but zinc condenses into the condenser to give zinc spelter. Condensed zinc is drawn out periodically. The residual aches in the retort move down gradually and are worked out from the bottom with the help of screw conveyer. Fresh charge is added through charging door at the top as ashes move downwards. The process is continuous.

4. *Refining:* Spelter zinc is impure and contains 97–98% Zn with 1–2% Pb and smaller amounts of Fe, Cd and As. The impurities can be removed by the following methods:

(a) *Liquation:* Crude zinc is placed on the sloping floor of a reverberatory furnace Fig. 31.4 and heated in an atmosphere of CO. As zinc melts, it flows downwards leaving behind non-fusible impurities.

(b) *Distillation:* A high quality zinc metal can be obtained by fractional distillation of the crude metal. Zinc (b.p. 1081 K) and cadmium (b.p. 1040 K) distil out in the first fraction leaving behind high boiling Pb (b.p. 2034 K) and Fe (b.p. 3008 k) as non-volatile residue. The above distillate with traces of cadmium is redistilled and the distillate condensed, whereby, zinc liquifies as crown special 99.99% pure and the more volatile cadmium precipitates out as cadmium dust alongwith some zinc vapours. This dust contains almost 40% Cd.

(c) *Electrolysis:* The crude metal is made the anode of an electrolytic a cell containing $ZnSO_4$ solution acidified with dil. H_2SO_4 as electrolyte. The cathode consists of a sheet of pure Al. The electolysis is done at 310–320 K. Pure zinc deposits on the Al strip from where it is stripped off and melted to get a material of high purity (99.99% Zn).

Electrolytic Process

In places where electric power is cheap, electrolytic process of extraction of zinc is preferred because of its neatness, convenience and less expensive work out.

The sulphide ore is first powder and concentrated as in the reduction process described above. The concentrated ore is then reasted at about 970 K to get a mixture of zinc oxide and zinc sulphate.

$$2ZnS + 3O_2 \rightarrow 2ZnO + 2SO_2$$

$$ZnS + 2O_2 \rightarrow ZnSO_4$$

The roasted ore is then leached with dil. H_2SO_4 to obtain $ZnSO_4$.

$$ZnO + H_2SO_4 \rightarrow ZnSO_4 + H_2O$$

The resulting $ZnSO_4$ solution is freed from impurities like Fe, Al, As and SiO_2 by treatment with $Ca(OH)_2$. Copper and cadmium are then removed by precipitation with zinc dust. The filtered solution of $ZnSO_4$ is then electrolysed in a large concrete tank lined inside with Pb or rubber. Chemically pure Pb or carbon rod is used as anode and a pure Al sheet is used as cathode. Electrolysis is done at 300–310 K using a current density of 300–400 amp. m_2 and a cell voltage of 3.5 volts. Pure zinc deposits at the cathode from which it is stripped off and melted. The metal is 99.99% pure. Electrolytic process has certain advantages over reduction method :

1. It gives a very pure metal in the very first attempt and separate refining is not needed.
2. Low grade ores which are otherwise not suited for smelting are easily worked out by this process.
3. It allows the recovery of precious metals that may be present as impurity in zinc ore.
4. It is quite economical where electricity is cheap.
5. Unlike reduction process, it is neat and does not pollute the atmosphere with unwanted gases.

Properties of zinc

Zinc is a bluish white metal when pure but tarnishes to grey colour in moisture. It melts at 692.5 K and boils at 1181.5 K. It has a specific gravity of 7.14. Commercial zinc is brittle at room temperature but becomes malleable and ductile at temperature above 370 K. It becomes brittle again at 470 K and remains so upto 570 K. It is a good conductor of heat and electricity.

Zinc is not a very active metal chemically. It is not affected by air but in moist air it is coated by a thin film of $Zn(OH)_2.AnCO_3$ which inhibits further corrosion.

When heated in air at 1300 K, it burns with a bluish white flame giving a dense smoke of light voluminous zinc oxide formerly known as philosopher's wool.

Uses of zinc

1. It is used for protecting iron sheets against rusting. This process is known as galvanisation.
2. In laboratory it is used for the preparation of H_2 gas. Zn dust is also used as a reducing agent.
3. It is used in the manufacture of dyes, drugs and chemicals.
4. It is used for the extraction of Ag and Au.
5. Zinc rods and plates are used in electric cells and batteries.
6. It is used in the preparation of pints like lithophone.
7. It is used in the preparation of alloys like brass (Cu 80%, Zn 20%), german silver (Cu 55%, Zn 25%, Ni 20%), delta metal (Cu 60%, Zn 38.2%, Fe 1.2%), die.casting metal (Zn92.5%, Al 4%, Cu 3.5%), etc.

METALLURGY OF NICKEL

Occurrance

Nickel occurs mostly in the combined state. Nickel ores are found in India in small deposits in Cuttack, Keonjhar and Mayurbhanj districts of Orissa (ore reserves—16.5 crore tons). Chief nickel ores are:

1. Pentlandite $CuFeS_2.NiS$
2. Garnierite $(MgNi)_3(OH)_4(Si_2O_5)$
3. Kupfer-nickel or nickelite $NiAs$
4. Nickel-blende or millerite NiS
5. Nickel glance $NiAsS$
6. Ullmannite $NiSbS$
7. Breithauptite $NiSb$

Elemental nickel is also found in meteoric irons. Besides, it is said that large quantities of nickel are also contained in the earth's nucleus. The ore pentlandite contains Cu as $CuFeS_2$ and Ni as NiS. Small amounts of Ag, Au and platinum metals are also contained in this ore. Next important Ni ore is Garnierite which is found in New Caledonia an island of North East Australia.

Treatment of Ore

The metallurgy of nickel is quite complicated due to the presence of large quantities of Fe and Cu in the ore. The extracation of nickel from sulphide ores, e.g., pentalandite (or arsenide ore) involves the following operations:

Concentration

The ore is hand picked and separated from the country rock (gangue). The ores is then crushed and concentrated by froth flotation method. The powdered or is mixed with water and pine oil and agitated by compressed air in flotation cell. The froth containing sulphides or arsendies of metal rises to the top and removed. The gangue sinks to the bottom. The froth is dried and powdered (Fig. 31.3) for greater details on froth flotation method.

Roasting

The concentrated ore is then fed into the hearth of reverberatory furnace (Fig. 31.4) and heated in a free supply or air where the following reactions take place:

Most of sulphur and arsenic are oxidised as volatile oxides and their burning produces enough heat and saves lot of fuel.

$$S + O_2 \rightarrow SO_2$$
$$4As + 3O_2 \rightarrow 2As_2O_3$$

Iron sulphide is partially converted into oxide. Nickel and copper sulphides remain mostly unaffected but trace amounts of oxides may also be formed.

$$2FeS_2 + 5O_2 \rightarrow 2FeO + 4SO_2 \text{ (major)}$$
$$2Cu_2S + 3O_2 \rightarrow 2Cu_2O + 2SiO_2$$
$$2NiS + 3O_2 \rightarrow 2NiO + 2SO_2$$

Smelting

The roasted material is then introduced into a small blast furnace and smelted with coke and fluxing material such as silica and lime stone. Iron oxide, CaO combine with silica to form the fusible silicate slags which are separated

$$FeO + SiO_2 \rightarrow FeSiO_2 \text{ (salg)}$$
$$CaO + SiO_2 \rightarrow CaSiO_2 \text{ (slag)}$$

from the top of the molten mass called crude matter. The molten matte contains predominantly sulphides of Ni and Cu. This process of smelting removes most of iron and has to be supplemented by Bessemerisation in order to remove last traces of iron.

Bessemerisation

The molten matte, containing small amounts of iron impurity, is mixed with requisite amount os silica and fed into a Bessemer converter (Fig. 31.12) provided with a basic lining. Blasts of air are introduced into the melt and regulated so that only iron sulphide is converted into iron oxide (FeO) which combines with SiO_2 to from silicate (as above). Nickel and copper sulphides are not affected significantly. The converter is tilted and slag is drawn from the surface. The lower layer of fused sulphides called refined matte is dropped out and left to solidify. The resulting bessemerised matte contains about 55% Ni, 30% Cu, 14% of sulphur in combined form and traces of iron (0.2–0.5%).

Removal of copper

Removal of copper and subsequent isolation of nickel is done in two ways: (i) orford process; (ii) mond's process.

1. *Orford Process:* This method is based on the following facts:

   (a) The difference in the solubilities of copper and nickel sulphides in Na_2S; copper sulphide being more soluble; and

   (b) The difference in specific gravity between NiS and Cu, Dissolved in Na_2S. The latter is lighter.

   In this process, the solidified matte is broken into smaller pieces, mixed with coke and sodium sulphate and fused in a closed furnace. Sodium sulphate gets reduced to Na_2S which dissolved out Cu_2S and forms the upper layer. The lower layer consists mainly of NiS as shown in Fig. 31.20. The top layer ($Cu_2S + Na_2S$) is separated and sent to copper converters. The heavier bottom layer is treated for extraction of Ni.

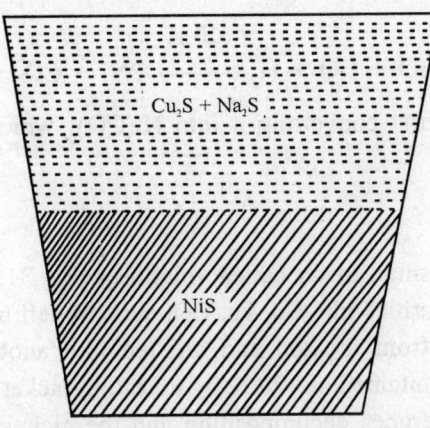

Fig. 31.20. Orford process.

The mass at the bottom is ground and Na_2S, if any, is leached out with hot water, The residual mass is again subjected to roasting in a reverberatory furnace. It is heated strongly in a current of air when NiS gets converted in NiO.

$$2NiS + 3O_2 \rightarrow 2NiO + SO_2m$$

The oxide is then purified by treatment with dil, H_2SO_4. the impurities due to copper and iron oxides get dissolved as $CuSO_4$ and $FeSO_4$, whereas, NiO remains unaffected.

The oxide is then reduced by heating in a current of water gas ($CO + H_2$).

$$NiO + H_2 \rightarrow Ni + H_2O$$
$$NiO + CO \rightarrow Ni + CO_2$$

The crude metal, thus obtained, contains a small percentage of Cu. The metal may then be refined as given in Mond's process.

2. Mond's Process. It is based on the fact that.

   (a) nickel (and not Cu and Fe) forms a volatile carbonayl $Ni(CO)_4$; and

   (b) $Ni(CO)_4$ decomposes at 450 K to form pure NI.

Process

The bessemerised matte is roasted in air in a special furnace to get oxides of Ni and Cu.

$$NiS + O_2 \rightarrow NiO + SO_2$$

$$Cu_2S + 2O_2 \rightarrow 2CuO + SO_2$$

The mixture of oxides is treated with 15% H_2SO_4 at 350 K when copper and iron oxides dissolve as sulphate leaving behind NiO.

$$CuO + H_2SO_4 \rightarrow CuSO_4 + H_2O$$

$$FeO + H_2SO_4 \rightarrow FeSO_4 + H_2O$$

The residual mass containing NiO is dried and them heated at 570–620 K in a " reducer" in a current of water gas (H_2 + CO). Under these conditions NiO is reduced.

$$2NiO + CO + H_2 \xrightarrow{620\ K} 2Ni + CO_2 + H_2O$$

Crude nickel is refined by first converting it into $Ni(CO)_4$ which is then decomposed as discussed below:

Refining

The crude metal is packed in a small tower called volatiliser Fig. 31.21 and a current of CO gas at 320-330 K is passed through it. Impurities like Cu, Fe, CO, etc., are left behind as unaffected but the vapours of $NI(COO_4)$ leave the tower from the top and are fed into another tower called decomposer. The decomposer is a small tower maintained at 470 K by a heating jacket (Fig. 31.21). It is filled with pellets of pure Ni. Here $Ni(CO)_4$ undergoes decomposition and the nickel form gets deposited on the pellets.

$$Ni(CO)_4 \xrightarrow{450\ K} Ni + 4CO \uparrow$$

Carbon monoxide is reused as shown in Fig. 31.21. The metal, thus formed, is 99.5–100% pure.

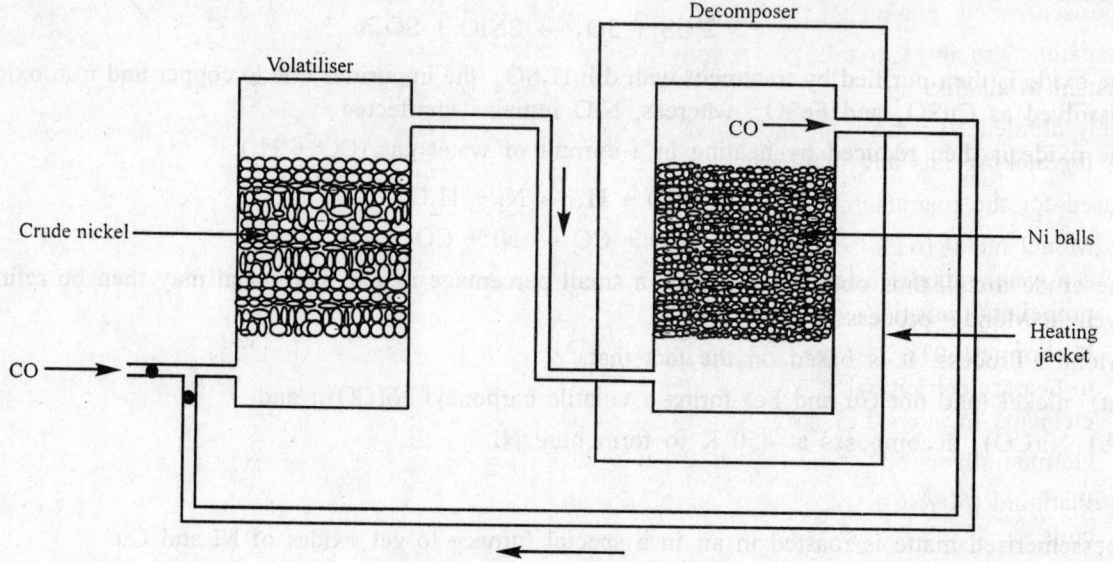

Fig. 31.21. Refining of nickel.

Electrorefining

Nickel is further refined by electolysis of a solution of nickel ammonium sulphate $NiSO_4(NH_4)_2SO_4.6H_2O$ at 293–298 K. The cell has a cast nickel block as anode and a thin polished sheet of pure Ni as cathode. Impurities like Hg, Au and Pb from anode mud.

Properties of nickel

Nickel is a soft silver white metal cable of taking a high polish. It is quite malleable and ductile. Its density is 8.8 g/cm^3, m.p. = 1725 K and b.p. = 3173 K. It is extremely resistant to corrosion and has excellent strength at low temperatures. It forms a number of useful alloys with Fe, Cu and Cr which are widely used in industry due to their resistance to corrosion.

It is resistant to air and moisture but finely divided metal can combine with O_3. Nickel wire burns in O_2 at 570 K.

Heated Ni catches fire in Cl_2 or Br_2 to form halides NiX_2. It combines with sulphur to form NiS but has no reaction with N_2.

Nickel decomposes steam slowly at red heat.

$$Ni + H_2O \rightarrow NiO + H_2$$

Dilute HCl and H_2SO_4 have a very slow action on Ni but dil. HNO_3 dissolves Ni to form $Ni(NO_3)_2$

$$3Ni + 8HNO_3 \rightarrow 3Ni(NO_3)_2 + 2NO + 4H_2O$$

The metal becomes passive when kept under fuming HNO_3 for some time.

Uses of nickel

Principal uses of nickel are as follows:

1. It is used for Ni plating on iron articles to avoid rusting and to give a good shining polish.

2. Ni spatula, crucibles, tonges and electrodes are used in laboratory and industry because metal is resistant to alkalis.

3. Finely divided Ni is an important catalyst used for the hydrogenation of vegetable oils and animal oils for making fats and margarine.

4. It used for the manufacture of very useful alloys.

   (a) Monel metal (67% Ni, 30% Cu, 3% Fe). It is strong and tough. It resists corrosion from air, sea-water, alkalis and most acids. It is, therefore, used in making of boilers, turbine blades, propellers, caustic soda evaporators, dye vats, food processing equipment, etc.

   (b) Nichrome (60%, 25% Cr, 15% Fe). It has very high meting point and shows marked resistance to heart, electricity and chemical action. Therefore, it is used for making resistant coils, heating elements in stoves, electric iron, toaster, etc.

   (c) German silver (55% Cu, 20% Ni, and 25 % Zn). It is used in ornaments, cutlery and utensil.

   (d) Platinoid (60% Cu, 24% Zn, 14% Ni and little W). It is used for thermocouples and resistant coils.

   (e) Platinite (46% Ni, 54% Fe). It is used in electric bulbs and radio valves due to its coefficient of expansion equal to that of glass.

Chapter 32

Pollution Control

INTRODUCTION

Pollution control is recognised as an important aspect of chemical engineering. Due to lack of development of a culture of pollution control, a heavy backlog of gaseous, liquid and solid pollution in our country has resulted. Thus pollution control in our country is a recent environmental concern. Pollution is a man-made problem, mainly of effluent countries. The developed countries have been in a mad race to exploit every bit of natural resource to convert them into goods for their comfort and to export them to needy developing world. In doing so, the industrialised countries dump lot of materials in their environment which becomes polluted. In a way, pollution in fact is being forced upon human beings.

Pollution is an undesirable change in the physical, chemical or biological characteristics of air, water and soil that may harmfully affect the life or create a potential health hazard of any living organism. Pollution is thus direct or indirect change in any component of the biosphere that is harmful to the living component(s), and in particular undesirable for man, affecting adversely the industrial progress, cultural and natural assets or general environment.

A social awareness of the pollution problem has arisen with a knowledge of its effects and the discharge of untreated effluent is rightly considered to be irresponsible behaviour. The problem has become more urgent on a global scale with the spread of industrialisation so that eventual dilution can no longer be accepted as a universal solution.

An ever present problem in legislation is in setting the permissible levels of emission of individual contaminants. These should be based on a sound knowledge of the effects of the contaminants on the environment and they must be changed from time to time as our knowledge of the effects of the contaminants changes. It must also be remembered that, without some pollution, modern society as we know it would not be able to exist. Social, political and economic pressures dictate that some pollution should be accepted. In consequence, a compromise is made—emissions are not permitted to exceed levels where they are considered to be harmful and it is required that the best practicable means are used to minimise them.

The cost of pollution abatement in the chemical industry is becoming an increasingly important factor as the trend continues towards tighter controls on emission levels. This has stimulated research and development activity in the field.

POLLUTION CONTROL AND ENVIRONMENTAL PROTECTION IN CHEMICAL INDUSTRY

Chemical industry occupies a major position in any developing nation. Be it plastics, fibres or refineries, there is some chemical process involved which generates pollutants. It is the duty of designers, operators or owners to run chemical industry so that pollutants are treated, disposed and possibly made use to generate useful products or energy and keep environment clean for public at large. Here some practical ways are enumerated to keep pollution under control meeting statutory regulations. Number of consultancy companies undertake detailed engineering and design work for large chemical projects. Number of small sector projects would benefit from these practical suggestions.

Dust Removal

In cement or calcined bauxite units, a venturi scrubber is installed to scrub dust. A powerful blower creates enough vacuum to suck air laden with dust. This air is made to go through venturi scrubber. Dust gets separated and pure air is vented to atmosphere. This method can be employed for meeting pollution control requirement for a calcined bauxite unit.

Phenol Removal

Phenol is harmful at a very low level in effluent and its reduction to few ppm is achieved by special algae which uses phenol for its growth. The resultant effluent is clarified or filtered to meet the statutory requirement.

Special Vegetable Plants

Many pollution control boards are modern and have encouraged research laboratories to develop special plants. One such plant is developed which have taken cyanide waste as its food. A good amount of work is done by Madhya Pradesh Pollution Control Board for various other toxic wastes which can be sprayed over controlled agricultural farms growing special vegetable plants. This method is utilised in handling acrylonitrile, SBR, ABS plant waste.

Solvent Vapour Removal

Activated carbon filters are available through which effluent gases are passed. Solvents get absorbed on activated carbon. These solvents can be removed and carbon can be regenerated for reuse by passing steam through bed of activated carbon. Gases can be further scrubbed in water tower. This method can be used for designing a grease plant where solvent vapours are generated in the reactor.

BOD Reduction

Surface aerators are put to induce fresh oxygen from air into effluent, to achieve desired BOD (Biological Oxygen Demand). Floating surface aerators are available with very low power requirement. The purpose of reducing BOD is to minimise the damage done by effluent stream going into river or sea as fish live on dissolved oxygen in water and any effluent which takes away dissolved oxygen from water is harmful to fish and other marine life.

Removal of Chlorine/Hydrochloric Acid Vapours

Water scrubbers are installed to absorb the hydrochloric acid vapours and make weak acid. This is used by various dyes, lube additives and bulk drug companies.

Removal of Hydrogen Sulphide

Hydrogen sulphide is a poisonous and dangerous gas. Exposure to small quantity of hydrogen sulphide gas can kill plant operator. It is best handled by sodium hypochlorite scrubber followed by burning in a flare unit or burning in incinerator. This method is used in organo-phosphorus based plants.

Incinerators

Certain waste cannot be handled except by burning at very high temperature of 1400°C to 1800°C in incinerator. This incinerators can be made profitable by generating steam or providing source of heat. Intelligent entrepreneur can be keep incinerators not only for their plant waste but offer services to neighbouring plant and earn enough money to pay for incinerator's cost in two years. Such cases are there in Andhra Pradesh in pesticides/bulk drug industry.

Chemical Treatment/Filtration

To reduce certain toxic elements, they are treated with other chemicals like alum or caustic soda or sulphuric acid or any other strong reactant to produce harmless chemicals, which can be separated by filtration. This can be used as a source of income also, like calcium chloride can be manufactured by treating lime and hydrochloric acid waste.

Special additives

Fuel additives have been developed which reduce pollution and increase efficiency of engine and offer better mileage.

Catalytic converters

Palladium converters are available to reduce emission of harmful vapours in fuel exhaust.

Membranes

Special membranes are available to selectively remove oxygen, nitrogen or carbon dioxide gases and purify the air.

Ultraviolet Rays

Sterile atmosphere is necessary for injectable area in pharmaceutical industry which uses of ultraviolet rays. Air is properly dehumidified, filtered and cleaned before exposing to ultraviolet rays.

Environment in Plant

Chemical plant area has to be made comfortable by proper air changes. The fresh air is to be sent. The vapours generated by handling of solvent drums should be sucked by downcomer pipes leading to scrubbing system. The dust is also contained by keeping crusher, pulveriser, storage silos under nitrogen or carbon dioxide. Later is preferred for phosphorus or sulphur pulverisers as it is heavier than air and descends when added from top of the system displacing air. Explosions have occurred in handling phosphorus pentasulphide, when they were not properly covered by inert gas.

Safety and environment protection

Both aspects are complimentary. A good clean plant will attain better safety standard as operators will be healthy and alert and exposed to minimum risk.

Spread of data

A number of publications are available giving dangerous property of every chemical, its TLV limit, and its effect on health when handled. Reputed manufacturers are benefitting from use of this data while handling chemicals, but common public is mostly unaware of dangerous properties of chemicals, particularly carried by tankers/trucks on roads.

Government/media—expectation

It is expected that Government educate public at large through TV, Radio and Press about the dangers posed by various chemicals, particularly transported via tanker/trucks. Even ICMA; FICCI or Tanker Owner Association should spend some money on TV advertisements or programmes to bring about awareness. Mobile foam fighters on highway should be kept for emergency fire.

Some budget must be kept by every industrialist for advertisement in educating the need of handling dangerous chemicals his factory produces. Government can take care through Directorate of Public Health, Safety & Factory Inspectorate. Public should not suffer for the mistake of tank lorry drivers or chemical goods carriers. A comprehensive insurance policy against such disaster should be made compulsory so that public exposed to chemical disaster is awarded suitable compensation.

Thus for the safety of mankind, our future generation and our own happy life, it is mandatory that we give attention to our rights and obligations to society and expect the same from manufacturers of chemicals. It is better to design a good plant and do preventive work to reduce pollution and safeguard environments. It is very costly to protect environment when factories are not designed properly. The initial investment for pollution control will go a long way in running the industry without fear or favour of government or public at large.

ENVIRONMENTAL POLLUTANTS

Any substance which causes pollution is called a pollutant. A pollutant may thus include any chemical or geochemical (dust, sediment, grit, etc.) substance, biotic component or its product, or physical factor (heat) that is released intentionally by man into the environment in such a concentration that may have adverse harmful or unpleasant effects on human health. A pollutant has also been defined as "any solid, liquid or gaseous substance present in such concentration as may be or tend to be injurious to the environment". Pollutants are the residues of things we make, use and throw away. There are many sources of such pollutants. The lakes and rivers are polluted by wastes from chemical and other factories, and the air by gases of automobile exhausts, industries, thermal power plants, etc.

The various principal pollutants which pollute air, water, land are as follows:

(1) Deposited matter–Soot, smoke, tar, dust, grit etc.

(2) Gases–Oxides of nitrogen (NO, NO_2), sulphur (SO_2), carbon monoxide, halogens (chlorine, bromine, iodine).

(3) Acids droplets–Sulphuric acid, nitric acid etc.

(4) Fluorides

(5) Metals–Mercury, lead, iron, zinc, nickel, tin, cadmium, chromium etc.

(6) Agrochemicals–Biocides (pesticides, herbicides, fungicides, nematicides bactericides, weedicides, etc.) and fertilisers.

(7) Complex organic substances–Benzene, ether, acetic acid, benzpyrenes etc.

(8) Photochemical oxidants–Photochemical smog, ozone, peroxyacetyl nitrate (PAN), peroxybenzoil nitrate (PB_zN), nitrogen oxides, aldehydes, ethylene, etc.

(9) Solid wastes.

(10) Radioactive waste.

(11) Noise.

Kinds of Pollutants

On the basis of the kind of pollutant involved, we may have sulphur dioxide pollution, fluoride pollution, carbon monoxide pollution, smoke pollution, lead pollution, mercury pollution, solid waste pollution, radioactive pollution, noise pollution, etc. Of the variety of pollutants, the following two basic types of pollutants non-degradable and biodegradable are recognised.

Non-degradable pollutants

These are the materials and poisonous substances like aluminium cans, mercuric salts, long-chain phenolics, DDT etc. that either do not degrade or degrade only very slowly in nature. They are cycled in ecosystem naturally. They not only accumulate but are often biologically magnified with their subsequent movement in food chains and biogeochemical cycles.

Biodegradable pollutants

They are the domestic wastes that can be rapidly decomposed under natural conditions. They may create problems when they accumulate (i.e., their input into the environment exceeds their decomposition).

Various types of pollution are classified in different ways on the basis of the types of environment being polluted, air pollution, water pollution, land (soil) pollution, marine pollution etc. are orcognised.

AIR POLLUTION

Air pollution may be defined as any atmospheric condition in which certain substances are present in such concentration that they can produce undesirable effects on man and his environment. These substances include gases (sulphur oxides, nitrogen oxides, carbon monoxides, hydrocarbons, etc.), particulate matter (smoke, dust, fumes, aerosols), radioactive materials and many others. Most of these substances are naturally present in the atmosphere in low (background) concentrations and are usually considered to be harmless.

Air Quality

The atmosphere is an insulating blanket around the earth. It is source of essential gases, maintains a narrow difference of day and night temperatures and provides a medium for long distance radiocommunication. It also acts as shield around the earth against lethal uv radiations. Without atmosphere, there will be no lightening, no wind, no clouds, no rains, no snow and no fire. Normal composition of clean air at or near sea is as follows:

Gases	Per cent (by volume)
Nitrogen	78.084
Oxygen	20.9476
Argon	0.934
Carbon dioxide	0.0314
Methane	0.0002
Hydrogen	0.00005
Other gases	minute

Although, there are several parameters to judge air quality, generally three–SO_2, NO_x and SPM (suspended particulate matter) are used which give a fair idea of pollutional load carried by the air.

Unfortulately, the concentration levels of different gases show variations due to pollutions. Moreover, there are also added a number of harmful gases to atmosphere. These adversely affect air quality and make it unfit for living beings.

Industrial air pollution makes an immediate impact on public opinion since it is generally visible and its effects on human health, vegetation and the general environment are well known.

Sources of Air Pollution

Air pollution results from gaseous emissions from mainly industry, thermal power stations, automobiles, domestic combustion, etc.

Industrial chimney wastes

There are a number of industries which are source of air pollution. Petroleum refineries are the major source of gaseous pollutants. The chief gases are SO_2 and NO_x. Mathura-based petroleum refinery is posing threat to Taj Mahal in Agra and other monuments at Fatehpur Sikri complex. Cement factories emit plenty of dust, which is potential health hazard. Stone crushers and hot mix plants also create a menace. The SPM levels in such areas of stone crushing are more than five times the industrial safety limits. There are many food and fertilisers industries which emit gaseous pollutants. There are also chemical manufacturing industries which emit acid vapours in air.

Thermal power stations

There are a number of thermal power stations and super thermal power stations in the country. The coal consumption of thermal plants is several millions tonnes. The chief pollutants are fly ash, SO_2 and other gases and hydrocarbons. Table 32.1 shows various gaseous pollutants from a 200 MW thermal power plant.

Table 32.1. Gaseous pollutants from a 200 MW thermal power plant.

Components	Emission factor kg/ton of coal	Emitted quantity (tons a day)
Aldehydes	0.0025	0.0035
Carbon monoxide	0.25	0.35
Hydrocarbons	0.10	0.14
NO_x	10.00	14.00
Oxides of sulphur (0.5% S)	19^s	13.30
Particulate matter (33% ash) Ash	8^a	369.60
	2^a	92.40

[a] Ash content in coal in per cent.

[s] Sulphur content in coal in per cent.

Automobiles

The toxic vehicular exhausts are a source of considerable air pollution, next only to thermal power plants. The ever increasing vehicular traffic density posed continued threat to the ambient air quality. In the major metropolitan cities, vehicular exhaust accounts for 70% of all CO, 50% of all hydrocarbons, 30-40% of all oxides and 30% of all SPM. The sources of emission in automobiles are:

(1) Exhaust system

(2) Fuel tank and carburettor

(3) Crankcase

The exhaust produces many air pollutants including unburnt hydrocarbons, CO, NO_x and lead oxides. There are also traces of aldehydes, esters, ethers, peroxides and ketones which are chemically active and combine to form smog in presence of light. Evaporation from fuel tank goes on constantly due to volatile nature of petrol, causing emission of hydrocarbons. The evaporation through carburettor occurs when engine is stopped and heat builds up and as much as 12 to 40 ml of fuel is lost during each long stop causing emission of hydrocarbons. Some gas vapour escapes between walls and the piston, which enters the crankcase and then discharges into the atmosphere. This accounts for 25% of the total hydrocarbon emissions of an engine.

Air Pollutants

From the different sources of air pollution, a variety of pollutants are released into atmosphere. The principal air pollutants emitted from these different sources are as follows:

(1) *Carbon compounds:* These are mainly CO_2 and CO, the former released by complete combustion of fossil fuels and the latter by automobile exhausts.

(2) *Sulphur compounds:* These include SO_2, H_2S and H_2SO_4, mostly released by fossil fuel (coal etc.) based power generating plants (thermal plants) and industrial units as refineries.

(3) *Nitrogen oxides:* These include chiefly NO, NO_2, HNO_3, mostly released by automobiles, power plants and industries.

(4) *Ozone:* Its level may rise in atmosphere due to human activities.

(5) *Fluorocarbons:* These come from industries, insecticides spray, etc.

(6) *Hydrocarbons:* These are chiefly benzene, benzpyrene, etc., which are mostly discharged by automobiles and industries.

(7) *Metals:* These include chiefly lead, nickel, arsenic, beryllium, tin, vanadium, titanium, cadmium etc., present in air as solid particles or liquid droplets or gases. They are produced mostly by metallurgical processes, automobiles, seaspray etc.

(8) *Photochemical products:* These are the photochemical smog, PAN, PB_2N, etc., released mostly by automobiles.

(9) *Particulate matter:* These are fly ash, dust, grit and other suspended particulate matter (SPM) released from power plants and industries (stone crushers etc). There are also bacterial cells, fungel spores and pollens in air as biological particulate pollutants.

(10) *Toxicants other than heavy metals:* These are complex chemical substances released during manufacture of other goods.

A brief account of harmful effects of individual air pollutant on biota of the system is given below.

Carbon compounds

The two important pollutants are carbon dioxide and carbon monoxide.

Carbon dioxide

Major amount of carbon dioxide is released in the atmosphere from burning of fossil fuel (coal, oil etc.) for domestic cooking, heating etc., and the fuel consumed in furnaces of power plants, industries, hot-mix plants etc. From fossil fuels alone more than 18×10^{12} tons of CO_2 is being released into atmosphere each year.

To some extent an increase in CO_2 level in atmosphere increases the photosynthesis rate and consequently plant growth, acting as fertiliser especially in hot tropical climates. This potential of fertiliser effect may be exploited by using modified crop varieties and agricultural practices. However, an increase in CO_2 concentration in atmosphere may result into disastrous effects also this is described below—greenhouse effect.

Greenhouse effect: The heat trap provided by atmospheric CO_2 probably helped to create the conditions necessary for the evolution of life and the greening of earth. Compared to moderately warm planet, Mars, with too little CO_2 in its atmosphere is frozen cold and Venus with too much is a dry furnace. The excess CO_2 to some extent is absorbed by the oceans.

Since CO_2 is confined exclusively to the troposphere, its higher concentration may act a serious pollutant. Under normal conditions (with normal CO_2 concentration) the temperature at the surface of the earth is maintained by the energy balance of the sun rays that strike the planet and heat that is radiated back into space. However, when there is an increase in CO_2 concentration, the thick layer of this gas prevents the heat from being re-radiated out. This thick CO_2 layer thus functions like the glass panels of a greenhouse (or the glass windows of a motor car), allowing the sunlight to filter through but preventing the heat from being re-radiated in outer space. This is the so-called *greenhouse effect.* Thus most heat is absorbed by CO_2 layer and water vapours in the atmosphere, which adds to the heat that is already present. The net results is the heating up of the earth's atmosphere. Thus increasing CO_2 levels tend to warm the air in the lower layers of atmosphere on a global scale. Nearly 100 years ago the CO_2 level was 275 ppm. Today it is 350 ppm and by the year 2035 and 2040 it is expected to reach 450 ppm. Imagine the earth's temperature. CO_2 increases the earth temperature by 50% while CFCs are responsible for another 20% increase. There are enough CFCs up there to last 120 years.

Carbon monoxide

The chief source are automobiles, though other involving a combustion process as stoves, furnaces, open fires, forests and bush fires, burning coal mines, factories, power plants etc. also give off CO. The principal source of this pollutant are the exhaust products from motor vehicles. The smoke of automobiles and thermal power and hot-mix plants, stone crushers etc. also contribute to CO level in air. CO comprises for as much as 80% of all automobile emissions and for more than 60% of all major pollutants added to the atmosphere.

Carbon monoxide is very harmful to those persons exposed to congested highways to a level of about 100 ppm. Thus drivers are the most affected people. CO causes difficulty in breathing, causes headache and irritation of mucous membranes. Inhaled CO combines with blood haemoglobin to form carboxy haemoglobin about 210 times faster than O_2 does. The gas is fatal over 1000 ppm, causing unconsciousness in an hour and death in four hours. If this gas is inhaled for few hours at even a low concentration of 200 ppm, it causes symptoms of poisoning.

Sulphur compounds

From amongst several other major sulphur compounds in the atmosphere, the oxides of sulphur are the most serious pollutants. The other S-compounds are carbonyl sulphide (COS), carbon disulphide (CS_2), dimethyl sulphide [$(CH_3)_3S$] and sulphates. The chief source of oxides of sulphur are the combustion of coal and petroleum. Thus, most oxides come from thermal power plants and other coal-based plants and smelting complexes. Automobiles also release SO_2 in air.

Sulphur dioxide

The major sources of SO_2 emission are burning of fossil fuels (coal) in thermal power plants, smelting industries (smelting sulphur containing metal ores) and other processes as manufacture of sulphuric acid and fertilisers. These account for about 75% of the total SO_2 emission. Most of the rest 25% emission is from petroleum refineries and automobiles.

SO_2 causes intense irritation to eyes and respiratory tract. It is absorbed in the moist passage of upper respiratory tract, leading to swelling and stimulated mucus secretion. Exposure to 1 ppm level of SO_2 causes a constriction of the air passage and causes significant broncho-constriction in asthmatics at even low (0.25-0.50 ppm) concentrations. Moist air and fogs increase the SO_2 dangers due to formation of H_2SO_4 and sulphate ions; H_2SO_4 is a strong irritant (4-20 times) than SO_2.

SO_2 is also involved in the erosion of building materials as limestone marble, the slate used in roofing, mortar and deterioration of statues. Petroleum refineries, smellers, kraft paper mills deteriorate the adjoining historic monuments.

Hydrogen sulphide

The chief source of H_2S are decaying vegetation and animal matter, especially in aquatic habitats. Sulphur springs, volcanic eruptions, coal pits and sewers also give of this gas. At a lower concentration, H_2S causes headache, nausea, collapse, coma and final death. Unpleasant odour may destroy the appetite at

5 ppm level in some people. A concentration of 150 ppm may cause conjunctivitis and irritation of mucus membranes. Exposure at 500 ppm for 15-30 min. may cause colic diarrhoea and bronchial pneumonia. Thus gas readily passes through alveolar membranes of the lung and enters the blood stream. Death occurs due to respiratory failures. The MAC for an 8 hours day is 20 ppm.

Nitrogen oxides (NO_x)

The measurable amounts of nitrous oxide, nitric oxide and nitrogen dioxide are present even in unpolluted atmosphere. Of these nitric oxide (NO) is the pivot compound. It is produced by combustion of O_2 and N_2 during lightening discharges and by bacterial oxidation of NH_3 in soil. NO contacts with air and combines with O_2 or even more readily with O_3 to form the more poisonous nitrogen dioxide (NO_2). NO_2 may react with water vapour in air to form HNO_3. This acid combines with NH_3 to form ammonium nitrate. Fossil fuel combustion also contributes to oxides of nitrogen. About 95% of the nitrogen oxide is emitted as NO and remaining 5% as NO_2. In urban areas about 46% of oxides of nitrogen in air come from vehicles and 25% from electric generation and the rest from other sources. In metropolitan cities, vehicular exhaust is the most important source of nitrogen oxides.

Nitric oxide (NO)

The chief source of this gas are the industries manufacturing HNO_3 and other chemicals and the automobile exhausts. At high temperature, combustion of gasoline produces this gas. A number of chemical reactions serially convert a large amount of NO to more toxic NO_2 in the atmosphere.

NO is responsible for several photochemical reactions in the atmosphere, particularly in the formation of several secondary pollutants like PAN, O_3, carbonyl compounds etc. in the presence of other organic substances. There is little evidence of the direct role of this gas is causing a health hazard, at the levels found in urban air.

Nitrogen dioxide (NO_2)

A deep reddish brown gas, nitrogen dioxide is the only widely prevalent coloured pollutant in the atmosphere. This gas is the chief constituent of photochemical smog in metropolitan areas. NO_2 causes irritation of alveoli, leading to symptoms resembling emphysema (inflammation) upon prolonged exposure to 1 ppm level. Lung inflammation may be followed by oedema and final death. The MAC for occupational exposure are set at 5 ppm for an 8 hour period. Smokers may readily develop lung diseases as the cigarettes and cigars contain 330-1,500 ppm nitrogen oxides. NO_2 is highly injurious to plants. Their growth is suppressed when exposed to 0.3-0.5 ppm for 10-20 days. Sensitive plants show visible leaf injury when exposed to 4-8 ppm for 1-4 hours.

Acid Rains—An Invisible Threat

The oxides of sulphur and nitrogen are important gaseous pollutants of air. These oxides are produced mainly by combustion of fossil fuels, smelters, power plants, automobile exhausts, domestic fires etc. These oxides are swept up into the atmosphere and can travel thousands of kilometres. The longer they stay in the atmosphere, the more likely they are to be oxidised into acids. Sulphuric acid and nitric acid are the two main acids, which then dissolve in the water in the atmosphere and fall to the ground as acid rain or may remain in atmosphere in clouds and fogs.

Acidification of environment is a man-made phenomenon. The acid rain is infact a cocktail of H_2SO_4 and HNO_3 and the ratio of the two may vary depending on the relative quantities of oxides of sulphur and nitrogen emitted. On an average 60-70% of the acidity is ascribed to H_2SO_4 and 30-40% to HNO_3. The acid rain problem has dramatically increased due to industrialisation. Burning of fossil fuels for power generation contributes to almost 60-70% of total SO_2 emitted globally. Emission of NO_x from anthropogenic sources ranges between 20-90 million tons annually over the globe. Acid rains have assumed global ecological problem, because oxides travel a long distance and during their journey in atmosphere they may undergo physical and chemical transformations to produce more hazardous products.

Acid rains create complex problems and their impacts are far reaching. They increase soil acidity, thus affecting land flora and fauna; cause acidification of lakes and streams thus affecting aquatic life, affects crop productivity and human health. Besides these they also corrodes buildings, monuments, statues, bridges, fences, railings etc. British parliament building also suffered damage due to H_2SO_4 rains. Due to acidity, levels of heavy metals as aluminium, manganese, zinc, cadmium, lead and copper in water increases beyond the safe limits.

Ozone (O₃)

It is universally accepted that the ozone layer in the stratosphere protects us from the harmful uv radiations from sun. The depletion of this O_3 layer by human activities may have serious implications and this has become a subject of much concern over the last few years. On the other hand, ozone is also formed in the atmosphere through chemical reactions involving certain pollutants (SO_2, NO_2, aldehydes) on absorption of uv radiations. The atmospheric ozone is now being regarded as potential danger to human health and crop growth. What makes ozone a *destroyer* as well as a *protector* needs to be elaborated to have a clear picture of its biopotency from human welfare view point.

Ozone—The destroyer

The temperature decreases with increasing altitude in the troposphere (8–16 km from earth surface), while it increases with increasing altitude in the stratosphere (above 16 km up to 50 km). This rise in temperature in stratosphere is caused by the ozone layer. The ozone layer has two important and interrelated effects. Firstly, it absorbs uv light and thus protects all life on earth from harmful effects of radiation. Second, by absorbing the uv radiation the ozone layer heats the stratosphere, causing temperature inversion. The effect of this temperature inversion is very interesting. It limits the vertical mixing of pollutants, thereby causing the dispersal of pollutants over larger areas and near the earth's surface. That is why a dense cloud of pollutants usually hangs over the atmosphere in highly industrialised areas causing several unpleasant effects. The wastes spread horizontally relatively fast (than slow mixing vertically), reaching all longitudes of the world in about a week and all latitudes within months. There is, therefore, very little that a country can do protect the ozone layer above it. The ozone problem is thus global in scope. In spite of slow vertical mixing, some of the pollutants (CFCs) enter the stratosphere and remain there for years until they are converted to other products or are transported back to the stratosphere. The stratosphere could be regarded as a sink, but unfortunately, these pollutants (CFCs) react with the ozone and deplete it.

The ozone near the earth's surface in the troposphere creates pollution problems. Ozone and other oxidants such as peroxyacetyl nitrate (PAN) and hydrogen peroxide are formed by light dependent reactions between NO_2 and hydrocarbons. Ozone may also be formed by NO_2 under uv-radiations effect. These pollutants cause photochemical smog.

Increase in O_3 concentration near the earth's surface reduces crop yields significantly. It also has adverse effect on human health. Thus, while higher levels of O_3 in the atmosphere protects us, it is harmful when it comes in direct contact with us and plants at earth's surface.

Ozone—The protector

Ozone protects us from the harmful uv radiations from the sun. In spite of being in such a small proportion (0.02–0.07 ppm), it plays a major role in climatology and biology of the earth. It filters out all radiations below 3000 Å. Thus, O_3 is intimately connected with the life-sustaining process. Any depletion of ozone would, therefore, have catastrophic effects on life systems of the earth. Over the last few years, it could be realised that the O_3 concentration of earth's atmosphere is thinning out. What has caused this depletion?

It has already been pointed out that O_3 layer by absorption of uv radiation heats the stratosphere, causing the temperature inversion. This temperature inversion limits the vertical mixing of pollutants. However, in spite of this slow vertical mixing, some pollutants enter the stratosphere and remain there for years until they react with ozone and are converted to other products. These pollutants thus deplete ozone in the stratosphere. Major pollutants responsible for this depletion are chlorofluorocarbons (CFCs), nitrogen oxides (coming from fertilisers) and hydrocarbons. CFCs are widely used as coolants in air conditioners and refrigerators, cleaning solvents, aerosol propellants and in foam insulation. CFC is also used in fire extinguishing equipment. They escape as aerosol in the stratosphere. Jet engines, motor vehicles, nitrogen fertilisers and other industrial activities are responsible for emission of CFCs, NO_x, etc. The supersonic aircrafts flying at stratosphere heights cause major disturbances in O_3 levels. The threat to O_3 is mainly from CFCs which are known to deplete O_3 by 14% at the current emission rate. On the other hand NO_x would reduce O_3 by 3.5%. The nitrogen fertilisers release nitrous oxide during dentrification. Depletion of O_3 would lead to serious temperature changes on the earth and consequent damage to life-support systems.

The reduction in ozone would lead to temperature changes and rainfall failures on earth. Moreover, 1% reduction in O_3 increases uv radiation on earth by 2%. A series of harmful effects are caused by an increase in uv radiation. Cancer is the best established threat to man. When the O_3 layer becomes thinner or has holes, it causes cancers, especially relating to skin like melanoma. A 10% decrease in stratospheric ozone appears likely to lead a 20-30% increase in skin cancer.

Apart from direct effects, there are also indirect effects. Under greenhouse effect conditions, plants exposed to uv radiation showed a 20–50% reduction in growth reduction in chlorophyll content and increase in harmful mutations. Enhanced uv radiation also impairs fish productivity.

Fluorocarbons

In minute amounts, fluorocarbons are beneficial helping prevention of tooth-decay in man. However, higher levels become toxic. Fluorides in atmosphere come from industrial processes of phosphate fertilisers, ceramics, aluminium, fluorinated hydrocarbons (refrigerants, aerosol propellants etc.), fluorinated plastic, uranium and other metals. The pollutant is in gaseous or particulate state.

In air, fluoride chiefly comes from smoke of industries, volcanic eruptions and insecticide sprays. Fluorides enter plant leaves through stomata. In plants it causes tip burn due to accumulation in leaves of conifers. Fluoride pollution in man and animals is mainly through water.

Hydrocarbons

From amongst many hydrocarbons the chief air pollutants are benzene, benzpyrene and methane. Their chief sources are the motor vehicles, being emitted by evaporation of gasoline through carburettors, crankcase etc. About 40% of the vehicular exhaust hydrocarbons are unburnt fuel components, the rest are the product of combustion.

They have carcinogenic effects on lung. They combine with NO_x under uv component of light to form other pollutants like PAN and O_3 (photochemical smog) which cause irritation of eye, nose and throat and respiratory distress.

Benzene, a liquid pollutant is emitted from gasoline. It causes lung cancer. Benzpyrene is most potent cancer inducing hydrocarbon pollutant. It is also present in small amounts in smoke, tobacco, charcoal boiled stakes and gasoline exhaust. *Methane* (marsh gas) is a gaseous pollutant, in minute quantity in air, about 0.0002% by volume. In nature this is produced during decay of garbage, aquatic vegetation etc. This is also released due to burning of natural gas and from factories. Higher concentrations may cause explosions. The excess of water seepage in filled up well and pits may lead to excess production of methane which bursts with high sound and may cause local destruction. At high levels in absence of oxygen, methane may be narcotic on man.

Metals

In air, the common metals present are mercury, lead, zinc and cadmium. They are released from industries and human activities in the atmosphere.

(1) *Mercury:* A liquid volatile metal (found in rocks and soils) is present in air as a result of human activities as the use of mercury compounds in production of fungicides, paints, cosmetics, paper pulp etc.

(2) *Lead:* Lead compounds added to gasoline to reduce knocking are emitted into the air with the exhaust as volatile lead halides (bromides and chlorides). About 75% of lead burnt in gasoline comes out as lead halides through tail pipe in exhaust gases. Of this about 40% settles immediately on the ground and the rest (60%) gets into air.

(3) *Zinc:* Zinc, not a natural component of air, occurs around zinc smelters and scrap zinc refineries. Copper, lead and steel refineries also release some zinc in the air. Open hearth furnaces emit 20-25 g zinc/hr in refining the galvanised iron scrap. Zinc in air occurs mostly as white zinc oxide fumes and is toxic to man.

(4) *Cadmium:* Cadmium occurs in air due to industries and human activities. Industries engaged in extraction, refining, electroplating and welding of cadmium-containing material and those in refining of copper, lead and zinc are the major source of cadmium in air. Production of some pesticides and phosphate fertilisers also emits cadmium to air. This metal is emitted as vapour and in this state it quickly reacts to form oxide, sulphate or chloride compounds. Cadmium is poisonous at very low levels and is known to accumulate in human liver and kidney. It causes hypertension, emphysema and kidney damage. It may turn to be carcinogenic in mammals.

Photochemical products

There is much interlinking of NO_x, hydrocarbons and O_3 in the atmosphere. These individually are recognised air pollutants. However, at the same time in presence of light as a result of photochemical reactions those may react with each other and/or may undergo transformations to produce even more toxic secondary pollutants in the air. There are also some other pollutants. The principal photochemical products are olefins, aldehydes, ozone, PAN, PB_zN and photochemical smog.

Olefins are produced directly from the exhaust and in the atmosphere from ethylene. At very low concentrations of few ppb, they affect plants seriously. Aldehydes as HCHO and olefin, acrolein irritate the skin, eyes and upper respiratory tract. Ozone, also a photochemical product, has already been touched. The aromatics are the most potent pollutants among the photochemical products. These are benzpyrene, peroxyacetyl nitrate (PAN) and peroxybenzoil nitrate (PB_zN).

Photochemical smog is highly oxidising polluted atmosphere comprising largely of O_3, NO_x, H_2O_2, organic peroxides, PAN and PB_zN. This is produced as a result of photochemical reaction among NO_x; hydrocarbons and oxygen and causes eye irritation and reduces visibility. The photochemical smog formation occurred only during night or cloudy days.

Some sulphates and nitrates can also be formed in photochemical smog due to oxidation of sulphur containing components (SO_2, H_2S) and NO_x (N_2O_5, NO_2). HNO_3 and nitrates are important toxicants of smog. They cause damage to plants, corrosion problems and are human health hazards.

Particulate matter (PM)

This is a discrete mass of any material, except pure water, that exists as liquid or solid in the atmosphere and of microscopic or sub-microscopic dimensions. Air borne matter results not only from direct emission of particles but also from emissions of some gases that condense as particles directly or undergo transformations to form particle. Thus PM may be primary or secondary. Primary PM includes dust, as a result of wind, or smoke particles emitted from some factory. Atmospheric PM ranges in size from 0.001 mm to several hundred mm. Particulate matter in atmosphere arises from natural as well as man-made sources. Natural sources are soil and rock debris (dust), volcanic emission, seaspray, forest fires and reactions between natural gas emissions. There are four types of sources of particulate matter.

(1) Fuel combustion and industrial operations (mining, smelting, polishing, furnaces and textiles, pesticides, fertilisers and chemical production).

(2) Industrial fugitive processes (materials handling, loading and transfer operations).

(3) Non-industrial fugitive processes (roadway dust, agricultural operations, construction, fire, etc.).

(4) Transportation sources (vehicles exhaust and related particles from fire, clutch and break wear).

The particulate matter is injurious to health. Soot, lead particles from exhaust, asbestos, fly ash, volcanic emission, pesticides, H_2SO_4, mist, metallic dust, cotton and cement dust etc. when inhaled by man cause respiratory diseases such as tuberculosis and cancer. Cotton dust causes an occupational disease *Byssinosis*, very common in India.

In addition to the above there are also many kinds of biological particulate matter that remain suspended in atmosphere. These are bacterial cells, spores, fungal spores, pollen grains. These cause bronchial disorders, allergy and many other diseases in man, animals and plants.

Toxicants

There is a wide variety of toxic substances, besides air pollutants, which have been shown to be implicated in human health hazards. Some of the principal toxicants are described below:

(1) *Arsenic* is produced as a by-product of metal refining process. In industrial areas its concentration may reach nearly 20–90 mg/m^3. It is found to cause cancer.

(2) *Asbestos* is a mineral fibre used in asbestos cement pipes, flooring products, paper, roofing products, asbestos cement sheets, packing and gaskets, insulation, textiles, etc. Asbestos fibres are non-degradable. They cause cancer in man.

(3) *Carbon tetrachloride* and *chloroform* are used for making fluorocarbons for refrigerants and propellants etc. Chloroform degrade slowly into phosgene, HCl and chlorine monoxide. Both have carcinogenic effects in rat, mouse and other animals. Chromium is used in stainless tool and alloy steel, heat and corrosion-resistant materials, alloy cast iron, pigments, metal plating, leather tanning, etc. Chromium components have carcinogenic effects.

(4) *Nickel* is used in chemicals, petroleum and metal products, electrical goods, household appliances, machinery, etc. Inorganic nickel is strongly carcinogenic in man. Nitrosamines are mostly used in rubber processing organic chemical manufacturing and in rocket fuel manufacturing. They are also considered carcinogenic.

(5) *Vinyl chloride* is the prime compound for polyvinyl chloride (PVC), a widely used plastic resin. It is a known carcinogen in man and also suspected to induce brain and lung cancer.

There are also several polycyclic aromatic hydrocarbons (PAH$_s$), which arrive in the atmosphere from coal production, vehicle disposal, wood burning, municipal incineration, petroleum refining and coal furnaces. In general, they do not produce adverse effect in their parent forms. However, if metabolised by enzymes of the body, they produce intermediates which are capable of inducing cancer.

PREVENTION AND CONTROL OF AIR POLLUTION

Steps are to be taken to control pollution at source (prevention) as well as after the release of pollutants in the atmosphere. Ideally, the engineering designer's strategy with regard to air pollution control should be to:

(1) Prevent the formation of the pollutant.

(2) Reduce emission by the removal of the pollutant from the process gas stream.

(3) Ensure efficient dispersal in the atmosphere.

Vehicular Pollution

Following are some of the measures to be adopted for control of air pollution by motor vehicles.

(1) *To check pollutant emission from vehicular exhaust.*

This can be achieved by the following practices:

(a) Using new proportion of gasoline and air.

(b) More exact timing of fuel feeding.

(c) Using gas additives to improve combustion.

(d) By injecting air into the exhaust to convert exhaust compounds to less toxic materials.

(e) Updating of engine design and/or install abatement equipment (device) to improve combustion with the existing engine design.

Carbon monoxide results from low air content of the fuel mixture, whereas NO_x production is promoted by high combustion temperatures. Hydrocarbons follows more or less the pattern of CO. The complete elimination of these three pollutants may be achieved by either updating the present design of engines or by making appropriate changes in devices for improving combustion.

(2) *To control evaporation from fuel tank and carburettor.*

This can be done by following methods:

(a) Collection of vapours with activated charcoal when the engine is turned off and its ignition when the engine is started.

(b) Subjecting the gasoline in the tank to slight pressure to prevent the gas from evaporation.

(c) Developing low-volatile gasoline that does not evaporate easily.

(3) *Use of filters.*

Some gas vapours escape between walls and the piston which enter the crankcase and then discharge into atmosphere. Hydrocarbons (about 25%) are released in this way. Thus use of filters that capture and recycle these escaped gases in the engine should help control the emission of these hydrocarbons.

Industrial Pollution

To check air pollution by industrial and power plant chimney wastes, designer must device measures for removal of the particulate matter and gaseous pollutants from the wastes. Before a pollution control system can be designed or assessed, the nature and quantity of the effluent must be determined. Normally this is done by withdrawing a representative sample from the atmosphere or process and analysing for chemical composition and/or particulate size distribution and concentration.

Particulates

The particle size distribution of a pollutant is important not only in assessing its potential as a hazard (the respirable size range is 0.3-5 μm) but in specifying the correct gas cleaning equipment since this must be used in conjunction with the grade efficiency curve to relate the outlet to the inlet pollutant concentrations.

With regularly shaped particles, it is easy to describe their size–the obvious examples is a sphere in which case the diameter is an adequate description.

However, when the shape is irregular the problem is extremely difficult and numerous definitions are used. The most widely used of these is probably the Stokes diameter—that of a sphere of the same density which settles in a fluid with the same speed. Most methods are related to the technique of analysis used. However, it is sound practice to examine a sample of the particulates by microscope as a preliminary step to a complete size distribution analysis.

Where the particulates are precipitated from the gas stream for subsequent analysis, further size classification can occur. Methods which have been introduced to obviate this include electrostatic precipitation. Notable amongst electrostatic precipitators are the Rochester instrument, Lovelace E.S.P. design, and pulsed voltage precipitator. Thermophoresis has also been used as a precipitation mechanism, but perhaps the simplest and most direct method is to draw the air through an absolute master filter. Membrane type filters which are manufactured with extremely small pore sizes are widely used for this purpose.

Particle size analysis

A very wide range of methods of analysis is now available. Much of the equipment can be obtained commercially and is automatic, requiring very little operation.

The principle of operation of particle counting instruments is shown in Fig. 32.1. A beam is light is focused onto a sensing zone through which the particles are passed individually in a sheath of clear air. The resultant scattered light pulse, which is a function of particle size, is picked up by a photomultiplier tube or solid state detector and fed to a pulse height analyser. The lower limit on particle size which can be measured is imposed by the signal to noise ratio which can be obtained in practice. Use of the laser cavity system in which the particle is fed through the cavity of the laser itself and in which a very high intensity of incident, and hence scattered, beams is achieved has led to an extension of the lower limit on particle size from about 0.3 to 0.1 μm. Counting methods which employ hydrodynamic particle focusing are restricted by particle coincidence errors for use in relatively low concentration situations. A recent development which allows *in situ* operation at higher concentrations is optical focusing in which a very small sensing volume is used.

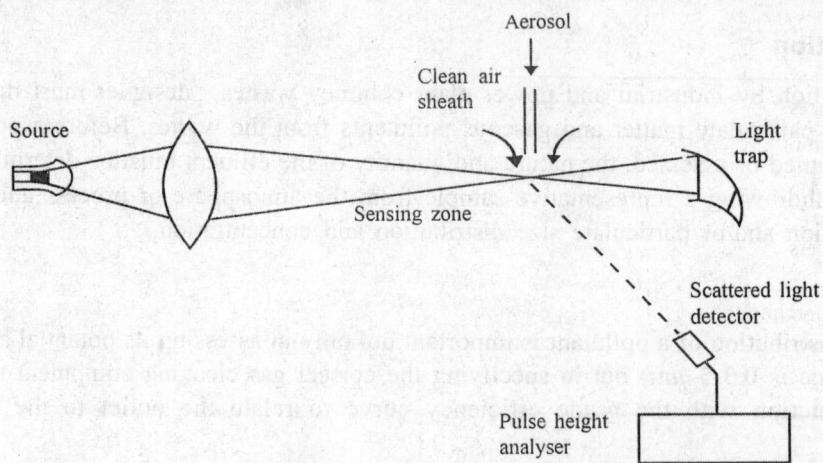

Fig. 32.1. Light scattering particle counter.

Gaseous contaminants

Representative samples of gaseous contaminants are easier to obtain than particulates since inertial classification does not take place. However, precautions may have to be taken to ensure that selective reaction does not occur with components of the sampling system or other gaseous components. Removal of the required contaminants may be achieved by the use of condensation traps, or by gas washers in which chemically selective absorbents are used or by adsorption on, for example, activated carbon, silica gel or molecular sieves. The sample may be removed for total volumetric analysis in the laboratory for it may be passed continuously to a monitor. Recently, there have been a number of developments in the application of remote monitoring.

The methods of analysis can be classified as chemical or physical. In the former the gas is trapped in a suitable medium and analysed chemically. The latter involves the direct measurement of some physical property, perhaps following chromatographic separation. Methods of chemical analysis used include colorimetric, acidimetric and particularly Coulometric methods. Of the physical methods, chemiluminescence and flame photometry, fluorescence and adsorption spectroscopy are all widely used. Both gas chromatography with flame ionisation detection and spectroscopic techniques are used in commercially available instruments for hydrocarbon analysis.

Control of emission of particulates

The selection of equipment for a specific application is dependent on a wide range of factors including the permissible pressure drop, dust loading, gas throughput and temperature, all of which are imposed by the process conditions. An extremely important parameter, however, is the size distribution of the particulates to be removed. The equipment is best characterised with regard to this parameter by its grade efficiency which is a graph of fractional collection efficiency drawn against particle size. There is a wide range of equipment available to the engineer and this is very simply classified in Table 32.2.

The mode of operation of these units is best considered by recourse to the particle removal mechanisms involved.

Table 32.2. Methods of gas/solid separation.

Equipment	Main collection mechanisms	Normally applicable in particle size range (mm)
Settling chambers	Gravity settling	> 100
Inertial separators	Inertia	> 50
Cyclones	Centrifugal settling	> 5
Scrubbers	Inertia	> 3
Venturi scrubbers	Inertia	> 0.3
Fabric filters	Sieving	> 0.1
Electrostatic precipitators	Electrostatic migration	> 0.001
Fibrous filtration	Inertial interception,	> 0.5
	Diffusion, eletrostatic precipitation	> 0.5

Gravity settlers

Typical gravity settlers (Fig. 32.2) and inertial collectors (Fig. 32.3) are used to collect fairly coarse dusts and may be used as prefilters. The gravity settler, in particular, works only for large particles the limiting factor being their settling speed in air. That is only particles in excess of 50 μm can be reasonably removed on an industrial scale using gravity separation. The units are, however, widely used as prefilters and their efficiency is enhanced by the inclusion of horizontally placed trays which reduce the sedimentation height and allow greater capture in the same residence time.

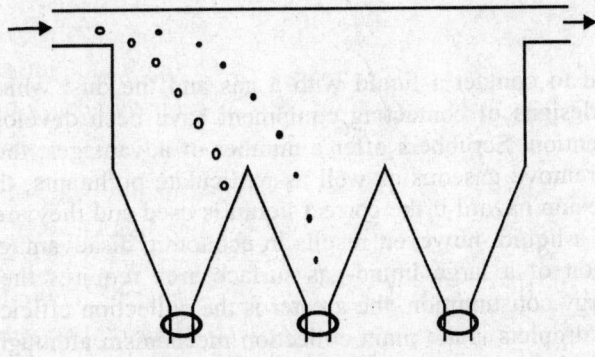

Fig. 32.2. Gravity settler.

Inertial Collectors

In inertial collectors (Fig. 32.3) an object is placed in the path of the gas. While the gas passes round the target, particles with sufficiently high inertia impinge on it and are removed from the stream. The inertial mechanism increases in significance with particle size, density and velocity and is reduced as the target size is increased, i.e. with decreasing gas streamline tortuosity. From Fig. 32.3 it can be seen that in the case of inertial separators, an attempt is made to achieve a high gas tortuosity but a pressure drop is incurred as a result of this.

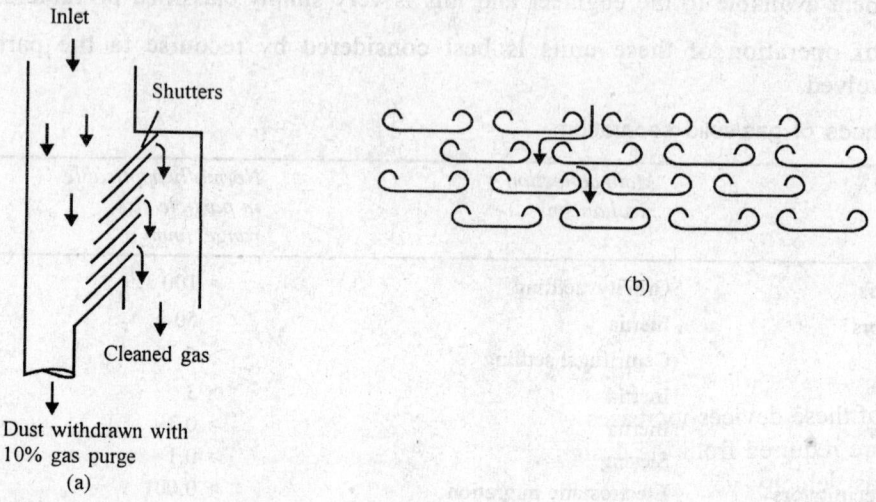

Fig. 32.3. Typical inertial separators: (a) Shutter collector; (b) Curtain collector.

Cyclones

Cyclones are also primarily used as precollectors. In these units centrifugal separation is applied and since this is, strictly speaking, an inertial mechanism, the Stokes number can again be used as a characterisation and scale-up parameter. By far the most commonly used type is the return flow cyclone whose operation is shown in Fig. 32.4. The dust laden gas enters tangentially with a velocity in the region of 20 ms^{-1} and spins downwards and inwards leaving by the vortex finder shown at the top of the unit. The strong vortex core in the centre has a diameter of about half that of the outlet thimble. Dust particles are thrown radially outwards by the centrifugal force to impinge on the wall of the cyclone and are then swept downwards and collected in the hopper.

Wet scrubbers

Wet scrubbers are designed to contact a liquid with a gas and the dust which the gas may contain. A large number of different designs of contacting equipment have been developed and the technique has recently required much attention. Scrubbers offer a number of advantages, the most obvious ones being that they can be uxsed to remove gaseous as well as particulate pollutants; the gas stream is cooled so that there is no fire or explosion hazard if the correct liquid is used and they are normally small compared with dry collectors. Use of a liquid, however, results in economic disadvantages and the gas stream will leave saturated. The creation of a large liquid-gas surface area requires the use of energy so that in general, the greater the energy consumption, the greater is the collection efficiency of the process. Inertial interception of particles by droplets is the main collection mechanism although in some cases, Brownian diffusion and condensation play a part.

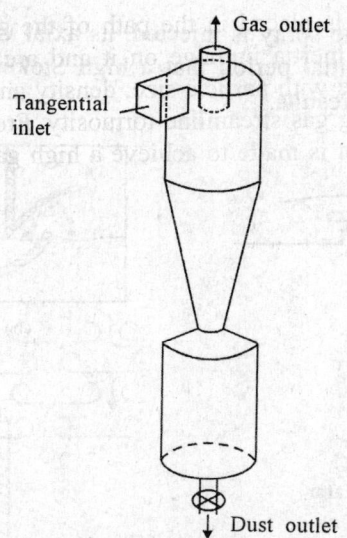

Gas outlet

Tangential
inlet

Dust outlet

Fig. 32.4. Cyclone separator.

The efficiency of these devices increases with particle size and density and small target sizes and high relative velocities are required from the design and operation point of view. Unfortunately, small droplets accelerate to the gas velocity very quickly, so that a compromise is often made between the two latter requirements. Some of the more common types of scrubbers are described below; in each case a different technique is used to achieve a high relative velocity between the gas and the small droplets.

In the wet impingement scrubber, water flows across an orifice plate through which the gas is passed (Fig. 32.5a). The main mechanism of collection is inertial impaction from the gas stream with the water or the plate located above the orifice. In the self induced spray type (Fig. 32.5b), droplets are sheared from the water surface and inertial collection takes place as they are accelerated by the gas stream. Simple water recirculation takes place and the low water consumption with these units is an advantage.

Packed towers are very widely used in gas absorption but particulates are also removed in the process. The liquid flows downwards through the packing and is contacted with the upward moving gas (Fig. 32.5c). A wide range of packings is available and the liquid flow for a given gas rate is critical. It must exceed the wetting point at which the flow of water is sufficient to wet the packing surface, but must be less than the flooding point at which water down-flow is prevented by the upward motion of the gas. It is most likely that both the droplet collection and impingement referred to above take place in these units. For a given particle size, penetration must fall off exponentially with depth of packing. Fluidised-bed scrubbers were originally designed to absorb fluorides from process gas that contained tarry substances but are now used for particulate removal. The fluidised-bed is retained between two grids and the action ensures continuous cleaning.

In centrifugal scrubbers, an attempt is made to increase the relative velocity of particles and droplets by centrifuging the droplets in an outward direction (Fig. 32.5d). This overcomes the objections to the spray tower in which the relative velocity is limited to the terminal settling speed of the droplets under gravity.

A considerable reduction in particle cut size at the expense of increased pressure drop is achieved using the venturi scrubber in which a high relative droplet velocity is obtained by injection of water into the throat of a venturi (Fig. 32.5e). When the spray is injected, its axial velocity resolute is close to zero but it rapidly increases. It is during this initial period that a high Stokes number is achieved even for submicron particles and a high efficiency results.

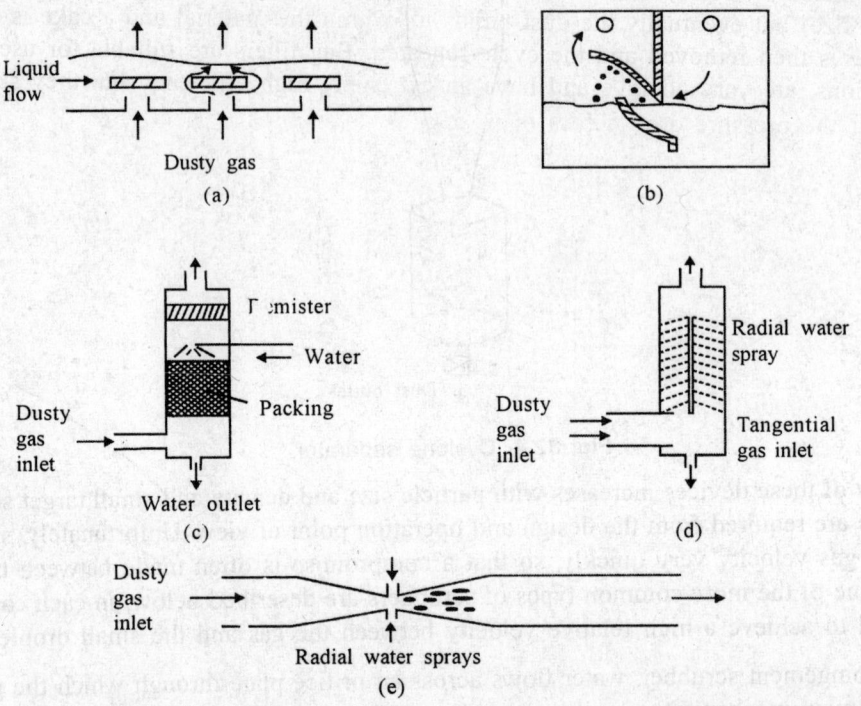

Fig. 32.5. Principles of operation of typical scrubbers: (a) Jet impingement; (b) Self-induced scrubber; (c) Packed tower; (d) Centrifugal Scrubber; (e) Venturi scrubber.

Recent developments which have been made in an attempt to reduce power requirements include the use of electrostatic enhancement. In some devices, the incoming gas is ionised by a corona discharge so that the particles become charged, usually negatively, and are attracted to droplets or to an irrigated surface. In other devices the droplets are charged. The pressure drop required to collect submicron particles is significantly reduced and a few such devices are now available commercially. Another development is the use of 'flux force condensation' scrubbing. In processes in which sufficient steam is available condensation contributes significantly towards collection. The 'particle size', and hence its inertia, is increased by condensate and both diffusphoresis and thermophoresis assist deposition during the condensation process.

Electrostatic precipitators (ESP)

Electrostatic precipitators are used where collection of fine particles at a high efficiency coupled with a low pressure drop is necessary. The principle of operation is shown in Fig. 32.6 in which dust laden gas is shown to enter a number of tubes or pass between parallel plates. The particulates are charged by corona discharge at the entrance and are subsequently deposited on the earthed plates or tube walls. The walls are 'rapped' periodically to remove the accumulated dust layer. These units are normally used in large installations such as power stations.

Bag filters

Bag filters comprise the simplest and probably the most common class of dust cleaners. Here a cloth or felt filter material which is impervious to the particles is used. Initially, particles are collected within the depth of the media by collision with fibres to which they adhere (the pore size being much greater than the particle size) but eventually the dust builds up within the material and a cake is formed on its surface; this cake is then removed and the cycle repeated. Bag filters are suitable for use in very high dust load conditions, are very positive and have an extremely high efficiency but they suffer from the disadvantage that the pressure drop across them may be high.

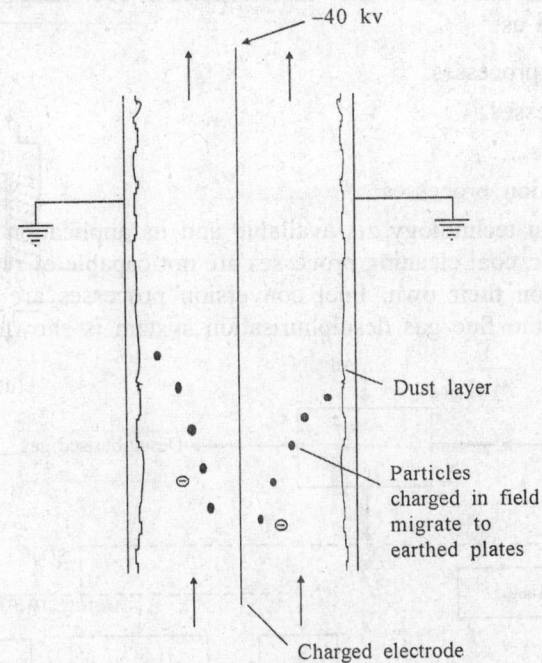

Fig. 32.6. Principle of electrostatic precipitation.

Fibrous filters

Fibrous filters are used to create very clean conditions, such as in clean room installations, to provide a sterile environment for the removal of possible radioactive dusts from nuclear exhausts and in air conditioning systems. They are used in situations in which the dust load is very low but they can be extremely efficient while offering a very low pressure drop.

Control of emission of gaseous contaminants

The major primary gaseous pollutants arising from industrial sources are those containing sulphur (SO_2, SO_3, H_2S), nitrogen (NO_x, NH_3), halogen compounds, hydrocarbons and carbon monoxide and dioxide. The most important to these from the point of view of volume produced are the products of combustion. CO_2 is not normally considered a pollutant from the industrial arrestment point of view and CO is released in minimum concentrations in the interests of efficiency. SO_2, however, is released in large quantities and presents one of the most difficult problems. The effects of cynogerism in which two relatively inoffensive emissions may combine in the atmosphere to cause an offensive haze must be considered and the production of odours is particularly obnoxious. The major methods of control of gaseous pollutants are absorption, adsorption, combustion and condensation.

Gas absorption is mainly used for the removal of SO_2, HF, Cl_2, HCl, H_2S and NO_x. The gas is brought into intimate contact with a liquid in which the pollutant is preferentially absorbed. The most common equipment is the packed tower scrubber described earlier and the design and operation of such units for gas absorption is a well established chemical engineering operation. The alternative scrubber systems described earlier are also used frequently as they serve the dual purpose of removing both gases and particulates. In particular applications, such as in the removal of SO_2 which has a very low solubility in water, the selection of a suitable solvent is most important. Technical processes for the control of SO_2 emissions can be classified as:

(1) Fuel desulphurisation processes.

(2) Fuel conversion processes.

(3) Combustion techniques.

(4) Flue gas desulphurisation processes.

Fuel oil desulphurisation technology zis available and its application could significantly reduce the pollution problem. However, coal cleaning processes are not capable of removing sufficient sulphur from the feedstock to be used on their own. Fuel conversion processes are not yet feasible on economic grounds. The principle of the flue gas desulphurisation system is shown schematically in Fig. 32.7.

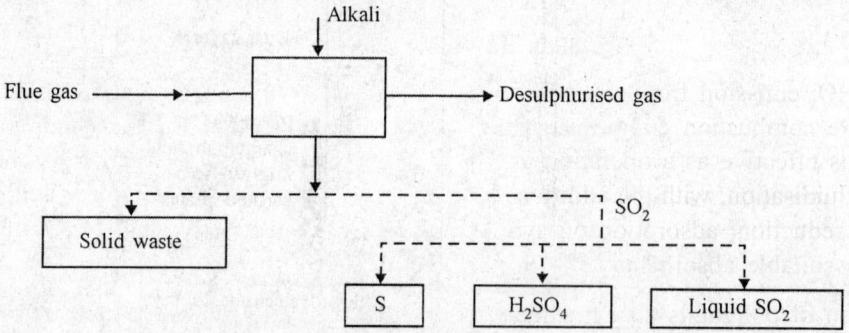

Fig. 32.7. Principle of flue gas desulphurisation.

The most popular process in the U.S., lime-limestone scrubbing, is shown in Fig. 32.8. Particulates are first removed and the flue gas passed to a scrubber where it is contacted with a 5–6% lime or limestone slurry. The following reactions take place:

$$CaCO_3 + SO_2 + \tfrac{1}{2}H_2O \longrightarrow CaSO_3 . \tfrac{1}{2}H_2O + CO_2$$
$$\text{or}$$
$$CaO + SO_2 + \tfrac{1}{2}H_2O \longrightarrow CaSO_3 . \tfrac{1}{2}H_2O$$

Other processes include MgO scrubbing in which the MgO is regenerated by calcination of the $MgSO_3$ and $MgSO_4$ precipitates and the Wellman-Lord process in which the SO_2 is again recovered as a by-product. Catalytic oxidation of the SO_2 to SO_3 using a V_2O_5 catalyst at about 450°C has been used but the dilute H_2SO_4 product has limited commercial value. Other control methods include simple packed tower scrubbing with dimethylaniline from which the SO_2 is distilled off.

H_2S is removed by scrubbing with diethanolamine which is then distilled and recycled. HCl, HF and other fluorides are soluble in water and a dilute solution of NaOH can be used for the absorption of fluorine. Scrubbing with $KMnO_4$ solution is used for industrial odour control.

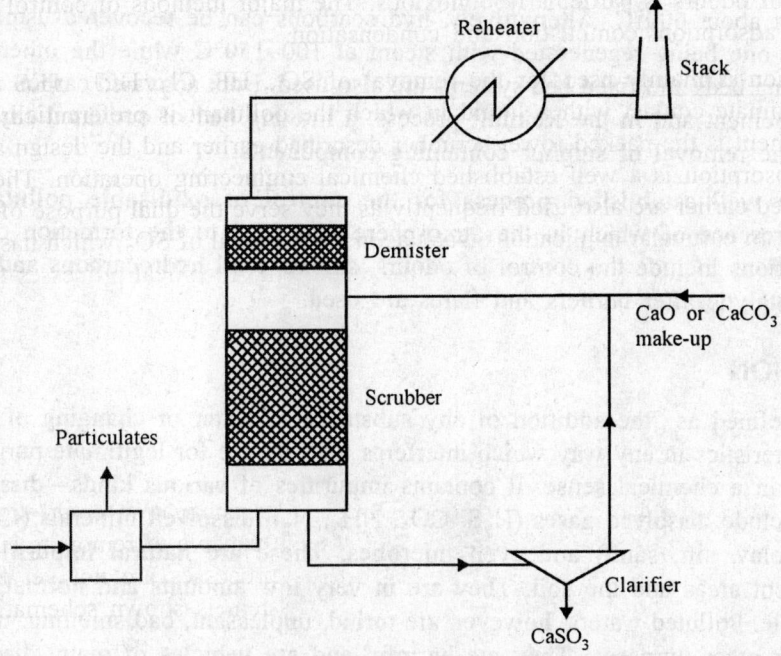

Fig. 32.8. Lime-limestone scrubbing process.

98% of NO_x emission from stationary sources is a result of combustion so this can be minimised by control of the combustion conditions. Operation at close to stoichiometric conditions using two stage combustion is effective as a operation at low combustion temperatures, which has been achieved using pressurised fluidisation, with the addition of limestone. NO_x emission from nitric acid plants is controlled by catalytic reduction, adsorption on synthetic zeolites and wet scrubbing. HNO_3 (30%) and hydrogen peroxide are suitable absorbents.

Absorption of pollutants as a method of abatement is well established. The gas is passed through a bed of solid adsorbent on the surface of which the pollutant molecules are preferentially adsorbed and retained by either physical or chemical forces. When the surface is saturated, regeneration of the adsorbent is necessary. Adsorbents can be classified as follows:

(1) Non-polar solids where adsorption is mainly physical. The most important of these is activated carbon which is very effective in retaining non-polar molecules and is widely used for the removal of hydrocarbons, odours and other trace impurities.

(2) Polar solids, where adsorption is chemical but in which no chemical change in either the surface or adsorbed molecules takes place. These preferentially adsorb polar molecules. Oxides of silica such as synthetic zeolites and metal oxides such as activated alumina are widely used. Applications include the removal of water vapour, NH_3, H_2S and SO_2.

(3) Chemically adsorbing solids which either catalytically or non-catalyt ically cause a chemical change in the adsorbed species and subsequently release them. This introduces the enormous subject of catalysis.

Since regeneration is necessary, adsorption is most suitable where the concentration of pollutant is very low. Thin beds of activated carbon used for odour control may have a lifetime of several months before reactivation at about 600°C. Alternatively, hydrocarbons can be recovered using two deep beds arranged in parallel, one being regenerated with steam at 100–150°C while the other is in operation. Fluidised beds, rotating beds and fixed bed systems are also used. Both activated carbon and copper oxide are used in SO_2 abatement and in the Reinluft process, a moving bed of activated carbon is used as a contact catalyst in the removal of sulphur containing components.

Combustion is a well established process for the control of oxidisable pollutants particularly hydrocarbons, the presence of which in the atmosphere may lead to the formation of photochemical smog. Other applications include the control of odours, smoke, total hydrocarbons and carbon dioxide. Both thermal and catalytic after burners and flares are used.

WATER POLLUTION

Water pollution is defined as "the addition of any substance to water or changing of water's physical and chemical characteristics in any way which interferes with its use for legitimate purposes". Normally water is never pure in a chemical sense. It contains impurities of various kinds—dissolved as well as suspended. These include dissolved gases (H_2S, CO_2, NH_3, N_2), dissolved minerals (Ca, Mg, Na salts), suspended matter (clay, silt, sand) and even microbes. These are natural impurities derived from atmosphere, catchment areas and the soil. They are in very low amounts and normally do not pollute water and it is potable. Polluted waters, however, are turbid, unpleasant, bad smelling, unfit for drinking, bath and washing or other purposes. They are harmful and are vehicles of many diseases as cholera, dysentry, typhoid, etc.

The major pollutants in industrial waste-waters include suspended solids and BOD (Biological Oxygen Demand) or COD (Chemical Oxygen Demand) which are measures of the concentration of oxygen consuming materials. The level of specific toxic materials is of course strictly controlled as are the pH, colour, odour etc. Clearly the quantity of flow and subsequent dilution are also important.

Sources of Pollution and the Pollutants

The chief sources of water pollution are as follows :

(1) Sewage and other waste.

(2) Industrial effluents.

(3) Agricultural discharges.

(4) Industrial wastes from chemical industries, fossil fuel plants (thermal power plants) and nuclear power plants.

Each of these sources of pollution carries a variety of pollutants that enter our water bodies. Following are the sources of water pollution and kind of the pollutants carried by them.

Sewage and other waste

Sewage is the waterborne waste derived from home (domestic waste) and animal or food processing plants. It includes human excreta, paper, cloth, soap, detergents etc. These are a major proportion of the pollutants entering our water. There is uncontrolled dumping of wastes of rural areas, towns and cities into ponds, lakes, stream or rivers. Due to accumulation of sewage and other wastes in these bodies, they are not able to recycle them and their self-regulatory capability is lost. The decomposition of these wastes by aerobic microbes decreases due to higher levels of pollution. The self-purifying ability of the water is lost and water becomes unfit for drinking and other domestic uses. Since decomposition of sewage and other wastes is largely an aerobic process, accumulation of these in water increases its oxygen requirements (BOD).

Phosphates are the major ingredient of most detergents. They favour the luxuriant growth of algae which form water blooms. This extensive algal growth also consumes most of the available oxygen from water. This decrease in O_2 level becomes detrimental to growth of other organisms which produces a foul s..ell upon decay. Some decomposing plants are known to produce toxins as *strychnine* which kills animals including cattle.

One of the most common primary sources of water pollution is the discharge of untreated or partly treated sewage in water bodies, sometimes due to improper sewage-handling processes of municipal bodies. This is not uncommon in major cities. Such a discharge of sewage and other wastes in water results into:

(1) Depletion of oxygen levels of water.
(2) Stimulation of algal growth.

Biological oxygen demand (BOD)

BOD is the amount of oxygen required for biological oxidation by microbes in any unit volume of water. The test is done at 20°C for at least five days. BOD values generally approximates the amount of oxidisable organic matter and is, therefore, used as a measure of degree of water pollution and waste level. Thus, mostly BOD value is proportional to the amount of organic waste present in water.

Thus due to addition of sewage and waste, oxygen levels are depleted, which are reflected in terms of BOD values of water. The number of microbes as *Escherichia coli* (bacterium) also increases tremendously and these also consume most of the oxygen. The number of bacteria as *E. coli* in unit volume of water is also taken (*E. coli* index) as a parameter of water pollution.

BOD values are thus useful in evaluation of self-purification capacity of a water body and for possible control measures of pollution. The quantity of oxygen in water (Dissolved Oxygen–DO) along with BOD is indicated by the kind of organisms present in water. Thus fish become rare at DO value of 4–5 ppm of water. Further decrease in DO value may lead to increase in anaerobic bacteria.

Eutrophication

Due to addition of domestic waste (sewage), phosphates, nitrates etc., from wastes or their decomposition products in water bodies, they become rich in nutrients, especially phosphates and nitrate ions. With the passage of these nutrients through such organic wastes, the water bodies become highly productive or eutrophic and the phenomenon is called as eutrophication. It must be remembered that ponds, lakes etc. during their early stages of formation are relatively barren and nutrient-deficient, thus supporting no or very poor aquatic life. This state of these bodies is known as *oligotrophic*. With the addition of nutrients, there is stimulated luxuriant growth of algae in water. There is also generally a shift in algal flora, blue-green algae begin to predominate. These start forming algal blooms, floating scums or blankets of algae. Blooms of algae are generally not utilised by zooplanktons. The algal blooms compete with other aquatic plants for light for photosynthesis. Moreover, these blooms also release some toxic chemicals which kill fish, birds and other animals thus water begins to stink. Decomposition of blooms also leads to oxygen depletion in water. Thus, in a poorly oxygenated water with higher CO_2 levels, fish and other animals begin to die and clean water body is turned into a stinking drain.

Eutrophication is thus limiting factor in supply of clean water for drinking, fishing and navigation etc. Following are the methods to stop or reverse eutrophication.

(1) The waste-water must be treated before its discharge into lake or river. This would limit its nutrient input.

(2) Bacterial multiplication should be stimulated in order to reduce the amount of nutrients solubilised in water. This would help disruption of algal food-web.

(3) Recycling of nutrients into the water should be checked through harvest and removal of algal blooms upon their death and decomposition.

(4) Dissolved nutrients should be removed from water by physical or chemical methods. For instance, phosphorus can be removed by precipitation; nitrogen by biological nitrification and dentrification or by air stripping of NH_3 from an alkalised waste-water, or by ion exchange, electrodialysis of reverse-osmosis.

Many pathogenic microbes (viruses, bacteria, protozoa etc.) may begin to grow on products coming from tanneries, slaughter houses, sewage disposal plants, etc., in the water bodies under anaerobic conditions. These may result into spread of fatal water-borne diseases, some of which may assume on epidermic state. These are viral hepatitis, polio (viral), cholera, typhoid, dysentery, diarrhoea (bacterial), amoebiasis (protozoal), etc.

Industrial Pollutants

A wide variety of both, inorganic and organic pollutants are present in effluents from breweries, tanneries, dying textiles, paper and pulp mills, steel industries, mining operations etc. The pollutants include oils, greases, plastics, plasticisers, metallic wastes, suspended solids, phenols, toxins, acids, salts, dyes, cyanides, DDT etc. many of which are not readily susceptible to degradation and cause serious pollution problems. H_2SO_4 as acid waste from coal mines is a serious pollutant that increases the hardness of water, has disastrous effect on live organisms and corrodes concrete etc. Na, Cu, Cr, Cd, Hg, Pb, etc. are the heavy metal effluents, discharged from industries.

Agricultural discharges

These include chiefly the chemicals used as fertilisers and the pesticides (biocides) used in disease control. Their discharges reach into the water bodies. As compared to developed nations, India has relatively a low use of these chemicals, thus discharges into water are still low.

Artificial fertilisers

Modern agriculture rely heavily on a wide range of synthetic chemicals which include different types of fertilisers and biocides (pesticides, herbicides or weedicides). These chemicals along with waste are wasted off lands through irrigation, rainfall, drainage etc. reaching into the rivers, lakes, streams etc. where they disturb the natural ecosystem. Artificial fertilisers crowd out useful minerals naturally present in the top soil. The microbes (bacteria, fungi, worms etc.) in top soil enrich the humus and help to produce nutrients to be taken up by the plant and later by animals. But fertiliser-enriched soil cannot support microbial life and hence there is less humus and less nutrients and the soil can easily become poor and eroded by wind and rain.

Chemical fertilisers are made up of only a few minerals. They impede the uptake of other minerals and imbalance the whole mineral pattern of plant body. Many crops today lack potassium due to excessive use of nitrogenous fertilisers. Excessive potash treatment decreases valuable nutrients in foods, such as ascorbic acid (vitamin C) and carotene. Liming can prevent the release and uptake into the plants of cobalt, nickel, manganese and zinc. Superphosphate may lead to a copper and zinc deficiency.

Nitrate fertilisers used on soil enter our wells and ponds. These waters thus are very rich in nitrates. It not only makes water unfit for drinking but also causes diseases. This water when taken by us, the nitrates are converted to nitrites by microbial flora of intestine. These nitrites then combine with the haemoglobin of blood to form methaemoglobin, which interferes with the O_2 carrying capacity of the blood. The disease produced is called methaemoglobinaemia. This leads to various ailments as damage to respiratory and vascular system, blue colouration of skin and even cancer. A healthy person contains 0.8% of methaemoglobin, whereas in methaneonoglobinaemia this level reaches to 10% in the blood. At above 20% there occurs headache, giddiness and above 60%, there begin consciousness, stiffness, ocular problem etc. At 80% death occurs.

Pesticides and biocides

Pesticides are the chemicals used for killing the plant and animal pests. It is a general term that includes bactericides, fungicides, nematicides, insecticides and also the herbicides or weedicides. Since weeds (herbs) are not pests like bacteria, fungi, nematodes, insects, the spectrum of activity of these chemicals is extended beyond the pests; and thus a broader term biocide, is used to include herbicides as well.

There is a wide range of chemicals used as biocides. But the most harmful are those which either do not degrade or degrade very slowly in nature. Such chemical substances are distinguished as hazardous substances or toxicants. These are highly potent chemicals that enter our food chain and then begin to increase in their concentrations at successive trophic levels in the food chain. Equally hazardous substances are the radionuclides. The hazardous biocides cause considerable harm since their effects are cumulative. Most nations have banned the use of some of these biocides and in India they are gradually being phased out.

The long range effects of such biocides are, infact, a threat to our ecological security.

Biomagnification

Some of the most toxic biocides are DDT (dichloro-diphenyl trichloroethane), BHC (benzene hexachloride), chlordane, heptachlor, methoxychlor, toxaphene, aldrin, endrin and PCBs (polychlorinated biphenyls). Indiscriminate use of the biocides could make them an integral part of our biological, geological and chemical cycles of the earth. They are everywhere in same form. Measurable amounts of DDT residues may be found in air, soil, water and at several thousand of kilometres from the point where it had originally entered the ecosystem. For instance if DDT enters a pond, lake, it is taken as such by the plants of the pond, then reaches to zooplankton feeding on plants, then to minnows eating the zooplanktons, then to fish which eats the minnows and finally in the body of birds who eat the fish. Not only DDT as such in its original form keeps on moving from water to different living components of the pond system but more threatening is that–DDT concentration continuously increases in successive trophic levels (various form of living organisms) in a food chain. This phenomenon is known as *biological magnification* or *biological amplification*. This is the reason why our food grains as wheat and rice and vegetables and fruits today contain varying amounts of pesticide residues which have become their integral part. They can not be removed by washing or other means.

Industrial wastes (physical pollutants)

The two chief pollutants are heat and radioactive substances. These are the wastes chiefly from power plants—thermal and nuclear—which use large quantities of water. Some other industries also give of waste-water after use. Nuclear power plants are the source of radionuclides.

The quantity of waste-water is highest in the thermal power plants in our country. This waste-water is returned after use at very high temperatures to the streams—a river, lake—affecting the aquatic life in these water bodies. This is also called thermal pollution since heat acts as a pollutant. Similarly, nuclear power plants besides causing fallout problems, also release waste heat. This also contributes to thermal pollution. Some plants and animals are killed outright by the very hot water. Though waste-water from nuclear power plant not as hot, but still has adverse effects on aquatic life. These are:

(1) Early hatching of fish eggs.

(2) Failure of trout eggs to hatch.

(3) Failure of salmon to spawn.

(4) Increase in BOD.

(5) Change in diurnal and seasonal behaviour and metabolic response of organisms.

(6) Significant shift in algal forms and other organisms towards more heat-tolerant forms leading to decrease in species diversity.

(7) Effect changes in macrophytes.

(7) Migration of some aquatic forms.

Ground Water Pollution

Ground water is threatened with pollution from seepage pits, refuse dumps, septic tanks, branyard manures, transport accidents and different pollutants. Important sources of ground water pollution are sewage and other wastes otherwise. Raw sewage is dumped in shallow soakpits. This gives birth to cholera, hepatitis, dysentry etc., especially in areas with high water table. The industries of woollens, bicycles contribute high amounts of Ni, Fe, Cu, Cr and cyanides to ground waters.

Marine Pollution

All the what is carried by rivers ultimately ends up in the seas. On their way to sea, rivers receive huge amounts of sewage, garbage, agricultural discharge, biocides, including heavy metals. These all are added to sea. Besides these discharge of oils and petroleum products and dumping of radionuclides waste into sea also cause marine pollution. Huge quantity of plastic is being added to sea and oceans.

The pollutants in sea may become dispersed by turbulence and ocean currents or concentrated in the food chain. They may sediment at the bottom by processes like adsorption, precipitation and accumulation. Bioaccumulation in food chain may result into loss of species diversity.

In marine water the most serious pollutant is oil, particularly when float on sea. An spill in oil or petroleum product due to accidents or a deliberate discharge of oil polluted waste brings about pollution.

Mercury Pollution

Mercury enters water naturally as well as through industrial effluents. It is a potent hazardous substance. Both, inorganic and organic forms are highly poisonous. Methyl mercury gives off vapours. Mercury was responsible for the Minamata epidemic that caused several deaths in Japan and Sweden.

From the effluents, mercury compounds enter the water body and at their bottom these are metabolically converted into methyl mercury compounds by anaerobic microbes. Methyl mercury is highly persistent, soluble in lipids and after being taken by animals enters into food chain and accumulates in fatty tissues. Fish may accumulate the methyl mercury ions directly. There may be nearly 3000 times more mercury in fish than in water.

Mercury poisoning is caused due to inactivation of several sulphydral enzymes by replacement of hydrogen atoms in sulphydral groups. The antidote, BAL (dimercaptol) is used for mercury poisoning.

Lead Pollution

Lead poisoning is common in adults. The chief source of lead to water are the effluents of lead and lead processing industries. Lead toys may be chewed by children. Painters also have a risk of lead consumption. In some plastic pipes lead is used as stabiliser. The water may become contaminated in these pipes. Lead is also used in insecticides, food, beverages, ointments and medicinal concoctions for flavouring and sweetening.

Lead pollution causes damage to liver and kidney, reduction in haemoglobin formation, mental retardation and abnormalities in fertility and pregnancy. Chronic lead poisoning may cause three general disease syndromes:

(1) Gastrointestinal disorders.

(2) Neuromuscular effects (leadlapsy)—weakness, fatigue muscular atrophy.

(3) Central nervous system effects or CNS syndrome—that may result in coma and death.

Lead poisoning also causes constipation, abdominal pain, etc.

Fluoride Pollution

Fluoride is also regularly present in water and soil besides air. In nature it is found as fluoride. The crop plants grown in high-fluoride soils in agricultural, non-industrial areas had a fluoride content as high as 300 ppm.

Yamuna—An Open Sewer

A dip in the Yamuna in Delhi no longer cleans the body or uplifts the soul; it only saps the body and sags the soul. A pure holy river has been turned into an open sewer. The river has been affected by the dumping of toxic effluents by various industries of Delhi. The Yamuna water near Delhi is not fit even for irrigation, what to say of drinking, bathing, swimming, fisheries, industrial cooling etc.

Ganga—The Polluted Purifier

Ganga along with its tributaries is the largest and a very important river basin of the country. It has been a symbol of purity but today it is grossly polluted, that is in utter disregard to its sanctity.

Essentially there are three types of pollutants—silt, biological and chemical. Sedimentation may be reduced by rehabilitation of catchment area through tree plantation. The present emphasis is on biological and chemical pollution.

Biological Pollution

There are three main sources:

(1) Urban liquid and solid waste (largest amount of pollution).
(2) Dead bodies of animals and humans.
(3) Wallowing of cattle.

The urban waste is mostly in the form of sheer faecal matter that also goes into the river as surface run off. Since the river is used for mass bathing at Rishikesh, Hardwar and Allahabad, there is problem of water-borne diseases.

Chemical Pollution

There are various chemical industries in India. But only very few have effluent treatment plants in operation. The control of urban waste is indeed a difficult task since most cities and town are on the bank of river which do not have proper sewer and treatment plant facilities. Therefore, first phase in control strategy should include diverting waste-water so as not to flow in the river and then put the same to beneficial use. Among other things the programme must include:

(1) Renovation of sewage pumps, treatment plants and augmentive digester capacity.
(2) Renovation of trunk sewer and other cognate elements.
(3) Extended sewerage in unsewered areas.
(4) Bring waste from unsewered areas to treatment plants.
(5) Provide pumping stations and intercepting sewer and lead to existing treatment.
(6) Install new treatment plants.

Wherever possible, sewage/sullage needs to be recycled for harvesting energy, feed for poultry and animal husbandry, manure and raising aquaculture and supply irrigation water.

The foregoing methods aim the pollution control at point sources. Also there is distributive pollution that reduces self-purifying capability of the river. Point sources are easy to control, distributive pollution cannot be controlled by the orthodox sewage systems. There are plenty of rains during monsoon. The extra water is not only a problem but also goes to sea without any use. There is a need to increase seewage and storing of the surplus monsoon water somewhere. One way is to create big reservoirs at ground level or in underground cavities. This surplus water may be used for different purposes. It also needs to pump water in rabi season and use in kharif. The monsoon water will sink and recharge underground systems.

PREVENTION AND CONTROL OF WATER POLLUTION

Biodegradable pollutants alone are not responsible for water pollution, though these indicate level of pollution (through BOD values). Besides these, a substantial pollution load is contributed by non-degradable or slow degrading pollutants, such as heavy metals, mineral oils, biocides, plastic materials etc. that are dumped into water. For biodegradable pollutants, pollution may be controlled at source by their treatment for reuse and recycling. The non-degradable toxic substances can be removed from water by suitable methods.

In addition to these methods, some standards, conditions and requirements are to be legally enforced by the government through various Acts. The various ways/techniques suggested for control of water pollution is given as under.

Stabilisation of the Ecosystem

This is the most scientific way to control water pollution. The basic principles involved are the reduction in waste input (thus control at source), harvesting and removal of biomass, trapping of nutrients, fish management and aeration. Various methods may be used (biological as well as physical) to restore species diversity and ecological balance in the water body to prevent pollution.

Reutilisation and Recycling of Waste

Various kinds of wastes which include industrial effluents (as paper pulp or other industrial chemicals), sewage/sullage of municipal and other systems and thermal pollutants (waste-water etc.) may be recycled to beneficial use. For instance, urban waste (sewage/sullage) may be recycled to generate cheaper fuel gas and electricity.

Removal of Pollutants

Various pollutants (radioactive, chemical, biological) present in water body can be removed by appropriate methods such as adsorption, electrodialysis, ion-exchange, reverse-osmosis etc. Reverse-osmosis is based on the removal of salts and other substances by forcing the water through a semi-permeable membrane under a pressure exceeding the osmotic pressure. Due to this, flow occurs in reverse direction. For this, we use a power membrane that attracts the solvent and repulses the solute. Reverse-osmosis is commonly used to desalinate the brackish water and can also be used for purifying water from sewage.

Fibrous filters are used to create very clean conditions, such as in clean room installations, to provide a sterile environment for the removal for the possible radio active dusts from nuclear exhausts and in air conditioning systems. They are used in situations in which the dust load is very low but they can be extremely efficient while offering a very low pressure drop.

Industrial Waste-water Treatment

The techniques applied in industrial waste-water treatment are summarised in Table 32.3 where they are generally classified as primary, secondary and tertiary. Primary treatment normally involves the removal of settleable material where this is present. Secondary, or biological, treatment is often sufficient on its own but with tighter effluent control levels tertiary treatment or 'polishing' is being widely applied. These physical chemical methods are also needed for the treatment of non-biodegradeable materials and for specific toxics. Obviously the most suitable effluent control process for any situation depends on the specific nature of the waste materials. For example considerable treatment which may be termed primary is required in the cleaning of refinery waste-waters. The broad classification of 'biological' and 'physical and chemical' processes is, however, convenient and the subject is dealt with under these headings below.

Table 32.3. Waste-water treatment processes.

Primary	Secondary	Tertiary
Sedimentation	Biological	Physical-chemical
	suspended growth	coagulation and filtration
	film	adsorption
	others (lagoon,	chemical oxidation
	anaerobic etc.)	membrane filtration (ultrafiltration,
		osmosis, electrodyialysis)
		ion exchange
		solvent extraction

Biological Treatment

This is the most common waste-water treatment process. A concentrated mass of micro-organisms is used to break down organic matter into stabilised wastes. These organisms can be classified as aerobic, which require molecular oxygen for synthesis, and anaerobic, which function without molecular oxygen and derive energy from organic compounds in the waste. Facultative organisms may operate in either environment. The most widely applied industrial techniques employ aerobic micro-organisms and these techniques can be classified according to the method of accumulating and storing the massed organisms. In the suspended growth method (activated sludge), the organisms are stored in the reacting liquid as a flocculated mass and in the film methods, of which the trickle filter is the most common, the biological mass is fixed as a film to an inert solid material over which the waste-water is passed.

Suspended growth method

A detailed and comprehensive review of this method, is shown in Fig. 32.9. Biological treatment takes place in the aeration tank where the waste is mixed with the flocculated biological sludge and oxygen. Conventionally, air but also pure oxygen and enriched air, are applied to the tank by surface and diffused aeration or by using static mixers. Sludge separation is normally achieved by gravity sedimentation.

The biological culture is in the form of large flocs comprised of a very large number of active bacterial cells each of which is about 1-2 μm diameter. A well flocculated system is necessary for efficient separation of the sludge from the treated water. The mechanisms of cell synthesis and oxidation may be depicted as follows.

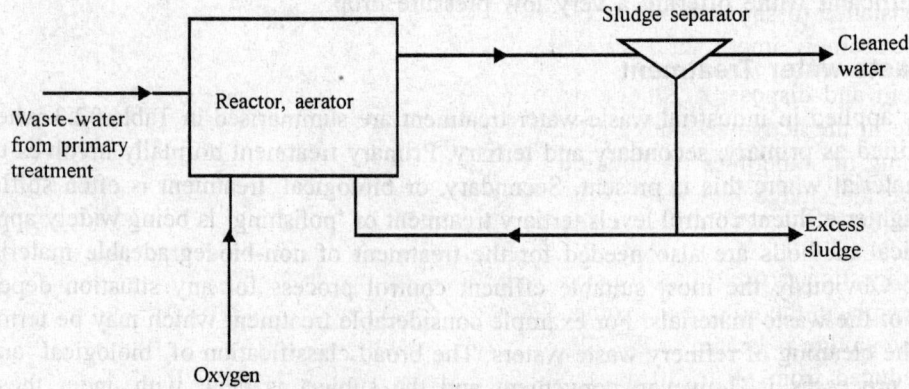

Fig. 32.9. Activated sludge process.

The degradable organic matter in the influent is used as food by the micro-organisms, new cells are formed, the organic waste is rendered inert and CO_2 and water are produced. The rate increases rapidly during the initial stages of the process when the food supply is unlimited and bacteria multiply rapidly. When the food supply runs short, bacterial reproduction ceases and eventually an endogenous phase is reached in which the mass uses up its stored food supply to remain alive. The net increase in biological mass is equal to that produced by food consumption less that lost by endogenous respiration. Clearly this is a very much simplified picture of an extremely complex situation.

The biochemical population can be specifically adapted to particular pollutants. However, in the majority of cases, a wide range of organics must be dealt with and mixed cultures are used. This also adds flexibility to the process since an increase in load of a specific organic will result in an increase in the culture which feeds on it and hence a more rapid removal rate of that material.

Some influent streams may be hydrolysed using caustic soda, carefully controlling the pH level, to render them more easily biodegradable. Excessively high organic loads can be reduced by steam or solvent stripping prior to biological treatment, a technique which is commonly used in the treatment of sour waters in petroleum refineries. Potentially inhibitory and toxic materials can also be removed using this method. Removal of solids prior to treatment and operation at high sludge concentrations (upto 10,000 mg/l) are also considered to enhance performance.

It is frequently required to remove the nutrients, nitrogen and phosphorus, from a waste stream. Nitrification is accomplished using suitable bacteria which can convert reduced nitrogen (as NH_3) to the nitrite and nitrate in the activated sludge process. However, the growth rate of nitrifiers is significantly less than that of the competing bacteria used in the breakdown of organics so where the organic content of the influent stream is high they are unable to compete. In this case, a two stage process can be used, the organics being substantially removed in the first stage and nitrification taking place in the second; the two types of bacteria are separated by an intermediate clarifier.

The effect of nitrites and nitrates on the environment is attracting increasing attention. Dentrification is an anaerobic process by which these are reduced to nitrogen gas. Since this follows nitrification, methanol is often added as a source of energy and carbon but this is expensive and careful control is needed. Soluble phosphorus is removed by the direct addition of lime, aluminium salts and ferric-iron salts.

The addition of powdered activated carbon to the aeration tanks results in the removal of non-biodegradable matter by adsorption. This adds a degree of tertiary treatment and it has been suggested that it adds resistance to shock loading and acts as a toxic sink but the economics of activated carbon and reuse are extremely important.

The treatment and disposal of sludge waste is a major problem. The flow can be as high as 2% of the effluent rate in the activated sludge process and handling and treatment are costly. Concentration by gravity thickening or sometimes by flotation reduces the volume flow rate. Anaerobic digestion is often used to stabilise the sludge into a less offensive material. This is followed by conditioning and finally dewatering.

Anaerobic digestion is often used to stabilise the sludge. In the absence of organic feed, endogenous respiration takes place, the culture consumes itself and gases, primarily methane, are released. It has been found that if the sludge is heated to 24–32°C the rate is increased and the residence time in the digester tank can be reduced from 60 to 30 days. The process is sensitive to the presence of contaminants so some pretreatment may be necessary.

The material is in a highly flocculated state and further chemical or physical conditioning is necessary before the dewatering stage. Heat treatment under pressure at 180–205°C for a period of 15–45 minutes breaks down the gel structure. Alternatively a mixture of sludge and air may be heated. Pressure filtration and rotary vacuum filtration are used for dewatering; an alternative is centrifugal separation. Despite much research activity, the gulf between filtration theory and practice is still too wide and the filterability of a sludge is best determined by pilot plant tests. Work on slurry characterisation using the compression permeability cell and capillary suction methods is promising.

The sludge may finally be used for agricultural purposes, composed by drying, or burnt to an ash using open hearth fluidised bed incinerators.

Fixed film method

The best known fixed film method is the trickle filter in which the waste-water is trickled over a packed bed in which the medium is covered with biological slime. Oxygen and biological matter diffuse into the slime, where oxidation and synthesis take place and the products appear in the filter effluent. Removal rates are normally low for soluble industrial wastes so the method is generally used only for pretreatment.

In rotating biological contactors, which are now widely used, the film is attached to the surface of large plastic discs which are rotated slowly with the bottom 1/2–1/4 submerged in the waste-water. Aeration is achieved by exposing the film of waste-water to the air and by air entrainment as the disc re-enters the tank. Since the rate process is mass transfer controlled, its advantages over the trickle filter are obvious. It is also claimed to have stable operation and low maintenance and power costs.

However, with high influent organic concentrations, many discs are required and the capital cost may be high. The system is readily adapted to situations where nitrification and dentrification are needed because the bacteria can be easily separated.

In the biological fluidised bed process, the waste-water fluidised particulate media, usually activated carbon or sand, to which the film is attached. The mass transfer rate is again high and the agitation prevents slugging.

Physical and chemical methods

Solids separation

Methods of particulate separation include flocculation or coagulation which may be required as a first step. Gravity separation and thickening are often used. Recent development include rapid deep bed granular filtration. Here, the water is passed through a granular bed, usually of sand, which is about 1 m in depth and suspended particles, which are much smaller than the pores in the bed, migrate to the media surface where they are retained. A range of washing procedures are used to regenerate the media, including fluidisation and air scouring. The use of graded media in which the grains become progressively finer in the direction of flow results in a higher bed capacity before regeneration and the use of a range of materials of varying density which settle with the largest grains at the top of the bed after washing and up-flow filtration are attracting much interest in this respect. Continuous media recycling-washing processes have also been developed. Precoat filters which use a thin particulate bed laid on cloth media are also widely used. The rotary vacuum filter is probably the most popular of these.

Adsorption

Adsorption, primarily on activated carbon, is being increasingly used for the removal of toxic materials, refractory organics and colour and typically the BOD content may be reduced to 3–10 mg/l. The system consists of carbon handling and storage facilities, reactivation facilities and the adsorption equipment. Both multiple-hearth and rotary-kiln furnaces are used for reactivation. In downflow fixed bed adsorbers, the water passes down through the bed in which organics are adsorbed. The beds which will also remove some particulate material may be arranged in parallel or in series, depending on process requirements and one bed at a time is removed for reactivation. In packed moving bed absorbers, the fluid passes upwards and the bed downwards, portions being removed periodically from the bottom; upflow expanded bed systems are also used.

In general, molecules of low solubility and polarity tend to be preferentially adsorbed, branched chain compounds are preferred to straight chain compounds and adsorptivity increases with molecular weight for members of a homologous series until the molecules physically block the pores. pH can strongly effect selectivity. Many classes of organics, particularly the oxygenated organics, are not readily adsorbed by activated carbon and much work remains to be done on multi-component systems and on process optimisation.

Chemical oxidation

Chemical oxidation can be used for the oxidation of non-biodegradable organics; the most common agents are chlorine, ozone and hydrogen peroxide.

The use of hydrogen peroxide is particularly effective with sulphur compounds and has been used in the treatment of wastes containing free chlorine, phenols and photographic chemicals. When mixed with formaldehyde, it is used in the treatment of plating wastes. In common with ozone it is a source of dissolved oxygen but the cost of the process has restricted its use.

A widespread example of chemical precipitation followed by solids separation is the lime-soda ash water softening process in which the bivalent cations which are the cause of hardness in water are precipitated on the addition of slaked lime. Many other ions may be precipitated by adjusting the pH and the method is useful in effluent control.

Membrane processes

Ultrafiltration, reverse osmosis and electrodialysis have recently been considered as tertiary effluent control and product recovery methods.

In ultrafiltration the effluent is passed across a semi-permeable membrane. Water passes through the membrane while submicron particles and large molecules are retained in the effluent and concentrated by recycling. The degree of separation and water flow rate depend largely on the pore size of the membrane which is available in a range of materials such as cellulose acetate, triacetate and combinations of these. Membranes based on polyamides can be used over a wide pH range and inorganic membranes can be used in the concentration of oily and latex waste-waters. The membrane is supported on a porous medium for strength and the system arranged geometrically to expose a large filtration surface per unit volume. Configurations used include tubes (1–2.5 cm dia.), plate and frame arrangements and spiral wound modules and hollow fibres. Relatively low pressure differentials (1–10 atmos) are used. These systems are primarily applied where product recovery is desirable such as in the recovery of electrophoretic paint, protein, starch and enzymes and in the recovery of latex concentrated product streams of 20–40% solids can be obtained. They are also used in oil separation and in the treatment of laundry wastes and in PVC recovery.

Reverse osmosis is a high pressure membrane separation process (20–100 atm) which can be used to reject dissolved inorganic salts. It is used industrially as a roughing demineraliser for water with a high total dissolved solids content. Pretreatment is often advisable to prevent membrane fouling. It is used in both product recovery and in water reuse schemes but its major impact is in desalination systems.

Electrodialysis is currently used in water desalination but because of its action in concentrating ionic materials, its potential as a waste-water treatment process is receiving attention and its use in cooling water systems is particularly interesting.

Ion exchange

Ion exchange is used for selective ion removal and is extensively applied in water treatment processes and finds some application in the recovery of specific materials from waste-waters. As with adsorption processes regeneration of the media is necessary. A continuous process incorporating removal of ion exchange resin from the primary tower and its regeneration with hot water is also carried out. It can be used to partially desalinate brackish water in the recovery of chromate and chromic acid from waste-water and in ammonia and dye removal.

Solvent extraction

In this conventional chemical engineering operation the effluent is contacted with a solvent in which organic waste is more soluble. The waste is then separated from the solvent which is recycled. Chemical extractants which are dissolved in a carrier solvent are used to effect the separation of both organic and inorganic chemicals from waste-waters that are difficult to separate using conventional solvents.

SOLID WASTE POLLUTION

The solid waste includes glass containers as bottles, crockeries, plastic containers, polythene and other packing materials that are used and then thrown away as garbage. These pile up at public places and cause obstruction in daily life. Besides these there are also other used things like automobile spares, machines, cycle parts etc. that are thrown as junk. The wastes from building material (during construction and demolition), sludge, dead animal skeletons, heaps of crop residues also contribute to solid waste.

To solve these problems technologies have been developed to recycle most of the solid waste items. Thus paper cans, newspapers and other waste paper may be easily recycled. Metallic components of vehicle spares may be recycled by cheap methods. However, there is problem of recycle of heavy metals, plastics, nylon, polythene, etc. To control solid waste pollution, following methods have been suggested:

(1) Recycling.
(2) Burning of waste and utilising heat to warm residential units, generation of electricity.
(3) Composting of organic waste for preparation of manures and biogas.

RADIOACTIVE POLLUTION

Discharge of potentially noxious materials into the environment involves some risk, which may or may not be measurable. Within very broad limits research in nuclear hazards enables us to forecast the effects of exposure of large groups of people, for extended periods, to low doses of radiation.

Among the hazardous substances, radioactive substances are most toxic. As compared to organic poisons, injurious effects of radio-nuclides are exceedingly high. For instance, radium is 25,000 times more lethal than arsenic. Nuclear war materials, test explosions, mad rush for power plants and radioisotope use in medicine, industry and research are the main sources of radioactive pollution that could threaten our environmental security.

Sources of Wastes or Nuclear Pollutants

Some of the sources which can cause radioactive pollution and radioactive wastes are listed below:

(1) Uranium mining and milling
(2) Processing of uranium oxide
(3) Fuel fabrication
(4) Reactor wastes
(5) Spent fuel processing
(6) Research and development
(7) Hospitals and biological laboratories
(8) Industrial application

Disposal Principles

If wastes are truly confined, in the sense that in no credible circumstances could they be liberated into the environment, then the only additional requirement is "perceptual custody" to ensure that the confinement is never broken. This is easier said than done. Few private firms go back for 100 years, political regimes have seldom lasted for as long as 500 years, and there are few civilisations that have survived for 2000 years. What then can be done to perpetual custody of wastes containing for example, plutonium with a half-life of 24,000 years?

Dilution and dispersion is the traditional method than men have always used for dealing with their wastes. It depends upon the capacity of the environment to dilute or detoxify the wastes to a level that is innocuous to man and to organisms of interest to man. Safety in discharge to the environment depends upon three factors:

(1) Dispersion by such means as atmospheric dilution, mixing into big bodies of water, or spreading through large volumes of soil.
(2) Fixation of radionuclides on soil minerals and organic detritus.
(3) Decay of radionuclides, dispersed or fixed, before they are able to affect man.

Disposal Practices

Gases

Radioactive gases arise mainly in reactors, spent fuel processing isotope production and research and development facilities. The general principles are the same for all procedures that depend upon dispersion into the atmosphere.

The maximum permissible emission rate—or in some cases the MPC at the stack mouth—is given in the regulations governing the plant or laboratory. It is then the responsibility of the operator to ensure that emissions are kept as for below the permissible level as may be practicable. Numerous methods are available:

(1) Filtration

(2) Electrostatic precipitators

(3) Incinerator off-gas

Liquids

(1) Storage

(2) Evaporation

(3) Flocculation and precipitation

(4) Ion exchange

Solids

(1) Conditioning

(2) Ground disposal

NOISE POLLUTION

The human ear is constantly being assailed by man-made sounds from all sides and there remain few places in populous areas where relative quiet prevails.

Aeroplanes, trains, cars and numatic drills and radio and television sets, all produce noise, the most dangerous pollutant of man's environment. Noise has become a permanent part of our lives these days because of the development of machinery, industry and technology. Noise harms the body and mind. It not only causes irritation or annoyance but is constricts arteries, increases the flow of adrenaline and forces the heart to work faster.

The word noise is usually defined as unwanted or unpleasant sound that causes discomfort. Noise is also defined as "wrong sound, in the wrong place at the wrong time". Noise pollution means "the unwanted sound dumped into the atmosphere leading to health hazards".

Formerly noise was limited only to the industry. This too was not much as there were only few industries. These days there has been rapid industrial growth. Moreover, there has been population explosion, due to which there is heavy traffic, urban crowd and electric equipment (luxury items and entertainment). All these have added to the noise nuisence in environment. In our country, besides these the two other factors are the religious and social functions which increase the gravity of situation.

Sources of Noise

The main contributors to noise are factories and industries, transportation (air, rail and road) and community and religious activities. There should not been exaggeration to say that we Indians are a noisy people and every sentiment and occasion is manifested in a noisy manner—be it a religious occasion, elections or a family celebration.

The chief man-made sources in urban areas are automobiles, factories, industries, trains, aeroplanes. Noise makers are horns, sirens, lawn mover, musical instruments, TV, radio, transistor, telephone, dogs, loudspeakers, washing machines, vaccum cleaner, food mixers, pressure cookers, fans, air conditioners, coolers, etc. Ever since the industrial revolution, there has been doubling every 10 years of environmental noise.

Properties

There are two basic properties of sound:

(1) Loudness; and

(2) Frequency.

Loudness is the strength of sensation of sound perceived by the individual. It is measured in terms of decibels. Just audible sound is about 10 dB, a whisper about 20 dB, library place 30 dB, normal conversation 35–60 dB, heavy street traffic 60–80 dB, boiler factories 120 dB, jet planes (take off) about 150 dB, rocket engine 180 dB. The loudest sound a person can stand without much discomfort is about 80 dB. Sounds beyond 80 dB can be safety regarded as pollutant as it harms hearing system. Loudness is also expressed in *sones*. One *sone* equals the loudness of 40 dB sound pressure at 1000 Hz.

Frequency is defined as the number of vibrations per second. It is denoted as Hertz (Hz). One Hz equals to one vibration per second. People can hear sound from 16 (infra-audible) to 20,000 (ultrasonic) Hz.

Effects of Noise Pollution

The different effects are categorised as:

(1) Auditory effects (affecting hearing faculty).

(2) Non-auditory effects (other than auditory ones).

Auditory Effects

These include auditory fatigue, and deafness. Auditory fatigue appears in the 90 dB and may be associated with side effects as whistling and buzzing in ears. Deafness can be caused due to continuous noise exposure. Temporary deafness occurs at 4000–6000 Hz. Permanent loss of hearing occurs at 100 dB.

Non-auditory Effects

These are discussed below:

(1) Interference with speech communication

(2) Annoyance

(3) Loss of working efficiency

(4) Physiological disorders

Interference with speech communication

A noise of 50–60 dB commonly interferes with speech; sound of warning (signal) may be misunderstood.

Annoyance

Balanced persons express great annoyance at even low level of noise as crowd, highway, radio etc. The effects are ill temper, brickering etc.

Loss in working efficiency

There develops tiredness and those doing mental work may put to deterioration in their efficiency or even complete loss of ability to work.

Physiological disorders

A number of physiological disorders develop due to imbalance in functioning of the body. These are neurosis, anxiety, insomnia, hypertension, hepatic diseases, behavioural and emotional stress, increase in sweating, giddiness, nausea, fatigue etc. Noise also causes visual disturbance, and reduces depth and quality of sleep thus affecting overall mental and physical health. Other effects are undesirable changes in respiration, circulation of blood in skin and gastrointestinal activity. Noise pollution also causes incidence of peptic ulcers.

Continuous noise causes an increase in cholesterol level resulting in the constriction of blood vessels making you prone to heart attack and strokes. There may be still births and usually low weight children born to mothers living near airports.

CONTROL OF NOISE

There are following ways to control and reduce the noise menace.

Source Control

The noise can be reduced and controlled by the following ways:

(1) Designing and fabricating silencing devices in aircraft engines, automobiles, industrial machines and home appliances.
(2) By segregating the noisy machines.

There could be developed gadgets to control noise at source.

Transmission Control

This can be achieved by covering the room walls with sound absorbers as acoustic tiles and construction of enclosures around industrial machinery.

Protect Exposed Person

The workers exposed to noise can be provided with wearing devices as ear plugs and ear muffs.

Create Vegetation Cover

Plants absorb and dissipate sound energy and thus act as buffer zone. Trees should be planted along with high ways, streets and other places. Ashok, Neem, Tamarind etc. are good for this purpose.

Noise Pollution Control through Law

Silence zones must be created near schools, hospitals and indiscriminate use of loudspeakers at public places may be done by laws. Adequate restrictions must be put on unnecessary use of horns and vehicles plying without silencers.

Public must be made aware and educated about noise nuisance through adequate news media, lectures and other programmes. The movement against noise pollution is very weak in India. The main reason being that most of us do not consider noise as a pollution but as a part of routine life.

Example 32.1. Find stoichiometric AFR for methyl mercaptan.

Solution.

$$CH_4S + x\ O_2 + 3.76\ x\ N_2 \longrightarrow yCO_2 + zH_2O + kSO_2 + 3.76\ x\ N_2$$

$1=y;\ 4=2z \Rightarrow z=2;\ 1=k;\ 2x=2y + z + 2k = 2 + 2 + 2 = 6\ \therefore x = 3$

$\therefore$ Stoichiometric eqn. is : $CH_4S + 3O_2 + 11.3N_2 \longrightarrow CO_2 + 2H_2O + SO_2 + 11.3\ N_2$

$$AFR = 28.97\ (3 + 11.3)/48 \times 1 = 8.631$$

Example 32.2. Find AFR if 150% excess air is used for combustion of diethyl ketone, $C_5H_{10}O$.

Solution.

$C_5H_{10}O + xO_2 + 3.76\ xN_2 \longrightarrow yCO_2 + zH_2O + 3.76\ xN_2$

$\therefore 5 = y;\ 10 = 2z;\ 2x + 1 = 2y + z \qquad \therefore 2x = 10 + 5 - 1 = 14 \qquad \therefore x = 7$

$\therefore C_5H_{10}O + 7O_2 + 26.3\ N_2 \longrightarrow 5\ CO_2 + 5H_2 + 26.3\ N_2$

$\therefore$ Stoichiometric AFR $= 28.97\ (7 + 26.3)/86\ (1) = 11.22$

$\therefore$ If 150% excess air is used, AFR $= 2.5 \times 11.22 = 28.05$

GLOSSARY

Abrasive. A hard substance with sharp edges, used for polishing and grinding.

Absolute temperature. Temperature measured on the absolute scale and is the Centigrade temperature plus 273. A = C + 273.

Absolute zero. Lowest possible temperature at which all molecular motion will probably cease. It is equal to −273°C.

Acid. A chemical compound, which in water solution forms no positive ions except H^+ ions.

Acid anhydride. The oxide of a non-metal which combines with water to form an acid.

Acid salt. A salt which is formed by an incomplete neutralisation of an acid and still contains replaceable hydrogen.

Alcohol. An organic compound containing hydroxyl group and derived from a hydrocarbon.

Alkali. The water soluble base.

Allotropy. Phenomenon of existence of an element in more than one form.

Alloy. A solid solution of one of more metals in another metal.

Alpha particle. A helium atom carrying two units of positive charge, emitted during radioactive disintegrations.

Alum. A double sulphate of monovalent metal (X) and a trivalent metal (Y) with general formula, $X_2SO_4.Y_2(SO)_3.24H_2O$.

Amalgam. An alloy of mercury with one or more metals.

Amorphous. Without definite shape.

Ampere. Strength of the current which deposits 0.001118 gm. of silver per second or sends one coulomb of electricity per second.

Amphoteric. Substance which acts both as an acid as well as a base.

Analysis. Breaking apart of chemical compound in order to determine its composition.

Anaesthetic. Anything that causes unconsciousness.

Anhydrous. Free from water.

Anion. Negatively-charged ion which moves towards the anode (positive electrode) during electrolysis.

Annealing. A regulated heat treatment used for toughening materials, e.g., glass.

Antichlor. A substance used for removing excess of chlorine during bleaching, e.g., sodium sulphide or hypo.

Antidote. A substance used to counteract the effect of a posion.

Antiseptic. Substances which prevent the growth of harmful micro-organisms in living tissues.

Aqua regia. A mixture of concentrated nitric and hydrochloric acid in the ration 1:3.

Atmosphere. The envelope of air completely surrounding the earth to a depth of many miles and exerting an average pressure of 14.7 pounds per square inch or 760 mm. column of mercury.

Atom. The smallest particle of an element which may or may not be capable of independent existence.

Atomic number. The net positive charge on the nucleus of an atom.

Atomic weight. The average, relative weight of an atom compared with that of carbon atom taken as 12.

Avogadro's number. The number of molecules present in one gram molecular weight of any gas, liquid or solid (= 6.023×10^{23}).

Baking powder. A leavening mixture (e.g., starch + sodium bicarbonate + potassium acid tartrate).

Baking soda. Sodium bicarbonate.

Base. A substance which in aqueous solution forms no negative ions other than OH^- ions.

Basic anhydride. The oxide of a metal which unites with water to form a base.

Basic salt. A salt which is produced by an incomplete neutralisation of a base and still contains some replaceable hydroxyl groups.

Beta rays. A stream of electrons emitted during radioactive disintegrations.

Bleaching. Changing naturally coloured substance to colourless.

Binary compounds. Compounds made up of only two elements.

Boiler scale. A deposit of calcium or magnesium salts formed in boilers using hard water.

Carat. Unit of weight (= 200 mgm.) used in the sale of diamonds. Also a unit of expression the composition of gold, e.g., 14 carat gold contains 14 parts of pure gold in 24 parts of the mixture.

Catalysis. The action of substances in changing the rate at which a chemical reaction proceeds.

Catalyst or catalytic agent. A substance which by its more presence alters the rate of a chemical reaction without itself being permanently changed.

Cathode. The negative electrode.

Cathode rays. A stream of negatively charged electrons.

Cation. Positively charged ion which moves towards the cathode during electrolysis.

Chemical change. A change in which new substances with different properties are formed.

Chemical action. A chemical change in which two or more substances are involved.

Chemical equation. The shorthand method of expressing a chemical reaction with the help of formulae. It is both a qualitative and a quantitative expression.

Chemistry. The exact science dealing with the changes in the composition of matter and our attempts to control or modify such changes.

Colloids. Substances with particle size intermediate between the size of molecules and the particles in suspension.

Combining weights. The proportions by weight in which elements combine with each other to form compounds.

Combustion. Any chemical action producing noticeable light and heat.

Compound. A substance composed of two or more elements combined in unvarying proportions.

Concentration. The quantity of a substance, expressed in weight, gram moles or gram equivalents per unit volume.

Cosmetics. Substance used for external applications to beautify or improve complexion, skin or hair.

Coulomb. The quantity of electricity which is transferred by one ampere current in one second or which deposits 0.001118 gm. of silver.

Covalency. The property by which linkages are set up by mutual sharing of electrons.

Crystal. Solid in which the atoms, ions or molecules of a substance are arranged in a definite orderly pattern.

Decomposition. The process of breaking up of a complex substance into simple ones.

Degree of dissociation. Fraction of the total number of molecules dissociated.

Dehydrating agent. A substance which removes water from another substance.

Deliquescent substance. A substance which takes moisture form the air to become wet.

Density. Mass per unit volume.

Destructive distillation. Distillation during which the original compound decomposes to give new products.

Deodorant. Substance used to prevent or mask odours.

Depilatory. A chemical used to remove hair form skin.

Deuterium. Heavy hydrogen, an isotope with atomic weight = 2.

Distillation. Process of vapourising a liquid and condensing the vapour produced.

Double salt. A salt containing two basic radicals combined with one acid radical or vice versa.

Dry ice. Solid carbon dioxide.

Ductility. Property of metals or their alloys to be drawn into wires.

Effervescence. A brisk escape of gases form a liquid or solution.

Efflorescent. The property of losing water of crystallisation by a hydrate on exposure to air.

Electrode. A pole in an electric cell.

Electrolysis. The process of decomposition of an electrolyte by means of an electric current.

Electrolytic dissociation. The process of splitting up of the solution of an electrolyte to form positive and negative ions.

Electron. The unit of negative charge of electricity and having a mass—1/1845 that of the proton.

Element. A substance which has not been broken up into simpler substances.

Emulsion. A colloidal dispersion of one liquid in another.

Endothermic reaction. A chemical reaction which proceeds with absorption of heat.

Energy. The capacity of doing work.

Equilibrium (Chemical). A chemical state in which the two opposing reactions in a reversible reaction proceed at equal velocities so that the concentrations of various constituents remain constant.

Equivalent weight. The number of parts by weight of a substance which combine with or displace directly or indirectly 1.008 parts by weight of hydrogen or 8 parts by weight of oxygen or 35.5 parts by weight of chlorine.

Evaporation. Change of liquid to vapour.

Exothermic reaction. A reaction which proceeds with an evolution of heat.

Faraday. 96,500 coulombs, the quantity of electricity which liberates one gm. equivalent of an ion.

Fermentation. Slow decomposition of organic compounds brought about by living organisms or enzymes.

Fission. Breaking up of complex atomic nuclei into simpler ones.

Fixation of nitrogen. Conversion of free atmospheric nitrogen into useful compounds.

Fixing. Removal of unchanged sensitive emulsion from sensitive plate with hypo—a process in photography.

Flux. The material added to the ore during smelting. It combines with a portion of the impurities to form slag.

Formula. Symbolic expression for a molecule.

Fractional distillation. The process of separating by distillation two or more liquids having different boiling points.

Fuel.. Any combustible substance which may be burnt to supply heat energy with the production of excessively objectionable by-products.

Fusion. The process of coating iron with zinc.

Galvanizing. The process of coating iron with zinc.

Gamma rays. Electromagnetic waves with high penetrating power, given off by a radioactive substance.

Gangue. Rocky matter that is not a part of the ore.

Germicide. A compound which kills harmful micro-organisms.

Gram molecular weight. The mass in grams of a substance numerically equal to its molecular weight.

Halogen. The name applied to salt forming non-metals belonging to the chlorine family of elements.

Hard water. Water containing dissolved calcium or magnesium salts which precipitate or curdle soap.

Heat of combustion. The quantity of heat evolved during the complete combustion of a gram molecule of the substance.

Heat of formation. The quantity of heat evolved or absorbed during the formation of a gram molecule of a compound from its elements.

Heat of neutralisation. The quantity of heat liberated when a gram equivalent of an acid completely neutralises a gram equivalent of base in dilute solution.

Heat of reaction. The quantity of heat evolved or absorbed during the chemical change as indicated by chemical equation.

Heat of solution. The quantity of heat evolved or absorbed when a gram molecule of the substance is dissolved in excess of the solvent.

Heavy water. Deuterium oxide written as D_2O, or H^2_2O.

Hydrate. A crystalline substance containing water of crystallization.

Hydride. A binary compound of an element with hydrogen.

Hydrocarbons. Compounds of carbon and hydrogen only.

Hydrolysis. The chemical reaction of a salt with water to form an acid and a base of unequal strengths– reverse of neutralisation.

Hypothesis. A scientific guess used to explain scientific phenomena.

Ignition temperature. The lowest temperature at which a substance catches fire and continues to burn in air.

Indicator. A substance used to indicate the exact stage at which the chemical reaction involved in a titration is just complete.

Insecticide. A chemical substance used to kill insects.

Ion. An atom or a radical carrying an electrical charge due to loss or gain of electrons.

Ionisation. The process of splitting up an electrolyte into ions in solution.

Isobars. Two atoms having the same atomic weight but different atomic numbers.

Isomerism. The condition where two or more molecules are made up of the same number and kind of atoms but differ in their structural formulae.

Isotopes. Two or more atoms of the same element, having the same atomic number but different atomic weights.

Lake. The combination of a dye adsorbed in a mordant in indirect dyeing.

Laws. The generalisations which sum up in themselves a multitude of facts based on laboratory experiments.

Leavening agents. A chemical compound, e.g., baking powder, used to make dough more digestible by producing gas, (CO_2) which permeates the mixture.

Malleability. The property of many metals and alloys of being hammered into thin sheets.

Matter. Anything that occupies space and has weight.

Metallurgy. The process of extraction of a metal from its ore.

Metals. Elements having a shiny lustre which are generally malleable, ductile and good conductors of heat and electricity.

Miscibility. The solubility of one liquid in another.

Mixture. A physical combination of two or more substances.

Molality. The strength of a solution expressed in terms of gram moles of the solute per 1,000 grams of the solvent.

Molarity. The strength of a solution expressed in terms of gram moles of the solute per litre of the solution.

Mole. The weight of a substance in grams, numerically equal to its molecular weight.

Molecular. The smallest particle of a substance which is capable of independent existence and has all the properties of the substance.

Monocular volume. The volume occupied by a gram-mole of the substance.

Molecular weight. The sum-total of a the atomic weights for all the atoms that make up a molecule of a substance.

Molecule acid. An acid having one replaceable hydrogen atom in each molecule, e.g., HCl.

Mordant. Substance used in dyeing to fix a dye on cloth; it is usually an insoluble metallic hydroxide.

Nascent. (Newly freed or new-born). The term applied to a gas that has just been released from chemical union, e.g., nascent hydrogen.

Neutralisation. The union of hydrogen ions of an acid with hydroxyl ions of a base to form water.

Neutron. A nuclear atomic particle having mass equal to that of a hydrogen atom and carrying no electric charge.

Non-electrolyte. A substance, the solution of which does not conduct electric current and is not decomposed by it.

Non-metal. An element which forms an acidic oxide.

Normal salt. A salt formed by the complete neutralisation of an acid by a base.

Normal solution. A solution containing one gram-equivalent weight of the solute per litre.

Nucleus. The central dense and positively charged part of an atom containing protons and neutrons.

Occlusion. Adsorption of gases by a metal.

Octet. The term applied to a group of eight electrons.

Ore. The mineral from which a metal can be profitably extracted.

Osmosis. The movement of a solvent through a semi-permeable membrane from a less concentrated solution into a more concentrated solution.

Osmotic pressure. Pressure exerted by the solute particles and responsible for osmosis.

Oxidation. A chemical change involving loss of electrons.

Oxide. A binary compound of an element with oxygen.

Oxidising agent. A substance which brings about oxidation.

Photosynthesis. Absorption of carbon dioxide by plant in presence of chlorophyll and sunlight to produce glucose, starch and cellulose.

Physical change. A change involving change in properties only, not in composition.

Precipitate. An insoluble solid which separates as a result of chemical reaction between two or more solutions.

Properties. The characteristics of a substance by which it is recognised.

Proton. A nuclear atomic particle having a mass equal to that of the hydrogen atom and carrying a unit positive charge.

Radical. An atom or a group of atoms forming part of a compound and behaves like a single atom in a chemical change.

Radioactivity. The process of spontaneous disintegration of the atomic nucleus to give alpha, beta and gamma rays.

Rayon. A synthetic fibre resembling natural silk to some extent and derived from cellulose (wood, paper, etc.)

Reducing agent. A substance which brings about reduction by supplying electrons.

Reduction. A chemical reaction involving gain of electrons.

Reversible reaction. A reaction in which the products can interact under a different set of conditions to re-form the original substances.

Salt. A compound whose aqueous solution contains a positive ions other than H^+ and negative ion other than OH^-.

Saturated solution. A solution containing as much dissolved solute as it can retain, at a given temperature.

Slag. Product of combination of flux and impurities present in the ore in metallurgy.

Smelting. Reduction of oxide obtained by roasting or calcination of the ore with carbon in the presence of flux.

Soap. The sodium or potassium salt of a higher fatty acid.

Solubility product. The ionic product in a saturated solution of an electrolyte.

Solute. The constituent of a solution which is dissolved or dispersed in a solvent.

Solution. A homogeneous mixture, the proportions of whose constituents can vary within certain limits.

Solvent. The constituent of a solution which forms the continuous medium.

Spontaneous combustion. Combustion set up by the accumulation of heat from slow oxidation.

Standard conditions. N.T.P. or S.T.P. (0°C and 1 atmosphere or 760 mm. pressure).

Sublimation. The process of changing of a solid to the gaseous state and vice versa without liquefaction.

Super-saturated solution. A solution which contains more of the solute than a saturated solution would contain at the same temperature.

Symbol. An abbreviation of the atom of an element and a definite weight of it.

Synthesis. The building up of complex compounds from simpler ones.

Tempering. A regulated heat treatment to bring a metal, e.g., steel, into required state of hardness.

Theory. An explanation of scientific phenomena supported by sufficient evidence but not capable of laboratory demonstration.

Tincture. A Solution with alcohol as solvent.

Titration. The process of adding one solution form the burette to a known volume of another in the conical flask, in order to complete the chemical reaction involved.

Transmutation. Changing of one atom into another.

Valency. The number of electrons an atom or a radical can lend, borrow or share.

Vapour density. Relative density of a vapour as compared with that of hydrogen.

Welding. The process of joining two metals by heating and pressure.

Zeolite. A natural mineral employed for softening hard water.

References

Adams J.A., and **Rogers D.F.**, *Industrial Chemistry*, Mcgraw-Hill, New York.

Baum B., and **Parker C.H.**, *Solid Waste Disposal Recycle and Pyrolysis*, Reinhold, New York.

Berenson C., *Inorganic Chemistry*, Wiley-Interscience, New York.

Boynton R.S., *Chemistry and Technology of Paints* and Synthetic Resins John-Willey, USA.

Budnikov P.O., *Technology of Ceramics and Refractories*, Prentice-Hall, UK.

Burke J.E., *Chemistry and Technology of Dyes*, Macmillan-Pergamon, New York.

Camp, T.R., *Water and Its Impurities*, Reinhold, New York.

Chanlett E.T., *Environmental Protection*, Mcgraw-Hill, New York.

Clauser H.R., *Organic Chemistry*, Reinhold, USA.

Cooper H.B.H., and **Rossano A.T.**, *Testing for Air Pollution Control*, Mcgraw-Hill, New York.

Conover M.S., *Pesticides and Insecticides*, Wiley-Interscience, New York.

DeGarmo E.P., *Industrial Chemistry*, John Willey-Interscience, New York.

Doremus R.H., *Chemistry of Coal and Coal Chemicals*, Wiley-Interscience, New York.

Eckenfelder W.W., and **O'Connor D.J.**, *Biological Waste Treatment*, Pergamon, UK.

Ewing G.W., *Handbook of Chemistry*, Mcgraw-Hill, New York.

Hampel B.H., *Electrochemical Encyclopaedia*, Reinhold, UK.

Harvey B.G., *Nuclear Physics and Chemistry*, Prentice-Hall, UK.

Jacobs M.B., *Chemical Analysis of Air Pollutants*, Wiley-Interscience, New York.

James A.N., *Industrial Chemistry*, Noyes, UK.

Jolles Z.E., *Chemistry of Polymerisation*, Noyes, UK.

Jurah J.M., *Chemistry of Petroleum*, Mcgraw-Hill, New York.

Kent J.A., *Riegel's Industrial Chemistry*, Reinhold, USA.

Kingzett M.J., *Chemical Encyclopaedia*, Van Nostrand, USA.

Kirk-Ohmer., *Encyclopaedia of Chemical Technology*, Wiley-Interscience, New York.

Kit B., and **Evered D.S.,** *Chemistry and Technology of Plastic and Rubber,* Macmillan, New York.

Laitinen H.A., and **W.E., Harris,** *Chemical Analysis,* Mcgraw-Hill, New York.

Morton M., *Rubber Technology,* Van Nostrand, Reinhold, USA.

Perry R.H., *Chemical Engineer's Handbook,* Mcgraw-Hill, New York.

R. Norris Shreve., and **Joseph A. Brink.,** *Chemical Process Industries,* Mcgraw-Hill, New York.

Rhodes T., and **Carroll G.,** *Organic Chemistry,* Mcgraw-Hill, New York.

Singer F.S., *Agro-Chemical Industries,* Chemical Publishing, UK.

Siting M., *Chemistry of Fermentation,* Noyes, UK.

Stephenson R.N., *Introduction to Chemical Process Industries,* Reinhold, USA.

Swern W., *Environmental Engineering,* Reinhold Publishing Company, USA.

Urbanski T., *Chemistry and Technology of Explosives,* Pergamon, UK.

Van Wazer J.R., *Industrial Chemistry,* Wiley-Interscience, New York.

Yaffe L., *Nuclear Chemistry,* Mcgraw-Hill, New York.

Index

89